Prediction of Protein Structure and the Principles of Protein Conformation

Prediction of Protein Structure and the Principles of Protein Conformation

Edited by
Gerald D. Fasman
Brandeis University
Waltham, Massachusetts

Plenum Press • New York and London

Library of Congress Cataloging in Publication Data

Prediction of protein structure and the principles of protein conformation / edited by Gerald D. Fasman.

p. cm.

Includes bibliographies and index.

ISBN 0-306-43131-9

1. Proteins—Structure—Mathematical models. 2. Proteins—Conformation. I. Fasman, Gerald D.

QP551.P655 1989 89-8555

574.19′296′011—dc20 CIP

Cover illustration: Ribbon drawing of triose phosphate isomerase (end view) as the classic example of a singly wound parallel β barrel. Figure 44 in Chapter 1, courtesy of Drs. Jane S. Richardson and David C. Richardson.

First Printing—October 1989
Second Printing—February 1990

A Division of Plenum Publishing Corporation
233 Spring Street, New York, N.Y. 10013

Printed in the United States of America

Contributors

Tom Alber • Institute of Molecular Biology, University of Oregon, Eugene, Oregon 97403. Present Address: Department of Biochemistry, University of Utah School of Medicine, Salt Lake City, Utah 84132

J. Fernando Bazan • Department of Biochemistry and Biophysics, University of California, San Francisco, California 94143

Peter Y. Chou • 3893 Ross Road, Palo Alto, California 94303

Fred E. Cohen • Department of Pharmaceutical Chemistry, School of Pharmacy, University of California, San Francisco, California 94143–04460

‡A. J. Cross • Biosym Technologies, Inc., San Diego, California 92121

Johann Deisenhofer • Department of Biochemistry, and Howard Hughes Medical Institute, University of Texas Southwestern Medical Center, Dallas, Texas 75235

Gilbert Deléage • Laboratory of Biological Chemistry, Claude Bernard University of Lyon I, 69622 Villeurbanne Cedex, France

Russell F. Doolittle • Center for Molecular Genetics, University of California, San Diego, La Jolla, California 92093

Jonathan E. Dworkin • Department of Biological Chemistry, Milton S. Hershey Medical Center, Pennsylvania State University, Hershey, Pennsylvania 17033

David Eisenberg • Molecular Biology Institute, and Departments of Chemistry and Biochemistry, University of California, Los Angeles, Los Angeles, California 90024

Gerald D. Fasman • Graduate Department of Biochemistry, Brandeis University, Waltham, Massachusetts 02254

Janet Finer-Moore • Department of Biochemistry and Biophysics, University of California, San Francisco, California 94143

J. Garnier • Laboratory of Physical Biochemistry, University of Paris-Sud, 91405 Orsay, France. Present Address: Protein Engineering Unit, INRA, 78350 Jouy-en-Josas, France

‡Deceased.

A. T. Hagler • Biosym Technologies, Inc., San Diego, California 92121; and Agouron Institute, La Jolla, California 92137.

Robert Huber • Division for Structural Research, Max Planck Institute for Biochemistry, D-8033 Martinsried, West Germany

Fritz Jähnig • Max Planck Institute for Biology, D-7400 Tübingen, West Germany

‡Emil T. Kaiser • Laboratory of Bioorganic Chemistry and Biochemistry, Rockefeller University, New York, New York 10021

Irwin D. Kuntz • Department of Pharmaceutical Chemistry, School of Pharmacy, University of California, San Francisco, California 94143–04460

Terry P. Lybrand • Department of Chemistry, University of Houston, Houston, Texas 77004. Present Address: Department of Medicinal Chemistry, University of Minnesota, Minneapolis, Minnesota 55455

D. H. J. Mackay • Biosym Technologies, Inc., San Diego, California 92121

J. Andrew McCammon • Department of Chemistry, University of Houston, Houston, Texas 77004

Harmut Michel • Division for Molecular Membrane Biochemistry, Max Planck Institute for Biophysics, D-6000 Frankfurt 71, West Germany

Kozo Nagano • Faculty of Pharmaceutical Sciences, University of Tokyo, Tokyo 113, Japan

Peter Prevelige, Jr. • Department of Biology, Massachusetts Institute of Technology, Cambridge, Massachusetts 02139

David C. Richardson • Department of Biochemistry, Duke University, Durham, North Carolina 27710

Jane S. Richardson • Department of Biochemistry, Duke University, Durham, North Carolina 27710

B. Robson • Proteus Biotechnology Ltd., Marple, Cheshire SK6 6AB, England; and Epistron Peptide and Protein Engineering Research Unit, University of Manchester, Manchester M13 9PT, England

Neil K. Rogers • Laboratory of Molecular Biophysics, Department of Zoology, University of Oxford, Oxford OX1 3QU, United Kingdom

George D. Rose • Department of Biological Chemistry, Milton S. Hershey Medical Center, Pennsylvania State University, Hershey, Pennsylvania 17033

Bernard Roux • Laboratory of Biological Chemistry, Claude Bernard University of Lyon I, 69622 Villeurbanne Cedex, France

John Rubin • Genentech, Inc., South San Francisco, California 94080

Robert M. Stroud • Department of Biochemistry and Biophysics, University of California, San Francisco, California 94143

Gerald Stubbs • Department of Molecular Biology, Vanderbilt University, Nashville, Tennessee 37235

‡Deceased.

Morgan Wesson • Molecular Biology Institute, and Departments of Chemistry and Biochemistry, University of California, Los Angeles, Los Angeles, California 90024

William Wilcox • Molecular Biology Institute, and Departments of Chemistry and Biochemistry, University of California, Los Angeles, Los Angeles, California 90024

Chung F. Wong • Department of Chemistry, University of Houston, Houston, Texas 77004

Preface

The prediction of the conformation of proteins has developed from an intellectual exercise into a serious practical endeavor that has great promise to yield new stable enzymes, products of pharmacological significance, and catalysts of great potential. With the application of prediction gaining momentum in various fields, such as enzymology and immunology, it was deemed time that a volume be published to make available a thorough evaluation of present methods, for researchers in this field to expound fully the virtues of various algorithms, to open the field to a wider audience, and to offer the scientific public an opportunity to examine carefully its successes and failures. In this manner the practitioners of the art could better evaluate the tools and the output so that their expectations and applications could be more realistic.

The editor has assembled chapters by many of the main contributors to this area and simultaneously placed their programs at three national resources so that they are readily available to those who wish to apply them to their personal interests. These algorithms, written by their originators, when utilized on PCs or larger computers, can instantaneously take a primary amino acid sequence and produce a two- or three-dimensional artistic image that gives satisfaction to one's esthetic sensibilities and food for thought concerning the structure and function of proteins. It is in this spirit that this volume was envisaged.

Thanks are due to Pamela Gailey for her assistance in the handling of the manuscripts and to Mary Born of Plenum Press, whose editorial help was always forthcoming. I owe a great debt to the staff of the Gerstenzang Science Library at Brandeis University for being so generous with their time in assisting me. Of course, the main credit for this volume belongs to the authors of the various chapters, who have labored to make their work available to the public.

The time is approaching when a new protein will be designed on the drawing board, using predictive algorithms, and its subsequent synthesis, via cloning or peptide coupling, will offer new and interesting challenges for biochemists and molecular biologists.

Gerald D. Fasman

Waltham, Massachusetts

Contents

1

Principles and Patterns of Protein Conformation

Jane S. Richardson and David C. Richardson

Jane S. Richardson and David C. Richardson • Department of Biochemistry, Duke University, Durham, North Carolina 27710.

I. INTRODUCTION

The raw materials of protein structure are the detailed geometry and chemistry of the polypeptide and side chains plus the solvent environment. The end result is a complex tapestry of details organized into a biologically meaningful whole: a variation on one of a few harmonious themes of three-dimensional structure. For the purposes of prediction we are not concerned primarily with either of the endpoints of this process but with the logical connection between the two. Therefore, we summarize what is known of that logical connection into a set of guiding principles: hydrophobicity, hydrogen bonding, handedness, history, and the tension between hierarchy and interrelatedness. In addition, we consider particularly relevant features of the starting and ending states. However, one should bear in mind, as cartooned in Fig. 1, that our abilities to follow a protein through this remarkable transition are still rather limited in both the experimental and the theoretical realms.

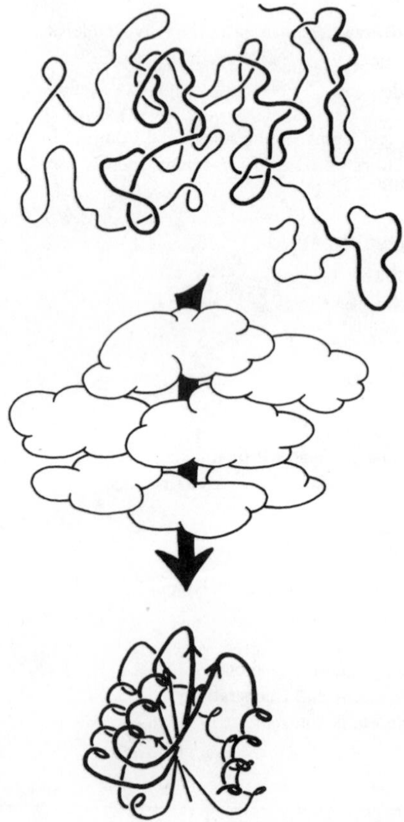

Figure 1. Protein folding.

Fortunately, there is a large and detailed body of evidence about the final folded state, from x-ray crystallography and now increasingly from two-dimensional NMR, most of which is available from the Brookhaven Protein Data Bank (Bernstein *et al.*, 1977). Appendix 1 lists the proteins used in assembling the data and figures for this chapter with their Brookhaven identification codes.

II. THE CONSTRAINTS AND OPPORTUNITIES INHERENT IN THE POLYPEPTIDE CHAIN

Because of the remarkable self-assembly capabilities shown by many proteins for refolding *in vitro,* we know that most aspects of protein configuration derive, in the final analysis, from the properties of the particular sequence of amino acids that make up its polypeptide chain. These properties include the characteristics of both the individual side chains and the polypeptide backbone, which exerts its influence in ubiquitous and sometimes rather subtle ways.

The geometric parameters of the peptide unit and the α carbon are illustrated in Fig. 2 (Momany *et al.*, 1975), and the two are strung together into the familiar polypeptide backbone in Fig. 3. The bond lengths and angles shown are the local minimum-energy values around which the structure fluctuates, both as a function of time for a given bond or angle and also statistically among the total set of such bonds or angles. For bond lengths the restoring forces are large, the time scale is very short, and the range of variation is quite small. Bond angles are somewhat looser, and dihedral or torsion angles are looser still.

Aside from its linear connectivity and steric volume, the most pervasive and significant influences of the polypeptide chain itself on protein conformation are the hydrogen-bonding capabilities of the peptide (treated in Section V) and the handedness imposed by the asymmetry of the α carbon. The convention shown in Fig. 4 (the "CORN crib") allows one to recognize the correct L-amino-acid handedness when dealing with physical models, stereo figures, or molecular-graphics displays: if one looks down on the α carbon from the direction of the hydrogen, the other substituents should read "CO-R-N" in clockwise order, where R stands for the R group of the side chain, CO for the peptide carbonyl, and N for the peptide nitrogen. Later we will investigate the unique role glycine plays because of its lack of handedness (Section V.A) and the influence L-amino acids exert on large-scale handedness phenomena at all levels of protein structure (Section VII).

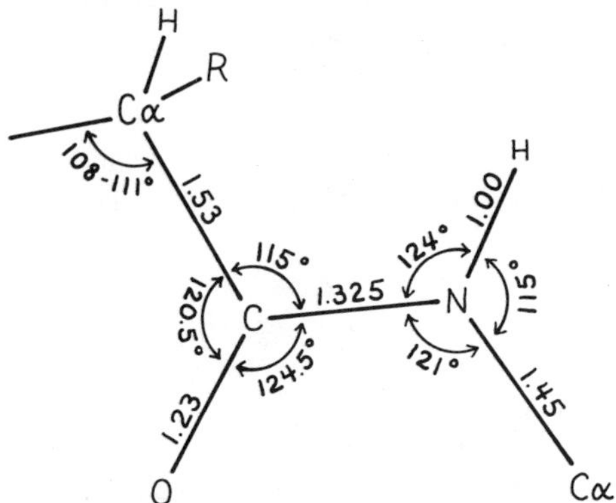

Figure 2. Numerical values for the bond lengths and angles of a peptide and an α carbon.

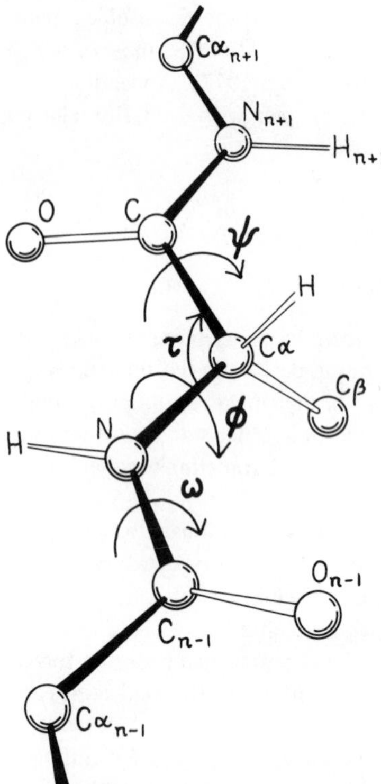

Figure 3. A key to nomenclature for the atoms of the polypeptide chain, the tetrahedral bond angle τ, and the backbone dihedral angles ϕ, ψ, and ω.

The other convention needed for understanding conformational details is the rule for assigning direction and numerical values to dihedral angles, as shown in Fig. 5. A dihedral angle involves four successive atoms—A, B, C, and D—and the three bonds joining them. If you look directly down the length of the central bond joining atoms B and C (fortunately, the answer is the same as viewed from either end of this bond) and put the atom nearest you (A) at 12 on the clock face, then the clock position of the far atom (D) reads out the angle. By convention, dihedral angles are assigned in the range $-180°$ to $+180°$ with the clockwise direction being positive. Thus, the dihedral angle shown in Fig. 5 is about $+35°$.

Assuming ideality for the rest of the geometry, then the three backbone dihedral angles per residue (ϕ, ψ, and ω) plus the dihedral angles χ_1, χ_2, . . . out the side chain provide a complete description of the local conformation. This description is ideal for comparing short pieces of structure, since it is independent of reference frame, but it is not workable for specifying global conformation because even very small round-off errors accumulate drastically. In practice, just ϕ and ψ suffice for the main chain, because the partial-double-bond

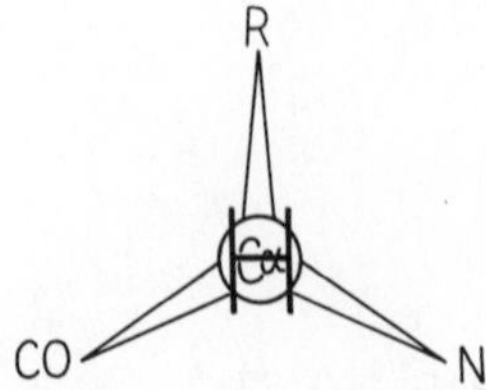

Figure 4. The "CORN crib": a mnemonic for determining the handedness of an amino acid. Looking at the α carbon from the direction of the hydrogen, the other substituents should read CO (carbonyl), R (side chain), and N (backbone NH) in clockwise order for a biologically appropriate L-amino acid.

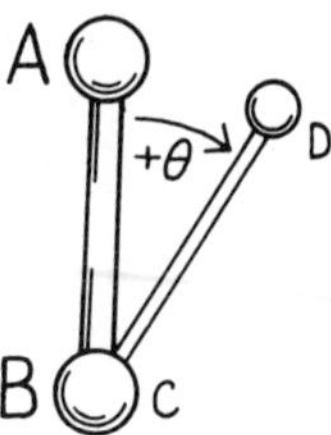

Figure 5. Illustration of the standard convention (IUPAC, 1970) for measuring dihedral angles. The dihedral, or torsion, angle around a bond B–C is defined by the relative positions of the four atoms A, B, C, and D. Looking down the B–C bond, atom A is placed at 12 o'clock, and atom D measures the dihedral angle: plus if clockwise, as in this example (about +35°), and minus if counterclockwise.

character of the peptide keeps ω very close to flat, with the two successive α carbons and the C, O, N, and H between them all lying in one plane. ω is almost always within about 10° of 180°, which is the fully extended or "*trans*" conformation. The curled-up "*cis*" conformation of ω at or near 0° is observed about 10% of the time for proline (see Section V.B for details) and extremely rarely for any other amino acid.

Since φ and ψ form a virtually complete description of backbone conformation, a two-dimensional plot of them (known as a Ramachandran plot) is an important type of representation. We use such plots to illustrate properties of repeating conformations, single residues, or two successive residues.

Regions of φ,ψ space are generally named after the conformation that results if they are repeated. In Fig. 6, the major regions are the right-handed α-helical cluster in the lower left, near −60°,−40°; the broad region of extended β strands in the upper left quadrant (centered around −120°,140°); and the sparsely populated left-handed α-helical region in the upper right near +60°,+40°. Vacant areas are conformations that place atoms unfavorably close together within the dipeptide unit: near 0°,0° the oxygen of residue $n - 1$ bumps the carbonyl C of residue n. The asymmetry of the plot results from collisions with C_β. Within each conformational region there can be significant differences, such as parallel versus antiparallel β, widely varying degrees of β-sheet twist, and extended collagen-type helix all lying within the "β" region. The "bridge" area across ψ = 0° between the α and β regions should be unfavorable based on a hard-sphere model, and yet it is in fact rather well populated; since the bump involved is between successive amide groups, it could be relieved either if the bond angle τ at the α carbon can stretch wider than tetrahedral or if the amide hydrogen is a bit "soft."

A more sophisticated energy calculation for the dipeptide around a central C_α (usually Ala) can do a fairly good job of matching the observed φ,ψ distribution (Anderson and Hermans, 1988). In a way this is rather surprising, because such a calculation leaves out both the favorable and the unfavorable effects of long-range interactions of the backbone as well as specific side-chain effects. One of the more remarkable properties of the repetitive secondary structures observed in proteins is that the optimum φ,ψ values and the permissible range for good long-range H-bonding and steric fit are so close to the optimum and range for favorable dipeptide conformations. Figure 7 shows a φ,ψ plot for residues in nonrepetitive loops, and one sees that the match to Fig. 6 is rather close in spite of the absence of helix and sheet. The presumption is that this neat match is what has, for instance, so strongly selected for the occurrence of right-handed α helices rather than for any of the slightly different versions such as 3_{10}, π, or left-handed α helices.

One should recognize that this fit of local and long-range preferences assumes that the influence of side-chain interactions is so variable as to cancel out on the average. This would not necessarily be true for highly repetitive amino-acid sequences, and there is every reason to suppose that these produce a different range of conformations. This has been convincingly demonstrated for elastin (poly VPGVG: Cook *et al.*, 1980) and collagen (Rich and Crick, 1961), for instance, and has been proposed for a number of other cases. Rather surprisingly, in

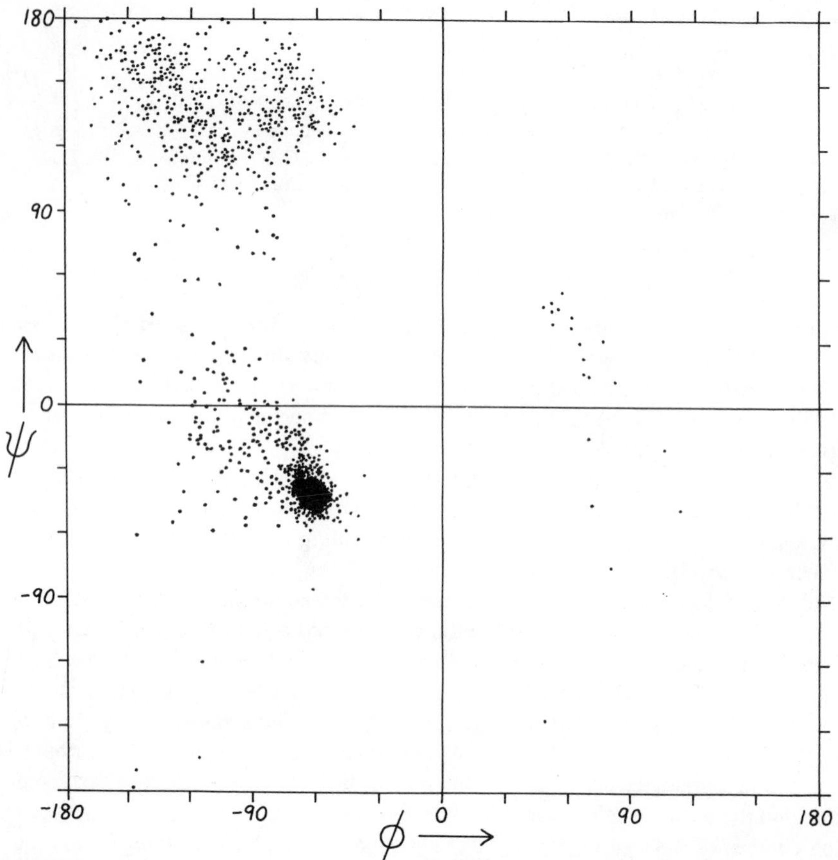

Figure 6. Plot of φ,ψ values found for all residues (excluding Gly and Pro) in a representative sample of highly refined x-ray structures at resolutions of 1.2 to 1.8 Å: 4DFR, 1ECA, 4FXN, 1INS, 2MHR, 1OVO, 1PCY, 5PTI, 2RHE, 5RXN, 2SGA, 1SN3.

the known globular protein structures there are no recognized examples of structural features that involve a cyclic repeat of two, three, or four φ,ψ values, except for the case of small alternating φ,ψ perturbations producing bends in β ribbons (see Section III.B).

If the broad conformational regions (such as "β" or "extended" taken to include most of the upper left quadrant of the φ,ψ plot) are considered in more detail, then some systematic differences emerge between repetitive and nonrepetitive conformations. Figure 7 shows a strong clustering around the polyproline conformation (near $\phi = -60°$, $\psi = 140°$) in spite of the fact that both Pro and Gly were omitted from the plot; there are also somewhat more points in the bridge region near $\phi = -90°$, $\psi = 0°$. Why are these conformations so commonly observed when they are neither at the local minimum for single amino acids nor very favorable for repeating structures? The answer appears to be that poly-Pro and 3_{10} conformations place successive carbonyls approximately perpendicular to each other rather than parallel as in α helix or antiparallel as in β strands. Figure 8 shows the angle between CO_i and CO_{i+1} as a function of φ and ψ. (Note: this function can also be expressed as the angle between successive peptide NH groups, which is a more useful construct for NMR.) Many of the common transitions between pieces of secondary structure (such as α–β connections) involve a 90°

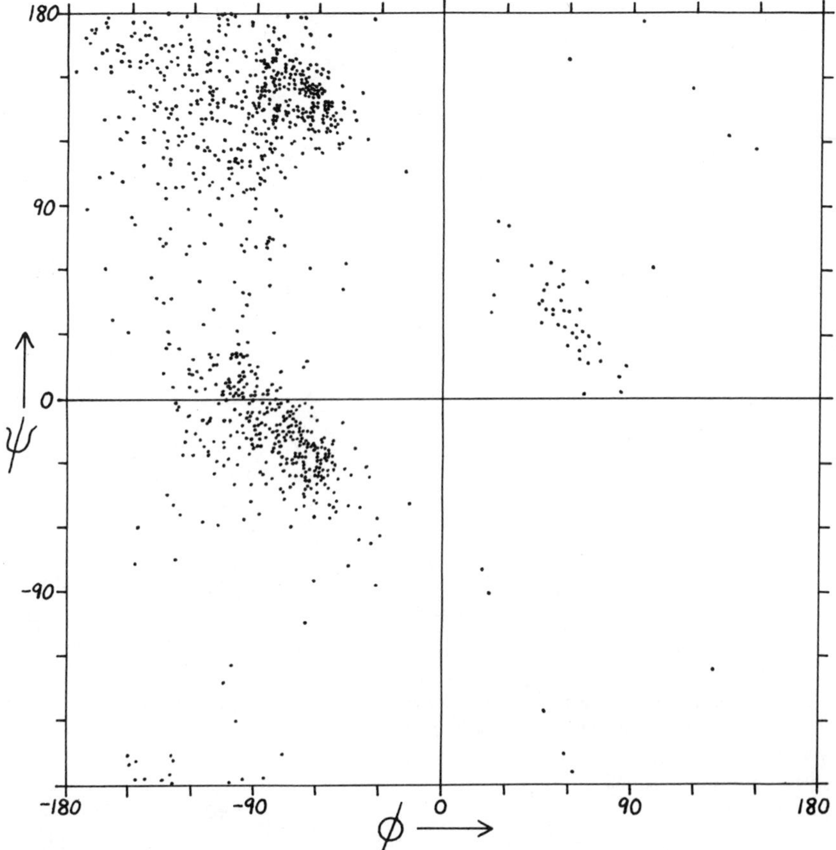

Figure 7. Plot of φ,ψ values for about 1000 residues in nonrepetitive structure (again excluding Gly and Pro) in protein crystal structures at about 2 Å resolution or better. Compared with Fig. 6, the same regions are populated, but there is relatively more emphasis of the polyproline, 3_{10}, Lα, and "bridge" conformations and less of α and β.

change in carbonyl direction, so if they are accomplished in a small number of residues these will usually include at least one of these "perpendicular" residues. This same distinction holds, of course, for the "left-handed" conformations with positive φ angles (which, unfortunately for terminological clarity, are on the right side of the standard φ,ψ plot). More than half of the "Lα" glycines are actually $L3_{10}$, so that their surrounding carbonyls are nearly perpendicular. Most helices and β strands begin and end with a residue in one of the perpendicular conformations (e.g., Fig. 9), and a tight turn requires two of them. The various perpendicular conformations act as the punctuation between secondary structures or as the creases that fold a polypeptide into the elaborate origami construction of a globular protein.

A more subtle influence of backbone geometry arises from the nonorthogonality of its rotations. If the bond angles were 90° and the bond lengths such that two bonds separated in the sequence could be collinear with each other, then it would be possible for some external subsections of the chain to move completely independently of the rest of the protein. As it is, such conditions can only be satisfied very approximately. The simplest example is coupled rotation of the angles at both ends of a single peptide, which leaves the net chain direction

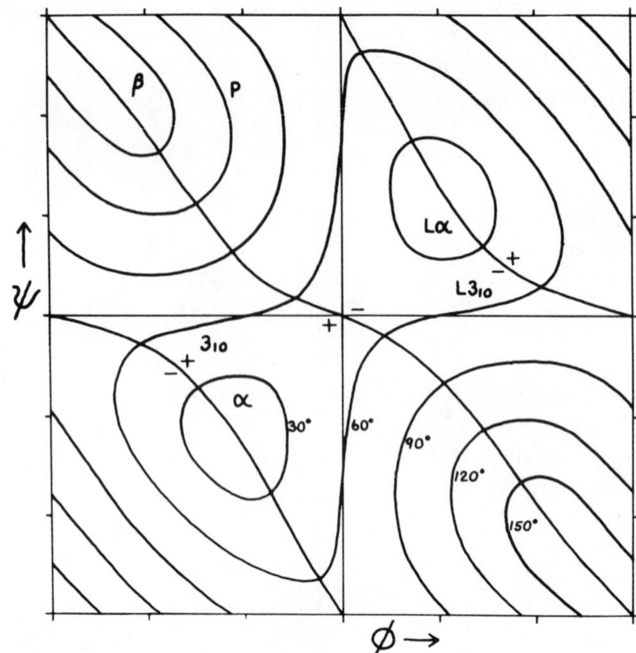

Figure 8. Plot of the angle between successive carbonyls as a function of φ and ψ angles. Contours are drawn every 30° for angle between carbonyls; the diagonal lines mark 0° and 180° in dihedral angle between successive carbonyls. The symbols α, β, P, 3_{10}, Lα, and L3_{10} represent α helix, β sheet, polyproline, 3_{10} helix, left-handed α helix, and left-handed 3_{10} helix.

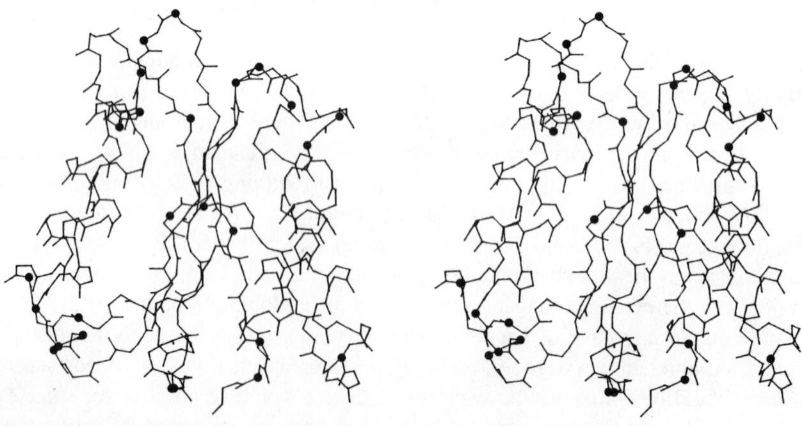

Figure 9. Stereo view of flavodoxin (4FXN), with circles around those residues with approximately perpendicular carbonyls.

nearly the same but produces an offset. Thus, main-chain hydrogen bonding cannot quite be optimized independently of the surrounding conformation, which means, conversely, that H-bonding details can help to determine the surrounding conformation. The overall result is that permissible conformational changes are usually distributed and complexly coupled, and they are sensed to some degree by the rest of the protein, at least within a single folding domain. The nonorthogonal degrees of freedom presumably are important to the cooperative nature of protein folding and conformational transitions. This cooperativity and the resulting interrelatedness among all parts of a protein are almost certainly very useful in engineering complex biological functions, but they make the task of structure prediction a difficult one.

III. HYDROGEN BONDING

The dual hydrogen-bonding capacity of the backbone peptide group is a pervasive influence on protein structure, especially striking in repeating secondary structures like the vast β sheet of bacteriochlorophyll protein (Fig. 10). Although H-bonds are weak, noncovalent interactions, they are fairly directional and specific. Since each peptide can form a bond in

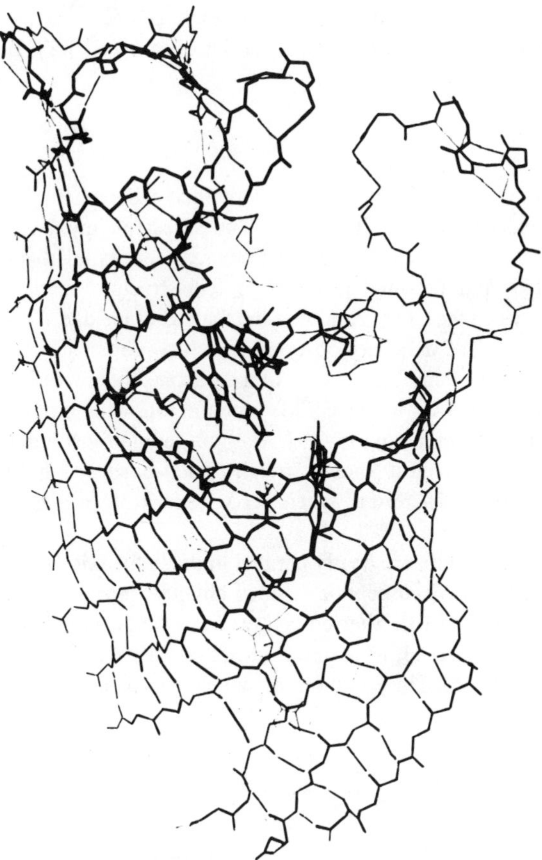

Figure 10. A view of part of the large β sheet in bacteriochlorophyll protein (3BCL), showing backbone and hydrogen bonds.

both directions, the cooperative effect of a network of such interactions can hold the polypeptide together in a strong and specific framework.

Hydrogen bonds involve an electrostatic attraction, either between actual charges (as, for instance, in a Glu-Lys salt link) or between dipoles (such as the peptide dipole, which puts a partial + charge on the NH and a partial − charge on the CO); and they also involve the sharing of a proton. The group on one side of the H-bond is the "donor," D (usually, in proteins, a nitrogen or a water but sometimes an OH), which has a hydrogen it can contribute to the bond. The other group is the "acceptor," A, with accessible unpaired electrons (usually a CO or water but sometimes an unprotonated N or the back side of an OH). The optimum distance for a strong H-bond is about 3 Å between D and A or 2 Å between H and A (Baker and Hubbard, 1984; see Fig. 11). This is a bit closer than the van der Waals bump distance and can be even shorter for a charged H-bond. The electrostatic part of the interaction only falls off as $1/d$ (Sheridan *et al.*, 1979), so there is still an effect at much greater separations, but beyond a certain point other atoms begin to intervene. Angles also matter, but again the energy falls off gradually. The D-H-A angle is fairly critical, with an optimum at 180° and falling to no interaction energy somewhere near 120° (Baker and Hubbard, 1984). For the H-A-C angle (where C is the carbon to which A is attached), there are usually optima in the 120°–150° range, but the interaction is still quite strong at either 180° or 90°. It should be clear that any sharp cutoff between the presence and absence of an H-bond is necessarily arbitrary. On the surface of a protein only an H-bond with very good geometry is useful because of the competition with solvent H-bonds, but in the interior even a very long H-bond is better than none at all.

A. Helices

Patterns of main-chain hydrogen bonding, combined with repeating values of ϕ,ψ angles, are what define secondary structures in proteins. The β structure involves repeating patterns of H-bonds between distant parts of the backbone, whereas helices involve repeating patterns of local H-bonding. The quintessential piece of protein structure is the α helix, formed by repeated H-bonds between the CO of residue n and the NH of residue $n + 4$, with repeated ϕ,ψ values near −60°,−40°. The α helices observed in actual protein structures are always right-handed both because of the cumulative effect of a moderate energy difference for each residue and even more because each C_β would collide with the following turn of a left-handed helix. Isolated residues in Lα conformation (ϕ = 60°, ψ = 40°) are fairly common, especially for Gly (Section V.A,G), but the longest known stretch of left-handed α helix found so far is at residues 226–229 of TLN. Therefore, whenever we speak of an α helix, we mean a right-handed one.

The H-bonds in an α helix are nearly parallel to the helix axis, with the COs all pointing toward the C-terminal end. Each peptide is tilted slightly, however, so that all the oxygens point a bit outward (see Fig. 12). Viewed end-on from the N terminus (e.g., Fig. 13), an α helix has a distinctive appearance, with almost square corners at the α carbons but considerable offset at each successive turn. The β carbons do not extend radially out from the α carbons but

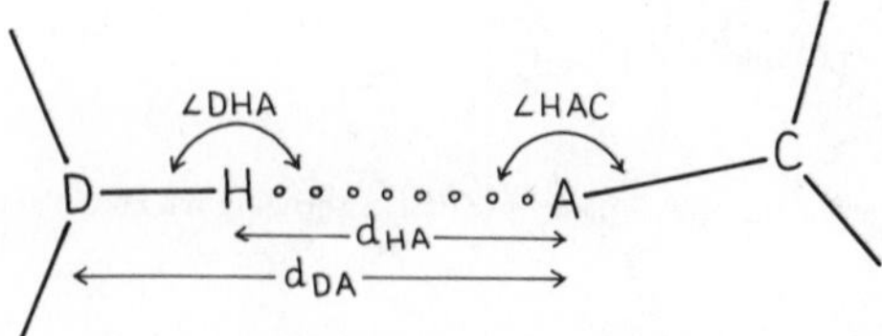

Figure 11. Diagram of some of the distances and angles involved in describing the geometry of a hydrogen bond, where D is the proton donor, A the acceptor, and C the carbon to which A (usually an oxygen for proteins) is attached.

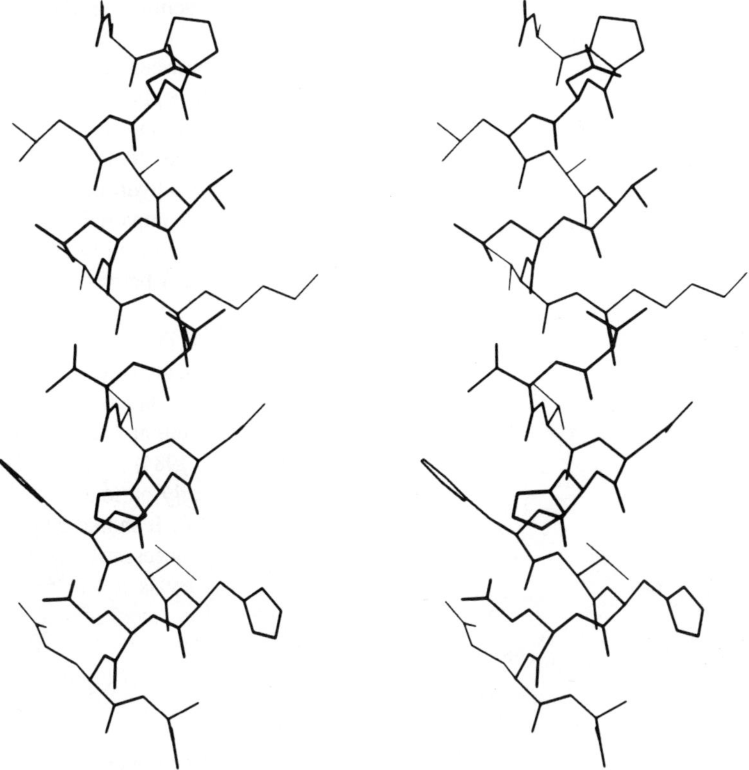

Figure 12. Stereo view of an α helix (2MHR 40–60) from the side. The N terminus is at the top, and the carbonyl groups point down.

make a clockwise pinwheel shape with C_β nearly in the plane of the preceding peptide. In side view (e.g., Fig. 12), the C_α–C_β bonds slant somewhat in the N-terminal direction and are rather close to the backbone of the preceding turn.

The pitch, or repeat, of an ideal α helix is 3.6 residues per turn. For that pitch, the rise per residue along the helix axis is 1.5 Å, or 5.4 Å per turn. Real helices match this value quite

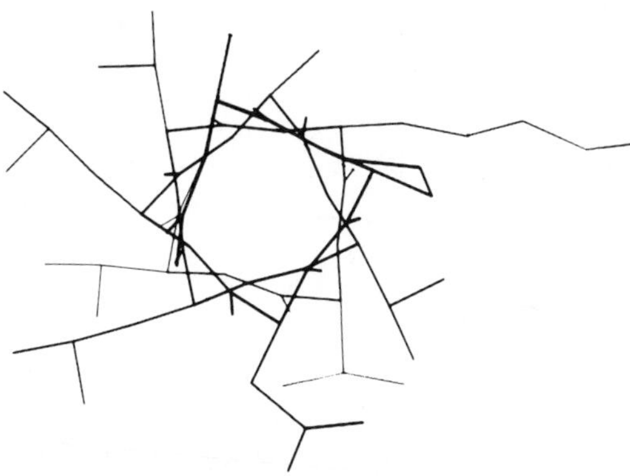

Figure 13. End view of an α helix (2MHR 40–60).

well; however, a difference in average pitch of 5% (between, say, 3.5 and 3.7 residues per turn, which is well within the common range of variation) produces an offset of an entire residue by the end of a typical four- or five-turn helix. That 5% difference makes a trivial change in ϕ,ψ angles but has a substantial effect on side-chain packing. For instance, in ROP protein (see Fig. 14), the neatly repeated layer arrangement of short side chains among the four helices necessitates a 3.5-residue/turn pitch in the α helices plus a noticeable amount of supercoiling. In the straight 3.6-pitch helices of most four-helix clusters, the ROP type of packing can be maintained for only one or two layers (see Section IV.B).

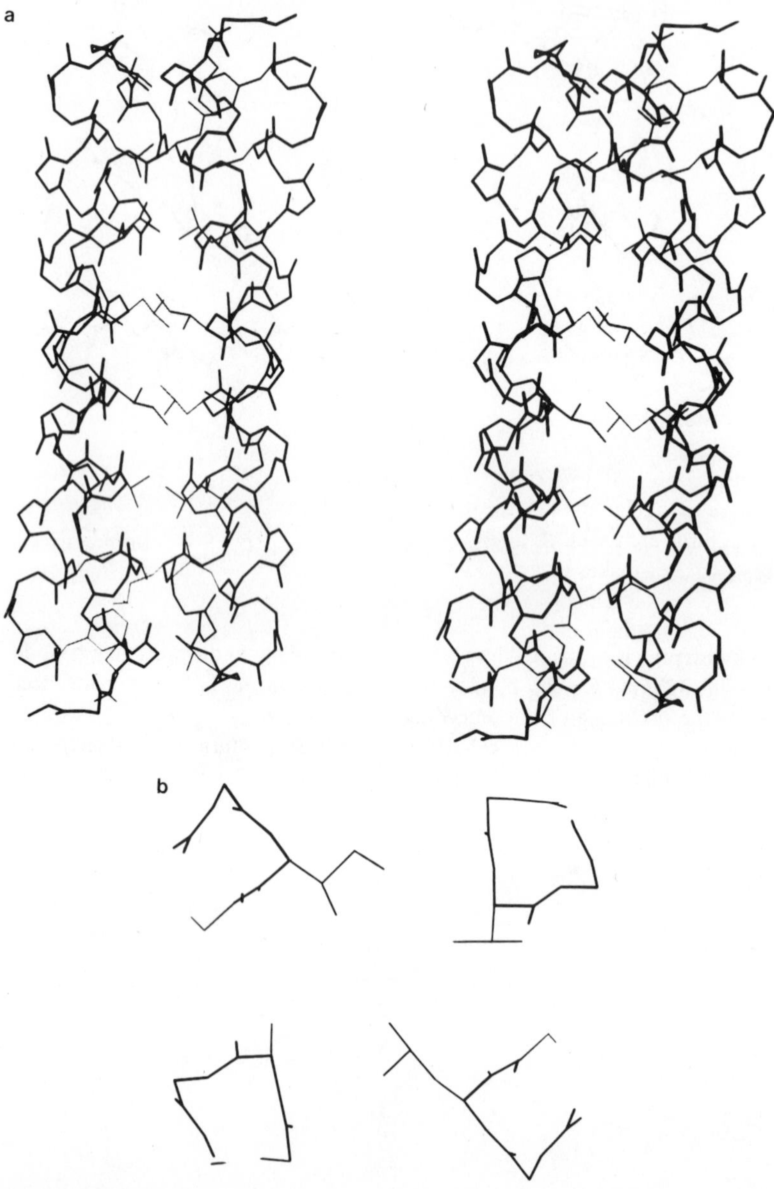

Figure 14. Side-chain packing in the four-helix bundle of ROP protein (Banner *et al.*, 1987). (a) Side view, in stereo. (b) End view of one thin slice through a layer of four contacting side chains.

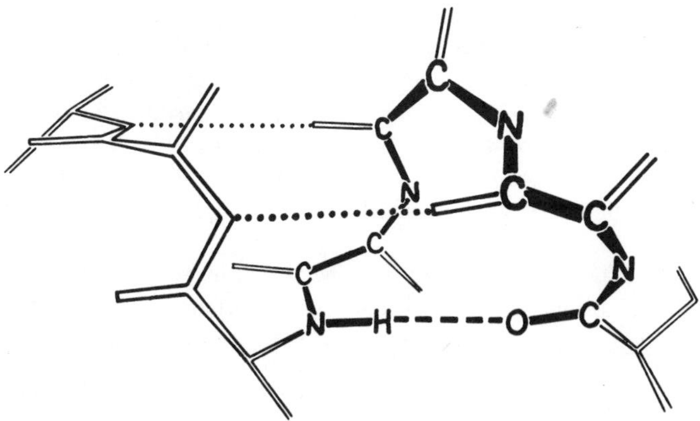

Figure 15. A piece of α helix, with dark lines and labels tracing the 13-atom loop joining the ends of one H-bond.

The H-bond in a helix can be thought of as closing a loop, which for the α helix contains 11 backbone atoms (or 3.6 residues worth) plus the carbonyl O on one end and the H on the other for a total of 13 atoms (see Fig. 15). An alternative name for the α helix is the 3.6_{13} helix, defined by its pitch and H-bond loop size.

The only other type of helix that occurs at all often is the 3_{10} helix, which has an H-bond between the CO of residue n and the NH of residue $n + 3$ (as opposed to $n + 4$ for the α helix) and ϕ,ψ values near $-70°, -5°$. The 3_{10} helix is more tightly wound than the α, and it has a very distinctive triangular appearance in end view (see Fig. 16). In the 3_{10} helix the α carbons on successive turns are exactly in line with one another since there are an integral number of residues per turn; this makes the H-bonds quite tilted relative to the helix axis. In contrast, the nonintegral pitch of an α helix lines up a CO on one turn with an NH on the next to make parallel H-bonds, and the α carbons do not line up. The H-bond geometry and van der Waals interactions between successive turns are not quite as favorable in 3_{10} helix, and long stretches of it are rare (two in thymidylate synthase, one in phosphorylase). The major importance of 3_{10} helix is that it very frequently forms the last turn at the C terminus of an α helix.

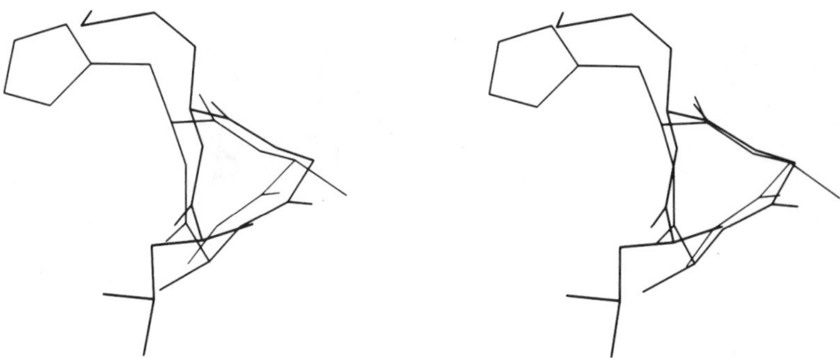

Figure 16. End view of a 3_{10} helix (2CYP 171–177), in stereo, showing the distinctively triangular pattern of the backbone.

In both α and 3_{10} helices all main-chain H-bonding groups within the body of the helix are satisfied by the secondary-structure formation. Each end produces three unsatisfied groups that often H-bond to solvent, especially the open carbonyls at the C terminus. Very frequently, one of the free NHs near the N-terminus H-bonds to the side chain of the N-cap residue (see Section V.G). Relatively few of the other helical side chains make hydrogen bonds. Alpha helices very commonly terminate with a residue in "αII" conformation and then a Gly C-cap in $\mathrm{L}3_{10}$ conformation, thus producing a distinctive pattern of H-bonds (Schellman, 1980; Fig. 17).

One trivial but difficult problem that must be settled in order to accumulate the quantitative information needed for predictions is the issue of defining exactly where a helix (or any other piece of structure) is considered to begin and end. The three major criteria one can use are hydrogen bonding (Kabsch and Sander, 1983; IUPAC, 1970), ϕ,ψ angles (Schellman, 1980; IUPAC, 1970), or α-carbon positions (Levitt and Greer, 1977); for each of these one must decide whether or not to include the interface residue, which fits the helical pattern on one side but not the other. There is no one "correct" answer to this problem, but for the present purpose the most suitable definition is the one that gives the strongest correlations with amino acid sequence. Preliminary trials indicate that using approximate C_α position gives the strongest correlations for helices and also that it is important to separate the preferences according to individual residue positions rather than by "first turn," "last turn," and "middle" (single-position statistics were first compiled by Argos and Palau, 1982, on an earlier data sample, which showed trends similar to those in Table I). The interface residue is of course treated individually, like each of the others, so that it becomes irrelevant whether or not it is called part of the helix.

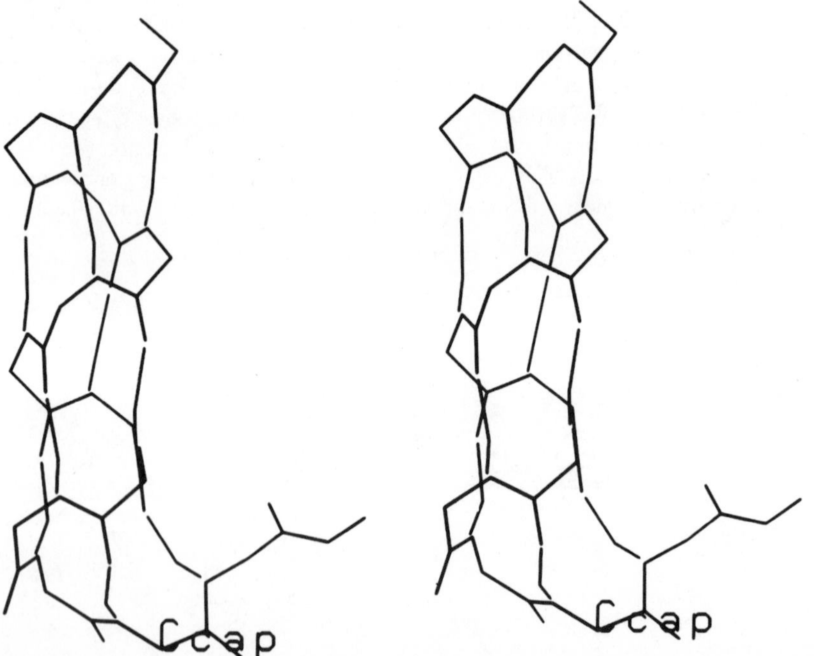

Figure 17. Stereo of the most common ending for an α helix (1TIM 109–122), with a Gly in $\mathrm{L}3_{10}$ conformation as the "C-cap" residue forming the last two hydrogen bonds, but in reverse order.

Table I. Position-Specific Amino Acid Preferences in Helices[a]

%		N″	N′	N-cap	N_1	N_2	N_3	N_4	N_5	Mid	C_5	C_4	C_3	C_2	C_1	C-cap	C′	C″
4.5	Pro	6	9	8	**25**	5	4	1	2	10	2	0	0	2	0	7	17	13
		0.7	1.0	0.8	**2.6**	0.5	0.4	0.1	0.3	0.3	0.3	0	0	0.2	0	0.7	1.9	1.5
8.7	Gly	29	25	**33**	6	16	12	6	8	29	6	8	2	3	4	**72**	21	10
		1.6	1.4	**1.8**	0.3	0.9	0.6	0.4	0.6	0.5	0.5	0.5	0.1	0.2	0.2	**3.9**	1.2	0.6
6.9	Ser	22	17	**34**	11	12	6	6	12	27	5	6	8	23	25	11	9	12
		1.6	1.2	**2.3**	0.7	0.8	0.4	0.4	1.1	0.6	0.5	0.5	0.6	1.6	1.7	0.8	0.7	0.9
6.1	Thr	4	10	21	11	9	13	6	6	43	4	6	9	12	13	4	10	**25**
		0.3	0.8	1.6	0.8	0.7	1.0	0.5	0.6	1.0	0.5	0.6	0.7	0.9	1.0	0.3	0.8	**2.1**
4.5	Asn	11	9	**34**	7	7	7	0	9	28	4	5	8	9	11	15	8	13
		1.2	1.0	**3.5**	0.7	0.7	0.7	0	1.2	0.9	0.6	0.6	0.8	.9	1.2	1.6	0.9	1.5
3.6	Gln	7	7	3	5	6	12	9	8	30	5	1	10	12	**16**	7	10	8
		1.0	0.9	0.4	0.7	0.8	1.5	1.3	1.4	1.3	1.0	0.2	1.3	1.6	**2.1**	0.9	1.4	1.1
6.0	Asp	17	10	**27**	10	**33**	**28**	4	6	38	6	5	7	9	5	9	17	17
		1.4	0.8	**2.1**	0.8	**2.6**	**2.2**	0.3	0.6	1.0	0.7	0.5	0.6	0.7	0.4	0.7	1.4	1.4
5.6	Glu	11	8	5	**27**	**24**	**40**	6	8	30	9	7	19	20	10	4	9	8
		1.0	0.7	0.4	**2.2**	**2.0**	**3.3**	0.5	0.9	0.8	1.1	0.7	1.6	1.7	0.8	0.3	0.8	0.7
6.7	Lys	14	16	10	9	14	12	11	**21**	49	13	12	**31**	**27**	**28**	18	16	17
		1.0	1.1	0.7	0.6	1.0	0.8	0.8	**1.9**	1.1	1.4	1.0	**2.2**	**1.9**	**2.0**	1.3	1.2	1.3
3.8	Arg	3	4	3	6	7	3	11	7	33	5	5	6	**17**	8	7	11	8
		0.4	0.5	0.4	0.7	0.9	0.4	1.5	1.2	1.3	1.0	0.8	0.8	**2.1**	1.0	0.9	1.4	1.1
2.1	His	5	4	5	3	3	3	6	3	14	7	7	5	8	**15**	6	5	4
		1.2	0.9	1.1	0.7	0.7	0.7	1.5	0.9	1.0	2.4	1.9	1.1	1.8	**3.4**	1.3	1.2	1.0
9.0	Ala	13	16	10	24	31	19	20	20	**106**	**23**	21	13	27	21	15	18	12
		0.7	0.8	0.5	1.2	1.6	1.0	1.1	1.4	**1.8**	**1.8**	1.3	0.7	1.4	1.1	0.8	1.0	0.7
7.7	Leu	15	20	3	15	5	10	**39**	13	63	13	19	**31**	13	11	11	14	8
		0.9	1.2	0.2	0.9	0.3	0.6	**2.6**	1.1	1.2	1.2	1.4	**1.9**	0.8	0.7	0.7	0.9	0.5
7.1	Val	10	15	1	17	9	16	21	9	56	4	17	19	10	10	3	9	14
		0.7	1.0	0.1	1.1	0.6	1.1	1.5	0.8	1.2	0.4	1.4	1.3	0.7	0.7	0.2	0.6	1.0
5.1	Ile	9	14	2	10	8	4	11	7	41	9	14	15	4	8	8	11	7
		0.9	1.3	0.2	0.9	0.7	0.4	1.1	0.9	1.2	1.3	1.6	1.4	0.4	0.7	0.7	1.1	0.7

(continued)

Table I. (Continued)

%		N″	N′	N-cap	N_1	N_2	N_3	N_4	N_5	Mid	C_5	C_4	C_3	C_2	C_1	C-cap	C′	C″
3.8	Phe	9	9	2	4	7	5	14	6	34	10	**19**	14	2	6	4	1	9
		1.2	1.1	0.2	0.5	0.9	0.6	1.9	1.0	1.3	1.9	**2.9**	1.8	0.3	0.7	0.5	0.1	1.2
3.4	Tyr	13	3	6	13	3	9	4	1	19	6	5	8	2	9	6	6	3
		1.9	0.4	0.8	1.8	0.4	1.2	0.6	0.2	0.8	1.3	0.8	1.1	0.3	1.2	0.8	0.9	0.5
1.8	Met	1	7	3	1	4	4	6	5	18	7	**9**	4	5	4	3	3	0
		0.3	1.8	0.8	0.3	1.0	1.0	1.7	1.7	1.5	2.7	**2.8**	1.0	1.3	1.0	0.8	0.8	0
1.3	Trp	3	2	1	6	5	4	**8**	3	14	2	5	1	1	0	0	1	7
		1.1	0.7	0.3	2.1	1.7	1.4	**3.1**	1.4	1.5	1.1	2.1	0.4	0.4	0	0	0.4	2.7
2.4	Cys	3	4	3	4	6	3	5	6	11	0	3	1	6	8	2	4	2
		0.6	0.8	0.6	0.8	1.2	0.6	1.1	1.6	0.7	0	0.7	0.2	1.2	1.6	0.4	0.8	0.4
Total		205	209	214	214	214	214	194	160	650	140	174	211	212	212	212	200	197
Chain termini		9	5														12	15

[a]The upper entry is the observed number of occurrences out of a total number listed at the bottom. The lower entry is the relative preference value: the ratio of observed occurrences to the expected number based on average percentage composition (column 1: adapted from Nakashima *et al.*, 1986). Boldface values are higher and underlined values are lower than expectation by more than 3 standard deviations. Out of 215 helices, one NH_2 terminus and three COOH termini were omitted because their location could not be specified unambiguously. This table is reproduced from Richardson and Richardson (1988a).

Table I summarizes statistics for a sample of 215 helices from 45 crystal structures (Richardson and Richardson, 1988a). The statistics were initially separated by tertiary-structure category (see Section IV), but there were no striking differences except that the preferences are a little weaker and less position-specific for α/β proteins than for the other categories. The ends of the helices, called the N-cap and C-cap residues, were defined as the first (or last) whose α carbon is still close to its correct spiral position on the helical cylinder; usually this residue makes one helical H-bond and may have a helical value either for ϕ or for ψ, but those features are not required. It is fairly common for one peptide in a helix to be rotated enough to break both H-bonds and considerably change ϕ,ψ values but to leave the α carbons approximately in place. This does not disrupt the rest of the helix at all and may, in fact, sometimes be an error caused by incomplete refinement. Such local distortions are ignored in the present definition. Bends of up to 30° or 40° in the helix axis (usually at Pro) are permitted if there is still at least one helical H-bond between the two segments. Location of the N-cap and the C-cap is usually quite unambiguous: of the 430 termini in the sample, only one N and three C termini were omitted from the table because they could not be clearly assigned. Five unique positions in from the cap are tabulated at each end, and also the next two positions outside the helix. No helix shorter than eight residues (including the caps) was tabulated. All residues in by more than five from both ends are collected as the "middle" group. Several of the preferences for specific positions are nearly 4 to 1 (normalized by average amino acid composition), which is higher than anything previously reported. Their significance is discussed further in Section V under the individual amino acids.

B. β Sheets

The only other common type of repeating secondary structure is the extended β strand with ϕ,ψ values in the upper left quadrant of the plot near $-120°,140°$. The backbone H-bonding groups are again completely satisfied within the body of a β sheet, but since the H-bonds go from one strand to another, β structure is inherently less local and modular than helices. As a result, the primitive unit of β structure is not the individual β strand but the β strand pair, which can be hydrogen bonded in either a parallel or an antiparallel arrangement. The almost fully extended β conformation is characterized by three mutually perpendicular directions: (1) the N-to-C direction of the backbone; (2) the plane of the peptides (the plane of the page in Figs. 18 and 19), in which close pairs of H-bonds are formed alternately to the left and to the right; and (3) the direction in which the side chains extend, alternating one up and one down. If one looks in the N-to-C direction along a strand of amino acids and the side chain is up, the close pair of H-bonds will be to the right.

For antiparallel β sheet, the relationship between adjacent strands is a twofold axis perpendicular to the sheet, with the H-bonds perpendicular to the strands and alternating between a closely spaced pair and a widely spaced pair (Fig. 18). In parallel β sheet, the H-bonds are evenly spaced but alternately slant forward and backward (Fig. 19), and the relationship between adjacent strands is a translation.

The side chains on β strands extend approximately perpendicular to the plane of H-bonding. Along the strand they alternate from one side to the other, but on adjacent strands they are in register. For antiparallel β sheets typically one side is buried in the interior and the other side is exposed to solvent, so that the amino acid types tend to alternate hydrophobic and hydrophilic. Parallel sheets, on the other hand, are usually buried on both sides, so that their central sequences are highly hydrophobic, and hydrophilics concentrate at the ends. For both types of β structure, edge strands can be much more hydrophilic than central strands. Distinguishing these characteristic patterns can be of some help in secondary-structure prediction and is clearly important for working out probable tertiary structures. The usefulness of this

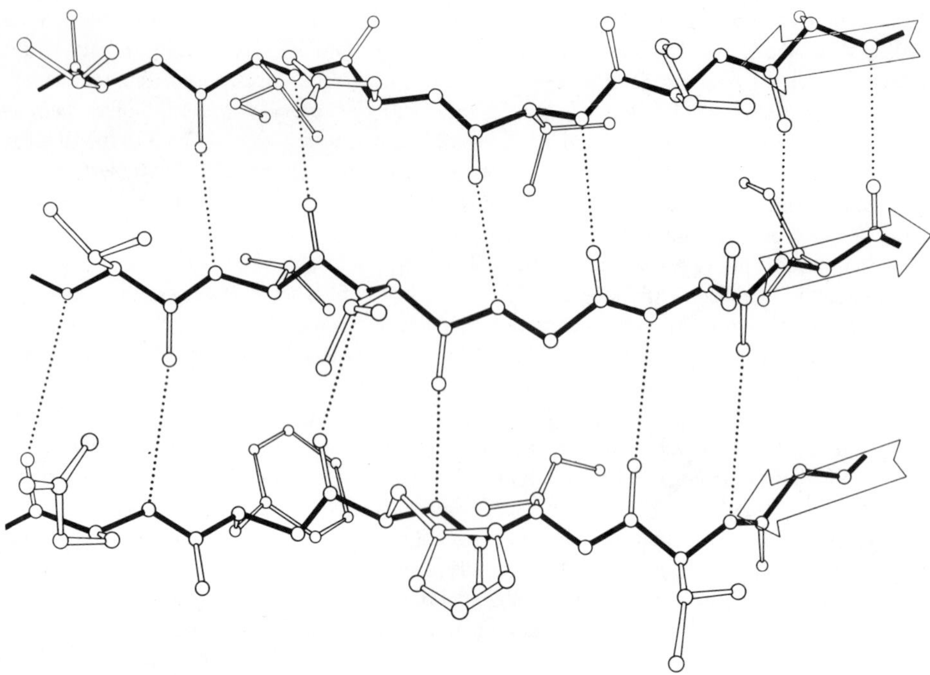

Figure 18. Drawing of a small piece of antiparallel β sheet (from SOD), illustrating the alternately narrow and wide pairing of H-bonds and the side-chain alternation above and below the plane of the sheet.

results from a strong tendency for β sheets to be either pure parallel or pure antiparallel. Mixed sheets occur, but not at anything like random expectation.

One of the most characteristic and interesting features of β structure is the twist both of strands and of whole sheets (Chothia, 1973). The planes of the peptides (and therefore also of H-bonding) twist in a right-handed sense along the strand direction. The twist can be almost absent in the center of some large sheets, like the BCL in Fig. 10, or it can be as much as 30° per residue in isolated two-stranded ribbons like the one in Fig. 23; most cases are intermediate. Right-handed twist puts the ϕ,ψ values to the right of the diagonal "$n = 2$" line on the Ramachandran plot. Twist values for individual residues vary quite a lot, especially in antiparallel sheets. Twist produces the flaring, saddle-shaped swirl of parallel β sheets (see Fig. 20) and makes the strands spiral around the axis of a β barrel (Fig. 21). Twist somewhat relieves the close contact of side chains directly opposite one another on neighboring β strands. Strong twist also affects pairs of side chains that are diagonally offset from each other by two residues on neighboring strands: if the strands go left and right (as in Figs. 18, 19, or 24), then upper-right-to-lower-left diagonal pairs facing the observer can contact one another, while pairs on the other diagonal are widely separated.

The quantity $\phi + \psi$ is a rough measure of twist and increases in the direction shown in Fig. 22, whereas $360° + \phi - \psi$ is a rough measure of the extent to which the β strand is "pleated" up and down out of the plane of H-bonding. Pleating would be zero for a hypothetical β sheet in fully extended conformation ($\phi = -180°$, $\psi = 180°$) but is fairly pronounced for real antiparallel β sheets and a bit more so, on average, for parallel ones. Pleating does not change the overall appearance of β strands very dramatically, but it must vary to accommodate the variation in H-bond spacing needed near the center versus the edges of twisted β sheets (Salemme and Weatherford, 1981).

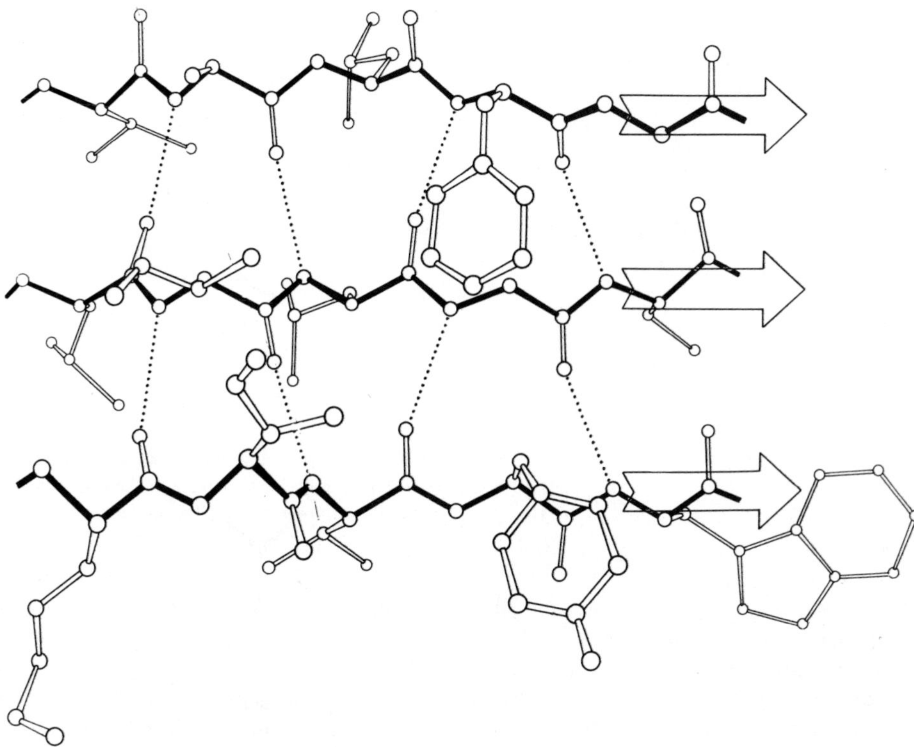

Figure 19. Drawing of a small piece of parallel β sheet (from FXN), showing the evenly spaced but alternately slanting H-bonds.

In addition to the ubiquitous right-handed twist, β sheets can either be "curled" along the strand direction or "arched" perpendicular to it. Arching is mainly a matter of nonlinear H-bonds and does not necessitate any changes in strand conformation; it must be fairly pronounced to produce a closed β barrel (Fig. 21), and it may be present to some extent even in isolated sheets (e.g., the 4 FXN sheet is arched slightly around the A and D helices, as seen on

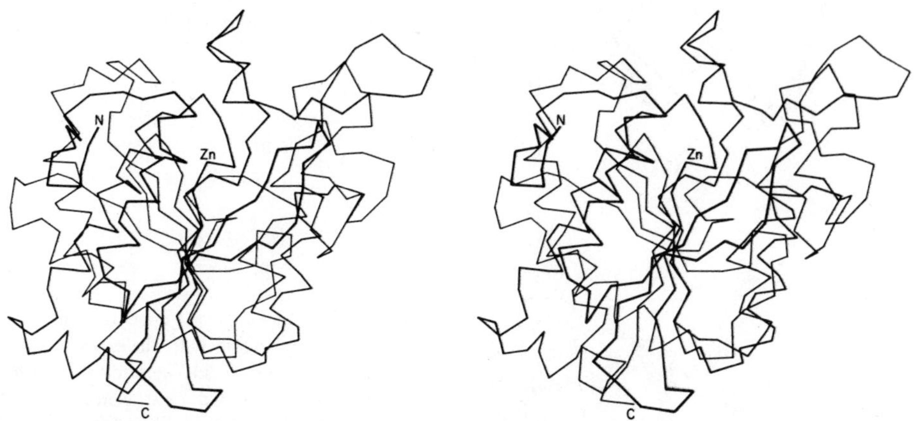

Figure 20. Backbone stereo of CPA, showing an edge view of the large, twisted β sheet.

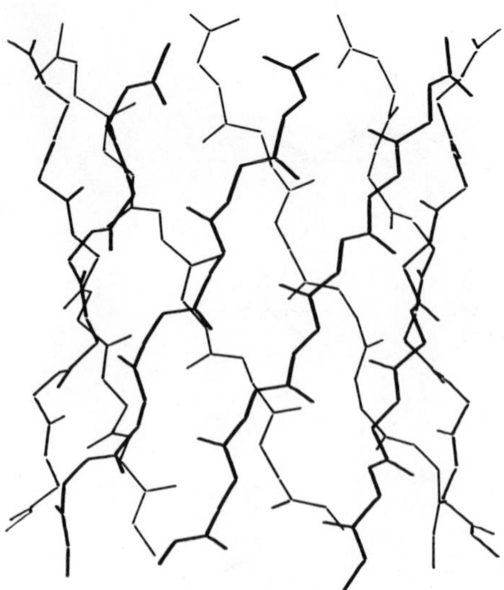

Figure 21. An idealized eight-strand parallel β barrel viewed from the side, showing how the strands spiral around the approximate cylinder of the barrel.

the right in Fig. 9). Moderate arching can quite readily be accommodated by the hydrogen bonds, but extreme arching produces collisions between the carbonyl oxygens on one strand and the β carbons on the next strand.

Curling, on the other hand, involves no change in H-bond geometry but an alternating perturbation of ϕ,ψ values (Chothia, 1983). Curl is easiest to visualize, and in fact also most common, for two-stranded antiparallel ribbons like the one in Fig. 23. If the strands were completely untwisted, then a curled β ribbon would return back on itself in a circle; the real cases are always quite twisted, however, so they form a large, open spiral. The probable cause of curl is the "sidedness" of two-stranded antiparallel ribbons. Since side-chain direction alternates in register on the two strands, the side chains of one side of the β ribbon are between close pairs of H-bonds (the "close-pair" side), and those of the other are between wide pairs (the "wide-pair" side). Figure 24a illustrates this geometry, and Fig. 24b shows the resulting

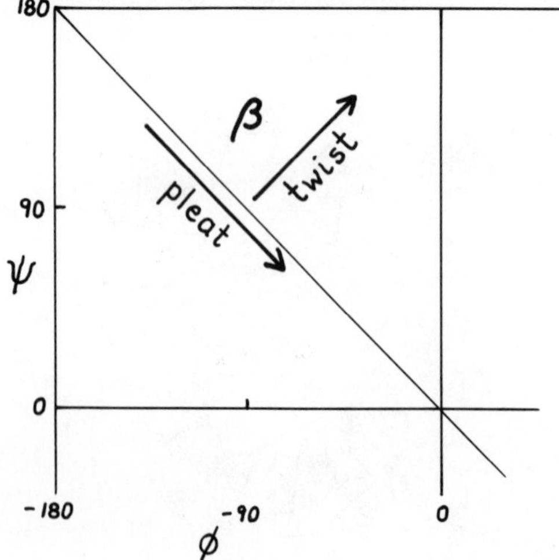

Figure 22. Drawing of the upper left quadrant of the ϕ,ψ plot, which includes the extended β conformations. Arrows mark the direction in which strand pleat and strand twist increase, respectively, parallel and perpendicular to the diagonal.

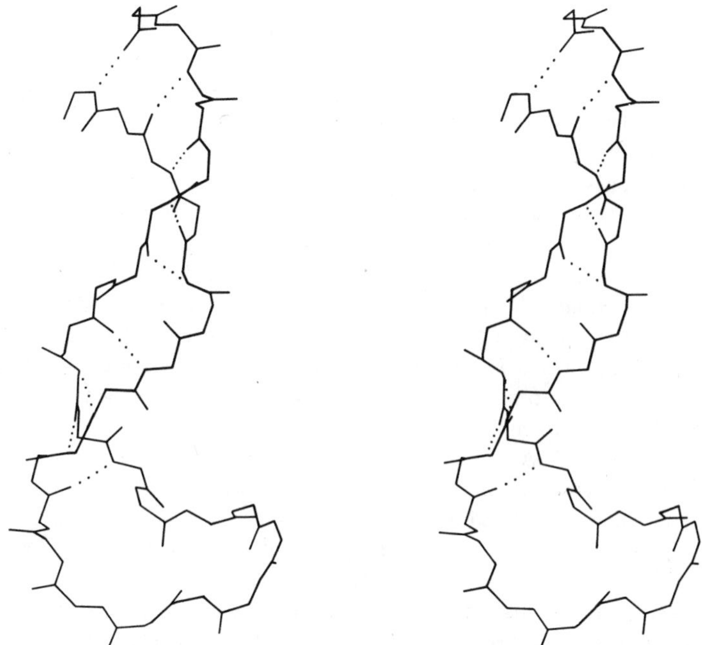

Figure 23. A strongly twisted, two-strand antiparallel β ribbon (LDH 263–294).

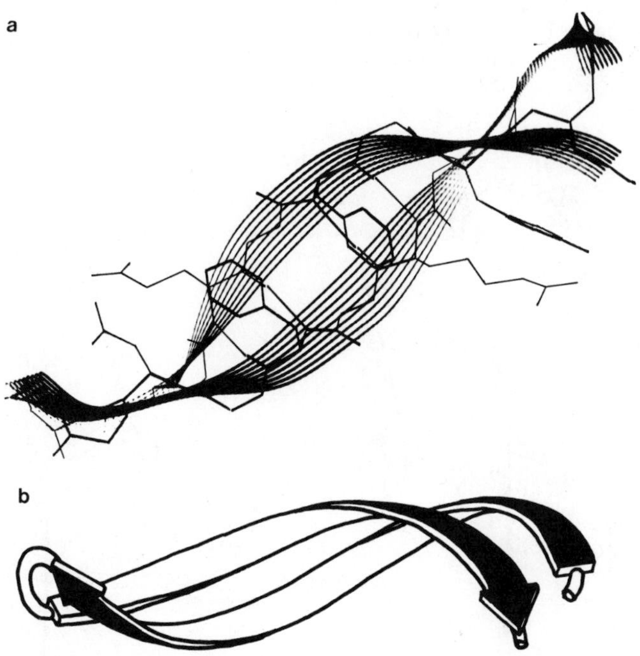

Figure 24. The "sidedness" of a two-strand β ribbon. (a) A detailed view (from PTI), in which it can be seen that the side chains on one side are between a narrow pair of H-bonds and on the other side between a wide pair. (b) A schematic of the same twisted pair with the hydrophobic, narrow-pair, concave side shown white and the exposed, wide-pair, convex side shown black. Note that the view is from the concave side and that the arrow directions are clockwise.

sidedness of the ribbon, with the wide-pair side darkened. The constraints of close versus wide pairs of H-bonds are a bit different, and the ϕ,ψ preferences of the residues between them differ slightly in such a way as to curl the ribbon toward the close-pair side (Salemme, 1983). Therefore, such a β ribbon promotes compact, stable structures if its hydrophobic side chains (especially large ones that interact strongly with each other) are on the concave, close-pair side. This restriction means that for β ribbons that either are isolated in the final structure or form an isolated pair during folding, their topology must be such that strand direction is up on the left and down on the right when viewed from the hydrophobic interior (see Fig. 24b). In Section IV.D we see how this rule affects overall tertiary structure in all-β proteins.

The parameters we have been discussing (twist, pleat, arch, and curl) can all vary significantly and still be accommodated within a regular network of β-sheet hydrogen bonding. However, as with a proline in α helix, there can be more violent local disruptions of hydrogen bonding in β structure. By far the most prevalent such local disruption is the β bulge (Richardson *et al.*, 1978). A β bulge can be thought of as the insertion of an extra residue into one strand, so that between a pair of H-bonds there is now one residue on the normal strand but two residues on the bulged strand (see Fig. 25). Bulges are common in antiparallel β structure (more than one per sheet, on average) but rare in parallel β. Usually they are located between a close pair of H-bonds rather than between a wide pair. The extra residue puts the hydrophobic–hydrophilic side-chain alternation out of register across the bulge, an effect that is sometimes recognizable in the sequence (e.g., TFTI(SS)LQ in REI V_L). To accommodate the surrounding H-bond pattern, usually one of the two bulge-strand residues stays close to normal β conformation while the other is close either to α-helical conformation (a "classic" bulge) or close to left-handed 3_{10} conformation (a "G-1" bulge). The single residue on the opposite strand is usually near polyproline conformation in order to match the greatly accentuated right-handed twist produced by a β bulge (see Fig. 26). That strong local twist is used to help close small β barrels or is found at the base of a compact loop or β ribbon to help it change direction.

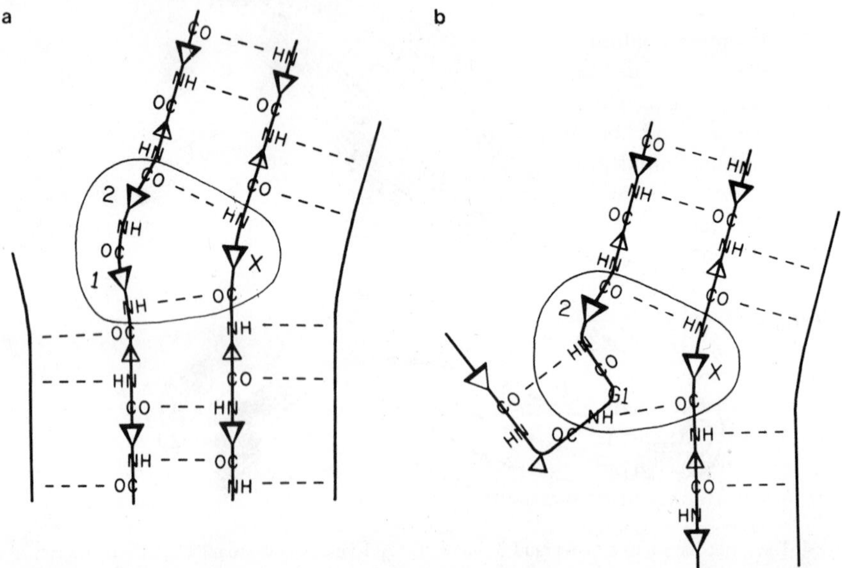

Figure 25. Diagrammatic illustrations of (a) a classic β bulge and (b) a $G_1\beta$ bulge. Small triangles represent side chains below the sheet, and larger triangles those above it.

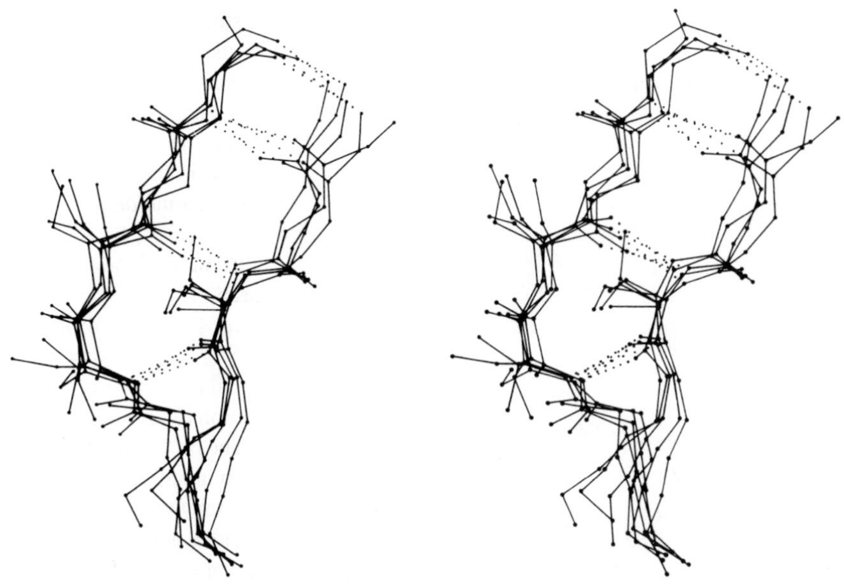

Figure 26. Five superimposed examples of classic β bulges (CHA 41–2, 33; 86–7, 107; CNA 107–8, 196; CAC 90–1, 120; SNS 15–6, 24). Side chains are shown (out to C_γ) only for the three positions in the bulge.

Bulges can mitigate the damage done by single-residue insertions or deletions in β strands, at least when they occur near an end or an edge of the β sheet.

C. Nonrepetitive Structure: Turns, Connections, and Compact Loops

The only protein conformations in which the same ϕ,ψ angles repeat for each consecutive residue are the right-handed helices and extended β strands. Large portions of protein structure, however, are made up of well-ordered but nonrepeating conformations. These have often been referred to as "coil" or even "random coil," which unfortunately has the connotation of disordered, mobile, unfolded chain. Disordered structure is often important (see Section VIII), but it is quite different from the nonrepetitive regions, which can be just as well-ordered, compact, and stable as pieces of repetitive secondary structure. Nonrepetitive structure includes both backbone H-bonds (mainly in turns or bulges) and frequent side-chain-to-main-chain H-bonds. It is of two fairly distinct general types: connections, or straps, that get from one piece of secondary structure to another and compact loops. A turn is both the shortest possible connection and the shortest possible compact loop, but turns also occur within larger loops or connections.

Turns can be defined in many different ways, and it is likely that they can best be predicted by a nonspecific definition such as direction changes or breaks between other pieces of structure. However, in order to describe turns in detail, we consider specific patterns of H-bonding and ϕ,ψ conformations.

The narrowest, classic definition of a tight turn (Venkatachalam, 1968) is in terms of a four-residue corner with a hydrogen bond between the CO of the first residue and the NH of the fourth residue. There are at least five distinct combinations of ϕ and ψ for the two central residues that can allow formation of that H-bond, and approximately those same five combinations also characterize most of the non-H-bonded turns. The only turn type that has both ϕ,ψ

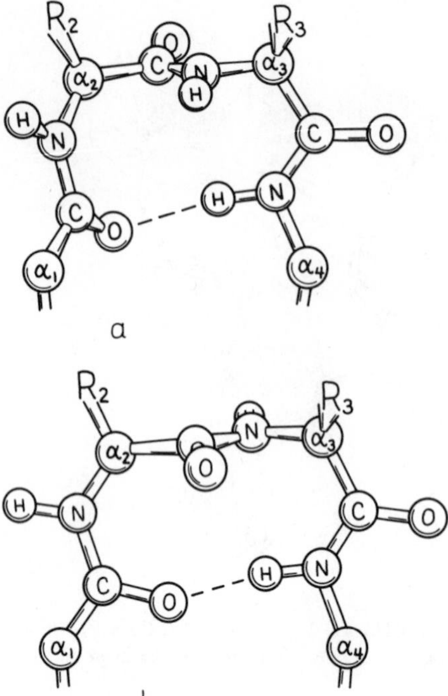

Figure 27. (a) Common turn (type I). (b) Glycine turn (type II), in which R_3 is most often Gly.

pairs with freely permissible values is the "common" turn* with residue 2 α-helical and residue 3 in 3_{10} conformation (Fig. 27a). This is indeed very much the most common type of turn, since any residues are permitted. The second most common type is the "glycine" turn (Fig. 27b), with residue 2 in poly-Pro conformation and residue 3 in left-handed 3_{10} conformation (near +90°,0°). An $\text{L}3_{10}$ conformation is favorable only for glycine, since a C_β bumps the carbonyl oxygen of the central peptide; about 60% of the glycine turns have Gly in position 3. Glycine turns have all four α-carbons and the H-bond very nearly in a plane. The common turn, on the contrary, is significantly nonplanar, with its four α-carbons making a dihedral angle of about +45° and the carbonyl group of residue 1 almost perpendicular to the average turn plane.

The first and fourth residues of a turn are in approximately the right relative position to be on antiparallel β strands, and one classic role of turns is to connect two consecutive antiparallel β strands in a hairpin connection. However, there is an additional complication for tight β hairpins (Sibanda and Thornton, 1985). The right-hand twist of β strands would form a negative dihedral angle at the end of the hairpin; for an isolated two-strand antiparallel ribbon, the twist would typically give a dihedral angle of about −45° for the end four α-carbons. Unfortunately, the common turn has a dihedral angle of about +45°, and the opposite twists would fight each other. This means that the typical textbook picture of a hairpin turn as shown before the Thornton and Sibanda paper is actually impossible.

*Turn conformations have usually been referred to by numbered "types" (Lewis *et al.*, 1973; Chou and Fasman, 1977); however, as we have found that few people remember the numbers, we use descriptive terms instead. The translation is as follows: "common," type I; "glycine," type II; "inverse common," type I′; "inverse glycine," type II′; "miscellaneous," type IV; "*cis*-Pro," type VI. We have argued (Richardson, 1981) that types III, V, and V′ do not form distinct clusters.

In the antiparallel sheets of real proteins, a majority of the connections are topologically hairpins, but relatively few of these are tight hairpins, very seldom more than one per sheet. Tight hairpins solve the problem of incompatible twist in one of several ways. The inverse common and inverse glycine types of turns have backbone conformations that are the mirror images of the common and the glycine turn, respectively (see Fig. 28), and therefore also have opposite C_α dihedral angles (around $-45°$ for inverse common and $-10°$ for inverse glycine). These inverse turns are extremely rare elsewhere, but they are essentially the only way for a tight β hairpin to form an n to $n + 3$ H-bond (Fig. 29). Bulged hairpins are also common, either with one extra residue forming first a common turn and then a G1 bulge to correct the twist (Fig. 30) or else with two extra residues forming one α-helical H-bond and then a G1 bulge (see Richardson, 1981, Fig. 55). There are also cases of common turns that break the H-bond and are thus somewhat less "tight" hairpins.

Nonrepetitive structure in the general sense is made up of a very wide variety of loops and connections. These range from large, compact "omega loops" (Leszczynski and Rose, 1986), which have close termini, good internal packing interactions, and many side-chain-to-main-chain H-bonds, all the way down to relatively extended "straps" for getting from one piece of repetitive structure to another. Figure 31 shows a large compact loop from Cu,Zn superoxide dismutase, which binds the structural Zn and has more H-bonds and more internal contact surface than the same number of residues would have if arranged in either an α helix or a β hairpin. As opposed to secondary structures, which are mainly defined and stabilized by their backbone interactions, it is the side chains that define and stabilize compact loops. It is the side chains that make them compact, and it is the side chains that form most of their H-bonds. As tabulated by Baker and Hubbard (1984) and mentioned in Section V, a given side-chain type has strong preferences for H-bonding at a certain distance and direction along the backbone, and these patterns influence loop configurations. Although compact loops are usually on the

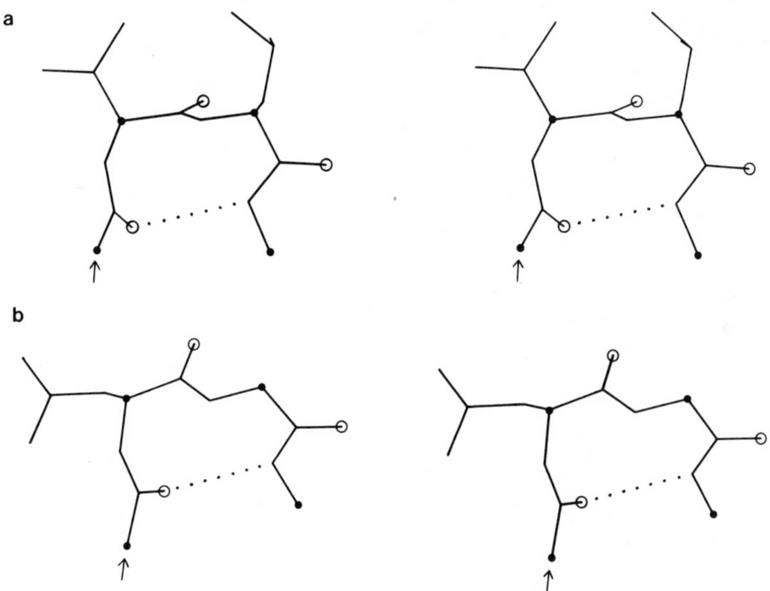

Figure 28. Stereo drawings of particular examples of (a) common turn (TLN 12–15) and (b) inverse common turn (PAP 183–6), or type I′.

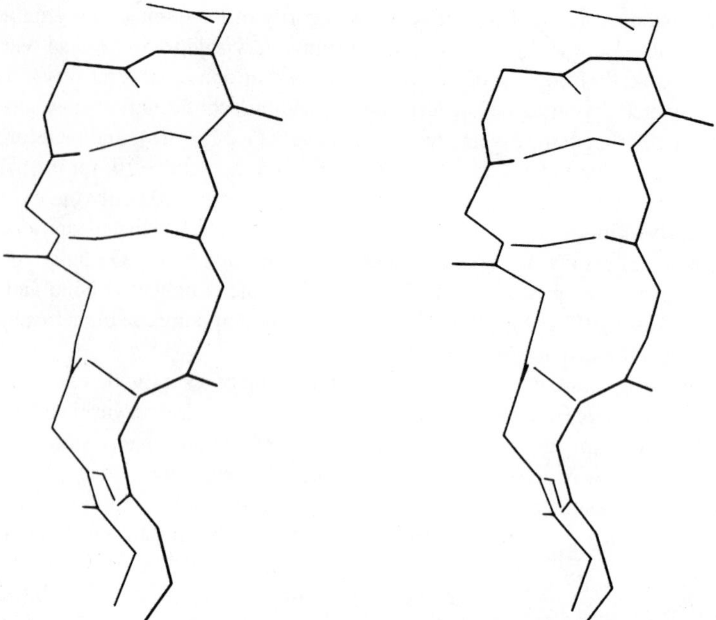

Figure 29. Stereo of a tight β hairpin between adjacent antiparallel β strands (2SOD 20–29), utilizing an inverse glycine turn.

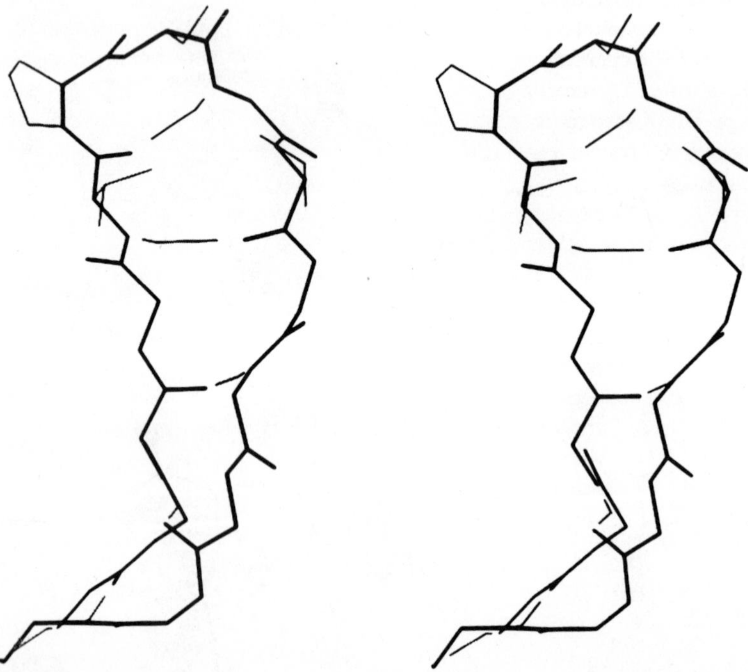

Figure 30. Stereo of a bulged three-residue hairpin (1ACX 68–81), which uses a common turn (P,S) followed by a G_1 bulge.

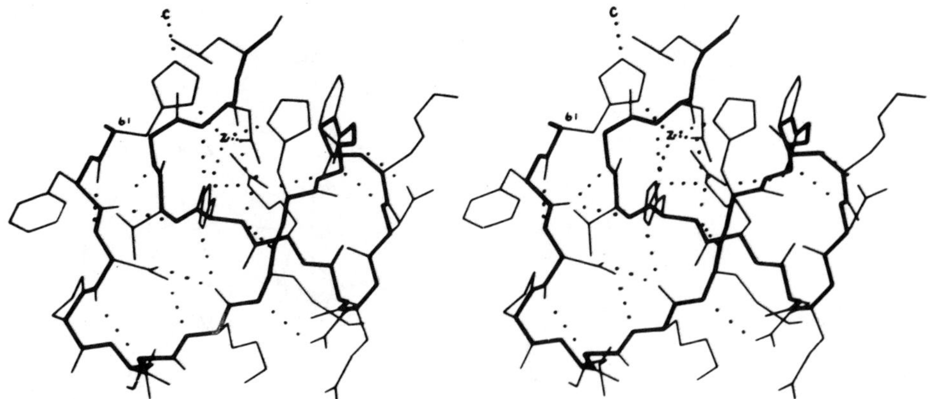

Figure 31. Stereo of a large compact loop (from SOD) showing the internal H-bonds and the bound Zn.

surface and sometimes disordered, they can also be among the most highly ordered parts of a protein and are certainly not "random" in any reasonable sense of the word. Their non-repetitive conformations and reliance on specific side-chain interactions, however, have made them resistant to either description or prediction.

The various sorts of "straps" can be categorized by the types and relative positions of secondary structure they connect. Short antiparallel and short perpendicular α–α connections have been surveyed by Efimov (1982) in terms of hydrophobic residue positions and geometrical packing constraints. A spacing of i to $i + 6$ between large hydrophobic residues is optimal for an antiparallel α connection that turns left when viewed from the outside, whereas $i + 5$, $i + 7$, or $i + 8$ connections are more likely to turn right. This asymmetry originates in the handed pinwheel of C_β directions around an α helix (Fig. 13). Short α–α connections have an "S" shape when viewed from the end (Fig. 32). They most often show the helix cap conformations and residue types that are most common for helices in general, that is, Gly in $L3_{10}$ conformation at the C-cap (see Section V.A) and Asn, Asp, Ser, or Thr in β or poly-Pro conformation at the N-cap (see Section V.G). The C-cap and N-cap residues can be adjacent, can have a single extended residue between them, or can have more complex loops between them.

Both β–α and α–β connections have been surveyed by Edwards *et al.* (1987). They found considerable variety of length and conformation in general but identified four families of short β–α or α–β connections, each with reproducible length, conformation, and residue preferences. The most common of these families, the adjacent β–α_1, is illustrated in Fig. 33. Again, these connections make use of the common types of helix terminations.

We have already discussed β hairpins, for which the short examples are tight turns or turns plus bulges, and which have an extremely high content of glycine. The other type of strap found in antiparallel β structure is the Greek key connection, or β arch, which joins nonadja-

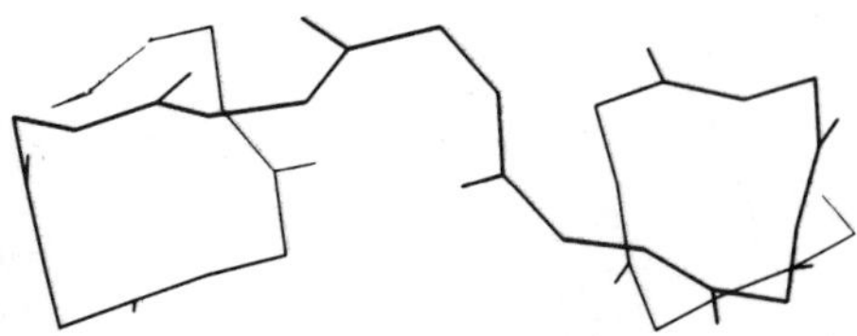

Figure 32. A short helix–helix connection (1HMQ 33–45), showing an overall "S" shape.

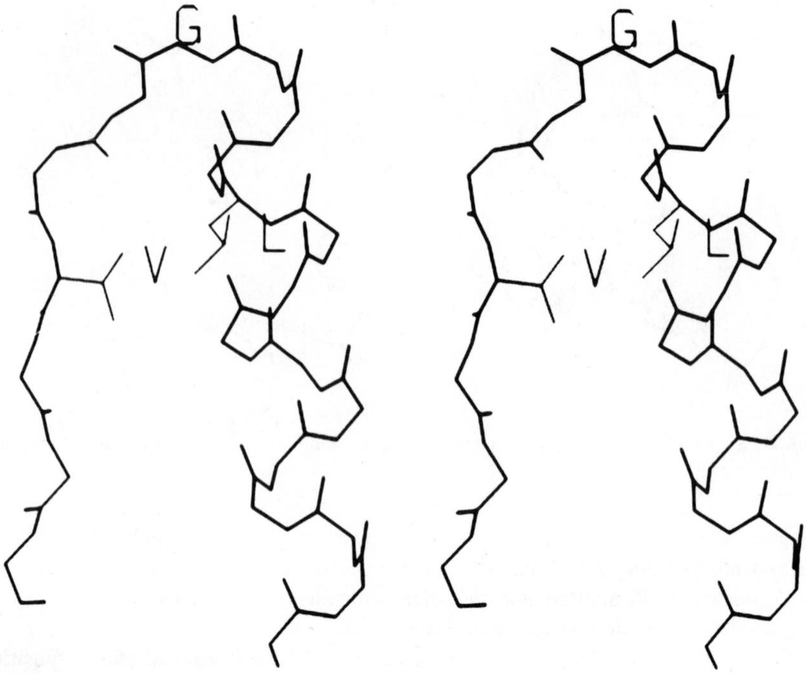

Figure 33. Stereo of a one-residue β–α connection (1CTF 100–120), with labels for the semiinvariant Gly in the connection and hydrophobics at G − 4 and G + 3 (see Edwards *et al.*, 1987).

cent β strands at the same end of the sheet and is usually found arching across one end of a Greek key β barrel (see Section IV.D). Taking as a reference point the last inward-pointing β strand residue at each end of the Greek key connection, the most common length is five residues between those reference points. The central portion of the Greek key connection is often in extended conformation with peptide planes parallel to the protein surface and side chains perpendicular to it; sometimes there are even β-type H-bonds to a neighboring connec-

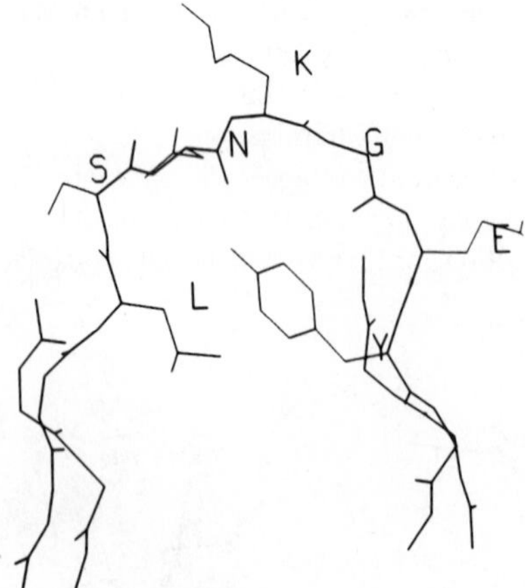

Figure 34. A Greek key connection with five residues between the inward-pointing reference residues in the β barrel (here L74 and Y80 of PCY) and with the most common starting-corner conformation (near α) and the most common ending-corner conformation (Lβ Gly).

tion (e.g., in the TBSV "tip" domain). Although the carbonyls do not need to change direction in this case, there is usually a relatively sudden change of backbone direction at each end where it joins the β strand. Such a sharp corner cannot easily be formed at the inward-pointing reference residue but typically is either one or two residues up from there. At the beginning of the connection the corner is almost always a sharp one, with more than half of the examples having the corner residue (most often a Ser or Pro) in α or near-α conformation and located one past the inward-pointing reference. At the end of the Greek key connection about half the cases have a Gly in Lβ conformation, making a relatively smooth corner at the position two before the reference residue. Not surprisingly, the most frequent Greek key connection type combines these starting and ending conformations and is either four or five residues long, as shown in Fig. 34. However, "strap"-type connections often seem to have relatively independent mixtures of ending conformations, and perhaps they can best be considered as separate halves with a variable-length strap across the middle.

IV. TERTIARY STRUCTURE: VARIATIONS ON A FEW HARMONIOUS THEMES

In a real sense, overall tertiary structure is the ultimate goal of prediction. One would be extremely pleased with a prediction that was accurate at the level of a schematic ribbon drawing, and a necessary, although hardly a sufficient, condition of such success is knowing what types of structure have been found in the past and therefore are the most likely possibilities for a new protein.

A. Domains

A complication that must be dealt with first is the issue of how the protein should be divided into domains. Since the different domains often have quite different types of tertiary structure (e.g., Fig. 35), they need to be described and classed separately and, if possible, should be predicted separately. Even with domains of the same structure type (e.g., Fig. 109), the rules for tertiary organization (such as number and placement of β strands) apply at the level of the individual domain and not the entire protein. Multiple domains should be considered for any sequence that suggests an internal repetition or that is longer than about 200

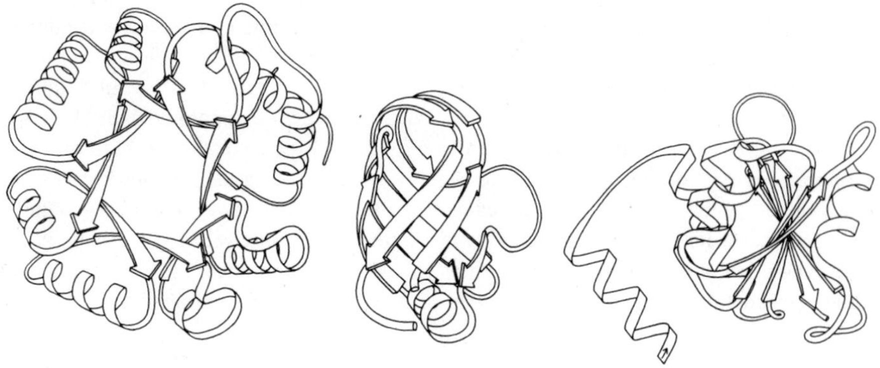

Figure 35. The three domains of pyruvate kinase (PYK), as an example of a protein whose domains are very different.

residues (250 for α/β structures or 100 for a disulfide-rich sequence) or if a preliminary prediction of secondary structure shows strong segregation into 100-residue or larger segments, each of which is either all α, all β, or alternating α/β.

Domains have been defined in a number of significantly different ways (e.g., Wetlaufer, 1973; Richardson, 1981; Rose, 1978), and in particular there is some disagreement on how small they can be. For the purposes of prediction this is not too important, since one clearly wants to deal together with all parts of a single sheet or a single barrel, and therefore the domains should be fairly large. However if spatially contiguous domains are made up from pieces separated in the sequence, a serious problem exists; fortunately this is fairly rare. In prediction one has no choice but to deal with sequence-contiguous groupings but should bear in mind any rules about with pieces are most likely to deviate: for instance, "arms" at the chain termini often form part of an adjacent domain. In the following subsections, we describe types of tertiary structure at the level of single domains, classified into four major groups. A more detailed description of the types, and many additional examples, can be found in Richardson (1981).

B. All α

The first major category of tertiary structure consists of α-helical proteins. A priori, it would seem possible to construct proteins that had their α helices arranged in parallel layers or groups, but in fact real all-helical proteins are always either antiparallel or nearly perpendicular. This can perhaps be explained by a preference for near-neighbor helix dipoles (see Section V.H or Hol *et al.*, 1978) or else by a preference for near-neighbor helix interactions during folding (Richardson, 1981).

The simplest type of antiparallel α structure is an approximately cylindrical bundle, or cluster, of helices with an up-and-down, near-neighbor connectivity (Argos *et al.*, 1977; Weber and Salemme, 1980). Since helices are rather stiff and weighty, four of them make a good, stable structure, and four-helix clusters are quite common (Fig. 36); 80–90% of them have simple up-and-down connectivity, with two well-documented counterexamples so far (ferritin: Rice *et al.*, 1983; growth factor: Abdel-Meguid *et al.*, 1987). It is theoretically somewhat easier to make a neat hairpin connection between adjacent antiparallel helices if the chain turns right at the top of the first helix, as seen from the solvent direction (Efimov, 1982). However, our only example so far of a dimer formed from subunits that are simple helix pairs is the ROP protein (Fig. 37b; Banner *et al.*, 1987), which does not have the favored type of hairpin. The ROP subunits are related to each other by an exact twofold axis that is crosswise to the long direction of the cluster; alternatively, such a dimer might be related by a lengthwise twofold axis (Fig. 37a). If there were a gene fusion in such a dimer, one with a lengthwise twofold axis could easily be joined with a short covalent connection into the up-and-down, +1,+1,+1 type of four-helix bundle (Fig. 36), whereas a ROP-type dimer would require a long crossover connection* and would end up with $+1,-2x,-1$ topology like the ferritin subunit (Fig. 38b). The growth hormone bundle (Fig. 38a), by the positioning both of its two long connections and of introns in related gene sequences, suggests yet another possible history in which the central pair of helices came first. The prevalence of up-and-down bundles may indicate that a lengthwise symmetry works better or, alternatively, just that helix-pair dimers that could be joined by short connections have predominated over the course of protein evolution. Presumably because of the preferred direction of hairpin turning, there is a prefer-

*Whereas a hairpin connection joins two antiparallel helices or two antiparallel β strands at their close ends, a crossover connection is between parallel elements and goes the long way around to the other end.

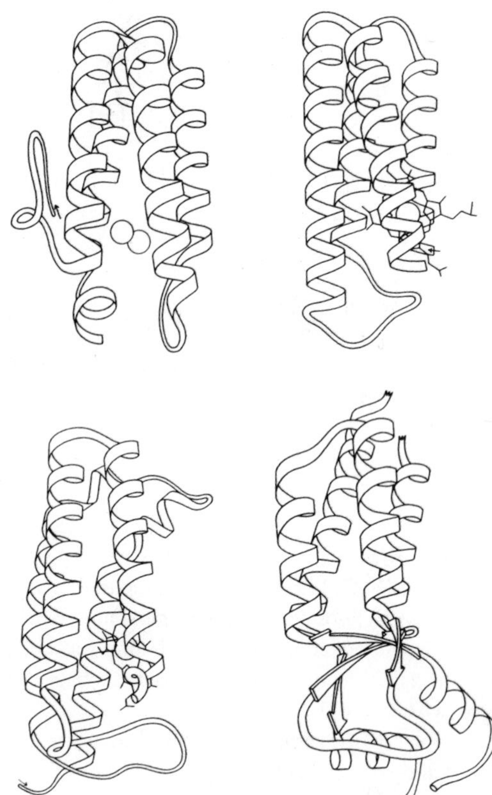

Figure 36. Ribbon drawings of four examples of up-and-down four-helix bundles (MHR, 56B, CCY, TMV).

ence for helix bundles to turn right at the top of the first helix (Fig. 39), but there are several counterexamples [e.g., ROP, cytochrome b5 (1B5C)].

There are a few three-helix bundles (e.g., protein A fragment; Deisenhofer *et al.*, 1978) and quite a few larger helix clusters. All known antiparallel clusters with five or more helices have at least one non-near-neighbor connection except for the seven helices in bacteriorhodopsin, which probably have an up-and-down topology (Engelman *et al.*, 1980). The bundles with a non-near-neighbor connection are typified by the "globin fold" in which the chain skips over the H helix in a +2 connection between helices F and G (Fig. 40). These structures can be

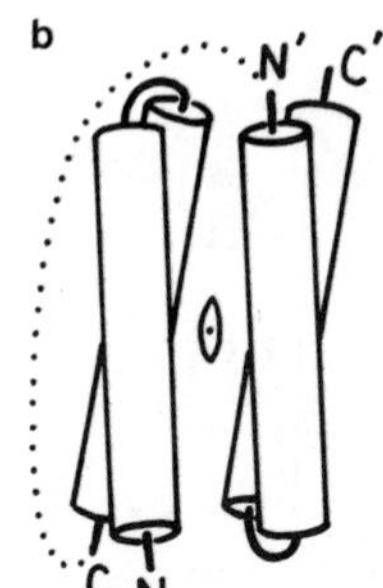

Figure 37. Diagram of two alternative arrangements for helix-pair dimers, with dotted lines to show the type of connection needed to join them into a single covalent chain. (a) The dimers are related by a lengthwise twofold axis, and they can be joined by a short connection into an up-and-down helix bundle (like those in Fig. 36). (b) The dimers are related by a crosswise twofold axis (as is the case for the ROP dimer), and they can only be joined by a long crossover connection of the type seen in ferretin and growth hormone.

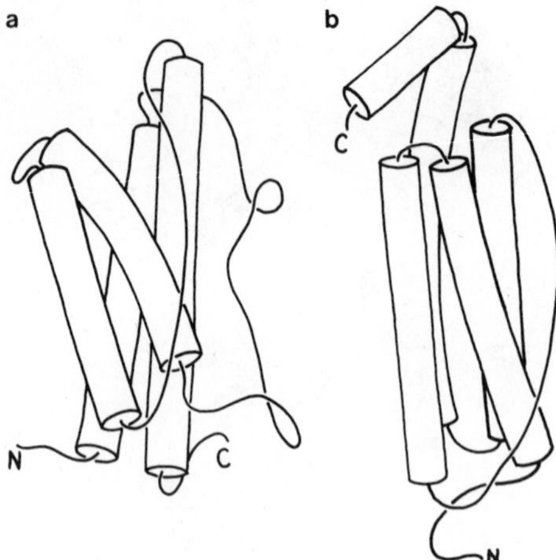

Figure 38. Layouts for two four-helix bundles with crossover connections: (a) porcine growth hormone; (b) apoferritin.

called ''Greek key helix bundles'' since they are formally analogous to the Greek key β barrels of Section IV.C, but they differ in having no strong handedness preferences and only a single non-near-neighbor connection (almost always a +2 connection). The Greek key helix bundles have almost no crossover connections [except for thermolysin (TLN)] of the sort seen in ferritin and growth hormone, which probably means they did not evolve from dimers with crosswise twofold symmetry.

Our set of examples now includes some integral membrane proteins, which apparently favor antiparallel helix bundles even more than globular proteins do. They are organized in remarkably similar ways, except that they are more accurately antiparallel in order for each helix to span the membrane, and their uniformly hydrophobic surroundings let them be much less compact in the crosswise direction (see Section VI).

Another type of organization seen in all-α proteins is based on successive helix pairs that are close to perpendicular and touch at their sequence-adjacent ends. The great majority of adjacent helix pairs are arranged with the two helix axes intersecting near their common ends, as though they had formed by bending a single, longer helix. The angle between the axes varies from nearly 180° for bent helices to nearly 0° for antiparallel pairs, but with numerous examples at intermediate angles. The major exceptions to the intersecting-axis geometry form a distinctive type of slightly offset and overlapped right-angled corner, which we call a helix lap joint (see Fig. 41; Richardson and Richardson, 1988b). These were first described for the ''E–F hands'' of reversible Ca-binding proteins (Kretsinger and Nockolds, 1973) and then discovered again in their sequence-reversed form as DNA-binding helix pairs (e.g., Steitz *et*

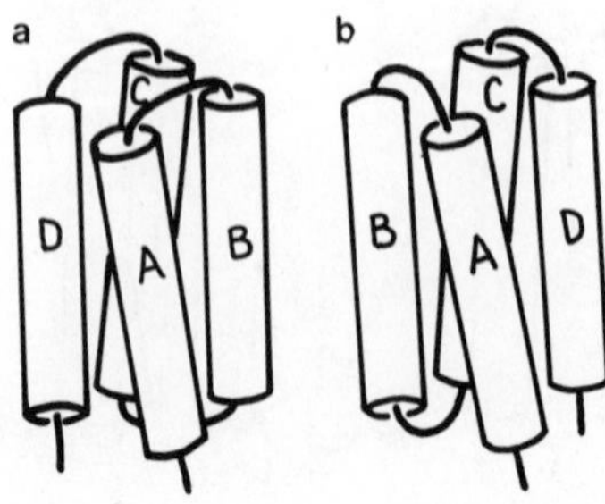

Figure 39. Diagrams of the two directions in which an up-and-down helix bundle can pack: (a) the more common version turns right at the top of the first helix, as in the proteins of Fig. 36; (b) the alternative version turns left at the top of the first helix, as seen in cytochrome b_5 and ROP protein.

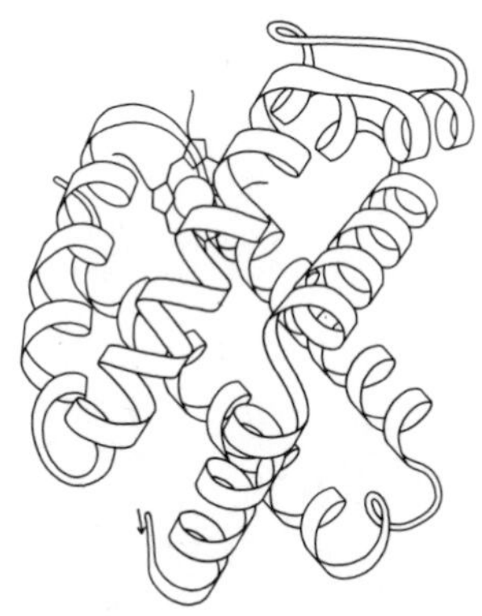

Figure 40. Ribbon drawing of the hemoglobin β subunit, an example of a Greek key helix bundle with one non-near-neighbor connection (between helices F and G).

al., 1982). This particular relationship of adjacent helices is seldom encountered except in these ion-binding sites, presumably because high charge at the loop is needed to stabilize the arrangement. The sequence reversal, which of course reverses the direction of the helix dipoles, is accompanied by reversal of the side-chain charges at the corner: E–F hands use Glu and Asp to bind Ca^{2+}, whereas DNA-binding motifs use Lys and Arg in binding the negative

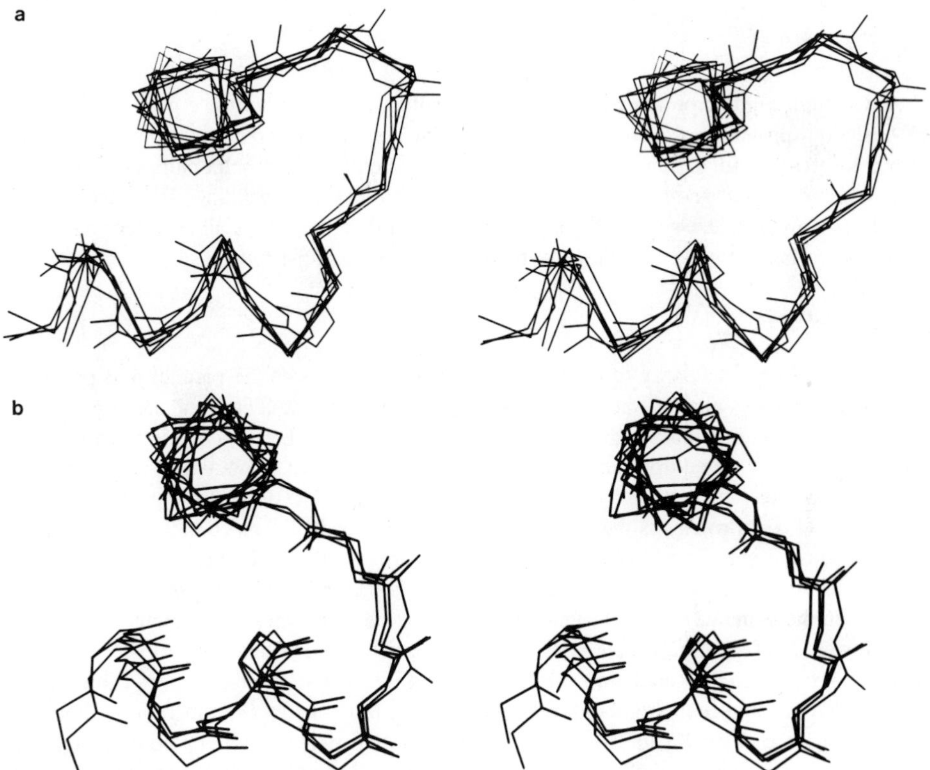

Figure 41. Offset perpendicular helix corners (''lap joints''): (a) six superimposed examples of Ca-binding ''E–F hands'' (from CPV, TNC, and CLN) and (b) five superimposed examples of DNA-binding two-helix motifs (from CRO, GAP, LRP, 1WRP, and 3WRP).

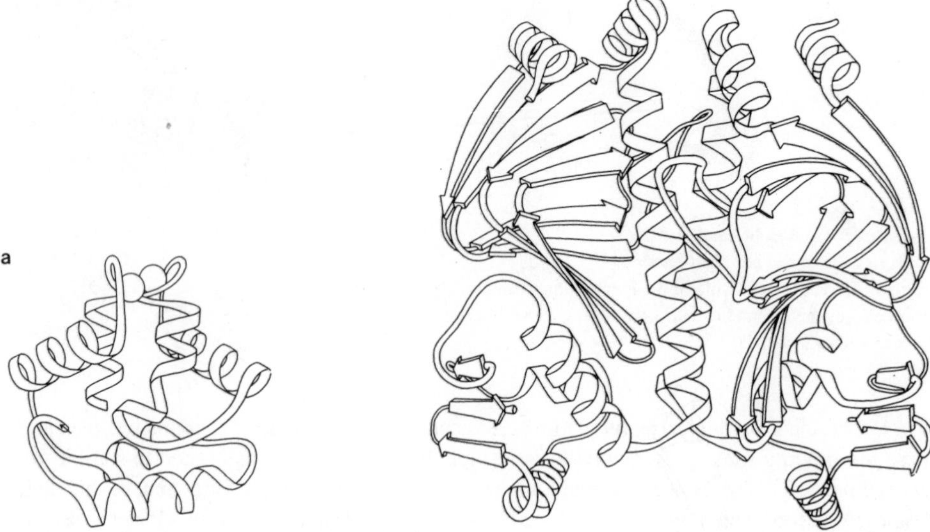

Figure 42. Examples of two proteins containing offset perpendicular helix corners: (a) carp Ca-binding protein (CPV) and (b) CAP protein (GAP).

DNA. Calcium-binding proteins consist predominantly of helix lap joints (e.g., Fig. 42a), but DNA-binding proteins so far use only one lap joint per subunit, combining that either with other α helices or with antiparallel β sheet (Fig. 42b). Perpendicular helix pairs with the more usual intersecting-axis geometry can combine with each other or with antiparallel helix pairs. It should certainly not be assumed that any small protein that predicts as all α is necessarily an up-and-down helix bundle, but they are the most common single type.

C. Parallel α/β

The next major category of overall tertiary structure includes the parallel α/β proteins, which approximately alternate helices and β strands along their sequences and are organized in multiple layers, with parallel β sheets on the inside and layers of helices covering them. These structures are constructed from another of the important primitive combinations or "supersecondary" structures: the β–α–β loop or crossover connection. The dominant organizing principle of parallel α/β proteins is the overwhelming right-handedness (much better than 99%) of β–α–β loops (Richardson, 1976; Sternberg and Thornton, 1977). Figure 43 shows the two possible forms schematically. In the known protein structures, these crossover connections never lie in the plane of the sheet (which would make the handedness ambiguous), and out of many hundreds of examples there is so far only one unassailably documented case of a left-handed β–α–β loop (in subtilisin, 1SBT, residues 45–95). For short crossover connections, the β-sheet twist makes the right-handed connection fit more neatly (Sternberg and Thornton, 1977; Chou *et al.*, 1989), and there are possible folding mechanisms that could couple the handedness to the right-handedness of the central α helix (Richardson, 1976); however, neither of these effects seems adequate to account for a 99% regularity. Perhaps the very long and convoluted crossover connections (a few of which encompass an entire domain) are right-handed primarily because they evolved from short β–α–β loops.

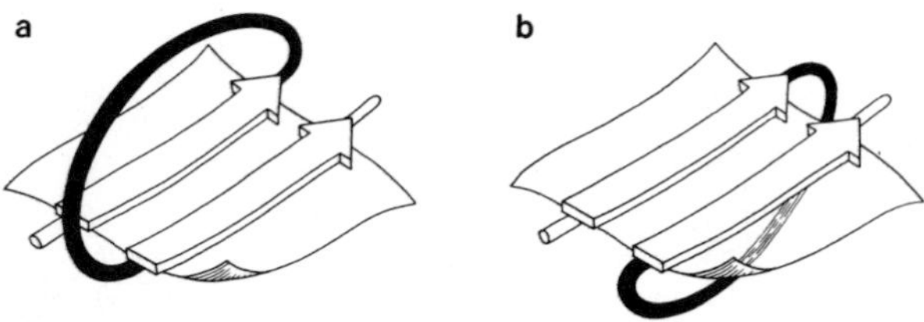

Figure 43. (a) A right-handed crossover connection. (b) A left-handed crossover connection. Note that virtually all actual examples are right-handed.

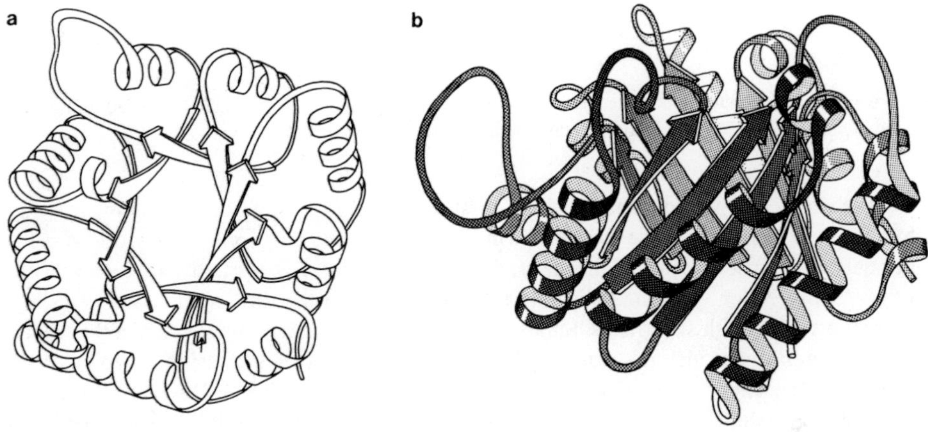

Figure 44. Ribbon drawings of triose phosphate isomerase (TIM) as the classic example of a singly wound parallel β barrel: (a) end view; (b) side view.

The simplest specific arrangement for parallel α/β structures is the singly wound parallel β barrel, typified by triose phosphate isomerase (Fig. 44; Richardson, 1977). More than a dozen of these structures have been found so far, all with eight parallel strands in the central β barrel and a concentric cylinder of eight α helices formed by right-handed crossover connections. The topology is invariant*, with each connection moving over by one strand, and always in the same direction (right-handed as referred to the strand direction), which is why they are called singly wound barrels. The β-sheet twist makes the strands, and also the helices, follow a right-handed spiral around the barrel axis.

A somewhat different pattern of parallel α/β organization is the doubly wound parallel β sheet, or nucleotide-binding domain (Rao and Rossmann, 1973), which has turned out to be the most common "fold" found for globular proteins. The doubly wound sheets are also twisted, but instead of arching over into a cylinder their β sheets form saddle-shaped wings. Again, all of their crossover connections are right-handed. The preference for near-neighbor $+1x$ connections is strong but not nearly universal, as it was for the singly wound barrels

*An exception has recently been reported for enolase (Lebioda, personal communication), for which the β-barrel starts with an antiparallel hairpin and an up-and-down helix pair, and then finishes with six strands in the usual singly-wound arrangement.

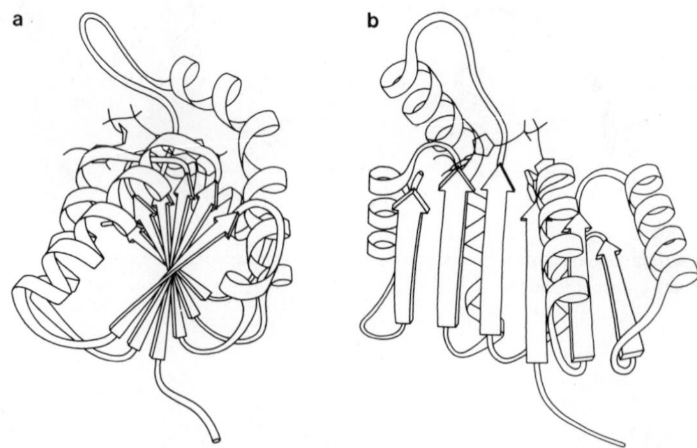

Figure 45. Ribbon drawings of lactate dehydrogenase (LDH) domain 1, as the classic example of a doubly wound parallel β sheet or "nucleotide-binding domain": (a) edge view; (b) side view.

The classical doubly wound topology, first seen in lactate dehydrogenase domain 1 (Fig. 45), begins in the center of the sheet and winds out toward one edge, with $+1x$ crossovers that pack helices against one side of the sheet; then the chain crosses back to the center and winds in the other direction, packing helices over the other side of the sheet. This classic topology occurs most often with six strands but also with five or seven. Spreading out from this central prototype is a whole hierarchy of parallel β-sheet structures of increasing complexity and irregularity (see Fig. 46); however, the great majority of them change "winding" direction once and only once. The most peripheral examples even incorporate some antiparallel strands, but they still present the same general appearance from a distance: a twisted central β sheet (predominantly parallel) flanked on each side by a layer of helices and loops. Parallel β sheet seems to require the protection of other main chain on both sides, and this is usually provided within the domain; therefore, parallel β strands are buried on both sides. Instead of the hydrophobic–hydrophilic alternation shown by antiparallel β strands, parallel strands are unrelievedly hydrophobic in their central portions and hydrophilic only at their ends (see Fig. 103), except for edge strands.

A majority of the doubly wound β sheets bind mono- or dinucleotides at a site near the middle of the C-terminal end of the sheet, at the position where the windings change direction. This is one of the few cases for which a type of catalytic site is correlated in a specific way with the global folding pattern rather than just with the local structure immediately surrounding it. Most categories of biological function include examples of proteins from very different tertiary-structure categories, giving the impression that evolution finds it easier to reinvent an active site than to reinvent a folded protein.

D. *Antiparallel* β

The third major category of tertiary structure consists of antiparallel β-sheet proteins. Like the α/β group, they can form either barrels or single sheets, but their topologies and relationships to other layers of structure are different from those of α/β proteins.

The simplest organization for an antiparallel β barrel would have all +1 near-neighbor hairpin connections between the strands, in analogy to the up-and-down helix bundles of Section IV.B. Such up-and-down β barrels do occur but are relatively rare: retinol-binding protein (Fig. 47; Newcomer *et al.*, 1984), catalase, and soybean trypsin inhibitor are the

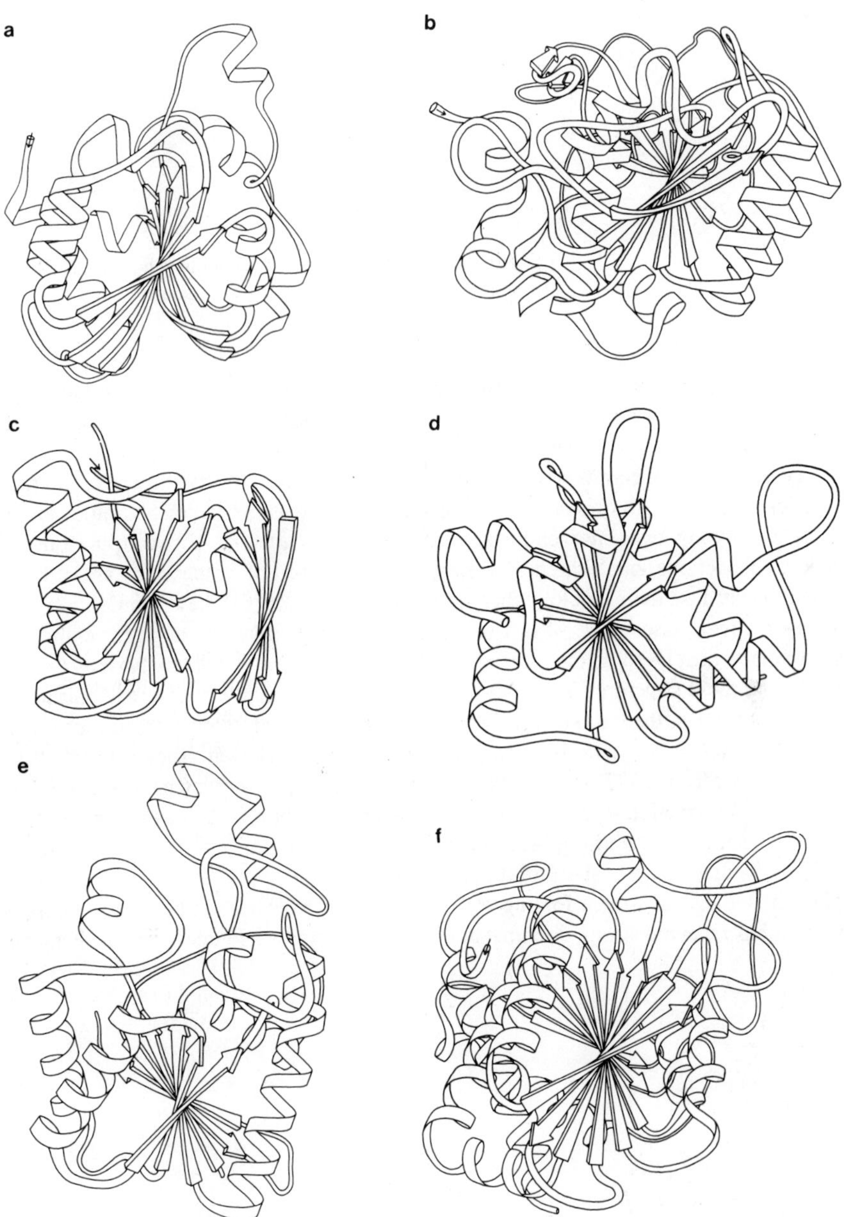

Figure 46. Examples of doubly wound parallel β-sheet structures with various overall shapes and specific connection topologies: (a) ATC catalytic chain, domain 2; (b) SBT; (c) GRS, domain 2; (d) RHD, domain 1; (e) PGM; (f) CPA.

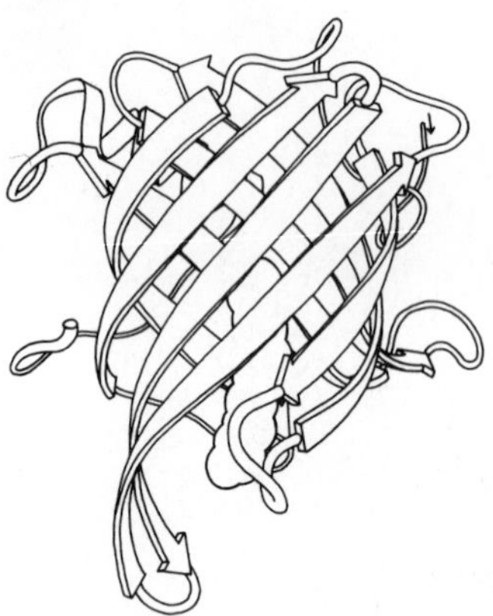

Figure 47. Ribbon drawing of retinol-binding protein (RBP), an example of an up-and-down β-barrel structure.

clearest cases. The barrels in the related enzymes papain and actinidin (and probably also the one in R67 DHFR) can be interpreted as having an up-and-down topology but are extremely irregular on one side.

Very much more common (the second most common tertiary structure, in fact) are antiparallel β barrels with one or more connections that cross the top or bottom of the barrel, most often skipping two intervening strands (therefore called "+3" hairpin connections, Greek key connections, or "β arches"). Barrels with this sort of connectivity are called Greek key β barrels because their topology diagrams resemble the "Greek key" border decoration so common on Greek vases (Fig. 48). The +3 hairpin connections are not very common elsewhere, so there must be some other reason for the prevalence of Greek key barrels. One way of characterizing these structures is that here, in contrast to the strictly local strand pairings in up-and-down topologies, there is a long pair of "superstrands" that follow next to each other. The swirl of this superstrand pair is always counterclockwise viewed from solvent. We have proposed (Richardson, 1977) that Greek keys are common because they have a favored folding pathway: initial formation of a very long two-stranded β ribbon, which can then curl up into a Greek key swirl, as shown in Fig. 49. The handedness of the final swirl is then a natural result of the right-handed twist of the initial two-stranded ribbon.

This hypothesis about the folding of Greek key β barrels has not been tested experimentally, but it has received strong circumstantial support from two sources. One is that a number

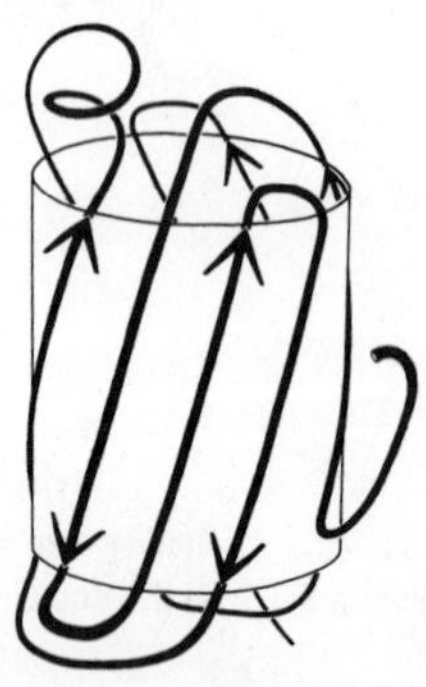

Figure 48. Simplified layout of a subunit of prealbumin (PAB), a Greek key β barrel. The border motif across the bottom has the same Greek key pattern as the five β strands on the front of the PAB barrel.

Figure 49. Hypothetical folding scheme for Greek key β structures, in which an initial twisted two-strand antiparallel β ribbon curls up into a barrel with the observed long-range handedness.

of proteins have been found to have "perfect" Greek keys in which all six strands (pyruvate kinase domain 2) or all eight strands (e.g., GAP, HMG, STV, TBV, SBV, MEV, RHV) take part in a clear superstrand pairing, and those pairs stay in register with each other as they swirl around the barrel (see Fig. 50). The second factor is an observation by Salemme (1983) that two-strand antiparallel β ribbons are different on their two sides (one has side chains extending from between narrow pairs of H-bonds, and the other from between wide pairs), which produces a preference for the direction of "curl": the concave side of the curl is the narrow-pair side. The property of sidedness is lost for β sheets of three or more strands because alternate strand pairs face in alternate directions; therefore, this effect can only apply to the initial β ribbon and not to the folded barrel. If a two-strand ribbon is to curl and so become compact, then its hydrophobic side chains must be on the narrow-pair side that will be on the inside of the curl. The sides can be distinguished even in a schematic representation: if a ribbon has its chain direction up on the right strand and down on the left, then you are looking at it from the wide-pair, or potentially convex, side (i.e., the outside of the final structure). In all the unambiguous known examples of Greek keys (e.g., Fig. 50), the preferred ribbon follows "Ray's rule," suggesting that the direction of curl around the hydrophobic core residues was indeed determined while that strand pair existed as an isolated ribbon.

The significance of the above analysis for prediction of tertiary structure is not only that Greek key topologies are favored for antiparallel β barrels but that among Greek keys only one out of four topologies is used (those with a counterclockwise swirl and with the central hairpin up on the right, when viewed from outside). The chief features that vary from one Greek key topology to another are the number and arrangement of extra up-and-down strands added on either side of the Greek key core and the number of strands (so far, four, six, or eight) used in that swirled core.

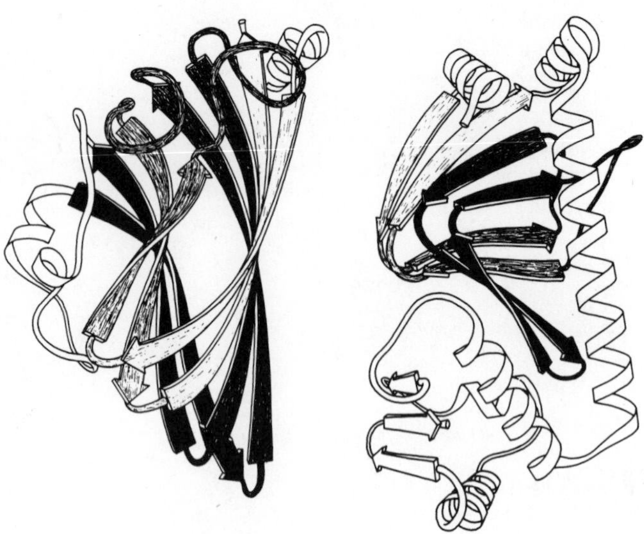

Figure 50. Two examples of Greek key proteins, shaded to indicate the two-strand pairing from which they could have folded up: (a) STV; (b) GAP.

Barrels with more than six strands have a flattened or elliptical cross section unless, as in retinol-binding protein, there is something bound in that central hydrophobic core. This is presumably a simple result of the average size of side chains, which pack well between opposite β sheets only in the range of about 10 to 13 Å apart. Because the strands twist, the direction of flattening twists, as shown in Fig. 51, often by close to 90° from top to bottom. For a flattened barrel, the local curvature (or arch) between strands is very strong around the edges, so that quite often there are few or no H-bonds in those positions. For that reason, these

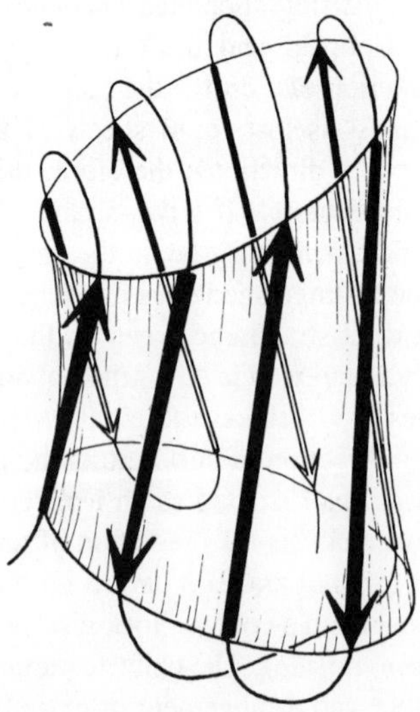

Figure 51. Illustration of the way in which the direction of barrel flattening (the elliptical cross section) twists from top to bottom.

structures are often described as β sandwiches or β-sheet-to-β-sheet proteins (Chothia and Janin, 1981). The sandwich description is the most appropriate for studying sheet-to-sheet packing in the native structures, but the barrel description is more appropriate for describing topologies and probably for folding and evolutionary relationships, since the chain repeatedly crosses back and forth between the two sheets.

The other common organization seen for antiparallel β structure is a single sheet (twisted but not closed into a cylinder) with helices and/or loops packed against one side (Fig. 52). In small proteins the other side of the sheet is exposed to solvent, whereas in larger proteins it may contact other domains or subunits. These "open-faced sandwich" structures have so far been found with from three-stranded (e.g., *cro* repressor) up to 15-stranded sheets (bacteriochlorophyll protein). Their topologies appear quite varied, but essentially all of them provide the covering helices or loops either within the β-sheet region of the sequence or immediately adjacent to it, perhaps because of the hydrophobicity of isolated β-sheet peptides. The covering is sometimes made up of right-handed crossover connections (like the βB–βC connection in L7/L12 ribosomal protein 1CTF) and sometimes from long extension of a hairpin connection (such as the βA–βB connection in 1CTF).

The topologies of small open-faced sandwiches show definite regularities, which are then combined into the more complex arrangements of the larger sheets. Of the 27 theoretically possible topologies for a three-stranded sheet, only three are actually observed: the up-and-down +1,+1; the "N-centered overhand" $+1,-2x$; and the "C-centered overhand", $+2x, -1$ (see Fig. 53). Restriction to these three topologies can be derived by ruling out parallel strands and left-handed crossovers. The N-centered overhands are somewhat more common than C-centered in isolated three-strand sheets, but they occur equally often as part of larger sheets. Overhands are just as numerous as up-and-down topologies. These three arrangements are apparently combined to form the bigger sheets, along with occasional +3 connections or longer crossovers, and with increasing occurrence of a few parallel strands as the sheets get larger. If there is any long-range regularity comparable to the Greek key organization of β barrels, then no one has yet found it.

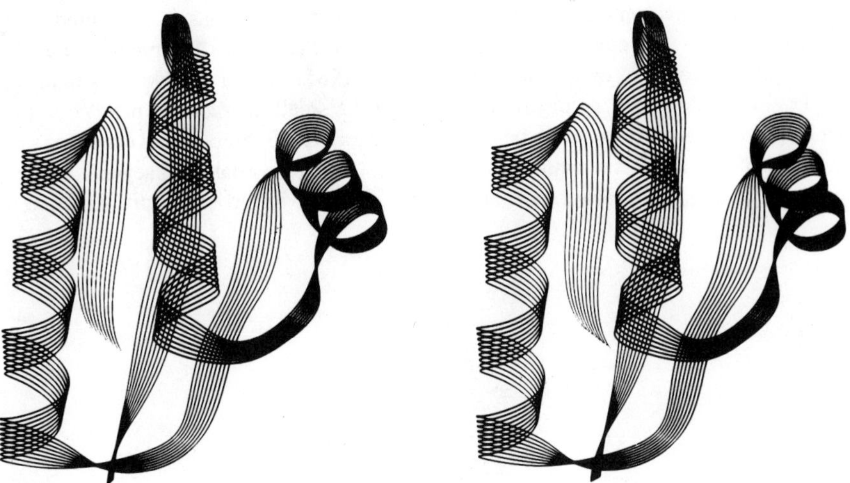

Figure 52. Stereo ribbon of ribosomal protein L7/L12 fragment (CTF), a small open-face β sheet with an N-centered overhand topology.

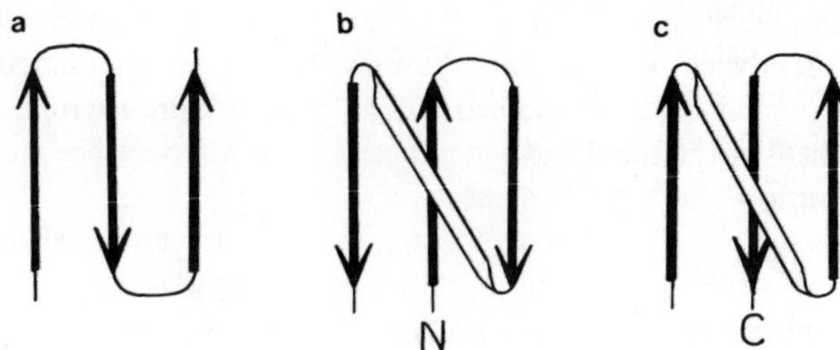

Figure 53. The three topologies observed for three-stranded β sheets: (a) up-and-down; (b) N-centered overhand; (c) C-centered overhand.

E. Small Irregular

Many of the smallest proteins are extremely difficult to fit into any overall set of tertiary-structure categories. They have relatively small amounts of secondary structure; what they do have is often irregular and not organized neatly into larger arrays. They are usually either SS-rich, with a high percentage of disulfide Cys, or else metal-rich, with bound metals or metal-containing prosthetic groups (but not both metals and disulfides). These additional nonpolypeptide interactions may substitute for the role of extensive secondary structure in stabilizing the folded protein. The current champion of small irregular structures is metallothionein, with 20 Cys-SH in 62 residues binding seven zincs or cadmiums. Although the x-ray (2MT2) and NMR (Braun *et al.*, 1986) structures disagree in detail, they both show two small domains with a series of short loops between ligands to the metal clusters.

The dividing line between small irregular structures and the other types of tertiary structure is sometimes gradual and, therefore, rather arbitrary, especially for the small open-faced β sheets discussed in the preceding section. There is a continuum all the way from bacteriochlorophyll protein down to the tiny C-centered overhand of potato carboxypeptidase inhibitor (Fig. 54), with only four β-sheet hydrogen bonds and one turn of helix in the crossover connection. There are also small irregular proteins that look like distorted all-α structures (e.g., cytochrome c, cytochrome C3) or distorted β barrels (e.g., rubredoxin; Fig. 55). It is difficult to find examples that resemble small, distorted α/β structures; perhaps the nearest candidate would be eglin C (McPhalen *et al.*, 1985), with a mixed four-strand sheet of $+3x, -1x, -1$ topology and no disulfides. Many of the small SS-rich and metal-rich proteins cannot be placed plausibly by this type of analogy, however. Examples like crambin, scorpion toxin, and pancreatic trypsin inhibitor (Fig. 56) are distantly related to several of the other categories if to any, and occasionally a protein is found with no secondary structure at all.

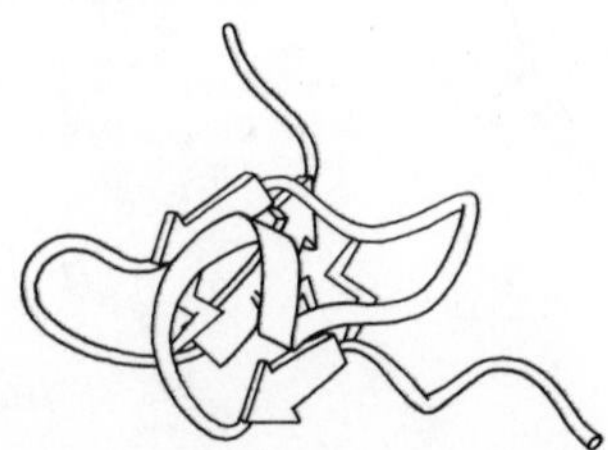

Figure 54. Backbone schematic of potato carboxypeptidase inhibitor (PCI), a small irregular structure with three disulfides and a C-centered overhand topology in its small β sheet.

The overall classification of tertiary structure can be visualized as an inverted pyramid, with the α, β, and α/β categories near the top corners and the small irregular structures near the base. Relative α and β contents vary in the horizontal dimensions, and total content of regular secondary structure increases in the vertical dimension.

V. ROLES OF THE INDIVIDUAL AMINO ACIDS

At the heart of the determination of structure by sequence lie the distinctive characteristics of each of the 20 different amino acids. When doing site-directed mutagenesis, one often wishes for additional possibilities, but natural proteins have done extremely well with the set that is available. They do a good job of spanning the reasonable range of such variables as size, shape, hydrophobicity, charge, and hydrogen bonding, although they avoid the extremes of high chemical reactivity (which would be too nonspecific) and also, except for proline, of groups that strongly restrict individual degrees of freedom. The 20 choices are not nearly enough to provide all combinations of all variables, but there are several pairs, such as Val and Ile or Asp and Glu, that seem highly similar. What is most relevant is how these individual side chains interact with the backbone and with one another and what roles they each play within particular types of secondary or tertiary structure. Figure 57 shows the amino acid side chains, arranged by the purely geometrical properties of size and shape. Glycine, proline, and cysteine have the most distinctive geometries and are considered first.

A. Glycine

Glycine, with only hydrogen for a side chain, is both the smallest residue and also the only one with a symmetrical α carbon. The lack of C_β means that a symmetrical, and also broader, region of the ϕ,ψ plot is accessible to glycine (see Fig. 58). The Lα local energy minimum is both small and rather shallow for other residues, but for glycine it is, of course, just as good as the Rα minimum. As with the nonrepetitive structures discussed in Section II, the ϕ,ψ values for glycine show a strong predilection for conformations with perpendicular carbonyl groups. In particular, the "left-handed" glycines (with positive ϕ) are predominantly L3_{10} or L-poly-Pro rather than Lα or Lβ. The L3_{10} conformation (or sometimes Lα) is necessary to form one of the most common endings of an α helix (Schellman, 1980); about one-third of all helices finish with Gly (see Fig. 59 and Table I). As seen in Fig. 17, this helix C-cap satisfies the same two H-bonds that a residue in α-helical conformation would, but in the reverse order, and sends the backbone off in a new direction. An L3_{10} residue is also essential

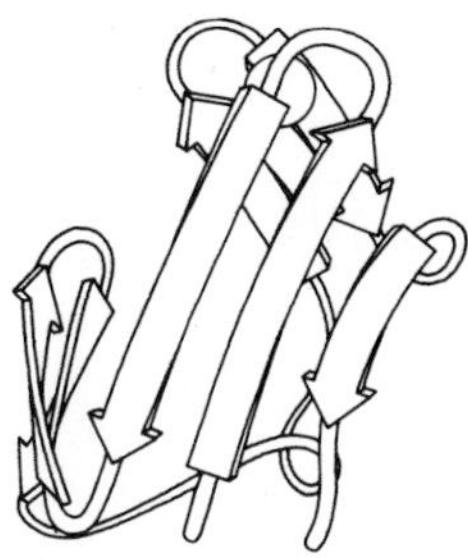

Figure 55. Backbone schematic of rubredoxin (RXN), a small irregular protein that binds a nonheme iron. Its β strands are H-bonded in pairs but not between the pairs.

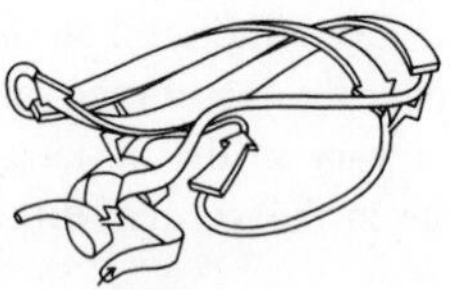

Figure 56. Backbone schematic of basic pancreatic trypsin inhibitor (PTI), a small irregular protein with three disulfides that is very stable and has been the subject of very extensive experimental and theoretical study.

in glycine, inverse glycine, and inverse common turns, and again those positions are overwhelmingly glycines. In many turns the Gly in $\text{L}3_{10}$ conformation is not strongly invariant because a common turn with unrestricted side chains does just as well. However, a tight hairpin between successive β strands requires an inverse turn because of the twist of the β strands (Sibanda and Thornton, 1985; Section III.C); therefore, there is at least one Gly in 17 out of 25 tight β hairpins (e.g., Fig. 60).

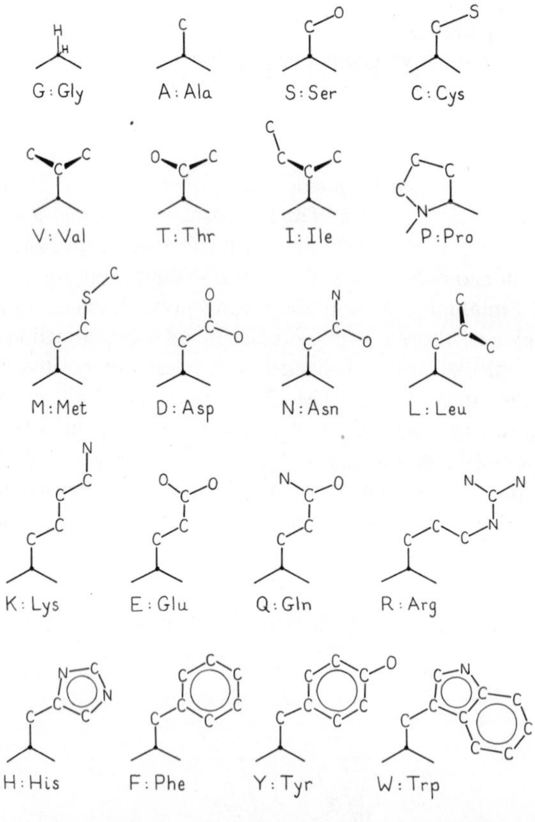

G: Gly A: Ala S: Ser C: Cys

V: Val T: Thr I: Ile P: Pro

M: Met D: Asp N: Asn L: Leu

K: Lys E: Glu Q: Gln R: Arg

H: His F: Phe Y: Tyr W: Trp

Figure 57. Structures of the 20 naturally occurring amino acid residues laid out according to size and shape of side chain, with three-letter and one-letter abbreviations: glycine, alanine, serine, cysteine; valine, threonine, isoleucine, proline; methionine, aspartate, asparagine, leucine; lysine, glutamate, glutamine, arginine; histidine, phenylalanine, tyrosine, tryptophan. Hydrogens are shown only for glycine. The α carbons are black dots; tapered bonds indicate the directionality at tetrahedral carbons; circles indicate aromatic rings.

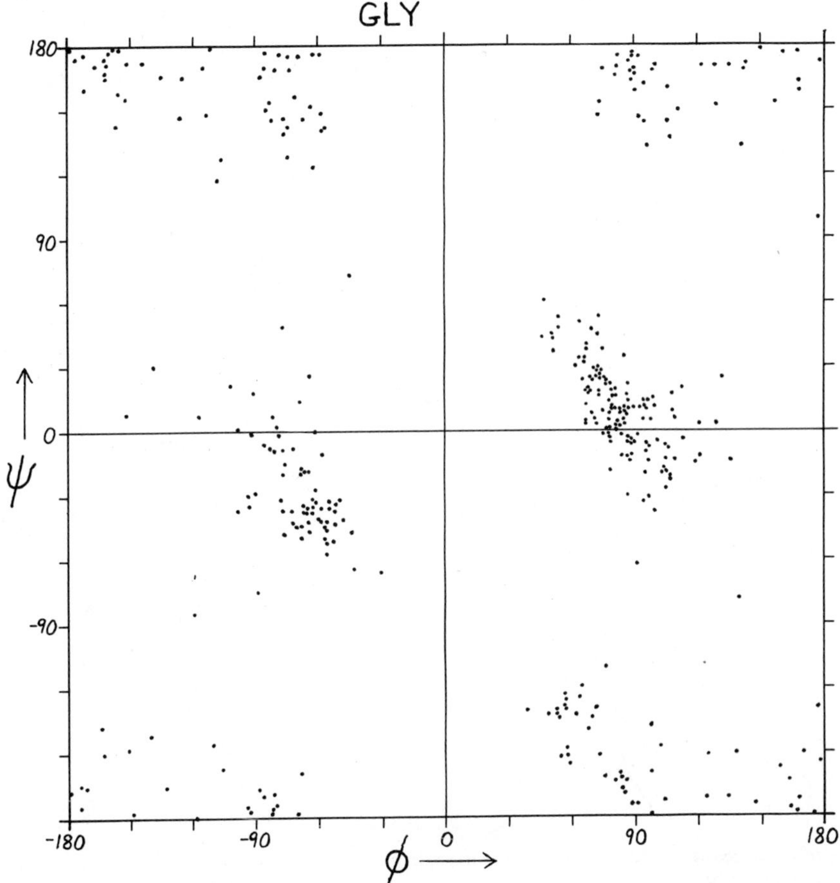

Figure 58. A φ,ψ plot for the glycine residues in a set of refined, high-resolution protein structures. Note the approximate mirror symmetry and the different distribution in the β region than for Fig. 6 or 7.

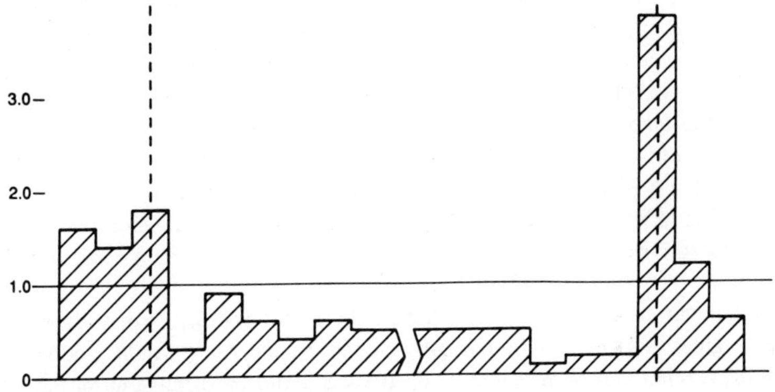

Figure 59. Plot of the relative preference values of Gly for occurrence in specific positions in α helix (from Table I, Section III). Dotted lines indicate the ends of the helix (N-cap and C-cap). Glycine is the best C-cap residue.

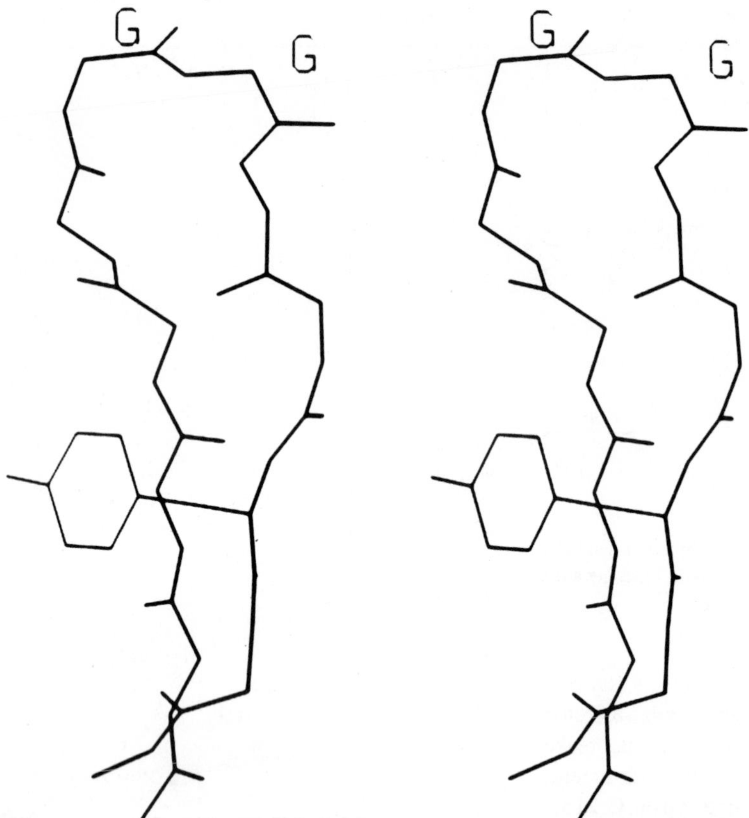

Figure 60. Glycines at the end of a tight β hairpin, in positions 2 and 3 of an inverse common turn (see Section III.C): ACT 168–179.

The Lβ region (the lower right quadrants of the Ramachandran plot) is rather similar to normal β conformation and can form both hydrogen bonds as part of a β sheet; however, it curves the sheet locally so that it is sightly concave where the side chain would otherwise have been (Fig. 61). The L-poly-Pro conformation, however, is likely to end a strand.

The broader range of accessible conformations for glycine produces another related, but logically distinct, role for Gly—that of facilitating movement. Glycine is frequently found in hinge regions (see also Section VIII). In fact, because of entropy considerations, glycine actually favors the unfolded state. It has been shown (Hecht *et al.*, 1986) that replacing a Gly with Ala, where there is no conformational or steric restriction against the change, increases stability of the folded protein, almost certainly partly by lowering the entropy of the unfolded state. For this reason, one would expect that when Gly occurs there is a particular need for its presence.

Glycine has yet another set of roles that can be important even when it adopts perfectly ordinary ϕ,ψ values: all of these are basically cases in which a C_β would produce long-range bumps with some other part of the protein, as opposed to the short-range bumps that determine permissible ϕ,ψ values. Many of these cases, of course, are idiosyncratically dependent on the details of a particular tertiary (or even quaternary) structure, but some are situations that recur fairly frequently. For instance, the close, high-angle contact of two α helices requires a very

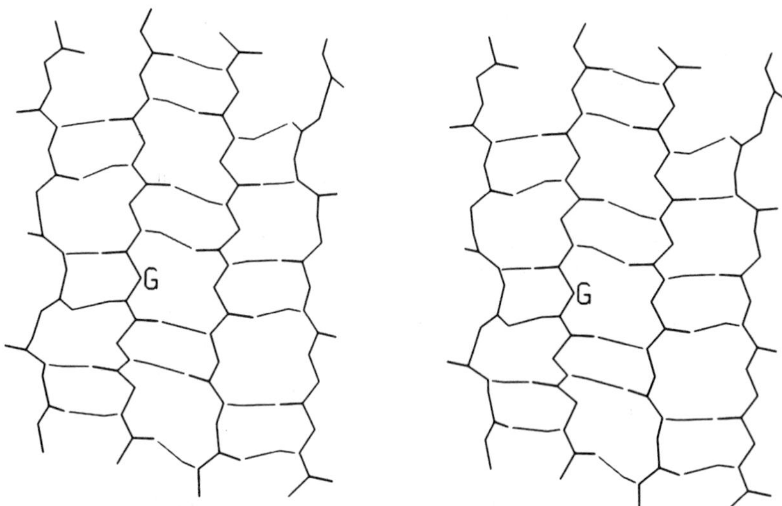

Figure 61. A Gly in left-handed β conformation as part of an antiparallel β sheet (SOD G31). A residue in β conformation would pleat inward at this position, but the Lβ Gly makes a convexity in the sheet.

small side chain like Gly or Ala at the center of each contact patch (Richmond and Richards, 1978). In an α helix, therefore, one Gly is quite plausible somewhere near the middle, as well as being an excellent breaker at the beginning or end. A tight, specific binding pocket is sometimes lined with extended main chain, necessitating Gly on that side of the backbone (e.g., Chymotrypsin G 216, G 226).

In β sheet, glycine serves a rather subtle role in modulating the right-handed twist and the curvature of the sheet. It has been known for some time that sharp points of especially high twist, such as curling over to close a small β barrel, can be produced either by a single Gly (Chothia *et al.*, 1977) or by a β bulge (Richardson *et al.*, 1978). However, even when the unusual twist is more uniform and widespread, and even when it is an especially low rather than especially high twist, glycines are usually involved, for steric rather than conformational reasons. For instance, in the parallel β barrels such as triose phosphate isomerase (see Section IV), not only the twist but also the curvature from one strand to the next must be very high in order to fold around into a cylinder that is smoothly H-bonded all the way around. If this is modeled using β strands with uniform φ,ψ angles, the shape can be fit very well; but near the "waist" of the flared cylinder there are very bad contacts between the inward-pointing β carbons and the main-chain carbonyl oxygens of adjacent strands. The proteins seem to solve this problem partly by having very irregular φ,ψ angles for parallel β strands and partly by sprinkling glycines (typically three or four per barrel) in inward-pointing positions near the narrowest part of the barrel. This is apparently important enough to be worth leaving internal holes above most of those glycines, where the other side chains cannot quite fill in (see Fig. 62).

At the opposite extreme, unusually many glycines are found in large antiparallel β sheets that are especially flat, such as bacteriochlorophyll protein or concanavalin A. This seems rather unexpected at first, but after all the only possible ultimate cause of a strong handedness preference is the handed position of a C_β; therefore, any such effect must be diminished by the presence of glycines.

The glycines that help allow flat sheets or TIM barrels, or the ones at close helix contacts,

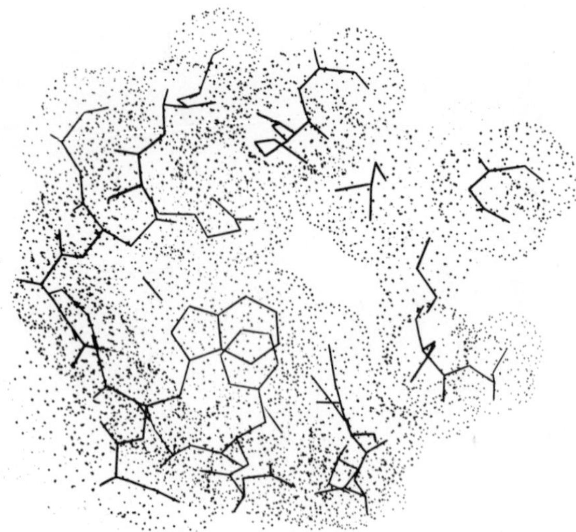

Figure 62. A cross section of the parallel β barrel in taka-amylase (TAA) with dots outlining the van der Waals surface of the strand residues and internal side chains. The large hole is next to an inward-pointing Gly.

illustrate amino-acid roles that are useful only after the protein structure is largely complete. They would presumably hinder the early stages of folding, since they would disfavor formation of the secondary structure in which they occur. In contrast, the glycines in turns would actively promote early turn formation, although only some of those would have specific roles in the final tertiary structure. Obviously, in the best cases an amino acid contributes in both of these ways, but we will often see examples of the logical distinction between these two sorts of roles.

B. Proline

Proline, in direct contrast with glycine, has a high additional constraint on the backbone relative to any other amino acid, since the ring closure keeps ϕ quite near $-60°$ (see Fig. 63). This range includes two well-populated areas of the ϕ,ψ plot: the α-helical region and the polyproline area at one side of the β region. The purely conformational restrictions are not really great except in lowering entropy. More influential is the fact that proline cannot fit into the regular, internal parts of either helix or β sheet, both because it does not have an NH available to take part in the H-bond network and also because the ring occupies some of the space where the neighboring piece of chain would be. Much of proline's importance is negative: as a breaker of regular secondary structures. Partly by default, then, it occurs in turns, nonrepetitive structure, and at the ends of strands and helices. Two prolines in a row favor the polyproline, or collagen, conformation, which is extended but cannot form a β sheet; short proline-rich sequences are useful as stiff, noninteracting connections, as in the hinge regions between the F_{AB} and F_C portions of immunoglobulins.

Proline is not entirely prohibited in helices. It works quite well in the first turn, where the ring and the unavailable nitrogen are harmlessly exposed. A closer examination shows that this positioning is surprisingly specific, with Pro the second most common residue in position 1 and fairly unusual in positions 2 or 3. Its preference values are 2.6, 0.5, and 0.4 there, compared with 0.2 for the rest of the helix (see Fig. 64). Figure 65 shows examples of prolines

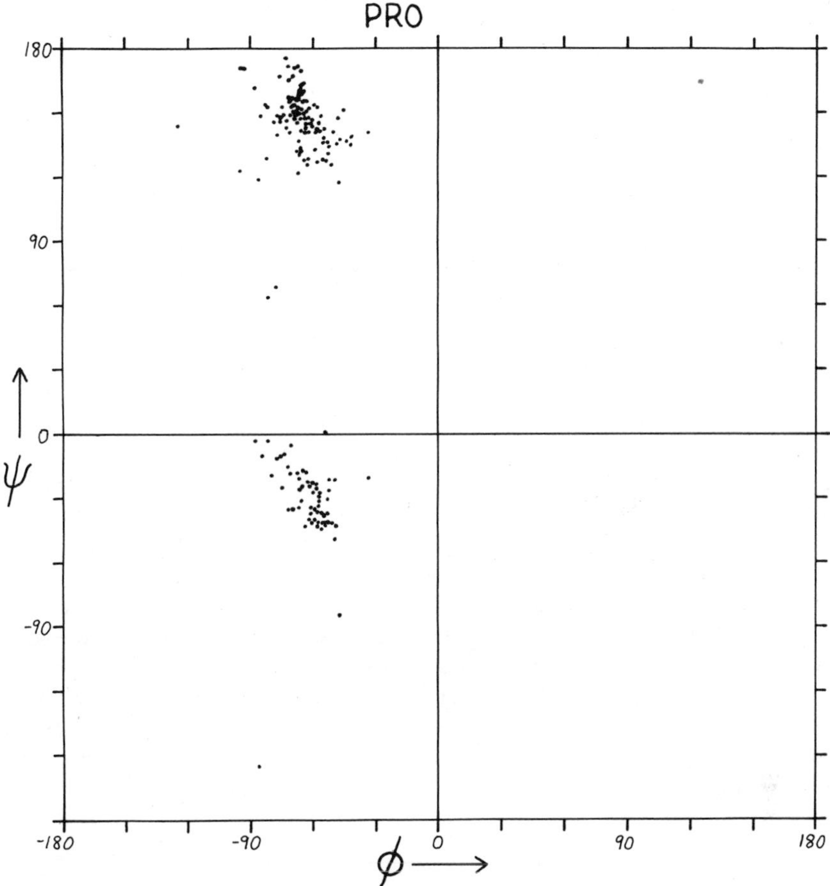

Figure 63. A φ,ψ plot for the proline residues in a set of refined, high-resolution protein structures. Note that the closed ring constrains φ to be near −60°.

in this favored position, where perhaps it should be thought of as a helix "starter" rather than a helix "breaker."

Even in the central portion, although Pro is the least common residue, it does occur.* The ring pushes away the preceding turn of the helix by approximately 1 Å on that side, producing a bend of about 30° in the helix axis and breaking the next hydrogen bond as well as the one at the proline N (see Fig. 66). Most of these cases seem best described as one bent helix rather than two separate helices, since there are typically one H-bond and two residues in common between the two sections, and the changes in φ,ψ values are often very small. It is unclear whether the bends serve some purpose or are merely tolerated. In some examples there are related proteins without the Pro (and without the bend); and in melittin, where it seems as though the bend might help in packing the group of four helices, it has been replaced synthetically with a slight improvement in activity (De Grado *et al.*, 1982). However, if there were ever a need for a bent helix, an internal proline would be the way to produce one.

*List of Pro in helices: CCY P25, P55 (56); MBD P88; ECA P89; LHB P25; MLT P14; CYP P94; LZM P86; DFR P105; FXN P68; ADK P159; LDH P125; ADH P329; SBT P225; CAT P161, P358; TLN P69.

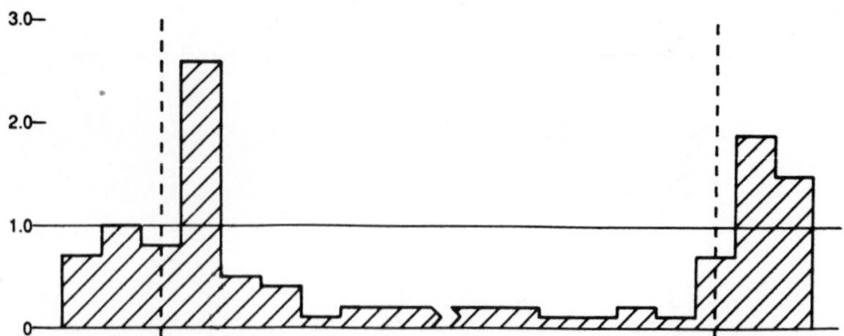

Figure 64. Relative preference of Pro for occurrence in specific positions along an α helix (see Table I, Section III). Proline is especially good in position N_1 and very bad within the body of the helix.

Proline is also very rare in the interior of β sheets, although it is common as a "breaker" residue at the end of a strand, on edge strands, and in β bulges. Even on an edge strand, with the NH not needed and the ring out of the way, a Pro bends the strand significantly because its polyproline ϕ,ψ values are fairly far from typical β-sheet values (see Fig. 67).

What proline really excels at is making turns, for which it has the highest specific positional preference of any residue (Chou and Fasman, 1977). It is very common in the second position of either common or glycine turns, both of which require ϕ close to $-60°$ (Fig. 68). As a result of its occurrence in turns and its virtual absence from the interior of regular secondary structures, Pro is almost always exposed on the protein surface. Thus, although its side chain is chemically hydrophobic, its typical location is that of a hydrophilic one.

Another unique property of proline is that it can occur with a *cis* rather than *trans* peptide bond (compare Fig. 69 and Fig. 68). *Cis* peptides for other amino acids are possible but almost

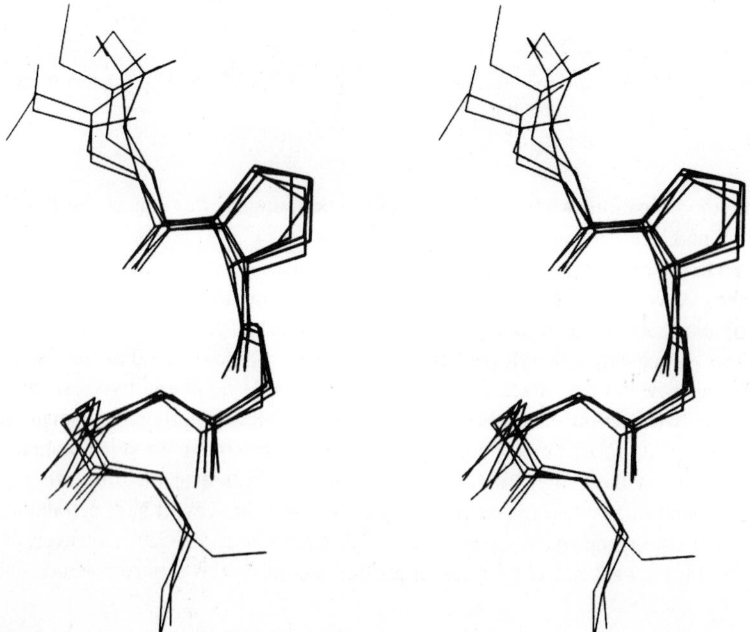

Figure 65. Five superimposed examples of helix N termini with Pro in position N1.

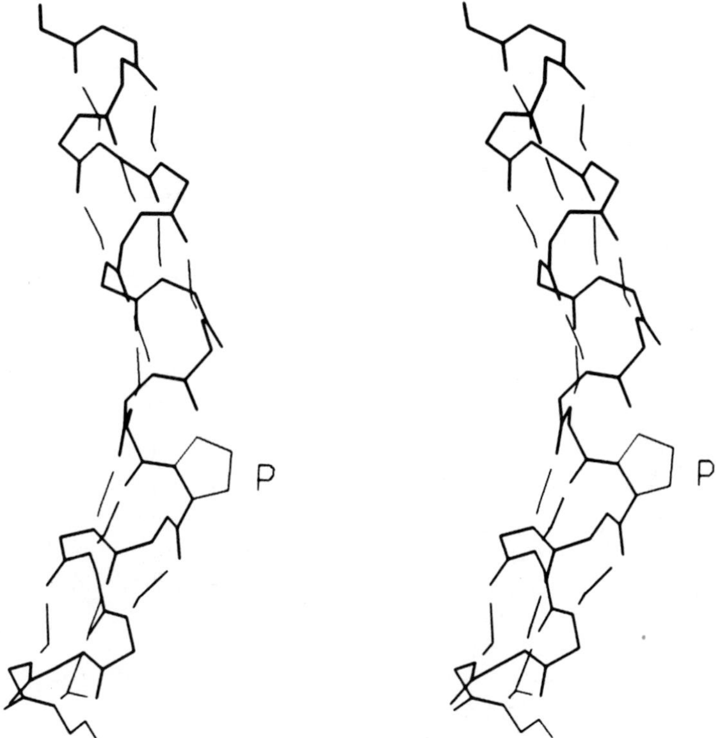

Figure 66. The effect of a Pro in the middle of an α helix (CCY 13–32), showing the bend produced and the two H-bonds broken.

vanishingly rare (5CPA Y 198, Y 206, and D 273; 4DFR G 97; Subtilisin Novo T 211; McPhelan, 1986). The energy barrier to rotation of the prolyl peptide bond is substantial, but isomerization occurs with a halftime on the order of a few tens of seconds (Brandts *et al.*, 1975). This process is of particular interest because it has been shown for several proteins to be responsible for a slow step in protein-folding kinetics.

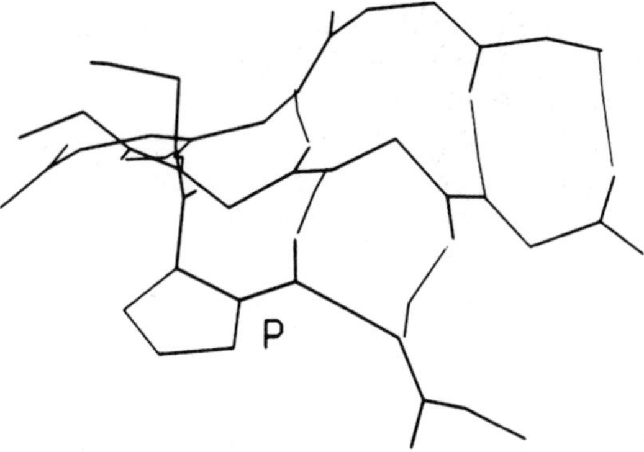

Figure 67. Proline on the edge strand of a β sheet (BCL P50). H-bonds can be formed on both sides of the Pro, but the backbone conformation is significantly distorted.

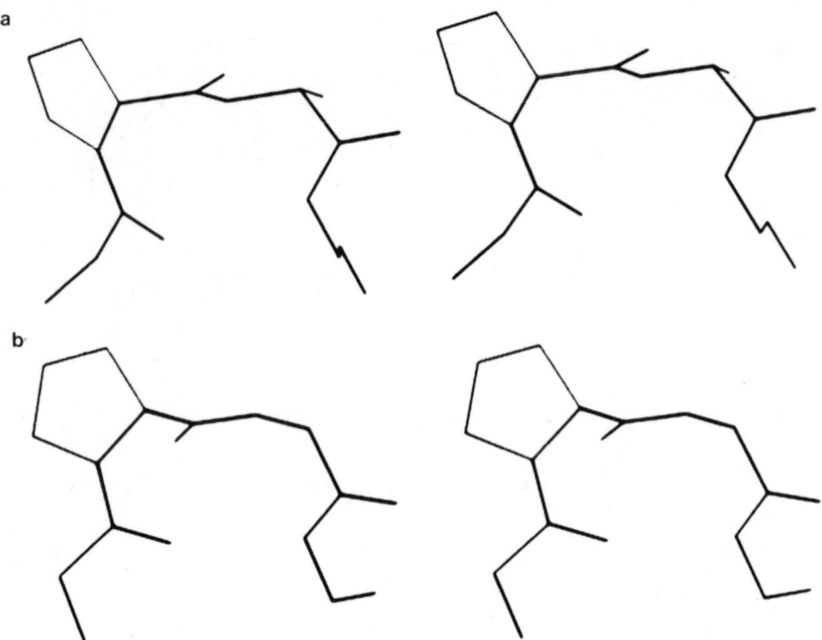

Figure 68. Proline in position 2 of tight turns: (a) a common turn (SGA P111); (b) a glycine turn (SGA P192). These prolines are preceded by *trans* peptides.

The geometry is such that a *cis* proline always forms a turn, which is known as a type VI or *cis*-Pro turn. The residue preceding the *cis*-Pro is always in extended conformation (it can be Lβ if it is a Gly), but there are two possible conformations for the proline. If ψ for the *cis*-Pro is near 0° (type VIa), then the NH of the third peptide is toward the center of the turn and is usually H-bonded to the CO of the first peptide (Fig. 69a), whereas with ψ near +150° (type VIb), the CO of the third peptide is toward the center of the turn, and there is no H-bond (Richardson, 1981; Fig. 69b). Although it is difficult to choose an unbiased statistical sample, since *cis* prolines are often not recognized until fairly late in refinement of a protein structure, a conservative estimate is that 10% of prolines occur as *cis* (28* out of 285 Pro in high-resolution structures surveyed). Seven of these are type VIa turns, 18 are type VIb, and three have an Lβ Gly and the proline ψ near 150°. The most surprising feature of *cis*-Pro distribution is that all-α, parallel α/β, and small irregular proteins as groups each have only 6% of their prolines *cis*, whereas antiparallel β proteins contain 15% *cis* prolines. Also in terms of surrounding local structure, essentially all of the *cis* prolines occur either in loops or at the end of an antiparallel β strand; there is only one case in a connection between helices. It seems likely that whatever contortions are necessary in a folded or mostly folded protein to allow isomerization of Pro, they can be more easily accommodated by loops and antiparallel β strands than by helices or parallel β.

*List of *cis*-Pro. Type VIa: SNS 117; TLN 51; RSA 93; APP 134; ECA 74; SBT 168; CAC 202. Type VIb: FAB h151; REI 8, 95; PCY 16, 36; GRS 375, 468; TBV 359; RSA 114; HMQ 7; SGA 99a; OVO 12; SN3 59; ADH 62; CAC 30; CRO 59; BCL 39, 320. With left-handed Gly: ACT 155; TBV 290; APP 315.

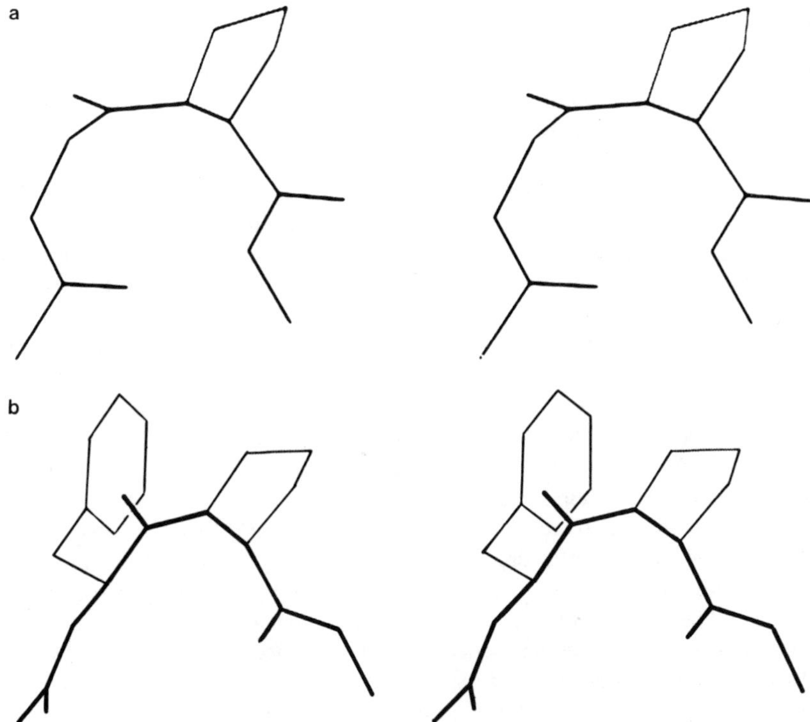

Figure 69. Prolines preceded by *cis* peptides: (a) a type VIa turn (SNS P117); (b) a type VIb turn (SGA P99a).

There are also some strong sequence preferences for *cis* versus *trans* prolines. The position immediately before the *cis*-Pro is heavily enriched in F, Y, and L (50% occurrence, compared with 16% preceding *trans*-Pro or 14% global) and very unfavorable for branched β carbons (3% I, V, and T compared with 25% before *trans*-Pro and 19% overall). Both before and after *cis*-Pro, charges are very unusual (7% D, E, K, and R in either place as opposed to 21% before *trans*-Pro, 19% after, and 20% global). None of the 28 *cis*-Pro are nearer than seven residues from a chain terminus. It seems quite plausible that a branched β carbon would interfere during isomerization (Brandts *et al.*, 1975), and a more charged environment could add to stabilization of the additional dipole in the *trans* form. The role of F, Y, or L is unclear; perhaps they provide hydrophobic shielding during isomerization, although only some of them are positioned over the peptide in the final structure (see Fig. 69). The percentage of *cis* versus *trans* forms varies widely for different prolines even in small peptides (Grathwohl and Wüthrich, 1974), to which effect these sequence preferences must contribute.

In places where it can be tolerated, Pro makes a positive contribution to protein stability purely by entropy effects. Since one degree of freedom is already frozen out even in the denatured state, Pro has a smaller loss of entropy on folding than any other residue. This has been experimentally verified by replacing Ala with Pro where the structure permitted in T4 lysozyme (Matthews *et al.*, 1987), and stability was indeed increased. This means that Pro can be expected sometimes to occur in otherwise permissive positions but in the absence of any structure-specific need for its presence.

C. Cysteine

Cysteine is notable both for its ability to bind Fe, Fe–S clusters, and various other prosthetic groups and also for its ability to form disulfides, the only common polypeptide linkage with a nonlinear topology. There are three very different types of Cys, then: (1) free SH groups, (2) ligand SH groups, and (3) disulfides (cystines). Each of these has distinct conformational preferences and distinct clustering properties in three dimensions.

Free SH groups are relatively uncommon; they occur predominantly where a reducing environment makes them less harmful, such as in bacterial and viral proteins. Although they look so much like serines, they really behave extremely differently. Free Cys SHs do not become phosphorylated or attach carbohydrates like Ser OHs, but they are more reactive to many external reagents (for instance, the mercurial compounds used as heavy-atom derivatives for x-ray crystallography). Cysteine is much poorer at hydrogen bonding than Ser. Substitution of one of these amino acids for the other is a frequent exercise for site-directed mutagenesis; many of those experiments have been extremely interesting (Ghosh *et al.*, 1986, and references therein), but they have never produced an equivalent functionality. Another difference is that Ser is fairly hydrophilic and is typically exposed, whereas Cys (even as the free-SH form) is the most thoroughly buried amino acid (Rose *et al.*, 1985; G. Rose, personal communication). Experimental measures of hydrophobicity (e.g., Wolfenden *et al.*, 1981) show Cys to be neutral or only mildly hydrophobic, which seems chemically reasonable. Perhaps its extreme buriedness could be a result of the deleterious effect of having such a reactive group exposed on the surface; however, the groups surrounding those buried SHs are almost always highly hydrophobic. Free-SH cysteines prefer regular secondary structures to turns or coil, and their favorite location (close to half of them) is on a β strand, pointing inward and buried (see Fig. 70).

Cysteines can serve as ligands to an incredible variety of different metals and prosthetic groups, including Fe, Zn, Cu, Fe–S clusters, and hemes. Typically two to four Cys are clustered together in three dimensions to bind a group, and often some or all of the cluster are also close together in the sequence (see Fig. 71). It is very common for two liganded Cys to be separated by only one or two intervening residues at or near the end of a loop (e.g., RXN) or on successive turns of distorted helix (e.g., CDV). Liganded cysteines are most common in coil regions, next in helix (often distorted), and quite rare in β sheet.

Disulfides, of course, are Cys pairs in three dimensions. Unfortunately, there is rather little strong preference for sequence separation or topology to help predict the correct pairings if they are not known chemically (as they less commonly are nowadays). Disulfides have one

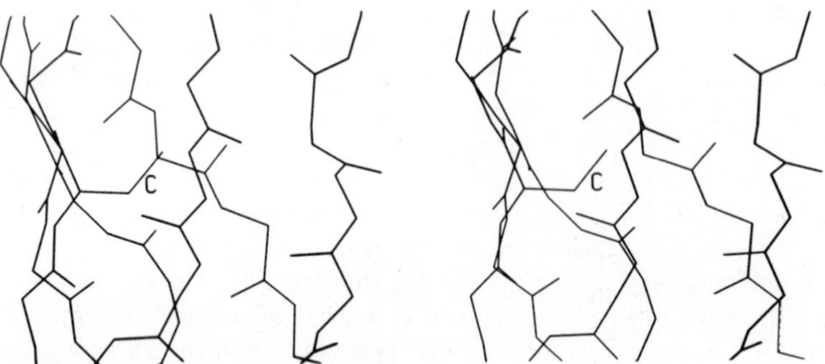

Figure 70. A free-SH Cys in its preferred location, pointing inward on a β strand (SOD C6).

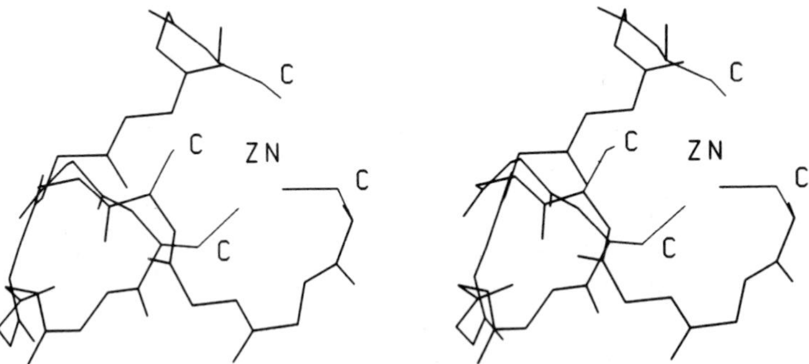

Figure 71. A group of four Cys as ligands to a Zn (ADH 97–111).

of the strongest and most distinctive roles in stabilizing final native protein structures, but they probably have very little role during early folding when they are in the reduced form. They fasten together neighboring parts of the polypeptide, but they do not themselves determine which parts will end up as neighbors. They are most useful, then, in a tertiary-structure prediction, particularly if the correct pairings have been established chemically or by homology, but to some extent even if they have not been. To make a disulfide, not only must the α carbons of the two half-cysteines be between 4 and 7.5 Å apart, but the C_α–C_β vectors must have the right sort of relationship. The most common disulfide conformations (Richardson, 1981) are the left-handed spiral (Fig. 72a), which has the C_α–C_β vectors almost antiparallel but offset, and the right-handed hook (Fig. 72b), which has the C_α–C_β vectors approximately perpendicular. There is no good way to join them if the vectors are close to parallel, which is why a disulfide is very seldom formed between H-bonded residues on neighboring β strands (exceptions: 1ACX, Karle *et al.*, 1988).

With a strong twist, a disulfide can form with a diagonal offset of two residues between adjacent antiparallel β strands (e.g., tendamistat, Kline *et al.*, 1986). Disulfides are quite common just past the ends of a β-strand pair (e.g., Fig. 73), in which position they are also very likely to be offset diagonally. Similarly, a disulfide cannot form between two Cys in the same α helix without extreme distortion (e.g., GRS active site). It is also surprisingly difficult, and unusual, to form a disulfide between two α helices. The only examples, CRN and BP2, both have the same asymmetric disulfide conformation and rather unusual interhelical angles. Half-cystines have relatively weak conformational preferences; however, they most often occur on backbone that is extended but not part of a β sheet.

For the purposes of prediction, then, it would be useful to distinguish the three different kinds of Cys. This may actually be possible. For the most part, proteins take one path or the other with their cysteines: there are very few cases with both disulfides and free SHs (e.g., SOD). Most such cases are either disulfide-type proteins, which use an active-site SH (e.g., PAP and ACT), or else SH-type proteins, which have an active-site disulfide (e.g., thioredoxin and GRS). From studies of related proteins with different numbers of disulfides, it seems that the stage with an unattached Cys must be brief; typically these related proteins differ by both ends of a disulfide when they differ. Similarly, proteins that use Cys as a ligand only occasionally contain unused free SH groups (e.g., 4ADH) and almost never disulfides (exception, 1AZA). There are no disulfides in α/β proteins (except at active sites) or in *E. coli* or viral proteins, but secreted proteins have disulfides if they contain Cys at all. Proteins that bind suitable metals or prosthetic groups are likely to have ligand Cys and perhaps some free-SH Cys but no disulfides. Most small proteins (fewer than 100 residues) without metals or prosthetic groups contain disulfides. Of course, if, as is increasingly common these days, the

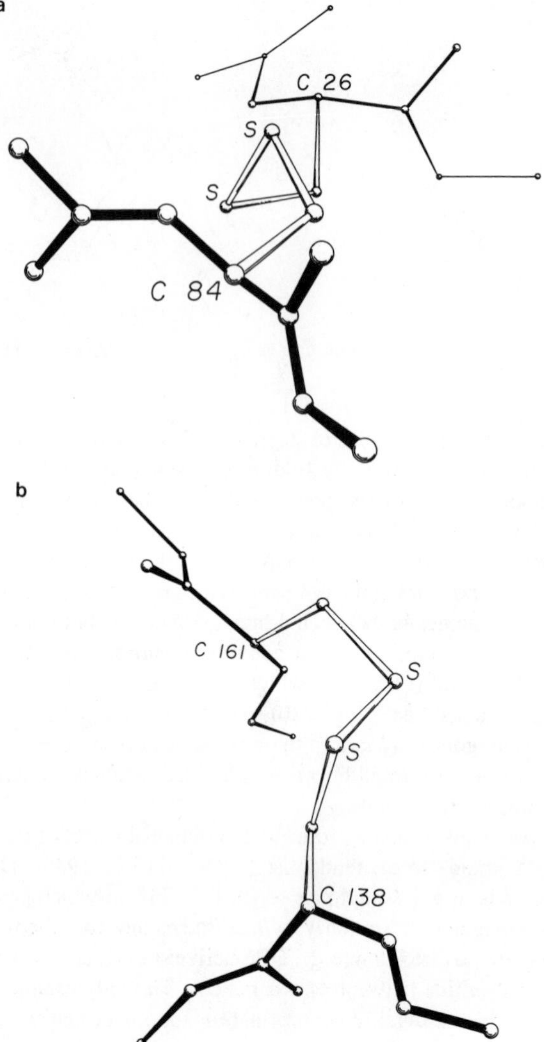

Figure 72. Two common disulfide conformations: (a) left-handed spiral (RNS C26–C84); (b) right-handed hook (CPA C138–C161).

protein is known only from its DNA sequence, one must keep in mind that a very high Cys content is possible with ligand SH groups as well as with disulfides; the extreme example is probably metallothionien, whose 20 Cys in 62 residues bind seven metals.

D. Alanine

Alanine is the default amino acid: no long side chain, no chemical reactivity, no unusual conformational properties, and fairly happy on either interior or surface. Likewise, α helix is the default protein conformation: the stablest compact, H-bonded structure in the absence of any particular disruptive influences. It seems natural, therefore, that Ala should be so prevalent in α helices (it has the strongest preference of any residue for a middle-helix location). Turns

Figure 73. A disulfide just past the end of β structure (RSA C65–C72) with the two ends diagonally offset by two residues relative to the β-strand pairing.

prefer more hydrophilic residues, and nonrepetitive structure in general favors amino acids with side-chain H-bonding capability. A β sheet prefers to be covered by larger side chains. An α helix, however, is well satisfied with Ala, which is just large enough to set the handedness but small enough that nothing special must be done to keep it happy while the backbone is completely satisfied by the helical H-bonds. In contrast with any other helix-forming residue, Ala is relatively indifferent to its location, with only a slight preference for either end relative to the middle or for outside versus inside (see Table I, Section III). Polyalanine forms α helices, and some helices in globular proteins have a remarkably high local concentration of Ala (e.g., Cyt C′ α78–103, with a total of eight Ala including the sequence AAAAKA shown in Fig. 74). One guesses that Ala helps encourage helix formation early in folding, in concert with other factors (Shoemaker *et al.*, 1987). Figure 75 shows a φ,ψ plot for Ala. In addition to the numerous, tightly clustered α-helical residues, an interesting feature is that extended alanines seem to have a tendency to cluster around the polyproline conformation at −60°,140°. There are very few points in the Lα region.

Inside antiparallel β barrels, Ala has a limited but specific role in the final packing. Sixty percent of those interior positions are occupied by the large, awkwardly shaped side chains of Val, Ile, Leu, and Phe (Chothia and Janin, 1981), which cannot be closely packed without a sprinkling of very short side chains. Thus, in spite of the fact that Ala strongly favors α over β conformation, 10% of those interior β barrel positions are alanines. Alanine is presumably less

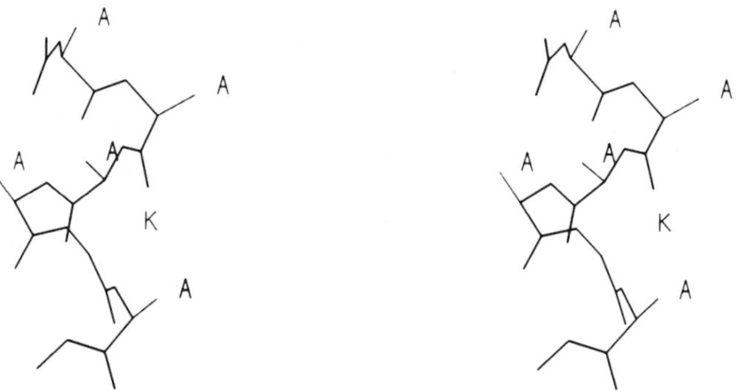

Figure 74. A set of five alanine residues in their classic location on an α helix (CCY 96–103).

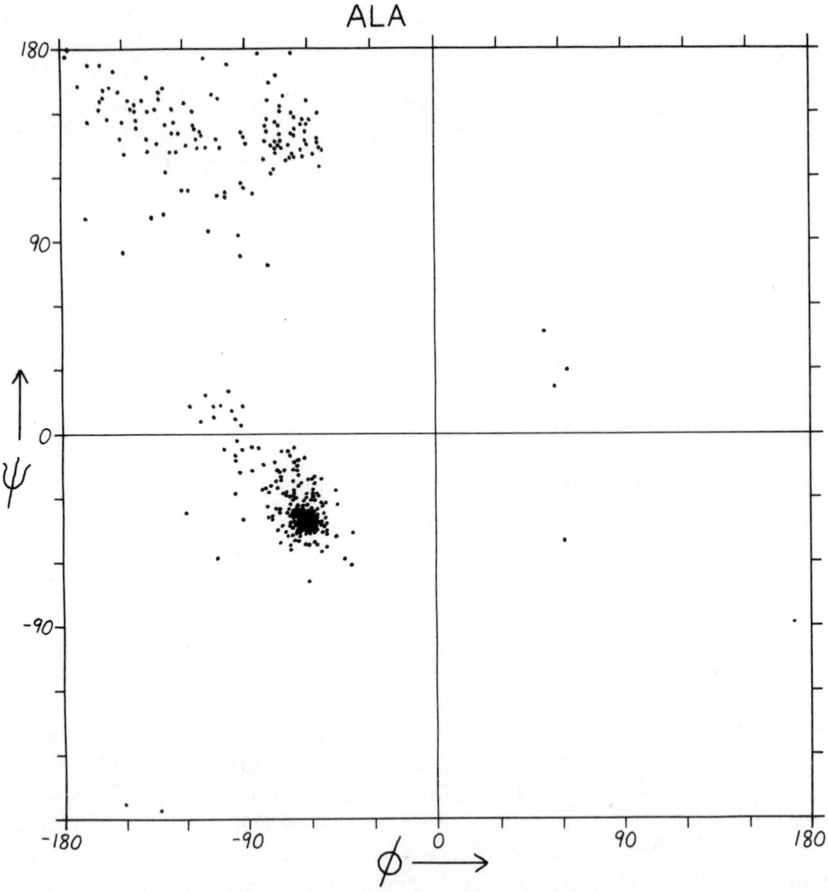

Figure 75. A ϕ,ψ plot for Ala showing a major concentration in α helix and a small cluster in the polyproline region.

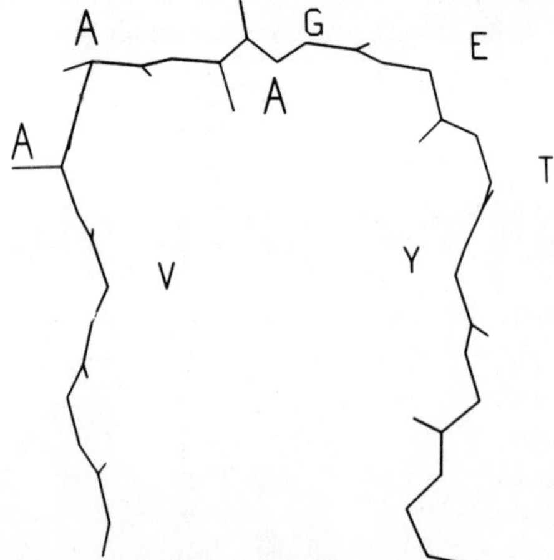

Figure 76. Three Ala in a Greek key connection (ACX 20–31).

dangerous than Gly, because a single Ala would probably not nucleate an α helix, whereas a single Gly could perhaps induce a turn, especially with hydrophilic residues on both sides of it, as would usually be the case in antiparallel β.

Alanine is also quite common in short, extended, straplike connections such as Greek key "β arches" (Fig. 76) or α/β connections. It is good at 90° corners, and it is fairly indifferent to solvent exposure, which would be useful in a strap that may change conformation during folding.

E. Valine, Isoleucine, Leucine, and Methionine

Valine, Ile, Leu, and Met (VILM) are the aliphatic, variously shaped jigsaw puzzle pieces provided for fitting together the hydrophobic interiors of proteins. This group of residues undoubtedly is important both early and late in protein folding. The water dancing unenthusiastically over their waxy expanses provides a major force drawing together separate parts of the polypeptide chain and orienting pieces of secondary structure. Although their eventual exclusion from the surface is not as perfect as the exclusion of charges from the interior (a small protein usually has three or four VILMs exposed), it is 85–90% (Rose *et al.*, 1985). A major requirement in the final structure is good fit in the interior. Hydrophobic cores in proteins are not much like oil drops in liquid motion; there is constant, rapid jiggling and occasional larger motions, but these are all excursions around a well-defined equilibrium structure with good packing. The variety of size and shape in hydrophobic side chains allows good fit for a great variety of backbone frameworks. For instance, although the sulfur may have some H-bonding capability, the major role of methionine is to provide a very flexible hydrophobic side chain. Valine and Ile are widest right next to the backbone, Leu a bit farther out, and the aromatic rings even farther.

In general, the packing arrangement has different rules on α helices and on β sheets. On β sheets the side chains that are neighbors, both alternate ones along a strand and those on adjacent strands, touch each other extensively and form a nearly continuous layer. Side chains coming from elsewhere pack against the outside of that layer or penetrate it partially but only rarely interdigitate as far as the backbone (Fig. 77). On the solvent-exposed face of an antiparallel β sheet, the layer is not continuous, since the hydrophilic side chains are less bulky and the sheet is often convex, so that solvent penetrates to the backbone in between them; however, on buried faces of β sheet the backbone is quite thoroughly covered by its own side chains. On α helices, in contrast, the contacts between neighboring side chains (n to $n+1$, $n+3$, $n+4$) are much sparser; solvent typically gets to the backbone on exposed faces, and in helix–helix contacts the side chains interdigitate extensively, and their ends often touch the backbone of the other helix (see Fig. 78). This is one reason why short, wide side chains like Val and Ile are most common on β sheets and long ones on helices. Aside from their average overall helix preferences, all four of these hydrophobics, plus Phe, show a pronounced periodicity in helix preference, with sharp maxima at N-cap + 4 and C-cap − 3.5 (see Fig. 79) and go in order of their length (Met > Leu > Ile > Val). Those two positions prefer hydrophobics because the side chain is usually covered by the connection and points approximately toward the rest of the protein.

Valine and Ile (and also Thr) are atypical for another reason: they branch at the β carbon, which puts some additional constraint on their conformations. Although they occur fairly frequently in α-helical conformation, it is not as favorable for them as extended. Branched C_β residues prefer α or β conformation strongly over 3_{10} or polyproline and almost never have a positive ϕ (see Fig. 80). Even outside of secondary structures, the conformational preferences for Val and Ile are different from those of Leu and Met.

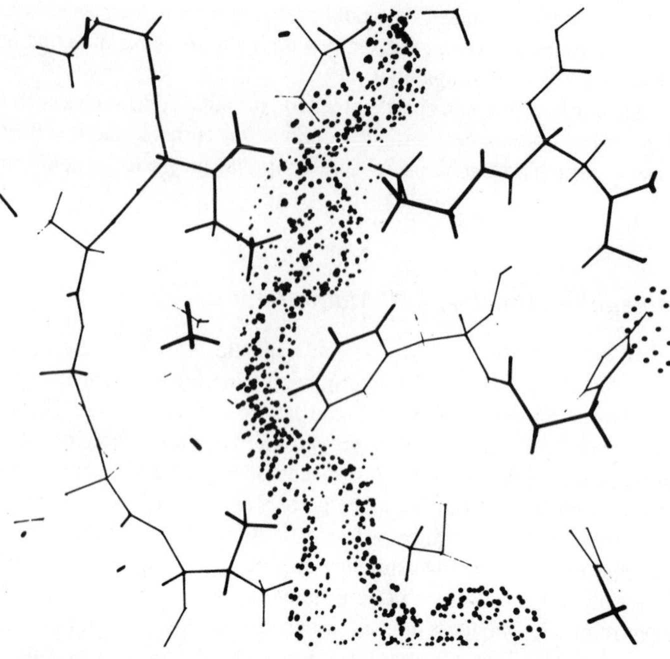

Figure 77. Dots outlining the two opposing contact surfaces of a pair of β sheets (SOD), one above and one below. The two surfaces have ridges and valleys that follow each other fairly closely, but all contact is out near the ends of the side chains.

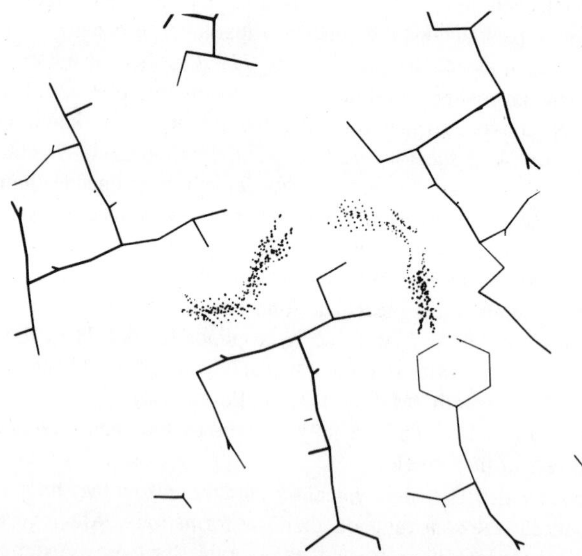

Figure 78. A thin slice through "small-probe" dots showing patches whenever an atom of one helix (MHR helix A) is within 0.5 Å of van der Waals contact of an atom in any of the other helices. Note that many of the contacts are between backbone and side chain, in contrast to Fig. 77.

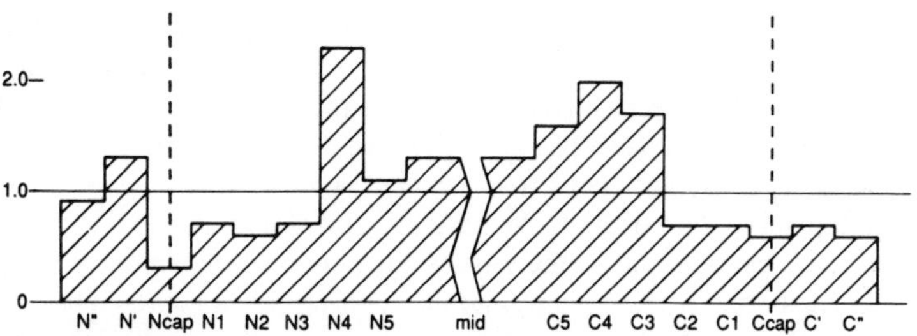

Figure 79. Relative preference values for Leu, Met, and Phe to occur at specific positions in α helix. Preference peaks at N_4 and C_4, one helical turn in from either end.

The branch at C_β puts stronger constraints on χ_1. All side chains have χ_1 near one of the three staggered positions −60° (opposite the CO), +60° (opposite the H), or 180° (opposite the NH), with the most common single value near −70° (Janin and Wodak, 1978). Valine, Ile, and Thr strongly prefer the χ_1 with one side-chain branch opposite NH and the other opposite CO; because of the vagaries of angle-naming convention, this same conformation is called

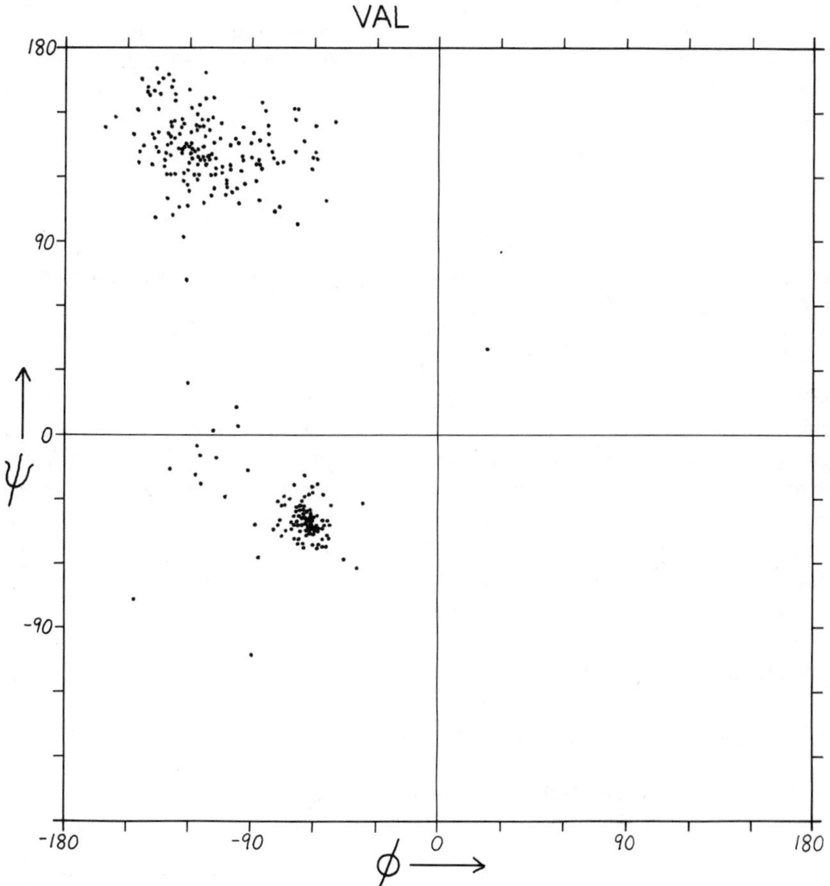

Figure 80. A φ,ψ plot for Val showing mainly α and β conformations.

$-60°$ for Ile and Thr but $180°$ for Val.* In α helices, the χ_1 constraints are even stronger than for the general case, with 90% of the Val and Ile opposite both NH and CO (see Fig. 82). This χ_1 preference has different effects for parallel and antiparallel β sheets. In parallel β, the relationship between adjacent strands is a translation, so that β-branched side chains with the preferred χ_1 can "cup" neatly against one another; indeed, whole strips of β-branched residues are commonplace on parallel β sheets (e.g., TIM, GPD: Lifson and Sander, 1980). For antiparallel β, on the other hand, the relationship between adjacent strands is a twofold rotation, so that neighboring β-branched residues with the preferred χ_1 would be either back to back, leaving unfilled space, or else face to face, producing a collision unless their backbone is adjusted. Therefore, on antiparallel β sheets the typical pattern is an alternation of branched and unbranched C_βs along strips perpendicular to the strands (e.g., EST or CNA).

F. Serine and Threonine

Serine and Thr are the short-chain OH residues. Both of them, but especially Ser, are chemically reactive, so that they have important roles as active site residues (e.g., Ser proteases), can be phosphorylated (e.g., glycogen phosphorylase), and can be carbohydrate attachment sites (e.g., flu hemagglutinin).

Since the OH can be either a hydrogen-bond donor or acceptor, Ser and Thr are quite versatile in H-bonding to backbone groups or solvent, but they are somewhat restricted by their short length and single variable side-chain angle. Their most classic structural role, noted even in the original myoglobin x-ray structure (Kendrew *et al.*, 1961), is to accept a hydrogen bond from one of the free NHs in the first turn of an α helix (see Fig. 81). In fact, they are sometimes found even in the middle of an α helix, making a second shared H-bond to the peptide CO of residue $n - 3$ in the same sort of arrangement that waters often occupy just next to a helical H-bond. This is presumably a destabilizing force on the α helix, since Ser is one of the most unfavorable residues in helix. The shared H-bond to the helical backbone means that the OH can readily occupy a position that is essentially forbidden for a methyl group. This gives the χ_1 angles for Ser and Thr in helices a strikingly different distribution from that of any other residue (see Fig. 82).

Serine is very common in tight turns and other nonrepetitive structures, where it may also make H-bonds to neighboring backbone NH or CO groups. However, even more often it is just interacting with solvent in the typically exposed location of turns and loops. Serine is also common in disordered regions, presumably because it has no hydrophobic character, is relatively unrestricted in backbone conformation, and can readily interact with solvent. Threonine, on the other hand, prefers β structure, especially antiparallel β, which typically has one side exposed to solvent. The geometry in β sheet is such that the Thr OH does not make an H-bond to a peptide, but it may interact indirectly through a water molecule. Serine also occurs in β but less commonly, since the branched β carbon of Thr favors extended conformation and packs well on the surface of β sheet. The extra methyl group makes Thr enough more hydrophobic than Ser that it occurs significantly more often in the interior and less often in turns, nonrepetitive loops, and disordered structure.

G. Asparagine and Glutamine

Asparagine and Gln are interesting because their side chains end in amide groups that are like the backbone peptide groups. Although the two are so similar, they fill surprisingly

*Ile and Thr have handed side chains: if their feet are on the C_α and their arms (representing the C_β–C_α are each slightly forward, then the left arm is the long branch for Ile and the OH for Thr.

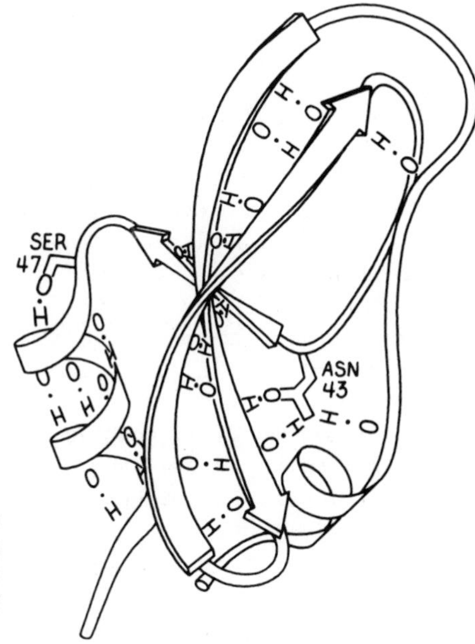

Figure 81. Ribbon diagram for basic pancreatic trypsin inhibitor (PTI) and its backbone H-bonds showing Ser47 in the N-cap position of the α helix, where it forms an H-bond with a backbone NH in the first turn of the helix.

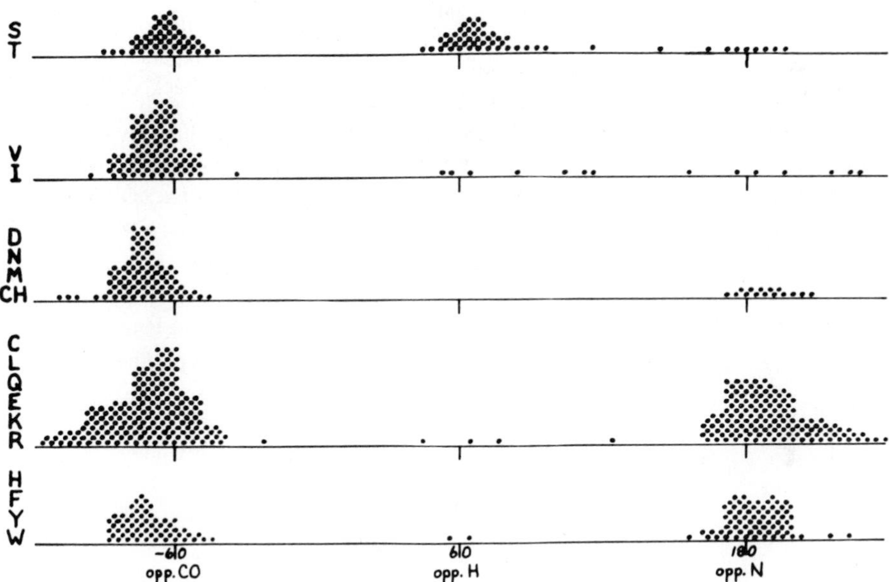

Figure 82. Distribution of χ_1 angles for side chains in the regular portions of α helices (excluding the first turn) in well-determined x-ray structures. Most examples are close to one of the three staggered conformations, but $\chi_1 = 60°$ (opposite the C_α hydrogen) almost never occurs except for Ser and Thr (see text). The other two conformations occur in ratios ranging from more than 10 : 1 for Val and Ile down to nearly 1 : 1 for the bottom two categories, which differ only in showing tighter distributions for the aromatics. Each residue type was compiled separately, and then they were gathered into groups. Note that SH cysteine and disulfide cysteine behave differently, as do Asn/Asp versus Gln/Glu. Somewhat surprisingly, Met looks more like the short than the long side chains.

distinct roles because of the extra length and additional degree of freedom of the Gln side chain. Although both residues end in "pseudopeptides," only Asn has a geometry that can connect it back to the main chain in a complete mimic of polypeptide geometry, and in only one direction, from the Asn backbone CO to its side-chain CO, with the Asn C_β playing the part (only geometrically, of course; not electronically) of a backbone NH (see Fig. 83). Of course, for Asn, there is rotation around the C_β–C_γ bond, but when $\chi_2 = 180°$ it mimics peptide geometry extremely closely. Therefore, in some situations the side chain O_δ of an Asn can make whatever H-bonds would have been made by the O_{n-1} carbonyl. One case of this behavior has previously been described: a pseudocommon turn (Tainer *et al.*, 1982) or "Asx turn" (Rees *et al.*, 1983) in which an Asn or Asp side chain takes the place of the first peptide in the turn (Fig. 84a). This configuration is very common, since the most frequent Asn H-bond is from O_δ to the backbone NH of residue $n + 2$ (Richardson, 1981; Baker and Hubbard, 1984). Neither an Asn N_δ nor the O_ϵ or N_ϵ of a Gln can perform this exact mimicry, since too many carbons separate them from the relevant adjoining peptide groups. Therefore, although those groups frequently form H-bonds, they do not have such a disruptive effect on the pieces of secondary structure they could replace. The ability to make pseudoturns probably explains why Asn is relatively infrequent in position 2 of tight turns (Chou and Fasman, 1977), where it competes with the backbone CO_{n-1} for the turn H-bond. In general, hydrophilic residues are strong turn formers, and Asn is very common in positions 1 and 3. In position 1 of a type 1 turn it can form an n + 2 H-bond to the NH of the central peptide, so stabilizing the turn (Richardson, 1981).

A second important way in which Asn mimics backbone is at the beginning of α helices. It can allow the main chain to depart in another direction while the side chain plays the part of one more helical residue (see Fig. 85), including an H-bond from the O_δ to the NH_{n+3} (equivalent to the n to $n + 4$ main chain H-bond of α helix). In that position, the Asn acts as the N-terminal "cap" of the helix: its main-chain CO makes a helical H-bond, but it has nonhelical (usually extended) ϕ,ψ angles that put the previous α carbon well outside of the cylinder of the helical backbone. Residue preferences for that specific position are very strong, and Asn is the strongest (Fig. 86), followed by Asp, Gly, and Ser. All of these except Asp are also very position-specific, since Asn, Ser, and Gly are among the least frequent residues in the next three positions at the beginning of an α helix. Asparagine, Asp, and Ser cannot make

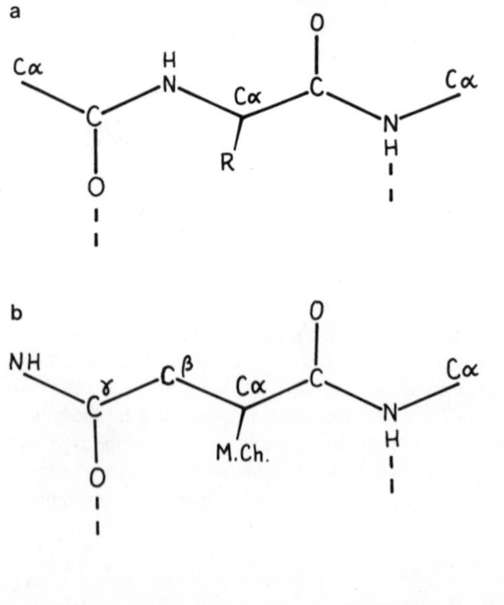

Figure 83. Illustration of the way in which the geometry and H-bonding of a dipeptide (a) can be mimicked by an Asn side chain plus the peptide following it (b).

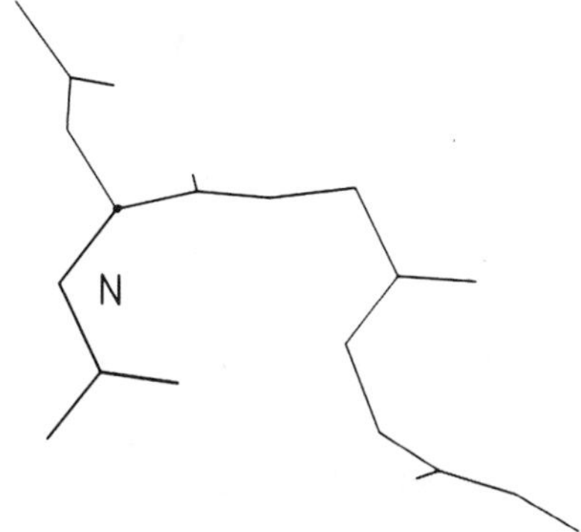

Figure 84. An Asn "pseudoturn" (SGA N170).

the described interaction in helical conformation but only as N-cap; however, Asp has good charge effects anywhere in the first turn, so that it is not position-specific. Fully half of the α helices begin with N, D, or S as initiator.

Another remarkable property of Asn (again, in distinction from Gln) is that it is the only residue besides Gly that is often found to have Lα conformation or, in general, positive values of ϕ. This is evident in a ϕ,ψ plot (compare Fig. 87 and Fig. 88), where Asn has about as many residues in Lα as in Rα as well as an unusually large number of "bridge" residues

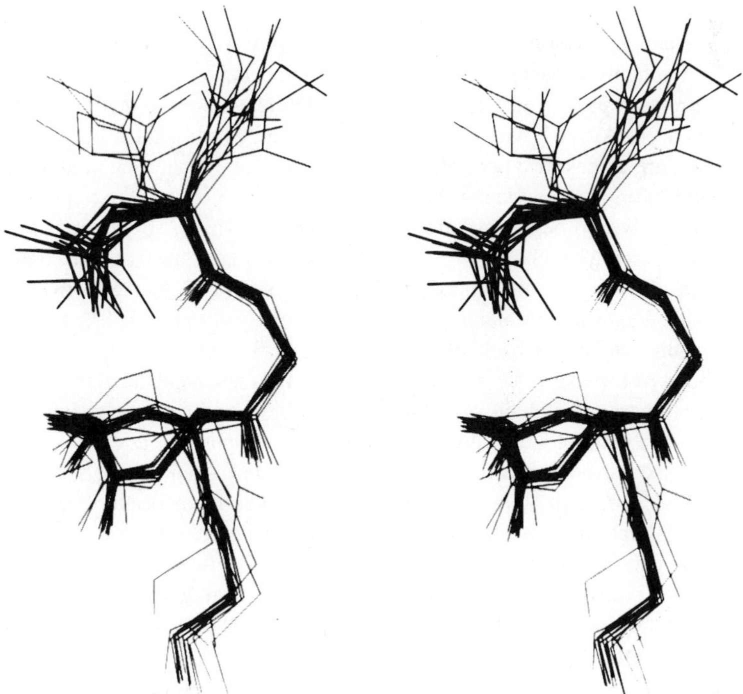

Figure 85. Stereo of 16 superimposed examples of Asn in the N-cap position of an α helix and forming an H-bond from Asn O_δ to a backbone NH in the first helical turn.

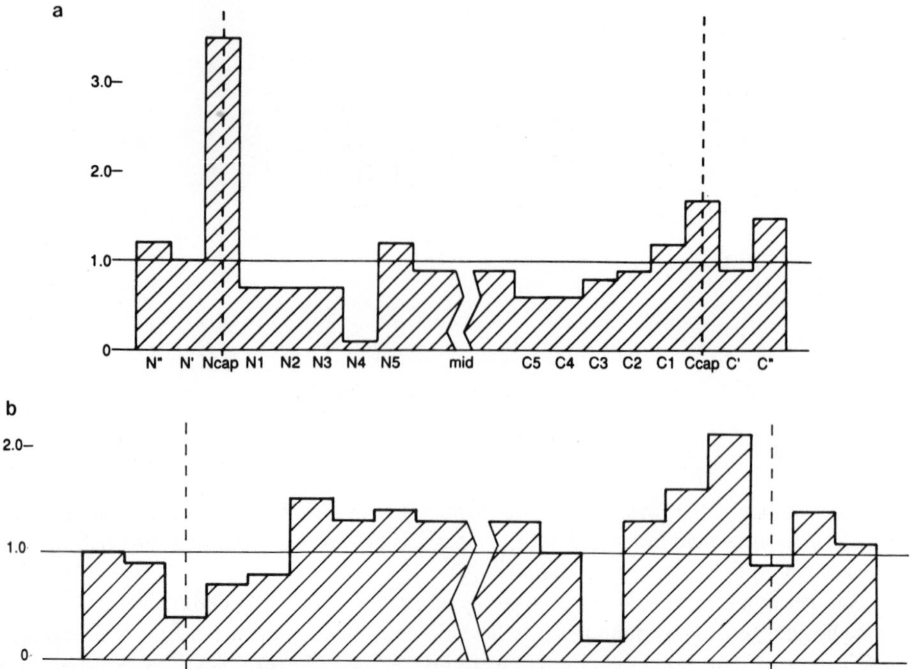

Figure 86. Relative preference values for Asn (a) and Gln (b) for specific positions in α helix. Asn is the strongest N-cap residue, while Gln has very different preferences.

between the Rα and Rβ regions and a spread-out and lopsided distribution in the upper left quadrant, which actually appears to avoid somewhat the actual β area near −120°,140°. In a sample of well-refined protein structures there were 326 residues with positive φ. Of the 101 nonglycines, 37 were Asn, 13 Asp, nine Ser, and six Gln; every residue but Pro occurred at least once. There are at least two possible ways to understand this unusual predilection of Asn for left-handed conformations. One is that it has a very directional hydrogen-bonding group (the amide) with only two degrees of freedom out the side chain, so that it can exert more leverage than most on the backbone. The other viewpoint is that if the Asn side chain is a close geometric mimic of the preceding peptide, as illustrated in Fig. 83, then effectively its C_α has two almost identical substituents and is therefore only weakly asymmetric. As a helix initiator, an Asn side chain can mimic backbone by adopting a β conformation, whereas at the C terminal of a helix or the end of a β strand, it must take up a left-handed conformation in order to put side chain where main chain would have been. These two cases, plus inverse common turns, are the most common situations for asparagines with positive φ.

Asparagine and Gln share a capacity for local side-chain-to-main-chain hydrogen bonding (also true for Ser, Thr, Asp, Glu, and His), which makes them occur frequently in non-repetitive loops. Glutamine, however, has a relatively good preference for α helix, since it is long and flexible enough to pack nicely on the outside of the helix rather than disrupting or competing with the backbone as does Asn. Glutamine is a relatively indifferent, plain vanilla residue that goes reasonably well with almost anything and has no extreme properties or violent preferences or aversions.

In marked contrast, Asn is an interesting, quirky, opinionated residue with many unique properties. The single methylene group by which they differ has remarkably profound consequences. In fact, it takes rather unusual circumstances for Gln to be the most conservative

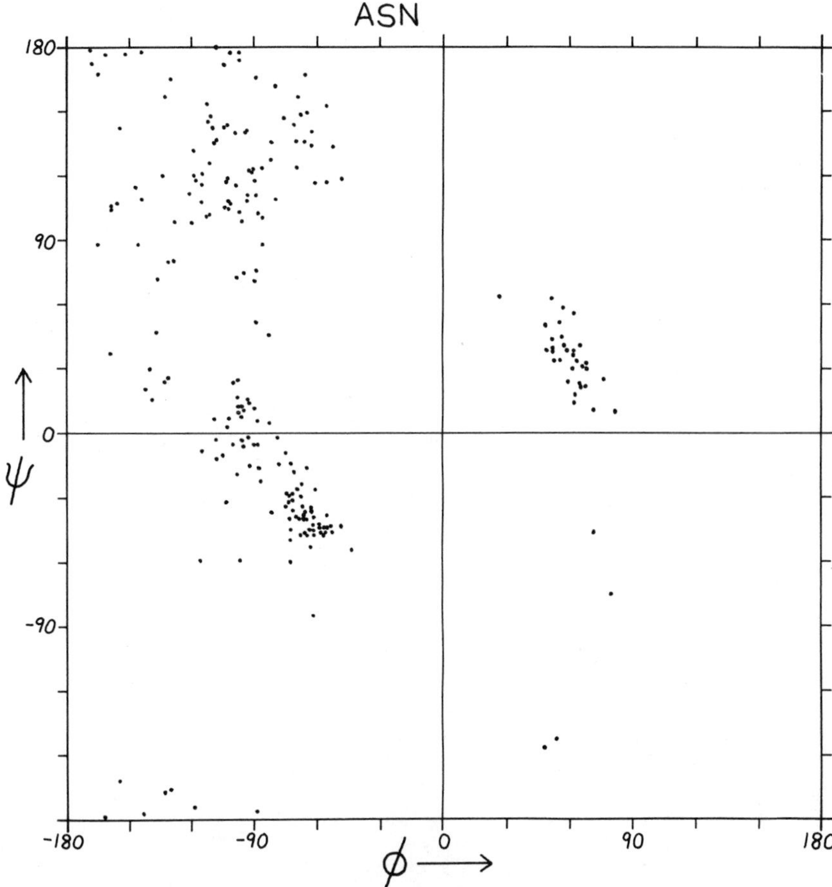

Figure 87. φ,ψ plots for Asn with many points in the Lα region.

substitution for Asn: only, in fact, where the role is amide H-bonding of a geometrically nonspecific sort. As a helix starter, Ser would be the most conservative replacement for Asn, since it shares the position specificity as well as the preference for that position; Ser would also be the conservative replacement for an Asn that bound carbohydrate (e.g., Fig. 84b). In the first or third position of a turn, Asp would be the most conservative, whereas in inverse turns or other left-handed conformations of Asn, Gly would be the conservative substitution. Conversely, in all roles involving unusual conformations, Asn is the conservative replacement for Gly, whereas for the purely steric roles of Gly, Ala would be the best available alternative (although often not very good). From this sort of analysis, one could perhaps figure out the role played by a particular instance of a residue by examining its replacements in related sequences.

H. Aspartate and Glutamate

Aspartate and Glu duplicate the Asn, Gln pair in every respect except that they are charged and their far ends are symmetrical to rotation. As one might expect, they are analogous to Asn and Gln in ways that depend on hydrogen bonding and geometry, but they have an

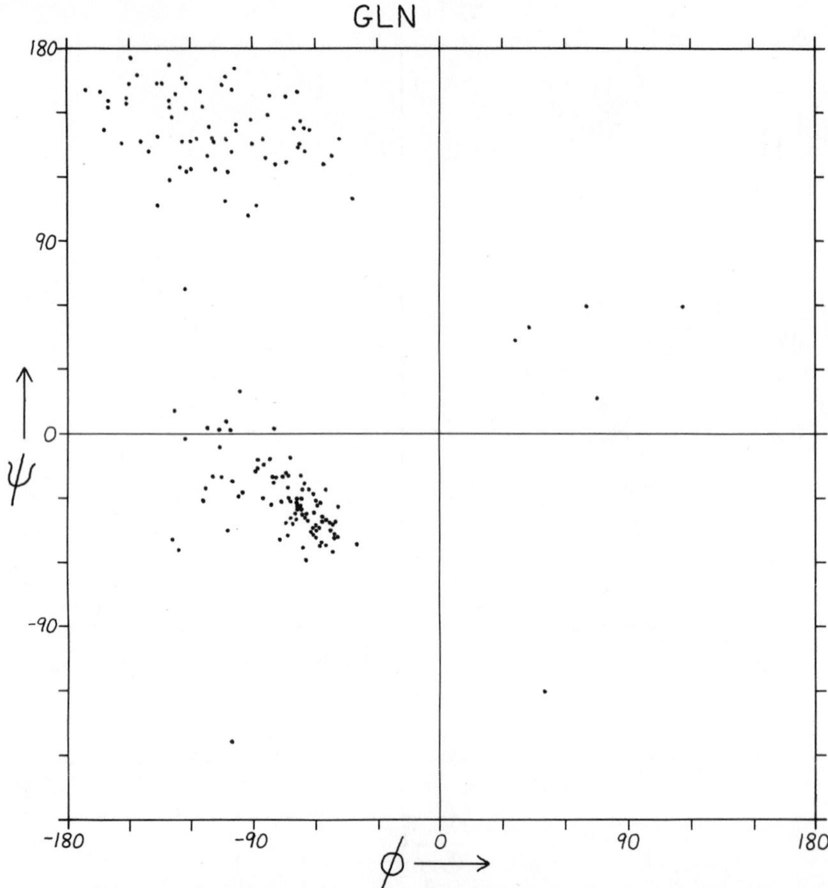

Figure 88. φ,ψ plot for Gln, which resembles the general distribution (Fig. 6) rather than the Asn distribution.

entire new set of roles that depend purely on their negative charge. The net effect is that they do not differ from each other as strongly as Asn and Gln, presumably because charge has such a dominant effect. Since the amide δ oxygen is more important than the δ nitrogen in asparagine's unusual roles (see previous section), it should not be surprising that aspartate can also do similar things with one of its carboxyl δ oxygens. However, in this set of roles the charge must be a slight detriment, since Asp is less prominently represented than Asn. For instance, Asp is the next most frequent residue after Asn to have positive φ values, but it is down by a factor of 3. Aspartate is common as a helix starter (N-cap); it can form the classic H-bond of O_δ to NH of residue $n + 3$, and it also has a favorable charge interaction (see below), but the preference is a bit lower and less position-specific than for Asn (see Table I). Aspartate is a prominent turn-forming residue, and its positional preferences are similar to those of Asn but less pronounced (Chou and Fasman, 1977). Glutamate is also a reasonable turn-forming residue (as are all hydrophilic residues), but it is even less position-specific than Asp.

However, in their charge-dependent roles, Asp and Glu are very similar and show very strong preferences. The classic case is their extremely frequent occurrence in the first turn of α helices, whereas Lys, Arg, and His are more common toward the C terminus (Fig. 89). This

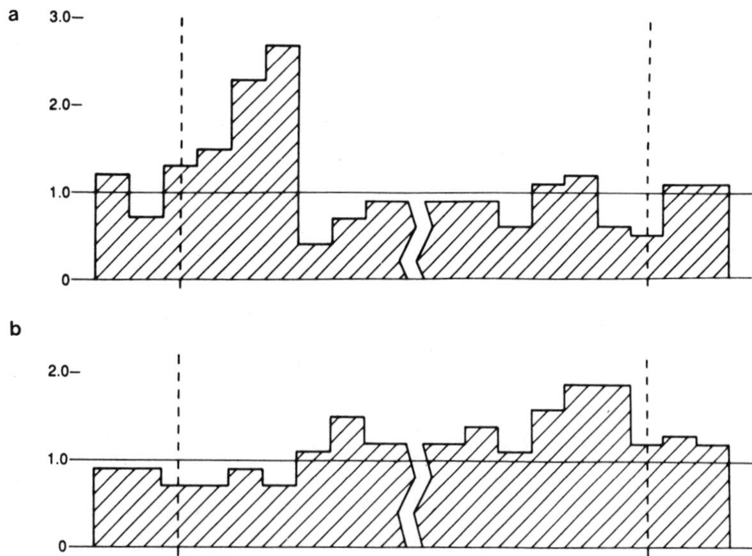

Figure 89. Relative preference values of (a) Glu and Asp and (b) Lys, Arg, and His for specific positions in α helix. The negatively charged side chains strongly prefer the first turn, and the positively charged ones the last turn.

can be explained mainly as a charge interaction with the helix dipole (Ptitsyn, 1969; Hol *et al.*, 1978). An individual peptide has a small dipole, with a partial positive charge on the H and a partial negative charge on the O; in a helix all of the peptide dipoles are lined up, giving an overall net dipole with a partial positive charge at the helix N terminus and a partial negative charge at its C terminus. Although there is considerable controversy about the quantitation of this effect, there is good evidence that it is indeed significant: the empirical distribution of charged side chains is such as to neutralize the helix dipoles; there is a tendency for binding of charged ligands (such as phosphates) to occur at the appropriate end of helices, sometimes with no other charges nearby (Pflugrath and Quiocho, 1985); for analogues of the S-peptide helix of RNAse, the percentage helix formation in solution correlates almost perfectly with the net amount and sign of charge at each terminus (Shoemaker *et al.*, 1987). Glutamate has the deserved reputation of being a very strong helix-forming residue, but that preference is almost entirely for the first turn; Glu is indifferent in the central regions of helices.

Aspartate and Glu, especially Glu, bind calcium, usually with about six oxygens as ligands. Often these sites are on loops and clustered tightly in the sequence; the classic case is the "E–F hand" sequence in reversible calcium-binding proteins (Fig. 90; Kretsinger and Nockolds, 1973). An even more extreme case of clustered binding is γ-carboxyl-Glu, the modified amino acid with two carboxyls that is used to bind Ca^{2+} on the activation peptides of some zymogens. Asymmetric charge distributions on the surface of a protein can form the site of interaction with another macromolecule. Aspartate and Glu can be active site residues also, such as in lysozyme or in acid proteases. Although Glu and Asp are charged at most biologically relevant pHs, a few enzymes operate in environments acid enough (e.g., stomach acid) for carboxyls to lose their charge. In addition to effects on activity, uncharged COOHs have been observed to H-bond with one another (James and Sielecki, 1983).

The two most general roles for charged residues are to stay on the outside (thereby orienting the neighboring backbone) and to form salt links (H-bonds with oppositely charged

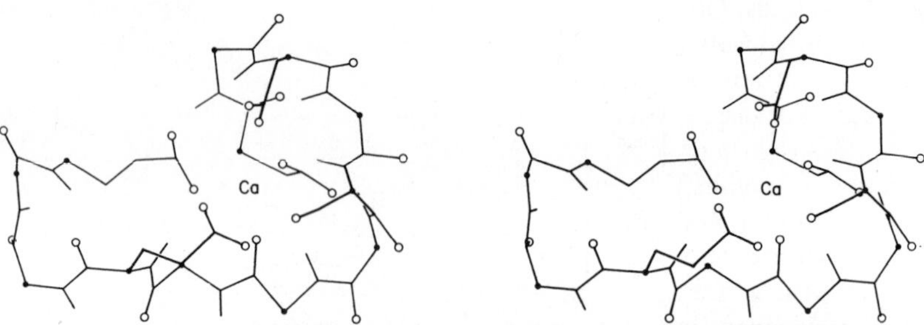

Figure 90. Stereo drawing of a Ca-binding "E–F hand" (CPV E–F loop) in which two Glu, two Asp, a Ser, and a backbone CO bind a Ca^{2+} ion.

groups). Charges are very rare on the inside, and those that do occur are almost always paired (Honig *et al.*, 1986). The paired internal charges make well-ordered salt links, often with a double H-bond between an Asp or Glu and an Arg (e.g., Fig. 91). Neighboring charges on the protein surface may or may not interact, either directly or through ordered water; salt links are common but not mandatory. They can occur between side chains n and $n + 3$ or n and $n + 4$ on an α helix, but most often they join residues distant from one another in the sequence. In most types of charge interactions, the polypeptide C terminus acts as an Asp or Glu, and the N terminus as a Lys.

I. Lysine and Arginine

Lysine and Arg are the opposite facet of the various charged roles described for Asp and Glu. As mentioned above, Arg and Lys (and also His) have a preference for the last turn of α helices (see Fig. 89), although it is not quite as strong as the comparable preference of Glu and Arg for the first turn. They stay on the protein surface with only rare exceptions. Lysine and Arg form salt links with negatively charged macromolecules, most notably nucleic acids. For nonspecific interactions, such as those of viral coat proteins, the positively charged residues can be on disordered loops or chain ends, which gives them freedom to interact with the negatively charged phosphates of RNA or DNA irrespective of position, conformation, or base sequence (e.g., "tails" of TBV, SBV). For sequence-specific or conformation-specific interaction with nucleic acids, such as required by repressor proteins or restriction enzymes, the positive charges must be accurately positioned, at least in the complex, so that their phosphate interactions can cooperate effectively with binding of the bases. Both amino acids are fairly often involved at active sites, Lys more often in a binding role and Arg more often as a

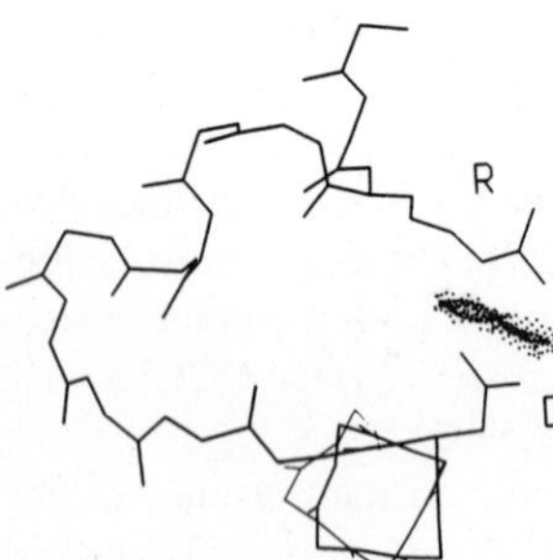

Figure 91. Small-probe contact dots illustrating the double H-bond of a salt link between Arg151 and Asp165 of cytochrome C peroxidase (CYP).

participant in catalysis. Often, in fact, the arginines most reactive to test reagents are those in an active site, presumably because other guanidinium groups are thoroughly H-bonded (see below) whereas an active-site Arg is held open for work.

In spite of their shared charge and long side chains, however, in many ways Lys and Arg are very different from one another. Lysine is by a very wide margin the most mobile of all the side chains, even when its backbone is well ordered. It is often invisible in electron-density maps beyond C_β or C_γ, and it buries the smallest fractional side chain surface area on folding of any residue (Rose *et al.*, 1985; Fig. 101) in spite of the considerable hydrophobic nature of much of its side chain. The Lys ϵ-amino group is charged at any reasonable pH and has three freely rotating hydrogens in tetrahedral geometry at the end of the side chain with the most possible conformations (Fig. 92). It is undoubtedly the part of a protein best able to interact with water molecules on their own terms, that is, with great flexibility, rapid rearrangement, and multiple interactions in tetrahedral geometry. Lysines are common on disordered loops or tails, but even in their typical position on ordered backbone, their side chains keep some of the high entropy of an unfolded state. They are the great solubilizers of globular proteins.

Arginine side chains, on the other hand, are only occasionally disordered and are more buried than lysines on the average (see Fig. 101). Both Lys and Arg have positive charges at the ends of long, flexible, locally hydrophobic side chains, but Lys seems to emphasize the flexibility, whereas Arg emphasizes the hydrophobicity relatively more. It is extremely rare for an Arg to be totally buried, but they usually make extensive van der Waals interactions along the side chain, and they can even curl around to produce a flat hydrophobic surface capable of conservatively replacing an Ile (Low *et al.*, 1976).

The guanidinium group at the end of an Arg has, in addition to its charge, five hydrogen-bond donors held in a large, rigid, planar array (see Fig. 93). They often interact with waters, which are usually seen as well-ordered peaks in electron-density maps if the Arg itself is well ordered. The spacings between optimal oxygen positions are quite good, with pairs bonded to neighboring quanidinium NHs about the right distance apart for near-neighbor waters and pairs bonded to the same nitrogen about right for next-nearest neighbors. The bond angles in these arrangements are not optimal for the waters, however, and of course the entropy is unfavorable

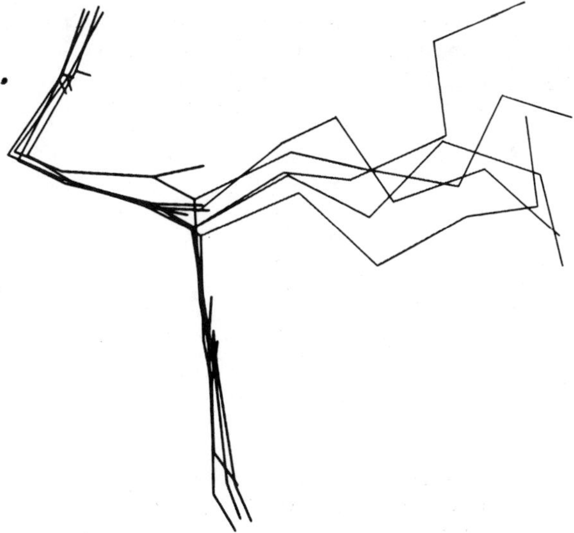

Figure 92. Multiple possible conformations for a Lys side chain.

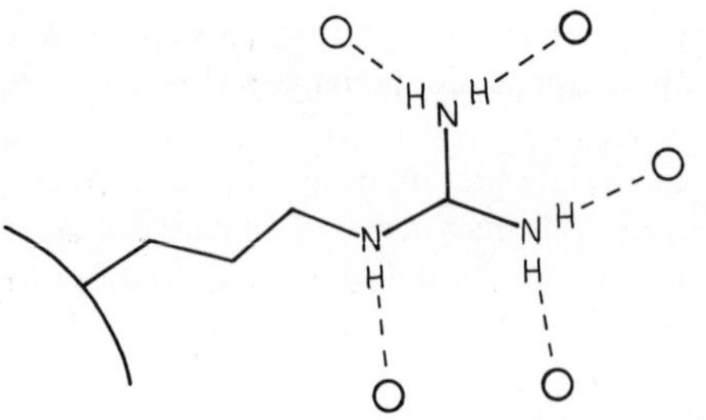

Figure 93. Diagram of the guanidinium group of an Arg, with the planar array of five oxygen atoms (either from protein or from solvent) that can H-bond with it.

for such a static single arrangement. It is not too surprising, therefore, that Arg typically satisfies several of its H-bonds with protein oxygens and only some with waters. The NH groups on neighboring Ns have very good geometry for making a double salt link to the two carboxyl oxygens of an Asp, a Glu, or a C terminus (see Fig. 91 or 94). No reasonable backbone conformation puts two consecutive carbonyl oxygens in the right position to interact with one Arg, but very often a single one does so. However, surprisingly, neither Arg nor Lys bonds at all often to backbone COs exposed at the ends of α helices. Serine, Thr, or Tyr OH groups are another good candidate for H-bonds with Arg (as acceptors, not donors), as also are Asn or Gln COs. Figure 31 shows an Arg in crambin, which illustrates the sort of complex, ordered network apt to surround an Arg: it makes a double salt link to the C terminus and an H-bond to the OH of Thr 2 and bonds two well-ordered waters. Arginine 17 holds together the two lobes of the small L-shaped crambin molecule and helps hold the N and C termini to the middle of the sequence and to each other. Replacement of this Arg with a Lys would presumably have disastrous consequences on stability. Incidentally, crambin, an insoluble seed protein, has no lysine.

J. Histidine, Phenylalanine, Tyrosine, and Tryptophan

Histidine, Phe, Tyr, and Trp are the aromatic rings. Phenylalanine is completely hydrophobic and has a strong preference for regular secondary structures. Tyrosine and Trp each have a single H-bond, which has only a small net effect on the hydrophobicity of Trp but makes Tyr almost indifferent to inside versus outside. Histidine has a charge group that is unique among amino acids in having a p*K* near neutrality, so that it can be made to gain or lose charge by quite feasible changes in its surroundings. Therefore, the overwhelmingly major role

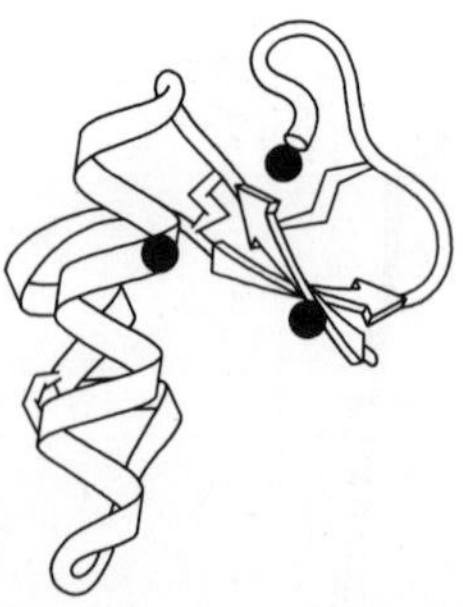

Figure 94. The Arg10 of crambin (CRN), which uses its array of H-bonded oxygens to tie together the two lobes and the middle with both termini of this small protein. The black dots show the C_α locations of Arg10 (on the helix at left), the C terminus, and Thr2.

of His is as an active-site residue or as a controllable element in conformational changes; it is rare elsewhere. Histidine is relatively buried on the Rose scale (Fig. 101), presumably because it is so often binding the buried charge of a reactive group, as in Fig. 95. The geometry of such binding can be quite finely controlled: there are only two variable side chain angles; the choice of proximal or distal N changes both distance and angle relative to the backbone; and the ring N not used as a ligand is typically H-bonded to some other group in the protein, giving further rigidity.

Phenylalanine, Tyr, and Trp are major constituents of the hydrophobic cores inside proteins. They pack against the peptides of the backbone, against the methyls of aliphatic side chains or the sulfurs of Met or disulfides, or against one another. This sort of packing is important in NMR, because ring currents produce large shifts in the resonances of, for instance, methyl protons, which allows them to be identified. In contrast to the larger aromatic systems of paired bases in nucleic acid structures, where there is a large energetic advantage to parallel stacking, the ring–ring orientations of protein side chains are more nearly random; in practice, that leads to frequent cases of perpendicular, herringbone-pattern stacking of aromatic side chains (Blundell *et al.*, 1986; Fig. 96). Clusters of aromatics are usually of mixed types, but occasionally Trp dominates (e.g., egg lysozyme). The most intriguing, and so far unexplained, such case is the dramatic layer of Trp just inside the level of the outer membrane surface in the photosynthetic reaction center structures (Deisenhofer *et al.*, 1985). Multiple tyrosines can be used for interaction with single-stranded nucleic acids, as, for instance, in gene 5 protein. Aromatic side chains probably form part of the pathway for some long-range electron transfer reactions.

In structural terms, aromatic side chains primarily provide interactions between separate pieces of secondary structure. The geometry is such that aromatics cannot pack well against their own local backbone in either a helical or an extended-chain conformation, but they can pack neatly against the backbone of a neighboring β strand (Fig. 60), or they can pack against the backbone of a neighboring helix at medium separations (see Fig. 19) or its side chains at wide separations (Fig. 97).

The aromatics are large, rigid structures with relatively few conformational degrees of

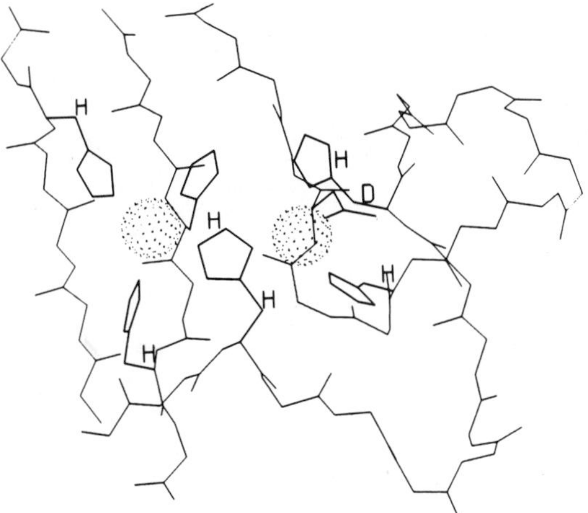

Figure 95. Six His acting as ligands to the Cu and Zn in Cu,Zn superoxide dismutase (SOD).

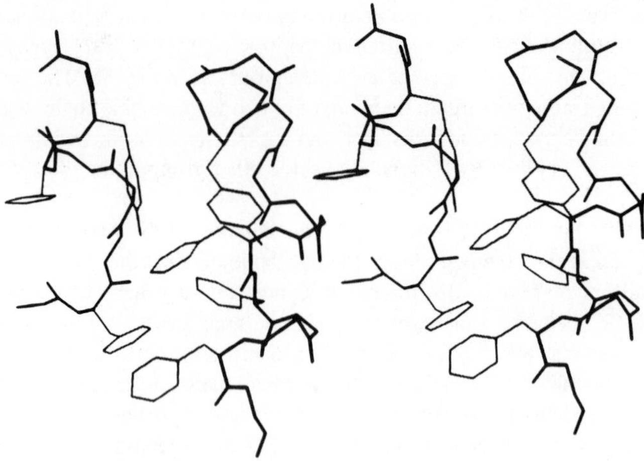

Figure 96. An internal cluster of aromatic side chains (CYP F198–F262), almost all of which show perpendicular, "herringbone" stacking with one another.

freedom, yet they do in fact stay very close to the optimal χ_1 and χ_2 angles in well-determined protein structures. On α helices, for instance (Fig. 82), the aromatic χ_1 angles are very tightly clustered around the two populated optima. Unfavorable torsion angles for aromatics usually have a functional rationale rather than just representing a structural compromise, and they are therefore most common in active-site histidines (e.g., His^{57} of the serine proteases; Sprang *et al.*, 1987).

Basically there are only three good values for χ_1 and two values for χ_2 in His and Trp (+90° or −90°), reducing to one for the symmetrical rings of Phe and Tyr. For side chains in general, the χ_1 with C_γ opposite the backbone CO (in most cases this is called −60°) is the lowest-energy state of the three, χ_1 with C_γ opposite the backbone NH is a close second, and χ_1 with $C_\gamma$opposite the C_α hydrogen is significantly less favorable. The minimum-energy conformation is actually a bit displaced, to −70° (Janin and Wodak, 1978). In α helix,

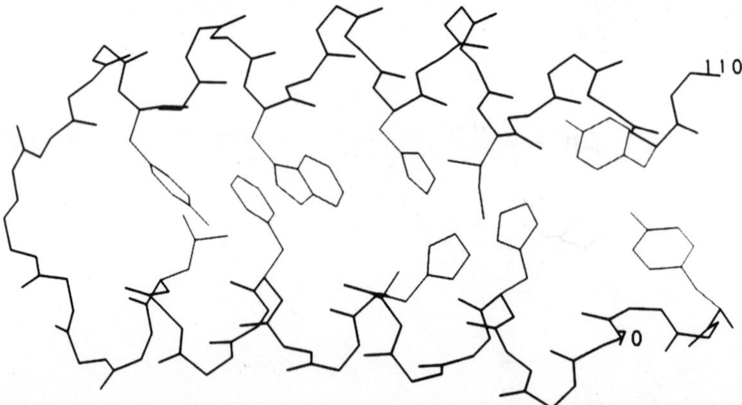

Figure 97. Aromatic side chains forming the contact between two rather widely spaced α helices (HMQ helices C and D).

aromatics populate the two best χ_1 optima (Fig. 82), with the only odd feature being that they are almost exactly equally populated. The third optimum is entirely missing because the ring would intersect the helix. On loops or on the exposed parts of β sheet, there is an extremely strong preference for the χ_1 opposite CO. However, for aromatic rings on the inside surface of antiparallel β sheet or on parallel β (which is buried on both sides, see Richardson, 1981), there is a mixture of all three conformations, with a preference for the "forbidden" χ_1 opposite H. The reason for this is almost certainly that it produces a tight packing of the aromatic ring against the neighboring β strand on the side with the narrow H-bond pair (see Fig. 98). Depending on the degree of strand twist, this interaction can be directly with backbone (Fig. 60), with the C_β of the neighbor-pair sidechain, or even with a neighbor-pair residue as bulky as Val (Fig. 98).

Aromatic rings are large enough so that their presence and local conformation have a significant effect on the surrounding structure, even in the absence of specific covalent or hydrogen bonds. They appear, for instance, to have a strong influence on two-stranded antiparallel β ribbons, occurring far more often than random just below a β hairpin (Fig. 60) and encouraging both twist and curl by their interactions with backbone and with one another on the concave inner surface of a β ribbon (Salemme, 1983; Fig. 24). As for many other examples we have seen in this section, their presence influences both secondary and tertiary structures in an extremely specific, but context-dependent, fashion.

VI. HYDROPHOBICITY, CHARGE, AND SOLVENT

Probably the two most classic and general statements about protein folding are that it is determined by the amino acid sequence (Epstein *et al.*, 1963) and that it is driven by the

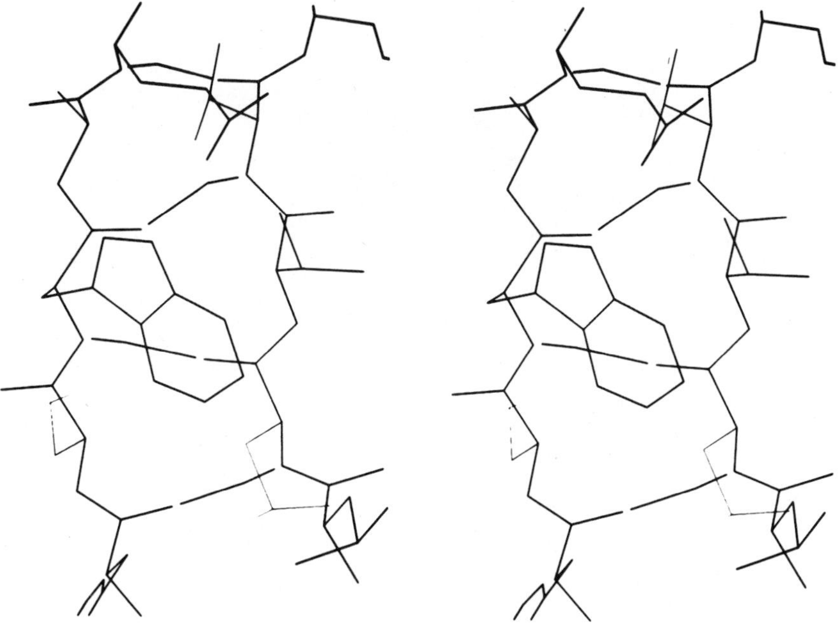

Figure 98. An example of an aromatic side chain on a buried β strand (DFR W5) in the normally disallowed χ_1 conformation opposite the C_α hydrogen. This conformation also occurs in Figs. 19 and 60.

entropy of removing hydrophobic groups from contact with the solvent (Kauzmann, 1959). Now that we have examined some of the conformational roles of the individual amino acids, we turn to the effects of hydrophobicity, charge, and solvent, some of which have already been mentioned.

The most basic level to be considered is the nature of the solvent environment in which the protein folds. Most of the proteins we describe here are soluble, globular proteins that operate in a physiological medium: aqueous solution of near-neutral pH, fairly high ionic strength, and with a sizable population of other small and macromolecules. Even globular proteins, however, may inhabit significantly unusual environments, such as the reducing conditions inside *E. coli* or the acid of the stomach. For the purposes of structure prediction, it pays to consider the organism, cell type, and compartment in which the protein occurs and any peculiarities of its genetics or its function, since any of these properties can influence the probabilities of some conformational features.

The most extreme case of such an importantly relevant environment, of course, is incorporation into a membrane. As can be appreciated in Fig. 99, the membrane has a profound

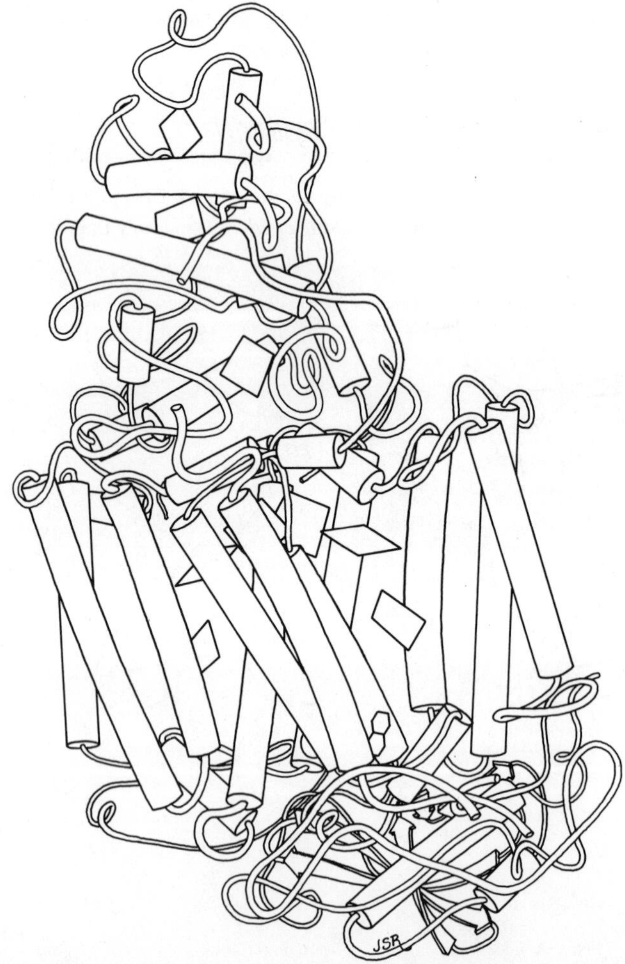

Figure 99. Simplified schematic drawing of the bacterial photosynthetic reaction center, the only membrane protein structure that has so far been determined at high resolution. The membrane portion is in the center, made up exclusively of helices. Prosthetic groups are shown as flat shapes.

effect on the permissible conformations, orientations, and topologies as well as on the hydrophobicity of the outer surface that interacts with the membrane lipids. Complete main-chain hydrogen bonding is stringently required, so that α helices, clusters of helices, or cylinders of β sheet are usual, whereas loops, other nonrepetitive structures, and isolated extended strands seem to be forbidden. Orientations documented so far are either along the membrane, as for amphiphilic helices, which lie along the surface, or else closely perpendicular to the membrane and extending all the way through it. Of course, membrane insertion logically requires some stage with either a free end or a nonrepetitive connection inside the membrane, but such features are probably only transiently permissible. Rather surprisingly, the membrane protein structures so far substantiated are a restricted subset of what is possible in solution rather than totally new organizations. Requirements for compactness are much lower (see Fig. 100), and optimism suggests that at some point we will encounter a truly unexpected surprise among these membrane structures.

In general the prospects seem excellent for prediction of membrane protein structures, since they are under much stronger and simpler constraints than globular proteins. Starting with the reaction center structure (Deisenhofer *et al.*, 1985), there is now accumulating a data base from which to determine residue preferences for lipid contact versus protein–protein contact versus hydrophilic channel surfaces versus interaction with lipid head groups at the membrane surface. Quite probably within a few years, membrane protein structures will be more reliably predictable than globular protein structures.

In globular proteins, the hydrophobic effect is evident as a partitioning of side-chain types between the protein interior and its exposed surface. Although far from perfect, that partitioning is one of the dominant principles of structure determination, with important effects on every type of conformation.

Although there is no doubt whatsoever of the overall importance of hydrophobicity, its detailed consideration has been complicated by the multiplicity of scales for measuring it. Hydrophobicity is clearly related to electrostatic charge, to hydrogen-bonding capability, and to surface area of aliphatic and aromatic carbons, but there is no consensus about how to weight these various factors. Hydrophobicity can be considered a property of entire residues, of groups of atoms, or of individual atom types; it can be measured experimentally by various solvent-partitioning effects (e.g., Wolfenden *et al.*, 1981) or estimated empirically as a function of position or accessibility in known protein structures. Almost certainly, the correct weighting of parameters should be different for the purposes of judging membrane insertion, antigenic potential, solubilizing ability, or globular protein conformation. Several hydrophobicity scales are compared in Fig. 101. The disagreements among them are about such things as whether the hydrogen bond in tryptophan makes it moderately hydrophilic, whether the aliphatic character or the surface location of proline is more important, and whether

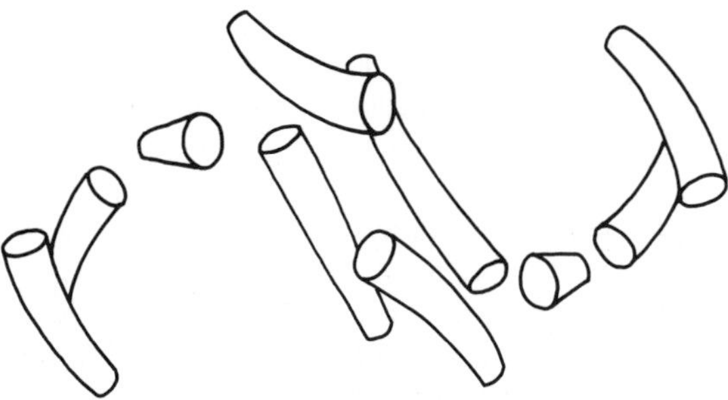

Figure 100. Cross section through the membrane portion of the photosynthetic reaction center, showing the relatively noncompact arrangement of the helices.

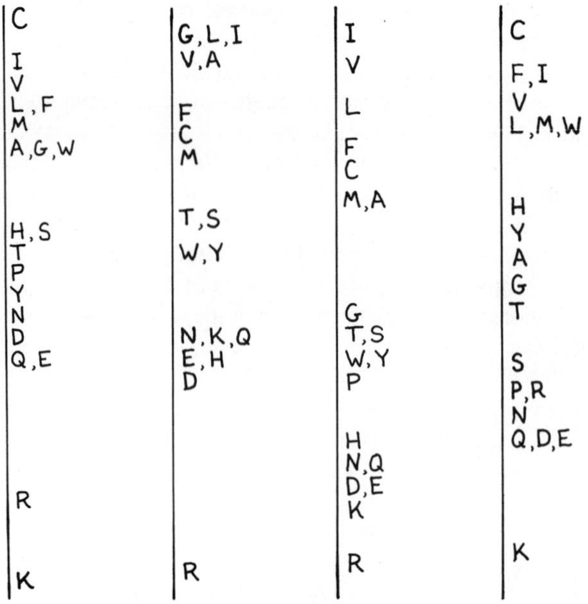

(a)	(b)	(c)	(d)
C	G, L, I	I	C
I	V, A	V	F, I
V	F	L	V
L, F	C	F	L, M, W
M	M	C	H
A, G, W	T, S	M, A	Y
H, S	W, Y	G	A
T	N, K, Q	T, S	G
P	E, H	W, Y	T
Y	D	P	S
N	R	H	P, R
D		N, Q	N
Q, E		D, E	Q, D, E
R		K	K
K		R	

Figure 101. Several hydrophobicity scales, approximately normalized to each other and increasing from bottom to top: from (a) Janin (1979), (b) Wolfenden *et al.* (1981), (c) Kyte and Doolittle (1982), and (d) Rose *et al.* (1985).

cysteine sulfurs have any hydrophilic character. In all probability there is no single correct answer to this set of questions, but for our present purposes we can rely on the general qualitative agreement of all the scales.

Patterns of hydrophobic versus hydrophilic side chains are very important for prediction of secondary and tertiary structure simply by virtue of their preferential occurrence on the inside versus the outside of various structural features. Those hydrophobicity patterns provide a rough translation of the three-dimensional structural features into the one-dimensional domain of the amino acid sequence. It is an inexact translation, unfortunately, because charged side chains are long enough so that their α carbons can be well inside the surface, buried hydrophilics can interact with each other or with hydrophilic prosthetic groups, and occasional large hydrophobics are often found on the surface. If the translation were more exact, structure prediction would no longer be referred to as a "problem." Even as things stand, hydrophobicity and charge repay consideration of their distribution as well as their overall occurrence frequencies.

Because of the limited size of protein domains and subunits, the great majority of α helices in globular proteins are at the surface, approximately half exposed and half buried. This puts a hydrophobic periodicity of about 3.6 residues into their sequences, which can be seen on a "helix wheel" representation such as Fig. 102 (Schiffer and Edmundson, 1967). The same periodicity is seen in the positional helix preferences of Table I, since the need for the end of a helix to connect back to the rest of the protein puts it approximately in register with the hydrophobic periodicity. There is a discontinuity at the ends, however, and both N-cap and C-cap are hydrophilic. We know from the work of Eisenberg (Eisenberg *et al.*, 1982) that the hydrophobic periodicity on a helix should not be too extreme, or else the helix will bind to membrane surfaces instead of forming part of a globular protein.

Charges on helices share the 3.6-residue periodicity of hydrophilics in general, but they

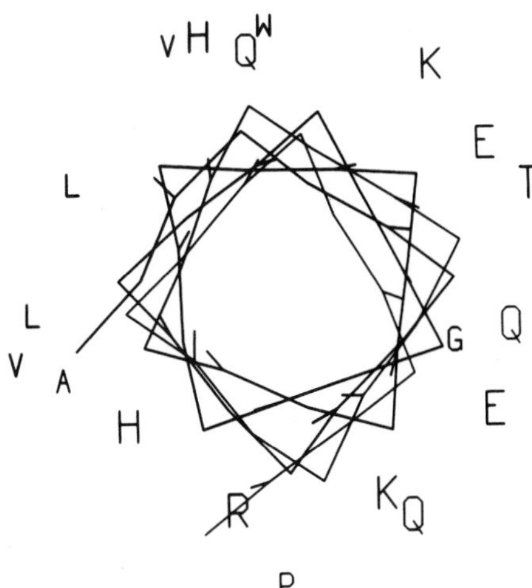

Figure 102. Example of a helix-wheel layout for an actual globular protein helix (TIM 178–196). Hydrophobic residues, mainly on the left, are in small letters, and hydrophilic residues in large letters.

show an even more significant distribution relative to the two ends of the helix. Charges are somewhat more common near the ends than in the middle, and there is a very strong preference for negative charges in the first turn and positive charges in the last turn. This effect was discussed in Sections V.H and V.I. Of particular interest is the fact that such asymmetric charge distributions are apparently effective in the early stages of protein folding, so that they can have a fairly direct influence on the probability of α-helix formation in a local region of the sequence.

The β strands alternate side-chain direction toward the opposite sides of the β sheet, so that for antiparallel β structure, which usually has one side buried and one side exposed to solvent, there is a hydrophobic–hydrophilic periodicity with a repeat of about two residues (see Fig. 103a). This periodicity is often quite apparent in the amino acid sequence, although it is nearly absent for strands at the edge of a sheet and can be thrown out of register by a β bulge (see Section III.B). Parallel β strands are typically buried on both sides, so that they show an entirely different pattern of quite solid hydrophobics near the center of a strand (except for edge strands) and hydrophilics, even charges, near each end of the strand (Fig. 103b). Most of the characteristic differences of residue preferences between parallel and antiparallel β structure result from this effect, such as the fact that Thr is much more common in antiparallel β and Val and Ile in parallel β (Lifson and Sander, 1979). However, it should be much more effective to apply these rules as distribution patterns rather than differences in overall β preferences.

Turns have a very striking tendency to occur at local maxima of hydrophilicity (Rose and Seltzer, 1977) because by their nature they are almost always exposed at the surface. They also are encouraged by residues with side-chain H-bonding capability, which can interact with the central peptide that is isolated from any main-chain H-bonding possibilities (see Section III.C). Nonrepetitive loops in general are rather hydrophilic as well, for much the same reasons: they are almost always exposed on the surface (Leczczynski and Rose, 1986), and they rely heavily on side-chain-to-main-chain H-bonding (Richardson, 1981), which implies hydrophilic side chains. Completely disordered structure is the most hydrophilic of all because it requires a fairly thorough absence of large hydrophobics as well as favoring groups that interact well with solvent. Tails at the N or C termini of a protein, in particular, are very likely to be disordered if they include no large aliphatic or aromatic hydrophobics and a large proportion of charges, H-bonding hydrophilic Ser, and mobile Gly.

The solvent, of course, is indirectly responsible for all the effects we have discussed in

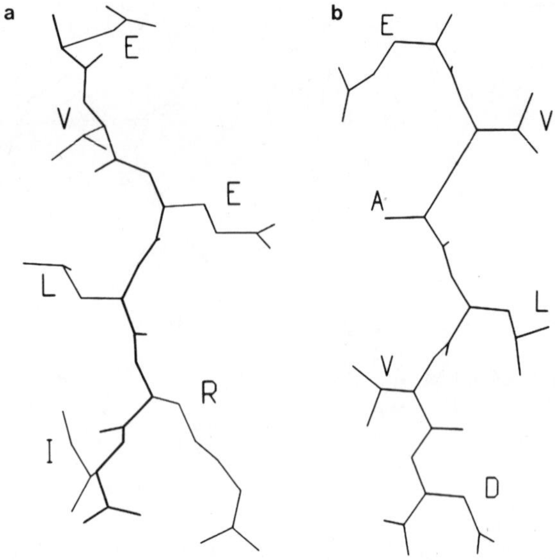

Figure 103. (a) An example of the hydrophobic–hydrophilic alternation usually seen on antiparallel β strands (RP2 157–162); (b) an example of the hydrophobic center and hydrophilic ends often seen for parallel β strands (LDH 47–53).

this section so far. Water is statistically more ordered next to hydrophobic side chains than in bulk solvent (an effect that can be visualized by realizing that the waters cannot H-bond with the hydrophobics and are thereby restricted in orientation) without any enthalpic compensation such as they get by H-bonded ordering next to hydrophilics. Therefore, it is more favorable for hydrophobic groups to interact with each other and waters with waters than for hydrophobics to interact with waters. This, as we have mentioned before, is the dominant effect that drives proteins toward compact, folded conformations in solution and is a recurrent theme in all these discussions. There are, however, two other types of more specific solvent effects to consider here. One is that in fact not all proteins live in, or are designed for, the same solvent conditions. In particular, the overall folding constraints are totally different for membrane than for soluble proteins and may vary in less severe but important ways for different solvent environments.

The other type of structurally relevant solvent effect is caused by the atomic details of specific interactions between individual water molecules and specific side-chain or peptide groups in particular positions and conformations. We have seen one example of this in Section V.I in the contrast between Lys and Arg. Lysine interacts with waters very strongly, but in a highly mobile and disordered mode, whereas Arg has a well-ordered, nearly planar array of five oxygens H-bonded to its guanidinium group, made up of a combination of waters and protein oxygens (Fig. 104). Glutamate, Asp, Gln, and Asn have lesser but similar H-bonding arrays, of which some portion is generally contributed by water. Hydroxyl groups can have very specific interactions with water, but they do not always result in visible electron-density peaks for ordered water, since the OH groups themselves are often rotationally disordered. Such interactions, however, can have a profound effect on protein structure even when they are not crystallographically ordered. Charged side chains on the surface, for instance, may not form direct, specific salt links with their opposite counterparts but may still prefer strongly to be near opposite rather than like charges and interact by a statistical network of indirect salt links through shared water molecules. The ordering around hydrophobic groups is also a

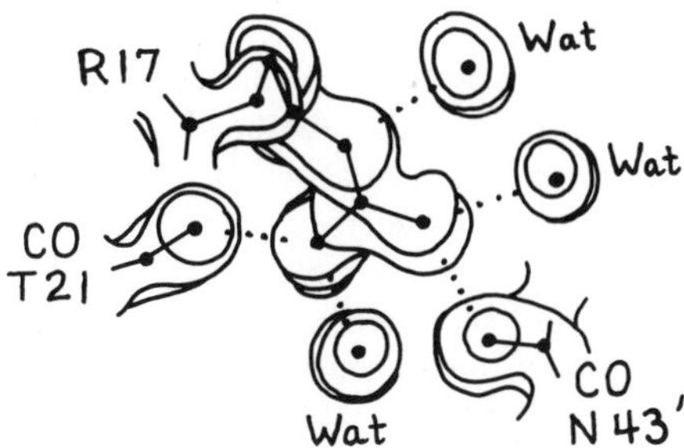

Figure 104. The arrangement of ordered water molecules and protein oxygens seen around an Arg side chain (CRN R17).

statistical and nonspecific one. In electron-density maps, the water molecules next to hydrophobic surfaces are seldom visible except when they can be held in place by neighboring polar groups or by adjacent protein molecules in the crystal, or after refinement down to low levels of occupancy (Teeter, 1984).

At a higher level of organization, solvent is by far the dominant player in determining supramolecular interactions. Association of subunits or macromolecular complexes is almost entirely driven by the change in surface area buried from solvent (Chothia, 1976). These buried surfaces are often as tightly fitted as the protein interior. However, in many domain interfaces, in some of the more mobile subunit contacts, and especially in crystal-packing contacts, a rather different sort of interface is seen, consisting of a single layer of ordered waters (Fig. 105; Richardson and Richardson, 1987). These contacts are less hydrophobic,

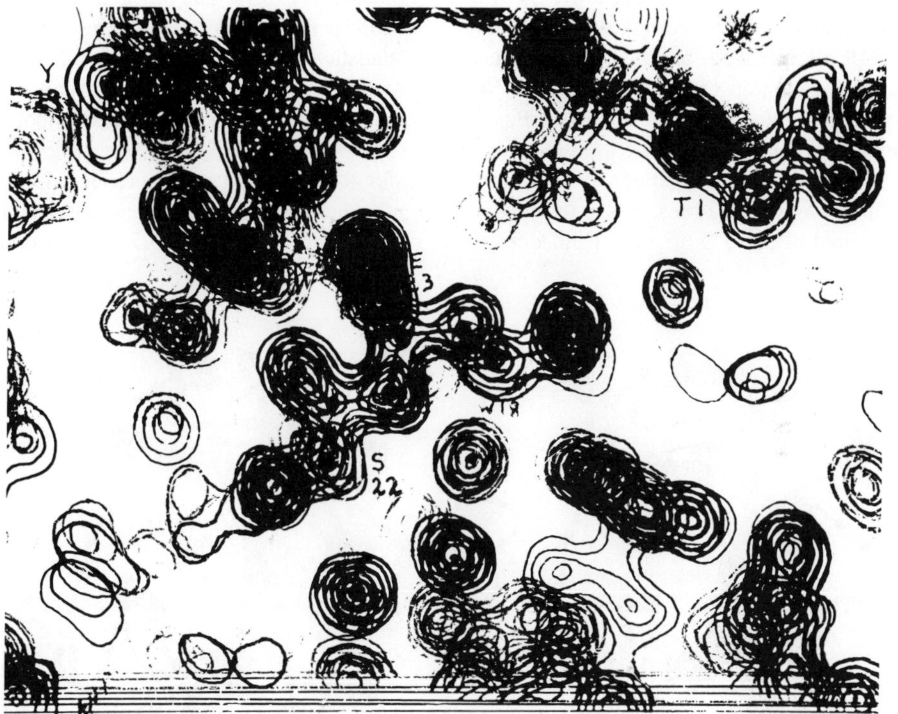

Figure 105. A crystal packing contact in the structure of crambin. One molecule has a helix starting near the center and running to the top left corner, and part of a second molecule shows at the bottom. In between are several round, dark peaks of electron density that represent ordered water bridging between the two protein molecules.

have relatively few direct contacts between protein atoms, and depend more on hydrogen bonds that bridge through the waters. These interactions are also driven by exclusion of the waters that were previously bound to the two protein surfaces, but in this case only about half of them are excluded as opposed to almost all of them in the more classic subunit contact. These water-layer interfaces bind more weakly, and they are more forgiving: the two surfaces need not be closely complementary in either shape or H-bonding pattern. However, they may be important biologically in systems requiring motion or reversible assembly or in the early evolution of tighter contacts.

VII. HANDEDNESS

One of the general principles of protein structure is a preference for one handedness over the other, seen at all levels of structure. The ultimate cause of all the handedness preferences is of course the asymmetric α carbon of the naturally occurring L-amino acids (see Section II), and the mechanism by which that creates larger-scale handedness is almost always through interactions of the β carbon. Helices are right-handed because left-handed ones would have a slightly unfavorable contact of C_β with the preceding CO, repeated at every turn. Left-handed α helices have been successfully produced by peptide synthesis using D-amino acids, for which C_β and H_α are reversed (Blanc and Kaiser, 1984). In tight turns (Section III.C) the differences among the common types involve bumps of the oxygen of the central peptide with the C_β of residue 3. Not only is this a less severe bump than the one in the left-handed α helix, but the turn involves only one such awkward position rather than a recurrence at every residue; therefore, a single glycine can solve the problem quite satisfactorily. The detailed mechanism is less clear for the preferred twist of β strands (Section III.B; Fig. 106); again, close contacts of C_βs are involved, but they act in a much more statistical fashion since they are distant in the sequence and can be compensated by a combination of minor readjustments in other parameters. However, the effects of this strong statistical bias are pervasive and important (Fig. 106).

The handedness preferences described above for secondary structures produce, in their turn, strongly handed features in the supersecondary and tertiary structure. Probably the most important of these is the right-handedness of crossover connections (Fig. 107), which domi-

Figure 106. Stereo ribbon display of carbonic anhydrase C (CAC), with the active site Zn and nearby histidines. The size and twist of the β sheet can be appreciated in this view.

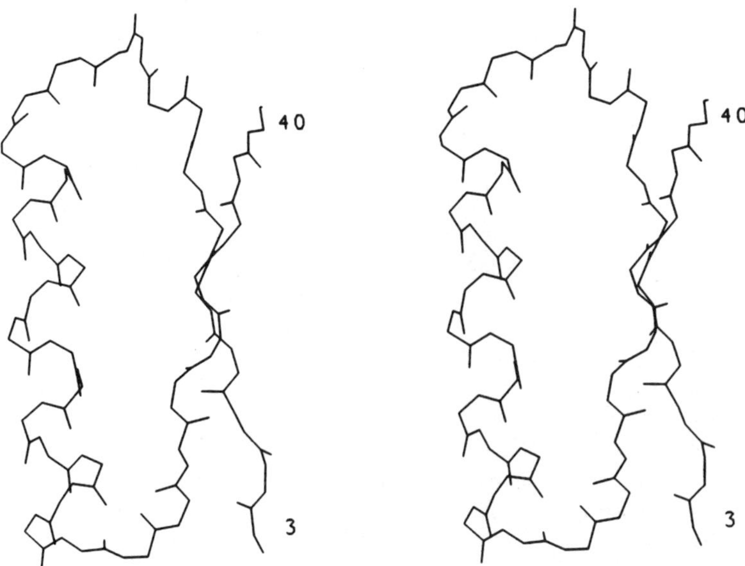

Figure 107. An example of a right-handed crossover connection (ABP 3–40) joining two parallel β strands with a right-handed overall spiral.

nates the organization of α/β proteins, including the most frequently encountered type of tertiary structure, the doubly wound parallel β sheet or nucleotide-binding domain (see Section IV.C). All-α proteins reflect the handed nature of helix–helix packing, which in turn reflects the handed spiral of side chains on the surface of right-handed α helices; antiparallel helix clusters always show a left-handed superhelical twist (Section IV.B). Antiparallel β barrels are handed at every possible level, with twisted β strands, twisted barrel cross-sections, and, most notably, handedness in the swirl direction of their Greek key motifs (Section IV.D). Disulfide bonds have a strong local handedness of conformation (Section V.C), but there are no recognized long-range handedness features of small disulfide-rich proteins. The chances are that such regularities exist but we have not yet learned to see them.

One of the best things about handedness preferences is that they are overwhelmingly strong. Local amino-acid preferences and local conformational regularities are always fairly weak, in the range from 1.5 : 1 to at most 20 : 1 (for prolines in the middle of helices). Most of the long-range handedness preferences, in contrast, are 10 : 1 to 100 : 1 or better. Since prediction is inherently a matter of playing the odds, we should concentrate on those factors that promise the best return.

VIII. HISTORY: EVOLUTION, FOLDING, AND FLUCTUATIONS

Informative and remarkable as the techniques are that show us the actual three-dimensional conformations of protein molecules, they do not deal explicitly with the proteins as historical objects in time. In addition to the energetics of the equilibrium structure, biology puts kinetic constraints on proteins at three very different time scales: over the evolution of the amino acid sequence; during the expression and folding of the actual molecule; and in the course of the rapid, atomic-level events of binding and catalysis. These historical processes have more in common than one might suppose from the fact that their time scales may differ by as much as 20 orders of magnitude.

Probably the most evident effect of evolutionary history on protein structure is the prevalence of approximately duplicated parts. Domains, and perhaps smaller sequences as well, are apparently interchanged within and between proteins. In multidomain proteins, about half the time the domains are totally different from each other, as in pyruvate kinase (Fig. 36); but in the other half of the cases they are more like each other than like any other protein, as for the γ-crystallin domains shown in Fig. 108. This presumably means that about half of the multiple domains are a result of internal gene duplication (Richardson, 1981). Unfortunately, that relationship is seldom evident in the raw amino acid sequence, but it can sometimes be picked up by similar patterns of secondary-structure predictions, and it is sometimes reflected in similarity of intron–exon patterns.

Within domains, duplicated structural patterns also seem much more frequent than random: the four most common tertiary-structure patterns (doubly wound parallel β sheet; as in Fig. 109, Greek key β barrel, up-and-down four-helix bundle, and singly wound parallel β barrel) are each built of two equivalent halves, and many other proteins have multiple parts, such as the four heme-binding sites in the reaction center cytochrome (Deisenhofer *et al.*, 1985; Fig. 99), the E–F hands of Ca-binding proteins, the halves of ferrodoxin, or the half-domain units in γ-crystallin (Fig. 110). Many of these repeats may merely reflect the fact that simple organization patterns are apt to include possible repeats. However, it is also quite possible that they are relics of the early history of proteins, when rather short sequences self-associated into domains. Later, after gene duplication had joined two such identical pieces with a workable connection, they could then mutate independently to optimize the whole structure but would still retain the approximate repeat in their tertiary structure. For γ-crystallin, evidence for this sort of history can still be seen in the amino acid sequence (Blundell *et al.*, 1981), with a primitive four-strand Greek key motif first duplicating to make an eight-stranded domain and then duplicating again into the two-domain protein.

Within closely related groups of proteins one can watch the moderate readjustments that preserve overall structure in the face of the mutations caused either by functional selection or by random drift. Subunit relationships and domain relationships often change, but the alterations within a domain are more subtle. Entire sections may be added or deleted, especially at

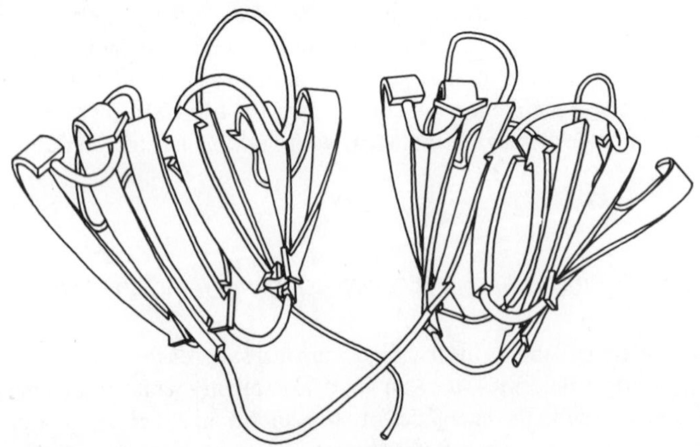

Figure 108. Schematic drawing of γ-crystallin (GCR) showing the two very similar domains into which it is divided.

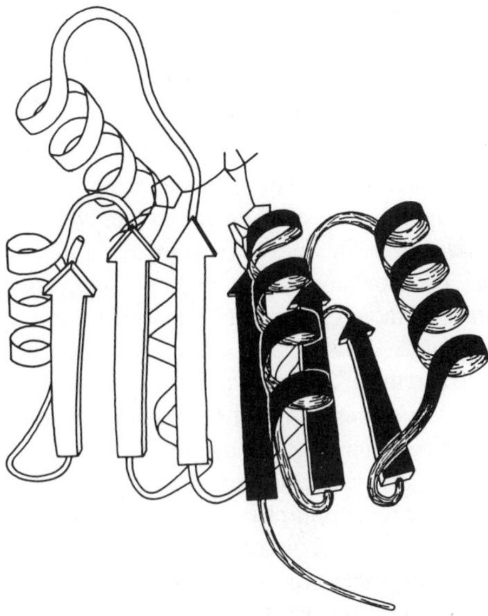

Figure 109. Schematic drawing of one domain from LDH showing the first of two similar halves in black and the second in white.

the N terminus but also internally; loops may change position on the surface; and secondary-structure elements may change in length or regularity; but we have no documented cases of secondary-structure elements rearranging their topological relationships.

When there is a mutation in an internal-contact residue, then there may, of course, be a compensating mutation in a neighboring residue that preserves packing geometry; however, it is more common for the backbone to shift slightly to accommodate the change (Lesk and Chothia, 1980). These backbone shifts often include relative rotation and/or translation at a helix contact or changes in local ϕ,ψ and β-strand spacing (Bryan *et al.*, 1986). If the local shifts are in a critical location, their effects may have to be neutralized by mutation elsewhere to provide compensating shifts (Lesk and Chothia, 1980).

Interestingly, very nearly these same types of changes occur during protein folding and during functional motions. Wholesale deletions occur during processing from precursor molecules, and some proteins are thought to have more regular secondary structure early in the

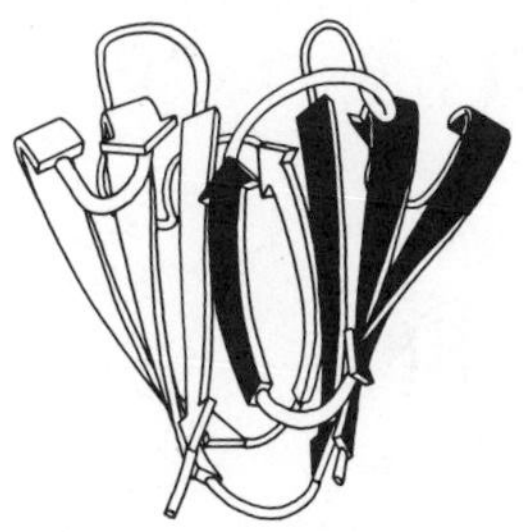

Figure 110. One domain of γ-crystallin (GCR) showing the first of its similar halves in black and the second in white.

folding process than at its end. At least for the case of pancreatic trypsin inhibitor (Creighton, 1977), however, folding has been shown to involve more drastic rearrangement than any that has yet been documented between evolutionarily related proteins. Of course, the primary step of condensation from an unfolded chain is also a drastic one, presumably similar to the very earliest events in evolution.

Functional motions in proteins most often involve either the movement of individual side chains or of loops (e.g., Fig. 111) or else the relative movement of subunits or domains. Hinging motions between domains can be well described in a first approximation as a relative pivoting of rigid bodies. The domains cannot remain rigorously rigid, however, and still maintain their large contact surfaces (see Fig. 112); instead, they use the same sort of subtle rearrangements that accommodate mutations. For hexokinase, a β sheet increase its "arch" and twist to help swing one end toward the active site (Anderson *et al.*, 1978). Citrate synthase uses small relative shifts between contacting helices (Lesk and Chothia, 1984), orchestrated so that their effects are cumulative and produce an overall bending of the domain. Another type of functionally important motion is a transition from disorder to order, such as the mobile loop in triose phosphate isomerase that folds into place over the substrate once it binds (Alber *et al.*, 1981); that sort of process is directly similar to protein folding.

The relevance of these historical processes to structure prediction is indirect but sometimes quite important. The most reliable predictions that can be made are those based on strong homology with another protein, or a large part of one, with a known three-dimensional structure. In such cases, one needs to pay careful attention to the rules for allowable types of change among related structures (e.g., Blundell *et al.*, 1987). The most frequent change of all is probably shortening or lengthening at either the N terminus or the C terminus. Also, between relatively distantly related proteins there is apt to be a change in the number or geometry of subunit contacts. Insertions or deletions are far more likely to occur between secondary

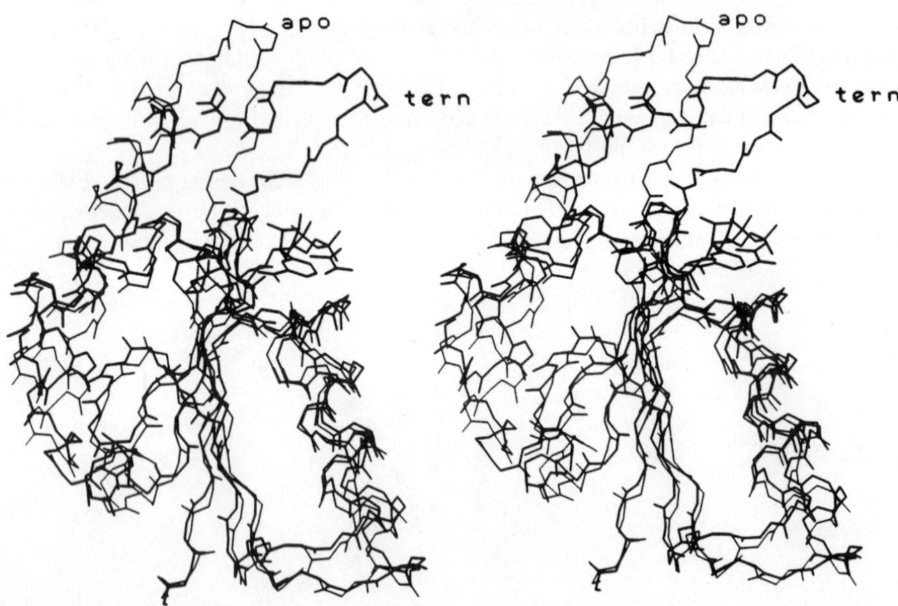

Figure 111. Stereo of the first domain of LDH in the apo form superimposed on the ternary complex structure to show that there is little change in general but a very large movement of the loop at the top.

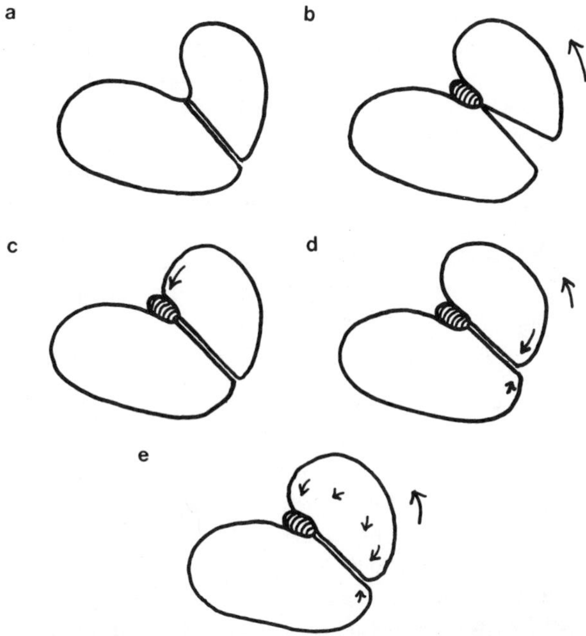

Figure 112. Possible scenarios for a "hinge" motion of a two-domain protein when the open form (a) closes down around a substrate or inhibitor. Rigid hinging would tear the domain contact apart (b). Proteins can accommodate by local motions near the substrate (c) (as shown also in Fig. 111), by hinging open and then conforming locally on the back side to close the gap (d), or else by many small, cumulative rearrangements throughout the domain as in (e).

structure elements than within them, but even in nonrepetitive loops, we have seen (Section III.C) that there can be fairly stringent limitations on allowable possibilities that mesh properly with the surrounding structure.

Folding is involved largely because it apparently restricts the possible tertiary arrangements to those that can be reached by a folding chain. In Section IV we saw a number of heuristic rules about such limitations for the various tertiary-structure categories. One very general limitation produced by folding constraints is that the polypeptide chain should not be knotted; this greatly limits, for instance, the allowable β-sheet topologies to 75% of those theoretically possible for five-stranded β sheets and to 55% for six-stranded ones (Richardson, 1977). The right-handedness of crossover connections, the swirl direction and "Ray's rule" for Greek keys, and the fact that doubly wound parallel β sheets always have the N rather than the C terminus in the center (see Sections IV.B and IV.C) are higher-level restrictions that are probably also imposed by the nature of the protein-folding process. Very fortunately for the enterprise of prediction, the functional motions seen in native proteins almost never involve much change in the "core" framework of a domain.

IX. THE TENSION BETWEEN HIERARCHY AND COOPERATIVITY

Descriptions of protein folding center around one of two complementary principles. Experimentalists generally emphasize the highly cooperative nature of the process, which in many cases consists of a pure two-state transition with no observable intermediates. Theorists and observers of the final folded state tend to emphasize hierarchies and intermediate steps along the way, such as helices, helix pairs, supersecondary structures, and domains. Both views have received new support from recent experiments. In favor of cooperativity, mutagenesis studies (e.g., Shortle and Lin, 1985) have found that stability changes are additive and that a single stabilizing mutant can compensate for destabilizing mutants almost anywhere throughout a protein. On the other hand, hierarchy is supported by peptide studies (e.g., Shoemaker *et al.*, 1987) that demonstrate formation of isolated secondary structures in short sequences.

Domains provide the one indisputable example of hierarchical folding at the tertiary-structure level. They have proven enormously fruitful in the description of native structures in spite of some disagreement over their exact definition. They are also often experimentally observed as folding intermediates, each domain showing its own cooperative transition (Pabo *et al.*, 1979). It has also become startlingly commonplace for geneticists to analyze protein function by deleting or rearranging large sections of the sequence, which are also usually referred to as domains.

It seems probable that both hierarchy and cooperativity are important principles in protein folding, and, in fact, the tension between these two extremes is much of what gives proteins their versatility, responsiveness, and functionality. A rigorously cooperative structure with the size and complexity of a protein would undoubtedly be unable to fold on anything approaching a biological time scale, whereas a completely hierarchical protein would be too rigidly and provincially structured to achieve long-range effects such as allosteric control.

On this view cooperativity and hierarchy are coexistent and complementary, and they form a continuum, not a dichotomy. Some pieces of a protein presumably are stable enough to form in isolation or to start the folding process, whereas others are marginal and only become uniquely ordered as part of a later cooperative transition. It is even possible that such marginally stable pieces of structure, frustrating as they are for prediction, are actually necessary to achieve successful folding rather than random bits of incompletely optimized design. An important point to keep in mind, however, is our present lack of proof that the most strongly predicted secondary structures are indeed the ones that form first. For all we know to the contrary, compact loops may form first rather than helices. In the near future, such hypotheses will be testable. For the time being, we must simply bear in mind that proteins are genuinely cooperative as well as hierarchical; hierarchy makes folding possible (and, with luck, perhaps prediction as well), but cooperativity and internal interrelatedness are crucial to biological function.

X. IMPLICATIONS FOR STRUCTURE PREDICTION

In addition to the specific facts that have been presented in this survey of protein conformation, there are several general ways in which these ideas bear on the methodology of structure prediction.

The most general lesson is one that has been pointed out by almost everyone in this field, namely, the general problem that local interactions in the sequence can only partially determine even the local levels of structure. Proteins are hierarchical, or modular, only to a limited extent, and the rest of the protein has an influence even on local secondary structure (Wilson *et al.*, 1985). One cure for this, which is being tried with some degree of success, is to determine the overall tertiary-structure type and then work within its rules for doing the more detailed prediction.

Perhaps the most promising implication of this present analysis is a strategy for making use of the greatly expanded and increasingly accurate data base of known three-dimensional structures. The improvements obtainable by recalculating conformational preferences from a much larger data base have been very disappointing. Plausibly, the problems with standard prediction methods did not result from statistical or sampling errors except perhaps for a few of the rarest amino acids, which by definition do not greatly affect an overall prediction. The primary problem, of course, is the one noted above of cooperative effects from the rest of the structure. However, there is an easier source for partial improvement in prediction simply by collecting statistics for more narrowly defined cases. With the relatively small number of known three-dimensional structures available when most types of prediction strategies were first explored, statistics needed to be collected for quite broadly defined categories in order to obtain useful levels of accuracy. Those pioneers understood the properties of statistics well enough to define their samples sensibly, so that the improvements from using those same sample categories on the larger data base are only marginal. However, every time the data base (total number of not-too-clearly homologous residues of known structures) doubles in size, one can obtain reasonable statistics for twice as many subcategories. It is our contention, which we have tried to substantiate throughout this chapter, that useful improvements in prediction are now available from this subdivision strategy and will continue to improve for one or two more data-base doublings even when the expansion does not involve novel types of structure. (Where new structure types are involved, such as for membrane proteins, then the improvements should be truly dramatic for that type of protein.) The following is a summary of the types of sample subdivision we believe could be useful in prediction:

1. Separation by tertiary-structure class. This type of subdivision is one that has already been tried with some success (e.g., Ptitsyn and Finkelstein, 1980; Taylor and Thornton, 1983), usually by restricting a prediction algorithm to one category or subcategory, such as Greek key β barrels or doubly wound α/β domains. Perhaps such strategies can evolve to a multistage procedure that uses differences in secondary-structure preferences to assign the tertiary-structure category and then uses somewhat different prediction algorithms based on that assignment. Alternatively, several parallel predictions can be done for different structure types if there are good criteria for ranking their relative success.

One promising area for such a multistage procedure would be in distinguishing parallel from antiparallel β sheets. Single-residue preferences are somewhat different (Lifson and Sander, 1979), but distribution along the strand is very strikingly different (see Section VI). A preliminary survey suggests that those differences are much more highly correlated with the overall organization of the sheet (that is, with tertiary-structure type if this is the principal sheet in that domain) than it is with the local H-bonding of the individual strand. In other words, an antiparallel or mixed strand in an otherwise parallel sheet has residue distribution more like parallel, whereas a parallel or mixed strand in an otherwise antiparallel sheet has a sequence more like typical antiparallel ones. This means that the overall organization type of a β sheet can probably be evaluated prior to determining any of its topological details. For both antiparallel and parallel sheets, edge strands show distinct (and less pronounced) residue distributions from interior strands (Sternberg and Thornton, 1977), and they also include bulges more frequently.

2. Separate prediction of the conformations with perpendicular carbonyls. Polyproline, 3_{10}, and $\text{L}3_{10}$ conformations could profitably be considered as separate categories, since they have distinctive amino acid preferences and are an essential component of corners between one piece of secondary structure and the next.

3. Inclusion of disordered structure as a separate conformational category, distinct from well-ordered, nonrepetitive loops. The use of the term "coil" or even "random coil" for anything not otherwise described has perhaps obscured the distinction between nonrepetitive loops, which have a single well-defined but not easily describable conformation, and truly disordered structure, which statistically samples multiple conformations. The amino acid preferences are rather distinctive, and the functional consequences even more so.

4. *Recognition that most amino acids play more than one structural role, which in some cases may be distinguishable from nonstructural evidence*. For instance, we saw in Section V.C that S-S, free-SH, and ligand-SH Cys have very different conformational preferences and clustering properties and that the three types can often be assigned from properties of the entire protein. Some roles could be distinguished by context in the sequence, such as whether a Gly is conferring flexibility (e.g., in a long stretch of all charges, hydroxyls, Pros, and Glys), is allowing a left-handed conformation (e.g., as a helix C-cap), or is permitting a tight tertiary contact (e.g., in the middle of a helix). Other roles could be distinguished by patterns of amino acid replacement if a large number of related sequences are available; e.g., an Asn replaced by Gly is probably in a left-handed conformation, whereas one replaced by Asp is making an O_δ H-bond.

5. *Separate consideration of categories of proteins different from the soluble, globular ones that have been easiest to crystallize*. At present, one primarily needs to be on the lookout for sequence characteristics or known behavior that indicates a membrane protein, a disordered protein, a fibrous protein, or a protein with many repeats of a short sequence and to avoid in such cases taking very seriously the results of a classical prediction. Of these other types, membrane proteins are in the best situation now, since a good deal of work has been done on the significance of long hydrophobic stretches for membrane spanning, and a data base is just now becoming available from which to judge more specific structural roles. As more information becomes available for the other categories, they can be considered separately as well. There is already a good deal of circumstantial evidence concerning disordered proteins, and more could be obtained by hydrodynamics, hydrogen exchange, and NMR. Repeating sequences (only recently found in large numbers, because they are much more amenable to DNA than to peptide sequencing) are of considerable interest, and one can expect much more structural information about them in the near future. For each of these cases, there is good reason to expect many of the predictional rules to be quite different from those for soluble globular proteins.

6. *Use of only those structures of accuracy good enough for whatever purpose is in question.* In spite of the lure of a big data base, caution must by exercised. Existence and approximate location of elements of secondary structure are quite reliable even in medium resolution (2.5–3 Å), unrefined structures (but not necessarily their *non*-existence). However, if H-bonds or ϕ,ψ angles are needed, then only x-ray structures at better than 2 Å resolution and under 20% residual, or the very best of the two-dimensional NMR structures, should be used. Even when helix ends are defined by Cα position, as in Section III.A, they may change after refinement.

In summary, the most general implication for prediction is a plea to go and look at three-dimensional protein structures and especially to choose which features to predict based on their suitability for prediction rather than simply to attempt correlations with someone else's definition of features made for another purpose.

XI. APPENDIXES

Appendix 1. Glossary of Proteins Listed by Brookhaven Data Bank File Code

Code	Protein	Reference
156B	Cytochrome b_{562}	Lederer, F., Glatigny, A., Bethge, P., Bellamy, H., and Mathews, F., 1981, *J. Mol. Biol.* 148:427
1ABP	Arabinose-binding protein	Gilliland, G., and Quiocho, F., 1981, *J. Mol. Biol.* 146:341
2ACT	Actinidin	Baker, E., and Dodson, E., 1980, *Acta Crystallogr.*, Section A, 36:559
1ACX	Actinoxanthin	Pletnev, V., Kuzin, A., Trakhanov, S., and Kostetsky, P., 1982, *Biopolymers* 21:287
4ADH	Alcohol deydrogenase	Eklund, H., Nordstrom, B., Zeppezauer, E., Soderlund, G., Ohlsson, I., Boiwe, T., Soderberg, B.-O., Tapia, O., Branden, C.-I., and Akeson, A., 1976, *J. Mol. Biol.* 102:27
3ADK	Adenyl kinase	Druesicke, D., Karplus, P., and Schulz, G., 1988, *J. Mol. Biol.* 199:359
5API	α_1-Antitrypsin	Loebermann, H., Tokuoka, R., Deisenhofer, J., and Huber, R., 1984, *J. Mol. Biol.* 177:531
2APP	Penecillopepsin	James, M., and Sielecki, A., 1983, *J. Mol. Biol.* 163:299
7ATC	Aspartate carbamoyl transferase	Kim, K., Pan, Z., Honzatko, R., Ke, H., and Lipscomb, W., 1987, *J. Mol. Biol.* 196:853
2AZA	Azurin	Norris, G., Anderson, B., and Baker, E., 1986, *J. Am. Chem. Soc.* 108:2784
2B5C	Cytochrome b_5	Mathews, F., Argos, P., and Levine, M., 1972, *Cold Spring Harbor Symp. Quant. Biol.* 36:387
3BCL	Bacterochlorophyll protein	Tronrud, E., Schmid, M., and Matthews, B., 1986, *J. Med. Biol.* 188:443
1BP2	Bovine phospholipase A_2	Dijkstra, B., Kalk, K., Hol, W., and Drenth, J., 1981, *J. Mol. Biol.* 147:97
1CAC	Carbonic anhydrase C	Liljas, A., Kannan, K., Bergsten, P.-C., Waara, I., Fridborg, K., Strandberg, B., Carlbom, U., Jarup, L., Lovgren, S., and Petef, M., 1972, *Nature [New Biol.]* 235:131
7CAT	Catalase	Fita, I., and Rossman, M., 1985, *Proc. Natl. Acad. Sci. U.S.A.* 82:1604
2CCY	Cytochrome C′	Finzel, B., Weber, P., Hardman, K., and Salemme, F., 1985, *J. Mol. Biol.* 186:627
2CDV	Cytochrome C_3	Higuchi, V., Kusunoki, M., Matsuura, V., Vasuoka, N., and Kakudo, M., 1984, *J. Mol. Biol.* 172:109

(*continued*)

Appendix 1. (Continued)

Code	Protein	Reference
5CHA	α-Chymotrypsin	Blevins, R., and Tulinsky, A., 1985, *J. Biol. Chem.* 260:4264
1CLN	Calmodulin	Babu, V., Sack, J., Greenhough, T., Bugg, C., Means, A., and Cook, W., 1985, *Nature* 315:37
3CNA	Concanavalin A	Hardman, K., and Ainsworth, C., 1972, *Biochemistry* 11:4910
5CPA	Carboxypeptidase A	Rees, D., Lewis, M., and Lipscomb, W., 1983, *J. Mol. Biol.* 168:367
1CPV	Carp Ca-binding protein	Kretsinger, R., and Nockolds, C., 1973, *J. Biol. Chem.* 248: 3313
1CRO	*cro* Repressor	Ohlendorf, D., Anderson, W., Takeda, Y., and Matthews, B., 1983, *J. Biomol. Struct. Dyn.* 1:553
1CRN	Crambin	Hendrickson, W., and Teeter, M., 1981, *Nature* 290:107
1CTF	L7/L12 ribosomal protein	Leijonmarck, M., and Liljas, A., 1987, *J. Mol. Biol.* 195:555
3CTS	Citrate synthase	Remington, S. Wiegand, G., and Huber, R., 1982, *J. Mol. Biol.* 158:111
2CYP	Cytochrome *C* reductase	Finzel, B., Poulos, T., and Kraut, J., 1984, *J. Biol. Chem.* 259:13027
4CYT	Cytochrome *C*	Takano, T., and Dickerson, R., 1981, *J. Mol. Biol.* 153:79
4DFR	Dihydrofolate reductase, *E. coli*	Bolin, J., Filman, D., Matthews, D., Hamlin, R., and Kraut, J., 1982, *J. Biol. Chem.* 257:13650
1ECA	Erythrocruorin	Steigemann, W., and Weber, E., 1979, *J. Mol. Biol.* 127:309
3FAB	Immunoglobulin Fab NEW	Saul, F., Amzel, L., and Poljak, 1978, *J. Biol. Chem.* 253:585
4FXN	Flavodoxin	Smith, W., Burnett, R., Darling, G., and Ludwig, M., 1977, *J. Mol. Biol.* 117:195
3GAP	Catabolite gene-activator	McKay, D., Weber, I., and Steitz, T., 1982, *J. Biol. Chem.* 257:9518
1GCR	γ-Crystallin	Wistow, G., Turnell, B., Summers, L., Slingsby, C., Moss, D., Miller, L., Lindley, P., and Blundell, T., 1983, *J. Mol. Biol.* 170:175
2GN5	Gene 5 protein, phage Fd	Brayer, G., and McPherson, A., 1983, *J. Mol. Biol.* 169:565
3GRS	Glutathione reductase	Karplus, P., and Schulz, G., 1987, *J. Mol. Biol.* 195:701
3HHB	Hemoglobin, deoxy	Fermi, G., Perutz, M., Shaanan, B., and Fourme, R., 1984, *J. Mol. Biol.* 175:159
1HMG	Haemagglutinin	Wilson, I., Skehel, J., and Wiley, D., 1981, *Nature* 289:366

Appendix 1. (*Continued*)

Code	Protein	Reference
1HMQ	Hemerythrin	Stenkamp, R., Sieker, L., and Jensen, L., 1983, *Acta Crystallogr.*, Section B, 39:697
1INS	Insulin	Dodson, E., Dodson, G., Hodgkin, D., and Reynolds, C., 1979, *Can. J. Biochem.* 57:469
4LDH	Lactate dehydrogenase	White, J., Hackert, M., Buehner, M., Adams, M., Ford, G., Lentz, P. Jr., Smiley, I., Steindel, S., and Rossman, M., 1976, *J. Mol. Biol.* 102:759
2LYM	Lysozyme (hen egg-white)	Kundrot, C., and Richards, F., 1987, *J. Mol. Biol.* 193:157
2LZM	Lysozyme (T_4 phage)	Matthews, B., and Remington, S., 1974, *Proc. Natl. Acad. Sci. USA* 71:4178
1MBD	Myoglobin, deoxy	Phillips, D., and Schoenborn, B., 1981, *Nature* 292:81
2MHR	Myohemerythrin	Sheriff, S., Hendrickson, W., and Smith, J.L., 1987, *J. Mol. Biol.* 197:273
1MLT	Melittin	Terwilliger, T., and Eisenberg, D., 1982, *J. Biol. Chem.* 257:6010
2MT2	Metallothionein	Furey, W., Robbins, A., Clancy, L., Winge, D., Wang, B., and Stout, C., 1986, *Science* 231:704
1NXB	Neurotoxin B	Tsernoglou, D., Petsko, G., and Hudson, R., 1978, *Mol. Pharmacol.* 14:710
1OVO	Ovomucoid, third domain	Papamokos, E., Weber, E., Bode, W., Huber, R., Empie, M., Kato, I., and Askowski, M. Jr., 1982, *J. Mol. Biol.* 158:515
2PAB	Prealbumin	Blake, C., Geisow, M., Oatley, S., Rerat, B., and Rerat, C., 1978, *J. Mol. Biol.* 121:339
PCI (in 4CPA)	Potato carboxypeptidase inhibitor	Rees, D., and Lipscomb, W., 1982, *J. Mol. Biol.* 160:475
1PCY	Plastocyanin	Colman, P., Freeman, H., Guss, J., Murata, M., Norris, V., Ramshaw, J., and Venkatappa, M., 1978, *Nature* 272:319
1PHH	p-Hydroxybenzoate hydroxylase	Schreuder, H., Van Der Laan, J., Hol, W., and Drenth, J., 1988, *J. Mol. Biol.* 199:637
5PTI	Pancreatic trypsin inhibitor	Wlodawer, A., Walter, J., Huber, R., and Sjolin, L., 1984, *J. Mol. Biol.* 180:307
1PYK	Pyruvate kinase	Stuart, D., Levine, M., Muirhead, H., and Stammers, D., 1979. *J. Mol. Biol.* 134:109
2PTN	Trypsin	Marquart, M., Walter, J., Deisenhofer, J., Bode, W., and Huber, R., 1983, *Acta Crystallogr.*, Section B, 39:480
1REI	Immunoglobulin REI, Bence-Jones	Epp, O., Lattman, E., Schiffer, M., Huber, R., and Palm, W., 1975, *Biochemistry* 14:4943
1RHD	Rhodanese	Ploegman, J., Drent, G., Kalk, K., and Hol, W., 1978, *J. Mol. Biol.* 123:557

(*continued*)

Appendix 1. (Continued)

Code	Protein	Reference
2RHE	Immunoglobin, RHE Bence-Jones protein	Furey, W., Jr., Wang, B., Yoo, C., and Sax, M., 1983, *J. Mol. Biol.* 167:661
5RSA	Ribonuclease A	Wlodawer, A., and Sjolin, L., 1983, *Biochemistry* 22:2720
5RXN	Rubredoxin	Watenpaugh, K., Sieker, L., and Jensen, L., 1980, *J. Mol. Biol.* 138:615
1SBT	Subtilisin	Wright, C., Alden, R., and Kraut, J., 1969, *Nature* 221:235
2SGA	*Strp.* griseus protease A	James, M., Sielecki, A., Brayer, G., Delbaere, L., and Bauer, C.-A., 1980, *J. Mol. Biol.* 144:43
1SN3	Scorpion neurotoxin	Almassy, R., Fontecilla-Camps, J., Suddath, F., and Bugg, C., 1983, *J. Mol. Biol.* 170:497
2SNS	Staphylococcal nuclease	Cotton, F., Hazen, E., Jr., and Legg, M., 1979, *Proc. Natl. Acad. Sci. USA* 76:2551
2SOD	Cu, Zn superoxide dismutase	Tainer, J., Getzoff, E., Richardson, J., and Richardson, D., 1972, *J. Mol. Biol.* 160:181
2SSI	*Strep.* subtilisin inhibitor	Mitsui, Y., Satow, Y., Watanabe, Y., and Iitaka, Y., 1979, *J. Mol. Biol.* 131:697
2STV	Satellite tobacco necrosis virus	Liljas, L., Unge, T., Jones, T.A., Fridborg, K., Lovgren, S., Skoglund, U., and Strandberg, B., 1982, *J. Mol. Biol.* 159:93
2TAA	Taka-amylase	Matsuura, V., Kusunoki, M., Harada, W., and Kakudo, M., 1984, *J. Biochem. Tokyo* 95:697
2TBV	Tomato bushy stunt virus	Hopper, P., Harrison, S., and Sauer, R., 1984, *J. Mol. Biol.* 177:701
1TIM	Triose phosphate isomerase	Banner, D., Bloomer, A., Petsko, G., Phillips, D., Pogson, C., and Wilson, I., 1975, *Nature* 255:609
3TLN	Thermolysin	Holmes, M., and Mathews, B., 1982, *J. Mol. Biol.* 160:623
2TNC	Troponin C	Herzberg, O., and James, M., 1985, *Nature* 313:653
1WRP	*trp* Repressor	Schevitz, R., Otwinowski, Z., Joachimiak, A., Lawson, C., and Sigler, P., 1985, *Nature* 317:782
2YHX	Hexokinase	Anderson, C., Stenkamp, R., and Steitz, T., 1978, *J. Mol. Biol.* 123:15

Appendix 2: Notes on Methodology

The protein structures were inspected on an Evans and Sutherland PS 330 display with a Tektronix stereo window on a UNIX MassComp workstation. The display program, CHAOS (David Richardson and Michael Zalis), runs entirely inside the PS 330 processor and is fed by utility programs on the MassComp that calculate vectors, ribbons, H-bonds, and dot surfaces. Superpositions were done with the INSIGHT program from Biosym, for which purpose the PS 330 was run over a serial line to the VMS VAX at the Duke Biomedical Simulations Resource.

Photographs were taken from the Evans and Sutherland PS 390 display, which is part of the Duke Comprehensive Cancer Center Shared Resource for Macromolecular Graphics. An even, "white" line was obtained with a color in the range between 114 (0.8) and 118 (0.5), depending on the black-and-white film used. Schematic drawings were made from stereo plots, using methods described by Richardson (1985). The drawings in this chapter are copyrighted by D. C. and J. S. Richardson. The coordinates of the protein structures are from the Brookhaven Data Bank (Bernstein *et al.*, 1977) unless otherwise noted.

ACKNOWLEDGMENTS. We would like to thank Brian Matthews, David Banner, Paul Sigler, Carl Pabo, Wayne Hendrickson, Kurt Wüthrich, and Allen Kline for the use of unpublished coordinates, many of which have since been deposited in the Brookhaven Data Bank, Robert Richardson for much of the photographic work, Caroline Usher for preparation of the manuscript, and Michael Hecht for suggesting the idea of Section V. This work was supported by NIH grant GM-15000 and in part by ONR research contract N00014-83-K-0596 and by the MacArthur Fellows Program.

XII. REFERENCES

Abdel-Meguid, S., Shieh, H.-S., Smith, W., Dayringer, H., Violand, B., and Bentle, L., 1987, Three-dimensional structure of a genetically engineered variant of porcine growth hormone, *Proc. Natl. Acad. Sci. U.S.A.* **84:**6434–6437.

Alber, T., Banner, D., Bloomer, A., Petsko, G., Phillips, D., Rivers, P., and Wilson, I., 1981, On the three-dimensional structure and catalytic mechanism of triose phosphate isomerase, *Phil. Trans. R. Soc. Lond. [Biol.]* **293:**159–171.

Anderson, A., and Hermans, J., 1988, Microfolding: Conformational probability map for the alanine dipeptide in water from molecular dynamics simulations, *Proteins* **3:**262–265.

Anderson, C., Stenkamp, R., McDonald, R., and Steitz, T., 1978, A refined model of the sugar binding site of yeast hexokinase B, *J. Mol. Biol.* **123:**207–210.

Argos, P., and Palau, J., 1982, Amino acid distribution in protein secondary structures, *Int. J. Peptide Protein Res.* **19:**380–393.

Argos, P., Rossman, M., and Johnson, J., 1977, A four-helical super-secondary structure, *Biochem. Biophy. Res. Commun.* **75:**83–86.

Baker, E., and Hubbard, R., 1984, Hydrogen bonding in globular proteins, *Prog. Biophys. Mol. Biol.* **44:**97–179.

Banner, D., Kokkinidis, M., and Tsernoglou, D., 1987, The structure of the ColE1 ROP protein at 1.7Å resolution, *J. Mol. Biol.* **196:**657–675.

Bernstein, F., Koetzle, T., Williams, G., Meyer, E., Jr., Brice, M., Rodgers, J., Kennard, O., Shimanouchi, T., and Tasumi, M., 1977, The Protein Data Bank: A computer-based archival file for macromolecular structures, *J. Mol. Biol.* **112:**535–542.

Blanc, J., and Kaiser, E., 1984, Biological and physical properties of a β-endorphin analog containing only D-amino acids in the amphiphilic helical segment 13–31, *J. Biol. Chem.* **259:**9549–9556.

Blundell, T., Lindley, P., Miller, L., Moss, D., Slingsby, C., Tickle, I., Turnell, B., and Wistow, G., 1981, The molecular structure and stability of the eye lens: X-ray analysis of gamma-crystallin II, *Nature* **289:** 771–777.

Blundell, T., Singh, J., Thornton, J., Burley, S., and Petsko, G., 1986, Aromatic interactions, *Science* **234:** 1005.

Blundell, T., Sibanda, B., Sternberg, M., and Thornton, J., 1987, Knowledge-based prediction of protein structures and the design of novel molecules, *Nature* **326:**347–352.

Brandts, J., Halvorson, H., and Brennan, M., 1975, Consideration of the possibility that the slow step in protein denaturation reactions is due to *cis–trans* isomerism of proline residues, *Biochemistry* **14:**4953–4963.

Braun, W., Wagner, G., Wörgötter, E., Vašák, M., Kägi, J., and Wüthrich, K., 1986, Polypeptide fold in the two metal clusters of metallothionein-2 by nuclear magnetic resonance in solution, *J. Mol. Biol.* **187:**125–129.

Bryan, P., Rollence, M., Pantoliano, M., Wood, J., Finzel, B., Gilliland, G., Howard, A., and Poulos, T.,

1986, Proteases of enhanced stability: Characterization of a thermostable variant of subtilisin, *Proteins* **1:** 326–334.

Chothia, C., 1973, Conformation of twisted β-pleated sheets in proteins, *J. Mol. Biol.* **75:**295–302.

Chothia, C., 1976, The nature of the accessible and buried surfaces in proteins, *J. Mol. Biol.* **105:**1–14.

Chothia, C., 1983, Coiling of β-pleated sheets, *J. Mol. Biol.* **163:**107–117.

Chothia, C., and Janin, J., 1981, Relative orientation of close-packed β-pleated sheets in proteins, *Proc. Natl. Acad. Sci. U.S.A.* **78:**4146–4150.

Chothia, C., Levitt, M., and Richardson, D., 1977, Structure of proteins: Packing of α-helices and pleated sheets, *Proc. Natl. Acad. Sci. U.S.A.* **74:**4130–4134.

Chou, P., and Fasman, G., 1977, β-Turns in proteins, *J. Mol. Biol.* **115:**135–175.

Chou, K.-C., Némethy, G., Pottle, M., and Scheraga, H., 1989, *J. Mol. Biol.* **205:**241–249.

Cook, W., Einspahr, H., Trapane, T., Urry, D., and Bugg, C., 1980, The crystal structure and conformation of the cyclic trimer of a repeat pentapeptide of elastin: Cyclo LVal-LPro-Gly-LVal-Gly, *J. Am. Chem. Soc.* **102:**5502–5505.

Creighton, T., 1977, Conformational restrictions on the pathway of folding and unfolding of the pancreatic trypsin inhibitor, *J. Mol. Biol.* **113:**275–293.

DeGrado, W., Musso, G., Lieber, M., Kaiser, E., and Kezdy, F., 1982, Kinetics and mechanism of hemolysis induced by melittin and by a synthetic melittin analogue, *Biophys. J.* **37:**329–338.

Deisenhofer, J., Jones, T., Huber, R., Sjodahl, J., and Sjoquist, J., 1978, Crystallization, crystal structure analysis and atomic model of the complex formed by a human Fc fragment and fragment B of protein A from *Staphylococcus aureus, Hoppe Zeylers Z. Physiol. Chem.* **359:**975–985.

Deisenhofer, J., Epp, O., Miki, K., Huber, R., and Michel, H., 1985, Structure of the protein subunits in the photosynthetic reaction center of *Rhodopseudomonas viridis* at 3Å resolution, *Nature* **318:**618–624.

Edwards, M., Sternberg, M., and Thornton, J., 1987, Structural and sequence patterns in the loops of βαβ units, *Protein Eng.* **1:**173–181.

Efimov, A., 1982, Role of constrictions in formation of protein structures containing four helical regions, *Mol. Biol. (Mosk.)* **16:**271–281.

Eisenberg, D., Weiss, R., and Terwilliger, T., 1982, The helical hydrophobic moment: A measure of the amphiphilicity of a helix, *Nature* **299:**371–374.

Engelman, D., Henderson, R., McLachlan, A., and Wallace, B., 1980, Path of the polypeptide in bacteriorhodopsin, *Proc. Natl. Acad. Sci. U.S.A.* **77:**2023–2027.

Epstein, C., Goldberger, R., and Anfinsen, C., 1963, The genetic control of tertiary protein structure: Studies with model systems, *Cold Spring Harbor Symp. Quant. Biol.* **28:**439–449.

Ghosh, S., Bock, S., Rokita, S., and Kaiser, E., 1986, Modification of the active site of alkaline phosphatase by site-directed mutagenesis, *Science* **231:**145–148.

Grathwohl, C., and Wüthrich, K., 1974, Carbon-13 NMR of the protected tetrapeptides TFA-Gly-Gly-L-X-L-Ala-OCH_3, where X stands for the 20 common amino acids, *J. Magnet. Res.* **13:**217–225.

Hecht, M., Sturtevant, J., and Sauer, R., 1986, Stabilization of lambda repressor against thermal denaturation by site-directed Gly → Ala changes in α-helix 3, *Proteins* **1:**43–46.

Hol, W., van Duijnen, P., and Berendsen, H., 1978, The α-helix dipole and the properties of proteins, *Nature* **273:**443–446.

Honig, B., Hubbell, W., and Flewelling, R., 1986, Electrostatic interactions in membranes and proteins, *Annu. Rev. Biophys. Bioeng.* **15:**163–193.

IUPAC–IUB Commission on Biochemical Nomenclature, 1970, Abbreviations and symbols for the description of the conformation of polypeptide chains, *J. Biol. Chem.* **245:**6489–6497.

James, M., and Sielecki, A., 1983, Structure and refinement of penicillopepsin at 1.8Å resolution, *J. Mol. Biol.* **163:**299–361.

Janin, J., 1979, Surface and inside volumes in globular proteins, *Nature* **277:**491–492.

Janin, J., and Wodak, S., 1978, Conformation of amino acid side-chains in proteins, *J. Mol. Biol.* **125:**357–386.

Kabsch, W., and Sander, C., 1983, Dictionary of protein secondary structure: Pattern recognition of hydrogen-bonded and geometrical features, *Biopolymers* **22:**2577–2637.

Karle, I., Kishore, R., Raghothama, S., and Balaram, P., 1988, Cyclic cystine peptides: Antiparallel β-sheet conformation for the 20-membered ring in BOC-Cys-Val-Aib-Ala-Leu-Cys-NHMe, *J. Am. Chem. Soc.* **110:**1958–1963.

Kauzmann, W., 1959, Some factors in the interpretation of protein denaturation, *Adv. Protein Chem.* **14:**1–63.

Kendrew, J., Watson, H., Strandberg, B., and Dickerson, R., 1961, The amino-acid sequence of sperm whale myoglobin: A partial determination by x-ray methods, and its correlation with chemical data, *Nature* **190:** 666–670.

Kline, A., Braun, W., and Wüthrich, K., 1986, Studies by ^{1}H nuclear magnetic resonance and distance geometry of the solution conformation of the α-amylase inhibitor Tendamistat, *J. Mol. Biol.* **189:** 377–382.

Kretsinger, R., and Nockolds, C., 1973, Carp muscle calcium-binding protein, *J. Biol. Chem.* **248:**3313–3326.

Kyte, J., and Doolittle, R., 1982, A simple method for displaying the hydropathic character of a protein, *J. Mol. Biol.* **157:**105–132.

Lesk, A., and Chothia, C., 1980, How different amino acids determine similar protein structures: The structure and evolutionary dynamics of the globins, *J. Mol. Biol.* **136:**225–270.

Lesk, A., and Chothia, C., 1984, Mechanisms of domain closure in proteins, *J. Mol. Biol.* **174:**175–191.

Leszczynski, J., and Rose, G., 1986, Loops in globular proteins: A novel category of secondary structure, *Science* **234:**849–855.

Levitt, M., and Greer, J., 1977, Automatic identification of secondary structure in globular proteins, *J. Mol. Biol.* **114:**181–239.

Lewis, P., Momany, F., and Scheraga, H., 1973, Chain reversals in proteins, *Biochim. Biophys. Acta* **303:**211–229.

Lifson, S., and Sander, C., 1979, Antiparallel and parallel β-strands differ in amino acid residue preferences, *Nature* **282:**109–111.

Lifson, S., and Sander, C., 1980, Specific recognition in the tertiary structure of β-sheets of proteins, *J. Mol. Biol.* **139:**627–639.

Low, B., Preston, H., Sato, A., Rosen, L., Searl, J., Rudko, A., and Richardson, J., 1976, Three dimensional structure of erabutoxin b neurotoxic protein: Inhibitor of acetylcholine receptor, *Proc. Natl. Acad. Sci. U.S.A.* **73:**2991–2994.

Matthews, B., Nicholson, H., and Becktel, W., 1987, Enhanced protein thermostability from site-directed mutations that decrease the entropy of unfolding, *Proc. Natl. Acad. Sci. U.S.A.* **84:**6663–6667.

McPhalen, C., 1986, *X-ray Crystallographic Studies on Subtilisins and Their Protein Inhibitors,* Ph. D. thesis, University of Alberta, Edmonton.

McPhalen, C., Schnebli, H., and James, M., 1985, Crystal and molecular structure of the inhibitor eglin from leeches in complex with subtilisin Carlsberg, *FEBS Lett.* **188:**55–58.

Momany, F., McGuire, R., Burgess, A., and Scheraga, H., 1975, Energy parameters in polypeptides. VII. Geometric parameters, partial atomic charges, nonbonded interactions, and intrinsic torsional potentials for the naturally occurring amino acids, *J. Phys. Chem.* **79:**2361–2380.

Nakashima, H., Nishikawa, K., and Ooi, T., 1986, The folding type of a protein is relevant to the amino acid composition, *J. Biochem. (Tokyo)* **99:**153–162.

Newcomer, M., Jones, T., Åqvist, J., Sundelin, J., Eriksson, U., Rask, L., and Peterson, P., 1984, The three-dimensional structure of retinol-binding protein, *EMBO J.* **3:**1451–1454.

Pabo, C., Sauer, R., Sturtevant, J., and Ptashne, M., 1979, The lambda repressor contains two domains, *Proc. Natl. Acad. Sci. U.S.A.* **76:**1608–1612.

Pflugrath, J., and Quiocho, F., 1985, Sulphate sequestered in the sulphate-binding protein of *Salmonella typhimurium* is bound solely by hydrogen bonds, *Nature* **314:**257–260.

Ptitsyn, O., 1969, Statistical analysis of the distribution of amino acid residues among helical and non-helical regions in globular proteins, *J. Mol. Biol.* **42:**501–510.

Ptitsyn, O., and Finkelstein, A., 1980, Self-organization of proteins and the problem of their three-dimensional structure prediction, in: *Protein Folding* (R. Jaenicke, ed.), Elsevier, Amsterdam, pp. 101–115.

Rao, S., and Rossmann, M., 1973, Comparison of super-secondary structures in proteins, *J. Mol. Biol.* **76:**241–256.

Rees, D., Lewis, M., and Lipscomb, W., 1983, Refined crystal structure of carboxypeptidase A at 1.54Å resolution, *J. Mol. Biol.* **168:**367–387.

Rice, D., Ford, G., White, J., Smith, J., and Harrison, P., 1983, The spatial structure of horse spleen apoferritin, *Adv. Inorg. Biochem.* **5:**39–50.

Rich, A., and Crick, F., 1961, The molecular structure of collagen, *J. Mol. Biol.* **3:**483–506.

Richardson, J., 1976, Handedness of crossover connections in β sheets, *Proc. Natl. Acad. Sci.* **73:**2619–2623.

Richardson, J., 1977, β-sheet topology and the relatedness of proteins, *Nature* **268:**495–500.

Richardson, J., 1981, The anatomy and taxonomy of protein structure, *Adv. Protein Chem.* **34:**167–339.

Richardson, J., 1985, Schematic drawings of protein structures, in: *Methods in Enzymology,* Vol. 115B, Chapter 24 (H. Wyckoff, C. Hirs, and S. Timasheff, eds.), Academic Press, Orlando, pp. 359–380.

Richardson, J., and Richardson, D., 1987, Some design principles: Betabellin, in: *Protein Engineering*, Chapter 12 (D. Oxender and C. Fox, eds.), Alan R. Liss, New York, pp. 149–163, 340–341.

Richardson, J., and Richardson, D., 1988a, Amino acid preferences for specific locations at the end of α-helices, *Science* **240:**1648–1652.

Richardson, J., and Richardson, D., 1988b, Helix lap-joints as ion-bonding sites, *Proteins* **4:**229–239.

Richmond, T., and Richards, F., 1978, Packing of α-helices: Geometrical constraints and contact areas, *J. Mol. Biol.* **119:**537–555.

Richardson, J., Getzoff, E., and Richardson, D., 1978, The β bulge: A common small unit of nonrepetitive protein structure, *Proc. Natl. Acad. Sci. U.S.A.* **75:**2574–2578.

Rose, G., 1978, Prediction of chain turns in globular proteins on a hydrophobic basis, *Nature* **272:**586–590.

Rose, G., 1979, Hierarchic organization of domains in globular proteins, *J. Mol. Biol.* **134:**447–470.

Rose, G., and Seltzer, J., 1977, A new algorithm for finding the peptide chain turns in a globular protein, *J. Mol. Biol.* **113:**153–164.

Rose, G., Geselowitz, A., Lesser, G., Lee, R., and Zehfus, M., 1985, Hydrophobicity of amino acid residues in globular proteins, *Science* **229:**834–838.

Salemme, F., 1983, Structural properties of protein β-sheets, *Prog. Biophys. Mol. Biol.* **42:**95–133.

Salemme, F., and Weatherford, D., 1981, Conformational and geometrical properties of β-sheets in proteins: II. Antiparallel and mixed β-sheets, *J. Mol. Biol.* **146:**119–141.

Schellman, C., 1980, The α_L conformation at the ends of helices, in: *Protein Folding* (R. Jaenicke, ed.), Elsevier, Amsterdam, pp. 53–61.

Schiffer, M., and Edmundson, A., 1967, Use of helical wheels to represent the structure of proteins and to identify segments of helical potential, *Biophys. J.* **7:**121–135.

Sheridan, R., Lee, R., Peters, N., and Allen, L., 1979, Hydrogen-bond cooperativity in protein secondary structure, *Biopolymers* **18:**2451–2458.

Shoemaker, K., Kim, P., York, E., and Baldwin, R., 1987, Tests of the helix dipole model for stabilization of α-helices, *Nature* **326:**563–567.

Shortle, D., and Lin, B., 1985, Genetic analysis of staphylococcal nuclease: Identification of three intragenic "global" suppressors of nuclease-minus mutations, *Genetics* **110:**539–555.

Sibanda, B., and Thornton, J., 1985, β-Hairpin families in globular proteins, *Nature* **316:**170–174.

Sprang, S., Standing, T., Fletterick, R., Stroud, R., Finer-Moore, J., Xuong, N.-H., Hamlin, R., Rutter, W., and Craik, C., 1987, The three-dimensional structure of Asn^{102} mutant of trypsin: Role of Asp^{102} in serine protease catalysis, *Science* **237:**905–909.

Steitz, T., Ohlendorf, D., McKay, D., Anderson, W., and Matthews, B., 1982, Structural similarity in the DNA-binding domains of catabolite gene activator and *cro* repressor proteins, *Proc. Natl. Acad. Sci. U.S.A.* **79:**3097–3100.

Sternberg, M., and Thornton, J., 1977, On the conformation of proteins: The handedness of the connection between parallel β-strands, *J. Mol. Biol.* **110:**269–283.

Tainer, J., Getzoff, E., Beem, K., Richardson, J., and Richardson, D., 1982, Determination and analysis of the 2Å structure of copper, zinc superoxide dismutase, *J. Mol. Biol.* **160:**181–217.

Taylor, W., and Thornton, J., 1983, Prediction of super-secondary structure in proteins, *Nature* **301:**540–542.

Teeter, M. M., 1984, Water structure of a hydrophobic protein at atomic resolution: Pentagon rings of water molecules in crystals of crambin, *Proc. Natl. Acad. Sci. U.S.A.* **81:**6014–6018.

Venkatachalam, M., 1968, Stereochemical criteria for polypeptides and proteins: Conformation of a system of three linked peptide units, *Biopolymers* **6:**1425–1436.

Weber, P., and Salemme, F., 1980, Structural and functional diversity in 4-α-helical proteins, *Nature* **287:**82–84.

Wetlaufer, D. B., 1973, Nucleation, rapid folding, and globular intrachain regions in proteins, *Proc. Natl. Acad. Sci. U.S.A.* **70:**697–701.

Wilson, I., Haft, D., Getzoff, E., Tainer, J., Lerner, R., and Brenner, S., 1985, Identical short peptide sequences in unrelated proteins can have different conformations: A testing ground for theories of immune recognition, *Proc. Natl. Acad. Sci. U.S.A.* **82:**5255–5259.

Wolfenden, R., Andersson, L., Cullis, P., and Southgate, C., 1981, Affinities of amino acid side chains for solvent water, *Biochemistry* **20:**849–855.

2

The Structure of the Photochemical Reaction Center of Rhodopseudomonas viridis and Its Implications for Function

Johann Deisenhofer, Robert Huber, and Harmut Michel

I. INTRODUCTION

Reaction centers (RC) in green and purple bacteria are protein–pigment complexes that are integrated in the photosynthetic membrane. In the RC charge separation across the membrane occurs on absorption of light. Electrons are transferred from a primary donor to a primary acceptor and finally to a secondary acceptor. The reduced secondary acceptor diffuses from the

Johann Deisenhofer • Department of Biochemistry, and Howard Hughes Medical Institute, University of Texas Southwestern Medical Center, Dallas, Texas 75235. *Robert Huber* • Division for Structural Research, Max Planck Institute for Biochemistry, D-8033 Martinsried, West Germany. *Hartmut Michel* • Division for Molecular Membrane Biochemistry, Max Planck Institute for Biophysics, D-6000 Frankfurt 71, West Germany.

RC and interacts with and transfers electrons to other electron carriers. After a complex series of reactions, an electron is returned to the oxidized primary donor. The cyclic electron flow is associated with vectorial proton transport across the membrane and coupled to adenosine triphosphate (ATP) synthesis.

This blunt characterization of the essential function of the RC is based on a series of fascinating scientific discoveries as reviewed by Parson (1978). With the elucidation of the functional properties, the stage was set to analyze the chemical and very recently also the three-dimensional structure of the RC.

In this short review we describe some general aspects of the three-dimensional structure of the RC of the purple nonsulfur bacterium *Rhodopseudomonas viridis* obtained by x-ray crystallography and amino acid sequence analysis, its implications for function, and its relevance for membrane proteins in general.

The original studies were done by Michel (1982), Deisenhofer *et al.* (1984, 1985a), Allen *et al.*, (1986), and Michel *et al.* (1985, 1986a,b). Short reviews are found in Deisenhofer *et al.* (1985b) and Michel and Deisenhofer (1986).

II. PHOTOSYNTHETIC APPARATUS IN BACTERIA

Figure 1a shows the principal arrangement of components of the photosynthetic apparatus in photosynthetic bacteria. The RC is surrounded by light-harvesting complexes (LHC). Light energy absorbed by the LHC is transmitted to the RC, where it is used to transfer electrons. The complex ensemble of pigments associated with the protein components is excited by light, but the different types of pigments follow different main paths of deexcitation. Quenching processes may follow two general pathways, energy and electron transfer (Kavarnos and Turro, 1986). The LHC and the RC follow almost exclusively and very efficiently one or the other pathway: energy transfer in LHC and electron transfer in RC.

The LHC are protein–pigment complexes far in excess of RC on a molar basis. In purple bacteria the LHC are aggregates of small peptides with BChl pigments that are integrated in the membrane. A plausible model suggests that BChl is bound to α-helical segments spatially close to the periplasmic side of the membrane (for reviews see Zuber, 1985; Drews, 1985). Photoreceptor units consisting of RC and LHC (three different polypeptides) of *R. viridis* in apparently regular arrangement have been prepared and investigated by electron microscopy (Jay *et al.*, 1984; Stark *et al.*, 1984) and optical spectroscopy (Breton *et al.*, 1985).

Some bacterial species possess LHC that are peripheral to the photosynthetic membrane (PLHC in Fig. 1a), and these have been studied in some detail. A three-dimensional structure of a BChl-containing protein from green bacteria has been determined (Tronrud *et al.*, 1986), but little information about its function is available.

The light-harvesting organelles of cyanobacteria (phycobilisomes) have been studied in very great detail (Schirmer *et al.*, 1985, 1986; Glazer, 1985), so that conclusions about the physical principles of their function may be drawn. In the context of this review we note that light energy is transferred very rapidly (in the picosecond time range) by pigments that are widely separated and are rigidly held in the protein matrix by predominantly ionic interactions. This arrangement is very different from the electron-conducting pigments in the RC, which are

Abbreviations used in this chapter: RC, reaction center; LHC, light-harvesting complex; L,M,H, L, M, and H subunits of the reaction center; A,B,C,D,E, sequential transmembrane α helices of the L and M subunits; BcHl-b, bacteriochlorophyll b; BPh-b, bacteriopheophytin b; P or P_{960}, primary electron donor; P*, primary electron donor in its lowest excited singlet state; P^+, primary electron donor in its oxidized state; Q_a, Q_b, Q_a^-, Q_b^-, primary and secondary electron-accepting quinones in oxidized and reduced states, respectively.

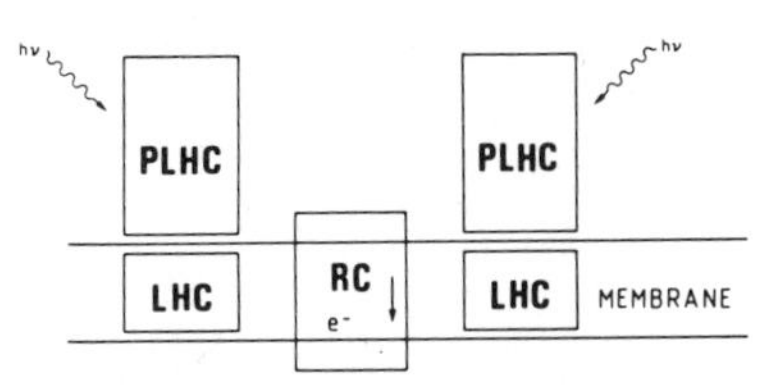

RC : REACTION CENTER OF PHOTOSYSTEM
PLHC : PERIPHERAL LIGHT HARVESTING COMPLEX
LHC : MEMBRANOUS LIGHT HARVESTING COMPLEX
MEMBRANE : PHOTOSYNTHETIC MEMBRANE

a

Periplasm
He He He He C
BC BC
BC BC
M L-U BP BP M-U M
Fe
$Q_b \cdot Q_a$
Cytoplasm H-U

C = Cytochrome
L-U, M-U, H-U = 'Light, Medium, Heavy' - Proteinsubunits

b

BC = BChl = Bacteriochlorophyll B
BP = BPh = Bacteriopheophytin
(Bacteriochlorophyll without Mg)

R = Phytyl: $-CH_2-CH{=}C(CH_3)-(CH_2)_3-CH(CH_3)-(CH_2)_3-CH(CH_3)-(CH_2)_3-CH(CH_3)-CH_3$

He = Protoheme

Q_a = Menachinone-9

Q_b = Ubichinone-9

c

Figure 1. (a) Principal arrangement of the reaction center and light-harvesting complexes in the photosynthetic membrane of bacteria. (b) Scheme of the arrangement of the subunits in the reaction center of *R. viridis*. C, cytochrome; L-U, M-U, H-U, L, M, and H subunits; M, membrane. (c) Chemical formulas of the pigments involved.

close in space and in a largely apolar environment. Energy transfer in phycobilisomes occurs by a mechanism of induced dipole interaction (Förster mechanism), which may operate over wide distances and does not involve charged species. Electron transfer has different structural requirements, as we describe below.

III. THE RC COMPOSITION

The RC of *R. viridis* is a complex of four protein subunits, the L, M, and H subunits and the cytochrome and various pigments, as shown schematically in Fig. 1b. Figure 1c displays the chemical formulas of the pigments. Amino acid sequence analysis on the basis of the nucleotide sequence of the gene has shown that the L, M, and H subunits consist of 273, 323, and 258 residues, respectively (Michel *et al.*, 1985, 1986a). The cytochrome has recently also been completely sequenced (K. A. Weyer, F. Lottspeich, and H. Michel, unpublished data). It is a polypeptide chain of 336 residues. The genes encoding L, M, and cytochrome subunits are in tandem. The cytochrome is of the c type with four heme groups covalently attached to the protein moiety via thioether linkages. The complex contains in addition four BChl-b, two BPh-b, one menaquinone-9 (Q_a), and a ferrous iron. These groups have well-defined electron density in the crystal structure, but the presence of other prosthetic groups that may be less well localized is not excluded. A second quinone (ubiquinone-9) (Q_b), which is a component of the functional complex, is lost during preparation and crystallization of the reaction center.

IV. THE RC CRYSTALS AND THE CRYSTAL STRUCTURE ANALYSIS

The RC complex of *R. viridis* was solubilized with detergent and precipitated with salt in the presence of a small amphiphilic molecule with the idea that the bulky detergent covering the hydrophobic surface of the protein might interfere with proper packing in the crystal lattice. Addition of an excess of a small amphiphilic molecule promotes crystallization (Michel, 1982). It is thought to replace those detergent molecules that interfere with crystal packing. The crystallographic analysis shows that the contacts in the crystal lattice of the RC are between parts of the complex molecule that are exposed to water and thus resemble the lattice contacts of water-soluble proteins. The RC of *R. sphaeroides* as determined in a preliminary crystal structure analysis (Allen *et al.*, 1986) shows similar contacts.

The *R. viridis* RC crystallizes in space group $P4_3\,2_12$ with cell lengths of $a = b = 223.5$ Å and $c = 113.6$ Å. The crystals contain one complex molecule in the asymmetric unit. Although the packing is quite loose, with about 70% of the cell volume occupied by solvent, i.e., water, salt, and detergent, the crystalline order is very good for a protein complex of this size. Significant diffraction intensities have been measured to about 2.5-Å resolution. The lifetime of the crystals in the x-ray beam was improved when the crystals were cooled to about 0°C. The diffraction experiments were carried out in a darkened cold room to avoid photochemical reactions. The analysis of the crystal structure proceeded by multiple isomorphous replacement on the basis of five derivatives, four different organic mercury (in part polymercury) compounds and a uranyl compound (Deisenhofer *et al.*, 1984).

The molecular boundary was clearly visible in the electron density map and allowed the application of a density modification procedure in which the density in the solvent region is set to zero (Wang, 1985) and density ripples around the main heavy atom sites are eliminated. Electron density modification, Fourier transformation, and combination with the isomorphous phases were carried out. The resulting map was the basis of the complete interpretation of pigment arrangement and polypeptide conformation. The amino acid sequence was identified, and the polypeptide chain was traced in the electron density map, which, except for a short segment of the H subunit, is very well defined (Deisenhofer *et al.*, 1985a). The crystallographic R value of the model was 0.36 for data from 7- to 3-Å resolution, a surprisingly good value. The model was subjected to repeated energy-restrained refinement and model building. The current *R* value is 0.23 for data between 7- and 2.9-Å resolution.

V. CHROMOPHORE ARRANGEMENT AND FUNCTIONAL PROPERTIES

The BChl-b and BPh-b chromophores are arranged in two branches related by an axis of approximate twofold symmetry and meet at a pair of BChl-b, the special pair (Fig. 2). This local symmetry axis is most likely normal to the plane of the membrane.

The two BChl-b of the special pair interact strongly by their pyrrole rings I, including the acetyl substituents and the central magnesium ions. The alignment of the transition dipole moments and the close approach of both BChl-b cause excitonic coupling, which partially explains the long-wavelength absorption band at 960 nm (P_{960}). Although many of the optical properties of the pigment system are rather well understood on the basis of the spatial structure (Knapp *et al.*, 1985), the electron transfer is less well understood.

The excited special pair is quenched by electron transfer to the BPh-b in 3 psec and further on to the primary acceptor Q_a in about 200 psec, driven by the redox potential gradient between P^*/P^+ (−760 mV) and Q_A/Q_A^- (−110 mV) (Cogdell and Crofts, 1972; Prince *et al.*, 1976; Bolton, 1978; Holten *et al.*, 1978; Netzel *et al.*, 1977; Breton *et al.*, 1986; Woodbury *et al.*, 1985). The redox potential of BPh-b/BPh-b$^-$ is intermediate with about −400 mV (Carithers and Parson, 1975).

The free energy change of electron transfer from the primary donor via the intermediate

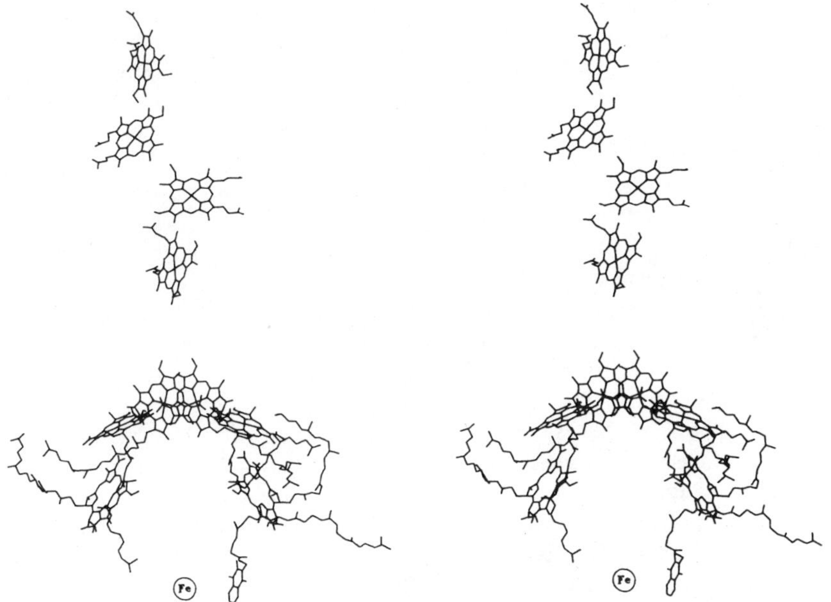

Figure 2. Stereo pair of the prosthetic groups of the RC showing four BChl-b (BC), two BPh-b (BP), one ferrous iron (Fe), one menaquinone-9 (MQ), and four heme groups (HE).

acceptor to the primary acceptor is governed by the chemical nature of the components the special pair P, BPh-b, and Q_a (I), but also by geometric factors (distance) (II) and environment (solvent polarity) (III). Specifically, I depends on the ionization potential of the donor in its excited state and the electron affinity of the acceptor. The coulombic interaction of the radical ion pair is small, as donor and acceptor are far apart (II). The effect of the environment (III) may be very substantial by stabilizing the radical ion pair. In general, ions are stabilized in polar media. Ionic interactions and hydrogen bonds of the special pair and the acceptors are therefore of great importance in a quantitative evaluation of the observed redox potential differences.

We must bear in mind, however, that a very strong stabilization of the radical ion pair would indeed favor electron transfer but oppose efficient turnover of the RC, as P^+ has to be reduced and Q_a^- oxidized to complete the catalytic cycle. This is analogous to enzyme–substrate interaction, where product release is crucial for efficient turnover. Strongly binding substrates act as inhibitors.

A negative free energy change is a necessary but not sufficient condition for electron transfer. Efficient electron transfer occurs only if it is fast enough to compete with other quenching processes: energy transfer, deactivation of the excited state by fluorescence emission, radiationless transitions, and chemical reactions.

Energy transfer from the excited special pair to pigments of the RC or the LHC may be very fast from a mechanistic point of view but is disfavored for energetic reasons. The special pair usually absorbs (but not in *R. viridis,* where the maximal apsorption of the RC and the LHC are at 960 and 1020 nm, respectively) at longer wavelengths than other pigments of the photosynthetic apparatus and represents the light energy sink. The natural radiative lifetime of the excited singlet state P* is around 20 nsec (Slooten, 1972; Parson and Cogdell, 1975) and may serve as a measure of the times involved in the other wasteful quenching processes. Clearly electron transfer is much faster and explains the very high quantum efficiency of electron transfer in the RC (Bolton, 1978).

Fast electron transfer requires effective overlap of the molecular orbitals by close approach. The orbital interaction decreases exponentially with the edge-to-edge distance of donor and acceptor and is insignificant at distances larger than about 10 Å (Kavarnos and Turro, 1986). In the RC the distance between the special pair and the electron acceptor Q_a is far too large to allow collision and direct electron transfer; instead the electron migrates via accessory BChl-b and BPh-b (Fig. 2). The BPh-b$^-$ is a spectroscopically and kinetically well-defined intermediate. Whether accessory BChl$^-$ is an intermediate is under debate (Breton *et al.*, 1986; Woodbury *et al.*, 1985). The distance between PHb-b and Q_a seems large for a fast transfer. Indeed, the gap is bridged by the aromatic side chain of a tryptophan residue M250 in the L branch of the pigment system (Deisenhofer *et al.*, 1985a; Michel *et al.*, 1986a,b, 1988). In addition, the isoprenoid side chain of Q_a is close to BPh-b. Electron transfer via long connecting chains by through-bond coupling of donor and acceptor orbitals has been observed (Kavarnos and Turro, 1986; Fasman *et al.*, 1982; Moore *et al.*, 1984), but there is only Van der Waals contact between BPh-b and the isoprenoid side chain of Q_a.

Electronic orbital overlap is one factor that determines electron transfer rates; nuclear rearrangements of the electron carriers and the environment are others. As the charge on donor and acceptor or intermediate acceptors develops, the nuclear configuration changes. This process has a certain free energy barrier related to the changes in molecular geometry. These are likely to be small in the RC, as the change is distributed over the electron systems of the BChl-b macrocycles. More important is the effect of the environment given by its polarizability. The polarization of the protein matrix must follow the electron transfer. The free energy barrier for electron transfer is low in a matrix with high electronic polarizability, as its polarization follows in phase and stabilizes the developing charge in the transition state of the reaction. Aromatic compounds have these characteristics.

The electron transfer from P* to Q_a occurs with very low activation energy (Bolton, 1978; Arnold and Clayton, 1960; Parson, 1974; Parson and Cogdell, 1975; Carithers and Parson, 1975; Kirmaier *et al.*, 1985a,b; Woodbury *et al.*, 1985) and proceeds readily at 1°K. Thermally activated processes, nuclear motions, and collisions are therefore unimportant. There is even a slight increase in rate with temperature decrease. An explanation might lie in structural changes accompanied by a closer approach of the pigments at low temperature, but the RC is a tightly packed protein pigment complex, which makes large changes of conformation unlikely. Low-temperature studies of some other protein crystals showed no evidence for large structural changes with temperature (Walter *et al.*, 1982). In general, the compressibility of proteins is low, and changes in linear dimensions with temperature are expected to be small (Kundrot and Richards, 1987). Changes with temperature of the vibrational spectra may be more important to explain the inverse temperature rate relationship, as they can lead to a faster electron transfer rate if the Franck–Condon factor increases with decreasing temperature.

The back reaction $P^+Q_a^-$ to PQ_a or P^*Q_a is slow and insignificant under physiological conditions but can be measured if the reduction of P^+ by cytochrome and the secondary electron transfer from Q_a to Q_b are blocked. It is a process proceeding with zero activation energy (Bolton, 1978).

The profound influence of the protein matrix on electron transfer in the RC is particularly obvious in the observed asymmetry of electron transfer in the two branches of the BChl-b and BPh-b pigments. Only the L branch is active. An explanation is offered by the fact that the protein environments of the two branches, although provided by homologous proteins (L and M), are rather different. Any of the observed differences in the pigment environment described by Michel *et al.* (1986b) may contribute to the functional asymmetry. The presence of the Trp M250 between BPh-b and Q_a is particularly conspicuous, as are the numerous differences in the Q_a and Q_b binding sites (see Fig. 4 in Michel *et al.*, 1988). Also, the special pair is asymmetric and may have an asymmetric electron density distribution as a result of different geometric distortions of the macrocycles and different hydrogen bonding to facilitate electron release into the L branch.

The protein matrix also serves to dissipate the excess energy of about 650 mV (Prince *et al.*, 1976) of the excited special pair (P^*/Q_a) over the radical ion pair $P^+Q_a^-$. These processes are probably very fast.

We now turn to the electron transfer between Q_a and Q_b. In *R. viridis*, Q_a is a menaquinone-9 and Q_b a ubiquinone-9, which differ in their redox potentials in solution by about 100 mV. In other purple bacteria both Q_a and Q_b are ubiquinones. The redox potential difference necessary for efficient Q_a–Q_b electron transfer in these cases must therefore be generated by the protein matrix, which in *R. viridis* and by homology in other purple bacteria is clearly rather different for both binding sites (see Fig. 4 in Michel *et al.*, 1986b). The asymmetric protein matrix must also be responsible for the quite different functional properties of Q_a and Q_b. The primary acceptor Q_a can accept one electron only, leading to the semiquinone anion. An electron has to be transferred to the secondary acceptor to allow the next electron transfer to Q_a to occur. Q_b, however, accepts two electrons and is protonated to form a hydroquinone to be able to diffuse from the RC. Q_b is close to glutamic acid L212, which opens a path to the H subunit and may protonate Q_b. The functional differences of Q_a and Q_b are obviously governed by the differences of their binding sites and possibly very importantly by the different potentials of protonation.

The electron transfer between Q_a and Q_b is rather different from the previous processes because it is much slower, about 6 μsec (at pH7, derived from Carithers and Parson, 1975), and has a substantial activation energy of about 8 kcal mole^{-1} (Carithers and Parson, 1975). The path between Q_a and Q_b is indeed very different from the environment of the other components of the electron transfer chain. The line connecting Q_a and Q_b [the Q_b binding site has been inferred from the binding mode of competitive inhibitors and ubiquinone-1 determined in *R. viridis* crystals (Deisenhofer *et al.*, 1985a)] is occupied by the ferrous iron and its five coordinating ligands, four histidine and one glutamic acid residues. Histidine M217 forms a hydrogen bond to Q_a. Histidine L190 is close to Q_b. Q_a and Q_b have an edge-to-edge distance of about 15 Å, which might explain the slow transfer. If electron transfer and protonation are coupled reactions, nuclear motions of the protein environment required for proton transfer may contribute to the activation energy barrier and would explain the observed pH dependence of the electron transfer rate of Q_a to Q_b (Kleinfeld *et al.*, 1985).

The role of the charged Fe-His_4-Glu complex in the Q_a to Q_b electron transfer is poorly understood at present, as it also occurs in the absence of the iron. The role of the metal seems to be predominantly structural (Debus *et al.*, 1986).

In summary, the electron transfer chain from the special pair P to Q_a is characterized by short distances between the aromatic macrocyles. The larger separation of BPh-b to Q_a is bridged by a tryptophan residue. The protein matrix in which the pigments are embedded and tightly held is lined predominantly with apolar amino acid side chains with a high proportion of aromatic residues. Electron transfer in RC occurs in the solid state via suitably spaced electron-conducting elements and does not require collisions or large nuclear motions. The electron path is far removed from the water phase, which by the necessary rearrangement of water dipoles would slow down electron transfer.

VI. THE SPATIAL STRUCTURE OF THE COMPLEX AND THE SUBUNITS

A. The Complex

Figures 3a–e are stereo drawings of the C_α carbon position of the RC complex and the four separate subunits: the cytochrome and the L, M, and H subunits. The crystal structure analysis provides no direct evidence for the location of the RC complex in the photosynthetic membrane, as detergent phase and water phase are not ordered.

But it follows from the functional properties of the RC, i.e., the vectorial electron transport by the pigments and their location, that the H subunit is at the cytoplasmic side of the membrane, the L–M complex spans the membrane, and the cytochrome is located at the periplasmic space. The local symmetry axis of the pigment system, which also relates L and M, is most likely normal to the membrane plane (Deisenhofer *et al.*, 1985a; Breton, 1985).

The central part of the RC complex is made up of subunits L and M, which bind the BChl-b, the BPh-b, the quinone, and the iron. The cytochrome is bound to the periplasmic flat surface of L–M and held tightly by a multitude of protein–protein contacts including the C-terminal arm of M. A major part of the H subunit is bound to the cytoplasmic surface of the L–M module. Its N-terminal helical segment, however, extends along the M subunit to reach the cytochrome. The central part of the reaction center (L and M subunits and N-terminal segment of H) is devoid of charged residues on its surface (Fig. 5) in accord with its presumed integration in the membrane.

Before we discuss structural details, the relationship in spatial structure of the RC of *R. viridis* with those of other purple bacteria and green plants deserves mentioning. Indirect evidence for a close relationship of purple bacterial RC comes from the amino acid sequence homology of L, M, and H subunits from several species (Youvan *et al.*, 1984; Williams *et al.*, 1984; Michel *et al.*, 1985, 1986a). Recent preliminary crystallographic analysis shows directly that the RCs of *R. viridis* and *R. sphaeroides* are very similar (Allen *et al.*, 1986), although the latter lacks the firmly bound cytochrome. Its functional role is played by a soluble cytochrome. Its functional role is played by a soluble cytochrome c_2. The interaction between the L–M complex and the soluble cytochrome c_2 in the encounter complex may be similar at the level of

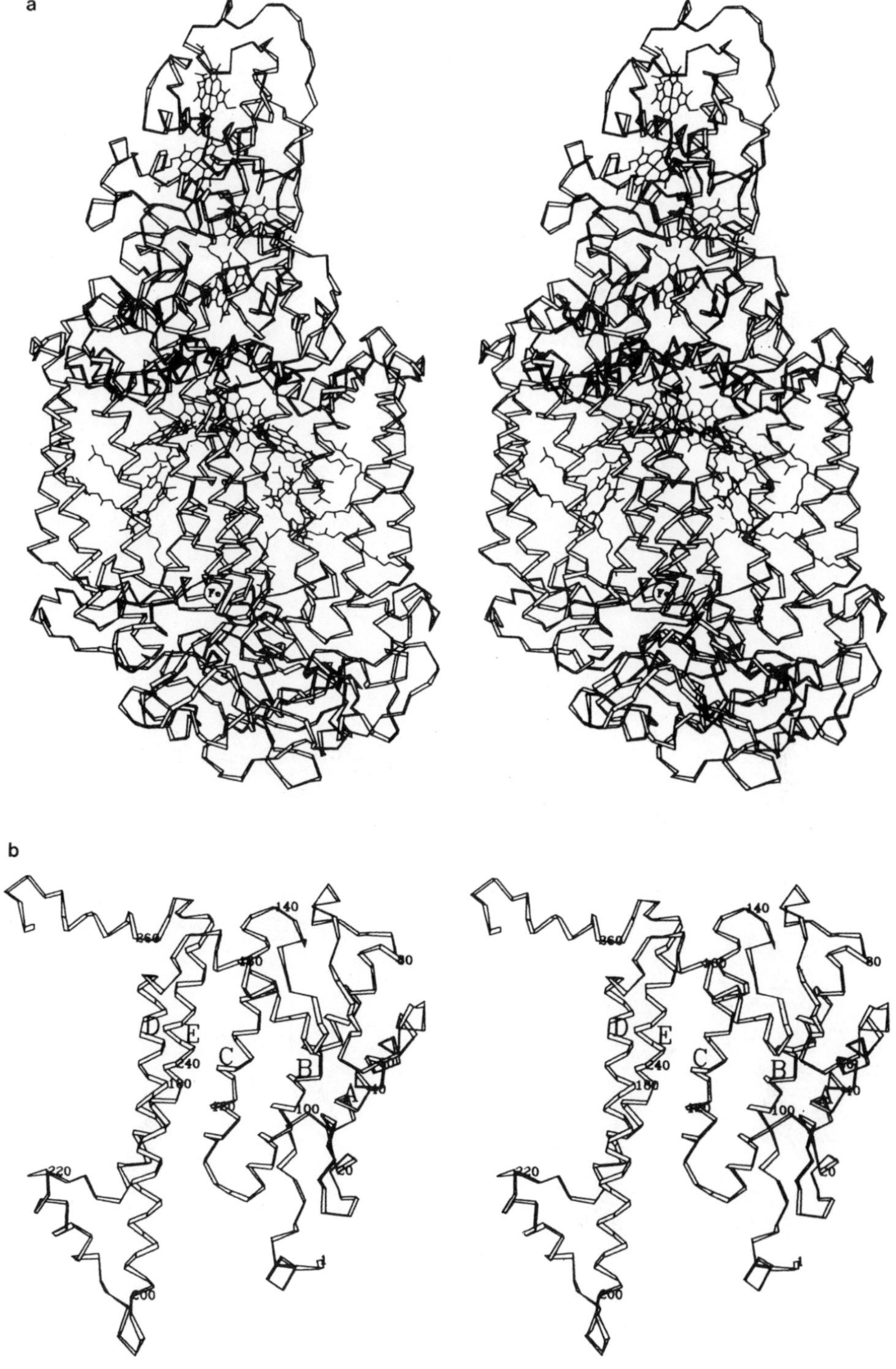

Figure 3. Ribbon drawings in stereo of the RC center including the chromophores (a), the L subunit (b), the M subunit (c), the H subunit (d), and the cytochrome (e).

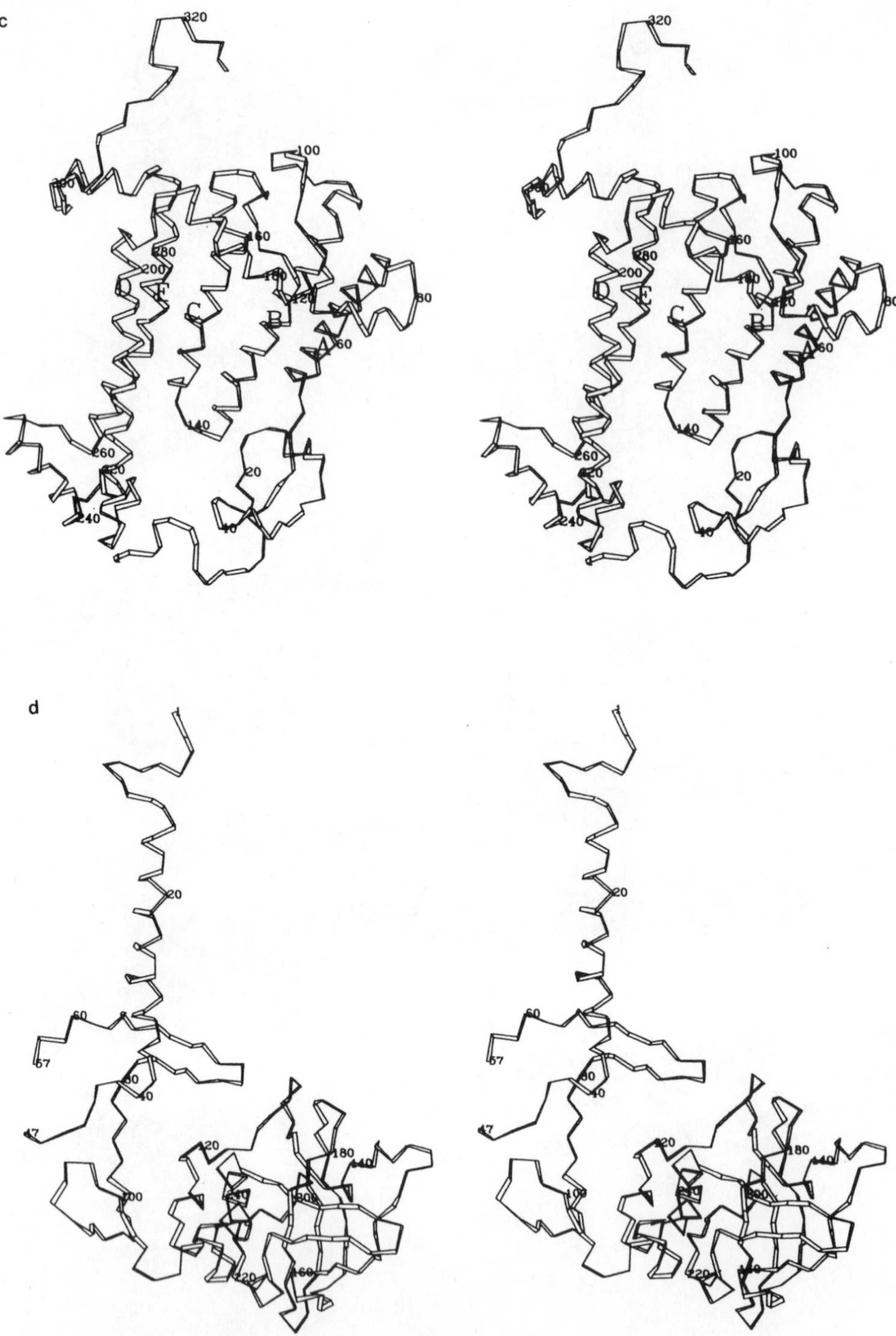

Figure 3 (cont.)

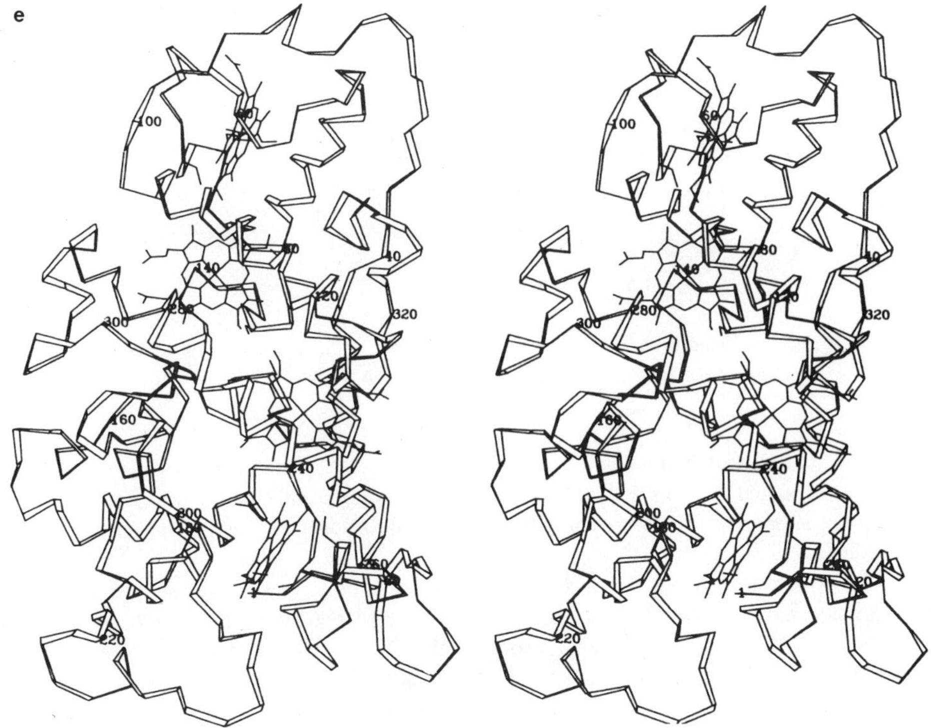

Figure 3 (cont.)

the prosthetic group arrangement as suggested by the conservation of the strategic Tyr L162 discussed later (Deisenhofer *et al.*, 1985a; Michel *et al.*, 1986).

A more distant relationship in amino acid sequence exists between L and M of purple bacteria and components D1 and D2 of photosystem II of chloroplasts. Many structurally and functionally important residues are conserved, in particular the histidine residues coordinated to the magnesium atoms of the special pair BChls and the iron. For a discussion, see Michel and Deisenhofer (1986), Trebst (1986), and Deisenhofer *et al.* (1985a).

B. The Cytochrome

According to its spectroscopic properties, the cytochrome was characterized as c-type. This is confirmed by the structural features. The cytochrome shows a novel polypeptide chain fold with four recurring motifs of α-helix–turn–irregular-segment, which wrap around the four heme groups (Fig. 4). Heme groups and heme-binding motifs are related by an approximate dyad axis of symmetry that is approximately perpendicular to the plane of projection in Fig. 3a and relate motifs 1 and 2 to 3 and 4. Residues 37 to 143 constitute an upper domain, and 144 to 318 a lower domain. The approximate symmetry between parts of these domains is not obeyed by the linking segment 145 to 215 and the N-terminus 1 to 25, which both form extensive contacts with the L–M module. The hemes are covalently attached to the protein via thioether linkages to Cys-X-Y-Cys-His segments located in the irregular segments of the heme-binding motifs. Histidines are the fifth ligand to the central iron atoms. All heme irons appear to be six-coordinated, with the sixth ligand projecting from the helix of the heme-binding motif except

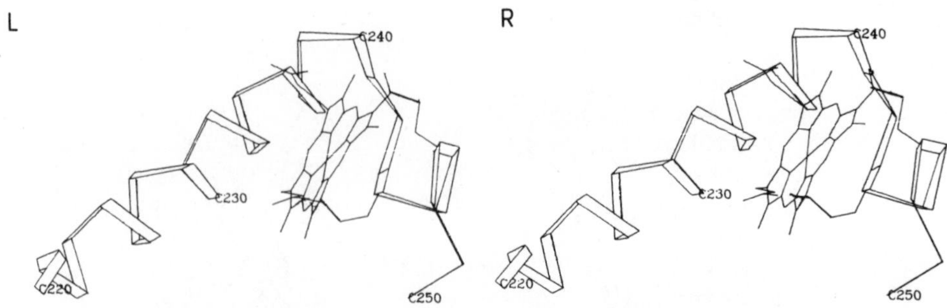

Figure 4. A ribbon drawing in stereo of the heme-binding motif of the *R. viridis* cytochrome. It comprises residues 220–250 and heme 3. This motif is repeated four times in the molecular structure.

in heme 4, which is linked to the turn segment of the heme 2 binding motif. The four heme groups are arranged in a row with close distances suggesting serial electron transfer from heme 1 to 3, but support from functional and spectroscopic studies for such a mechanism is not available.

The cytochrome displays two redox potentials of about 0 mV (c_{553}) and 330 mV (c_{558}) and can reduce the oxidized special pair, which has a redox potential P/P^+ of 480 mV (Case *et al.*, 1970; Cogdell and Crofts, 1972; Thornber and Olson, 1971; Dutton and Prince, 1978). These have not been assigned to the four hemes, but it is known that histidine is usually the sixth ligand in low-potential and methionine in high-potential cytochromes. However, electrostatic interactions also control the redox potential in cytochromes, as shown by a change of 65 mV on ionization of a heme propionic acid in mitochondrial cytochrome c (for a review see Matthews, 1985). For functional reasons, heme 3, which has a methionine ligand, should be one of the high-potential cytochromes.

The electron transfer from heme 3 to the special pair bridges a long distance of about 11 Å between the *n*-electron systems. The electron transfer time from cytochrome to the special pair is 270 nsec (Holten *et al.*, 1978) and about 1000 times slower than the electron transfer from the primary donor to the primary acceptor, perhaps a reflection of insufficient orbital overlap. Tyrosine L162 is located midway and may mediate the transfer by electron hopping. Simple rotations allow its aromatic side chain to approach pyrrole ring II of the BChl-b linked to the L subunit and pyrrole ring I of heme 3. The electron may therefore be transported by a thermally activated process of collisions at high temperatures. At low temperatures, when nuclear motions are frozen, electron tunneling may prevail to explain the biphasic temperature dependence (Hopfield, 1974; De Vault and Chance, 1966; Dutton and Prince, 1978).

C. The L and M Subunits

The L and M subunits, although of different lengths, have very similar folds with five transmembrane α helices. In addition, the L and M subunits have α helices in common, which connect helices C with D, D with E, and E with the C-terminal segment. L and M subunits differ in particular in the N termini, the A–B and D–E connections, and the C termini, which are all longer in M. The transmembrane α helices are arranged in the order ABCED. The α helices are consequently antiparallel except C and E, which are parallel.

The α helices show a right-handed relative rotation of the directions of their axes, which adds to about +80° between the most distant α helices A and D. The angles between the axes of contacting helices are around +20° to +30°, a preferred value for packing of α helices to allow interlocking of amino acid side chains (Chothia, 1984). The α-helical arrangement has a left-handed twist when viewed along the helical axes, as has also been observed in the low-resolution structure of bacteriorhodopsin (Henderson and Unwin, 1975; Eisenberg, 1984). In contrast, however, to the closed seven-α-helical barrel of bacteriorhodopsin, the five α helices of the L and M subunits form open shells.

The D–E α-helical submotifs of the L and M subunits are associated around the local dyad axis and form the center of the L–M module that binds the ferrous iron and the special pair of BChl-b. The four D and E α helices of the L and M subunits are arranged as a bundle with a left-handed twist. They are tied together by the iron and splay out toward the cytoplasmic side to accommodate the large special pair.

The four-α-helix bundle is a motif quite common in electron-transfer proteins (Weber and Salemme, 1980) such as *Escherichia coli* cytochrome b_{562} (Mathews *et al.*, 1979), cytochrome c′ from *Rhodospirillum molischianum* (Weber *et al.*, 1980), and nonheme iron-containing oxygen transport proteins from marine worms haemerythrin and myohaemerythrin (Hendrickson *et al.*, 1975; Stenkamp *et al.*, 1978). In all the molecules the four α-helix bundles diverge from the point of closest approach so that a binding site for the prosthetic group is created: the iron pair in haemerythrin, the heme group in the cytochromes, and the special pair of bacteriochlorophylls in the RC. In the RC the closest approach is near the iron that joins the four α helices by binding four histidine residues, one from each of the four α helices. The D–E connecting loop of the M subunit provides the fifth ligand to the iron, Glu M232. The α helices in the RC are splayed out to accommodate the very large bacteriochlorophyll pair, and in addition, α helices C and E are also considerably bent.

D. The H Subunit

The H subunit, which forms the cytoplasmic cover of the reaction center complex, is clearly divided into two domains, a loosely folded N-terminal part of residues 1 to 105 and a globular C-terminal part. The globular domain has two twisted β sheets to form a barrel (residues 150 to 222) filled with hydrophobic residues. In the front sheet (Fig. 3d) the two middle strands are parallel; all other strands are antiparallel. The lower strand (161 to 176) participates in and links both sheets. An irregular loop around residue 150 and the C-terminal helix (231 to 249) shield the open ends of the barrel. In contrast to the compact C-terminal domain of 153 residues, the N-terminal segment is very loosely folded into a prominent transmembrane helix (residues 12 to 37) and two two-stranded sheets (residues 64 to 79 and 89 to 104, respectively) that form flaps that interact with the L and M subunits.

E. Subunit Association in the RC Complex

The L and M subunits cohere mainly near the H contact surface. Further contacts between L and M are at the N-terminal segments, the D–E connecting loops, at the C-terminal segments, and the A–B connections. Indirect contacts between L and M are mediated by the accessory BCHl-b and BPh-b. Further contributions are the D and E α helices of the α-helical bundle, which are connected by the iron and the special pair. The protrusion connecting the D and E α helices in M (residues 240 to 255) forms the binding site for Q_a. The counterpart in the L subunit (residues 208 to 221) is the binding site of Q_b. Thus, it is remarkable that the four-α-helical motif forms the scaffold for and contributes the ligands to essential elements of the electron transfer chain. It is also noteworthy that these ligands are conserved in chloroplast proteins D1 and D2, to indicate that a similar structural motif may occur in green plants (Michel and Deisenhofer, 1986).

The quasisymmetry of the L–M complex is not obeyed by the peripheral subunits H and cytochrome, which even have their centers of gravity shifted from the dyad axis of symmetry. Main areas of contact of L with H are at the N-terminal region, the B–C connection, and the D–E connection. H and M have contacts at the N-terminal region of M and, particularly noteworthy, at the E transmembrane α helix, which is close to and almost parallel to the N-terminal transmembrane α helix of H. This α helix of H highlights the asymmetry of the L–M complex and may contribute to the functional asymmetry of the two branches of pigments, as it is close to the L branch, in particular to Q_a.

The cytochrome covers the periplasmic surface of the L–M complex and is in contact with most of the connecting segments of the transmembrane α helices of both L and M subunits. The specific contacts of the C-terminus of M and the N-terminus of H with the cytochrome are particularly noteworthy.

VII. IMPLICATIONS FOR MEMBRANE PROTEINS IN GENERAL

The RC is composed of components that are quite differently arranged with respect to the membrane. The cytochrome is located on the periplasmic side. The H subunit is folded into a globular part, which is located on the cytoplasmic side, and a transmembrane α helix. The L and M subunits are incorporated into the phospholipid bilayer. Consequently, the cytochrome has to be completely translocated across the membrane from the intracellular site of synthesis to its binding site in the RC. In the H, L, and M subunits, the transmembrane α helices are embedded in the bilayer. Only the N-terminal segment of H and the C-termini and connecting segments of the α helices located at the periplasmic side of L and M (A–B, C–D) require transfer.

It is interesting to note that only the cytochrome gene possesses a prokaryotic signal sequence as indicated by the sequence of the gene (Michel *et al.*, 1986a). Transfer of the large hydrophilic cytochrome may require a complex translocation system affording a signal sequence, whereas H, L, and M may spontaneously insert into the bilayer because of the affinity of the contiguous hydrophobic segments with the phospholipids (for a review of this and related problems see Rapoport, 1986).

A "simple" process of dissolution still requires transfer across the membrane of those charged residues that are located at the periplasmic side (Deisenhofer *et al.*, 1985a,b; Michel *et al.*, 1986a). The activation energy for this process may be lowered by he developing favorable protein lipid interaction. M and L have considerably more charged residues at the cytoplasmic side (41) than at the periplasmic side (24), providing a lower free energy barrier for correct insertion. The net charge distribution of the L–M complex is asymmetric, with a net distribution (counting Asp, Glu, and carboxy termini as negatively charged and Lys, Arg, and amino termini as positively charged) of six positive charges at the cytoplasmic side and eight negative charges at the periplasmic side. Because the intracellular membrane potential is negative, the observed orientation of the L–M complex is also thermodynamically favored by its electric dipole.

The H subunit has a very polar amino acid sequence at the C terminus of the transmembrane α helix with a stretch of seven consecutive charged residues (H33–H39) (Deisenhofer *et al.*, 1985a; Michel *et al.*, 1985), which may efficiently stop membrane insertion. There are no comparable sequences in the L and M subunits, but there are 3 to 11 charged residues in each of the connecting segments of the α helices at the cytoplasmic side, which might stop the transfer of helices or α-helical pairs (Engelman *et al.*, 1986). As an alternative to sequential insertion of α-helical elements, the L and M subunits may be inserted into the membrane as assembled protein–pigment complexes, because they cohere tightly by protein–protein interactions and additional linkages mediated by the iron, the special pair, the accessory BChl-b, and the BPh-b.

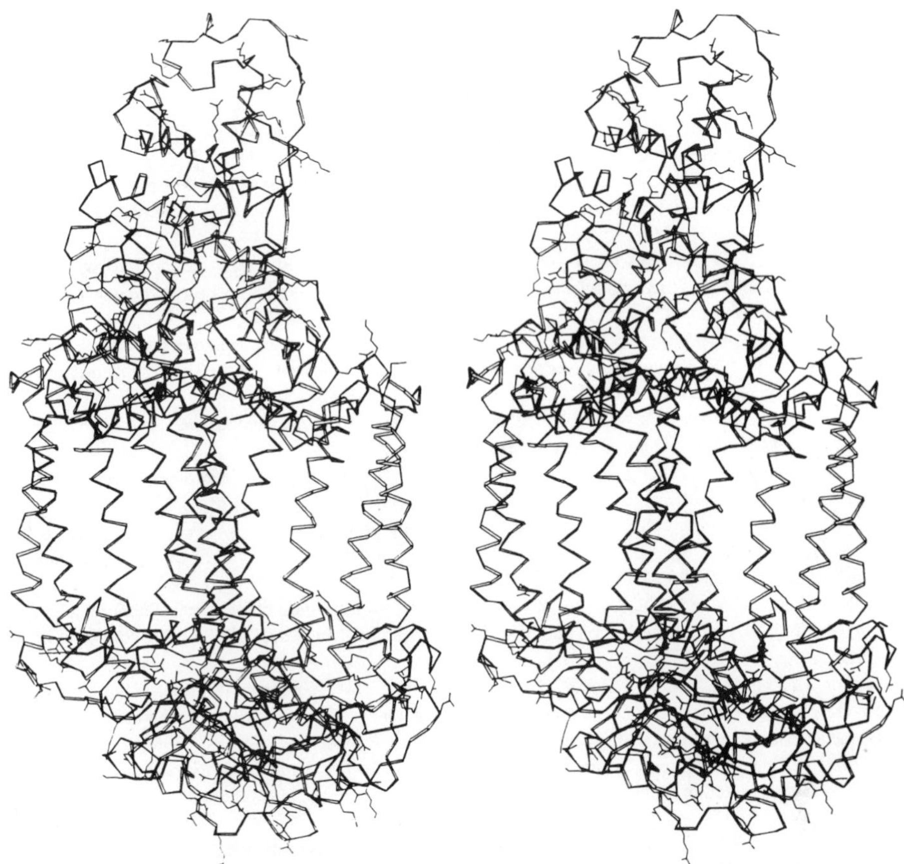

Figure 5. A ribbon drawing in stereo of the reaction center with the side chains of charged residues (Lys, Arg, Glu, Asp) added. The view corresponds to Fig. 3a.

The 11 transmembrane α-helices of the complex consist of 26 residues (H subunit) to 24–30 residues (L and M subunits), appropriate lengths to span the membrane. The amino acid sequences of these segments are devoid of charged residues (Fig. 5) and are characterized by maxima in the hydropathy index (Michel *et al.*, 1986a). Few charged residues occur close to the ends of the α helices.

Rather characteristically, glycine residues initiate and terminate almost all α-helical segments, both the transmembrane and the connecting α helices. It is well known from soluble proteins that glycine residues are abundant in turns. A different aspect concerns flexibility often associated with glycine residues (Huber and Bennett, 1983; Bennett and Huber, 1984) and possibly important for the insertion into the membrane by allowing rearrangements of α-helical segments.

VIII. REFERENCES

Allen, J. P., Feher, G., Yeats, T. O., Rees, D. C., Eisenberg, D. C., Deisenhofer, J., Michel, H., and Huber, R., 1986, Preliminary electron density mapping of the reaction centers from *R. sphaeroides* using the molecular replacement method, *Biophys. J.* **49:**583.

Arnold, W., and Clayton, R. D., 1960, The first step in photosynthesis: Evidence for its electronic nature, *Proc. Natl. Acad. Sci. U.S.A.* **46:**769–776.

Bennett, W. S., and Huber, R., 1984, Structural and functional aspects of domain motions in proteins, *CRC Crit. Rev. Biochem.* **15:**291–384.

Bolton, J. R., 1978, Primary electron acceptors, in: *The Photosynthetic Bacteria* (R. K. Clayton and W. R. Sistrom, eds.), Plenum Press, New York, London, pp. 419–442.

Breton, J., 1985, Orientation of the chromophores in the reaction center of *Rhodopseudomonas viridis.* Comparison of low-temperature linear dichroism spectra with a model derived from x-ray crystallography, *Biochim. Biophys. Acta* **810:**235–245.

Breton, J., Farkas, D. L., and Parson, W. W., 1985, Organization of the antenna bacteriochlorophylls around the reaction centre of *Rhodopseudomonas viridis* investigated by photoselection techniques, *Biochim. Biophys. Acta* **808:**421–427.

Breton, J., Martin, J.-L., Migus, A., Antonetti, A., and Orszag, A., 1986, Femtosecond spectroscopy of excitation energy transfer and initial charge separation in the reaction centre of the photosynthetic bacterium *Rhodopseudomonas viridis, Proc. Natl. Acad. Sci. U.S.A.* **83:**5121–5125.

Carithers, R. P., and Parson, W. W., 1975, Delayed fluorescence from *Rhodopseudomonas viridis* following single flashes, *Biochim. Biophys. Acta* **387:**194–211.

Case, G. D., Parson, W. W., and Thornber, J. P., 1970, Photooxidation of cytochromes in reaction center preparations from *Rhodopseudomonas viridis, Biochim. Biophys. Acta* **223:**122–128.

Chothia, C., 1984, Principles that determine the structure of proteins, *Annu. Rev. Biochem.* **53:**537–572.

Cogdell, R. J., and Crofts, A. R., 1972, Some observations on the primary acceptor of *Rhodopseudomonas viridis, FEBS Lett.* **27:**176–178.

Debus, R. J., Feher, G., and Okamura, M. Y., 1986, Iron-depleted reaction centers from *Rhodopseudomonas sphaeroides* R26.1. Characterization and reconstitution with Fe, Mn, Co, Ni, Cu and Zn, *Biochemistry* **25:** 2276–2287.

Deisenhofer, J., Epp, O., Miki, K., Huber, R., and Michel, H., 1984, X-ray structure analysis of a membrane protein complex electron density map at 3 Å resolution and a model of the chromophores of the photosynthetic reaction center from *Rhodopseudomonas viridis, J. Mol. Biol.* **180:**385–398.

Deisenhofer, J., Epp, O., Miki, K., Huber, R., and Michel, H., 1985a, Structure of the protein subunits in the photosynthetic reaction centre of *Rhodopseudomonas viridis* at 3 Å resolution, *Nature* **318:**618–624.

Deisenhofer, J., Michel, H., and Huber, R., 1985b, The structural basis of photosynthetic light reactions in bacteria, *Trends Biochem. Sci.* **10:**243–248.

DeVault D., and Chance, B., 1966, Studies of photosynthesis using a pulsed laser, *Biophys. J.* **6:**825–847.

Drews, G., 1985, Structure and functional organization of light-harvesting complexes and photochemical reaction centres in membranes of phototropic bacteria, *Microbiol. Rev.* **49:**59–70.

Dutton, P. L., and Prince, R. C., 1978, Reaction center-driven cytochrome interactions in electron and protein translocation and energy coupling, in: *The Photosynthetic Bacteria* (R. K. Clayton and W. R. Sistrom, eds.), Plenum Press, New York, pp. 525–565.

Eisenberg, D., 1984, Three-dimensional structure of membrane and surface proteins, *Annu. Rev. Biochem.* **53:** 595–623.

Engelman, D. M., Steitz, T. A., and Goldman, A., 1986, Identifying nonpolar transbilayer helices in amino acid sequences of membrane proteins, *Annu. Rev. Biophys. Chem.* **15:**321–353.

Fasman, P., Rob, F., and Verhoeven, J. W., 1982, Intramolecular charge-transfer absorption and emission resulting from through-bond interaction in bichromophoric molecules, *J. Am. Chem. Soc.* **104:**5127–5133.

Glazer, A. N., 1985, Light harvesting by phycobilisomes, *Annu. Rev. Biophys. Biophys. Chem.* **14:**47–77.

Henderson, R., and Unwin, P. N. T., 1975, Three-dimensional model of purple membrane obtained by electron microscopy, *Nature* **257:**28–32.

Hendrickson, W. A., Klippenstein, G. L., and Ward, K. B., 1975, Tertiary structure of myohemerythrin at low resolution, *Proc. Natl. Acad. Sci. U.S.A.* **72:**2160–2164.

Holten, D., Windsor, M. W., Parson, W. W., and Thornber, J. P., 1978, Primary photochemical processes in isolated reaction centers of *Rhodopseudomonas viridis, Biochim. Biophys. Acta* **501:**112–126.

Hopfield, J. J., 1974, Electron transfer between biologically molecules by thermally activated tunneling, *Proc. Natl. Acad. Sci. U.S.A.* **71:**3640–3644.

Huber, R., and Bennett, W. S., Jr., 1983, Functional significance of flexibility in proteins, *Biopolymers* **22:** 261–279.

Jay, F., Lambillotte, F., Stark, W., and Mühlethaler, K., 1984, The preparation and characterization of native photoreceptor units from the thylacoid of *Rhodopseudomonas viridis, EMBO J.* **3:**773–776.

Kavarncs, G. J., and Turro, N. J., 1986, Photosensitization by reversible electron transfer: Theories, experimental evidence, and examples, *Chem. Rev.* **86:**401–449.

Kirmaier, C., Holten, D., and Parson, W. W., 1985a, Temperature and detection-wavelength dependence of the picosecond electron transfer kinetics measured in *Rhodopseudomonas sphaeroides* reaction centers. Resolution of new spectral and kinetic components in the primary charge separation process, *Biochim. Biophys. Acta* **810:**33–48.

Kirmaier, C., Holten, D., and Parson, W. W., 1985b, Picosecond-photodichroism studies of the transient states in *Rhodopseudomonas sphaeroides* reaction centers at 5°K, *Biochim. Biophys. Acta* **810:**49–61.

Kleinfeld, D., Okamura, M. Y., and Feher, G., 1985, Electron transfer in reaction centers of *Rhodopseudomonas sphaeroides*. II. Free energy and kinetic relations between the acceptor state $O_a^- O_b^-$ and $O_aO_b^{2-}$, *Biochim. Biophys. Acta* **809:**291–310.

Knapp, E. W., Fischer, S. F., Zinth, W., Sander, M., Kaiser, W., Deisenhofer, J., and Michel, H., 1985, Analysis of optical spectra from single crystals of *Rhodopseudomonas viridis* reaction centers, *Proc. Natl. Acad. Sci. U.S.A.* **82:**8463–8467.

Kundrot, C. E., and Richards, F. M., 1987, The crystal structure of egg-white lysozyme at a hydrostatic pressure of 1000 atmospheres, *J. Mol. Biol.* **193:**157–170.

Matthews, F. S., 1985, The structure, function and evolution of cytochromes, *Prog. Biophys. Mol. Biol.* **45:**1–56.

Matthews, F. S., Bethge, P. H., and Czerwinski, E. W., 1979, The structure of cytochrome b_{562} from *Escherichia coli* at 2.5 Å resolution, *J. Biol. Chem.* **254:**1699–1706.

Michel, H., 1982, Three-dimensional crystals of a membrane protein complex. The photosynthetic reaction centre from *Rhodopseudomonas viridis, J. Mol. Biol.* **158:**567–572.

Michel, H., and Deisenhofer, J., 1986, X-ray diffraction studies on a crystalline bacterial photosynthetic reaction center: A progress report and conclusions on the structure of photosystem II reaction centers, in: *Encyclopedia of Plant Physiology, Photosynthesis III,* Volume 19 (A. C. Staehelin and C. J. Arntzen, eds.), Springer, Berlin, pp. 371–381.

Michel, H., Weyer, K. A., Gruenberg, H., and Lottspeich, F., 1985, The "heavy" subunit of the photosynthetic reaction centre from *Rhodopseudomonas viridis*. Isolation of the gene, nucleotide and amino acid sequence, *EMBO J.* **4:**1667–1672.

Michel, H., Weyer, K. A., Gruenberg, H., Dunger, I., Oesterhelt, D., and Lottspeich, F., 1986a, The "light" and "medium" subunits of the photosynthetic reaction centre from *Rhodopseudomonas viridis*. Isolation of the genes, nucleotide and amino acid sequence, *EMBO J.* **5:**1149–1158.

Michel, H., Epp, O., and Deisenhofer, J., 1986b, Pigment–protein interactions in the photosynthetic reaction centre from *Rhodopseudomonas viridis, EMBO J.* **5:**2445–2451.

Moore, T. A., Gust, D., Mathis, P., Bialocq, J.-C., Chachaty, C., Benasson, R. V., Land, E. J., Doizi, D., Liddell, P. A., Lehman, W. R., Nemeth, G. A., and Moore, A. L., 1984, Photodriven charge separation in a carotenoporphyrin–quinone triad, *Nature* **307:**630–632.

Netzel, T. L., Rentzepis, P. M., Tiede, D. M., Prince, R. C., and Dutton, P. L., 1977, Effect of reduction of the reaction center intermediate upon picosecond oxidation reaction of the bacteriochlorophyll dimer in *Chromatium vinosum* and *Rhodopseudomonas viridis, Biochim. Biophys. Acta* **460:**467–479.

Parson, W. W., 1974, Bacterial photosynthesis, *Annu. Rev. Microbiol.* **28:**41–59.

Parson, W. W., 1978, Role of reaction in photosynthesis, in: *The Photosynthetic Bacteria* (R. K. Clayton and W. R. Sistrom, eds.), Plenum Press, New York, pp. 317–322.

Parson, W. W., and Cogdell, R. J., 1975, The primary photochemical reaction or bacterial photosynthesis, *Biochim. Biophys. Acta* **416:**105–149.

Prince, R. C., Leigh, J. S., and Dutton, P. L., 1976, Thermodynamic properties of the reaction center of *Rhodopseudomonas viridis, Biochim. Biophys. Acta* **440:**622–636.

Rapoport, T. A., 1986, Protein translocation across and integration into membranes, *CRC Crit. Rev. Biochem.* **20:**73–137.

Schirmer, T., Huber, R., Schneider, M., Bode, W., Miller, M., and Hackert, M. L., 1986, Crystal structure analysis and refinement at 2.5 Å resolution of hexameric C-phycocyanin from the cyanobacterium *Agmenellum quadruplicatum, J. Mol. Biol.* **188:**651–676.

Slooten, L., 1972, Reaction center preparations of *Rhodopseudomonas sphaeroides, Biochim. Biophys. Acta* **256:**452–466.

Stark, W., Kühlbrandt, W., Wildhaber, I., Wehrli, E., and Mühlethaler, K., 1984, The structure of the photoreceptor unit of *Rhodopseudomonas viridis, EMBO J.* **3:**777–783.

Stenkamp, R. E., Sieker, L. C., Jensen, L. H. and McQueen, J. E., Jr., 1978, Structure of methemerythrin at 2.8 Å resolution: Computer graphics fit of an averaged electron density map, *Biochemistry* **17:**2499–2504.

Thornber, J. P., and Olson, J. M., 1971, Chlorophyll-proteins and reaction-center preparations from photosynthetic bacteria, algae, and higher plants, *Photochem. Photobiol.* **14:**329–341.

Trebst, A., 1986, The topology of the plastoquinone and herbicide binding peptides of photosystem II in the thylokoid membrane, *Z. Naturforsch.* **C41:**240–245.

Tronrud, D. E., Schmid, M. F., and Matthews, B. W., 1986, Structure and x-ray amino acid sequence of a bacteriochlorophyll: A protein from *Prosthecochloris aestuarii* refined at 1.9 Å resolution, *J. Mol. Biol.* **188:**443–454.

Walter, J., Steigemann, W., Sing, T. P., Bartunik, H., Bode, W., and Huber, R., 1982, On the disordered activation domain in trypsinogen: Chemical labelling and low-temperature crystallography, *Acta Cryst.* **B38:**1462–1472.

Wang, B. C., 1986, Resolution of phase ambiguity in macromolecular crystallography, in: *Methods in Enzymology,* Volume 115 (H. W. Wyckoff, C. H. W. Hirs, and S. N. Timasheff, eds.), Academic Press, London, pp. 90–117.

Weber, P. C., and Salemme, F. R., 1980, Structural and functional diversity of 4-α-helical proteins, *Nature* **287:**82–84.

Weber, P., Bartsch, R. G., Cusanovich, M. A., Hamlin, R. C., Howard, A., Jordan, S. R., Kamen, M. D., Meyer, T. E., Weatherford, D. W., Xuong, N., and Salemme, F. R., 1980, Structure of cytochrome c′: A dimeric, high-spin haem protein, *Nature* **286:**302–304.

Williams, J. C., Steiner, L. A., Feher, G., and Simon, M. I., 1984, Primary structure of the L subunit of the reaction center from *Rhodopseudomonas sphaeroides, Proc. Natl. Acad. Sci. U.S.A.,* **81:**7303–7307.

Woodbury, N. W., Becker, M., Middendorf, D., and Parson, W. W., 1985, Picosecond kinetics of the initial photochemical electron-transfer reaction in bacterial photosynthetic reaction centers, *Biochemistry* **24:** 7516–7521.

Youvan, D. C., Bylina, E. J., Alberti, M., Begusch, H., and Hearst, J. E., 1984, Nucleodide and deduced polypeptide sequences of the photosynthetic reaction center, B870 antenna and flanking polypeptides from *R. capsulata, Cell* **37:**949–995.

Zuber, H., 1985, Structure and function of light-harvesting complexes and their polypeptides, *Photochem. Photobiol.* **42:**821–844.

3

Virus Structure

Gerald Stubbs

I. INTRODUCTION

It is scarcely more than ten years since we first learned of the fold of a viral coat protein. The structures of tobacco mosaic virus, by fiber diffraction at 4-Å resolution (Stubbs *et al.*, 1977), and tomato bushy stunt virus by crystallography at 2.9 Å (Harrison *et al.*, 1978) were very different from each other, and at that time it might have been expected that further work would lead only to an ever-increasing array of unrelated structures. In fact, virus structures deter-

Abbreviations used in this chapter: AMV, alfalfa mosaic virus; BBV, black beetle virus; BSMV, barley stripe mosaic virus; CpMV, cowpea mosaic virus; CGMMV-W, cucumber green mottle mosaic virus, watermelon strain; HRV, human rhinovirus; SBMV, southern bean mosaic virus; STNV, satellite tobacco necrosis virus; TBSV, tomato bushy stunt virus; TCV, turnip crinkle virus; TMV, tobacco mosaic virus; TRV, tobacco rattle virus.

Gerald Stubbs • Department of Molecular Biology, Vanderbilt University, Nashville, Tennessee 37235.

mined since then have shown a considerable range of variation, but on a surprisingly limited number of themes. In particular, all the spherical viruses whose structures have been determined by crystallography until now have been shown to be related to the original example of bushy stunt virus, even in cases where there is no similarity in amino acid sequence, serology, or host range. At present, there are only three known basic virus coat protein folds: the β barrel of the spherical viruses, the α-helix bundle of the rod-shaped tobamoviruses, and the interleaved α helices of the filamentous bacteriophages. This is not to imply that no more will be found; it does, however, suggest that a few viruses have shown a remarkable capacity to adapt and evolve.

Structural studies of viruses began in the late 1930s, particularly in the laboratory of J. D. Bernal, where the first x-ray diffraction patterns were obtained from oriented gels of TMV (Bawden *et al.*, 1936; Bernal and Fankuchen, 1941) and crystals of TBSV (Bernal *et al.*, 1938). Wyckoff and Corey (1936) had looked earlier at solution scattering from TMV. Yet it was to be 40 years before these studies were to bear fruit in the form of interpretable electron-density maps. In the case of TBSV, this was because of the sheer size of the problem. The molecular weight of the virion is almost 10 million, and of the asymmetric unit of the crystallographic unit cell (cell edge 383 Å) over 600,000. TMV has not been crystallized in any form suitable for structural analysis, and therefore its structure had to be determined by the methods of fiber diffraction. Fiber diffraction data are cylindrically averaged because of the random orientation of the rod-shaped particles about their long axes; there is, therefore, a loss of information, amounting to a factor of about 2.5 in the case of TMV at 2.9-Å resolution (Makowski, 1982; Stubbs *et al.*, 1986).

Structure determinations such as these relied heavily on technical developments in accurate data collection. Well-focused, high-intensity x-ray beams were needed to resolve the diffraction data from such large assemblies, and considerable computing resources were required, particularly in the case of the crystalline viruses. The 2.9-Å data set for TBSV contained 198,000 reflections (Harrison *et al.*, 1978). It was necessary to make use of the considerable redundancy in the data stemming from the noncrystallographic icosahedral symmetry of the virus; this was done by cycles of symmetry averaging (Bricogne, 1976). The problem of cylindrical averaging in TMV was overcome using a multidimensional analogue of the isomorphous replacement method of protein crystallography (Stubbs and Diamond, 1975; Namba and Stubbs, 1985).

Other plant virus structures followed TBSV. The structure of southern bean mosaic virus was determined by Rossmann and his colleagues (Abad-Zapatero *et al.*, 1980), and that of a smaller virus, satellite tobacco necrosis virus, by Strandberg and his colleagues (Liljas *et al.*, 1982). Both of these viruses, unrelated to each other or to TBSV by the usual criteria of virology (Argos and Johnson, 1984), were found to have a β-barrel domain as the core of the coat protein subunit, very similar to one of the two domains of TBSV.

Several small animal viruses, in particular picornaviruses, were crystallized many years ago (Finch and Klug, 1959). The protein coats of these viruses include four types of subunit rather than one as in the simple plant viruses. Further crystallographic work was deferred until the inherent technical problems had been at least partially overcome in the plant virus studies. Three picornavirus structures have recently been determined: human rhinovirus 14 (Rossmann *et al.*, 1985), poliovirus (Hogle *et al.*, 1985), and Mengo virus (Luo *et al.*, 1987). Three of the viral coat proteins are folded in the same β-barrel topology as that of the plant viruses; the fourth, which is much smaller, is the N-terminal extension of one of the others. These structures thus provide a fascinating and largely unexpected insight into virus evolution. They are also adding significantly to our understanding of immunology and receptor binding as well as to the store of information about assembly and disassembly derived from the plant viruses.

There have been other structural studies of both plant and animal viruses, many of which are mentioned below. One noteworthy series of structure determinations is discussed only in passing in this chapter, which is otherwise confined to the proteins required for the structural integrity of viruses. These determinations, of the influenza surface antigens hemagglutinin at 3.0-Å resolution (Wilson *et al.*, 1981) and of neuraminidase at 2.9 Å (Varghese *et al.*, 1983), represent major achievements in crystallography and virology and have had considerable bearing on the elucidation of mechanisms of antigenicity (Wiley *et al.*, 1981; Colman *et al.*, 1983).

One more category of viruses has yielded to diffraction methods: the filamentous bacteriophages. In many cases, well-oriented partially dried fibers of these phages can be made, and high-quality diffraction patterns obtained (Marvin, 1966; Makowski, 1984). Orientation can often be improved by using strong magnetic fields during the preparation of the fibers (Torbet and Maret, 1979). Heavy-atom derivatives have been of only limited use in the study of filamentous bacteriophages because the heavy atoms induce small but significant changes in the symmetry of the virion (Nave *et al.*, 1981). Fortunately, however, these viruses, which have only about 50 amino acids in their coat protein subunits, have a relatively simple α-helical structure (Marvin, 1966), and this has allowed model-building approaches to structure determination. The structure of the bacteriophage Pf1 has been described at 7-Å resolution (Makowski *et al.*, 1980) as a double layer of α helices running approximately longitudinally.

II. SPHERICAL VIRUSES

A. Subunit Packing

All viruses use many copies of relatively small protein subunits to protect their nucleic acids. Arrangements having cubic symmetry, such as icosahedra or dodecahedra, are well suited to enclose the nucleic acid in an isometric particle (Hodgkin, 1949; Crick and Watson, 1956), and Caspar (1956) showed that the x-ray diffraction pattern from TBSV was consistent with the particle's having icosahedral symmetry.

An icosahedron has 20 triangular faces and thus can accommodate 60 protein subunits in identical environments (Fig. 1a). Most viruses, however, have far more than 60 subunits in their coat protein shells. Caspar and Klug (1962) put forward a theory in which multiples of 60 subunits could be accommodated in almost equivalent environments within a particle of icosahedral symmetry. The surface of the icosahedron can be subdivided into triangles, introducing approximate local sixfold axes of symmetry (Fig. 1). It was suggested that subunits clustering about the sixfold axes could have almost identical intersubunit interactions to those clustered about the fivefold axes; only a slight distortion of the intersubunit bonds would be required. Although strict equivalence would not be maintained, the subunits were said to be quasiequivalent. In this way, $60T$ subunits could be accommodated, where $T = H^2 + HK + K^2$, and H and K are any integers. The first four possible values for T are 1, 3, 4, and 7. The theory includes elements derived from the geodesic domes of Buckminster Fuller. These artificial constructions have in the end proved to be too limited to continue serving as models for virus shells; nature remains far more inventive, even devious, than art. Nonetheless, the Caspar and Klug theory guided understanding of the packing principles of viruses for 20 years, and many of the concepts behind it have been verified by structure determinations.

The structures of TBSV and SBMV partially confirmed the quasisymmetry theory in that 180 chemically identical subunits were found to be arranged on a $T = 3$ surface lattice with icosahedral symmetry (Fig. 1b). In some of the subunit interfaces, small bond deformations are sufficient to accommodate the different types of subunit packing; the quasithreefold axes in SBMV are a good example (Rossmann, 1984). In most of the interfaces, however, completely different bonding patterns are required. Perhaps the most dramatic example is the 35-residue

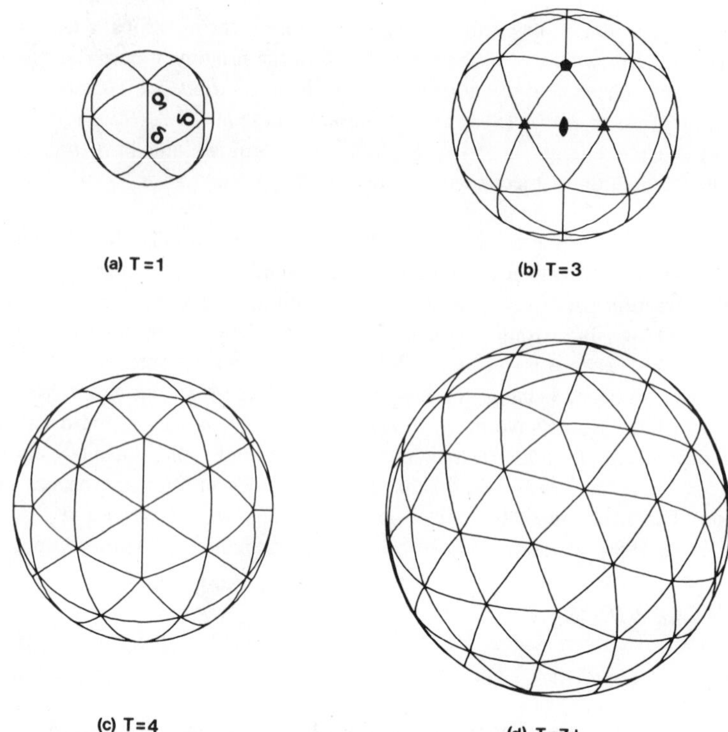

Figure 1. Orthogonal projections of triangulated icosahedral surface lattices for $T = 1, 3, 4$, and 7. One of two possible arrangements for $T = 7$ is shown. The symbol δ shows the relationship between three protein subunits in the $T = 1$ lattice. The symbols ⬟, ▲, and ◆ in the $T = 3$ lattice represent the icosahedral fivefold, threefold, and twofold axes, respectively, and correspond to the same symbols in Fig. 2. (From Rayment, 1984.)

''N-terminal arm'' found in the C subunits (Fig. 2) of TBSV. In both TBSV (Harrison *et al.*, 1978) and SBMV (Abad-Zapatero *et al.*, 1980), these arms from three equivalent subunits form a β annulus (Section II.C) about the icosahedral threefold axis, but the corresponding peptides in the two quasiequivalent subunits A and B are completely disordered.

An even greater departure from the Caspar and Klug principles is found in the case of the papovavirus polyoma. This virus was known from electron microscopy to contain 72 morphological units, located at the 12 fivefold and 60 quasisixfold axes of a $T = 7$ icosahedral lattice (Klug, 1965). It was therefore assumed for many years to have 420 subunits, arranged according to the principles of quasisymmetry. Refinement of models from electron microscopy against x-ray diffraction data at 22.5-Å resolution showed conclusively, however, that all the morphological units were pentamers: 60 pentamers lie on points corresponding to sixfold axes in the surface lattice and have six nearest-neighbor pentamers; 12 lie on fivefold axes (Rayment *et al.*, 1982). There are three distinctly different types of pentamer–pentamer interaction. Baker *et al.* (1983) showed by electron microscopy that another assembly of polyoma capsid proteins, the ''hexamer tube,'' is also made up of hexavalent pentamers. Frozen hydrated electron microscopic studies (Baker *et al.*, 1988) have shown that the papovavirus SV40 has the same coat protein packing as polyoma.

A number of viruses, including the picornaviruses, have been classified as $T = 1$ viruses, with each of the 60 icosahedral asymmetric units accommodating four nonidentical subunits. Crystallographic structure determinations (Rossmann *et al.*, 1985; Hogle *et al.*, 1985; Luo *et al.*, 1987) have shown, however, that three of the four coat proteins have extremely similar

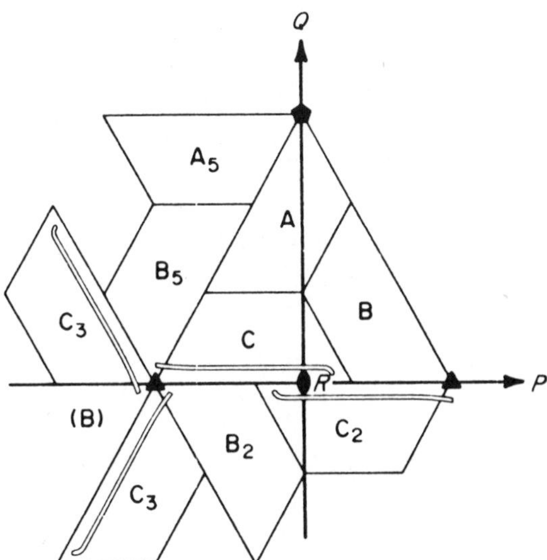

Figure 2. Arrangement of the subunits A, B, and C in the $T = 3$ viruses with respect to the icosahedral fivefold (⬟), threefold (▲), and twofold (◆) axes, looking toward the center of the virus. The subunits may be oriented in the complete surface lattice by reference to Fig. 1b. (From Rossmann, 1984.)

folds and are arranged at the lattice points of a $T = 3$ icosahedron. In this way, approximate equivalence is achieved, but the deviations from strict equivalence are accommodated by using different proteins rather than by conformational changes in one protein. Other genetic ways of approximating equivalence, discussed below, are used by the comoviruses, typified by cowpea mosaic virus, and by adenovirus.

B. The β-Barrel Fold: Satellite Tobacco Necrosis Virus

The simplest virus whose structure has been solved is satellite tobacco necrosis virus. STNV has 60 identical coat protein subunits, arranged with icosahedral symmetry on a $T = 1$ lattice. They encapsidate a single strand of RNA, of 1239 bases (Ysebaert *et al.*, 1980). The virus structure was determined at 3.0-Å resolution (Liljas *et al.*, 1982) and further refined at 2.5 Å (Jones and Liljas, 1984). The structural studies have been reviewed by Liljas and Strandberg (1984).

The protein subunits form a shell about 170 Å in diameter, having an average thickness of about 20 Å. The shell is only about 10 Å thick around the fivefold axes. Most of the protein (residues 26–195) is in a domain built around two four-stranded β sheets folded into a β barrel with the "jellyroll" (Richardson, 1981) topology (Fig. 3). The subunit is wedge-shaped, being narrower at the end nearer the fivefold axis, with the β strands at this end connected by very short peptide loops (Fig. 4). The tapered end permits extended intersubunit contacts about the fivefold axes. At the other end of the subunit, two longer loops connect the sheets, and a particularly long loop connects the E and F strands (see Fig. 4 for strand nomenclature) of one sheet. This loop includes short α-helical segments at residues 101–106 and 118–123. At the N-terminal end of the protein, an α helix between residues 12 and 24 extends toward the center of the virus, between radii 58 Å and 43 Å.

The protein subunits are closely packed at all interfaces, leaving no large holes in the shell. The most extensive contacts are between subunits related by the fivefold axes; these interactions are mostly polar and include two salt bridges. Most of the threefold contact is also polar, except where the three N-terminal helices form a short right-handed superhelix; the interacting side chains here are hydrophobic. The twofold contact includes both hydrophobic and hydrophilic interactions, with a salt bridge connecting Arg^{96} of one subunit with Glu^{104} and Asp^{92} of the other.

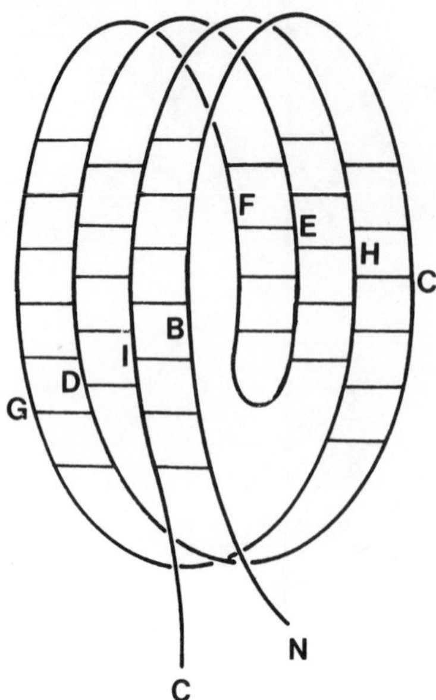

Figure 3. The jellyroll topology. The N and C termini are marked; the labeling of the strands corresponds to that used for the S domains of the simple icosahedral viruses.

The virus contains three types of calcium-binding sites. There is a site on the fivefold axis, coordinated by five equivalent carbonyl groups from residue 138 and two water molecules, one on the threefold axis, coordinated by three Asp^{55} residues, and one in a general position, coordinated by Glu^{25} from one subunit and Asp^{194} from another, two carbonyl groups, and two water molecules. As is characteristic of both icosahedral and rod-shaped viruses (Stubbs, 1984), all of these sites are in subunit interfaces.

No electron density attributable to the RNA is visible in the high-resolution map, although low-resolution x-ray and neutron scattering does show a preferred general distribution of the RNA (Liljas, 1986). The lack of high-resolution detail is because the RNA lacks precise icosahedral symmetry; the protein shell does not impose its symmetry on the nucleic acid. Amino acid residues 1–11, including four basic residues and the positively charged N terminus, are disordered. Three more basic residues are on the N-terminal helix; their side chains are also disordered. Six other basic residues are exposed to the RNA on the inside of the protein shell, and all except one of these have disordered side chains. Together these residues account for about two thirds of the positive charges that would be required to neutralize the RNA phosphates.

C. Simple Plant Viruses: One Coat Protein

Structural studies of a wide variety of small spherical plant viruses, including representatives of several different virus groups, have been reported. A number of these have been reviewed by Argos and Johnson (1984). Of those whose coats consist of only one type of protein subunit, STNV (Jones and Liljas, 1984; Liljas and Strandberg, 1984), TBSV (Olson *et al.*, 1983), SBMV (Abad-Zapatero *et al.*, 1980; Rossmann, 1984; Silva and Rossmann, 1987), and turnip crinkle virus (Hogle *et al.*, 1986) have been described in molecular detail. All of these except STNV are $T = 3$ viruses with diameters between 280 and 310 Å. Amino acid sequences are available for STNV (Henriksson *et al.*, 1981), TBSV (Hopper *et al.*, 1984), and SBMV (Hermodson *et al.*, 1982). The structure of the empty $T = 1$ capsid of alfalfa mosaic virus has been determined at 4.5-Å resolution (Fukuyama *et al.*, 1983). Preliminary reports of

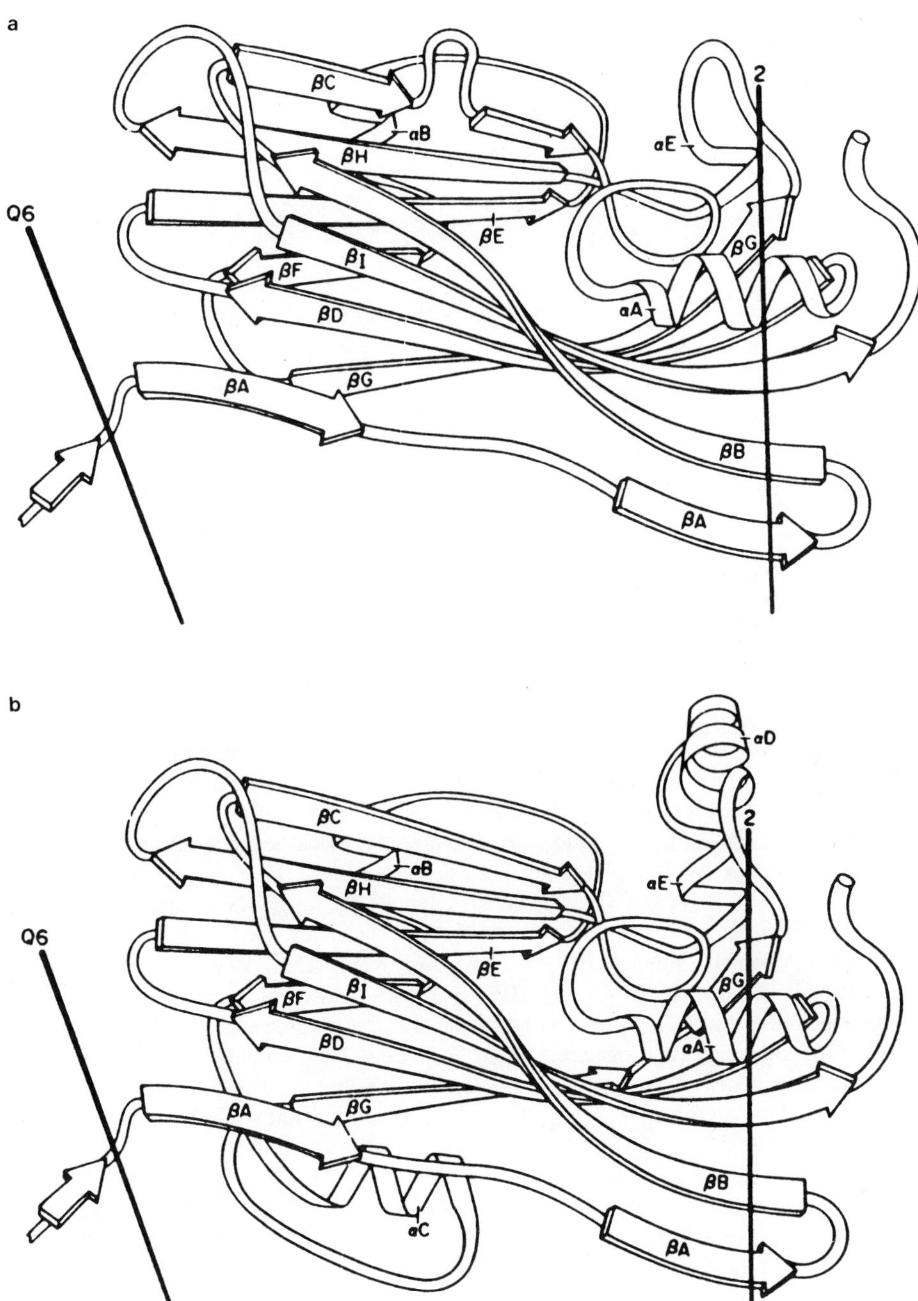

Figure 4. The S domains of (a) tomato bushy stunt virus, (b) southern bean mosaic virus, and (c) satellite tobacco necrosis virus. Icosahedral axes and the quasisixfold (Q6) axes are shown. (From Rossmann *et al.*, 1983a.)

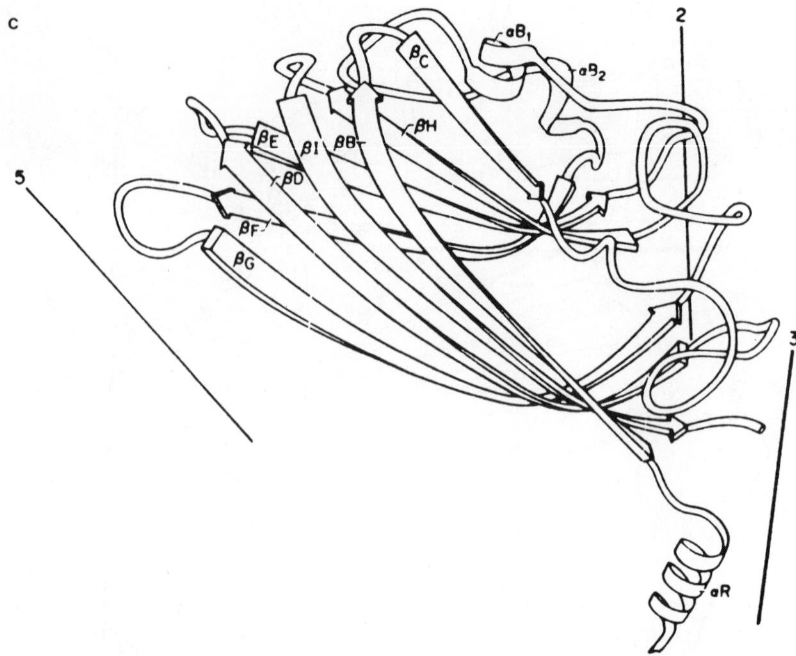

Figure 4 (cont.)

crystals that diffract to high resolution have appeared for erysimum latent virus (Colman *et al.*, 1980), belladonna mottle virus (Heuss *et al.*, 1981a), cowpea chlorotic mottle virus (Heuss *et al.*, 1981b), tobacco ringspot virus (Heuss *et al.*, 1981c), and the $T = 3$ bacteriophage MS2 (Valegård *et al.*, 1986). In this section, we also consider black beetle virus, the only insect virus whose structure has been determined (Hosur *et al.*, 1984, 1987).

TBSV, SBMV, TCV, and BBV all contain a domain like the STNV β barrel, the shell (S) domain, which makes the intersubunit contacts to form the capsid shell. AMV also appears to have a β-barrel domain, although the lower resolution of the AMV data makes unambiguous chain tracing impossible. In SBMV and BBV, as in STNV, this domain makes up most of the coat protein; TBSV and TCV contain an additional domain, the protruding (P) domain. Within each $T = 3$ virus, the three quasiequivalent S domains have virtually identical backbone conformations except in the β annulus, discussed further below. The S domains of STNV, SBMV, and TBSV are shown in Fig. 4; these three viruses and AMV have been compared by Rossmann *et al.* (1983a).

The $T = 3$ viruses are much more like each other than like STNV. This is not unexpected in view of the significantly different subunit packing in STNV. The β barrel in STNV is more regular than in the other viruses: TBSV and SBMV have a single, short α helix in the loop between strands E and F where STNV has two short helices, and another helix between strands C and D where STNV has irregular structure. These two helices, together with the β barrel, are the most common secondary structure features of spherical viruses. The backbone conformation of TCV in both S and P domains is almost identical to that of TBSV except for small differences in the loops between the β strands. The C strand of TBSV (although not TCV) is interrupted to form a loop, which interacts with the P domain, but most of the differences between the S domains of the spherical viruses are in the loops between the β strands of the barrel. SBMV has two extra helices between strands G and H as well as a helix between F and G that lies on the inner surface of the virus, close to the RNA. BBV has large inserts between

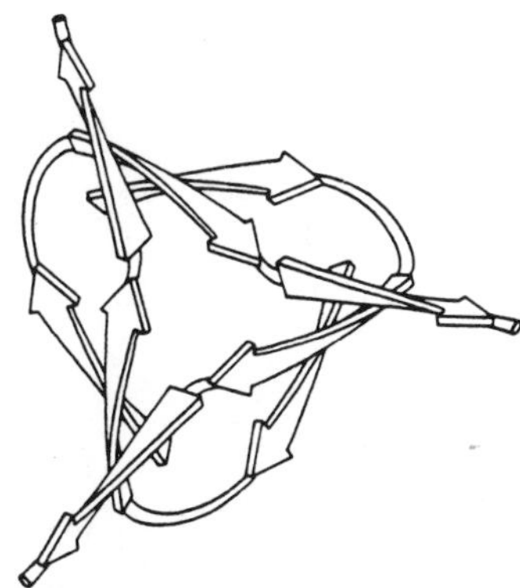

Figure 5. The β annulus found in several icosahedral plant viruses. Strands from three symmetrically related subunits form three short β sheets. A similar fivefold structure is found in the picornaviruses. (From Richardson, 1981.)

strands C and D, E and F, and G and H; the CD and GH loops together include four β strands, which extend the CHEF sheet of the barrel and interact with quasiequivalent inserts from two other subunits to form a prominent surface protrusion. BBV also has a long C-terminal appendage consisting of two α helices close to the inner surface of the S domain. The N-terminal helix of STNV is not seen in the other viruses.

The N-terminal peptides of these viruses, called the R (random) domains, do not conform to the icosahedral symmetry and are therefore not seen in the electron-density maps. Nuclear magnetic resonance experiments (McCain *et al.*, 1982a; Rossmann, 1984) suggest, however, that they do have fixed structures. It has been proposed on the basis of predictions from sequence that these structures may be α helical (Argos, 1981). The R domains vary considerably in length: whereas STNV has only 11 disordered residues, the A and B subunits of SBMV have 62 and 64, respectively, and those of TBSV 101. The subunits of the $T = 3$ viruses are not equivalent, and this is most evident in the N-terminal sections of the C subunits. In both SBMV and TBSV, these subunits include an additional ordered strand, from residue 39 in SBMV and 67 in TBSV. Part of this strand extends one of the β sheets in the barrel, and part interacts with equivalent strands from two other C subunits to form a "β annulus" (Harrison *et al.*, 1978) about the quasisixfold axis. In this structure (Fig. 5), chains from three symmetrically related subunits interact to form three short (three-residue) β sheets, each containing three strands in TBSV and two strands in SBMV and TCV. BBV does not have a β annulus, although it does have a peptide chain between the twofold related C subunits that is not seen at the quasitwofold AB interface. This strand is not, however, the N-terminal extension of strand B of the C subunit (Hosur *et al.*, 1987); all three subunits in BBV are ordered beginning at residue 65.

As in STNV, no RNA is visible in the electron-density maps of these viruses, although nuclear magnetic resonance has established that the nucleic acid does have a discrete structure (Munowitz *et al.*, 1980; McCain *et al.*, 1982b). Many basic residues, almost enough to balance the negative charge of the RNA, lie on the inner surface of the S domain or in the R domain (Rossmann, 1984). Rossmann *et al.* (1983b) have proposed, on the basis of a model-building study, that a double-helical RNA structure could bind to the inner surface of the S domain in SBMV.

In TBSV and TCV, the peptide chain continues beyond the S domain to form the P (protruding) domain. The P domains interact with each other across the twofold and quasitwofold axes, conferring the characteristic twofold clustering appearance of the tombusviruses. Here again the protein is folded into a jellyroll, although not in such a regular way as in the S domain. There are ten strands in the β barrel, with strands 7, 8, 5, 10, 1, and 2 (numbering from the N terminus) forming a sheet that provides the twofold interaction, while strands 3, 4, 9, and 6 form another sheet facing the first (Fig. 6). Strands 3 to 10 have the same topology as

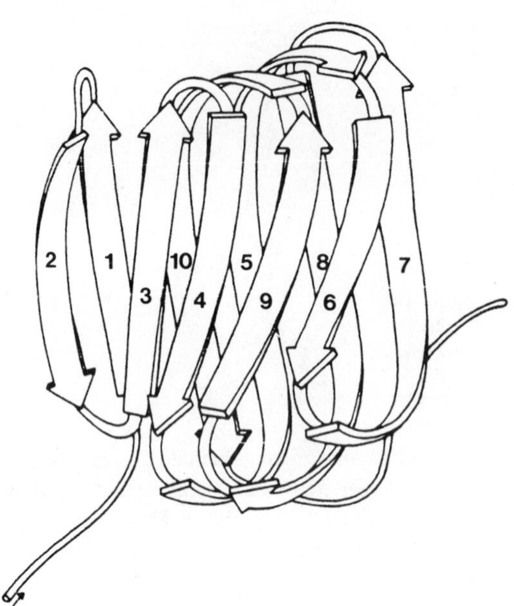

Figure 6. The P domain in tomato bushy stunt virus. Ten strands form a β barrel; strands 3 to 10 have the same topology as strands B to I of the S domain. (From Richardson, 1981.)

the strands of the S domain barrel (Argos *et al.*, 1980; Richardson, 1981), although the hydrogen-bonding pattern does not correspond; strands 4, 9, 6, and 7 correspond topologically to the strands of the CHEF sheet, but strand 7 is in a different sheet from the other three. If the jellyroll is seen as a folded hairpin (Fig. 4), the end loop is between strands E and F in the S domain, but to maintain the integrity of the sheets in the P domain, it must be placed between strands 7 and 8, which correspond topologically to F and G. The loops connecting the β strands in the P domain are very short, often containing only the residues of a β bend.

Residues 269–273 in TBSV form a hinge between the S and P domains. The backbones of the two domains have almost identical conformations in all three subunit positions, but the hinge adopts two distinct conformations, described as "down" in the A and B positions and "up" in C (Olson *et al.*, 1983). Other than through the hinge, there are very few contacts between the domains.

Protein–protein interactions have been discussed in detail for SBMV by Rossmann *et al.* (1983c) and in less detail for TBSV by Olson *et al.* (1983). Despite the similarity of protein folds, the interactions in the $T = 3$ viruses are very different from those in STNV, even about the icosahedral fivefold axes (Rossmann *et al.*, 1983a). Contacts are extensive about the fivefold, quasisixfold, and quasithreefold axes but very sparse about the twofold axes (except in viruses with a P domain). Interactions at quasisymmetrically related interfaces are similar except where the N-terminal arm of the C subunit is involved; in those cases, they are very different. For example, there are two types of interface between subunits related by quasisixfold axes, depending on whether or not the arm is interposed between them. If the arm is not present, the subunits can be superimposed by a rotation of 68°, very close to the 72° of the icosahedral fivefold axes. If the arm is present, however, the required rotation is only 57°. The two types of interaction are related by a rotation of one subunit about a vector in the subunit interface: 39° when comparing the quasisixfold interaction with the icosahedral fivefold and 36° between the quasitwofold and the icosahedral twofold. Thus, the vectors of rotation act as hinges between two states of contact. Contacts at these vectors tend to involve α helices, which may be particularly well suited structurally to function as hinges. Among the interfaces related by the quasithreefold axes, AB and AC are very similar, but BC, closer to the arm, is slightly but significantly different.

Almost all the protein residues in the S domains make some sort of intersubunit contact (Rossmann *et al.*, 1983c). There are extensive charge clusters, large hydrophobic areas of contact, and numerous possible hydrogen-bonding interactions. The twofold interaction in SBMV is stabilized by the stacking of Trp^{99} and Trp^{107} in a hydrophobic pocket. The threefold interaction of the N-terminal arms is through hydrophobic contacts. Contact between the P domains occurs only at the twofold axes and consists of hydrophobic interactions surrounded by surface hydrophilic interactions (Olson *et al.*, 1983). These interactions are identical about the icosahedral twofold and quasitwofold axes, the differences being accommodated by the hinge between S and P. The P domain is more tilted toward the viral surface in TCV than in TBSV, so that the twofold interface is somewhat altered.

The charge clusters in the subunit interfaces are of particular interest, as they are believed to be important in viral assembly. Rossmann *et al.* (1983c) found three such clusters in the quasithreefold interfaces and two in the fivefold interface. One of these two was also found in the quasisixfold interface. One of the clusters is on the quasithreefold axis, but it is not the site found on the threefold axis in STNV. Calcium is bound to one of the most extensive charge clusters (Abdel-Meguid *et al.*, 1981; Silva and Rossmann, 1987), in the quasithreefold interface at the viral surface. Hogle *et al.* (1983) found a site binding two calcium ions in TBSV, involving five aspartate groups and a lysine; one of these aspartate groups corresponds in the sequence to Asp^{138} in the second SBMV calcium-binding site. This site has been implicated in viral expansion by Robinson and Harrison (1982), as it is located in the interface that dissociates in that process and includes a ligand from the hinge between the S and P domains (Olson *et al.*, 1983). The other SBMV site appears to help determine the T number of the capsid (Savithri and Erickson, 1983). Charge clusters of carboxylic acids alone, or of carboxylic acids and histidine (as in one of the fivefold clusters), are particularly suited to the control of assembly under physiological conditions; because of their near-neutral p*K*s, they are sensitive to changes in environmental factors such as pH or calcium concentration and thus serve as electrostatic switches to control assembly and disassembly (Caspar, 1963, 1976; Stubbs, 1984). The charge clusters are not conserved between SBMV and TBSV or between the cowpea and bean strains of SBMV; nor are any of the calcium-binding sites of STNV found in the $T = 3$ viruses. A similar circumstance is found in the rod-shaped viruses; one of the charge clusters that controls assembly of TMV is not conserved in cucumber green mottle mosaic virus (watermelon strain) (Lobert *et al.*, 1987). Electrostatic switches, being structurally flexible and nondirectional, appear to migrate easily within subunit interfaces during evolution, thus reducing evolutionary constraints on the developing structure.

D. Simple Animal Viruses: Four Coat Proteins

Three picornavirus structures have recently been determined. These small RNA viruses make up a very large family, responsible for such diverse diseases as the common cold, poliomyelitis, hepatitis, and foot-and-mouth disease. They have diameters of about 300 Å, comparable to the $T = 3$ plant viruses. The structure of human rhinovirus 14 has been determined at 3.0-Å resolution (Rossmann *et al.*, 1985), and that of poliovirus type 1 (Mahoney strain) at 2.9 Å (Hogle *et al.*, 1985). Both of these structures were initially determined at 5-Å resolution by isomorphous replacement and noncrystallographic symmetry averaging. Resolution was extended in further cycles of symmetry averaging. The structure of Mengo virus has been determined at 3.0 Å by phase extension from the 8-Å model of HRV14 (Luo *et al.*, 1987). The three structures are similar, although HRV14 and poliovirus are more like each other than Mengo virus. A preliminary report of crystals of foot-and-mouth disease virus (Fox *et al.*, 1987) has been published.

Each picornavirion contains 60 copies of each of four proteins, VP1, VP2, VP3, and VP4, and one molecule of RNA containing 7200 to 8500 nucleotides (for comparison, the

RNA of SBMV contains 4195 nucleotides; Wu *et al.*, 1987). The RNA is covalently linked to a single copy of a small protein, VPg. The viral proteins are first synthesized as one large polyprotein, from which is cleaved a capsid protein precursor. During viral assembly, this protein is cleaved into three chains, VP0, VP1, and VP3. VP0 is cleaved to yield VP2 and VP4 at the final stage of assembly. In HRV14, the molecular weights of VP1, VP2, VP3, and VP4 are 32,000, 29,000, 26,000, and 7000, respectively. The amino acid sequences of rhinovirus and poliovirus proteins appear to be more similar to each other than to the other picornaviruses. The properties of the picornaviruses have been reviewed by Rueckert (1985).

The structures of VP1, VP2, and VP3 are similar to each other and to the coat proteins of the $T = 3$ plant viruses in that each contains a wedge-shaped jellyroll β barrel with various loops and extensions and two short α helices, one between strands C and D of the barrel and one between strands E and F. Between strand C and the CD helix of VP2 and VP3, there is an additional short helix, of two turns in poliovirus and one turn in HRV14 and Mengo. The lengths of the strands in the barrels vary, but the positions of the barrels correspond very closely to those of the $T = 3$ plant viruses, with VP1 at position A, VP2 at position C, and VP3 at position B. The N termini of VP1, VP2, and VP3 are on the inner surface of each virus, and the C termini are on the outer surface. This spatial separation confirms that proteolytic processing must take place prior to viral assembly (Hogle *et al.*, 1985). VP4 is located on the inner surface of the virus. The proteins of HRV14, poliovirus, and Mengo virus are compared in Fig. 7.

Although the regular secondary-structure elements of the picornaviruses closely resemble those of the plant viruses, there are considerable differences in the connecting loops. The largest differences are in the loop connecting strands G and H in VP1 and in the major excursion (the "puff"; Rossmann *et al.*, 1985) between strands E and F in VP2. Mengo virus has a considerably attenuated puff and two large loops (I and II) between strands C and D of VP1. The N- and C-terminal extensions also vary significantly; unlike the plant viruses, only very short lengths of the peptide chains are disordered at the N termini. There is a short piece of α helix before strand B in all the picornavirus barrels; in VP3 of poliovirus, the N-terminal part of strand B is disrupted to form a significant excursion between this helix and the remainder of the strand, and in HRV14 and Mengo, the corresponding excursion (which is much smaller in Mengo) interrupts the strand.

The N-terminal residues of five VP3 subunits form a β barrel about the icosahedral fivefold axis, analogous to the threefold β annulus of TBSV and SBMV. These residues lie under the VP1 barrel in the virus, and the N-terminal residues of VP1 lie under VP3. The N-terminal arms are in contact with the RNA, but in contrast to the plant viruses, they are more acidic than basic.

VP4 is the N-terminal extension of VP2. Its N terminus is near the fivefold axis, and in HRV14 and poliovirus, it runs as an extended chain toward the threefold axis, near the N terminus of VP2. In Mengo virus, the path of the chain is rather different; it interacts extensively with VP3 and ends near the N terminus of the neighboring VP2, related by a fivefold rotation to the corresponding VP2 in HRV14 and poliovirus. VP4 is much more exposed to the RNA core in Mengo virus than in HRV14 or poliovirus. A serine from VP2 in HRV14 and Mengo virus is so located that, activated by a base from the RNA, it could be responsible for the autocatalytic cleavage of VP4 and VP2 at the final stage of assembly (Rossmann *et al.*, 1985; Luo *et al.*, 1987).

In overall shape, the picornaviruses present a much more uneven outer surface than, for example, SBMV because of protrusions formed by the various excursions of the peptide chain. A large peak at the fivefold axis, formed by sequence insertions in VP1, is surrounded by a ridge of VP2 and VP3 subunits at the twofold and threefold axes and separated from them by a broad valley. In HRV14, the valley is more pronounced than in poliovirus, giving the ap-

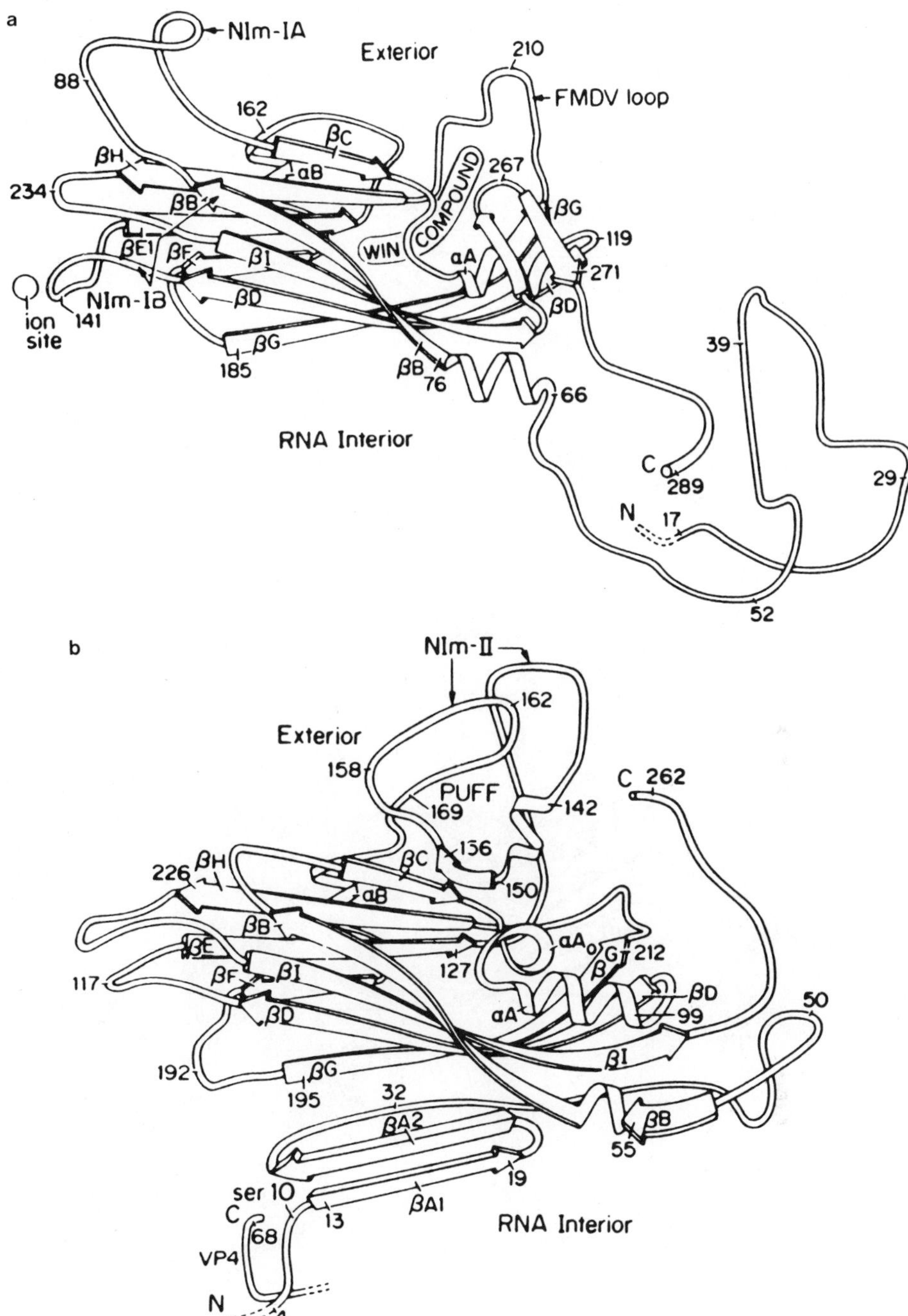

Figure 7. The S domains of HRV14: (a) VP1, (b) VP2, and (c) VP3; Mengo virus: (d) VP1, (e) VP2, and (f) VP3; and poliovirus: (g) VP1, (h) VP2, and (i) VP3. (From Luo *et al.*, 1987, and Hogle *et al.*, 1985.)

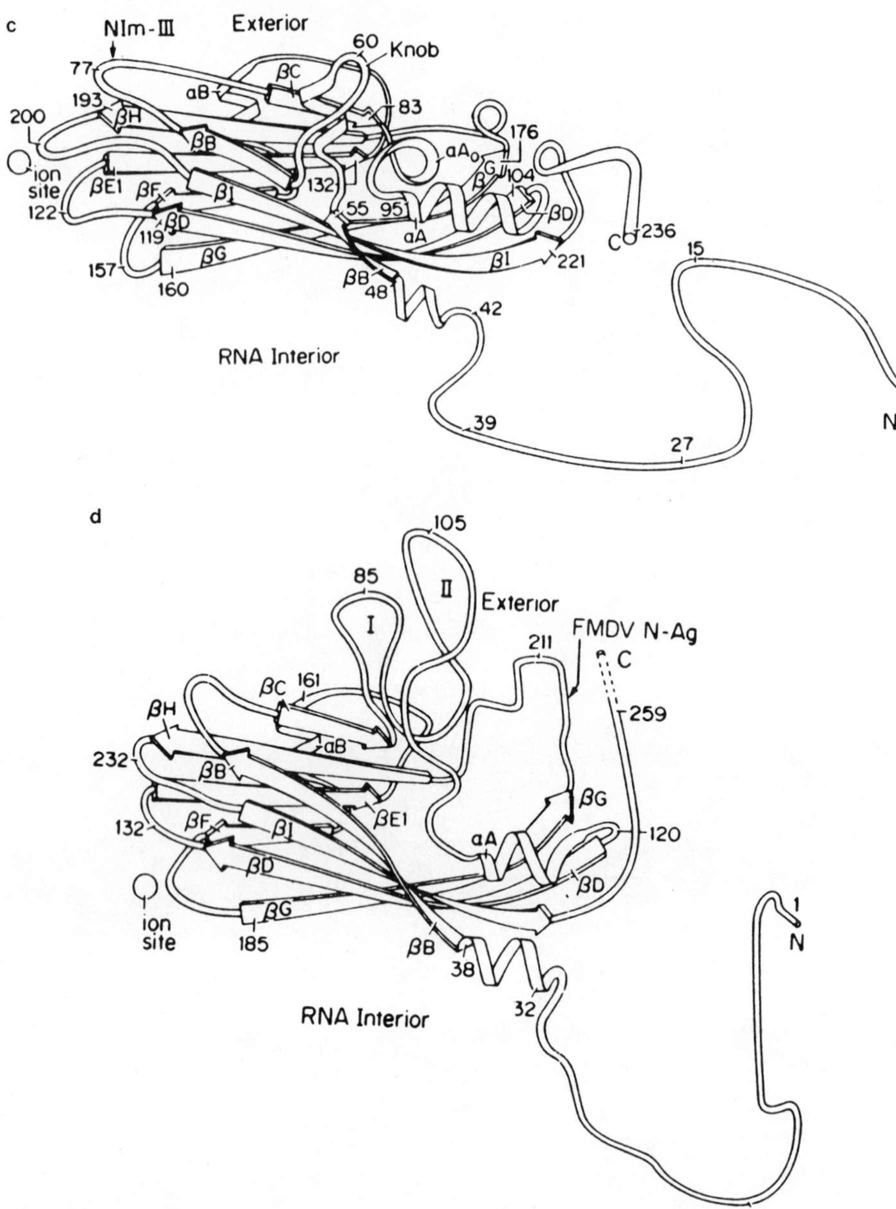

Figure 7 (cont.)

e

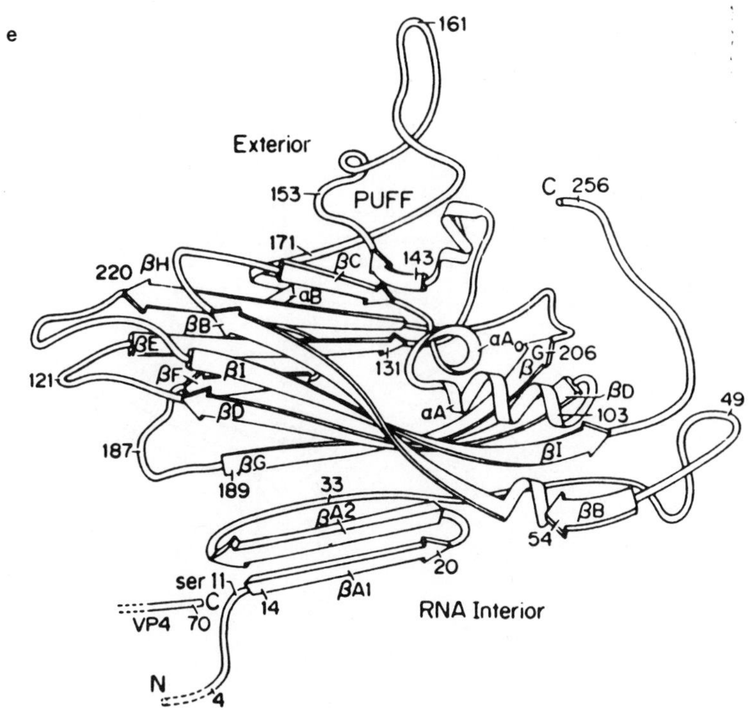

f

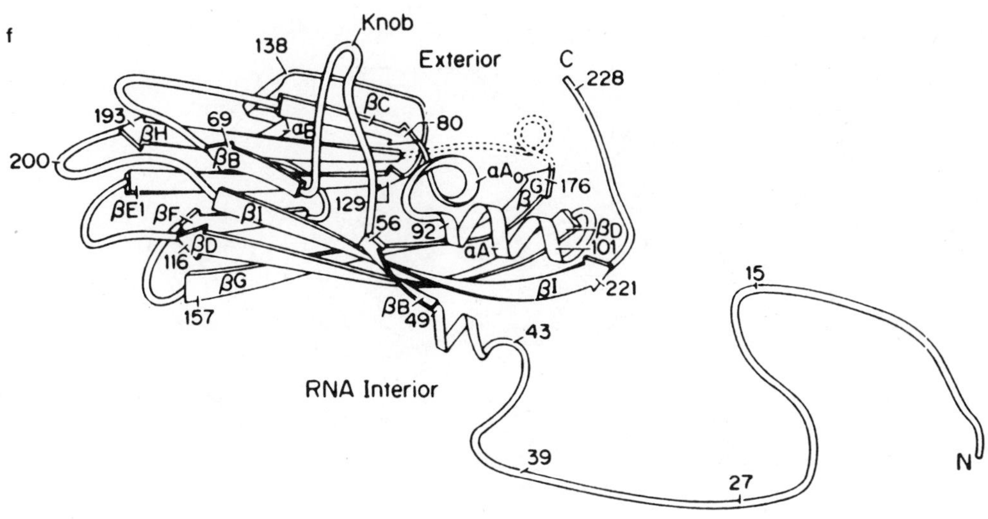

Figure 7 (cont.)

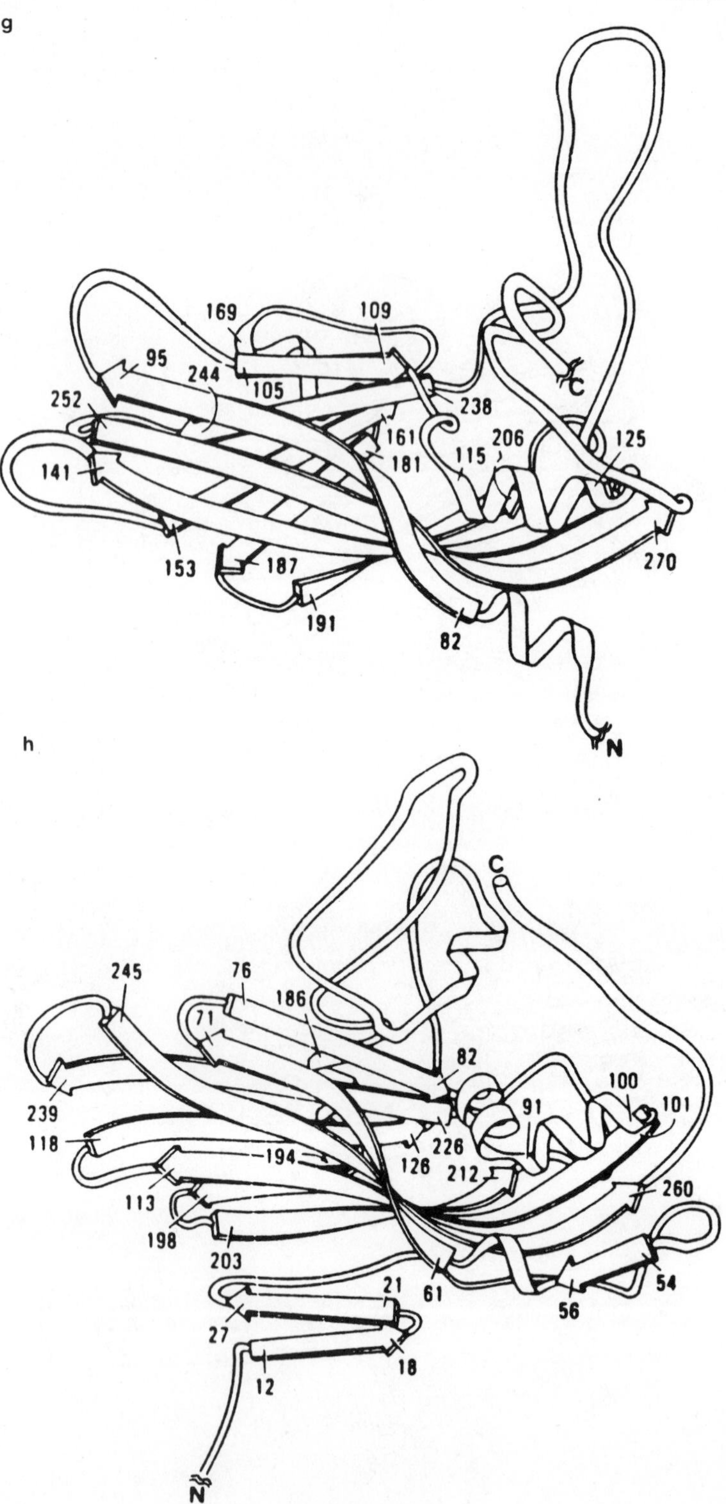

Figure 7 (cont.)

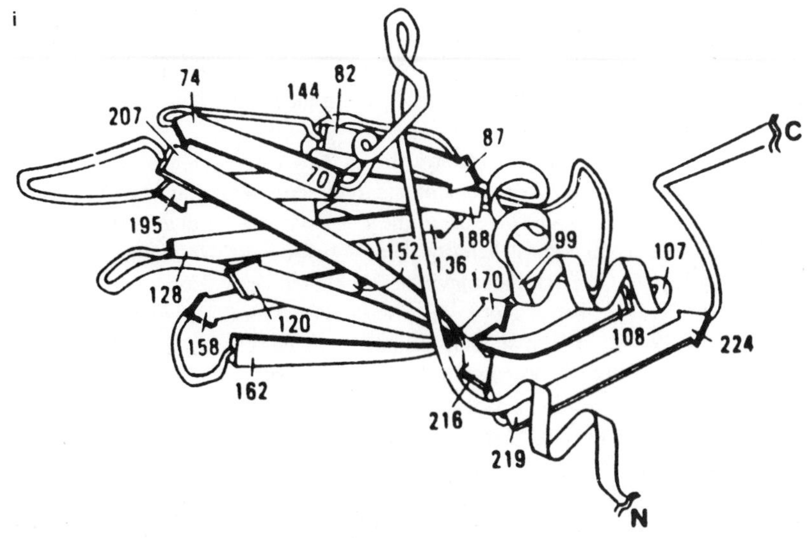

Figure 7 (cont.)

pearance of a canyon or moat, and in Mengo virus, loops I and II interact with the modified puff, filling in part of the canyon to leave a deep pit. Most of the identified antigenic binding sites are on the protrusions, and Rossmann *et al.* (1985) have suggested that the receptor binding site for HRV is on the floor of the canyon, protected from antibodies by the steep walls and thus freed from the need to mutate under immune selection pressure.

The packing of the barrels in the picornaviruses is very similar to that in the $T = 3$ plant viruses. Luo *et al.* (1987) compared the positions of the proteins of HRV14 and SBMV and found that the largest rotation that was needed to superimpose corresponding subunits was only 19°, for VP2 and subunit C. Some differences in subunit contacts caused by the chemical nonequivalence of the animal virus subunits, in particular the packing of α helices from neighboring subunits against each other, were discussed by Hogle *et al.* (1985). The internal surfaces of the viruses are characterized by the complex interactions between the N-terminal peptide chains and the surfaces of different subunits discussed above. There appears to be a calcium-binding site on the fivefold axis of HRV14, created by the juxtaposition of carbonyl groups from five equivalent Asn^{141} residues of VP1, and a similar site in Mengo virus at Pro^{178}. The negative charge concentrations found in the plant viruses have not been reported for the picornaviruses, but the interactions of the acidic N-terminal arms with the RNA could provide a similar energy balance.

The binding of an antiviral compound to HRV has been described by Smith *et al.* (1986). The compound inserts itself into a hydrophobic pocket in VP1, near the floor of the canyon, and prevents viral uncoating, either directly, by stabilizing the virion, or indirectly, by preventing ions from reaching the virus interior. Weak electron density has been observed at the corresponding site in poliovirus (J. M. Hogle, personal communication). This hydrophobic cavity is present in all three viruses studied, although it is not accessible in Mengo virus. Luo *et al.* (1987) have suggested that the cavity may be required for viral disassembly.

E. Multiple Barrels: Variations on a Hexagonal Theme

β barrels are formed into spherical shells in a variety of ways, all based on some distortion of hexagonal packing: by strict pentameric equivalence in STNV, by quasihexagonal symmetry of chemically identical subunits in the $T = 3$ plant viruses, and by pseudohexagonal symmetry of different subunits in the picornaviruses. As new virus structures are determined, even stranger solutions to the problem of constructing approximately hexagonal packing are coming to light. In the two structures described in this section, cowpea mosaic virus and adenovirus, nonequivalent barrels are again used, but in these viruses, more than one barrel is found in a single protomer.

The structure of CpMV has been determined at 3.5-Å resolution by Stauffacher *et al.* (1987), using single isomorphous replacement and symmetry averaging at 5.5 Å followed by phase extension. This virus is about the size of the $T = 3$ viruses, with an average diameter of 280 Å; the capsid contains 60 copies each of two proteins, one of 374 amino acids and one of 213. It is thus technically (like the picornaviruses) a $T = 1$ virus. Nonetheless, the structure reveals the widespread three-barrel motif, distributed on a $T = 3$ icosahedral net.

The smaller of the two protein subunits contains one barrel, arranged about the fivefold axes and corresponding to the A position in the $T = 3$ viruses. In addition to the usual eight strands and the CD and EF helices, this barrel contains two extra β strands in a loop between strand C and the CD helix. The larger subunit contains two barrels. The N-terminal part of the chain forms a barrel that occupies the C position and has very short connecting loops, with little more than the minimal eight β strands and two helices other than a long, trailing C-terminal chain. The C-terminal barrel, in the B position, is very similar to the SBMV barrel except that it does not have the FG helix. CpMV does not have a β annulus; the N termini of the C domains interact about the twofold axes. The A barrel is tilted out of the surface relative to the B and C barrels, forming a significant bulge at the fivefold axis. There is consequently not a good pseudothreefold axis relating the barrels.

Adenovirus is much larger and more complex than any of the viruses so far discussed. Structural studies have been reviewed by Burnett (1984). The icosahedral virion has a diameter of about 900 Å and contains more than ten different proteins. Like CpMV, adenovirus uses separate proteins for icosahedral vertices and faces, although the relative numbers of the two types of protein are very different. There are 252 morphological units on the surface, consisting of 240 hexons (Fig. 8) and, at the vertices, 12 penton complexes. Each penton consists of a base and a projecting fiber. There is also an inner icosahedral protein core enveloping a protein–DNA complex. The architecture of adenovirus, in particular the modification of the

Figure 8. A model adenovirus capsid showing the organization of 12 hexons on each icosahedral facet. The vacant position at each vertex is filled by a penton complex. (From Roberts *et al.*, 1986.)

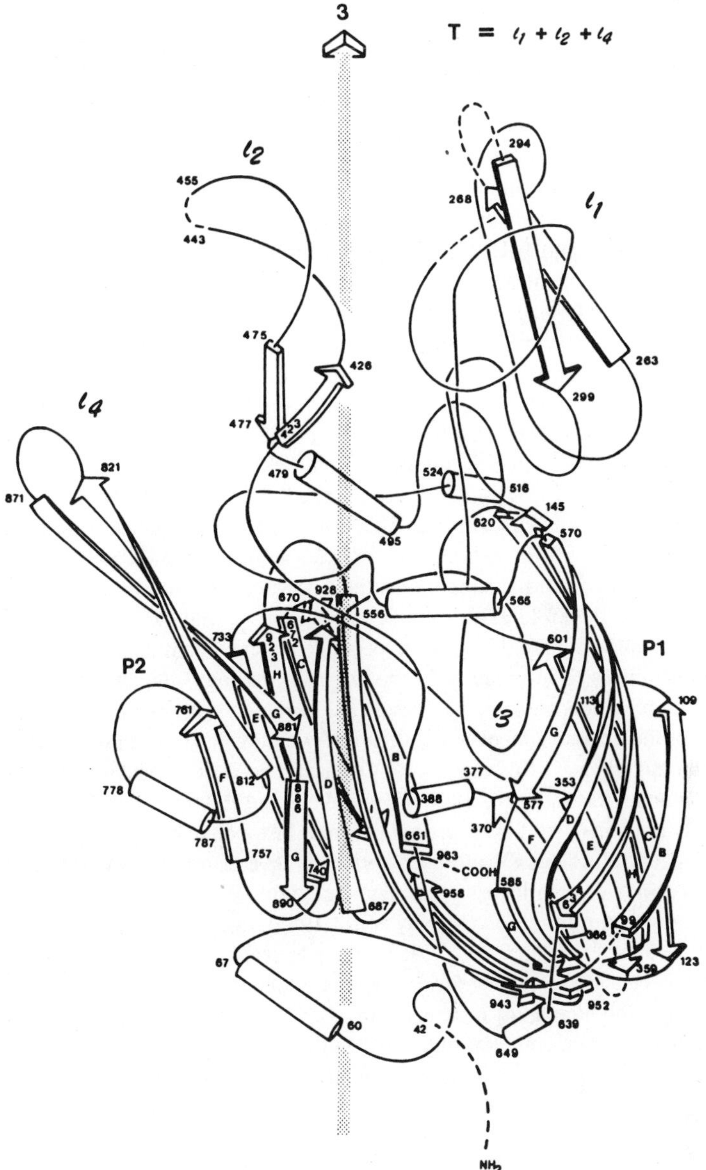

Figure 9. The hexon subunit of adenovirus, showing domains P1 and P2 and the tower (T). The threefold axis of the hexon is marked. (From Roberts *et al.*, 1986.)

concepts of quasisymmetry to treat this and other large viruses, has been discussed by Burnett (1985).

The hexon of adenovirus type 2 has been crystallized, and its structure determined at 2.9-Å resolution (Roberts *et al.*, 1986). The amino acid sequence is known from the gene sequence (Akusjärvi *et al.*, 1984). The hexon is a trimer of identical subunits, each containing 967 amino acids, 859 of which have been located. The N-terminal peptide is disordered, as in the RNA viruses, although in the hexon, this peptide interacts with another protein rather than with nucleic acid. The structure of one subunit is shown in Fig. 9. At the base of the subunit are two jellyroll β barrels, P1 and P2. Many of the interstrand loops are very short, in four cases utilizing a Pro-X-X-X-Pro sequence. Prolines are used to form short loops in many places in the structure, with 19 examples of proline pairs separated by five or fewer residues. In general, the hexon contains very little regular secondary structure, being only 8% helix and 22% β sheet.

From the base, three loops form the "towers," protrusions from the viral surface. One of these loops, l_2, includes two antiparallel β strands. These strands interact with their symmetry equivalents from the other two subunits to form the "β constriction," a six-stranded tube that is similar to the three- and five-stranded β annuli found in many of the small icosahedral viruses. The loops that form the towers are extensively intertwined, with each of the three towers in the trimer containing one loop from each of the three subunits. Roberts *et al.* (1986) have suggested that this complexity of structure may account for the unusual requirement of the hexon for another protein as part of its folding mechanism (Cepko and Sharp, 1982). One of the strands leading to l_1 includes a sequence of 16 consecutive acidic residues; it has been suggested (Roberts *et al.*, 1986) that these may be involved in a pH-dependent conformational change as part of the mechanism of dissociation of hexon shells.

The barrels are distributed on a $T = 25$ icosahedral net, oriented with their axes approximately perpendicular to the viral surface. This orientation is in contrast to the parallel orientation found in the small viruses. Although there is no sequence homology between the barrels, they superimpose very well; the superimposition, however, requires rotation about an axis at 25° to the molecular threefold axis, so there is no good sixfold axis in the virion.

Many of the authors quoted here have speculated on the evolutionary relationships among the icosahedral viruses. The ubiquity of the jellyroll topology suggests that they are indeed derived from a common ancestor, although the complete absence of sequence homology among many of the viruses, and even between the two domains of adenovirus hexon, suggests a very ancient divergence. Rossmann (1984) has proposed a probable set of evolutionary relationships among AMV, STNV, SBMV, and the S and P domains of TBSV and suggested also that the R domains could be descended from a histonelike protein.

Jellyroll β-barrel domains have been found in several proteins in addition to the viral capsid proteins, including influenza virus hemagglutinin (Wilson *et al.*, 1981), the catabolite gene activator protein from *E. coli* (McKay *et al.*, 1982), and lectins including concanavalin A (Reeke *et al.*, 1975; Einspahr *et al.*, 1986). Concanavalin A competes with HRV and poliovirus for polysaccharide cell surface receptors (Lonberg-Holm, 1975), and Argos *et al.* (1980) found that the strands of concanavalin A and TBSV superimposed better than other pairs of proteins of known common function. On these grounds it has been speculated (Argos *et al.*, 1980; Rossmann, 1984; Rossmann *et al.*, 1985) that the gene for a protein such as concanavalin A could have been ancestral to the family of icosahedral virus coat protein genes.

F. Equivalence, Quasiequivalence, and Pseudoequivalence

Determination of new structures has made it necessary to go beyond the original description of quasiequivalence by Caspar and Klug (1962), based on distortions of similar bonds between identical subunits. The inadequacy of the quasiequivalence theory stems at least partly from the fact that it fails to take into account the significant thicknesses of viral coat protein shells (Rossmann *et al.*, 1983c; Rayment, 1984). Approximate equivalence is reached in practice by conformational switches (Caspar, 1976) leading to completely new arrangements of bonds and by genetic variation leading to protein subunits or domains specialized to occupy one of the nonequivalent positions in the surface lattice. The principle stands, however, that by small deviations from perfectly symmetrical arrangements, protein molecules are able to form much larger icosahedral particles than they could were rigid symmetry to be the rule.

III. HELICAL VIRUSES

The helical viruses include the rod-shaped plant viruses, the filamentous bacteriophages, and the enveloped nucleocapsids found in several families of animal viruses and a few plant viruses. It has not been possible to crystallize these viruses, so their structures can only be visualized by x-ray fiber diffraction or electron microscopy. Only two helical viruses, TMV and the bacteriophage Pf1, have so far been described in molecular detail, and only TMV at a resolution comparable to that of the small icosahedral viruses.

Some structural studies have been made of the enveloped nucleocapsid viruses, which include several families of animal viruses, in particular, the paramyxoviruses (mumps, measles), the orthomyxoviruses (influenza), and the rhabdoviruses (vesicular stomatitis, rabies, and some plant viruses). A brief review is given by Casjens (1985). Single-stranded RNA is bound to subunits of one major coat protein, the N protein, which is rich in basic amino acids. Small amounts of other proteins are present, but they are not required for the structural integrity of the nucleocapsid. There are generally six to nine nucleotides bound to each protein subunit, but the orthomyxovirus subunits appear to bind about 20 nucleotides (Compans *et al.*, 1972). Finch and Gibbs (1970) reconstructed a projection of the paramyxovirus Sendai virus from electron micrographs at about 30-Å resolution and found the virus to be hollow, with inner and outer radii of 20 and 100 Å. The subunits are arranged in a helix of pitch about 50 Å, probably with 11 subunits per turn of the helix.

A. Plant Viruses

1. Tobacco Mosaic Virus

The tobamoviruses are rod-shaped, 3000 Å long and 180 Å in diameter, with a central hole of diameter 40 Å. Approximately 2130 identical protein subunits of molecular weight 17,500 form a right-handed helix of pitch 23 Å with 49 subunits in three turns, protecting a single strand of RNA that follows the basic helix between the protein subunits at a radius of 40 Å (Fig. 10). There are three nucleotides bound to each protein subunit. The type member,

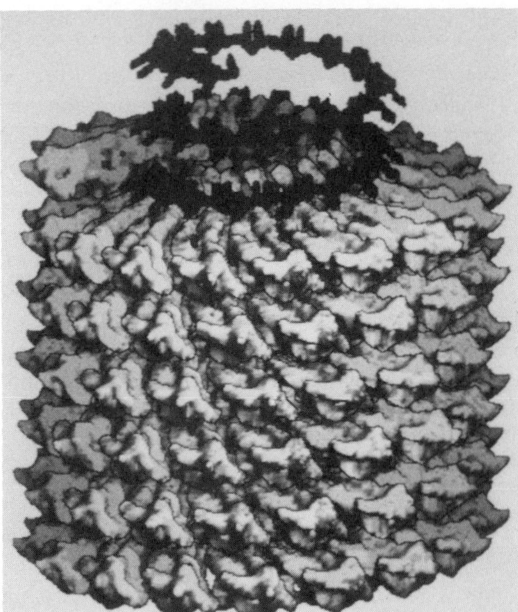

Figure 10. Packing of the subunits in TMV. About $\frac{1}{20}$ of the particle is shown. RNA (dark) is shown beyond the end of the protein helix (light) for clarity.

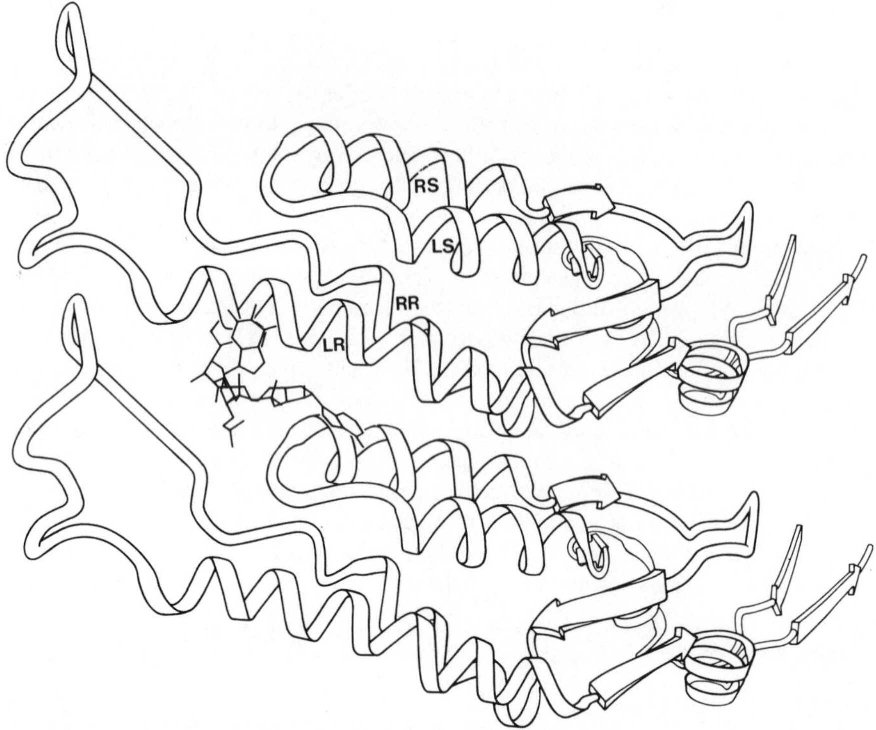

Figure 11. The folding of the polypeptide chain and the axial packing of two subunits in tobacco mosaic virus. The virus axis runs vertically, to the left of the figure, and the outer surface is on the right. The four closely packed α helices are termed left slewed (LS, top and toward the reader), right slewed (RS, top and away), left radial (LR, bottom and toward the reader), and right radial (RR, bottom and away).

TMV, has been a model system for both fiber diffraction and structural studies of viruses for many years. The amino acid sequence of the coat protein was determined by Wittmann-Liebold and Wittmann (1967) and confirmed by the complete sequence of the viral genome (Goelet *et al.*, 1982). Using x-ray fiber diffraction from oriented gels, the virus structure was determined at 3.6-Å resolution by Namba and Stubbs (1986) and has been extended to 2.9 Å using a restrained least-squares refinement modified (Stubbs *et al.*, 1986) for fiber diffraction data. The TMV is the only virus in which protein–nucleic acid interactions have been visualized.

In contrast with the spherical viruses, the secondary structure of the TMV coat protein is largely α helical. The chain fold of two adjacent subunits is shown in Fig. 11. The central part of the molecule consists of four closely packed α helices, which form a hydrophobic core extending radially from 45 Å to 65 Å. This α-helical core is functionally analogous to the β barrel of the spherical viruses, dividing the amino acid side chains into an inner hydrophobic layer that stabilizes the protein folding and an outer layer of mixed hydrophobic and hydrophilic character that stabilizes the subunit packing. There are two other short helical segments and a small region of β structure connecting the outer ends of the four core helices. Beyond this region, aromatic and hydrophobic residues interact with each other and across subunit boundaries to form a continuous hydrophobic sheet beneath the outer surface of the virus. Observing that the overall protein structure contains a twice-repeated pattern of two core α helices and two strands of β structure, McLachlan *et al.* (1980) have suggested that TMV protein evolved by gene duplication of a primitive dimeric protein.

The long, thin subunits of TMV are packed closely together, creating extensive protein–protein interactions, including close packing between the α helices of adjacent subunits as well as interactions between the connecting loops. There is a continuous cavity between the subunits and along the basic viral helix, which accomodates the RNA. The vertical chains at low radius pack closely together, each filling a space 9 Å wide and 23 Å high and forming a dense wall around the central hole of the virus particle.

Like the spherical plant viruses, TMV has charge clusters in the subunit interfaces, which are believed to control viral assembly and disassembly. One cluster is in the side-to-side interface at low radius, and another, at about 60 Å radius, is in the top-to-bottom interface. In both cases carboxylate groups, partially stabilized by nearby arginines, are clustered at sites believed to bind calcium (Namba and Stubbs, 1986). These sites are also believed to bind protons that titrate with anomalous p*K*s near 7. Asp^{116} is very close to a phosphate group, and the resulting electrostatic repulsion is also believed to be important in viral disassembly.

The RNA in TMV is well ordered, and the protein–RNA interactions have been described by Namba and Stubbs (1986). Some of these interactions are shown in Fig. 12. The RNA conformation has been described by Stubbs and Stauffacher (1981). The phosphate groups in the RNA are neutralized by Arg^{41}, Arg^{90}, and Arg^{92}, but they do not all form simple ion pairs. The concentration of negative charge at Asp^{115}, Asp^{116}, and one of the phosphates is partially neutralized by Arg^{112}. All three bases lie flat against the LR helix, making hydrophobic interactions that are not RNA sequence-specific. Two bases, one in the *syn* conformation and one *anti,* stack together and point up into a cavity between the intersubunit salt bridges Arg^{113}-Asp^{115} and Arg^{122}-Asp^{88}. The third base, in the *anti* conformation, lies along the LR helix, between this helix and the connecting LS–RS loop from the subunit below. Side chains from subunits above and below the RNA provide a hydrophilic environment for the polar parts of the

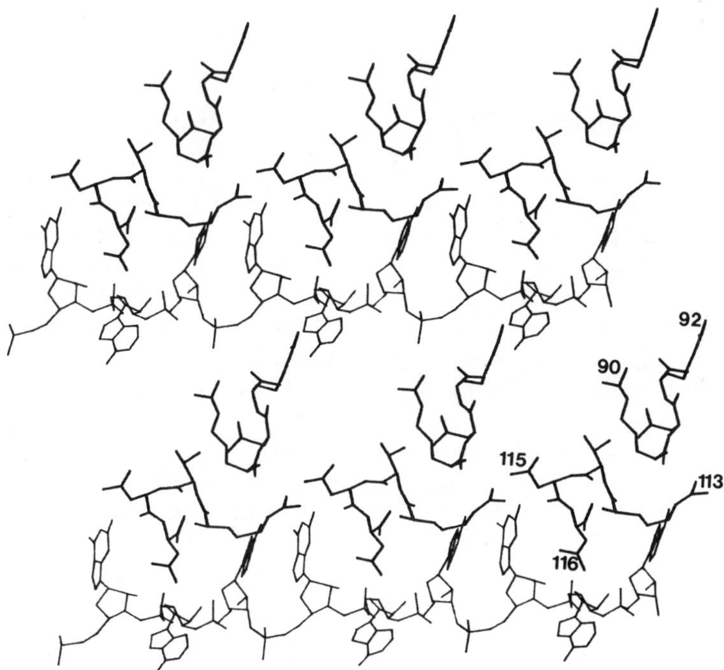

Figure 12. Protein–nucleic acid interactions in six subunits of TMV, looking along a radius toward the viral axis. Heavy lines: protein, residues 90–92 and 113–116. Light lines: RNA. Two bases extend vertically into the intersubunit space between two equivalent α helices, and one base extends toward the reader, along the α helix.

bases. Although most of the protein–nucleic acid interactions are not base specific, several hydrogen bonds have been tentatively identified that would specifically favor the AAG base sequence, known to be important in recognition for the early stages of viral assembly (Zimmern, 1977), and to bind strongly as a trinucleotide to TMV protein (Steckert and Schuster, 1982).

2. Other Helical Plant Viruses

Although detailed structures are not yet available for any helical plant viruses other than TMV, preliminary diffraction results have been reported in several cases. Franklin (1956) obtained fiber diffraction patterns for three tobamoviruses, and Holmes and Franklin (1958) showed that the radial density distributions for four tobamoviruses were similar. The similarities in the diffraction patterns of about ten tobamoviruses (Franklin, 1956; G. Stubbs, S. Warren, and S. Morris, unpublished results) clearly imply a generally similar chain fold. Lobert *et al.* (1987) found that the radial density distribution of CGMMV-W resembles that of TMV and is particularly similar to that of cucumber virus 4 (Holmes and Franklin, 1958). CGMMV-W does not contain Glu^{50} or Asp^{77}, but it was suggested on the basis of lead-binding studies that other carboxylate side chains not present in TMV could have the same function as these two groups.

Fiber diffraction studies have been reported for viruses from three other groups: several potexviruses (Tollin *et al.*, 1979, 1980, 1981), the CAM strain of tobacco rattle virus (Finch, 1965; Tollin and Wilson, 1971), and barley stripe mosaic virus. Symmetries and axial repeat distances have been reported in each case. All of these viruses diffract strongly in both equatorial and meridional directions at about 10-Å resolution, which would be consistent with α helices packed approximately at right angles to the viral axis, like TMV. The helical packing of the potexviruses suggests a subunit thickness of about 35 Å, which would be too much for a simple double layer of α helices, but the corresponding dimensions for TRV and BSMV are 25 Å and 26 Å, quite close to the 23 Å of TMV.

B. Filamentous Bacteriophages

Structural studies of the filamentous bacteriophages have been reviewed by Makowski (1984). They are flexible viruses about 60 Å in diameter and of variable length. Depending on the environment, fd is about 8900 Å long, and Pf1 is about 19,400 Å long. They contain one molecule of single-stranded circular DNA running the length of the virion, several thousand copies of a major coat protein, and a few copies of several minor proteins at the ends of the filament. The number of nucleotides per coat protein subunit varies considerably, being one for Pf1, two for Xf, and 2.3 for fd. The amino acid sequence for the coat protein is known in many cases; references are given by Makowski (1984). These phages have been divided into two structural classes on the basis of the symmetry of the major coat protein (Marvin *et al.*, 1974a,b). In class I, including the phages fd and M13, the viral axis is a fivefold axis of rotation; that is, the subunits are arranged on a set of five coaxial helices. In class II, including Pf1, there is no rotation axis, and the subunits are arranged on a simple helix. The subunit packing in class II can, however, be described as a small perturbation of the packing in class I (Makowski, 1984). The precise symmetry of Pf1 depends on temperature, heavy atom binding, and other factors and varies from 5.40 to 5.46 subunits per turn of the basic helix.

The structure of Pf1 has been determined at 7-Å resolution for high-temperature (Makowski *et al.*, 1980) and low-temperature (Nave *et al.*, 1981) forms, in each case by refining models of α helices. Although the types of model and methods of analysis were very different, the electron-density maps are very similar. A map of the low-temperature form at 3-Å resolution has been reported (Marvin *et al.*, 1987).

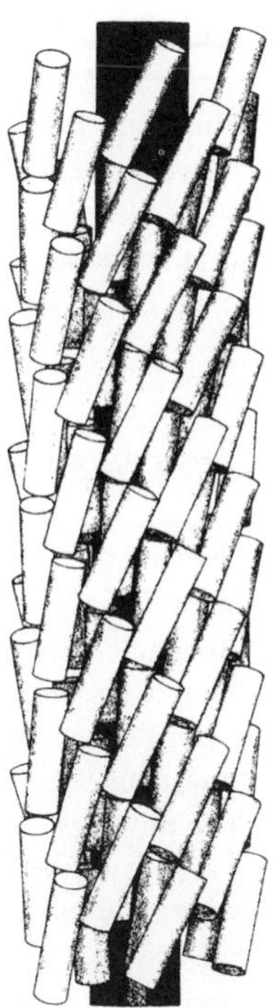

Figure 13. The arrangement of the α helices in Pf1. The α helices are represented by cylinders about 30 Å long and 9 Å in diameter. The central core, occupied by DNA, is represented by a cylinder 20 Å in diameter. (From Makowski, 1984.)

The DNA is encased in a double layer of α helices, stabilized by hydrophobic interactions between the layers. Figure 13 shows the helices as described by Makowski (1984). Each molecule contains two straight segments of α helix, one nearer the axis and one at the outer viral surface. The inner helices make an angle of 6° to the viral axis, and the outer helices 26°. The interaction between helices of different layers is, unexpectedly, right-handed (Bryan *et al.*, 1983; Stark *et al.*, 1988). The peptide chain runs from the N terminus at high radius and axial coordinate (z) to the C terminus at low radius and z, with a disruption of the secondary structure at the connection between the two helices. The connectivity was not firmly established by the 7-Å maps; another possible interpretation could have been that a hairpin bend connected the helices. The elongated structure described here is more consistent with the 3-Å map of Marvin *et al.* (1987), however, and has now been firmly established by neutron-scattering experiments at 8-Å resolution in which five selectively deuterated valine residues were located (Stark *et al.*, 1988). The position of Tyr^{25} is known from an iodinated derivative (Nave *et al.*, 1981), and other residues have been located from the constraints of the helical structure and considerations of side-chain interactions. Asp^{14} and Asp^{18} are not readily accessible to solvent and may be buried in the hydrophobic part of the molecule; they may function in a similar way to the carboxyl clusters of other viruses (Makowski, 1984).

An alternative model for Pf1 (Marvin *et al.*, 1987) contains one continuous curved helix rather than two helical segments, but solid-state NMR experiments have shown that the orientations of the amide bonds relative to the viral axis in Pf1 fall into two narrowly defined populations, corresponding to the two helical segments. In fd, in contrast, the orientations of the amide bonds vary continuously over a small range (Cross *et al.*, 1983), in agreement with a curved-helix model based on fiber diffraction data (Banner *et al.*, 1981). This difference between the class I and class II phages is also indicated by Patterson maps from fiber diffraction data for Pf1 and M13 (Glucksman *et al.*, 1987), which are consistent with straight helices in Pf1 and curved or in some way continuously perturbed helices in M13. The segmented nature of the Pf1 helices has been confirmed by neutron scattering (Stark *et al.*, 1988).

Density interpreted as DNA is seen near the viral axis in the map of the high-temperature form of Pf1 (Makowski *et al.*, 1980) and in the preliminary map of the low-temperature form (Bryan *et al.*, 1983). Solid-state NMR (Cross *et al.*, 1983) suggests that the DNA is ordered in Pf1 but disordered in fd.

IV. STRUCTURAL CHANGES IN VIRUSES

Virus capsid proteins are actively involved in viral assembly and disassembly. The need to control these processes requires that the proteins be flexible and versatile. Flexibility is also necessary to allow the different types of subunit contact found in icosahedral viruses that have only one type of coat protein subunit. Multiple structural states and the conditions for switching between these states have been characterized in many viral systems, and in several cases structural information is available for more than one state.

Many icosahedral viruses swell on removal of divalent metal ions at alkaline pH (reviewed by Argos and Johnson, 1984). The swelling process is generally believed to be functionally related to assembly and disassembly of the viruses (Bancroft, 1970; Durham *et al.*, 1977; Robinson and Harrison, 1982; Stubbs, 1984). TMV is also destabilized by the removal of cations. In cases where the structure is known, swelling (or, for TMV, disassembly) appears to be mediated by removal of the cations from binding sites at intersubunit carboxylate clusters. Robinson and Harrison (1982) determined the structure of swollen TBSV. Resolution was limited to 8 Å because of disorder in the expanded form of the virus, but the secondary structures of the domains, including the β annulus, appeared to be well preserved. A 20-Å hole through the capsid shell opens up between the subunits near the quasithreefold axis, consistent with charge repulsion between the carboxylate groups clustered about this axis. The hole is large enough to account for the increased susceptibility of the N-terminal arm of the protein to proteolysis in the swollen form of the virus. The protruding P-domain dimers experience large rotations about their twofold axes, but because these domains are not present in many viruses, the rotation was presumed to be a consequence rather than an integral part of the structural change in the S domains.

Capsid proteins from a number of $T = 3$ viruses can form $T = 1$ particles, particularly at low pH and in the presence of divalent ions. The $T = 1$ particles usually form only if the N-terminal arm of the protein is removed, but they can be made from intact SBMV protein in the presence of RNA of suitable size (Savithri and Erickson, 1983). The $T = 1$ particles of SBMV (Erickson and Rossmann, 1982) and TCV (Leberman and Longley, 1970) have been crystallized, and the structure of the SBMV particles has been described at 6-Å resolution (Erickson *et al.*, 1985). The fold of the protein was apparently that of the A subunit in the $T = 3$ particle, and the intersubunit contacts of the A subunit were maintained. On the basis of the stability of the fivefold packing demonstrated in this structure, and after examining the patterns of intersubunit contacts in the native virus, Rossmann *et al.* (1983c) proposed a model of assembly in which five AB dimers associate first, and C_2 dimers add to this complex, ordering the βA arm and forming the β annulus. The picornaviruses are known to assemble via intermediates with fivefold symmetry (Rueckert, 1985; Rossmann *et al.*, 1985; Hogle *et al.*, 1985). Harrison and his colleagues (Sorger *et al.*, 1986) preferred a model in which the C subunits associated first, through the β annulus. Both models emphasize the importance of the structural difference between the AB and the C subunits in forming the $T = 3$ capsid.

Tobacco mosaic virus protein forms a variety of aggregates, depending particularly on pH and ionic strength. At low pH, two rodlike assemblies have been observed, one isomorphous to the virus and one having $17\frac{1}{3}$ rather than $16\frac{1}{3}$ subunits per turn of the viral helix. The structures of both of these forms were studied by Mandelkow *et al.* (1981), who found that the protein structure was very similar to that of the virus. The RNA was replaced by bound anions. Much greater structural changes are found at high pH and ionic strength, where the protein crystallizes as a 34-subunit two-layered disk. The structure of the disk has been determined at 2.8 Å by Bloomer *et al.* (1978). The top-to-bottom protein–protein interactions are completely different in the disk and the virus, but the side-to-side interactions involve only minor rearrangements of side chains (Namba and Stubbs, 1986). Much of the structure in the disk is essentially that of the virus, but a 25-residue loop of the protein, ordered in the virus, is disordered in the disk and therefore not visible in the electron-density map. This loop contains most of the RNA binding site and the low-radius carboxyl cluster responsible for control of assembly. Experiments using NMR (Jardetzky *et al.*, 1978) showed that the two-turn helical aggregate that nucleates viral assembly (Correia *et al.*, 1985; Namba and Stubbs, 1986) also has this disordered loop; it appears to be stabilized either by binding RNA or by lowering the pH, protonating the carboxylate groups and overcoming their mutual electrostatic repulsion.

A common motif in all the structural transitions observed in virus coat proteins is a transition between order and disorder, mediated by electrostatic interactions. Although there is no reason to suppose that this pattern will be universal, it is easy to see that it is sufficiently useful to have been independently arrived at even by such unrelated viruses as the α-helical tobamoviruses and the β-barrel icosahedral viruses. Disordered structures may favor nonspecific protein–nucleic acid interactions in the early stages of assembly and prevent the formation of abortive protein assemblies in the absence of nucleic acid. Ordered structures are essential in stable, protective nucleocapsids. Electrostatic interactions are responsive to the dramatic differences in the ionic composition of intra- and extracellular environments. New structures will reveal new variations on these principles; the picornaviruses, for example, appear to make use of charge interactions between protein and nucleic acid as part of the delicate balance between assembly and disassembly. New folding patterns may emerge to use the same principles. The variations are limited only by our capacity to discover them.

ACKNOWLEDGMENTS. I thank Peter Heil and Cynthia Stauffacher for comments on this manuscript, and Jack Johnson, Lee Makowski, and Cynthia Stauffacher for providing manuscripts prior to publication. This work was supported by National Institutes of Health grant GM33265.

V. REFERENCES

Abad-Zapatero, C., Abdel-Meguid, S. S., Johnson, J. E., Leslie, A. G. W., Rayment, I., Rossmann, M. G., Suck, D., and Tsukihara, T., 1980, Structure of southern bean mosaic virus at 2.8Å resolution, *Nature* **286:**33–39.

Abdel-Meguid, S. S., Yamane, T., Fukuyama, K., and Rossmann, M. G., 1981, The location of calcium ions in southern bean mosaic virus, *Virology* **114:**81–85.

Akusjärvi, G., Aleström, P., Pettersson, M., Lager, M., and Jörvall, H., 1984, The gene for the adenovirus 2 hexon polypeptide, *J. Biol. Chem.* **259:**13976–13979.

Argos, P., 1981, Secondary structure prediction of plant virus coat proteins, *Virology* **110:**55–62.

Argos, P., and Johnson, J. E., 1984, Chemical stability in simple plant viruses, in: *Biological Macromolecules and Assemblies,* Volume I: *Virus Structures* (F. A. Jurnak and A. McPherson, eds.), pp. 1–43, Wiley, New York.

Argos, P., Tsukihara, T., and Rossmann, M. G., 1980, A structural comparison of concanavalin A and tomato bushy stunt virus protein, *J. Mol. Evol.* **15:**169–179.

Baker, T. S., Caspar, D. L. D., and Murakami, W. T., 1983, Polyoma virus "hexamer" tubes consist of paired pentamers, *Nature* **303:**446–448.

Baker, T. S., Drak, J., and Bina, M., 1988, Reconstruction of the three-dimensional structure of simian virus 40 and visualization of the chromatin core, *Proc. Natl. Acad. Sci. U.S.A.* **85:**422–426.

Bancroft, J. B., 1970, The self-assembly of spherical plant viruses, *Adv. Virus Res.* **16:**99–134.

Banner, D. W., Nave, C., and Marvin, D. A., 1981, Structure of the protein and DNA in fd filamentous bacterial virus, *Nature* **289:**814–816.

Bawden, F. C., Pirie, N. W., Bernal, J. D., and Fankuchen, I., 1936, Liquid crystalline substances from virus-infected plants, *Nature* **138:**1051–1052.

Bernal, J. D., and Fankuchen, I., 1941, X-ray and crystallographic studies of plant virus preparations, *J. Gen. Physiol.* **25:**111–165.

Bernal, J. D., Fankuchen, I., and Riley, D. P., 1938, Structure of the crystals of tomato bushy stunt virus preparations, *Nature* **142:**1075.

Bloomer, A. C., Champness, J. N., Bricogne, G., Staden, R., and Klug, A., 1978, Protein disk of tobacco mosaic virus at 2.8Å resolution showing the interactions within and between subunits, *Nature* **276:**362–368.

Bricogne, G., 1976, Methods and programs for direct-space exploitation of geometric redundancies, *Acta Crystallogr.* **A32:**832–847.

Bryan, R. K., Bansal, M., Folkhard, W., Nave, C., and Marvin, D. A., 1983, Maximum-entropy calculation of the electron density at 4Å resolution of Pf1 filamentous bacteriophage, *Proc. Natl. Acad. Sci. U.S.A.* **80:** 4728–4731.

Burnett, R. M., 1984, Structural investigations of hexon, the major coat protein of adenovirus, in: *Biological Macromolecules and Assemblies,* Volume I: *Virus Structures* (F. A. Jurnak and A. McPherson, eds.), pp. 337–385, Wiley, New York.

Burnett, R. M., 1985, The structure of the adenovirus capsid. II. The packing symmetry of hexon and its implications for viral architecture, *J. Mol. Biol.* **185:**125–143.

Casjens, S., 1985, Nucleic acid packaging by viruses, in: *Virus Structure and Assembly* (S. Casjens, ed.), pp. 75–147, Jones and Bartlett, Boston.

Caspar, D. L. D., 1956, Structure of bushy stunt virus, *Nature* **177:**475–476.

Caspar, D. L. D., 1963, Assembly and stability of the tobacco mosaic virus particle, *Adv. Protein Chem.* **18:** 37–121.

Caspar, D. L. D., 1976, Switching in the self-control of self-assembly, in: *Structure–Function Relationships of Proteins* (R. Markham and R. W. Horne, eds.), pp. 85–89, North Holland, Amsterdam.

Caspar, D. L. D., and Klug, A., 1962, Physical principles in the construction of regular viruses, *Cold Spring Harbor Symp. Quant. Biol.* **27:**1–24.

Cepko, C. L., and Sharp, P. A., 1982, Assembly of adenovirus major capsid protein is mediated by a nonviral protein, *Cell* **31:**407–415.

Colman, P. M., Tulloch, P. A., Shukla, D. D., and Gough, K. H., 1980, Particle and crystal symmetry of erysimum latent virus, *J. Mol. Biol.* **142:**263–268.

Colman, P. M., Varghese, J. N., and Laver, W. G., 1983, Structure of the catalytic and antigenic sites in influenza virus neuraminidase, *Nature* **303:**41–44.

Compans, R. W., Content, J., and Duesberg, P. H., 1972, Structure of the ribonucleoprotein of influenza virus, *J. Virol.* **10:**795–800.

Correia, J. J., Shire, S., Yphantis, D. A., and Schuster, T. M., 1985, Sedimentation equilibrium measurements of the intermediate-size tobacco mosaic virus protein polymers, *Biochemistry* **24:**3292–3297.

Crick, F. H. C., and Watson, J. D., 1956, Structure of small viruses, *Nature* **177:**473–475.

Cross, T. A., Tsang, P., and Opella, S. J., 1983, Comparison of protein and deoxyribonucleic acid backbone structures in fd and Pf1 bacteriophages, *Biochemistry* **22:**721–726.

Durham, A. C. H., Hendry, D. A., and Von Wechmar, M. B., 1977, Does calcium ion binding control plant virus disassembly? *Virology* **77:**524–533.

Einspahr, H., Parks, E. H., Suguna, K., Subramanian, E., and Suddath, F. L., 1986, The crystal structure of pea lectin at 3.0Å resolution, *J. Biol. Chem.* **261:**16518–16527.

Erickson, J. W., and Rossmann, M. G., 1982, Assembly and crystallization of a $T = 1$ icosahedral particle from trypsinized southern bean mosaic virus coat protein, *Virology* **116:**128–136.

Erickson, J. W., Silva, A. M., Murthy, M. R. N., Fita, I., and Rossmann, M. G., 1985, The structure of a $T = 1$ icosahedral empty particle from southern bean mosaic virus, *Science* **229:**625–629.

Finch, J. T., 1965, Preliminary x-ray diffraction studies on tobacco rattle and barley stripe mosaic viruses, *J. Mol. Biol.* **12:**612–619.

Finch, J. T., and Gibbs, A. J., 1970, Image reconstruction in projection of the nucleocapsid from Sendai, *J. Gen. Virol.* **6:**141–150.

Finch, J. T., and Klug, A., 1959, Structure of poliomyelitis virus, *Nature* **183:**1709–1714.

Fox, G., Stuart, D., Acharya, K. R., Fry, E., Rowlands, D., and Brown, F., 1987, Crystallization and preliminary x-ray diffraction analysis of foot-and-mouth disease virus, *J. Mol. Biol.* **196:**591–597.

Franklin, R. E., 1956, X-ray diffraction studies of cucumber virus 4 and three strains of tobacco mosaic virus, *Biochim. Biophys. Acta* **19:**203–211.

Fukuyama, K., Abdel-Meguid, S. S., Johnson, J. E., and Rossmann, M. G., 1983, Structure of a $T = 1$ aggregate of alfalfa mosaic virus coat protein seen at 4.5Å resolution, *J. Mol. Biol.* **167:**873–894.

Glucksman, M. J., Stubbs, G., and Makowski, L., 1987, X-ray fiber diffraction studies of filamentous bacteriophage M13 protein structure, *Ann. N.Y. Acad. Sci.* **494:**403–406.

Goelet, P., Lomonossoff, G. P., Butler, P. J. G., Akam, M. E., Gait, M. J., and Karn, J., 1982, Nucleotide sequence of tobacco mosaic virus RNA, *Proc. Natl. Acad. Sci. U.S.A.* **79:**5818–5822.

Harrison, S. C., Olson, A. J., Schutt, C. E., Winkler, F. K., and Bricogne, G., 1978, Tomato bushy stunt virus at 2.9Å resolution, *Nature* **276:**368–373.

Henriksson, D., Tanis, R. J., Tashian, R. E., and Nyman, P. O., 1981, Amino acid sequence of the coat protein subunit in satellite tobacco necrosis virus, *J. Mol. Biol.* **152:**171–179.

Hermodson, M. A., Abad-Zapatero, C., Abdel-Meguid, S. S., Pundak, S., Rossmann, M. G., and Tremaine, J. H., 1982, Amino acid sequence of southern bean mosaic virus coat protein and its relation to the three-dimensional structure of the virus, *Virology* **119:**133–149.

Heuss, K. L., MohanaRao, J. K., and Argos, P., 1981a, Crystallization of belladonna mottle virus, *J. Mol. Biol.* **146:**629–633.

Heuss, K. L., MohanaRao, J. K., and Argos, P., 1981b, Crystallization of cowpea chlorotic mottle virus, *J. Mol. Biol.* **146:**635–640.

Heuss, K. L., Murthy, M. R. N., and Argos, P., 1981c, Crystallization of tobacco ringspot virus, *J. Mol. Biol.* **153:**1161–1168.

Hodgkin, D. C., 1949, X-ray analysis and protein structure, *Cold Spring Harbor Symp. Quant. Biol.* **14:**65–78.

Hogle, J., Kirchhausen, T., and Harrison, S. C., 1983, Divalent cation sites in tomato bushy stunt virus, *J. Mol. Biol.* **171:**95–100.

Hogle, J. M., Chow, M., and Filman, D. J., 1985, Three-dimensional structure of poliovirus at 2.9Å resolution, *Science* **229:**1358–1365.

Hogle, J. M., Maeda, A., and Harrison, S. C., 1986, Structure and assembly of turnip crinkle virus. I. X-ray crystallographic structure analysis at 3.2Å resolution, *J. Mol. Biol.* **191:**625–638.

Holmes, K. C., and Franklin, R. E., 1958, The radial density distribution in some strains of tobacco mosaic virus, *Virology* **6:**328–336.

Hopper, P., Harrison, S. C., and Sauer, R. T., 1984, Structure of tomato bushy stunt virus: V. Coat protein sequence determination and its structural implication, *J. Mol. Biol.* **177:**701–713.

Hosur, M. V., Schmidt, T., Tucker, R. C., Johnson, J. E., Selling, B. H., and Rueckert, R. R., 1984, Black beetle virus—crystallization and particle symmetry, *Virology* **133:**119–127.

Hosur, M. V., Schmidt, T., Tucker, R. C., Johnson, J. E., Gallagher, T. M., Selling, B. H., and Rueckert, R. R., 1987, Structure of an insect virus at 3.0Å resolution, *Proteins Struct. Funct. Genet.* **2:**167–176.

Jardetzky, O., Akasaka, K., Vogel, D., Morris, S., and Holmes, K. C., 1978, Unusual segmental flexibility in a region of tobacco mosaic virus coat protein, *Nature* **273:**564–566.

Jones, T. A., and Liljas, L., 1984, Structure of satellite tobacco necrosis virus after crystallographic refinement at 2.5Å resolution, *J. Mol. Biol.* **177:**735–767.

Klug, A., 1965, Structure of viruses of the papilloma–polyoma type. II. Comments on other work, *J. Mol. Biol.* **11:**424–431.

Leberman, R., and Longley, W., 1970, X-ray diffraction studies on crystals of the small particles, *J. Mol. Biol.* **50:**213.

Liljas, L., 1986, The structure of spherical viruses, *Prog. Biophys. Mol. Biol.* **48:**1–36.

Liljas, L., and Strandberg, B., 1984, The structure of satellite tobacco necrosis virus, in: *Biological Macromolecules and Assemblies,* Volume I: *Virus Structures* (F. A. Jurnak and A. McPherson, eds.), pp. 97–119, Wiley, New York.

Liljas, L., Unge, T., Jones, T. A., Fridborg, K., Lövgren, S., Skoglund, U., and Strandberg, B., 1982, Structure of satellite tobacco necrosis virus at 3.0Å resolution, *J. Mol. Biol.* **159:**93–108.

Lobert, S., Heil, P., Namba, K., and Stubbs, G., 1987, Preliminary x-ray fiber diffraction studies of cucumber green mottle mosaic virus, watermelon strain, *J. Mol. Biol.* **196:**935–938.

Lonberg-Holm, K., 1975, The effects of concanavalin A on the early events of infection by rhinovirus type 2 and poliovirus type 2, *J. Gen. Virol.* **28:**313–327.

Luo, M., Vriend, G., Kamer, G., Minor, I., Arnold, E., Rossmann, M. G., Boege, U., Scraba, D. G., Duke, G. M., and Palmenberg, A. C., 1987, The atomic structure of Mengo virus at 3.0Å resolution, *Science* **235:**182–191.

Makowski, L., 1982, The use of continuous diffraction as a phase constraint. II. Application to fiber diffraction data, *J. Appl. Crystallogr.* **15:**546–557.

Makowski, L., 1984, Structural diversity in filamentous bacteriophages, in: *Biological Macromolecules and Assemblies,* Volume I: *Virus Structures* (F. A. Jurnak and A. McPherson, eds.), pp. 203–253, Wiley, New York.

Makowski, L., Caspar, D. L. D., and Marvin, D. A., 1980, Filamentous bacteriophage Pf1 structure determined at 7Å resolution by refinement of models for the α-helical subunit, *J. Mol. Biol.* **140:**149–181.

Mandelkow, E., Stubbs, G., and Warren, S., 1981, Structures of the helical aggregates of tobacco mosaic virus protein, *J. Mol. Biol.* **152:**375–386.

Marvin, D. A., 1966, X-ray diffraction and electron microscope studies on the structure of the small filamentous bacteriophage fd, *J. Mol. Biol.* **15:**8–17.

Marvin, D. A., Wiseman, R. L., and Wachtel, E. J., 1974a, Filamentous bacterial viruses. XI. Molecular architecture if the class II (Pf1, Xf) virion, *J. Mol. Biol.* **82:**121–138.

Marvin, D. A., Pigram, W. J., Wiseman, R. L., Wachtel, E. J., and Marvin, F. J., 1974b, Filamentous bacterial viruses. XII. Molecular architecture of the class I (fd, If1, IKe) virion, *J. Mol. Biol.* **88:**581–600.

Marvin, D. A., Bryan, R. K., and Nave, C., 1987, Pf1 Inovirus. Electron density distribution calculated by a maximum entropy algorithm from native fibre diffraction data to 3Å resolution and single isomorphous replacement data to 5Å resolution, *J. Mol. Biol.* **193:**315–343.

McCain, D. C., Virudachalum, R., Markley, J. L., Abdel-Meguid, S. S., and Rossmann, M. G., 1982a, Carbon-13 NMR study of southern bean mosaic virus, *Virology* **117:**501–503.

McCain, D. C., Virudachalum, R., Santini, R. E., Abdel-Meguid, S. S., and Markley, J. L., 1982b, Phosphorus-31 nuclear magnetic resonance study of internal motion in ribonucleic acid of southern bean mosaic virus, *Biochemistry* **21:**5390–5397.

McKay, D. B., Weber, I. T., and Steitz, T. A., 1982, Structure of catabolite gene activator protein at 2.9Å resolution, *J. Biol. Chem.* **257:**9518–9524.

McLachlan, A. D., Bloomer, A. C., and Butler, P. J. G., 1980, Structural repeats and evolution of tobacco mosaic virus coat protein and RNA, *J. Mol. Biol.* **136:**203–224.

Munowitz, M. G., Dobson, C. M., Griffin, R. G., and Harrison, S. C., 1980, On the rigidity of RNA in tomato bushy stunt virus, *J. Mol. Biol.* **141:**327–333.

Namba, K., and Stubbs, G., 1985, Solving the phase problem in fiber diffraction. Application to tobacco mosaic virus at 3.6Å resolution, *Acta Crystallog.* **A41:**252–262.

Namba, K., and Stubbs, G., 1986, Structure of tobacco mosaic virus at 3.6Å resolution: Implications for assembly, *Science* **231:**1401–1406.

Nave, C., Brown, R. S., Fowler, A. G., Ladner, J. E., Marvin, D. A., Provencher, S. W., Tsugita, A., Armstrong, J., and Perham, R. N., 1981, Pf1 filamentous bacterial virus. X-ray fibre diffraction analysis of two heavy-atom derivatives, *J. Mol. Biol.* **149:**675–707.

Olson, A. J., Bricogne, G., and Harrison, S. C., 1983, Structure of tomato bushy stunt virus. IV. The virus particle at 2.9Å resolution, *J. Mol. Biol.* **171:**61–93.

Rayment, I., 1984, Animal virus structure, in: *Biological Macromolecules and Assemblies,* Volume I: *Virus Structures* (F. A. Jurnak and A. McPherson, eds.), pp. 255–298, Wiley, New York.

Rayment, I., Baker, T. S., Caspar, D. L. D., and Murakami, W. T., 1982, Polyoma virus capsid structure at 22.5Å resolution, *Nature* **295:**110–115.

Reeke, G. N., Becker, J. W., and Edelman, G. M., 1975, The covalent and three-dimensional structure of concanavalin A. IV. Atomic coordinates, hydrogen bonding, and quaternary structure, *J. Biol. Chem.* **250:** 1525–1547.

Richardson, J. S., 1981, The anatomy and taxonomy of protein structure, *Adv. Protein Chem.* **34:**167–339.

Roberts, M. M., White, J. L., Grütter, M. G., and Burnett, R. M., 1986, Three-dimensional structure of the adenovirus major coat protein hexon, *Science* **232:**1148–1151.

Robinson, I. K., and Harrison, S. C., 1982, Structure of the expanded state of tomato bushy stunt virus, *Nature* **297:**563–568.

Rossmann, M. G., 1984, The structure of southern bean mosaic virus, in: *Biological Macromolecules and*

Assemblies, Volume I: *Virus Structures* (F. A. Jurnak and A. McPherson, eds.), pp. 45–96, Wiley, New York.

Rossmann, M. G., Abad-Zapatero, C., Murthy, M. R. N., Liljas, L., Jones, T. A., and Strandberg, B., 1983a, Structural comparisons of some small spherical plant viruses, *J. Mol. Biol.* **165:**711–736.

Rossmann, M. G., Chandrasekaran, R., Abad-Zapatero, C., Erickson, J. W., and Arnott, S., 1983b, RNA–protein binding in southern bean mosaic virus, *J. Mol. Biol.* **166:**73–80.

Rossmann, M. G., Abad-Zapatero, C., Hermodson, M. A., and Erickson, J. W., 1983c, Subunit interactions in southern bean mosaic virus, *J. Mol. Biol.* **166:**37–83.

Rossmann, M. G., Arnold, E., Erickson, J. W., Frankenberger, E. A., Griffith, J. P., Hecht, H.-J., Johnson, J. E., Kamer, G., Luo, M., Mosser, A., Rueckert, R. R., Sherry, B., and Vriend, G., 1985, Structure of a human common cold virus and functional relationship to other picornaviruses, *Nature* **317:**145–153.

Rueckert, R. R., 1985, Picornaviruses and their replication, in: *Virology* (B. N. Fields, D. M. Knipe, R. M. Chanock, J. L. Melnick, B. Roizman, and R. E. Shope, eds.), pp. 705–738, Raven Press, New York.

Savithri, H. S., and Erickson, J. W., 1983, The self-assembly of the cowpea strain of southern bean mosaic virus: Formation of $T = 1$ and $T = 3$ nucleoprotein particles, *Virology* **126:**328–335.

Silva, A. M., and Rossman, M. G., 1987, Refined structure of southern bean mosaic virus at 2.9Å resolution, *J. Mol. Biol.* **197:**69–97.

Smith, T. J., Kremer, M. J., Luo, M., Vriend, G., Arnold, E., Kamer, G., Rossmann, M. G., McKinlay, M. A., Diana, G. D., and Otto, M. J., 1986, The site of attachment in human rhinovirus 14 for antiviral agents that inhibit uncoating, *Science* **233:**1286–1293.

Sorger, P. K., Stockley, P. G., and Harrison, S. C., 1986, Structure and assembly of turnip crinkle virus. II. Mechanism of reassembly *in vitro, J. Mol. Biol.* **191:**639–658.

Stark, W., Glucksman, M. J., and Makowski, L., 1988, Conformation of the coat protein of filamentous bacteriophage Pf1 determined by neutron diffraction from magnetically oriented gels of specifically deuterated virions, *J. Mol. Biol.* **199:**171–182.

Stauffacher, C. V., Usha, R., Harrington, M., Schmidt, T., Hosur, M. V., and Johnson, J. E., 1987, The structure of cowpea mosaic virus at 3.5Å resolution, in: *Crystallography in Molecular Biology* (D. Moras, J. Drenth, B. Strandberg, D. Suck, and K. Wilson, eds.), pp. 293–308, Plenum, New York.

Steckert, J. J., and Schuster, T. M., 1982, Sequence specificity of trinucleoside diphosphate binding to polymerized tobacco mosaic virus protein, *Nature* **299:**32–36.

Stubbs, G., 1984, Macromolecular interactions in tobacco mosaic virus, in: *Biological Macromolecules and Assemblies,* Volume I: *Virus Structures* (F. A. Jurnak and A. McPherson, eds.), pp. 149–202, Wiley, New York.

Stubbs, G., and Diamond, R., 1975, The phase problem for cylindrically averaged diffraction patterns. Solution by isomorphous replacement and application to tobacco mosaic virus, *Acta Crystallogr.* **A31:**709–718.

Stubbs, G., and Stauffacher, C., 1981, Structure of the RNA in tobacco mosaic virus, *J. Mol. Biol.* **152:**387–396.

Stubbs, G., Warren, S., and Holmes, K., 1977, Structure of RNA and RNA binding site in tobacco mosaic virus from a 4Å map calculated from x-ray fibre diagrams, *Nature* **267:**216–221.

Stubbs, G., Namba, K., and Makowski, L., 1986, Application of restrained least-squares refinement to fiber diffraction from macromolecular assemblies, *Biophys. J.* **49:**58–60.

Tollin, P., and Wilson, H. R., 1971, Some observations on the structure of the Campinas strain of tobacco rattle virus, *J. Gen. Virol.* **13:**433–440.

Tollin, P., Bancroft, J. B., Richardson, J. F., Payne, N. C., and Beveridge, T. J., 1979, Diffraction studies of papaya mosaic virus, *Virology* **98:**108–115.

Tollin, P., Wilson, H. R., and Bancroft, J. B., 1980, Further observations on the structure of particles of potato virus X, *J. Gen Virol.* **49:**407–410.

Tollin, P., Wilson, H. R., Bancroft, J. B., Richardson, J. F., Payne, N. C., and Alford, W. P., 1981, Diffraction studies of clover yellow mosaic virus, *J. Gen Virol.* **52:**205–208.

Torbet, J., and Maret, G., 1979, Fibres of highly oriented Pf1 bacteriophage produced in a strong magnetic field, *J. Mol. Biol.* **134:**843–845.

Valegård, K., Unge, T., Montelius, I., and Strandberg, B., 1986, Purification, crystallization and preliminary x-ray data of the bacteriophage MS2, *J. Mol. Biol.* **190:**587–591.

Varghese, J. N., Laver, W. G., and Colman, P. M., 1983, Structure of the influenza virus glycoprotein antigen neuraminidase at 2.9Å resolution, *Nature* **303:**35–40.

Wiley, D. C., Wilson, I. A., and Skehel, J. J., 1981, Structural identification of the antibody-binding sites of Hong Kong influenza haemagglutinin and their involvement in antigenic variation, *Nature* **289:**373–378.

Wilson, I. A., Skehel, J. J., and Wiley, D. C., 1981, Structure of the haemagglutinin membrane glycoprotein of influenza virus at 3Å resolution, *Nature* **289**:366–373.

Wittmann-Liebold, B., and Wittmann, H. G., 1967, Coat proteins of strains of two RNA viruses: Comparison of their amino acid sequences, *Mol. Gen. Genet.* **100**:358–363.

Wu, S., Rinehart, C., and Kaesberg, P., 1987, Sequence and organization of southern bean mosaic virus genomic RNA, *Virology* **161**:73–80.

Wyckoff, R. W. G., and Corey, R. B., 1936, X-ray diffraction patterns of crystalline tobacco mosaic proteins, *J. Biol. Chem.* **116**:51–55.

Ysebaert, M., van Emmelo, J., and Fiers, W., 1980, Total nucleotide sequence of a nearly full-size DNA copy of satellite tobacco necrosis virus RNA, *J. Mol. Biol.* **143**:273–287.

Zimmern, D., 1977, The nucleotide sequence at the origin for assembly on tobacco mosaic virus RNA, *Cell* **11**:463–482.

4

Protein Stability and Function

Theoretical Studies

J. Andrew McCammon, Chung F. Wong, and Terry P. Lybrand

I. INTRODUCTION

The convergence of several lines of development in chemistry and molecular biology has created major new needs and opportunities for theoretical studies of proteins. The traditional approaches of organic synthesis have been supplemented by methods for automated chemical synthesis and genetic engineering that allow the preparation of a wide variety of polypeptides, specifically altered enzymes, and other complex molecules. The choice of molecules to be synthesized for a

J. Andrew McCammon, Chung F. Wong, and Terry P. Lybrand • Department of Chemistry, University of Houston, Houston, Texas 77004. *Present address for T.P.L.:* Department of Medicinal Chemistry, University of Minnesota, Minneapolis, Minnesota 55455.

given application is increasingly guided by structural information in addition to traditional methods such as chemical intuition and empirical correlation (quantitative structure–activity relationships, or QSAR). X-ray area detectors and new methods in NMR spectroscopy, combined with the improvements in our ability to synthesize and purify samples, are increasing the rate at which high-resolution structures of proteins are becoming available.

Theoretical studies can contribute to this work in several different ways (McCammon and Harvey, 1987). Energy refinement and molecular dynamics calculations can be used to develop or refine structural models from x-ray or NMR data. Calculations can also be used to help provide molecular interpretations of experimental data on protein function. For example, molecular dynamics simulations are useful in determining the roles of steric, electrostatic, solvation, and other effects in protein conformational transitions or binding phenomena. Finally, calculations can be used to design new proteins or new molecules that influence protein function. In particular, calculations based on the high-resolution structure of a protein or protein–ligand complex can be used to predict how the thermodynamic or kinetic properties of these systems will change as the result of changes in chemical composition.

A. The Thermodynamic Cycle-Perturbation Method

The present chapter focuses on certain new methods for predicting thermodynamic properties of proteins. These methods are designed to answer questions such as the following. Given the structure of a protein–ligand complex, how will the free energy of ligand binding change if one group on the ligand is replaced by another? If the complex is that of an enzyme and substrate, how will changes in the substrate affect the free energy of activation for the reaction steps that follow binding? How might possible changes in the protein (such as might be produced by site-directed mutation) alter the thermodynamic stability of the protein? And how would these changes alter ligand-binding properties and subsequent reactivity?

The methods considered here are the thermodynamic cycle-perturbation methods (Tembe and McCammon, 1984; Lybrand *et al.*, 1986; Wong and McCammon, 1986a,b, 1987). In principle, these methods can be used to answer any of the questions given above. In practice, these methods have been used successfully to predict changes in equilibrium constants for noncovalent associations when small but significant chemical changes are made in a ligand or a protein. The range and power of these methods is increasing very rapidly as a result of developments in fundamental theory, computational algorithms, and computer hardware (McCammon and Harvey, 1987).

The general idea of the thermodynamic cycle-perturbation approach is quite simple. Suppose that one is interested in computing the relative free energy change for two different processes:

$$\mathrm{A + B \rightarrow AB} \qquad \Delta G_1 \tag{1}$$

$$\mathrm{A' + B \rightarrow A'B} \qquad \Delta G_2 \tag{2}$$

The desired quantity, $\Delta\Delta G = \Delta G_2 - \Delta G_1$, can in principle be obtained from molecular dynamics (or Monte Carlo) simulations in which processes 1 and 2 are caused to occur sufficiently slowly in the appropriate solvent surroundings. Analysis of the simulations using standard methods yields ΔG_1 and ΔG_2 (Berendsen, 1985; Beveridge and Mezei, 1985; Jorgensen *et al.*, 1986; McCammon *et al.*, 1986; McCammon and Harvey, 1987). Although free energies of association of simple molecules in solution have been obtained by such procedures, this direct approach is unworkable for most problems involving proteins. The difficulty is that each process must be carried out slowly enough that the system remains in thermodynamic equilibrium. Many factors (e.g., complicated conformation or solvation changes requiring more than a few tens of picoseconds) can make it difficult to insure that representative configurations of the system develop during the simulations, because these simulations themselves typically cover only 10 to 100 psec. In the thermodynamic cycle-perturbation approach, one considers instead the nonphysical processes

$$A + B \rightarrow A' + B \qquad \Delta G_3 \tag{3}$$

$$AB \rightarrow A'B \qquad \Delta G_4 \tag{4}$$

Because processes 1 through 4 form a thermodynamic cycle, the desired relative free energy change $\Delta\Delta G = \Delta G_2 - \Delta G_1 = \Delta G_4 - \Delta G_3$. The latter quantities are again obtained by molecular dynamics simulations. Because the changes in 3 and 4 are typically much smaller and more localized than those in 1 and 2, the calculations are greatly simplified.

B. Range of Application

A wide variety of transformations can be mapped formally onto processes 1 through 4. To study the binding of two drugs with different substituents to a given protein, one would let A and A′ represent the drugs and B the protein. To study the effects of a mutation on ligand binding, A and A′ would be the different mutants, and B would be the ligand. To study reaction rates, AB and A′B could be taken to be activated complexes, in which case ΔG_1 and ΔG_2 are free energies of activation. Special methods are required in this case to allow for the fact that changes in reactant composition will generally lead to differences in location of the complexes along the reaction coordinate (Madura *et al.*, 1987). Unimolecular processes such as protein folding can be treated by a corresponding analysis of

$$A_1 \rightarrow A_2 \qquad \Delta G_5 \tag{5}$$

$$A_1' \rightarrow A_2' \qquad \Delta G_6 \tag{6}$$

where the subscripts on A indicate states (e.g., unfolded and folded). The analysis then involves

$$A_1 \rightarrow A_1' \qquad \Delta G_7 \tag{7}$$

$$A_2 \rightarrow A_2' \qquad \Delta G_8 \tag{8}$$

where, e.g., processes 7 and 8 might correspond to replacing one amino acid side chain by another in the unfolded and folded states, respectively (Wong and McCammon, 1987). The utility of this approach for folding problems depends on the quality of assumptions concerning the structure of the unfolded state.

In the following section, we consider the use of thermodynamic cycle-perturbation methods in studies of ligand binding. Several examples based on a very simple technique for calculating free energy changes such as those in 3 and 4 are described. In Section III we briefly consider the less-developed area of predicting protein-folding stability. Some remarks on ongoing and future work are provided in Section IV.

II. LIGAND BINDING

A. Thermodynamics

Recognition and binding are key elements of protein function. The ability to discriminate markedly in binding is a characteristic of enzymes, antibodies, polypeptide hormones, and other proteins. Recognition is usually assessed quantitatively in terms of equilibrium constants. A familiar example is the inhibitor dissociation constant K_I. For the process EI $\rightleftharpoons$ E +

I, $K_I = [E][I]/[EI]$. In theoretical studies, one usually works with free energies. The dissociation constant K_I is related to free energy by the familiar equation

$$\Delta G^0 = +RT\ln K_I \tag{9}$$

where ΔG^0 is the standard-state free energy change for the inhibitor binding process (Cantor and Schimmel, 1980). The sign of the right-hand side of equation 9 reflects the fact that the equilibrium constant corresponding to the binding process is K_I^{-1}. For the binding of two different inhibitors I and I′, one has

$$\Delta G_2^0 - \Delta G_1^0 = +RT\ln(K_I'/K_I) \tag{10}$$

where ΔG_1^0 and K_I correspond to the binding and dissociation of I, and ΔG_2^0 and K'_I correspond to the binding and dissociation of I′. The theoretical studies discussed here deal with model systems in which solutes (E and I, for example) are effectively at infinite dilution. Thus, the theoretical result $\Delta\Delta G = \Delta G_4 - \Delta G_3$ from calculations for processes 3 and 4 can be identified with the standard-state difference, $\Delta\Delta G = \Delta G_2^0 - \Delta G_1^0$, if the customary biochemical definition of standard states is used (Van Holde, 1971).

B. Perturbation Methods

Each of the ΔG_3 and ΔG_4 values can be evaluated by using the statistical mechanical perturbation theory (Postma *et al.*, 1982; Berendsen, 1985; Beveridge and Mezei, 1985; Friedman, 1985; McCammon and Harvey, 1987). For instance, ΔG_4 can be obtained as

$$\Delta G_4 = -k_BT \ln\langle\exp(\Delta V/k_BT)\rangle_{AB} + \Delta G_4^m \tag{11}$$

for small perturbations. In equation 11, $\Delta V = V_{AB} - V_{A'B}$ in which V_{AB} and $V_{A'B}$ are the potential energies of the systems containing AB and A′B, respectively. T is the absolute temperature, k_B is the Boltzmann constant, and $\langle\ \rangle_{AB}$ denotes an ensemble average over the states of the system containing AB. ΔG_4^m is the contribution of the difference in masses of AB and A′B. It is usually not included in the calculation because it cancels with a similar term in the other half of the thermodynamic cycle; that is, $\Delta G_4^m = \Delta G_3^m$ (Lybrand *et al.*, 1985).

The ensemble average $\langle\exp(\Delta V/k_BT)\rangle_{AB}$ can be obtained from molecular dynamics simulations of AB and the surrounding solvent under isothermal–isobaric conditions in order to obtain ΔG_4. However, the results usually converge slowly except for very small perturbations. A more general way to obtain a value of ΔG is to express this quantity as

$$\Delta G = \sum_{i=1}^{n} \Delta G_i \tag{12}$$

where

$$\Delta G_i = -k_BT \ln\langle\exp[(V_i - V_{i+1})/k_BT]\rangle_i \tag{13}$$

In the case of ΔG_4, one has $V_1 = V_{AB}$, $V_{n+1} = V_{A'B}$, and V_i changes gradually (linearly or nonlinearly) from V_{AB} to $V_{A'B}$ when i changes from 1 to $n + 1$. (Mass terms are not shown explicitly.) Other technical improvements are discussed in Section IV.

C. Trypsin–Benzamidine Binding

The thermodynamic cycle-perturbation method has been applied to compute the relative affinity of two different benzamidine inhibitors for trypsin and the relative affinity of benzamidine for wild-type and a mutant trypsin (Wong and McCammon, 1986a,b, 1987). The agreement with experimental data is quite encouraging. For the binding of different inhibitors to trypsin, the thermodynamic cycle involved can be written as

$$\begin{array}{ccc} \text{Tr} + \text{BH} & \xrightarrow{\Delta A_1} & \text{Tr:BH} \\ \Delta A_3 \downarrow & & \downarrow \Delta A_4 \\ \text{Tr} + \text{BF} & \xrightarrow[\Delta A_2]{} & \text{Tr:BF} \end{array}$$

where Tr, BH, and BF represent trypsin, benzamidinium ion, and parafluorobenzamidinium ion, respectively. The symbol ":" indicates formation of a complex. Helmholtz free energies (ΔAs) are shown here because canonical ensemble (constant temperature, constant volume) simulations were performed. Equation 11 instead of 12 was used for the initial studies. Two simulations were required: one of benzamidinium ion in water and one of trypsin–benzamidinium ion in water.

For the trypsin–benzamidine–water simulation, SPC water molecules (Hermans *et al.*, 1984) were put in a rectangular box of dimensions 49.2 Å × 54.5 Å × 64.3 Å by starting with a cubic box (18.6 Å × 18.6 Å × 18.6 Å) of well-equilibrated bulk water and applying translational symmetry in the *x*, *y*, and *z* directions. Benzamidine-inhibited trypsin and 62 associated water molecules were then put in the center of the box. The inhibitor was in its protonated (amidinium) state. The initial coordinates of the protein–inhibitor complex and associated waters were taken from x-ray results (Bode and Schwager, 1975). Bulk water molecules having a distance of 2.3 Å or less from the nonhydrogen atoms of the x-ray structure and any bulk water molecules lying inside cavities of the protein were then deleted. The resulting system consists of 4785 water molecules and the protein–benzamidine complex. Hydrogen atoms that can form hydrogen bonds were treated explicitly, whereas other hydrogen atoms (such as the methyl hydrogens of the protein) were treated as extended atoms (Hermans *et al.*, 1984). The total number of explicit atoms included in the dynamics calculations was 16,384. The simulations were performed using periodic boundary conditions.

The empirical potential energy function used was that from the GROMOS program (Hermans *et al.*, 1984). It comprises terms for bond angles, dihedral angles, improper dihedrals, and Lennard-Jones and electrostatic interactions. Bond lengths were fixed at ideal values by using the SHAKE algorithm (Ryckaert *et al.*, 1977). The nonbonded interactions were calculated within cutoff distances of 8 Å for the Lennard-Jones terms and 20 Å for the electrostatic terms.

The whole protein–benzamidine–water system was equilibrated for ~16 psec, followed by a 28.8-psec production run for the free energy calculations. In the dynamics calculations, the equations of motion were solved by using the leap-frog algorithm (Hockney and Eastwood, 1981) with a time step of 2 fsec. The nonbonded pair list was updated every 10 steps. During the simulation, the temperature of the system was kept close to 300 K by coupling it to a heat bath with a relaxation time of 0.1 psec (Berendsen *et al.*, 1984).

The benzamidinium–water simulation was performed by taking initial benzamidinium ion coordinates from the x-ray structure of the trypsin–benzamidine complex. The molecule was put in a cubic box (18.6 Å × 18.6 Å × 18.6 Å) of well-equilibrated water. Water molecules that overlapped the benzamidine were deleted as before. The system consists of 212 water molecules and the ion, for a total of 650 atoms. The system was then equilibrated for 32 psec and simulated for 64 psec at 300 K.

The $\Delta\Delta A = \Delta A_2 - \Delta A_1 = \Delta A_4 - \Delta A_3$ value obtained from the thermodynamic cycle-perturbation method is 0.9 ± 0.5 kcal mole^{-1} at 300 K. The experimental result, approximately corrected from 288 K to 300 K, is 0.5 ± 0.3 kcal mole^{-1} (Mares-Guia *et al.*, 1977). The error estimate in the predicted value is ± 1 standard deviation for results obtained by separate analyses of 8-psec segments of each simulation. That calculations and experiment agree in indicating that the unsubstituted benzamidinium binds slightly more strongly is encouraging. The theoretical studies help in the interpretation of the experimental results as well. For instance, the values of ΔA_3 and ΔA_4 are -0.8 ± 0.1 kcal mole^{-1} and 0.1 ± 0.5 kcal mole^{-1}, respectively, which suggests that benzamidine binds more strongly than its parafluoro analogue because of the unfavorable desolvation of the latter.

To predict the relative free energy of binding of benzamidine to wild-type trypsin and one of its mutant forms in which glycine 216 is replaced by alanine, an additional simulation of the uninhibited trypsin in water is required, as indicated from the appropriate thermodynamic cycle:

$$\begin{array}{ccc} \text{Tr} + \text{BH} & \xrightarrow{\Delta A_1} & \text{Tr:BH} \\ \Delta A_3 \downarrow & & \downarrow \Delta A_4 \\ \text{Tm} + \text{BH} & \xrightarrow[\Delta A_2]{} & \text{Tm:BH} \end{array}$$

where now Tm represents the mutant protein. The simulation was performed by taking the 16-psec equilibrated structure of the inhibited complex and replacing the inhibitor by three water molecules. To improve the distribution of water in the binding pocket, the 25 waters in or nearest the pocket were dynamically equilibrated for 26 psec while the rest of the system was held fixed. The whole system was equilibrated for an additional 0.6 psec and then allowed to evolve without intervention for 1 psec. Finally, a 22.4-psec simulation was performed as for the trypsin–inhibitor complex.

The $\Delta\Delta A$ value obtained is 1.3 kcal mole^{-1} at 300 K. An experimental estimate, $\Delta\Delta A = 2.0$ kcal mole^{-1}, can be obtained from the Michaelis constants for a homologous system, namely, wild-type and mutant rat pancreatic trypsin II acting on arginine substrates (Craik *et al.*, 1985). Experiment and theory appear to agree that amidinium (or guanidinium) ligands bind more strongly to the native enzyme than to the mutant. The theoretical components ΔA_3

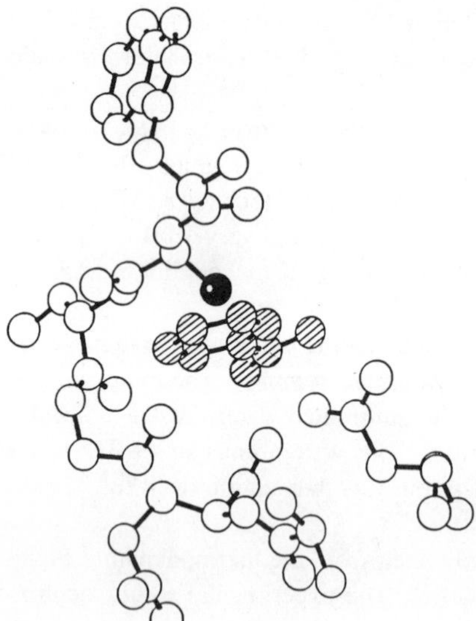

Figure 1. View into the trypsin binding site, showing how the methyl group (solid sphere) added in the glycine 216 → alanine 216 mutation projects toward the amidinium group of the benzamidine inhibitor (hatched spheres) (Wong and McCammon, 1986b).

= 0.6 kcal mole^{-1} and ΔA_4 = 1.9 kcal mole^{-1} suggest that steric crowding caused by the added methyl group in the mutant enzyme is the primary factor in determining binding preference (Fig. 1).

III. PROTEIN STABILITY

In the previous section, we described a prototype calculation of the change in enzyme–substrate affinity produced by a single-site mutation in an enzyme. That calculation rests on the implicit assumption that the conformation of the mutant enzyme is similar to the native conformation of the wild-type enzyme. X-ray studies show that the substitution of one amino acid for another generally results in small conformational changes that are restricted to the immediate vicinity of the substitution site (Lesk and Chothia, 1986). The thermodynamic stability of the single-site mutant can, however, be changed significantly from that of the corresponding wild-type protein (Hawkes *et al.,* 1984). In this section, we first indicate how one can quantitatively describe the effects of altered stability on net ligand binding by a protein and on the net catalytic rate of an enzyme. We then indicate how theory might be used to predict changes in protein stability and describe some initial applications of these ideas to the enzyme trypsin.

A. Stability–Activity Relationships

To analyze the effects of protein stability on ligand binding, consider the processes

$$P_u + L \overset{K_f}{\rightleftharpoons} P_f + L \underset{k_{-1}}{\overset{k_1}{\rightleftharpoons}} P_fL \tag{14}$$

Here, P_u and P_f represent the unfolded and folded proteins, respectively, and L represents a ligand. The equilibrium constants for folding and for binding to the folded protein are given by

$$K_f = [P_f]/[P_u] \tag{15}$$

$$K_b = k_1/k_{-1} = [P_fL]/[P_f][L] \tag{16}$$

An effective equilibrium constant for ligand binding can be defined as

$$K = [P_fL]/([P_T] - [P_fL])([L_T] - [P_fL]) \tag{17}$$

where $[P_T]$ and $[L_T]$ are the total concentrations of protein and ligand, respectively. With a few lines of algebra, one can show that

$$K = K_b/(1 + K_f^{-1}) \tag{18}$$

For a protein whose folded conformation is quite stable, $K_f >> 1$ and $K \simeq K_b$. In the opposite case of a very unstable protein, $K_f \simeq 0$ and $K \simeq 0$. The theory described in Section 2 can be used to predict the effects of mutations on K_b. An approach for predicting the effects of mutations on K_f is given in Section III.B.

If P and L represent an enzyme and its substrate, the kinetic effects of folding stability can be analyzed by extending equation 14 to include the process

$$P_fL \underset{k_{-2}}{\overset{k_2}{\rightleftharpoons}} P_f + \text{product} \tag{19}$$

With the usual assumptions of excess substrate and steady-state conditions, one can derive a modified Lineweaver–Burk equation,

$$v^{-1} = (K_M/\bar{K}_f k_2[P_T])\,(1/[L]) + (1/k_2[P_T]) \quad (20)$$

Here, v is the initial rate of formation of product, $K_M = (k_{-1} + k_2)/k_1$ is the usual Michaelis constant, and $\bar{K}_f = K_f(1 + K_f)^{-1}$. If the folded form of the enzyme is very stable, $K_f >> 1$, $\bar{K}_f \simeq 1$, and one recovers the conventional form of the Lineweaver–Burk equation. For less stable enzymes, $\bar{K}_f < 1$, and the rate v is lower than in the case of a stable enzyme. It is often the case that $K_M \simeq K_b^{-1}$, so that the methods of Section II can be used directly to predict changes in K_M that would result from mutations. An approach for predicting changes in K_f is described below, and a possible approach for predicting changes in k_2 is mentioned in Section IV.

B. Trypsin Perturbations

Formally, changes in the free energy of folding that result from mutations or other chemical changes in a protein can be expressed as

$$\Delta\Delta G_f = \Delta G_6 - \Delta G_5 = \Delta G_8 - \Delta G_7 \quad (21)$$

where ΔG_5 through ΔG_8 are defined by equations 5 through 8 (Wong and McCammon, 1987). The calculation of ΔG_8 requires replacing one side chain or other chemical group by another in the folded conformation of the protein. The assumptions and procedures are essentially identical to what was described in Section II. The calculation of ΔG_7 requires a corresponding replacement in the unfolded state. At present, too little is known about the nature of the unfolded state to allow accurate calculations of ΔG_7. In particular, unfolded proteins may have more or less long-lived elements of secondary and tertiary structure on a local scale and may be fairly compact on a global scale (Ahmad and Bigelow, 1986; Baldwin, 1986). It is therefore uncertain what solvation and electrostatic effects operate at any given site of an unfolded protein.

To illustrate the kinds of calculations that can be done, however, we have carried out simplified calculations for the effects of a single-site mutation in trypsin. The unfolded state was assumed to have an extended, highly exposed nature at the site of substitution and was modeled by a short segment of the protein immersed in water. Here, we tried to model the effects of changing glycine 216 and glycine 226 into alanines, and thus a glycine "dipeptide" (N-methylglycylacetamide) was simulated. The ϕ,ψ angles were constrained to about $-130°$ and $100°$, respectively, by means of two harmonic potentials to account for part of the effects of neglecting the rest of the protein. After careful equilibration, a 40-psec simulation was used to estimate the difference in free energy between the unfolded wild-type enzyme and the unfolded mutant enzyme in which either Gly^{216} or Gly^{226} was mutated to alanine. The value obtained for ΔA_7 is -1.6 kcal/mole. Using the simulation of uninhibited trypsin described in Section II, perturbation theory gives a ΔA_8 value of 0.6 kcal/mole when Gly^{216} is changed into alanine and a value of -0.6 kcal/mole when Gly^{226} is changed into alanine. The resulting $\Delta\Delta A$ values for changing Gly^{216} to Ala^{216} is $+2.2$ kcal/mole. The corresponding value for changing Gly^{226} to Ala^{226} is $+1.0$ kcal/mole. This model calculation suggests that the enzyme trypsin unfolds more easily when either Gly^{216} or Gly^{226} is changed into alanine. Although this model is oversimplified, it clearly indicates the potential of the thermodynamic cycle-perturbation method for the study of protein folding/unfolding problems.

IV. PROSPECTS AND LIMITATIONS

The potential range of applicability of the thermodynamic cycle-perturbation approach is wide. The approach can be used to study virtually any problem involving substituent effects in molecular recognition. In our research group, this approach is being applied to study the binding of antiviral agents to viruses and of antigens to antibodies, the effects of mismatches in DNA double-strand formation, and nonbiological problems such as the binding between alkali metal cations and modified crown ethers in liquid methanol. The utility of this approach is increasing as a result of different technical improvements.

A. Local Site Simulations

Local site simulations can often be performed to save computing time. In such calculations, the free energy changes are obtained by simulating only that part of a system where the effects of a chemical modification are most noticeable. This can be achieved by performing molecular dynamics simulations with stochastic boundary conditions (Berkowitz and McCammon, 1982; Brunger *et al.*, 1985). Because the dynamic response of the region surrounding the local site is less important than the thermodynamic response in these calculations, an alternative method is to perform molecular dynamics simulations with harmonic constraints imposed on the outer atoms. This method is now being used to study the trypsin–inhibitor complex. Initial results show that computing time can be reduced by one half to one third of the time required for full simulation of this complex in water. In another alternative method, the outer atoms are kept rigidly fixed, and only the local site atoms are allowed to move. This approach is being used in initial simulations of the drug–rhinovirus complex, as described elsewhere (Lybrand *et al.*, 1987). The local site simulation methods are certainly adequate for many problems, but they may have to be modified or abandoned if long-range effects are important. Such effects can arise, for example, when substantial changes in electrostatic charge or in the global conformation of a receptor would result from the chemical modification that is being studied.

B. Perturbation Methods

As mentioned in Section II, the perturbation formula (equation 11) can be used a number of times (equation 12) to enable better sampling. In the limit of infinite n, the free energy changes can be expressed as

$$\Delta G = \int_a^b \langle \partial V(\lambda)/\partial\lambda \rangle_\lambda \, d\lambda \tag{22}$$

where the potential energy V is now expressed as a function of some continuous coupling parameter λ instead of using discrete labels i. The potential energy function $V(\lambda)$ changes from one system ($\lambda = a$) to another ($\lambda = b$) smoothly when λ changes from a to b. Equation 22 can be derived by expanding the logarithmic and exponential functions in Taylor series and taking limits as $n \to \infty$ in equations 12 and 13. Equation 22 is the familiar Kirkwood thermodynamic integration formula (Friedman, 1985).

In practical calculations, it is hard to evaluate the ensemble averages $\langle \partial V/\partial\lambda \rangle_\lambda$ for a large number of λ values and then perform the integration numerically. An approximate method to evaluate the free energy changes from equation 22 is to approximate $\langle \partial V(\lambda)/\partial\lambda \rangle_\lambda$ by $\partial V/\partial\lambda|_{\lambda(t)}$. Here $\partial V/\partial\lambda|_{\lambda(t)}$ is obtained from a molecular dynamics simulation in which $\lambda(t)$, a function of time t, is changed gradually from a to b during the simulation. This method has been applied successfully to some simple systems, and the extension to study protein molecules would be useful (Straatsma *et al.*, 1986).

C. Limitations

The major limitations of the thermodynamic cycle-perturbation approach are the requirements that an initial structure be known to fairly high resolution and that the structural effects of perturbations be predictable. The former requirement limits our present ability to treat protein-folding phenomena, as described in Section III. This limitation should be relaxed as more experimental information on unfolded structures becomes available. Also, integral equation techniques that have been used for some time to study the structure of liquids have recently been extended to small peptides; these methods may eventually provide a theoretical guide to the structures of unfolded proteins (Pettitt and Karplus, 1985). Improvements in our ability to predict the structural effects of perturbations will be needed for cases in which local chemical modification would lead to a conformational change that does not represent a typical thermal fluctuation of the unperturbed state. Such predictive methods are being developed (Madura *et al.*, 1987). One application will be to calculate relative free energies of activation of chemical reactions. Such methods should prove useful in studies of the catalytic steps in enzyme action.

ACKNOWLEDGMENTS. Professor Peter Kollman and Dr. Wilfred van Gunsteren provided the AMBER and GROMOS programs, respectively. This work was supported in part by grants from the National Science Foundation, the Robert A. Welch Foundation, the National Center for Supercomputing Applications, the John von Neumann Center, and HNS Supercomputers, Inc. Computing equipment was provided in part by the Texas Advanced Technology Research Program and the Texas Advanced Research Program. J.A.M. is the recipient of the George H. Hitchings Award from the Burroughs Wellcome Fund. T.P.L. is the recipient of a Presidential Young Investigator Award from NSF.

V. REFERENCES

Ahmad, F., and Bigelow, C. C., 1986, Estimation of the stability of globular proteins, *Biopolymers* **25:**1623.

Baldwin, R. L., 1986, Seeding protein folding, *Trends Biochem. Sci.* **11:**6.

Berendsen, H. J. C., 1985, Statistical mechanics and molecular dynamics: The calculation of free energy, in: *Molecular Dynamics and Protein Structure* (J. Hermans, ed.), University of North Carolina, Chapel Hill, pp. 43–46.

Berendsen, H. J. C., Postma, J. P. M., van Gunsteren, W. F., DiNola, A., and Haak, J. R., 1984, Molecular dynamics with coupling to an external bath, *J. Chem. Phys.* **81:**3684.

Berkowitz, M., and McCammon, J. A., 1982, Molecular dynamics with stochastic boundary conditions, *Chem. Phys. Lett.* **90:**215.

Beveridge, D., and Mezei, M., 1985, Free energy simulations: The coupling parameter approach and topographical transition coordinates, in: *Molecular Dynamics and Protein Structure* (J. Hermans, ed.), University of North Carolina, Chapel Hill, pp. 53–57.

Bode, W., and Schwager, P., 1975, The refined crystal structure of bovine β-trypsin at 1.8 angstroms resolution, *J. Mol. Biol.* **98:**693.

Brunger, A. T., Brooks, C. L., and Karplus, M., 1985, Active site dynamics of ribonuclease, *Proc. Natl. Acad. Sci. U.S.A.* **82:**8458.

Cantor, C. R., and Schimmel, P. R., 1980, *Biophysical Chemistry,* W. H. Freeman, San Francisco.

Craik, C. S., Largman, C., Fletcher, T., Roczniak, S., Barr, P. J., Fletterick, R., and Rutter, W. J., 1985, Redesigning trypsin: Alteration of substrate specificity, *Science* **228:**291.

Friedman, H. L., 1985, *A Course in Statistical Mechanics,* Prentice-Hall, Englewood Cliffs, N.J.

Hawkes, R., Grutter, M. G., and Schellman, J., 1984, Thermodynamic stability and point mutations of bacteriophage T4 lysozyme, *J. Mol. Biol.* **175:**195.

Hermans, J., Berendsen, H. J. C., van Gunsteren, W. F., and Postma, J. P. M., 1984, A consistent empirical potential for water–protein interactions, *Biopolymers* **23:**1513.

Hockney, R. N., and Eastwood, J. W., 1981, *Computer Simulation Using Particles,* McGraw-Hill, New York.

Jorgensen, W. L., Chandrasekhar, J., Buckner, J. K., and Madura, J. D., 1986, Computer simulations of organic reactions in solution, *Ann. N.Y. Acad. Sci.* **482:**198.

Lesk, A. M., and Chothia, C. H., 1986, The response of protein structures to amino-acid sequence changes, *Phil. Trans. R. Soc. Lond.* **A317:**345.

Lybrand, T. P., Ghosh, I., and McCammon, J. A., 1985, Hydration of chloride and bromide anions: Determination of relative free energy by computer simulation, *J. Am. Chem. Soc.* **107:**7793.

Lybrand, T. P., McCammon, J. A., and Wipff, G., 1986, Theoretical calculation of relative binding affinity in host–guest systems, *Proc. Natl. Acad. Sci. U.S.A.* **83:**833.

Lybrand, T. P., Lau, W. F., McCammon, J. A., and Pettitt, B. M., 1987, Molecular dynamics studies on antiviral agents: Thermodynamics of solvation and binding, in: *Protein Structure and Design* (D. Oxender, ed.), p. 227, Alan R. Liss, New York.

Madura, J. D., Pettitt, B. M., and McCammon, J. A., 1987, Geometric considerations in the calculation of relative free energies of activation, *Chem. Phys. Lett.* **140:**83.

Mares-Guia, M., Nelson, D. L., and Rogana, E., 1977, Electronic effects in the interaction of para-substituted benzamidines with trypsin: The involvement of the π-electronic density at the central atom of the substituent in binding, *J. Am. Chem. Soc.* **99:**2331.

McCammon, J. A., and Harvey, S. C., 1987, *Dynamics of Proteins and Nucleic Acids,* Cambridge University Press, Cambridge.

McCammon, J. A., Karim, O. A., Lybrand, T. P., and Wong, C. F., 1986, Ionic association in water: From atoms to enzymes, *Ann. N.Y. Acad. Sci.* **482:**210.

Pettitt, B. M., and Karplus, M., 1985, The potential of mean force surface for the alanine dipeptide in aqueous solution: A theoretical approach, *Chem. Phys. Lett.* **121:**194.

Postma, J. P. M., Berendsen, H. J. C., and Haak, J. R., 1982, Thermodynamics of cavity formation in water, *Faraday Symp. Chem. Soc.* **17:**55.

Ryckaert, J. P., Ciccotti, G., and Berendsen, H. J. C., 1977, Numerical integration of Cartesian equations of motion of a system with constraints: Molecular dynamics of *n*-alkanes, *J. Comp. Phys.* **23:**327.

Straatsma, T. P., Berendsen, H. J. C., and Postma, J. P. M., 1986, Free energy of hydrophobic hydration: A molecular dynamics study of noble gases in water, *J. Chem. Phys.* **85:**6720.

Tembe, B. L., and McCammon, J. A., 1984, Ligand–receptor interactions, *Comput. Chem.* **8:**281.

Van Holde, K. E., 1971, *Physical Biochemistry,* Prentice-Hall, Englewood Cliffs, N.J.

Weiner, P. K., and Kollman, P. A., 1981, AMBER: Assisted model building with energy refinement. A general program for modeling molecules and their interactions, *J. Comput. Chem.* **2:**287.

Wong, C. F., and McCammon, J. A., 1986a, Dynamics and design of enzymes and inhibitors, *J. Am. Chem. Soc.* **108:**3830.

Wong, C. F., and McCammon, J. A., 1986b, Computer simulation and the design of new biological molecules, *Isr. J. Chem.* **27:**211.

Wong, C. F., and McCammon, J. A., 1987, Thermodynamics of enzyme folding and activity: Theory and experiment, in: *Structure, Dynamics and Function of Biomolecules* (A. Ehrenberg and R. Rigler, eds.), Springer-Verlag, Berlin, pp. 51–55.

5

Stabilization Energies of Protein Conformation

Tom Alber

I. INTRODUCTION

The study of protein stability is currently undergoing a dramatic change. Early work, especially after Kauzmann (1959), centered on the analysis of simple chemical model systems to determine the contributions of different types of interactions to stability. Amino acids that are sequestered from solvent in the folded protein were emphasized, since these groups presumably undergo the largest changes in environment during folding (Tanford, 1968, 1970; Pace,

Tom Alber • Institute of Molecular Biology, University of Oregon, Eugene, Oregon 97403; *present address:* Department of Biochemistry, University of Utah School of Medicine, Salt Lake City, Utah 84132.

1975; Privalov, 1979). From an inventory of interactions derived from the x-ray crystal structure of a protein, the stabilizing contribution of each amino acid could be calculated using the free energies determined from thermodynamic studies of model systems. Implicit in this approach is the assumption that each interaction of a specific type—each hydrogen bond, for example, or each square angstrom of buried hydrophobic surface area—contributes equally to stability regardless of its unique structural context.

In the last decade, improved methods of peptide synthesis, chemical modification, and genetic manipulation have allowed workers to ask if results from model systems can be used to understand the roles of specific substituents in proteins. Thus far, it appears that the contribution of each interaction depends strongly on its context. Although this finding raises the disquieting possibility that each interaction will have to be understood independently, it is hoped that structural patterns that distinguish stabilizing contacts will be discovered. Indeed, organizing principles are already beginning to emerge.

At the same time, a conceptual framework that accounts for the disparate contributions of specific interactions has been proposed by Creighton (1983a,b, 1985; Creighton and Goldenberg, 1984; Goldenberg, 1985). Creighton's approach is based on the idea of "effective concentration," the ratio of the intramolecular and intermolecular association constants for two groups. Substituents will contribute to protein stability to the extent that they have higher effective concentrations (their interactions are more favored) in the folded state than in the unfolded state. This means that groups that are constrained to interact productively in the folded structure will contribute to stability. Amino acids and water molecules that are solvated better in the unfolded state will detract from stability, even if they participate in observable interactions in the x-ray crystal structure of the folded protein.

The concept of effective concentration includes all the determinants of association constants, such as entropic effects, orientation effects, and interaction potentials. As a result, to say that two groups have higher effective concentrations in the folded state does not necessarily provide a mechanistic understanding of their contribution to stability. Nonetheless, this framework has several virtues in addition to allowing interactions of each type to make a wide range of contributions. (1) The cooperativity of folding is an integral part of the approach. (2) In favorable circumstances, effective concentrations of specific groups in a protein can be measured experimentally in both the folded and unfolded states. (3) Several testable predictions follow from the theory. One of these that has recently been lent experimental support is that atoms that are relatively rigid in the folded structure make important contributions to stability (Alber *et al.*, 1987a).

This chapter outlines some of the recent progress toward understanding the structural basis of protein thermodynamic stability. Because of the rapidly growing literature in this area, I have found it impossible to be comprehensive. A number of other reviews have been written recently (Ackers and Smith, 1985; Dill, 1987; Baldwin and Eisenberg, 1987; Matthews, 1987; Schellman, 1987a; Goldenberg, 1988). Several important topics are not treated here, including recent progress on the mechanisms of irreversible protein inactivation and the sources of ligand-binding energy. Discussions of the methods of thermodynamics, kinetics, x-ray crystallography, genetics, and calculations of free energy are also excluded. A theme of this review is the comparison of studies of proteins with studies of model systems. The emphasis is on the search for structural patterns associated with stability, which, with luck, may provide rigorous criteria for assessing the plausibility of predicted structures.

A. The Problem

The central problem of protein stability is to account quantitatively for the small differences in free energy between the relatively small ensemble of folded conformations and the immense ensemble of unfolded, rapidly interconverting alternatives. Qualitatively, the conformational entropy and hydration of the unfolded state are thought to be balanced by specific stabilizing interactions in the folded molecule. The physical forces that underlie this balance—such as the hydrophobic effect, van der Waals forces, electrostatic interactions, hydrogen bonds, covalent cross links—were largely identified in the 1950s (Kauzmann, 1959). A current challenge is to determine the contributions of individual amino acids to the stability of a specific protein. Which residues are essential for stability, and what are the contributions of specific physical forces at essential sites? How do the contributions of critical amino acids depend on environmental conditions? How are the interactions integrated to produce the observed thermodynamic stability? How does the amino acid sequence destabilize alternative nonfunctional conformations?

Though the problem is simply stated, major barriers to solving it are presented by the complexity of the folded and unfolded states as well as the transition between them. Even under the most favorable conditions, the folded state is only stabilized by 5–15 kcal/mole. This narrow range of stabilization free energy is independent of molecular weight (Privalov, 1979). Larger proteins gain less stabilization per residue than small ones, in part by folding in distinct domains. The observed stability is the small difference between the free energies of the folded and unfolded states, which are each on the order of 10^7 kcal/mole including covalent bonds (Baldwin and Eisenberg, 1987). The need to evaluate energies very accurately, coupled with the inherent structural complexities, makes it impossible to calculate protein stability from first principles.

B. The Folded State

The first high-resolution x-ray crystal structure of a protein, that of myoglobin (Kendrew *et al.*, 1958, 1960, 1961), dashed hopes that a small number of simple rules could account for the distinctive fold of the molecule. The words of Kendrew and co-workers (1958) still ring true for each new structure determined: "Perhaps the most remarkable features of the molecule are its complexity and lack of symmetry. The arrangement seems to be almost totally lacking in the kind of regularities which one instinctively anticipates."

In his succinct description of the low-resolution model of myoglobin, Perutz (1964) captured the disappointment that the structure did not provide a molecular Rosetta stone for translating the sequence of a protein into its three-dimensional conformation: "Could the search for ultimate truth really have revealed so hideous and visceral looking an object?"

By now the structural complexity of folded proteins is better appreciated. Analysis of structural patterns has led to the discovery of architectural rules (Richardson, 1981; Chothia, 1984) and to proposals for potentially stabilizing features such as the α-helix dipole (Arridge and Cannon, 1963; Brant and Flory, 1965; Hol *et al.*, 1978) and polar interactions between aromatic groups (Burley and Petsko, 1985). The flexibility of the folded state has been identified as a well of vibrational and conformational entropy (Sturtevant, 1977). Alternative folded conformations have been detected in a number of proteins (Privalov, 1979; Dlott *et al.*, 1983; Svensson *et al.*, 1986; Smith *et al.*, 1986; Evans *et al.*, 1987). Interactions of the polypeptide chain with solvent and counterions are fluid and hard to model rigorously. Because of the modest resolution of most protein x-ray crystal structures, an extended atom representation that does not include hydrogens is commonly used. Yet in terms of energetics, incorrect placement of even a single hydrogen atom can thwart the computational analysis of stability (Ponder and Richards, 1987; Grütter *et al.*, 1987).

C. The Unfolded State

Since each amino acid influences the free energy of both the folded and unfolded states, insight into denatured proteins is crucial for understanding protein stability. Information about the denatured state also constrains models for the early events of the folding process. Major issues include the number of accessible conformations, the amount and nature of preferred structure, and the relationship between structure and free energy. It is also not known if short peptides in solution provide appropriate models for the same sequences in unfolded proteins.

On unfolding, the polypeptide chain becomes less compact, more highly solvated, and much more flexible (Tanford, 1968). Intrinsic viscosity measurements made at identical temperatures suggest that the unfolded chains produced by heat and by guanidinium hydrochloride (GuHCl) are very nearly random coils (Privalov, 1979). Whatever residual structures may exist in unfolded states, changes in heat capacity, enthalpy, and free energy associated with denaturation by heat and by GuHCl are very similar when measurements made at the same temperature and pH are compared (Privalov, 1979). Differences in local order produced by different denaturants do not result in a change in thermodynamic state.

Of course, the unfolded polypeptide cannot be an ideal mathematically random chain. Dill (1985, 1987) has shown that excluded volume effects alone reduce the estimated number of allowed backbone conformations of a 100-residue chain from about 10^{60} in a "random coil" to about 10^{16}. There is considerable experimental evidence for local order in proteins unfolded by different denaturants. Addition of GuHCl to several acid-denatured proteins resulted in small but cooperative changes in optical rotation (Aune *et al.*, 1967). Differences in the average structures of ribonuclease denatured by heat and GuHCl have been detected by NMR (Matthews and Westmoreland, 1975) and circular dichroism (Labhart, 1982). The NMR data suggested that ribonuclease denatured by GuHCl adopts preferred conformations near His^{12} (Bierzynski and Baldwin, 1982). Using fluorescence energy transfer, Haas *et al.* (Amir and Haas, 1987; Haas and Amir, 1987) observed a nonrandom distribution of conformations in reduced bovine pancreatic trypsin inhibitor denatured by GuHCl. Distinguishable unfolded forms of staphylococcal nuclease were detected by NMR methods (Evans *et al.*, 1987). Kinetic studies of folding have also provided evidence for slowly interconverting conformational distributions in the unfolded state that differ in isomerization of prolines (Kim and Baldwin, 1982; Hurle and Matthews, 1987).

Preferred conformations in isolated short peptides in water have been detected under conditions that favor structure formation (Brown and Klee, 1971; Bundi and Wüthrich, 1979; Dyson *et al.*, 1985). Such local interactions may influence the unfolded state (Bierzynski and Baldwin, 1982; Labhart, 1982). For adjacent residues, effective concentrations may exceed several molar (Richards, 1977). For groups distant in the amino acid sequence, effective concentrations of 10^{-4} to 10^{-2} M have been measured (Mutter, 1977; Illuminati and Mandolini, 1981). As a result, dilute aqueous solutions of model compounds may not adequately represent the unfolded state (Privalov, 1979).

Whatever structure forms in the unfolded state under a given set of conditions, it is likely to be very transient. Bond rotations are slowed up to 100-fold relative to free amino acids (Glushko *et al.*, 1972), and NMR spectra of denatured proteins generally show that side chains are found in similar average environments (Roder *et al.*, 1985; States *et al.*, 1987). Unfolded proteins undergo rapid hydrogen-exchange reactions (Roder *et al.*, 1985; Loftus *et al.*, 1986), and the rates of helix propagation in amino acid homopolymers have been estimated at 10^{6}–10^{7} sec^{-1} (Barksdale and Stuehr, 1971). Proline residues undergo slow isomerization in the denatured state (Kim and Baldwin, 1982; Evans *et al.*, 1987). These data suggest the presence of a large number of rapidly interconverting isoenergetic conformations. This immense conformational variability and flexibility presents severe problems for establishing simple structural models of the unfolded state whose interactions can be evaluated analytically or experimentally.

D. Thermodynamics of Protein Stability

The thermodynamics of the transition between the folded and unfolded states is also proving to be complex. Thermodynamic studies provide the data that must be explained by structural theories of protein stability. This topic has been developed in a number of excellent papers (Pace, 1975; Privalov, 1979; Schellman *et al.*, 1981; Privalov *et al.*, 1986; Schellman, 1987a,b; Becktel and Schellman, 1987) and are only summarized here.

Protein denaturation is a highly cooperative reaction. For single-domain proteins, intermediates are rarely detected at equilibrium. This is the basis for the simplifying assumption that the polypeptide chain occurs in two states, folded and unfolded.

Protein stability depends on environmental conditions such as temperature, pressure, pH, ionic strength, and the concentration of specific ligands, stabilizers, and denaturants. Each of these environmental factors provides an experimental probe of the magnitudes of specific stabilizing forces.

A central feature of the energetics of protein denaturation is that changes in enthalpy and entropy are strongly dependent on temperature. In particular:

$$\Delta H_T = \Delta H_0 + \Delta C_p(T - T_0)$$

$$\Delta S_T = \Delta S_0 + \Delta C_p \ln(T/T_0)$$

where ΔH_T and ΔS_T are the changes in enthalpy and entropy at temperature T, ΔH_0 and ΔS_0 are the changes in enthalpy and entropy at a reference temperature T_0, and ΔC_p is the difference in heat capacity between the folded and unfolded states at constant pressure. ΔC_p for unfolding is large and positive (~1–2 kcal mole^{-1} K^{-1}), and ΔC_p per gram of protein is proportional to the number of nonpolar contacts per gram in the folded state (Privalov, 1979). ΔC_p is also constant or nearly constant in the temperature range studied (0–80°C). This behavior has been taken as a confirmation of the central role of the hydrophobic effect in protein stabilization.

The large value of ΔC_p means that ΔH_T and ΔS_T are steep functions of temperature. A temperature change of 1°C, for example, causes changes in ΔH_T and in $T\Delta S_T$ of approximately 1–2 kcal/mole. ΔH_T and ΔS_T are zero near room temperature, and they have large compensating values at temperatures at which proteins denature. The mechanisms of this compensation and of the integration of stabilizing contributions are not well understood (Go, 1975; Ueda and Go, 1976).

The relationship between free energy of stabilization and temperature, $\Delta G_T = \Delta H_T - T\Delta S_T$, defines the protein stability curve (Privalov, 1979; Schellman *et al.*, 1981; Becktel and Schellman, 1987). The free energy of stabilization reaches a maximum near room temperature, where $\Delta S_T = 0$, and the equilibrium constant for folding ([F]/[U]) reaches a maximum at a slightly lower temperature, where $\Delta H_T = 0$. Denaturation can be induced by heat and, surprisingly, by cold as well (Privalov *et al.*, 1986). At high temperature, unfolding results in a large increase in entropy, presumably because of the added flexibility of the polypeptide chain, and a compensating increase in enthalpy, attributed to changes in interactions in the protein and solvent. In the cold, the system actually loses entropy (becomes more ordered) and releases heat on unfolding.

E. Thermodynamics of Amino Acid Substitutions

Major barriers to understanding how the amino acid sequence gives rise to this behavior have been described above. They include the complexities of folded and unfolded structures and the critical role of the solvent in determining protein conformation. A fundamental practical problem is that thermodynamic measurements only provide information about the sum of all interactions in the system. The measurements do not directly identify features of a complicated structure that determine its stability.

Traditional methods of assigning the relative magnitudes of specific interactions are to characterize simple chemical model systems and to study the behavior of protein variants. Comparisons of closely related proteins offer the possibility of ascribing differences in stability to the small number of atoms that are altered. Modern methods of directed mutagenesis have significantly increased the power of this approach. Despite these technical advances, at least three problems in interpreting the data complicate the genetic approach. (1) Even chemically simple alterations can simultaneously change many different interactions in the folded protein (Grütter *et al.*, 1979, 1983, 1987; Luisi and Nagai, 1986; Howell *et al.*, 1986; Nagai *et al.*, 1987; Villafranca *et al.*, 1987; Alber *et al.*, 1987b). This is true even when shifts in the three-dimensional structure are localized to the site of the amino acid substitution. This emphasizes the importance of detailed structural studies for understanding the properties of mutant proteins. (2) Amino acid substitutions can also alter the distribution of unfolded conformations (Shortle and Meeker, 1986; Evans *et al.*, 1987). (3) Thermodynamic measurements only evaluate the differences between the folded and unfolded states, not the absolute effect of the mutation on either state. This represents a fundamental limitation of the genetic approach, because the magnitudes of the energetic effects that must be accounted for in each state are unknown.

This problem is illustrated by the following thermodynamic cycle:

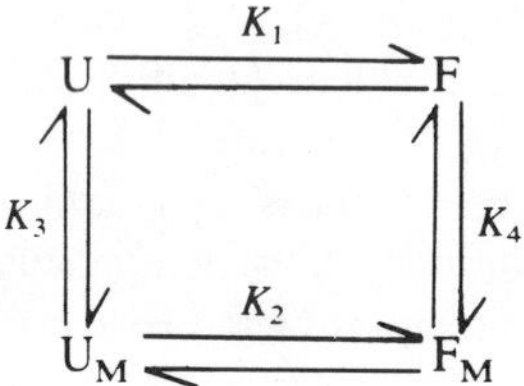

where K_1 and K_2 are the equilibrium constants for folding a wild-type (U,F) and a mutant protein (U_M,F_M), respectively. K_1 and K_2 cannot be related directly because the two proteins are different compounds. They have different numbers of atoms. K_3 and K_4 are the hypothetical equilibrium constants for making the amino acid substitution reversibly in the unfolded and folded states. Although free energy perturbation methods are being used to try to calculate K_3 and K_4, no general method of measuring the free energy cost of a substitution in either state is available. As a result, it is impossible to tell whether a change in stability from a mutation is caused by effects on the folded state, on the unfolded state, or on a combination of the two.

Five approaches—measurement of disulfide stability, a kinetic test, a thermodynamic test, an empirical test, and a trick to equate the free energies of the unfolded states—have been put forward to overcome this problem. None of these methods is completely satisfactory. Partitioning the free energy of mutation remains a primary limitation to interpretation of genetic experiments in structural terms. The five approaches are reviewed here before we go on to discuss the insights gained from studying protein variants.

For the specific case of the stability of disulfide bonds, the effects on the unfolded and folded states can be determined relative to an added disulfide exchange reagent (Creighton, 1983a, 1985; Creighton and Goldenberg, 1984; Goldenberg, 1985). The analysis is based on the following thermodynamic cycle:

$$
\begin{array}{ccc}
\text{RSSR} + \text{U}\begin{smallmatrix}\text{SH}\\ \text{SH}\end{smallmatrix} & \underset{}{\overset{K_1}{\rightleftharpoons}} & \text{F}\begin{smallmatrix}\text{SH}\\ \text{SH}\end{smallmatrix} + \text{RSSR} \\
K_3 \updownarrow & & \updownarrow K_4 \\
2\text{RSH} + \text{U}\begin{smallmatrix}\text{S}\\ | \\ \text{S}\end{smallmatrix} & \overset{K_2}{\rightleftharpoons} & \text{F}\begin{smallmatrix}\text{S}\\ | \\ \text{S}\end{smallmatrix} + 2\text{RSH}
\end{array}
$$

The equilibrium constants K_1 and K_2 describe the stability of the reduced and oxidized proteins, and K_3 and K_4 reflect the stability of the disulfide bond in the unfolded and folded states. Linkage between these equilibria requires that $K_1K_4 = K_2K_3$. Measured effective concentrations for disulfide bonds in unfolded proteins range from 10^{-3} to 10^{-2} M, and the effective concentrations of the three disulfides in the folded conformation of bovine pancreatic trypsin inhibitor range between 10^2 M and 5×10^5 M (Creighton, 1983a).

Relating these values to theoretical estimates of the reduction in conformational entropy caused by forming a cross link in the unfolded state is not straightforward. Such calculations assume a value (10–10^3 M) for the effective concentration of the cross link in the folded state, and they also assume that no entropy is lost on forming the cross link in the folded state (Goldenberg, 1985). Nonetheless, the possibility of using the stabilities of disulfides to determine the effects of amino acid substitutions on the folded (K_4) and unfolded (K_3) states is also being explored (Goldenberg and Creighton, 1984). This involves using a thiol reagent as a common reference to relate thermodynamic cycles for disulfide formation in two proteins that differ by a single amino acid.

A second approach to assigning the effects of amino acid substitutions relies on the analysis of kinetic data. Using concepts from transition state theory, C. R. Matthews and co-workers (Beasty *et al.*, 1986) have proposed that mutations that selectively alter the rate of folding shift the free energy of the unfolded state, mutations that alter only the rate of unfolding shift the free energy of the folded state, and mutations that alter both the rates of unfolding and folding shift the free energy of the transition state. This formalism describes the simplest interpretation of changes in the kinetics of folding and denaturation, but other effects are not ruled out. Mutations that selectively alter folding rates, for example, could reasonably affect the energies of both the folded state and the rate-limiting transition state. Nonetheless, the parsimonius interpretation of kinetic data has led to testable structural models for the folding process (Beasty *et al.*, 1986, 1987; Matthews and Hurle, 1987; Perry *et al.*, 1987).

Shortle and Meeker (1986) have suggested a thermodynamic test for assigning selective effects of an amino acid substitution to the unfolded state. They proposed that a change in the average structure of the unfolded state can be inferred from a change in the dependence of the logarithm of the equilibrium constant for folding on denaturant concentration. This idea is based on an interpretation of the thermodynamics of solvent denaturation in terms of the amount of exposed surface area available for binding in the unfolded state. This physical model, however, does not reflect all the interactions of denaturants with proteins and water (Schellman, 1978, 1987b). As a result, the thermodynamic test does not rule out alternative interpretations. An increase in the slope of ln K versus denaturant concentration could reflect increased association of a component of the solvent with the native state. A decrease in slope could reflect a higher population of intermediates in the transition region. The relationship between structure and free energy in the unfolded state is also not well understood. Despite these ambiguities, the analysis of Shortle and co-workers has emphasized the critical importance of understanding more about changes in the unfolded state caused by mutations.

An empirical test for analyzing the effects of mutations has been used by several laboratories. This relies on measuring the stabilities of a series of mutant proteins, each containing a different amino acid at a given position. Correlations between stability and structure identify critical interactions (Ackers and Smith, 1985; Alber *et al.*, 1987a). A correspondence between transfer free energy and protein stability, for example, was used to support the importance of the hydrophobic effect at specific sites in the α subunit of trp synthase (Yutani *et al.*, 1984, 1987) and kanamycin nucleotidyl transferase (Matsumura *et al.*, 1988). The importance of a particular hydrogen bond in the folded state of phage T4 lysozyme was inferred from the presence of this interaction in the x-ray crystal structures of the most stable of 14 variants with different amino acids at position 157 (Alber *et al.*, 1987b). This approach requires considerable effort, and it is only relevant if a correlation is found. Interpretation of the results is complicated by the possibility that an observed correlation arises by coincidence due to effects that are unique to each of the mutant proteins studied.

The final approach to partitioning the effects of mutations is to try to equate the free energies of the unfolded states by comparing proteins with identical amino acid compositions (L. McIntosh, W. Becktel, W. Baase, D. Muchmore, and F. W. Dahlquist, unpublished data). Such so-called "permutants" contain reciprocal substitutions at two positions in the amino acid sequence. Differences in stability are ascribed to the folded state on the assumption that the free energy of the unfolded state is more sensitive to composition than to sequence. This approach suffers from the questionable validity of this assumption and from the need to account simultaneously for the effects of two amino acid substitutions (L. McIntosh, W. Becktel, and F. W. Dahlquist, personal communication).

Despite the difficulty of accounting for the effects of mutations on the folded and unfolded states, the genetic approach to the protein stability problem is proving to be a rich source of new information. Recent progress in this area is highlighted in the following section.

II. GENETIC STUDIES OF PROTEIN STABILITY

Classical genetic studies of protein function have suggested that proteins can be very tolerant of amino acid substitutions. In the *E. coli lac* repressor, for example, a wild-type phenotype was produced by 58% of 323 substitutions generated by suppression of nonsense mutations at 90 sites (Miller *et al.*, 1979). Repressor activity was drastically reduced by 15% of the substitutions, and only 11% of the sequence changes resulted in a temperature-sensitive (*ts*) phenotype. Analysis of almost 2000 missense mutants of the *lac* repressor showed that *ts* and inactivating lesions are clustered (Miller and Schmeissner, 1979; Miller, 1984). The protein apparently adapts to changes in large regions of the sequence.

Analyses of protein families emphasized similar conclusions. The globins, for example, contain fewer than ten absolutely conserved residues (Perutz *et al.*, 1965; Perutz and Lehmann, 1968). In several families, the most variable residues are on the protein surface (Perutz and Lehmann, 1968; Go and Miyazawa, 1980).

Recent studies of mutant proteins have revealed two reasons for this pattern of variation. (1) Only a fraction of the residues in a protein make large contributions to stability (Alber *et al.*, 1987a). This limits the size of the target for destabilizing mutations. (2) Structural adjustments can mitigate the intrinsic effects of amino acid substitutions, even at critical sites (Alber *et al.*, 1987b, 1988).

"Randomly" induced mutations that alter protein stability provide a relatively unbiased experimental identification of amino acids that make essential contributions. A large collection of mutant hemoglobins has been obtained by random screening of the human population (Fermi and Perutz, 1981). Collections of mutations have also been obtained by screening for temperature sensitivity of phage T4 lysozyme (Streisinger *et al.*, 1961; Alber *et al.*, 1987a) and for reduced activity of staphylococcal nuclease (Shortle and Lin, 1985), cytochrome *c* (Hampsey *et al.*, 1986), and the *cI* and *cro* repressors of phage λ (Hecht *et al.*, 1983, 1984a; Pakula *et al.*, 1986). Several methods for isolating thermostable variants have also been reported. These include screening for resistance to heat inactivation (Alber and Wozniak, 1985; Bryan *et al.*, 1986), isolating pseudorevertants of destabilizing lesions (Hecht and Sauer, 1985; Shortle and Lin, 1985), and selecting for gene function at elevated temperatures in a thermophilic bacterium (Matsumura and Aiba, 1985; Hendrix and Welker, 1985; Liao *et al.*, 1986). Although the characterizations of stability differ from system to system, a number of general conclusions have been drawn.

The most obvious finding is that amino acid substitutions that alter stability are chemically varied. Changes in charge, size, polarity, hydrophobicity, or hydrogen bonding capacity are found. This indicates that many different types of noncovalent interactions—including hydrogen bonds, van der Waals contacts, hydrophobic contacts, and ionic interactions—can make quantitatively comparable contributions to stability. No single type of interaction dominates; all the forces proposed to play a role can be important at specific sites.

It is also apparent that the effect of a substitution depends on the nature of the change and on its structural context. Analysis of destabilizing lesions in the amino terminal domain of λ *cI* repressor (Hecht *et al.*, 1983, 1984a) and in phage T4 lysozyme (Alber *et al.*, 1987a) in terms of the x-ray crystal structures of these proteins showed that most critical amino acids are both relatively rigid and inaccessible to solvent in the folded protein. Substitutions at mobile and exposed sites generally have little effect on stability. Analysis of a large collection of inactivating substitutions in the λ *cro* protein (Pakula *et al.*, 1986) pointed to similar conclusions and highlighted several interesting exceptions. Destabilizing mutations can alter residues with relatively high mobility if the sites are charged or are undergoing segmental, rather than local, motion. Substitutions that would be expected to cause propagating shifts in the folded structure, such as insertion of prolines at inappropriate locations or replacement of conformationally special glycines, can also be destabilizing.

Calculated solvent accessibility and crystallographic thermal factors are highly correlated, making it difficult to determine if one of these parameters is more important for stability than the other. Further studies of rare sites that are either buried and mobile or exposed and rigid may clarify this issue. One example is provided by Cys^{54} in phage T4 lysozyme, which is buried and relatively mobile. Substitutions at this position have small effects on stability (Wetzel *et al.*, 1988), consistent with the high mobility of this residue. More examples are needed to draw general conclusions.

Genetic studies suggest that only some of the amino acids in a protein make large contributions to stability. This implies that many interactions inferred from the x-ray crystal structure of a protein are not stabilizing. Critical amino acids are generally rigid and buried in the folded protein. This pattern provides simple rules for using crystallographic data to predict the effects of amino acid substitutions on protein stability (Alber *et al.*, 1987a) and accounts for much of the silent variation observed in protein families and in genetic studies of protein function.

The correlations among low mobility, low solvent accessibility, and high sensitivity to destabilizing lesions also bear on current theories of protein stability. Side chains with a large amount of buried nonpolar surface area are sensitive to destabilizing substitutions, emphasizing the importance of van der Waals contacts and the hydrophobic effect. Even small buried residues, though, form part of the target for *ts* mutations. This is hard to rationalize on the basis of a constant linear relationship between surface area and hydrophobic stabilization, because small side chains have limited surface area. Instead, it supports the view that groups that become surrounded by protein atoms during folding can form specific stabilizing interactions. Substituents that are exposed to solvent in the folded and unfolded states may contribute less to stability, because their average environments in the two states are similar.

The correlation between rigidity and sensitivity to destabilizing substitutions was predicted by Creighton's hypothesis (1983a,b) that the relative effective concentrations of groups in the folded and unfolded states determine their contribution to stability. According to this view, rigid amino acids can make the largest contributions because they are constrained to interact productively. Their entropy is reduced by the folded structure *in toto*, so a greater fraction of the enthalpy of interaction is expressed in the free energy of stabilization. More flexible substituents may interact productively for a smaller fraction of time and lose more entropy in forming a stable contact.

A final intriguing conclusion drawn from the structural patterns of destabilizing substitutions is that severe mutations that primarily affect the free energy of the denatured state may be rare (Alber *et al.*, 1987a). In principle, mutations that affect the unfolded state could occur in any part of the sequence. Their positions should not correlate with features of the folded structure. Since a tight correlation is observed between the positions of severe destabilizing mutations and properties of the folded structure, large changes in thermodynamic stability may generally involve significant effects on the folded protein. The rigid and buried residues in the folded state may comprise the part of the system that is least able to relax structurally to compensate for amino acid substitutions. The conformational diversity of the unfolded state may allow it to compensate more completely for changes in sequence. As a result, the free energy of the unfolded state may be less sensitive to amino acid substitutions. Although the unfolded state cannot be ignored, analysis of mutant proteins would be simplified if the effects on the folded state are dominant. Effects on the unfolded state would be more important for understanding mutations that cause small changes in stability.

This argument notwithstanding, the possible magnitude of the effects of substitutions on the free energy of the unfolded state remains controversial. Shortle and Meeker (1986) have suggested that changes in interactions in the unfolded state can fully account for large changes in stability. The kinetic formalism adopted by C. R. Matthews and co-workers (Beasty *et al.*, 1986; Matthews and Hurle, 1987) also suggests that some mutations may differentially alter the free energy of unfolded forms.

A surprising and unexplained finding from studies of thermal denaturation of protein variants is that multiple amino acid substitutions generally have additive effects on stability (Shortle and Lin, 1985; Matsumura *et al.*, 1986; Becktel and Schellman, 1987; L. McIntosh, W. Becktel, W. Baase, D. Muchmore, and F. W. Dahlquist, unpublished data). Nonadditive effects can occur if one of the substitutions alters ΔC_p (Wetzel *et al.*, 1988). Solvent denaturation studies carried out near room temperature also generally reflect the additivity of the effects of substitutions (Shortle and Meeker, 1986), although nonadditive effects have also been reported (Hurle *et al.*, 1986). The additivity of changes in free energy of stabilization is unexpected because of the cooperativity of protein folding. The theoretical treatment of Creighton (1983a,b) discussed above also implies that substitutions should synergistically alter the contributions of neighboring interactions and lead to nonadditive effects even in the absence of structural shifts (Goldenberg and Creighton, 1985). The apparent additivity of amino acid substitutions implies that cooperative interactions are highly localized. This supports the simplifying assumption that protein stability may be understood as a sum of local interactions. Practically speaking, this means that multiple substitutions might be engineered to set the stability of a protein at any desired value.

III. STRUCTURAL STUDIES OF PROTEIN VARIANTS

The growing number of structural studies of protein variants using x-ray crystallography and NMR emphasize that the native structure (and the unfolded state as well) can adjust to accommodate changes in sequence. The effects of a given substitution depend on its intrinsic impact on the free energies of the folded and unfolded states and on the relative abilities of the two states to compensate structurally for the new amino acid (Alber *et al.*, 1988). Flexibility provides a mechanism of compensation. If proteins were completely rigid, changes in sequence would have larger effects on thermodynamic stability.

A range of conformational adjustments allow proteins to tolerate substitutions. Replacements of surface residues often result in small localized shifts. Positional changes of less than 1 Å in nearby side-chain atoms and 0.5 Å in main-chain atoms are common. Changes in the distribution of bound solvent have also been observed (Howell *et al.*, 1986; Grütter *et al.*, 1987; Alber *et al.*, 1987b). Even surface substitutions, however, can cause propagating structural adjustments: NMR data show that replacement of Pro^{117} in staphylococcal nuclease changes the equilibrium between similar alternative conformations in both the folded and unfolded states (Evans *et al.*, 1987). Positional adjustments propagate extensively through the folded conformation. In phage T4 lysozyme, replacement of Gly^{156} by Asp (Gray and Matthews, 1987) or of Pro^{86} by any of seven amino acids (Alber *et al.*, 1988) causes crystallographically observable shifts up to 20 Å apart on the protein surface. Surface substitutions generally do not cause structural changes to propagate into the protein interior.

Substitutions of interior residues sometimes cause only localized adjustments (Grütter *et al.*, 1984), but propagating shifts are more common. In phage T4 lysozyme, replacement of Ala^{98} by Val in an interhelical contact causes structural changes in a 20 Å × 20 Å × 10 Å slice through the protein (T. Alber and B. W. Matthews, unpublished). The high-resolution x-ray crystal structure of the mutant protein slows extensive subtle repacking of the enzyme's hydrophobic core. This change is associated with a reduction in stability of ~3 kcal/mole at pH 2.0 and 42°C, but the activity of the protein is preserved. Replacements of Ala^{146} by Thr, Val, Cys, Gly, and Ile also cause shifts to propagate over 10 Å into the interior, where a cavity in the wild-type protein becomes filled (Grütter *et al.*, 1983; K. Wilson, D. Maki, T. Alber, D. Tronrud, M. Karpusas, J. Mendel-Hartvig, L. McIntosh, and B. W. Matthews, unpublished results). These substitutions reduce stability even though the mutant proteins bury additional hydrophobic surface area in the folded state. In the collection of over 80 purified mutants of phage lysozyme, proteins with interior substitutions have proven to be the most difficult to crystallize and the most likely to produce new crystal forms (Matthews, 1987). This is an indication that structural changes propagate from the altered site to regions that make intermolecular contacts in the crystals of the wild-type enzyme.

Several other generalizations are emerging from structural studies of protein variants. (1) Positional shifts are channeled in specific directions and tend to be damped with distance from the site of the substitution. (2) Even chemically and structurally simple substitutions can simultaneously alter several different kinds of interactions. As a result, careful comparisons of multiple substitutions may be required to evaluate the contributions of specific noncovalent bonds. (3) New side chains generally adopt commonly observed rotamer conformations (Ponder and Richards, 1987; Villafranca *et al.*, 1987; Alber *et al.*, 1987b, 1988), indicating an absence of torsional strain. Side chains in rare unstrained rotamer conformations can also be stabilizing. (4) Examples of dramatic changes in solvent binding on the surface of mutant proteins have been reported (Howell *et al.*, 1986; Alber *et al.*, 1987b). Binding of specific water molecules can be stabilizing. (5) Unexpected features are often found, and proteins can respond differently to seemingly conservative replacements. For example, functionally important differences in the position and mobility of leucine and isoleucine side chains at specific sites in hemoglobin and phage lysozyme have been reported (Nagai *et al.*, 1987; Alber *et al.*, 1987b). At position 157 in phage lysozyme, Glu, Asp, and Asn are found in different average conformations, and Arg forms a new ion pair with a neighboring residue. These findings emphasize the importance of structural studies for understanding the properties of mutant proteins. Predicting structural shifts in protein variants is not yet a simple proposition (Alber *et al.*, 1986; Snow and Amzel, 1986).

IV. FORCES THAT STABILIZE PROTEINS

As noted at the outset, it has been traditional to estimate the magnitudes of the interactions that stabilize proteins by considering simpler model systems. A theme of recent studies has been to use directed mutagenesis and chemical modification to determine if data from model systems quantitatively describe the contributions of specific interactions. Although such studies are in their infancy, early indications are that proteins harbor complexities that are not accounted for by traditional model systems. Data on several classes of stabilizing interactions are summarized in this section.

A. Effects on Conformational Entropy

Covalent cross links are thought to stabilize proteins by reducing the conformational entropy of the unfolded chain. Statistical treatments have suggested that the destabilization of the unfolded state depends on the length of the loop formed by a single cross link. Recent studies of proteins containing natural or engineered cross links indicate that effects on the folded state can also be crucial.

Chemical methods were used to join Gly^{35} and Trp^{108} in hen lysozyme (Johnson *et al.*, 1978), Lys^{7} and Lys^{41} in ribonuclease A (Lin *et al.*, 1984), and the N and C termini of bovine pancreatic trypsin inhibitor (BPTI) (Goldenberg and Creighton, 1984). Crystallographic and NMR studies indicate that the cross links did not cause large changes in the folded conformations of ribonuclease (Weber *et al.*, 1985) and BPTI (Chazin *et al.*, 1985). The cross links in hen lysozyme and ribonuclease increased the melting temperature by 29°C and 25°C, respectively, at pH 2.0. Analysis of thermodynamic data suggested that the Lys^{7}–Lys^{41} cross link stabilizes ribonuclease by ~5 kcal/mole at 53°C. This effect is almost entirely entropic, consistent with a destabilization of the unfolded state (Lin *et al.*, 1984). Kinetic data, however, show that the cross link specifically reduces the rate of unfolding (Lin *et al.*, 1985). This led Matthews and Hurle (1987) to suggest that the cross link may differentially affect the folded state.

Studies of cross links in BPTI emphasize the importance of effects on the folded state. A peptide bond joining the natural chain termini provided as much stabilization as the ion pair between the termini in the wild type protein (Goldenberg and Creighton, 1984; Goldenberg, 1985). This suggested that the cross link destabilized both the folded and the unfolded states equally. Strain and loss of entropy in the folded conformation were proposed as potential destabilizing effects.

Creighton (1983a) has analyzed the stabilities of the three natural disulfides in BPTI relative to an added disulfide exchange reagent. The protein disulfides have similar stabilities in the unfolded state, with effective concentrations of about 10^{-2} M. In the folded state, however, the measured effective concentrations ranged from 2.3×10^{2} M to 4.6×10^{5} M. The effective concentration of each disulfide bond in the folded state correlated qualitatively with the change in the melting temperature of the protein caused by reduction of the disulfide.

Overall, these data suggest that the stabilizing contribution of a disulfide bond (or other cross link) is determined by the size of the loop formed in the chain and by the compatibility of the cross link with the folded structure. For a given loop size, the most effective cross link may be formed between groups that are rigidly held in an optimum orientation by the folded structure (Creighton, 1983a; Goldenberg, 1985). Cross links that disrupt interactions or reduce the entropy of the folded state by reducing its flexibility may be proportionately less stabilizing.

Attempts to stabilize proteins with engineered disulfide bonds have produced mixed results. Sites for engineered disulfides have generally been selected to satisfy conformational restrictions observed in naturally occurring proteins (Richardson, 1981). In the absence of shifts in the folded structure, very few pairs of α carbons in any given protein can be joined by a disulfide bond with standard geometry (Pabo and Suchanek, 1986). Single added disulfides in subtilisin BPN′ and dihydrofolate reductase (DHFR) produce complex effects. Observation of an increase in stability depends on solution conditions (Wells and Powers, 1986; Pantoliano *et al.*, 1987) or on the method of extrapolation to zero denaturant concentration (Villafranca *et al.*, 1987). Crystallographic studies show that the S-S bridge between residues 39 and 85 in DHFR reduces van der Waals contacts among nearby residues. The mobility of residues 80 to 90 also increases, perhaps reflecting destabilization of the native state (Villafranca *et al.*, 1987).

An intermolecular disulfide bond joining residue 88 in adjacent subunits of λ repressor increases the T_m of the N-terminal domain by approximately 10°C (Pabo and Suchanek, 1986; Sauer *et al.*, 1986). Formation of an engineered disulfide between residues 3 and 97 of phage T4 lysozyme increases T_m by 7°C at pH 2 (Perry and Wetzel, 1984; Wetzel *et al.*, 1988). The free energy of stabilization increases about 1.8 kcal/mole, which is less than expected from loop formation. It has not been possible to crystallize this mutant protein, perhaps because of conformational shifts that alter intermolecular contacts (J. A. Wozniak and B. W. Matthews, personal communication).

In most cases reported to date, introduction of the Cys residue(s) required for an engineered disulfide was destabilizing. This reduced the net stabilization afforded by the S-S bridge. The phage lysozyme containing a disulfide bond between residues 3 and 97, for example, is 1.8 kcal/mole more stable than the reduced mutant protein but only 1.2 kcal/mole more stable than the wild-type protein at 42°C and pH 2.0 (Wetzel *et al.*, 1988).

Residues that do not introduce cross links may also influence the conformational entropy of the unfolded state. Simple estimates of the degrees of freedom of the polypeptide backbone suggest that removal of glycine (Nemethy *et al.*, 1966) or introduction of proline (Matthews *et al.*, 1987) could destabilize the unfolded state by 0.8 and 1.4 kcal/mole, respectively, at 65°C. Substitutions that introduce β-branched or bulkier side chains may also restrict main chain rotations in the unfolded state (Nemethy *et al.*, 1966). Increases in stability of <1 kcal/mole are caused by Gly-to-*X* and *X*-to-Pro substitutions that are compatible with the folded structures of phage lysozyme (Matthews *et al.*, 1987), phage λ repressor (Hecht *et al.*, 1984, 1985), and the neutral protease of *B. stearothermophilus* (Imanaka *et al.*, 1986). In λ repressor and neutral protease, increases in stability were attributed to enhanced "helical propensity" at helical sites. Entropic effects, however, may contribute to the observed tendency of glycine to destabilize model helices. In addition to possible effects on the unfolded state, these substitutions could enhance stability by increasing the rigidity, and consequently the effective concentration, of neighboring groups in the folded structure (Alber *et al.*, 1987a,b).

B. The Hydrophobic Effect

One of the most important contributions to protein stability is made by the hydrophobic effect. Chain segments that are disordered in folded proteins contain few hydrophobic residues (G. A. Petsko, unpublished observation). Current estimates of amino acid hydrophobicity are based on the measured free energies of transferring side chains from water to organic solvents (Tanford, 1962; Nozaki and Tanford, 1971; Kyte and Doolittle, 1982; Abraham and Leo, 1987), to SDS micelles (Wishnia, 1963), and to the vapor phase (Wolfenden *et al.*, 1981). On the basis of these data, scaler multipliers of 22–25 cal per Å^2 have been proposed to estimate the free energy of stabilization from buried hydrophobic surface area (Richards, 1977). More sophisticated approaches based on summing the different contributions of individual atoms have also been developed (Eisenberg and McLachlan, 1986; Abraham and Leo, 1987). The free energy of transferring side chains from water to ethanol is correlated with their average extent of burial in known protein structures, supporting the applicability of the solvent transfer model to the problem of protein stability (Rose *et al.*, 1985; Miller *et al.*, 1987; Lawrence *et al.*, 1987).

Baldwin (1986) has analyzed the temperature dependence of the transfer of six liquid hydrocarbons to water and used the results to interpret thermodynamic data on the unfolding of hen lysozyme. For the unfolding of hen lysozyme, the temperature dependence of ΔH_T and ΔS_T could be largely accounted for by the hydrophobic effect. The fraction of ΔH_T and ΔS_T arising from other contributions was independent of temperature. The temperature-independent part of ΔH_T was large and favored folding, presumably because of noncovalent interactions in the folded state. The temperature-independent part of ΔS_T was large and favored unfolding, presumably because of conformational entropy. From 10 to 100°C, the free energy of stabilization from the hydrophobic effect increased with temperature, and it reached a maximum at ~112°C if ΔC_p is constant. This analysis suggested that cold denaturation is caused by the weakening of the hydrophobic effect. Unfolding occurs at high temperatures because the destabilizing contribution of the conformational entropy of the unfolded state ($T\Delta S_{\text{conf}}$) increases more rapidly with temperature than the stabilizing contribution of the hydrophobic effect. Baldwin noted that his analysis was limited by assumptions made to determine ΔC_p and by the possibility that the hydrocarbon transfer experiments may not be directly applicable to protein folding.

The relevance of transfer experiments has been questioned on theoretical and experimental grounds. (1) Measured hydrophobicities depend on the nature of the nonaqueous phase (Wolfenden *et al.*, 1981), and organic solvents, detergent micelles, and gases are poor models of the protein interior. These models differ dramatically from proteins in packing density (Klapper, 1971; Richards, 1977), polarity (Chothia, 1976), absolute size (Lee, 1985), and in the energy of cavity formation. (2) The loss of entropy associated with fixing the position of a side chain during folding is likely to be significantly larger than the entropy change of transferring a side chain from water to another bulk phase (Tanford, 1980). (3) Cooperativity may play a less significant role in solvent transfer (Tanford, 1980; Creighton, 1983b). (4) The transfer data may not account for the effects of pressure on protein stability (Brandts *et al.*, 1970; Zipp and Kauzmann, 1973; Hvidt, 1975). These criticisms suggest that the relationship between hydrocarbon transfer free energy in model systems and hydrophobic stabilization of proteins is undetermined. Because of differences in the energy of cavity formation (Lee, 1985), hydrophobicity may even vary from site to site and from protein to protein.

This problem has begun to be addressed in recent studies of multiple amino acid substitutions in the α subunit of trp synthase (Yutani *et al.*, 1984, 1987) and kanamycin nucleotidyl transferase (Matsumura *et al.*, 1988). Yutani *et al.* (1987) constructed 20 variants of the α subunit of trp synthase, each with a different amino acid at position 49. The kinetics and thermodynamics of unfolding by guanidine hydrochloride were studied. With the exception of the bulky aromatic residues Trp, Tyr, Phe, and His and the negatively charged residues Asp and Glu, the stability of the protein was proportional to the free energy of transfer of side chains from water to ethanol. This result supports the qualitative validity of the transfer model.

Surprisingly, the constant of proportionality between protein stabilization and transfer free energy depended on the concentration of denaturant and on the pH. Based on extrapolations to zero denaturant concentration, the protein was apparently more hydrophobic than the ethanol model by factors of 2 at pH 5.5, 3.7 at pH 7, and 1.3 at pH 9. Yutani and co-workers (1987) proposed that this variation may result from differences in electrostatic interactions arising from differences in the local dielectric constant. The difficulty of extrapolating to zero denaturant concentration and the different degrees of extrapolation at each pH could also influence the apparent hydrophobicities.

For eight replacements of Asp^{80} in kanamycin nucleotidyl transferase (KNTase), Matsumura and co-workers (1988) discovered a qualitatively similar relationship between transfer free energy and resistance to irreversible inactivation at 58°C. Although the structure of

KNTase is not known, the observed correlation was taken as evidence that residue 80 is at least partially buried and that the solvent transfer model at least qualitatively applies to the hydrophobic stabilization of proteins.

C. Hydrogen Bonds

One of the major contributions to the largely temperature-independent part of the enthalpy of stabilization is thought to come from hydrogen bonds. Hydrogen-bonding groups are part of the target for destabilizing amino acid substitutions (Fermi and Perutz, 1981; Alber *et al.*, 1987a). A survey of highly refined protein structures indicates that hydrogen-bonding geometry is quite variable, and unpaired H-bond donors and acceptors are rare (Baker and Hubbard, 1984).

Hydrogen bond partners are exchanged during folding. Intramolecular bonds are formed at the expense of intermolecular hydrogen bonds with water. Water molecules are released to interact with each other, and specific water-binding sites are formed on the protein.

Model compounds that form detectable intermolecular hydrogen bonds in water have been used to estimate the contribution of hydrogen bonding to protein stability. Because of competition with water, model compounds generally associate weakly in aqueous solution (Schellman, 1955a,b; Klotz and Franzen, 1962; Susi *et al.*, 1964; Stahl and Jencks, 1986; Rebek, 1987). Different model systems have provided different estimates of the enthalpy of hydrogen-bond formation. Studies of the dimerization of urea and δ-valerolactam in water led Schellman (1955a,b) and Susi and co-workers (1964) to conclude that the heat of formation of hydrogen bonds may be -1.5 kcal/mole or more. Klotz and Franzen (1962), using aqueous N-methylacetamide as a model for the peptide bond, estimated that ΔH for hydrogen-bond formation is close to zero.

The latter result has supported the popular view that as long as equal numbers of hydrogen bonds are formed in the folded and unfolded states, hydrogen bonds do not contribute to the difference between the free energies of the states. Hydrogen bonds can help establish the specificity of folding, since unpaired donors or acceptors sequestered from solvent destabilize the folded conformation by reducing the relative number of hydrogen bonds. A central feature of this analysis is that hydrogen bonds in the folded and unfolded states are considered to be energetically equivalent. The concentration of donors and acceptors in water, ~110 M, is assumed to be the effective concentration of all hydrogen-bonding groups in both the folded and unfolded states.

Several authors have pointed out, however, that differences in the average geometry, the entropy of formation, and the average number of interacting partners can lead to significant energetic differences between hydrogen bonds (and other noncovalent interactions) in the folded and unfolded states (Page and Jencks, 1971; Creighton, 1983a,b; Goldenberg, 1985; Stahl and Jencks, 1986). Hydrogen-bonding groups can have effective concentrations higher or lower than 110 M. Groups with higher effective concentrations in the folded state will be stabilizing; those with lower effective concentrations will be destabilizing. All possible hydrogen bonds are not made in liquid water, so an increase or a decrease in the average number of partners can influence the contribution of a particular substituent. Obviously, if hydrogen bonds in a folded protein make a wide range of contributions to stability, data obtained from model systems cannot provide reliable estimates of the contributions of specific interactions.

To determine the range of potential contributions, the consequences of adding or removing specific hydrogen-bonding groups have been analyzed. Studies of protein–ligand complexes and of nucleic acids suggest that hydrogen bonds between uncharged donors and acceptors can contribute 0.5 to 1.8 kcal/mole to the observed association energy (Page, 1984; Fersht *et al.*, 1985; Fersht, 1987). Hydrogen bonds between charged groups can contribute up to 6 kcal/mole. Unpaired donors or acceptors in a protein–ligand complex are destabilizing, reflecting a reduction in the relative number of hydrogen bonds in the bound state.

Evidence for intramolecular hydrogen bonds in the unfolded state has been obtained in NMR studies of short peptides. Based on the pH dependence of amide proton chemical shifts, Bundi and Wüthrich (1979) concluded that ionized carboxylates in short peptides can hydrogen bond to nearby amides. Dyson and co-workers (1985) found that a nonapeptide fragment of influenza virus hemagglutinin can form a hydrogen-bonded β turn in water. Surprisingly, this sequence is part of a strand of β sheet in the folded protein, suggesting that the conformational preference of the fragment must be overcome during folding.

Several hydrogen bonds in the folded state of phage T4 lysozyme have been studied using oligonucleotide-directed mutagenesis, x-ray crystallography, and thermodynamic methods (Grütter *et al.*, 1987; Alber *et al.*, 1987b, 1988). These studies confirmed that unpaired donors or acceptors sequestered from solvent in the folded state are destabilizing. Conclusive evidence for excess free energy of stabilization from specific hydrogen bonds was harder to obtain, however, because the amino acid replacements simultaneously changed several interactions in the folded structure.

The replacement of Thr^{157} by Ile, for example, reduced the stability of phage lysozyme by about 3 kcal/mole at 42°C and pH 2.0 (Hawkes *et al.*, 1984). The x-ray crystal structure of the mutant protein showed differences in hydrogen bonds, electrostatic interactions, hydration, and van der Waals contacts at the altered site (Grütter *et al.*, 1987). Twelve additional substitutions were engineered at position 157 to evaluate systematically the contributions of each of the atoms in the Thr and Ile side chains (Alber *et al.*, 1987b). The mutant proteins were intermediate in stability between the wild-type and the Ile^{157} variant. Crystallographic and thermodynamic studies showed that the five most stable lysozymes contained a hydrogen bond to the buried main-chain amide of Asp^{159}. This correlation between structure and stability suggested that the hydrogen bond in the folded state dominates the effects of all 13 amino acid substitutions on the free energies of both the folded and unfolded states.

Though the hydrogen bonds to residue 157 made the largest contribution to stability, loss of the Thr γ-methyl group in the Ser^{157} mutant protein reduced stability by ≈0.7 kcal/mole. The methyl group could stabilize the protein directly, through van der Waals and hydrophobic contacts, or indirectly, by increasing the probability of forming the hydrogen bonds to the Thr^{157} hydroxyl group (Alber *et al.*, 1987b).

One striking finding from this study is that bound water can stabilize proteins. A water molecule is localized by three hydrogen bonds in the crevice formed by Gly^{157}, and this mutant protein is almost as stable as the Ser^{157} variant.

Compensating structural changes in the 13 mutant lysozymes make it impossible to determine the stabilizing contributions of specific hydrogen bonds. Nonetheless, it is clear that different hydrogen bonds are associated with different contributions to stability. Among the most stabilizing amino acids at position 157, differences in the number of donor and acceptor groups allow Asn, Asp, and Gly H_2O to form different numbers of hydrogen bonds than Thr and Ser (Alber *et al.*, 1987b). A new ion pair between Arg^{157} and Asp^{159} contains two hydrogen bonds but does not fully compensate for the unpaired amide of Asp^{159}. Some of the least stabilizing substitutions—Ile, His, Val, and Ala—showed evidence of additional hydrogen bonds between the protein and ordered solvent. These results leave open the possibility that some hydrogen bonds preferentially stabilize the folded protein.

Similar methods were used to analyze changes in structure and thermal stability caused by ten substitutions for the surface-exposed Pro^{86} in phage T4 lysozyme (Alber *et al.*, 1988). Comparison of the x-ray crystal structures of seven of the mutant proteins showed that the side chains of Ser^{86} and Cys^{86} formed a unique tertiary hydrogen bond to the side chain of Gln^{122}.

Nonetheless, all the mutant proteins had essentially the same thermodynamic stability; the additional hydrogen bond did not correlate with added stability. Residues at position 86 have higher than average mobility and surface accessibility. As a result, the intramolecular hydrogen bond may be equivalent to the local interactions with water. The high mobility of the side chains at this site may reflect a significant loss of entropy in the formation of the intramolecular bond.

Further studies will undoubtedly clarify the structural features that determine the contributions of specific hydrogen bonds to protein stability. Obvious parameters to consider include surface accessibility, mobility, average geometry, and number of partners in the folded state. In addition, the effects of competing structures in the unfolded state have not been quantified. Such information will help distinguish hydrogen bonds that are essential for protein stability from those that are neutral or destabilizing.

D. Dispersion Forces

The importance of dispersion or van der Waals forces for protein stability hinges on differences in packing in the folded and unfolded states. Unfolding usually involves very small changes in the volume of the system, so the average interatomic distance remains constant. This has been interpreted to imply that van der Waals forces do not change appreciably between the two states. According to this view, cavities in proteins are destabilizing because they reduce the relative contact surface of the folded state. Steric overlap is avoided in both states. This implies that van der Waals forces can contribute to the specificity of folding but not to the excess free energy of stabilization.

The overall volume of the system, however, is not an adequate measure of the sum of the interactions. During folding, water is transferred to the relatively open bulk phase, and the atoms that form the protein interior become as densely packed as molecules in crystals of small organic molecules (Klapper, 1971; Chothia, 1975; Richards, 1977). This changes the neighbors of solvent and protein atoms (the unfolded chain is more hydrated), and it changes the distribution of interatomic distances. Both of these effects can alter van der Waals interactions. A likely source of an increase in van der Waals forces is the interior of the folded protein, since this is the most densely packed phase of the system. Bello (1977, 1978) has argued from the similarity in densities that the average van der Waals contribution in the folded state can be estimated from the heat of fusion of small organic solids (~30 cal/g). This value, however, may not be appropriate because it does not take into account the changes in packing in the solvent, the greater flexibility of proteins, or the irregularities in the packing of individual side chains with different neighbors (Hvidt, 1975; Narayana and Argos, 1984).

Analysis of the packing of proteins provides circumstantial evidence for the importance of dispersion forces (Richards, 1977). Contact surfaces of secondary structures are generally nonpolar (Chothia, 1976), and cavities that are large enough are often filled by water molecules or nonpolar solutes. Short nonbonded contacts are generally absent.

Amino acid substitutions that alter interior packing can reduce protein stability, but quantitative correlations have not yet been established. A reduction in tertiary contact surface is associated with the lack of stabilization of an engineered disulfide bond in dihydrofolate reductase (Villafranca *et al.*, 1987). In phage T4 lysozyme, van der Waals and hydrophobic contacts of the γ-methyl group of Thr^{157} may contribute up to 0.7 kcal/mole to stability at pH 2 and 42°C (Alber *et al.*, 1987b). Several destabilizing substitutions in phage lysozyme that alter interior packing have been analyzed crystallographically. The replacement of Ala^{98} by Val in a helix–helix contact causes extensive subtle rearrangements in packing and destabilizes the protein by over 3 kcal/mole at pH 2 and 42°C (T. Alber, S. Cook, and B. W. Matthews, unpublished results). At residue 146, Gly, Val, Thr, Ile, and Phe are equally poor substitutes for the wild-type alanine. Stability is not correlated with side-chain volume at this largely buried site. A cavity in the interior of phage lysozyme is eliminated by the replacement of Ala^{146} by Thr, and the same cavity is expanded by the replacement of Met^{102} by Thr (Grütter *et al.*, 1983; D. E. Tronrud and B. W. Matthews, personal communication). Both amino acid substitutions destabilize the protein.

An algorithm for identifying combinations of amino acids that could pack similarly has been developed by Ponder and Richards (1987). This list of sequences defines what they have called the "tertiary template" for a given region of main-chain conformation. The approach is based on the requirement that side chains fill space without steric overlap. Possible side-chain conformations are taken from a rotamer library derived from 19 highly refined protein crystal structures. Surprisingly, the conformations of 17 of the 20 amino acids can be represented by only 67 rotamers. The dihedral angles in the rotamer library cluster around minima of torsional energy, indicating that side chains generally adopt unstrained conformations (Janin *et al.*, 1978).

Determining the stabilization afforded by different members of a tertiary template provides a systematic approach to quantifying the effect of packing on stability. Modern genetic methods may also allow proteins with all possible combinations of amino acids in a given tertiary cluster to be compared.

E. Electrostatic Interactions

The associations of acetate ions with guanidinium or *n*-butylammonium cations in water have been used to model the contributions of ion pairs to protein stability (Tanford, 1954; Springs and Haake, 1977). Interactions between these ions are weak, fueling early skepticism that electrostatic interactions could be stabilizing. Like other simple model systems, however, these salt solutions do not account for unique changes in the solvation of specific groups during folding or for the differences in the entropy of formation of specific interactions. Folding alters the effective dielectric constant around each ion and creates a boundary between the high-dielectric solvent and the lower dielectric of the protein interior. These effects can increase the interactions of ions and dipoles.

Data on protein variants suggest that ionic interactions can be important for folding and stability. Even though charged residues are often mobile and exposed to solvent, they are sensitive to destabilizing mutations (Fermi and Perutz, 1981; Hecht *et al.*, 1983, 1984; Pakula *et al.*, 1986; Alber *et al.*, 1987a). Additional ion pairs distinguish several thermophilic proteins from their mesophilic counterparts (Perutz and Raidt, 1975; Perutz, 1978). In the α subunit of trp synthase, replacement of Arg^{211} by Glu increases the rate of folding by a factor of ~70 (Matthews *et al.*, 1983; Beasty *et al.*, 1987; C. R. Matthews, unpublished results). Charged groups can also have dramatic effects on the stability of isolated helices in solution (Ihara *et al.*, 1982; Shoemaker *et al.*, 1985; Mitchinson and Baldwin, 1986; Marqusee and Baldwin, 1987).

Ionizable groups are not distributed randomly over protein surfaces, reflecting structural and functional roles. Including the peptide dipoles, charges are, on average, surrounded by charges of the opposite sign (Wada and Nakamura, 1981; Thornton, 1982; Barlow and Thornton, 1983). Examples of highly asymmetric charge distributions are found, and the resulting macrodipoles often point in the right direction to facilitate the binding of charged ligands (Barlow and Thornton, 1986). On average, only a third of the charged residues in proteins are involved in ion pairs, and 76% of these are between residues in different elements of secondary structure (Barlow and Thornton, 1983). Surprisingly, 17% of the ion pairs are buried, and these generally play clearly identifiable functional roles. Overall, however, ion pairs are poorly conserved in protein families, suggesting that at least some of them are not critical for folding and stability. Most unexpected was the finding that 20% of the buried charged groups do not form ion pairs and are solvated, instead, by hydrogen bonds (Rashin and Honig, 1984). This highlights the importance of the self-energy of ionic groups (Russell and Warshel, 1985; Gilson *et al.*, 1985; Warshel, 1987).

The stabilizing contributions of several specific ion pairs have been measured. Analysis of the pH dependence of the activity of α- and δ-chymotrypsins suggested that the stabilization energy of the buried ion pair between Asp^{194} and the α-amino group of Ile^{16} is 2.9 kcal/mole (Fersht, 1971). A solvent-exposed salt bridge between the termini of bovine pancreatic trypsin inhibitor stabilizes the protein by ~1 kcal/mole in 6 M guanidine hydrochloride at pH 6.0 and 75°C (Brown *et al.*, 1978). In phage T4 lysozyme, a new ion pair is formed when Thr^{157} is replaced by Arg (Alber *et al.*, 1987b). Comparison of the denaturation behavior of this variant at pH 6.0 and pH 2.0 suggests that the ion pair contributes about 1 kcal/mole to the stability of the mutant protein at 65°C (T. Alber, S. Cook, and B. W. Matthews, unpublished results). Based on studies of protein–ligand complexes, Fersht and co-workers (1985, 1987) have concluded that hydrogen-bonded ion pairs can contribute 3 to 6 kcal/mole to ligand–binding energies.

The stabilizing contributions of charged groups that do not form ion pairs are also dependent on position. Increases in the positive charge density of BPTI and phage T4 lysozyme (*pI*s > 10) at solvent-exposed positions decrease stability by 0.25–1 kcal/mole (Wagner *et al.*, 1979; Grütter *et al.*, 1982; Alber *et al.*, 1988). In the α subunit of trp synthase, a more neutral protein, replacement of Gly^{211} by either Glu or Arg increases stability (Matthews *et al.*, 1980). Isolated charges that cannot be well solvated in the folded protein are apparently very destabilizing (Yutani *et al.*, 1987; Ahern *et al.*, 1987).

Ionic groups at helix termini also influence stability. The stability of ribonuclease S increases with increasing negative charge on the N terminus of the S-peptide helix (Mitchinson and Baldwin, 1986). Replacement of Arg^{96} by His at the C terminus of a helix in phage lysozyme reduces the free energy of stabilization by ~3.5 kcal/mole at low ionic strength (Sturtevant, 1987). The guanidium group of Arg^{96} forms hydrogen bonds to two carbonyl groups at the helix terminus, and the histidine side chain, although positively charged at low pH, is not long enough to make these interactions in the folded structure (Grütter *et al.*, 1979).

Electrostatic interactions can have very long-range effects in the folded state. Shifts of up to one unit in the pK_a of His^{64} in *B. amyloliquefaciens* subtilisin were induced by substitutions of six charged residues 12–20 Å from the imidazole nitrogen (Thomas *et al.*, 1985; Sternberg *et al.*, 1987; Gilson and Honig, 1987). In phage T4 lysozyme, several replacements that increased positive charge over 20 Å from the catalytic Glu^{11} carboxyl group caused a marked decrease in enzyme activity at neutral pH (Grütter and Matthews, 1982; Alber *et al.*, 1988). Such long-range interactions may account for the finding that mobile and solvent-exposed charged residues can make crucial contributions to protein stability.

Interactions of ionizable groups vary with pH, and the stability of a protein depends on the number of protons bound or released on denaturation (Tanford, 1968, 1970; Privalov, 1979; Becktel and Schellman, 1987). Surprisingly, many proteins are maximally stable near pH 6, regardless of their isoelectric points (Becktel and Schellman, 1987). Stability varies smoothly with pH, suggesting the presence of compensating interactions in the constellation of ionizable groups. Consistent with this behavior, a calculation of the contribution of electrostatic energy to the stability of sperm whale myoglobin showed that stabilizing and destabilizing interactions were redistributed over the charge pairs as the pH was varied (Garcia-Moreno *et al.*, 1985).

The strength of ionic interactions can be affected by changes in local effective dielectric constants. As a result, amino acid substitutions that alter neutral residues may alter electrostatic contributions to stability (Flanagan *et al.*, 1983; Yutani *et al.*, 1984, 1987). The local dielectric within a protein can be quite high because of the distribution of polar groups (Macgregor and Weber, 1986). Local differences in polarity have been proposed (Warshel, 1987) to contribute to the surprising 2.7 kcal/mole difference in the stability of an Asp-Arg ion pair compared to the inverted Arg-Asp pair in aspartate amino transferase (Cronin *et al.*, 1987).

Nonpolar amino acids may also participate in "weakly polar interactions" (Burley and Petsko, 1988). For example, aromatic groups in proteins tend to pack edge to face. This arrangement may be favored by short-range dispersion forces and by a longer-range electrostatic attraction between the electron-poor aromatic hydrogens and the electron rich π orbitals (Burley and Petsko, 1985, 1988).

The factors that determine the electrostatic contribution to protein stability are not quantitatively understood. Theoretical approaches to calculating electrostatic energies are under rapid development (reviewed by Warshel and Russell, 1984; Matthew, 1985; Warshel, 1987). Critical questions include how to model the protein structure and the solvent, how to distribute charge in the model, which partial charges to consider, which interactions to include, and how to account for polarization effects and for local variations in dielectric constant. Whatever the computational method used, it is clear that experimental studies of mutant proteins can provide accurate benchmarks of electrostatic contributions in stability. Changes in stability related to pH and ionic strength as well as titrations of folded and unfolded proteins can be used to evaluate electrostatic effects. As a result, ionic interactions may soon become the best understood contribution to protein stability.

F. Helix Stabilization

The helix–coil transition has long served as a model of protein folding, and considerable effort has been invested in understanding the balance of forces that promote helix formation. One aim of this approach is to understand the stabilizing contribution of protein helices in terms of their intrinsic stability and their affinity for the remainder of the protein. The validity of this concept was directly demonstrated by Mitchinson and Baldwin (1986). They found that addition of the ribonuclease S protein to S peptides of increasing intrinsic helical stability leads to concomitant stabilization of the resulting ribonuclease S variants.

The relative abilities of each of the 20 common amino acids to stabilize model α helices in water has been investigated by Scheraga and co-workers by measuring the effect of "guest" amino acids on the thermodynamic stability of "host" α helices (Scheraga, 1978; Sueki *et al.*, 1984). Copolymers of hydroxypropyl- and hydrobutyl-L-glutamine randomly "doped" with the guest amino acids were studied. Equilibrium data on helix stability were analyzed using the theory of Zimm and Bragg (1959) to derive nucleation constants, σ, and propagation constants, s, for each residue as a function of temperature. Helix formation in this system is highly cooperative. Values of σ range from 10^{-5} to 10^{-2}. The amino acids show small differences in helical propensity, with s values ranging from 0.59 to 1.35 at 20°C (Sueki *et al.*, 1984). Helix formation is favored when $s > 1$; helix termination is favored when $s < 1$. The values of σ and s indicate that isolated short helices are likely to be unstable because both nucleation and termination are unfavorable. The relative helical propensities obtained from the analysis of random copolymers generally parallel the relative frequencies of occurrence of each amino acid in protein helices (Lewis *et al.*, 1970; Scheraga, 1978).

Although the thermodynamic and statistical preferences for helix formation are weak, these measures of helix stability correlate qualitatively with the effects of amino acid substitutions on the thermodynamic stability of several proteins (Alber *et al.*, 1986). In α helices in T4 lysozyme, for example, most mutations that reduce the stability of the protein also reduce the helical propensity at the altered site. An increase in helix propensity is associated with substitutions in two helices that increase the stability of staphylococcal nuclease (Shortle and Lin, 1985). Similar correlation has been found among variants of λ repressor (Hecht *et al.*, 1984b, 1985). These results are consistent with the idea that the host–guest data can be used to describe both the stability of individual protein helices and the contributions of helices to the overall stability of a protein.

This conclusion has been challenged by investigations of the effects of amino acid substitutions on isolated short helices (Kim and Baldwin, 1984; Shoemaker *et al.*, 1985) and in a helix in phage T4 lysozyme (Alber *et al.*, 1988). These studies indicate that the intrinsic effects of changes in σ and s can be small or can be easily dominated by sequence-specific interactions. The decrease in helical propensity often associated with destabilizing substitutions may be a fortuitous consequence of the relatively high helical propensities of residues in actual protein helices.

To examine the relationship between protein stabilization and helical propensity, ten substitutions were made for Pro^{86} in a helix in phage T4 lysozyme (Alber *et al.*, 1988). The x-ray crystal structures of seven of the mutant lysozymes were determined at high resolution, and in each case, the helix was extended by two residues. The amino acids at position 86 spanned a wide range of helical propensity. Surprisingly, the amino acid substitutions had little effect on protein stability. The strongest helix former studied, Ala, was no more stabilizing than the strongest helix breaker, Gly. This insensitivity of protein stability to the residue at position 86 is not simply explained by thermodynamic studies of model helices.

Also contrary to expectations from the host–guest data, several short peptides have been found to form isolated helices in water. The peptides include block copolymers of Ala_{20} and Glu_{20} (Ihara *et al.*, 1982), a designed 17-residue sequence with three Glu-Lys pairs (Marqusee and Baldwin, 1987) and variants of the S peptide (residues 1–20) and C peptide (residues 1–13) of ribonuclease (Brown and Klee, 1971; Bierzynski *et al.*, 1982).

Helix formation in S-peptide and C-peptide analogues is promoted by low temperature and moderately acidic pH. The bell-shaped dependence of helix stability on pH suggested a crucial role for charged residues (Bierzynski *et al.*, 1982). An initial hypothesis that a salt bridge between Glu^{9} and His^{12} stabilizes the helix was ruled out, because replacement of Glu^{9} by Leu increased helix stability (Shoemaker *et al.*, 1985). On the other hand, variants of C peptide lacking Glu^{2} and His^{12} formed less helix under optimal conditions, and helix stability depended on the ionization of the remaining residue (Shoemaker *et al.*, 1985, 1987). On this basis, the "charged group effect" was assigned to electrostatic interactions of Glu^{2} and His^{12} with the macrodipole resulting from the partial charges on the peptide groups at the helix termini. This explanation for the charged group effect is consistent with the finding that isolated helices are stabilized by increasing N-terminal negative charge and destabilized by increasing N-terminal negative charge and destabilized by increasing N-terminal positive charge (Ihara *et al.*, 1982; Mitchinson and Baldwin, 1986; Marqusee and Baldwin, 1987). Other residues in the S peptide, including Phe^{8} and Arg^{10}, may also be important for helix stability (Rico *et al.*, 1986; Shoemaker *et al.*, 1987).

Helix termination in short peptides is also not expected on the basis of host–guest data on model helices (Scheraga, 1978). Nonetheless, the helix in the isolated S peptide stops near Met^{13}, the last helical residue in this region of folded ribonuclease S (Kim and Baldwin, 1984). Helix termination in the S peptide does not require tertiary interactions. These studies indicate that specific interactions of the side chains can be as important for helix formation as the intrinsic helical propensities of the individual residues (Kim and Baldwin, 1984; Shoemaker *et al.*, 1985, 1987; Scheraga, 1985).

The α-helix macrodipole has been proposed to account for the charged group effect on the stability of isolated helices (Shoemaker *et al.*, 1985, 1987) and for several structural features of proteins. These include (1) binding sites for ions and other ligands at helix termini (Hol *et al.*, 1978, 1981), (2) antiparallel packing of helices (Sheridan *et al.*, 1982), and (3) the preponderance of acidic amino acids at helix N termini and basic amino acids at helix C termini (Chou and Fasman, 1974; Blagdon and Goodman, 1975). Although the importance of electrostatic interactions involving the partial charges on polarized helical peptide groups seems well established, the use of the dipole model is controversial.

The macrodipole model was developed to estimate the effects of the parallel alignment of individual peptide dipoles (Arridge and Cannon, 1963; Brant and Flory, 1965) and the polarization of peptide groups (Wada, 1976; Hol *et al.*, 1978) in a helix. Calculations assuming a low dielectric constant indicate that the effective charges at the helix termini may be as high as ± 0.5 electron (Wada, 1976; Hol *et al.*, 1978). Sheridan and co-workers (1982) compared the calculated potential of an α helix modeled as a macrodipole and as an array of discrete charges. As expected, the models were essentially identical at distances that are large compared to the helix length. Long-range interactions of the helix dipole (such as those between secondary structural elements), however, may be very weak when the ends of the helices are in a high-dielectric medium (Rogers and Sternberg, 1984). The dipole model broke down near the helix termini (Sheridan *et al.*, 1982), resulting in large differences in the calculated potentials where binding sites are commonly located. In addition, the dipolar potential is cylindrically symmetric, whereas the calculated potential of the "all atom" model reflects the asymmetric positions of the amide and carbonyl groups in the helix.

A striking example of ion binding at helix termini is provided by the sulfate-binding protein (SBP) of *S. typhimurium* (Pflugrath and Quiocho, 1985). The SBP has a dissociation constant for sulfate of $\sim 10^{-6}$ and does not bind phosphate. X-ray crystallographic analysis showed that the bound sulfate ion is sequestered from bulk solvent without being neutralized by positively charged side chains. The SO_4^{2-} is coordinated by seven charge-stabilized hydrogen bonds near the N termini of three helices. Quiocho and co-workers (1987; Pflugrath and Quiocho, 1985) have rationalized this structural arrangement without reference to the helix macrodipole. The sulfate ion may be neutralized by the partial charges on the three amide hydrogens at each of the three helix termini.

It can also be argued that the sulfate ion is, in fact, not completely neutralized when it is bound to the protein. Seven charge-stabilized hydrogen bonds would be expected to provide much tighter binding than is observed. Lack of charge neutralization may provide the mechanism for dissipating some of the intrinsic binding energy of the hydrogen bonds. This would allow the protein to have both extreme specificity for sulfate and, at the same time, a dissociation constant that is high enough to allow the ion to be released for transport into the cell.

Determining the contributions of side-chain interactions and electrostatic effects to the stability of isolated helices are active areas of inquiry.

V. APPLICATIONS TO PREDICTING PROTEIN STRUCTURE

There is no shortage of difficulties in predicting protein conformations from amino acid sequence data. Extended sequence homology remains the best indication of structural similarity, but this approach is limited by the data base of known structures. The success of approaches based on building up and combining local structures is hampered by the difficulty of determining which long-range interactions are critical (Cohen *et al.*, 1986). Computer simulations of protein folding are limited by the lack of a quantitative understanding of stabilizing forces and by the problem of escaping local minima in potential energy (Levitt and Warshel, 1975; Lesk and Chothia, 1980).

These problems, at face value, would not seem to inspire optimism. However, even an empirical and semiquantitative understanding of protein stability can be used to develop stringent criteria for assessing the plausibility of predicted structures. Whatever the method of prediction, a proposed structure that does not contain sufficient stabilizing interactions could be identified and rejected.

The power of this capability was demonstrated by Cohen and Sternberg (1980), who used conformational criteria to identify the myoglobin fold among a set of $\sim 10^{32}$ randomly generated conformations. The criteria used to "filter out" implausible conformations included requirements for compact globular structure, correct helix content, correct helical boundaries, reasonable packing of connected helices, and, finally, a requirement that the distance between the proximal and distal histidines be within the value measured in solution. The root mean square deviation of the predicted structure from the x-ray crystal structure was 4.5 Å. The structures of human growth hormone (Cohen and Kuntz, 1987) and the α subunit of tryptophan synthase (Hurle *et al.*, 1987) have also been predicted using this general approach.

A number of additional conformational filters based on studies of protein stability and structure have been tested. It is worth noting that minimization of the calculated global potential energy has not provided a useful filter. Novotny, Bruccoleri, and Karplus (1984) interchanged the amino acid sequences of hemerythrin and the variable domain of an immunoglobulin light chain and compared the minimized energies of the correctly and incorrectly folded models. They found that small structural adjustments were sufficient to produce equivalent calculated potential energies, even though the two proteins have no sequence homology or structural similarity. However, a comparison of the total solvation energies of the models by Eisenberg and McLachlan (1986) showed that atom hydrophobicities could be used to distinguish the correct and incorrect folds. The correct folds also had lower solvation energies than models in which the chain was fully extended.

Residue hydrophobicity has proven to be a powerful filter in the method of Kyte and Doolittle (1982) for identifying potential transmembrane α helices. Filters that depend on structural arrangements include the method of Ponder and Richards (1987) to identify combinations of amino acids that fill space without steric overlap. Connolly (1986a,b) has developed algorithms for docking elements of structure based on packing criteria. The presence of unpaired hydrogen-bonding groups or unsolvated charges in the protein interior would also be expected to reduce the likelihood that a predicted structure is correct. The correlation between low mobility and sensitivity to destabilizing mutations suggests that it may be possible to identify correct structural elements by their behavior in a molecular dynamics simulation. The usefulness of these structural criteria will be inversely proportional to the computer time required to evaluate proposed conformations.

This approach is illustrated by a recent paper by Moult and James (1986), who combined several structural filters in an algorithm for predicting the conformations of loops of up to six residues. The criteria implemented include requirements for standard bond lengths and angles, reasonable main-chain torsion angles, side chains with little torsional strain, the absence of steric overlap, minimum electrostatic energy, and minimum exposed hydrophobic surface. The loop conformations predicted by the algorithm have root mean square deviations from their x-ray structures of about 1 Å.

This degree of accuracy is modest. It is not sufficient to support mechanistic conclusions or, as Moult and James point out, to serve as a basis for rational drug design. Only short sections of the polypeptide chain were considered, because randomly generated conformations were used as the starting point of the analysis. The approach, however, might also be used to analyze more complete predicted structures obtained by other means.

Improvements in the accuracy of the approach may depend on the development of new structural filters. Filters, in turn, will surely be refined with advances in understanding the structural basis of protein stability. Whatever the ultimate utility of this understanding may be, the use of physical methods to determine the properties of carefully chosen protein variants has clearly brought our appreciation of protein stability into a period of rapid change.

VI. SUMMARY

The thermodynamic stability of proteins is modest and depends on environmental conditions in a complex way. The magnitudes of specific interactions that favor the folded state, such as disulfide bonds, hydrophobic forces, hydrogen bonds, electrostatic interactions, and dispersion forces, have traditionally been estimated using simple chemical model systems. Recently a combination of genetic techniques, chemical modification, thermodynamic studies, and structural analysis has been used to measure the stabilizing contributions of specific interactions in proteins. Despite the difficulty of partitioning the effects of amino acid substitutions between the folded and unfolded states, these studies have led to several important conclusions:

1. Many different types of interactions, including those listed above, make quantitatively comparable contributions to stability.
2. Specific interactions of each type can make a wide range of contributions to stability.
3. The observed range of contributions is not adequately described by the behavior of simple chemical model systems. Model systems generally do not account for the unique environments of each residue in the folded and unfolded states or for the entropy changes associated with forming specific interactions.
4. The role of each amino acid depends on its structural context. Sensitivity to severe destabilizing mutations is qualitatively correlated with features of the folded state. With the exception of charged residues, critical interactions generally involve groups that are rigid or buried in the folded structure.
5. Many amino acid replacements do not have large effects on stability. Proteins tolerate substitutions because (a) many interactions do not make large contributions to stability and (b) protein structures adjust to compensate for changes in sequence.
6. Physical studies of protein variants are leading to rapid advances in understanding the essential features of stabilizing and destabilizing interactions. These phenomenological data on the structural basis of stability provide stringent tests for mechanistic theories of protein stability and for the plausibility of predicted protein structures.

ACKNOWLEDGMENTS. I am deeply grateful to Brian Matthews for my experiences in his laboratory working on the structural basis of the stability of phage T4 lysozyme. I would like to thank W. A. Baase, W. J. Becktel, J. Bell, B.-L. Chen, S. Cook, F. W. Dahlquist, D. Goldenberg, T. Gray, E. Haas, P. Kim, S. Marqusee, B. W. Matthews, C. R. Matthews, L. McIntosh, H. Nicholson, J. Nye, G. A. Petsko, C. Schellman, J. Schellman, G. Streisinger, D.-P. Sun, D. Tronrud, L. H. Weaver, K. Wilson, and J. A. Wozniak for many wonderful collaborations and discussions on the problem of protein stability and for communications of unpublished results. Wayne Becktel, Peter Kim, Lawrence McIntosh, Brian Matthews, Marty Rechsteiner, and Larry Weaver provided critical comments on this manuscript. This work was supported in part by grants from the National Institutes of Health (to B.W. Matthews) and the Helen Hay Whitney Foundation.

VII. REFERENCES

Abraham, D. J., and Leo, A. J., 1987, Extension of the fragment method to calculate amino acid zwitterion and side chain partition coefficients, *Proteins* **2:**130–152.

Ackers, G. K., and Smith, F. R., 1985, Effects of site-specific amino acid modification on protein interactions and biological function, *Annu. Rev. Biochem.* **54:**597–629.

Ahern, T. J., Casal, J. I., Petsko, J. A., and Klibanov, A. M., 1987, Control of oligomeric enzyme thermostability by protein engineering, *Proc. Natl. Acad. Sci. U.S.A.* **84:**675–679.

Alber, T., and Wozniak, J. A., 1985, A genetic screen for mutations that increase the thermal stability of phage T4 lysozyme, *Proc. Natl. Acad. Sci. U.S.A.* **82:**747–750.

Alber, T., Grütter, M. G., Gray, T. M., Wozniak, J. A., Weaver, L. H., Chen, B.-L., Baker, E. N., and Matthews, B. W., 1986, Structure and stability of mutant lysozymes from bacteriophage T4, in: *UCLA Symposia on Molecular and Cellular Biology, New Series,* Volume 39, *Protein Structure Folding and Design* (D. L. Oxender, ed.), Alan R. Liss, New York, pp. 307–318.

Alber, T., Sun, D.-P., Nye, J. A., Muchmore, D. C., and Matthews, B. W., 1987a, Temperature-sensitive mutations of bacteriophage T4 lysozyme occur at sites of low mobility and low solvent accessibility in the folded protein, *Biochemistry* **26:**3754–3758.

Alber, T., Sun, D.-P., Wilson, K., Wozniak, J. A., Cook, S. P., and Matthews, B. W., 1987b, Contributions of hydrogen bonds of Thr 157 to the thermodynamic stability of phage T4 lysozyme, *Nature* **330:**41–46.

Alber, T., Bell, J. A., Sun, D.-P., Nicholson, H., Wozniak, J. A., Cook, S. P., and Matthews, B. W., 1988, Replacements of Pro 86 in phage T4 lysozyme extend an α-helix but do not alter protein stability, *Science* **239:**631–635.

Amir, D., and Haas, E., 1987, Estimation of intramolecular distance distributions in bovine pancreatic trypsin inhibitor by site-specific labeling and nonradiative excitation energy-transfer measurements, *Biochemistry* **26:**2162–2175.

Arridge, G. C., and Cannon, C. G., 1963, Calculation of the CONH dipole contribution to lattice energies of amides, polyamides, and polypeptides, *Proc. R. Soc. Lond. [A]* **278:**91–109.

Aune, K. C., Salahuddin, A., Zarlengo, M. H., and Tanford, C., 1967, Evidence for residual structure in acid- and heat-denatured proteins, *J. Biol. Chem.* **242:**4486–4489.

Baker, E. N., and Hubbard, R. E., 1984, Hydrogen bonding in globular proteins, *Prog. Biophys. Mol. Biol.* **44:** 97–179.

Baldwin, R. L., 1986, Temperature dependence of the hydrophobic interaction in protein folding, *Proc. Natl. Acad. Sci. U.S.A.* **83:**8069–8072.

Baldwin, R. L., and Eisenberg, D., 1987, Protein stability, in: *Protein Engineering* (D. L. Oxender and C. F. Fox, eds.), Alan R. Liss, New York, pp. 127–148.

Barksdale, A. D., and Stuehr, J. E., 1971, Kinetics of the helix–coil transition in aqueous poly(-glutamic acid), *J. Am. Chem. Soc.* **94:**3334–3338.

Barlow, D. J., and Thornton, J. M., 1983, Ion-pairs in proteins, *J. Mol. Biol.* **168:**867–885.

Barlow, D. J., and Thornton, J. M., 1986, The distribution of charged groups in proteins, *Biopolymers* **25:** 1717–1733.

Beasty, A. M., Hurle, M. R., Manz, J. T., Stackhouse, T., Onuffer, J. J., and Matthews, C. R., 1986, Effects of the Phe 22 to Leu, Glu 49 to Met, Gly 234 to Asp, and Gly 234 to Lys mutations on the folding and stability of the α-subunit of tryptophan synthase from *Escherichia coli, Biochemistry* **25:**2965–2974.

Beasty, A. M., Hurle, M., Manz, J. T., Stackhouse, T., and Matthews, C. R., 1987, Mutagenesis as a probe of protein folding and stability in: *Protein Engineering* (D. L. Oxender and C. F. Fox, eds.), Alan R. Liss, New York, pp. 91–102.

Becktel, W. J., and Schellman, J. A., 1987, Protein stability curves, *Biopolymers* **26:**1859–1877.

Bello, J., 1977, Stability of protein conformation: Internal packing and enthalpy of fusion of model compounds, *J. Theor. Biol.* **68:**139–142.

Bello, J., 1978, Tight packing of protein cores and interfaces, *Int. J. Peptides* **12:**38–41.

Bierzynski, A., and Baldwin, R. L., 1982, Local secondary structure in ribonuclease A denatured by guanidine-Hcl near 1°C, *J. Mol. Biol.* **162:**173–186.

Bierzynski, A., Kim, P. S., and Baldwin, R. L., 1982, A salt bridge stabilizes the helix formed by isolated C-peptide of RNAse A, *Proc. Natl. Acad. Sci. U.S.A.* **79:**2470–2474.

Blagdon, D. E., and Goodman, M., 1975, Mechanisms of protein and polypeptide helix initiation, *Biopolymers* **14:**241–245.

Brandts, J. F., Oliveira, R. J., and Westort, C., 1970, Thermodynamics of protein denaturation. Effect of pressure on the denaturation of ribonuclease A, *Biochemistry* **9:**1038–1048.

Brant, D. A., and Flory, P. J., 1965, The role of dipole interactions in determining polypeptide configurations, *J. Am. Chem. Soc.* **87:**663–664.

Brown, J. E., and Klee, W. A., 1971, Helix–coil transition of the isolated amino terminus of ribonuclease, *Biochemistry* **10:**470–476.

Brown, L. R., DeMarco, A., Richarz, R., Wagner, G., and Wüthrich, K., 1978, The influence of a single salt bridge on static and dynamic features of the globular solution conformation of the bovine pancreatic trypsin inhibitor, *Eur. J. Biochem.* **88:**87–95.

Bryan, P. N., Rollence, M. L., Pantoliano, M. W., Wood, J., Finzel, B. C., Gilliland, G. L., Howard, A. J., and Poulos, T. L., 1986, Proteases of enhanced thermostability: Characterization of a thermostable variant of subtilisin, *Proteins* **1:**326–334.

Bundi, A., and Wüthrich, K., 1979, Use of amide ^{1}H-NMR titration shifts for studies of polypeptide conformation, *Biopolymers* **18:**299–311.

Burley, S. K., and Petsko, G. A., 1985, Aromatic–aromatic interaction: A mechanism of protein structure stabilization, *Science* **229:**23–28.

Burley, S. K., and Petsko, G. A., 1988, Weakly polar interactions in proteins, *Adv. Protein Chem.* **39:**125–189.

Chazin, W. J., Goldenberg, D. P., Creighton, T. E., and Wüthrich, K., 1985, Comparative studies of conformation and internal mobility in native and circular basic pancreatic trypsin inhibitor by ^{1}H nuclear magnetic resonance in solution, *Eur. J. Biochem.* **152:**429–437.

Chothia, C., 1975, Structural invariants in protein folding, *Nature* **254:**304–308.

Chothia, C., 1976, The nature of the accessible and buried surfaces in proteins, *J. Mol. Biol.* **105:**1–14.

Chothia, C. 1984, Principles that determine the structure of proteins, *Annu. Rev. Biochem.* **53:**537–572.

Chou, P. Y., and Fasman, G. D., 1974, Conformational parameters for amino acids in helical, β-sheet, and random coil regions calculated for proteins, *Biochemistry* **13:**211–222.

Cohen, F. E., and Kuntz, I. D., 1987, Prediction of the three-dimensional structure of human growth hormone, *Proteins* **2:**162–167.

Cohen, F. E., and Sternberg, M. J. E., 1980, On the use of chemically derived distance constraints in the prediction of protein structure with myoglobin as an example, *J. Mol. Biol.* **137:**9–22.

Cohen, F. E., Abarbanel, R. M., Kuntz, I. D., and Fletterick, R. J., 1986, Turn prediction in proteins using a pattern-matching approach, *Biochemistry* **25:**266–275.

Connolly, M. L., 1986a, Shape complementarity at the hemoglobin α-1-β-1 subunit interface, *Biopolymers* **25:** 1229–1247.

Connolly, M. L., 1986b, Atomic size packing defects in proteins, *Int. J. Peptides* **28:**360–363.

Creighton, T. E., 1983a, An empirical approach to protein conformation, stability and flexibility, *Biopolymers* **22:**49–58.

Creighton, T. E., 1983b, *Proteins,* W. H. Freeman, New York.

Creighton, T. E., 1985, The problem of how and why proteins adopt folded conformations, *J. Phys. Chem.* **89:** 2452–2459.

Creighton, T. E., and Goldenberg, D. P., 1984, Kinetic role of a metastable native-like two-disulphide species in the folding transition of bovine pancreatic trypsin inhibitor, *J. Mol. Biol.* **179:**497–526.

Cronin, C. N., Malcolm, B. A., and Kirsh, J. F., 1987, Reversal of substrate specificity by site directed mutagenesis of aspartate amino transferase, *J. Am. Chem. Soc.* **109:**2222–2223.

Dill, K. A., 1985, Theory for the folding and stability of globular proteins, *Biochemistry* **24:**1501–1509.

Dill, K. A., 1987, The stabilities of globular proteins, in: *Protein Engineering* (D. L. Oxender and C. F. Fox, eds.), Alan R. Liss, New York, pp. 187–192.

Dlott, D. D., Frauenfelder, H., Langer, P., Roder, H., and DiIorio, E. E., 1983, Nanosecond flash photolysis study of carbon monoxide binding to the β chain of hemoglobin Zurich [β63(E7)His to Arg], *Proc. Natl. Acad. Sci. U.S.A.* **80:**6239–6243.

Dyson, H. J., Cross, K. J., Houghten, R. A., Wilson, I. A., Wright, P. E., and Lerner, R. A., 1985, The immunodominant site of a synthetic immunogen has a conformational preference in water for a type-II reverse turn, *Nature* **318:**480–483.

Eisenberg, D., and McLachlan, 1986, Solvation energy in protein folding and binding, *Nature* **319:**199–203.

Evans, P. A., Dobson, C. M., Kautz, R. A., Hatfull, G., and Fox, R. O., 1987, Proline isomerism in staphylococcal nuclease characterized by NMR and site-directed mutagenesis, *Nature* **329:**266–268.

Fermi, G., and Perutz, M. F., 1981, *Haemoglobin and Myoglobin,* Clarenden Press, Oxford.

Fersht, A. R., 1971, Conformational equilibria and the salt bridge in chymotrypsin, *Cold Spring Harbor Symp. Quant. Biol.* **36:**71–73.

Fersht, A. R., 1987, The hydrogen bond in molecular recognition, *Trends Biochem. Sci.* **12:**301–304.

Fersht, A. R., Shi, J.-P., Knill-Jones, J., Lowe, D. M., Wilkinson, A. J., Blow, D. M., Brick, P., Carter, P., Waye, M. M. Y., and Winter, G., 1985, Hydrogen bonding and biological specificity analyzed by protein engineering, *Nature* **314:**235–238.

Flanagan, M. A., Garcia-Moreno, E. B., Friend, S. H., Feldmann, R. J., Scouloudi, H., and Gurd, F. R. N.,

1983, Contributions of individual amino acid residues to the structural stability of cetacean myoglobins, *Biochemistry* **22:**6027–6037.

Garcia-Moreno, E. G., Chen, L. X., March, K. L., Gurd, R. S., and Gurd, F. R. N., 1985, Electrostatic interactions in sperm whale myoglobin, *J. Biol. Chem.* **260:**14070–14082.

Gilson, M. K., and Honig, B. H., 1987, Calculation of electrostatic potentials in an enzyme active site, *Nature* **330:**84–86.

Gilson, M. K., Rashin, A., Fine, R., and Honig, B., 1985, On the calculation of electrostatic interactions in proteins, *J. Mol. Biol.* **183:**503–516.

Glushko, V., Lawson, P. J., and Gurd, F. R. N., 1972, Conformational states of bovine pancreatic ribonuclease A observed by normal and partially relaxed carbon-13 nuclear magnetic resonance, *J. Biol. Chem.* **247:** 3176–3185.

Go, M., and Miyazawa, S., 1980, Relationship between mutability, polarity and exteriority of amino acid residues in protein evolution, *Int. J. Peptides* **15:**211–224.

Go, N., 1975, Theory of reversible denaturation of globular proteins, *Int. J. Peptides* **7:**313–323.

Goldenberg, D. P., 1985, Dissecting the roles of individual interactions in protein stability: Lessons from a circularized protein, *J. Cell Biochem.* **29:**321–335.

Goldenberg, D. P., 1988, Genetic studies of protein stability and mechanisms of folding, *Annu. Rev. Biophys. Biophys. Chem.* **17:** 481–507.

Goldenberg, D. P., and Creighton, T. E., 1984, Folding pathway of a circular form of bovine pancreatic trypsin inhibitor, *J. Mol. Biol.* **179:**527–545.

Goldenberg, D. P., and Creighton, T. E., 1985, Energetics of protein structure and folding, *Biopolymers* **24:** 167–182.

Gray, T. M., and Matthews, B. W., 1987, Structural analysis of the temperature-sensitive mutant of bacteriophage T4 lysozyme, glycine 156 to aspartic acid, *J. Biol. Chem.* **262:**16858–16864.

Grütter, M. G., and Matthews, B. W., 1982, Amino acid substitutions far from the active site of bacteriophage T4 lysozyme reduce catalytic activity and suggest that the C-terminal lobe of the enzyme participates in substrate binding, *J. Mol. Biol.* **154:**525–535.

Grütter, M. G., Hawkes, R. B., and Matthews, B. W., 1979, Molecular basis of thermostability in the lysozyme of bacteriophage T4, *Nature* **277:**667–668.

Grütter, M. G., Weaver, L. H., Gray, T. M., and Matthews, B. W., 1983, Structure, function, and evolution of the lysozyme from bacteriophage T4 lysozyme, in: *Bacteriophage T4* (C. K. Matthews, E. M. Kutter, G. Mosig, and P. M. Berget, eds.), American Society for Microbiology, Washington, pp. 356–360.

Grütter, M. G., Gray, T. M., Weaver, L. H., Alber, T., Wilson, K., and Matthews, B. W., 1987, Structural studies of mutants of the lysozyme of bacteriophage T4: The temperature-sensitive mutant protein Thr 157 to Ile, *J. Mol. Biol.* **197:**315–329.

Haas, E., and Amir, D., 1987, BPTI has a compact structure when the disulfide bonds are reduced, *J. Cell. Biochem.* **11C:**214.

Hampsey, D. M., Das, G., and Sherman, F., 1986, Amino acid replacements in yeast iso-1-cytochrome c: Comparisons with the phylogenetic series and the tertiary structure of related cytochromes c, *J. Biol. Chem.* **261:**3259–3271.

Hawkes, R., Grütter, M. G., and Schellman, J., 1984, Thermodynamic stability and point mutations of bacteriophage T4 lysozyme, *J. Mol. Biol.* **175:**195–212.

Hecht, M. H., and Sauer, R. T., 1985, Phage λ repressor revertants, *J. Mol. Biol.* **186:**53–63.

Hecht, M. H., Nelson, H. C. M., and Sauer, R. T., 1983, Mutations in λ repressor's amino-terminal domain: Implications for protein stability and DNA binding, *Proc. Natl. Acad. Sci. U.S.A.* **80:**2676–2680.

Hecht, M. H., Sturtevant, J. M., and Sauer, R. T., 1984a, Effect of amino acid replacements on the thermal stability of the NH_2-terminal domain of phage λ repressor, *Proc. Nat. Acad. Sci. U.S.A.* **81:**5685–5689.

Hecht, M. H., Sturtevant, J. M., and Sauer, R. T., 1984b, Stabilization of λ repressor against thermal denaturation by site-directed Gly to Ala changes in α-helix three, *Proc. Nat. Acad. Sci. U.S.A.* **81:**5685–5689.

Hecht, M. H., Hehir, K. M., Nelson, H. C. M., Sturtevant, J. M., and Sauer, R., T., 1985, Increasing and decreasing protein stability: Effects of revertant substitutions on the thermal denaturation of phage λ repressor, *J. Cell. Biochem.* **29:**217–224.

Hendrix, J. D., and Welker, N. E., 1985, Isolation of a *Bacillus stearothermophilus* mutant exhibiting increased thermostability in its restriction endonuclease, *J. Bacteriol.* **162:**682–692.

Hol, W. G. J., van Duijnen, P. T., and Berendsen, H. J. C., 1978, The α-helix dipole and the properties of proteins, *Nature* **273:**443–446.

Hol, W. G. J., Halie, L. M., and Sander, C., 1981, Dipoles of the α-helix and β-sheet: Their role in protein folding, *Nature* **294**:532–536.

Howell, E. E., Villafranca, J. E., Warren, M. S., Oatley, S. J., and Kraut, J., 1986, Functional role of aspartic acid-27 in dihydrofolate reductase revealed by mutagenesis, *Science* **231**:1123–1128.

Hurle, M. R., and Matthews, C. R., 1987, Proline isomerization and the slow folding reactions of the α subunit of tryptophan synthase from *Escherichia coli, Biochim. Biophys. Acta* **913**:179–184.

Hurle, M. R., Tweedy, N. B., and Matthews, C. R., 1986, Synergism in folding of a double mutant of the α subunit of tryptophan synthase, *Biochemistry* **25**:6356–6360.

Hurle, M. R., Matthews, C. R., Cohen, F. E., Kuntz, I. D., Toumadje, A., and Johnson, W. C., Jr., 1987, Prediction of the tertiary structure of the α-subunit of tryptophan synthase, *Proteins* **2**:210–224.

Hvidt, A., 1975, A discussion of pressure–volume effects in aqueous protein solutions, *J. Theor. Biol.* **50**:245–252.

Ihara, S., Ooi, T., and Takahashi, S., 1982, Effects of salts on the nonequivalent stability of the α-helices of isomeric block copolypeptides, *Biopolymers* **21**:131–145.

Illuminati, G., and Mandolini, L., 1981, Ring closure reactions of bifunctional chain molecules, *Acc. Chem. Res.* **14**:95–102.

Imanaka, T., Shibazaki, M., and Takagi, M., 1986, A new way of enhancing the thermostability of proteins, *Nature* **324**:695–697.

Janin, J., Wodak, S., Levitt, M., and Maigret, B., 1978, Conformation of amino acid side-chains in proteins, *J. Mol. Biol.* **125**:357–386.

Johnson, R. E., Adams, P., and Rupley, J. A., 1978, Thermodynamics of protein crosslinks, *Biochemistry* **17**:1479–1484.

Kauzmann, W., 1959, Some factors in the interpretation of protein denaturation, *Adv. Protein Chem.* **14**:1–63.

Kendrew, J. C., Bodo, G., Dintzis, H. M., Parrish, R. G., Wyckoff, H., and Phillips, D. C., 1958, A three-dimensional model of the myoglobin molecule obtained by x-ray analysis, *Nature* **181**:662–666.

Kendrew, J. C., Dickerson, R. E., Strandberg, B. E., Hart, R. G., Davies, D. R., Phillips, D. C., and Shore, V. C., 1960, Structure of myoglobin, *Nature* **185**:422–437.

Kendrew, J. C., Watson, H. C., Strandberg, B. E., Dickerson, R. E., Phillips, D. C., and Shore, V. C., 1961, A partial determination by x-ray methods, and its correlation with chemical data, *Nature* **190**:666–670.

Kim, P. S., and Baldwin, R. L., 1982, Specific intermediates in the folding reactions of small proteins and the mechanism of protein folding, *Annu. Rev. Biochem.* **51**:459–489.

Kim, P. S., and Baldwin, R. L., 1984, A helix stop signal in the isolated S-peptide of ribonuclease A, *Nature* **307**:329–334.

Klapper, M. H., 1971, The nature of the protein interior, *Biochim. Biophys. Acta* **229**:557–566.

Klotz, I. M., and Franzen, J. S., 1962, Hydrogen bonds between model peptide groups in solution, *J. Am. Chem. Soc.* **84**:3461–3466.

Kyte, J., and Doolittle, R. F., 1982, A simple method for displaying the hydropathic character of a protein, *J. Mol. Biol.* **157**:105–132.

Labhart, A. M., 1982, Secondary structure in ribonuclease I. Equilibrium folding transitions seen by amide circular dichroism, *J. Mol. Biol.* **157**:331–355.

Lawrence, C., Auger, I., and Mannella, C., 1987, Distribution of accessible surface areas of amino acids in globular proteins, *Proteins* **2**:153–167.

Lee, B., 1985, The physical origin of the low solubility of nonpolar solutes in water, *Biopolymers* **24**:813–823.

Lesk, A. M., and Chothia, C., 1980, Solvent accessibility, protein surfaces, and protein folding, *Biophys. J.* **32**:35–37.

Levitt, M., and Warshel, A., 1975, Computer simulation of protein folding, *Nature* **253**:694–698.

Lewis, P. N., Go, N., Go, M., Kotelchuck, D., and Scheraga, H. A., 1970, Helix probability profiles of denatured proteins and their correlation with native structures, *Proc. Nat. Acad. Sci. U.S.A.* **65**:810–815.

Liao, H., McKenzie, T., and Hageman, R., 1986, Isolation of a thermostable enzyme variant by cloning and selection in a thermophile, *Proc. Natl. Acad. Sci. U.S.A.* **83**:576–580.

Lin, S. H., Konishi, Y., Denton, M. E., and Scheraga, H. A., 1984, Influence of an extrinsic crosslink on the folding pathway of ribonuclease A. Conformational and thermodynamic analysis of cross-linked (lysine^{7}-lysine^{41})-ribonuclease A, *Biochemistry* **23**:5504–5512.

Lin, S. H., Konishi, Y., Nall, B. T., and Scheraga, H. A., 1985, Influence of an extrinsic cross-link on the folding pathway of ribonuclease A. Kinetics of folding–unfolding, *Biochemistry* **24**:2680–2686.

Loftus, D., Gbenle, G. O., Kim, P. S., and Baldwin, R. L., 1986, Effects of denaturants on amide protein exchange rates: A test for structures in protein fragments and folding intermediates, *Biochemistry* **25**:1428–1436.

Luisi, B. F., and Nagai, K., 1986, Crystallographic analysis of mutant haemoglobins made in *Escherichia coli*, *Nature* **320:**555–556.

Macgregor, R. B., and Weber, G., 1986, Estimation of the polarity of the protein interior by optical spectroscopy, *Nature* **319:**70–73.

Marqusee, S., and Baldwin, R. L., 1987, Helix stabilization by Glu-Lys salt bridges in short peptides of *de novo* design, *Proc. Natl. Acad. Sci. U.S.A.* **84:**8898–8902.

Matsumura, M., and Aiba, S., 1985, Screening for thermostable mutant of kanamycin nucleotidyl transferase by the use of a transformation system for a thermophile, *Bacillus stearothermophilus, J. Biol. Chem.* **260:** 15298–15303.

Matsumura, M., Yasumura, S., and Aiba, S., 1986, Cumulative effect of intragenic amino acid replacements on the thermostability of a protein, *Nature* **323:**356–358.

Matsumura, M., Yahanda, S., and Aiba, S., 1988, Site-directed mutagenesis: Role of tyrosine 80 in thermal stabilization of kanamycin nucleotidyltransferase, *Eur. J. Biochem.* **171:**715–720.

Matthew, J. B., 1985, Electrostatic effects in proteins, *Annu. Rev. Biophys. Biophys. Chem.* **14:**387–417.

Matthews, B. W., 1987, Genetic and structural analysis of the protein stability problem, *Biochemistry* **26:**6885–6888.

Matthews, B. W., Nicholson, H., and Becktel, W. J., 1987, Enhanced protein thermostability from site-directed mutations that decrease the entropy of unfolding, *Proc. Natl. Acad. Sci. U.S.A.* **84:**6663–6667.

Matthews, C. R., and Hurle, M. R., 1987, Mutant sequences as probes of protein folding mechanisms, *Bioessays* **6:**254–257.

Matthews, C. R., and Westmoreland, D. G., 1975, Nuclear magnetic resonance studies of residual structure in thermally unfolded ribonuclease A, *Biochemistry* **14:**4532–4538.

Matthews, C. R., Crisanti, M. M., Gepner, G. L., Velicelebi, G., and Sturtevant, J. M., 1980, Effect of single amino acid substitutions on the thermal stability of the α subunit of tryptophan synthase, *Biochemistry* **19:** 1290–1293.

Matthews, C. R., Crisanti, M. M., Manz, J. T., and Gepner, G. L., 1983, Effect of a single amino acid substitution on the folding of the α subunit of tryptophan synthase, *Biochemistry* **22:**1445–1452.

Miller, J. H., 1984, Genetic studies of the *lac* repressor XII. Amino acid replacements in the DNA binding domain of the *Escherichia coli lac* repressor, *J. Mol. Biol.* **180:**205–212.

Miller, J. H., and Schmeissner, V., 1979, Genetic studies of the *lac* repressor X. Analysis of missense mutations in the *lacI* gene, *J. Mol. Biol.* **131:**223–248.

Miller, J. H., Coulondre, C., Hofer, M., Schmeissner, V., Sommer, H., Schmitz, A., and Lu, P., 1979, Genetic studies of the *lac* repressor IX. Generation of altered proteins by the suppression of nonsense mutations, *J. Mol. Biol.* **131:**191–222.

Miller, S., Lesk, A. M., Janin, J., and Chothia, C., 1987, The accessible surface area and stability of oligomeric proteins, *Nature* **328:**834–836.

Mitchinson, C., and Baldwin, R. L., 1986, The design and production of semisynthetic ribonucleases with increased thermostability by incorporation of S-peptide analogues with enhanced helical stability, *Proteins* **1:**23–33.

Moult, J., and James, M. N. G., 1986, An algorithm for determining the conformation of polypeptide segments in proteins by systematic search, *Proteins* **1:**146–163.

Mutter, M., 1977, Macrocyclization equilibria of polypeptides, *J. Am. Chem. Soc.* **99:**8307–8314.

Nagai, K., Luisi, B., Shih, D., Miyazaki, G., Imai, K., Poyart, C., DeYoung, A., Kwiatkowski, L., Noble, R. W., Lin, S.-H., and Yu, N.-T., 1987, Distal residues in the oxygen binding site of haemoglobin studied by protein engineering, *Nature* **329:**858–860.

Narayana, S. V. L., and Argos, P., 1984, Residue contacts in protein structures and implications for protein folding, *Int. J. Peptides* **24:**25–39.

Nemethy, G., Leach, S. J., and Scheraga, H. A., 1966, The influence of amino acid side chains on the free energy of helix–coil transitions, *J. Phys. Chem.* **70:**998–1004.

Novotny, J., Bruccoleri, R., and Karplus, M., 1984, An analysis of incorrectly folded protein models, *J. Mol. Biol.* **177:**787–818.

Nozaki, Y., and Tanford, C., 1971, The solubility of amino acids and two glycine peptides in aqueous ethanol and dioxane solutions, *J. Biol. Chem.* **246:**2211–2217.

Pabo, C. O., and Suchanek, E. G., 1986, Computer-aided model-building strategies for protein design, *Biochemistry* **25:**5987–5991.

Pace, C. N., 1975, The stability of globular proteins, *CRC Crit. Rev. Biochem.* **3:**1–43.

Page, M. I., 1984, The energetics and specificity of enzyme–substrate interactions, in: *The Chemistry of Enzyme Action* (M. I. Page, ed.), Elsevier, Amsterdam, pp. 1–54.

Page, M. I., and Jencks, W. P., 1971, Entropic contributions to rate accelerations in enzymic and intramolecular reactions and the chelate effect, *Proc. Natl. Acad. Sci. U.S.A.* **68:**1678–1683.

Pakula, A. A., Young, V. B., and Sauer, R. T., 1986, Bacteriophage λ *cro* mutations: Effects on activity and intracellular degradation, *Proc. Natl. Acad. Sci. U.S.A.* **83:**8829–8833.

Pantoliano, M. W., Ladner, R. C., Bryan, P. N., Rollence, M. L., Wood, J. F., and Poulos, T. L., 1987, Protein engineering of subtilisin BPN′: Enhanced stabilization through the introduction of two cysteines to form a disulfide bond, *Biochemistry* **26:**2077–2082.

Perry, K. M., Onuffer, J. J., Touchette, N. A., Herndon, C. S., Gittelman, M. S., Matthews, C. R., Chan, J.-T., Mayer, R. J., Taira, K., Benkovic, S. J., Howell, E. E., and Kraut, J., 1987, Effect of single amino acid replacements on the folding and stability of dihydrofolate reductase from *Escherichia coli, Biochemistry* **26:**2674–2682.

Perry, L. J., and Wetzel, R., 1984, Disulfide bond engineered into T4 lysozyme: Stabilization of the protein toward thermal inactivation, *Science* **226:**555–557.

Perutz, M. F., 1964, The hemoglobin molecule, *Sci. Am.* **211**(5):64–76.

Perutz, M. F., 1978, Electrostatic effects in proteins, *Science* **201:**1187–1191.

Perutz, M. F., and Lehmann, H., 1968, Molecular pathology of human haemoglobin, *Nature* **219:**902–909.

Perutz, M. F., and Raidt, H., 1975, Stereochemical basis of heat stability in bacterial ferredoxins and in haemoglobin A_2, *Nature* **255:**256–259.

Perutz, M. F., Kendrew, J. C., and Watson, H. C., 1965, Structure and function of haemoglobin 2. Some relations between polypeptide chain configuration and amino acid sequence, *J. Mol. Biol.* **13:**669–678.

Petsko, G. A., and Ringe, D., 1984, Fluctuations in protein structure from X-ray diffraction, *Annu. Rev. Biophys. Bioeng.* **13:**331–371.

Pflugrath, J. W., and Quiocho, F. A., 1985, Sulphate sequestered in the sulphate-binding protein of *Salmonella typhimurium* is bound solely by hydrogen bonds, *Nature* **341:**257–260.

Ponder, J. W., and Richards, 1987, Tertiary templates for proteins, *J. Mol. Biol.* **193:**775–791.

Privalov, P. L., 1979, Stability of proteins, *Adv. Protein Chem.* **33:**167–241.

Privalov, P. L., Griko, Y. V., Venyaminov, S. Y., and Kutyshenko, V. P., 1986, Cold denaturation of myoglobin, *J. Mol. Biol.* **190:**487–498.

Quiocho, F. A., Sack, J. S., and Vyas, N. K., 1987, Stabilization of charges on isolated ionic groups sequestered in proteins by polarized peptide units, *Nature* **329:**561–564.

Rashin, A. A., and Honig, B., 1984, On the environment of ionizable groups in globular proteins, *J. Mol. Biol.* **173:**515–521.

Rebek, J., Jr., 1987, Model studies in molecular recognition, *Science* **235:**1478–1484.

Richards, F. M., 1977, Areas, volumes, packing, and protein structure, *Ann. Rev. Biophys. Bioeng.* **6:**151–176.

Richardson, J. S., 1981, The anatomy and taxonomy of protein structure, *Adv. Protein Chem.* **34:**167–339.

Rico, M., Santoro, J., Bermejo, F. J., Herranz, J., Nieto, J. L., Gallego, E., and Jimenez, M. A., 1986, Thermodynamic parameters for the helix–coil thermal transition of ribonuclease S-peptide and derivatives from ^{1}H-NMR data, *Biopolymers* **25:**1031–1053.

Roder, H., Wagner, G., and Wüthrich, K., 1985, Individual amide proton exchange rates in thermally unfolded basic pancreatic trypsin inhibitor, *Biochemistry* **24:**7407–7411.

Rogers, N. K., and Sternberg, M. J. E., 1984, Electrostatic interactions in globular proteins: Different dielectric models applied to the packing of α-helices, *J. Mol. Biol.* **174:**527–542.

Rose, G. D., Geselowitz, A. R., Lesser, G. J., Lee, R. H., and Zehfus, M. H., 1985, Hydrophobicity of amino-acid residues in globular proteins, *Science* **229:**834–838.

Russell, S. T., and Warshel, A., 1985, Calculations of electrostatic energies in proteins, *J. Mol. Biol.* **185:**389–404.

Sauer, R. T., Hehir, K., Stearman, R. S., Weiss, M. A., Jeitler-Nilsson, A., Suchanek, E. G., and Pabo, C. O., 1986, An engineered intersubunit disulfide enhances the stability and DNA binding of the N-terminal domain of λ repressor, *Biochemistry* **25:**5992–5998,

Schellman, J. A., 1955a, The thermodynamics of urea solutions and the heat of formation of the peptide hydrogen bonds, *C. R. Lab. Carlsberg Ser. Chim.* **29:**223–229.

Schellman, J. A., 1955b, The stability of hydrogen-bonded peptide structures in aqueous solution, *C. R. Lab. Carlsberg Ser. Chim.* **29:**230–259.

Schellman, J. A., 1978, Solvent denaturation, *Biopolymers* **17:**1305–1322.

Schellman, J. A., 1987a, The thermodynamic stability of proteins, *Annu. Rev. Biophys. Chem.* **16:**115–137.

Schellman, J. A., 1987b, Selective binding and solvent denaturation, *Biopolymers* **26:**549–559.

Schellman, J. A., Lindorfer, M., Hawkes, R., and Grütter, M., 1981, Mutations and protein stability, *Biopolymers* **20:**1989–1999.

Scheraga, H. A., 1978, Use of random copolymers to determine the helix–coil stability constants of the naturally occurring amino acids, *Pure Appl. Chem.* **50:**315–324.

Scheraga, H. A., 1985, Effect of side chain–backbone electrostatic interactions on the stability of α-helices, *Proc. Natl. Acad. Sci. U.S.A.* **82:**5585–5587.

Sheridan, R. P., Levy, R. M., and Salemme, F. R., 1982, α-Helix dipole model and electrostatic stabilization of 4-α-helical proteins, *Proc. Natl. Acad. Sci. U.S.A.* **79:**4545–4549.

Shoemaker, K. R., Kim, P. S., Brems, D. N., Marqusee, S., York, E. J., Chaiken, I. M., Stewart, J. M., and Baldwin, R. L., 1985, Nature of the charged group effect on the stability of the C-peptide helix, *Proc. Natl. Acad. Sci. U.S.A.* **82:**2349–2353.

Shoemaker, K. R., Kim, P. S., York, E. J., Stewart, J. M., and Baldwin, R. L., 1987, Tests of the helix dipole model for stabilization of α-helices, *Nature* **326:**563–567.

Shortle, D., and Lin, B., 1985, Genetic analysis of staphylococcal nuclease: Identification of three intragenic ''global'' suppressors of nucelase-minus mutations, *Genetics* **110:**539–555.

Shortle, D., and Meeker, A. K., 1986, Mutant forms of staphylococcal nuclease with altered patterns of guanidine hydrochloride and urea denaturation, *Proteins* **1:**81–89.

Smith, J. L., Hendrickson, W. A., Honzatko, R. B., and Sheriff, S., 1986, Structural heterogeneity in protein crystals, *Biochemistry* **25:**5018–5027.

Snow, M. E., and Amzel, L. M., 1986, Calculating three-dimensional changes in protein structure due to amino acid substitutions: The variable region of immunoglobulins, *Proteins* **1:**267–279.

Springs, B., and Haake, P., 1977, Equilibrium constants for association of guanidinium and ammonium ions with oxyanions, *Bioorg. Chem.* **6:**181–190.

Stahl, N., and Jencks, W. P., 1986, Hydrogen bonding between solutes in aqueous solution, *J. Am. Chem. Soc.* **108:**4196–4205.

States, D. J., Creighton, T. E., Dobson, C. M., and Karplus, M., 1987, Conformations of intermediates in the folding of the pancreatic trypsin inhibitor, *J. Mol. Biol.* **195:**731–739.

Sternberg, M. J. E., Hayes, F. R. F., Russell, A. J., Thomas, P. G., and Fersht, A. R., 1987, Prediction of electrostatic effects of engineering of protein charges, *Nature* **330:**86–88.

Streisinger, G., Mukai, F., Dreyer, W. J., Miller, B., and Horiuchi, S., 1961, Mutations affecting the lysozyme of phage T4, *Cold Spring Harbor Symp. Quant. Biol.* **26:**25–30.

Sturtevant, J. M., 1977, Heat capacity and entropy changes in processes involving proteins, *Proc. Natl. Acad. Sci. U.S.A.* **74:**2236–2240.

Sturtevant, J. M., 1987, Biochemical applications of differential scanning calorimetry, *Annu. Rev. Phys. Chem.* **38:**463–488.

Sueki, M., Lee, S., Powers, S. P., Denton, J. B., Konishi, Y., and Scheraga, H. A., 1984, Helix–coil stability constants for the naturally occurring amino acids in water, *Macromolecules* **17:**148–155.

Susi, H., Timasheff, S. N., and Ard, J. S., 1964, Near infrared investigation of interamide hydrogen bonding in aqueous solution, *J. Biol. Chem.* **239:**3051–3054.

Svensson, L. A., Sjölin, L., Gilliland, G. L., Finzel, B. C., and Wlodawer, A., 1986, Multiple conformations of amino acid residues in ribonuclease A, *Proteins* **1:**370–375.

Tanford, C., 1954, The association of acetate with ammonium and guanidinium ions, *J. Am. Chem. Soc.* **76:** 945–946.

Tanford, C., 1962, Contribution of hydrophobic interactions to the stability of the globular conformation of proteins, *J. Am. Chem. Soc.* **84:**4240–4247.

Tanford, C., 1968, Protein denaturation, *Adv. Protein Chem.* **23:**121–282.

Tanford, C., 1970, Protein denaturation Part C. Theoretical models for the mechanism of denaturation, *Adv. Protein Chem.* **24:**1–95.

Tanford, C., 1980, *The Hydrophobic Effect: Formation of Micelles and Biological Membranes,* John Wiley & Sons, New York.

Thomas, P. G., Russell, A. J., and Fersht, A. R., 1985, Tailoring the pH dependence of enzyme catalysis using protein engineering, *Nature* **318:**375–376.

Thornton, J. M., 1982, Electrostatic interactions in proteins, *Nature* **295:**13–14.

Ueda, Y., and Go, N., 1976, Theory of large-amplitude conformational fluctuations in native globular proteins, *Int. J. Peptides* **8:**551–558.

Villafranca, J. E., Howell, E. E., Oatley, S. J., Xuong, N.-H., and Kraut, J., 1987, An engineered disulfide bond in dihydrofolate reductase, *Biochemistry* **26:**2182–2189.

Wada, A., 1976, The α-helix as an electric macrodipole, *Adv. Biophys.* **9:**1–63.

Wada, A., and Nakamura, H., 1981, Nature of the charge distribution in proteins, *Nature* **293:**757–758.

Wagner, G., Kalb, A. J., and Wüthrich, K., 1979, Conformational studies by ^{1}H nuclear magnetic resonance of the basic pancreatic trypsin inhibitor after reduction of the disulfide bond between Cys-14 and Cys-38, *Eur. J. Biochem.* **95:**249–253.

Warshel, A., 1987, What about protein polarity? *Nature* **330:**15–16.

Warshel, A., and Russell, S. T., 1984, Calculations of electrostatic interactions in biological systems and in solutions, *Q. Rev. Biophys.* **17:**282–422.

Weber, P. C., Sheriff, S., Ohlendorf, D. H., Finzel, B. C., and Salemme, F. R., 1985, The 2Å resolution structure of a thermostable ribonuclease A chemically cross-linked between lysine residues 7 and 41, *Proc. Natl. Acad. Sci. U.S.A.* **82:**8473–8477.

Wells, J. A., and Powers, D. B., 1986, *In vivo* formation and stability of engineered disulfide bonds in subtilisin, *J. Biol. Chem.* **261:**6564–6570.

Wetzel, R., Perry, L. J., Baase, W. A., and Becktel, W. J., 1988, Disulfide bonds and thermal stability in T4 lysozyme, *Proc. Natl. Acad. Sci. U.S.A.* **85:**401–405.

Wishnia, A., 1963, The hydrophobic contribution to micelle formation: The solubility of ethane, propane, butane, and pentane in sodium dodecyl sulfate solution, *J. Phys. Chem.* **67:**2079–2082.

Wolfenden, R., Andersson, L., Cullis, P. M., and Southgate, C. C. B., 1981, Affinities of amino acid side chains for solvent water, *Biochemistry* **20:**849–855.

Yutani, K., Ogasahara, K., Aoki, K., Kakuno, T., and Sugino, Y., 1984, Effect of amino acid residues on conformational stability in eight mutant proteins variously substituted at a unique position of the trp synthase α-subunit, *J. Biol. Chem.* **259:**14076–14081.

Yutani, K., Ogasahara, K., Tsujita, T., and Sugino, T., 1987, Dependence of conformational stability on hydrophobicity of the amino acid residue in a series of variant proteins substituted at a unique position of tryptophan synthase α subunit, *Proc. Natl. Acad. Sci. U.S.A.* **84:**4441–4444.

Zimm, B. H., and Bragg, J. K., 1959, Theory of the phase transition between helix and random coil in polypeptide chains, *J. Chem. Phys.* **31:**526–535.

Zipp, A., and Kauzmann, W., 1973, Pressure denaturation of metmyoglobin, *Biochemistry* **12:**4217–4228.

6

The Development of the Prediction of Protein Structure

Gerald D. Fasman

Gerald D. Fasman • Graduate Department of Biochemistry, Brandeis University, Waltham, Massachusetts 02254.

I. INTRODUCTION

The tenet of structural biology that function follows form had its seeds in the monograph by C. B. Anfinsen, *The Molecular Basis of Evolution* (Anfinsen, 1959), wherein he stated "Protein chemists naturally feel that the most likely approach to the understanding of cellular behavior lies in the study of structure and function of protein molecules." The achievement of protein crystallography over the past 30 years has confirmed this view whereby the description of the structure and function of proteins is now frequently understood at the atomic level.

The classical experiments of Anfinsen and co-workers (Anfinsen *et al.*, 1961; Anfinsen, 1973) proved that ribonuclease could be denatured and refolded without loss of enzymatic activity. This implied that the amino acid sequence contains sufficient information to define the three-dimensional structure of a protein in a particular environment. The acceptance of this tenet has led to multifarious efforts to predict the conformation of proteins based only on the consideration of sequence alone, which has been termed the protein-folding problem. Both theoreticians (e.g., Levitt and Warshel, 1975; Nemethy and Scheraga, 1977; Karplus and Weaver, 1979; Schulz and Schirmer, 1979; Sternberg, 1983) and experimentalists (e.g., Shoemaker *et al.*, 1987; Creighton, 1978) have tackled the chain-folding problem with very limited success.

It is thought that the native structure of a protein will lie near the minimum of free energy; however, it will not fold by sampling every possible conformation (Levinthal, 1968; Wetlaufer, 1973), but there will be one or more pathways along which the protein folds (Creighton, 1979). Anfinsen (1973) had proposed that one or more regions of secondary structure, e.g., α helices or a two-stranded antiparallel β sheet, having marginal stability would act as nucleation sites and direct the refolding.

The advent of recombinant DNA techniques has led to an explosion of information concerning the sequences of receptors and enzymes that will be important for drug, herbicide, and pesticide design. Technological developments of industrial, clinical, and agricultural importance may be achieved in the coming years by imitation of the interactions between macromolecules and ligands that occur naturally in the living cell (Blundell *et al.*, 1987). Engineered hormones may have more desirable properties than their native counterparts in terms of stability or activity. All these promising prospects have fired the desire to predict the conformation of and design protein molecules with a high degree of sophistication.

The theoretical efforts could be categorized into three main areas: energetic, heuristic, and statistical. The school following the assumption that a protein folds so as to minimize the free energy of the system has had many contributors (e.g., Levitt and Warshel, 1975; Nemethy and Scheraga, 1977; McCammon *et al.*, 1977; Weiner *et al.*, 1984). These researchers have developed potential functions to describe the energy surface of a polypeptide chain. Chain folding is simulated computationally, directed by surface gradients to find the energy minimum. Alternatively, conformation space is probed from a starting point by integrating the equations of motion over time. These theoretical predictions of structure have been influenced by both thermodynamic and kinetic arguments. The thermodynamic properties of the polypeptide chain were the first to be considered. Liquori and co-workers (DeSantis *et al.*, 1965) and Ramachandran *et al.* (1963) first demonstrated that the peptide unit can adopt only certain allowed conformations. They constructed ϕ, ψ plots, which were subsequently improved by semiempirical energy calculations (Lewis *et al.*, 1973b), to predict preferred regions for the various secondary structures of polypeptides. Recent energy calculations have been refined, either by choosing parameters to fit crystal structures (Hagler *et al.*, 1974) or by performing complex molecular orbital calculations (Pullman and Pullman, 1974). On the basis of the thermodynamic hypothesis, by calculating the total free energy of a protein and finding the global minimum, it should be possible to predict the native structure. However, minimization schemes have failed to predict chain folding accurately (Hagler and Honig, 1978; Cohen and Sternberg, 1980a,b). The lack of success presumably stems from the difficulties in modeling protein–solvent interactions, the use of analytical functions to approximate the chemical potential, and the compounding of these errors in the computed gradient; in addition, the energy surface has multiple minima, which make it nearly impossible to locate a global minimum. Attempts at solving these problems are underway. Molecular dynamics offers solutions to these problems but remains computationally limited as a technique for studying chain folding (McCammon *et al.*, 1977). Chain folding is thought to take place in the millisecond time scale (Baldwin, 1980). Elaborate computing resources must be applied to sample 100 nsec in the life of a small protein (Post *et al.*, 1986).

One of the first attempts to predict the conformation of a protein, ribonuclease, was that of Scheraga (1960). This model used the available chemical information, deuterium–hydrogen exchange data, and the knowledge of the α helix. It later was shown to have little similarity to the native molecule.

The prediction of the secondary structure of protein conformation was pioneered by Guzzo (1965). By analysis of the known sequences and structures of myoglobin and the α- and β-hemoglobins, he found that groups of amino acids were helix disrupters. He also emphasized the role of hydrophobic interactions. He noted that Blout (1962) had pointed out that the known poly(α-amino acids) fall into two categories, helix formers and breakers. However, earlier, Davies (1964) had detected a correlation between amino acid composition and protein structure. By using optical properties for 15 proteins and their known sequences, it was possible to compare estimates of helicity with amino acid composition. Other early researchers in this area included Prothero (1966, 1968), Cook (1967), Periti *et al.* (1967), and Low *et al.* (1968), who all examined the few x-ray structures available and attempted to ferret out patterns of recognition. Schiffer and Edmundson (1967) adopted an innovative way of locating hydrophobic arcs by the "helical wheel" method. Dunhill (1968) slightly improved this approach by constructing helical net diagrams to locate hydrophobic clusters. Ptitsyn (1969) made a statistical analysis of the distribution of different amino acid residues among helical and nonhelical regions of seven globular proteins: myoglobin, α- and β-hemoglobin, lysozyme, ribonuclease, α-chymotrypsin, and papain. It was found that the distribution of a number of amino acid residues differed essentially from the distribution averaged over all amino acids. Hydrophobic amino acids, Ala and Leu, showed a tendency to occupy internal turns of helical regions. Negatively and positively charged amino acids have a tendency to concentrate, correspondingly, at the amino and carboxyl ends of helical regions. Amino acids with heteroatoms (O or S atoms) near the main chain (Ser, Thr, Cys, Asn) have a tendency to concentrate in nonhelical regions (including the regions with β structure). Proline can be located either on the amino ends of helical regions or in nonhelical regions. On examination it was found that Guzzo's (1965) empirical rules concerning the relationship between the amino acid composition of the given regions of the polypeptide chain and its structure was in disagreement with the results of the statistical analysis, whereas Prothero's (1966) rule agreed with them. Lewis *et al.* (1970, 1971), using the Zimm–Bragg (1959) helix–coil theory, suggested that better predictions may result when σ and s parameters for all 20 amino acids were experimentally determined.

Kotelchuck and Scheraga (1968, 1969) were among the first to attempt energy-minimization calculations to test the hypothesis that short-range interactions seem to play an important role in helix formation. Calculations were carried out to obtain the energy of interactions of individual amino acids along the backbone. A series of papers followed (see review by Scheraga, 1985), but these energy-minimization schemes have failed to predict chain folding accurately (Hagler and Honig, 1978; Cohen and Sternberg, 1980a,b). The reasons for this have been discussed above.

There have been between 15 and 20 proposals for the prediction of the secondary structure of protein conformation from the amino acid sequence. The number of methods counted depends on the degree of difference one defines between methods.

To predict the various secondary structures first requires a precise definition of the various secondary structures. Various criteria have been used by x-ray crystallographers, and thus secondary structures are delineated in different ways in various publications (see Blundell and Johnson, 1976). The crystallographers' assignments of secondary structure in the Brookhaven Protein Data Bank are often subjective and often incomplete. In an attempt to overcome this problem, Levitt and Greer (1977) described a computer program to analyze automatically and objectively the atomic coordinates of a large number of globular proteins (62) in order to identify the regions of α-helix, β-sheet, and reverse-turn secondary structure. The most successful criterion was based on the patterns of peptide hydrogen bonds, inter-C^{α} distances, and inter-C^{α} torsion angles. A more recent compilation of secondary structures has been published by Kabsch and Sander (1983b). This is a pattern-recognition process of hydrogen-bonded and geometrical features extracted from x-ray coordinates. Cooperative secondary structure is recognized as repeats of the elementary hydrogen-bonding patterns "turn" and "bridge." Repeating turns are "helices," repeating bridges are "ladders," and connecting ladders are "sheets." Geometric structure is defined in terms of the concepts of torsion and curvature of differential geometry. Local chain "chirality" is the torsional handedness of four consecutive C^{α} position and is positive for right-handed helices and negative for ideal twisted β sheets. Curved pieces are defined as bends. Solvent exposure is given as the number of water molecules in possible contact with residues. This dictionary is available in computer-readable form for protein structure prediction work.

II. PROTEIN TOPOLOGY

Proteins may be classified into three main categories: (1) globular proteins, (2) membrane proteins, and (3) fibrous proteins. Although fibrous proteins play very important roles in the anatomy and physiology of animals, they are not considered in this chapter.

The globular proteins pack tightly to form distinct tertiary conformations, usually producing a hydrophobic core. The packing involves the association of the various secondary structures, α helices and β strands with intervening β turns, into various domains. Often disulfide cross links between cysteine residues further stabilize the structure. These domains are formed from compact, contiguous chain structures (Wetlaufer, 1973; Liljas and Rossmann, 1974). These units have been proposed as intermediates in the folding process, as functional units within the native structure (Richardson, 1981; Rossmann and Argos, 1981; Wetlaufer, 1981; Kim and Baldwin, 1982), and possibly even as fundamental genetic units (Gilbert, 1978; Blake, 1979). Schulz (1977) outlined a conceptual, partially hierarchic framework that emphasized the role of domains within the known protein structures. Levitt and Chothia (1976) had earlier observed that assemblies of a few adjacent secondary structural units often pack and serve as structural building blocks. Lewis *et al.* (1971) and Kuntz (1972), in defining turns, laid the foundation for the concept of domains. Turns constitute recognizable structural units in proteins and are situated at the solvent-accessible surface of the molecule.

The domains have been classified into five classes according to their secondary structural arrangement (Levitt and Clothia, 1976; Richardson, 1981). These classes are (1) all α, with α helices and no β strands; (2) all β, with only β strands and no α helices; (3) α/β, in which the chain alternates between α helices and β strands; (4) α and β, in which the α and β regions tend to segregate into separate regions; and (5) coil, in which there is little or no regular secondary structure. The excellent review article by Richardson (1981) is amply illustrated with schematic diagrams, which have made the visualization of protein structure more aesthetic and a much simpler task. Levitt and Greer (1977) developed a computer program to analyze, automatically and objectively, the atomic coordinates of a large number of globular proteins in order to identify regions of α helix, β sheets, and β turns (reverse turns). The pattern of hydrogen bonds, inter-C^{α} distances, and inter-C^{α} torsion angles was used to find the secondary structure. Rose (1979) devised an automatic procedure for the identification of domains in globular proteins. The known protein structures were shown to be iteratively subdivisible into a hierarchy of disjunct contiguous-chain regions ranging in size from whole-protein monomers down to individual helices and strands. This analysis was later expanded as a control study to a set of 1000 simulated chain folds, and their organization was found to be similar to that of authentic molecules (Yuschok and Rose, 1983). The hierarchic nature of proteins was also observed by Crippen (1978) using a different algorithm. Rashin (1981) has proposed an algorithm to calculate domains defined as globular fragments, and Wodak and Janin (1980, 1981) have also shown that the presence of domains is easily detected by an automatic procedure based on surface areas only. More recently, Kabsch and Sander (1983b) described a program similar to Levitt and Greer's (1977), a pattern-recognition process of hydrogen-bonded and geometrical features plus solvent exposure of each residue, to produce a *Dictionary of Protein Secondary Structure* for 62 different globular proteins.

In addition to the important hydrophobic interactions and the hydrogen-bonding networks, there are several other interactions that play a role in protein conformation. The role that electrostatic forces play in the structure of proteins has been frequently discussed (see Chapter 8 by Rogers). Paul (1982) presented an interesting proposition that it is the electrostatic interactions that have a major influence on the low-resolution features of protein tertiary structure. Barlow and Thornton (1983) examined the role of ion pairs in proteins. Their "working definition" for an ion pair has been derived on the basis of an analysis of the distance distributions for like- and oppositely charged groups in 38 proteins. Ion pairs defined according to this criterion (≤4 Å between charged groups) have been analyzed in respect to (1) the frequencies of different pair types, (2) the residue separations and secondary structural locations of the residues involved, (3) the flexibility of the side chains involved, (4) their conformation, (5) their environmental accessibility to solvent and proximity to active site or ligand-binding regions, and (6) their conservation in related proteins. On the average, one third of the charged residues in a protein are involved in ion pairs, and 76% of these are concerned with stabilizing the tertiary structure. Only 17% of ion pairs are buried.

Rashin and Honig (1984) examined the environment of ionizable groups in 36 proteins and characterized them in terms of solvent accessibility, salt-bridge formation, and hydrogen bonding. An interesting finding was that there are on the average two completely buried ionizable groups per protein, of which at least 20% do not form salt bridges. However, all buried ionizable groups form hydrogen bonds with neutral polar groups. In a recent paper Sundaralingam *et al.*(1987) surveyed 47 globular proteins to determine the probability of occurrence of ion pairs separated by various numbers of residues in α helices. Ion pairs of the type $i,i + 3$ and $i,i + 4$ were the most predominant. The normalized frequencies of occurrence of ion pairs were also found to increase generally with helix length. These results indicate that ion pairs may contribute to the stability of the solvent-exposed α helices.

Sawyer and James (1982) reported finding carboxyl–carboxylate interactions important in stabilizing the structures of protein crystals and multisubunit complexes at low pH. These were found in protease A and protease B found in *Streptomyses griseus* and in penicillopepsin. Leszczynski and Rose (1986) introduced the proposition that loops in globular proteins constitute an additional secondary structure. These so-called "omega" loops were usually previously classified as "random coil." The segment length of the omega loop must be between six and 16 residues, have their backbone groups packed closely together, have a distance between segment termini of less than 10 Å, and may not exceed two thirds the maximum distance between any two α carbons within the segment under consideration. The frequencies of amino acid residues in loops were calculated and normalized in 67 proteins analyzed. The overall conformation was distributed as follows: 26% in the α helix, 19% in the β, 26% in turns, and 21% in loops. Thus, a loop may be described as a continuous-chain segment that adopts a "loop-shaped" conformation in three-dimensional space with a small distance between its termini. Morgan and McAdon (1980) reported finding that sulfur–aromatic interactions in globular proteins are not random events and are promoted in the presence of positively charged side chains. It has been pointed out (Burley and Petsko, 1985; Singh and Thornton, 1985) that aromatic–aromatic interactions frequently occur in proteins and that about 60% of the aromatic side chains are involved in such pairs, 80% of which form networks of three or more interacting side chains. Phenylalanine–phenylalanine interactions occur frequently with dihedral angles of 90° to each other. They stabilize tertiary structure (80%) and quaternary structure (20%). Amino–aromatic interactions also occur frequently, with the positively charged or δ(+) amino groups of Lys, Arg, Asn, Gln, and His preferentially located within 6 Å of the ring centroids of Phe, Tyr, and Try, where they make van der Waals contact with the δ(−) π electrons and avoid the δ(+) ring edge (Burley and Petsko, 1986).

Blundell *et al.* (1986b) comment on the paper of Burley and Petsko (1986) concerning the distribution of dihedral angles between two aromatic residues in globular proteins. It can provide evidence of specific interactions between aromatic rings only if it differs significantly from a random distribution. The expected distribution can be considered in terms of the angles between the normals to the two ring planes. Singh and Thornton (1985) have pointed out that the distribution of this angle that would arise by chance varies as the sine of the angle and has a mean value of ~57°. The overall distribution in the paper of Burley and Petsko (1986) thus closely approximates a random distribution. However, it is also necessary to consider the spatial displacement of the two aromatic rings between the ring centers. When this is examined, it shows significant deviations from random arrangement. Thus, the importance of taking into consideration the available three-dimensional space and expected "random" distribution for such side-chain interactions is stressed. A striking preference for perpendicular packing of aromatic rings is observed for a small subgroup in a special spatial displacement.

III. TECHNIQUES OF PROTEIN PREDICTION

The approach termed "knowledge-based" prediction (Blundell *et al.*, 1987) depends on analogies between a protein that is to be modeled and other proteins of known three-dimensional structure at all levels in the hierarchy of protein organization: secondary structure, motifs, domains, and quaternary or ligand interactions.

The first requirement is to have available the amino acid sequence. Protein sequences are collected and made available from the Protein Information Resource (PIR) databank at the National Biomedical Research Foundation, Maryland (Georgetown University Medical Center, 3900 Reservoir Road, N.W., Washington, DC 20007), where over 7000 entries are available (April, 1988, with about 1000 new entries/month), or the NEWAT data bank in the United States (for review of sequence data banks, see Kneale and Bishop, 1985). The second source is from the DNA data banks, which can be easily converted into the primary amino acid structure. The European Molecular Biology Laboratory (EMBL, Heidelberg), the National Biomedical Research Foundation, Maryland, and GenBank (Intelligenetics, 700 East El Camino Real, Mountain View, California 94040-2216) have vast collections of DNA sequences (see list at end of chapter).

A. Sequence Alignment

One of the earliest attempts to determine whether the relationships existing between protein sequences resulted from homology or chance was by the sequence alignment algorithm of Needleman and Wunsch (1970). The maximum match is a number dependent on the similarity of the sequences. Comparisons were made from the smallest unit of significance, a pair of amino acids, one from each protein. All possible pairs are represented by a two-dimensional array, and all possible comparisons were represented by pathways through the array. The use of dynamic programming methods to carry out global comparison of two sequences was soon amplified by many techniques (e.g., Sellers, 1974; Waterman *et al.*, 1976). Their proficiency depends on the degree of similarity and may give variable results depending on the gap-penalty parameters chosen, even for closely related sequences (Fitch and Smith, 1983; Barton and Sternberg, 1987a,b). For sequences with greater than 25% similarity, such automatic procedures will identify the homology above the background of randomized sequences. Because insertions and deletions often occur at the loop regions of proteins between secondary structures, improvements can be made by introducing penalties for insertions/deletions in α helices or β strands (Barton and Sternberg, 1987a; Lesk *et al.*, 1986).

For very distantly related sequences, homology may be restricted to a few key residues or sequence segments whose separation along the chain may vary considerably between proteins. There are several newer algorithms to identify such local homologies (Goad and Kanehisa, 1982; Sellers, 1979; Boswell McLachlan, 1984; Taylor, 1986a; Gribskov *et al.*, 1987). McLachlan (McLachlan, 1971; McLachlan and Stewart, 1976) was an early contributor to sequence alignment and published an improved method based on those of Fitch (1966a–c), Cantor and Jukes (1966), Needleman and Blair (1969), and Haber and Koshland (1970).

Crippen (1977a,b) examined the x-ray crystal structures of 19 selected proteins for correlations between amino acid sequences and long-range tertiary conformation. He found clear evidence for preferential association between certain types of amino acids, particularly among hydrophobic alphatic, aromatic, and cysteine residues. Because the likelihoods of forming these residue pair contacts are all less than 12%, packing and geometric requirements must take precedence over energetic considerations.

An algorithm for secondary structure determination based on sequence similarity (Levin *et al.*, 1986) claimed a prediction accuracy of 62.2% over three states for 61 proteins for a new set of seven proteins not in the original database. An example of the use of sequence analysis can be found in the work of de Groot *et al.* (1987). The amino acid sequences of the spike proteins from three distantly related coronaviruses were aligned initially by FASTP analysis (Lipman and Pearson, 1985). These alignments were further extended by reiterating FASTP with nonaligned parts as query sequences and by DIAGON comparison (Staden, 1982). DIAGON plots revealed two repetitious regions in the C-terminal domains with a seven-residue periodicity. Further analysis showed the presence of so-called "heptad repeats" (Cohen and Parry, 1986). Because heptad repeats are indicative of a coiled-coiled structure, it was suggested that these spike proteins of coronaviruses had this supersecondary structure.

The following have published algorithms based on various sequence alignment techniques: Low *et al.* (1968), Remington and Matthews (1978), Lesk and Chothia (1980), Maizel and Lenk (1981), Staden (1982), Sippl (1982), Wilbur and Lipman (1983), Kabsch and Sander (1984), Murata *et al.*, (1985), Lipman and Pearson (1985), Sweet (1986), Lesk *et al.* (1986), Chothia and Lesk (1986), Taylor (1986a), Nishikawa and Ooi (1986), Bacon and Anderson (1986), Gribskov *et al.* (1987), Argos (1987a,b), Barton and Sternberg (1987b), Zvelebil *et al.* (1987), and Bashford *et al.* (1987).

B. Hydrophobicity

The hydrophobic effect, as first expounded by Kauzmann (1959), is primarily entropic, resulting from the unfavorable ordering of water molecules that associate with exposed nonpolar atoms. Since the publication of this report, the nature of the hydrophobic core in globular proteins has been a central focus in the studies of protein folding, self-assembly, and conformation. The first experimental values for this effect on the individual amino acids were published by Tanford (1962) (Nozaki and Tanford, 1971; also see Tanford, 1980). Since that date the controversy over the relative magnitude of the hydrophobicity of each of the amino acid residues in globular proteins had produced a vast literature without consensus. Cornette *et al.* (1987) discuss 38 published hydrophobicity scales, which are compared for their ability to identify the characteristic period of α helices, and also computed an optimum scale for this purpose using a new Eigenvector method. Scales have previously been compared and discussed (Rose *et al.*, 1985; Meirovitch *et al.*, 1980).

Several papers are discussed to demonstrate the differences in principle involved in developing these scales. Such scales can be classified as solution measurements, empirical calculations, or some combination of both. Solution scales are based on distribution coefficients between an aqueous phase and a suitably chosen organic phase, whereas empirical scales are based on partitioning between the solvent-accessible surface and the buried interior in proteins of known structure. Significant differences exist between scales. Residues that are strongly hydrophobic on one scale appear to be strongly hydrophilic on another scale [e.g., Try and Tyr were found to be hydrophobic by Nozaki and Tanford (1971) but found to be hydrophilic by Wolfenden *et al.* (Wolfenden *et al.*, 1981; Wolfenden, 1983)]. Five significant and different approaches are briefly discussed.

It has been pointed out that the hydrophobic indices of the amino acids have a poor correlation with the extent to which the residues are buried in the native folded-protein matrix (Chothia, 1976). Although the hydrophobic index is a measure of the preference of the nonpolar environment by a residue, it does not necessarily reflect to the same extent the environment in protein crystals. Ponnuswamy and co-workers (Manavalan and Ponnuswamy, 1977, 1978; Ponnuswamy *et al.*, 1980) have defined a parameter called "the surrounding hydrophobicity" for residues, which reflects realistically the preferred nonpolar environment of the residue in protein crystals. This environment is represented by a sphere of 8 Å radius around the residue. This preferred environment is obtained by each residue associating itself with a set of specific surrounding residues, and this requirement of each residue in the protein molecule drives the linear chain to fold into a specific compact globular shape. The hydrophobicity of a residue in a native protein is the same as the product of the surrounding residues and their hydrophobic indices (using either values of Tanford, 1962, or Jones, 1975). This scale provides valuable information with regard to hydrophobic domains, nucleation sites, surface domains, loop sites, and the spatial positions of residues in protein molecules.

Wolfenden (Wolfenden, 1983; Wolfenden *et al.*, 1979, 1981, 1983) determined the equilibria of distribution of amino acid side chains between their dilute aqueous solutions and the vapor phase at 25° by dynamic vapor pressure measurements. The resulting scale of "hydration potentials" or free energies of transfer from the vapor phase to neutral aqueous solutions spans a range of ~22 kcal/mole. These hydration potentials are closely correlated with the relative tendencies of various amino acids to appear at the surface of globular proteins. Guy (1985) calculated the energies required to transfer amino acid side chains from water to less polar environments, and these were compared with several statistical analyses of residue distributions in soluble proteins. He found that an analysis that divides proteins into layers parallel with their surface was more informative than those that simply classify each residue as exposed or buried. Most residues were found to be distributed as a function of the distances from the protein–water interface in a manner consistent with partition energies calculated from partitioning of amino acids between water and octanol phases rather than from solubilities of amino acids in water ethanol and methanol.

Kyte and Doolittle (1982) described a computer program that progressively evaluates the hydrophilicity and hydrophobicity of a protein along with its amino acid sequence. The hydrophobic properties of each of the 20 amino acid side chains are taken into consideration. The scale is based on an amalgam of experimental observations derived from the literature. They draw attention to the fact that the extent to which residues are buried depends not only on the strict hydrophobicity but also on steric effects that determine packing between the secondary structure in the crowded interior of the macromolecule. An excellent discussion of the literature bearing on the hydropathicity of proteins can be found in this paper.

Eisenberg *et al.* (1982a) analyzed the structure of proteins in terms of the "hydrophobic moments" (1) of the entire molecule and (2) of the segment of secondary structure that makes up the polypeptide chain. The zeroeth moment is defined as the sum of the hydrophobicities of the amino acid residues (an analogue of the net charge of a cluster of point charges), and the first moment, or hydrophobic dipole moment, is the analogue of the electric dipole moment of a cluster of charges. The hydrophobic dipole can be used to measure the amphilicity of the structure and can be applied to relating the function and secondary structure of a region to its amino acid sequence (Eisenberg *et al.*, 1982b). It can also be useful in the analysis of interactions of a segment or domain of a protein with neighboring regions in the protein (the "hydrophobic field") and to detect the periodicity in protein hydrophobicity (Eisenberg *et al.*, 1984a,b) (see Chapter 16 by Eisenberg *et al.*).

Direct measurements have been made, on long water-soluble double-chained alkylammonium acetate surfactants adsorbed onto sheets of muscovite mica, of the forces between electrically neutral planar hydrophobic surfaces in aqueous solution (Pashley *et al.*, 1985). Such forces reflect interactions caused by surface-induced water structure and are long-ranged, with an exponential decay length of about 1.4 nm. Over 0 to 8 nm, the forces are 10 to 100 times stronger that the van der Waals forces that would operate in the absence of any surface-induced order in water. It was concluded that there is available to biological systems a hierarchy of attractive forces that operate between hydrophobic moieties. These forces depend on the dimensions and geometry of the surfaces and are much stronger, longer-ranged, and more variable than classical colloid science previously indicated.

Rose *et al.* (1985) derived two new scales that are based on accessibility to solvent for residues within proteins of known structure. These two scales measure two quantities that can be distinguished: (1) the area lost when a residue is transferred from a defined standard state to a folded protein—the area a residue buries on folding is proportional to the conformational free energy ΔG_{conf} (Richards, 1977)—and (2) the fractional accessibility of a residue, defined as its mean accessible area in protein molecules divided by the standard-state area. The fractional accessibility is an intrinsic measure of hydrophobicity. These results revealed a strong correlation between hydrophobicity and the surface area residues buried on folding (see Chapter 15 by Rose and Dworkin). Wolfenden *et al.* (Wolfenden *et al.*, 1979, 1981, 1983; Wolfenden, 1983) found that $\Delta G^0{}_h$, the free energy of transfer of side-chain analogues between water and the dilute vapor phase, correlated with the empirical tendency of residues to be buried within proteins. Conversely, Chothia (1976, 1984) and Janin (1979), in an empirical analysis, and Wolfenden *et al.* (1981, 1983) noted a lack of correlation between the degree to which residues are buried and the Nozaki–Tanford $\Delta G_t{}^0$.

C. Minimum Energy Calculations

At present, it is still impossible to predict the three-dimensional structure of a protein by the minimization of the free energy of an all-atom representation. There have been numerous quantum-mechanical studies of peptide systems. Some of these have been *ab initio* studies (e.g., Shipman and Christoffersen, 1973, and references herein), but the majority have used more approximate methods, mostly semiempirical (Scheraga, 1985). Many investigators have developed potential functions to describe the energy surface of a polypeptide chain. However, to date energy-minimization schemes have failed to predict chain folding accurately (Hagler and Honig, 1978; Cohen and Sternberg, 1980a,b). Scheraga and co-workers were among the first to tackle this exceedingly complex task (see review by Nemethy, 1974). The main problems involved are (1) the large number of variables and interaction energy terms, (2) uncertainties about the function form of the potential energy terms, (3) the existence of many conformations corresponding to a minimum in potential energy, (4) interaction of the protein with the solvent, and (5) vibrations and other free energy contributions. A basic question that can be asked is whether the native conformation of a protein molecule is its thermodynamically most stable state. Many investigators recognized that the kinetic question of the folding pathway was significantly more complex than the thermodynamic issue of predicting the optimal folded conformation. One major problem in this approach is that the contributions to the free energy cannot be modeled adequately. Many workers (e.g., Weiner *et al.*, 1984) have developed a new force field simulation for proteins. They have used the powerful Cartesian-coordinate energy refinement of Lifson and Warshel (1968) and developed empirical force fields within this context. The related force field for proteins was similar and contained minor modifications of the parameters used by Gelin and Karplus (1979). The most important changes concerned the explicit inclusion of H-bonding hydrogens parameters and the use of partial charges taken from Mulliken populations of *ab initio* calculations. They focused on ϕ, ψ maps of glycyl and alanyl dipeptides, hydrogen-bonding groups, and energy-refinement calculations on insulin. They emphasize that the single crudest aspect of this work in the application of the force field is the way solvation effects are modeled, and this is the area that requires the most refinement in the near future.

Another approach has been to predict the tertiary structure on the basis of sequence homology and regularize a predicted structure by energy calculations. Lewis and Scheraga (1971) carried this out for α-lactalbumin. Following the first energy calculations on known protein conformations (Levitt and Lifson, 1969), more attention has been given to the energy refinement of x-ray coordinates (e.g., Levitt, 1974; Burgess and Scheraga, 1975), but with little more success. Further work (Levitt, 1976; Levitt and Warshel, 1975) introduced a simplified representation of protein conformation for rapid simulation of protein folding using the concept of time-averaged forces, but this did not produce significant improvements.

Conformational energy calculations were used to analyze the interactions of structural substructures in subtilisin BPN (Honig *et al.*, 1976). These substructures are kept fixed or "rigid" so that the only variables in the calculations are the backbone segments that separate them. The flexible segments are assumed to be free turns. By using this representation of the protein, it is possible to predict both a likely order of events along the folding pathway and preferred modes of conformational changes of the native protein. These results suggest an approach to the folding problem based on the piecemeal formation of tertiary structure from smaller prefolded fragments. The major improvement of the Robson and Osguthorpe (1979) procedure over previous methods was to retain a more realistic and complete representation of the protein backbone and alternately to reduce the number of variables by coupling their behavior. They attempted to satisfy the criteria essential for folding as discussed by Nemethy and Scheraga (1977). This "preliminary investigation" left much work to be done before such computer simulation could become a realistic approach to protein folding. Finney *et al.* (1980), using detailed hydrogen bonding, surface exposure, internal environment, and solvent interaction calculations, in conjunction with data from quantum mechanical hydrogen-bonding studies, estimated various contributions to the free energy of folding. Their conclusion stated that "a picture emerges of globular proteins as extremely well-fitting jigsaw puzzles, in which no single driving force dominates the marginal stability of the native conformation. Rather, the folded structure is seen as the result of a complex global minimization of several strongly interacting driving forces." The necessity to maintain a very efficient internal hydrogen bonding and the role of solvent as a hydrogen-bond sink are stressed as strong constraints on the (incomplete) maximization of hydrophobic effects.

Robson and Platt (1986) reappraised the interatomic potential functions for protein structure calculations, using the all-atom approximation (except CH, CH_2, and CH_3, which were treated as "united atoms"). This produced a more efficient and robust folding algorithm. The potentials were calibrated for the rigid geometry approximation, since use of fixed standard bond lengths and valence angles (and fixed transplanar peptide groups) reduces the number of conformational variables and saves a great deal of computer time. These potentials do not generally give the nativelike form as the least energy form, in common with other similar methods. A statistical approach to the calculation of conformation of proteins was published by Crippen (1977a,b) and applied to the reduced pancreatic trypsin inhibitor. The theory applies to both thermodynamically and kinetically determined processes, so that not only can the equilibrium result be calculated but also the time course of folding. The results look promising.

Bash *et al.* (1987) have applied recent advances in statistical mechanical theory and molecular dynamics to the understanding of the role of solvation in determining molecular properties. The free energies of solvation of all the chemical classes of amino acids side chains were calculated using supercomputers. The effect of a site-specific mutation on the stability of trypsin was predicted, which yielded results in agreement with available experiments.

IV. APPROACHES TO PROTEIN CONFORMATION

The x-ray crystallographic study of proteins has opened up the opportunity to evaluate protein conformation from many points of view. Each of these opens a different window to view the architecture of protein structure. These various windows are briefly reviewed.

A. Solvent Accessibility

The counterparts of the buried hydrophobic areas are those that are accessible to solvent, usually water. Lee and Richards (1971) developed a program that permits the accessibility of atoms, or groups of atoms, to solvent or solute molecules of specified size to be quantitatively assessed. This accessibility was found to be proportional to the surface area. This approach can also define internal cavities. It was found that about 40–50% of the surface area of each protein is occupied by nonpolar atoms. The numerical results are sensitive to the choice of the van der Waals radii of the various groups. For the atoms in ribonuclease S, lysozyme, and myoglobin, the average change in accessibility for the atoms in going from a hypothetical chain to the folded conformation of the native protein is about a factor of 3. Another alternative would be to consider the contact surface, those parts of the molecular van der Waals surface that can actually be in contact with a probe examining the surface (Richards, 1977). Richards and co-workers (Richards, 1974, 1977; Richmond and Richards, 1978; Richmond, 1984) have examined the total volume, group volume distributions, and packing densities of proteins. Richmond and Richards (1978) discussed the packing of α helices in sperm whale myoblobin and proposed an algorithm for picking potentially strong helix–helix interaction sites in peptides of known sequence. Richmond (1984) has expanded on this idea. A computer program, using the equations for area, has been tested and has had limited application to the docking of protein α helices. Chothia (1974, 1975, 1976) found that the accessible surface area is proportional to the hydrophobicities of residues in the protein. The loss of accessible surface area by monomeric proteins on folding is proportional to the hydrophobic energy and is simply proportional to the two-thirds power of their molecular weight. Wodak and Janin (1980) proposed an analytical substitute to the protein surface area that is accessible to solvent. A statistical approach leads to an expression of accessible surface areas as a function of distance between pairs of atoms or of residues in the protein structure, assuming only that these atoms or residues are randomly distributed in space and not penetrating each other. More recently, Zehfus *et al.* (1985) published equations that approximate the accessible area of a continuous protein segment using the surface area of an inertial ellipsoid that approximates the molecular volume from the number of nonhydrogen atoms in a segment.

B. Packing of Residues

Solvent accessibility is the result of failure to achieve maximum close packing (Richards, 1974) of the atoms in the folding of the polypeptide chain. Although similar in principle to solvent accessibility, its approach is based on a different outlook. The result of maximum packing is to produce a center of residues, which has been classified as the hydrophobic core, that has maximum shielding from solvent (Richards, 1977; Chothia, 1975). This core consists mainly of nonpolar atoms or polar atoms that form either hydrogen bonds or salt bridges. Thus, the degree to which ideal packing occurs determines the degree of relative stability of the particular protein. Crippen and Kuntz (1978) surveyed the atom packing in high-resolution x-ray crystal structures of 21 proteins and concluded that the atom density around a given central atom is determined primarily by its covalently bonded neighbors and proximity to the surface of the protein. Long-range hydrophobic, hydrogen-bonding, and electrostatic interactions are strictly of secondary importance.

The interior of globular proteins has very significant density inhomogeneities on a scale of 100–1000 Å^3. The interior densities range from less than 0.5 g/cm^3 to over 3 g/cm^3. The low local densities are primarily associated with clusters of nonpolar side chains, and the high-local-density regions arise from the protein backbone secondary structures: helices and β sheets (Crippen and Kuntz, 1978). Lesk and Rose (1981) developed a method to identify all compact, contiguous-chain structural units in a globular protein from x-ray coordinates. These units were then used to describe a complete set of hierarchic folding pathways for the molecule. The analysis showed that the larger units are combinations of smaller units, giving rise to structural hierarchy ranging from the whole protein monomer through supersecondary structures down to individual helices and strands. Thus, there is hierarchic condensation. In this model, neighboring hydrophobic chain sites interact to form folding clusters, with further stepwise cluster association giving rise to a population of folding intermediates.

C. Distance Geometry

The mathematics of distance geometry constitutes the basis of a group of algorithms for revealing the structural consequences of diverse forms of information about the macromolecule's conformation. An excellent review of this approach was published by Havel *et al.* (1983). The article presents the basic theorems of distance geometry in Euclidean space and gives formal proofs of the correctness and, where possible, of the complexity of the algorithms. Crippen (1977a,b) originally conceived the idea as a means of circumventing the longstanding local minimum problem of molecular minimization, but since then it has led to the development of a data structure for representing classes of related conformations consistent with experimental, energetic, and functional information that is compact and succinct.

Rackovsky and Goldstein (1987) extended a previous differential geometric analysis of the conformational properties of the various amino acids to study their influence on folding over a larger backbone interval. In addition, statistical effects associated with the variation in the number of individual amino acids in the data base were treated in greater detail, using a simulation method. It was found that the amino acids could be divided into three groups on the basis of their conformational influence over four-C^{α} units in the interval $i - 6 \leq j \leq i + 6$. Group Ia is composed of seven amino acids (His, Leu, Ala, Met, Lys, Glu, Ile) that encourage the formation of a right-handed α-helical structure. Group Ib (Glu, Phe, Trp, Val, Asp) is composed of amino acids with some helix-forming tendency but that also showed a positive extended strand formation tendency. They therefore act as a bridge between group Ia and group II (Cys, Gly, Asn, Pro, Arg, Ser, Thr, Tyr), which contains amino acids that encourage the formation of extended structure and bends. It was shown that, in general, such influences extend further in the N-terminal direction than in the C-terminal direction. Insofar as comparison is possible, these results correlate with those of Robson and Pain (Robson, 1974; Robson and Pain, 1974a–c) and Robson and Suzuki (1976) and are complementary to earlier work of Maxfield and Scheraga (1976, 1979). In general, they also agree with the propensity parameters of Chou and Fasman (1974a,b).

D. Amino Acid Physicochemical Properties

A large number of physical-chemical properties, manifest in the amino acid side chains, have been thoroughly examined by many investigators. Attempts have been made to correlate these properties with their relatedness between protein sequences. Sneath (1966) sought the relationship between chemical structure and biological function in peptide hormones using a new measure of difference in chemical structure. A large number of physicochemical properties (32) were investigated, and a principal component analysis yielded four vectors, which represent major composite chemical factors. Zimmerman *et al.* (1968) performed three different but related comprehensive statistical analyses of amino acid sequences in proteins. The goal was to find evidence of significant sequence structure related to a purely random arrangement of amino acid residues and to attempt to relate any significant structure uncovered to secondary and/or tertiary conformation of the protein. A continuous physical scale property for six scales was employed. These were bulkiness, polarity, R_F, pI, pK, and hydrophobicity. No striking conclusions could be made. Grantham (1974) developed a formula for the difference between amino acids that combines the properties that correlated best with protein residue substitution frequencies. The properties that yielded the best correlations were composition, polarity, and molecular volume.

Jones (1975) replaced the amino acid sequence of a protein by a numerical sequence of values representing a physical or chemical property of the amino acids, and the resulting numerical sequence was amenable to autocorrelation analysis. Similarly, certain geometrical parameters, calculated from the three-dimensional structure of a protein to form a configurational series, can be analyzed by cross-correlation techniques. On the basis of ten proteins, it was found that the hydrophobicity of an amino acid residue in a protein influences the orientation angle of the amino acid side chain. Padlan (1977) used a technique to compare the dissimilarity of physicochemical properties of the amino acids in immunoglobulin sequences. Exterior residues showed greater structural variability than the interior residues. The dissimilarities were taken from the work of Sneath (1966) and Grantham (1974). Yockey (1977) utilized Grantham's (1974) method of representing amino acid residues to present a prescription that predicts all functionally equivalent residues expected at a given site in a protein sequence if at least two such residues are known. The prescription established a conceptual framework wherein the validity of Grantham's hypothesis can be tested objectively and, if necessary, generalized.

Levitt (1978) analyzed the secondary structure of 50 different globular proteins to give the frequency of occurrence of the 20 naturally occurring amino acids in the α helix, β sheet, and reverse turn. These preferences correlate well with the chemical structure and stereochemistry of the particular amino acid. The rules that emerged from this study can be summarized as follows: bulky amino acids, namely, those that are branched at the β carbon or have a large aromatic side chain, prefer β sheet. The shorter, polar side chains prefer reverse turns, as do Gly and Pro, the special side chains. All other side chains prefer α helix except Arg, which has no preference. The polar side chains with hydroxyl groups disrupt α helix; the other polar side chains disrupt reverse turns. This conclusion agrees well with previous analyses of a similar nature (Nagano, 1973; Chou and Fasman, 1974a,b; Robson and Suzuki, 1976; Chou and Fasman, 1977a–c; Chou *et al.*, 1975). Similar conclusions were also previously drawn from studies on poly-α-amino acids (for review see Blout, 1962; Fasman, 1987). Wertz and Scheraga (1978) examined the x-ray structures of 20 proteins to determine which amino acid residues reside in the inside or outside of the molecule and to assign a conformational state. The data confirm that polar groups are generally found on the outside of proteins and that nonpolar residues are generally found on the inside. Seven amino acids (Ala, Arg, Cys, His, Pro, Ser, Tyr) have inside/outside preferences that are not consistent with their usual assignment as either polar or nonpolar residues. It is suggested that differences in entropy play an important part in the inside/outside preferences of backbone structures.

There are generally significant changes in the conformational preferences of the residues in going from the inside to the outside of proteins; environmental (rather than local) solute–solvent interactions seem to be the predominant cause of these changes in conformational preferences. Kubota *et al.* (1981) examined internal homologies in an amino acid sequence and in amino acid sequences of two different proteins using correlation coefficients calculated from the sequences when residues are replaced by various quantitative properties such as hydrophobicity. These quantities inherent in amino acids were those previously discussed by Zimmerman *et al.* (1968). The sequences of α-tropomyosin, calmodulin, troponin C, and L2 light chain of myosin were evaluated in this manner.

Argos and Palau (1982) probed the compositional distribution of the 20 amino acids for particular positions within the secondary structures (α helices, β strands, and turns) in a 44-protein sample. Correlation coefficients between positional composition of the amino acids and various of their physicochemical characteristics indicated considerable asymmetry in the properties of the residues comprising regions within and adjacent to secondary structures, modes of helix formation, physical parameters most sensitive to the buriedness of residues in β strands, and possible improvements in the accuracy of secondary structural prediction methodologies. The physical parameters used were bulkiness and polarity (Jones, 1975), hydrophobic index (Nozaki and Tanford, 1971), conformational propensity (Levitt, 1978), the "residue surrounding" hydrophobicity (Manavalan and Ponnuswamy, 1978), hydration potential (Wolfenden *et al.*, 1979), and free energy transfer (Janin, 1979). Taylor (1986b) classified the amino acid type based on a synthesis of physicochemical and mutation data. The major sets group the amino acids by size and hydrophobicity. These relationships were displayed as a Venn diagram, from which subsets are derived that include groups of amino acids likely to be conserved for similar structural reasons.

There have been several previous evaluations of amino acid mutations relative to evolution (e.g., Grantham, 1974; Sellers, 1974). The assessment of the structural effect of introducing a new amino acid into a known structure is often based on the likehood matrix of amino acid mutabilities derived by Dayhoff (1972, 1978). The prediction of secondary structure by evolutionary comparison was made by Crawford *et al.* (1987) for the α subunit of tryptophan synthase from ten different microorganisms. Both the Chou and Fasman (1974a,b) and Garnier *et al.* (1978) methods were applied, as well as profiles of hydropathy (Kyte and Doolittle, 1982) and chain flexibility values (Karplus and Schulz, 1985), to give a joint prediction. There was good agreement (1) among predicted β strands, maximal hydropathy, and minimal flexibility and (2) among predicted loops, great chain flexibility, and protein segments that accept insertions of various lengths of individual sequences.

Hellberg *et al.* (1986) predicted a set of bradykinin-potentiating potencies of pentapeptides with a variation in amino acid sequences. This was performed by varying three parameters per amino acid position. The variables were derived from a principal-component analysis of a property matrix for the 20 amino acids. The resulting structure descriptor describes the observed activity of the peptides to 97% by means of a multivariable partial least-squares (PLS) model. It was demonstrated that this quantitative structure–activity relationship (QSAR) can be used to predict the activity of new peptide analogues. The relationship is based on the same principle of quantitative analogy models that previously have been shown to apply to structure–activity relationships in organic chemistry, e.g., the Hammet (1970) relationship. Three scales are used to characterize each individual amino acid. These scales are derived by a principal-component analysis of a matrix of 20 properties of each of the amino acids. Vondervisztz *et al.* (1986) studied the occurrence of all di- and tripeptide segments of proteins from a large data base containing 119,000 residues. It was found that the abundance of the amino acids does not determine the frequencies of the various di- and tripeptide segments. The pair-frequency distribution of amino acids is highly asymmetric, pairs formed from identical residues are generally preferred, and amino acids cannot be clustered on the basis of their first-neighbor preferences. These data indicate the existence of general short-range regularities in the primary structure of proteins. The consequences of these short-range regularities were studied by comparing Chou–Fasman parameters with analogous parameters determined from the results of conformational energy calculations (Zimmerman *et al.*, 1977) of single amino acids. This comparison showed that the Chou–Fasman parameters carry significant information about the environment of each amino acid. The success of Chou and Fasman's (1974a,b) prediction and the properties of the pair and triplet distributions of the amino acid residues suggests that every amino acid has a characteristic sequential residue environment in proteins.

Kelly and Holladay (1987) compared the scales of amino acid side-chain properties by conservation during evolution of four proteins. As the amino acid sequence of a given protein changes along the phylogenetic tree, enough of the overall folding pattern must be conserved to ensure that the protein fulfills its biological function. Eighteen published scales that tabulate various side-chain properties were compared by computing the variance of each scale when applied to each of several protein families. The conservation of each scale of side-chain properties was examined for the 20,627 residues in 60 mammalian myoglobins, 31 mammalian ribonucleases, insulin A and B chains (29 sequences each), and 29 vertebrate and 28 plant cytochrome *c*s. Those scales that are the most highly conserved may well be the best prediction of protein folding patterns. The mean-area-buried scale (Rose *et al.*, 1985) and the optimized matching hydrophobicities scale (Sweet and Eisenberg, 1983) are more conserved than other scales. An additional result is the relatively poor conservation of the Chou–Fasman (1974a,b) secondary structure predictions.

V. PREDICTION OF THE SECONDARY STRUCTURE OF PROTEINS: α HELIX, β STRANDS, AND β TURN

The origins of the prediction of the secondary structure of proteins have been previously outlined (Section I). Over 20 different methods have been proposed in addition to variations of several of these. These predictive schemes mostly assume that the local sequence (short-range interactions) determines local structure. The following authors have proposed the main schemes representing significantly different approaches: Kotelchuck and Scheraga (1968a,b), Lewis *et al.* (1970), Ptitsyn and Finkelstein (1970a,b), Toitskii and Zav'yalov (1972), Kabat and Wu (1973a,b), Chou and Fasman (1974a,b, 1978a,b), Burgess *et al.* (1974), Lim (1974a,b), Robson and Suzuki (1976), Maxfield and Scheraga (1976), Nagano (1977), McLachlan (1977), Barkovsky and Bandarin (1979), and Barkovsky (1982).

These methods can be categorized into two broad classes. The empirical statistical methods use parameters obtained from the analysis of known sequences and structures (e.g., Chou and Fasman, 1974a,b). The second method is based on stereochemical criteria (e.g., Lim, 1974a,b). The most frequently used methods to date have been the empirical approaches of Chou and Fasman (1974a,b) and of Robson and co-workers (Garnier *et al.*, 1978) and the stereochemical method of Lim (1974a,b).

The Chou and Fasman (1974a,b) method has been widely used because of the simplicity of its application and the ease of understanding its premise. The statistical derivation of the propensity of residues to be in α helices, β strands, or turns (P_α, P_β, P_t) is straightforward. The relative magnitudes of these values assist in classifying residues as α formers, α indifferent, and α breakers. The β-strand residues are likewise categorized. The β-turn parameters (frequencies) are also used in a direct manner. There are rules for utilizing these parameters for nucleation, propagation, and termination of α helices and β sheets. A problem of interpretation of regions of equal potential for α helices and β strands has dismayed some users (Sternberg, 1983), and the lack of clear-cut rules has prevented programmers from writing computer algorithms that yield the same results. One factor, unfortunately, has been overlooked that gives this method an advantage over others. This ambiguity that arises in delineating sequences of equal potential for α and β structures is not a detrimental factor, as is so frequently pointed out, but rather gives insight into regions of sequence that can undergo conformational change. This has been illustrated for glucagon (Chou and Fasman, 1975) and for the preproparathyroid hormone (Rosenblatt *et al.*, 1980). The percentage accuracy of the method was stated to be 75–80%. The Chou and Fasman method is discussed in detail in Chapter 9.

The method of Robson and co-workers (Robson and Suzuki, 1976; Garnier *et al.*, 1978) is based on information theory and can be programmed easily and unambiguously. It considers the effects of residues from positions $i - 8$ to $i + 8$ on the conformation of position i. The Garnier *et al.* (1978) method has recently been reevaluated (Gibrat *et al.*, 1987), and the validity of the approximations drawn from the theory examined. It was concluded that the existing data base does not allow evaluation of parameters required for an exact treatment of the problem. It was shown that the first-level approximation, involving single-residue parameters, is only marginally improved by an increase in the data base. The second-level approximation, involving pairs of residues, provides a better model. However, in this case, the data base is not big enough, and this method leads to parameters with deficiencies. This new version of the Garnier–Osguthorpe–Robson (GOR) method increases the accuracy of prediction by 7%, giving a 63% correctly predicted residues count for three states in 68 proteins, each protein to be predicted being removed from the data base and the parameters derived from the other proteins. If the protein to be predicted is kept in the data base, the accuracy goes up to 69.7%. The GOR method is discussed in detail in Chapter 10.

The method of Lim (1974a,b) is based on a stereochemical theory of globular protein secondary structure. This approach considers the "architectural" principles of packing of polypeptide chains and the interactions of proteins with water molecules. Thus, this procedure takes into account long-range interactions, as was first applied by Schiffer and Edmundson (1967). The most salient structural features utilized are compactness of form, the presence of a tightly packed hydrophobic core (cores), and a polar shell. These structural features impose the following requirements for possible conformations of any region of the polypeptide chain:

> (1a) All the helical, β-structural and irregular regions must be attached by non-covalent interactions to the main part of the globule. (1b) The linear dimensions of regular and irregular regions cannot exceed the linear dimensions of the globule. The following requirements result from the second structural principle: (2a) The overwhelming majority of massive hydrophobic side groups must be fully or partially immersed in hydrophobic cores, while hydrophilic groups must not penetrate into these hydrophobic cores; NH and CO groups of the backbone also must not penetrate into these cores without formation of a hydrogen bond. (2b) The conformation of the backbone of the region having hydrophobic side groups must be compatible with the tight packing of these hydrophobic groups in the hydrophobic core or cores. (2c) The hydrophilic shell must at least shield each hydrophobic core from water molecules; hydrophilic groups must, if possible, form salt and hydrogen bonds which together with hydrophobic interactions will attach the separate parts of the globule together (Lim, 1974a).

Lim (1974b) published an algorithm that was stated to predict accurately ~80% and 85% of the α-helical and β-structural regions, respectively, for 25 proteins.

Pongor and Szaley (1985) have described a quantitative procedure for the comparison of predicted secondary structures of homologous proteins. The predictive procedures of Chou and Fasman (1978a) and Garnier *et al.* (1978) and the hydrophilicity values of Hopp and Woods (1981) show that correlation coefficients of structural profiles can be used to describe similarity in secondary structures. The method is potentially useful to describe evolutionary changes in protein secondary structure as well as in the design of peptide analogues.

A. β Turns

A proposed mechanism for folding of polypeptide chains in proteins was put forth by Lewis *et al.* (1971). It was suggested that secondary structures (e.g., α helix or other ordered regions) could be stabilized only by long-range interactions arising from the proximity of two such ordered regions. These regions would be brought near each other by the directing influence of certain other amino acid sequences that have a high probability of forming β bends or variants thereof and also on the basis of short-range interactions. An analysis was made, on three proteins of known structure, of the tendency of various amino acids to occur in β bends, and it was found that it was possible to predict the regions of the chain in which a β bend will occur with a high degree of reliability (~80%). The β bends were defined in the manner previously described by Venkatachalan (1968). Dickerson *et al.* (1971) determined the structure of horse heart ferricytochrome *c* to a resolution of 2.8 Å. They carefully analyzed the folding of the polypeptide chain and carefully delineated the β turns (calling them 3_{10} bends). The 3_{10} bend occurred six times in cytochrome *c*. Two variants of the 3_{10} bend, which were previously classified by Venkatachalan (1968) as type I and type II, were found: three type I and three type II bends, with Gly in the third α-carbon position in the latter.

Kuntz (1972) developed simple computer routines for locating regions in which the peptide backbone of a globular protein folds back on itself. These programs were used to locate "turns" in carboxypeptidase and α-chymotrypsin. The following generalizations were made: (1) turns occur on the surface of the protein; (2) turn segments are less hydrophobic than the protein as a whole; (3) uninterrupted sequences of three to eight hydrophilic residues are frequently associated with folding of the peptide chain. It was found that Ser, Thr, Gly, Pro, Gln, and Asn frequencies were about 20% enhanced in turns. Bunting *et al.* (1972) predicted the β bends in the immunoglobulin light and heavy chains using the method of Lewis *et al.* (1971). They concluded "that there was an overall common design in light chains, and in heavy chains, which, in the absence of significant quantities of other recognizable conformational features, may be at least partially influenced by the position of β-bends."

It is now known that immunoglobulins contain a large amount of β-sheet structure, and the β strands are frequently connected by β turns. Thus, the above conclusions have been verified by x-ray analysis. Esipova and Tumanyan (1972) suggested a model for the tertiary structure of proteins whereby relatively straight segments are interrupted by a folding point between them. Turns in proteins are regarded either as flexible hinge regions that permit the chain to bend in this region because of forces acting outside the regions of folding or as rigid segments that alter the course of the polypeptide chain through local interactions in the bending region. Folding regions were determined for seven proteins. These globular proteins were analyzed by means of stereo projections and special curvature functions. The suggested model encompasses not only regions with the secondary structure of a protein globule but also those regions that do not enter into secondary structure and previously have been negatively described as "unordered." It is primarily these latter segments that in the different portions of the protein determine the most important properties of the tertiary structure. It was shown that certain amino acids—Gly, Asn, Ser, etc.—have a specific significance in these regions of transition, forming hydrogen bonds between their side groups and the backbone of the chain. These transitions agree with those found by Lewis *et al.* (1971). Lewis *et al.* (1973a) analyzed eight proteins for the presence of chain reversals and found 135 bends. Of these, 129 belong to a set of ten types.

Conformational energy calculations were carried out on the pentapeptide N-acetyl-N′-methyl-Ala-Ala-Ala-Ala-amide in order to determine the backbone conformations of minimum energy for bend and nonbend structures. The conformations of lowest energy did not correspond to the conformations obtained from a minimization starting from the x-ray structural data, although in all three cases, one type of bend or another was found to be of lowest energy compared to other conformations studied. Crawford *et al.* (1973) defined a reverse turn as a tertiary conformation in globular proteins and defined it in terms of the dihedral angles, the C^{α}_{1}–C^{α}_{4} distance (less than 5.7 Å), and the $O_1 \ldots H\text{-}N_4$ hydrogen-bond distance (less than 3.2 Å). In seven proteins, 125 examples of turns were found, comprising 33% of the amino acids in these proteins as compared with 34% of the residues forming helices and only 17% forming β sheets. The amino acid compositions of turns, helices, and β sheets were analyzed in detail.

They found Asn and Gly mainly in turns, Pro in turns (and at the beginning of helices), and Glu in helices. In these turns, 19% of Asp was found in the first position, 33% of Pro in the second position, 24% of Asn in the third position, and 26% of Trp in the fourth position.

Chou and Fasman's (1974b) algorithm for the prediction of the secondary structure of proteins predicted α helices, β sheets on the basis of the conformational parameters P_α and P_β (Chou and Fasman, 1974a). In order to compare the β-turn potentials for all the 20 amino acids, with their α and β potentials, the conformational parameter for the β turns, $P_t = f_t/\langle f \rangle$ ($f_t = n_t/n$, where n_t and n are, respectively, the total occurrence of each residue in the β turns and in all the regions of the 12 proteins; $\langle f_t \rangle = n_t/n$ is the frequency of all residues in the β-turn regions), was obtained in the same manner as P_α and P_β (Chou and Fasman, 1974a) by the process of normalization (see Chapter 9). These turns were chosen in part on stereo diagrams inasmuch as atomic coordinates were not available. There is almost an inverse relationship between the β-turn potential and the α potential for the 20 amino acids. The relative probability that a tetrapeptide will form a β turn (Lewis *et al.*, 1971), $p_t = f_i \times f_{i+1} \times f_{i+2} \times f_{i+3}$, where f_i, f_{i+1}, f_{i+2}, and f_{i+3} are, respectively, the frequency of occurrence for a certain residue at the first, second, third, and fourth positions of a β turn. The cutoff value of $p_t = 0.5 \times 10^{-4}$ was found to be a reasonable value in predicting the β turns of the 12 proteins studied herein.

Burgess *et al.* (1974) described methodology for describing a discrete number of conformational states of amino acids in proteins and used these to examine the relative importance of short-range, medium-range, and long-range interactions in light proteins of known structure. A prediction algorithm that assigns four states to each residue of a protein chain (α helix, extended structures, bend, and coil) was developed from consideration of both short- and medium-range interactions and was applied to 13 proteins of known structure. First the frequencies of occurrence of the various conformations of each of the 20 amino acids in several proteins are examined and represented on φ,ψ conformational maps. Second, these empirical frequencies are compared to those deduced (Lewis *et al.*, 1970) from the statistical weights of various conformations of these residues (obtained from conformational energy calculations on the N-acetyl-N′-methylamides of each residue; Lewis *et al.*, 1973b). Third, the empirical frequency data are used to develop an algorithm to predict simultaneously the location of α-helical, extended, bend, and coil regions of a protein, taking into account the conformational states of four residues on each side of the given one (the prediction of β turns was not as good as those for the α helix). The authors state that their method is more reliable than other predictive schemes, but in relative trials (Schulz and Schirmer, 1974) this was not borne out.

Nishikawa *et al.* (1974) examined the low-energy conformations of two dipeptides using "empirical" energy calculations and the virtual-bond method. The complete conformational space of the dipeptide N-acetyl-N′-methyl-L-alanineamide was searched systematically, and all conformations of minimum energy were found. The low-energy conformation of the alanine dipeptide, which closedly approximates a type II bend, does not correspond to a combination of single-residue minima. Conformations similar to type I bends are not found to be of minimum energy for either the glycine or alanine dipeptide. Chou and Fasman (1977b) expanded their sample and utilized the x-ray atomic coordinates from 29 proteins of known sequence and structure to elucidate 459 β turns in regions of chain reversal. Tetrapeptides whose $C^\alpha{}_i$–$C^\alpha{}_{i+3}$ distances were below 7 Å and not in a helical region were characterized as β turns. In addition, β turns were considered to have hydrogen bonding if their computed $O_{(i)}$–$N_{(i+3)}$ distances were ≤3.5 Å. The torsion angles of 26 proteins containing 421 β turns were classified in 11 bend types based on (φ,ψ) dihedral angles of the $i + 1$ and $i + 2$ bend residues. The average frequency of β turns was found to be 32% as compared to 38% helices and 20% β sheets.

The relative frequencies for all the amino acids in the four positions were established (see Tables XI, XII, and XIII, Chapter 9). Residues with the highest β-turn potential in all four positions are Pro, Gly, Asn, Asp, and Ser, with the most hydrophobic residues (i.e., Val, Ile, and Leu) showing the lowest bend potential. An environmental analysis of β-turn neighbors shows that reverse chain folding is stabilized by antiparallel β sheets as well as helix–helix and α–β interactions. The β-turn potential at the 12 positions adjacent to and including the bend were plotted for the 20 amino acids and showed dramatic positional preference, which may be classified according to the nature of the side chains. Eight type VI bends were found with a *cis*-Pro at the third position. Chou and Fasman (1979a) then presented an automated computer prediction of the chain-reversal regions in globular proteins, using the bend frequencies and β-turn conformational parameters (P_t) determined from 408 β turns in 29 proteins calculated from x-ray atomic coordinates. The probability of bend occurrence at residue i is $p_t = f_i \times f_{i+1} \times f_{i+2} \times f_{i+3}$, with the average bend probability $\langle p_t \rangle = 0.55 \times 10^{-4}$. Therefore, tetrapeptides with $p_t > 0.75 \times 10^{-4}$ ($\simeq 1.5 \times \langle p_t \rangle$) as well as at $\langle P_t \rangle > 1.00$ and $\langle P_\alpha \rangle < \langle P_t \rangle > \langle P_\beta \rangle$ were selected by the computer as probable bends. When adjacent probable bends or overlapping bends occur, the tetrapeptide with the higher p_t value is predicted as the β turn. The percentage of bend and nonbend residues predicted correctly for 29 proteins by this computer algorithm is 70%, whereas 78% of β turns were localized correctly within ±2 residues. The average β-turn content in the 20 proteins is 32%, with helical proteins having fewer bends (17%) than the β-sheet proteins (41%). The accuracy of β-turn prediction using the probability of bend occurrence at residue i obtained from the frequencies of 2, 4, 8, and 12 residues were investigated, and very little change occurred beyond four residues.

Tanaka and Scheraga (1976d) developed a statistical mechanical treatment of protein conformation wherein chain-reversal regions were predicted in 23 proteins after predictions of helical and β-sheet regions. Their percentage of bends localized (±2 residues) was 59%. Lenstra (1977) has compared the accuracy of bend predictions in 25 proteins according to the computer methods of Nagano (1977) and Argos *et al.* (1976) and obtained correlating coefficients of $C_t = 0.40$ and 0.44, respectively. Although the automatic bend prediction (Chou and Fasman, 1979a) using four residues has a smaller correlation coefficient, $C_t = 0.38$, its prediction of turns is 70.4%

Rose and Seltzer (1977) presented an algorithm to identify peptide chain turns from x-ray-elucidated coordinate data. This algorithm uses only the C^α coordinate for every residue in the protein. No other information is required, and notions about hydrogen bonding at these loci are irrelevant to the geometric nature of the argument. The radius of curvature for each amino acid along the chain is calculated, and this is the basis of detecting turns. A turn corresponds to a locus where the chain direction vector is changing rapidly and the value of the radius of curvature is at a local minimum. The results are compared to data of Kuntz (1972), Crawford *et al.* (1973), and Lewis *et al.* (1971). Both Kuntz (1972) and Rose and Seltzer (1977) use visual parsing to detect β turns, and their results are quite similar. The authors stated that their algorithm yields a more comprehensive set of turns than the other three. Because the sample is relatively small, it is difficult to evaluate accurately these different sets of results.

Zimmerman and Scheraga (1977) calculated probabilities of bend formation in 47 amino acid sequences of the N-acetyl-N′-methylamide dipeptides with a statistical mechanical analysis using empirical conformational energies and compared these results with the fraction of bends formed in the same 47 dipeptide sequences in the x-ray structures of 20 globular proteins. Agreement was found for 26 dipeptides, suggesting that for those particular dipeptide sequences, local interactions dominate over long-range interactions in determining conformational preferences. Rose and Wetlaufer (1977), using the algorithm of Rose and Seltzer (1977) for determining turns, showed that the number of turns in a protein is a linear function of the number of amino acid residues in the protein. In this sequence-dependent model, turns are a linear function of the molecular weight, whereas in a shape-dependent model, they are a function of the two-thirds power of the molecular weight. The number of turns (T) is a linear function of the number of amino acids (R) as given by $T = 0.125R + 2.28$.

Rose (1978) hypothesized that turns occur at those sites in the polypeptide chain where the hydrophobicity is at a local minimum. The measure of hydrophobicity of the amino acid side chains is taken to be the Nozaki and Tanford (1971) free energy of transfer from water to an organic solvent. One problem arises because one or more local minima in hydrophobicity are always found to be associated with helical secondary structure. An agreement of 78% between measure and predicted turns was claimed. Conformational energy calculations were carried out on the two terminally blocked tetrapeptides N-acetyl-Thr-Asp-Gly-Lys-N′-methylamide and N-acetyl-Ala-Asp-Gly-N′-methylamide (Simon *et al.*, 1978). The latter occurs as a bend at residues 94–97 in staphylococcal nuclease. Several groups of low-energy conformations were found. They were compactly folded structures, but they differed from the "standard" chain reversals. One group, containing Thr peptides, was stabilized by a network of hydrogen bonds involving polar atoms of both the backbone and side chains of Thr, Asp, and Lys.

Anderson *et al.* (1979) optimized energies of seven β bends, repeating C5 and C7 and right- and left-handed α-helical conformations for each of eight tetrapeptides have been computed using empirical methods. Eight tetramers were selected: four helix-forming sequences with hydrophobic residues such as Val, Leu, Ile, and Trp and four helix-breaking sequences with hydrophilic residues such as Asp, Asn, and Ser, as determined by their frequency of occurrence in β turns in proteins. Analysis of the optimized conformations with energies ≤2.1 kcal/mole from absolute minimum energy conformer for each tetramer revealed a correlation between low-energy conformations and those predicted from observed protein structure. These results indicated that energy calculations, performed with the Empirical Conformational Energy Program for Peptides (ECEPP) developed by Scheraga and co-workers (Momany *et al.*, 1975), on small peptide fragments may be useful in predicting protein structure. From the bend frequencies based on 29 proteins (Chou and Fasman, 1979a), the β-turn probability profiles were calculated for three sets of proteins: ten mammalian proinsulins, seven proteinase inhibitors, and 12 species of pancreatic ribonucleases. Despite relatively low sequence conservation in these three sets of proteins, β turns were predicted to be highly conserved: 33% sequence versus 78% bend for proinsulins, 20% sequence versus 85% bend for proteinase inhibitors, and 65% sequence versus 92% bend for ribonucleases.

It was suggested that chain-reversal regions play an essential role in keeping the active structural domains in hormones and enzymes intact for their specific biological function. The concept of β bends [nonhelical dipeptide sequences in which the distance $R_3(i,i + 3)$ between C^{α} of residues i and $i + 3$ is ≤7.0 Å] has been extended to define double bends as tripeptide sequences not in an α helix in which two successive distances $R_3(i,i + 3)$ and $R_3(i + 1,i + 4)$ are both ≤7.0 Å, with analogous definitions for higher-order multiple bends, as examined by Isogai *et al.* (1980). A sample of 23 proteins consisting of 4050 residues contains 235 single, 58 double, and 11 higher-order multiple bends. Multiple bends may occur as combinations of the "standard" type I, II, and III chain reversals (as well as their mirror images), but usually they require distortions from these well-defined conformations. The frequencies of occurrence of amino acids often differ significantly between single and multiple bends. The probability of distribution of R_3 distances does not differ in single and multiple bends. However, R_4 (the distance between the C^{α} atoms of residues i and $i + 4$) in multiple bends is generally shorter than in tripeptide sequences containing single bends. The value of R_4 in many multiple bends is near those for α helices. Double bends in which the signs of two successive virtual-bond dihedral angles differ have conformations that are very different from an α helix. They act as chain reversals occurring over three residues. Multiple bends may play an important role in protein folding because they occur fairly frequently in proteins and cause major changes in the direction of the polypeptide chain. Such multiple β bends have been frequently predicted by the Chou and Fasman (1977b) method (G. D. Fasman, unpublished data).

A simple algorithm was developed by Kolasker *et al.* (1980) to detect β bends and "loop" chain reversals containing five amino acid residues, using only coordinates of C^{α} atoms from crystal structure data of globular proteins. Analysis of bends showed that the total number of bends in each protein (T_B) is linearly related to the total number of nonhydrophobic residues in that protein, which in turn is related linearly to the total number of amino acid residues. This result is similar to that reported by Rose and Wetlaufer (1977). They also reported that a large number of consecutive bends occur in each protein, which give rise to, on an average, only three independent residues per turn. Positional preferences of amino acid residues in chain reversals were stressed.

Nemethy and Scheraga (1980) characterized β bends in proteins by a range of dihedral angles. These were classified into eight groups according to the orientation of the three peptide groups comprising the bend. The possibility of formation of intrabend hydrogen bonds involving N-H and C-O groups depends on the relative orientation of the peptide groups and hence differs for various types of bends. On occasion, β turns were buried in the hydrophobic interior of the molecule, although most turns are always situated at the surface of the protein in contact with solvent water (Rose *et al.*, 1983). In every instance of a buried turn, one or more solvent molecules were also found in a hydrogen-bonded complex with main-chain atoms of the turn residues. These bound water molecules appear to function as an integral part of the protein structure.

Cohen *et al.* (1983) described an algorithm for assigning secondary structure of α/β proteins. Turns were identified very accurately (98%) by simultaneously considering hydrophilicity and the ideal spacing of turns throughout the sequence. The segments bounded by these turns are labeled by a pattern recognition scheme based on the physical properties of α helices and β strands in this class of proteins. Long-range as well as local information is incorporated to enhance the quality of the assignments. The algorithm successfully divides proteins into two classes: α/β and non-α/β. This method is discussed in detail in Section VI.

Tetrapeptide sequences of the type *Z*-Pro-*Y*-*X* were obtained from the crystal structure data on 34 globular proteins and used in an analysis of the positional preferences of the individual amino acids in the β-turn conformation (Ananthanarayanan *et al.*, 1984). This work is an extension of the work of Chou and Fasman (1977b). The effect of fixing proline as the second-position residue in the tetrapeptide sequence was studied. Differences were found in positional preferences for the two sequences *Z*-*R*-*Y*-*X* and *Z*-Pro-*Y*-*Y*. Murakami (1985, 1987) made an attempt to predict the conformations, particularly β turns, around the mutation regions of the p21 protein and its cancer-associated variants by using the prediction method of Chou and Fasman (1978a). Mutations affecting the 12th and 61st amino acid of p21 resulted in a decreased probability of β-turn occurrence. Point mutations at residues 12, 13, or 61 are involved in malignant activation of *ras* protooncogenes. Probabilities of β-turn occurrence at residues 10–13 or 58–61 of the p21 proteins are high. Thus, these critical amino acids lie within β turns.

β Hairpins are widespread in globular proteins and have often been suggested as possible sites for nucleation (Ptitsyn, 1981). Sibanda and Thornton (1985) examined the loop regions of β hairpins in proteins of known structure and found that the "tight" β hairpins, classified by the length and conformation of their loop regions, form distinct families, and the loop regions of the family members have sequences that are characteristic of that family. The two-residue hairpin loops include entirely I′ or II′ β turns, in contrast to the general preferences for type 1 and type 2 turns (Chou and Fasman, 1977b; Lewis *et al.*, 1973a). Milner-White and Poet (1986, 1987) demonstrated that β hairpins can be divided into four classes, each with a number of members. Hairpins from a single class are readily interconverted by loss or gain of hydrogen bonds, but interconversion between classes requires unzippering and reformation of the entire β hairpin.

Milner-White (1988) determined that there is a recurring loop motif in proteins that occurs in both right-handed and left-handed forms, found frequently at the C-terminal end of α helices with a characteristic hydrogen bond pattern, which is called a paperclip. It was also noted that several loops with the same structural features occur independently of α helices; e.g., two are situated at the loop end of β hairpins. These paperclips exist in two classes depending on the number of residues at the loop end. Two such loops belong to the common class except that the main-chain conformation is the mirror image of that normally found. The majority of paperclips were shown to have tightly clustered sets of main-chain dihedral angles.

These are somewhat similar to, but distinct from, a subgroup of another common family of loops that have been called β bulge loops, whose dihedral angles are also tightly clustered. The high degree of clustering in both cases is likely to be a result of steric constraints associated with hydrogen-bond patterns at the ends of loops. Sibanda and Thornton (1985) have classified β hairpins in terms of the number of "loop residues" as either two-residue, three-residue, four-residue, etc., loops. This leads to ambiguity, but if the class (one of the four mentioned above) is also specified, the description of β hairpins becomes straightforward.

Cohen *et al.* (1986a,b) extended the use of amino acid sequence patterns (Cohen *et al.*, 1983) to the identification of turns in globular proteins. The approach uses a conservative strategy combined with a hierarchic search and length-dependent masking to achieve high accuracy (95%) on a test of proteins of known structure. Applying the same procedure to homologous families gives a 90% success rate. The loops in globular proteins were considered in detail on 67 proteins of known structure by Leszczynski and Rose (1986) were categorized as a novel secondary structure. The protein loop, a novel category of nonregular secondary structure, is a segment of contiguous polypeptide chain that traces a "loop-shaped" path in three-dimensional space; the main chain of an idealized loop resembles a Greek omega (Ω). A systematic study revealed 270 omega loops. Although such loops are typically regarded as "random coil," they are, in fact, highly compact substructures and may also be independent folding units. Loops are almost invariably situated at the protein surface, where they can assume important roles in molecular function and biological recognition.

To identify a loop from x-ray coordinates, the following rules are applied. (1) The segment length must be between six and 16 residues. Loops of this length allow their side-chain atoms to pack within the loop's own core. (2) The distance between segment termini, that is, end-to-end distance, is measured as the distance from the first α carbon to the last α carbon in the segment, must be less than 10 Å, and may not exceed two-thirds the maximum distance between any two α carbons within the segment under consideration. The residue composition of loops was assessed by calculating the normalized frequency of occurrence, f, for each residue type, X, such that $f = X_L X_T / N_L N_T$, where X_L is the number of the residues of type X in loops, X_T is the total number of residues of type X, N_L is the total number of residues in loops, and N_T is the total number of residues in the data base. These frequencies, when compared to the p_t values of Chou and Fasman (1977c), reveal that the residues most often found in reverse turns are also found most often in loops (Gly, Pro, Asp, Asn, and Ser). Parrilla *et al.* (1986) described a simple PASCAL microcomputer program based on the Chou and Fasman (1974a,b) algorithm for the prediction of protein secondary structure. The program also performs an analysis of the hydrophobic character of the residues. The authors compare their results with the original results of Chou and Fasman (1978a) for the prediction of β turns and report very good agreement.

Edwards *et al.* (1987) analyzed 129 loops of 70 βαβ units from 17 α/β proteins for patterns. There were many different conformations of the loop regions, but 18 of the loops could be classified into one of four loop families with distinctive conformations and sequence patterns: (I) adjacent α/β loops with one residue between the α helix and β strand—the residue is a Gly with conformationally restricted φ,ψ angles; (II) adjacent αβ loops of three residues with a conformationally restricted Gly as the first of the loops followed by an Ala or His and a third residue with helical φ,ψ angles; (III) adjacent βα loops of three or four residues previously reported to bind nucleotides and that have three Gly residues in the loop region; (IV) nonadjacent βα loops of zero residues with a Ser or Thr as the last residue of the β strand. The analysis provides information for model building of loops and prediction of secondary structure from amino acid sequences.

In a series of papers, Efimov (1985, 1986a,b) performed a stereochemical analysis of regions previously termed irregular in known protein structures and found new regions of "unknown standard conformations." Using Ramachandran maps (Ramachandran *et al.*, 1963), he assigned six regions of "condensation," designated by the symbols α, α_L, β, γ, δ, and ϵ. The conformation of an irregular region can be written by enumerating the conformations of its constituent residues, starting from the N terminus. The α, α_L, and β have the previous ψ,φ values of Ramachandran *et al.* (1963), whereas γ, δ, and ϵ are newly found regions of high occupancy in protein structures. For example, the two β strands in a β sheet can be linked by the structures $\beta\beta\alpha_L\beta$, $\beta\alpha\gamma\alpha_L\beta$, $\beta\alpha\alpha\alpha_L\beta$, etc., which are referred to as turns. The structures of α-helix–β-strand type examined were those in which the α helix and β strand were packed approximately antiparallel with formation of α–β hairpins. It was shown that for α–β hairpins with connector length no more than five or six peptide units, a limited number of standard conformations was found in proteins (Efimov, 1986a). Each of these standard α–β hairpins must have its own strictly determined alternation of hydrophobic, hydrophilic, and glycine residues. In general, two conditions are required for formation of any structure in globular proteins: (1) the prohibition of dehydration of polar groups and (2) the prohibition of strong steric tension. The simplest transition from an α into a β structure via an α–β hairpin is observed with an $\alpha_m\gamma\alpha_L\beta$–$\beta_n$ conformation where m is the number of residues in the α helix and n is the number of residues in the β structure. Efimov categorizes all possible connectors (or gaps) up to six residues. The β–α hairpin is also analyzed in the same manner (Efimov, 1986b). The standard β–α hairpins are two-layered structures in which hydrophobic clusters of α helices and β strands are oriented approximately toward one another and are located at different layers. They also differ primarily in the length and the conformation of the connectors and also by the degree of twisting of the β strands. A series of proteins with various numbers of amino acids in the connectors were examined, and a defined set of parameters was obtained for these regions previously termed "irregular" (e.g., $\beta_m\beta\alpha\beta\alpha_n$). Thus, it is not obligatory that each β–α hairpin in a protein be unequivocally governed by primary structure, but a certain pattern of residues can cause formation of β turns or other hairpins.

Wilmot and Thornton (1988) examined 59 nonidentical proteins whose x-ray structure had been determined with a resolution of ≤2 Å, and extracted 735 β turns. Using φ,ψ angles, these β turns were classified into seven conventional types (I, I′, II, II′, IV, VIa, VIb) and a new class turn, designated VIII, in which the central residues ($i + 1$, $i + 2$) adopt an $\alpha_R\beta$ conformation. Type I′ and II′ turns were found in 83% and 53%, respectively, of β hairpins. These two turn types were shown to be strikingly different in their sequence preference. Type I turns favor Asp, Asn, Ser, or Cys at i; Asp, Ser, Thr, or Pro at $i + 1$; Asp, Ser, Asn, or Arg at $i + 2$; and Gly, Trp, or Met at $i + 3$. Type II turns prefer Pro at $i + 1$; Gly or Asn at $i + 2$; and

Gln or Arg at $i + 3$. The positional trends for types I and II were incorporated into a simple empirical predictive algorithm originally developed by Lewis *et al.* (1971). With these new parameters, 72% of β turns were predicted within ±2 residues, compared to 41% using the original Chou and Fasman parameters.

Aubert *et al.* (1976) calculated the structure of the peptide segment around the carbohydrate–peptide linkage in glycopeptides by the Chou and Fasman (1974a,b) algorithm. Sequences around Thr and/or Ser for O-glycosidic and Asn for N-glycosidic linkages were studied. Nine O-glycosidically linked glycans and 28 N-glycosidically linked glycans were predicted. All O-glycosidic links were within β turns. The Asn residue was frequently found (19 out of 28) either in or near a β turn. A conformational study of α_1-acid glycoprotein (Aubert and Loucheaux-Lefebrve, 1976) and a prediction of its conformation showed that four of five glycan chains are linked to Asn residues that are situated either in a reverse β turn or in regions where charged residues are numerous. Beeley (1976) located the four carbohydrate groups of ovomucoid, a family of glycoproteins with antitryptic activity that have been isolated from the egg white of several avian species, on Asn groups. These Asn groups were very close to groups of amino acids that occur with high frequency in β turns (Chou and Fasman, 1974b). The sequence Asn(CHO)*X*-Ser/Thr has previously been proposed as a necessary requirement for glycosylation of Asn (Neuberger and Marshall, 1968). Beeley (1977) applied the Chou and Fasman (1974b) predictive scheme to predict the peptide chain conformation of the amino acid sequences adjacent to carbohydrate attachment sites of glycoproteins containing N-glucosylamine-type protein–carbohydrate linkages. Of 31 glycosylated residues examined, 30 occur in sequences favoring β-turn structures.

Small *et al.* (1977) examined 14 different proteins that were highly phosphorylated and found that 24 out of 30 phosphorylated residues (80%) existed within regions predicted to be β turns. Phosphorylated sites not predicted within turns were found to be adjacent to predicted turns (±2 residues) in four other cases. It was suggested that these β turns play a more active role in biological function in addition to their directional effect on the folding of globular proteins. Previously Mercier and Chobert (1976) had pointed out that in the caseinomacropeptides, phosphorylated hydroxy amino acid residues are located in a tripeptide sequence Ser-*X*-Glu. A more detailed predictive study of the K-caseins by Loucheux-Lefebrve *et al.* (1978) demonstrated that both glycosylation and phosphorylation sites were attached to β turns (Chou and Fasman, 1974a). Loucheux-Lefebrve (1978) also predicted β turns in different regions of peptide and glycopeptide antifreezes. Ricard *et al.* (1983) examined the structural requirement of the Asn-*X*-Thr(Ser) sequence for the N-glycosylation of proteins as a local conformation acting as a signal for the enzymatic process. Energy calculations were performed on the substrate Ac-Asn-Ala-Thr-NH_2, and the lowest-energy conformers have been characterized as β-bend structures. The structural comparison of protein sequences around potential N-glycosylation sites was made by Mononen and Karjalainen (1984). One hundred and five proteins with 139 glycosylated and 57 nonglycosylated Asn-*X*-Ser/Thr sites were analyzed. The primary sequence data indicated a lack of glycosylation in structures containing either Pro or Glu at position *X*. The study of predicted structures (Chou and Fasman, 1974b) showed that in both glycosylated and nonglycosylated acceptor sites, 70% of the Asn occurred in β turns, approximately 20% in β sheets, and 10% in helical conformations. The Asn had no preference for a particular position in the β turn. Statistical tests did not show significant differences in the predicted secondary structure, contradicting claims of the importance of the β turns. Parrilla *et al.* (1986) described a simple PASCAL microcomputer program for predicting secondary structure according to the Chou and Fasman (1978a,b) algorithm. They predicted the β turns in four proteins—myothemerythrin, superoxide dismutase, thioredoxin, and trypsin inhibitor—and compared the results to the x-ray-determined structures. The results were very satisfactory.

B. Evaluation of Predictive Methodologies

The three most frequently used predictive methods of the secondary structure of proteins—those of Chou and Fasman (1974a,b), Garnier *et al.* (1978), and Lim (1974a,b)—have come under considerable scrutiny and evaluation. These three methods are compared in this section.

The first comparison of various predictive schemes for predicting α helices, performed before the three above named methods were published, was that reported by Dickerson *et al.* (1971), who had determined the structure of both horse and bonito ferricytochrome *c* at 2.8-Å resolution. The methods of Prothero (1966), Schiffer and Edmundson (1967), Low *et al.* (1968), Kotelchuck and Scheraga (1969), and Lewis *et al.* (1970) were compared. All the methods except that of Kotelchuck and Scheraga (1969) identified the α-helical 91 to 101 region correctly. Most of the methods predicted helix in regions of residues up to 21 and 80–88. The first region, up to 21, has a helical entity in the crystal structure, but the predicted region from 80 to 88 is completely wrong. The helical segment in the sequence in the region of 60–70 is approximately correct for most schemes. Thus, for cytochrome *c*, the helix-predicting methods appeared to that date (1971) to have considerable validity, their chief defect being that they were not sufficiently discriminating, with a tendency to predict more helix than was actually present. The partition function of Lewis *et al.* (1973b) was particularly poor in this respect. Thus, the methods available in 1971 did not give satisfactory predictions.

The second comparison of various predictive schemes was carried out by Schulz in 1974 (Schulz *et al.*, 1974a) for adenylate kinase, whose 3-Å x-ray crystallographic structure he and co-workers (Schulz *et al.*, 1974a) had determined. He invited the authors of 11 predictive methods to use the known sequence and contribute to this joint effort. Figure 1 shows the results of this comparison of the predicted and the experimentally determined α helices, strands of β sheets, and β bends in adenylate kinase. The highest percentage of α helices correctly predicted was 82% (Lim, 1974a,b), with 22% β sheets and 38% turns; Chou and Fasman (1974a,b) predicted correctly 28% of bends, 20% of β sheets, and 70% of α helix. Matthews (1975) compared the predicted and observed secondary structure of T4 phage lysozyme. Within the amino-terminal half of the molecule, the location of helices, predicted by a number of methods, agrees moderately well with the observed structure; however, within the carboxyl half of the molecule, the overall agreement was poor. For 11 different helix predictions, the coefficients giving the correlation between prediction and observation ranged from 0.14 to 0.42 (Lim, 1974a,b, 60% α helix, 42% β sheet; Chou and Fasman, 1974a,b, 53% α helix, 42% β sheet, 36% β turn). This result is less successful than that reported by Schulz *et al.* (1974a) above. Burgess and Scheraga (1975) emphasized that these predictive algorithms gave little information about the three-dimensional structure of a protein. They stated, "However, starting from the results of a *perfect* prediction algorithm, it appears that conformational energy minimization (with long interactions included) can lead to a structure having the general features of the native protein."

Argos *et al.* (1976) computerized five secondary predictive methods (Nagano, 1973, 1974; Barry and Friedman, as quoted in Schulz *et al.* 1974a; Chou and Fasman, 1974a,b; Kabat and Wu, 1973a,b) based on protein amino acid sequence and combined them to calculate a joint prediction. Forty known structures were predicted. It was found, as previously, that accuracy of prediction of α helices is better than that for β sheets or turns and that the amino-terminal-half prediction is superior to that for the carboxyl-terminal half. However, there was a little improvement in this joint predictive scheme.

Lenstra *et al.* (1977) compared the primary structures of 24 homologous ribonucleases that differ in up to 34% of their amino acids. The results using the Chou and Fasman (1974a,b)

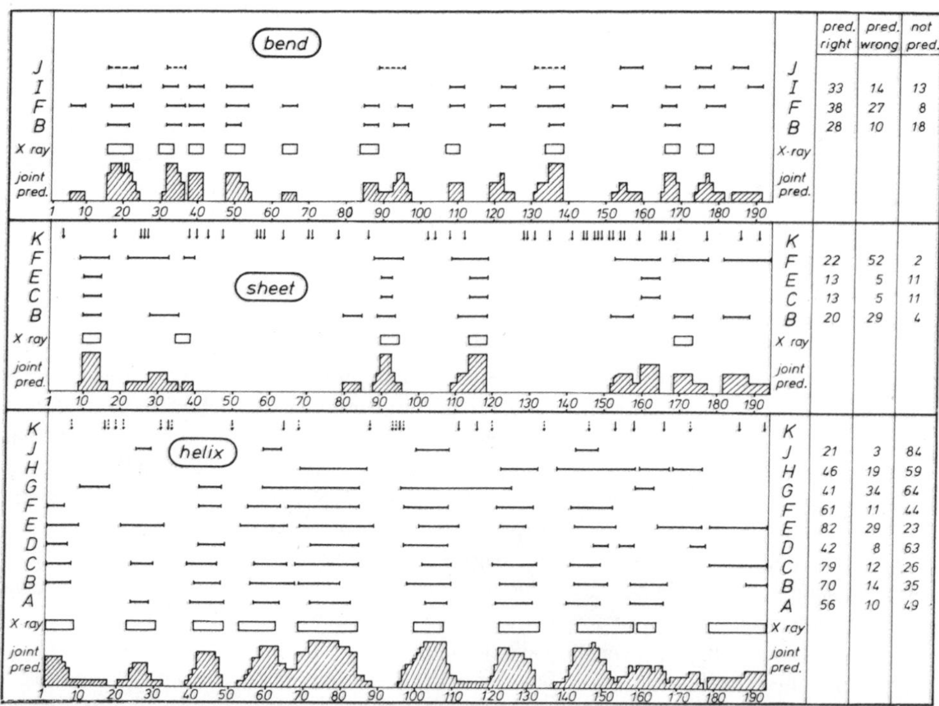

Figure 1. Comparison of predicted and experimentally determined α helices, strands of β pleated sheets, and bends in adenylate kinase. The experimental data (x-ray) have been derived from a 3-Å electron density map. At this resolution the exact geometry of bends cannot be evaluated with certainty. Therefore, the experimentally determined bends are defined as changes of more than 120° in the overall direction of the polypeptide chain and without reference to any hydrogen-bonding scheme. None of the predictions is biased by previous structural information about the enzyme. The predictions *A* to *K* have been supplied by *A,* Barry and Friedman; *B,* Chou and Fasman, 1974a,b; *C,* Ptitsyn and Finkelstein, 1970a,b; *D,* Levitt and Robson; *E,* Lim, 1974a,b; *F,* Nagano, 1973; *G,* Kotelchuck and Scheraga, 1969; *H,* Lewis *et al.*, 1971; *I,* Burgess and Scheraga, 1975; *J,* Burgess *et al.*, 1974; and *K,* Kabat and Wu, 1973a. They are based on prediction schemes that have been described in the corresponding references. (The method applied by Barry and Friedman has not been published yet. It uses an analysis of the distribution of amino acids occurring at both ends of helices to define potential starting and termination points of helices in a sequence. In this method the residues three before and three after, as well as the residue defined as the beginning (or end) of an helix, are considered to be structurally significant. The predictions are based on data for 90 helices from 23 distinct three-dimensional protein structures.) Although predictions are usually made as probability profiles, all groups converted their profiles beforehand to yes-or-no decisions for each residue by comparing the probabilities with a given threshold value. Thus, every prediction depends on the threshold value applied[32]. This procedure simplified the comparison appreciably. But concomitantly, it reduced the amount of information contained in any of the predictions. Kabat and Wu predicted helix and sheet-breaking residues, which are indicated by vertical arrows. Dashed arrows point to residues that are helix breaking but with a lower probability. Dashed lines in the bend prediction *J* indicate "multiple bend regions." A bend is predicted to be anywhere within such a region. Three joint prediction histograms have been produced by adding predictions *A* to *J* for helix, predictions *B, C, E,* and *F* for sheet, and predictions, *B, F, I,* and *J* for bends, respectively. Scores of correctly and incorrectly predicted residues as well as residues not predicted are listed on the right side of the figure. From Schulz *et al.* (1974a).

method were in better agreement with the x-ray structure of bovine RNAse than the method of Burgess *et al.* (1974). The method of Lim (1974a,b) gave the most satisfactory results. Lenstra (1977) also evaluated the predictive accuracy of the histogram method of Argos *et al.* (1976), the statistical method of Nagano (1973), and the stereochemical method of Lim (1974a,b). The method of Nagano yielded the best prediction of β structure, whereas the β-structure predictions of Lim and Argos *et al.* were not significantly different. The results of the α-helix and β-structure predictions according to the statistical mechanical method of Tanaka and Scheraga (1976a–d) were inferior to those obtained by the other three methods. For the prediction of turns, there was no significant difference between the methods of Nagano (1974) and Argos *et al.* (1976). The method of Chou and Fasman (1974a,b) was not utilized, as it was stated that their rules were ambiguous and difficult to program.

Kabsch and Sander (1984, 1985) pointed out the lack of structural significance of short sequence homologies. In 62 proteins with 10,000 residues, they found the longest isolated homologies between correlated proteins to be five residues long. In six out of the 25 cases, they found structural adaptability; the same five residues are part of an α helix in one protein and part of a β strand in another. These examples show quantitatively that pentapeptide structure within a protein is strongly dependent on sequence context. However, they state that this fact is essentially ignored in most protein structure prediction methods. This statement is misleading, because it does not apply to several predictive methods (e.g., Chou-Fasman, 1978a,b), where this is taken into account by the methodologies used. Unfortunately, this paper has been quoted to demonstrate the unreliability of predictive schemes (e.g., Harrision, 1985). Wilson *et al.* (1985) have used this observation, that the same pentapeptides can have different conformations in different proteins, to test the theories of immune recognition.

Two sequences that have been found to have different conformations (by x-ray diffraction studies) have been examined by the Chou–Fasman (1978a,b) predictive scheme to investigate this conformational duplicity. The sequence VELIRG, found in influenza neuraminidase (Elleman *et al.*, 1982), was found to have an α-helix structure (Varghese *et al.*, 1983). The same sequence found in the protein disk of tobacco mosaic virus (Anderer, 1963) was shown to have the β-sheet conformation (Bloomer *et al.*, 1978). On prediction of the secondary structure (Chou and Fasman, 1978a,b), $\langle P_\alpha \rangle = 1.07$ and $\langle P_\beta \rangle = 1.20$ were obtained, indicating a β-sheet conformation, which is an incorrect prediction. However, the prediction for the TMV coat protein, $\langle P_\beta \rangle = 1.21 > \langle P_\alpha \rangle = 1.12$, indicated a β-sheet conformation, which agrees with the x-ray structure.

Another sequence, NAAIRS, found in phosphofructokinase (Kolb *et al.*, 1980) was located in an α-helical segment. This same sequence in thermolysine (Titani *et al.*, 1982) was present in a β-sheet conformation (Holmes and Matthews, 1982). On predicting this sequence, with its contiguous residues on either side, the phosphofructokinase gave values of $\langle P_\beta \rangle = 1.14$, $\langle P_\alpha \rangle = 1.06$, an incorrect prediction. However, this sequence in thermolysine, with contiguous sequences, had $\langle P_\alpha \rangle = 1.03$ and $\langle P_\beta \rangle = 1.05$. Thus, the prediction could not distinguish between these structures. The most likely explanation for these questionable predictions is that the secondary structure of a sequence in an intact protein is strongly influenced by its neighboring amino acids as well as by the tertiary interactions encompassed in the domain in which it resides. Attempts are now being made to include the latter factor in secondary structural predictions (e.g., Chou–Fasman).

Schulz and Schirmer (1979) have discussed the merits of various predictive algorithms in great detail and have determined the relationship between the various quality indices used by the individual authors.

Busetta and Hospital (1982) analyzed the prediction of the secondary structures of 38 proteins of known structure, using the methods of Chou and Fasman (1974a,b) and Garnier *et al.* (1978). The percentage of correctly predicted residues in the three states helix, extended, or turn (H, E, T) of the basic Garnier *et al.* (1978) method was 56.8%, whereas for the Chou and Fasman (1974a,b) model it was found to be 47%. By varying the decision constant DC_H, which corresponds to a proportion of H states, less than 0.2 or greater than 0.5, the percentage of correctly predicted states remains important at, respectively, 49.1% and 56.6%. The use of hydrophobic reinforcements (with $k_E1 = 1.5$) improves the basic procedure of Garnier *et al.* (58.4% correctly predicted states). The prediction level is clearly better for proteins with a single type of secondary structure (all α or all β) than for proteins of a mixed type (α + β or α/β). A previous knowledge of the protein type greatly improves the prediction level. They suggest that the efficiency of prediction could be improved by the use of the distribution of hydrophobic residues.

Kabsch and Sander (1983a) tested the three most widely used methods (Chou and Fasman 1974a,b; Garnier *et al.*, 1978; Lim, 1974a,b,c) for prediction of protein secondary structure from the amino acid sequence on 62 proteins of known structure using a new program package and data collection. They state that they have overcome the ambiguities in two of the best-known methods, the Chou and Fasman (1978a) and Lim (1974a–c) methods, and have also removed the variation in the definition of secondary structure given by crystallographers by their objective and accurate assignment of secondary structure by a pattern recognition algorithm (Kabsch and Sander, 1983b). They have now applied these "improved" methods to more than 10,000 residues. For the three-state definition of secondary structure (helix, sheet, loop/turn), the overall prediction accuracy for these newly defined secondary structures did not exceed 56% for the best of these (Lim, 1974a–c; Garnier *et al.*, 1978) and was only 50% for the most widely used method of Chou and Fasman (1978a). The Chou and Fasman ambiguities were overcome by selecting possible secondary structure segments such that the sum of the preference parameters over all chosen segments was maximal. The β-turn prediction was done separately using the Chou and Fasman (1979a) method. Ambiguities in the method of Lim (1974a) were overcome by a simplified iterative procedure for segment selection written by Lenstra (1977). The method of Robson and co-workers (Garnier *et al.*, 1978) was used as programmed by the authors.

Nishikawa (1983) also assessed the predictive accuracy of the same three methods. The predictive abilities of the three methods turn out to be almost at the same level but unexpectedly low, less than 55% measured by the three-state assessment (α, β, and coil) or less than 45% measured by the four-state assessment (α, β, turn, and coil). Wallace *et al.* (1986) have evaluated the validity of using predictive schemes developed for soluble proteins (Chou and Fasman, 1974a,b; Garnier *et al.*, 1978; Burgess *et al.*, 1974) for membrane proteins and have concluded that they are inappropriate for predicting the structure of membranes (15 examined). Only two of these membrane proteins, crambin and the reaction center from *Rhodopseudomonas viridis,* have had their structure determined by x-ray crystallography. The other membrane protein conformations were determined by physical-chemical techniques, and the interpretation of the data is open to question. Thus, the sample is very limited to draw such a definitive conclusion. Further discussion of this point can be found in Section VII.

Scheraga and co-workers (Burgess *et al.*, 1974; Tanaka and Scheraga, 1976a–c; Maxfield and Scheraga, 1976) have been severely critical of the Chou and Fasman (1974a,b) method. They have pointed out apparent ambiguities in some of the predictive rules, have suggested that the scheme is incomplete, have pointed out perceived misconceptions in their mathematical treatment, and have demonstrated that they cannot reproduce the claimed results.

In conclusion, it can be stated unequivocally that the original claims of accuracy in the predictability of the various methods of the secondary structure of proteins have not been found to be maintained in the laboratories of others.

C. Other Predictive Algorithms

Other predictive algorithms that have been published but not used extensively because of either their lesser predictability score or their newness are those of Scheraga and co-workers (Burgess *et al.*, 1974; Tanaka and Scheraga, 1976a,b,c; Maxfield and Scheraga, 1976) and Troitskii and Zav'yalov (1972). McLachlan (1977) developed a purely statistical theory that uses the observed tendencies of single amino acids and the lengths of typical helices and β sheets. Ptitsyn and Finkelstein (1983) developed a molecular theory of protein secondary structure that takes into account both local interactions inside each chain region and long-range interactions between different regions, incorporating all these interactions in a single Ising-like model. Local interactions are evaluated from the stereochemical theory describing the relative stabilities of α and β structures for different residues in synthetic polypeptides, and long-range effects are approximated by the interactions of each chain region with the averaged hydrophobic template (also see Ptitsyn, 1985). Busetta and Hospital (1982) studied ways of introducing different kinds of interactions that lead to protein folding into secondary structure predictions. The secondary structure prediction was based on the Garnier *et al.* (1978) method with hydrophobic reinforcement (Busetta and Hospital, 1982; Lim, 1974a–c). A previous knowledge of the protein type (e.g., α, α/β) greatly improves the accuracy of prediction.

Moult and James (1987) examined the feasibility of determining the conformation of segments of a polypeptide chain up to six residues in length in globular proteins by means of a systematic search through the possible conformations. Trial conformations are generated by using representative sets of φ, ψ, and χ angles that have been derived from an examination of the distribution of these angles in refined protein structures. A set of filters based on simple rules that protein structures obey was used to reduce the number of conformations to a manageable total. The most important filters are the maintenance of chain integrity and the avoidance of too-short van der Waals contacts with the rest of the protein and with other portions of the segment under construction. The electrostatic energy, including a solvent-screening term, and the exposed hydrophobic area are evaluated for each accepted conformation. The method was tested on two segments of chain in the trypsinlike enzyme from *Streptomyces griseus*. It was found that there is a wide spread of energies among the accepted conformations, and the lowest-energy ones have satisfactorily small root mean square deviations from the x-ray structure.

Bashford *et al.* (1987) examined 226 globin amino acid sequences for any unique features. These sequences, which only have two residues absolutely conserved and the residue identities of some pairs of sequences only 16%, were aligned by the use of structural data and analyzed by a new procedure. Although individual chains vary in size between 132 and 157 residues, deletions and insertions result in there being only 102 residue sites in common in all globins, forming six separate regions. Within the conserved regions 32 sites are highly hydrophobic. Another 32 sites are almost always occupied by charged polar or small nonpolar (Gly or Ala) residues, which occur on the protein surface. These six conserved regions and the residue restrictions that occur at the 66 sites within these regions were encoded into two "templates." One was based only on the sequences so far determined; the other was extended to include yet unobserved substitutions that seemed plausible on the basis of size, hydrophobicity, and polarity. Each of the 3286 nonglobin sequences in the data bank was examined by a computer program to see how closely it could be matched to these templates, and it was found that no nonglobin made an exact match to either template. Thus, the features of the globin sequences that are conserved and define its fold are essentially unique to that family.

Klein, DeLisi, and co-workers (Klein *et al.*, 1984; Klein, 1986), using the National Biomedical Research Foundation sequence data, found by discriminant analysis that the protein superfamilies cluster into six groups that can be distinguished on the basis of four variables characterizing amino acid composition and local sequence data. These variables are average hydrophobicity, net charge, sequence length, and periodic variation in hydrophobic residues along the chain. The clusters they distinguished were (1) globins, (2) chromosomal proteins, (3) contractile system proteins and respiratory proteins other than cytochromes, (4) enzyme inhibitors and toxins, (5) enzymes except hydrolases, and (6) all other proteins. The overall probability of correctly allocating a given protein to one of these functional groups was 0.76, with the allocation reliability being highest for globins (0.97) and for chromosomal proteins (0.93). They also found that approximately 53% of these protein sequences can be allocated to one of 26 functional classes, each of which can be characterized by the joint occurrence of four or fewer attributes. The attributes reflect collective physicochemical properties of the sequences in a class, ranging from simple characteristics of composition, such as average hydrophobicity and net charge, to amphilicity and the propensities of various residues to be in certain preferred conformations. These attributes permit 17 of the 26 groups to be filtered from all other proteins in the data base with a miscalculation error of less than 2%, and the remaining nine groups can be filtered with errors not exceeding 13%.

Argos *et al.* (1982) developed a prediction algorithm for membrane-bound proteins based on physical characteristics of the 20 amino acids and refined by comparison to the proposed bacteriorhodopsin structure. The scheme was devised to delineate likely membrane-buried regions in the primary sequences of protein known to interact with the lipid bilayer. Using the thus-calculated lipid-buried segments in several membrane-bound proteins allowed a hierarchic ranking of the 20 amino acids in their preference to be in lipid contact. A helical wheel analysis of the predicted regions suggested which helical faces are within the protein interior and which were in contact with the lipid bilayer. Such analyses should be viewed with caution, as the original bacteriorhodopsin structure used is a highly speculative one.

An interesting approach was developed by Manavalan, Ponnuswamy, and co-workers (Manavalan and Ponnuswamy, 1977, 1978; Ponnuswamy *et al.*, 1981) based on the "preferred environment" associated with each of the 20 amino acids. Previously one of the best parameters exploited in studies on protein structure has been the hydrophobic index of amino acids (Tanford, 1962; Jones, 1975; Rose, 1978; see Section III.B). However, Chothia (1976) noted that this parameter had a very poor correlation with the extent to which the residue is buried in the native folded-protein matrix. Although the hydrophobic index is a measure of the preference by a residue in a nonpolar environment, it does not reflect to the same extent the environment in protein crystals. This is because of the difference in the environment within the protein molecule and in the nonpolar solvent used in deriving the parameter. Manavalan and Ponnuswamy (1977) defined a parameter called "the surrounding hydrophobicity" for residues, a set of modified hydrophobic indices. This new set of parameters was found to correlate in a much more significant way with the buried/exposed behavior of the residues found in the protein matrix. They extended their studies to define hydrophobic domains, nucleation sites, loop sections, characteristic directionality of the chain (segments), and the depth of the residue from the surface. The phenomenon of protein folding was interpreted in terms of the enrichment of the hydrophobic environment, and an attempt was made to assign spatial positions for the residues from the centroid of the molecule, using statistical parameters.

Argos (1987b) utilized certain residue physical characteristics and the Dayhoff relatedness-odds amino acid exchange matrix (Dayhoff *et al.*, 1983) to provide sensitive criteria for detection of weak sequence homologies. The search procedure uses several residue probe lengths in comparing all possible segments of two protein sequences, and search plots are shown with peak values displayed over the entire search length. Alignments are automatically effected using the highest search matrix values without the necessity of gap penalties.

D. Chou–Fasman Algorithm

The Chou–Fasman algorithm for the prediction of the secondary structure of proteins and its application by the authors (Chou and Fasman, 1974a,b, 1975a,b, 1977a–c, 1978a,b, 1979a,b; Fasman and Chou, 1974; Schulz *et al.*, 1974a,b; Fasman *et al.*, 1976, 1977; Small *et al.*, 1977; Fasman, 1980, 1982, 1985, 1987; Rosenblatt *et al.*, 1980, 1981; Heber-Katz *et al.*, 1985; Nussbaum *et al.*, 1985; Ötvös *et al.*, 1988; Murphy *et al.*, 1988) have been employed in studies of conformational problems in the following areas: a guide to x-ray crystallographic studies; segments with potential for conformational changes; rationale for amino acid substitution in peptide synthesis of biologically active polypeptides; recognition of homologous conformations in analogous proteins from various species with sequence differences; conformational dependence of protein-binding sites to membranes, nucleic acids, etc.; and the understanding of the loss of biological activity through enzyme cleavage (Fasman and Chou, 1974c).

The Chou and Fasman (1974a,b; 1978a,b) method has attracted much attention because it is seemingly simple to use and can be utilized without a computer. A recent survey by ISI, Philadelphia, showed that it had been referenced over 1061 times; however, popularity does not guarantee efficacy. Although its simplicity has been attractive, it has been frequently severely criticized (Matthews, 1975; Kabsch and Sander, 1983a; Burgess and Scheraga, 1975; Maxfield and Scheraga, 1976; Lenstra, 1977; Busetta and Hospital, 1982; Nishikawa, 1983; Wallace *et al.*, 1986). Many of the criticisms are well founded, and many others seek perfection comparable to x-ray crystallography. The original intent (see above) was to aid in our understanding of structure–function relationships and not to substitute for x-ray crystallography. In early comparative evaluations with 11 other methods (Schultz *et al.*, 1974a,b; Schultz and Schirmer, 1979), the Chou–Fasman algorithm compared favorably and ranked near the most accurate in all secondary structure predictions, while the other methods had poorer records. The detailed comparison with other methods is discussed below.

Because the object of this predictive scheme was to open up new approaches to biochemical and molecular biological problems, it can be rated as having contributed significantly in this area (see reviews Fasman, 1985, 1987). The following examples show its wide utility and contributions to interesting biological problems. One of the early predictions was that made for the pancreatic trypsin inhibitor (Chou and Fasman, 1974b). This 58-amino-acid-residue polypeptide was predicted with 87% of the α-helical and 95% of the β residue agreeing with the x-ray-determined structure (Huber *et al.*, 1971). This also illustrated the fact that predictions can be made on shorter sequences than those used to assemble the parameters.

The predictive algorithm not only can locate where the secondary structural regions are but will detect those regions in proteins having both helical and β-forming potentials and therefore the potential for conformational changes. For example, the x-ray crystallographic studies of concanavalin A (Edelman *et al.*, 1972; Hardman and Ainsworth, 1972) showed only 2% helix in the native structure. However, 55% helicity can be induced in concanavalan A with 70% chloroethanol (McCubbin *et al.*, 1971). The predictive scheme (Chou and Fasman, 1974b) correctly located all 12 β-sheet regions in concanavalin A with only one overpredicted β region. In addition, it showed that a total of 47% of its residues in 13 regions also have α potential, although many of these had still higher β potentials. Similarly, elastase has 7% helicity as shown from x-ray diffraction (Shotton and Watson, 1970), but circular dichroism studies showed 35% helicity in sodium dodecyl sulfate (Visser and Blout, 1971). The predictive method (Chou and Fasman, 1974b) showed that there are 79 residues in 15 regions with helical potential, accounting for 33% helicity. Hence, the easily computed $\langle P_\alpha \rangle$ and $\langle P_\beta \rangle$ values for the α and β segments in proteins may assist in elucidating the regions potentially capable of undergoing conformational change. It is interesting to note that the B1–7 region of insulin was predicted as β sheet (Chou and Fasman, 1974b) with $\langle P_\beta \rangle = 1.15 > \langle P_\alpha \rangle = 1.07$, in agreement with the x-ray data. Since B1–7 also has α potential, it is not surprising that this region was found to be helical in 4-Zn insulin in 6% NaCl (Bentley *et al.*, 1976).

The predictions of the conformation of glucagon, a hormone containing 29 amino acid residues, offers an excellent example of the potential of the method. Utilizing the conformational parameters for helix, β sheet, β turns, and random coil, Chou and Fasman (1975) predicted two conformational states for glucagon. They showed that the conformational sensitivity of glucagon may reside in residues 19–27, which have both α-helical potential ($\langle P_\alpha \rangle = 1.19$) and β-sheet potential ($\langle P_\beta \rangle = 1.25$). Thus, in predicted form *a*, residues 5–10 with $\langle P_\beta \rangle = 1.08 > \langle P_\alpha \rangle = 0.86$ adopt a β conformation, while residues 19–27 form a helical region (32% α, 21% β). In predicted form *b*, both regions, residues 5–10 and residues 19–27, are β sheets (0% α, 52% β). Circular dichroism studies (Srere and Brooks, 1969) of glucagon solutions (12·6 mg/ml) yield 33% α and 20% β, supportive of form *a*. Infrared studies of glucagon gels and fibrils (Gratzer *et al.*, 1967; Epand, 1971) have a predominant β conformation consistent with form *b*. In addition, three reverse β turns were predicted at 2–5, 10–13, and 15–18, suggesting that this small polypeptide has the potential to fold into a relatively compact structure. Thus, it appears that glucagon has different α and β conformations under different concentration conditions. Hence, residues 19–27 may be involved in an α → β transition.

The *in vivo* concentration of glucagon is probably too small to elicit the β conformation, but this conformational state may be induced on binding of glucagon to its receptor site. Because the conformational state of region 19–27 is sensitively balanced between α and β states, it was predicted that replacement of one or more residues of high β potential in this region with strong α formers would lock the conformation in the helical state. It is also feasible to lock the β conformation by suitable substitutions. Thus, the predictive scheme offers a working hypothesis whereby the structure of the biologically active hormone may be arrived at. The α → β transition of glucagon in solution has been followed by means of circular dichroism (Moran *et al.*, 1977), providing evidence for the potential conformational change predicted. Hruby *et al.* (1986) tested this hypothesis and studied the conformational aspects of glucagon agonists and antagonists by using synthetic analogues. Full analogues of glucagon designed to increase α-helical probabilities in the C-terminal region of glucagon lead to highly potent analogues. [Lys^{17}, Lys^{18}, Glu^{21}]Glucagon was 500% and 700% more potent than glucagon in the receptor-binding and adenylate cyclase assays, respectively. [Phe^{13}, Lys^{17}, Lys^{18}, Glu^{21}]Glucagon was also more potent than the native hormone.

Merrifield and co-workers (Murphy *et al.*, 1988) synthesized six new analogues of glucagon containing replacements at positions 19, 22, and 23. They were designed to study the correlation between predicted conformation in the 19–27 segment of the hormone and the conformation calculated from circular dichroism measurements and observed activation of adenylate cyclase in the liver membrane. The analogues were [Val^{19}]glucagon, [Val^{22}]glucagon, [Glu^{23}]glucagon, [Val^{19}, Glu^{23}]glucagon, [Glu^{22}, Glu^{23}]glucagon, and [Ala^{22}, Ala^{23}]-glucagon. The structures predicted for the 19–27 segment ranged from strongly α helical to weakly β sheet. The observed conformations varied as functions of amino acid composition, solvent, concentrations, pH, and temperature but did not correlate with prediction. There was, however, a correlation between predicted structure and activation of adenylate cyclase in rat liver membranes.

The conformational parameters P_α, P_β, and P_t are expedient for detecting regions in proteins with potential for conformational changes caused by mutations or changes in solvent conditions. The *lac* repressor–*lac* operator interaction of *Escherichia coli* provides an excellent example of the specificity of protein binding to DNA (Bourgeois and Pfahl, 1976). The amino acid sequence of the *lac* repressor, a polypeptide subunit containing 347 amino acid residues, has been determined. Its secondary structure was predicted to contain 37% α helix and 35% β sheet (Chou *et al.*, 1975). The extensive β sheets predicted in the 215–324 region may be responsible for tetramer stabilization found in both the *lac* repressor and the core. These β sheets are almost devoid of charge and would have an extremely hydrophobic nucleus. There are 23 predicted β turns in the *lac* repressor, made up of 50% charged and polar residues (serine and threonine), which would be found on the surface, conferring solubility.

Examination of five *lac* repressor mutants yields significant information regarding conformational requirements for repressor function. Mutant AP46 has an $Ala^{53} \rightarrow Val^{53}$ replacement, causing a predicted $\alpha \rightarrow \beta$ transition at residues 52–57 and a loss of repressor activity. Several amber mutants at Gln^{26} (Leu, Ser, Tyr) still cause repression, and no conformational change was predicted, in agreement with this observation. In mutant AP309 a $Ser^{16} \rightarrow Pro^{16}$ change is incurred with loss of biological activity, and a predicted β turn at 14–17 is lost by this mutation. Thus, in these examples it is possible to correlate biological activity with definite secondary structures, and a loss of activity results from induction of a conformational change by a mutation.

The β-turn conformation can now be predicted with the same degree of accuracy as the α-helical and β-sheet regions in proteins (Chou and Fasman, 1979a). The three-dimensional structure of proinsulin has been predicted (Snell and Smyth, 1975). The C-peptide sequences of ten mammalian species show a remarkable conservation of predicted conformation, with a β turn at residues 15–18 flanked by two helices. Data from 29 proteins (Chou and Fasman, 1978a) have shown that a high β-turn potential exists in the 12–17 regions for this series and, more importantly, in none outside of it (Chou and Fasman, 1979b). Although no biological role has been assigned to the C-peptide, the present prediction shows that the β-turn conservation in proinsulin is probably necessary for directing the proper folding of C-peptide helices, which possibly masks the receptor binding region of the hormone, thus making the precursor, proinsulin, inactive.

The designs of several biologically active polypeptides, based on predicted secondary structures, have been reported (Chakravarty *et al.*, 1973; Gutte *et al.*, 1979; Fukushima *et al.*, 1979; Moser *et al.*, 1983, 1987).

Gutte *et al.* (1979) published an interesting and bold application of the predictive scheme. A model of a neutral artificial 34-residue polypeptide with potential nucleic acid binding activity was synthesized. The conformation was designed to bind the nucleotide sequence of the anticodon of yeast $tRNA^{Phe}$, $m^{2\prime}GAA$. The backbone of this peptide chain was to fold around this ligand. The structure chosen for the 34-residue peptide contained a β strand, a reverse turn, an antiparallel β strand, a second reverse turn, and an α helix running parallel to the second β strand. The residues chosen were based on the parameters of the Chou and Fasman method (1978a). Specific amino acid side chains were to form salt bridges with the phosphate moieties, other hydrophobic aromatic rings allowed stacking interactions or intercalation with the bases of the trinucleotide ligand, and specific hydrogen bonds were also projected. The product and its covalent dimer showed strong interaction with cytidine phosphates and single-strand DNA. The dimer had considerable ribonuclease activity.

Kaiser and co-workers (Fukushima *et al.*, 1979) designed a synthetic amphiphilic helical docosapeptide with the surface properties of plasma apolipoprotein A-1 based on the Chou and Fasman parameters (1978a).

Gutte and co-workers (Moser *et al.*, 1983) designed and synthesized a hydrophobic 24-residue polypeptide that could potentially form a four-stranded antiparallel β sheet and bind the insecticide DDT. More recently, this group (Moser *et al.*, 1987) expressed the synthetic gene for this artificial DDT-binding polypeptide (DBP) in *E. coli*. Recombinant and chemically synthesized DBP showed identical properties.

Recently there has been considerable interest in the manner in which precursor proteins are synthesized as prepro sequences and the manner in which they are transported across the membrane (Davies and Tai, 1980). Potts *et al.* (1982) have carefully studied the physiology, biosynthesis, and mechanism of action of the parathyroid hormone (PTH). Rosenblatt *et al.* (1979) have chemically synthesized the peptide representing the NH_2-terminal extension of prepro-PTH. The 30-amino-acid single-chain peptide contains the pre region (24 residues) and the prohormone specific hexapeptide of PTH. The pre sequence has also been called the signal or leader peptide, and it has been found to contain a common hydrophobic core in many proteins. This core may aid it in penetrating the membrane (Wickner, 1979). By the predictive method of Chou and Fasman (1978a), two secondary structures were found to be highly probable for the precursor sequence of prepro-PTH: (1) a form with high β-sheet content ($\alpha = 20\%$, $\beta = 57\%$), corresponding experimentally to the aqueous conformation, and (2) a form with high α-helix content and no β sheet ($\alpha = 83\%$, $\beta = 0\%$), corresponding to the observed conformation in a nonpolar environment. The circular dichroism spectrum of the prepro-PTH was examined in an aqueous buffer (pH 7.0) and was found to contain 27% α-helix and 43% β, whereas in a nonpolar solvent, hexafluoro-2-propanol, similar to an intramembrane environment, the polypeptide was found to contain 46% α helix and 0% β. This would support the mechanism by which leader sequences may facilitate passage of nascent preproteins across the lipid layer of the rough endoplasmic reticulum (RER) and into the RER cisternal space by assuming an α-helical conformation to aid in passing through the membrane (Wickner, 1979). The prediction also indicates the presence of a β turn at the end of the leader sequence, which may delineate the COOH terminus and the cleavage site of the leader sequence.

A closer examination of the conformational roles of signal peptides was made by Briggs and Gierasch (1984). The λ receptor protein wild-type and mutant signal peptides were synthesized, and conformational analyses were performed. Secretion of the *E. coli* λ receptor protein (LamB protein) appears from genetic evidence to be correlated with the predicted tendency of its signal sequence to adopt a α-helical conformation (Emr and Silhavy, 1983). They have isolated a strain of *E. coli* with 12 base-pair deletions in the region for the signal sequence of the λ receptor protein. This mutant synthesizes the LamB protein but is unable to export it to its normal location in the outer membrane. Two pseudorevertants containing point mutations that lead to amino acid substitutions in the shortened signal sequences are able to export and process the LamB protein. Emr and Silhavy (1983) report the results of the Chou and Fasman calculations on all four of the signal peptide sequences. The wild-type signal sequence is predicted to adopt a α-helical conformation. The deletion mutant (missing four amino acids) cannot nucleate a helix and is random. The two revertants were predicted to be capable of assuming an α helix. Briggs and Gierasch (1984) synthesized the 12- to 16-residue portions of the four signal peptides and examined their conformations by CD in both aqueous solvents and micellar systems (40 mM aqueous SDS). In aqueous media all signal peptides were largely in the random conformation. In 40 mM SDS the wild-type and pseudorevertant peptides contained considerable helical content (≃25%), whereas the deletion mutant signal peptide had little change in structure between SDS and water. The predictive method (Chou and Fasman, 1978a) had suggested that the signal sequences capable of assuming an α helix would be found to be exported, whereas nonhelical signal sequences would not. This hypothesis has been further confirmed by the synthesis of the signal peptides and analysis of their conformation by CD. This elegant work also illustrates the fact that the predictive scheme is applicable to membrane proteins. The internal milieu of a membrane is not significantly different from that of the hydrophobic core of a globular protein. Thus, the conformational parameters are applicable to both environments.

An interesting application of the predictive method was the successful demonstration of the structural homology between proteins from widely differing sources that have been suspected of having a possible ancestral relationship. Goodman, Rivier, and colleagues (Pallai *et al.*, 1983) examined corticotropin-releasing factor (CRF) from the hypothalamus, sauvagine from the skin of a frog, and urotensin I from a teleost fish. These polypeptides have approximately 50% homology in amino acid sequences. The circular dichroism spectra in trifluoroethanol all have approximately the same structure, ≃70–80% α helix. The predicted conformation is in good agreement with the measured value (e.g., CRF, CD 78% α; predicted, 78%). All three polypeptides possess a long internal helix, spanning about 25 residues, connected to a turn region to a COOH-terminal element that is an α-helix in CRF and urotensin I and a β sheet in sauvagine. Thus, based on secondary structure, the similarities and differences in biological activity could be rationalized.

The secondary and tertiary structures of interferon were predicted from four homologous amino acid sequences (Sternberg and Cohen, 1982). Several predictive methods were used, and although satisfactory agreement was lacking, four α helices were found to be important in the tertiary fold. A possible tertiary model for interferon was proposed in which the four α helices pack into a right-handed bundle similar to that observed in several other protein structures. This tertiary structure was obtained by using a helix-docking algorithm (Cohen *et al.*, 1979; Cohen and Sternberg, 1980a,b).

The three-dimensional structure of interleukin-2 has recently been solved to 3.0-Å resolution (Brandhuber *et al.*, 1987). Helices B, C, D, and F form an apparent antiparallel α-helical bundle that differs significantly from the classical four-helix bundle represented by cytochrome *c*′, cytochrome b_{562}, and myohemerythrin (Richardson, 1981). The packing regions of the helices are shorter, involving only three to four turns of helix, whereas classical four-helix bundles usually have at least five turns in each helix. There are also three other helical regional (B′, A, and E). Therefore, the predictive result is encouraging but requires further modification and refinement.

Kaiser and co-workers (De Grado *et al.*, 1981) designed and synthesized a cytotoxic peptide with mellitinlike activity. The desired polypeptide of 26 residues was to possess an amphiphilic α helix, bind to phospholipid layers, and form monolayers as well as have a basic C-terminal hexapeptide. A synthetic peptide with sequence residues based on P_α and P_β values that was homologous to mellitin was produced, and its CD spectrum in aqueous media had a 69% α-helical content and formed a tetramer in solution as mellitin did. This synthetic polypeptide formed stable monolayers, caused hemolysis of erythrocytes, and disrupted phospholipid bilayers. Thus, this amphiphilic synthetic peptide appears to possess some of the properties necessary for cytotoxic behavior.

An approach to the design of peptide-hormone analogues in which amino acid substitutions are based on predicted effects on secondary structure was investigated. The structural requirements for analogues of the parathyroid hormone's binding domain in the region 25–34 were investigated for bioactivity, and their solution conformation was determined by circular dichroism (Nussbaum *et al.*, 1985). Biological activity of these analogues in the rat renal adenylate cyclase assay *in vitro* and binding affinity in a radioreceptor assay were threefold those of the unsubstituted PTH_{1-34}.

Hruby *et al.* (1986) and Merrifield and co-workers (Murphy *et al.*, 1988) studied the conformational dependence of the bioactivity of glucagon and found striking effects. Hruby *et al.* (1986) prepared a number of analogues of glucagon by total synthesis using the solid-phase method of peptide synthesis. Analogues of the 29-amino-acid hormone designed to increase α-helical probabilities in the C-terminal region of glucagon led to highly potent analogues. [Lys^{17}, Lys^{18}, Glu^{21}]Glucagon was 500% and 700% more potent than native glucagon in the receptor-binding and adenylate cyclase assays, respectively. Glucagon has a $\langle P_\alpha \rangle_{19-27} = 1.18$, whereas [Lys^{17}, Lys^{18}, Glu^{21}]glucagon has a $\langle P_\alpha \rangle_{19-27} = 1.23$, thus having higher helical potential. [Phe^{13}, Lys^{18}, Glu^{21}]Glucagon was also more potent than the native hormone. The importance of the 10–13 residues of glucagon as a "hinge region" for correlating binding and transduction regions of glucagon was investigated. It was found that [Phe^{13}]glucagonamide is a potent analogue with full biological activity; [Phe^{10}]glucagon, although quite potent, is a partial agonist. Other derivatives were investigated for their agonist and antagonist potentials. Merrifield and co-workers (Murphy *et al.*, 1988) synthesized analogues of glucagon with amino acid replacements at positions 19, 22, and 23. They were designed to study the correlation among predicted conformation in the 19–27 segment of the hormone, the conformation calculated from circular dichroism measurements, and the observed activation of adenylate cyclase in the rat liver membrane. The observed conformations did not correlate well with prediction, but the predicted conformation and activation of adenylate cyclase in the rat liver membrane did correlate well.

Because a computer program was not published by the authors for the Chou–Fasman predictive method for the secondary structure of proteins (Chou and Fasman, 1974a,b), there have been many published procedures (approximately 20) as well as methods that have alterations and additions that are claimed to have improved the method. The first computerized method to appear was that of Argos *et al.* (1976). The known structures of about 40 proteins were compared to other predictive methods (see below). The accuracy of the predicted helices was found to be better than for β-sheet regions and turns. The amino-terminal half of the protein molecule was predicted with higher accuracy than was the carboxyl half. Dufton and Hider (1977) applied a modified Chou–Fasman algorithm to the prediction of the secondary structure of 57 snake venom toxins. A common distribution of secondary structure was detected throughout these toxins with a distinctive pattern of β sheet, α helix, and β bend for each toxin. The basic difference utilized in their scheme, relative to the original algorithm, was to employ straightforward multiplication instead of the arithmetic mean to find favorable segments.

Argos *et al.* (1978) applied their computer program (Argos *et al.*, 1977) with an expanded data base but found no improvement in the correctness of the predictions. The data base used was that obtained by Levitt (1978), who expanded the data base from 1939 residues (Chou and Fasman, 1974a) to a 5523-residue sample. Busetta and Hospital (1982) attempted to improve the efficiency of the Chou and Fasman (1974a,b) and the Garnier *et al.* (1978) procedures for the prediction of secondary structures by the introduction of the use of the distribution of hydrophobic residues. A set of 38 proteins was examined, and the ratio of residues in the helix versus helix plus extended was utilized, as previously demonstrated by Garnier *et al.* (1978), which produced slightly improved predictions. These authors noted that the prediction level is better for proteins with a single type of secondary structure (all α or all β) than for a mixed type (α + β or α/β). A BASIC microcomputer program for plotting the secondary structure of proteins was developed by Corrigan and Huang (1982) for use on an Apple II+. A modified version of the Chou and Fasman method (1974a,b) was constructed that included nucleation site determination through multiplication of conformational preferences as well as weighing factors to represent structurally stabilizing short-range interactions. They concluded that this method has nearly achieved its upper limit of prediction accuracy, although slight improvement through the use of stereochemical weighing factors and conformational parameters might be possible. On analyzing the conformational parameters of Chou and Fasman (1978a), Charton and Charton (1983) found that for the α helix, coil, and turn parameters, steric effects were predominant, whereas for β-sheet parameters, intramolecular forces are dominant.

The need for a computer program to apply the Chou and Fasman (1978a, 1979a) prediction method was pointed out by Rawlings *et al.* (1983), and a DEC System 10 FORTRAN computer program was published and made available. Comparisons of their results for seven proteins were also made available with the predictive schemes of Nagano (1977) and Burgess *et al.* (1974). The results achieved for the seven proteins studied were excellent (between 80 and 96% correct). Novotny and Auffray (1984) described a computer program (written in FORTRAN, which runs on a Digital VAX 11/780 computer) that, given a nucleotide or amino acid sequence, outputs protein secondary structure prediction curves as well as hydrophobicity and charge-residue profiles (graphics output on several terminals, VT125, Tektronix 4010, GIGI). The program allows for cumulative averaging of properties (secondary structure propensities, hydrophobicity, and charge profiles) of several homologous primary structures, a concept shown to improve the predictive accuracy. The smoothing of hydrophobicity profiles used in this procedure was that of Rose and Roy (1980). This method was applied to human and murine histocompatability antigens of class I and II. A PASCAL microcomputer program for prediction of protein secondary structure, written for use on an Apple IIe or IIc, was published by Parrilla *et al.* (1986). This uses the Chou and Fasman (1974a,b) method with minor modifications and, in addition, performs an analysis of the hydropathic character of the residues for predicting of external/internal regions of the polypeptide chain. It also can search for probable glycosylation and phosphorylation sites. The hydropathic character is similar to that chosen by Kyte and Doolittle (1982). The sequence chosen for N-glycosylation is Asn-*X*-Thr/Ser (Mononen and Karjalainen, 1984), and the target for phosphorylation is Gly/Ile-Ser/Thr-Gly/Ala/Val-Lys/Arg (Schulz and Schirmer, 1979).

MESQ, a versatile and "user-friendly" software package dedicated to the analysis, display, and prediction of protein structure was published by Black and Glorioso (1986). Protein secondary structure is predicted using the parameters of Chou and Fasman (1978a,b). Zero-order sequence hydrophobicity is calculated with the use of four user-selectable sets of 20 amino acid side-chain polarity values obtained from Argos *et al.* (1982), von Heijne (1981a), Hopp and Woods (1981), and Kyte and Doolittle (1982). Both the α and β-strand hydrophobic moments are also calculated according to the method of Eisenberg *et al.* (1982a,b, 1984a,b). The program is written for the IBM microcomputer family with the extensive use of color graphics, graphic output, and structural "cartoons."

Gribskov *et al.* (1986) described a computer program for the analysis of protein secondary structure that produces both graphic and printed output. Structural predictions are based on the Chou and Fasman (1978a,b) and Garnier *et al.* (1978) methods, hydropathy analysis by the method of Kyte and Doolittle (1982), and a simplified method of hydrophobic moment analysis (Eisenberg *et al.*, 1984a,b) is included.

An interactive analysis of protein structure using a microcomputer spreadsheet, Lotus 1-2-3, was used to predict the α-helix, β-sheet, and hydrophobicity profiles of protein sequences (Vickery, 1987). The Chou and Fasman (1978a,b) empirical algorithm was used to predict secondary structure, and the Kyte and Doolittle (1982) method for the analysis of hydrophilicity/hydrophobicity was applied, using an IBM personal computer. Williams *et al.* (1987) incorporated information about neighboring residues participating in short- and medium-range interactions into predictions of protein secondary structure and, contrary to proposals by several authors, found that no improvement resulted. However, they obtained an 8% improvement for predictions of secondary structure based on the algorithm of Chou and Fasman (1974a,b) by eliminating many rules that include neighboring residue interactions and choosing the best decision constants for structure assignments. Specifically, all rules described by Chou and Fasman (1978a,b) regarding α-helix and β-sheet nucleation, propagation, termination, breakers, boundaries, and overlapping regions were not used. This simplified method is claimed to yield a 57% correct assignment for three states α helix, β strand, and β turn.

Krchnak *et al.* (1987) describe a computer program (written in BASIC for an IBM personal computer) that predicts protein immunogenic determinants on the basis of the probability of β turns as specified in the Chou and Fasman (1978a,b) procedure.

Deléage *et al.* (1987) developed a computerized program (written for an Apple IIe microcomputer) for predicting the secondary structure of proteins from their amino acid sequence by closely following the scheme of Chou and Fasman (1978a,b). Some of the qualitative rules (Chou and Fasman, 1978a,b) have been converted to numerical scales to obtain unambiguous predictions. On testing 21 proteins with known three-dimensional structure, the percentage of correctly predicted amino acids was between 41 and 66% for a three-state (α helix, β sheet, and coil) description of protein secondary structure. Spectroscopic and structural properties of three homologous dimeric inhibitors of microbial origin, *Streptomyces* subtilisin inhibitor, alkaline proteinase inhibitor, and plasminostreptin, were examined by comparing hydropathy maps (method of Kyte and Doolittle, 1982).

Deléage and Roux (1987) described an algorithm for the prediction of protein secondary structure with a marked improvement of accuracy by taking into account the predicted class of the proteins. This "double predictive method" consists of first predicting the secondary structure from a new algorithm that uses parameters of the type described by Chou and Fasman (1978a,b) and then predicting the class of proteins from their amino acid composition (Nakashima *et al.*, 1986). The parameters were obtained from a 59-protein data base (Kabsch and Sander, 1983b). These two independent predictions allow one to optimize the parameters calculated over the secondary structure data base to provide a final improved prediction of secondary structure. When the procedure was tested on the 59 proteins in the data base, a 72% success in class prediction was obtained, and 61.3% of residues were correctly predicted for three states (α helix, β strand, and coil). This method shows great promise.

Ralph *et al.* (1987) have written a FORTRAN program, called PRSTRC for the prediction of the secondary structure of proteins. This modified Chou and Fasman (1978a,b) analysis carries out a running average of amino acid structure occurrence frequencies, utilizes a simple set of nucleation conditions, and in addition, allows the user control over nucleation threshold and cutoff parameters. The algorithm has also included the prediction of omega loops (Leszczynski and Rose, 1986) and includes a profile of charge distribution and a hydropathy profile (Rose *et al.*, 1985). The authors claim an overall improvement in the average correlation coefficient for α helices and β strands of 12% and 24%, respectively, over the Chou–Fasman algorithm as written by IntelliGenetics (1981). PRSTRC allows the user control of all threshold and cutoff values in the prediction scheme. Thus, if the x-ray structure is known, one may manipulate the values to optimize the accuracy. They state that this flexibility allows for optimization of secondary structure prediction for homologous proteins. Therefore, their figures for increased accuracy must be taken with caution, as it has not been demonstrated on a blind test with many proteins.

Lathrop *et al.* (1987) developed a computer program, ARIADNE, as a hierarchic pattern-directed inference system for protein structural analysis. Input to ARIADNE consists of the primary sequence, any secondary structure predictions, and patterns describing the structure of interest. The secondary structure, α helices, β strands, and β turns, were predicted by the program PRSTC (Ralph *et al.*, 1987) on the basis of the Chou and Fasman (1974b) conformational probabilities. The system identifies the optimal match between a given complex pattern descriptor and protein sequences annotated with various inferred properties by abstracting

Table I. Conformational Parameters for α-Helical and β-Sheet Residues

α-Residues	P_α		β-Residues	P_β	
		64 Proteins			
Glu	1.44 ± 0.06		Val	1.64 ± 0.07	
Ala	1.39 ± 0.05		Ile	1.57 ± 0.08	
Met	1.32 ± 0.11	H_α	Thr	1.33 ± 0.07	
Leu	1.30 ± 0.05		Tyr	1.31 ± 0.09	
Lys	1.21 ± 0.05		Trp	1.24 ± 0.14	
His	1.12 ± 0.08		Phe	1.23 ± 0.09	
Gln	1.12 ± 0.07	h_α	Leu	1.17 ± 0.06	
Phe	1.11 ± 0.07		Cys	1.07 ± 0.12	
Asp	1.06 ± 0.06		Met	1.01 ± 0.13	
Trp	1.03 ± 0.10	I_α	Gln	1.00 ± 0.09	
Arg	1.00 ± 0.07		Ser	0.94 ± 0.06	
Ile	0.99 ± 0.06		Arg	0.94 ± 0.09	
Val	0.97 ± 0.05		Gly	0.87 ± 0.05	
Cys	0.95 ± 0.09	i_α	His	0.83 ± 0.09	
Thr	0.78 ± 0.05		Ala	0.79 ± 0.05	
Asn	0.78 ± 0.06		Lys	0.73 ± 0.06	
Tyr	0.73 ± 0.06	b_α	Asp	0.66 ± 0.06	
Ser	0.72 ± 0.04		Asn	0.66 ± 0.06	
Gly	0.63 ± 0.04	B_α	Pro	0.62 ± 0.07	
Pro	0.55 ± 0.05		Glu	0.51 ± 0.06	
		24 Proteins			
Glu	1.51		Val	1.70	
Met	1.45		Ile	1.60	H_β
Ala	1.42		Tyr	1.47	
Leu	1.21	H_α	Phe	1.38	
Lys	1.16		Trp	1.37	
Phe	1.13		Leu	1.30	
Gln	1.11		Cys	1.19	h_β
Trp	1.08		Thr	1.19	
Ile	1.08	h_α	Gln	1.10	
Val	1.06		Met	1.05	
Asp	1.01	I_α	Arg	0.93	
His	1.00		Asn	0.89	
Arg	0.98		His	0.87	i_β
Thr	0.83	i_α	Ala	0.83	
Ser	0.77		Ser	0.75	
Cys	0.70		Gly	0.75	b_β
Tyr	0.69	b_α	Lys	0.74	
Asn	0.67		Pro	0.55	
Pro	0.57	B_α	Asp	0.54	B_β
Gly	0.57		Glu	0.37	

intermediate levels of structural organization. This method was applied to predict a common structural domain in amino acyl-tRNA synthetases (Webster *et al.*, 1987). The general utility and reliability of ARIADNE will have to await further tests on known structures.

E. Class Prediction

Chou (1979, 1980) analyzed the x-ray structures of 64 different proteins containing 11,444 residues in terms of their α-helical and β-sheet regions as well as their amino acid compositions. This sample represented a larger number of nonhomologous proteins than the Levitt (1978) sample (49 different proteins out of 66). It was found that the four distinct classes of proteins—α, β, α + β, and α/β (Levitt and Chothia, 1976; Schulz and Schirmer, 1979; Richardson, 1981)—had significantly different amino acid compositions. The sample had 19 α proteins (69% α, 1% β), 15 β proteins (15% β, 5% α), 16 α/β proteins (35% α, 23% β), and 14 α + β proteins (34% α, 21% β). The average residue chain length for the four classes of proteins varied in decreasing order: $\langle N_{\alpha/\beta}\rangle = 271 > \langle N_{\beta}\rangle = 185 > \langle N_{\alpha+\beta}\rangle = 135 > \langle N_{\alpha}\rangle > 129$. The overall secondary structural content in the 64 proteins was 35% α helical and 25% β sheet, which is very similar to the α and β content in α + β as well as α/β proteins. An earlier analysis based on 29 proteins (4741 residues) showed an average 38% helicity and 20% β content (Chou and Fasman, 1978b), whereas the more recent analysis (Levitt, 1978) found 31% α and 28% β in 66 proteins. It should be noted that of the 66 proteins analyzed by Levitt (1978), 17 were redundant in the sense that some proteins were independently solved halves of dimers and some proteins were the same but solved by different laboratories. Compared to the conformational parameters of Chou and Fasman (1978b), there were a few significant changes (Table I), and the relative order of the amino acids was relatively constant. A computerized algorithm was developed that could assign proteins to the correct structural class based on their amino acid compositions with 80% accuracy. Chou (1979) developed a new set of conformational parameters, P_{α} and P_{β}, for the amino acids in the four different classes of proteins (see Chapter 12 by Chou). With this new methodology, some previously incorrect predictions (Chou and Fasman, 1978b) were corrected (for greater details of the method, see Chapter 12).

Geisow and Roberts (1980) also reported that the amino acid preferences for secondary structure vary with the protein class. The classes were also determined by the algorithm described by Levitt and Greer (1977). The *P* parameters for the classes α, β, mixed α/β, and the global values were calculated. The global values found by Geisow and Roberts (1980) were similar but not identical to those of Chou and Fasman (1974a,b). Structurally important residues (hydrophobic, disulfide-bond-forming, and charged amino acids) have the largest index changes between classes. Previously it had been reported (Lifson and Sander, 1979) that the preference of amino acids for parallel or antiparallel β-strand arrangements differed substantially. The Chou and Fasman (1974a,b) procedure was slightly modified by Geisow and Roberts (1980) to incorporate the information provided by a periodic structure, and predictions were made with the new indices; significant improvements in the α proteins were obtained, but not with the β-proteins. Busett and Hospital (1982) incorporated the possible use of the distribution of the hydrophobic residues in both the Chou and Fasman (1978a,b) and Garnier *et al.* (1978) methods. The prediction level was improved for proteins of a single type of secondary structure (all α or all β) more than for the mixed type (α + β or α/β).

Klein and DeLisi (1986) used the multidimensional statistical technique of discriminant analysis to allocate amino acid sequences to one of four secondary structural classes: high α content, high β content, mixed α and β, and low content of ordered structure. The discrimination was based on four attributes: estimates of percentages of α and β structures (Chou and Fasman, 1974b, as modified by Dufton and Hider, 1977; and the method of Garnier *et al.*, 1978) and regular variations in the hydrophobic values of residues along the sequence occurring with periods of 2 and 3.6 residues. The accuracy of the method, classifying 138 sequences, is 80% with no misallocations between α-rich and β-rich classes. Nakashima *et al.* (1986) analyzed the folding type of 135 proteins of known three-dimensional structure in terms of their amino acid composition. The amino acid composition of a protein was expressed as a point in a multidimensional space spanned with 20 axes, on which the corresponding contents of 20 amino acids in the protein are represented. The distribution pattern of proteins in this composition space was examined in relation to five folding types: α, β, α/β, α + β, and irregular type. The α, β, and α/β types are distinctly separated; however, the points representing proteins of the α + β and irregular types cannot be easily classified. The assignment of the folding type to five classes by this method gave an accuracy of 70%, and to four folding classes a reliability of 79%. The authors found ambiguity in assigning a unique type of some proteins by the method of Levitt and Chothia (1976). Therefore, a quantitative measure to define a folding type was adopted: α-type proteins, $\alpha > 15\%$ and $\beta < 10\%$; β-type proteins, $\alpha < 15\%$ and $\beta > 10\%$; α/β-type proteins, $\alpha > 15\%$ and $\beta > 10\%$ with dominantly parallel β sheets; α + β-type proteins, $\alpha > 15\%$ and $\beta > 10\%$ with predominantly antiparallel β sheets; and irregular-type proteins, $\alpha < 15\%$ and $\beta < 10\%$. The success of this procedure has an important implication, namely, that properties of single amino acids are, as an approximation, additive regardless of the sequence. The nature of a protein as a whole is, therefore, approximately determined by its amino acid composition.

Klein (1986) improved on the Nakashima *et al.* (1986) analysis by the use of discriminant analysis. Analysis by class—α, β, mixed (α/β or α + β), and irregular (four classes)—can be predicted with an 83% reliability, and to one of five classes (separate α/β and α + β) with 78% reliability. It was shown that optimal linear combinations of amino acid frequencies used as attributes give slightly better accuracy than attributes using secondary prediction methods for each residue.

Deléage and Roux (1987) have used the technique of Nakashima *et al.* (1986) for class prediction and utilized this in a modified Chou and Fasman (1974a, 1978a) algorithm, called a "double prediction method," for the prediction of secondary structure of proteins from their amino acid sequence (see Chapter 12). When tested on 59 proteins, the methods yielded a 72% success in class prediction, with 61.3% of residues correctly predicted for three states (α helix, β sheet, and coil). Ponder and Richards (1987) approached the folding problem, i.e., structural class prediction, from another point of view, that previously expressed by Drexler (1980): "What sequences are compatible with a given structure?" Ponder and Richards (1987) have elaborated on the Drexler suggestion through the development of tertiary templates for the various classes of protein domains. Blundell and Sternberg (1985) had previously suggested the phrase "tertiary template" to describe a peptide that is characteristic of a particular tertiary structure or fold and in which selected sequence positions have a specified composition.

Ponder and Richards' (1987) use of the term tertiary template is an extension of the above definition to the entire core of a protein domain and to the development of the templates through stereochemical considerations rather than statistical inference. They assume that each class of protein has a core structure that is defined by internal residues and that external, solvent-contacting residues contributing to the stability of the structure are of primary importance to function but do not determine the architecture of the core portions of the polypeptide chain. An algorithm has been developed to supply a list of permitted sequences of internal residues compatible with a known core structure. This list is referred to as the tertiary template for that structure. In general, the positions in the template are not sequentially adjacent and are distributed through the polypeptide chain. The template is derived using the fixed positions of the main-chain and β-carbon atoms in the test structure and selected stereochemical rules. Two packing criteria were used: avoidance of steric overlap and complete filling of available space. The program also notes potential polar group interactions and disulfide bonds as well as possible burial of formal charges. Central to the algorithm is the 64-side-chain rotamer library. Preliminary tests make it appear likely that templates prepared from the currently known core structures will be able to discriminate between these structures and be useful in deciding whether a sequence of unknown tertiary structure fits any of the known core classes and, if a fit is found, how the sequence should be aligned in three dimensions to fit the core of that class.

Sheridan *et al.* (1985) found that the amino acid composition and hydrophobicity patterns of protein domains correlate with structures. They correlated the sequence and tertiary structure for 212 domains from globular proteins and polypeptides. The sequence of each domain is described as a set of 25 features: the mole percentage of 20 amino acids, the number of residues in a domain, and the abundance of four simple patterns in the hydrophobicity profile of the sequence. Pattern recognition methods were applied to find the two axes through the 35-dimensional sequence–feature space that best discriminate, respectively, predominantly α-helix domains from predominantly β-strand domains and parallel α/β domains from other domains. The domains were further divided into two categories based on whether the Cys content is above (Cys-rich) or below (normal) 4.5%. They found the secondary structure vector for the subset of Cys-rich domains points in a significantly different direction than the equivalent vector for the normal domains. Thus, Cys-rich and normal domains are best treated separately. On projection of the secondary structure vectors onto the plane containing the origin of the feature space, it is seen that α, β and parallel domains cluster in a plane, with the β cluster partially overlapping the parallel cluster. They could correctly predict the structural class with 83% accuracy.

McGregor *et al.* (1987) studied the relationship between side-chain conformation (dihedral angles) and secondary structure in globular proteins. Sixty-one proteins solved to a resolution of 2 Å or better were analyzed. The strongest feature observed was that the χ_1 distribution (rotation around the C_α–C_β bond) for most side chains in an α helix showed the absence of the g^-(χ_1 = 60°) conformation and a shift towards the *t* (180°) conformation when compared to the non-α/β structures. The exceptions to this tendency were for short polar side chains that form hydrogen bonds with the main chain, which prefers g^+ (300°). Shifts in the χ_1 preferences for residues in the β sheet were observed.

VI. PREDICTION OF TERTIARY STRUCTURE

The future prospects for the prediction of the tertiary structure of proteins are unlimited. The opportunity for the design and construction of new proteins, e.g., vaccines (Mutter, 1985), drugs, herbicides, and pesticides (Blundell *et al.*, 1986b, 1987), await a better understanding of the protein-folding problem (Ghelis and Yon, 1982; Jaenicke, 1984). The advent of recombinant DNA techniques has led to an explosion of information and techniques for the construction of these new hybrid molecules. Protein engineering will be added to the armament of the protein chemist, making possible studies, such as of enzyme mechanisms, at a level previously thought impossible.

There are three broad approaches presently utilized for the prediction of the tertiary structure of a polypeptide (Ponder and Richards, 1987): (1) use of sequence homology with peptides of known three-dimensional structure; (2) prediction of secondary structure units followed by the assembly of these units into a compact structure; and (3) use of empirical energy functions *ab initio* to derive the tertiary structure of minimum potential energy. The first two rely heavily on the data base of structures determined by x-ray crystallography, but the third, in principle, does not. The first procedure uses rules of structure only by implication. The second uses them quantitatively, with the goal of a rough outline of the structure. The third relies on quantitative calculations for evaluation of trial structures.

In 1975, the characterization of the tertiary structure in globular proteins was still elementary (Kuntz, 1975). The approach that showed promise had been the identification of "folding domains" with larger proteins (Wetlaufer and Ristow, 1973; Liljas and Rossman, 1974; Rossman and Liljas, 1974). The "distance plots" method for the location of domains showed promise (Phillips, 1970; Rossman and Liljas, 1974; Liljas and Rossman, 1974; Kuntz, 1975). These "distance plots" are graphs of C_α–C_α distances plotted against residue number, with contour lines drawn at fixed interatomic distances. Kuntz (1975) pointed out that these maps lead to an ordering of tertiary structural features such as distorted three-dimensional "superhelical" structures that are principal constituents of folding domains.

By 1980, with the structures of ~80 globular proteins known to atomic resolution, it became evident that the tertiary fold is largely determined by the packing of α helices and/or β strands (Levitt and Chothia, 1976; Chothia and Janin, 1982; Richardson, 1976, 1977; Sternberg and Thornton, 1976, 1977a–c; Chothia *et al.*, 1977; Richmond and Richards, 1978). One of three motifs was generally found: a stacked pair of β sheets (β/β); α helices packed against a predominantly parallel β sheet (α/β); or an assembly of α helices (α/α). These motifs satisfy the hydrogen-bonding requirements of buried main-chain nitrogen and oxygen atoms while shielding a substantial fraction of the nonpolar atoms from solvent (Chothia and Janin, 1975). Taylor and Thornton (1984) have devised a procedure that further assists in recognizing these supersecondary structures in proteins.

The supersecondary structures are the main components of the domains that, when assembled, constitute the three-dimensional conformations of proteins. An interesting paper by Dill (1985) discusses the theory of the folding and stability of globular proteins. A theory was developed for the folding of proteins to the globular and soluble state using lattice statistical mechanics. Folding is assumed to be driven by the association of solvophobic monomers to avoid solvent and is opposed by the chain configurational entropy. The theory predicts a phase transition as a function of temperature and solvent character. Molecules that are too short or too long or that have too few solvophobic residues are predicted not to fold. Globular molecules should have a largely solvophobic core, but there is an entropic tendency for some residues to be "out of place," particularly in small molecules. For long chains, molecules comprised of globular domains are predicted to be thermodynamically more stable than spherical molecules. The number of accessible conformations in the globular state is calculated to be an exceedingly small fraction of the number available to the random coil. Thus, as the molecular weights of proteins increase, there is a high probability that they will become divided into domains rather than increasing in size of a single unit.

A. Combinatorial Approach

Cohen, Sternberg, and co-workers (Cohen *et al.*, 1979, 1980, 1981, 1983; Cohen and Sternberg, 1980a; Edwards *et al.*, 1987) have developed a stepwise method, termed the combinatorial approach, for predicting the three-dimensional structure of a protein from its amino acid sequence (for reviews see Nemethy and Scheraga, 1979; Sternberg and Thornton, 1978; Schulz and Schirmer, 1979). There are three stages in this procedure: (1) predict the regular secondary structures, now possible with up to 80% accuracy; (2) pack the α helices and β strands into an approximate native fold; (3) use simplified energy calculations (Levitt, 1976; Kuntz *et al.*, 1976; Robson and Osguthorpe, 1979) to refine the fold into the native structure. Because it has been shown (Hagler and Honig, 1978; Cohen and Sternberg, 1980b) that structures predicted solely by simplified energy calculations are not significantly better than random models for a compact globular protein, the combinatorial approach appears to have many advantages. The failure of the energy calculations presumably stems from the difficulties in modeling protein–solvent interactions, the use of analytical functions to approximate the chemical potential and compounding of these errors in the computed gradient, and because the energy surface has multiple minima, which makes it nearly impossible to locate the global minimum.

In the combinatorial approach it is generally assumed that step 1, the secondary structure prediction, has been successful, and attention is mainly directed to step 2, the docking of secondary structures into a nativelike three-dimensional structure. Initially, a list of trial structures is generated by packing all combinations of the α helices and β strands. The native fold will be in this list, and structures are eliminated that violate stereochemical rules when applied to the packing of α helices in myoglobin (Richmond and Richards, 1978; Cohen *et al.*, 1979; Cohen and Sternberg, 1980a). In predicting the structure of myoglobin, over 10^8 trial structures were generated by docking together hydrophobic patches on the surface of the α helices. However, only 20 folds did not violate the steric and connectivity constraints. The addition of two distance constraints obtained from experimental data on heme binding further restricted the number of allowed structures to two. The relative arrangements of α helices in one of these structures closely resembled that in the native structure. The root-mean-square deviation (Cohen and Sternberg, 1980b) between predicted structure and the native protein was 4.3 Å. With this same combinatorial approach, an algorithm has been written to predict β-sheet structures in proteins, e.g., β_2-microglobulin and an HLA-B7 antigen fragment (Cohen *et al.*, 1980).

The analysis and prediction of the structural motifs in the glycolytic enzymes were reviewed by Sternberg *et al.* (1981). Ten of the 13 enzymes in this pathway have been studied by x-ray crystallography. It is shown that all the enzyme structures are variations and extensions of a basic theme of a many-stranded (four to nine), predominantly or totally parallel β sheet that is shielded from solvent by α helices (i.e., α/β structure). There were strong structural similarities between the domains of some but not all enzymes. In particular, the dinucleotide-bonding fold of lactate dehydrogenase and the β barrel of triose phosphate isomerase are found in other domains. General rules governing the topology and packing of α helices against β sheet provided a basis for the combinatorial prediction of the tertiary fold of glycolytic domains from their amino acid sequence and observed secondary structure. The prediction algorithm demonstrates that there are severe restrictions on the number of possible structures. However, these restrictions do not fully explain some of the remarkable structural similarities between different enzymes that probably result from evolution from a common ancestor.

Taylor and Thornton (1984) devised a procedure to recognize supersecondary structure in protein sequences. The term supersecondary structure was introduced by Rao and Rossmann (1973) to describe larger continuous folds found in proteins. An idealized template derived from known supersecondary structure was used to locate probable sites by matching secondary structure probability profiles. The method was applied to the identification of βαβ units in β/α type proteins with 75% accuracy. The location of supersecondary structure was then used to refine the original Garnier *et al.* (1978) secondary structure prediction, resulting in an 8.8% improvement, which correctly assigned 83% of secondary structure elements in 14 proteins. Slight modifications of the Garnier *et al.* (1978) method were suggested, producing a more accurate identification of protein class and a better prediction of β/α-type proteins. A method for the incorporation of hydrophobic information into the prediction was also described. Taylor (1984) published an algorithm to compare secondary structure predictions rather than percentages of all residues correctly assigned. This procedure is more important if the tertiary structural prediction is the subsequent goal.

Vonderviszt *et al.* (1986) developed a simple approach to domain border prediction, relying only on the amino acid sequence. Statistically determined sequential and association preference data of amino acids were combined to generate short-range preference profiles along the polypeptide chains. Domain boundaries correlate with the minima of preference profiles, but some false minima also exist. Preferences of pairs of amino acids and amino acids separated from each other by one residue were obtained. Statistically determined association potentials (E_{ij}) of Narayana and Argos (1984) were used to describe interaction between amino acids.

Hones *et al.* (1987) tried a manual prediction of glucose dehydrogenase based on the hydropathic nature of the internal β sheet and the amphilic character of external helices. The overall homology of primary structure between this enzyme and lactate dehyrogenase was low, and independent predictions of secondary structure produced different patterns of β strands and α helices. However, studies on physicochemical and chemical modification indicated similarity of structure. This method led to the identification of analogues of all the β strands present in lactate dehydrogenase with one exception.

Cohen and co-workers (Hurle *et al.*, 1987) have applied their combinatorial approach to tertiary structural prediction and combined it with circular dichroism measurements to establish the class of the protein. Hurle *et al.* (1987) used vacuum circular dichroism to assign the α subunit of tryptophan synthetase to the α/β class of supersecondary structure. The two-domain structure of the α subunit was assumed based on the work of Miles *et al.* (1982) and Beasty and Matthews (1985), which eliminated consideration of a barrel structure and focused attention on a β-sheet structure. With the algorithm of Cohen *et al.* (1983) a secondary structure was predicted. By the use of other algorithms, the final structure was predicted to have a parallel β sheet flanked on both sides by α helices. This did not agree with the subsequently obtained x-ray structure, which was shown to be a α/β barrel (D. Davies and C. Hyde, quoted in Hurle *et al.*, 1987). Cohen *et al.* (1986b) applied their scheme to predict the three-dimensional structure (Cohen *et al.*, 1979; Cohen and Sternberg, 1979) of interleukin-2. This combinatorial approach generated 3.9×10^4 structures. Of these, 27 satisfied steric constraints and maintained the connectivity bridge between Cys^{58} and Cys^{105}. This allowed the structure to be placed into five structural categories, which contained the right-handed cylinders for a fourfold α-helical bundle. Circular dichroism measurements placed the protein into the α-helical class. Thus, the most plausible structure for interleukin-2 is a right-handed fourfold α-helical bundle.

The three-dimensional structure of interleukin-2 has subsequently been solved to a 3.0-Å resolution (Brandhuber *et al.*, 1987). Four helices form an apparent antiparallel helical bundle that differs significantly from the classical four-helix bundle represented by several proteins (Richardson, 1981). The packing of regions of the helices is shorter, involving only three to four turns of the helix, whereas classical four-helix bundles usually have at least five turns in each helix. There are also three other helical regions.

Billeter *et al.* (1987b) have applied the new method of constrained optimization, known as the "ellipsoid algorithm" (Shor, 1977), to the problem of the docking of two molecules. This method is efficient with respect to computer time and is robust when dealing with nonconvex problems. Its ability to make large steps, especially at the beginning of the optimization, and the fact that only one violated constraint is used in any iteration makes it a powerful tool for problems that involve many local minima and for which no good starting points are available (e.g., Ecker and Kupferschmid, 1982). Although the maximum number of variables that can be used has not yet been determined, problems with up to 55 torsion-angle variables have been solved (Billeter *et al.*, 1987a). Billeter *et al.* (1987b) used this method to explore it as a tool for determining sterically acceptable interactions between two molecules. These interactions are described by constraints on intermolecular distances. To specify the relative orientation of the two molecules, a new set of variables had to be introduced, which represents the set of all possible rotations in a vector space as required by the ellipsoid algorithm. Applications discussed include the docking of two macromolecules and the formation of an enzyme–inhibitor complex. Previously, Kuntz *et al.* (1982) had designed an algorithm that matches the geometry of a binding site to that of a ligand, docking a rigid ligand to a rigid receptor. This method was recently adapted (DesJarlais *et al.*, 1986) to allow for some flexibility in the ligand by docking fragments of the ligand and screening the docked fragments for subsets that can be rejoined to yield the complete molecule with acceptable stereochemistry.

B. β Sheets

The β sheets observed in globular proteins exhibit an extraordinary diversity of structural forms. Salemme (1983) has reviewed the structural properties of protein β sheets and has presented a clear and incisive review of this secondary protein structure. The classical flat β-sheet arrangement originally described by Pauling and Corey (1951) is in contrast to the great variety of twisted and curved surfaces found in protein β sheets. Salemme (1983) reviewed the operative forces and constraints that produce different twisted β-sheet geometries. Briefly, the following factors contribute to the complexity of β sheets. Because the repeating subunits of the chains are chiral, they tend to assume minimum-energy conformations whose effect is to twist the polypeptide chains away from the twofold helical conformation that produces flat sheets. This in turn results in the introduction of energetically unfavorable distortions into the interchain hydrogen bonds. The final configuration of the sheet represents an energetic compromise between optimizing the conformational energy of the polypeptide chains and preserving the interchain hydrogen bonds (Weatherford and Salemme, 1979). The final geometry of the sheet therefore depends on the specific features of interchain hydrogen bonding that constrain the possible ways in which the polypeptide chain can twist. The great variety of sheet geometries observed in globular proteins reflect alternate ways of reaching this compromise. Because of differences in parallel and antiparellel sheet symmetry properties, the effects of chain twisting can produce quite different structural geometries under the corresponding different constraints imposed by hydrogen bonds.

Chothia (1973) showed that, in general, the β sheet with a right-handed twist, when viewed along the polypeptide chain direction, has a lower free energy than sheets that are straight or have a left-handed twist. Ananthanarayanan and Bandekar (1976) attempted to predict the β regions in 16 globular proteins by applying the one-dimensional Ising model of Lifson and Roig (1961). The parameters for the theory were derived from the statistical data on globular proteins given by Chou and Fasman (1974a,b). Results obtained compared favorably to those from other methods, but it was pointed out that not considering the long-range interactions in their method and in other methods based on short-range interactions would make these methods incomplete and incapable of being uniformly applicable to all proteins. Finkelstein *et al.* (1970) showed that the domains constructed from antiparallel β structures can have only an extremely limited set of topologies depending primarily on the number of β portions and the localization of the β hairpin initiating the formation of the protein structure. Ptitsyn *et al.* (1979) proposed a folding mechanism for β proteins on physically reasonable assumptions. According to this mechanism, the folding pathway and final protein topology depend only on the total number of β strands and on the localization of the initiating complex in the given protein chain. This reduces the number of possible topologies for β protein from 10^2 or 10^8 (depending on the number of β strands) to only a few. Lifson and Sander (1979) determined the frequencies for the 20 amino acids separately for antiparallel and parallel arrangements of strands.

Parallel (β_P) and antiparallel (β_A) arrangements of strands in a sheet differ in the hydrogen pattern between strands and in the type of chain connectivity they allow: short reverse-turn connections for β_A and longer crossover connections for β_P (Levitt and Chothia, 1976; Richardson, 1977; Sternberg and Thornton, 1977a–c). The distinction between the two arrangements results in strikingly different sets of preference parameters, including some of the largest values reported so far for any substructure. These authors suggested the use of these new parameters; beyond secondary structure prediction, the different preferences for β_A and β_P may aid in predicting the tertiary interaction between strands. Lifson and Sander (1980a,b) also determined the frequency of occurrence of nearest-neighbor residue pairs on adjacent antiparallel and parallel strands in 30 known protein structures. These were studied by statistical methods. The largest and most significant correlations were: Ser/Thr (1.9 ± 0.3), Ile/Val (1.7 ± 0.3), and Lys/Arg/Asp-Gln (1.8 ± 0.3) in β_A and Ile/Leu (1.9 ± 0.4) in β_P. The pair Gly/Gly never occurs in any β sheet.

Previous attempts at analyzing the tertiary structure of β sheets by statistical means have been of two types: (1) analysis of sheet topology and (2) analysis of the amino acid content of β sheets. Schulz and Schirmer (1974), Richardson (1976, 1977), Levitt and Chothia (1976), and Sternberg and Thornton (1977a,b) have analyzed known β sheets as to length, direction, and number of strands, ordering of strands within the sheet, length, type, and handedness of crossover conditions, and statistical significance of the occurrences of folding units containing β strands. Sternberg and Thornton (1977c) observed that the most hydrophobic strands tend to occur at the center of the sheet and put forward the hypothesis of the hydrophobic ordering of strands. Von Heijne and Blomberg (1977, 1978) analyzed pair correlations among hydrophobic neutral and polar classes of residues. They found that interstrand pairs between residues of the same class occur more often than expected by random chance. They concluded from their observations "that inter- and intrastrand nearest neighbor interactions of a rather unspecific character are responsible for the main stabilizing forces in β sheets." Lifson and Sander (1980a) challenged this conclusion and report pair correlations to the level of specific individual recognition among the most frequent amino acid residues. Less frequent residues are grouped according to their resemblance in size, structure, polarity, and genetic exchangeability (Sander and Schulz, 1979). For antiparallel strands, with a database of 788 residue pairs, the statistical analysis of pair correlations involves seven individual residues, five groups of two residues each, and one group of three residues. For parallel strands, with a data base of only 263 pairs, the groups are more extensive.

Chothia and Janin (1982) studied the packing of β-sheet-to-β-sheet in a family of proteins that are formed by two β-pleated sheets packed face to face. Concanavalin A, plastocyanin, α-crystallin, superoxide dismutase, prealbumin, and the immunoglobulin domains are representative members of this family. They found that when β-pleated sheets pack face to face in proteins, the angle between the strand directions of the two β sheets is near −30°. They propose a simple model that shows how the relative orientation of two packed β sheets is a consequence of (1) the rows of side chains at the interface being approximately aligned and (2) the β sheet having a right-handed twist. The special amino acid composition of residues at the β-sheet-to-β-sheet interfaces makes the contact surfaces essentially smooth and hydrophobic. Cohen *et al.* (1981) examined the tertiary structure of ten β sandwiches formed by face-to-face packing of two primarily antiparallel β sheets. They formed a well-defined structural class with the following features: (1) a standard packing geometry with the two twisted β sheets separated by 8.3 to 10.3 Å and rotated counterclockwise by 20° to 50° and (2) common values and positions for changes in solvent-accessible contact area during the condensation of β strands → β sheets → β sandwich. Sheet packing produces anticomplementary patterns of area changes. (3) The sheet–sheet interface has a bilayer structure when medium-sized residues in the two sheets stack, but there is interdigitation of large and small residues in different sheets; and (4) the twisted nature of the β sheet explains both the left-handed rotation between the two sheets and the observed anticomplementary pattern of area changes.

These observations have been incorporated into a computer algorithm to predict the tertiary fold of β sandwiches from primary and secondary structure. Salemme and Weatherford (1981a,b; Salemme, 1981) studied the conformational and geometric properties of idealized β sheets. In a series of three papers, the parallel, antiparallel and mixed, and isotropically stressed β sheets were studied. Comparison of observed parallel β structure with the idealized models showed that to the extent that the observed structures are regularly hydrogen bonded, they are closely approximated by the models. The long-range geometric configuration of twisted sheets was shown to be primarily an equilibrium between the forces that cause the individual polypeptide chains to twist in order to minimize their local conformational potential energies and the requirements for interchain hydrogen bonding, which generally resist the introduction of twist into the sheet. Antiparallel structures possess conformational degrees of freedom that allow them to assume a greater diversity of spatial configurations than occur in parallel sheets.

Hol *et al.* (1981) demonstrated that as a result of the regular arrangement of peptide dipoles in secondary structure segments and low effective dielectric constant in hydrophobic cores, the electrostatic energy of a protein is very sensitive to the relative orientations of the segments. Evidence was provided that the alignment of secondary structure dipoles is significant in determining the three-dimensional structure of globular proteins. In globular proteins containing only β sheet there is an overwhelming tendency to align in an antiparallel manner, and all peptide dipole moments cancel each other, with no residual dipole in either direction. In a parallel arrangement, the components of the dipole moment parallel to the strand direction interact unfavorably with each other. This effect in the antiparallel strand may only be ~0.4 kcal/mole for a pair of short β strands containing four residues. This gives a difference in electrostatic energy of ~0.8 kcal/mole between parallel and antiparallel arrangements in three strands, which is smaller than the energy difference for α helices, but it may nevertheless be significant. Chou *et al.* (1982) minimized the energy of two- and three-chain antiparallel and parallel β sheets. All computed minimum-energy β sheets were found to have a right-handed twist, as observed in proteins. As in the case of the right-handed α helices, it is the intrastrand nonbonded interaction energy that plays a key role in forcing β sheets of L-amino acids to adopt a right-handed twist. The minimized energies of parallel β sheets are considerably higher than those of the corresponding antiparallel β sheet, indicating that parallel β sheets are intrinsically less stable. The energy difference between antiparallel and parallel β sheets arises from closer packing of the chains and a more favorable alignment of the peptide dipoles in the antiparallel structures.

Chothia and Lesk (1982a,b) have studied the family of small copper-containing proteins plastocyanin and azurin, which are active in the electron transport systems of plants and bacteria. Both proteins contain two β sheets packed face to face. Using computed superpositions of the structures, they aligned the sequences, identified homologous positions, and studied how the structures have changed as a result of mutations. The residues in the vicinity of the copper-bonding site show minimal amino acid substitution and form almost identical structures, whereas other portions are more variable in sequence and structure. Buried residues tend to maintain their hydrophobic character, but mutations change their values. Eleven immunoglobulin domains whose x-ray structures were known were also studied. Mutations caused (1) displacements and rotations of the β sheets relative to each other up to 2 Å and 20°, (2) lateral insertions of side chains from extended loops into the interface regions to compensate for reductions in the volume of β-sheet residues, (3) insertion of a residue into a strand to form a β bulge, (4) local changes in conformation, and, only rarely, (5) complementarity in adjacent mutations. The mutations of interior residues are accommodated through substantial structural changes consistent with the preservation of their function.

Chothia and Janin (1982) distinguished two classes of β-sheet-to-β-sheet packing in globular proteins. Both classes have β sheets with the usual right-handed twist packed face to face. In orthogonal β-sheet packings, the strand directions of the different β sheets are 90° to each other. Twisted sheets in this orientation have anticomplementary surfaces: one pair of diagonally opposite corners in β sheets is very close, and the other pairs of corners splay apart. At the close corner, the β sheets are usually covalently connected, connected by a β bend. Contacts between the β sheets occur along the diagonal joining the close corners. They involve about one quarter of the β-sheet residues, and two-thirds of them are Val, Ile, or Leu. In aligned β-sheet packings, the angle between the strand's directions of the packed β-sheets is ~30°. In this orientation, the twisted β-sheet surfaces are complementary. Novotny *et al.* (1984) presented a twisted hyperboloid (strophoid) as a model of β barrels in proteins. With a least-squares fitting procedure, polypeptide backbones of one parallel and seven antiparallel β barrels were approximated with various curved surfaces. Although the hyperboloid gave better approximations to all the β-barrel backbones than the ellipsoid, elliptical cylinder, or catenoid, the best approximations were obtained with a novel surface, a twisted hyperboloid (strophoid).

Chothia *et al.* (1985) studied the domain association in immunoglobulin molecules, which have variable domains. Approximately three quarters of the interfaces between VL and VH domains are formed by packing of VL (variable light chains) and VH (variable heavy chains) β sheets in the conserved "framework," and one quarter from contacts between the hypervariable regions. The β sheets that form the interface have edge strands that are strongly twisted (coiled) by β bulges. As a result, the edge strands fold back over their own β sheet at two diagonally opposite corners. When the VL and VH domains pack together, residues from these edge strands form the central part of the interface and give what Chothia *et al.* called a three-layer packing; i.e., there is a third layer composed of side chains inserted between the two backbone side-chain layers that are usually in contact. This three-layer packing is different from previously described β-sheet packings. The 12 residues that form the central part of the three observed VL–VH packings are absolutely or very strongly conserved in all immunoglobulin sequences. This strongly suggests that this structure is a general mode for the association of VL and VH domains and that three-layer packing plays a central role in forming the antibody-combining site.

Garratt *et al.* (1985) applied the prediction algorithm of Garnier *et al.* (1978) to 16 proteins whose structures are dominated by β sheet. Comparisons of the predicted structures with those defined by Kabsch and Sander (1983a,b) showed that for β-sheet residues, the quality of prediction falls off markedly with increasing residue accessibility. Two subclasses (internal and external) of β residues have been distinguished on the basis of hydrogen-bonding patterns, and the distribution of amino acid types within each subclass was found to be different. Thus, the Chou and Fasman (1974a,b) P_β-type parameters for these previously indistinguished states have been derived.

Getzoff *et al.* (1986) presented a qualitative computer graphics approach to the characterization of forces important to the assembly of β domains that should have general utility for examining protein interactions and assembly. In their approach, the nature of the molecular surface buried by the domain contacts, the specificity of the residue-to-residue interactions, and the identity of electrostatic, hydrophobic, and hydrophilic interactions were elucidated. These techniques were applied to the β-barrel domains of Cu,Zn-superoxide dismutase, immunoglobulin Fab, and tomato bushy stunt virus protein. Strong β-domain interactions (identified by their biochemical integrity) apparently result from chemical, electrostatic, and shape complementarity of the molecular surfaces buried from interaction with solvent molecules. Electrostatic forces appear to be important in both stabilizing and destabilizing specific contacts.

Taylor (1986a) described a pattern-matching procedure, based on fitting templates to the sequence, that allows general structural constraints to be imposed on the patterns identified. The templates correspond to structurally conserved regions of the sequence and were initially derived from a small number of related sequences whose tertiary structures are known. The templates are then made more representative by aligning other sequences of unknown structure. Two alignments were built up containing 100 immunoglobulin variable-domain sequences and 85 constant-domain sequences, respectively. From each of these extended alignments, templates were generated to represent features conserved in all the sequences. These consisted mainly of patterns of hydrophobicity associated with β structure. For structurally conserved β strands with no conserved features, templates based on general secondary structure prediction principles were used to identify their possible locations. The specificity of the templates was demonstrated by their ability to identify the conserved features in known immunoglobulin and immunoglobulin-related sequences but not in other nonimmunologic sequences.

C. Packing of α Helices (α/α)

One of the first attempts to fold secondary structures into the native tertiary fold was that of Ptitsyn and Rashin (1975), who predicted the folding of the α helices of myoglobin. Using the crystallographic assignments of α helices, they docked the α helices to bury the hydrophobic residues maximally. One of the most favorable structures obtained in this manner coincided in rough resolution with the native tertiary fold. Richmond and Richards (1978) also examined the helical packing in sperm whale myoglobin. The approach of two helices along the contact normal connecting their axes produced solvent-exclusive effects at a distance of about 6 Å from the final position. The solvent-excluded area formed in such an interaction site is equivalent to a large hydrophobic contribution to the free energy of association. Helices of close-packed spheres form useful approximations to actual peptide helices. On the basis of the geometry implied by the close-packed-sphere helix, an algorithm was proposed for picking potentially strong helix–helix sites in the peptides of known sequence. When combined with preliminary secondary structure predictions, this algorithm might usefully restructure the search for these specific types of contact in the docking portion of a general fold program.

Efimov (1979) likewise considered the packing of α helices in globins. It was shown that close packing of hydrophobic side chains on the surface of an individual α helix or on the surface of the bihelical structure is obtained at a certain combination of rotational isomers. This allowed the prediction of rotational isomers in α-helical regions of proteins. Cohen *et al.* (1979) devised a computer program to fold a peptide chain consisting solely of helical segments and connecting links of known length using myoglobin as the example. This was an extension of the earlier work of Ptitsyn and Rashin (1975) and Richmond and Richards (1978). The helices are paired according to the list of potential sites, with each helix paired at least once. The list of pairs is then examined geometrically, with two filters being used: (1) lengths of connecting links must be equal to or greater than the end-to-end distances of helices, and (2) nonpaired helices must not collide. Adjustment of parameters reduced the final number of possible structures to 20, one of which closely resembles the actual distribution of helices in myoglobin. Cohen and Sternberg (1980a) added two new filters to their predictive method (Richmond and Richards, 1979). The use of chemical information to constrain the distal–proximal histidine separation aided in the prediction of the structure of myoglobin. Out of 20 structures, only two very similar structures satisfied this additional filter. The two remaining structures had root-mean-square deviations of 4.48 and 4.53 Å from the crystal structure. Cohen and Sternberg (1980b) reported a method and quantified the significance of obtaining a specific root-mean-square deviation when folding proteins of different molecular weight. The average root-mean-square deviation was found to be proportional to the number of residues, and this correlation was explained by a mathematical model.

The analysis of the pattern of residue-to-residue contacts at the interface of 50 helix-to-helix packings observed in ten proteins of known structure was carried out by Chothia *et al.* (1981). This analysis supported a model for helix-to-helix packing in which the ridges and grooves on the helix surface intercalate. These ridges are formed by rows of residues whose separation in sequence is usually four, occasionally three, and rarely one. The model explains the observed phenomenon of packings whose interhelical angle is ~50°. Of the 50 packings, 38 agree with the model, and the general features of another ten packings were described by an extension to the model in which ridges can pack across each other if a small side chain occurs at the place where they cross. Before this paper there were only three other detailed models for helix-to-helix packing, those described by Crick (1953), by Efimov (1977, 1979), and by Richmond and Richards (1978). Crick's (1953) knobs-into-holes model for helix packing was developed as part of the coiled-coil model for α-keratin. Efimov (1977, 1979) has described two models for helix packing: the polar and the apolar. In the apolar model, side chains are clustered on one face of a helix by residues i, $i + 4$, and $i + 8$ having a *trans* conformation, $\chi \simeq 180°$, and residues $i + 1$ and $i + 5$ having a *gauche* conformation, $\chi_1 \simeq -60°$. If side chains are represented by spheres, he shows that the packing of two such helices gives the torsion angle between helices, Ω, values of +30°, −30°, and 90°. Efimov (1979) states that this model (Efimov, 1977) coincides in principle with that described by Crick (1953), by Chothia *et al.* (1977), and by Richmond and Richards (1978).

A domain of four α helices packed into a right-handed bundle has been proposed to be a recurrent motif in protein structure (Sternberg and Cohen, 1982; Sheridan *et al.*, 1982; Efimov, 1982a). The fourfold α-helical structure had been previously suggested (Argos *et al.*, 1977; Weber and Salemme, 1980) to be similar to that observed in several proteins, e.g., the disk of tobacco mosaic virus (Bloomer *et al.*, 1978). Sternberg and Cohen (1982) predicted the secondary and tertiary structures of four interferons from homologous amino sequences. Three methods of predicting the secondary structure were used (Lim, 1974a–c; Chou and Fasman, 1978a; Garnier *et al.*, 1978) and gave differing results, but all of them suggested four α helices. The prediction of Hayes (1980) for the interferons also agreed in principle with that of Sternberg and Cohen (1982). The algorithm for docking α helices into a tertiary fold (Richmond and Richards, 1978; Cohen *et al.*, 1979; Cohen and Sternberg, 1980a,b), which was slightly modified, substantiated the proposed right-handed four-helical bundle model. One model was shown to be compatible with the known disulfide linkages in interferon. It should be noted that there are instances of left-handed bundles formed from four α helices (Blow *et al.*, 1977; Matthews *et al.*, 1971; Remington *et al.*, 1978; McLachlan *et al.*, 1980).

Sheridan *et al.* (1982) developed a simple dipole model for estimation of the electrostatic interactions between α helices in the protein tertiary structural motif of an array of four closely packed α helices. It was found that, for the proteins examined (cytochrome *c*′, hemerythrin, myohemerythrin, cytochrome b_{562}, and a T4 phage lysozyme domain), their common antiparallel arrangement of adjacent helices confers a stabilization of 5–7 kcal/mole. In contrast, a similarly packed array of parallel helices is relatively destabilized by 20 kcal/mol. Hol *et al.* (1981) has calculated the electrostatic interactions between the peptide groups of both α helices and β sheet in various proteins and demonstrated that secondary structures pack in a manner to provide significantly favorable electrostatic energy. Efimov (1982a,b) demonstrated that four-helix complexes can form two "mirror-symmetric" structures. A consideration of hydrophobic interactions, hydrogen bonds, and salt bonds is not sufficient for an unequivocal choice of one of these two structures. It was shown that severe stereochemical restrictions on the packing of the α helices are imposed by the length of the interhelical regions, i.e., the constrictions.

Murzin and Finkelstein (1983) considered the packing of α-helical portions in globules with a single monobound nucleus. A scheme for the description of α-helical complexes by a system of "longitudinal" and "transverse" interhelical contacts was proposed. It was shown that the closed helical globules are well described by quasispherical polyhedra with the ends of the α helices serving as their apices and the axes of the α helices and the lines of contacts between their ends as ribs. It was found that the cross pieces between the helical portions run along certain ribs of such polyhedra. A novel supersecondary structure, referred to as an αα corner, was described by Efimov (1984). The αα corner is formed by two consecutive α helices packed approximately crosswise and connected by two or more peptide units. It was shown that the amino acid sequences coding for the αα corners have a strictly definite order of hydrophobic, hydrophilic, and Gly residues. A hypothesis was suggested that the αα corner can be an embryo of protein folding.

Cohen and Kuntz (1987) applied a series of heuristic algorithms to the sequence of human growth hormone. A family of five structures that are genetically right-handed fourfold α-helical bundles was found from an investigation of 10^8 structures. A plausible receptor binding site was suggested. Circular dichroism studies showed a secondary structure consisting of 45–50% α helix and no β structure, thus placing this protein in the α/α class of proteins. The turn algorithm of Cohen *et al.* (1986a,b) using the patterns for α/α domains (AA-TURNS; Cohen *et al.*, 1982) identified 12 turns. The key positions for hydrophobic residues that favor helicity (Cohen *et al.*, 1982; Richmond and Richards, 1978) were also obtained. Helical boundaries were defined by the delimit methods (Cohen *et al.*, 1983). Helix–helix interactions, which were divided into three classes by the helix-packing algorithm of Cohen *et al.* (1979), and helix docking sites were located by the method of Richmond and Richards (1978). This list was further processed using the steric restriction of Cohen *et al.* (1979). The interaction site central residue, interaction class, and secondary structure provide the input for the combinatorial helix-packing algorithm of Cohen *et al.* (1979). Distance restraints were used to further eliminate possible structures (Havel *et al.*, 1983; Cohen and Sternberg, 1980a,b). The surface and volume characteristics of proteins have indicated that proteins do not have a high surface-area-to-volume ratio (Richards, 1977). Thus, 119 of the structures generated have high surface-area-to-volume ratios and long loop excursions, so these were eliminated. The remaining 67 structures fall into the fourfold helical bundle motif. Five of the remaining structures have the right-handed topological preference (Weber and Salemme, 1980) and were kept and have small variations. The application of the helix dipole for stability (Shoemaker *et al.*, 1987) further assisted in establishing the connectivity of the four helices.

D. Amphipathic α Helices and β Strands: Dipole Moments and Electrostatic Interactions

Segrest and Feldman (1977) presented a computer analysis of the general occurrence of amphipathic helix patterns in proteins with known sequence. An α helix is considered amphipathic when it can be divided into separate polar and nonpolar faces. A specific distribution of charged residues along the polar face is frequently observed. A minimal residue length of 11 was set for selection of amphipathic helices. The *Protein Sequence Data Tape* (1972) plus *Supplemental* (1973), National Biomedical Research Foundation, Georgetown University

Medical Center, Washington, D.C. 20007, was used. A total of 649 amino acid sequences had two distinct properties that could be quantified: the number of ion pairs and the degree of hydrophobicity of the potentially nonpolar face. Positively charged lysyl and arginyl residues occurred at the interface between the polar and nonpolar "faces" of each amphipathic sequence. Amphipathic helices were found to occur with much greater frequency in known lipid-binding proteins than in proteins in general.

Fourier analysis of the hydrophobicities (Kyte and Doolittle, 1982; Eisenberg *et al.*, 1982a,b) of the acetylcholine receptor subunits by Finer-Moore and Stroud (1984) revealed regions of amphipathic secondary structure. Prediction of a consensus secondary structure based on this analysis and on empirical prediction methods (Garnier *et al.*, 1978) for the sequence external to the bilayer ($\simeq$75%) and separate evidence for oriented helices in the transmembrane regions led to a testable hypothesis about how the ion channel is formed and might function. Previously McLachlan and Karn (1983) had discussed the use of Fourier analysis as a means of identifying periodic properties of amino acid sequences. The periodicity of the amphipathicity was found to be exactly that expected for an α helix at least 30, and up to 50, amino acids in length. The predicted overall secondary structure of the acetylcholine receptor has 27% β sheet and 44% α helix, in rough agreement with both circular dichroism studies [29% β, 34% α (Moore *et al.*, 1974)] and Raman spectroscopy studies [34% β, 25% α helix, 14% disordered helix (Chang *et al.*, 1983)]. They proposed a mode of assembly of the acetylcholine receptor complex from its subunits in which the transmembrane helices, H1–H3, are hydrophobic and could be inserted into the bilayer during synthesis. As the subunits associate, the COOH-terminal hydrophobic helix H4 and amphipathic helix A could fold into the bilayer looped together. Thus, a hydrophilic lipid-free channel could be formed between subunits from residues that would be unstable in the individual subunit's interface with lipids. Correct ion pairing could encode correct multisubunit assembly.

Krebs and Phillips (1984) showed that the amphiphilicity of an α-helical segment in a protein may be quantitated by calculating its mean helical hydrophobic moment (μ_H) (Eisenberg *et al.*, 1982a,b). For proteins whose hydrophobic interactions with interfaces are mediated by α helices, the surface pressure exerted at the air–water interface correlates with the product ($\mu_H \cdot F$), where μ_H is the mean helical hydrophobic moment averaged over all helices in the entire molecule and F is the fraction of α helix in the protein. Knowledge of μ_H permits a description of amphipathic α helices and their surface activities at the air–water interface of serum apolipoproteins, surface-seeking peptides, and globular water-soluble proteins. Jähnig and co-workers (Vogel and Jähnig, 1986; Vogel *et al.*, 1985) estimated the secondary structure of the lactose permease of *Escherichia coli* reconstituted in lipid membranes by Raman spectroscopy. They obtained an estimate of ~70% α-helix content, β-strand content below 10%, and a β-turn contribution of 15%. About one third of the residues in α helices and most other residues are exposed to water. Then, by a method for structural prediction that accounts for amphipathic helices (Eisenberg *et al.*, 1984a,b; Finer-Moore and Stroud, 1984), ten membrane-spanning helices were predicted that are either hydrophobic (Kyte and Doolittle, 1982) or amphipathic. These are expected to form an outer ring of helices in the membrane, the interior of which would be made of residues that are predominantly hydrophilic and, by analogy to sugar-binding proteins, suited to provide the sugar-binding site. The β-turn profile (Chou and Fasman, 1978a) was also utilized.

Vogel and Jähnig (1986) also determined the secondary structure of porin, maltoporin, and OmpA protein reconstituted in lipid membranes by Raman spectroscopy. The three proteins have similar structures consisting of 50 to 60% β strand, about 20% β turn, and less that 15% α helix. By a method for structural prediction that accounts for amphipathic β strands, folding models were developed for porin and for a segment of OmpA protein incorporated into a membrane. In the model, the OmpA fragment consists of eight amphipathic membrane-spanning β strands that form a β barrel. Similarly, porin is folded into ten amphipathic membrane-spanning β strands that are located at the surface of the trimer towards the lipids and eight predominantly hydrophilic strands in the interior. The search for stretches of 20 amino acid residues that would form a membrane-spanning α helix used the Kyte and Doolittle (1982) method. To ascertain β-strand amphipathicity, the method for elaboration of amphipathic helices (Eisenberg *et al.*, 1984a,b; Finer-Moore and Stroud, 1984; Vogel *et al.*, 1985) was modified. The secondary structure of residues exposed to water and the β turns was predicted by the Chou and Fasman (1978a) algorithm.

Cornette *et al.* (1987) developed computational techniques for detecting amphipathic structures in proteins. Protein segments that form amphipathic α helices have periodic variation in the hydrophobicity value of the residues along the segment, with a 3.6-residue-per-cycle period characteristic of an α helix. They compared the usual method for detecting periodicity based on a discrete Fourier transform with a method based on a least-squares fit of a harmonic sequence to a sequence of hydrophobicity values. The analogue to the usual Fourier transform power spectrum is the "least-squares power spectrum," the sum of squares accounted for in fitting a sinusoid of given frequency to a sequence of hydrophobicity values. The sum of the spectra of the α helices in their data base peaked at 97.5°, and approximately 50% of the helices can be accounted for in this peak. Thus, approximately 50% of the α helices appear to be amphipathic, and, of those that are, the dominant frequency at 97.5° rather than 100° indicates that the helix is slightly more open than previously thought, with the number of residues per turn closer to 3.7 than 3.6. This extra openness was examined in crystallographic data and was shown to be associated with the C terminus of the helix. The α amphipathic index, the key quantity in their analysis, measures the fraction of the total spectral area that is under the 97.5 peak and is characteristic of hydrophobicity scales that are consistent for different sets of helices. They developed an optimized hydrophobicity scale that maximizes the amphipathicity index and has a correlation of 0.85 or higher with nine previously published scales. Although the scale is optimal only for predicting α amphipathicity, it also ranks high in identifying β amphipathicity and in distinguishing interior from exterior residues in proteins.

The possibility of certain residues exerting helix-forming influence in either the COOH-terminal or NH_2-terminal direction preferentially was suggested by the nonrandom distribution of amino acid residues between the two ends of helical regions in globular proteins (Cook, 1967; Ptitsyn, 1969). Robson and Pain (1972) confirmed that this directional effect exists and characterized this effect for individual residues and its dependence on distance along the polypeptide chain. A study of primary sequence–conformation relationships by information theory analysis involved the estimation of an information function, $I(S_j; R_{j+m})R_j$ (Robson and Pain, 1971). This describes the information in residue R_{j+m} at position $j + m$ (where $-8 \leq m \leq +8$) concerning the conformation S_j of residue R_j and will include any directional effect, since the value of the function will be dependent on the sign of m. The function $i(S_j; \mathrm{R}_{j+m})$ is the information in the residue at position $j + m$ concerning the configuration per average-type residue, which could be at position i. The analysis of 11 proteins showed which residues showed well-defined characteristics: i.e., Ala, Leu, and Ile exhibit a tendency to support helix formation in both directions; Pro disrupts helices in the NH_2-terminal direction and, interestingly, has a definite tendency to potentiate helix formation in the COOH-terminal direction; Asp and Glu show strong tendencies to helix disruption in the NH_2-terminal direction; Glu exhibits a strong directional influence in that it supports helix formation in the COOH-terminal direction, whereas the influence of Asp is of marginal significance; basic amino acids Lys and marginally His and Arg show tendencies to support helix formation in the NH_2-terminal direction. Other groups have also studied the tendency of certain side chains to cluster at the end of helices (Lewis *et al.*, 1971; Crawford *et al.*, 1973).

Lewis and Bradbury (1974) found the attractive and repulsive electrostatic interactions of the *i*th residue with its neighbors $i \pm 1, 2, 3, 4$, and 7 to be helix-breaking if there was more than one net repulsion. Maxfield and Scheraga (1975) studied the effect of neighboring charges on the helix-forming ability of charged amino acids in proteins. They found helix-disruptive effects of some charged residues at the $i \pm 4$ position, whereas helix-stabilizing effects were found with the oppositely charged residues at $i \pm 2$ and $i \pm 3$.

Chou and Fasman (1974a) further characterized this directional propensity and evaluated the frequency of helical boundary residues in 15 proteins. Their results were in general agreement with those of Robson and Pain (1972); however, there were several major differences: Arg, His, and Lys were all found with the highest frequency at the COOH terminus of helices, and Pro, Asp, and Glu were found to have the highest frequency at the NH_2 terminus. Blagdon and Goodman (1975) proposed that helix initiation could occur at the end of helices, i.e., terminal initiation by polar groups and by turns, and this would have a directionality because of the distribution of charge discussed above. The significance of the fairly strong electric fields produced by electric macrodipoles of the α helix and by ionic charges in stabilizing proteins has been recognized for some years (Wada, 1976; Perutz, 1980; Hol *et al.*, 1978; van Duijen *et al.*, 1979). These ideas have been incorporated into the α-helix dipole hypothesis. Wada (1976) discussed the α helix as an electric macrodipole and its importance in protein structure. Hol *et al.* (1978) discussed in detail the role of the α-helix dipole for the properties of proteins. In an α helix the peptide dipoles are aligned nearly parallel to the helix axis, and the axial component is 97% of the dipole moment. Theoretical considerations indicate an increase in the peptide dipole moment by polarization related to hydrogen bonds in the α helix, yielding a value of up to 5 D. The α helix has a considerable electric field, which runs from the C terminus to the N terminus. The strength of the field increases up to a helix length of about 10 Å (two turns), whereafter further elongation has only a marginal effect.

Hol *et al.* (1981) applied the concept of peptide dipoles to protein folding. As a result of the regular arrangement of peptide dipoles in secondary structure segments and the low effective dielectric constant in hydrophobic cores, the electrostatic energy of a protein is very sensitive to the relative orientation of the segments. Evidence was provided that the alignment of secondary structure dipoles is significant in determining the three-dimensional structure of globular proteins. Model calculation of helix–helix interactions showed that electrostatic energy between the backbone dipoles of helices was maximum for antiparallel helices rather that parallel helices. In proteins with alternating α and β structures the α–β dipolar interaction is favorable when helices and strands are antiparallel.

Sheridan and Allen (1980) presented electrostatic potential maps of α helices of various lengths using a point-charge model. It was shown that the potential field of the helix can be mimicked by two equal and opposite charges, one at each terminus. The magnitude of these equivalent charges reaches its limiting value of ±0.2 to 0.3 electron at a helix length of approximately seven to ten residues. Hydration, the presence of counterions, and solvent dielectric would no doubt reduce the influence of both charged residues and the helix dipole, the former probably to a greater extent. In order to have the helix dipole potential as the major electrostatic influence on substrate or coenzyme binding site, a protein must sequester these sites within crevices shielded from solvent and at the same time keep any charged residues at a reasonably large distance from the site. If these conditions are met, anions will find stability at the N termini of helices. On the other hand, at points somewhat removed from binding-site crevices, the greater electrostatic influence is very likely to be the charged residues on the protein surface. That is, helices will stabilize the presence of anions already in the binding site but will not draw them in from a distance.

Wada and Nakamura (1981) analyzed the distribution of the distance separating the ionic charges (ionized groups and the apparent charges at the termini of the α helices) for more that

44,000 charge pairs in 14 proteins. Their results show that charges in the proteins are, on average, surrounded by charges of the opposite sign. Previous calculations of the electric potential near the helix termini have shown that the electrostatic effect of the α-helix dipole is equivalent to the effect of one half of a positive unit charge at the N terminus of the α helix and one half of a negative charge at the C terminus (Hol *et al.*, 1978; Sheridan and Allen, 1980). The macrodipole of the α helix has been found to have the same order of contribution to stabilizing the native protein conformation as ionized groups. These electrostatic interactions have been stated to be involved at active sites of functional proteins in binding substrate or a coenzyme or in enhancing enzymatic reaction rates (Perutz, 1980; Hol *et al.*, 1978; van Duijen *et al.*, 1979; Sheridan and Allen, 1980). Perutz *et al.* (1985), in discussing the pK_a values of two His residues in human hemoglobin, invoke the dipole of the helices to explain the abnormal values. In HbCO A, His, FG(97)β has a pK_a of 7.8 compared to the pK_a value of about 6.6 characteristic of free His at the surface of proteins. This high pK_a is accounted for by its interaction with the negative pole at the C terminus of helices F and FG. It corresponds to a free energy change of the same order as that observed in the interaction of histidines with carboxylate ions and confirms the strongly dipolar character of α helices, which manifests itself even when they lie on the surface of the protein.

Thornton and Sibanda (1983) analyzed the structural, dynamic, and functional aspects of amino- and carboxyl-terminal regions of proteins of known structure. Terminal regions are usually located on the surfaces of the protein, accessible to solvent, and are often flexible. There is a significant preference for terminal regions in single-domain proteins to be in close proximity. This partially reflects the compact globular nature of proteins, but the preference for spatial proximity is stronger in native proteins than in randomly generated structures. In addition, in multidomain and multisubunit proteins, it was found that the terminal regions are commonly involved in the interface between domains and subdomains.

The role of electrostatic interactions is complicated and not readily described analytically. However, efforts have been made to analyze and calculate their contribution to protein folding and stability. As an example, Rogers and Sternberg (1984) have discussed the different dielectric models applied to the packing of α helices. The effect of the α-helix dipole in stabilizing the tertiary structure of globular proteins was examined using three of the commonly used dielectric models. These are (1) the uniform dielectric model, (2) the distance-dependent dielectric model, and (3) the cavity dielectric model. They found the cavity model to be the most reasonable since it attempts to describe the markedly different dielectric responses of the solvent and the protein. The protein is set at a low continuous dielectric value, and the solvent is set at a high continuous dielectric value. It was found that for the cavity model of the dielectric, the calculated interaction energy between two helices is strongly dependent on how exposed the helix termini are to solvent. For helices with exposed termini, the calculations using the cavity model yielded electrostatic interaction energies that were lower by an order of magnitude than those using the uniform dielectric model.

Several authors have attempted to calculate the contribution of electrostatic potentials in an enzyme active site. Gilson and Honig (1987) used experimental studies on the serine protease subtilisin to compare calculated electrostatic interactions between two specific sites on a protein (Klapper *et al.*, 1986). The extent of agreement between the theoretical and experimental results suggested that the continuum solvent model used in the calculation reproduces the essential feature of the interaction. Sternberg *et al.* (1987) have shown that the algorithm of Warwicker and Watson (1982), which uses classical electrostatics and considers both the charge position and the shape of the molecule, can be used to model several pK_a shifts in subtilisin. These pK_a shifts were produced by site-directed mutagenesis. Quiocho *et al.* (1987) examined three highly refined atomic structures of periplasmic binding proteins. Hydrogen bonds, acting primarily through backbone peptide units, are mainly responsible for the involvement of the positively charged Arg^{151} residue in the ligand site of the arabinose-binding protein, for the association between the sulfate-binding protein and the completely buried sulfate dianion, and the formation of the complex of the Leu/Ile/Val binding protein with the Leu zwitterion. They propose a general mechanism in which the isolated charges on various buried, desolvated ionic groups are stabilized by the polarized peptide units.

Van Belle *et al.* (1987) described calculations implementing molecular mechanics and molecular dynamics simulation procedures on crystal structures of crambin, liver alcohol dehydrogenase, and ribonuclease T_1. Evaluation of the contribution of polarizability of the protein matrix to electrostatic energies, local fields, dipole moments of peptide groups, and secondary structure elements in the polypeptide chain was carried out. The results confirmed that induced dipole moments in proteins provide important stabilizing contributions to electrostatic energies and that these contributions cannot be mimicked by the usual approximations where either a continuum dielectric constant or a distance-dependent dielectric function is used. It was found that an induced protein dipole appreciably affects the magnitude and direction of local electrostatic fields in a manner that is strongly influenced by the microscopic environment in the protein. The induction effects from surrounding protein atoms tend on average to increase peptide dipoles and helix macrodipoles by about 16%, again reflecting electrostatic stabilization by the protein matrix, and show that (at least in the α/β domain of alcohol dehydrogenase) the contributions of side chains to this stabilization is significant.

The role of electrostatics and calculations of electrostatic interactions have been reviewed in depth (Rogers and Sternberg, 1984; Warshel and Russell, 1984; Matthew, 1985; Matthew and Gurd, 1986a,b; Honig *et al.*, 1986; Rogers, 1986).

The overall stability of a protein structure is considered to be the sum of contributions from electrostatic interactions, hydrogen bonding, and van der Waals and hydrophobic interactions (solvent exclusion). The role of solvent accessibility and the effective p*K* values for ionizable groups has been studied for many years and plays a role in enzyme catalysis, etc. The generation and utilization of electrical potentials is a central function of biological membranes. These topics are not discussed in this chapter, and the reader is referred to Chapter 8 by N. K. Rogers.

E. Packing of α Helices and β Pleated Sheets (α/β)

Chothia *et al.* (1977) presented simple models that describe the rules for almost all the packing that occurs between and among α helices and pleated sheets. These packing rules, together with the primary and secondary structure, are the major determinants of the three-dimensional structure of proteins. The two principles that dominate the way secondary structures associate are (1) that residues that become buried in the interior of a protein close-pack and occupy a volume similar to that which they occupy in crystals of their amino acids and (2) that associated secondary structures retain a conformation close to the minimum-free-energy conformations of the isolated secondary structures. Efimov (1977) also considered the stereochemistry of α helices and the β structure in globular proteins. Side-chain packing of hydrophobic amino acids was represented in a rough approximation in the form of spheres. Principles of close packing were applied for formation of close-packed layers of hydrophobic side chains on the surface of α helices and β structure.

Janin and Chothia (1980) proposed a model for the packing of α helices on β sheets in α/β proteins. It involved the association of two smooth surfaces with complementary twists: the surface of a regular β sheet with a right-handed twist and the helix face formed by two rows of residues, i, $i + 4$, $i + 8$, etc., and $i + 1$, $i + 5$, $i + 9$, etc. The model required the helix areas to be parallel to the β strands and contacts to be limited to the residues mentioned above. It also follows from the model that when two helices pack against each other as well, their axes should be at an angle of about $-40°$, compatible with only one class of helix–helix interaction. To substantiate their model, they examined helix–sheet and helix–helix contacts in eight proteins of known three-dimensional structure. The contact regions have a typical amino composition, especially in the β sheets, that is very rich in Val and Ile, two residues that contribute to the formation of a smooth surface. Thus, the requirement of helix–sheet packing sets severe restrictions on the configuration and on the amino acid composition of the secondary structures in α/β proteins.

Sternberg and Cohen (1982) and Sternberg *et al.* (1982) applied their combinatorial predictive approach to the docking of α helices and β strands into a tertiary fold. They point out that several problems still remain before a general predictive scheme using their combinatorial docking of secondary structure can be developed. (1) Secondary prediction is not yet accurate enough to provide the starting assignment required by the scheme (Schulz and Schirmer, 1979), and it was suggested that an improvement of secondary structure prediction might be to have some feedback from a tertiary docking algorithm. (2) Another problem is that many proteins are divided into spatially distinct regions, known as domains (Wodak and Janin, 1981), and an algorithm is required to locate the link between domains starting from the amino acid sequence. At present the combinatorial approach is a reduced list of possible structures, and methods are required to identify the correct fold and discard the alternatives.

Sternberg *et al.* (1982) studied the packing of α helices and β strands in six α/β proteins (e.g., flavodoxin) and developed a computer algorithm to predict the tertiary structure of an α/β protein from its amino acid sequence and actual assignment of secondary structure. The packing of an individual α helix against a β sheet generally involves two adjacent rows ±4 rows on nonpolar residues on the α helix. The pattern of interacting β-sheet residues results from the twisted nature of the sheet surface and the attendant rotation of side chains. In general, the α helix sits 10 Å above the sheet and lies parallel to the strand direction. The prediction follows a combinatorial approach. First, a list of possible β-sheet structures (10^6 to 10^{14}) is constructed by the generation of all β-sheet topologies and β-strand assignments. This list is reduced by constraints of topology and the location of nonpolar residues to mediate the sheet–helix packing and then rank ordered on the extent of hydrogen bonding. This algorithm was uniformly applied to 16 α/β domains in 13 proteins.

Thornton and Chakauya (1982) analyzed a sample of 70 nonhomologous proteins whose terminal residues had been classified either as α or β. It was found that the resulting distribution of the helical, β-strand, and coil conformations is significantly different for the amino and carboxy terminals. The amino terminal preferentially adopts an extended β strand, whereas the carboxyl terminal is usually helical. The observed difference derives from the α/β proteins in which the helix and strand alternate along the sequence, suggesting that the origin of this preference lies in the special structural topology of α/β proteins and the βα unit. Taylor and Thornton (1983) analyzed 62 βαβ units of known sequence and structure from 18 different proteins, which allowed them to calculate the average lengths of helix, strand, and coil regions and constructed an "ideal" βαβ unit. The observed βαβ sequences were scaled (expanded or contracted) to produce a maximum correspondence with the ideal sequence. The βαβ unit having its two strands lying adjacent in the β sheet was more precisely defined than the βαβ unit having nonadjacent strands.

The secondary structure was predicted by the method of Garnier *et al.* (1978). The predicted scheme is based on finding the best position along the protein sequence for a scaled ideal βαβ template. The application of this basic method to the location of βαβs and the refinement of secondary structure prediction is as follows. (1) A standard secondary structure prediction method (e.g., Chou and Fasman, 1978a; Garnier *et al.*, 1978) is used to generate the helical (P_α), extended (P_β), or turn/coil (P_t) probability profiles from the sequence. (2) Additional information on the distribution of hydrophobic residues, as specifically required in the βαβ unit, is added to these probabilities (Cohen *et al.*, 1982). Typically, this contribution is the same order of magnitude as the original P_α and P_β (Schulz and Schirmer, 1979). (3) The ideal βαβ template, which can expand or contract, is fitted to the β, α, and turn profiles for each position, on the sequence and the maximum F is found (the F value, the goodness of fit, is obtained as a product of the area under the predicted secondary structure probability curves for each type of structure in the region where it corresponds to the idealized sequence). The fits having the highest F that do not cause β and α regions to overlap are taken as the most likely βαβ locations. (4) For the refinement of the secondary structure prediction, the P_α, P_β, and P_t profiles are weighed by F (in the Garnier *et al.*, 1978, method, this is equivalent to a local change in decision constant), and a new secondary structure prediction is obtained. When this approach was applied to 16 βαβ proteins of known sequence and structure, 80% of adjacent-stranded βαβ units and 50% of the nonadjacent βαβs were correctly located. An average improvement of 7.5% was observed. The overall accuracy of 67% can be analyzed for each structural component.

Cohen *et al.* (1983) attempted to suggest a route to more productive secondary structure assignments by means of a physical-model approach. The general strategy was adopted from the "expert systems" formulation of artificial intelligence, which consists of stating explicit hypotheses or rules about the system, providing a way of evaluating the validity of the rules, and developing a set of higher-order rules to resolve conflicts. The rules were simple lists of generalized amino acid sequences or patterns that are associated via the physical model with specific secondary structures. The model is a natural evolution of the suggestions of Lim (1974a–c), Nagano (1973), and others. The use of artificial intelligence technology offers novel departures from earlier efforts: (1) the algorithms are completely defined; (2) useful results can be obtained even when a unique assignment is not possible; and (3) the entire process is refinable so that new information or insights can be added and tested at any stage. The method was applied to α/β proteins. Turns were identified very accurately (98%) by simultaneously considering hydrophobicity and the ideal spacing of turns throughout the sequence. Although the assignment for any one sequence is not unique, at least one of the assignments bears a close resemblance to the native structure. The algorithm successfully divided proteins into two classes: α/β and non-α/β. The accuracy of the secondary structure assignments in the α/β class is sufficient to provide useful input for tertiary structure assignments.

Taylor and Thornton (1984) described a procedure to recognize supersecondary structure in protein sequences (see above). When applied to the identification of βαβ units in β/α type proteins, it was found to have 75% accuracy. Wierenga *et al.* (1986) derived an amino acid sequence "fingerprint" that has been used to test if a particular sequence will fold into a βαβ unit with ADP-binding properties. The fingerprint, constructed from analysis of five proteins, is a set of 11 rules describing the type of amino acid that should occur at a specific position in a peptide fragment. The total length of this fingerprint varies between 29 and 31 residues. When the data were checked against all possible sequences in a data base (PIR: Barker *et al.*, 1984), it appeared that every peptide that exactly follows this fingerprint does indeed fold into an ADP-binding βαβ unit.

Edwards *et al.* (1987) examined the structural and sequence patterns in the loops of βαβ units (See Section V.A.).

F. Prediction of Protein Conformation by Minimum-Energy Calculations

The earliest attempts to determine the structure of proteins by means of minimum-energy calculations were those of Ramachandran *et al.* (1963) and Liquori and co-workers (DeSantis *et al.*, 1965), who showed that the peptide unit can only adopt certain conformations. Nonbonding interactions were considered, and ϕ,ψ plots were constructed, which were later improved by semiempirical calculations (Lewis *et al.*, 1973b), to predict preferred regions. Scheraga and co-workers have contributed significantly to this area and have frequently reviewed the field (Scheraga, 1968, 1971, 1985; Nemethy and Scheraga, 1977). The assumption in all these calculations is that a protein folds so as to minimize the free energy of the system, and many investigators (e.g., Levitt and Warshel, 1975; Nemethy and Scheraga, 1977; McCammon *et al.*, 1977) have developed potential functions to describe the energy surface of the polypeptide chain. Chain folding is simulated computationally to find the energy minimum. Alternatively, conformational space is probed from a starting point by integrating the equations of motion over time.

There are two major problems in this approach that have yet to be effectively overcome: (1) decisions on the appropriate parameterization of the energy, including derivation of the values of the parameters, and (2) the multiple-minimum problem on the global energy surface (Ponder and Richards, 1987). Much research is presently directed toward solving these problems, and substantial progress is being made (e.g., Weiner *et al.*, 1984). However, the approximations required to model the computationally complex electrostatic and solvent effects properly are major stumbling blocks. This problem of the free energy of folding must be solved if we are to understand protein structure based on first principles. It must be noted that it has not yet been determined unambiguously whether the minimum corresponding to the native protein is the one of lowest energy. If the native conformation were not the global minimum, this would imply that certain conformations are kinetically inaccessible because of very high potential energy barriers. To date, no evidence is available that contradicts the hypothesis that the native conformation is the one of the lowest free energy. Although progress is being made, energy-minimization schemes have failed to predict chain folding accurately (Hagler and Honig, 1978; Cohen and Sternberg, 1980a,b). Because the literature covering this area is vast, no attempt is made to summarize it, but only significant trends are mentioned. An excellent in-depth review can be found in Chapter 7 by Mackay, Cross, and Hagler.

The methods of attempting to predict tertiary structure have been of two types. (1) The Monte Carlo approach of the Scheraga school, discussed in detail by Nemethy and Scheraga (1977), samples the conformational energy surface at random, and the emphasis is on the prediction of long-range residue-to-residue contact when the secondary structure is well predicted in advance or is constrained closely to the observed secondary structure. Further, it assumes that the native structure of a protein is that of least energy, whereas it might be the lowest-energy conformation that the protein can reach from an open conformation in reasonable time. (2) Folding procedures, on the other hand, take some account of the latter possibility but, in order to reduce the computation time to manageable proportions, use a very simplified representation of the molecular structure of the polypeptide chain (Levitt and Warshel, 1975; Levitt, 1976; Kuntz *et al.*, 1976). This simplification and other assumptions made in the latter methods have left them open to the criticism that they may have led to fortuitous approximate agreement with the observed structure (Robson, 1974; Nemethy and Scheraga, 1977; Hagler and Honig, 1978). Further, they simplify the backbone structure to such a degree that they cannot be used to provide detailed information about changes in secondary structure during folding (Robson and Osguthorpe, 1979).

Scheraga and co-workers (Ponnuswamy *et al.*, 1973) considered the role of medium-range interactions in proteins. The energies of oligopeptide segments of lysozyme were minimized with respect to the dihedral angles of the central residue. As the length of the oligopeptide segment increased up to a nonapeptide, the low-energy conformation became that observed in the x-ray structure in most cases. This finding suggests that although short-range interactions appear to play the dominant role in determining the conformation of an amino acid residue in a protein, the additional interactions required to stabilize the conformation uniquely may be only of medium range, i.e., those within a nonapeptide, and longer-range interactions may be of considerably less importance.

Levitt and Warshel (1975; Levitt, 1976) developed a simple representation of protein conformations that was used together with energy minimization and normal-mode thermalization to simulate protein folding. Under certain conditions, the method succeeded in "renaturing" bovine pancreatic trypsin inhibitor from an open-chain conformation into a folded conformation close to that of the native molecule. Levitt (1976) described the representation in some detail and tested the methods extensively under a variety of different conditions. The use of the concept of time-averaged forces, introduced by Levitt and Warshel (1975), was shown to simplify conformational energy calculations on globular proteins. A detailed description is given of the simplified molecular geometry, the parameterization of suitable force fields, the best energy-minimization procedure, and the techniques for escaping from local minima. Extensive tests of the method on the native conformation of pancreatic trypsin inhibitor showed that this simplification worked well in representing the stable native conformation of this globular protein. The original early optimism for the application of minimization of the total potential energy of a polypeptide to lead to the observed equilibrium structure has not overcome the two major problems of (1) choice of appropriate parameterization of the energy and (2) the multiple-minimum problem on the global potential energy surface (Ponder and Richards, 1987).

Kuntz *et al.* (1976) described a method for calculating the tertiary structure of proteins given their amino acid sequence. The algorithm involves locally minimizing an energylike expression as a function of the Cartesian coordinates of the C_β of all residues. Although the approximation to the true polypeptide geometry and conformational energies is extremely approximate, quite respectable results were obtained for the small proteins subredoxin and trypsin inhibitor, where the root-mean-square errors were as low as 4.0 Å and 4.7 Å, respectively. This paper is illustrative of the then-developing interest in extremely simplified models to predict the structure of small globular proteins (e.g., Levitt and Warshel, 1975; Ptitsyn and Rashin, 1975; Burgess and Scheraga, 1975).

Robson and Osguthorpe (1979) proposed a new model and parameters for the computer simulation of protein folding that satisfied requirements for a fully automatic simulation of protein folding. The major improvement over previous methods was to retain a more realistic and complete representation of the protein backbone and to reduce the number of variables by coupling their behavior. When applied to the folding of pancreatic trypsin inhibitor, a root-mean-square fit of 6 Å with good secondary structure was obtained. This method allowed a more detailed examination of secondary structure transitions during protein folding. By observing changes in secondary structure during the simulated folding, the authors showed that α helices and extended chain regions predicted at the outset or formed early in the simulation are conserved and that certain residues are crucial as flexible hinge points to bring the secondary structure together in order to achieve tertiary packing. Hinge points were found in general not to contain Gly. This made an important distinction between a "reverse-turn region," for which Gly is statistically a strong candidate, and a hinge point in the protein backbone.

Finney *et al.* (1980) used detailed hydrogen-bonding, surface exposure, internal environment, and solvent interaction calculations on several proteins in conjunction with data from quantum mechanical hydrogen-bonding studies to estimate various contributions to the free energy of folding and discussed their likely significance. A picture emerged of globular proteins as extremely well-fitted jigsaw puzzles in which no single driving force dominated the marginal stability of the native conformation. Rather, the folded structure is seen as the result of a complex global maximization of several strongly interacting driving forces. In particular, the need to maintain very efficient internal hydrogen bonding and the role of the solvent as a hydrogen-bond sink were stressed as strong constraints on the (incomplete) maximization of hydrophobic effects. The possible significance of internal-dipole–induced-dipole interaction was tentatively discussed.

Pincus and Klausner (1982) predicted the three-dimensional structure of the signal sequence for murine κ light chain using conformational energy calculations. The strategies employed to calculate the conformation involved judicious combinations of the local minima for simple peptides to construct longer peptide chains. A simple method was presented for combining the local minima for the component di- and tripeptides of the leader sequence that allows construction of long peptide chains. These calculations, based on tested and reliable potential energy functions, employ a novel global search technique to identify the lowest-energy structures for a hexadecapeptide signal sequence, Glu-Thr-Asp-Thr-(Leu$_3$-Trp-Val)$_2$-Pro-Gly. It has been found that the core hydrophobic sequence, Leu$_3$-Trp-Val-Leu, adopts an α-helical conformation that is terminated by chain-reversal conformations for the four residues Trp-Val-Pro-Gly. The amino-terminal four residues adopt a low-energy conformation that is fully compatible with the succeeding α helix. The immediate neighboring sequence, Asp-Thr, exists in a single lowest-energy, double-equatorial conformation, whereas the first two residues, Glu-Thr, can adopt a variety of low-energy conformations. The calculations arrive at a highly structured and specific model for the conformation of the leader sequence that is compatible with experimental data.

Levitt (1983) calculated the nativelike folded conformations of bovine pancreatic trypsin inhibitor protein by searching for conformations with the lowest possible potential energy. Twenty-five random starting structures were subjected to soft-atom-restrained energy minimization with respect to both torsion angles and the atomic Cartesian coordinates. The restraints used to limit the search included the three disulfide bridges and the 16 main-chain hydrogen bonds that define the secondary structure. Novel features of the methods used included soft atoms to make restrained energy minimization work, writing numbers to classify chain threadings, and molecular dynamics followed by energy minimization to anneal the conformations and reduce the energies further. The best conformation calculated had a root-mean-square deviation of only 3 Å and showed the same special threading found in the x-ray structure.

Weiner *et al.* (1984) presented the development of a force field for simulation of nucleic acids and proteins. The approach began by obtaining equilibrium bond lengths and angles from microwave, neutron diffraction, and prior molecular mechanical calculations, torsional constants from microwave, NMR, and molecular mechanical studies, nonbonded parameters from crystal-packing calculations, and atomic charges from the fit of a partial-charge model to electrostatic potentials calculated by *ab initio* quantum mechanical theory. For proteins they focused on ϕ,ψ maps of Gly and Ala dipeptides, hydrogen-bonding interactions involving various protein polar groups, and energy refinement calculations on insulin. The authors discuss the improvements over previous attempts and have delineated areas for further improvement.

Robson and Platt (1986) made a reappraisal of the interatomic potential functions for protein structure calculations using the all-atom approximation (except CH, CH_2, and CH_3, which are treated as "united atoms"). The potential functions are somewhat novel in form and consistent with more efficient and robust folding algorithms. In addition, the potentials are calibrated for the rigid geometry approximation, since use of fixed standard bond lengths and valence angles (and fixed transplanar peptide groups) reduced the number of conformational variables and saves a great deal of computer time. Although these algorithms demand the use of potential functions of this special type, these functions can be readily implemented in more classical programs for the conformational analysis of proteins. They were calibrated or tested against a large body of experimental data, including (1) extended-basis-set *ab initio* quantum mechanical calculations, (2) nuclear magnetic resonance spectroscopic data and dipole moment data for di- and oligopeptides, (3) characteristic ratio data for random-coil homopolypeptides, (4) extensive data from peptide solubility studies, and (5) experimental structures of polyalanine fibers and globular proteins.

Brucceroleri and Karplus (1987) developed a procedure, CONGEN, for uniformly sampling the conformational space of short polypeptide segments in proteins. This method of homology modeling greatly simplifies the problem of tertiary structure prediction. Because the time required for this sampling grows exponentially with the number of residues, parameters were introduced to limit the conformational space that has to be explored. This was done by using the empirical energy function of CHARMM (Brooks *et al.*, 1983) and truncating the search when conformations of grossly unfavorable energies were sampled. Tests were made to determine control parameters that optimize the search without excluding important configurations. When applied to known protein structures, the resulting procedure is generally capable of generating conformations where the lowest-energy conformation matches the known structure with a root-mean-square deviation of 1 Å.

Bash *et al.* (1987) tackled the fundamental problem in chemistry and biochemistry of understanding the role of solvation in determining molecular properties. Recent advances in statistical mechanical theory and molecular dynamics can be used to solve this problem with the aid of supercomputers. These advances permit the free energies of solvation of all chemical classes of amino acid side chains, four nucleic acid bases, and other organic molecules to be calculated. The effect of a site-specific mutation on the stability of trypsin was predicted. The results are in good agreement with available experiments.

Although progress is being made, the formidable task of calculating the tertiary structure of proteins requires a great effort in future research.

G. Expert Systems

Robson and co-workers (Robson *et al.*, 1987; Fishleigh *et al.*, 1987) have studied rationales for an expert-system approach to the interpretation of protein sequence data. Fishleigh *et al.* (1987) analyzed the sequence of the epidermal growth factor receptor to identify regions of potential structural and functional importance. This receptor protein was analyzed using a variety of established methods and novel procedures developed for the study of weak internal and external homologies and for the use of homologous sequences in the prediction of secondary and supersecondary structures. First, a composition scan of sequences was done to locate consensus sets for posttranslational modification, identifying features such as transmembrane sections and analyzing regions of unusual amino acid composition. This was followed by searches for internal and external homologues and the prediction of the secondary structure of the protein of interest, with any homologous sequences identified. Information gained from the composition scan and the study of homologues can then be used to refine the secondary structure prediction, which, in conjunction with any homology data, may permit the development of a model for the gross structures of the protein.

An additional technique enables the division of the protein sequence into sections that may represent current or ancestral exons, collectively referred to as paraexonic fragments. The Cys-rich regions were analyzed in terms of their separations along the sequence. The methods of Garnier *et al.* (1978) and Lim (1974a–c) were used for secondary structure prediction.

Robson *et al.* (1987) developed a suite of programs, named LUCIFER (Logical Use of Conformational Information and Fast Energy Routines) for the conformational study of drugs, proteins, and other biomolecules. The suite explicitly seeks to carry out global energy minimization as rapidly as possible (Robson and Platt, 1986) but yet puts emphasis on the rational use of external data to overcome any deficiencies. The external data are of three basic types. The first come from the scanning of data bases to identify primary, secondary, supersecondary, and tertiary structural homologies. Second, experimental data from physical chemists and pharmacologists are analyzed and exploited. Such data are exemplified by intergroup distances from NMR spectroscopy, infrared spectral data, characteristic ratios of polymers, net dipole moments, and pharmacological potencies of structural analogues. Third, the Imman graphics system developed by their group provides a vehicle for more qualitative judgments.

The value of this system has been illustrated in the modeling of immunoglobulins against electron microscopic and other experimental data (Pumphrey, 1986a,b; Robson and Garnier, 1986). This method was described for chloramphenicol acetyltransferase (CAT). The first step, homology detection, utilized the method of Fishleigh *et al.* (1987), that of Garnier *et al.* (1978), and a computerized version of the Lim method (Protein Resource Identification, National Biomedical Research Foundation, Georgetown University Medical Center, Washington, DC, 1986) for the secondary structure prediction. New procedures were also described for identification of secondary structure template similarity on the basis of theorem-proof algebra and for rapid evaluation of the hydrophobic packing of a protein conformation. These methods are discussed in relation to the molecular-graphics modeling of CAT, where it is assumed that there was a tertiary structure relationship between CAT and cat muscle pyruvate kinases. An attractive feature of the structure was that its hydrophobic radius of gyration was very close to the value that would have been expected for a protein of this size. The hydrophobic radius of gyration of a protein is a criterion by which nonnativelike conformations may be rapidly identified. Thirty-four proteins in the range of 40 to 500 residues were analyzed to obtain the hydrophobic radii of gyration for use in comparative studies. Although the authors state that LUCIFER is of proven worth for the study of biological peptides and for protein modeling against homologues, its applicability to *de novo* protein structure prediction is untested. A preliminary study of an avian pancreatic polypeptide is described and classified as both informational and promising.

In an excellent review on "Folding and Association of Proteins," Jaenicke (1987) gives a somber evaluation of the prospects for a more accurate prediction of the tertiary structure of proteins. There is still much work to be done.

VII. PREDICTION OF MEMBRANE STRUCTURE: METHODS OF PREDICTION

Membrane proteins are a diverse group of proteins that play an important role in modulating the activity of many of the cell's functions. With the advent of cloning methodologies and the ease of DNA sequencing, the availability of membrane proteins, previously obtainable in minute amounts, has made their study feasible. Of great interest is their mode of function. The diversity of proteins residing in the lipid bilayer of the membrane is immense, as is their function from receptors to active enzymes. Those presently under study include bacteriorhodopsin, acetylcholine receptor, ATPases, T-cell receptor, and the sodium channel.

The method of prediction for membrane proteins that is utilized most frequently is that of Kyte and Doolittle (1982). This simple method for displaying the hydropathic character of a protein has been used to define the sequences that have a high probability of being embedded in the nonpolar membrane environment. This agrees with the first report by Capaldi and Vanderkooi (1972) of the low polarity of many membrane proteins. With membrane-bound proteins, the portions of their sequences that are located within the lipid bilayer are clearly delineated by large uninterrupted areas on the hydrophilic side of the midpoint line in the well-known plot of the hydropathic index versus sequence number. Kyte and Doolittle (1982) devised their own hydropathy scale, and a computer program, SOAP, assigns the appropriate hydropathic value to each residue in a given amino acid sequence and then successively sums those values, starting at the amino terminal, within overlapping segments displayed from each other by one residue. Although a segment of any size can be chosen, ordinarily spans of 7, 9, 11, or 13 were employed, odd numbers being used so that a given sum could be plotted above the middle residue of the segment. One of the novel features of the approach is that membrane-spanning segments can be identified and distinguished from sequences that merely pass through the interior of a protein. Kyte and Doolittle (1982) have carefully avoided associating the hydrophobic segments with any particular secondary structure, i.e., α helix or β strand. However, unfortunately, in the literature, it has been generally accepted and stated that these segments represent α-helical segments. It is possible, of course, that they do represent helical sections, and they probably are helical, but the Kyte and Doolittle method cannot be used to verify the presence of a helical segment in a bilayer. The β sheets are also extremely hydrophobic. On examination of P_α and P_β values (Chou and Fasman, 1974a) together with the hydropathy scale of Kyte and Doolittle (1982), it can be seen that β sheets on average have a higher average hydrophobicity that do α helices. As was recently shown, porin, which spans the outer membranes of *Escherichia coli* and forms voltage-dependent transmembrane channels, has a conformation dominated by β structure (Paul and Rosenbusch, 1985).

Paul and Rosenbusch (1985) have pointed out that conventional methods of secondary structure prediction for bacteriorhodopsin have been ambiguous; however, when they applied the Chou and Fasman (1978a,b, 1979a,b) method of utilizing β turns as well, they could rationalize the hydrophobic segments as being α helices immersed in the bilayer. This subject is discussed in detail later in this chapter.

Argos *et al.* (1982) developed an algorithm based on physical characteristics of the 20 amino acids and refined by comparison to the proposed bacteriorhodopsin structure to delineate likely membrane-buried regions in the primary sequences of proteins known to interact with the lipid bilayer. Application of the method to the sequence of the carboxyl-terminal one third of bovine rhodopsin predicted a membrane-buried helical hairpin structure. With the use of lipid-buried segments in bacteriorhodopsin as well as regions predicted by the algorithm in other membrane-bound proteins, a hierarchic ranking of the 20 amino acids in order of their preferences to be in lipid contact was calculated. A helical wheel analysis of the predicted regions suggested which helical faces are within the protein interior and which are in contact with the lipid bilayer. Nine arbitrarily chosen physical parameters that may have some relevance to conformational preference were used. However, a more serious criticism of the method is that it is based on a highly speculative structure of bacteriorhodopsin and is therefore suspect. Several other membrane proteins—glycophorin A, cytochrome b_5, cytochrome *c*, ATP synthase, M13 coat protein, precasein, proteolipid, and porin—were also predicted by this approach by Argos *et al.* (1982).

Mohana-Rao *et al.* (1983), using the algorithm of Argos *et al.* (1982), predicted the hydrophobic helical spans in the thylakoid membrane protein (TMP). A second method was used to indicate the number of possible helices in TMP as well as the N terminus of each helix. Curve segments from the helical spans of the vertebrate photoreceptor protein rhodopsin (RHO), a five-parameter plot, assumed to be another seven-helical protein like bacteriorhodopsin, were compared to all segments of the comparable TMP plot, and correlation coefficients were calculated at each value. The helical wheels were calculated using a table of membrane-buried helical preference values. Thus, it was concluded that the three proteins, thylakoid protein, bovine rhodopsin, and bacteriorhodopsin, are all seven-helical bundles, and it was hypothesized that this uniquely stable arrangement was evolved by convergent evolution and will be found frequently in membrane proteins. The consecutive assumptions leading to this conclusion cast doubts on this conclusion. Flinta *et al.* (1983) challenged the often-claimed proposition that transmembrane helices show ''sidedness'' in the distribution of polar and hydrophobic residues. However, an analysis of the statistical distribution of polar residues in randomly generated helices shows that the degree of bias commonly observed in real helices is far from statistically significant. Thus, it is concluded that a ''patchy'' distribution of residues in such helices should be interpreted with great care.

Eisenberg *et al.* (1984a,b; Eisenberg, 1984) published an algorithm that identifies α helices involved in the interactions of membrane proteins with lipid bilayers and distinguishes them from helices in soluble proteins. The membrane-associated helices are then classified with the aid of the hydrophobic moment plot, on which the hydrophobic moment of each helix is plotted as a function of its hydrophobicity. The magnitude of the hydrophobic moment measures the amphilicity of the helix (and hence its tendency to seek a surface between hydrophobic and hydrophilic phases), and the hydrophobicity measures its affinity for the membrane interior. Segments of membrane proteins in α helices tend to fall in one of three regions of a hydrophobic moment plot: (1) monomeric transmembrane anchors (class I HCA transmembrane sequences) lie in a region of highest hydrophobicity and smallest hydrophobic moment; (2) helices presumed to be paired (such as the transmembrane M segments of surface immunoglobulins) and helices that are bundled together in membranes (such as bacteriorhodopsin) fall in the adjacent region with higher hydrophobic moment and smaller hydrophobicity; and (3) helices from surface-seeking proteins (such as melittin) fall in the region with still higher hydrophobic moment. The α helices from globular proteins mainly fall in a region of lower mean hydrophobicity and hydrophobic moment. This procedure demonstrated that the sequences of diphtheria toxin may have four transmembrane helices and a surface-seeking helix in fragment B, the moiety known to have a transmembrane function.

The method of Finer-Moore and Stroud (1984), based on a Fourier analysis of hydrophobicities, revealed that the subunit sequences of the acetylcholine receptor have sequences of amphipathic secondary structures. Prediction of a consensus secondary structure based on this analysis and on an empirical prediction method leads to a testable hypothesis about how the ion channel is formed and might function. The acetylcholine receptor (AcChR), is a ≃250-kDa complex of five homologous glycoprotein subunits in stoichiometry $\alpha_2\beta\delta\gamma$. The local periodicity in the hydrophobicities of the primary sequences of all subunits was quantitated by Fourier analysis in the manner of McLachlan and Karn (1983). Fourier analysis was used for amphipathic secondary structure correlation together with a secondary structure prediction method for the ≃75% sequence external to the bilayer and separate evidence for oriented helices in the transmembrane regions to develop an overall structural scheme. Amphipathic spectra $I_k(\nu)$ (power spectra of hydrophobicities) were computed for stretches of 25 residues in length according to

$$I_k(\nu) = \Sigma_j = {}^{k+12}_{k-12} (h_j - h_k) \exp(2\pi \cdot j \cdot \nu)^2$$

where h_j is the hydrophobicity of residue j (kcal/mole) from a consensus set of values developed by Eisenberg *et al.* (1982a,b), and h_k is the average hydrophobicity of the 25 residues from $k - 12$ to $k + 12$. Hydrophobicity plots (Kyte and Doolittle, 1982) H_i were generated by

$$H_i = (1/7)\, \Sigma_j = {}^{i+3}_{i-3} h_j$$

Secondary structure prediction for extramembrane regions used the Garnier *et al.* (1978) algorithm. Sequence alignment of the AcChR subunits was carried out by hydrophobicity correlation by the comparison matrix method of McLachlan (1971). The amphipathic power spectrum $I_k(\nu)$ for each sequence was plotted as a two-dimensional contoured map with frequency ν (residues^{-1}) as abscissa and residue number k as ordinate. A prominent feature is the intense peak at $\nu = 1/3.5$ residues between $k = 412$ and 470. This periodicity is exactly that expected for an α helix, and the length and intensity of the peak strongly suggest an amphipathic α helix up to at least 30, possibly up to 58, amino acids in length. The overall secondary structure has 27% β sheet and 44% α helix, in agreement with circular dichroism studies (Moore *et al.*, 1974) and Raman spectroscopy (Chang *et al.*, 1983). A detailed discussion of the Finer-Moore and Stroud predictive method can be found in Chapter 19 by Finer-Moore, Bazan, Rubin, and Stroud.

Gray and Matthews (1984) surveyed the known protein structures and found that approximately 70% of serine residues and at least 85% (potentially 100%) of threonine residues in helices make hydrogen bonds to carboxyl oxygen atoms in the preceding turn of the helix. The high frequency of intrahelical hydrogen bonding is of particular significance for intrinsic membrane-bound proteins that form transmembrane helices. Hydrogen bonding within a helix provides a way for serine, threonine, and cysteine to satisfy their hydrogen-bonding potential, permitting such residues to occur in helices buried within a hydrophobic milieu. Honig and Hubbell (1984) estimated the free energies of transfer of ionized amino acid side chains in water to both their ion-paired and neutral hydrogen-bonded states in low-dielectric media. The difference between the two free energies corresponds to the proton transfer free energy in a "salt bridge" formed between acidic and basic groups (i.e., Lys and Glu residues). Dielectrics of 80 (H_2O) and 1 (vacuum) were used in the calculation. Their results suggest that it costs approximately 10–16 kcal/mole to transfer a salt bridge from water to a medium of $\epsilon = 2$–4 in ionized or neutral form. The proton transfer energy is thus approximately zero. The tendency of salt bridges to form additional hydrogen bonds in real proteins suggests that the ion pair will be present in most biological systems. Kuhn and Leigh (1985) developed a statistical technique for predicting transmembrane segments of membrane proteins from their amino acid sequences. A propensity scale was derived from the frequency of occurrence of amino acids in transmembrane fragments. Those values were compared to the hydropathy scale of Kyte and Doolittle (1982) and the signal sequence helical potential scales of Argos *et al.* (1982). The difficulty of such an approach is that the sequences termed transmembrane are rough estimates, as these are not definitely known.

Klein *et al.* (1986) used discriminant analysis to classify membrane proteins precisely as integral or peripheral and to estimate the odds that the classification is correct. On 102 membrane proteins from the National Biomedical Research Foundation, it was found that the discrimination between integral and peripheral membrane proteins can be achieved with 99% reliability. Hydrophobic segments of integral membranes can also be distinguished from interior segments of globular soluble proteins with better that 95% reliability. A procedure was also proposed for determining boundaries of membrane-spanning segments, and it was applied to several integral membrane proteins. From the limited data available, the residues at the boundaries of a membrane-spanning segment are predictable to within the error in the concept of a boundary. As a specific indication of resolution, seven membrane-spanning segments of bacteriorhodopsin were resolved with no information other than sequence, and the predicted boundary residues agree with the experimental data on proteolytic cleavage sites. A computer program in FORTRAN for prediction of membrane-spanning segments is available from the sources listed in the appendices at the end of the text.

Argos and Mohana-Rao (1985) applied their predictive scheme (Argos *et al.*, 1982; Mohana-Rao *et al.*, 1983) to five functionally distinct lipid-bound proteins whose exonic structure is known in an attempt to shed light on the mechanism and etiology of splice junctions. They found that the splice junctions largely map to the predicted surface segments and that the number of junctions correlates with the length of the surface spans in the five proteins.

Jähnig and co-workers (Vogel *et al.*, 1985; Vogel and Jähnig, 1986) employed a structural predictive method that takes into account the amphipathic helices. The method of Kyte and Doolittle (1982), based only on hydrophobicity, underestimates the number of membrane-spanning helices. To predict amphipathic α helices, the authors represent the amphipathy as real space instead of Fourier space as in the analysis of the hydrophobicity, $H_{\alpha}(i)$ (Eisenberg *et al.*, 1984a,b; Finer-Moore and Stroud, 1984). Turns are also predicted by the Chou and Fasman (1978a) method. A structural model of lactose permease is proposed in which the ten membrane-spanning helices are expected to form an outer ring of helices in the membrane. The interior of the ring would be made up of residues that are predominantly hydrophilic and, by analogy to sugar-binding proteins, suited to provide the sugar-binding site. Vogel and Jähnig (1986) predicted the outer-membrane proteins of *Escherichia coli*, porin, maltoporin, and OmpA protein. Using an adaption of their method for structural prediction of amphipathic α helices, they predicted the amphipathic β strands in the membrane. In the model suggested, the OmpA fragment consists of eight amphipathic membrane-spanning β strands that form a β barrel. This agrees with the structure determined by Raman spectroscopy, which estimated 50 to 60% β strands, about 20% β turn, and less than 15% α helix. Similarly porin is folded into ten amphipathic membrane-spanning β strands that are located at the surface of the trimer towards the lipids and eight predominantly hydrophobic strands in the interior.

Edmonds (1985) has calculated the interaction energies between hydrophobic α-helical sections that span membranes, which are known to possess large electric dipole moments. These interaction energies, which include screening effects, remain comparable with a typical thermal energy of kT up to a separation of 20 Å. In addition, it is shown that, solely because of its dipole moment, an α helix that completely spans the membrane has an energy up to $5kT$ lower than one that terminates within the membrane width. The paper also describes the electrical interaction of the charge structure of a membrane channel and the protein helices that surround the pore. The gating charge transfer that is measured when a voltage-sensitive ion channel switches means that the dipole moment of the ion channel changes. This in turn results in a change in the radial forces that act between the pore and helices that surround it. A change in these radial forces, which tend to open or close the pore, constitutes an electrically silent gating mechanism that must necessarily act subsequent to the gating charge transfer. The gating mechanism could consist of a radial translation of the neighboring proteins or of their axial rotation under the influence of the torque that would act on a pair of approximately equidistant but oppositely directed α helices. An attempt to calculate the interaction energy of a typical pore and a single α helix spanning a membrane results in an energy of many times kT.

Engelman *et al.* (1986) reviewed the literature on identifying nonpolar transbilayer helices in amino acid sequences of membrane proteins. They consider the arguments that support the notion that helical structure will be a dominant motif in integral membrane protein organization, introduce the problem of suitable scaling of amino acids in terms of their polar and nonpolar characteristics, and discuss further the use of such scales in prediction of protein structure. They conclude that a suitable scale and protocol can lead to the successful identification of transmembrane helical structures in integral membrane proteins. The authors discuss the hydrophobicity scale they have developed over the past several years (Engelman *et al.*, 1981; Engelman and Steitz, 1981, 1984; Steitz *et al.*, 1982) in which the nonpolar properties of the amino acids, as they exist in a helix, were calculated using a semitheoretical approach that combines separate experimental values for the polar and nonpolar characteristics of groups in the amino acid chain. The development is similar to von Heijne's earlier work (von Heijne, 1981a,b), but with important differences. Their table of transfer energies for amino acid side chains in α-helical polypeptides includes hydrophobic and hydrophilic components of the transfer of amino acid side chains from water to a nonaqueous environment of dielectric 2 (GSE scale, Goldman, Engelman, Steitz). The GSE scale is used as a point of reference to compare with other scales (e.g., von Heijne, 1981b; von Heijne and Blomberg, 1979; Kyte and Doolittle, 1982; Rose *et al.*, 1985; Guy, 1985).

Using the GSE scale in sequence analysis, they tested their prediction of the photosynthetic reaction center of *Rhodopseudomonas viridis,* whose high-resolution structure is known (Deisenhofer *et al.*, 1984, 1985; Chapter 2 herein). The macromolecular assembly consists of four polypeptide chains: two of these (L and M) are globular integral membrane proteins, and one (H) is an anchored membrane protein. The crystal structure shows a region in which bundles of helices transverse an apparently nonpolar region. All putative membrane-spanning helices observed in the crystal are predicted from the hydrophobicity analyses of the sequences. Four helices each were suggested in both the L and M chains, and a fifth helix is possible, but of marginal significance and not corresponding to a transmembrane helix in the structure. The H-subunit profile suggests a single transmembrane helix. Eleven helices were assigned and are somewhat shorter that those actually observed (e.g., subunit M, helix A, predicted 52–71, observed 52–78). This method fails to predict the β-sheet structure found in porin (Kleffel *et al.*, 1985; Paul and Rosenbusch, 1985).

Engelman and Steitz (1981) proposed that the initial event for either secretion of protein across or insertion into membranes is the spontaneous penetration of the hydrophobic portion of the bilayer by a helical hairpin. The major proposals of this model are the following. (1) Energetic considerations of polypeptide structures in a nonpolar lipid environment as compared with an aqueous environment have led to the conclusion that only α and 3_{10} helices will be observed in the hydrophobic interiors of membranes. (2) During protein synthesis, the nascent polypeptide chain folds in the aqueous environment to form an antiparallel pair of helices, each of which is ~20 residues long. (3) The helical hairpin partitions into the membrane if the free energy arising from burying hydrophobic helical surfaces exceeds the free energy cost of burying potentially charged and hydrogen-bonding side groups. (4) Globular membrane proteins will be formed by the insertion of several pairs of helical hairpins, which are expected to be the fundamental unit of membrane protein folding. (5) In secreted proteins, the hydrophobic leader peptide forms one of these two helices and functions to pull polar portions of the secreted protein into the membrane as the second helix of the hairpin. (6) Insertion of the helical hairpin into the bilayer initiates secretion if the second helix is polar, and secretion of the newly synthesized protein continues until or unless a hydrophobic segment is encountered. (7) Alternatively, if both helices are hydrophobic, the hairpin will simply remain inserted in the membrane.

Steitz *et al.* (1982) wrote a computer program to analyze the amino acid sequence of secreted and membrane proteins in order to estimate quantitatively whether insertion of membrane into lipid bilayers can be expected to be spontaneous on thermodynamic grounds and also to establish the probable topology of membrane proteins. Amino acid sequences were analyzed for the free energy of burying an α helix of definable length, usually 21 amino acid residues long. A 21-residue probe helix was removed down the sequence, and the free energy of burying each successive 21-residue helix was calculated and plotted. Bacteriorhodopsin and glycophorin were calculated in this manner. Approximately seven helices were found for bacteriorhodopsin.

Wallace *et al.* (1986) evaluated the accuracy of several predictive schemes in predicting the secondary structure of 15 integral membrane proteins and membrane-spanning polypeptides. Statistical analyses (χ^2) indicated a less than 0.5% correlation between the net predicted secondary structures and experimental results. The authors conclude that predictive schemes using soluble protein bases are inappropriate for the prediction of membrane protein folding. The methods evaluated were the Chou and Fasman (1974a,b), Garnier *et al.* (1978), and Burgess *et al.* (1974) algorithms, and a smaller number of proteins were analyzed by the Kyte and Doolittle (1982) and Engelman–Goldman–Steitz (1986) methods. This conclusion can be serious questioned, as only two of the 15 proteins evaluated have had their structure determined by x-ray crystallography. The remaining proteins' structures were determined by a combination of electron microscopy, Raman spectroscopy, infrared spectroscopy, and circular dichroism, methods whose accuracy has been frequently questioned.

Mohana-Rao and Argos (1986) developed a conformational preference parameter to predict helices in integral membrane proteins. Five parameters were found to be most suitable for predicting the presence or absence of hydrophobic stretches: hydration potential (Wolfenden *et al.*, 1979), the free energy of transfer for a given residue in a helix in aqueous medium to a helix in a nonpolar phase (von Heijne, 1981a), polarity (Trewhella *et al.*, 1983), bulk (Trewhella *et al.*, 1983), and turn conformational preferences (Levitt, 1978). The method was described in detail in an earlier paper (Argos *et al.*, 1982). The proteins chosen were those having at least two transmembrane helices. A total of 256 membrane-buried helices were predicted in 49 integral membrane proteins. The Chou–Fasman membrane-buried helix preference parameter for a particular amino acid is defined as the ratio of its composition in predicted helices to its composition in all sequence regions of the integral membrane proteins. When applied to the L and M subunits of *Rhodopseudomonas sphaeroides,* five helices were predicted, in agreement with the three-dimensional x-ray crystal structure. Data on signal sequences and amino acid exchanges in membrane proteins were also analyzed.

Cornette *et al.* (1987) developed a computational technique for detecting amphipathic structures in proteins. They optimized hydrophobicity scales and used these with a method based on a least-squares fit of a harmonic sequence to a sequence of hydrophobicity values. They termed this the "least-squares power spectrum." The sum of the spectra of the α helices in their data base peaked at 97.5°, and approximately 50% of the helices can account for this peak. Amphipathic α helices in their natural state have a periodic variation in the hydrophobicity values of the residues along the segment, with a 3.6-residue-per-cycle period characteristic of an α helix. Thus, approximately 50% of the α helices appear to be amphipathic, and of those that are, the dominant frequency at 97.5° rather than 100° indicates that the helix is slightly more open than previously thought, with the number of residues per turn closer to 3.7 than 3.6. Although the scale is optimal only for predicting α amphipathicity, it also ranks high in identifying β amphipathicity and distinguishing interior from exterior residues in a protein (see further discussion on amphiphilic helices in Section VI.D).

Furois-Corbin and Pullman (1987) undertook theoretical studies on the packing properties of α helices and on their ability to form conducting bundles using minimization techniques. Packages of two and five α helices containing leucines on their faces of contact and made otherwise of alanine were studied. Such bundles were compared to pure poly(L-alanine) packages. They concluded that the essential packing properties were conserved, with near antiparallelism and a preponderance of nonbonded interactions. Helical packing is different when leucine is included, and substitution of serines for the alanine lying on the inner wall has little effect on interhelix packing.

VIII. PREDICTED MEMBRANE STRUCTURES

A. Bacteriorhodopsin

Bacteriorhodopsin is the transmembrane protein found in *Halobacterium halobium*. The purple membrane, which contains bacteriorhodopsin, is present in a number of extremely halophilic bacteria and catalyzes the light-driven proton translocation from the inside to the outside of the cell membrane. This generates an electrochemical gradient, which is used by the cell for the synthesis of ATP. Bacteriorhodopsin contains one retinal molecule per protein molecule, and this is linked as a Schiff base to the ε-amino group of Lys^{216}. This protein is probably the most studied membrane protein to date. Henderson and Unwin (1975) were able to deduce an electron density map at 7-Å resolution by an electron diffraction technique. From this is was concluded that bacteriorhodopsin forms a continuum of seven α helices, each of which spans the membrane and is largely embedded in it. Ovchinnikov *et al.* (1979) proposed the first model of bacteriorhodopsin based on studies of limited proteolysis. There were seven transmembrane helices accounting for 207 residues, which yielded a total of 83% α helix. This model was slightly altered when the antigenic structure and topography of bacteriorhodopsin were probed (Ovchinnikov *et al.*, 1985).

Engelman *et al.* (1980) attempted to fit the amino acid sequence of bacteriorhodopsin to the three-dimensional map of the molecule. Seven segments of the sequence were selected as being probable transmembrane α helices. The complete amino acid sequence had been previously determined (Ovchinnikov *et al.*, 1979; Gerber *et al.*, 1979; Walker *et al.*, 1979). There were 5040 ways of fitting these seven segments into the seven regions of helical density in the map, and these were evaluated on the basis of the criteria of connectivity of the nonhelical link regions, charge neutralization, and total scattering density per helix. The seven helices, with between 24 and 28 residues each, contained a total of 178 amino acids, which gave a helical content of 72%. Steitz *et al.* (1982) further confirmed the analysis of Engelman *et al.* (1980). The α helices appear to have their hydrophobic sides facing the nonpolar lipid moiety, whereas more hydrophilic portions are in the interior of the molecule (Engelman and Zaccai, 1980). Ovchinnikov (1982) combined the sites of enzyme proteolysis with the hydrophobicity profile (Rose and Roy, 1980) and the β-turn prediction (Chou and Fasman, 1979a) to revise the probable disposition of the polypeptide in the bilayer.

Huang *et al.* (1982), using a photosensitive *m*-diazirinophenyl analogue of retinal bound to bacterioopsin at Lys^{216}, regenerated a chromphore with λ_{max} at 470 nm. Photolysis of the complex at 365 nm resulted in covalent cross linking of the retinal analogue to the bacterioopsin in a >30% yield. Investigation of the sites of cross linking between the ^3H-labeled retinal analogue and the protein showed the peptide fragment (amino acid residues 190–248) to be the main radioactively labeled product. Stepwise Edman degradation showed Ser^{193} and Glu^{194} to be the predominant sites of cross linking. These results show that the chromophore in bacteriorhodopsin is inclined towards helix 6 and towards the exterior of the cell. These data provided information on the approximate angle that the chromophore makes with the plane of the membrane and required a modification of the then-current secondary structural model for

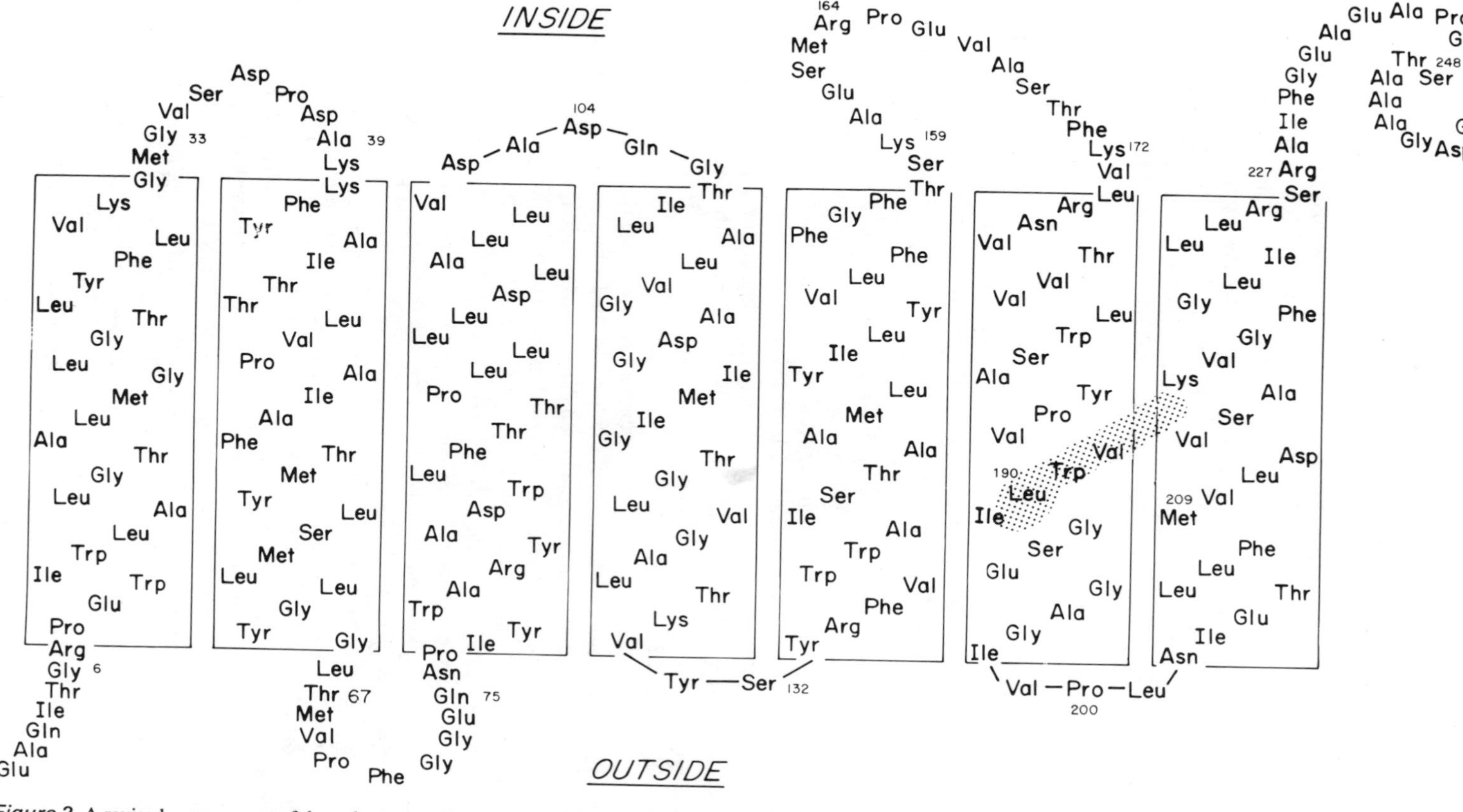

Figure 2. A revised arrangement of the polypeptide chain of bacteriorhodopsin across the membrane based on the work of Huang *et al.* (1982) and the proposal of Steitz *et al.* (1982). The shaded area represents the retinal analogue II forming a Schiff base with Lys^{216} at one end and cross links with amino acids 193 and 194 at the distant end.

bacteriorhodopsin (Engelman *et al.*, 1980). Modifications of the structures of helices 6 and 7 and the loop connecting helices 5 and 6 were required. Thus, if the overall size and structure of helix 7 in the original model of Engelman *et al.* (1980) is assumed to be correct, then Lys^{216}, the site of attachment of retinal, is positioned close to the center of the bilayer. To span the full length of the retinal analogue II (15 Å) between the Schiff base linkage and the cross-linking sites, amino acids 193 and 194, the latter amino acids must be placed within the layer, as shown in Fig. 2, rather than in the loop connecting helices 6 and 7, as in the original model. These seven helices vary in length from 23 to 29 residues, having a total of 173 residues in helices, or a 72% helical structure. Trewhella *et al.* (1983, 1984), using neutron diffraction data, assigned specific amino acid sequences to specific regions in the sequences in α helices. The seven helices had been shortened to contain either 20 or 21 amino acid residues, for a total of 57% helical content.

Paul and Rosenbusch (1985) found that the available predictive methods utilized for membrane proteins did not do well for the protein porin, which spans the outer membrane of *Escherichia coli*. By other criteria, this protein had a large β-strand component. They proposed to use the β turn as an indication of chain reversal and to avoid the hydrophobicity parameters. Such a model for porin was in good agreement with all experimental data. Application to the paradigm of hydrophobic membrane proteins, bacteriorhodopsin, revealed a pattern consistent with its α-helical folding, yielding seven helical segments. With the Chou–Fasman (1979a,b) algorithm for predicting β turns, it was found that the β turns appear between the proposed α-helical segments.

Wallace *et al.* (1986) evaluated three methods for the prediction of membrane protein secondary structure: the Chou–Fasman (1974a,b, 1978a,b) (CF), Garnier *et al.* (1978) (G), and Burgess *et al.* (1974) (B) algorithms. For bacteriorhodopsin, the experimental values of 74–83% α, 0–9% β sheet were quoted, having been determined by image reconstruction, infrared spectroscopy, x-ray diffraction, and circular dichroism. The predicted values obtained were: CF, 39–53% α, 22–28% β; G, 53% α, 41% β; B, 9% α, 26% β. It was thus concluded that these methods inadequately predict membrane protein structure. However, the methodologies used, because of the degree of resolution obtained with them, cannot be accepted with any degree of certainty, and this conclusion can be questioned. The predicted values for the photochemical reaction center of *Rhodopseudomonas viridis,* whose structure has been determined by x-ray crystallography (see Deisenhofer *et al.*, Chapter 2), were not satisfactory: L chain, experimental, 61% α, 21% β; CF, 33–44% α, 12–20% β, G, 19% α, 34% β; B, 3% α, 33% β; M chain, experimental, 65% α; CF, 16–33% α, 21–32% β; B, 11% α, 18% β; H chain, membrane, experimental, 100% α; CF, 8–19% α, 62–73% β; G, 28% α, 56% β; B, 54% α, 35% β (Wallace *et al.*, 1986).

Wu *et al.* (1982) examined lipid-induced conformations of peptide hormones and concluded that the α helix is the preferred conformation, as determined by circular dichroism, when the predicted conformation has potential for both α and β forms. These authors applied the Chou–Fasman (1974a,b, 1978a,b) algorithm to predict the secondary structure of bacteriorhodopsin and found that 60% β would predominate, with 20% α and 10% β turn. The six potential turns occurred at the seven ordered segments. Thus, it was concluded that the potential parameters for the amino acids might vary in a lipid environment.

The above predictions of bacteriorhodopsin used the conformational parameters based on 29 proteins. On the basis of the 29 data base and the 64 data base (Chou, Chapter 12), the results obtained were as follows: 29 data base, 16% α, 57% β; 64 data base, 21% α, 48% β (G. D. Fasman, unpublished data). Thus, there was a small improvement from using the larger data base.

The conformation of bacteriorhodopsin is as yet undetermined. Extensive studies using circular dichroism to determine the secondary structure of bacteriorhodopsin have been carried out. Studies of Dencher and Heyn (1978) showed that nonionic detergents caused dissociation of bacteriorhodopsin into monomers without a change in the retinal environment. They also showed that light and dark adaptation causes large absorbance changes. Octyglucoside and Triton 100-X caused changes in the circular dichroism spectrum with time, with a disappearance of the exciton bands at 318 and 560 nm. However, no change in secondary structure of the protein was observed in the 200 to 250-nm region. Nabedryk *et al.* (1985) summarized the many circular dichroism studies on bacteriorhodopsin. The helical content is reported as 45–83% (Beecher and Cassim, 1976; Long *et al.*, 1977; Jap *et al.*, 1983; Mao and Wallace, 1984). The value obtained depends on the state of the bacteriorhodopsin. The intact purple membrane, the solubilized bacteriorhodopsin, and bacteriorhodopsin reconstituted in lipid vesicles all yield different helical contents. This is probably because of both difference in secondary structure and, in large part, absorption flattening and scattering, to various degrees, in each of these preparations.

Jap *et al.* (1983) considered the electron microscopy study of Hayward and Stroud (1981), which reported possible β structure in bacteriorhodopsin, and interpreted their circular dichroism and infrared data as having β structure. They suggested a model for bacteriorhodopsin consisting of only five α-helical rods (50%) and four β sheets of 11 amino acids (20% β). Nabedryk *et al.* (1985) interpreted their circular dichroism and polarized infrared studies to indicate 74 ± 5% transmembrane helices and detected no significant contribution of β strands running perpendicular to the membrane plane. Their data agree with the reports of Mao and Wallace (1984) and the model of Trewhella *et al.* (1983). The studies of Glaeser and Jap (Jap and Kong, 1986; Glaeser and Jap, 1985; Jap *et al.*, 1983) find that by using the theoretical analysis and empirical estimation of Gordon and Holzworth (1971) and the data of Mao and Wallace (1984), there is no major absorption flattening for submicrometer-sized membranes. In agreement with the reports of Mao and Wallace (1984), Khorana's laboratory, using circular dichroism, reports that bleaching of delipidated bacteriorhodopsin dissolved in deoxycholate caused a 25% drop in the $[\theta]_{222}$ (London and Khorana, 1982), thus indicating a significant conformational change in protein structure. Huang *et al.* (1981) report a helical content of 50% in SDS. Popot *et al.* (1987) report a helical content of 58% by circular dichroism measurements of a reconstituted, cleaved, two-chain assembly of bacteriorhodopsin, whereas in lipid vesicles a helical content of 75–80% was found. Thus, a wide divergence in helical content of bacteriorhodopsin is reported. Differences are probably related to the preparative methods, particle size, the degree of association of the puple membrane, and the methodology used to deconvolute the circular dichroism curves to obtain the secondary structure.

B. Acetylcholine Receptor

Another thoroughly studied membrane protein is the nicotinic acetylcholine receptor protein (AcChR). The AcChR complex is comprised of five structurally similar subunits (two α, one β, one γ, and one δ) that stack next to each other so that a channel forms between the subunits when the AcChR is in the open conformation. This assembly contains both the binding site for the neurotransmitter and the cation-gating unit. Smythies (1980) applied the Chou–Fasman (1978a) predictive scheme to the amino acid sequence of the N-terminal segment of the α subunit. Two α helices cross linked by four ionically bound complementary amino acids (Arg/Lys to Glu) were suggested as the secondary structure.

Guy (1981) modeled the protein structure of agonist, competitive antagonist, and snake-neurotoxin-binding sites using the first 54 residues of AcChR α subunit from *Torpedo califor-*

nica. These models were based on the premise that the N-terminal portion of the subunits form the outermost extracellular surface of the AcChR and that agonists bind to this portion. The structure proposed was based on a comparison of the secondary structure predicted by the methods of Chou and Fasman (1978a) (using the Levitt, 1978, data base) and Lim (1974a–c). The tertiary structures of the α subunit were developed by requiring that it bind strongly to the crystal structures of the snake neurotoxins α cobra toxin and erabutoxin b. Finally, the quaternary structures were developed on the bases of interactions among the proposed tertiary structure (Devillers-Thiery *et al.*, 1983). A 1350-base-pair-long cDNA was cloned by Devillers-Thiery *et al.* (1983) and sequenced to yield the amino acid sequence (437 residues) of the α subunit of the receptor. Four regions of high hydrophobicity were assumed to be helical regions, and a very approximate model was constructed. Kosower (1983a,b, 1987) predicted six transmembrane segments of the α subunit of the AcChR from the amino acid sequence as determined by Toda *et al.* (1982) on the basis of hydrophobic sequences (Kyte and Doolittle, 1982). The choice of ion channel elements is based on the theory of single-group rotation (Kosower, 1982, 1983a).

Noda *et al.* (1983) reported the primary structure of the γ-subunit precursor of the AcChR. Comparison with the four subunits ($\alpha_2\beta\gamma\delta$) revealed marked homology. The hydrophobicity profiles (Hopp and Woods, 1981) as well as the predicted secondary structures (Chou and Fasman, 1978a,b) were used. As with the predicted α subunit (Noda *et al.*, 1982), based on both the hydrophilicity profile (Hopp and Woods, 1981) and the predicted secondary structure (Chou and Fasman, 1978a,b), each subunit has four strongly hydrophobic regions, which may represent regions with secondary structure that may be transmembrane segments or may be involved in intersubunit interaction. Claudio *et al.* (1983) also obtained the sequence of the γ subunit and predicted the secondary structure on the basis of hydropathicity profiles (Kyte and Doolittle, 1982; Hopp and Woods, 1981). Four stretches of hydrophobic residues were located and were attributed to membrane-spanning regions. Finer-Moore and Stroud (1984) developed an amphipathic analysis of membrane proteins and applied it to acetylcholine (see discussion of method in Section VI.D). On the basis of the prediction of amphipathic α helices, hydrophobicity plots, and the secondary structure as predicted by the Garnier *et al.* (1978) method, a model of the subunits of acetylcholine receptor subunits was devised. Stroud and Finer-Moore (1985) have reviewed the work on the acetylcholine receptor.

In summary, the predictive scheme of Claudio *et al.* (1983), Noda *et al.* (1983), and Devillers-Thiery *et al.* (1983) proposed four hydrophobic transmembrane domains. A fifth amphipathic transmembrane domain was proposed by Finer-Moore and Stroud (1984) and Guy (1981). Immunologic techniques suggest two additional amphipathic domains (Criado *et al.*, 1985) to account for the transmembrane orientation of various sequences.

Brisson and Unwin (1985) resolved an electron microscopic image and determined that the acetylcholine receptor had five membrane-spanning subunits, which were shown to be at pentagonally symmetrical positions around a channel over a large fraction of their length. The channel consists of a wide synaptic portion and a narrow portion extending through the membrane into the interior of the cell. Thus, the predictive schemes of Finer-Moore and Stroud (1984) and of Guy (1981) appear to give the most accurate representation of the structure.

Dixon *et al.* (1986) cloned the gene and cDNA for the mammalian β-adrenergic receptor (βAR). This receptor is known to modulate adenylate cyclase activity and consists of a catalytic moiety and regulatory guanine nucleotide-binding proteins, which provide the effector mechanism for the intracellular actions of many hormones and drugs. Analysis of the amino acid sequence of βAR indicates a significant amino acid homology with bovine rhodopsin and suggests that, like rhodopsin, βAR possesses multiple membrane-spanning regions.

Hydropathicity profiles of the βAR amino acid sequences were produced using the analysis of Hopp and Woods (1981) and were similar to the rhodopsins, of which bacteriorhodopsin is known to contain seven membrane-spanning helices (Engelman *et al.*, 1982).

Numa and co-workers (Kubo *et al.*, 1986) cloned, sequenced, and expressed the complementary cDNA encoding the muscarinic acetylcholine receptor. The muscarinic receptor was found to be homologous with the β-adrenergic receptor and rhodopsin in both amino acid sequences and suggested transmembrane topography. The Kyte and Doolittle (1982) method of expressing the hydropathicity was utilized.

Peralta *et al.* (1987) determined the amino acid sequence of the 466 amino acids comprising the M_2 muscarinic acetylcholine receptor. This receptor is predicted to have seven membrane-spanning regions distinguished by the disposition of a large cytoplasmic domain. The atrial muscarinic receptor is distinct from the cerebral muscarinic receptor gene product, sharing only 38% overall amino acid homology and possessing a completely nonhomologous large cytoplasmic domain. The hydrophobicity, as determined by the Kyte and Doolittle (1982) method, suggested six transmembrane regions with a seventh region of less pronounced hydropathicity. Bonner *et al.* (1987) isolated cDNAs for three different muscarinic acetylcholine receptors from a rat cerebral cortex library, and the cloned receptors were expressed in mammalian cells. This gene family provided a new basis for evaluating the diversity of muscarinic mechanisms in the nervous system. Hydropathicity profiles of these receptor sequences indicated the presence of seven transmembrane domains.

Greeningloh *et al.* (1987) have demonstrated that the ligand-binding subunit of an inhibitory central neurotransmitter receptor, the glycine receptor, shares homology with the nicotinic acetylcholine receptor polypeptide family. The primary structure of the 48K glycine receptor was deduced from its cDNA. Analysis for regional hydropathicity (Hopp and Woods, 1981) revealed four hydrophobic segments long enough to form transmembrane α helices, which is similar to that of the nicotinic acetylcholine receptor subunits. Recent reviews covering the acetylcholine receptor area have appeared (McCarthy *et al.*, 1986; Hucho, 1986), and a discussion of the family of receptors coupled to guanine nucleotide regulatory proteins, which include the β-adrenergic receptor, covers the area in great detail.

C. ATPases

Sodium- and potassium-dependent ATPase [$(NA^+ + K^+)$-ATPase], which is responsible for the active transport of Na^+ and K^+, is distributed universally among animal cell membranes and consists of two subunits, α and β. The larger α subunit with a relative molecular mass (M_r) of 84,000–120,000 is thought to have the catalytic role. Numa and co-workers (Kawakami *et al.*, 1985) determined the primary sequence of the α subunit of the *Torpedo californica* $(Na^+ + K^+)$-ATPase. Analysis by Kawakami *et al.* (1985) of the amino acid sequence for local hydrophobicity (Kyte and Doolittle, 1982) and the predicted secondary structure (Chou and Fasman, 1978b) suggests the presence of at least six transmembrane segments, presumably α-helical structures. Shull *et al.* (1985) characterized the complementary DNA for the catalytic subunit of the sheep kidney $(Na^+ + K^+)$-ATPase. The 1016-amino-acid sequence was analyzed by the Kyte and Doolittle (1982) procedure to establish the hydrophobic sequences, and eight major hydrophobic sequences (H1–H8), ranging in size from 17 to 29 amino acids, were found, which are the likely transverse membrane regions.

MacLennon *et al.* (1985) deduced the amino acid sequence from the cDNA sequence of the $(Ca^+ + Mg^+)$-dependent ATPase from rabbit muscle sarcoplasmic reticulum. They propose that the protein has three cytoplasmic domains joined to a set of ten transmembrane helices by a narrow pentahelical stalk. The homology between the amino acid sequence of the extramembranous segments of the Ca^{+}-ATPase and the K^{+}-ATPase of *Escherichia coli* was previously pointed out by Shull *et al.* (1985). There was ~50% homology in some limiting regions. MacLennan *et al.* (1985) identified the transmembrane sequences by their hydrophobicity and applied standard methods only to the extramembranous regions. These regions are mainly on the cytoplasmic regions of the membrane and are separated by transmembrane hydrophobic hairpins, which are located by the polarity plot. This plot gives results similar to those of the Kyte and Doolittle (1982) plot. An earlier prediction of the Ca^{+}-ATPase (Allen *et al.*, 1980) produced an ambiguous answer using the method of McLachlan (1978); therefore, the present authors utilized the procedure of Taylor and Thornton (1984), which resolved some of the ambiguities in priority of strand and helix in the earlier prediction.

The segments connecting the globular region to the membrane are of particular interest, as they are likely to contain the Ca^{2+}-binding region and to transmit movements from the phosphorylation and nucleotide domains that bring about ion translocation. Brandl *et al.* (1986) demonstrated that rabbit genomic DNA contains two genes that encode Ca^{2+}-ATPase of fast twitch and slow twitch (and cardiac) sarcoplasmic reticulum, respectively. The deduced amino acid sequences of the products of the two genes are highly conserved in putative Ca^{2+}-binding regions, in sectors leading from cytoplasmic domains into transmembrane domains, and in transmembrane helices. The assignment of the secondary structure was that previously described by Brandl *et al.* (1986). Ovchinnikov *et al.* (1986) determined the primary structure and spatial organization of pig kidney (Na^{+} + K^{+})-ATPase. The cDNA complementary to pig kidney RNAs coding for α and β subunits of (Na^{+} + K^{+})-ATPase were cloned and sequenced. Selected tryptic hydrolysis of the α subunit within the membrane-bound enzyme and trypsin hydrolysis of the immobilized isolated β subunit were also performed. The mature α and β subunits contain 1016 and 302 amino acids, respectively. Hydropathicity profiles were determined by two methods (Ovchinnikov, 1982; Capaldi and Vanderkooi, 1972; Kyte and Doolittle, 1982). Eleven regions were found that could serve as intramembrane segments. However, comparison of the above data and those on amino acid sequences of the hydrophilic peptides of the α subunit (Arzamazova *et al.*, 1985) allowed a selection of transmembrane segments. They assumed that there are five probable transmembrane fragments localized inside the membrane, with the remaining two, shorter and less hydrophobic, being exposed outside.

The plasma membrane ATPase of *Neurospora crassa* was deduced from genomic and cDNA sequences by Hagler *et al.* (1986). It contains a protein of 920 amino acids possessing as many as eight transmembrane segments. The *Neurospora* ATPase shows a significant amino acid sequence homology with the (Na^{+} + K^{+})- and Ca^{2+}-transporting ATPases of animal cells, particularly in regions that appear to be involved in ATP binding and hydrolysis. The method of Engelman *et al.* (1986) was used to determine the hydropathicity, and eight transmembrane segments were predicted.

Vogel *et al.* (1986) mapped the ATP substrate site in the epidermal growth factor (EGF-K). Sequence alignment of four protein kinases, cGMP-dependent protein kinase (cGK), cAMP-dependent protein kinase (cAK), the γ-subunit of phosphorylase b kinase (γ), and the EGF receptor (EGF-K) indicated that only a few residues are strictly conserved. The segment 1–11 contains three invariant Gly residues and was predicted to adopt an α-helical structure (Nagano, 1974; Maxfield and Scheraga, 1976; Robson and Suzuki, 1976; Chou and Fasman, 1978a). Such an element is consistently found in all nucleotide-binding sites, including various dehydrogenases (Sternberg and Taylor, 1984). The secondary structure of segment 2, including Cys^{157}, was analyzed by various predictive models (Nagano, 1974; Maxfield and Scheraga, 1976; Robson and Suzuki, 1976; Chou and Fasman, 1978a; von Heijne, 1981a,b), including a hydropathy plot using a similar approach to that outlined by Taylor and Thornton

(1984). Turns for the EGF receptor were predicted using several methods (Maxfield and Scheraga, 1976; Robson and Suzuki, 1976; Chou and Fasman, 1978a; Sternberg and Taylor, 1984). Kanazawa *et al.* (1985) cloned a mutant gene of the γ subunit of H^+-tranlocating ATPase from *Escherichia coli,* mutant NR70, and found that there were seven amino acids deleted from the amino-terminal portion. This deletion resulted in the loss of a hydrophobic domain (23–28) as determined by a Kyte and Doolittle (1982) hydrophobicity plot. This caused total loss of the assembly of F_1 on the membrane.

The plasma membrane ATPase of plants and fungi is a hydrogen pump. The protein gradient generated by the enzyme drives the active transport of nutrients by H^+–symport. In addition, the external acidification in plants and the internal alkalinization in fungi, both resulting from activation of the H^+ pump, have been proposed to mediate growth responses. The ATPase has a relative molecular mass similar to those of the (Na^+ + K^+)- and Ca^{2+}-ATPases of animal cells and, like those proteins, forms an aspartylphosphate intermediate. Serrano *et al.* (1986) cloned, mapped, and sequenced the gene encoding the yeast plasma membrane ATPase (PMAI). The strong homology between the amino acid sequences encoded by PMAI and those of (Na^+ + K^+)-, Na^+-, K^+-, and Ca^{2+}-ATPases is consistent with the notion that the family of cation pumps that form a phosphorylated intermediate evolved from a common ancestral ATPase. The hydrophobicity profiles of (Na^+ + K^+)-ATPase, Ca^{2+}-ATPase, and K^+-ATPase are very similar to that of the H^+-ATPase as determined by the Kyte and Doolittle (1982) and Eisenberg *et al.* (1984a,b) methods. A possible ten-trans-membrane-segment structure was proposed.

D. T-Cell Receptor

Hedrick *et al.* (1984a) have isolated a cDNA clone, TM86, that represents a species of mRNA that is expressed in T cells but not in B cells, encodes a membrane-associated protein, and is rearranged in T cells. Hedrick *et al.* (1984b) have analyzed clone TM86 and also several cross-reacting clones isolated from a thymocyte cDNA library by nucleotide sequencing. The data reveal a remarkable resemblance between the amino acid sequences of the protein encoded by the mRNAs represented by the TM86 and related cDNA clones and the immunoglobulin proteins of B cells. An analysis of the amino acid sequences was carried out to predict the secondary structure by the method of Kyte and Doolittle (1982). The hydrophobicity plot indicates a number of interesting features. (1) The 86T1 protein has the alternating hydrophobic–hydrophilic stretches characteristic of globular proteins (Kyte and Doolittle, 1982). (2) The predicted leader polypeptide occurs in a very adequate hydrophobic environment, and, most importantly, it predicts a possible transmembrane region spanning the end of the 86T1 sequence, followed by a string of positive charges (Lys-Arg-Lys), which are characteristic of the cystoplasmic portion of a number of lymphocyte cell-surface markers (Kabat *et al.,* 1983).

Barth *et al.* (1985) found ten different V_β gene segments when the sequences of 15 variable (V_β) genes of the mouse T-cell receptor were examined. Both the T-cell receptor and the immunoglobulins are heterogeneous cell-surface glycoproteins that can recognize many antigens (e.g., Allison *et al.,* 1982). T-cell receptor molecules are composed of α and β chains, each of which, like the immunoglobulin chains, is divided into variable (V) and constant (C) regions (e.g., Allison *et al.,* 1982). The V regions together form the antigen-binding domain. Like the immunoglobulin gene, they are divided into separate V_β, diversity (D_β), and joining (J_β) gene segments that are assembled by recombination during T-cell development to form a V_β that is associated with either of two constant ($C_\beta 1$ and $C_\beta 2$) genes. Barth *et al.* (1985) have compared the V_β and immunoglobulin V segments by analyzing them for two properties believed to reflect important structural features of these molecules: the distribution of β-pleated-sheet-forming potential (Chou and Fasman, 1978b) and the predicted hydrophobicity profile (Kyte and Doolittle, 1982). The analysis showed nearly identical results for mouse V_β, V_H, and V_κ segments. This agrees with work of Patten *et al.* (1984). The authors conclude that the T-cell receptor and immunoglobulin molecules fold into comparable tertiary structure.

Arden *et al.* (1985) analyzed 19 complementary DNA clones encoding the α chain of the T-cell antigen receptor derived from thymic transcripts. The primary sequence was analyzed by several methods for primary structural patterns. The variability plot of Wu and Kabat (1970) examines the variation of each residue position between members of a set of similar sequences. In immunoglobulin and β-chain V regions, two regions of relative hypervariability were noted, with a third hypervariability region positioned at the junction of the V, D, and J gene segments. The Kyte and Doolittle (1982) hydropathicity and the Chou and Fasman (1974a,b) method, which determines the relative potential for β-pleated-sheet formation were used. Both methods showed that the V_α, V_β, V_H, and V_κ sequences are very similar. These data suggest that the T-cell antigen receptor and immunoglobulin molecules have similar structure.

Tillinghast *et al.* (1986) characterized the variability of the expressed T-cell receptor β-chain repertoire and compared this variability to the known murine β-chain repertoire. No features were found that distinguish human V_β genes from their murine counterparts. Variability among the 15 human V_β genes was analyzed by the Wu and Kabat (Kabat *et al.*, 1983) variability plot. Twelve human V_β genes were also analyzed with the hydrophobicity algorithm of Hopp and Woods (1981). The areas of highest variability on the Wu–Kabat analysis correspond, in general, with the hydrophilic regions of the molecule, consistent with these regions being available to interact with ligand. The hydrophilic plot shows conservation of several hydrophobic regions that are thought to be involved in the formation of the common tertiary structure shared among all T-cell receptor chains as well as members of the immunoglobulin supergene family (e.g., Novotny and Haber, 1985). Schiffer *et al.* (1986) examined the β-chain variable regions of human, mouse, and rabbit T-cell receptors for antigen and were able to classify them into two subgroups by the application of the variability plots of Wu and Kabat (1970).

E. Sodium Channel

The sodium channel is a membrane protein that mediates the voltage-dependent modulation of the sodium ion permeability of electrically excitable membranes (Catterall, 1984). Numa and co-workers (Noda *et al.*, 1984) cloned and sequenced the cDNA for the *Electrophorus electricus* electroplax sodium channel. Analysis of the derived protein sequence indicated a protein consisting of 1820 amino acids that exhibited four repeated homologous units, determined by homology matrix comparison (Toh *et al.*, 1983), that are presumably oriented in a pseudosymmetric fashion across the membrane. Each homology unit contains a unique segment with clustered positively charged residues, which may be involved in the gating structure, possibly in conjunction with negatively charged residues clustered elsewhere. The amino acid sequence of the sodium channel protein was analyzed for predicted secondary structure (Chou and Fasman, 1978b) and for local hydropathy (Kyte and Doolittle, 1982). The segments that can form α helix or β sheet extend over a wide range of the polypeptide chain.

The hydropathy profile along the polypeptide chain indicates that each repetitive homology unit contains five hydrophobic segments (S1, S2, S3, S5, and S6) at equivalent positions. In addition, each unit contains a characteristic segment with strong positive charge (S4) located between segments S3 and S5. Segments S5 and S6 in each repeat correspond to highly hydrophobic regions with predicted secondary structures that comprise a continuous stretch of 24–38 uncharged amino acid residues including many nonpolar residues. These hydrophobic regions are flanked on both sides by charged residues. Because such a sequence is characteristic of transmembrane protein segments, segments S5 and S6 most probably transverse the membrane, forming α-helical structures. Segments S1, S2, and S3 in each repeat represent hydrophobic regions with predicted secondary structure consisting of 18–28 amino acid residues that are largely nonpolar but include a few charged residues. Negatively charged residues predominate in segments S1 and S3, whereas both positively charged and negatively charged residues are present in segment S2, so that this segment generally has no net charge. If an α-helical structure is assumed for these segments, the charged side chains in them are clustered largely on one side of the helix, the opposite side being occupied mostly by nonpolar side chains. Thus, segments S1, S2, and S3 can also span the membrane, forming amphipathic α-helical structures. Segment S4 in each repeat represents a positively charged region with predicted secondary structure comprising 21 amino acid residues, including five to seven Arg or Lys residues, Arg being predominant. This segment exhibits a unique structural feature in that the Arg or Lys residues are located every third position. The residues intervening between these basic residues are mostly nonpolar. If an α-helical structure is assumed for this segment, the positively charged side chains would be distributed along a spiral line, making a three quarters to one turn. Alternatively, if a 3_{10} helical structure is assumed, they would lie on one side of the helix so that the segment would be strongly amphipathic. On the other hand, if a β-sheet structure is assumed, they would extend alternately toward both sides of the peptide backbone.

It seems reasonable to postulate that the four repeated homology units are oriented in a pseudosymmetric fashion across the membrane. If this is the case, each unit should contain an even number of transmembrane segments because no additional hydrophobic segments are predicted outside the homology units. The transmembrane segments are most likely to be involved directly or indirectly in the formation of the ionic channel.

Noda *et al.* (1986) isolated complementary DNA clones derived from two distinct rat brain mRNAs encoding sodium channel large polypeptides and have determined the complete amino acid sequences of the two polypeptides (designated sodium channels I and II) as deduced from the cDNA sequences. A partial DNA sequence complementary to a third homologous mRNA from a rat brain has also been cloned. The degree of homology is 87%, 62%, and 62% for the rat I/rat II, rat I/*Electrophorus,* and rat II/*Electrophorus* pairs, respectively. Rat I and II sodium channel also contain four homologous repeats (deletions or insertions). The regions corresponding to these repeats are highly conserved among all three sodium channels, whereas the remaining regions, all of which are assigned to the cytoplasmic side of the membrane, are less well conserved except for the region between repeats III and IV. Rat sodium channel I and II have hydropathicity profiles similar to that of the *Electrophorus* sodium channel (Noda *et al.,* 1984). Each internal repeat has five hydrophobic segments (S1, S2, S3, S5, and S6) and one positively charged segment (S4), all of which exhibit predicted secondary structure (Chou and Fasman, 1978b). The distribution of charge is similar to the *Electrophorus* channel. However, based on other considerations, a different arrangement of segments is proposed, namely, that all of the six segments S1–S6 span the membrane, presumably as α helices, forming an ion channel.

Guy and Seetharamulu (1986) have presented alternative models of the sodium channel in *Electrophorus electricus*. The favored model has the same general folding pattern as postulated by Noda *et al.* (1984) except that each domain has four additional transmembrane segments, S3, S4, S6, and S7. The postulated activation gating mechanism involves a screwlike motion of the S4 helices. The sequence was analyzed with a method that predicts which portions of α helices and β structures are exposed to water, buried inside the protein, or exposed to lipid (Guy, 1984, 1985). Interactions among transmembrane segments were examined by constructing Nicholson molecular models of alternative conformations and by using computer graphics. Primary features of the models are that the sodium channel lining is formed by positively and negatively charged segments and that movement of the positively charged segments underlies voltage-dependent activation.

Table II contains a list of predicted structures of membrane proteins.

Table II. Predicted Structure of Membrane Proteins

Protein	Method used[a]	Reference
Coat protein	C–F	Green and Flanagan (1976)
Filamentous bacteriophages	L	
Pf1; fd; ZJ-2		
Outer envelope of *E. coli*		
Erythrocyte glycophorin		
Bovine liver microsomal	C–F	Fleming *et al.* (1978)
Cytochrome b_5		
Signal sequences	ΔG_{trans}	von Heijne and Blomberg (1979)
Prelysozyme	Chothia	
M-41 κ L chain		
Prealbumin		
β-Lactamase		
Phage fl-coat		
E. coli lipoprotein		
M-104E λ, L chain		
Preconalbumin		
p-450 M_2		
Mitochondrial ADP/ATP translocase	K–D	Saraste and Walker (1982)
	S	
	M	
α and β subunits of ATP synthase myosin; kinases and other ATP-requiring enzymes and a common nucleotide fold	M	Walker *et al.* (1982)
Bacteriorhodopsin	C–F	Senior (1983)
E. coli H^+-ATPase	V–H	
Cytochrome oxidase subunit III		
UNC E protein		
Lac permease of *E. coli*	K–D	Foster *et al.* (1983)
	C–F	Menick *et al.* (1986)
UNC operon	K–D	Walker *et al.* (1984)
ATP synthetase	W	
	S–M	
	M	
	S	
Photosynthetic reaction center, B870 antenna and flanking polypeptides from *R. capsulta*	K–D	Youvan *et al.* (1984)

Table II. (*Continued*)

Protein	Method used[a]	Reference
ATP binding site of oncogene product v-*src*	T–T	Sternberg and Thorton (1984)
Epidermal growth factor receptor, cAMP-dependent protein kinase		
Bovine brain myelin proteolipid	K–D	Lauresen *et al.* (1984)
	C–F	
	R	
Protein disulfide isomerase of the lumen of the endoplasmic reticulum	F–M–S	Edman *et al.* (1985)
GTPase of bovine rod outer segments	C–F	Yatsunami and Khorana (1985)
3-Hydroxymethylglutaryl–coenzyme A reductase (glycoprotein of the endoplasmic reticulum)	F–M–S	Liscum *et al.* (1985)
Phosphocarrier protein factor III^{lac} of lactose-specific phosphotransferase system of *Staphylococcus aureus*	C–F	Stuber *et al.* (1985)
	K–D	
	St.	
Colicin A	G–O–R	Pattus *et al.* (1985)
	C–F	
	F–M–S	
	E(1)	
Human insulin receptor	K–D	Ullrich *et al.* (1985)
Murine anion exchange protein	K–D	Kopito and Lodish (1985)
	S–E	
Wheat chloroplast gene for CF_0 subunit of ATP synthase	K–D	Bird *et al.* (1985)
	C–F	
	G–O–R	
Human glucose transporter	K–D	Mueckler *et al.* (1985)
	E(2)	
	C–F	
Glucose permease of bacterial phosphotransferase	K–D	Erni and Zanolari (1986)
	E(3)	
	S–E	
32-kd Qb-binding chloroplast thylakoid membrane protein	R–H–A	Sayre *et al.* (1986)
		Rao *et al.* (1983)
Potential membrane proteins of Epstein–Barr virus	C–F	Modrow and Wolf (1986)
	G–O–R	
	H–W	
Nucleotide-binding domain in the β subunit of *E. coli* F_1-ATPase	W–S–G	Duncan *et al.* (1986)
Mannitol permease of *E. coli*	K–D	Stephan and Jacobson (1986)
		Lee and Saier (1983)
Protein 4.1, human erythrocyte membrane skeleton	F–M–S	Conboy *et al.* (1986)
Ca^{2+}-dependent membrane-binding proteins (colelectron) in *Torpedo marmorata* and mammalian cells	C–F	Geisow *et al.* (1986)
Amelogenin from bovine tooth enamel	C–F	Renugopalakrishnan *et al.* (1986)
Coronavirus protein E1	C–F	Rottier *et al.* (1986)
	E(2)	
	L	
	H	
$GABA_A$ benzodiazepine receptor from bovine brain	K–D	Schofield *et al.* (1987)
	H–W	
Dihydropyridine calcium channel blocker receptor from skeletal muscle (DHP receptor)	K–D	Tanabe *et al.* (1987)
	C–F	

(*continued*)

Table II. (*Continued*)

Protein	Method used[a]	Reference
Family of receptors coupled to guanine nucleotide regulatory proteins	K–D	Dohlman *et al.* (1987)
Human platelet glycoprotein 1b	K–D	Lopez *et al.* (1987)
Willebrand factor-binding domain of platelet membrane glycoprotein 1b	H–W	Titani *et al.* (1987)
	K–D	
	C–F	
Human cation-dependent mannose-G-phosphate-specific receptor	N–A (C–F)	Dohlman *et al.* (1987)
	R–R	
Bovine retinal S antigen	G–O–R	Shinohara *et al.* (1987)
	N–T	
Murine band 3 gene of erythrocytes	K–D	Kopito *et al.* (1987)
	C–F	
Insulinlike growth factor II receptor	K–D	Morgan *et al.* (1987)
Growth hormone receptor and serum binding protein	K–D	Leung *et al.* (1987)
Myelin P_2 protein	C–F	Shin and McFarlane (1987)
	B–B	
Calcium-dependent membrane-binding proteins p35, p36, and p32	G–O–R	Taylor and Geisow (1987)
Scrapie prion protein	E	Bazan *et al.* (1987)
	K–D	
	C–F	
	G–O–R	
	F–M–S	
Bovine substance-K receptor (SKR)	K–D	Masu *et al.* (1987)

[a]Sources of references: B–B, Barkovsky and Bandarin (1979); C–F, Chou-Fasman (1978a); E(1), Eisenberg *et al.* (1982a); E(2), Eisenberg *et al.* (1984a); E(3), Eisenberg (1982b); F–M–S, Finer-Moore and Stroud (1984); G–O–R, Garnier *et al.* (1978); H, von Heijne (1981b); H–W, Hopp and Woods (1981); K–D, Kyte and Doolittle (1982); L, Lim (1974a,b,c); M, McLachlan (1977); N–A, Novotny and Auffray (1984); N–T, Nozaki and Tanford (1971); R, Rose (1978); R–H–A, Mohana-Rao *et al.* (1983); R–R, Rose and Roy (1980); S, Staden (1982); S–E, Schiffer and Edmunson (1967); S–M, Staden (1982); St., Stuber (1982); T–T, Taylor and Thornton (1983, 1984); V–H, von Heijne (1981a,b); W, Walker *et al.* (1984).

ACKNOWLEDGMENTS. This work was supported by a National Science Foundation grant, DMB-8713193. I would like to thank Pamela Gailey for the preparation of the manuscript.

IX. REFERENCES

Abercrombie, D. M., and Khorana, H. G., 1986, Regeneration of native bacteriorhodopsin following acetylation of ϵ-amino groups of Lys-30, -40 and -41, *J. Mol. Biol.* **261:**4875–4880.

Allen, G., Trinnaman, B. J., and Green, N. M., 1980, The primary structure of the calcium ion-transporting adenosine triphosphatase protein of rabbit skeletal sarcoplasmic reticulum, *Biochem. J.* **187:** 591–616.

Allison, J. P., McIntyre, B. W., and Bloch, D., 1982, Tumor-specific antigen of murine T-lymphoma defined with monoclonal antibody, *J. Immunol.* **129:**2293–2300.

Ananthanaryanan, V. S., and Bandekar, J., 1976, Application of one-dimensional Isling model to the secondary structure in globular proteins: Predicted β-regions, *Int. J. Peptide Protein Res.* **8:**615–623.

Ananthanaryanan, V. S., Brahmachari, S. K., and Paltabiraman, N., 1984, Proline-containing β-turns in peptides and proteins: Analysis of structural data on globular proteins, *Arch. Biochem. Biophys.* **232:**482–495.

Anderer, F. A., 1963, Recent studies on the structure of tobacco mosaic virus, *Adv. Protein Chem.* **13:**1–35.

Anderson, W., Burt, S., and Loew, G., 1979, Energy-conformation studies of frequency of β-turns in tetrapeptide sequences, *Int. J. Peptide Protein Res.* **14:**402–408.

Anfinsen, C. B., 1959, *The Molecular Basis of Evolution,* John Wiley & Sons, New York.

Anfinsen, C. B., 1973, Principles that govern the folding of protein chains, *Science* **181:**233–230.

Anfinsen, C. B., Haber, E., Sela, M., and White, F. H., 1961, The kinetics of formation of native ribonuclease during oxidation of the reduced polypeptide chain, *Proc. Natl. Acad. Sci. U.S.A.* **47:**1309–1314.

Arden, B., Klotz, J. L., Siu, G., and Hood, L., 1985, Diversity and structure of genes of the α family of mouse T-cell antigen receptor, *Nature* **316:**783–787.

Argos, P., 1987a, Analysis of sequence-similar pentapeptides in unrelated protein tertiary structures. Strategies for protein folding and a guide for site-directed mutagenesis, *J. Mol. Biol.* **197:**331–348.

Argos, P., 1987b, A sensitive procedure to compare amino acid sequences, *J. Mol. Biol.* **193:**385–396.

Argos, P., and Mohana-Rao, J. K., 1985, Relationships between exons and the predicted structure of membrane-bound proteins, *Biochim. Biophys. Acta* **827:**283–297.

Argos, P., and Palau, J., 1982, Amino acid distribution in protein secondary structures, *Int. J. Peptide Protein Res.* **19:**380–393.

Argos, P., Schwarz, J., and Schwarz, J., 1976, An assessment of protein secondary structure prediction methods based on amino acid sequence, *Biochem. Biophys. Acta* **439:**261–273.

Argos, P., Rossman, M., and Johnson, J. E., 1977, A four-helical super-secondary structure, *Biochem. Biophys. Res. Commun.* **75:**83–86.

Argos, P., Hanei, M., and Garavito, R. M., 1978, The Chou-Fasman secondary structure prediction method with an extended data base, *FEBS Lett.* **93:**19–24.

Argos, P., Mohana-Rao, J. K., and Hargrave, P. A., 1982, Structural prediction of membrane-bound proteins, *Eur. J. Biochem.* **128:**565–575.

Arzamazova, N. M., Arystarkhova, E. A., Shafieva, G. I., Nazimov, I. V., Aldanova, N. A., and Modyanov, N. N., 1985, Primary structure of the α-subunit of $Na^+ + K^+$-ATPase. I. Analysis of hydrophilic fragments of the polypeptide chains, *Bioorg. Khim.* **11:**1598–1601.

Aubert, J.-P., and Loucheaux-Lefebvre, M.-H., 1976, Conformational study of α_1-acid glycoprotein, *Arch. Biochem. Biophys.* **175:**400–409.

Aubert, J.-P., Biserte, G., and Loucheux-Lefebvre, M. H., 1976, Carbohydrate–peptide linkage in glycoproteins, *Arch. Biochem. Biophys.* **175:**410–418.

Aubert, J.-P., Helbecque, N., and Loucheux-Lefebvre, M.-H., 1981, Circular dichroism studies of synthetic Asn-X-Ser/Thr-containing peptides: Structure of glycosylation relationship, *Arch. Biochem. Biophys.* **208:** 20–29.

Bacon, D. J., and Anderson, W. F., 1986, Multiple sequence alignment, *J. Mol. Biol.* **191:**153–161.

Baldwin, R. L., 1980, The mechanism of folding of ribonucleases A and S, in *Protein Folding* (R. Jaenicke, ed.), Elsevier Amsterdam, pp. 369–384.

Barker, W. C., Hunt, L. T., Orcutt, B. C., George, D. G., Yeh, L. S., Chen, H. R., Blomquist, M. C.,

Johnson, G. C., Seibel-Ross, E. I., and Ledky, R. S., 1984, *Report of the National Biomedical Research Foundation,* Georgetown University, Washington.

Barkovsky, E. V., 1982, Prediction of the secondary structure of globular proteins by their amino acid sequence, *Acta Biol. Med. Germ.* **41:**751–758.

Barkovsky, E. V., and Bandarin, V. A., 1979, Secondary structure prediction of globular proteins from their amino acid sequence, *Bioorg. Khim.* **5:**24–34.

Barlow, D. J., and Thornton, J. M., 1983, Ion-pairs in proteins, *J. Mol. Biol.* **168:**867–885.

Barth, R. K., Kim, B. S., Lan, N. C., Hunkapiller, T., Sobieck, N., Winoto, A., Gershenfeld, H., Okada, C., Hansburg, D., Weissman, I. L., and Hood, L., 1985, The murine T-cell receptor uses a limited repertoire of expressed V_β gene segments, *Nature* **316:**517–523.

Barton, G. J., and Sternberg, M. J. E., 1987a, Evaluation and improvements in the automatic alignment of protein sequences, *Protein Eng.* **1:**89–94.

Barton, G. J., and Sternberg, M. J. E., 1987b, A strategy for the rapid multiple alignment of protein sequences. Confidence levels from tertiary structure comparison, *J. Mol. Biol.* **198:**327–337.

Bash, P. A., Singh, V. C., Langridge, R., and Kollman, P. A., 1987, Free energy calculations by computer simulation, *Science* **236:**564–568.

Bashford, D., Chothia, C., and Lesk, A. M., 1987, Determinants of a protein fold. Unique features of the globin amino acid sequences, *J. Mol. Biol.* **196:**199–216.

Bazan, J. F., Fletterick, R. J., McKinley, M. P., and Pruisner, S. B., 1987, Predicted secondary structure and membrane topology of the scrapie prion protein, *Protein Eng.* **1:**125–135.

Beasty, A. M., and Matthews, C. R., 1985, Characterization of an early intermediate in the folding of the α-subunit of tryptophan synthase by a hydrogen exchange measurement, *Biochemistry* **24:**3547–3553.

Beecher, B., and Cassim, J. Y., 1976, Effects of light adaption on the purple membrane of *Halobacterium halobium, Biophys. J.* **16:**1183–1200.

Beeley, J. G., 1976, Location of the carbohydrate groups on ovomucoid, *Biochem. J.* **159:**335–345.

Beeley, J. G., 1977, Peptide chain conformation and the glycosylation of glycoproteins, *Biochem. Biophys. Res. Commun.* **76:**1051–1055.

Bentley, G., Dodson, E., Dodson, G., Hodgkin, D., and Mercola, D., 1976, Structure of insulin in 4-zinc insulin, *Nature* **261:**166–168.

Billeter, M., Havel, T. F., and Kuntz, I. D., 1987a, A new approach to the problem of docking two molecules: The ellipsoid algorithm, *Biopolymers* **26:**777–793.

Billeter, M., Havel, T. F., and Wuthrich, K., 1987b, The ellipsoid algorithm as a method for the determination of polypeptide conformations from experimental distance constraints and energy minimization, *J. Comput. Chem.* **8:**132–141.

Bird, C. R., Koller, B., Auffret, A. D., Huttly, A. K., Howe, C. J., Dyer, T. A., and Gray, J. C., 1985, The wheat chloroplast gene for CF_0 subunit I of ATP synthase contains a large intron, *EMBO J.* **4:**1381–1388.

Black, S. D., and Glorioso, J. C., 1986, MSEQ: A microcomputer-based approach to the analysis, display, and prediction of protein structure, *Biol. Techniques* **4:**448–460.

Blagdon, D. E., and Goodman, G., 1975, Mechanisms of protein and polypeptide helix initiation, *Biopolymers* **14:**241–245.

Blake, C. C. F., 1979, Exons encode protein functional units, *Nature* **277:**598.

Bloomer, A. C., Champness, J. N., Bricogne, G., Staden, R., and Klug, A., 1978, Protein disk of tobacco mosaic virus at 2.8 Å resolution showing the interactions within and between units, *Nature* **276:**362–368.

Blout, E. R., 1962, The dependence of the conformation of polypeptides and proteins upon amino acid composition, in: *Polyamino Acids, Polypeptides, and Proteins* (M. Stahman, ed.), University of Wisconsin Press, Madison, pp. 275–279.

Blow, D. M., Irwin, M. J., and Nyborg, J., 1977, The peptide chain of tyrosyl t-RNA synthetase: No evidence for a super-secondary structure of four-α-helices, *Biochem. Biophys. Res. Commun.* **76:**728–734.

Blundell, T. L., and Johnson, L. N., 1976, *Protein Crystallography,* Academic Press, New York.

Blundell, T. L., and Sternberg, M. J. E., 1985, Computer-aided design in protein engineering, *Trends Biotechnol.* **3:**228–235.

Blundell, T., Singh, J., Thornton, J., Burley, S. K., and Petsko, G. A., 1986a, Aromatic interactions, *Science* **234:**1005.

Blundell, T. L., Barlow, D., Sibanda, B. L., Thornton, T. M., Taylor, W., Tickle, I. J., Sternberg, M. J. E., Pitts, J. E., Haneef, I., and Hemmings, A. M., 1986b, Three-dimensional aspects of the design of new protein molecules, *Phil. Trans. R. Soc. Lond. [A]* **317:**333–344.

Blundell, T. L., Sibanda, B. L., Sternberg, M. J. E., and Thornton, J. M., 1987, Knowledge-based prediction of protein structures and the design of novel molecules, *Nature* **326:**347–352.

Bonner, T. I., Buckley, N. J., Young, A. C., and Brann, M. R., 1987, Identification of a family of muscarinic acetylcholine receptor genes, *Science* **237:**527–532.

Boswell, D. R., and McLachlan, A. D., 1984, Sequence comparison by exponentially-damped alignment, *Nucleic Acids Res.* **12:**457–464.

Bourgeois, S., and Pfahl, M., 1976, Repressors, *Adv. Protein Chem.* **30:**1–99.

Brandhuber, B. J., Boone, T., Kenney, W. C., and McKay, D. B., 1987, Three-dimensional structure of interleukin-2, *Science* **283:**1707–1709.

Brandl, C., Green, N. M., Korczak, B., and MacLennan, D. H., 1986, Two Ca^{2+} ATPase genes: Homologies and mechanistic implications of deduced amino acid sequences, *Cell* **44:**597–607.

Briggs, M. S., and Gierasch, L. M., 1984, Exploring the conformational role of signal sequences: Synthesis and conformational analysis of λ receptor protein wild type and mutant signal peptides, *Biochemistry* **23:**3111–3114.

Brisson, A., and Unwin, P. N. T., 1985, Quaternary structure of the acetylcholine receptor, *Nature* **315:**474–477.

Brooks, B. R., Bruccoleri, R. E., Olafson, B. D., States, D. J., Swaminathan, S., and Karplus, M., 1983, CHARMM: A program for macromolecular energy minimization and dynamics calculations, *J. Comput. Chem.* **4:**187–217.

Bruccoleri, R. E., and Karplus, M., 1987, Prediction of the folding of short polypeptide segments by uniform conformational sampling, *Biopolymers* **26:**137–168.

Bunting, J. R., Athey, T. W., and Cathou, R. E., 1972, Backbone folding of immunoglobulin light and heavy chains: A comparison of predicted β-bend positions, *Biochim. Biophys. Acta* **285:**60–71.

Burgess, A. W., and Scheraga, H. A., 1975, Assessment of some problems associated with the prediction of the three-dimensional structure of a protein from its amino acid sequence, *Proc. Nat. Acad. Sci. U.S.A.* **72:** 1221–1225.

Burgess, A. W., Ponnuswamy, P. K., and Scheraga, H. A., 1974, Analysis of conformations of amino acid residues and prediction of backbone topography in proteins, *Israel J. Chem.* **12:**239–286.

Burley, S. K., and Petsko, G. A., 1985, Aromatic–aromatic interaction: A mechanism of protein stabilization, *Science* **229:**23–28.

Burley, S. K., and Petsko, G. A., 1986, Amino–aromatic interactions in proteins, *FEBS Lett.* **203:**139–143.

Busetta, B., and Hospital, M., 1981, Improving the accuracy of secondary structure predictions, *Biochemie* **63:** 951–954.

Busetta, B., and Hospital, M., 1982, An analysis of the prediction of secondary structures, *Biochim. Biophys. Acta* **701:**111–118.

Cantor, C. R., and Jukes, T. H., 1966, The repetition of homologous sequences in the polypeptide chains of certain cytochromes and globins, *Proc. Natl. Acad. Sci. U.S.A.* **56:**177–184.

Capaldi, R. A., and Vanderkooi, G., 1972, The low polarity of many membrane proteins, *Proc. Natl. Acad. Sci. U.S.A.* **69:**930–932.

Catterall, W. A., 1984, The molecular basis of neuronal excitability, *Science* **223:**653–661.

Chakravarty, P. K., Mathur, K. B., and Dhar, M. M., 1973, The synthesis of a decapeptide with glycosidase activity, *Experientia* **29:**786–788.

Chang, E. L., Yager, P., Williams, R. W., and Dalziel, A. W., 1983, The secondary structure of reconstituted acetylcholine receptors as determined by Raman spectroscopy, *Biophys. J.* **41:**65a.

Charton, M., and Charton, B. I., 1983, The dependence of the Chou–Fasman parameters on amino acid side-chain structure, *J. Theor. Biol.* **102:**121–134.

Chothia, C., 1973, Conformation of twisted β-pleated sheets in proteins, *J. Mol. Biol.* **75:**295–302.

Chothia, C., 1974, Hydrophobic bonding and accessible surface area in proteins, *Nature* **248:**338–339.

Chothia, C., 1975, Structural invariants in protein folding, *Nature* **254:**303–308.

Chothia, C., 1976, The nature of accessible and buried surfaces in proteins, *J. Mol. Biol.* **105:**1–14.

Chothia, C., 1984, Principles that determine the structure of proteins, *Annu. Rev. Biochem.* **53:**537–572.

Chothia, C., and Janin, J., 1975, Principles of protein-protein recognition, *Nature* **256:**705–708.

Chothia, C., and Janin, J., 1982, Orthogonal packing of β-sheets in proteins, *Biochemistry* **21:**3955–3965.

Chothia, C., and Lesk, A. M., 1982a, Evolution of proteins formed by β-sheets. I. The core of the immunoglobulin domains, *J. Mol. Biol.* **160:**325–342.

Chothia, C., and Lesk, A. M., 1982b, Evolution of proteins formed by β-sheets. II. Plastocyanin and azurin, *J. Mol. Biol.* **160**:303–323.

Chothia, C., and Lesk, A. M., 1986, The relation between the divergence of sequence and structure in proteins, *EMBO J.* **5**:823–826.

Chothia, C., Levitt, M., and Richardson, D., 1977, Structure of proteins: Packing of α-helices and pleated sheets, *Proc. Natl. Acad. Sci. U.S.A.* **74**:4130–4134.

Chothia, C., Levitt, M., and Richardson, D., 1981, Helix to helix packings in proteins, *J. Mol. Biol.* **145**:215–250.

Chothia, C., Novotny, J., Bruccoleri, R., and Karplus, M., 1985, Domain association in immunoglobulin molecules. the packing of variable domains, *J. Mol. Biol.* **186**:651–663.

Chou, P. Y., 1979, New approaches to protein structural analysis and conformational predictions, in: *CECM Protein Folding Workshop,* Université de Paris-Sud, Orsay, France.

Chou, P. Y., 1980, Amino acid compositions of four structural classes of proteins, in: *Abstracts, Second Chemical Congress of the North American Continent,* Las Vegas.

Chou, P. Y., and Fasman, G. D., 1973, Structural and functional role of Leu residues in proteins, *J. Mol. Biol.* **74**:263–281.

Chou, P. Y., and Fasman, G. D., 1974a, Conformational parameters for amino acids in helical, β-sheet, and random coil regions calculated from proteins, *Biochemistry* **13**:211–222.

Chou, P. Y., and Fasman, G. D., 1974b, Prediction of protein conformation, *Biochemistry* **13**:222–245.

Chou, P. Y., and Fasman, G. D., 1975, The conformation of glucagon: Predictions and consequences, *Biochemistry* **14**:2536–2541.

Chou, P. Y., and Fasman, G. D., 1977a, Secondary structural prediction of proteins from their amino acid sequence, *Trends Biochem. Sci.* **2**:128–132.

Chou, P. Y., and Fasman, G. D., 1977b, β-Turns in proteins, *J. Mol. Biol.* **115**:135–175.

Chou, P. Y., and Fasman, G. D., 1977c, Prediction of protein secondary structure, in: *Fifth American Peptide Symposium* (M. Goodman and J. Meienhofer, eds.), John Wiley & Sons, New York, pp. 284–287.

Chou, P. Y., and Fasman, G. D., 1978a, Prediction of the secondary structure of proteins from their amino acid sequence, *Adv. Enzymol.* **47**:45–148.

Chou, P. Y., and Fasman, G. D., 1978b, Empirical predictions of protein conformation, *Annu. Rev. Biochem.* **47**:251–276.

Chou, P. Y., and Fasman, G. D., 1979a, Prediction of β-turns, *Biophys. J.* **26**:367–384.

Chou, P. Y., and Fasman, G. D., 1979b, Conservation of chain reversal regions in proteins, *Biophys. J.* **26**: 385–400.

Chou, P. Y., Adler, A. J., and Fasman, G. D., 1975, Conformational prediction and circular dichroism studies on the *lac* repressor, *J. Mol. Biol.* **96**:29–45.

Chou, K.-C., Pottle, M., Nemethy, G., Veda, Y., and Scheraga, H. A., 1982, Structure of β-sheets, *J. Mol. Biol.* **162**:89–112.

Claudio, T., Ballivet, M. Patrick, J., and Heinemann, S., 1983, Nucleotide and deduced amino acid sequences of *Torpedo californica* acetylcholine receptor γ-subunit, *Proc. Natl. Acad. Sci. U.S.A.* **80**:1111–1115.

Cohen, C., and Parry, D. A. D., 1986, α-Helical coiled-coils: A widespread motif in proteins, *Trends Biochem. Sci.* **11**:245–248.

Cohen, F. E., and Kuntz, I. D., 1987, Prediction of the three-dimensional structure of human growth hormone, *Proteins* **1**:162–166.

Cohen, F. E., and Sternberg, M. J. E., 1980a, The use of chemically derived distant constants in the prediction of protein structure with myoglobin as an example, *J. Mol. Biol.* **137**:9–22.

Cohen, F. E., and Sternberg, M. J. E., 1980b, On the prediction of protein structure: The significance of the root-mean-square deviation, *J. Mol. Biol.* **138**:321–333.

Cohen, F. E., Richmond, J. T., and Richards, F. M. J., 1979, Protein folding: Evaluation of some simple rules for the assembly of helices into tertiary structure with myoglobin as an example, *J. Mol. Biol.* **132**:275–288.

Cohen, F. E., Sternberg, M. J. E., and Taylor, W. R., 1980, Analysis and prediction of protein β-sheet structures by a combinatorial approach, *Nature* **285**:378–382.

Cohen, F. E., Sternberg, M. J. E., and Taylor, W. R., 1981, Analysis of the tertiary structure of protein sandwiches, *J. Mol. Biol.* **148**:253–272.

Cohen, F. E., Sternberg, M. J. E., and Taylor, W. R., 1982, Analysis and prediction of the packing of α-helices against a β-sheet in the tertiary structure of globular proteins, *J. Mol. Biol.* **156**:821–862.

Cohen, F. E., Abarbanel, R. M., Kuntz, I. D., and Fletterick, R. J., 1983, Secondary structure assignment for α/β proteins by a combinatorial approach, *Biochemistry* **22:**4894–4904.

Cohen, F. E., Abarbanel, R. M., Kuntz, I. D., and Fletterick, R. J., 1986a, Turn prediction in proteins using a pattern matching approach, *Biochemistry* **25:**266–275.

Cohen, F. E., Kosen, P. A., Kuntz, I. D., Epstein, C. B., Cardelli, T. C., and Smith, K. A., 1986b, Structure activity studies of interleukin-2, *Science* **234:**349–356.

Conboy, J., Kan, Y. W., Shohet, S. B., and Mohandas, N., 1986, Molecular cloning of protein 4.1, a major structural element of the human erythrocyte membrane skeleton, *Proc. Natl. Acad. Sci. U.S.A.* **83:**9512–9516.

Cook, D. A., 1967, The relation between amino acid sequence and protein conformation, *J. Mol. Biol.* **29:**167–171.

Cornette, J. L., Cease, K. B., Margalit, J. H., Spouge, J. L., Berzofsky, J. A., and DeLisi, C., 1987, Hydrophobicity scales and computational techniques for detecting amphipathic structures in proteins, *J. Mol. Biol.* **195:**659–685.

Corrigan, A. J., and Huang, P. C., 1982, A BASIC microcomputer program for plotting the secondary structure of proteins, *Comput. Prog. Biomed.* **15:**163–168.

Crawford, I. P., Niermann, T., and Kirshner, K., 1987, Predictions of secondary structure by evolutionary comparison: Application to the α-subunit of tryptophan synthase, *Proteins* **1:**118–129.

Crawford, J. L., Lipscomb, W. N., and Schellman, C. G., 1973, The reverse turn as a polypeptide conformation in globular proteins, *Proc. Natl. Acad. Sci. U.S.A.* **70:**538–542.

Creighton, T. E., 1978, Experimental studies of protein folding and unfolding, *Prog. Biophys. Mol. Biol.* **33:** 231–297.

Creighton, T. E., 1979, Electrophoretic analysis of the unfolding of proteins by urea, *J. Mol. Biol.* **129:**235–264.

Criado, M., Hochschwender, S., Sarin, V., Fox, J. L., and Lindstrom, J., 1985, Evidence for unpredicted transmembrane domains in acetylcholine receptor subunits, *Proc. Nat. Acad. Sci. U.S.A.* **82:**2004–2008.

Crick, F. H. C., 1953, The packing of α-helices: Simple coiled coils, *Acta Crystallogr.* **6:**689–697.

Crippen, G. M., 1977a, A statistical approach to the calculation of conformation of proteins. 1. Theory, *Macromolecules* **10:**21–25.

Crippen, G. M., 1977b, A statistical approach to the calculation of conformation of proteins. 2. The reoxidation of reduced trypsin inhibitor, *Macromolecules* **10:**25–28.

Crippen, G., M., 1977c, A novel approach to the calculation of conformation: Distance geometry, *J. Comp. Physiol.* **26:**449–452.

Crippen, G. M., 1978, The tree structural organization of proteins, *J. Mol. Biol.* **126:**315–332.

Crippen, G. M., and Kuntz, I. D., 1978, A survey of atom packing in globular proteins, *Int. J. Peptide Protein Res.* **12:**47–56.

Davies, B. D., and Tai, R. C., 1980, The mechanism of protein secretion across membranes, *Nature* **283:**433–438.

Davies, D. R., 1964, A correlation between amino acid composition and protein structure *J. Mol. Biol.* **9:**605–609.

Dayhoff, M. O., 1972, *Atlas of Protein Sequence and Structure,* National Biomedical Research Foundation, Washington.

Dayhoff, M. O., 1978, *Atlas of Protein Sequence and Structure. Supplement 3,* National Biomedical Research Foundation, Washington.

Dayhoff, M. O., Barker, W. C., and Hunt, L. T., 1983, Establishing homologies in protein sequences, *Methods Enzymol.* **91:**524–545.

DeGrado, W. F., Kezdy, E. J., and Kaiser, E. T., 1981, Design, synthesis and characterization of a cytotoxic peptide with melittin-like activity, *J. Am. Chem. Soc.* **103:**679–681.

de Groot, R. J., Luytjes, W., Horzinek, M. C., van der Zeijst, B. A. M., Spaan, W. J. M., and Lenstra, J. A., 1987, Evidence for a coiled-coil structure in spike proteins of coronaviruses, *J. Mol. Biol.* **196:**963–966.

Deisenhofer, J., Epp, O., Mikki, K., Huber, R., and Michel, H., 1984, X-ray structural analysis of a membrane protein complex electron density map at 3 Å resolution and a model of the chromophores of the photosynthetic reaction centre from *Rhodopseudomonas viridis, J. Mol. Biol.* **180:**385–398.

Deisenhofer, J., Epp, O., Mikki, K., Huber, R., and Michel, H., 1985, Structure of the protein subunits in the photoreaction centre of *Rhodopseudomonas viridis* at 3Å resolution, *Nature* **318:**618–624.

Deléage, G., and Roux, B., 1987, An algorithm for protein secondary structure prediction based on class prediction, *Protein Eng.* **1:**289–294.

Deléage, G., Tinland, B., and Roux, B., 1987, A computerized version of the Chou and Fasman method for predicting the secondary structure of proteins, *Anal. Biochem.* **165:**200–207.

Dencher, N. A., and Heyn, M. P., 1978, Formation and properties of bacteriorhodopsin monomers in the non-ionic detergents octyl-β-glucoside and Triton X-100, *FEBS Lett.* **96:**322–396.

DeSantis, P., Giglio, E., Liquori, A. M., and Ripamonti, A., 1965, Van der Waals interaction and the stability of helical polypeptide chains, *Nature* **206:**456–458.

DesJarlais, R. L., Sheridan, R. P., Dixon, J. S., Kuntz, I. D., and Venkataraghavan, R., 1986, Docking flexible ligands to macromolecular receptors by molecular shape, *J. Med. Chem.* **29:**2149–2153.

Devereaux, J., Haeben, P., and Smithies, O., 1984, A comprehensive set of sequence analysis programs for the VAX, *Nucleic Acids Res.* **17:**387–395.

Devillers-Thiery, A., Giraudet, J., Bentaboulet, M., and Changeux, J.-P., 1983, Complete mRNA coding sequence of the acetylcholine binding α-subunit of *Torpedo marmorata* acetylcholine receptor: A model for transmembrane organization of the polypeptide chain, *Proc. Natl. Acad. Sci. U.S.A.* **80:** 2067–2071.

Dickerson, R. E., Takano, T., Eisenberg, D., Kallai, O. B., Samson, L., and Cooper, A., 1971, Ferricytochrome c: General features of the horse and bonito proteins at 2.8 Å resolution, *J. Biol. Chem.* **246:** 1511–1535.

Dill, K. A., 1985, Theory for the folding and stability of globular proteins, *Biochemistry* **24:**1501–1509.

Dixon, R. A., Kobilka, B. K., Strader, D. J., Benovic, J. L., Dohlman, H. G., Frielle, T., Bolanowski, M. A., Bennet, C. D., Rands, E., Diehl, R. E., Mumford, R. A., Slater, E. E., Sigal, I. S., Caron, M. G., Lefkowitz, R. J., and Strader, C. D., 1986, Cloning of the gene and cDNA mammalian β-adrenergic receptor and homology with rhodopsin, *Nature* **321:**75–79.

Dohlman, H. G., Caron, M. G., and Lefkowitz, R. L., 1987, A family of receptors coupled to guanine nucleotide regulatory proteins, *Biochemistry* **26:**2657–2664.

Drexler, K. E., 1980, Molecular engineering: An approach to the development of general capabilities for molecular manipulation, *Proc. Natl. Acad. Sci. U.S.A.* **78:**5275–5278.

Dufton, M. J., and Hider, R. C., 1977, Snake toxin secondary structure predictions. Structure activity relationships, *J. Mol. Biol.* **115:**177–193.

Duncan, T. M., Parsonage, D., and Senior, A. E., 1986, Structure of the nucleotide-binding domain in the β-subunit of *Escherichia coli* F_1-ATPase, *FEBS Lett.* **208:**1–6.

Dunhill, P., 1968, The use of helical net-diagrams to represent protein structures, *Biophys. J.* **8:**865–875.

Ecker, J. G., and Kupferschmid, M., 1982, *Report OR,* Rensselaer Polytechnic, Troy, NY.

Edelman, G. M., Cunningham, B. A., Reeke, G. N., Jr., Becker, J. W., Waxdall, M. J., and Wang, J. L., 1972, The covalent and three-dimensional structure of concanavalin A, *Proc. Natl. Acad. Sci. U.S.A.* **69:** 2580–2584.

Edman, J. C., Ellis, L., Blacher, R. W., Roth, R. A., and Rutter, W. J., 1985, Sequence of protein disulphide isomerase and implications of its relation to thioredoxin, *Nature* **317:**267–270.

Edmonds, D. T., 1985, The α-helix dipole in membranes: A new gating mechanism for ion channels, *Eur. Biophys. J.* **13:**31–35.

Edsall, J. Y., and McKenzie, H. A., 1983, Water and proteins II. The location and dynamics of water in protein systems and its relation to their stability and properties, *Adv. Biophysics* **16:**53–183.

Edwards, M. S., Sternberg, M. J. E., and Thornton, J. M., 1987, Structure and sequence patterns in the loops of βαβ units, *Protein Eng.* **1:**173–181.

Efimov, A. V., 1977, Stereochemistry of the packing of α-helices and the β-structure in a compact globule, *Dokl. Akad. Nauk SSSR* **235:**699–702.

Efimov, A. V., 1979, Packing of α-helices in globular proteins. Layer-structure of globular hydrophobic cores, *J. Mol. Biol.* **134:**23–46.

Efimov, A. V., 1982a, Role of constrictions in formation of protein structures containing four helical regions, *Mol. Biol.* **16:**271–281.

Efimov, A. V., 1982b, Super-secondary structures of β-proteins, *Mol. Biol.* **16:**799–806.

Efimov, A. V., 1984, A novel super-secondary structure of proteins and the relation between the structure and amino acid sequence, *FEBS Lett.* **166:**33–38.

Efimov, A. V., 1985, Standard conformations of polypeptide chains in irregular regions of proteins, *Mol. Biol.* **20:**350–360.

Efimov, A. V., 1986a, Standard structures in protein molecules. I. α–β Hairpins, *Mol. Biol.* **20:**329–339.

Efimov, A. V., 1986b, Standard structures in protein molecules. II. β–α Hairpins, *Mol. Biol.* **20:**340–345.

Eisenberg, D., 1984, Three-dimensional structure of membrane surface proteins, *Annu. Rev. Biochem.* **53:**595–623.

Eisenberg, D., Weiss, R. M., Terwilliger, T. C., and Wilcox, W., 1982a, Hydrophobic moments and protein structure, *Faraday Symp. Chem. Soc.* **17:**109–120.

Eisenberg, D., Weiss, R. M., and Terwilliger, T. C., 1982b, The helical hydrophobic moment: A measure of the amphilicity of a helix, *Nature* **299:**371–374.

Eisenberg, D., Schwartz, E., Komaromy, M., and Wall, R., 1984a, Analysis of membrane and surface protein sequences with the hydrophobic moment plot, *J. Mol. Biol.* **179:**125–142.

Eisenberg, D., Weiss, R. M., and Terwilliger, T. C., 1984b, The hydrophobic moment detects periodicity in protein hydrophobicity, *Proc. Natl. Acad. Sci. U.S.A.* **81:**140–144.

Elleman, T. C., Azad, A. A., and Ward, C. W., 1982, Neuraminidase gene from early Asian strain human influenza virus, A/RI/5$^-$/57(H2N2), *Nucleic Acids Res.* **10:**7005–7015.

Emr, S. D., and Silhavy, T. J., 1983, Importance of secondary structure in the signal sequence for protein secretion, *Proc. Natl. Acad. Sci. U.S.A.* **80:**4599–4603.

Engelman, D. M., and Steitz, T., 1981, The spontaneous insertion of proteins into and across membranes: The helical hairpin hypothesis, *Cell* **23:**411–422.

Engelman, D. M., and Steitz, T. A., 1984, On the folding and insertion of globular membrane proteins, in: *The Protein Folding Problem* (D. Wetlaufer, ed.), Westview Press, Boulder, CO, pp. 87–113.

Engelman, D. M., and Zaccai, G., 1980, Bacteriorhopdopsin is an inside-out protein, *Proc. Natl. Acad. Sci. U.S.A.* **77:**5894–5898.

Engelman, D. M., Henderson, R., McLachlan, A. D., and Wallace, B. A., 1980, Path of the polypeptide in bacteriorhodopsin, *Proc. Natl. Acad. Sci. U.S.A.* **77:**2023–2027.

Engelman, D. M., Goldman, A., and Steitz, T., 1986, The identification of helical segments in the polypeptide chain of bacteriorhodopsin, *Methods Enzymol.* **88:**81–88.

Engelman, D. M., Steitz, T. A., and Goldman, A., 1986, Identifying non-polar transbilayer helices in amino acid sequences of membrane proteins, *Annu. Rev. Biophys. Biophys. Chem.* **15:**321–353.

Epand, R. M., 1971, Studies on the conformation of glucagon, *Can. J. Biol. Chem.* **49:**166–169.

Erni, B., and Zanolari, B., 1986, Glucose-permease of the bacterial phosphotransferase system, *J. Biol. Chem.* **261:**16398–16403.

Esipova, N. G., and Tumanyan, V. G., 1972, Factors determining the formation of the tertiary structure of globular protein, *Mol. Biol.* **6:**840–850.

Fasman, G. D., 1980, Prediction of protein conformation from the primary structure, *Ann. N.Y. Acad. Sci.* **348:** 147–159.

Fasman, G. D., 1982, Prediction of the secondary structure of proteins, in: *From Cyclotrons to Cytochromes. Essays in Molecular Biology and Chemistry* (N. O. Kaplan and A. Robinson, eds.), Academic Press, New York, pp. 455–468.

Fasman, G. D., 1985, A critique of the utility of the prediction of protein secondary structure. International Symposium on Biomolecular Structure and Interactions, *J. Biosci.* **8:**15–23.

Fasman, G. D., 1987, The road from poly-α-amino acids to the prediction of protein conformation. Biopolymers and biotechnology symposium in honor of Prof. Ephraim Katzir on his 70th birthday, *Biopolymers* **26:** 559–579.

Fasman, G. D., and Chou, P. Y., 1974, Prediction of protein conformation: Consequences and aspirations, in: *Peptides, Polypeptides, and Proteins* (E. R. Blout, F. A. Bovey, M. Goodman, and N. Lotan, eds.), John Wiley & Sons, New York, pp. 114–125.

Fasman, G. D., Chou, P. Y., and Adler, A. J., 1976, Prediction of the conformation of the histones, *Biophys. J.* **16:**1201–1238.

Fasman, G. D., Chou, P. Y., and Adler, A. J., 1977, Histone conformation: Predictions and experimental studies, in: *The Molecular Biology of the Mammalian Genetic Apparatus*—I, (P. O. P. Ts'o, ed.), Elsevier–Exerpta Medica/North Holland, Amsterdam, pp. 1–52.

Finer-Moore, J., and Stroud, R. M., 1984, Amphipathic analysis and possible formation of the ion channel in acetylcholine receptor, *Proc. Natl. Acad. Sci. U.S.A.* **81:**155–159.

Finkelstein, A. V., Ptitsyn, O. B., and Bendsko, P., (1970), Coiling and topology on the anti-parallel β-structure, *Biofisika* **24:**21–26.

Finney, J. L., Gellatly, B. J., Golton, I. C., and Goodfellow, J., 1980, Solvent effects and polar interactions in the structural stability and dynamics of globular proteins, *Biophys. J.* **32:**17–23.

Fishleigh, R. V., Robson, B., Garnier, J., and Finn, P. W., 1987, Studies on rationales for an expert system approach to the interpretation of protein sequence data. Preliminary analysis of the human epidermal growth factor receptor, *FEBS Lett.* **214:**219–225.

Fitch, W. M., 1966a, The relation between frequencies of amino acids and ordered trinucleotides, *J. Mol. Biol.* **16:**1–8.

Fitch, W. M., 1966b, An improved method of testing for evolutionary homology, *J. Mol. Biol.* **16:**9–16.

Fitch, W. M., 1966c, Evidence suggesting a partial, internal duplication in the ancestral gene for heme-containing globins, *J. Mol. Biol.* **16:**17–27.

Fitch, W. M., and Smith, T. F., 1983, Optimal sequence alignments, *Proc. Natl. Acad. Sci. U.S.A.* **80:**1382–1386.

Fleming, P. J., Dailey, H. A., Corcoran, D., and Strittmatter, P., 1978, The primary structure of the non-polar segment of bovine cytochrome b_5, *J. Biol. Chem.* **253:**5369–5372.

Flinta, C., von Heijne, G., and Johansson, J., 1983, Helical sidedness and the distribution of polar residues in trans-membrane helices, *J. Mol. Biol.* **168:**193–196.

Foster, D. L., Boublik, M., and Kaback, H. R., 1983, Structure of the *lac* carrier protein of *Escherichia coli*, *J. Biol. Chem.* **258:**31–34.

Froimowitz, M., and Fasman, G. D., 1974, Prediction of secondary structure of proteins using the helix-coil transition theory, *Macromolecules* **7:**583–589.

Fukushima, D. Kupferberg, J. P., Yokoyama, S., Kroon, D. J., Kaiser, E. T., and Kezdy, F. J., 1979, A synthetic amphiphilic helical docosapeptide with the surface properties of plasma apolipoprotein A-I, *J. Am. Chem. Soc.* **101:**3703–3704.

Furois-Corbin, S., and Pullman, A., 1987, Theoretical studies of the packing of α-helices into possible transmembrane bundles: Sequences including alanines, leucines and serines, *Biochim. Biophys. Acta* **902:**31–45.

Garnier, J., Osguthorpe, D. J., and Robson, B., 1978, Analysis of the accuracy and implications of simple methods for predicting the secondary structure of globular proteins, *J. Mol. Biol.* **120:**97–120.

Garratt, R. C., Taylor, W. R., and Thornton, J. M., 1985, The influence of tertiary structure on secondary structure prediction. Accessibility versus predictability for β-structure, *FEBS Lett.* **188:**59–62.

Geisow, M. J., and Roberts, R. D. B., 1980, Amino acid preferences for secondary structure vary with protein class, *Int. J. Biol. Macromol.* **2:**387–389.

Geisow, M. J., Fritsche, U., Hexham, J. M., Dash, B., and Johnson, T., 1986, A consensus amino acid sequence repeat in *Torpedo* and mammalian Ca^{2+}-dependent membrane-binding proteins, *Nature* **320:**636–638.

Gelin, B., and Karplus, M., 1979, Side-chain torsional potentials: Effect of dipeptide, protein, and solvent environment, *Biochemistry* **18:**1256–1268.

Gerber, G. E., Anderegg, R. J., Herlihy, W. C., Gray, C. P., Bieman, K., and Khorana, H. G., 1979, Partial primary structure of bacteriorhodopsin: Sequencing methods for membrane proteins, *Proc. Natl. Acad. Sci. U.S.A.* **76:**227–231.

Getzoff, E. D., Tainer, J. A., and Olson, A. J., 1986, Recognition and interactions controlling the assemblies of β-barrel domains, *Biophys. J.* **49:**191–206.

Ghelis, C., and Yon, J., 1982, *Protein Folding*, Academic Press, New York.

Gibrat, J.-F., Garnier, J., and Robson, B., 1987, Further developments of protein secondary structure prediction using information theory: New parameters and consideration of residue pairs, *J. Mol. Biol.* **198:**425–443.

Gilbert, W., 1978, Why genes in pieces? *Nature* **271:**501.

Gilson, M. K., and Honig, B. H., 1987, Calculation of electrostatic potentials in an enzyme active site, *Nature* **330:**84–86.

Glaeser, R. M., and Jap, B. K., 1985, Absorption flattening in the circular dichroism spectra of small membrane fragments, *Biochemistry* **24:**6398–6401.

Goad, W. B., and Kanehisa, M. I., 1982, Pattern recognition in nucleic acid sequences I. A general method for finding local homologies and symmetries, *Nucleic Acids Res.* **10:**247–263.

Gordon, D. J., and Holzworth, G., 1971, Artifacts in the measure of optical activity of membrane suspensions, *Arch. Biochem. Biophys.* **142:**481–488.

Grantham, R., 1974, Amino acid difference formula to help explain protein evolution, *Science* **185:**862–864.

Gratzer, W. B., Bailey, E., and Beaven, G. H., 1967, Conformational states of glucagon, *Biochem. Biophys. Res. Commun.* **28:**914–919.

Gray, T. M., and Matthews, B. W., 1984, Intrahelical hydrogen bonding of serine, threonine, and cysteine residues within α-helices and its relevance to membrane-bound proteins, *J. Mol. Biol.* **175:**75–81.

Green, N. M., and Flanagan, M. T., 1976, The prediction of the conformation of membrane proteins from the sequence of amino acids, *Biochem. J.* **153:**729–732.

Grenningloh, G., Rienitz, A., Schmitt, B., Methfessel, C., Zensen, M., Beyreuther, K., Grundelfinger, E. D., and Betz, H., 1987, The strychnine-binding subunit of the glycine receptor shows homology with nicotinic acetylcholine receptors, *Nature* **328:**215–220.

Gribskov, M., Burgess, R. R., and Devereaux, J., 1986, PEPPLOT, a protein secondary analysis program for the UWGCG sequence analysis software package, *Nucleic Acids Res.* **14:**327–334.

Gribskov, M., McLachlan, A. D., and Eisenberg, D., 1987, Profile analysis: Detection of distantly related proteins, *Proc. Natl. Acad. Sci. U.S.A.* **84:**4355–4358.

Gutte, B., Daumigen, M., and Wittschieber, E., 1979, Design, synthesis and characteristics of a 34-residue polypeptide that interacts with nucleic acids, *Nature* **281:**650–655.

Guy, H. R., 1981, Structural models of the nicotinic acetylcholine receptor and its toxin-binding sites, *Cell. Mol. Neurobiol.* **1:**231–258.

Guy, H. R., 1984, A structural model of the acetylcholine receptor channel based on partition energy and helix packing calculations, *Biophys. J.* **45:**249–261.

Guy, H. R., 1985, Amino acid side-chain partition energies and distribution of residues in soluble proteins, *Biophys. J.* **47:**61–70.

Guy, H. R., and Seetharamulu, P., 1986, Molecular model of the action potential sodium channel, *Proc. Natl. Acad. Sci. U.S.A.* **83:**508–512.

Guzzo, A. V., 1965, The influence of amino acid sequence on protein structure, *Biophys. J.* **5:**809–822.

Haber, J. E., and Koshland, Jr., D. E., 1970, An evaluation of the relatedness of proteins based on comparison of amino acid sequences, *J. Mol. Biol.* **50:**617–639.

Hager, K. M., Mandala, S. M., Davenport, J. W., Speicher, D. W., Benz, Jr., E. J., and Slayman, C. W., 1986, Amino acid sequences of the plasma membrane ATPase of *Neurospora crassa:* Deduction from genomic and c-DNA sequences, *Proc. Natl. Acad. Sci. U.S.A.* **83:**7693–7697.

Hagler, A. T., and Honig, B., 1978, On the formation of protein tertiary structure on a computer. *Proc. Natl. Acad. Sci. U.S.A.* **75:**554–558.

Hagler, A. T., Huler, E., and Lifson, S., 1974, Energy functions for peptides and proteins. I. Derivation of a consistent force field including the hydrogen bond from amide crystals, *J. Am. Chem. Soc.* **96:**5319–5327.

Hammett, L. P., 1970, *Physical Organic Chemistry,* 2nd ed., McGraw-Hill, New York.

Hardman, K. D., and Ainsworth, C. F., 1972, Structure of concanavalin A at 2.4 Å resolution, *Biochemistry* **11:**4910–4919.

Harrison, S. C., 1985, Two for the price of one, *Nature* **313:**736–737.

Havel, T. F., Kuntz, I. D., and Crippen, G. M., 1983, The theory and practice of distance geometry, *Bull. Math. Biol.* **45:**665–720.

Hayes, T. G., 1980, Chou–Fasman analysis of the secondary structure of F and LE interferons, *Biochem. Biophys. Res. Commun.* **95:**872–879.

Hayward, S. B., and Stroud, R. M., 1981, Projected structure of purple membrane determined to 3.7 Å resolution by low temperature electron microscopy, *J. Mol. Biol.* **151:**491–517.

Heber-Katz, E., Hollosi, M., Dietzschold, B., Hudecz, F., and Fasman, G. D., 1985, The T cell response to the glycoprotein D of the herpes simplex virus: The significance of antigen conformation, *J. Immunol.* **135:** 1385–1390.

Hedrick, S. M., Cohen, D. I., Nielsen, E. A., and Davis, M. M., 1984a, Isolation of cDNA clones encoding T-cell-specific membrane associated proteins, *Nature* **308:**149–153.

Hedrick, S. M., Nielsen, E. A., Kavaler, J., Cohen, D. I., and Davis, M. M., 1984b, Sequence relationships between putative T-cell receptor polypeptides and immunoglobulins, *Nature* **308:**153–158.

Hellberg, S., Sjostrom, M., and Wold, S., 1986, The prediction of bradykinin potentiating potency of pentapeptides. An example of a peptide quantitative structure–activity relationship, *Acta Chem. Scand.* **B40:**135–140.

Henderson, R., and Unwin, P. N. T., 1975, Three-dimensional model of purple membrane obtained by electron microscopy, *Nature* **257:**28–32.

Hol, W. G. J., van Duijnen, P. T., and Berendsen, H. J. C., 1978, The α-helix dipole and the properties of proteins, *Nature* **273:**443–446.

Hol, W. G. J., Halie, L. M., and Sander, C., 1981, Dipoles of the α-helix and β-sheet: Their role in protein folding, *Nature* **294:**532–536.

Holmes, M. A., and Matthews, B. W., 1982, Structure of thermolysin refined at 1.6 Å resolution, *J. Mol. Biol.* **160:**623–639.

Hones, J., Jany, K.-D., Pfleider, G., and Wagner, A. F. V., 1987, An integrated prediction of secondary, tertiary, and quaternary structure of glucose dehydrogenase, *FEBS Lett.* **212:**193–198.

Honig, B. H., and Hubbell, W. L., 1984, Stability of "salt bridges" in membrane proteins, *Proc. Natl. Acad. Sci. U.S.A.* **81:**5412–5416.

Honig, B. H., Ray, A., and Levinthal, C., 1976, Conformational flexibility and protein folding: Rigid structural fragments connected by flexible joints in subtilisn BPN, *Proc. Natl. Acad. Sci. U.S.A.* **73:**1974–1978.

Honig, B. H., Hubbell, W. L., and Flewelling, R. F., 1986, Electrostatic interactions in membranes and proteins, *Annu. Rev. Biophys. Biophys. Chem.* **15:**163–193.

Hopp, T. P., and Woods, K. R., 1981, Prediction of protein antigenic determinants from amino acid sequence, *Proc. Natl. Acad. Sci. U.S.A.* **78:**3824–3828.

Hruby, W., Krstenansky, J., Gysin, B., Pelton, J. T., Trivedi, D., and McKee, R. L., 1986, Conformational considerations in the design of glucagon agonists and antagonists: Examination using synthetic analogs, *Biopolymers* **25:**S135–S155.

Huang, K.-S., Bayley, H., Liao, M.-J., London, E., and Khorana, H. G., 1981, Refolding of an integral membrane protein. Denaturation, renaturation, and reconstitution of intact bacteriorhodopsin and two proteolytic fragments, *J. Biol. Chem.* **256:**3802–3809.

Huang, K.-S., Radhakrishnan, R., Bayley, H., and Khorana, H. G., 1982, Orientation of retinal in bacteriorhodopsin as studied by cross-linking using a photosensitive analog of retinal, *J. Biol. Chem.* **257:** 13616–13623.

Huber, R., Kulka, D., Ruhlman, A., and Steigman, W., 1971, Pancreatic trypsin inhibitor (Kunitz) Part 1. Structure and function, *Cold Spring Harbor Symp. Quant. Biol.* **36:**141–150.

Hucho, F., 1986, The nicotinic acetylcholine receptor and its ion channel, *Eur. J. Biochem.* **158:**211–256.

Hurle, M. R., Matthews, C. R., Cohen, F. E., Kuntz, I. D., Toumadje, A., and Johnson, Jr., W. C., 1987, Prediction of the tertiary structure of the α-subunit of tryptophan synthetase, *Proteins* **2:**210–224.

IntelliGenetics, Inc., 1981–1985. *PER References Manual,* IntelliGenetics, Mountain View, CA.

Isogai, Y., Nemethy, G., Rackovsky, S., Leach, S. J., and Scheraga, H. A., 1980, Characterization of multiple bends in proteins, *Biopolymers* **19:**1183–1210.

Jaenicke, R., 1984, Protein folding and protein association, *Angew. Chem. [Engl.]* **23:**395–413.

Jaenicke, R., 1987, Folding and association of proteins, *Prog. Biophys. Mol. Biol.* **49:**117–237.

Janin, J., 1979, Surface and inside volumes in globular proteins, *Nature* **277:**491–492.

Janin, J., and Chothia, C., 1980, Packing of α-helices onto β-pleated sheets and anatomy of α/β proteins, *J. Mol. Biol.* **143:**95–128.

Jap, B. K., and Kong, S. H., 1986, Secondary structure of halorhodopsin, *Biochemistry* **25:**502–505.

Jap, B. K., Maestre, M. F., Hayward, S. B., and Glaeser, R. M., 1983, Peptide-chain secondary structure of bacteriorhodopsin, *Biophys. J.* **43:**81–89.

Jones, D. D., 1975, Amino acid properties and side-chain orientation in proteins: A cross correlation approach, *J. Theor. Biol.* **50:**167–183.

Kabat, E. A., and Wu, T. T., 1973a, The influence of nearest-neighbor amino acids on the conformation of the middle amino acid in proteins: Comparison of predicted and experimental determination of β-sheets in conconavalin A, *Proc. Natl. Acad. Sci. U.S.A.* **70:**1473–1477.

Kabat, E. A., and Wu, T. T., 1973b, The influence of nearest-neighboring amino acid residues on aspects of secondary structure of proteins. Attempts to locate α-helices and β-sheets, *Biopolymers* **12:**751–774.

Kabat, E. A., Wu, T. T., Bilofsky, H., Reid-Miller, M., and Perry, H., 1983, *Sequences of Immunological Interest,* U.S. Dept. of Health and Human Services, Washington.

Kabsch, W., and Sander, C., 1983a, How good are predictions of protein secondary structure? *FEBS Lett.* **155:** 179–182.

Kabsch, W., and Sander, C., 1983b, Dictionary of protein secondary structure: Pattern recognition of hydrogen-bonded and geometric features, *Biopolymers* **22:**2577–2637.

Kabsch, W., and Sander, C., 1984, On the use of sequence homologies to predict protein structure: Identical pentapeptides can have completely different conformations, *Proc. Natl. Acad. Sci. U.S.A.* **81:**1075–1078.

Kabsch, W., and Sander, C., 1985, Identical pentapeptides with different backbones, *Nature* **317:**207.

Kanazawa, H., Hama, H., Rosen, B. P., and Futai, M., 1985, Deletion of seven amino acids from the γ subunit of *Escherichia coli* H^+-ATPases causes total loss of F_1 assembly on membrane, *Arch. Biochem. Biophys.* **241:**364–370.

Karplus, M., and Weaver, D. L., 1979, Diffusion–collision model for protein folding, *Biopolymers* **18:**1421–1437.

Karplus, P. A., and Schulz, G. E., 1985, Prediction of chain flexibility in proteins, *Naturwissenschaften* **72:** 212–213.

Kauzmann, W., 1959, Some factors in the interpretation of protein denaturation, *Adv. Protein Chem.* **14:**1–63.

Kawakami, K., Noguchi, S., Noda, M., Takahashi, H., Ohta, T., Kawamura, M., Nojima, H., Nagano, K., Hirose, T., Inayama, S., Hayashida, H., Miyata, T., and Numa, S., 1985, Primary structure of the α-subunit of *Torpedo californica* (Na^+ + K^+)ATPase deduced from cDNA sequence, *Nature* **316:** 733–736.

Kelly, L., and Holladay, L. A., 1987, Comparison of scales of amino acid side chain properties by conservation during evolution of four proteins, *Protein Eng.* **1:**137–140.

Kim, P. S., and Baldwin, R. L., 1982, Specific intermediates in the folding reactions of small proteins and the mechanism of protein folding, *Annu. Rev. Biochem.* **51:**459–489.

Klapper, I., Hagstrom, R., Fine, R., Sharp, K., and Honig, B., 1986, Focussing of electric fields in the active site of Cu–Zn superoxide dismutase: Effects of ionic strength and amino acid modification, *Proteins* **1:**47–59.

Kleffel, B., Garavito, R. M., Baumeister, W., and Rosenbusch, J. P., 1985, Secondary structure of a channel-forming protein: Porin from *E. coli* outer membrane, *EMBO J.* **4:**1589–1592.

Klein, P., 1986, Prediction of protein structural class by discriminant analysis, *Biochim. Biophys. Acta* **874:** 205–215.

Klein, P., and DeLisi, C., 1986, Prediction of protein structural class from amino acid sequence, *Biopolymers* **25:**1659–1672.

Klein, P., Kanehisa, M., and DeLisi, C., 1984, Prediction of protein function from sequence properties. Discriminant analysis of a data base, *Biochim. Biophys. Acta* **787:**221–226.

Klein, P., Kanehisa, M., and DeLisi, C., 1985, The detection of membrane-spanning proteins, *Biochim. Biophys. Acta* **815:**468–476.

Klein, P., Jacquez, J. A., and DeLisi, C., 1986, Prediction of protein function by discriminant analysis, *Math. Biosci.* **81:**177–189.

Kneale, G. G., and Bishop, M. J., 1985, Nucleic acid and protein sequence databases, *Cabios Rev.* **1:**11–17.

Kolaskar, A. S., Ramabrahmam, V., and Soman, K. V., 1980, Reversal of polypeptide chain in globular proteins, *Int. J. Peptide Protein Res.* **16:**1–11.

Kolb, E., Hudson, P. J., and Harris, J. I., 1980, Phosphofructokinase: Complete amino acid sequence of the enzyme from *Bacillus stearothermophilus, Eur. J. Biochem.* **108:**587–597.

Kopito, R. R., and Lodish, H. F., 1985, Primary structure and transmembrane orientation of the murine anion exchange protein, *Nature* **316:**234–238.

Kopito, R. R., Andersson, M., and Lodish, H. F., 1987, Structure and organization of the murine band 3 gene, *J. Biol. Chem.* **262:**8035–8040.

Kosower, E. M., 1982, in: *International Symposium on Structure and Dynamics of Nucleic Acids and Protein,* pp. 52–53.

Kosower, E. M., 1983a, Partial tertiary structure assignment for the acetylcholine receptor on the basis of the hydrophobicity of amino acid sequences and channel location using single group rotation theory, *Biochem. Biophys. Res. Commun.* **111:**1022–1024.

Kosower, E. M., 1983b, Partial tertiary structure assignments of the β_1-, γ-, and δ subunits of the acetylcholine receptor on the basis of the hydrophobicity of amino acid sequences and channel location using single group theory, *FEBS Lett.* **155:**245–247.

Kosower, E. M., 1987, A structural and dynamic model for the nicotinic acetylcholine receptor, *Eur. J. Biochem.* **168:**431–449.

Kotelchuck, D., and Scheraga, H. A., 1968, The influence of short-range interactions on protein conformation. I. Side-chain–backbone interactions with a single peptide unit, *Proc. Natl. Acad. Sci. U.S.A.* **61:**1163–1170.

Kotelchuck, D., and Scheraga, H. A., 1969, The influence of short-range interactions on protein conformation II. A model for predicting the α-helical regions of proteins, *Proc. Natl. Acad. Sci. U.S.A.* **62:**14–21.

Krchnak, V., Mach, O., and Maly, A., 1987, Computer prediction of potential immunogenic determinants from protein amino acid sequence, *Anal. Biochem.* **165:**200–207.

Krebs, K. E., and Phillips, M. C., 1984, The contribution of α-helices to the surface activities of proteins, *FEBS Lett.* **175:**263–266.

Kubo, T., Fukuda, K., Mikami, A., Maeda, A., Takahashi, H., Mishina, M., Hoga, T., Haga, K., Ichiyama, A., Kangawa, K., Kojima, M., Matsuo, H., Hirose, T., and Numa, S., 1986, Cloning, sequencing and expression of complementary DNA encoding the muscarinic acetylcholine receptor, *Nature* **323:**411–416.

Kubota, Y., Takahashi, S., Nishikawa, K., and Ooi, T., 1981, Homology in protein sequences expressed by correlation coefficients, *J. Theor. Biol.* **91:**347–361.

Kuhn, L. A., and Leigh, J. S., Jr., 1985, A statistical technique for predicting membrane protein structure, *Biochim. Biophys. Acta* **828:**351–361.

Kuntz, I. D., 1972, Protein folding, *J. Am. Chem. Soc.* **94:**4009–4012.

Kuntz, I. D., 1975, An approach to the tertiary structure of globular proteins, *J. Am. Chem. Soc.* **97:**4362–4366.

Kuntz, I. D., and Crippen, G. M., 1979, Protein densities, *Int. J. Peptide Protein Res.* **13:**223–228.

Kuntz, I. D., Crippen, G. M., Kollman, P. A., and Kimelman, D., 1976, Calculation of protein tertiary structure, *J. Mol. Biol.* **106:**983–994.

Kuntz, I. D., Blaney, J. M., Oatley, S. J., Langridge, R., and Ferrin, T. E., 1982, A geometric approach to macromolecule–ligand interactions, *J. Mol. Biol.* **161:**269–288.

Kyte, J., and Doolittle, R. F., 1982, A simple method for displaying the hydropathic character of a protein, *J. Mol. Biol.* **157:**105–132.

Lathrop, R. H., Webster, T. A., and Smith, T. F., 1987, ARIADNE: Pattern-directed inference and hierarchical abstraction in protein structure recognition, *Commun. ACM* **30:**909–921.

Laursen, R. A., Samiullah, M., and Lees, M. B., 1984, The structure of bovine brain myelin proteolipid and its organization in myelin, *Proc. Natl. Acad. Sci. U.S.A.* **81:**2912–2916.

Lee, B., and Richards, F. M., 1971, An interpretation of protein structures: Estimation of static accessibility, *J. Mol. Biol.* **55:**379–400.

Lee, C. A., and Saier, M. H., Jr., 1983, Mannitol-specific enzyme II of the bacterial phosphotransferase system, *J. Biol. Chem.* **258:**10761–10767.

Lenstra, J. A., 1977, Evolution of secondary structure prediction of proteins, *Biochim. Biophys. Acta* **491:**333–398.

Lenstra, J. A., Hofsteenge, J., and Beintema, J. J., 1977, Invariant features of the structure of pancreatic ribonuclease. A test of different predictive models, *J. Mol. Biol.* **109:**185–193.

Lesk, A. M., and Chothia, C., 1980, How different amino acid sequences determine similar protein structures: The structure and evolutionary dynamics of the globins, *J. Mol. Biol.* **136:**225–270.

Lesk, A. M., and Rose, G. D., 1981, Folding units in globular proteins. *Proc. Natl. Acad. Sci. U.S.A.* **78:** 4304–4308.

Lesk, A., Levitt, M., and Chothia, C., 1986, Alignment of the amino acids sequences of distantly related proteins using variable gap penalties, *Protein Eng.* **1:**77–78.

Leszczynski, J., and Rose, G. D., 1986, Loops in globular proteins: A novel category of secondary structure, *Science* **234:**849–855.

Leung, D. W., Spenser, S. A., Cachianes, G., Hammonds, R. G., Collins, C., Henzel, W. J., Barnard, R., Waters, M. J., and Wood, W. I., 1987, Growth hormone receptor and serum binding protein: Purification, cloning and expression, *Nature* **330:**537–543.

Levin, J. M., Robson, B., and Garnier, J., 1986, An algorithm for secondary structure determination in proteins based on sequence similarity, *FEBS Lett.* **205:**303–308.

Levinthal, C., 1968, Are there pathways for protein folding? *J. Chem. Phys.* **65:**44–45.

Levitt, M., 1974, On the nature of the binding of hexa-N-acetylglucosamine substrate to lysozyme, in: *Peptides, Polypeptides, and Proteins* (E. R. Blout, F. A. Bovey, M. Goodman, and N. Lotan, eds.), John Wiley & Sons, New York, pp. 99–113.

Levitt, M., 1976, A simplified representation of protein conformations for rapid simulation of protein folding, *J. Mol. Biol.* **104:**59–107.

Levitt, M., 1978, Conformational preferences of amino acids in globular proteins, *Biochemistry,* **17:**4277–4285.

Levitt, M., 1983, Protein folding by restrained energy minimization and molecular dynamics, *J. Mol. Biol.* **170:** 723–764.

Levitt, M., and Chothia, C., 1976, Structural patterns in globular proteins, *Nature* **261:**552–558.

Levitt, M., and Greer, J., 1977, Automatic identification of secondary structure in globular proteins, *J. Mol. Biol.* **114:**181–293.

Levitt M., and Lifson, S., 1969, Refinement of protein conformations using a macromolecular energy minimization procedure, *J. Mol. Biol.* **46:**269–279.

Levitt, M., and Warshel, A., 1975, Computer simulation of protein folding, *Nature* **253:**694–698.

Lewis, P. N., and Bradbury, E. M., 1974, Effect of electrostatic interactions on the prediction of helices in proteins, *Biochim. Biophys. Acta* **336:**153–164.

Lewis, P. N., and Scheraga, H. A., 1971, Predictions of structural homologies in cytochrome c proteins, *Arch. Biochem. Biophys.* **144:**576–583.

Lewis, P. N., Go, N., Go, M., Kotelchuck, D., and Scheraga, H. A., 1970, Helix probability profiles of denatured proteins and their correlation with native structures, *Proc. Natl. Acad. Sci. U.S.A.* **65:** 810–815.

Lewis, P. N., Momany, F. A., and Scheraga, H. A., 1971, Folding of polypeptide chains in proteins: A proposed mechanism for folding, *Proc. Natl. Acad. Sci. U.S.A.* **68:**2293–2297.

Lewis, P. N., Momany, F. A., and Scheraga, H. A., 1973a, Chain reversals in proteins, *Biochim. Biophys. Acta* **303:**211–229.

Lewis, P. N., Momany, F. A., and Scheraga, H. A., 1973b, Energy parameters in polypeptides. VI. Conformational energy analysis of the N-acetyl-N′-methyl amides of the twenty naturally occurring amino acids, *Israel J. Chem.* **11:**121–152.

Lifson, S., and Roig, A., 1961, On the theory of helix–coil transitions in polypeptides, *J. Chem. Phys.* **34:** 1963–1974.

Lifson, S., and Sander, C., 1979, Antiparallel and parallel β-strands differ in amino acid residue preferences, *Nature* **282:**109–111.

Lifson, S., and Sander, C., 1980a, Specific recognition in the tertiary structure of β-sheets in proteins, *J. Mol. Biol.* **139:**627–639.

Lifson, S., and Sander, C., 1980b, Composition, cooperativity and recognition in proteins, in: *Protein Folding* (R. Jaenicke, ed.), Elsevier/North-Holland Biomedical Press, Amsterdam, pp. 289–316.

Lifson, S., and Warshel, A., 1968, Consistent force field calculations, vibrational spectra, and enthaipies of cycloalkane and *n*-alkane molecules, *J. Chem. Phys.* **49:**5116–5129.

Liljas, A., and Rossman, M. G., 1974, X-ray studies of protein interactions, *Annu. Rev. Biochem.* **43:**475–507.

Lim, V. I., 1974a, Structural principles of the globular organization of protein chains: A stereochemical theory of globular protein secondary structure, *J. Mol. Biol.* **88:**857–872.

Lim, V. I., 1974b, Algorithms for prediction of α-helices and β-structural regions in globular proteins, *J. Mol. Biol.* **88:**873–894.

Lipman, D. J., and Pearson, W. R., 1985, Rapid and sensitive protein similarity searches, *Science* **227:**1435–1441.

Lisium, L., Finer-Moore, J., Stroud, R. M., Luskey, K. L., Brown, M. S., and Goldstein, J. L., 1985, Domain structure of 3-hydroxy-3-methylglutaryl coenzyme A reductase, a glycoprotein of the endoplasmic reticulum, *J. Biol. Chem.* **260:**522–530.

London, E., and Khorana, H. G., 1982, Denaturation and renaturation of bacteriorhodopsin in detergents and lipid–detergent mixtures, *J. Biol. Chem.* **257:**7003–7011.

Long, M. M., Urry, D. W., and Stoeckenius, W., 1977, Circular dichroism of biological membranes: Purple membrane of *Halobacterium halobium, Biochem. Biophys. Res. Commun.* **75:**725–731.

Lopez, J. A., Chung, D. W., Fujikawa, K., Hagen, F. S., Papayannopoulou, T., and Roth, G. J., 1987, Cloning of the α chain of human platlet glycoprotein 1b: A transmembrane protein with homology to leucine-rich α_2-glycoprotein, *Proc. Natl. Acad. Sci. U.S.A.* **84:**5615–5619.

Loucheux-Lefebvre, M.-H., 1978, Predicted β-turns in peptide and glycopeptide antifreezes, *Biochem. Biophys. Res. Commun.* **81:**1352–1356.

Loucheux-Lefebvre, M.-H., Aubert, J.-P., and Jolles, P., 1978, Prediction of the conformation of the cow and sheep k-caseins, *Biophys. J.* **23:**323–336.

Low, B. W., Lovell, F. M., and Rudko, A. D., 1968, Prediction of α-helical regions in proteins of known sequence, *Proc. Natl. Acad. Sci. U.S.A.* **60:**1519–1526.

MacLennan, D. H., Brandl, C. J., Korczak, B., and Green, N. M., 1985, Amino acid sequences of a Ca^{+} + Mg^{+}-dependent ATPase from rabbit muscle sarcoplasmic reticulum, deduced from its complementary DNA sequence, *Nature* **316:**696–700.

Maizel, J. V., Jr., and Lenk, R. P., 1981, Enhanced graphic matrix analysis of nucleic acid and protein sequence, *Proc. Natl. Acad. Sci. U.S.A.* **78:**7665–7669.

Manavalan, P., and Ponnuswamy, P. K., 1977, A study of the preferred environment of amino acid residues in globular proteins, *Arch. Biochem. Biophys.* **184:**476–487.

Manavalan, P., and Ponnuswamy, P. K., 1978, Hydrophobic character of amino acid residues in globular proteins, *Nature* **275:**673–674.

Mao, D., and Wallace, B. A., 1984, Differential light scattering and absorption flattening optical effects are minimal in circular dichroism spectra of small unilamellar vesicles, *Biochemistry* **23:**2667–2673.

Masu, Y., Nakayama, K., Tamaki, H., Harada, Y., Kuno, M., and Nakanishi, S., 1987, cDNA cloning of bovine substance-K receptor through oocyte expression system, *Nature* **329:**836–838.

Matthew, J. B., 1985, Electrostatic effects in proteins, *Annu. Rev. Biophys. Biophys. Chem.* **14:**387–417.

Matthew, J. B., and Gurd, F. R. N., 1986a, Calculation of electrostatic interactions in proteins, *Methods Enzymol.* **130:**413–436.

Matthew, J. B., and Gurd, F. R. N., 1986b, Stabilization and destabilization of protein structure by charge interactions, *Methods Enzymol.* **130:**437–453.

Matthews, B. W., 1975, Comparison of the predicted and observed secondary structure of T4 phage lysozyme, *Biochim. Biophys. Acta* **405:**442–451.

Matthews, F. S., Argos, P., and Levine, M., 1971, The structure of cytochrome b_5 at 2.0 Å resolution, *Cold Spring Harbor Symp. Quant. Biol.* **36:**387.

Maxfield, F. R., and Scheraga, H. A., 1975, The effect of neighboring charges on the helix forming ability of charged amino acids in proteins, *Macromolecules* **8:**491–493.

Maxfield, F. R., and Scheraga, H. A., 1976, Status of empirical methods for the prediction of protein backbone topography, *Biochemistry* **15:**5138–5153.

Maxfield, F. R., and Scheraga, H. A., 1979, Improvements in the prediction of protein backbone topography by reduction of statistical errors, *Biochemistry* **18:**697–704.

McCammon, J. A., Gelin, B. R., and Karplus, M., 1977, Dynamics of folded proteins. *Nature* **267:**585–590.

McCarthy, M. P., Earnest, J. P., Young, E. F., Choe, S., and Stroud, R. M., 1986, The molecular neurobiology of the acetylcholine receptor, 1986, *Annu. Rev. Neurosci.* **9:**383–413.

McCubbin, W. D., Oikawa, K., and Kay, C. M., 1971, Circular dichroism studies on concanavalin A, *Biochem. Biophys. Res. Commun.* **43:**666–674.

McGregor, M. J., Islam, S. A., and Sternberg, M. J. E., 1987, Analysis of the relationship between side-chain conformation and secondary structure in globulary proteins, *J. Mol. Biol.* **198:**295–310.

McLachlan, A. D., 1971, Tests for comparing related amino-acid sequences. Cytochrome c and cytochrome c_{551}, *J. Mol. Biol.* **61:**409–424.

McLachlan, A. D., 1977, Quantum chemistry and protein folding: The art of the possible, *Int. J. Quant. Chem.* **13**(Suppl. 1):371–385.

McLachlan, A. D., and Karn, J., 1983, Periodic features in the amino acid sequence of nemotode myosin rod, *J. Mol. Biol.* **164:**605–626.

McLachlan, A. D., and Stewart, M., 1976, The 14-fold periodicity in α-tropomyosin and the interaction with actin, *J. Mol. Biol.* **103:**271–298.

McLachlan, A. D., Bloomer, A. C., and Butler, P. J. G., 1980, Structural repeats and evolution of tobacco mosaic virus coat protein and RNA, *J. Mol. Biol.* **136:**203–224.

Meirovitch, S., Rackovsky, S., and Scheraga, H. A., 1980, Empirical studies of hydrophobicity. 1. Effect of protein size on the hydrophobic behavior of amino acids, *Macromolecules* **13:**1398–1405.

Menick, D. R., Carrasco, N., Antes, L., Patel, L., and Kaback, H. R., 1986, *Lac* permease of *Escherichia coli:* Arginine-302 as a component of the postulated proton relay, *Biochemistry* **26:**6638–6644.

Mercier, J.-C., and Chobert, J. M., 1976, Comparative study of the amino acid sequences of the caseinomacropeptides from seven species, *FEBS Lett.* **72:**208–214.

Mercier, J.-C., Uro, J., Ribadeau-Daumas, B., and Grosclaude, F., 1972, Structure primaire du caséinomacropeptide de la caséine k_{β_1} bovine, *Eur. J. Biochem.* **27:**535–547.

Miles, E. W., Yutani, K., and Ogarsahara, K., 1982, Guanidine hydrochloride-induced unfolding of the α-subunit of tryptophan synthetase and of the two α-proteolytic fragments. Evidence for stepwise unfolding of the two α domains, *Biochemistry* **21:**2586–2592.

Milner-White, E. J., 1988, Recurring loop motif in proteins that occurs in right-handed and left-handed forms. Its relationship with alpha-helices and beta-bulge loops, *J. Mol. Biol.* **199:**503–511.

Milner-White, E. J., and Poet, R., 1986, Four classes of β-hairpins in proteins, *Biochem. J.* **240:**289–292.

Milner-White, E. J., and Poet, R., 1987, Loops, bulges, turns and hairpins in proteins, *Trends Biochem. Sci.* **12:**189–192.

Modrow, S., and Wolf, H., 1986, Characterization of two related Epstein–Barr virus-encoded membrane proteins that are differentially expressed in Burkitt lymphoma and *in vitro*-transformed cell lines, *Proc. Natl. Acad. Sci. U.S.A.* **83:**5703–5707.

Mohana-Rao, J. K., and Argos, P., 1986, A conformational preference parameter to predict helices in integral membrane proteins, *Biochim. Biophys. Acta* **869:**197–214.

Mohana-Rao, J. K., Hargrave, P. A., and Argos, P., 1983, Will the seven-helix bundle be a common structure for integral membrane proteins? *FEBS Lett.* **156:**165–169.

Momany, F. A., MacGuire, R. F., Burgess, A. W., and Scheraga, H. A., 1975, Energy parameters in polypeptides. VII. Geometric parameters, partial atomic charges, nonbonded interactions, hydrogen bond interaction, and intrinsic torsion potentials for the naturally occurring amino acids, *J. Phys. Chem.* **79:** 2361–2381.

Mononen, I., and Karjalainen, E., 1984, Structural comparisons of protein sequences around potential N-glycosylation sites, *Biochim. Biophys. Acta* **788:**364–367.

Moore, W. M., Holladay, L. A., Puett, D., and Brody, R. N., 1974, On the conformation of the acetylcholine receptor protein from *Torpedo nobiliana, FEBS Lett.* **45:**145–149.

Moran, E. C., Chou, P. Y., and Fasman, G. D., 1977, Conformational transitions of glucagon in solution: The $\alpha \rightleftarrows \beta$ transition, *Biochem. Biophys. Res. Commun.* **77:**1300–1306.

Morgan, D. O., Edman, J. C., Standring, D. N., Fried, V. A., Smith, M. C., Roth, R. A., and Rutter, W. J., 1987, Insulin-like growth factor II receptor as a multifunctional binding protein, *Nature* **329:** 301–307.

Morgan, R. S., and McAdon, J. H., 1980, Predictor for sulfur–aromatic interactions in globular proteins, *Int. J. Peptide Protein Res.* **15:**177–180.

Moser, R., Thomas, R. M., and Gutte, B., 1983, An artificial crystalline DDT-binding polypeptide, *FEBS Lett.* **157:**247–251.

Moser, R., Frey, S., Münger, K., Hehlgans, T., Klausen, S., Langen, H., Winnacker, E.-L., Mertz, R., and Gutte, B., 1987, Expression of the synthetic gene of an artificial DDT-binding polypeptide of *E. coli, Protein Eng.* **1:**339–343.

Moult, J., and James, M. N. G., 1987, An algorithm for determining the conformation of polypeptide segments in proteins by systematic search, *Proteins* **1:**146–163.

Mueckler, M., Caruso, C., Baldwin, S. A., Panico, M., Blench, I., Morris, H. R., Ailard, W. J., Lienhard, G. E., and Lodish, H. F., 1985, Sequence and structure of a human glucose transporter, *Science* **299:**941–945.

Murakami, M., 1985, Mutation affecting the 12th and 61st amino acids of p21 protein result in decreased probability of β-turn occurrence around the mutation positions: A prediction, *J. Theor. Biol.* **114:**193–198.

Murakami, M., 1987, Critical amino acids of p21 protein are located within β-turns: Further evaluation, *J. Theor. Biol.* **128:**339–347.

Murata, M., Richardson, J. S., and Sussman, J., 1985, Simultaneous comparison of three protein sequences, *Proc. Natl. Acad. Sci. U.S.A.* **82:**3073–3077.

Murphy, J., Zhang, W.-J., Macaulay, W., Fasman, G., and Merrifield, R. B., 1988, The relation of predicted structure to observed conformation and activity of glucagon analogs containing replacements at positions 19, 22 and 23, *J. Biol. Chem.* **262:**17304–17312.

Murzin, A. G., and Finkelstein, A. V., 1983, Polyhedra describing the packing of helices in a protein globule, *Biofisika* **28:**905–911.

Mutter, M., 1985, The construction of new proteins and enzymes. A prospect for the future, *Angew. Chem. [Engl.]* **24:**639–653.

Nabedryk, E., Bardin, A. M., and Breton, J., 1985, Further characterization of protein secondary structures in purple membrane by circular dichroism and polarized infrared spectroscopies, *Biophys. J.* **48:**873–876.

Nagano, K., 1973, Logical analysis of the mechanism of protein folding. I. Prediction of helices, loops and β-structures from primary structure, *J. Mol. Biol.* **75:**401–420.

Nagano, K., 1974, Logical analysis of the mechanism of protein folding. II. The nucleation process, *J. Mol. Biol.* **84:**337–372.

Nagano, K., 1977, Triplet information in helix prediction applied to the analysis of super-secondary structures, *J. Mol. Biol.* **109:**251–274.

Nagarajan, M., and Rao, V. S. R., 1977, Conformational analysis of glycoproteins. Part I. Conformation of the protein segment at the site of peptide-sugar linkage, *Current Sci.* **46:**395–400.

Nakashima, H., Nishikawa, K., and Ooi, T., 1986, The folding type of a protein is relevant to the amino acid composition, *J. Biochem. (Tokyo)* **99:**153–162.

Narayana, S. V. L., and Argos, P., 1984, Residue contacts in protein structures and implications for protein folding, *Int. J. Peptide Protein Res.* **24:**25–39.

Needleman, S. B., and Blair, T. T., 1969, Homology of *Pseudomonas* cytochrome c-551 with eukaryotic c-cytochromes, *Proc. Natl. Acad. Sci. U.S.A.* **63:**1227–1233.

Needleman, S. B., and Wunsch, C. D., 1970, A general method applicable to the search for similarities in the amino acid sequences of two proteins, *J. Mol. Biol.* **48:**443–453.

Nemethy, G., 1974, Conformational energy calculations and the folding of proteins, *PAABS Rev.* **3:**51–61.

Nemethy, G., and Scheraga, H. A., 1977, Protein folding, *Q. Rev. Biophys.* **3:**239–352.

Nemethy, G., and Scheraga, H. A., 1980, Stereochemical requirements for the existence of hydrogen bonds in β-bends, *Biochem. Biophys. Res. Commun.* **95:**320–327.

Neuberger, A., and Marshall, R. D., 1968, Aspects of the structure of glycoproteins, in: *Carbohydrates and Their Roles* (H. W. Schultze, R. F. Cain, and R. W. Molstad, eds.), AVI, Westport, CT, p. 115.

Nishikawa, K., 1983, Assessment of secondary structure prediction of proteins: Comparisons of computerized Chou–Fasman method with others, *Biochim. Biophys. Acta* **748:**285–299.

Nishikawa, K., and Ooi, T., 1986, Amino acid sequence homology applied to the prediction of protein secondary structure, and joint prediction with existing methods, *Biochim. Biophys. Acta* **871:**45–54.

Nishikawa, K., Momany, F. A., and Scheraga, H. A., 1974, Low energy structure of two dipeptides and their relationship to bend conformation, *Macromolecules* **7:**797–806.

Noda, M., Takahashi, H., Tanabe, T., Toyosato, M., Furutani, Y., Hirose, T., Asai, M., Inayama, S., Miyata, T., and Numa, S., 1982, Primary structure of α-subunit precursor of *Torpedo californica* acetylcholine receptor deduced from c-DNA sequence, *Nature* **299:**793–797.

Noda, M., Takahashi, H., Tanabe, T., Toyosato, M., Kikyotani, S., Furutani, Y., Hirose, T., Takashima, G., Inayama, S., Miyata, T., and Numa, S., 1983, Structural homology of *Torpedo californica* receptor subunits, *Nature* **302:**528–532.

Noda, M., Shimizu, S., Tanabe, T., Takai, T., Kayano, T., Ikeda, T., Takahashi, H., Nakayama, H., Kanaoka, Y., Minamino, N., Kangawa, K., Matsuo, H., Raftery, M. A., Hirose, T., Inayama, S., Hayashida, H., Miyata, T., and Numa, S., 1984, Primary structure of *Electrophorus electricus* sodium channel deduced from c-DNA sequence, *Nature* **312:**121–127.

Noda, M., Ikeda, T., Kayano, T., Suzuki, H., Takeshima, J.-H., Kurasaki, M., Takahashi, T., and Numa, S., 1986, Existence of distinct sodium channel messenger RNAs in rat brain, *Nature* **320:**188–192.

Novotny, J., and Auffray, C., 1984, A program for prediction of protein secondary structure from nucleotide sequence data: Application to histocompatability antigens, *Nucleic Acids Res.* **12:**243–255.

Novotny, J., and Haber, E., 1985, Structural invariants of antigen binding: Comparison of immunoglobulin V_2–V_H and V_2–V_2 domain dimers, *Proc. Natl. Acad. Sci. U.S.A.* **82:**4592–4596.

Novotny, J., Bruccoleri, R. E., and Newell, J., 1984, Twisted hyperboloid (strophoid) as a model of β-barrels in proteins, *J. Mol. Biol.* **177:**567–573.

Nozaki, Y., and Tanford, C., 1971, The solubility of amino acids and two glycine peptides in aqueous ethanol and dioxane solutions, *J. Biol. Chem.* **246:**2211–2217.

Nussbaum, S., Beaudette, N. V., Fasman, G. D., Potts, J. T., Jr., and Rosenblatt, M., 1985, Design of analogues of parathyroid hormone: A conformational approach, *J. Protein Chem.* **4:**391–406.

Otvös, L., Jr., Hollosi, M., Perczel, A., Dietzschold, B., and Fasman, G. D., 1988, Phosphorylation loops in synthetic peptides of the human neurofilament protein middle-sized subunit, *J. Protein Chem.* **7:**365–376.

Ovchinnikov, Yu. A., 1982, Rhodopsin and bacteriorhodopsin: Structure-function relationships, *FEBS Lett.* **148:**179–191.

Ovchinnikov, Yu., A., Abdulaev, N. G., Feigina, M. Y., Kiselev, A. V., and Lobanov, N. A., 1979, The structural basis of the functioning of bacteriorhodopsin: An overview, *FEBS Lett.* **100:**219–224.

Ovchinnikov, Yu., A., Abdulaev, N. G., Vasilov, R. G., Vturina, I. Yu., Kuryatov, A. B., and Kiselev, A. V., 1985, The antigenic structure and topography of bacteriorhodopsin in purple membranes as determined by interaction with monoclonal antibodies, *FEBS Lett.* **179:**343–350.

Ovchinnikov, Yu., A., Modyanov, N. N., Brovde, N. E., Petrukhin, K. E., Grishin, A. V., Arzamazova, N. M., Aldanova, N. A., Monastyrskaya, G. S., and Sverdlov, E. D., 1986, Pig kidney, $Na^+ + K^+$ATPase, *FEBS Lett.* **201:**237–245.

Padlan, E. A., 1977, Structural implications of sequence variability in immunoglobulins, *Proc. Natl. Acad. Sci. U.S.A.* **74:**2551–2555.

Palau, J., Argos, P., and Puigdomenech, P., 1982, Protein secondary structure studies on the limits of prediction accuracy, *Int. J. Peptide Protein Res.* **19:**394–401.

Pallai, P. V., Mabilla, M., Goodman, M., Vale, W., and Rivier, J., 1983, Structural homology of corticotropin-releasing factor, sauvagine and urotensin I: Circular dichroism and prediction studies, *Proc. Natl. Acad. Sci. U.S.A.* **80:**6770–6774.

Parrilla, A., Domenech, A., and Querol, E., 1986, A PASCAL microcomputer program for prediction of protein secondary structure and hydropathic segments, *Cabios* **2:**211–215.

Pashley, R. M., McGuiggan, P. M., Ninham, B. W, and Evans, D. F., 1985, Attractive forces between uncharged hydrophobic surfaces: Direct measurements in aqueous solution, *Science* **229:**1088–1089.

Patten, P., Yokota, T., Rothbard, J., Chien, Y.-H., Arai, K.-I., and Davis, M. M., 1984, Structure, expression and divergence of T-cell receptor β-chain variable regions, *Nature* **312:**40–46.

Pattus, F., Heitz, F., Martinez, C., Provencher, S. W., and Lazdunski, C. L., 1985, Secondary structure of the pore-forming colicin A and its C-terminal fragment. Experimental fact and structure prediction, *Eur. J. Biochem.* **152:**681–689.

Paul, C., and Rosenbusch, J. P., 1985, Folding patterns of porin and bacteriorhodopsin, *EMBO J.* **4:**1593–1597.

Paul, C. H., 1982, Building models of globular proteins. Molecules from their amino acid sequences. I. Theory, *J. Mol. Biol.* **155:**53–62.

Pauling, L., and Corey, R. B., 1951, Configurations of polypeptide chains with favored orientations around single bonds: Two new pleated sheets, *Proc. Natl. Acad. Sci. U.S.A.* **37:**729–740.

Peralta, E. G., Winslow, J. W., Peterson, G. L., Smith, D. H., Ashkenazi, A., Ramachandran, J., Schimerlik, M. I., and Capon, D. J., 1987, Primary structure and biochemical properties of an M_2 muscarinic receptor, *Science* **236:**600–605.

Periti, P. F., Quagliarotti, G., and Liquori, A. M., 1967, Recognition of α-helical segments in proteins of known primary structure, *J. Mol. Biol.* **24:**313–322.

Perutz, M. F., 1980, Electrostatic effects in proteins, *Science* **201:**1187–1191.

Perutz, M. F., Gronenborn, A. M., Clore, G. M., Fogg, J. H., and Shih, D. T.-B, 1985, The pK_a values of two histidine residues in human haemoglobin, the Bohr effect, and the dipole moments of α-helices, *J. Mol. Biol.* **183:**491–498.

Phillips, D. C., 1970, in: *British Biochemistry, Past and Present* (T. W. Goodwin, ed.), Academic Press, London, pp. 11–28.

Pincus, M. R., and Klausner, R. D., 1982, Predictions of the three-dimensional structure of the leader sequence of pre-κ-light chain, a hexadecapeptide, *Proc. Natl. Acad. Sci. U.S.A.* **79:**3413–3417.

Pohlman, R., Nagel, G., Schmidt, B., Stein, M., Lorkowski, G., Krentler, C., Cully, J., Meyer, H. E., Grzeschik, K.-H., Mersmann, G., Hasilik, A., and von Figura, K., 1987, Cloning of a c-DNA encoding the human cation-dependent mannose-6-phosphate receptor, *Proc. Nat. Acad. Sci. U.S.A.* **84:**5575–5579.

Ponder, J. W., and Richards, F. M., 1987, Tertiary templates for proteins. Use of packing criteria in the enumeration of allowed sequences for different structural classes, *J. Mol. Biol.* **193:**775–791.

Pongor, S., and Szaley, A. A., 1985, Prediction of homology and divergence in the secondary structure of polypeptides, *Proc. Natl. Acad. Sci. U.S.A.* **82:**366–370.

Ponnuswamy, P. K., Worme, P. K., and Scheraga, H. A., 1973, Role of medium-range interactions in proteins, *Proc. Natl. Acad. Sci. U.S.A.* **70:**830–833.

Ponnuswamy, P. K., Prabhakran, M., and Manavalan, P., 1981, Hydrophobic packing and spatial arrangement of amino acid residues in globular proteins, *Biochim. Biophys. Acta* **623:**301–316.

Popot, J.-L., Gerchman, S.-E., and Engelman, D. M., 1987, Refolding of bacteriorhodopsin in lipid bilayers. A thermodynamically controlled two-stage process, *J. Mol. Biol.* **198:**655–676.

Post, C. B., Brooks, B. R., Karplus, M., Dobson, C. M., Artymiuk, P. C., Cheetham, J. C., and Phillips, D. C., 1986, Molecular dynamics. Simulations of native and substrate bound lysozyme. A study of the average structures and atomic fluctuations, *J. Mol. Biol.* **190:**455–479.

Potts, J. T., Jr., Kronenberg, H. M., and Rosenblatt, M., 1982, Parathyroid hormone: Chemistry, biosynthesis, and mode of action, *Adv. Protein Chem.* **35:**322–396.

Prothero, J. W., 1966, Correlation between the distribution of amino acids and alpha helices, *Biophys. J.* **6:** 367–370.

Prothero, J. W., 1968, A model of alpha-helical distribution in proteins, *Biophys. J.* **8:**1236–1255.

Ptitsyn, O. B., 1969, Statistical analyses of the distribution of amino acid residues among helical and nonhelical regions in globular proteins, *J. Mol. Biol.* **42:**501–510.

Ptitsyn, O. B., 1981, Protein folding: General physical model, *FEBS Lett.* **131:**197–202.

Ptitsyn, O. B., 1985, Physical principles of protein structure and protein folding, *J. Biosci.* **8:**1–13.

Ptitsyn, O. B., and Finkelstein, A. V., 1970a, Connection between the secondary and primary structures of globular proteins, *Biofisika* **15:**757–768.

Ptitsyn, O. B., and Finkelstein, A. V., 1970b, Prediction of helical portions of globular proteins according to their primary structure, *Dokl. Akad. Nauk. SSSR* **195:**221–224.

Ptitsyn, O. B., and Finkelstein, A. V., 1979, Coiling and topology of the parallel β-structure, *Biofisika* **24:**27–30.

Ptitsyn, O. B., and Finkelstein, A. V., 1983, Theory of protein secondary structure and algorithm of its prediction, *Biopolymers* **22:**15–25.

Ptitsyn, O. B., and Rashin, A. A., 1975, A model of myoglobin self-organization, *Biophys. Chem.* **3:**1–20.

Ptitsyn, O. B., Finkelstein, A. V., and Falk, P., 1979, Principal folding pathway and topology of all β-proteins, *FEBS Lett.* **101:**1–5.

Pullman, B., and Pullman, A., 1974, Molecular orbital calculations on the conformation of amino acid residues of proteins, *Adv. Protein Chem.* **28:**347–526.

Pumphrey, R. S. H., 1986a, Computer models of the human immunoglobulins. I. Shape and segmental flexibility, *Immunol. Today* **7:**174–178.

Pumphrey, R. S. H., 1986b, Computer models of the human immunoglobulins. II. Binding sites and molecular interactions, *Immunol. Today* **7:**206–211.

Quiocho, F. A., Sack, J. S., and Vyas, N. K., 1987, Stabilization of charges on isolated ionic groups sequestered in proteins by polarized peptide units, *Nature* **329:**561–564.

Rackovsky, S., and Goldstein, D. A., 1987, Differential geometry and protein conformation. V. Medium-range conformational influence of the individual amino acids, *Biopolymers* **26:**1163–1187.

Ralph, W. W., Webster, T., and Smith, T. F., 1987, A modified Chou and Fasman protein structure algorithm, *Cabios* **3:**211–216.

Ramachandran, G. N., Ramakrishnan, C., and Sasisekharan, V., 1963, Stereochemistry of polypeptide chain configurations, *J. Mol. Biol.* **7:**95–99.

Rao, S. T., and Rossman, M. G., 1973, Comparison of super-secondary structures in proteins, *J. Mol. Biol.* **76:** 241–250.

Rao, J. K. M., Hargrave, P. A., and Argos, P., 1983, Will the seven-helix bundle be a common structure for integral membrane proteins? *FEBS Lett.* **156:**165–169.

Rashin, A. A., 1981, Location of domains in globular proteins, *Nature* **291:**85–86.

Rashin, A. A., and Honig, B., 1984, On the environment of ionizable groups in globular proteins, *J. Mol. Biol.* **173:**515–521.

Rawlings, N., Ashman, K., and Wittman-Leibold, B., 1983, Computerized version of the Chou and Fasman protein secondary structure predictive method, *Int. J. Peptide Protein Res.* **22:**515–524.

Remington, S. J., and Matthews, B. W., 1978, A general method to assess similarity of protein structures, with applications to T4 bacteriophage lysozyme, *Proc. Natl. Acad. Sci. U.S.A.* **75:**2180–2184.

Remington, S. J., Anderson, W. F., Owen, J., Ten Eyck, L. F., Grainger, C. T., and Matthews, B. W., 1978, Structure of the lysozyme from bacteriophage T4: An electron density map at 2.4 Å resolution, *J. Mol. Biol.* **118:**81–98.

Renugopalakrishnan, V., Strawich, E. S., Horowitz, P. M., and Glimcher, M. J., 1986, Studies on the secondary structures of amelogenin from bovine tooth enamel, *Biochemistry* **25:**4879–4887.

Ricard, J. M., Perez, J. J., Pons, M., and Giralt, E., 1983, Conformational basis of N-glycosylation of proteins: Conformational analysis of Ac-Asn-Ala-Thr-NH_2, *Int. J. Biol. Macromol.* **5:**279–282.

Richards, F. M., 1974, The interpretation of protein structures: Total volume, group volume distributions and packing density, *J. Mol. Biol.* **82:**1–14.

Richards, F. M., 1977, Areas, volumes, packing, and protein structure, *Annu. Rev. Biophys. Bioeng.* **6:**151–176.

Richardson, J. S., 1976, Handedness of crossover connections in β sheets. *Proc. Natl. Acad. Sci. U.S.A.* **73:** 2619–2623.

Richardson, J. S., 1977, β-Sheet topology and the relatedness of proteins, *Nature* **268:**495–500.

Richardson, J. S., 1981, The anatomy and taxonomy of protein structure, *Adv. Protein Chem.* **34:** 167–339.

Richmond, T. J., 1984, Solvent accessible surface area and excluded volume in proteins, *J. Mol. Biol.* **178:**63–89.

Richmond, T. J., and Richards, F. M., 1978, Packing of α-helices: Geometrical constraints and contact areas, *J. Mol. Biol.* **119:**775–791.

Robson, B., 1974, Analysis of the code relating sequence to conformation in globular proteins: Theory and application of expected information, *Biochem. J.* **141:**853–867.

Robson, B., and Garnier, J., 1986, *Introduction to Proteins and Protein Engineering,* Elsevier, Amsterdam.

Robson, B., and Osguthorpe, D. J., 1979, Refined models for computer simulations of protein folding. Applications to the study of conserved secondary structure and flexible hinge points during the folding of pancreatic trypsin inhibitor, *J. Mol. Biol.* **132:**19–51.

Robson, B., and Pain, R. H., 1971, Analyses of the code relating sequence to conformation in proteins: Possible implications for the mechanism of formation of helical regions, *J. Mol. Biol.* **58:**237–259.

Robson, B., and Pain, R. H., 1972, Directional information transfer in protein helices, *Nature (New Biol.)* **238:** 107–108.

Robson, B., and Pain, R. H., 1974a, Analysis of the code relating sequence to conformation in globular proteins: Development of a stereochemical alphabet on the basis of intra-residue information, *Biochem. J.* **141:**869–882.

Robson, B., and Pain, R. H., 1974b, Analysis of the code relating sequence to conformation in globular proteins: An informational analysis of the role of the residue in determining the conformation of its neighbours in the primary sequence, *Biochem. J.* **141:**869–882.

Robson, B., and Pain, R. H., 1974c, Analysis of the code relating sequence to conformation in globular proteins: The distribution of residue pairs in turns and kinks in the backbone chain, *Biochem. J.* **141:**899–904.

Robson, B., and Platt, E., 1986, Refined models for computer calculations in protein engineering. Calibration and testing of atomic potential functions compatible with more efficient calculations, *J. Mol. Biol.* **188:** 259–281.

Robson, B., and Suzuki, E., 1976, Conformational properties of amino acid residues in globular proteins, *J. Mol. Biol.* **107:**327–356.

Robson, B., Platt, E., Fishleigh, R. V., Marsden, A., and Millard, P., 1987, Expert system for protein engineering: Its application in the study or chloroamphenicol acetyltransferase and avian pancreatic polypeptide, *J. Mol. Graphics* **5:**8–17.

Rogers, N. K., 1986, The modelling of electrostatic interactions in the function of globular proteins, *Prog. Biophys. Mol. Biol.* **48:**37–66.

Rogers, N. K., and Sternberg, M. J. E., 1984, Electrostatic interactions in globular proteins. Different dielectric models applied to the packing of α-helices, *J. Mol. Biol.* **174:**527–542.

Rose, G. D., 1978, Prediction of chain turns in globular proteins on a hydrophobic basis, *Nature* **272:**586–590.

Rose, G. D., 1979, Hierarchic organization of domains in globular proteins, *J. Mol. Biol.* **134:**447–470.

Rose, G. D., and Roy, S., 1980, Hydrophobic basis of packing in globular proteins, *Proc. Natl. Acad. Sci. U.S.A.* **77:**4643–4647.

Rose, G. D., and Seltzer, J. P., 1977, A new algorithm for finding the peptide chain turns in a globular proteins, *J. Mol. Biol.* **113:**153–164.

Rose, G. D., and Wetlaufer, D. B., 1977, The number of turns in globular proteins, *Nature* **268:**769–770.

Rose, G. D., Young, W. B., and Gierasch, L. M., 1983, Interior turns in globular proteins, *Nature* **304:**655–657.

Rose, G. D., Gierasch, L. M., and Smith, J. A., 1985, Turns in peptides and proteins, *Adv. Protein Chem.* **37:** 1–109.

Rosenblatt, M., Habener, J. F., Tyler, F. A., Shepard, G. L., and Potts, J. T., Jr., 1979, Chemical synthesis of the precursor-specific region of preproparathyroid hormone, *J. Biol. Chem.* **254:**1414–1421.

Rosenblatt, M., Beaudette, N. V., and Fasman, G. D., 1980, Conformational studies of the synthetic precursor-specific regions of pre-parathyroid hormone, *Proc. Natl. Acad. Sci. U.S.A.* **77:**3983–3987.

Rosenblatt, M., Majzoub, J. A., Beaudette, N. V., Kronenberg, H. M., Potts, J. T., Fasman, G. D., and Habener, J. F., 1981, Chemically synthesized precursor-specific fragment of preproparathyroid hormone: Conformational and biological properties, in: *Peptides 1980. Proceedings of the Sixteenth European Peptide Symposium* (K. Brunfeldt, ed.), Scriptor, Copenhagen, pp. 572–577.

Rossman, M. G., and Argos, P., 1981, Protein folding, *Annu. Rev. Biochem.* **50:**497–533.

Rossman, M. G., and Liljas, A., 1974, Recognition of structural domains in globular proteins, *J. Mol. Biol.* **85:** 177–181.

Rottier, P. J. M., Welling, G. W., Welling-Wester, S., Niesters, G. M., Lenstra, J. A., and van der Zeijst, B. A. M., 1986, Predicted membrane topology of the coronavirus E1, *Biochemistry* **25:**1335–1339.

Sack, G. H., Jr., 1983, Molecular cloning of human genes for serum amyloid A, *Gene 22:* 19–29.

Salemme, F. R., 1981, Conformational and geometrical properties of β-sheets in proteins: III. Isotropically stressed configurations, *J. Mol. Biol.* **146:**143–156.

Salemme, F. R., 1983, Structural properties of protein β-sheets, *Prog. Biophys. Mol. Biol.* **42:**95–133.

Salemme, F. R., and Weatherford, D. W., 1981a, Conformational and geometrical properties of β-sheets in proteins: I. Parallel β-sheets, *J. Mol. Biol.* **146:**101–117.

Salemme, F. R., and Weatherford, D. W., 1981b, Conformation and geometrical properties of β-sheets in proteins: II. Antiparallel and mixed β-sheets, *J. Mol. Biol.* **146:**119–141.

Sander, C., and Schulz, G. E., 1979, Degeneracy of the information contained in amino acid sequences: Evidence for overlaid genes, *J. Mol. Evol.* **13:**245–252.

Saraste, M., and Walker, J. E., 1982, Internal sequence repeats and the path of polypeptide in mitochondrial ADP/ATP translocase, *FEBS Lett.* **144:**250–254.

Sawyer, L., and James, M. N. G., 1982, Carboxyl–carboxylate interactions in proteins, *Nature* **295:**79–80.

Sayre, R., Anderson, B., and Bogorad, L., 1986, The topology of a membrane protein: The orientation of the 32 kd Qb-binding chloroplast thylakoid membrane protein, *Cell* **47:**601–608.

Scheraga, H. A., 1960, Structural studies of ribonuclease III. A model for the secondary and tertiary structure, *J. Am. Chem. Soc.* **82:**3847–3852.

Scheraga, H. A., 1968, Calculations of conformations of polypeptides, *Adv. Phys. Org. Chem.* **6:**103–184.

Scheraga, H. A., 1971, Theoretical and experimental studies of conformations of polypeptides, *Chem. Rev.* **71:** 195–217.

Scheraga, H. A., 1985, Calculations of the three-dimensional structures of proteins, *Ann. N.Y. Acad. Sci.* **439:** 170–194.

Schiffer, M., and Edmundson, A. B., 1967, Use of helical wheels to represent the structures of proteins and to identify segments with helical potential, *Biophys. J.* **7:**121.

Schiffer, M., Wu, T. T., and Kabat, E. A., 1986, Subgroups of variable regions genes of β-chains of T-cell receptors for antigen, *Proc. Natl. Acad. Sci. U.S.A.* **83:**4461–4463.

Schofield, P. R., Darlison, M. G., Fujita, N. Burt, D. R., Stephenson, F. A., Rodriguez, H., Rhee, L. M., Ramachandran, J., Reale, V., Glencorse, T. A., Seeburg, P. H., and Barnard, E. A., 1987, Sequence and functional expression of the $GABA_A$ receptor shows a ligand-gated receptor super-family, *Nature* **328:** 221–227.

Schulz, G. E., 1977, Structural rules for globular proteins, *Angew. Chem. [Engl.]* **16:**23–32.

Schulz, G. E., 1980, Gene duplication in glutathione reductase, *J. Mol. Biol.* **138:**335–347.

Schulz, G. E., and Schirmer, R. H., 1974, Topological comparison of adenyl kinase with other proteins, *Nature* **250:**142–144.

Schulz, G. E., and Schirmer, R. H., 1979, *Principles of Protein Structure,* Springer-Verlag, New York.

Schulz, G. E., Barry, C. D., Friedman, J., Chou, P. Y., Fasman, G. D., Finkelstein, A. V., Lim, V. I., Ptitsyn, O. B., Kabat, E. A., Wu., T. T., Levitt, M., Robson, B., and Nagano, K., 1974a, Comparison of predicted and experimentally determined secondary structure of adenylate kinase, *Nature* **250:** 140–142.

Schulz, G. E., Elzinga, M., Marx, F., and Schirmer, R. H., 1974b, Three-dimensional structure of adenyl kinase, *Nature* **250:**120–123.

Segrest, J. P., and Feldman, R. J., 1977, Amphipathic helices and plasma lipoproteins. A computer study, *Biopolymers* **16:**2053–2065.

Sellers, P., 1974, On the theory and computation of evolutionary distances, *J. Appl. Math.* **26:**787–793.

Sellers, P., 1979, Pattern recognition in genetic sequences, *Proc. Natl. Acad. Sci. U.S.A.* **76:**3041.

Senior, A. E., 1983, Secondary and tertiary structure of membrane proteins involved in proton translocation, *Biochim. Biophys. Acta* **726:**81–95.

Serrano, R., Kiedland-Brandt, M. C., and Fink, G. R., 1986, Yeast plasma membrane ATPase is essential for growth and has homology with ($Na^+ + K^+$), K^+- and Ca^{2+}-ATPases, *Nature* **319:**689–693.

Sheridan, R. P., Dixon, J. S., Venkataraghavan, R., Kuntz, I. D., and Scott, K. P., 1985, Amino acid composition and hydrophobicity patterns of protein domains correlate with their structure, *Biopolymers* **24:** 1995–2023.

Sheridan, R. P., and Allen, L. C., 1980, The electrostatic potential of the alpha helix (electrostatic potential/ α-helix/secondary structure/helix dipole), *Biophys. Chem.* **11:**133–136.

Sheridan, R. P., Levy, R. M., and Salemme, F. R., 1982, α-Helix dipole model and electrostatic stabilization of 4-α-helical proteins, *Proc. Natl. Acad. Sci. U.S.A.* **79:**4545–4549.

Shin, H.-C., and McFarlane, E. F., 1987, The secondary structure of myelin P_2 protein derived by secondary structure prediction methods, circular dichroism, and 400-MHz ^{1}H-NMR spectroscopy: Implications for tertiary structure, *Biochim. Biophys. Acta* **913:**155–162.

Shinohara, T., Dietzschold, B., Craft, C. M., Wistow, G., Early, J. J., Donoso, L. A., Horowitz, J., and Tao, R., 1987, Primary and secondary structure of bovine retinal S antigen (48 kDa protein), *Proc. Natl. Acad. Sci. U.S.A.* **84:**6975–6979.

Shipman, L. L., and Christoffersen, R. E., 1973, *Ab initio* calculations on large molecules using molecular fragments. Model peptide studies, *J. Am. Chem. Soc.* **95:**1408–1416.

Shoemaker, K. R., Kim, P. S., York, E. J., Stewart, J. M., and Baldwin, R. L., 1987, Tests of the helix dipole model for stabilization of α-helices, *Nature* **326:**563–567.

Shor, N. Z., 1977, Cut-off method with space extension in convex programming problems, *Cybernetics* **12:**94–96.

Shotton, D. M., and Watson, H. C., 1970, Three-dimensional structure of tosyl-elastase, *Nature* **235:**811–816.

Shull, G., Schwartz, A., and Lingrel, J. B., 1985, Amino-acid sequence of the catalytic subunit of the (Na^+ + K^+) ATPase deduced from a complementary DNA, *Nature* **316:**691–695.

Sibanda, B. L., and Thornton, J. M., 1985, β-Hairpin families in globular proteins, *Nature* **316:**170–174.

Simon, I., Nemethy, G., and Scheraga, H. A., 1978, Conformational energy calculations of the effects of sequence variation on the conformations of two tetrapeptides, *Macromolecules* **11:**797–804.

Singh, J., and Thornton, J. M., 1985, The interaction between phenylalanine rings in proteins, *FEBS Lett.* **191:** 1–6.

Sippl, M. J., 1982, On the problem of comparing protein structures. Development and application of a new method for the assessment of structural similarities of polypeptide conformations, *J. Mol. Biol.* **156:**359–388.

Small, D., Chou, P. Y., and Fasman, G. D., 1977, Occurrence of phosphorylated residues in predicted β-turns: Implications for β-turns participation in control mechanisms, *Biochem. Biophys. Res. Commun.* **79:**341–346.

Smith, T. F., and Waterman, M. S., 1981, Identification of common molecular subsequences, *J. Mol. Biol.* **147:**195–197.

Smythies, J. R., 1980, An hypothesis concerning the molecular structure of the nicotinic acetylcholine receptor, *Med. Hypothesis* **6:**943–950.

Sneath, P. H. A., 1966, Relations between chemical structure and biological activity in peptides, *J. Theor. Biol.* **12:**157–195.

Snell, C. R., and Smyth, D. G., 1975, Proinsulin: A proposed three-dimensional structure, *J. Biol. Chem.* **250:** 6291–6295.

Srere, P. A., and Brooks, G. C., 1969, The circular dichroism of glucagon solutions, *Arch. Biochem. Biophys.* **129:**708–710.

Staden, R., 1982, An interactive graphics program for comparing and aligning nucleic acid and amino acid sequences, *Nucleic Acids Res.* **10:**2951–2961.

Steitz, T. A., Goldman, A., and Engelman, D. M., 1982, Quantitative application of the helical hairpin hypothesis to membrane proteins, *Biophys. J.* **37:**124–125.

Stephan, M. M., and Jacobson, G. R., 1986, Membrane disposition of the *Escherichia coli* mannitol permease: Identification of membrane-bound and cytoplasmic domains, *Biochemistry* **25:**8230–8234.

Sternberg, M. J. E., 1983, The analysis and prediction of protein structure, in: *Computing in Biological Sciences* (M. Geisow and A. Barret, eds.), Elsevier, Amsterdam, pp. 143–177.

Sternberg, M. J. E., and Cohen, F. E., 1982, Prediction of the secondary and tertiary structures of interferon from four homologous amino acid sequences, *Int. J. Biol. Macromol.* **4:**137–144.

Sternberg, M. J. E., and Taylor, W. R., 1984, Modelling the ATP-binding site of oncogene products, the epidermal growth factor receptor and related proteins, *FEBS Lett.* **175:**387–392.

Sternberg, M. J. E., and Thornton, J. M., 1976, On the conformation of proteins: The handedness of the β-strand–α-helix–β-strand unit, *J. Mol. Biol.* **105:**367–382.

Sternberg, M. J. E., and Thornton, J. M., 1977a, On the conformation of proteins: The handedness of the connection between parallel β-strands, *J. Mol. Biol.* **110:**269–283.

Sternberg, M. J. E., and Thornton, J. M., 1977b, On the conformation of proteins: An analysis of β-pleated sheets, *J. Mol. Biol.* **110:**285–296.

Sternberg, M. J. E., and Thornton, J. M., 1977c, On the conformation of proteins: Hydrophobic ordering of strands in β-pleated sheets, *J. Mol. Biol.* **115:**1–17.

Sternberg, M. J. E., and Thornton, J. M., 1978, Prediction of protein structure from amino acid sequence, *Nature* **271:**15–20.

Sternberg, M. J. E., Cohen, F. E., Taylor, W. R., and Feldman, R. J., 1981, Analysis and prediction of structural motifs in the glycolytic enzymes, *Phil. Trans. R. Soc. Lond. [Biol.]* **293:**177–189.

Sternberg, M. J. E., Cohen, F. E., and Taylor, W. R., 1982, A combinatorial approach to the prediction of the tertiary fold of globular proteins, *Biochem. J.* **10:**299–301.

Sternberg, M. J. E., Hayes, F. R. F., Russell, A. J., Thomas, P. G., andFerscht, A. R., 1987, Prediction of electrostatic effects of engineering of protein charges, *Nature* **330:**86–88.

Stroud, R. M., and Finer-Moore, J., 1985, Acetylcholine receptor structure, function and evolution, *Annu. Rev. Cell. Biol.* **1:**317–351.

Stuber, K., Deutscher, J., Sobek, H. M., Hengstenberg, W., and Beyreuther, K., 1985, Amino acid sequence of the amphiphilic phosphocarrier protein factor III^{lac} of the lactose-specific phosphotransferase system of *Staphylococcus aureus, Biochemistry* **24:**1164–1168.

Stuber, M., 1982, Doctoral Thesis, University of Cologne, Cologne, Germany.

Sundaralingam, M., Sekharudu, Y. C., Yathindra, N., and Ravichandran, V., 1987, Ion pairs in alpha-helices, *Proteins: Structure, Function and Genetics* **2:**64–71.

Sweet, R. M., 1986, Evolutionary similarity among peptide segments is a basis for prediction of protein folding, *Biopolymers* **25:**1565–1577.

Sweet, R. M., and Eisenberg, D., 1983, Correlation of sequence hydrophobicities measures similarity in three-dimensional protein structure, *J. Mol. Biol.* **171:**479–488.

Tanabe, T., Takeshima, H., Mikami, A., Flockerzi, V., Takahashi, H., Kangawa, K., Kojima, M., Matsuo, H., Hirose, T., and Numa, S., 1987, Primary structure of the receptor for calcium channel blockers from skeletal muscle, *Nature* **328:**313–318.

Tanaka, S., and Scheraga, H. A., 1976a, Statistical mechanical treatment of protein conformation. I. Conformational properties of amino acids in proteins, *Macromolecules* **9:**142–159.

Tanaka, S., and Scheraga, H. A., 1976b, Statistical mechanical treatment of protein conformation. II. A three-state model for specific-sequence copolymers of amino acids, *Macromolecules* **9:**159–167.

Tanaka, S., and Scheraga, H. A., 1976c, Statistical mechanical treatment of protein conformation. III. Prediction of protein conformation based on a three-state model, *Macromolecules* **9:**168–182.

Tanaka, S., and Scheraga, H. A., 1976d, Statistical mechanical treatment of protein conformation. IV. A four-state model for specific-sequence copolymers of amino acids, *Macromolecules* **9:**812–833.

Tanford, C., 1962, Contributions of hydrophobic interactions to the stability of the globular conformation of proteins, *J. Am. Chem. Soc.* **84:**4240–4247.

Tanford, C., 1980, *The Hydrophobic Effect,* 2nd ed., John Wiley & Sons, New York.

Taylor, W. R., 1984, An algorithm to compare secondary structure predictions, *J. Mol. Biol.* **173:**512–514.

Taylor, W. R., 1986a, Identification of protein sequence homology by consensus template alignment, *J. Mol. Biol.* **188:**233–258.

Taylor, W. R., 1986b, The classification of amino acid conservation, *J. Theor. Biol.* **119:**205–218.

Taylor, W. R., and Geisow, M. J., 1987, Predicted structure for the calcium-dependent membrane-binding proteins, p35, p36, p32, *Protein Eng.* **1:**183–187.

Taylor, W. R., and Thornton, J. M., 1983, Prediction of super-secondary structure in proteins, *Nature* **301:** 540-542.

Taylor, W. R., and Thornton, J. M., 1984, Recognition of super-secondary structures in proteins, *J. Mol. Biol.* **173:**487–514.

Thornton, J. M., and Chakauya, B. L., 1982, Conformation of the terminal regions in proteins, *Nature* **298:** 296–297.

Thornton, J. M., and Sibanda, B. L., 1983, Amino and carboxyl-terminal regions in globular proteins, *J. Mol. Biol.* **167:**443–460.

Tillinghast, J. P., Behlke, M. A., and Loh, D. Y., 1986, Structure and diversity of the human T-cell receptor β-chain variable region genes, *Science* **233:**879–883.

Titani, K., Hermodson, M. A., Ericsson, C. H., Walsh, K. A., and Neurath, H., 1982, Amino acid sequence of thermolysin, *Nature [New Biol.]* **238:**35–37.

Titani, K., Takio, K., Handa, M., and Ruggeri, Z. M., 1987, Amino acid sequences of the von Willebrand factor-binding domain of platelet membrane glycoprotein Ib, *Proc. Natl. Acad. Sci. U.S.A.* **84:** 5610–5614.

Toda, M., Takahashi, H., Tanabe, T., Toyosato, M., Furutani, Y., Hirose, T., Asai, M., Inayama, S., Miyata, T., and Numa, S., 1982, Primary structure of α-subunit precursor of *Torpedo californica* acetylcholine receptor deduced from cDNA sequence, *Nature* **299:**793–797.

Toh, H., Hayashida, H., and Miyata, T., 1983, Sequence homology between retroviral reverse transcriptase and putative polymerases of hepatitis B virus and cauliflower mosaic virus, *Nature* **305:**827–829.

Toitskii, G. V., and Zav'yalov, V. P., 1972, Calculation of the conformations of proteins with the aid of a modified nonagram. Establishment of the interrelationship between the primary and secondary structures of the polypeptide chain, *J. Mol. Biol.* **6:**645–647.

Trewhella, H., Anderson, S., Fox, R., Gogol, R., Khan, S., Engelman, D., and Zaccai, G., 1983, Assignment of segments of the bacteriorhodopsin sequence to positions in the structural map, *Biophys. J.* **42:**233–241.

Trewhella, J., Gogol, E., Zaccai, G., and Engelman, D. M., 1984, Neutron diffraction studies of bacteriorhodopsin structure, in: *Neutrons in Biology* (B. P. Schoenborn, ed.), Plenum Press, New York, pp. 227–246.

Ullrich, A., Bell, J. R., Chen, E. Y., Herrera, R., Petruzzelli, L. M., Dull, T. J., Gray, A., Coussens, L., Liao, Y.-C., Tsubokawa, M., Mason, A., Seeburg, P. H., Grunfeld, C., Rosen, O. M., and Ramachandran, J., 1985, Human insulin receptor and its relationship to the tyrosine kinase family of oncogenes, *Nature* **313:**756–761.

Van Belle, D., Couplet, I., Prevost, M., and Wodak, S., 1987, Calculations of electrostatic properties in proteins. Analysis of contributions from induced protein dipoles, *J. Mol. Biol.* **198:**721–735.

van Duijnen, P. T., Thole, B. T., and Hol, W. G. J., 1979, On the role of the active site helix in papain. An *ab initio* molecular orbital study, *Biophys. Chem.* **9:**273–280.

Varghese, J. N., Laver, W. G., and Colman, P. M., 1983, Structure of the influenza virus glycoprotein antigen neuraminidase at 2.9Å resolution, *Nature* **303:**35–40.

Venkatachalan, C. M., 1968, Stereochemical criteria for polypeptides and proteins. V. Conformation of a system of three linked peptide units, *Biopolymers* **6:**1425–1436.

Vickery, L. E., 1987, Interactive analysis of protein structure using a microcomputer spread sheet, *Trends Biochem. Sci.* **12:**37–39.

Visser, L., and Blout, E. R., 1971, Elastase. II. Optical properties and the effects of sodium dodecyl sulfate, *Biochemistry* **10:**743–752.

Vogel, H., and Jähnig, F., 1986, Models for the structure of outer-membrane proteins of *Escherichia coli* derived from Raman spectroscopy and prediction methods, *J. Mol. Biol.* **190:**191–199.

Vogel, H., Wright, J. K., and Jähnig, F., 1985, The structure of the lactase permease derived from Raman spectroscopy and prediction methods, *EMBO J.* **4:**3625–3631.

Vogel, S., Freist, W., and Hoppe, J., 1986, Assignment of conserved amino acid residues to the ATP site in the protein kinase domain of the receptor for epidermal growth factor, *Eur. J. Biochem.* **154:**529–532.

Vonderviszt, F., and Simon, I., 1986, A possible way for prediction of domain boundaries in globular proteins from amino acid sequence, *Biochem. Biophys. Res. Commun.* **139:**11–17.

Vonderviszt, F., Matrai, G., and Simon, I., 1986, Characteristic sequential residue environment of amino acids in proteins, *Int. J. Peptide Protein Res.* **27:**483–492.

von Heijne, G., 1981a, On the hydrophobic nature of signal sequences, *Eur. J. Biochem.* **116:**419–422.

von Heijne, G., 1981b, Membrane proteins. The amino acid composition of membrane-penetrating segments, *Eur. J. Biochem.* **120:**275–278.

von Heijne, G., and Blomberg, C., 1977, The β-structure: Inter-strand correlations, *J. Mol. Biol.* **117:**821–824.

von Heijne, G., and Blomberg, C., 1978, Some global β-sheet characteristics, *Biopolymers* **7:**2033–2037.

von Heijne, G., and Blomberg, C., 1979, Trans-membrane translocation of proteins. The direct transfer model, *Eur. J. Biochem.* **97:**175–181.

Wada, A., 1976, The α-helix as an electric macro-dipole, *Adv. Biophysics* **9:**1–63.

Wada, A., and Nakamura, H., 1981, Nature of the charge distribution in proteins, *Nature* **293:**757–758.

Walker, J. E., Crane, A. F., and Schmitt, H., 1979, The topology of the purple membrane, *Nature* **278:**653–654.

Walker, J. E., Saraste, M., Runswick, M. J., and Gay, N. J., 1982, Distantly related sequences in the α- and β-subunits of ATP synthase, myosin, kinases, and other ATP-requiring enzymes and a common nucleotide binding fold, *EMBO J.* **1:**945–951.

Walker, J. E., Saraste, M., and Gay, N. J., 1984, The *unc* operon. Nucleotide sequence, regulation and structure of ATP-synthase, *Biochim. Biophys. Acta* **768:**164–200.

Wallace, B. A., Cascio, M., and Mielke, D. L., 1986, Evaluation of methods for the prediction of membrane protein secondary structure, *Proc. Natl. Acad. Sci. U.S.A.* **83:**9423–9427.

Warshel, A., and Russell, S. T., 1984, Calculations of electrostatic interactions in biological systems and in solution, *Quart. Rev. Biophys.* **17:**283–422.

Warwicker, J., and Watson, H. C., 1982, Calculation of the electric potential in the active site cleft due to α-helix dipoles, *J. Mol. Biol.* **157:**671–679.

Waterman, M. S., Smith, T. F., and Beyer, W. A., 1976, Some biological sequence metrics, *Adv. Math.* **20:** 367–387.

Weatherford, D. W., and Salemme, F. R., 1979, Conformations of twisted parallel β-sheets and the origin of chirality in protein structures, *Proc. Nat. Acad. Sci. U.S.A.* **76:**19–23.

Weber, P. C., and Salemme, F. R., 1980, Structural and functional diversity in 4-α-helical proteins, *Nature* **287:**82–84.

Webster, T. A., Lathrop, R. H., and Smith, T. F., 1987, Prediction of a common structural domain in amino acid-tRNA synthetases through use of a new pattern-directed inference system, *Biochemistry* **26:**6950–6957.

Weiner, S. J., Kollman, P. A., Case, D. A., Singh, U. C., Ghio, C., Alagona, G., Profeta, S., and Weiner, P., 1984, A new force field for molecular mechanical simulation of nucleic acids and proteins, *J. Am. Chem. Soc.* **106:**765–784.

Wertz, D. H., and Scheraga, H. A., 1978, Influence of water on protein structure. An analysis of the preferences of amino acids residues for the inside or outside and for specific conformations in a protein molecule, *Macromolecules* **11:**9–15.

Wetlaufer, D. B., 1973, Nucleation, rapid folding, and globular intrachain regions in proteins, *Proc. Natl. Acad. Sci. U.S.A.* **70:**697–701.

Wetlaufer, D. B., 1981, Folding of protein fragments, *Adv. Protein Chem.* **34:**61–92.

Wetlaufer, D. B., and Ristow, S., 1973, Acquisition of three-dimensional structure of proteins, *Annu. Rev. Biochem.* **42:**135–158.

Wickner, W., 1979, The assembly of proteins into biological membranes: The membrane trigger hypothesis, *Annu. Rev. Biochem.* **48:**23–45.

Wierenga, R. K., Terpstra, P., and Hol, W. G. J., 1986, Prediction of the occurrence of the ADP-binding βαβ-fold in proteins, using an amino acid sequence fingerprint, *J. Mol. Biol.* **187:**101–107.

Wilbur, W. J., and Lipman, D. J., 1983, Rapid similarity searches of nucleic acids and protein data banks, *Proc. Natl. Acad. Sci. U.S.A.* **80:**726–730.

Williams, R. W., Chang, A., Juretic, D., and Loughran, S., 1987, Secondary structure predictions and medium-range interactions, *Biochim. Biophys. Acta* **916:**200–204.

Wilmot, C. M., and Thornton, J. M., 1988, Analysis and prediction of the different types of β-turn in proteins, *J. Mol. Biol.* **203:**221–232.

Wilson, I. A., Haft, D. H., Getzoff, E. D., Tainer, J. A., Lerner, R. A., and Brenner, S., 1985, Identical short peptide sequences in unrelated proteins can have different conformations: A testing ground for theories of immune recognition, *Proc. Natl. Acad. Sci. U.S.A.* **82:**5255–5259.

Wodak, S. J., and Janin, J., 1980, Analytical approximation to the accessible surface area of proteins, *Proc. Natl. Acad. Sci. U.S.A.* **77:**1736–1740.

Wodak, S. J., and Janin, J., 1981, Location of structural domains in proteins, *Biochemistry* **20:**6544–6552.

Wolfenden, R., 1983, Waterlogged molecules, *Science* **222:**1087–1093.

Wolfenden, R., Anderson, L., Cullis, P. M., and Southgate, C. C. B., 1981, Affinities of amino acid side chains for solvent water, *Biochemistry* **20:**849–855.

Wolfenden, R. V., Cullis, P. M., and Southgate, C. C. F., 1979, Water, protein folding, and the genetic code, *Science* **206:**575–577.

Wu, C.-S. C., Hachimori, A., and Yang, J. T., 1982, Lipid induced ordered conformation of some peptide hormones and bioactive oligopeptides: Predominance of helix over β-form, *Biochemistry* **21:**4556–4562.

Wu, T. T., and Kabat, E. A., 1970, An analysis of the sequences of the variable regions of Bence Jones proteins and Myeloma light chains and their implications for antibody complementarity, *J. Expt. Med.* **132:**211–250.

Yatsunami, K., and Khorana, H. G., 1985, GTPase of bovine rod outer segments: The amino acid sequence of the α-subunit as derived from the c-DNA sequence, *Proc. Natl. Acad. Sci. U.S.A.* **82:**4316–4320.

Yockey, H. P., 1977, A prescription which predicts functionally equivalent residues at given sites in protein sequences, *J. Theor. Biol.* **67:**337–343.

Youvan, D. C., Bylina, E. J., Albert, M., Begusch, H., and Hearest, J., 1984, Nucleotide and deduced polypeptide sequence of the photosynthetic reaction center, B870 antenna, and flanking polypeptides from *R. capsulata, Cell* **37:**949–957.

Yuschok, T. J., and Rose, G. D., 1983, Hierarchic organization of globular proteins. A control study, *Int. J. Peptide Prot. Res.* **21:**479–484.

Zehfus, M. H., Seltzer, J. P., and Rose, G. D., 1985, Fast approximation for accessible surface area and molecular volume of protein segments, *Biopolymers* **24:**2511–2519.

Zimm, B. H., and Bragg, J. K., 1959, Theory of the phase transition between the helix and random chain in polypeptide chains, *J. Chem. Phys.* **31:**526–535.

Zimmerman, J. M., Eliezer, N., and Simha, R., 1968, The characterization of amino acid sequences in proteins by statistical methods, *J. Theor. Biol.* **21:**170–201.

Zimmerman, S. S., and Scheraga, H. A., 1977, Local interactions in bends of protein, *Proc. Natl. Acad. Sci. U.S.A.* **74:**4126–4129.

Zimmerman, S. S., Pottle, M. S., Nemethy, G., and Scheraga, H. A., 1977, Conformational analysis of the 20 naturally occurring amino acid residues using ECEPP, *Macromolecules* **10:**1–9.

Zvelebil, M. J., Barton, G. J., Taylor, W. R., and Sternberg, M. J. E., 1987, Prediction of protein secondary structure and active sites using alignment of homologous sequence, *J. Mol. Biol.* **195:**957–961.

X. APPENDIXES

Appendix 1: List of Reviews on Protein Folding and Prediction of Secondary and Tertiary Structure

Argos, P., and Mohana Rao, J. K., 1986, Prediction of protein structure, *Methods Enzymol.* **130:**185–207.

Bajaj, M., and Blundell, T., 1984, Evolution and the tertiary structure of proteins, *Annu. Rev. Biophys. Biophys. Chem.* **13:**453–492.

Blake, C. C. F., and Johnson, L. N., 1984, Protein structure, *Trends Biochem. Sci.* **9:**147–151.

Blundell, T., and Sternberg, M. J. E., 1985, Computer-aided design in protein engineering, *Trends Biotechnol.* **3:**228–235.

Blundell, T. L., Sibanda, B. L., Sternberg, M. J. E., and Thornton, J. M., 1987, Knowledge-based prediction of protein structures and the design of novel molecules, *Nature* **326:**347–352.

Cantor, C. R., and Schimmel, P. R., 1980, *Biophysical Chemistry,* Volume I, W. H. Freeman, San Francisco.

Chothia, C., 1984, Principles that determine the structure of proteins, *Annu. Rev. Biochem.* **53:**537–572.

Chou, P. Y., and Fasman, G. D., 1977, Secondary structural prediction of proteins from their amino acid sequence, *Trends Biochem. Sci.* **2:**128–132.

Chou, P. Y., and Fasman, G. D., 1978a, Empirical predictions of protein conformation, *Annu. Rev. Biochem.* **47:**251–276.

Chou, P. Y., and Fasman, G. D., 1978b, Prediction of the secondary structure of proteins from their amino acid sequence, *Adv. Enzymol.* **47:**45–108.

Creighton, T. E., 1983, *Proteins,* W. H. Freeman, New York.

Doolittle, R. F., 1986, *Of URFS and ORFS. A Primer on How to Analyse Derived Amino Acid Sequences,* University Science Books, Mill Hill, CA.

Edsall, J. T., and McKenzie, H. A., 1983, Water and proteins II. The location and dynamics of water in protein systems and its relation to their stability and properties, *Adv. Biophys.* **16:**53–183.

Eisenberg, D., 1984, Three-dimensional structure of membrane and surface proteins, *Annu. Rev. Biochem.* **53:** 595–623.

Engelman, D. M., Steitz, T. A., and Goldman, A., 1986, Identifying nonpolar transbilayer helices in amino acid sequences of membrane proteins, *Annu. Rev. Biophys. Biophys. Chem.* **15:**321–353.

Fasman, G. D., 1980, Prediction of protein conformation from the primary structure, *Ann. N.Y. Acad. Sci.* **348:** 147–159.

Fasman, G. D., 1985, A critique of the utility of the prediction of protein secondary structure, *J. Biosci.* **8:**15–23.

Fasman, G. D., 1987, The road from poly-α-amino acids to the prediction of protein conformation, *Biopolymers* **26:**S59–S79.

Fletterick, R., and Zoller, M., eds., 1986, Computer graphics and molecular modeling, in: *Current Communications in Molecular Biology,* Cold Spring Harbor Laboratory, Cold Spring Harbor, NY.

Ghelis, C., and Yon, J., 1982, *Protein Folding,* Academic Press, New York.

Go, N., 1983, Theoretical studies of protein folding, *Annu. Rev. Biophys. Biophys. Chem.* **12:**183–210.

Hohne, E., and Kretschmer, R. G., 1985, Description of secondary structure in proteins, *Stud. Biophys.* **108:** 165–186.

Honig, B. H., Hubbell, W. L., and Flewling, R. F., 1986, Electrostatic interactions in membranes and proteins, *Annu. Rev. Biophys. Biophys. Chem.* **15:**163–193.

Jaenicke, R., ed., 1984, *Protein Folding,* Elsevier/North-Holland Biomedical Press, Amsterdam.

Jaenicke, R., 1987, Folding and association of proteins, *Prog. Biophys. Mol. Biol.* **49:**117–237.

Janin, J., and Wodak, S. J., 1983, Structural domains in proteins and their role in the dynamics of protein function, *Prog. Biophys. Mol. Biol.* **42:**21–78.

Jungck, J. R., Friedman, R. M., 1984, Mathematical tools for molecular genetics data: An annotated bibliography, *Bull. Math. Biol.* **46:**699–744.

Kauzmann, W., 1959, Some factors in the interpretation of protein denaturation, *Adv. Protein Chem.* **14:**1–63.

Kneale, G. G., and Bishop, M. J., 1985, Nucleic acid and protein sequence databases, *Cabios* **1:**11–17.

Kollman, P., 1987, Molecular modeling, *Annu. Rev. Phys. Chem.* **38:**303–316.

Lesk, A. M., and Hardman, K. D., 1985, Computer-generated pictures of proteins, *Methods Enzymol.* **115:** 381–390.

Levitt, M., 1982, Protein conformation, dynamics, and folding by computer simulation. *Annu. Rev. Biophys. Bioeng.* **11:**251–271.

Matthew, J. B., and Gurd, F. R. N., 1986a, Stabilization and destabilization of protein structure by charge interactions, *Methods Enzymol.* **130:**437–453.

Matthew, J. B., and Gurd, F. R. N., 1986b, Calculation of electrostatic interactions in proteins, *Methods Enzymol.* **130:**413–436.

Matthews, J. B., 1985, Electrostatic effects in proteins, *Annu. Rev. Biophys. Biophys. Chem.* **14:**387–417.

Nagano, K. and Ponnuswamy, P. K., 1984, Prediction of packing of secondary structure, *Adv. Biophys.* **18:** 115–148.

Nemethy, G., and Scheraga, H. A., 1977, Protein folding, *Q. Rev. Biophys.* **10:**239–352.

Ptitsyn, O. B., and Finkelstein, A. V., 1980, Similarities of protein topologies: Evolutionary divergence, functional convergence or principles of folding, *Q. Rev. Biophys.* **13:**339–386.

Richards, F. M., 1977, Areas, volumes, packing, and protein structure, *Adv. Biophys. Bioeng.* **6:**151–176.

Richards, J. S., 1985, Schematic drawings of protein structures, *Methods Enzymol.* **115:**359–380.

Richardson, J., 1981, The anatomy and taxonomy of protein structure, *Adv. Protein Chem.* **34:**167–339.

Richardson, J. S., 1985, Describing patterns of protein tertiary structure, *Methods Enzymol.* **115:**341–358.

Robson, B., 1982, The prediction of molecular conformation, *Biochem. J.* **10:**297–298.

Robson, R., and Garnier, J., 1986, *Introduction to Proteins and Protein Engineering,* Elsevier, Amsterdam.

Rose, G. D., Gierasch, L. M., and Smith, J. A., 1985, Turns in peptides and proteins, *Adv. Protein Chem.* **37:** 1–109.

Rossman, M. G., and Argos, P., 1981, Protein folding, *Annu. Rev. Biochem.* **53:**497–533.

Salemme, F. R., 1983, Structural properties of protein β-sheets, *Prog. Biophys. Mol. Biol.* **42:**95–133.

Scheraga, H. A., 1985, Calculations of the three-dimensional structure of proteins, *Ann. N.Y. Acad. Sci.* **439:** 170–194.

Schulz, G. E., 1977, Structural rules for globular proteins, *Angew. Chem. [Engl.]* **16:**23–32.

Schulz, G. E., and Schirmer, R. H., 1979, *Principles of Protein Structure,* Springer-Verlag, Berlin.

Sternberg, M. J. E., 1983, The analysis and prediction of protein structure, in: *Computing in Biological Sciences* (M. S. Geisow and A. N. Barrett, eds.), Elsevier Biomedical Press, Amsterdam, pp. 143–177.

Sternberg, M. J. E., 1986, Prediction of protein structure from amino acid sequence, *Anticancer Drug Design* **1:** 169–178.

Sternberg, M. J. E., and Thornton, J. M., 1978, Prediction of protein structure from amino acid sequence, *Nature* **271:**15–20.

Taylor, W. R., 1987, Protein structure prediction in: *Nucleic Acid and Protein Sequence Analysis* (M. J. Bishop and G. J. Rawlings, eds.), IRL Press: Oxford.

von Heijne, G., 1987, *Sequence Analysis in Molecular Biology,* Academic Press, New York.

Warshel, A., and Russell, S. T., 1984, Calculations of electrostatic interactions in biological systems and in solutions, *Q. Rev. Biophys.* **17:**283–422.

Wetlaufer, D. B., 1981, Folding of protein fragments, *Adv. Protein Chem.* **34:**335–347.

Wetlaufer, D. B., ed., 1984, *The Protein Folding Problem,* AAAS, Washington.

Wetlaufer, D. B., and Ristow, S., 1973, Acquisition of three-dimensional structure of proteins, *Annu. Rev. Biochem.* **42:**135–158.

Appendix 2: Programs Available through This Publication for Protein Secondary Structure Prediction

Deposited at:

1. The Protein Identification Resource (PIR), National Biomedical Research Foundation, Georgetown University Medical Center, 3900 Reservoir Road, NW, Washington, D.C. 20007
2. Molecular Biology Computer Research Resource (MBCRR), Dana Farber Cancer Institute, 44 Binney Street, Boston, Massachusetts 02115
3. Whitehead Institute for Biochemical Research, 9 Cambridge Center, Cambridge, Massachusetts 02142, c/o W. Gilbert

1. Chou–Fasman–Prevelige derived from the original Chou–Fasman algorithm [C-F-P]:
   a. Chou, P. Y., and Fasman, G. D., 1974, Prediction of protein conformation, *Biochemistry* **13:**222–245.
   b. Chou, P. Y., and Fasman, G. D., 1978, Prediction of the secondary structure of proteins from their amino acid sequence, *Adv. Enzymol.* **47:**45–148.
   c. Chou, P. Y., and Fasman, G. D., 1979, Prediction of β-turns, *Biophys. J.* **26:**367–384.
   d. Chou, P. Y., Fasman, G. D., and Prevelige, P., Chapters 9 and 12, this volume.

   Written in C for IBM-PC-XT.

2. Deléage, F., Tinland, B., and Roux, B., 1987, A computerized version of the Chou and Fasman method for predicting the secondary structure of proteins, *Anal. Biochem.* **163:**292–297. [D-T-R]

   Some of the qualitative rules in the original rules have been converted to numeric scales to obtain unambiguous predictions.
   Written for an Apple IIe (128k) microcomputer.

3. Eisenberg, D., Wesson, M., and Wilcox, W., Chapter 16, this volume. [E]

   Written in FORTRAN to be used on a Vax computer.

4. Finer-Moore, J., and Stroud, R. M., 1984, Amphipathic analysis and possible formation of the ion channel in an acetylcholine receptor, *Proc. Natl. Acad. Sci. U.S.A.* **81:**155–159. [F-M-S]
   Finer-Moore, J., Bazan, F., Rubin, J., and Stroud, R. M., 1989, Identification of membrane proteins and soluble protein secondary structural elements, domain structure, and packing arrangements by Fourier-transform amphipathic analysis, Chapter 19, this volume.

   Written in FORTRAN for use on a VAX computer on a VMS operating system.

5. Garnier, J., Osguthorpe, D. G., and Robson, B., 1978, Analysis of the accuracy and implications of simple methods for predicting the secondary structure of globular proteins, *J. Mol. Biol.* **120:**97–120. [G-O-R]
   and updated:
   Gibrat, J.-F., Garnier, J., and Robson, B., 1987, Further developments of protein secondary structure prediction using information theory, new parameters, and consideration of residue pairs, *J. Mol. Biol.* **198:**425–443.
   Garnier, J., and Robson, B., 1989, The G-O-R method for predicting secondary structure in proteins, Chapter 10, this volume.

Written in FORTRAN for use on a Micro VAX II computer. Another program is available to be run on a microcomputer (e.g., IBM PC).

6. Vogel, H., Wright, J. K., and Jähnig, F., 1985, The structure of the lactose permease derived from raman spectroscopy and prediction methods, *EMBO J.* **4:**3625–3631. [J]

   Vogel, H., and Jähnig, F., 1986, Models for the structure of outer-membrane proteins of *Escherichia coli* derived from raman spectroscopy and prediction methods, *J. Mol. Biol.* **190:**191–199.

   Jähnig, F., 1989, Structure prediction for membrane proteins, Chapter 18, this volume.

   Written in FORTRAN for use on an IBM PC/AT computer.

7. Klein, P., 1986, Prediction of protein structural class by discriminant analysis, *Biochim. Biophys. Acta* **874:**205–215. [K]

   Program STRCLS, written for VAX/VMS in FORTRAN.

8. Klein, P., Kanehisa, M., and DeLisi, C., 1985, The detection and classification of membrane-spanning regions, *Biochim. Biophys. Acta* **815:**468–476. [K-K-D]

   Written in FORTRAN.

9. Kyte, J., and Doolittle, R. F., 1982, A simple method for displaying the hydropathic character of a protein, *J. Mol. Biol.* **157:**105–132.* [K-D]

   Program SOAP, written in language C for use in the software system Unix Vax with a C compiler (K-D: Program 8). Will send other programs on 1600 bpi tape.

   Doolittle Programs: Protein sequence alignment and phylogenetic tree construction. D.-Feng and R. F. Doolittle, 1987, Progressive sequence alignment as a prerequisite to correct phyogenetic trees, *J. Mol. Evol.* **23:**351.

   Seven programs: format.c–for DNA or protein
   Score.c–for nearest relationships
   prealign.c
   dfalign.c
   blen.c
   mulpub.c
   dfplot.c

   (The .c indicates that the programs are written in C language. All these programs are in their uncompiled form. Instructions are given to compile the C programs.)

10. Lim, V. I., 1974, Algorithms for prediction of α-helical and β-structural regions in globular proteins, *J. Mol. Biol.* **88:**873–894.

    Programs written by: Johannes A. Lenstra, Vakgroep Infectieziekten en Immunologie, Facultair Recombinant DNA Laboratorium Fakulteit Der Diergeneeskunde, Rijksuniversiteit Te, Utrecht, Yalelaan 1, Postbus 80-165, 3508 TD Utrecht, The Netherlands, and

    Kabsch, W. and Sander, C., Biophysics Department, Max Planck Institute of Medical Research, D-6900 Heidelberg, Federal Republic of Germany.

*Data-sieving program can be used. Based on a running median (between 5 and 19 amino acids) smooths the raw data, rendering the domain more visible. Bangham, J. A., 1988, Data-sieving hydrophobicity plots, *Anal. Biochem.* **174:**142–145.

11. Nagano, K., 1973, Logical analysis of the mechanism of protein folding. I. Prediction of helices, loops and β-structures from primary structure, *J. Mol. Biol.* **75:**401–420. [N]

    Nagano, K., 1974, Logical analysis of the mechanism of protein folding. II. The nucleation process, *J. Mol. Biol.* **84:**337–372.

    Nagano, K., 1977a, Logical analysis of the mechanism of protein folding. IV. Super-secondary structure, *J. Mol. Biol.* **109:**235–250.

    Nagano, K., 1977b, Triplet information in helix prediction applied to the analysis of super-secondary structures, *J. Mol. Biol.* **109:**251–274.

    Nagano, K., 1980, Logical analysis of the mechanism of protein folding. V. Packing game simulation of α/β proteins, *J. Mol. Biol.* **138:**797–832.

    Nagano, K., and Ponnuswamy, P. K., 1984, Prediction of packing of secondary structure, *Adv. Biophys.* **18:**115–148.

    Nagano, K., 1989, Prediction of packing of secondary structure, Chapter 11, this volume.

    Written in FORTRAN for use with an HITAC M-682H/680 computer system; compatible with the IBM 370 series computer.

12. Rose, G. D., Geselowitz, A. R., Lesser, G. J., Lee, R. H., and Zehfus, M. H., 1985, Hydrophobicity of amino acids in globular proteins, *Science* **229:**834–838. [R-D]

    Dworkin, J. E., and Rose, G. D., 1987, Hydrophobicity profiles revisited, in *Methods in Protein Sequence Analysis* (K. A. Walsh, ed.), Humana Press, Clinton, New Jersey, pp. 573–586.

    Rose, G. D., and Dworkin, J. E., 1989, The hydrophobicity profile, Chapter 15, this volume.

    Written in FORTRAN for use on a VAX or MICROVAX computer on a VMS operating system.

Appendix 3. Commercially Available Programs

1. HIBIO-PROSIS

   Secondary structure prediction:

   1. Chou, P. Y., and Fasman, G. D., 1978, *Adv. Enzymol.* **47**:45–148; 1978, *Annu. Rev. Biochem.* **47**:251–276.
   2. Garnier, J., Osguthorpe, D. J., and Robson, B., 1978, *J. Mol. Biol.* **120**:97–120.

   Hydrophobicity:

   1. Kyte, J., and Doolittle, R. F., 1982, *J. Mol. Biol.* **157**:105–132.
   2. Hopp, T. P., and Woods, K. R., 1981, *Proc. Natl. Acad. Sci. U.S.A.* **78**:3824–3828.
   3. Rose, G., 1978, *Nature* **272**:586–590.

   Written in C for use on any IBM-XT,AT microcomputer.

   Available from: Pharmacia LKB Biotechnology, 800 Centennial Avenue, P.O. Box 1327, Piscataway, New Jersey 08855-1327.

2. MSEQ: A Microcomputer-Based Approach to the Analysis, Display and Prediction of Protein Structure. Black, S. D., and Gloriso, J. C., 1986, *Bio Techniques* **4**:448–460.

   Chou, P. Y., and Fasman, G. D., 1978, Secondary structure prediction, *Annu. Rev. Biochem.* **47**:251–276.

   Hydrophobicity — 4 scales:

   1. Argos, P., and Palau, J., 1982, *Int. J. Peptide Prot. Res.* **19**:380–393.
   2. von Heijne, G., 1981, *Eur. J. Biochem.* **116**:419–422.
   3. Hopp, T. P., and Woods, K. R., 1981, *Proc. Natl. Acad. Sci. U.S.A.* **78**:3824–3828.
   4. Kyte, J., and Doolittle, R. F., 1982, *J. Mol. Bio.* **157**:105–132.

   Hydrophobic moments:

   Eisenberg, D., Weiss, R. M., and Terwilliger, T. C., 1982, *Nature* **299**:371–374; *Proc. Natl. Acad. Sci. U.S.A.* **81**:140–144.

   Graphic cartoons.

   Written in Basic for use on an IBM microcomputer family IBM-PC XT.

   Contact: Mr. Fred Reinhardt, University of Michigan Software, Intellectual Properties Office, 225W Engineering, Ann Arbor, Michigan 48109–1092.

3. NEWAT 85

   Protein sequence data base and programs.

   Categorized phylogenetically: *E. coli,* other prokaryotes, DNA viruses, eykaryotes (except vertebrate animals), vertebrates (except human), human and viruses.

   Software to enter new sequences: search for homologies between a sequence and the data base and to align a pair of sequences.

   Programs for displaying the hydropathy of a protein sequence; translating DNA sequences into putative amino acid sequences.

Written for IBM-PC.

Available from: Newat Distribution Co., Inc., P.O. Box 12822, La Jolla, California 92037.

4. PEP: Analyzing protein sequences.

   Reverse translates peptides and indicates ambiguity due to codon preferences or to the degeneracy of the genetic code.

   Identifies the least ambiguous regions of a peptide for making hybridization probes of different lengths.

   Simulates and maps protease digestion fingerprints.

   Determines hydropathicity to predict antigenic sites or membrane-binding regions.

   Predicts and maps protein secondary structure with the Chou-Fasman algorithm.

   Allows variably set amino acid equivalencies for similarity searches and alignments.

   Written in Mainsail for use on the VMS VAX, Microvax II comuters and Sunwork Stations.

   Available from: Intelligenetics, 1975 El Camino Road, Mountain View, California 94040-2216.

5. PEPPLOT

   Gribskov, M., Burgess, R. R., and Devereux, J., 1986, PEPPLOT: A protein secondary structure analysis program for the UWGCG sequence analysis software package.

   Structure Prediction:

   1. Chou, P. Y., and Fasman, G. D., 1978, Secondary structure prediction, *Adv. Enzymol.* **47**:45–147.
   2. Garnier, J., Osguthorpe, D. J., and Robson, B., 1978, Secondary structure prediction, *J. Mol. Biol.* **120**:97–120.
   3. Kyte, J., and Doolilttle, R. F., 1982, Hydropathy profile, *J. Mol. Biol.* **157**:105–132.
   4. Eisenberg, D., Sweet, R. M., and Terwilliger, T. C., 1984, Hydrophobic moment, *Proc. Natl. Acad. Sci. U.S.A.* **81**:140–144.

   Written in FORTRAN 77 for use on a Vax computer running version 3 or 4 of VMS.

   Available from: John Devereux, University of Wisconsin, Biotechnology Center, 1710 University Avenue, Madison, Wisconsin 53705.

Appendix 4: Relevant Programs Described in the Literature

Arnold, J., Eckerrode, U. K., Lemke, J., Phillips, G. J., and Schaeffer, S. W., 1986, A comprehensive package for DNA sequence analysis in FORTRAN IV for the PDP-11, *Nucleic Acids Res.* **14:**239–254.

Klein, P., and DeLisi, C., 1986, Prediction of protein structural class from the amino acid sequence, *Biopolymers* **25:**1659–1672.

Mount, D. W., 1986, Improved programs for DNA and protein sequence analysis on the IBM personal computer and other standard computer systems, *Nucleic Acids Res.* **14:**443–454.

Nakashima, H., Nishikawa, K., and Ooi, T., 1986, The folding type of a protein is relevant to the amino acid composition, *J. Biochem. (Tokyo)* **99:**153–162.

Nishikawa, K., and Ooi, T., 1986, Amino acids sequence homology applied to the prediction of protein secondary structures, and joint prediction with existing methods, *Biochim. Biophys. Acta* **871:**45–54.

Novotny, J., and Auffray, C., 1984, A program for prediction of protein secondary structure from nucleotide sequence data: Application to histocompatibility antigens, *Nucleic Acids Res.* **12:**243–253. (Editor's note: A combination of Chou, P. Y., and Fasman, G. D., 1978, *Adv. Enzymol.* **47:**45–148 and Rose, G. D., and Roy, S., 1980, *Proc. Natl. Acad. Sci. U.S.A.* **77:**4643–4647)

Peltola, H., Soderlund, H., and Ukkonen, E., 1986, Algorithms for the search of amino acid patterns in nucleic acid sequences, *Nucleic Acids Res.* **14:**99–107.

Reisner, A. H., and Bucholtz, C. A., 1986, The MTX package of computer programs for the comparison of sequences of nucleotides and amino acid residues, *Nucleic Acids Res.* **14:**233–238.

Staden, R., 1986, The current status and portability of our sequence handling software, *Nucleic Acids Res.* **14:** 217–231.

Taylor, P., 1986, A computer program for translating DNA sequences in protein, *Nucleic Acids Res.* **14:**437–441.

Trifonov, E. D., and Brendel, V., 1986, *GNOMIC. A Dictionary of Genetic Codes,* Balaban Publishers, Philadelphia.

van der Berg, J. A., and Osinga, M., 1986, A peptide to DNA conversion program, *Nucleic Acids Res.* **14:**137–140.

Review Articles

Moore, J., Engelberg, A., and Bairoch, A., 1988, Using PC/GENE for proteins and nucleic acid analysis, *Biotechniques* **6:**566–572.

Roe, B. A., 1988, Computer programs for molecular biology: An overview of DNA sequencing and protein analysis packages, *Biotechniques* **6:**560–565.

Appendix 5. National Resource Data Bases

1. BIONET™: National Computer Resource for Molecular Biology

Smith, D. H., Brutlag, D., Friedland, P., and Kedes, L. H., 1986, *Nucleic Acids Res.* **14:**17–20.

Kirstofferson, D., 1987, *Nature* **325:**555–556.

Available through: IntelliGenetics, 700 East El Camino Real, Mountain View, CA 94040.

IntelliGenetics Software

CLONER:	Recombinant DNA simulation system
DDMATRIX:	Dot matrix sequence similarity program
GEL:	DNA sequencing project management system
GENALIGN:	Multiple sequence alignment program
GENED:	Genetic sequence editor
IFIND:	Sequence similarity and alignment program
MAP:	Restriction map generator and editor
PEP:	Polypeptide sequence analysis system
QUEST:	Biological searching system
SEQ:	DNA sequence analysis system
SIZER:	Fragment length analysis system

2. The EMBL Data Library

Hamm, G. H., and Cameroa, G. N., 1986, The EMBL Data Library, *Nucleic Acids Res.* **14:**5–9.

The EMBL Data Library was the first internationally supported central resource for nucleic acid sequence data. Working in close collaboration with its American counterpart, GenBank, the library prepares and makes available to the scientific community a comprehensive collection of the published nucleic acid sequences.

Available through: European Molecular Biology Laboratory, Meyerhofstrasse 1, 6900 Heidelberg, Federal Republic of Germany.

3. The GenBank® Genetic Sequence Databank

Bilofsky, H. S., Burks, C., Fickett, J. W., Goad, W. B., Lewitter, F. I., Rindone, W., Swindell, C. D., and Tung, C.-S., 1986, The GenBank genetic sequence databank, *Nucleic Acids Res.* **14:**1–4.

The GenBank® Genetic Sequence Data Bank contains over 5700 entries for DNA and RNA sequences that have been reported since 1967. This paper briefly describes the contents of the data base, the forms in which the data base is distributed, and the services offered to scientists who use the GenBank data base.

Available through: GenBank, 700 East El Camino Real, Mountain View, CA 94040.

4. Molecular Biology Computer Research Resource (MBCRR)

Smith, T. F., Grushin, K., Tolman, S., and Faulkner, D., 1986, *Nucleic Acids Res.* **14:**25–29.

Analytic tools:

MASE: Full-screen multiple aligned sequence editor with regular expression highlighter.

LOCAL: Dynamic programming maximum local subsequence alignment algorithm, 1981, *J. Mol. Biol.* **147:**195–197.

PRSTRC: A modified Chou and Fasman protein structure algorithm, Ralph, W. W., Webster, T., and Smith, T. F., 1987, *Cabios* **3:**211–216.

DASHER: A high-speed hash-linked list sequence similarity search tool which ranks identified similarities by chi-square test on the occurrence distribution of common *n*-mers.

GGREP: Regular expression pattern search tool for GenBank and NBFR analogous to UNIX GREP. Employs the regular expression handler from GNU-EMACS.

ARIADNE: A pattern-directed inference and hierarchical abstraction in protein structure recognition, Lathrop, R. H., Webster, T. A., and Smith, T. F., 1987, *Commun. ACM* **30:**909–921.

RZMAP: A branch and bound algorithm to reconstruct restriction maps from double digest lengths, 1983, *Gene* **22:**19–29.

The MBCRR distributes the source, documentation, and MS-DOS executables of various utilities and programs for genetic sequence analyses to the noncommercial scientific community:

1. "Fristensky Package": Brian Fristensky's Cornell DNA sequence analysis programs.
2. "Mount Package": The Genetics PC-Software Center of the University of Arizona sequence analysis tools (developed by D. W. Mount, B. Conrad, and E. Myers).
3. "Lipman/Pearson Package": David Lipman and William Pearson's rapid biosequence similarity analysis code. *Science* **227:**1435–1441.
4. "Shalloway Package": David Shalloway's restriction/functional site data base management program (IBM-compatible executable code only).
5. "Zucker Package": Michael Zucker's RNA secondary structure software.
6. "Caltech Package": Alan Goldin's series of routines to analyze DNA or protein sequence data.

IntelliGenetics Programs (see above).

Available through: Molecular Biology Computer Research Resource, Dana Farber Cancer Institute, 44 Binney Street, Boston, MA 02115.

5. Protein Identification Resource (PIR). Sponsored by the National Biomedical Research Foundation (NBRF)

George, D. G., Barker, W. C., and Hunt, L. T., 1986, The protein identification resource, *Nucleic Acids Res.* **14:**11–15.

The NBRF–PIR Protein Sequence Database and the NBRF Nucleic Acid Sequence Database are distributed on magnetic tape in VAX/VMX and ASCII card image formats on a quarterly basis. The PSQ (Protein Sequence Query) and NAQ (Nucleic Acid Query) programs for browsing and information retrieval are distributed with the respective data bases in VAX/VMS format. The PIR sequence analysis software is updated approximately once a year. The data bases and software of the PIR are in the public domain and may be freely copied and redistributed provided the Protein Identification Resource is acknowledged as the source.

The NBRF–PIR Protein Sequence Database includes the following: all substantially sequenced proteins, including sequences translated from nucleic acid sequences; bibliographic citations for amino acid sequences, nucleic acid sequences, x-ray crystallography, active site determination, etc.; annotations identifying posttranslational modifications, active site, signal sequences, activation cleavages, disulfide bonds, intron locations, etc. Sequences translated from nucleotide sequences are checked against the author's translations and against sequences reported from protein sequencing. An auxiliary data base includes sequences in preparation as well as additional fragmentary and predicted sequences.

The NBRF Nucleic Acid Sequence Database contains entries annotated to show the locations of protein coding regions. The VAX/VMS nucleotide database tape also includes programs to reformat GenBank® and EMBL nucleotide databases to be accessible by the NAQ program.

The NBRF-PIR Sequence Analysis Software programs run on a VAX-11/780 operating under VAX/VMS version 4.2. All programs are written in VAX-11 FORTRAN (a superset of ANSI FORTRAN 77), with the exception of the Lipman–Pearson programs, which are written in VAX-11 C.

PSQ: Protein sequence query retrieval program.

NAQ: Nucleic acid query retrieval program.

FASTP and FASTN: Programs written by William Pearson of the University of Virginia that use the Lipman–Pearson algorithm (*Science* **227:**1435–1441, 1985) to search the protein and nucleic acid sequence data bases. The programs produce near-optimal alignments of segments identified in the search; they are extremely rapid.

RDF: Lipman–Pearson program for evaluation of RASTP results.

SEARCH: To compare a protein segment of, for instance, 25 residues with every 25-residue segment in the data base (gaps not permitted) using a scoring matrix.

ISEARCH: An interactive adaptation of our SEARCH program that includes a procedure for searching for ambiguous segments in which two or more amino acids may be found at some positions.

ALIGN: Uses a version of the Needleman–Wunsch algorithm to determine the best alignment of two sequences by computing a maximum match score; this score is compared with the average maximum match score from random permutations of the two sequences to derive an alignment score in standard deviation units. ALIGN uses a scoring matrix to assign a value to each pair of aligned amino acids in the sequences being compared; certain scoring matrices increase sensitivity for detecting similarity between distantly related sequences.

IALIGN: Interactive ALIGN program.

RELATE: To compare all segments of a given length from one sequence with all segments of a second one, using a scoring matrix. Statistics can be generated by comparing the results with those from permuted sequences.

DOTMATRIX: Graphic segment comparison display program similar to program RELATE. Output is specific for Printronix 300-line printer.

PRPLOT: A general-purpose program that plots values that can be used for Chou–Fasman-type predictions, such as hydrophilicity, hydrophobicity, β-turn-forming potential, etc., averaged over several contiguous amino acids.

CHOFAS: A secondary structure prediction program with a convenient and compact display developed by M. Kanehisa as part of the IDEAS package.

HYDRO: Hydrophobicity scoring matrix.

Levitt, M., 1976, *J. Mol. Biol.* **104:**59–109.

Available through: National Biomedical Research Foundation. Georgetown University Medical Center, 3900 Reservoir Road NW, Washington, D.C. 20007.

6. Protein Data Bank: A computer-based archival file for macromolecular structures.

Bernstein, F. C., Koeyzle, T. F., Williams, G. J. B., Meyer, J. E. F., Brice, M. D., Rodgers, J. R., Kennard, O., Shimanouchi, T., and Tatsumi, M., 1977, The Protein Data Bank: A computer-based archival file for macromolecular structures, *J. Mol. Biol.* **112:**525–542.

Available through: Protein Data Bank, Chemistry Department, Brookhaven National Laboratory, Upton, NY 11973; University Chemical Laboratory, Lensfield Road, Cambridge CB2 1EW, United Kingdom; and University of Tokyo, Hongo, Tokyo, Japan.

7

The Role of Energy Minimization in Simulation Strategies of Biomolecular Systems

D. H. J. Mackay, A. J. Cross, and A. T. Hagler

D. H. J. Mackay and A. J. Cross • Biosym Technologies, Inc., San Diego, California 92121. *A. T. Hagler* • Biosym Technologies, Inc., San Diego, California 92121; and Agouron Institute, La Jolla, California 92137.

I. INTRODUCTION

It is now possible to calculate the classical energy of a complex system such as a protein as a function of its coordinates. By making many such calculations for various coordinate values, one can explore multidimensional energy surfaces. These energy surfaces are the basis for molecular dynamics and Monte Carlo studies. Another important method for exploring these energy surfaces is to find configurations for which the energy is a minimum. By this, we mean finding a point in configuration space where all of the forces on the atoms are balanced. By simply minimizing the energy of a molecule, we can identify stable conformations. Perhaps more importantly, by adding external forces to the molecule in the form of restraints and constraints, a wide range of modeling strategies can be developed using minimization techniques as the foundation to answer specific questions. For example, by forcing specific atoms to overlap atoms in a template structure during a molecular geometry minimization, one can answer the question, "how much energy is required for one molecule to adopt the shape of another." In this chapter, we discuss how minimization techniques are used in a variety of molecular modeling strategies, focusing on the use of constraints and restraints to extend the scope and utility of traditional structure minimization.

Before we discuss the advanced minimization strategies, however, we will spend some time discussing the characteristics and challenges of the commonly used algorithms. For example, although the goal of a minimization can be simply stated, there may be difficulties in achieving it for a given system, particularly a large system such as a protein. There generally exist several points where the atomic forces are zero, although we usually are interested in the point of lowest energy, known as a global minimum. Convergence is critical in minimization studies (as well as in dynamics) and is, in theory, easy to define: all derivatives are zero (and the second-derivative matrix is positive definite). In practice, however, being "close" to the true minimum can be difficult to define and is also sometimes difficult to judge accurately. Finally, there are different algorithms for minimizing the energy of a system, and one must choose the algorithm that is appropriate. We address these issues as well as review the various algorithms that are appropriate for particular types of systems or at certain stages of the minimization procedure.

Minimization provides information that is complementary to that obtained from molecular dynamics or Monte Carlo. Ensembles of dynamics or Monte Carlo structures are useful for calculating thermodynamic averages and estimating entropy, but the large number of structures involved makes detailed microscopic analysis cumbersome. Minimized structures in some sense represent the underlying configurations about which fluctuations occur during dynamics and, as such, provide a convenient and meaningful basis for structural analysis (Hagler, 1985; Struthers *et al.*, 1984).

A minimized structure can be analyzed in a variety of ways to obtain diverse information about a molecular system. For example, a minimized enzyme–substrate complex provides detailed structural information, and the energies calculated at the minimum can be partitioned into contributions from specific interactions. If minimizations of similar complexes are sufficiently refined, one can obtain estimates of the relative enthalpies of binding as well as the strain energy introduced in the ligands on binding. In a recent study of the binding of the ligand trimethoprim (TMP) to bacterial and vertebrate dihydrofolate reductase (DHFR), it was found that although the strain induced in TMP on binding was nearly the same in both complexes, it arose from distortions in different regions of the ligand in the two cases (Roberts *et al.*, 1986).

Minimization also provides a straightforward way of studying the structural consequences of modifications to the primary sequence of a protein. Since the recent development of site-directed mutagenesis of proteins, it is nearly routine to incorporate desired amino acid substitution into a protein of known sequence. Such experiments may be complemented by computer modeling calculations from which a coherent understanding of the relevant molecular interactions may emerge.

A series of constrained minimizations can be used to model the energetics of small conformational changes in protein. For example, the crossing of a torsional barrier can be studied by fixing the torsion angle at a series of values and minimizing the energy of the surrounding system (i.e., all remaining degrees of freedom) at each point. This technique has been used successfully in small molecules (Stern *et al.*, 1983) and is easily extended to larger systems.

There are several limitations to such calculations. A fundamental limitation is the reliability of the mathematical models describing the molecular energies and interactions. Much progress has been made in the past 20 years to derive valid energy functions and parameters to simulate simple systems, beginning with hydrocarbons and more recently peptides (Lifson *et al.*, 1979; Hagler *et al.*, 1979a,b; Dinur *et al.*, 1986) and nucleic acids (Nilsson and Karplus, 1986; Weiner *et al.*, 1984). However, this work is incomplete. Solvent effects, polarizability, cross terms or coupling, and quantum effects are often ignored in biomolecular calculations. Until these aspects are included and the models are shown to be rigorous and robust, conclusions based solely on energy calculations must be critically examined. Often, these should be regarded as powerful enhanced modeling studies rather than as rigorous theoretical thermodynamic or mechanics studies of the systems of interest.

Insufficient computer resources also limit the scope of problems that can be addressed by energy calculations. For example, it is virtually impossible to amass enough computer time to fold proteins by brute force beginning only with the amino acid sequence. However, if additional clues are available (e.g., NMR data and/or structural templates from homologous proteins), one can model a reasonable protein structure from the linear sequence with some success. Nuclear Overhauser effect (NOE) constraints can be used in conjunction with "distance geometry" (Crippen, 1977; Havel *et al.*, 1983) to yield structural information, and this is discussed below. Also, known structures and structural components can be used as templates to model homologous sequences in unknown proteins (Greer, 1985). Advances in these modeling areas are typified by recently developed algorithms that can search a data base of known protein structures for loops that are structurally homologous to a template structure on either side of the loop but are not necessarily homologous in the loop region itself (Dayringer *et al.*, 1986; Jones, 1982). Since most insertions and deletions among homologous proteins occur in the loop regions, this data base can be used to model amino acid insertions and deletions when building unknown structures by homology. A multidisciplinary approach including such knowledge-based systems, experimentally determined constraints, and molecular modeling provides a viable strategy for addressing the formidable problem of protein folding.

The remainder of this chapter examines various optimization strategies commonly used in molecular modeling. A description of typical minimization algorithms is followed by a discussion and examples of how constraints and restraints expand the applications of minimization methods.

II. MINIMIZATION ALGORITHMS

A. The General Minimization Strategy

Minimization of a molecular structure occurs in two steps. First, an equation describing the energy of the system as a function of its coordinates must be defined and evaluated for a given conformation. Target functions may be constructed, which also include external restraining terms to bias the minimization in addition to the energy terms. Next, the conformation is adjusted to lower the value of the target function. A minimum may be found after one adjustment or may require many thousands of iterations, depending on the nature of the algorithm, the form of the target function, and the size of the molecule. The efficiency of the minimization is therefore judged by both the time to evaluate the target function and the number of structural adjustments iteration of the minimization. In steepest descent of the example given above, it would be along $(2x, 10y)$, the gradient. If our direction is the gradient, this new one-dimensional surface can be expressed parametrically in terms of a new one-dimensional coordinate, α (Fig. 3)

1. A Specific Minimization Example

To introduce the various minimization algorithms, we apply each to the minimization of a pure quadratic in two dimensions. Most minimization algorithms, as we will see, assume that the energy surface is approximately harmonic. Even for nonharmonic surfaces, the shape of the surface becomes harmonic in the limit as one converges on the minimum. Therefore, this is a useful approximation. The target function we will use is an elliptical surface in two dimensions described by the equation

$$E(x,y) = x^2 + 5y^2 \tag{1}$$

Although this is a simplified function, it illustrates the properties and captures the mathematical essence of the formulations. Every minimization begins with some equation analogous to equation 1. In addition to an equation defining the energy surface, one must provide a starting set of coordinates, an initial guess for (x,y). Figure 1 is a contour plot of the energy E in the (x,y) plane. Each ellipse is spaced two energy units apart and represents a locus of points with the same energy. (This is analogous to a contoured topographical map.) Of course, the minimum in this simple function is trivial and can be discerned by inspection to be $x = 0$, $y = 0$ (the origin).

Given a target function that defines the energy surface (such as shown in Fig. 1) and an initial starting point, a minimizer must determine both the direction toward a minimum and the distance to the minimum in that direction. A good initial direction is simply the slope or gradient of the function at the current point. The gradient of equation 1 is a two-dimensional vector:

$$\left(\frac{dE}{dx}, \frac{dE}{dy}\right) = \nabla E = (2x, 10y)$$

Note that the gradient is proportional to the coordinates in this case; thus, the farther we are from the minimum (the origin), the larger the gradient will be. This is an important property of a gradient: it includes information on how far away we are from the minimum. The magnitude of the gradient is also the most rigorous way to characterize the convergence of a minimization. A minimization is converged when the gradient is equal (or nearly equal) to zero.

The gradient, however, only points "downhill" and not necessarily to the minimum (see Fig. 2). Thus, as we follow in the direction of the initial gradient, it changes and points in yet a new direction. Predicting how the gradient changes and using this information to improve the efficiency of the minimization is the primary objective of both the conjugate gradients and the Newton–Raphson algorithms and are discussed below. First, however, we introduce the concept of a line search, which is an implicit component of most minimizers.

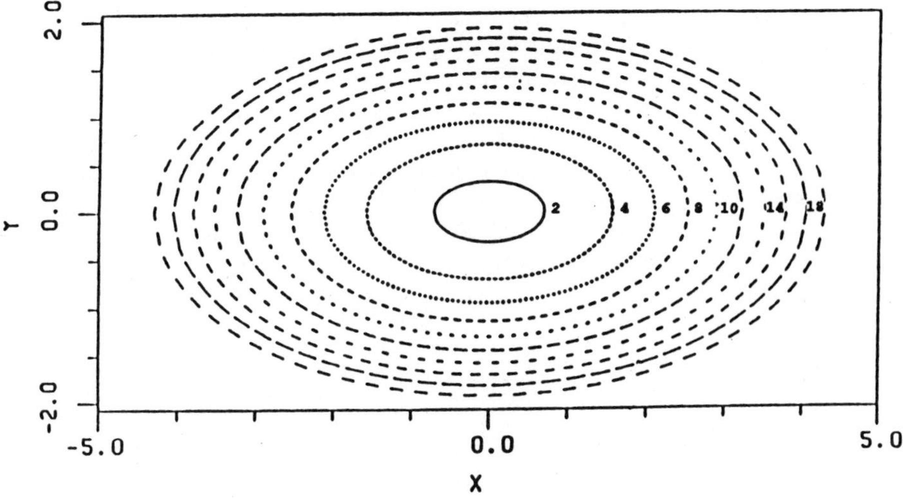

Figure 1. An energy contour surface for the function $x^2 + 5y^2$. Each contour represents an increase of two arbitrary energy units.

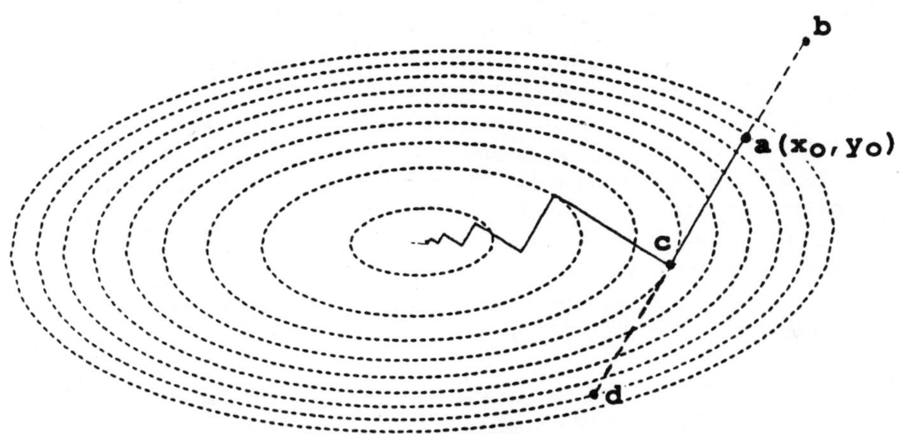

Figure 2. An energy surface for equation 1 with the gradient from the initial point a (x_0,y_0) defining the line search direction. Note that the gradient does not point directly to the minimum. Compare this representation to that of Fig. 3, where the line (b–a–c–d) is searched in one dimension for the minimum. Note that the minimum (point c) occurs precisely at the point where the gradient is tangent to the energy contours, thus implying that the subsequent gradient will be orthogonal to the previous gradient.

2. The "Line Search"

Minimizers usually have two major components. The more generic part is the so-called "line search," which actually changes the coordinates to a new lower-energy structure. Line searches can be critical to some algorithms such as conjugate gradients. To illustrate what is meant by a line search, consider Fig. 2 in which the gradient direction from an arbitrary starting point has been superimposed onto our elliptic function. We start at point a, defined by the coordinates (x_0,y_0). Simply put, a line search amounts to a one-dimensional minimization along a given direction. The line to be searched is taken along the direction vector determined at each iteration of the minimization. In steepest descent of the example given above, it would be along $(2x, 10y)$, the gradient. If our direction is the gradient, this new one-dimensional surface can be expressed parametrically in terms of a new one-dimensional coordinate, α (Fig. 3):

$$x' = x_0 + \alpha \left.\frac{\partial E}{\partial x}\right|_{x_0 y_0} \tag{2}$$

$$y' = y_0 + \alpha \left.\frac{\partial E}{\partial y}\right|_{x_0 y_0}$$

where (x',y') are coordinates along the line away from the current point, (x_0,y_0), in the direction of the gradient at (x_0,y_0), $(\partial E/\partial x,\ \partial E/\partial y)_{x_0,y_0}$. If we calculate the energy of these new points and plot them as a function of α, we obtain the curve in Fig. 3.

As one minimizes along this line from point a to point c, note that we have moved to a point on the line that is tangential to a contour. A general result of a line search is that the gradient at the minimum of the line search must be perpendicular to the previous direction. This is an important property of line searches, which we invoke in our discussion of the conjugate gradients algorithm (Section II.C).

Line searches do not depend on the algorithm that generated the direction vector. The general strategy is simply to bracket the one-dimensional minimum between two points higher in energy (for example, points b and d in Fig. 3) and converge on the actual minimum by successive iterations (e.g., starting at point a, the first step might take us to b, then reversing direction to d and finally to c, the minimum). Line searches are aesthetically attractive since they extract all the energy from one direction before moving on to the next. Also, since the new gradients are always perpendicular to the previous directions, one can hope that this would produce an efficient path to the minimum. In practice, however, line searches are costly in terms of the number of function evaluations that must be performed. The energy must be evaluated at three to ten points to locate precisely the one-dimensional minimum; thus, extensive line searches can be inefficient.

We now consider three algorithms for choosing directions that are commonly used in molecular mechanics minimizations: steepest descents, conjugate gradients, and Newton–Raphson. To be consistent in our discussions of efficiency, we must explicitly define what we mean by minimization iteration. An iteration is complete when the direction vector is updated. For minimizers using a line search, each completed line search is therefore an iteration. Iterations should not be confused with function evaluations. As we have seen, a single iteration may require several function evaluations during the line search. Therefore, the efficiency of a minimizer must be judged both by the number of iterations required to converge and the number of function evaluations needed per iteration.

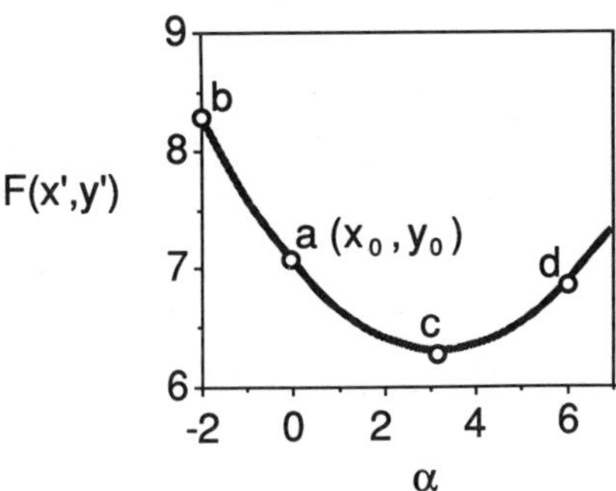

Figure 3. The cross section of the energy surface defined by the intersection of the gradient in Fig. 2 with the energy surface. The independent variable, α, is a one-dimensional parameter that is adjusted to minimize the value of the function $E(x',y')$, where x' and y' are parameterized in terms of α in equation 2. The point a corresponds to the initial point (when α is 0), and point c is the local one-dimensional minimum. Points b and d along with a bound the minimum and form the basis for an iterative search for the minimum.

B. Steepest Descents: A Robust but Slowly Converging Algorithm

In steepest descents, the line search direction is simply taken as the gradient. After each line search, the old direction is replaced with the gradient at the new point, and the line search is repeated. Figure 4 shows the minimization path followed by a steepest-descent approach on our simple quadratic function. As expected, each line search produces a new direction that is perpendicular to the previous gradient. However, using the current gradient as the next direction leads to oscillations on the way to the minimum. This inefficient behavior is characteristic of steepest descents, especially on energy surfaces with narrow valleys.

In Fig. 4, although each successive direction is orthogonal to the previous direction, the path to the minimum is not particularly efficient. What would happen if we eliminated the line search and simply updated the position any time the trial point along the gradient has a lower energy? The advantage is that the number of function evaluations performed per iteration would be dramatically decreased. Furthermore, by constantly changing the direction to match the current gradient, we might hope to damp out the oscillations along the minimization path. The result of such a minimization is given in Fig. 5. The minimization begins from the same point as in Fig. 4, but each "line search" uses at most two function evaluations (if the trial point has a higher energy, the step size is adjusted downward, and a new trial point generated). Note here that the steps are more erratic, but we arrive at the minimum in roughly the same number of iterations. The critical aspect, however, is that by avoiding comprehensive line searches, the total number of function evaluations is only 10–20% of that used by the rigorous line search method.

Steepest descents' exclusive reliance on gradients is both its weakness and its strength close to the minimum. Its convergence is slow near the minimum because the gradient approaches zero, but it is extremely robust even when systems are far from harmonic. It is the most likely method to generate a lower-energy structure no matter what the function is or where it begins. Therefore, steepest descents is often used initially when configurations are far from the minimum, i.e., when the gradients are large (and it is efficient in this region). This is commonly the case for initial relaxation of poorly refined crystallographic data or for molecules built graphically. In fact, as we see below, more advanced algorithms often begin by design with a steepest descents direction as the first step.

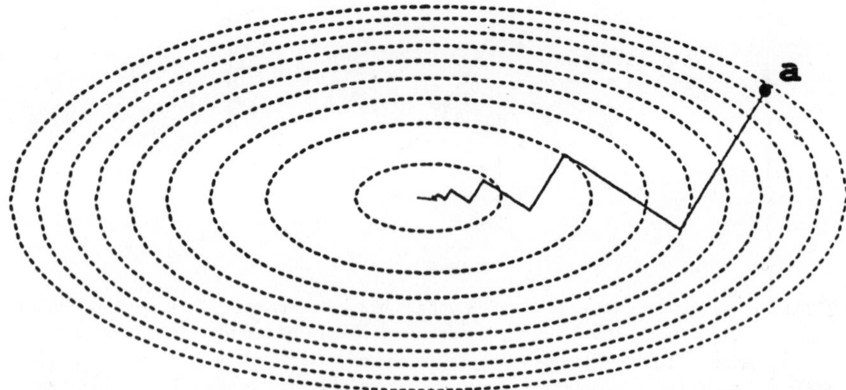

Figure 4. Minimization path following a steepest descent path using complete line searches starting from point a and converging on the minimum in about 12 iterations. In this case, where a rigorous line search is carried out, approximately eight function evaluations were needed for each line search using a quadratic interpolation scheme. Note how steepest descents consistently overshoots the best path to the minimum, resulting in an inefficient oscillating trajectory.

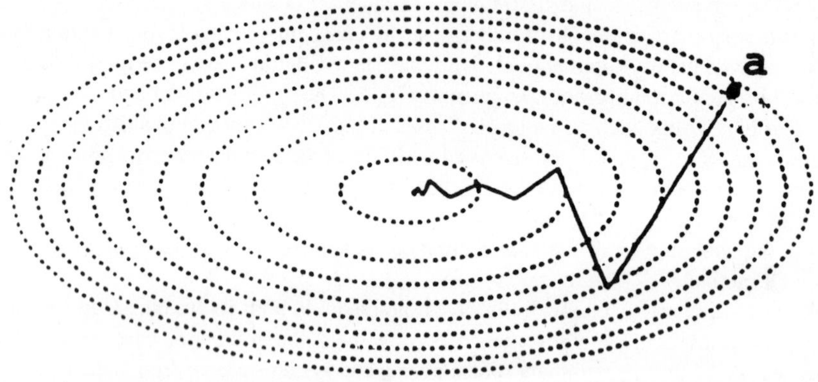

Figure 5. Minimization path following a steepest descent path with no line searches starting from point a and converging on the minimum in about 12 iterations. Although the number of iterations is comparable to a steepest descent path with line searches (Fig. 4), the total minimization was five times faster since, on the average, each iteration used only 1.3 function evaluations. In most applications to biological systems, the function evaluation is the most time-consuming portion of the calculation.

C. Conjugate Gradients

One reason why the minimization paths followed in steepest descents are inefficient is that each segment of the path tends to retrace progress made in an earlier iteration. For example, in Fig. 4, each line search deviates somewhat from the ideal direction to the minimum. Successive line searches correct this deviation, but because each direction must be orthogonal to the previous direction, the deviation is not completely corrected. Thus, the path oscillates and continually overcorrects for poor choices of directions in earlier steps.

A preferable scheme would be one that prevents the next direction vector from undoing earlier progress or, in other words, an algorithm that produces a complete set of mutually conjugate directions such that each successive step continually refines the direction toward the minimum. The conjugate gradients algorithm constructs and follows such a set of directions.

In conjugate gradients, $\boldsymbol{h}_{i+1}$, the new direction vector leading from point $i + 1$, is computed by adding the gradient at point $i + 1$, $\boldsymbol{g}_{i+1}$ to the previous direction $\boldsymbol{h}_i$ scaled by a constant γ_i:

$$\boldsymbol{h}_{i+1} = \boldsymbol{g}_{i+1} + \gamma_i \boldsymbol{h}_i \tag{3}$$

γ_i is a scalar defined as:

$$\gamma_i = \frac{\boldsymbol{g}_{i+1} \cdot \boldsymbol{g}_{i+1}}{\boldsymbol{g}_i \cdot \boldsymbol{g}_i} \tag{4}$$

This direction is then used in place of the gradient in equation 2, and a new line search is conducted. This construction has the remarkable property that the next gradient, $\boldsymbol{g}_{i+1}$, will be orthogonal to all previous gradients $(\boldsymbol{g}_0, \boldsymbol{g}_1, \boldsymbol{g}_2, \ldots, \boldsymbol{g}_i)$ and that the next direction, $\boldsymbol{h}_{i+1}$, will be conjugate to all previous directions $(\boldsymbol{h}_0, \boldsymbol{h}_1, \boldsymbol{h}_2, \ldots, \boldsymbol{h}_i)$.*

Several practical aspects of conjugate gradients are worth mentioning before we proceed. First, the conjugate gradients algorithm requires convergence along each line search before continuing in the next direction. The gradient at $(i + 1)$ *must* be perpendicular to $\boldsymbol{h}_i$ or the derivation guaranteeing a conjugate set of directions will break down. Secondly, to start conjugate gradients, one must choose an initial direction, $\boldsymbol{h}_0$, equal to the initial gradient. Finally, the only additional storage required is an extra vector of N elements to hold the N components of the "old" gradient. For minimizing molecular energy in Cartesian space, this would mean storing the derivatives of the energy with respect to the x, y, and z coordinates of each atom. This makes conjugate gradients the method of choice for systems too large to store and manipulate a second-derivative matrix (requiring $N \times N$ components), which is required by Newton-Raphson style minimizers, as discussed in Section II.D.

To illustrate the properties of conjugate gradients, let us return to our sample quadratic function, $E(x,y) = x^2 + 5\,y^2$. The minimization path followed by conjugate gradients on this surface is depicted in Fig. 6.

In the first iteration, a line search is conducted from a along the gradient identical to that of steepest descents. At point b, a value of γ can be calculated from:

$$\gamma = \frac{(\mathrm{d}E_\mathrm{b}/\mathrm{d}x)^2 + (\mathrm{d}E_\mathrm{b}/\mathrm{d}y)^2}{(\mathrm{d}E_\mathrm{a}/\mathrm{d}x)^2 + (\mathrm{d}E_\mathrm{a}/\mathrm{d}y)^2} = (4x_\mathrm{b}^2 + 25y_\mathrm{b}^2)/(4x_\mathrm{a}^2 + 25y_\mathrm{a}^2) \tag{5}$$

*Thus, the term "conjugate gradients" is somewhat of a misnomer. The algorithm produces a set of mutually orthogonal gradients and a set of mutually conjugate directions.

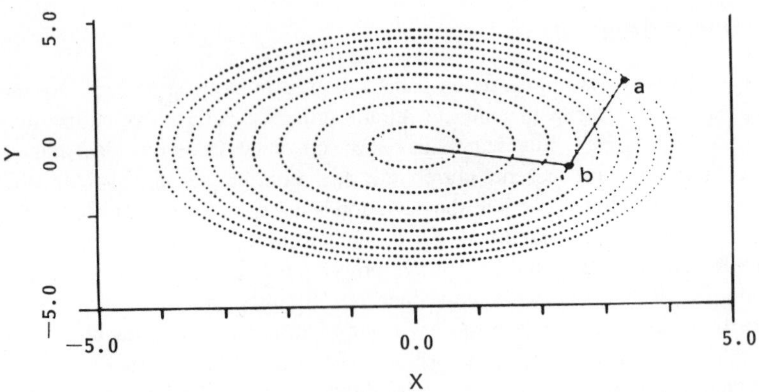

Figure 6. Minimization path following a conjugate gradients path with line searches starting from point a and converging on the minimum in two iterations. The total number of function evaluations needed was approximately half those needed for steepest descents without line searches and only 10% compared to steepest descents with line searches. As in steepest descents, successive line searches result in a set of mutually orthogonal gradients. Unlike steepest descents, however, successive directions are not orthogonal but rather conjugate. Constraining the second direction to be conjugate to the first results in a vector that passes through the minimum. The two directions used are thus a complete set of mutually conjugate directions for this two-dimensional system. Conjugate gradients will converge in N iterations for a harmonic system (where N is the dimensionality of the system) if the line search is exact. Anharmonic systems may require several passes of N steps each.

where dE_i/dx and dE_i/dy are the derivatives of E at point i. Then, the new direction vector at point b, $\boldsymbol{h}_b$, can be calculated:

$$\boldsymbol{h}_b = g_b + \gamma \boldsymbol{h}_a = (2\,x_b, 5\,y_b) + \gamma(2\,x_a, 5\,y_a)$$

No matter the what the initial point is, γ multiplied by the old direction will always provide the correction to the gradient to produce a new direction conjugate to all previous directions. In our example in two dimensions, we see that $\boldsymbol{h}_b$ places the minimum directly in the path of the next line search, and the minimum is found in just two iterations.

Conjugate gradients is the method of choice for large systems where storing and manipulating a large second-derivative matrix is impractical. Note, however, that since complete line search minimizations must be performed along each direction, and since line searches consume several function evaluations per search, conjugate gradients may, in practice, not be as efficient as a quasi-Newton–Raphson in terms of total function evaluations required. Also, although not discussed here, the derivation of equations 3 and 4 assumes that the function being minimized is quadratic. For nonharmonic systems, conjugate gradients can exhaustively minimize along its conjugate directions without converging. In these cases, one can restart the algorithm, and several iterations of N steps each may be required. For a detailed discussion of this algorithm, see the recent excellent text by Press *et al.* (1986) or the somewhat more formal treatments by Fletcher (1980).

D. Newton–Raphson: An Expensive but Rapidly Converging Algorithm

Steepest descents uses the gradient as the search direction directly. Conjugate gradients improves on steepest descents by minimizing only along directions that are mutually conjugate so that movement along one direction will not counteract progress made in earlier iterations. In each case, the first derivative or gradient of the function is used. In general, for a harmonic function, N^2 independent data points are required to solve an equation in N variables numerically. Since a gradient is a vector N long, the best one can hope for in a gradient-based minimizer is to converge in N steps (conjugate gradients achieves this goal for harmonic functions, but steepest descents is less efficient). However, if one can apply second-derivative information, a minimization could in principle converge in one step, since each second derivative is an $n \times n$ matrix. This is the principle behind the variable metric minimization algorithms, of which Newton–Raphson is perhaps the most common.

Another way of looking at Newton–Raphson is that in addition to using the gradient, which helps identify a direction, it also uses the curvature of the function (the second derivative) to predict where along the gradient the function will change directions (pass through a minimum). Since the complete second-derivative matrix defines the curvature in each gradient direction, we can multiply the inverse of the second-derivative matrix by the gradient to obtain a vector that will translate us directly to the nearest minimum. This is expressed mathematically as:

$$\boldsymbol{r}_{\min} = \boldsymbol{r}_0 - \mathrm{A}_0^{-1} \cdot \nabla V(\boldsymbol{r}_0) \tag{6}$$

where $\boldsymbol{r}_{\min}$ is the predicted minimum, $\boldsymbol{r}_0$ is an arbitrary starting point, A_0 is the matrix of second partial derivatives of the energy with respect to the coordinates at $\boldsymbol{r}_0$ (also known as the Hessian matrix), and $\nabla V(\boldsymbol{r}_0)$ is the gradient of the potential energy at $\boldsymbol{r}_0$. To illustrate how this expression works in practice, consider once again our elliptical function, $E(x,y) = x^2 + 5\,y^2$. The gradient and Hessian matrix at any given point, (x_0,y_0), are easily obtained as:

$$\left.\frac{\mathrm{d}\,E}{\mathrm{d}x}\right|_0 = 2\,x_0 \qquad \left.\frac{\mathrm{d}\,E}{\mathrm{d}y}\right|_0 = 10y_0 \tag{7}$$

$$A_0 = \begin{pmatrix} \dfrac{\mathrm{d}^2E}{\mathrm{d}x^2} & \dfrac{\mathrm{d}^2E}{\mathrm{d}y\mathrm{d}x} \\ \dfrac{\mathrm{d}^2E}{\mathrm{d}x\mathrm{d}y} & \dfrac{\mathrm{d}^2E}{\mathrm{d}y^2} \end{pmatrix} = \begin{pmatrix} 2 & 0 \\ 0 & 10 \end{pmatrix} \tag{8}$$

The inverse of the Hessian $\boldsymbol{A}^{-1}$, is trivial for this diagonal matrix and is given by

$$\boldsymbol{A}^{-1} = \begin{pmatrix} \frac{1}{2} & 0 \\ 0 & \frac{1}{10} \end{pmatrix} \tag{9}$$

Substituting these explicit terms into Equation 6, we may solve for

$$\begin{aligned} (x_{\min},y_{\min}) &= (x_0,y_0) - \begin{pmatrix} \frac{1}{2} & 0 \\ 0 & \frac{1}{10} \end{pmatrix}(2x_0, 10y_0) \\ &= (x_0,y_0) - \tfrac{1}{2}(2x_0) - \tfrac{1}{10}(10y_0) = (x_0,y_0) - (x_0,y_0) = (0,0) \end{aligned} \tag{10}$$

Thus, we have found the minimum (for any arbitrary initial point x_0,y_0) without line searches and with only a single evaluation of the gradient and the second-derivative matrix. Since, as was noted above, a molecular energy surface in general is not harmonic, the algorithm must be applied iteratively. In practice, however, convergence is extremely rapid near the minimum.

As elegant as this algorithm appears, there are several drawbacks to its use in molecular modeling. First, the terms in the Hessian matrix are difficult to derive and are computationally costly for molecular force fields. Furthermore, when a structure is far from the minimum (where the energy surface is anharmonic), the minimization can become unstable. For example, consider the case of two atoms within van der Waals contact of each other. The steep repulsive wall of the van der Waals potential results in a large "flat" gradient. For this case, the Newton–Raphson algorithm will compute a large step (a large gradient divided by the small curvature) that may overshoot the minimum and lead to a point even further from the minimum than where we started. Thus, the method can diverge rapidly if the initial forces are too high (or if the surface is too "flat"). Finally, calculating, inverting, and storing an $n \times n$ matrix for a large system can become unwieldy. Even taking into account that the Hessian is symmetric and that each of the tensor components is also symmetric, the storage requirements scale as ($3N^2$) for N atoms. Thus, for a 200-atom system, 120,000 words are required. The Hessian alone for a 1000-atom system already approaches the limits of a Cray-XMP super computer, and a 10,000-atom system is currently intractable.

A variation on the Newton–Raphson algorithm, known as "quasi-Newton–Raphson" or "variable metric" methods, was developed by Fletcher and Powell (Fletcher, 1980) and is made by calculating the Hessian matrix numerically using only first-derivative information rather than with analytical second derivatives. The energy function evaluation is faster since the second derivatives are not calculated analytically. Furthermore, since the second-derivative information is built up slowly, the algorithm behaves like steepest descents at the start when the structure is least harmonic and like Newton–Raphson at the end as the structure approaches the minimum. However, since the Hessian matrix is maintained, the size limitation remains. These "quasi-Newton–Raphson" or "variable metric" methods are in principle no faster at converging than conjugate gradients. However, they are less sensitive to line search convergence, and so in practice, a quasi-Newton–Raphson without line searches will converge faster than conjugate gradients.

For systems that fall within the requirements of the "quasi-Newton–Raphson" algorithm, it is the preferred method. Pure Newton–Raphson is primarily reserved for cases where rapid convergence to an extremely precise minimum is required, for example, from initial derivatives of 0.1 kcal/mole-Å to 10^{-8}. Such extreme convergence is necessary to perform vibrational normal mode analysis, where even small residual derivatives can lead to errors in calculated vibrational frequencies.

E. Dynamic Quenching: Minimization by Freezing

Molecular dynamics can be used to minimize a system if the integration algorithm used simulates contact with a temperature bath (Berendsen *et al.*, 1984). This algorithm constantly monitors and adjusts the average kinetic energy of the atoms to maintain a given temperature. By simply setting this temperature to a very low value (e.g., 1 K) or by decreasing it gradually during the course of the simulation, one can reduce both the kinetic and potential energy of the system. This is similar in spirit to steepest descents except the energy is allowed to increase through fluctuations in the exchange of potential and kinetic energy. It has been argued that this should allow the system to overcome small barriers during the relaxation procedure and finally settle in a lower energy minimum. In practice, the time scale during which the cooling must occur has not been well characterized. In at least one case we have found that dynamic quenching did not provide better convergence than steepest descents and did not lead to better quality minima for medium-sized peptides within 20 psec once the temperature dropped below 300 K (unpublished results). Dynamic annealing as described below does, however, appear to be a useful method to search conformational space in conjunction with high-temperature dynamics.

III. MODELING/DESIGN STRATEGIES USING MINIMIZATION

The "target function" is the term used to describe the function for which a minimum value is sought. As discussed in the introduction, by adding various biasing functions to the energy, we can construct target functions to carry out a variety of modeling strategies. Later in this section, we describe how to bias the energy for specific modeling tasks. First, however, we describe the fundamental energy function and give examples of what can be learned from detailed analysis of straightforward energy-minimization studies.

A. The Molecular Energy Expression

Every target function includes terms to describe the energy of a molecule: the strain present in bond lengths, angles, and dihedrals (internal coordinates), and between so-called "nonbonded" atoms (primarily, electrostatic and van der Waals). In energy minimization, as in molecular dynamics or Monte Carlo calculations, the potential energy of the system is expressed in terms of an analytical function and the internal coordinates of the molecules and the distances between atoms. Equation 11 gives a representation that has been used successfully to describe such a molecular force field; this expression reflects the energy necessary to stretch bonds (b), to distort bond angles (θ) from their unstrained geometries, and to rotate about torsion angles (ϕ) by twisting atoms about the bond axis that determines the torsion angle. The parameters D_b, α, H_θ, H_ϕ, and H_χ are the force constants for the corresponding deformations, and $F_{bb'}$, $F_{\theta\theta'}$, $F_{b\theta}$, $F_{\theta\theta'\phi}$, and $F_{\chi\chi'}$ are force constants for cross terms, which account for off-diagonal interactions of the internal coordinates. The final summation represents the nonbonded interactions by a sum of repulsions, attractive dispersion forces, and coulomblike electrostatics as a function of the distance, r_{ij}, between atom pairs.

$$
\begin{aligned}
E_{\text{pot}} = {} & \sum_b D_b\{1 - \exp[-\alpha(b - b_0)]\}^2 + \tfrac{1}{2}\sum_\theta H_\theta(\theta - \theta_0)^2 + \tfrac{1}{2}\sum_\phi H_\phi[1 + s\cos(n\phi)] + \\
& \tfrac{1}{2}\sum_\chi H\chi^2 + \sum_b\sum_b F_{bb'}(b - b_0)(b' - b_0') + \sum_\theta\sum_\theta F_{\theta\theta'}(\theta - \theta_0)(\theta' - \theta_0') + \\
& \sum_b\sum_\theta F_{b\theta'}(b - b_0)(\theta - \theta_0) + \sum_\phi F_{\theta\theta'\phi}\cos\phi(\theta - \theta_0)(\theta' - \theta_0') + \sum_\chi\sum_{\chi'} F_{\chi\chi'}\chi\chi' + \\
& \sum_i\sum_{j>i}[A_{ij}/r_{ij}^{12} - B_{ij}/r_{ij}^{6} + q_i q_j/r_{ij}]
\end{aligned}
\qquad (11)
$$

There is increasing evidence that off-diagonal terms are important for an accurate representation of the force field (Hagler, 1985; Ermer, 1976). Recent comparisons have been made between energies calculated from the analytical force field expression and *ab initio* calculations for a small molecule that is representative of the important interactions in polypeptides, N-methylacetamide (Dinur *et al.*, 1986). Preliminary results indicate that anharmonic terms in bond stretching (as accounted for by the Morse potential form), and particularly the bond-angle cross terms, are essential for accurate representation of the energy of a moderately distorted molecule. Several alternative force fields for modeling macromolecules are currently available (Lifson *et al.*, 1979; Hagler *et al.*, 1979a,b; Weiner *et al.*, 1984; Brooks *et al.*, 1983).

In general, the accuracy required in a force field depends on the properties of the system that are of interest. A less accurate force field may reproduce the minimum-energy structure of a molecule reasonably well, although the calculated energy may not be accurate enough for quantitative comparisons of similar structures. For example, minimization schemes are being increasingly included in real-time molecular graphics applications: relaxing a mutated side chain or docking a substrate in the field of a fixed protein (Katz and Levinthal, 1972; Pattabiraman *et al.*, 1985; Karfunkel, 1986). To achieve real-time performance, the force fields are reformulated, and major portions of the nonbonded interactions are precalculated for fast lookup. These minimizations are primarily used to generate and screen structures for subsequent more rigorous calculations.

Several approximations can be made to the force field to improve the computational efficiency, the most common of which is a nonbond cutoff, i.e., neglecting nonbonded interactions for pairs of atoms separated by distances greater than a cutoff value. Clearly this has the potential for being a rather drastic simplification. A review of several cutoff methods has been given by Brooks *et al.* (1985). To appreciate the impact of cutoffs on computational efficiency, consider a protein–substrate–solvent system with 5000 total atoms. This size system would be typical of a small protein (100–150 residues) surrounded by one or two layers of water.

Figure 7 shows how the number of nonbonded interactions increases with the cutoff distance. This calculation would run at least ten times faster with an 8.0 Å cutoff than with no cutoff (assuming the usual case that the nonbond term is rate limiting). The trade-off one makes, of course, is that interactions beyond the cutoff distance are not accounted for. Both van der Waals interactions and electrostatic interactions are significant up to 15 Å. For example, in a recent calculation of the nonbonded energy as a function of cutoff distance in the [Ala-Pro-D-Phe]$_2$ crystal, Kitson and Hagler have shown that the energy changes from 63% to 97% of the asymptotic value as the cutoff distance is increased from 8 to 15 Å (Kitson and Hagler, 1986). Figure 8 shows how the van der Waals component of the nonbond energy varies as a function of cutoff distance for this system. The van der Waals energy changes by 40% on going from 8 to 15 Å. The exact dependence of the energy on the cutoff distance depends on the system itself and should be calibrated for each new system of interest.

We now continue our discussion on the development of other approaches and tools that can expand the scope, flexibility, and efficiency of minimization strategies to address a variety of modeling objectives.

B. Constraints to the Target Function

Modifying the target function is a powerful way to customize a minimization to address specific modeling objectives. In this section, we discuss ways to constrain the target function, i.e., minimize it with respect to a subset of the total degrees of freedom, and comment on the implications constraints have for simulation strategies.

We define constraints as degrees of freedom that are fixed, i.e., not allowed to vary during the course of the minimization. This should not be confused with a restraint (discussed below), which biases the target function toward a specific value for a degree of freedom but does not fix it. The impact of constraints on computational efficiency is twofold. First, when constraints are in place, the target function may have fewer terms to evaluate and can usually be evaluated faster. In addition, since the number of iterations required to converge is roughly proportional to the number of degrees of freedom in most algorithms (especially those likely to be used on macromolecules), reducing the degrees of freedom translates directly into faster minimizations. The trade-off is that the resulting model may include unrealistic strain or other artifacts that could compromise structural and energetic analysis.

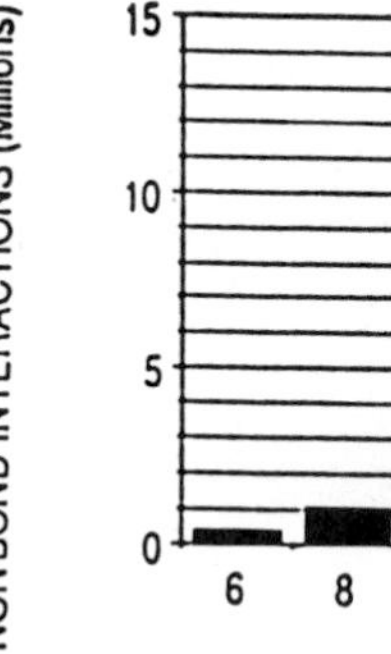

Figure 7. The number of nonbonded pairwise interactions (in millions) expected for a 5000-atom system as a function of a cutoff distance. The time required to evaluate the total energy of this system is approximately proportional to the number of nonbonded interactions.

Strategically, constrained minimizations are used in preliminary studies to set up systems quickly for subsequent more rigorous calculations, or when computer time for unconstrained calculations is unavailable. Three common types of constraints are discussed: treatment of molecules as rigid bodies, minimization in torsional space, and fixing of atoms in Cartesian space.

1. Rigid-Body Minimization

If all the internal motions are constrained, translation and rotation are the only remaining degrees of freedom. "Rigid-body" minimizations are useful for studying the docking of two (or more) relatively rigid molecules. Even for molecules that are not particularly rigid, such as the active site of an enzyme, rigid-body dockings can quickly position the substrate in a likely orientation in preparation for more rigorous minimizations or dynamics simulations. If the docking involves only two rigid molecules, the interaction of the larger molecule with an "atom probe" can be precalculated and mapped onto a three-dimensional lattice. Interaction with a second molecule can then be evaluated by simply summing the energies of each atom with the nearest lattice point. Pattabiraman *et al.* (1985) have shown that this approach can evaluate the intermolecular energy of a substrate and macromolecular receptor in "real time." With real-time energies, a user can receive immediate feedback during manual docking.

Another example of a strategy involving rigid-body minimization is the study of macromolecular packing. Several membrane-spanning channels such as the sodium channel, bacteriorhodopsin, or the acetylcholine receptor are suspected to be constructed of several α-helical bundles (Eisenberg, 1984; Brisson and Unwin, 1985). In order to predict structures for these channels, it is first necessary to understand how such structural components pack. Chou *et al.* (1983) and, more recently, Furois-Corbin and Pullman (1986) have used rigid-body minimization to identify the best chain–chain packing arrangements of polyalanine in α-helical structures. From these studies, Furois-Corbin and Pullman report that α helices tend to pack into three classes depending on the chain length and that attractions caused by dispersion forces dominate those from electrostatics for long chains. It is likely that such studies will ultimately involve detailed calculations of the energetics of interaction of the membrane with these helices, relaxing all degrees of freedom. Clearly, however, the "low-resolution" rigid-body packing is a first step in the modeling problem.

The advantage of rigid-body constraints is that minimizations can be performed quickly, even to the point of making them interactive when docking substrates into proteins. The

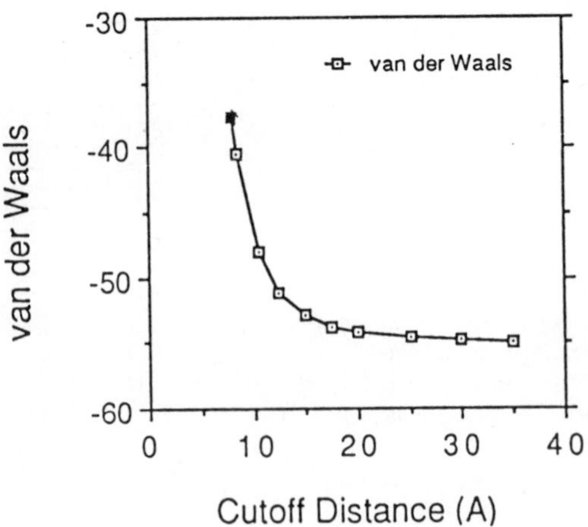

Figure 8. The van der Waals energy for the hexapeptide crystal $[\text{Ala-Pro-D-Phe}]_2$ as a function of energy cutoff distance. Note that the van der Waals energy does not converge until ≈20 Å.

primary restriction is that there is no way for the internal degrees of freedom to absorb any strain introduced in the docking process. Nevertheless, as noted, rigid-body docking is useful to prepare initial structures of flexible molecules, which can be subsequently refined with more rigorous minimizers.

2. Torsional Minimization

Since, in general, bond length and angle distortions require significant energy, most bond lengths and angles in amino acids vary only slightly from average values. Thus, to a first approximation, the conformational flexibility of proteins can be attributed to low-energy rotations about single bonds. By only allowing motion about dihedral angles, the number of degrees of freedom can be reduced from $3N$ to approximately $N/2$ (the number of non-hydrogen-containing bonds) or even lower if bonds with double-bond character are constrained.

The approximations implicit in torsional minimizers do lead to well-known quantitative problems. Consider the difference between the energy surface of an alanine amino acid as a function of its backbone angles (ϕ,ψ), as shown in Fig. 9. Panel A plots the contours of equal energy when the (ϕ,ψ) dihedrals are rotated rigidly with no relaxation of the bond lengths or angles. Panel B is for the same peptide but with flexible geometry, that is, minimizing all degrees of freedom for a given set of (ϕ,ψ) angles. The qualitative difference is that the flexible-geometry map has a larger region of accessible conformational space. This is because flexible geometry allows bad contacts resulting from torsional displacements to be reduced by small bond length and bond angle displacements, as indeed they would be in nature. Not surprisingly, restricted flexibility has a quantitative effect on both entropy and enthalpy. For example, the positions and energies of various extrema for Fig. 9 are given in Table I, which shows that the energy barriers to rotation are consistently overestimated by 4–14 kcal/mole by rigid torsional minimizers in this molecule.

A

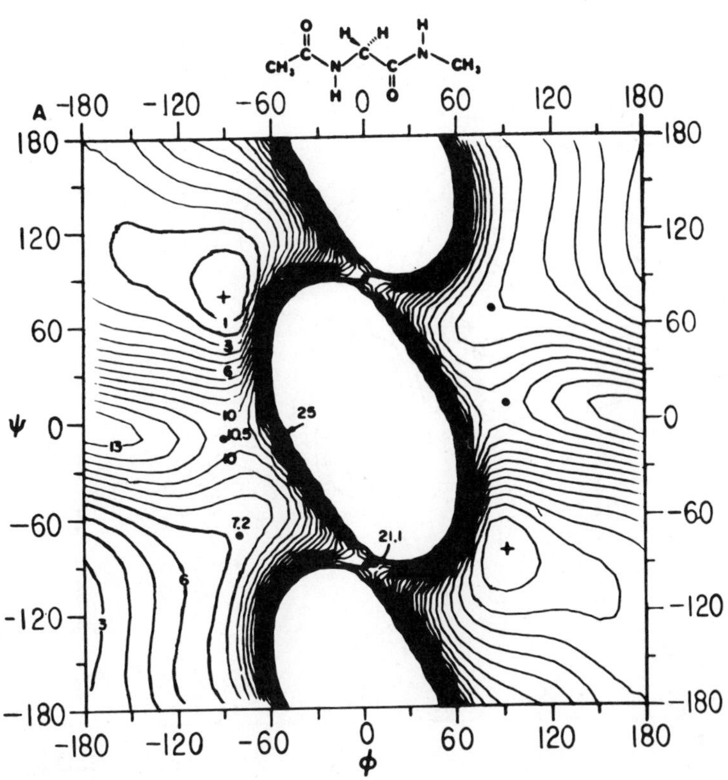

Figure 9. Energy maps for N-acetylglycine-N′-methylamide (taken from Stern *et al.*, 1983). (A) A rigid-geometry map produced by calculating the energy of the peptide after rotating in 10° increments about φ and ψ with fixed bond lengths and bond angles. (B) A flexible-geometry map produced by minimizing with respect to all degrees of freedom except the φ/ψ angles, which were rotated in 20° increments.

Table I. Local Minima and Energy Barriers (Saddle Points) for N-Acetylglycine-N′-methylamide[a]

Rigid geometry			Description of point	Flexible geometry		
φ	ψ	Energy		φ	ψ	Energy
−90	80	0.0	C_7^{eq} minimum	−80	80	0.0
−90	−10	10.5	Saddle point	−90	−70	6.7
−80	−70	7.2	α_R (no min)	−60	−60	6.3
0	−90	21.1	Saddle point	−20	−70	7.2
90	−80	0.0	C_7^{ax} minimum	80	−80	0.0

[a]From Stern *et al.* (1983).

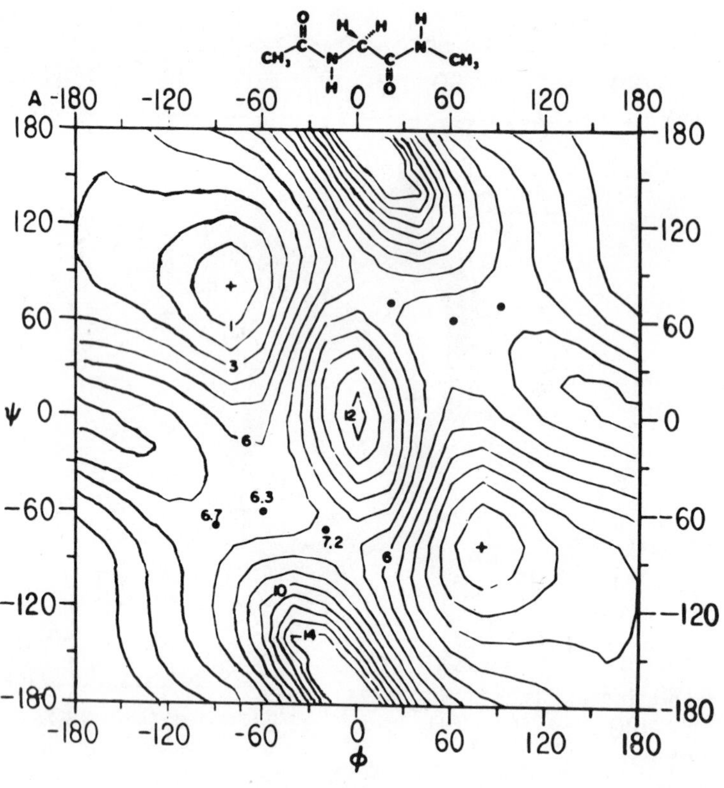

Figure 9 (cont.)

Torsional minimization has nevertheless proven to be a useful modeling tool in spite of the limitations of the quantitative results. Levinthal (Katz and Levinthal, 1972) and more recently Go (Noguti and Go, 1983; Abe *et al.*, 1984) have described minimization algorithms that rely on torsional degrees of freedom alone to search for and identify stable peptide and protein structures.

One of the most useful applications of torsional minimization is as an interactive tool so that portions of large structures can be relaxed in real time while being displayed in a molecular graphics system. Katz and Levinthal were among the first to describe a system, PACGRAF, that coupled fast torsional relaxation to graphics (Katz and Levinthal, 1972). More recently, Karfunkel (Karfunkel, 1986) has described a real-time energy calculation algorithm that, unlike the approach of Pattabiraman *et al.*, allows flexible geometry to relax during a docking exercise. First, a target region (e.g., the active site) of the macromolecule is selected. Then, from a normal-mode analysis in the torsional space of the protein, those low-energy deformations that move the target region atoms the most are selected as "docking coordinates." By restricting the motion to these relevant coupled sets of rotatable bonds, the effective degrees of freedom are reduced by two orders of magnitude. Together with efficient ways to calculate the nonbonded interactions, this method has the potential to reduce the computational requirements by three orders of magnitude compared to minimizations with all degrees of freedom. Although no results of this method have been published, it may prove to be an extremely valuable tool for interactive modeling.

Harvey and McCammon (1982) have described a rather different approach using pseudotorsional degrees of freedom. In an application to tRNA, which they treated as rigid domains connected by torsional hinge regions, they reported that large-scale motions can be simulated more quickly and converge on better-quality minima than through minimization in Cartesian space alone. In their approach, they define a pseudodihedral angle between unconnected atoms that are thought to form pivot points of hinge regions. Whereas Karfunkel uses the "docking coordinates" to describe local correlated deformations, Harvey and McCammon construct arbitrary dihedral angles that allow large-scale motions between otherwise rigid bodies. They have shown that when this method is used initially, followed by refinement with steepest descents, transfer RNA can be minimized to lower-energy structures in 60% of the time taken by steepest descents alone.

Thus, torsional minimization techniques are most useful as interactive tools where structures are being screened or prepared for more extensive calculations. It is not appropriate to use rigid body techniques to estimate energy barriers about bonds except for the most qualitative cases. Because torsional minimizers cannot distribute strain into the bond and angles of the system, they tend to overestimate the height of energy barriers and distort the energy surface.

3. Minimization with Fixed Atoms

Perhaps the simplest constraint conceptually is to fix atoms in Cartesian space. With every constrained atom, three degrees of freedom are removed. Depending on the modeling objective, holding a protein fixed (or at least that part of it that is far from the region of interest) may allow one to focus on a region of interest. Hagler and Moult (1978) modeled the water structure around the protein lysozyme applying Monte Carlo to rigid waters surrounding a fixed protein. More recently, Moult and James (1986) have successfully predicted flexible loop regions in trypsin from *Streptomyces griseus* with the remainder of the protein fixed by using energy minimization in conjunction with systematic search methods. Fine *et al.* (1986) have also minimized loops in a fixed immunoglobulin with some success. Structures of loops containing five and seven amino acids were correctly reproduced, but loops of nine and 11 amino acids had significant deviations. Both of these cases represent control studies in which the modeled loops were from the native sequence. One expects the fixed native protein to be an "excellent approximation" for such cases. The next step, which is still more challenging, is modeling a foreign sequence into a homologous framework.

C. Restraints to the Target Function

A restraint is a term added to the target function that biases the system toward adopting a certain value for a degree of freedom (but not requiring it absolutely as in the case of constraints). There is usually no computational advantage in adding restraints to a minimization; the number of degrees of freedom is not reduced, and the target function is actually more complex (though not significantly). The main advantage is in having more control over the minimization, exploring defined regions of conformation space, and testing conformational hypotheses, such as in template forcing (Struthers *et al.*, 1984). Four types of restraints are discussed: torsional restraints, distance restraints, template forcing, and tethering.

1. Torsional Restraints

A torsional restraint places a harmonic torque about a bond to force it to a new value:

$$E_{torque} = K_{torque} (\phi - \phi_{target})^2$$

By systematically changing the target angle, ϕ_{target}, and minimizing the entire structure at each step, one can produce an adiabatic energy surface similar to those in Fig. 9. These energy maps are useful to quantify low-energy pathways across a conformational barrier (Stern *et al.*, 1983). The torsion angles implied by NMR coupling constants can also be used as target angles using minimization to generate a conformation biased toward a structure consistent with experimental data.

2. Distance Restraints

A simple but powerful restraint is to add a term restraining the distance between two atoms. Distance restraints are very useful for incorporating NOE data from NMR experiments into a structural model. Including NOE distance restraints with the energy function can produce a family of representative structures of reasonable energy consistent with the given data and constitutes a powerful method for solving solution structures.

Various functional forms for the restraint are useful. A harmonic function amounts to placing a classical spring between atom pairs (Braun and Go, 1985; Van Gunsteren and Karplus, 1980).

$$E = K(R_{ij} - R_{target})^2$$

Often, a one-sided restraint is preferred. By turning off the restraint at distances greater than R_{target}, one can push the atoms apart. Conversely, if the restraint is turned off at distances less than R_{target}, the atoms are pulled together. Finally, constant rather than harmonic forces may be best when initial structures are far from the target value. For example, a constant force would gently close a ring when a harmonic force could add enough strain to invert chiral centers.

Most applications of NOE-type distance restraints to structure solving have used either "distance geometry" algorithms (Crippen, 1977; Havel *et al.*, 1983) or molecular dynamics to satisfy the initial restraints (Clore *et al.*, 1986; Havel and Wuthrich, 1984). Minimization is used only in the final refinement stages. A notable exception is the approach of Braun and Go (1985), where minimization was used exclusively in conjunction with novel perturbations to the target function to solve NOE restrained structures. The essential feature of their approach is that long-range nonbond interactions are ignored during the initial stages of restrained minimization. This allows atoms and residues to pass completely through each other as they seek to satisfy the experimental restraints. Gradually, the sphere of nonbonded interactions is increased as the model becomes more refined. In addition, the minimization is conducted in torsion space, which accelerates convergence and allows for larger-scale motions in this protocol.

3. Template Forcing

When testing conformational hypotheses for "binding conformations" in drug design, it is often useful to know if one molecule can adopt the conformation (or more generally, the shape) of another (Struthers *et al.*, 1984). By selecting corresponding atoms or functional groups between two molecules, one can force atoms from a flexible molecule to superimpose onto atoms of a rigid "template" molecule. Effectively, a spring is placed between the paired

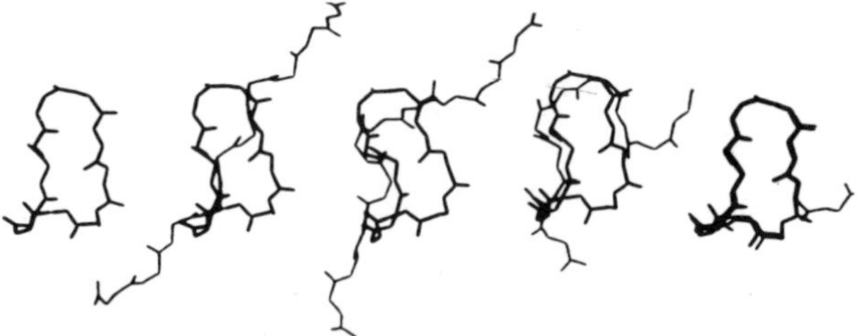

Figure 10. Template forcing of gonadotropin-releasing hormone onto a constrained active cyclic analogue. Starting from an extended conformation of a linear antagonist of GnRH (far left), template forcing is used to induce it to adopt a putative receptor-binding conformation deduced for a second, conformationally constrained antagonist. As the forcing proceeds (from left to right), the linear antagonist looks more and more like the template molecule, until at the end the two molecules are virtually identical. The energy of the final conformation is low, which is evidence that the putative receptor-binding conformation could be adopted.

sets of atoms, pulling the atoms in the flexible molecule toward the rigid molecule.

$$E_{\text{template}} = \Sigma_{\text{atom pairs}} K_{\text{template}} (R_i^{\text{analogue}} - R_i^{\text{template}})2$$

The restraining or forcing term is added to the energy to construct the target function to be minimized. By varying the spring force constant, one can force the flexible molecule into the conformation of the template to an arbitrarily small deviation at the lowest energy cost. The energy required to achieve a given "fit" can then be used to evaluate how easily an analogue can adopt the conformation of a given template. Figure 10 shows how template forcing was used to derive and test the putative binding conformations of peptide hormone analogues of GnRH and in the design of active analogues (Struthers *et al.*, 1984; Baniak *et al.*, 1988; Struthers *et al.*, 1987).

4. Tethering Atoms to Points in Space

Tethering is a special case of template forcing in which the "template" is a set of points in space, usually corresponding to a desired target conformation such as a crystal structure. Tethering is useful for "annealing" crude complex structures before carrying out unrestrained minimization or dynamics. For example, when minimizing solvent with a protein, one can tether the oxygens of the water to their initial positions while the hydrogens reorient (Roberts *et al.*, 1986). One may gradually reduce the force constant during the minimization and update the tethering points periodically to correspond to the current structure. Warme and Scheraga (1974) first used tethering to keep lysozyme close to the crystal coordinates during minimization. This is desirable since the crystal data may include disordered or poorly ordered regions resulting in strained geometries and overlapping atoms that would result in unrealistic long-range disruptions to the structure.

D. Exploring Conformational Space

A common limitation of classical minimization algorithms is that they locate a local minimum usually close to the starting configuration, not the global minimum. This is because the minimizers discussed above are specifically designed to ignore configurations if the energy increases. Therefore, they do not generally push a system over barriers but down into the nearest valley. Modified strategies are needed to search conformational space more thoroughly.

Most attempts to search conformation space can be classified into two approaches: (1) screen from a large number of systematically generated points in conformational space, "best" starting points from which to begin a minimization; (2) modify the target function and/or the minimization algorithm to allow the system to overcome local barriers en route to finding lower-energy minima.

1. Systematic Searching of Conformational Space

Systematic conformations are usually generated by rotating about all torsion angles in small increments (~2–3°). Such an exhaustive search, though computationally demanding, is the only method that can assure that a global minimum will be found. However, it is severely restricted since the number of initial starting points increases exponentially with the number of degrees of freedom. Even with sophisticated pruning algorithms that efficiently ignore large volumes of unlikely conformational space (Crippen and Scheraga, 1971a, 1973), the method is practically limited to 10–20 degrees of freedom. A more promising approach along these lines is the so-called "build-up" method of Scheraga (Vasquez and Scheraga, 1985; Gibson and Scheraga, 1986) applied to peptides. This method uses standard geometries of smaller peptide segments to build up dipeptides, and after culling out emerging high-energy structures, these are used to build all possible tetrapeptides, which in turn are used to build larger structures in an iterative process.

2. "Uphill" Minimizers

The second approach attacks the problem directly by overcoming the natural tendency of minimizers to get stuck in local wells. The "deflation" method minimizes a target function that is constructed by dividing the potential energy by the found "roots" of the energy surface each time a local minimum is found and uses the new target function (which, in principle, no longer has the previous minimum as a root) to continue the search. Crippen and Scheraga (1969) demonstrated that this method identified 27 of 37 local extrema and 10 of 12 minima for a dipeptide with two degrees of freedom. Another approach described by Crippen and Scheraga adds a term to the target function that forces the molecule up the path of "gentlest ascent" from a local minimum to a saddle point (Crippen and Scheraga, 1971b). Once past the saddle point, conventional minimization takes over until the next minimum is found. This continues until all the minima have been found. The disadvantage of these approaches is that one cannot be sure when the search of conformational space is sufficiently complete. There is no way of knowing *a priori* how many local extrema to expect on the energy surface of a molecule.

3. Minimization through Reduction of Dimensionality

A recent algorithm that is neither a systematic search nor what could be called an "uphill" minimizer uses Cayley–Menger determinants to search for the global minimum directly (Purisima and Scheraga, 1986). Instead of working in three-dimensional space, the minimization begins in a space with a dimension equal to the number of atoms squared (N^2). In N^2 space, the global energy minimum solution is trivial. Each atom's position (coordinates) can be defined in this higher dimension space so that its distance to every other atom is optimal for the pairwise interaction. Of course, this creates an "abstract" structure that is meaningless in three dimensions. Thus, the next step is to contract the dimensionality of the space while continuing the minimization until a three-dimensional structure is obtained. By starting from a global minimum and adding dimensional constraints, the final minima will be closer to the true global minimum. Purisima and Scheraga (1986) have used this approach for the minimization of four terminally blocked amino acids and were able to converge to the global minimum directly from a total of 326 possible local minima.

So far, all of these approaches to surmounting the local minima problem have been restricted to relatively small molecules. However, increased speed and the availability of supercomputers have encouraged researchers to speculate that such techniques could soon be applied to proteins (Vasquez and Scheraga, 1985).

4. Dynamic Searching

Real molecules find their conformational global minimum by fluctuating about an ensemble of configurations within energetic reach (as determined by the system temperature). In principle, then, if one can simulate the motions and fluctuations experienced by a molecule, the global minimum will be sampled by this simulation (eventually). What we have described is precisely the molecular dynamics approach. By solving Newton's equations of motion, $F = ma$, for a molecule, one can predict where atoms will move over a short time interval. By taking successive time steps, a time-dependent trajectory can be constructed for all the atoms representing the molecular motions.

By using the available thermal energy to climb and cross conformational energy barriers, dynamics provides insight into the accessible conformational states of the molecule. Molecular dynamics has been used in a variety of small peptide systems to find lower-energy conformational states across energy barriers that would be inaccessible to classical minimization. In particular, it has been applied to the study of vasopressin (Hagler *et al.*, 1985) and gonadotropin-releasing hormone (GnRH) (Hagler, 1985; Struthers *et al.*, 1984). In the case of GnRH, novel constrained analogues with 100-fold higher potency than the parent cyclic compounds were designed based on the conformations of GnRH and GnRH analogues identified with a combined strategy incorporating energy minimization, template forcing, and molecular dynamics.

One of the disadvantages of dynamics, however, is that many hundreds and even thousands of conformational states are sampled during a trajectory, thus compounding the analysis process. Furthermore, the transitions of interest may take place on a time scale of milliseconds to minutes or longer at 300°, whereas dynamic simulations at present are practically limited to tens of nanoseconds. Minimization periodically during a dynamics trajectory helps the analysis problem by identifying those structures that represent the local minimum about which the dynamics fluctuations take place. This is illustrated in Fig. 11, where the dark line superimposed on our elliptic energy surface represents a dynamic trajectory of a particle beginning at rest from point a. The initial gradient pulls the particle downhill, but since energy is conserved,

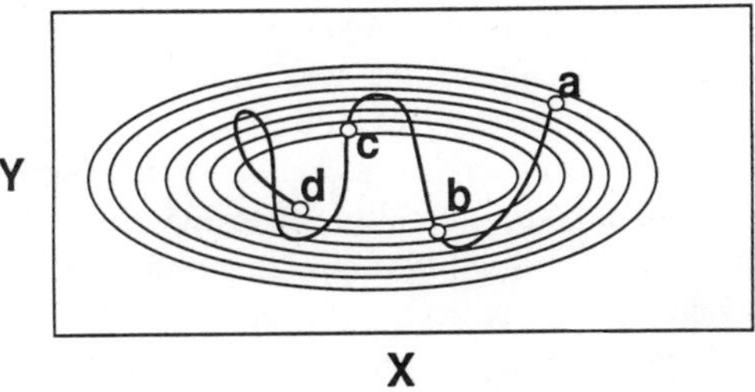

Figure 11. A dynamics trajectory for a particle beginning at rest at point a and continuing with constant total energy (kinetic plus potential) until point d. On this elliptic energy surface, minimization from any of the points, a, b, c, or d, will converge to the same point. Thus, minima can be used to characterize many nearby points sampled during dynamics. In molecules in which there are many more degrees of freedom, periodic minimization during a dynamic calculation will result in relatively fewer (compared to the number of discrete dynamic structures) structures for use in detailed structural and energetic studies.

its velocity increases as it gets closer to the minimum. Its momentum takes the particle back and forth, visiting many points on the energy surface during the trajectory. However, if we stop the particle at any point during the dynamics and minimize, only one point will be found. In a higher-dimensional system such as molecules, minimization from various dynamic points may result in the identification of several different local minima. However, the number of minima will be far smaller than the number of points sampled during dynamics, and these are useful as points of reference when analyzing molecule structures and their energies.

Of course, in a real molecular system, there are far too many degrees of freedom to plot the trajectory on a realistic energy surface. Any attempt to project a trajectory into one or two dimensions will result in a far more complex pattern that will obscure any existing points that are close to a minimum. In real calculations, then, one must resort to periodic (systematic) minimizations to discover what minima are being sampled by dynamics. A detailed discussion of how dynamic searching has been used to characterize the conformational states of the peptide hormone atrial natriuretic factor (ANF) is presented in Section V.B.

IV. PRACTICAL HINTS FOR PERFORMING MACROMOLECULAR SIMULATIONS

There are many issues involved in designing an appropriate simulation strategy for a given molecular system, some of which only have to do with minimization algorithms themselves. One of the most important steps in any simulation is properly preparing the system to be simulated. Calculations on the fastest computer running the most efficient minimization or dynamics algorithm may be worthless if the hydrogen is put on the wrong nitrogen of an active site histidine or a structural "water" is omitted from the active site. Unfortunately, it is impossible to provide a single recipe for a successful model; too much depends on the objectives and expectations of each calculation. Will energies be compared quantitatively? Are we expecting or interested in conformational changes leading to very different structures? What is the hypothesis being tested? The effects of tethering, fixing, energy cutoffs, etc., on the results may only be answered by controlled preliminary experiments. Several common considerations that are necessary to address in building almost all macromolecular systems are discussed in the following sections.

A. Theoretical Considerations

When dealing with macromolecular optimization calculations, it is important to keep in mind the theoretical significance of the minimum-energy structure and its calculated energy. For all force fields used in calculations of this type, the energy zero is arbitrary; thus, one cannot compare the total potential energy of different molecules directly. However, it is meaningful to make comparisons of energies calculated for different configurations of chemically identical systems. Also, although the calculated energy of a fully minimized structure is the classical enthalpy at absolute zero and thus ignores quantum effects for a molecule that is sufficiently small that its normal modes can be calculated, quantum corrections for zero-point energy and free energy at higher temperatures can be taken into account (Hagler *et al.*, 1979c).

The minimized energies calculated for enzyme–substrate complexes can be used to estimate relative binding enthalpies, but there are two caveats that should be noted. First, for a meaningful comparison of the relative binding of two different substrates, one must always consider a complete thermodynamic cycle (Kirkwood, 1935; Quirke and Jacucci, 1982; Tembe and McCammon, 1984; Mezei and Beveridge, 1985). In practical terms, this means that an enthalpy calculation for the various substrates in water must be made. In cases in which relative binding of two different enzymes to the same substrate is calculated, one must calculate the energy of each enzyme with solvent in the binding site.

A second consideration in using minimization results to estimate relative binding strengths is that the entropy is neglected in such calculations. Direct calculation of entropy differences is a computationally intensive process, and it is only recently that it has been taken into account correctly by calculations of relative free energies (Hagler *et al.*, 1979c; Hwang and Warshel, 1987; Warshel *et al.*, 1986; Singh *et al.*, 1987; Straatsma *et al.*, 1986). The extent of the errors introduced in neglecting entropic contributions in the simpler minimization calculations is difficult to estimate, although as with the zero-point energy, the entropy can be estimated for a system sufficiently small that its normal-mode frequencies can be calculated (Hagler *et al.*, 1979c).

The relative importance of these fundamental considerations depends on the objective of the calculation. When studying the relative binding of two substrates to an enzyme active site, one of which is flexible and the other rigid, entropic effects may be crucial to obtain even qualitative agreement with experimental binding constants. On the other hand, if a putative compound overlaps sterically with many active site atoms and results in hundreds of kilocalories of strain energy even in a minimized structure, one can confidently reject the compound. The bottom line is that one must not abandon common physical chemistry sense when setting up a calculation and interpreting the results.

B. System Definition

1. Adding Hydrogens

Most macromolecular simulations begin with a crystal structure. These structures generally do not include hydrogen coordinates. In general, defining coordinates for nonpolar hydrogens in a protein is straightforward; however, polar and ionizable hydrogens are more difficult. First and foremost, one must decide the protonation state of ionizable amino acid residues. If a realistic solvent environment with explicit waters and counterions is to be included, then it can be assumed that most amino acids have a protonation state consistent with their p*K*s at the experimental pH. However, for modeling carried out in the absence of a shielding solvent and counterions, charged side chains may be inappropriate. A tendency to form salt bridges between oppositely charged side chains may be overemphasized, and electrostatic repulsion between like charged groups can cause unrealistic conformational distortions.

Apart from charged amino acids, polar hydrogens on water, serine, threonine, and tyrosine not defined by x ray for protein structures must be oriented in an initial structure. Often, this can only be done by searching for specific hydrogen bonds that require certain orientations either manually or through "systematic search" algorithms. However, it may be possible to minimize with restraints (tethering the heavy atoms) and let the hydrogens find their own best orientation. In any case, one must consider hydrogen positions and protonation states carefully, especially in regions of particular interest such as the active site.

An option that may be available with respect to hydrogens is whether to include nonpolar hydrogens at all. The so-called "united-atom" approximation absorbs the hydrogens implicitly into the carbons to which they are bonded and can reduce the atom count by as much as a factor of 2. The computational savings are significant. However, detailed studies indicate that an all-hydrogen definition may fit experimental structures and energies more reliably, and one should consider including all hydrogens at least in critical regions such as the active site (Dauber-Osguthorpe *et al.*, 1988).

2. Solvent Effects

Solvent must be included if a result is suspected to depend on charged interactions (which are screened by water) or if hydrophobic effects are to be modeled. A rigorous protein simulation will include explicit water molecules at least two or three layers thick around the surface of a protein. For example, one of the most complete macromolecular simulations attempted to date includes all water molecules and counterions (1429 waters and 42 Na^+ and $H_2PO_4^-$ ions), and two independent proteins in an asymmetric unit used in replicating the crystal of protease A from *Streptomyces griseus* (F. Abvelg, D. Kitson, and A. T. Hagler, unpublished results). Such rigor can be computationally prohibitive for large systems and may not be necessary for limited studies where the results are not expected to depend on solvent effects.

Often, the screening effect of solvent is modeled implicitly by adjusting the behavior of the dielectric constant. The dielectric constant is either set to a value greater than unity or is replaced with a function, usually the distance to the first power (the so-called "distance-dependent dielectric"). The value of such models has been questioned (Warwicker *et al.*, 1985; Gilson *et al.*, 1985). They cannot account for the structural contribution of solvent introduced through hydrogen bonds. Furthermore, applying a macroscopic dielectric function to microscopic domains may be entirely inappropriate. Even different parts of the same protein may have completely different dielectric properties. A promising approach to this problem is a numerical solution of the Poisson–Boltzman equation (Warwicker, 1986; Warwicker *et al.*, 1985; Gilson *et al.*, 1985).

C. Minimization Strategies

1. When to Use Different Algorithms

The choice of which algorithm to use depends on two factors: the size of the system and the current state of optimization. Until the derivatives are well below 100 kcal/mole-Å, it is likely that the point is sufficiently away from a minimum that the energy surface is far from quadratic. Algorithms that assume the energy surface is quadratic (Newton–Raphson, quasi-Newton–Raphson, or conjugate gradients) can be unstable when the molecule is far from the quadratic limit. Newton–Raphson is particularly sensitive since it must invert the Hessian matrix. Therefore, as a general rule, steepest descents is often the first minimizer to use, after which conjugate gradients or a Newton–Raphson minimizer can be used to complete the minimization to convergence.

2. When to Use Constraints/Restraints

Constraints and restraints are often used to control and direct the minimization. For example, if a substrate is being docked into an enzyme and a specific hydrogen bond between the enzyme and the ligand is thought to be involved in binding, the donor and acceptor atoms can be pulled together to provide a docking coordinate. Thus, the results are not so dependent on the initial starting configuration, which may only have been a crude graphic alignment. In cases like this, the restraint is turned off at some point to make sure that the biased minimum is close to a true minimum.

Another example of the use of restraints is in modeling incomplete systems. Often, it is difficult or impossible to construct a realistic environment around parts of a model system. Only a partial structure of a large protein complex may be available, and one must restrain some atoms to stay near the initial crystal positions since they do not have the interactions with neighboring (missing) amino acids, membrane, or solvent to stabilize the structure. If the site of interest (for instance, a binding site for a competitive inhibitor) is well characterized but other parts of the enzyme are unknown or simply would require too much computation time if included, a limited study can still be carried out with the ends of the known structure tethered to their crystal coordinates. Usually, these restraints are permanent parts of the model. The results of such calculations must be critically evaluated but can be valid if the ligand binding does not depend on interactions with missing pieces of the model or on conformational flexibility in the tethered regions.

As a final example, tethering can be used to gently relax a protein crystal. Often, crystal coordinates of proteins, even if highly refined, have several strained interactions as a result of intrinsically disordered or poorly defined atomic positions, which, on minimization, give rise to large initial forces. If these forces are not restrained, they can result in artifactual movement away from the original structure. The general approach is to relax parts of the model progressively in stages, starting with the least-well-determined atoms, until the entire system can minimize freely. The restraints are ultimately removed so that the final minimum represents an unperturbed conformation. It is not usually necessary to minimize to convergence at each stage; the object is to relax the most strained parts of the system as quickly as possible without introducing artifacts. A typical approach to relaxing a protein system would be:

1. Fix the crystal coordinates of the protein heavy atoms to allow added hydrogens (and perhaps added solvent) to adjust to a static crystallographically defined environment. Only steepest descents is necessary, and this stage is complete once the derivatives are of the order of 10 (the next stages will allow us to proceed to convergence).
2. Fix or tether the main-chain atoms while allowing the side chains to adjust (this may occur in stages, with surface side chains being relaxed before interior residues). Again, use steepest descents until the derivatives are less than 10.
3. Gradually decrease the tethering constant for the backbone atoms until the system can be totally relaxed. Steepest descents can be used initially but should be replaced with a more efficient algorithm as soon as possible.

3. Convergence Criteria

There is in the literature a wide variety of criteria used to judge minimization convergence in molecular modeling. Mathematically, a minimum is defined by the point at which the derivative of the function is zero [and the second derivative (matrix) is positive (definite)]. Nongradient minimizers use the increment in the energy and/or coordinate. In gradient minimizers, derivatives are available analytically and should be used directly.

In a molecular minimization, the atomic derivatives may be summarized as either an average, a root-mean-square value, or the largest value. The average, of course, must be an average of absolute values of the derivatives, since the distribution of derivatives is symmetric about zero. A root-mean-square derivative is a better measure than the average: it weights larger derivatives more and therefore makes it less likely for a few large derivatives to escape detection, as can occur with simple averages. Nevertheless, regardless of whether one chooses to report convergence in terms of average or RMS values of the derivatives, one should always check that the maximum derivative is not unreasonable. There can be no ambiguity about the quality of the minimum if all derivatives are less than an acceptable value.

The more difficult question is what value of the average or RMS derivative constitutes convergence. The specific value depends on the objective of the minimization. If we simply want to relax overlapping atoms before beginning a molecular dynamics run, minimizing to a maximum derivative of 1.0 kcal/mole-Å is usually sufficient. However, to perform a normal-

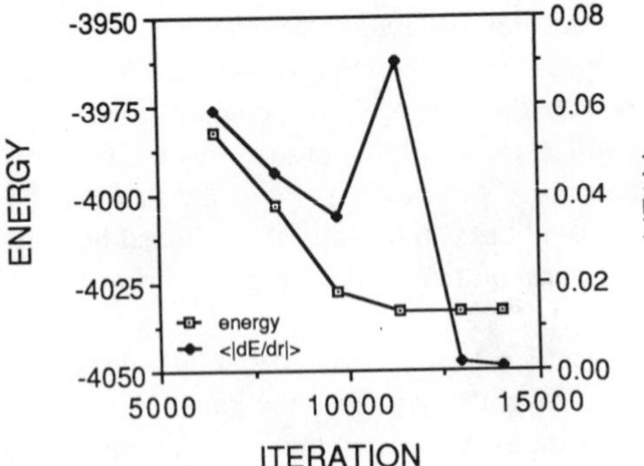

Figure 12. Minimization history of a dihydrofolate-reductase–trimethoprim complex. The left axis tracks the energy as a function of the iteration step (in kcal/mole), while the right axis is the average absolute derivative. Note that the energy changes by some 50 kcal/mole as the derivatives go from 0.06 to 0.001. Thus, convergence is only achieved once the derivatives are less than 0.001 (V. A. Roberts and A. T. Hagler, unpublished results).

mode analysis, the maximum derivative must be less than 10^{-5}, or the frequencies can be shifted by several wave numbers. Figure 12 shows the minimization history of the DHFR–trimethoprim protein–substrate system, which required some 14,000 iterations to converge to an average derivative of 0.0002 kcal/mole-Å (V. A. Roberts and A. T. Hagler, unpublished results). Often, protein-sized minimizations achieve average derivatives between 0.02 kcal/mole-Å and 0.5 kcal/mole-Å. If this simulation had been terminated when the average derivative had reached just 0.02, the structure would have been 30 kcal/mole higher in energy and approximately 0.3 Å RMS away in structure from the final minimum (V. A. Roberts and A. T. Hagler, unpublished results). This difference in RMS movement represents 25% of the total movement during minimization. Thus, if quantitative measurements and comparisons of macromolecular structures are to be reliable, it appears from this result that it may be necessary to allow minimizations to converge to average derivatives of the order of 0.002.

V. SPECIFIC EXAMPLES

We discuss applications involving a protein–ligand system, a medium-sized peptide, and a small peptide.

A. Substrate Binding Study of DHFR–Trimethoprim

Roberts *et al.* (1986) have recently reported a comprehensive energetic and structural analysis of the dihydrofolate reductase–trimethoprim system. The objective was to understand the molecular basis for trimethoprim binding to bacterial DHFR 100,000 times more strongly than to vertebrate DHFR. It is this selectivity that makes trimethoprim useful as an antibiotic. This study provides an excellent example of how to analyze minimized structures.

1. Strategic Approach to Studying an Enzyme–Ligand Complex

The central objective of the DHFR–trimethoprim study was to probe the energetic basis for trimethoprim selectivity between bacterial and vertebrate DHFR as well as to account for an observed cooperativity (enhanced trimethoprim binding) exhibited by the presence of NADPH in the bacterial system. The approach was to subject four different complexes to rigorous energy minimization beginning from crystal structures and then thoroughly analyze the energetic components of the minimized structures in order to discover the structural basis for the observed selectivity. The four systems minimized were:

1. *E. coli* DHFR + trimethoprim
2. *E. coli* DHFR + trimethoprim + NADPH
3. Chicken liver DHFR + trimethoprim
4. Chicken liver DHFR + trimethoprim + NADPH

Crystal structures were available only for the *E. coli* binary and chicken liver ternary complexes (1 and 4). The *E. coli* ternary complex (2) was constructed by using the NADPH coordinates from a known structure for *L. casei* DHFR with methotrexate and NADPH bound. The chicken liver binary system (3) was built by replacing the NADPH from system 4 with waters.

Binding of ligands to enzymes depends strongly on solvent and ionic interactions. Therefore, one of the first steps is choosing the ionization state of side chains and the treatment of solvent. In this study, the side chains of Glu, Asp, Arg, and Lys were charged, as were the N- and C-terminal amine and carboxylate groups. In addition, a His at position 45 is known to be protonated when NADPH is bound, and so this amino acid is treated as a special case. Waters that appeared to play a structural role in the protein structure and any other crystallographic waters within 7 Å of trimethoprim and NADPH were retained. Additional waters were added to fill the volume within 7 Å of trimethoprim and NADPH and within 3 Å of all charged residues. All hydrogens were included explicitly, and a dielectric constant of 1 was used throughout.

A common and important modeling objective is to relax poorly defined regions in a protein structure without disrupting well-defined regions. We have discussed earlier how template forcing and tethering can be used for this. In this study, the protein was tethered in stages to relax the least-well-refined parts of the model first, starting with the solvent. The backbone of the protein was restrained next to allow the side chains to make minor adjustments. Eventually, the entire system was relaxed unrestrained until the minimization was converged (an average absolute derivative of 0.0002 kcal/mole-Å).

2. Structural Analysis

Following minimization of structures begun from crystal coordinates, it is necessary to confirm that the model is sufficiently accurate to justify detailed energy analysis. Unrealistic deviations from the experimental structure may be indications that the model is improperly defined or that the energy expression is inaccurate and could compromise subsequent structural or energetic analysis. A common measure of the fit of two protein structures is the root-mean-square (RMS) deviation of corresponding atoms in the two superimposed structures. Minimized structures are superimposed onto the corresponding initial crystal structure using a least-squares fit of the heavy atoms of their secondary structures, i.e., the α helices and β strands. By only using atoms in the relatively rigid secondary structures, one avoids artifacts in the superposition arising from bias caused by poorly defined regions or loops at the surface (where the model is expected to be less rigorous, since bulk solvent, counterions, and crystal contacts are ignored). Given this preliminary superposition, the RMS deviation can be calculated for the entire protein or for local regions such as the active site to quantify the relative movement throughout the protein complex.

Table II summarizes the results of several RMS deviation analyses for each of the DHFR complexes. Figure 13 provides a graphic representation of the RMS deviations for the *E. coli*

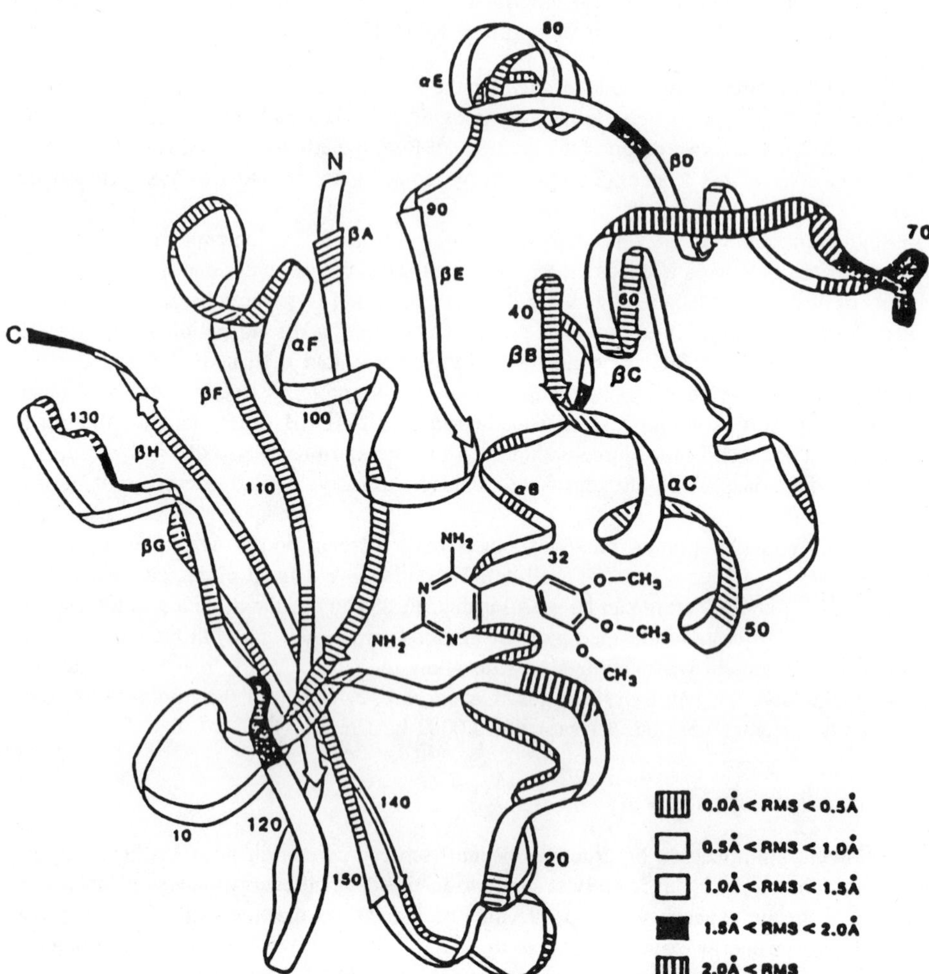

Figure 13. Schematic representation of *E. coli* DHFR showing relative positions of α helices, β strands, and the loops connecting them. The bound inhibitor trimethoprim in the active site is also indicated. The RMS deviation of C_α atoms between the experimental and minimized structure for *E. coli* DHFR + trimethoprim is superimposed on the ribbon representation. Note that the largest deviations occur in external loops. Residues in the active site region all have RMS deviations less than 1.0 Å. This figure is adapted from Matthews *et al.* (1977).

Table II. RMS Deviations (Å) for Heavy Atoms and C_α between the Minimized and the Observed Structures[a,b]

Region	*E. coli* binary	*E. coli* ternary	Chicken liver binary	Chicken liver ternary
All heavy atoms[c]	1.23 (1.07)	0.94 (0.72)	1.47 (1.10)	1.49 (1.10)
Trimethoprim	0.81 (0.00)	0.69 (0.00)	0.42 (0.00)	0.68 (0.00)
Pyrimidine ring	0.42 (0.00)	0.55 (0.00)	0.24 (0.00)	0.65 (0.00)
Trimethoxyphenyl ring	0.97 (0.00)	0.74 (0.00)	0.50 (0.00)	0.70 (0.00)
NADPH	—	0.95 (0.00)	—	0.67 (0.00)
Active site[d]	0.58 (0.54)	0.50 (0.41)	0.60 (0.51)	0.72 (0.57)
α Helices	1.10 (0.72)	0.81 (0.47)	0.99 (0.64)	1.02 (0.76)
β Strands	0.82 (0.66)	0.68 (0.47)	1.34 (1.00)	1.34 (0.89)

[a]From Roberts *et al.* (1986).
[b]Numbers in parentheses are the RMS deviations of C_α atoms.
[c]Deviations of water molecules are not included.
[d]Residues with any atom within 4 Å of trimethoprim.

+ trimethoprim system. As expected, internal regions of the protein, such as the active site, were found to fit the experimental structure more accurately than areas on the surface of the protein. The well-defined areas of the protein, such as the active site and the C_α carbons of the α helices and β strands, show RMS deviations substantially lower than that of the overall protein. The pyrimidine ring of trimethoprim, deep in the active site cleft, deviates from the experimentally observed structure less than does the trimethoxyphenyl moiety, which extends out towards the surface of the protein. Similarly, in the chicken liver ternary complex, the nicotinamide ring, which extends into the active site, has a much smaller RMS deviation (0.34 Å) than the NADPH molecule as a whole, which generally lies on the surface of the protein. Thus, the crucial area for this study, the active site, shows good agreement with the experimental structure and was judged adequate for further detailed structural and energetic analysis.

3. Structural and Energetic Analysis

Roberts *et al.* (1986) analyzed the energetics of the minimized structures by breaking down the total energy into various components including intramolecular strain (bond lengths and angles, etc.) and intermolecular energies (van der Waals and electrostatics) between and among the various molecules. The ability to consider the energetic contributions of various kinds of interactions is a crucial part of any energetic modeling strategy. For the purpose of illustration, we restrict our discussion to the analysis of how NADPH enhances the binding of trimethoprim and, in particular, to different degrees in the vertebrate and bacterial systems.

The selectivity of bacterial versus vertebrate DHFR for trimethoprim binding partly depends on enhanced cooperativity by NADPH in the two enzymes. Specifically, Baccanari *et al.* (1982) have shown that the *E. coli* DHFR dissociation constant for trimethoprim is decreased 40-fold in the presence of NADPH, whereas for rodent lymphoma DHFR the corresponding decrease is only a factor of 2.8. By examining the two DHFR systems both with and without NADPH, the energetic basis underlying any cooperativity was investigated. Two general mechanisms for cooperativity are possible: an indirect mechanism in which the presence of NADPH changes the protein to be conducive to the binding of trimethoprim or a direct mechanism in which favorable interactions between the NADPH and trimethoprim directly stabilize the complex. If the indirect mechanism is responsible, one expects the interaction energy between the protein and trimethoprim to improve when NADPH is present. The calculated intermolecular energy between trimethoprim and the protein in the binary complex was −214.4 kcal/mole compared to −188.7 kcal/mole with NADPH present. The major difference in binding energy arises because His[45] becomes protonated on binding of NADPH (as evidenced by proton magnetic resonance studies) (Poe *et al.*, 1979). The interaction of this protonated histidine with the positively charged trimethoprim accounts for 20.9 kcal/mole of the 25.7 kcal/mole difference in the interaction of energy. (Note the importance of correctly identifying the charged state of His[45].) Thus, the energetic results indicate that binding of NADPH does not predispose the enzyme to interact more favorably with trimethoprim, and protein–ligand interactions cannot explain the observed cooperativity in *E. coli* DHFR.

Comparison of the interactions of NADPH with trimethoprim in the *E. coli* ternary complex and the chicken liver ternary complex provides a clue to the source of the enhanced cooperativity of NADPH in the bacterial enzyme. The energy of this interaction in the minimized structures is 6.6 kcal/mole more favorable in the *E. coli* complex. This is not sufficient to overcome the 25.7 kcal/mole increase in energy experienced on binding of NADPH and alerts us that quantitative energy comparisons may be unreliable in this model. Nevertheless, the qualitative trend is still worth investigating more closely. By partitioning these interaction energies into contributions from the various functional groups, one can determine the specific interactions responsible for this difference (Table III). The favorable interactions arise predominantly from the spatial relationship between the trimethoxyphenyl ring of trimethoprim and NADPH. The trimethoxyphenyl ring occupies a different cleft in the chicken liver enzyme

Table III. Nonbonded Interactions of Trimethoprim with NADPH[a,b]

NADPH	Energy	*E. coli* ternary			Chicken liver ternary		
		PRMD[c]	CH_2	Phenyl[d]	PRMD[c]	CH_2	Phenyl[d]
Nicotinamide	VdW	− 3.7	−0.6	−3.7	− 3.6	−0.9	−1.9
	Coul	−14.6	−1.7	−1.3	−14.0	−1.9	−0.3
	Total	−18.3	−2.3	−5.0	−17.6	−2.7	−2.2
Nicotinamide ribose	VdW	− 0.1	0.0	−0.7	− 0.1	−0.1	−0.2
	Coul	5.7	0.6	0.2	5.7	0.7	0.1
	Total	5.6	0.5	−0.4	5.7	0.6	−0.1
Pyrophosphate linkage	VdW	0.0	0.0	0.0	0.0	0.0	0.0
	Coul	−48.1	−4.6	−2.5	−48.2	−4.4	−1.0
	Total	−48.2	−4.6	−2.5	−48.2	−4.4	−1.1
Adenine ribose	VdW	0.0	0.0	0.0	0.0	0.0	0.0
	Coul	−25.4	−2.1	−0.7	−24.9	−1.9	0.0
	Total	−25.4	−2.1	−0.7	−24.9	−1.9	0.0
Adenine	VdW	0.0	0.0	0.0	0.0	0.0	0.0
	Coul	− 5.4	−0.5	−0.2	− 5.4	−0.4	−0.1
	Total	− 5.4	−0.5	−0.2	− 5.4	−0.4	−0.1
Total	VdW	− 3.8	−0.6	−4.4	− 3.7	−0.9	−2.2
	Coul	−87.9	−8.3	−4.4	−86.7	−8.0	−1.3
	Total[e]	−91.7	−9.0	−8.8	−90.4	−8.9	−3.5

[a]From Roberts *et al.* (1986).
[b]Energies in kcal/mole.
[c]PRMD, 2,4-diaminopyrimidine.
[d]Phenyl, 3,4,5-trimethoxyphenyl.
[e]The total interaction of trimethoprim with NADPH in the *E. coli* ternary complex is −109.4 kcal/mole. The total energy of interaction of trimethoprim with NADPH in the chicken liver ternary complex is −102.8 kcal/mole.

so that this ring is closer to the nicotinamide of NADPH and its associated ribose in the *E. coli* system than in the chicken liver ternary complex. Table III shows that the van der Waals interactions between these groups of the NADPH and the trimethoprim phenyl ring are 2.2 kcal/mole more favorable in the *E. coli* complex. Coulombic interactions between the entire NADPH molecule and the trimethoxyphenyl ring are 3.1 kcal/mole more favorable in the *E. coli* ternary complex. This is interpreted to suggest that it is the difference in the trimethoprim–NADPH interactions that underlies the cooperativity in the *E. coli* complex. Specifically, the cooperativity appears to result from better van der Waals interactions between NADPH and the trimethoxyphenyl ring of trimethoprim in the *E. coli* complex compared to the vertebrate DHFR.

Minimizing macromolecular structures presents unique challenges. This example has illustrated the strategy involved in constructing appropriate molecular structures, minimizing the structures, and extracting and interpreting both structural and energetic information from the minimized systems.

B. Conformational Search of ANF Using Dynamics and Minimization

As our next example, we consider a strategy for searching the conformational space of a medium-sized peptide using a combination of dynamics and minimization steps.

Atrial natriuretic factor (ANF) is a polypeptide hormone isolated from the atrium with both natriuretic and vasodilator properties. Because of its implied role in the regulation of blood pressure, the pharmaceutical industry is actively engaged in determining the structural determinants responsible for the activity of this molecule. However, the structure of ANF when it is bound to the receptor is unknown. In fact, the molecule has resisted all attempts at crystallization, and NMR studies suggest that it does not have a single conformation in solution. Thus, molecular modeling may provide a valuable structure-directed design strategy.

We have performed a set of molecular dynamics calculations on ANF at elevated temperature to search for structural features that may be significant for the binding of ANF to its receptor (unpublished results). The goals of these calculations were (1) to search the conformational space of the molecule, (2) to find unique regions that may be cross linked to test conformational hypotheses, and (3) to compare different approaches using dynamics to search conformational space.

1. System Setup and Strategic Approach

The principal active form of ANF in the blood is a 28-amino-acid fragment with a disulfide bond between residues 7 and 23:

```
                 5                     10
Ser-Leu-Arg-Arg-Ser-Ser-Cys-Phe-Gly-Gly- Arg-Met-┐
                         /                      Asp
                        S                       Arg
                        S                       Ile  15
                       /                        Gly
Tyr-Arg-Phe-Ser-Asn-Cys-Gly-Leu-Gly-Ser-Gln-Ala-┘
              25                  20
```

An initial geometry was generated by first building the sequence into an extended conformation using standard amino acid bond lengths, angles, and side-chain dihedrals. The disulfide bridge was formed by imposing a β turn at Gly[16] and minimizing the resulting structure as a starting conformation for use in dynamics.

Solvent was not included in this system because we are not interested in the conformation of ANF in solution but rather when it is bound to the receptor. Lacking an explicit receptor model with which to interact, our objective is then to look for any conformations accessible to the ANF regardless of the environment. We will not ignore high-energy structures (since the receptor may stabilize these) and will compare the structures identified with structures of other analogues. Common structural features among ANF and ANF analogues may suggest motifs that can be used to construct a putative active conformation.

Since solvent is not included, all the amino acids were defined to be in their neutral states to keep interactions between charged side chains from dominating the structural search. This structure was then equilibrated by running dynamics at 300 K and 600 K for 20 psec each. Data were collected from a subsequent 100-psec run at 900 K. A total of 100,000 configurations (one every 10^{-15} sec) were sampled during this simulation. To reduce the volume of data to a more manageable level, instantaneous dynamic structures were minimized at 1-psec intervals, reducing the number of structures to be analyzed to 100.

2. Classification of Conformational States

Similarity of the structures minimized along the trajectory was quantified by the RMS deviation between backbone atoms of residues 7–23 (the ring residues). The similarity between structures was evaluated by comparing the RMS deviation for each possible pair of structures. The procedure is summarized in Fig. 14, where the RMS deviations are mapped into a gray scale and plotted in a 100 × 100 matrix. Elements near the diagonal representing structures that are close in time generally had low RMS values, whereas off-diagonal elements reflecting structures separated by long times generally had high RMS values. A number of boxlike areas appear along the diagonal, indicating that several distinct families of structures were encountered during the progress of the dynamics calculations. These areas were divided into 11 conformational groups, represented by the 11 boxed regions of Fig. 14. Thus, the 100 structures found during dynamics were classified into 11 conformational groups that had mutual RMS deviations less than a threshold value (3 Å) and were located sequentially in time.

A high temperature vastly increases the efficiency of producing conformational transitions. Consider a simple Arrhenius model for reaction rates:

$$k = A \exp(-\Delta E/RT)$$

where k is the reaction rate, A is a preexponential factor describing the frequency of thermal collisions, ΔE is the activation energy, R is the gas constant, and T is the temperature. Assuming a value of A of 10^{12} sec^{-1}, which is a typical rate of collisions for liquid rare gas atoms, the ΔE that will yield a reaction rate of once per 100 psec at 300 K is about 5 kcal/mole. At 900 K, this barrier would be crossed 250 times in 100 psec.

The conformational search of ANF at 900 K produced at least 11 distinct families in 100 psec. A similar control run at 300 K produced no significant new conformational states over the same period. Dynamics at 900 K was not without risk, however. For example, the peptide bond was observed to undergo *trans–cis* conversions in 42 of 100 minimized structures. This is clearly a result of the high temperature; no *trans*-to-*cis* transitions were observed for room-temperature simulations. To avoid formation of *cis* bonds, a torsional restraint of 5 kcal/mole-rad^2 was added to the peptide bond that effectively eliminated *trans–cis* conversions. This is another good example of how restraints can be incorporated into modeling strategies. Without a restraint to prevent *trans–cis* interconversion, high-temperature dynamics would not be as useful for conformational searching of peptides.

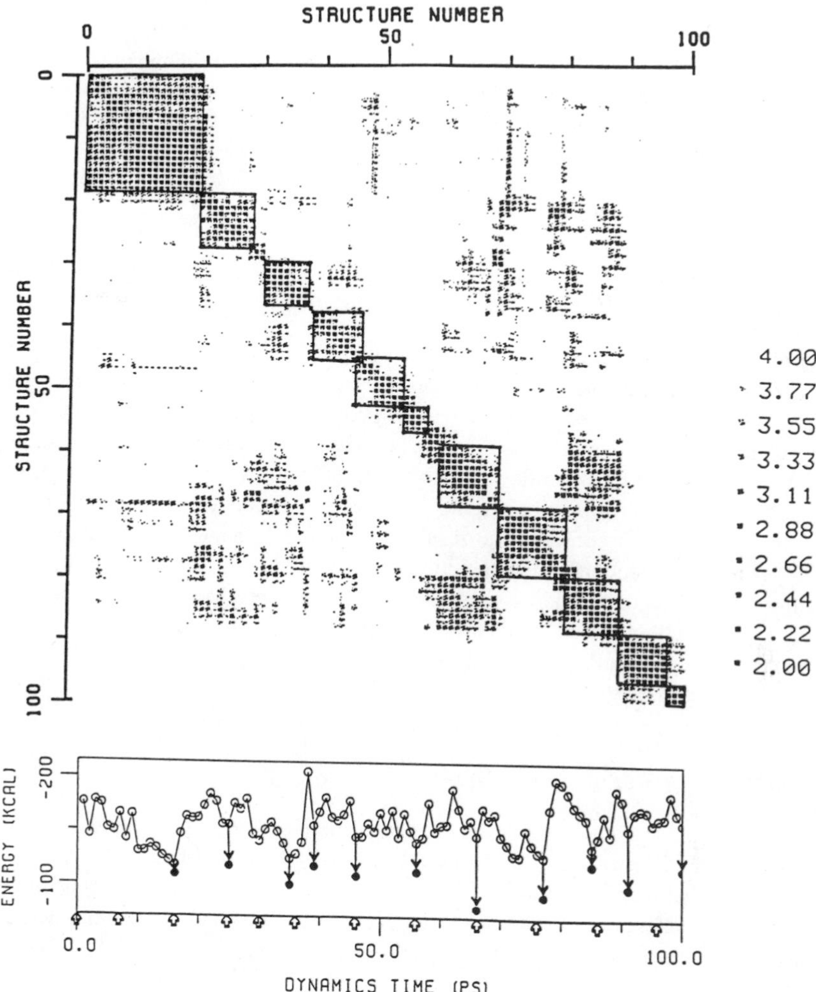

Figure 14. A matrix of RMS deviations between backbone atoms in the ring region of ANF (residues 7–23) for 100 minimized conformations of ANF with each other. The RMS deviation between the conformations minimized at time *I* and time *J* is mapped onto a gray scale from 2.0 Å to 4.0 Å and plotted as the matrix element, *IJ* for *I*,*J* = 1 to 100. Boxes are placed around groups of structures whose small RMS deviations and closeness in time suggest that they may be of the same conformational "family." The bottom graph contains the minimized energies for each of the 100 minimized structures (open circles). The filled circles represent the lowest energy found by annealing representative structures at 300 K for 20 psec.

3. Refinement of Strained Structures via Annealing

A second side effect of high-temperature dynamics is that it leads to higher-energy minima (by starting from a higher point on the energy surface, one falls into a higher-energy local minimum). By subsequently subjecting the minimized structures to lower-temperature dynamics at 300 K for 10–20 psec, up to 60 kcal/mole of strain could be removed. This annealing process also changed the relative ordering of the 11 conformational families and is therefore crucial in any energy-based selection of conformations.

Overall, then, minimization and structural classification strategies were used to refine 100,000 dynamic structures to 11 representative structures. This more manageable set of conformations was then analyzed in detail for important structural features to be incorporated into an analogue design strategy.

4. Applying Simulation Results to Specific Design Strategies

The biological significance of a structure can be best assessed through structure–activity relationships. Typically, the presence and location of specific functional groups are correlated to either decreased or enhanced activity to suggest requirements for activity. However, constraining conformational features should also be effective in distinguishing the biological significance of conformational families. Conformational features common to active analogues but inaccessible to inactives may be required for activity. Introducing a chemical cross link between two residues can constrain the resulting structure to remain in this conformation. Identifying a unique residue–residue interaction for each of the 11 identified ANF families and introducing a chemical cross link to restrain this residue–residue distance could provide analogues to test such conformational features. A matrix of C_α–C_α distances was calculated for each of the 11 representative annealed structures. These unique distances might be used as the basis for such a strategy.

C. Characterizing Vasopressin Dynamics

The final example considers a small peptide hormone system to demonstrate how minimization and dynamics strategies can be used to analyze small molecules.

Vasopressin is a peptide hormone that mediates such biological processes as milk ejection, uterine contraction, vasoconstriction, and antidiuretic functions. It is composed of nine amino acids, including two cysteines that form a disulfide bond between residues 1 and 6. Its sequence is:

S—S (disulfide bridge linking Cys 1 and Cys 6)

Cys-Tyr-Phe-Gln-Asn-Cys-Pro-Lys-Gly-NH_2

Thus, its structure includes a constrained ring of six residues and a flexible three-residue tail. We recently reported a dynamics study of vasopressin with the objective of characterizing the dynamic transitions and conformational equilibria of this flexible peptide (Hagler *et al.*, 1985). As in the ANF example, minimization techniques play a major role in understanding and analyzing dynamic results.

1. System Setup and Strategy

The dynamics of vasopressin were carried out *in vacuo* and with neutral side chains for the same reasons as in the ANF example. The initial structure was modeled after a postulated structure of the homologous peptide oxytocin. Dynamics were run at room temperature for approximately 8 psec, and the resulting trajectory was analyzed for transitions between conformational states. As for ANF, minimizations were performed periodically during the trajectory to characterize the conformational states being sampled.

2. Structural and Energetic Analysis

The flexibility of vasopressin was confirmed immediately by the dynamics. Surprisingly, even residues within the six-residue ring underwent several backbone transitions during the short simulation. The Phe^3, for example, started from a C_7^{eq} state and proceeded to a right-handed α helix before eventually finding a C_7^{ax}, all in just over 6 psec. As for the ANF example, structures during the dynamics were minimized to confirm that the dynamics trajectory was passing through legitimate minima and not just fluctuating about a single minimum. The instantaneous conformational states along with the resulting minimized states are given in Table IV for Phe^3. Note that dynamic structures as characterized by the (ϕ,ψ) values for Phe^3 can vary widely (by as much as 100°) but still minimize to the same conformational state. Nevertheless, there are three distinct states to which the dynamic structures minimize, which rigorously demonstrates that transitions are occurring.

One of the most significant results of this study was that the energy levels of conformational states for Phe^3 in vasopressin are more closely spaced than for isolated phenylalanine. For example, the difference between C_7^{eq} and α-helical state in the isolated blocked Phe

Table IV. Intraresidue Energies of the Conformations of the Phe^3 Residue of Vasopressin[a]

	Energy (kcal/mole)		(ϕ,ψ)(degrees)	
Time (psec)	Dynamics	Minimized	Dynamics	Minimized
		C_7 equatorial conformation		
0.001	45.1	32.8	(−49,54)	(−74,68)
0.2	36.9	32.7	(−81,30)	(−74,69)
1.1	40.5	33.3	(−112,78)	(−76,63)
		α-Helical conformation		
1.4	43.7	34.7	(−54,100)	(−57,−54)
1.6	44.8	35.3	(−77,83)	(−51,−56)
1.8	38.4	35.0	(−80,54)	(−55,−53)
2.7	42.0	35.4	(−58,−14)	(−46,−52)
3.2	41.5	35.0	(−59,−48)	(−50,−51)
3.9	44.7	35.4	(−31,−64)	(−50,−57)
		C_7 axial conformation		
5.1	42.1	34.9	(−51,−39)	(48,−81)
6.0	43.3	34.9	(35,−74)	(48,−81)
6.9	44.3	34.8	(48,−70)	(48,−81)

[a]The minimum energies and torsion angles for isolated residues in the three conformational states found are as follows: C_7 equatorial, 26.2 kcal/mole and (ϕ,ψ) (188°,97°); α-helical, 30.9 kcal/mole and (ϕ,ψ) (−78°,−55°); C_7 axial, 29.3 kcal/mole and (ϕ,ψ) (75°,−69°).

residue is calculated to be 4.7 kcal/mole, whereas in vasopressin, it is only 2.2 kcal/mole. This effective leveling of the energy levels imposed by constraints in the ring will contribute to an enhanced flexibility of this part of the vasopressin molecule.

3. Mapping an Adiabatic Torsion Potential Using Torsion Restraints

As a final example of how one may use minimization in conjunction with restraints to perform energy analysis, consider once again the rigid-geometry versus flexible-geometry torsion map. As pointed out above, energy barriers to rotation calculated by rotating about a bond with rigid bonds and angles can be misleading. What would appear to be very high barriers because of van der Waals overlap are much lower when small bond stretches and angle bends can absorb the strain. To illustrate this point, consider Fig. 15, which superimposes the Phe^3 (ϕ,ψ) trajectory onto an "adiabatic" (that is, flexible-geometry) contour surface. Note that the barrier, particularly from the α helix to C_7^{ax}, is only 3–4 kcal/mole. The flexible-geometry torsion map is particularly useful to compare with a dynamics trajectory, since it can demonstrate nicely the path the molecule takes on the energy surface.

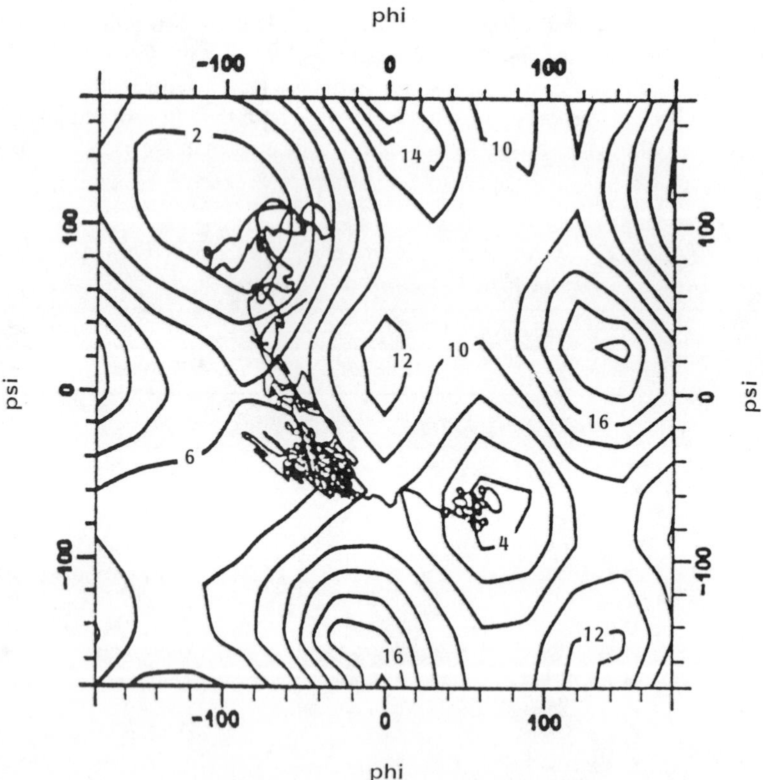

Figure 15. A "flexible geometry" contour map of the energy of the Phe^3 as a function of (ϕ,ψ) calculated by relaxing the entire phenylalanine while systematically sampling the ϕ and ψ angles. The energy is contoured in 2 kcal/mole. The dynamic trajectory of the residue in the vasopressin molecule is superimposed on the energy contour, showing the conformations assumed by this residue sequentially during the 8-psec run.

VI. CONCLUSIONS

We have discussed the significance and methodologies of minimization methods applied to macromolecular systems. We have seen that by introducing restraints and constraints into the classical minimization algorithms, modeling strategies can be formulated to answer a wide range of specific questions. Not only can minimization be used to refine a structure in preparation for detailed structural and energetic analysis (as we have seen for DHFR), but molecular mechanics algorithms can be directed to preserve structural features (such as peptide bonds), relax strain, prepare systems for dynamic simulations, test conformational hypotheses, search and categorize conformational space, and calculate energy barriers, among other applications.

The most common minimization algorithms in use include steepest descents, conjugate gradients, and Newton–Raphson. These differ mainly in the choice of the direction in which to move atoms to lower the energy. Steepest descents is the most robust and the method of choice for highly strained systems. Conjugate gradients converges more efficiently and is particularly well suited for large systems for which storage space is a constraint. Newton–Raphson-style minimizers are limited by their large memory requirements; therefore, they usually are practical only for small systems.

Finally, through several detailed examples, we have shown applications to a protein–ligand system and also described how minimization is a crucial component in characterizing and analyzing the results of conformational searches on small and medium peptides.

VII. REFERENCES

Abe, H., Braun, W., Noguti, T., and Go, N., 1984, Rapid calculation of first and second derivatives of conformational energy with respect to dihedral angles for proteins. General recurrent equations, *Comp. Chem.* **8:**239–247.

Baccanari, D. P., Daluge, S., and King, R. W., 1982, Inhibition of dihydrofolate reductase: Effect of reduced nicotinamide adenine dinucleotide phosphate on the selectivity and affinity of diaminobenzylpyrimidines, *Biochemistry* **21:**5068–5075.

Baniak, E. L., Rivier, J. E., Struthers, R. S., Hagler, A. T., and Gierasch, L. M., 1987, Nuclear magnetic resonance analysis and conformational characterization of a cyclic decapeptide antagonist of gonadotropin-releasing hormone, *Biochemistry* **26:** 2642–2656.

Berendsen, H. J. C., Postma, J. P. M., van Gunsteren, W. F., DiNola, A., and Haak, J. R., 1984, Molecular dynamics with coupling to an external bath, *J. Chem. Phys.* **81:**3684–3690.

Braun, W., and Go, N., 1985, Calculation of protein conformations by proton–proton distance constraints: A new efficient algorithm, *J. Mol. Biol.* **186:**611–626.

Brisson, A., and Unwin, P. N. T., 1985, Quaternary structure of the acetylcholine receptor, *Nature* **315:**474–477.

Brooks, B. R., Bruccoleri, R. E., Olafson, B. D., States, D. J., Swaminathan, S., and Karplus, M., 1983, CHARMM: A program for macromolecular energy, minimization, and dynamics calculations, *J. Comp. Chem.* **4:**187–217.

Brooks, C. L. III, Pettitt, B. M., and Karplus, M., 1985, Structural and energetic effects of truncating long ranged interactions in ionic and polar fluids, *J. Chem. Phys.* **83:**5897–5908.

Chou, K. C., Nemethy, G., and Scheraga, H. A., 1983, Energetic approach to the packing of α-helices. 1. Equivalent helices, *J. Phys. Chem.* **87:**2869–2881.

Clore, G. M., Bruenger, A. T., Karplus, M., and Gronenborn, A. M., 1986, Application of molecular dynamics with interproton distance restraints to three-dimensional protein structure determination. A model study of crambin, *J. Mol. Biol.* **191:**523–551.

Crippen, G. M., 1977, A novel approach to calculation of conformation: distance geometry, *J. Comp. Phys.* **24:** 96–107.

Crippen, G. M., and Scheraga, H. A., 1969, Minimization of polypeptide energy. VIII. Application of the deflation technique to a dipeptide, *Proc. Natl. Acad. Sci. U.S.A.* **64:**42–49.

Crippen, G. M., and Scheraga, H. A., 1971a, Minimization of polypeptide energy. X. A global search algorithm, *Arch. Biochem. Biophys.* **144:**453–461.

Crippen, G. M., and Scheraga, H. A., 1971b, Minimization of polypeptide energy. XI. Method of gentlest ascent, *Arch. Biochem. Biophys.* **144:**462–466.

Crippen, G. M., and Scheraga, H. A., 1973, Minimization of polypeptide energy. XII. Methods of partial energies and cubic subdivision, *J. Comput. Phys.* **12:**491–497.

Dauber-Osguthorpe, P., Roberts, V. A., Osguthorpe, D. J., Wolff, J., Genest, M., and Hagler, A. T., 1988, Structure and energetics of ligand binding to proteins: *E. coli* dihydrofolate reductase–trimethoprim, a drug-receptor system, *Proteins: Structure, Function and Genetics,* **4:** 31–47.

Dayringer, H. E., Tramontano, A., Sprang, S. R., and Fletterick, R. J., 1986, Interactive program for visualization and modelling of proteins, nucleic acids, and small molecules, *J. Mol. Graphics* **6:**82–87.

Eisenberg, D., 1984, Three-dimensional structure of membrane and surface proteins, *Annu. Rev. Biochem.* **53:** 595–623.

Ermer, O., 1976, Calculation of molecular properties using force fields. Applications in organic chemistry, *Structure Bonding* **27:**161–211.

Fine, R. M., Wang, H., Shenkin, P. S., Yarmush, D. L., and Levinthal, C., 1986, Predicting antibody hypervariable loop conformations II: Minimization and molecular dynamics studies of MCPC603 from many randomly generated loop conformations, *Proteins* **1:**342–362.

Fletcher, R., 1980, *Practical Methods of Optimization,* Volume 1, John Wiley & Sons, New York.

Furois-Corbin, S., and Pullman, A., 1986, Theoretical study of the packing of α-helices by energy minimization: Effect of the length of the helices on the packing energy and on the optimal configuration of a pair, *Chem. Phys. Lett.* **123:**305–310.

Gibson, K. D., and Scheraga, H. A., 1986, Predicted conformations for the immunodominant region of the circumsporozoite protein of the human malaria parasite *Plasmodium falciparum, Proc. Natl. Acad. Sci. U.S.A.* **83:**5649–5653.

Gilson, M. K., Rashin, A., Fine, R., Honig, B., Kline, A. D., and Wüthrich, K., 1985, Secondary structure of the α-amylase polypeptide inhibitor tendamistat from *Streptomyces tendae* determined in solution by ^{1}H nuclear magnetic resonance, *J. Mol. Biol.* **183:**503–507.

Greer, J., 1985, Protein structure and function by comparative model building, *Ann. N.Y. Acad. Sci.* **439:**44–63.

Hagler, A. T., 1985, Theoretical simulation of conformation, energetics, and dynamics of peptides, in: *The Peptides,* Volume 7 (V. J. Hruby and J. Meienhofer, eds.), Academic Press, New York, pp. 213–299.

Hagler, A. T., and Moult, J., 1978, Computer simulation of the solvent structure in biological macromolecules, *Nature* **272:**222–226.

Hagler, A. T., Lifson, S., and Dauber, P., 1979a, Consistent force field studies of intermolecular forces in hydrogen bonded crystals. 2. A benchmark for the objective comparison of alternative force fields, *J. Am. Chem. Soc.* **101:**5122–5130.

Hagler, A. T., Dauber, P., and Lifson, S., 1979b, Consistent force field studies of intermolecular forces in hydrogen bonded crystals. 3. The C═O···H-O hydrogen bond and the analysis of the energetics and packing of carboxylic acids, *J. Am. Chem. Soc.* **101:**5131–5141.

Hagler, A. T., Stern, P. S., Sharon, R., Becker, J. M., and Naider, F., 1979c, Computer simulation of the conformational properties of oligopeptides: Comparison of theoretical methods and analysis of experimental results, *J. Am. Chem. Soc.* **101:**6842–6852.

Hagler, A. T., Osguthorpe, D. J., Dauber-Osguthorpe, P., and Hempel, J. C., 1985, Dynamics and conformational energetics of a peptide hormone: Vasopressin, *Science* **227:**1309–1315.

Harvey, S. C., and McCammon, J. A., 1982, Macromolecular conformational energy minimization: An algorithm varying pseudodihedral angles, *Comput. Chem.* **6:**173–179.

Havel, T. F., and Wuthrich, K., 1984, A distance geometry program for determining the structures of small proteins and other macromolecules from nuclear magnetic resonance measurements of intramolecular ^{1}H–^{1}H proximities in solution, *Bull. Math. Biol.* **46:**673–698.

Havel, T. F., Kuntz, I. D., and Crippen, G. M., 1983, The theory and practice of distance geometry, *Bull. Math. Biol.* **45:**665–720.

Hwang, J. K., and Warshel, A., 1987, Semiquantitative calculations of catalytic free energies in genetically modified enzymes, *Biochemistry* **26:**2669–2273.

Jones, T. A., 1982, FRODO: A graphics fitting program for macromolecules, in: *Computational Crystallography* (D. Sayre, ed.), Clarendon Press, London, p. 303.

Karfunkel, H. R., 1986, A fast algorithm for the interactive docking maneuver with flexible macromolecules and probes, *J. Comput. Chem.* **7:**113–128.

Katz, L., and Levinthal, C., 1972, Interactive computer graphics and representation of complex biological structures, *Annu. Rev. Biophys. Bioeng.* **1:**465–504.

Kirkwood, J. G., 1935, Statistical mechanics of fluid mixtures, *J. Chem. Phys.* **3:**300–313.

Kitson, D. H., and Hagler, A. T., 1988, Theoretical studies of the structure and molecular dynamics of a peptide crystal, *Biochemistry* **27:** 5246–5257.

Lifson, S., Hagler, A. T., and Dauber, P., 1979, Consistent force field studies of intermolecular forces in hydrogen bonded crystals. 1. Carboxylic acids, amides, and the C=O···H-O hydrogen bonds, *J. Am. Chem. Soc.* **101:**5111–5121.

Maple, J. R., Dinur, U., and Hagler, A. T., Derivation of forcefields for molecular mechanics and dynamics from *ab initio* energy surfaces, *Proc. Natl. Acad. Sci. U.S.A.* **85:**5350–5354.

Matthews, D. A., Alden, R. A., Bolin, J. T., Freer, S. T., Hamlin, R., Xuong, N., Kraut, J., Poe, M., Williams, M., and Hoogsteen, K., 1977, Dihydrofolate reductase: X-ray structure of the binary complex with methotrexate, *Science* **197:**452–455.

Mezei, M., and Beveridge, D., 1986, Free energy simulations. *Ann. N.Y. Acad.Sci.* **482:** 1–23.

Moult, J., and James, M. N. G., 1986, An algorithm for determining the conformation of polypeptide segments in proteins by systematic search, *Proteins* **1:**146–163.

Nilsson, L., and Karplus, M., 1986, Empirical energy functions for energy minimization and dynamics of nucleic acids, *J. Comput. Chem.* **7:**591–616.

Noguti, T., and Go, N., 1983, A method of rapid calculation of a second derivative matrix of conformational energy for large molecules, *J. Phys. Soc. (Jpn.)* **52:**3685–3690.

Pattabiraman, N., Levitt, M., Ferrin, T. E., and Langridge, R., 1985, Computer graphics in real-time docking with energy calculation and minimization, *J. Comput. Chem.* **6:**432–436.

Poe, M., Hoogsteen, K., and Matthews, D. A., 1979, Proton magnetic resonance studies on *E. coli* dihydrofolate reductase, *J. Biol. Chem.* **254:**8143–8152.

Press, W. H., Flannery, B. P., Teukolsky, S. A., and Vetterling, W. T., 1986, *Numerical Recipes, The Art of Scientific Computing,* Cambridge University Press, Cambridge.

Purisima, E. O., and Scheraga, H. A., 1986, An approach to the multiple-minima problem by relaxing dimensionality, *Proc. Natl. Acad. Sci. U.S.A.* **83:**2782–2786.

Quirke, N., and Jacucci, G., 1982, Energy difference functions in Monte Carlo simulations: Application to (1) the calculation of free energy of liquid nitrogen, (2) the calculation of fluctuation in Monte Carlo averages, *Mol. Phys.* **45:**823–838.

Rivier, J., Kupryszewski, G., Varga, J., Porter, J., Rivier, C., Perrin, M., Hagler, A., Struthers, S., and Corrigan, A., Design of potent cyclic gonadotropin releasing hormone antagonists, *J. Med. Chem.* **31:** 677–682.

Roberts, V. A., Dauber-Osguthorpe, P., Osguthorpe, D. J., Levin, E., and Hagler, A. T., 1986, A comparison of the binding of the ligand trimethoprim to bacterial and vertebrate dihydrofolate reductases, *Isr. J. Chem* **27:**198–210.

Singh, U. C., Brown, F. K., Bash, P. A., and Kollman, P. A., 1987, An approach to the application of free energy perturbation methods using molecular dynamics: Applications to the transformations of methanol to ethane, oxonium to ammonium, glycine to alanine, and alanine to phenylalanine in aqueous solution and to $H_3O^+(H_2O)_3$ NH_4^+ $(H_2O)_3$ in the gas phase, *J. Am. Chem. Soc.* **109:**1607–1614.

Stern, P. S., Chorev, M., Goodman, M., and Hagler, A. T., 1983, Computer simulation of the conformational properties of retro–inverso peptides. I. Empirical force field calculations of rigid and flexible geometries of *N*-acetylglycine-*N'*-methylamide, bis(acetamido)methane, and *N,N'*-dimethylmalonamide and their corresponding C^{α}-methylated analogs, *Biopolymers* **22:**1885–1900.

Straatsma, T. P., Berendsen, H. J. C., and Postma, J. P. M., 1986, Free energy of hydrophobic hydration: A molecular dynamics study of noble gases in water, *J. Chem. Phys.* **85:**6720–6727.

Struthers, R. S., Rivier, J., and Hagler, A. T., 1984, Design of peptide analogs: Theoretical simulation of conformation, energetics, and dynamics, in: *Conformationally Directed Drug Design: Peptides and Nucleic Acids as Templates or Targets* (J. A. Vida and M. Gordon, eds.), American Chemical Society, Washington, pp. 239–261, American Chemical Society, Washington.

Tembe, B. L., and McCammon, J. A., 1984, Ligand–receptor interactions, *Comput. Chem.* **8:**281–283.

Van Gunsteren, W. F., and Karplus, M., 1980, A method for constrained energy minimization of macromolecules, *J. Comput. Chem.* **1:**266–274.

Vasquez, M., and Scheraga, H. A., 1985, Use of buildup and energy-minimization procedures to compute low-energy structures of the backbone of enkephalin, *Biopolymers* **24:**1437–1447.

Warme, P. K., and Scheraga, H. A., 1974, Refinement of the x-ray structure of lysozyme by complete energy minimization, *Biochemistry* **13:**757–767.

Warshel, A., Sussman, F., and King, G., 1986, Free energy of charges in solvated proteins: Microscopic calculations using a reversible charging process, *Biochemistry* **25:**8368–8372.

Warwicker, J., 1986, Continuum dielectric modelling of the protein-solvent system, and calculation of the long-range electrostatic field of the enzyme phosphoglycerate mutase, *J. Theor. Biol.* **121:**199–210.

Warwicker, J., Ollis, D., Richards, F. M., and Steitz, T. A., 1985, Electrostatic field of the large fragment of *Escherichia coli* DNA polymerase I, *J. Mol. Biol.* **186:**645–649.

Weiner, S. J., Kollman, P. A., Case, D. A., Singh, U. C., Ghio, C., Alagona, G., Profeta, S., Jr., and Weiner, P., 1984, A new force field for molecular mechanical simulation of nucleic acids and proteins, *J. Am. Chem. Soc.* **106:**765–784.

8

The Role of Electrostatic Interactions in the Structure of Globular Proteins

Neil K. Rogers

I. INTRODUCTION

The structure of a protein may be considered as a scaffolding that brings into the correct spatial arrangement a constellation of amino acid side chains to perform a given function. If the

Neil K. Rogers • Laboratory of Molecular Biophysics, Department of Zoology, University of Oxford, Oxford OX1 3QU, United Kingdom.

structure is disrupted, the function is impaired. The spectrum of functions performed by proteins in living systems makes an understanding of the factors that determine their structures highly desirable.

Since the renaturation experiments of Anfinsen (1973), it has been largely accepted as axiomatic that the three-dimensional structure of a protein is determined by its sequence of amino acids. Such a view is helpful but incomplete, since it ignores the caveats that Anfinsen included in that paper;

> . . . the three dimensional structure of a native protein in its normal physiological milieu. . . . is the one in which the Gibbs free energy of the whole system is lowest; that is, that the native conformation is determined by the totality of interatomic interactions and hence by the amino acid sequence, in a given environment.

The environment in which a protein finds itself is just as significant a factor as the sequence in determining the structure. When, for example, a protein is extracted from a membrane or removed from a ribosome, immersed in alcohol or titrated with acid, deprived of its prosthetic group or exposed to extremes of temperature, nothing happens to the primary sequence, but the three-dimensional structure is often changed or lost. Thus, the structure of a protein is determined by its amino acid sequence and its entire environment.

The thermodynamic hypothesis as given by Anfinsen (*vide supra*) is a useful view of protein structure because it means that if all the interactions can be described by some potential function, then it should be possible to minimize that expression for the free energy and so predict the native conformation of the protein. There are many contributions to the free energy of stabilization of a protein structure.

$$\begin{aligned}\text{Free energy of protein} = {} & \text{covalent bonds} \\ & + \text{torsional potentials} \\ & + \text{hydrogen bonds} \\ & + \text{Van der Waals interactions} \\ & + \text{electrostatic interactions} \\ & + \text{hydrophobic effect} \\ & + \text{dynamic energy}\end{aligned} \tag{1}$$

It is a matter of conjecture whether or not the native structure occurs at the global minimum on the free energy surface defined by this equation or whether it is a local minimum that is found by some kinetic folding pathway (Creighton, 1985). Each of the terms in the equation contains protein–protein, protein–solvent, and solvent–solvent interactions and makes both enthalpic and entropic contributions, which need not necessarily be of the same sign, to the free energy of stabilization.

A. The Importance of Electrostatic Interactions

The purpose of this chapter is to examine the role that electrostatic interactions play in the determination of protein structure. Each atom in the structure is a possible source of an electrostatic field arising from its partial charge and intrinsic dipole moment, and there are $n(n - 1)/2$ interactions to be accounted for (if n is the total number of sources in the system). The protein, the solvent, cofactors, prosthetic groups, and substrates are almost invariably charged or dipolar, and so the range of effects in which electrostatics are involved is broad. Unlike the Van der Waals interactions there is no effective cutoff to the interaction distance, and so the electrostatic energy is the sum of a large number of weak interactions, whereas other contributions are perhaps the sum of a smaller number of strong interactions. Thus, careful consideration must be given to electrostatic effects in any calculations on the structure and function of proteins.

B. Comparison with Other Factors

It is no trivial matter to evaluate the relative magnitudes of each of the energy terms in equation 1. However, it seems clear that the major reason why soluble proteins adopt a globular shape is to limit the extent of water–hydrophobic interactions. This is an entropy-driven effect, since the water structure around hydrophobic groups is thought to be more ordered than bulk water. The burial of hydrophobic groups leads to an increase in entropy of the solvent but to a decrease in entropy of the globule (Kauzmann, 1959). Kauzmann concludes, "hydrophobic bonds are much more important than salt linkages in maintaining the stability of most native proteins."

The charged amino acid side chains are usually found at the boundary between the hydrophobic interior and the solvent (at least in soluble proteins). This means that the environment of the charged groups in the native protein is not dissimilar to that in the unfolded state, i.e., exposed to solvent to some degree (Tanford, 1962). Charged groups of the protein order the water structure more than hydrophobic residues, but the difference in solvent entropy around charged groups between the folded and unfolded states is not large, and the enthalpic cost of burying charges is high (see Section IV.B); thus, the charges prefer to stay at the surface of the globule.

What are the charges doing at the surface? If the surface of proteins were to be extensively composed of hydrophobic residues, then the proteins would probably aggregate by the hydrophobic effect, and solubility would be lost; hence, charges maintain solubility. Secondly, the charged residues are often the ones responsible for the function, and they therefore need to be accessible to the solvent, so that substrates can encounter their active centers. Thirdly, the charged atoms may form stabilizing networks of salt bridges or ion pairs on the surface of the molecule and aid stability in that way.

Van der Waals interactions are largely responsible for the determination of secondary structure. The energetically favorable regions of conformational space correspond almost entirely with the allowed regions derived from considerations of hard spherical atoms (Ramachandran *et al.*, 1963) (see Section III.A). The later addition of other factors and potential functions did little to change the conformational possibilities that had already been discovered by the simple steric model. Edsall *et al.* (1966) established conventions for the representation of conformational geometries and refer to all the major contributions that were already made. There are also steric interactions in the packing of secondary structure units (Chothia *et al.*, 1977; Cohen *et al.*, 1982), the so-called "knobs-into-holes" idea of secondary structure packing, which Chou *et al.* (1983, 1984) have shown to be considerably more significant than electrostatic interactions between the secondary structures.

Hydrogen bonds are found extensively in proteins and are particularly prominent in secondary structure elements such as the α helix and the β sheet. The peptide linkage and the side chains of the 11 hydrophilic residues all have the capability of forming one or more hydrogen bonds, as has been reviewed by Baker and Hubbard (1984). In the folding process, the H-bonds between the polypeptide chain and the solvent will be largely replaced by intrachain bonds, and so the total number of H-bonds will not significantly alter. Thus, the contribution that H-bonds make to the stability of proteins may not be large. On the other hand, Perutz and Raidt (1975) have suggested that the stability of one mutant hemoglobin over another could be explained in terms of some extra hydrophobic interactions and an extra H-bond, the sum of which came to about 5 kJ/mole. This energy is not large, but it is sufficient to increase the denaturation rate constant sixfold.

The bending, stretching, and torsional potentials are likely to be the fine-tuning effects rather than the major determinants of the overall structure. However, they are numerous, and their contribution to the total free energy will not be negligible.

C. The Development of Models and Methods

The force of interaction between any two charges in any environment is given by Coulomb's Law:

$$\mathbf{F}_{12} = q_1 q_2 \mathbf{r}_{12}/4\pi\epsilon_0 r_{12}^3 \tag{2}$$

If there is other material present (as there always is in proteins), then the law still holds, but the other material is polarized by the presence of the charges, and the induced dipoles of the material interact with the original charges. Thus, the total resolved force (as opposed to the actual force between the charges) on each of the charges is altered, and in a uniform isotropic medium this can be accounted for by the use of a dielectric constant, so that Coulomb's law can be expressed as;

$$\mathbf{F}_{12} = q_1 q_2 \mathbf{r}_{12}/4\pi\epsilon_0\epsilon_r r_{12}^3 \tag{3}$$

This does not mean that the force between the original charges is diminished, but the forces between the charges and the polarized atoms of the media give rise to a net effect that apparently decreases the interaction by a factor that is designated the dielectric constant.

The major problem in studying the role of electrostatic interactions in globular proteins is the treatment of the dielectric constant. Ideally an approach that discarded the use of a dielectric constant and actually considered the individual atoms of the system and their polarizabilities and dynamic behavior would be the way to tackle the problem. This would be a "half-way house" between a full quantum mechanical treatment (which is not a feasible option for a system of this size) and the use of macroscopic dielectric constants. An attempt at such an approach has been made by Warshel (reviewed by Warshel and Russell, 1984), but there are yet problems. The more detailed the model is, the more detailed the information going into it should be; otherwise there is little point in going to such exotic lengths. However, this information is not always available, and, even when known, it is not always computationally convenient to incorporate it into the model. This discrepancy between the detail of the model and its implementation is dictated in part by the size of the system that is being modeled. Although there are problems, this approach is the most rigorous of those available at present and should be useful in the future, when the problems have to some extent been overcome.

The alternative to such a microscopic approach is to use the macroscopic concept of a dielectric constant. Three different models have been used that incorporate a dielectric constant: the uniform dielectric model, the distance-dependent dielectric model, and the cavity dielectric model. These have been discussed in a complementary review (Rogers, 1986), but an overview is given here.

The uniform dielectric model assumes that there is one value of the dielectric constant that may be used on the molecular level. There have been many suggestions as to what this value might be; since the inside of a protein may be considered as being somewhat similar to a polyamide, a value in the range 2–5 is often quoted (e.g., Ramachandran and Sasisekharan, 1968). At the other extreme, others have attempted to find values for an effective dielectric constant and found values in the rage 44–128 (e.g., Rees, 1980).

The lower ranges are fine as an estimate of the dielectric constant of the material of the protein interior but ignore the fact that the protein is surrounded by water, which has a high dielectric constant, and that the solvent has a substantial effect even within the protein. The upper estimates are not of an actual dielectric but of an effective dielectric constant. An effective dielectric constant is useful in describing a given interaction, but it should be noted that it may at best be misleading if it is used in calculations between any other charges. In 1956, Hill pointed out that each pair of interacting charges would possess its own effective dielectric constant and that this would be a unique property of those two charges. Rogers (1986) has gone on to point out that even when appropriate values for the dielectric constant of the protein and the solvent are employed, the properties of the effective dielectric constant are rather unusual:

1. Effective dielectric constants should have an angular dependence.
2. Effective dielectric constants are not symmetric.
3. Effective dielectric constants may be infinite.
4. Effective dielectric constants are not transferable between energy and force calculations.

Since the effective dielectric is a mathematical function of the actual dielectric values of the media, it too is a function of temperature in an unpredictable way.

The distance-dependent dielectric constant is a pragmatic rather than physical description of electrostatic interactions. It is assumed that at separations of the order of atomic dimensions the dielectric constant between two points is that of free space and that at much larger distance the bulk value for water may be used. For intermediate distances it is assumed that the dielectric varies with distance in some appropriate fashion. In certain well-defined geometries it is possible to approximate the effective dielectric constant by a distance-dependent function, as has been shown by Ramachandran and Srinivasan (1970), Warshel *et al.* (1984), Ehrensen (1984), and Hingerty *et al.* (1985), but as Warshel *et al.* (1984) aptly point out, "$e(R)$ is not a universal function, and its use may be quite misleading in analysis of many biological processes."

Rogers (1986), in a critique of the use of the distance-dependent model, has shown that the form of distance dependence is not the same in field and energy calculations and that whatever form of the distance-dependent dielectric constant is used, there is grave danger of contravening Gauss' law.

The cavity dielectric model has several subforms, but the model is a macroscopic one that considers the protein to be a medium of low polarizability with a dielectric of less than 5 embedded in a solvent of high dielectric (water, for which the dielectric constant is 80). The shape of the cavity is modeled as a sphere, an ellipsoid, or some nonanalytical shape, and the electrostatic potentials are calculated by solution of Poisson's equation. The method of solution may be analytical if the shape of the cavity is geometrically definable or by a numerical procedure if the cavity is irregular. In some geometries it is possible to circumnavigate the cumbersome solution to Poisson's equations by the method of images (Hammond, 1960); however, in anything but planar geometry the method is approximate, and the accuracy decreases the nearer the charges are to the dielectric interface (Friedman, 1975). This is unfortunate, since this is where nearly all the formal charges are to be found in globular proteins.

The best known of the cavity dielectric models is the formalism of Tanford and Kirkwood (1957) and the subsequent modifications to it (see Matthew *et al.*, 1985, for review). The finite difference procedure of Warwicker and Watson (1982) and the surface polarization method of Zauhar and Morgan (1985) are attempts to describe more accurately the three-dimensional structure of the protein and involve numerical solutions to the electrostatic equations.

D. The Energetics of Electrostatic Interactions

There are two fundamental approaches to the energetics of electrostatic interactions. The first takes no account of the individual atoms of the system and their polarizabilities, since it is assumed that the macroscopic assignments of the dielectric constant subsume all those details. The energy that is calculated by this method is strongly affirmed to be a free energy of interaction (Born, 1920; Abraham and Becker, 1950; Schellman, 1953; Tanford and Kirkwood, 1957; Linderstrom-Lang and Schellman, 1959; Tanford, 1961; Friedman, 1975; Matthew and Richards, 1982). It is found that the free energy is a function of the dielectric constant and the geometry of the system:

$$\Delta G^{\mathrm{ES}} = \Delta G^{\mathrm{ES}}(\text{geometry}, \epsilon_r) \tag{4}$$

Manipulations of the analytic expression of the free energy yield the entropic and enthalpic contributions that electrostatic interactions make to the overall free energy of the system:

$$\Delta S^{\mathrm{ES}} = \frac{-\partial}{\partial T}\Delta G^{\mathrm{ES}} = \Delta S^{\mathrm{ES}}\left(\frac{\partial \text{geometry}}{\partial T}, \frac{\partial \epsilon_r}{\partial T}\right) \tag{5}$$

$$\Delta H^{\mathrm{ES}} = \Delta G^{\mathrm{ES}} + T\Delta S^{\mathrm{ES}} = \Delta H^{\mathrm{ES}}\left((\text{geometry}, \frac{\partial \text{geometry}}{\partial T}, \epsilon_r, \frac{\partial \epsilon_r}{\partial T}\right) \tag{6}$$

It is found that the critical factor is the way in which the dielectric constants of the media vary as a function of temperature. This may be known macroscopically, and so a handle on the thermodynamics of the system may be gained. What is more difficult to incorporate into such an approach is how the geometry varies as a function of temperature, which may be just as significant but is almost impossible to consider in a macroscopic model.

The alternative approach is to consider all the atoms in the system and the environment. It is possible in principle to incorporate all the various interactions and to calculate a conformational potential, which will include electrostatic interactions, for all possible arrangements of the system. This conformational potential may be sampled over a range of states, and a partition function may be calculated:

$$Z = \Sigma_{\text{conformations}} \exp(-V_{\text{conf}}/kT) \tag{7}$$

The partition function may be manipulated to get all of the relevant thermodynamic parameters of the system by well-known statistical mechanical procedures.

This or similar procedures have been followed by Brant *et al.* (1967), Florey (1969), Warshel (1978a), Hagler *et al.* (1979), Karplus and Kushick (1981), and Warshel and Russell (1984). In such a method the enthalpy is the statistically averaged potential of the system, and the entropy and free energy of the system are derived from the partition function. The dielectric constant should not appear explicitly in this method, since ideally the treatment of every atom should mean that dielectric effects are contained implicitly.

There are serious questions about the validity and implementation of both the macro- and microscopic approaches (Rogers, 1986), but the fundamental point is that the type of energy that is calculated is a function of the model that is adopted for the calculation. Microscopic approaches calculate a conformational potential energy that may be manipulated to give the enthalpy, entropy, and free energy. Macroscopic models produce free energies from which the other thermodynamic functions may be derived (Fig. 1).

What actually happens in practice is neither of these. A compromise model is usually used that is neither truly microscopic nor macroscopic, and the energy so derived is neither an enthalpy nor a free energy but a theoreticians' fiction, as has been pointed out by McCammon (1984):

> . . . the adjustment of the energy function parameters to reflect the average effects of the solvent surroundings . . . . corresponds to replacing the potential energy function by a potential of mean force that is averaged with respect to the possible configurations of the water molecules that surround the protein. The potential of mean force is a type of free energy and is dependent on temperature.

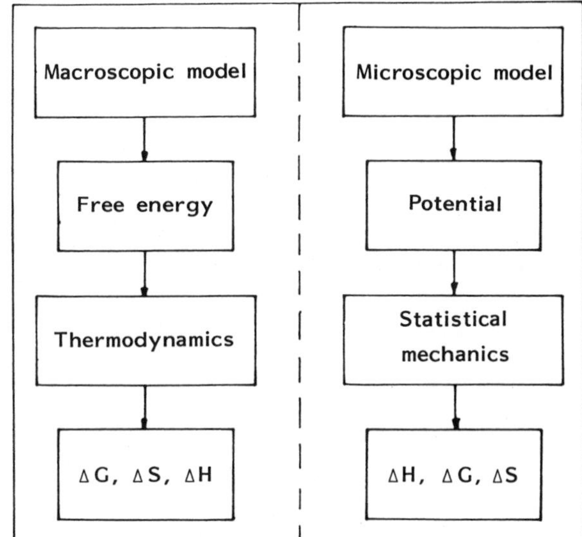

Figure 1. Schematic representation of the energetics of electrostatic interactions, indicating that the type of energy calculated is dependent on the dielectric model chosen.

A common example of this compromise is that of conformational analyses in which many of the atoms of the system are considered but the effects of the solvent are incorporated by the use of some effective dielectric constant. The energy that is calculated is neither a free energy nor an enthalpy but falls into a thermodynamic no-man's land. This highlights the confusion of the present situation. Indeed, this author was caught short by this problem, as can be seen in the paper by Moore and Rogers (1985a) and the subsequent erratum to it (Moore and Rogers, 1985b).

This is a serious setback to our understanding of the role of electrostatic interactions in the structure and function of globular proteins and indicates the need for theoreticians to consider carefully what they are actually calculating.

II. SOURCES OF ELECTROSTATIC FIELDS

A. Formal Charges of the Polypeptide Chain

Five of the side chains of the 20 commonly occurring amino acids may bear a full charge under normal physiological conditions. The amino group of lysine and the guanidinium group of arginine are protonated in all but the most extreme conditions of pH. The carboxyl groups of the glutamic acid and aspartic acid side chains carry a full charge at normal pH. Histidine is commonly but by no means always protonated and as such bears a full positive charge.

The p*K*s of the ionizable groups are shown in Fig. 2 (after Timasheff, 1970) along with their structures. The values quoted are for the idealized side chains, but these values may vary quite extensively in native proteins (and also between sources in the literature). The reason for this lies in the proximity of other charged groups to the side chain concerned, which may alter the electrostatic potential at the ionizable site and shift the p*K* according to the equation

$$\Delta pK = e\Delta\phi/2.303zkT \tag{8}$$

where e is the electronic charge and $\Delta\phi$ is the shift in electrostatic potential at the titrating site (Tanford and Roxby, 1972).

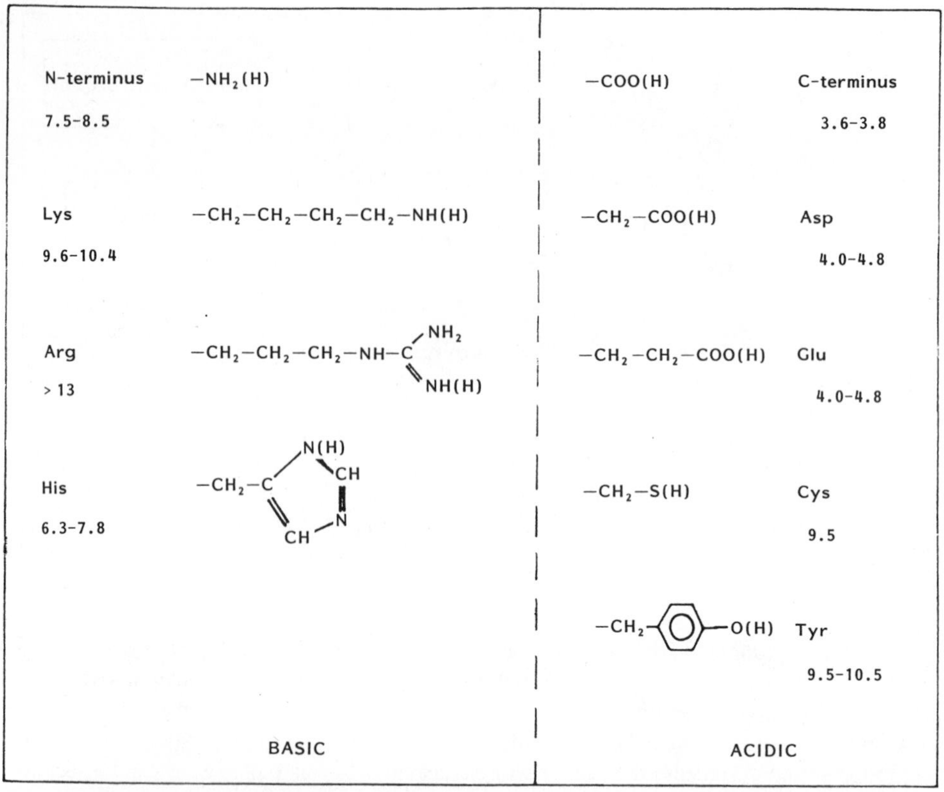

Basic	pK	Structure	Structure	Acidic	pK
N-terminus	7.5–8.5	$-NH_2(H)$	$-COO(H)$	C-terminus	3.6–3.8
Lys	9.6–10.4	$-CH_2-CH_2-CH_2-CH_2-NH(H)$	$-CH_2-COO(H)$	Asp	4.0–4.8
Arg	> 13	$-CH_2-CH_2-CH_2-NH-C(NH_2)=NH(H)$	$-CH_2-CH_2-COO(H)$	Glu	4.0–4.8
His	6.3–7.8	$-CH_2-C$ (ring: N(H), CH, N, CH)	$-CH_2-S(H)$	Cys	9.5
			$-CH_2-C_6H_4-O(H)$	Tyr	9.5–10.5
		BASIC	ACIDIC		

Figure 2. The structures and p*K* values of the ionizable side chains in globular proteins (after Timasheff, 1970).

The shape of the protein, irrespective of other charges in the proximity, may also contribute to shifted p*K*s of the ionizable side chains (Rogers, 1986). The precise values of the p*K*s of the ionizable side chains are of functional significance, and in the active sites of proteins, where charges are in close proximity to each other, the balance of p*K*s is critical.

In addition to these five side chains, others may also bear full formal charges in extremes of pH. Perutz (1974) suggested that the alkali denaturation of certain hemoglobins could be accounted for by the ionization of cysteine and tyrosine side chains at high pH. He proposed that they donated a proton from their sulfhydryl and hydroxyl groups, respectively, and that the presence of an unusually charged side chain in the closely packed interior of the protein or at the subunit interface attracted a solvation shell that led to the disruption of the structure as depicted in Fig. 3 (see Section IV.B).

The amino and carboxy termini of the polypeptide chain are also commonly charged, although the N terminus may be chemically modified by acetylation or some other process (Driessen *et al.*, 1985). Thornton and Sibanda (1983) have analyzed the crystallographic data on N and C termini in a sample of different proteins. They concluded that there is a preference for termini to lie much closer together than would be predicted by the generation of random structures and that this may have some role in the stabilization of protein structures.

B. Partial Charges of the Polypeptide Chain

Because of the differences in electronegativity of the atoms in the system, most carry an appreciable partial charge. Their values may be calculated by quantum mechanical methods

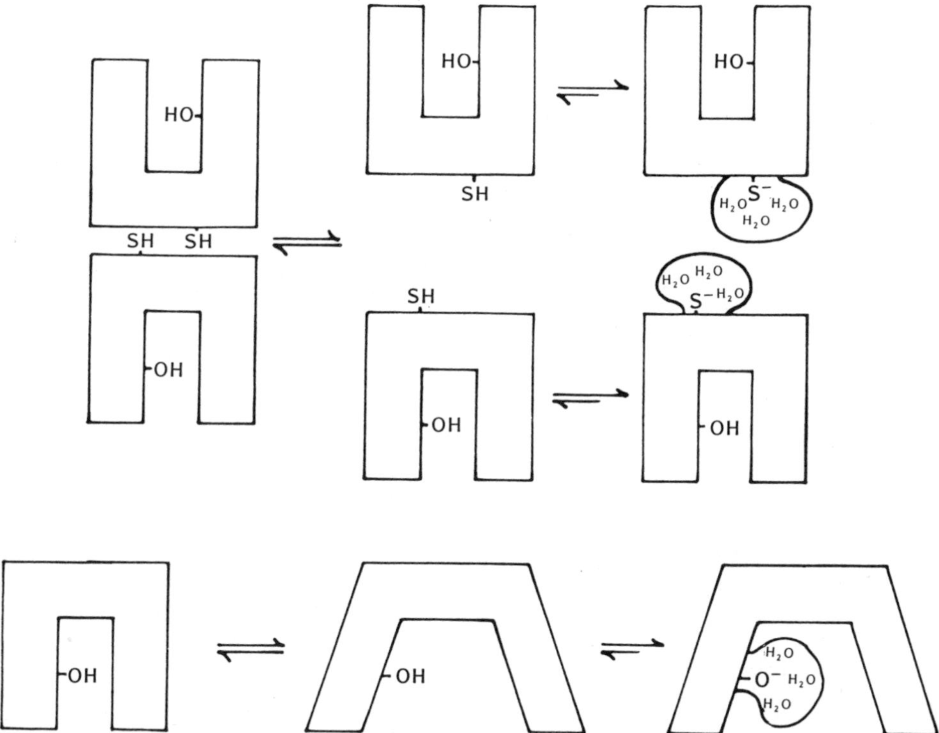

Figure 3. Cartoon of the mechanism of alkali denaturation of hemoglobin in which the solvation of buried or interfacial sulfhydryl and hydroxyl groups disrupts the protein's structure (after Perutz, 1974).

(e.g., Sheridan and Allen, 1981). By far the most significant of these partial charges are those distributed over the peptide bond (Fig. 4). Their importance is a result of their being distributed throughout the protein structure in ordered arrangements in the secondary structure units, whereas the other atomic partial charges are arranged on a somewhat more random basis. Rashin and Honig (1984) and Wada *et al.* (1985) have suggested that the atomic partial charges play an important role in ''solvating'' buried formal charges. The role of secondary structure dipole moments is discussed in Section III.

C. Cofactors, Regulators, Prosthetic Groups, and Substrates

Many proteins require the incorporation of non-amino-acid components for their function, and many of these carry formal charges that are of functional and structural import. For example, the porphyrins found in different forms in the globins, the cytochromes, photosynthetic proteins, and elsewhere have a metal ion, and some have ionizable substituents such as hydroxyl and propionyl groups. Nucleotide cofactors, phosphoryl groups, divalent metal ions, and vitamin coenzymes usually carry charges of some magnitude, and even if they do not the very presence of extra polarizable material in the vicinity of the protein will alter its electrostatic properties.

It is not difficult to see how these groups modify the function of proteins either directly or indirectly. Their presence can trigger conformational changes by covalent or noncovalent bonding, which may include electrostatic interactions. They can alter the dynamics of the system, which in itself could be a means of allosteric control without the need for conforma-

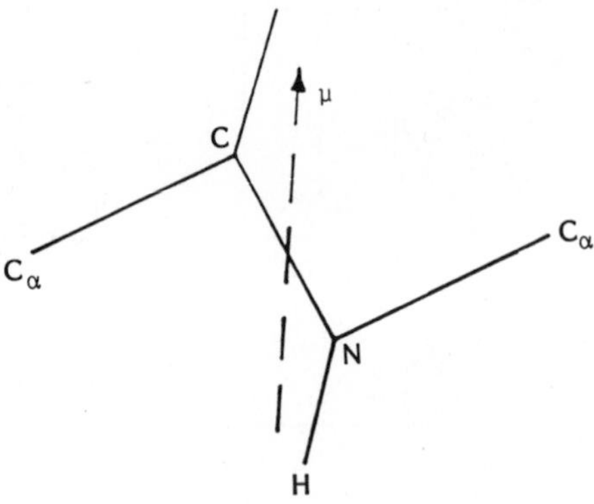

Partial Charges	N	H	C	O
Ramachandran and Sasisekharan, 1968	−0.3	0.3	0.4	−0.4
Momanay *et al.*, 1975	−0.344	0.176	0.450	−0.384
Hol *et al.*, 1978	−0.2	0.2	0.42	−0.42
Sheridan *et al.*, 1982	−0.378	0.201	0.310	−0.303

Figure 4. Distribution of partial charges in the peptide bond, which gives rise to the peptide dipole moment of 3.5 D.

tional change (Cooper, 1984) and could also be a result of electrostatic interactions between the cofactor and the protein. The presence of new charges and extra polarizable bulk can shift the p*K*s of the functional groups on the polypeptide chain or even form part of the functional constellation of groups. Thus, the electrostatics of any protein is likely to be incomplete unless it has some consideration of the contribution made by the non-amino-acid components of the system.

D. The Environment of Proteins

1. Water

All proteins have at least some degree of exposure to water even if they are membrane bound or part of a ribosome or some other macromolecular complex. The behavior of pure water is complex, and although much is known (Franks, 1972–1982), it is far from being fully understood. The solvent in biological systems is not pure water, and the presence of salt ions, lipids, and other components of the protein's milieu complicates the situation to no end. The properties of water molecules near a protein's surface are a matter of great interest and have been studied by detailed experimental and theoretical methods (see, for example, Finney, 1979; Edsall and McKenzie, 1983, for reviews).

The permanent orientable water molecule dipole is 1.85 D, and this is held to be the major contributor to the high dielectric constant of water (Warshel and Russell, 1984). There have also been suggestions that the polarizability of the water molecule is also an important factor in the determination of the high dielectric constant of water (Goodfellow *et al.*, 1982). Pethig

(1979) gives a helpful discussion of the properties of water in biological systems. From the point of view of calculations on proteins, the water is important because it gives rise to a reaction field. For any field set up by the protein, the water molecules orient and polarize to oppose the protein's field. This reaction field is ignored in many calculations.

2. Salt

The concentrations of salt ions in the environment of most proteins are carefully controlled. Within cells the concentrations of individual ionic species are regulated by the action of membrane ion pumps, channels, and various ion-exchange mechanisms. The same is true within the organelles of the cells. In multicellulate organisms the ionic balance in the extracellular fluid is also under close rein. The only circumstance in which proteins exist and function without control being exerted over their ionic environments is extracellular proteins that are also extraorganismal.

Many ions are involved in the regulation of protein structure and function. For example, intracellular calcium acts by specific binding to sites on specific proteins, whereas hydrogen ions modify the structure and function of all proteins by nonspecific binding. Every conceivable point between these two extremes is taken up. The salt ions can act in two major ways: by perturbing the structure of the protein directly or by altering the structure and properties of the solvent. This section deals with each of these in turn and then describes some of the methods of describing the ionic distribution.

Direct effects on the protein will be exerted by the presence of ions close to the protein surface. Perutz (1978) showed that it is not energetically favorable for ions to approach the surface of a low-dielectric medium from a high-dielectric solvent because the image charge repels the approaching ion. However, if a protein is highly charged or there are patches of high charge density on the surface of the protein (as is commonly the case), then the charge–charge interactions will overcome the charge–image charge interactions, and the protein will attract a preponderance of oppositely charged ions from the solution simply to maintain electroneutrality. This cloud of counterions will have a statistical distribution, and the most commonly used description of that is the analysis of Debye and Huckel (1923). The importance that the ions have can be seen in the effects that may be observed by altering the ionic strength. Solubility (Cohn and Ferry, 1943; Arakawa and Timasheff, 1985), stability (Von Hippel and Schleich, 1969), redox potential (George *et al.*, 1966), and hence all the kinetic properties (e.g., Frieden *et al.*, 1957; Stankovich *et al.*, 1978; Voet *et al.*, 1981) and almost every other physical attribute of a protein are dependent on the ionic strength and composition of the solution.

Specific ion binding to proteins occurs commonly, and often as a means of regulation of function. The nature of the binding will depend on the properties of the ion—whether the ion is simple or compound and if it has any other components such as organic groups, which may or may not have some hydrophobic character. Its overall charge, size, shape, and dipole moment will all be significant in determining specific ion binding (Tam and Williams, 1985).

Effects on the solvent are complex and involve several different and competing processes. The presence of a salt ion in the solution must mean that at least one water molecule is now absent from that site. The replacement of a polar water molecule by an ion of different dipole moment (if it has one) means that there will be a change in the dielectric properties of the solvent. Unless the ion is exactly the same shape and size as the water molecule it replaces, its presence in solution will also cause a steric perturbation of the solvent structure. There will be, of necessity, a rearrangement of coordination, density, and radial distribution in the vicinity of the ion, which will also have effects on the dielectric properties of the solution. The introduction of a charged ion into a dipolar solvent will cause electrostrictive effects, which will act as

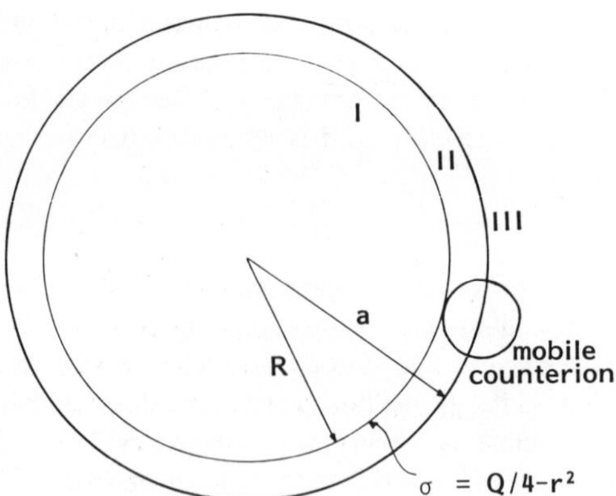

Figure 5. Debye–Huckel formalism applied to a spherical model of a globular protein. Region I is the protein interior defined by the radius R. Region II is the ionic exclusion radius and is defined by the radius a. Region I is the solvent. There is a smeared surface charge density as indicated.

a result of the charge attracting the water molecule dipoles. There will be substantial volume and entropy effects associated with this electrostriction and the ordering of the solvent in the neighborhood of the ion.

The description of ionic solutions is most frequently carried out by an extension of the formalism of Debye and Huckel (1923) (Gabler, 1978). The ion is treated as a cavity within a high-dielectric medium. The surrounding ions are unable to approach the ion under consideration closer than a given radius (see Fig. 5). The potential at any point is given by Poisson's equation:

$$\nabla(\epsilon_r \nabla \phi) = -\rho/\epsilon_0 \tag{9}$$

in terms of the charge on the ion and the distribution of other ions around it. The unknown in this equation is the counterion distribution, but this may be calculated by the Boltzmann distribution as a function of the potential at any point:

$$\rho = \rho_0[\exp(-e\phi/kT) - \exp(e\phi/kT)] \tag{10}$$

Thus, we have two coupled equations that need to be solved. By passing through the appropriate mathematical hoops of linearizing the Boltzmann distribution and observing the boundary equations, the following expressions may be found for the potential in each region.

$$\left.\begin{aligned} \phi_{\mathrm{I}} &= (q/\epsilon_r R)[1 - KR/(1 + Ka)] \\ \phi_{\mathrm{II}} &= (q/\epsilon_r R)[1 - Kr/(1 + Ka)] \\ \phi_{\mathrm{III}} &= (q/\epsilon_r r)\exp(-Kr)[\exp(Ka)/(1 + Ka)] \end{aligned}\right\} \tag{11}$$

where

$$K = (8\pi e^2/\epsilon_r kT)^{\frac{1}{2}}\, N^{\frac{1}{2}} \tag{12}$$

These expressions may be used for two purposes. First, the potential at any point may be used to give an indication of the effective dielectric constant of the ionic solution:

$$\phi_{\mathrm{III}} \simeq (q/\epsilon_r r)\exp(-Kr) = q/\epsilon_{\mathrm{eff}} r \tag{13}$$

thus,

$$\epsilon_{\mathrm{eff}} \approx \epsilon_r \exp(-Kr) \tag{14}$$

This expression for the effective dielectric constant of ionic solutions is often quoted (e.g., Matthew *et al.*, 1985) and indicates that the effective dielectric constant of an ionic solution should be in excess of that for pure water. Unfortunately, this is not commonly the case, and only recently has such an effect been observed, since normally the dielectric constant of an ionic solution is less than that of pure water and the decrement is a linear function of the concentration (Pethig, 1979).

Secondly, the expression for the potential can be used to calculate the free energy of charging the ion in the first place, which is equivalent to the electrostatic work of bringing charges from infinity to the ion:

$$W^{ES} = \int_0^q \phi \, dq \tag{15}$$

which for a spherical ion is given by

$$W^{ES} = (z^z e^2/2\epsilon_r R)[1 - KR/(1 + Ka)] \tag{16}$$

where z is the valence of the ion. A full review of geometries and systems to which this has been applied is given by Tanford (1961), with all the mathematical details.

The major problem in applying this method to proteins is that the procedure assumes that the charge of the protein is evenly smeared over the surface of a geometrically definable cavity. This is clearly not the case for proteins, since the charged sites are discrete and it is not possible to describe accurately the shape of the protein by an analytical mathematical expression. The problem of discrete charges was tackled by the Tanford–Kirkwood formalism (1957), but even then the shape of the protein was approximated by a sphere. The Tanford–Kirkwood procedure is the only major treatment of protein electrostatics that includes any account of the ionic properties of the solvent. Warwicker *et al.* (1985) refer to an ionic strength correction for the Warwicker and Watson procedure (1982). However, no details are given of how this is implemented. Klapper *et al.* (1986) have outlined an approach to incorporating an ionic strength correction to the finite difference procedure of Warwicker and Watson (1982) and have applied it to a calculation on the active site potential in superoxide dismutase.

A second approach to the description of ions in solution is that of Bjerrum (1926). Implicit in the Debye–Huckel formalism is that the counterions do not have defined positions but are statistically distributed. Bjerrum argued that if the ions do get close to each other, then they may become trapped within each other's coulombic shell and exist as an ion pair. The procedure, in outline, is to consider the probability of finding an ion in a shell of given radius around the central ion. The probability is proportional to the fraction of the total volume that is occupied by the shell and to the Boltzmann probability of having the associated electrostatic energy, U:

$$P(r) \propto 4\pi r^2 dr \exp(-U(r)/kT) \tag{17}$$

By integrating this probability between the limits of the closest approach and some arbitrary cutoff, at which the energy of interaction between the ions is $2kT$, the fraction of ions involved in ion pairs may be found. From this the association constant for the ions may be calculated. The bottom line of the calculation is the expression for the association constant:

$$k = (4 \pi N/1000)\,(z_1 z_2 e^2/\epsilon_r kT)^3 \int_2^b y^4 \exp(y) dy \tag{18}$$

where

$$y = z_1 z_2 e^2/\epsilon_r rkT \qquad b = z_1 z_2 e^2/\epsilon_r akT \tag{19}$$

where r is the interionic distance and a is the closest approach. The most notable use of this expression in the analysis of protein electrostatics is by Schellman (1953), and this is discussed in Section IV.A. Bockris and Reddy (1970) describe and compare the Debye–Huckel and Bjerrum approaches.

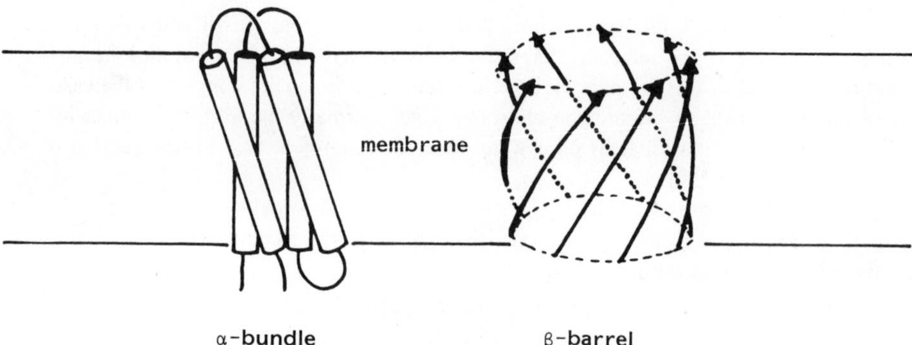

Figure 6. Possible models of membrane-bound proteins as envisaged by Henderson (1979). In the helical bundle, the peptide dipoles are all compensated for by intrahelical hydrogen bonds. In the β-barrel structure the compensation occurs by interstrand hydrogen bonds. The uncompensated charges and dipoles are envisaged as lying at the surfaces of the membrane.

3. Membranes

Membranes can influence the structure of proteins through electrostatic mechanisms in three distinct ways. First, the dielectric properties of the membrane are such that it presents a low-dielectric environment for any protein that is bound to it or in close proximity to it. This is a consequence of the membrane being largely composed of hydrophobic, nonpolarizable aliphatic chains. For a protein bound in a membrane, the structural motif of an oil drop surrounded by charges is no longer appropriate, since the charges can not happily rest in the membrane. Neither can the hydrophobic effect be the major determinant of the globular shape, since the predominant solvent for the protein is hydrophobic itself. Henderson (1979) considered the possible conformational properties of a membrane-bound protein and in particular the necessity of compensating for the buried hydrogen-bond donors and acceptors of the peptide chain. He proposed that the best way of saturating the hydrogen-bond potential was to have only regular secondary structure elements as components of the protein actually embedded in the membrane: α helices and β barrels (and combinations of the two) were his proposals, although one could also envisage other highly compensated structures such as the prealbumin tetramer acting in similar fashion. In such a model channels could be formed by the arrangement of helices or by the barrel. The turns in the protein structure, which disrupt the H-bond compensation, have to be at the surface of the membrane, and any charges have to be either at the surface or facing into the channel, where they are exposed to the aqueous environment (Fig. 6). Gray and Matthews (1984) consider the possibility of side-chain–main-chain hydrogen bonds as a means of compensating for buried charges or dipoles.

Engelman (1982) compared the strengths of the various structural interactions in a solvated and a membrane-bound protein and concluded that because hydrophobic factors are decreased and electrostatic factors are potentiated (because of the low-dielectric medium), the leading structural role must be played by electrostatic interactions. This was based on a consideration of the model of bacteriorhodopsin as an "inside-out" protein (Engelman and Zaccai, 1980) in which the charged residues were found clustered around a central channel where they were accessible to the aqueous solvent, as Henderson (1979) had suggested.

Kyte and Doolittle (1982) analyzed the hydrophobicity profiles of a number of proteins and found that long sequences of hydrophobicity corresponded to the membrane-bound portions of several proteins. Eisenberg *et al.* (1982) also noted that transmembrane helices normally had a high hydrophobicity and a small hydrophobic dipole moment, which would agree with Henderson's model (1979) except in the detail of charges facing an internal channel, which was later considered by Eisenberg *et al.* (1984).

Paul and Rosenbusch (1985) have disputed the above as a general model for membrane proteins. They consider the structure of porin, which appears to be predominantly composed of β strands and has a considerable number of apparently buried charges. They suggest that buried charges may be compensated for by an array of hydrogen bonds and other partial charges in a similar fashion to that of Rashin and Honig (1984). They present an algorithm to search for turns in membrane-bound proteins that appears to work satisfactorily.

Whatever the outcome of these various suggestions, it seems clear that the familiar models for aqueous proteins will not be reproduced in membrane-bound proteins. Particular care should be taken in the use of prediction algorithms for membrane proteins if the probability parameters for the algorithm have been derived from structures of soluble proteins.

Secondly, the presence of charged head groups on the surface of the membrane will also contribute to the electrostatic fields felt by the protein. Since the composition of the membrane and its head groups is a variable parameter, it may be most appropriate to consider this by some statistically averaged model. Phosphates, saccharides, and peptides with their various mobilities, charge densities, and polarizabilities will all have different effects.

Thirdly, there is a resting potential across cellular and organelle membranes. This is usually of the order of a few tens of millivolts and is a consequence of the maintained imbalance of ionic concentrations on each side of the membrane. The structure of many channels, pumps, and ports is affected by any change in the resting potential, and these structural changes are of functional importance in the propogation of action potentials, messenger reception and secretion, electron transport, and many other physiological processes.

Honig *et al.* (1986) give a much more detailed account of the role of membranes in protein electrostatics as part of an extensive review.

III. SECONDARY STRUCTURE

A. Conformational Analysis

The literature on conformational analysis of peptides is enormous, but it would not be far from the truth to say that most of the important questions were answered before 1964. The role of steric factors in determining the allowed values of the dihedral angles about the C_α atoms was largely dealt with by Ramachandran *et al.* (1963), and the role of the peptide dipole in the stabilization of secondary structure was thoroughly treated by Arridge and Cannon (1964). Subsequent conformational analyses have not really changed our overall perspective of what is and what is not allowed in protein secondary structure. The major advances since these early treatments have been in the area of supersecondary arrangements (e.g., Cohen *et al.*, 1980), the analysis and prediction of chain topologies and folding patterns (e.g., Sternberg and Thornton, 1978), and the establishment of criteria by which the accuracy of predictions may be assessed (e.g., Cohen and Sternberg, 1980; Nishikawa, 1983).

The early conformational analyses were reviewed by Ramachandran and Sasisekharan (1968). Other useful reviews are to be found in the *Proceedings of the Madras Conference on the Conformation of Biopolymers* (Ramachandran, 1967) and in the annual Specialist Periodical Reports of the Royal Society of Chemistry, *Amino Acids, Peptides and Proteins.*

Ramachandran *et al.* (1963) showed that the possible conformations that could be adopted by small peptides are severely limited by the steric constraints of the atoms actually coming into contact. Most of the energetically favorable regions of the Ramachandran map are predicted by hard-sphere analyses. Small changes in the coordinates of the dipeptide atoms can

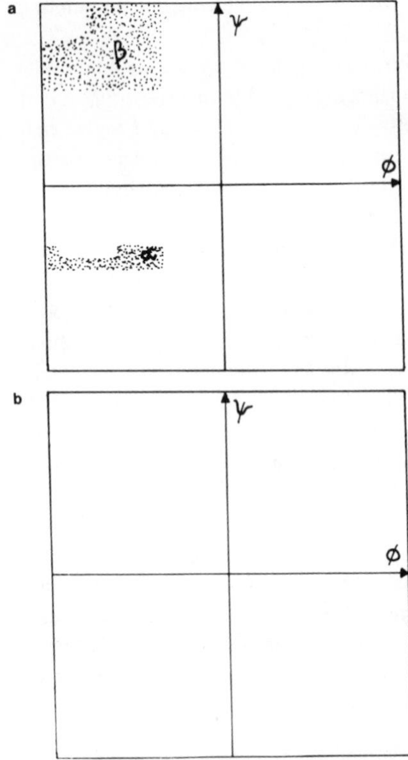

Figure 7. Hard-sphere conformational analysis of two Ala residues joined by a peptide bond. The allowed regions in a are very similar to those found in other studies—the coordinates were taken from Pauling and Corey (1951). The second plot (b) used exactly the same parameters and program, but the coordinates were taken from Dickerson and Geiss (1969). Thus, all allowed conformations are removed by just small changes of a few degrees and tenths of Angstrom units in the coordinates.

completely eliminate the possibility of any allowed regions; thus, it is fairly safe to say that local secondary structure is primarily determined by steric interactions as shown in Fig. 7. The influence of minor changes in the geometry of the peptide unit was also explored by Gibson and Scheraga (1966), who showed that allowing a degree of flexibility in the structure considerably increased the conformational possibilities.

The inclusion of electrostatic interactions (Brant *et al.*, 1967) does not significantly change the allowed conformations derived from steric analysis. There are minor alterations in the shape of the contours and in the positions of the absolute minima on the map, but only a seasoned connoisseur of Ramachandran plots would be able to detect whether he was viewing a potential map that included or excluded electrostatic interactions, because alterations in the parmeters of the energy function might cause more significant alterations. However, Brant and Flory (1965a,b) showed that the inclusion of dipole–dipole interactions between the peptide dipoles was an important factor in predicting the average end-to-end length of a polypeptide that was statistically coiling in solution.

B. The α Helix and the β Strand

The lattice energy of the array of dipoles in nylon was calculated by Rogers (1961) and then applied to polypeptides by Arridge and Cannon (1964). They calculated the interaction energy between the peptide dipoles in an α helix and a β strand. They showed that for the α helix the dominant term in the energy was the interaction between adjacent peptide dipoles, which is repulsive. Indeed, it is so repulsive that they predicted that an isolated helix would have to be longer than 14 residues to be stabilized by the other attractive interactions (Fig. 8).

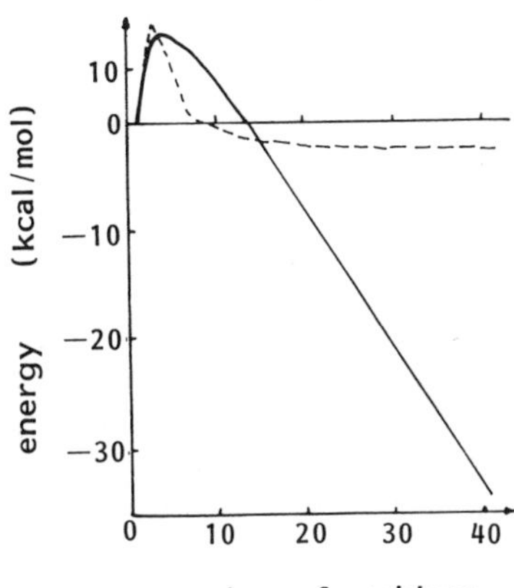

Figure 8. Energy of the array of dipoles in the α helix. The dotted line is the energy of the center dipole in the helix, and the solid line is the total lattice energy. Only a helix longer than 14 residues is electrostatically stable (after Arridge and Cannon, 1964).

For the β strand, the *in vacuo* energy is shown to be roughly twice as favorable as for the α helix, but perhaps more significant is that the energy of the β sheet is a function of the register of the component strands. These results are in keeping with those of Brant *et al.* (1967), who also showed that the β strand was electrostatically more favorable than the α helix (Fig. 9).

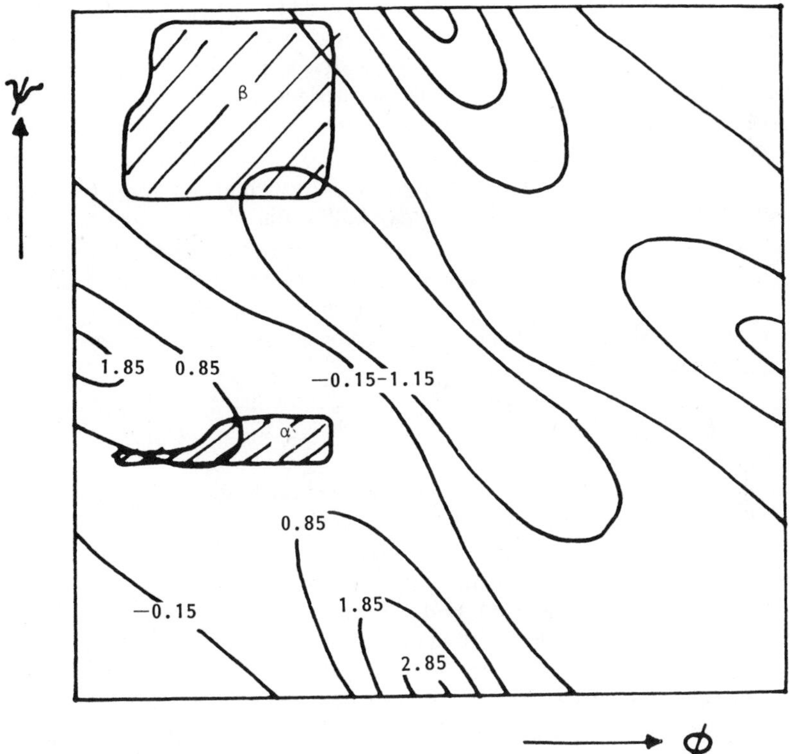

Figure 9. Electrostatic contribution to the conformational potential as derived by Brant *et al.* (1967) with the α and β allowed conformations superimposed. The isolated β strand is shown to be more electrostatically stable than the α helix. The energy levels are in kilocalories per mole.

It has been suggested that intrahelical salt bridges (Bierzynski *et al.*, 1982; Sundaralingam *et al.*, 1985) and side-chain–main-chain interactions (Gray and Matthews, 1984; Scheraga, 1985) may be involved in stabilizing helices in proteins. These charge–charge and charge–dipole interactions may be significant in maintaining structures that have been constructed under the direction of other factors. Barlow and Thornton (1983) suggest that charge–charge interactions are primarily concerned with the stabilization of tertiary rather than secondary structure but do not rule out the possibility that secondary structure stabilization may also occur.

IV. TERTIARY STRUCTURE

A. Charge–Charge Interactions

There is a long and involved history of ideas about charge–charge interactions in the stabilization of protein tertiary structure. In 1931, Speakman and Hirst proposed that salt bridges were formed between the long chains of the keratin molecule. They based their suggestion on two lines of evidence: (1) amino acid analysis showed that there were similar quantities of Glu and Arg residues, which indicated the possibility of pairing between them, and (2) pH titration experiments on the elastic properties of wool indicated that the amino groups played a significant role in the mechanical properties of the molecule. Their model was found to be consistent with the then available x-ray data (Astbury and Woods, 1933) on the separation of the chains.

On the basis of molal volumes of amino acids and proteins, Cohn *et al.* (1933) put forward the seemingly contrary proposal that salt bridges were actually instrumental in the denaturation of proteins. They argued that since denaturation was usually accompanied by a decrease in solubility, and because the formation of salt bridges would decrease the dipole moment of a protein (and hence its solubility), the formation of salt bridges was one of the steps involved in protein denaturation. This may well be true under circumstances of organic solvent denaturation, where the surface charges find themselves in an unfavorable environment, and the protein may be forced to turn itself "inside-out" in a similar fashion to the bacteriorhodopsin molecule. This would leave the hydrophobics on the exterior (and the likely prospect of aggregation and precipitation by the hydrophobic effect) and the charges forming a self-solvating network of salt bridges in the interior.

In a discussion of the factors involved in determining the structure of proteins, Mirsky and Pauling (1936) suggested that the so-called salt bridges were actually hydrogen bonds and not ion pairs at all.

It was to this confusing situation that Jacobsen and Linderstrom-Lang turned their attention in 1949. They collated these data and others from kinetic and thermodynamic experiments and made two fundamental conclusions: (1) charged groups must not be buried within the bulk of the protein but should be exposed to the solvent; (2) the contribution made to the stability of soluble proteins by the presence of salt bridges is of doubtful importance.

Since then the wealth of structural and experimental data on proteins has led to a not dissimilar state of confusion as existed prior to 1949. On the basis of experimental and theoretical considerations there have been suggestions that range from attributing prime importance to charge–charge interactions (e.g., Perutz, 1978) to neglecting them altogether (e.g., Wakana *et al.*, 1984; Henry *et al.*, 1985). On the basis of chemical modification experiments (Hollecker and Creighton, 1982; Aviram *et al.*, 1981), it appears that some charged groups may not only be neutralized but also reversed in sign without any significant change in the stability, whereas other charges may not be modified without spontaneous unfolding ensuing.

On the other hand, there are many reports in the literature of how salt bridges are crucial determinants of protein structure. Perutz and Raidt (1975) and Perutz (1978) have suggested that salt bridges act to stabilize the structures of enzymes from thermophilic organisms. In a model-building study they replaced the side chains on a ferredoxin of known three-dimensional structure by the side chains of a ferredoxin sequence (the three-dimensional structure of which was unknown) from a thermophile. They noticed that a favorable salt bridge could be formed, and from the kinetic data on unfolding they suggested that it would only have to contribute about 10 kcal/mole in order to cause an order of magnitude decrease in the rate of denaturation. Similar suggestions have been made by Walker *et al.* (1980) and by Hase *et al.* (1983). Although it might be the case that charge–charge interactions stabilize certain thermophilic proteins, Holmes and Matthews (1982) conclude:

> We do not believe that increased thermostability *per se* is always due exclusively to an increased number of ionic interactions. Rather we suggest that in a given instance a variety of stabilizing effects can contribute; these include hydrophobic interactions, ionic and hydrogen bonding, metal binding, disulphide bonds, stabilisation of secondary structure and so on (Holmes and Matthews, 1982).

There have been four major analyses of the distribution of charges in proteins, and the conclusions from these have by no means been consistent. Warme and Morgan (1978) analyzed the environments of the side chains in 21 proteins. They concluded that polar groups do aggregate (perhaps a reflection of their having to be on the surface) but that electrostatic interactions between them are probably quite well screened by the solvent. This they inferred from the fact that there are a substantial number of unfavorable electrostatic interactions, although favorable interactions do dominate. Wada and Nakamura (1981; Wada *et al.*, 1985) calculated a ''coulomblike'' interaction parameter between all the charges in a sample of 14 proteins. They concluded that there is a preponderance of positively charged residues around each negative group and vice versa, which is in agreement with the conclusions of Warme and Morgan (1978). However, Wada and Nakamura (1981) go on to try and turn the ''coulomblike'' interaction parameter into an energy, and this is where their analysis overstretches itself, since each interacting pair will have its own effective dielectric, and this will be a sensitive function of position (Hill, 1956; Rogers, 1986). Thus, the energies that they calculate are unlikely to be reliable estimates of the contribution that charge–charge interactions make to the stability of soluble proteins.

Barlow and Thornton (1983) carried out a qualitative analysis of ion pairs in proteins, from which they concluded that (1) oppositely charged groups surround each other preferentially, (2) a third of charged groups take part in the formation of stabilizing ion pairs, and (3) their contribution is to the stability of tertiary and not secondary structure.

Rashin and Honig (1984) have considered the environments of charged groups in a representative sample of known protein structures. They observe that on average there are about two buried charges in a protein and that these may be stabilized by the presence of compensating formally charged groups or simply by the formation of hydrogen bonds with partially charged atoms from the rest of the structure. This agrees well with the suggestion of Nakamura and Wada (1985) that there is an extensive and complete compensation among all the charges (formal and partial) in a protein. Of the fully buried ion pairs, they note that all correspond to functionally important groups in the active sites of proteins.

The energetics of salt-bridge formation is not immediately obvious. In 1953, Schellman proposed the following scheme in an investigation of the binding of ions to proteins. First, he calculated the interaction energy between an ion in solution and a charged group in the protein by use of the method of images. He assigned the protein as a low-dielectric medium and the solvent as a high-dielectric medium. From the interaction energy so found, he derived an effective dielectric by

$$\epsilon_{\text{eff}} = q_1q_2/4\pi\epsilon_0 r_{12}W_{12} \tag{20}$$

Now with expressions for the interaction energy and the effective dielectric constant, he substituted them into Bjerrum's expression for the association constant of two ions (equation 18). This expression was manipulated by the normal procedure:

$$\Delta H = -R\partial\, lnk/\partial(1/T) \tag{21}$$

to give the enthalpy of association and subsequently the entropy.

His conclusions were that the sign of the enthalpy change depends on the sign of the differential

$$\partial(\epsilon_{\text{eff}}T)/\partial T \tag{22}$$

Unfortunately, the effective dielectric constant is not a differentiable quantity (Linderstrom-Lang and Schellman, 1959). The differential is negative for water (e.g., Malmberg and Maryott, 1956) and positive for polyamides (Baker and Yager, 1942). The conclusion was that the enthalpy of ion association with proteins is small and that association is an entropy-driven process, the entropy being derived from the partial liberation of the water of solvation on association. He also considered the formation of ion pairs in proteins to be an entropy-driven process.

Kauzmann in 1959 used a Born (1920) solvation model to describe the formation of an ion pair. The difference of free energy between two separated charges and an ion pair is

$$\Delta G = -(q_1q_2/4\pi\epsilon_0\epsilon_r)\,(1/r_1 + 1/r_2) \tag{23}$$

where r_1 and r_2 are the ionic radii. [It might have been more accurate to include an Onsager (1936) dipole reaction field in this expression, but that was neglected.] The entropy and enthalpy changes can be calculated from this expression:

$$\Delta S = -\frac{\partial}{\partial T}\,\Delta \text{G} = \frac{1}{\epsilon_r}\frac{\partial\epsilon_r}{\partial \text{T}}\,\Delta \text{G} \tag{24}$$

$$\Delta H = -\frac{T^2\partial}{\partial T}\left(\frac{\Delta G}{T}\right) = \frac{1}{\epsilon_r}\,\frac{\partial(\epsilon_r T)}{\partial T}\,\Delta \text{G} \tag{25}$$

The conclusions were that the enthalpic and entropic contributions to the free energy of ion-pair formation are always in the opposite direction and that in an aqueous environment the association is always entropy driven. This conclusion again rests on the temperature dependence of the dielectric constant. Most ion pairs in protein exist at the dielectric interface between the protein and the solvent, and the likely dielectric behavior of the boundary zone of both the protein and the solvent is far from clear. The conclusion could then be tentatively drawn that for ion pairs that are well buried in the protein interior, enthalpy is the dominant force, and that for highly exposed groups entropy drives association (Fig. 10). For the majority of ion pairs that are positioned at the interface, the enthalpy and the entropy largely compensate, and hence the free energy of salt-bridge formation is small. Perhaps this explains why only 30% of charged residues are involved in ion pairs (Barlow and Thornton, 1983), the rest having no overriding entropic or enthalpic compulsion to associate.

These considerations of the energetics of ion-pair formation would not be anticipated by a simple-minded approach to the calculation of conformational potential as indicated in Section I.D. The microdielectric model has been implemented in an examination of this problem (Warshel, 1978a,b, 1979), and the explicit account of some of the solvent molecules yields the somewhat curious result that

$$\Delta H \approx \Delta G \tag{26}$$

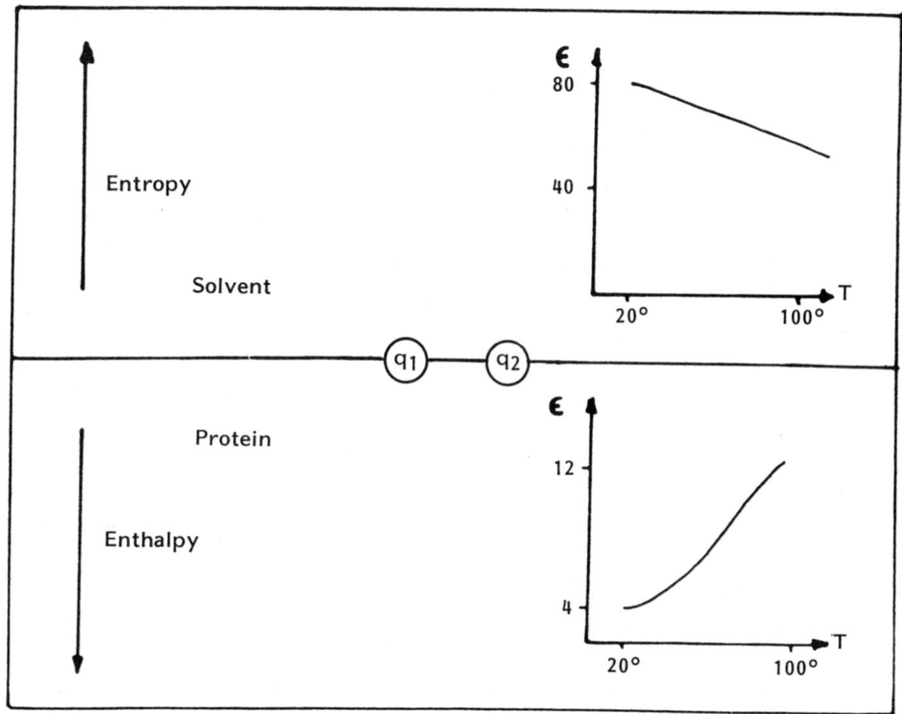

Figure 10. Speculative proposal concerning the driving force behind the formation of salt bridges in proteins. For highly exposed groups, entropy is the dominant factor, and for buried pairs enthalpy, as suggested by the temperature dependence of the dielectric constants.

The general applicability of this approach is yet to be demonstrated, given this disregard for entropic effects, which must be of significance in ionic processes in solution.

B. Why Are Charges Normally at the Surface?

It was suggested in Section I.B that the charges on the surface provide solubility and stability and may be involved in the function. It is indeed observed that the majority of charged residues are exposed to the solvent (Barlow and Thornton, 1983; Rashin and Honig, 1984; Wada *et al.*, 1985). Such a view has been held in at least vague terms since 1909 (Chibnall, 1967), but perhaps one of the first papers to crystallize the prevalent conception of a hydrophobic core surrounded by a polar coat was that of Jacobsen and Linderstrom-Lang (1949). This doctrine was subsequently adopted as dogma on the basis of theoretical and experimental evidence (e.g., Schellman, 1953; Tanford, 1957, 1962) and was later borne out in the structural analyses of the first proteins (e.g., Perutz *et al.*, 1965). The physical rationale behind the presence of charges at the solvent-exposed surface of proteins can be expressed in at least two ways.

Born's (1920) solvation model gives the free energy of solvation of an ion in a dielectric medium as

$$\Delta G = -q^2/4\pi\epsilon_0 r\ (1 - 1/\epsilon_r) \tag{27}$$

The free energy is more negative for large values of the dielectric constant than it is for low-dielectric media. Hence, given the option, a charge would prefer to be surrounded by a solvent of high-dielectric constant than one of low-dielectric constant, and so the charges are constrained by free energy to be as exposed to the aqueous environment as is possible.

The reaction field model considers the entire protein–solvent in a manner similar to that of Paul (1982). A charge in a low-dielectric medium experiences a reaction field from the polarization of the solvent. This can be visualized in terms of image charges (e.g., Friedman, 1975) or as a result of an interfacial surface polarization charge density (Zauhar and Morgan, 1985). The self energy is given by

$$(z^2/2\epsilon_p) \sum_{n=0}^{\infty} \frac{(n+1)(\epsilon_s - \epsilon_p)}{(n+1)\epsilon_s + n\epsilon_p} (r/b)^{2n} \tag{28}$$

where z is the valence of the charge, ϵ_s is the solvent dielectric, ϵ_p is the protein dielectric, b is the radius of the cavity, and r is the radial position of the charge (Fig. 11; Tanford and Kirkwood, 1957; Paul, 1982).

However one visualizes the effect, the result is that any charge in a medium of low dielectric constant is attracted toward the nearest interface with a material of high dielectric constant. It is only a matter of semantics, but this is the same effect that Perutz (1974) proposed for the disruption of mutant hemoglobins by buried charges. The only difference is that he saw the charges drawing water into the structure, whereas I have described the solvent drawing the charges out of the structure. The significant point is that to move a charge from the surface of a protein to its interior costs several tens of kilocalories per mole, as shown in Fig. 11 (Rogers, 1986).

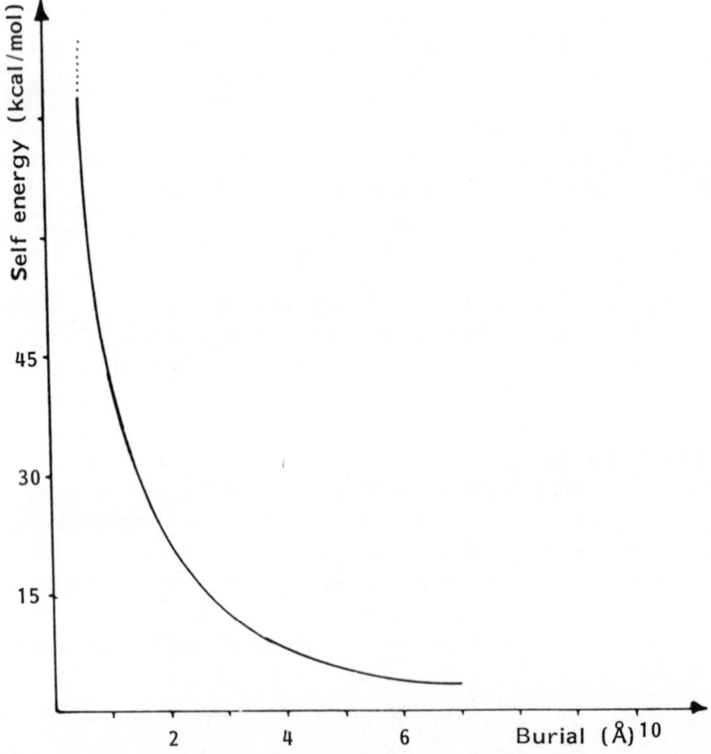

Figure 11. The self-energy of a charge in a sphere of radius 15 Å in which the charge is buried by a depth d beneath the surface of the dielectric interface. The internal dielectric is set to 3.5, and that of the exterior is 80.0.

C. What About Buried Charges?

It is observed in both aqueous (Rashin and Honig, 1984) and membrane-bound proteins (Honig and Hubbell, 1984) that charged groups may occur buried within the structure, inaccessible to solvent. This may indicate that the models presented above are less than accurate, which is not surprising given the use of a macroscopic dielectric constant on the molecular level. On the basis of the microdielectric model, Warshel (1978b) has argued that the protein may actually be a better solvent than water and as such more than adequately stabilize buried charges. The same idea is expressed in different form by Rashin and Honig (1984) and by Wada *et al.*, (1985), who note that buried charges may be compensated for by the permanent partial charges of the protein. Paul and Rosenbusch (1985) have used these ideas to explain the charge distribution in the membrane-bound protein porin, which appears to have a substantial number of charges buried within the membrane and inaccessible to the solvent. As a rationale for buried charges this may be correct, but it fails to explain why the majority of charges are exactly where one would expect them to be from macroscopic considerations, i.e., on the surface.

Buried charges are almost certain to have structural and functional importance, since the field of a charge in a low-dielectric medium is considerably more pervasive than one exposed to the high-dielectric solvent. Rashin and Honig (1984) concur with this in their analysis of the environment of charged groups.

D. The α-Helix Dipole

It has been suggested that the α-helix dipole (see Section III.C) is significant in protein function insofar as it is able to "bury" charge effectively (Warwicker and Watson, 1982). The degree of exposure to the termini has been shown to be the major factor that determines the interaction energy of two helices (Rogers and Sternberg, 1984; *vide infra*). Wada *et al.* (1985) have conducted a survey of exposures of charged groups including helix termini, and although there is a problem in defining the terminus of a helix, they conclude that virtually all helices terminate closer than 5 Å from the nearest solvent-accessible surface. For those that do not, one might expect that charged groups would be the intervening material as proposed by Cook (1967).

Cook's (1967) observation that there is a predilection for positively charged amino acid side chains to cluster at the C terminus of an α helix and for negatively charged residues to occur at the N terminus was the first intimation of a dipole moment associated with an α helix. He suggested that the charged groups competed for the hydrogen-bonding capacity of the terminal carbonyl and amide groups and so disrupted the helical structure. Commenting on this phenomenon, Ptitsyn (1969) proposed that the asymmetry of formal charges along an α helix could lead to ". . . electrostatic attraction between the ends of neighboring antiparallel helical regions. . . ." He noted that the asymmetry of formal charges opposed the dipole moment of the α helix, which arises from the orientation of the peptide dipoles along the helix axis.

Blagdon and Goodman (1975) also noticed these recurring features and explicitly attributed a dipole moment to the α helix arising from the distribution of peptide dipoles (Fig. 12). Wada (1976) extensively reviewed the properties of the α-helix dipole of polypeptides in solution, and this marks the start of extensive interest in the role that the α-helix dipole may play in the structure and function of proteins. The implementation of these ideas by Hol *et al.*, (1978, 1981) has popularized the notion and is widely cited in the literature.

The structural role for dipole–dipole interactions was investigated analytically by Weill and Andre (1978), but this was primarily concerned with solution aggregation effects. Hol *et al.* (1981) were the prime movers in suggesting that α-helix dipole moments could interact

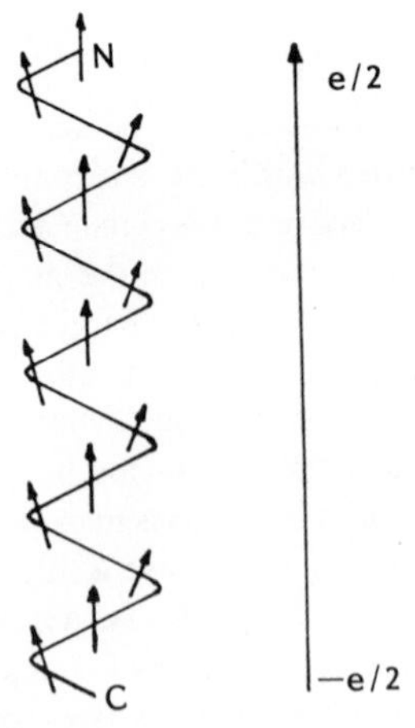

Figure 12. Schematic representation of the arrangement of peptide dipoles in the α helix and the equivalent line dipole, which may be mimicked by half an electronic charge at each terminus of the helix.

favorably and stabilize protein tertiary structure to the tune of 10 kcal/mole or more. The model they used is shown in Fig. 13, and they made the following assumptions:

1. The peptide dipoles polarize each other, and this increases their dipole moment and hence the interaction energy between helices. This increase is thought to be of the same order as the decrease in interaction produced by the dielectric constant of the protein (taken to be 2).
2. The reaction field of the solvent is assumed to be small in the interior and only of significance near the surface of the protein.

A condensation of these results and further considerations are given by Hol in two recent reviews (Hol, 1985a,b). Rogers and Sternberg (1984) have disputed the magnitudes of the interaction energies calculated by Hol *et al.* (1981). The question of polarization of the peptide dipoles is by no means settled. Arridge and Cannon (1964) and Arridge (1965) attempted to include polarization effects and showed that they are dependent on the geometry of the interacting dipoles and have a substantial angular component. Sheridan and Allen (1980) doubt it to be of any significance at all. Real helices are rarely as perfectly formed as those used in the model calculations (Blundell *et al.*, 1983; Rogers 1986), and the deviations from near-

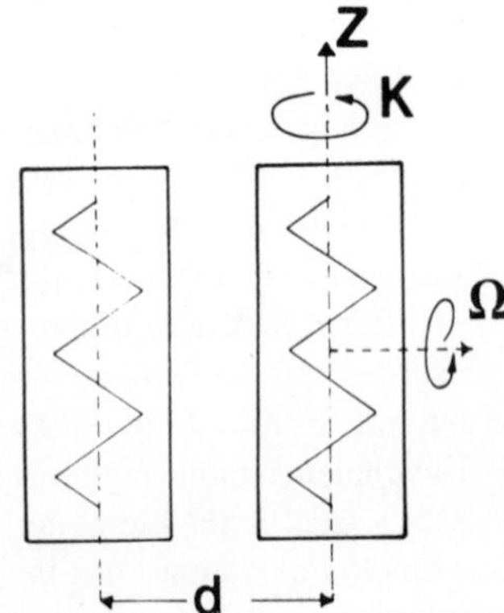

Figure 13. The model for helix dipole–dipole interactions used by Hol *et al.* (1981).

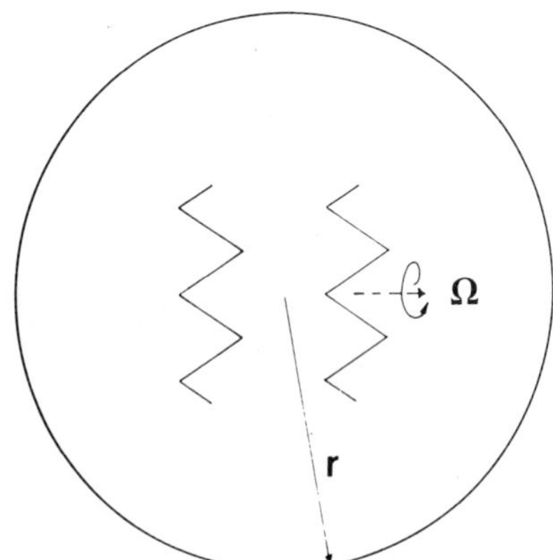

Figure 14. The model for helix dipole–dipole interactions used by Rogers and Sternberg (1984), in which the effect of the solvent is explicitly accounted for by assigning the sphere and the solvent with dielectric constants of 3.5 and 80.0, respectively.

perfect alignment will diminish the overall dipole and mitigate any polarization effect should it occur. The value of 2 for the internal dielectric of a protein has been sanctioned by Pethig (1979), but most authors have estimated a value between 2 and 5 (e.g., Ramachandran and Sasisekharan, 1968).

The major question to be resolved is to what extent the solvent modifies the interaction energy of two dipoles in the protein. Rogers and Sternberg (1984) attempted to model the effect of the solvent by calculating the interaction energy between two helices in a sphere of dielectric constant 3.5 embedded in a solvent of dielectric constant 80 (Fig. 14). This calculation was performed by an all-atom approach using the method of Warwicker and Watson (1982) and also by an image-charge analysis in which the helices were approximated by half-unit charges at their termini (see Section III.C). The calculation of interaction energy as a function of the rotation angle was carried out for helices in spheres of radii between 15 and 20 Å, and the results are plotted in Fig. 15.

The major conclusion of this analysis is that the solvent plays a significant role in

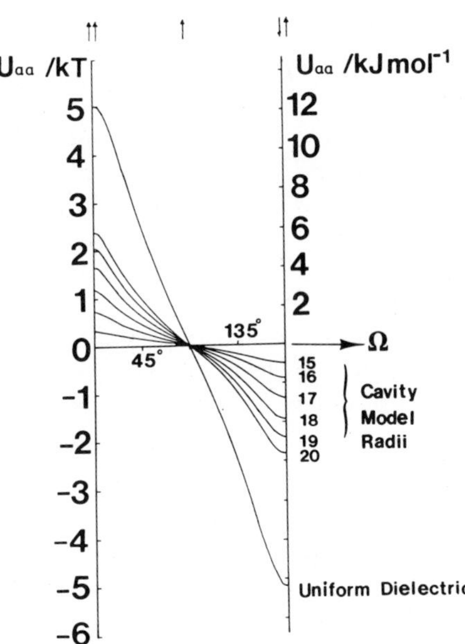

Figure 15. Results from the analysis of Rogers and Sternberg (1984) as outlined in Fig. 14. The uniform model calculated the interaction energy according to the model of Hol *et al.* (1981) (using a dielectric constant of 3.5). The results of the cavity model show that the nearer the helix termini are to the dielectric interface, the lower the energy of interaction becomes.

electrostatic interactions even within the globule and that the size of the solvent effect is principally a function of the proximity of the charged groups to the protein-solvent interface: the closer the charges are to the interface, the stronger the solvent reaction field. The most generous estimate of the interaction energy using this procedure was of the order of a few kilocalories per mole, which is significant compared with thermal energies but is about an order of magnitude less than the magnitudes proposed by Hol *et al.* (1981) or Sheridan *et al.* (1982) for the energy of forming a fourfold helical bundle.

There are questions that must be asked (even if they cannot be answered) about both approaches. The fundamental flaw in all cavity models for the dielectric constant is the abrupt discontinuity in the dielectric constant at the interface. The protein surface is mobile, polar, and highly unlikely to have the same dielectric properties as the hydrophobic, nonpolar interior. Similarly, the interfacial water is very unlikely to behave in a similar fashion to bulk solvent. Thus, the calculations for buried helices may produce reasonable estimates, but the results for exposed helices may be less than accurate. The question of just how exposed the helices are to solvent has been tackled by Wada *et al.* (1985), and they find that the answer is not the same for N and C termini. This may reflect the observation of Sheridan and Allen (1980) that the position of the effective terminal charge is different for the N and C termini.

The effect of formal charges near helix termini is the one large factor that is as yet completely unknown. Ptitsyn (1969) originally thought that the major interaction would be between the clusters of formal charges at the termini of the helices, not between the helix dipole moments themselves. Sheridan and Allen (1980) attempted to evaluate the relative importance of the helix dipole and the presence of formal charges on a helix. Their results seemed to indicate that dipolar effects are swamped by monopole effects, but this will have to be tempered by the proximity of each to the solvent interface, i.e., the degree of solvent accessibility.

Scheraga and co-workers (Silverman and Scheraga, 1972; Chou *et al.*, 1983, 1984, 1985) have carried out extensive calculations on the various factors involved in the assembly of secondary structure units in proteins. On the packing of α helices they conclude that although electrostatic interactions are important, they are not as significant as steric effects in bringing together helices into their energetically favored conformations.

V. CONCLUSION

It has been shown that electrostatic interactions are involved in the determination of protein structure. Their effect on secondary structure is of less importance than steric factors. The role of electrostatics in tertiary structure is dependent on position within the globule and the geometry of the protein's shape. The energetics of electrostatic interactions have been addressed and shown to be at best less than obvious and at worst a serious pitfall.

The role of electrostatics in protein–protein interactions and quaternary structure has not been considered and is a large area of further study that includes the association of multisubunit enzymes, the assembly of nucleosomes, ribosomes, and viruses, the formation of reaction intermediates, and the allosteric functioning of polymeric proteins. Many of the same features that have been discussed will apply, but a full discussion is beyond the scope of this chapter.

ACKNOWLEDGMENTS. I am grateful to Professor Sir David Phillips for a place to work in the Laboratory of Molecular Biophysics in Oxford. Professor R. J. P. Williams, Dr. D. T. Edmonds, Dr. G. R. Moore, and Dr. M. J. E. Sternberg have been encouraging in their comments and criticisms.

VI. REFERENCES

Abraham, M., and Becker, R., 1950, *The Classical Theory of Electricity and Magnetism,* 2nd ed., Blackie and Son, London.

Anfinsen, C. B., 1973, Principles that govern the folding of protein chains, *Science* **181:**223–230.

Arakawa, T., and Timasheff, S. N., 1985, Theory of protein solubility, *Methods Enzymol.* **114:**49–77.

Arridge, R. G. C., 1965, Calculation of the total dipole moment of a polarizable dipole in a lattice of similar dipoles, *Proc. Phys. Soc.* **85:**1157–1161.

Arridge, R. G. C., and Cannon, C. G., 1964, Calculation of the CONH dipole contribution to lattice energies of amides, polyamides and polypeptides, *Proc. R. Soc. Lond.* **A278:**91–109.

Astbury, W. T., and Woods, H. J., 1933, X-ray studies of the structure of hair, wool and related fibres. II. The molecular structure and elastic properties of hair keratin, *Phil. Trans. R. Soc. Lond.* **232A:**33–394.

Aviram, I., Myer, Y. P., and Schejter, A., 1981, Stepwise modification of the electrostatic charge of cytochrome c, *J. Biol. Chem.* **256:**5540–5544.

Baker, E. N., and Hubbard, R. E., 1984, Hydrogen bonding in globular proteins, *Prog. Biophys. Mol. Biol.* **44:** 97–179.

Baker, W. O., and Yager, W. A., 1942, The relation of dielectric properties to structure of crystalline polymers. II. Linear polyamides, *J. Am. Chem. Soc.* **64:**2171–2177.

Barlow, D. J., and Thornton, J. M., 1983, Ion pairs in proteins, *J. Mol. Biol.* **168:**867–885.

Bierzynski, A., Kim, P. S., and Baldwin, R. L., 1982, A salt bridge stabilizes the helix formed by isolated C-peptide of RNAse A, *Proc. Natl. Acad. Sci. U.S.A.* **79:**2470–2474.

Bjerrum, N., 1926, Ionic association. I. Influence of ionic association on the activity of ions at moderate degrees of association, *Kgl. Danske. Videnskab. Selskab. Math-Fys. Medd.* **7:**1–48.

Blagdon, D. E., and Goodman, M., 1975, Mechanisms of protein and polypeptide helix initiation, *Biopolymers* **14:**241–245.

Blundell, T. L., Barlow, D. J., Borkakoti, N., and Thornton, J. M., 1983, Solvent induced distortions and the curvature of alpha-helices, *Nature* **306:**281–283.

Bockris, J. O'M., and Reddy, A. K. N., 1970, *Modern Electrochemistry,* McDonald, London.

Born, M., 1920, Volumen und Hydratationswarme der Ionen, *Z. Phys.* **1:**45–48.

Brant, D. A., and Flory, P. J., 1965a, The role of dipole interactions in determining polypeptide configurations, *J. Am. Chem. Soc.* **87:**663–664.

Brant, D. A., and Flory, P. J., 1965b, The configuration of random polypeptide chains. II. Theory, *J. Am. Chem. Soc.* **87:**2791–2800.

Brant, D. A., Miller, W. G., and Florey, P. J., 1967, Conformational energy estimates for statistically coiling polypeptide chains, *J. Mol. Biol.* **23:**47–65.

Chibnall, A. C., 1967, The Armstrongs and the polypeptide chain in 1909, *Chem. Ind.* **1967:**1289–1290.

Chothia, C., Levitt, M., and Richardson, D., 1977, Structure of proteins: Packing of alpha-helices and pleated sheets, *Proc. Natl. Acad. Sci. U.S.A.* **74:**4130–4134.

Chou, K. C., Nemethy, G., and Scheraga, H. A., 1983, Energetic approach to the packing of alpha-helices. I. Equivalent helices, *J. Phys. Chem.* **87:**2869–2881.

Chou, K. C., Nemethy, G., and Scheraga, H. A., 1984, Energetic approach to the packing of alpha-helices. 2. General treatment of nonequivalent and nonregular helices, *J. Am. Chem. Soc.* **106:**3161–3170.

Chou, K. C., Nemethy, G., Rumsey, S., Tuttle, R. W., and Scheraga, H. A., 1985, Interactions between an alpha-helix and a beta-sheet. Energetics of alpha/beta packing in proteins, *J. Mol. Biol.* **186:**591–609.

Cohen, F. E., and Sternberg, M. J. E., 1980, On the prediction of protein structure: The significance of the root mean square deviation, *J. Mol. Biol.* **138:**321–333.

Cohen, F. E., Sternberg, M. J. E., and Taylor, W. R., 1980, Analysis and prediction of protein beta-sheet structures by a combinatorial approach, *Nature* **285:**378–382.

Cohen, F. E., Sternberg, M. J. E., and Taylor, W. R., 1982, Analysis and prediction of the packing of alpha-helices against a beta-sheet in the tertiary structure of globular proteins, *J. Mol. Biol.* **156:**821–862.

Cohn, E. J., and Ferry, J. D., 1943, The interactions of proteins with ions and dipolar ions, in: *Proteins, Amino-acids and Peptides* (E. J. Cohn and J. T. Edsall, eds.), Reinhold, New York, pp. 586–622.

Cohn, E. J., McMeekin, T. L., Edsall, J. T., and Blanchard, M. H., 1933, The electrical forces in systems containing biological components. II. Molal volumes of amino-acids, proteins and certain related substances, *J. Biol. Chem.* **100:**xxvii.

Cook, D. A., 1967, The relation between amino acid sequence and protein conformation, *J. Mol. Biol.* **29:**167–171.

Cooper, A., 1984, Protein fluctuations and the thermodynamic uncertainty principle, *Prog. Biophys. Mol. Biol.* **44:**181–214.

Creighton, T. E., 1985, The problem of how and why proteins adopt folded conformations, *J. Phys. Chem.* **89:** 2452–2459.

Debye, V. P., and Huckel, E., 1923, Zur Theorie der Elektrolyte. I. Gefrierpunktserniedrigung und verwandte Erscheinungen, *Z. Phys.* **24:**185–206.

Dickerson, R. E., and Geiss, I., 1969, *The Structure and Action of Proteins,* Harper and Row, New York.

Driessen, H. P. C., de Jong, W. W., Tesser, G. I., and Bloemendal, H., 1985, The mechanism of N-terminal acetylation of proteins, *Crit. Rev. Biochem.* **18:**281–325.

Edsall, J. T., and McKenzie, H. A., 1983, Water and proteins. II. The location and dynamics of water in protein systems and its relation to their stability and properties, *Adv. Biophys.* **16:**53–183.

Edsall, J. T., Flory, P. J., Kendrew, J. C., Liquori, A. M., Nemethy, G., Ramachandran, G. N., and Scheraga, H. A., 1966, A proposal of standard conventions and nomenclature for the description of polypeptide conformations, *J. Mol. Biol.* **15:**399–407; *J. Biol. Chem.* **241:**1004–1012; *Biopolymers* **4:**121–129.

Ehrenson, S., 1984, Solvent dielectric attenuation of substituent effects. Dependence on boundary representation in prolate spheroidal cavity models, *J. Comp. Chem.* **5:**56–63.

Eisenberg, D., Weiss, R. M., and Terwilliger, T. C., 1982, The helical hydrophobic moment: A measure of the amphiphilicity of a helix, *Nature* **299:**371–374.

Eisenberg, D., Schwarz, E., Komaromy, M., and Wall, R., 1984, Analysis of membrane and surface protein sequences with the hydrophobic moment plot, *J. Mol. Biol.* **179:**125–142.

Engelman, D. M., 1982, An implication of the structure of bacteriorhodopsin. Globular membrane proteins are stabilized by polar interactions, *Biophys. J.* **37:**187–188.

Engelman, D. M., and Zaccai, G., 1980, Bacteriorhodopsin is an inside-out protein, *Proc. Natl. Acad. Sci. U.S.A.* **77:**5894–5898.

Finney, J., 1979, The organisation and function of water in protein crystals, in: *Water—A Complete Treatise,* Volume 6 (F. Franks, ed.), Plenum Press, New York, pp. 47–122.

Florey, P. J., 1969, *Statistical Mechanics of Chain Molecules,* Wiley, New York.

Franks, F. (ed.), 1972–1982, *Water—A Complete Treatise,* Plenum Press, New York.

Frieden, C., Wolfe, R. G., and Alberty, R. A., 1957, Studies of the enzyme fumarase. IV. The dependence of the kinetic constants at 25C on buffer concentration, composition and pH, *J. Am. Chem. Soc.* **79:**1523–1525.

Friedman, H. L., 1975, Image approximation to the reaction field, *Mol. Physiol.* **29:**1533–1543.

Gabler, R., 1978, *Electrical Interactions in Molecular Biophysics—An Introduction,* Academic Press, New York.

George, P., Hanania, G. I. H., and Eaton, W. A., 1966, Effect of electrostatic environment of redox potentials, in: *Hemes and Hemoproteins* (B. Chance, R. W. Estabrook, and T. Yonetani, eds.), Academic Press, New York, pp. 267–271.

Gibson, K. D., and Scheraga, H. A., 1966, Influence of flexibility on the energy contours of dipole maps, *Biopolymers* **4:**709–712.

Goodfellow, J. M., Finney, J. L., and Barnes, P., 1982, Monte-Carlo computer simulation of water–amino acid interactions, *Proc. R. Soc. Lond. [Biol.]* **214:**213–228.

Gray, T. M., and Matthews, B. W., 1984, Intrahelical hydrogen bonding of serine, threonine and cysteine residues within alpha-helices and its relevance to membrane-bound proteins, *J. Mol. Biol.* **175:**75–81.

Hagler, A. T., Stern, P. S., Sharon, R., Becker, J. M., and Naider, F., 1979, Computer simulation of the conformational properties of oligopeptides. Comparison of theoretical methods and analysis of experimental results, *J. Am. Chem. Soc.* **101:**6842–6852.

Hammond, P., 1960, Electric and magnetic images, *Proc. IEEE* **107C:**306–313.

Hase, T., Matsubara, H., Koike, H., and Katoh, S., 1983, Amino acid sequence of ferredoxin from a thermophilic blue-green alga, *Synechococcus* sp. Salt bridges responsible for thermostability, *Biochim. Biophys. Acta* **744:**46–52.

Henderson, R., 1979, The structure of bacteriorhodopsin and its relevance to other membrane proteins, *Soc. Gen. Physiol.* **33:**3–15.

Henry, E. R., Levitt, M., and Eaton, W. A., 1985, Molecular dynamics simulation of photodissociation of carbon monoxide from hemoglobin, *Proc. Natl. Acad. Sci. U.S.A.* **82:**2034–2038.

Hill, T. L., 1956, Influence of electrolyte on effective dielectric constants in enzymes, proteins and other molecules, *J. Phys. Chem.* **60:**253–255.

Hingerty, B. E., Ritchie, R. H., Ferrell, T. L., and Turner, J. E., 1985, Dielectric effects in biopolymers: The theory of ionic saturation revisited, *Biopolymers* **24:**427–439.

Hol, W. G. J., 1985a, The role of the alpha-helix dipole in protein function and structure, *Prog. Biophys. Mol. Biol.* **45:**149–195.

Hol, W. G. J., 1985b, Effects of the alpha-helix dipole upon the functioning and structure of proteins and peptides, *Adv. Biophys.* **19:**133–165.

Hol, W. G. J., van Duijnen, P. T., and Berendsen, H. J. C., 1978, The alpha-helix dipole and the properties of proteins, *Nature* **273:**443–446.

Hol, W. G. J., Halie, L. M., and Sander, C., 1981, Dipoles of the alpha-helix and beta-sheet: Their role in protein folding, *Nature* **294:**532–536.

Hollecker, M., and Creighton, T. E., 1982, Effect on protein stability of reversing the charge of amino groups, *Biochim. Biophys. Acta* **701:**395–404.

Holmes, M. A., and Matthews, B. W., 1982, Structure of thermolysin refined at 1.6A resolution, *J. Mol. Biol.* **160:**623–639.

Honig, B. H., and Hubbell, W. L., 1984, Stability of salt-bridges in membrane proteins, *Proc. Natl. Acad. Sci. U.S.A.* **81:**5412–5416.

Honig, B. H., Hubbell, W. L., and Flewelling, R. F., 1986, Electrostatic interactions in membranes and proteins, *Annu. Rev. Biophys. Biophys. Chem.* **15:**163–193.

Jacobsen, C. F., and Linderstrom-Lang, K., 1949, Salt-linkages in proteins, *Nature* **164:**411–412.

Karplus, M., and Kushick, J. N., 1981, Method for estimating the configurational entropy of macromolecules, *Macromolecules* **14:**325–332.

Kauzmann, W., 1959, Some factors in the interpretation of protein denaturation, *Adv. Protein Chem.* **14:**1–63.

Klapper, I., Hagstrom, R., Fine, R., Sharp, K., and Honig, B., 1986, Focussing of electric fields in the active site of Cu–Zn superoxide dismutase: Effects of ionic strength and amino-acid modification, *Proteins* **1:**47–59.

Kyte, J., and Doolittle, R. F., 1982, A simple method for displaying the hydropathic character of a protein, *J. Mol. Biol.* **157:**105–132.

Linderstrom-Lang, K. U., and Schellman, J. A., 1959, Protein structure and enzyme activity, in: *The Enzymes*, 2nd ed. (P. D. Boyer, H. Hardy, and K. Myrbäck, eds.), pp. 443–510, Academic Press, New York.

Malmberg, C. G., and Maryott, A. A., 1956, Dielectric constant of water from 0 to 100 degrees C, *J. Res. Natl. Bur. Stand.* **56:**1–8.

Matthew, J. B., and Richards, F. M., 1982, Anion binding and pH dependent electrostatic effects in ribonuclease, *Biochemistry* **21:**4989–4999.

Matthew, J. B., Gurd, F. R. N., Garcia-Moreno, B. E., Flanagan, M. A., March, K. L., and Shire, S. J., 1985, pH dependent processes in proteins, *Crit. Rev. Biochem.* **18:**90–197.

McCammon, J. A., 1984, Protein dynamics, *Rep. Prog. Phys.* **47:**1–41.

Mirsky, A. E., and Pauling, L., 1936, On the structure of native, denatured and coagulated proteins, *Proc. Natl. Acad. Sci. U.S.A.* **22:**439–447.

Momany, F. A., McGuire, R. F., Burgess, A. W., and Scheraga, H. A., 1975, Energy parameters in polypeptides. VII. Geometric parameters, partial atomic charges, non-bonded interactions, hydrogen bond interactions and intrinsic torsional potentials for the naturally occurring amino acids, *J. Phys. Chem.* **79:**2361–2381.

Moore, G. R., and Rogers, N. K., 1985a, The influence of electrostatic interactions between buried charges on the properties of membrane proteins, *J. Inorg. Biochem.* **23:**219–226.

Moore, G. R., and Rogers, N. K., 1985b, Erratum to "The influence of electrostatic interactions between buried charges on the properties of membrane proteins," *J. Inorg. Biochem.* **25:**295.

Nakamura, H., and Wada, A., 1985, Nature of the charge distribution in proteins. III. Electric multipole structures, *J. Phys. Soc. (Jpn.)* **54:**4047–4052.

Nishikawa, K., 1983, Assessment of secondary structure prediction of proteins. Comparison of computerised Chou–Fasman method with others, *Biochim. Biophys. Acta* **748:**285–299.

Onsager, L., 1936, Electric moments of molecules in liquids, *J. Am. Chem. Soc.* **58:**1486–1493.

Paul, C. H., 1982, Building models of globular protein molecules from their amino acid sequences. I. Theory, *J. Mol. Biol.* **155:**53–62.

Paul, C. H., and Rosenbusch, J. P., 1985, Folding patterns of porin and bacteriorhodopsin, *EMBO J.* **4:**1593–1597.

Pauling, L., and Corey, R. B., 1951, Atomic coordinates and structure factors for two helical configurations of polypeptide chains, *Proc. Natl. Acad. Sci. U.S.A.* **37:**235–240.

Perutz, M. F., 1974, Mechanism of denaturation of haemoglobin by alkali, *Nature* **247:**341–344.

Perutz, M. F., 1978, Electrostatic effects in proteins, *Science* **201:**1187–1191.

Perutz, M. F., and Raidt, H., 1975, Stereochemical bases of heat stability in bacterial ferredoxins and in haemoglobin A_2, *Nature* **255:**256–259.

Perutz, M. F., Kendrew, J. C., and Watson, H. C., 1965, Structure and function of haemoglobin II. Some relations between polypeptide chain configuration and amino-acid sequence, *J. Mol. Biol.* **13:**669–678.

Pethig, R., 1979, *Dielectric and Electronic Properties of Biological Materials,* John Wiley & Sons, New York.

Ptitsyn, O. B., 1969, Statistical analysis of the distribution of amino acid residues among helical and non-helical regions in globular proteins, *J. Mol. Biol.* **42:**501–510.

Ramachandran, G. N., 1967, *Conformation of Biopolymers: Proceedings of the 1967 Conference in Madras,* Academic Press, New York.

Ramachandran, G. N., and Sasisekharan, V., 1968, Conformation of polypeptides and proteins, *Adv. Protein Chem.* **23:**283–437.

Ramachandran, G. N., and Srinivasan, R., 1970, Effective dielectric constant values to be used in biopolymer energy calculations, *Ind. J. Biochem.* **7:**95–97.

Ramachandran, G. N., Ramakrishnan, C., and Sasisekharan, V., 1963, Stereochemistry of polypeptide chain configurations, *J. Mol. Biol.* **7:**95–99.

Rashin, A. A., and Honig, B., 1984, On the environment of ionizable groups in globular proteins, *J. Mol. Biol.* **173:**515–521.

Rees, D. C., 1980, Experimental evaluation of the effective dielectric constant of proteins, *J. Mol. Biol.* **141:** 323–326.

Rogers, D., 1961, ZEBRA and the crystallographer, in: *Computing Method and the Phase Problem in X-Ray Crystal Analysis* (R. Pepinsky, J. M. Robertson, and J. C. Speakman, eds.), Pergamon Press, Oxford, pp. 1–8.

Rogers, N. K., 1986, The modelling of electrostatic interactions in the function of globular proteins, *Prog. Biophys. Mol. Biol.* **48:**37–66.

Rogers, N. K., and Sternberg, M. J. E., 1984, Electrostatic interactions in globular proteins: Different dielectric models applied to the packing of alpha-helices, *J. Mol. Biol.* **174:**527–542.

Schellman, J. A., 1953, The application of the Bjerrum ion association theory to the binding of anions by proteins, *J. Phys. Chem.* **57:**472–475.

Scheraga, H. A., 1985, Effect of side chain–backbone electrostatic interactions on the stability of alpha-helices, *Proc. Natl. Acad. Sci. U.S.A.* **82:**5585–5587.

Sheridan, R. P., and Allen, L. C., 1980, The electrostatic potential of the alpha-helix, *Biophys. Chem.* **11:**133–136.

Sheridan, R. P., and Allen, L. C., 1981, The active site potential of human carbonic anhydrase, *J. Am. Chem. Soc.* **103:**1544–1550.

Sheridan, R. P., Levy, R. M., and Salemme, F. R., 1982, Alpha-helix dipole moment and electrostatic stabilization of 4-alpha-helical proteins, *Proc. Natl. Acad. Sci. U.S.A.* **79:**4545–4549.

Silverman, D. N., and Scheraga, H. A., 1972, Hairpin bend and interhelical interactions in alpha-helical poly(L-alanine) in water, *Arch. Biochem. Biophys.* **153:**449–456.

Speakman, J. B., and Hirst, M. C., 1931, Constitution of the keratin molecule: Interchain salt linkages in wool, *Nature* **128:**1073–1074.

Stankovich, M. T., Schopfer, L. M., and Massey, V., 1978, Determination of glucose oxidase oxidation–reduction potentials and the oxygen reactivity of fully reduced and semi-quinoid forms, *J. Biol. Chem.* **253:** 4971–4979.

Sternberg, M. J. E., and Thornton, J. M., 1978, Prediction of protein structure from amino acid sequence, *Nature* **271:**15–20.

Sundaralingam, M., Drendel, W., and Greaser, M., 1985, Stabilization of the long central helix of troponin C by intrahelical salt bridges between charged amino acid side chains, *Proc. Natl. Acad. Sci. U.S.A.* **82:** 7944–7947.

Tam, S. C., and Williams, R. J. P., 1985, Electrostatic and biological systems, *Struct. Bonding* **63:**103–151.

Tanford, C., 1957, Theory of protein titration curves. II. Calculations for simple models at low ionic strength, *J. Am. Chem. Soc.* **79:**5340–5347.

Tanford, C., 1961, *Physical Chemistry of Macromolecules,* Wiley, New York.

Tanford, C., 1962, Contribution of hydrophobic interactions to the stability of the globular conformation of proteins, *J. Am. Chem. Soc.* **84:**4240–4247.

Tanford, C., and Kirkwood, J. G., 1957, Theory of protein titration curves. I. General equations for impenetrable spheres, *J. Am. Chem. Soc.* **79:**5333–5339.

Tanford, C., and Roxby, R., 1972, Interpretation of protein titration curves. Application to lysozyme, *Biochemistry* **11:**2192–2198.

Thornton, J. M., and Sibanda, B. L., 1983, Amino and carboxy-terminal regions in globular proteins, *J. Mol. Biol.* **167:**443–460.

Timasheff, S. N., 1970, Polyelectrolyte properties of globular proteins, *Biol. Macromol. Ser.* **3:**1–64.

Voet, J. G., Coe, J., Epstein, J., Matossian, V., and Shipley, T., 1981, Electrostatic control of enzyme reactions: Effects of ionic strength on the pKa of an essential acidic group on glucose oxidase, *Biochemistry* **20:**7182–7185.

Von Hippel, P. H., and Schleich, T., 1969, The effects of neutral salts on the structure and conformational stability of macromolecules in solution, *Biol. Macromol. Ser.* **2:**417–574.

Wada, A., 1976, The alpha-helix as an electric macro-dipole, *Adv. Biophys.* **9:**1–63.

Wada, A., and Nakamura, H., 1981, Nature of the charge distribution in proteins, *Nature* **293:**757–758.

Wada, A., Nakamura, H., and Sakamoto, T., 1985, Nature of the charge distribution in proteins. II. Effect of atomic partial charges on ionic charges, *J. Phys. Soc. (Jpn.)* **54:**4042–4046.

Wakana, H., Wako, H., and Saito, N., 1984, Monte-Carlo study on local and small amplitude conformational fluctuation in hen egg white lysozyme, *Int. J. Peptide Protein Res.* **23:**315–323.

Walker, J. E., Wonacott, A. J., and Harris, J. I., 1980, Heat stability of a tetrameric enzyme, D-glyceraldehyde-3-phosphate dehydrogenase, *Eur. J. Biochem.* **108:**581–586.

Warme, P. K., and Morgan, R. S., 1978, A survey of amino-acid side chain interactions in 21 proteins, *J. Mol. Biol.* **118:**289–304.

Warshel, A., 1978a, A microscopic model for calculations of chemical processes in aqueous solutions, *Chem. Phys. Lett.* **55:**454–458.

Warshel, A., 1978b, Energetics of enzyme catalysis, *Proc. Natl. Acad. Sci. U.S.A.* **75:**5250–5254.

Warshel, A., 1979, Calculations of chemical processes in solutions, *J. Phys. Chem.* **83:**1640–1652.

Warshel, A., and Russell, S. T., 1984, Calculations of electrostatic interactions in biological systems and in solutions, *Q. Rev. Biophys.* **17:**283–422.

Warshel, A., Russell, S. T., and Churg, A. K., 1984, Macroscopic models for studies of electrostatic interactions in proteins: Limitations and applicability, *Proc. Natl. Acad. Sci. U.S.A.* **81:**4785–4789.

Warwicker, J., and Watson, H. C., 1982, Calculation of the electric potential in the active site cleft due to alpha-helix dipoles, *J. Mol. Biol.* **157:**671–679.

Warwicker, J., Ollis, D., Richards, F. M., and Steitz, T. A., 1985, Electrostatic field of the large fragment of *Escherichia coli* DNA polymerase I, *J. Mol. Biol.* **186:**645–649.

Weill, G., and Andre, J. J., 1978, Dipole–dipole interactions and the relative stability of different types of aggregates of helical polypeptides, *Biopolymers* **17:**811–814.

Zauhar, R. J., and Morgan, R. S., 1985, A new method for computing the macromolecular electric potential, *J. Mol. Biol.* **186:**815–820.

9

Chou–Fasman Prediction of the Secondary Structure of Proteins

The Chou–Fasman–Prevelige Algorithm

Peter Prevelige, Jr., and Gerald D. Fasman

I. INTRODUCTION

The Chou–Fasman algorithm for the prediction of protein secondary structure is one of the most widely used predictive schemes. This is because of its relative simplicity and its reasonably high degree of accuracy.

Peter Prevelige, Jr. • Department of Biology, Massachusetts Institute of Technology, Cambridge, Massachusetts 02139. *Gerald D. Fasman* • Graduate Department of Biochemistry, Brandeis University, Waltham, Massachusetts 02254.

A number of modifications of the Chou–Fasman algorithm have been developed and published (see G. D. Fasman, Chapter 6, this volume, for a review). However, in general these suffer from one of two faults: either they are completely computerized and hide much of the decision-making process from the user or they leave the user to make decisions but do not adequately describe the decision-making process used by the authors.

This chapter attempts to outline the approach that has been successfully employed by the authors over the past several years. The approach is one in which a computer program is employed to perform the arithmetic calculations and then the data reduction is performed by hand. This approach utilizes the computer to reduce the tedious calculations while at the same time allowing the individual to bring his experience and intuition to bear. The computer program itself was developed from ideas in a program originally written by Dr. George Long and Jeff Siegel in 1979.

The first section of this chapter reviews the Chou–Fasman method for prediction of protein structure. This is followed by a section that lays out the mechanics of operating the program and then by a discussion of the process of data reduction. Finally, worked examples are provided in the hope that they will make more concrete the many considerations involved in predicting a protein secondary structure.

II. REVIEW OF THE METHOD AND RATIONALE OF THE CHOU–FASMAN ALGORITHM

The Chou-Fasman algorithm is an algorithm to predict the secondary structure of proteins from their amino acid sequence. It falls into the class of the statistical approach as discussed by Fasman (Chapter 6, this volume).

The x-ray-determined structures of 15 proteins containing 2473 amino acid residues were carefully examined, and the number of occurrences of a given amino acid in the α helix, β sheet, and coil was tabulated (Table I). From this, the conformational parameters for each amino acid were calculated by considering the relative frequency of a given amino acid within a protein, its occurrence in a given type of secondary structure, and the fraction of residues occurring in that type of structure (Chou and Fasman, 1974a). This conformational parameter is essentially a measure of a given amino acid's preference to be found in α helix, β sheet, or coil. These parameters, symbolized by P_α, P_β, and P_c, respectively, presumably contain information about the physical–chemical parameters defining protein stability, such as hydrophobicity, properly weighted for their relative importance. These parameters therefore should be useful for predicting a protein's secondary structure based on the amino acid sequence.

Having computed these conformational parameters, Chou and Fasman formulated a set of empirical rules for predicting secondary structure (Chou and Fasman, 1974b). The development of these empirical rules was guided by underlying considerations of protein structure. These rules, when applied by Chou and Fasman, resulted in a 70–80% predictive accuracy. The rules were never developed as a computer algorithm and hence lack the type of rigorous definition that a computer algorithm requires. This has led to a wide variety of implementations, which have an equally wide variety of accuracies.

Chou and Fasman later extended the analysis of α helix, β sheet, and coil to include 29 proteins of known x-ray structure. This increased the total number of residues classified to 4741, or approximately double the initial number (Chou and Fasman, 1978). The most pronounced change occurred for Met. This change resulted from an underrepresentation of Met in the initial 15 proteins examined. Less pronounced changes were also seen in Asn, Asp, Ala, His, Gly, Ile, Lys, and Tyr (Table II).

Table I. Assignment of Amino Acids as Formers, Breakers, and Indifferent for Helical and β-Sheet Regions in Proteins Based on P_α and P_β Values[a]

Helical residues[b]	P_α		β-Sheet residues[c]	P_β	
Glu(−)	1.53	H_α	Met	1.67	H_β
Ala	1.45		Val	1.65	
Leu	1.34		Ile	1.60	
His(+)	1.24	h_α	Cys	1.30	h_β
Met	1.20		Tyr	1.29	
Gln	1.17		Phe	1.28	
Trp	1.14		Gln	1.23	
Val	1.14		Leu	1.22	
Phe	1.12		Thr	1.20	
Lys(+)	1.07	I_α	Trp	1.19	
Ile	1.00		Ala	0.97	I_β
Asp(−)	0.98	i_α	Arg(+)	0.90	i_β
Thr	0.82		Gly	0.81	
Ser	0.79		Asp(−)	0.80	
Arg(+)	0.79		Lys(+)	0.74	b_β
Cys	0.77		Ser	0.72	
Asn	0.73	b_α	His(+)	0.71	
Tyr	0.61		Asn	0.65	
Pro	0.59	B_α	Pro	0.62	
Gly	0.53		Glu(−)	0.26	B_β

[a]Chou and Fasman (1974b).

[b]Helical assignments: H_α, strong α former; h_α, α former; I_α, weak α former; i_α, α indifferent; b_α, α breaker; B_α, strong α breaker. I_α assignments are also given to Pro and Asp (near the N-terminal helix) as well as Arg (near the C-terminal helix).

[c]β-Sheet assignments: H_β, strong β former; h_β, β former; I_β, weak β former; i_β, β indifferent; b_β, β breaker; B_β, strong β breaker; b_β assignment is also given to Trp (near the C-terminal β region).

A similar analysis, using the 29-protein data base, was performed for amino acid residues that were found in β turns (Chou and Fasman, 1977). The conformational parameter P_t was determined. In the case of turns, a significant difference was also observed in the frequency of residues in the first, second, third, and fourth positions of β turns for all residues (Table II). Some residues were found to have a dramatic positional preference, e.g., proline. Proline occurs 30% of the time in position number 2 of the β bend but less than 4% of the time in position number 3. Therefore, for the prediction of turns a method to factor in positional preference was devised (Chou and Fasman, 1979).

The Chou–Fasman algorithm is simple in principle. Using the conformational parameter, one finds nucleation sites within the sequence and extends them until a stretch of amino acids is encountered that is not disposed to occur in that type of structure or until a stretch is encountered that has a greater disposition for another type of structure. At that point, the structure is terminated. This process is repeated throughout the sequence until the entire sequence is predicted. The conformational parameters for coil are not employed; coil is predicted by default.

Table II. Conformational Parameters for α-Helical, β-Sheet, and β-Turn Residues in 29 Proteins[a]

P_α			P_β			P_i		f_i		f_{i+1}		f_{i+2}		f_{i+3}	
Glu	1·51		Val	1·70		Asn	1·56	Asn	0·161	Pro	0·301	Asn	0·191	Trp	0·167
Met	1·45	H_α	Ile	1·60	H_β	Gly	1·56	Cys	0·149	Ser	0·139	Gly	0·190	Gly	0·152
Ala	1·42		Tyr	1·47		Pro	1·52	Asp	0·147	Lys	0·115	Asp	0·179	Cys	0·128
Leu	1·21		Phe	1·38		Asp	1·46	His	0·140	Asp	0·110	Ser	0·125	Tyr	0·125
Lys	1·16		Trp	1·37		Ser	1·43	Ser	0·120	Thr	0·108	Cys	0·117	Ser	0·106
Phe	1·13		Leu	1·30		Cys	1·19	Pro	0·102	Arg	0·106	Tyr	0·114	Gln	0·098
Gln	1·11	h_α	Cys	1·19	h_β	Tyr	1·14	Gly	0·102	Gln	0·098	Arg	0·099	Lys	0·095
Trp	1·08		Thr	1·19		Lys	1·01	Thr	0·086	Gly	0·085	His	0·093	Asn	0·091
Ile	1·08		Gln	1·10		Gln	0·98	Tyr	0·082	Asn	0·083	Glu	0·077	Arg	0·085
Val	1·06		Met	1·05		Thr	0·96	Trp	0·077	Met	0·082	Lys	0·072	Asp	0·081
Asp	1·01	I_α	Arg	0·93		Trp	0·96	Gln	0·074	Ala	0·076	Thr	0·065	Thr	0·079
His	1·00		Asn	0·89	i_β	Arg	0·95	Arg	0·070	Tyr	0·065	Phe	0·065	Leu	0·070
Arg	0·98		His	0·87		His	0·95	Met	0·068	Glu	0·060	Trp	0·064	Pro	0·068
Thr	0·83	i_α	Ala	0·83		Glu	0·74	Val	0·062	Cys	0·053	Gln	0·037	Phe	0·065
Ser	0·77		Ser	0·75		Ala	0·66	Leu	0·061	Val	0·048	Leu	0·036	Glu	0·064
Cys	0·70		Gly	0·75	b_β	Met	0·60	Ala	0·060	His	0·047	Ala	0·035	Ala	0·058
Tyr	0·69	b_α	Lys	0·74		Phe	0·60	Phe	0·059	Phe	0·041	Pro	0·034	Ile	0·056
Asn	0·67		Pro	0·55		Leu	0·59	Glu	0·056	Ile	0·034	Val	0·028	Met	0·055
Pro	0·57	B_α	Asp	0·54	B_β	Val	0·50	Lys	0·055	Leu	0·025	Met	0·014	His	0·054
Gly	0·57		Glu	0·37		Ile	0·47	Ile	0·043	Trp	0·013	Ile	0·013	Val	0·053

[a] P_α, P_β, and P_i are conformational parameters of helix, β sheet, and β turns, $f_i, f_{i+1}, f_{i+2}, f_{i+3}$ are bend frequencies in the four positions of the β turn. H_α, H_β, etc., as defined previously (Chou and Fasman, 1974b). From Chou and Fasman (1977, 1978).

An abbreviated set of rules follows (Fasman, 1985).

1. A cluster of four helical residues (Hα or hα) out of six along the protein sequence will initiate a helix. The helical segment is extended in both directions until sets of tetrapeptide breakers ($\langle P_\alpha \rangle < 1.00$) are reached. Proline cannot occur in the inner helix or at the C-terminal helical end but can occur within the last three residues at the N-terminal end. The inner helix is defined as one omitting the three helical end residues at both the amino and carboxyl ends. Any segment that is at least six residues long with $\langle P_\alpha \rangle > 1.03$ and $\langle P_\alpha \rangle > \langle P_\beta \rangle$ is predicted as helical.

2. A cluster of three β formers or a cluster of three β formers out of five residues along the sequence will initiate a β sheet. The β sheet is propagated in both directions until terminated by a set of tetrapeptide breakers ($\langle P_\beta \rangle < 1.00$). Any segment with $\langle P_\beta \rangle > 1.05$ as well as $\langle P_\beta \rangle > \langle P_\alpha \rangle$ is predicted as β sheet.

3. The probability of a bend at residue i is calculated from $p_t = f_i \times f_{i+1} \times f_{i+2} \times f_{i+3}$. Tetrapeptides with $p_t > 0.75 \times 10^{-4}$ as well as $\langle P_t \rangle > 1.00$ and $\langle P_\alpha \rangle < \langle P_t \rangle > \langle P_\beta \rangle$ are predicted as β-turns.

4. Any segment containing overlapping α and β regions is helical if $\langle P_\alpha \rangle > \langle P_\beta \rangle$ or β sheet if $\langle P_\beta \rangle > \langle P_\alpha \rangle$.

III. OPERATION OF THE PREDICTION PROGRAM: THE INPUT FILE

The input file must be an ASCII file containing the single-letter codes for the amino acids. The format of the file does not matter—the letters can be upper or lower case, there can be one

```
> PREDICT

Enter name of input file: B:SUBTIL

Enter name of output file: B:SUBTIL.PRN

Enter the name of the protein ----> SUBTILISIN BPN'

Use database of 29 proteins (Chou-Fasman '78) ----> enter 29

Use database of 64 proteins (Chou '79) ----> enter 64

29

ASCII format enter A, or LOTUS format, enter L:

L

PROGRAM EXECUTING.

To obtain print-out enter:

PRINT SUBTIL.PRN (Filename OUTPUT.PRN)
```

Figure 1. Sample of screen during program execution. User input is shown in all capital letters and bold print.

per line or as many as desired—however, each letter must be separated by a "white space." White spaces are spaces, tabs, or carriage returns. The program will accept drive and path information up to a total of 128 characters. The input file can be written using any word processor program that has a nondocument or ASCII mode.

To run the program, type PREDICT. The program will query the user for the name of the input file followed by the name of the output file (Fig. 1).

The next question asked by the program concerns the data base to use in predicting the secondary structure. There are two choices: the Chou–Fasman original data set based on 29 proteins (Chou and Fasman, 1978) or the expanded data set based on 64 proteins (P. Chou, Chapter 12, this volume).

Finally, the program will query the user for the type of output file. There are two formats, one that can be read into Lotus-123 and a plain ASCII-type file. The only difference between the two is that the Lotus version has the characters surrounded by quotation marks. This enables it to be read into Lotus using the File Import function (see Section IV).

The program will then proceed to perform the prediction and write the output file. The output file requires that the data be reduced by hand. The program does not perform this function. This is a deliberate decision. The Chou–Fasman algorithm is really a set of guidelines, and in reducing the data to a single predicted structure significant information is lost. Frequently a region will display significant propensity for more than one type of structure and be "hard to call." These regions are actually very interesting and may be sites of conformational change (see Fasman, Chapter 6, this volume). It is hoped that by being forced to be in closer contact with the raw data, the user will become aware of these potentialities.

		Pa	Pb	Pt	a	b	<Pa>	<Pb>	<Pt>	<pt>
1	A	142	83	66	H	i	109 *	109 *	89	3.90e-005
2	Q	111	110	98	h	h	87	102 *	110	1.96e-005
3	S	77	75	143	i	b	77	111 *	114	2.45e-005
4	V	106	170	50	h	H	72	111 *	118	3.23e-004 *
5	P	57	55	152	B	B	72	111 *	118	6.68e-005
6	Y	69	147	114	b	H	77	116 *	115	2.07e-005
7	G	57	75	156	B	b	87	107 *	111	6.00e-005
8	V	106	170	50	h	H	100 *	128 *	84	1.79e-005
9	S	77	75	143	i	b	103 *	104 *	97	1.45e-005
10	Q	111	110	98	h	h	119 *	106 *	78	1.05e-005
11	I	108	160	47	h	H	105 *	93	91	1.18e-005
12	K	116	74	101	h	b	114 *	73	96	8.24e-006
13	A	142	83	66	H	i	115 *	87	85	4.42e-005
14	P	57	55	152	B	B	105 *	88	93	1.51e-005
15	A	142	83	66	H	i	110 *	93	90	1.48e-005
16	L	121	130	59	H	H	102 *	100 *	98	3.51e-005
17	H	100	87	95	I	i	86	86	123	1.09e-004 *
18	S	77	75	143	i	b	78	101 *	127	2.79e-004 *
19	Q	111	110	98	h	h	80	112 *	116	5.66e-005
20	G	57	75	156	B	b	66	104 *	130	6.55e-005
21	Y	69	147	114	b	H	71	104 *	127	1.78e-004 *
22	T	83	119	96	i	h	71	89	137	8.32e-005 *
23	G	57	75	156	B	b	76	102 *	126	1.44e-004 *
24	S	77	75	143	i	b	91	102 *	112	2.65e-005
25	N	67	89	156	b	i	98	125 *	89	2.95e-005
26	V	106	170	50	h	H	117 *	124 *	66	1.16e-005
27	K	116	74	101	h	b	117 *	124 *	66	4.90e-006
28	V	106	170	50	h	H	115 *	145 *	53	7.39e-006
29	A	142	83	66	H	i	114 *	116 *	77	3.03e-006
30	V	106	170	50	h	H	98	114 *	96	4.00e-005
31	I	108	160	47	h	H	85	91	123	8.99e-005 *
32	D	101	54	146	I	B	85	91	123	2.17e-004 *
33	S	77	75	143	i	b	85	91	123	1.07e-005
34	G	57	75	156	B	b	85	91	123	6.58e-005
35	I	108	160	47	h	H	90	91	119	6.27e-005
36	D	101	54	146	I	B	88	72	131	1.38e-004 *
37	S	77	75	143	i	b	77	73	133	1.05e-004 *
38	S	77	75	143	i	b	83	67	134	1.55e-005
39	H	100	87	95	I	i	94	81	113	5.28e-004 *
40	P	57	55	152	B	B	98	78	114	3.84e-005
41	D	101	54	146	I	B	111 *	107 *	89	1.40e-005
42	L	121	130	59	H	H	121 *	114 *	69	1.14e-005
43	K	116	74	101	h	b	105 *	100 *	93	1.40e-005

Figure 2. Sample of ASCII-mode output file for the Chou–Fasman algorithm. Input file, a:SUBTIL; protein name, SUBTILISIN BPN'; data base used was 29 proteins.

IV. DATA REDUCTION: THE OUTPUT FILE

A sample output file is presented in Fig. 2. The first column is the residue number, and the second column is the single-letter code for the amino acid at that position.

The next five columns are data base data; they are, in order, the P_α value, the P_β value, the P_t value, the helical assignment, and the β-sheet probability. It should be noted that since the helical and β-sheet assignments were made only for the original Chou–Fasman data set, these are the ones that are always listed. Likewise, all turn values are from the original Chou–

Fasman data set. All the data base values and all the output values have been multiplied by a factor of 100.

The four columns on the right are the calculated output data. The first three of these are the tetrapeptide averages calculated from the P_α, P_β, and P_t values. The asterisks indicate when the tetrapeptide average is greater than 100. The last column is the position-dependent turn calculation $\langle p_t \rangle$. This value is flagged with an asterisk when the calculated value is above 100.

An approach to data reduction is presented below in terms of the search for helical regions, the search for β-sheet regions, and the search for β-turn sites. In fact, the process is most easily performed in a single pass through the data, checking for all three types of structure, rather than three passes as might be implied by the presentation.

V. ANALYSIS OF OUTPUT

A. Search for Helical Regions

1. Helix Nucleation

Helical regions are nucleated by the presence of four sequential tetrapeptide averages with a value greater than 100. These are easily located by visually scanning for a cluster of asterisks.

2. Helix Propagation

The helix is extended towards the carboxyl terminus until the tetrapeptide average drops below 100. Thus, the following three residues are included in the helical region. At this point, the residues immediately following are examined. If they are of class H or h, they are generally included; if they are of class i, with high P_α values they are included. The $\langle P_\alpha \rangle$ for the entire region is then calculated and recalculated with the exclusion of the terminal class i residues (should any be present). If the calculated value is above the threshold value of 103, the region may be assigned as helical. In deciding whether to include a terminal residue, one should consider that the distribution of number of residues per helical segment has peaks at multiples of four residues (Fig. 3). This reflects the fact that the hydrogen-bonding scheme in an α helix has residue i hydrogen bonded to residue $i + 3$. Therefore, inclusion of unfavorable residues is more acceptable when they make the total number a multiple of four.

It will often be the case that the tetrapeptide average will drop below 100 for a residue while it is above 100 for the residues on either side. The somewhat low residue is generally included in the structural region.

3. Proline as a Helix Breaker

Proline cannot occur in the inner helix or at the C-terminal end. It can occupy the first turn in the N-terminal helix.

B. Search for β-Sheet Regions

1. β-Sheet Nucleation

β-Sheet regions are nucleated by the presence of three sequential tetrapeptide averages with a value greater than 100. These are easily located by visually scanning for a cluster of asterisks.

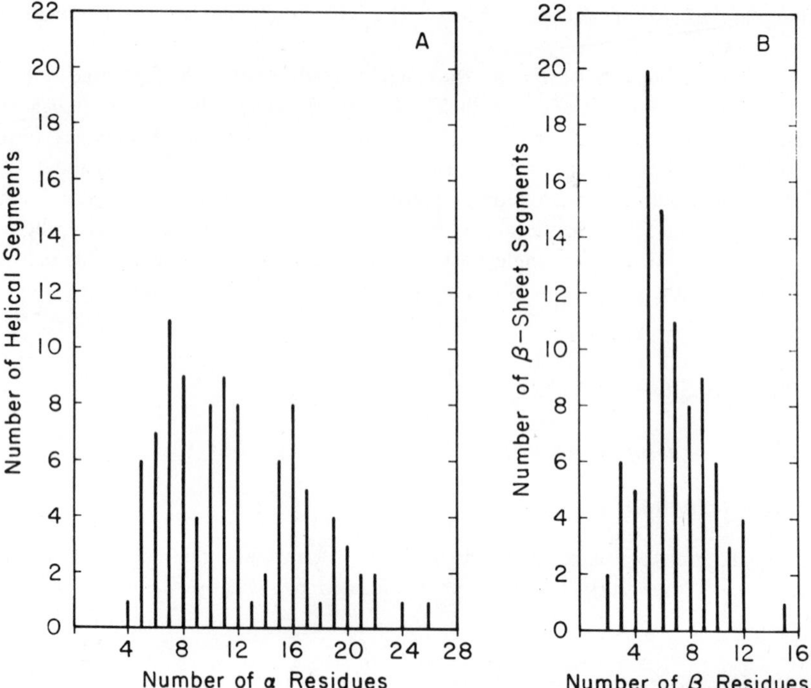

Figure 3. Distribution of the number of residues per helical segment (A) and per β-sheet segment (B) for 29 proteins determined from x-ray crystallography (Chou and Fasman, 1974b).

2. β-Sheet Propagation

The sheet is extended towards the carboxyl terminus until the tetrapeptide average drops below 100. The next three residues (the four of the tetrapeptide) are included in the β-sheet sequence. At this point, the residues immediately following are examined. If they are of class H or h, they are generally included; if they are of class i with high P_β values, they are included. The $\langle P_\beta \rangle$ for the entire region is then calculated and recalculated with the exclusion of the terminal class i residues (should any be present). If the calculated value is above the threshold value of 103, the region may be assigned as β sheet.

It will often be the case that the tetrapeptide average will drop below 100 for a residue while it is above 100 on either side. These residues are generally included in the structural region.

C. Search for β Turns

1. Definition of β Turns

Beta turns are defined by having $\langle P_t \rangle > 100$, $\langle P_\alpha \rangle < \langle P_t \rangle > \langle P_\beta \rangle$, and $p_t > 0.75 \times 10^{-4}$. In general, turns are assigned regardless of disruption of helical or sheet regions.

2. Resolving Overlapping Turns

Where a series of turns is found to overlap, the assignment is made to the turn with the higher local $\langle p_t \rangle$ value. One should then go back to make sure that no potential turns have been discounted. If a turn was discounted because of overlap with another turn, which finally was not assigned, the first turn should be reconsidered.

D. Resolving Overlapping Regions

Resolving the overlapping regions is the most difficult aspect of the analysis. It is at this point that using a spreadsheet program such as Lotus-123 greatly reduces the tedium of the process. In general the overlapping regions are compared for the calculated average value. If the $\langle P_\alpha \rangle$ is greater than the $\langle P_\beta \rangle$, the region is assigned as α helical; if the situation is reversed, the region is assigned as β sheet.

When one region of structure is wholely contained within another and the lengths are similar, the procedure is relatively straightforward. If, for example, there is a region of β sheet contained within a longer region of potential helix, the average value for the contained β-sheet region is calculated and compared to the average helical value for the same residues. If the average β-sheet value is greater, then the region is assigned as β sheet; otherwise it is α helix.

If one region of structure is significantly longer than another, it is handled as described above. Should the contained region have a larger average value, the assignment is made for the contained region. The user should then go back and reevaluate the remaining region to see if it will be stable on its own.

A problem arises when the contained region has a larger average value and the leftover residues are not long enough to form a stable structure. Should a small region of stable structure disrupt a longer region? Although generally, a small locally stable region is not allowed to disrupt a longer region of secondary structure, if the difference in average value is large it may do so.

VI. GRAPHIC DISPLAY

It is frequently useful to look at the tetrapeptide averages for α helix, β sheet, and β turns in a graphic format. If the Lotus output file has been generated and the file imported, this is a simple procedure using the graph option (see Fig. 4).

A program to display simultaneously the three tetrapeptide averages for the α helix, β sheet, and β turns, called PREVIEW, has been written by R. Lee for use on an IBM-PC. Such a plot is displayed in Fig. 5 for staphylococcal nuclease, which is discussed later in this chapter. As an aid in portraying the prediction of secondary structure, a diagramatic plot can be made of the secondary structure, as was demonstrated in the original paper on the prediction of secondary structure (Chou and Fasman, 1974b). A diagramatic representation of the bovine pancreatic trypsin inhibitor is shown in Fig. 6 (Chou and Fasman, 1974b).

VII. PORTABILITY

This program is written in the C programming language. The source code is reproduced in Appendix 1. The data set is contained in an Include file called Protein.Dat. The Include file

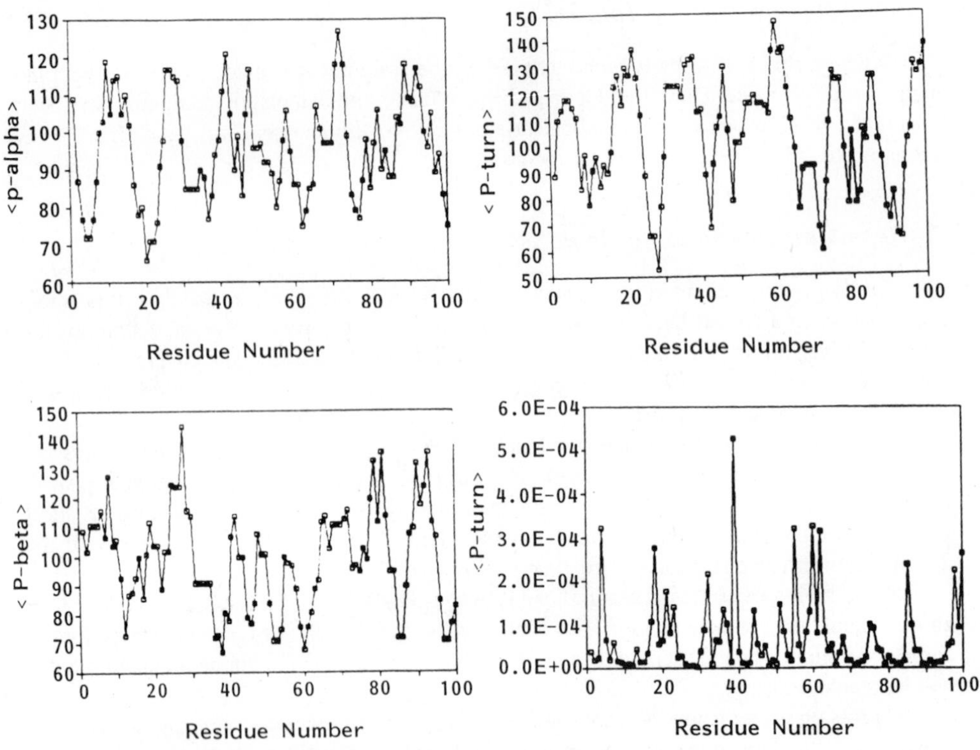

Figure 4. Graphic display of tetrapeptide averages produced using Lotus.

program must be available for compilation. The program uses only standard C functions and thus should be readily portable from machine to machine.

VIII. LOTUS HINTS

The Lotus-compatible version of the output file can be read into Lotus using the File Import function. The file should be imported as "numbers." Once imported it is useful to reformat the file so that it is displayed conveniently on the screen. In order to do this, select the Worksheet function, followed by the Global function. Then use the Column-Width feature and set the column width to 4. This takes care of all but the position-dependent turn data, which will be displayed as all zeros. This can be set as follows: select Range, then Format, and finally select Scientific with two decimal places. The range will now appear as all asterisks. Adjust the column width by selecting Worksheet and Column-Width: set the column width to 9.

In order to obtain the average of the P_α or P_β values for a range, use the @AVG function. A convenient place to locate the calculation is in the adjacent column where the asterisks are located. It is also possible to obtain a running average by fixing the starting point, using the average function, and copying the function through the range.

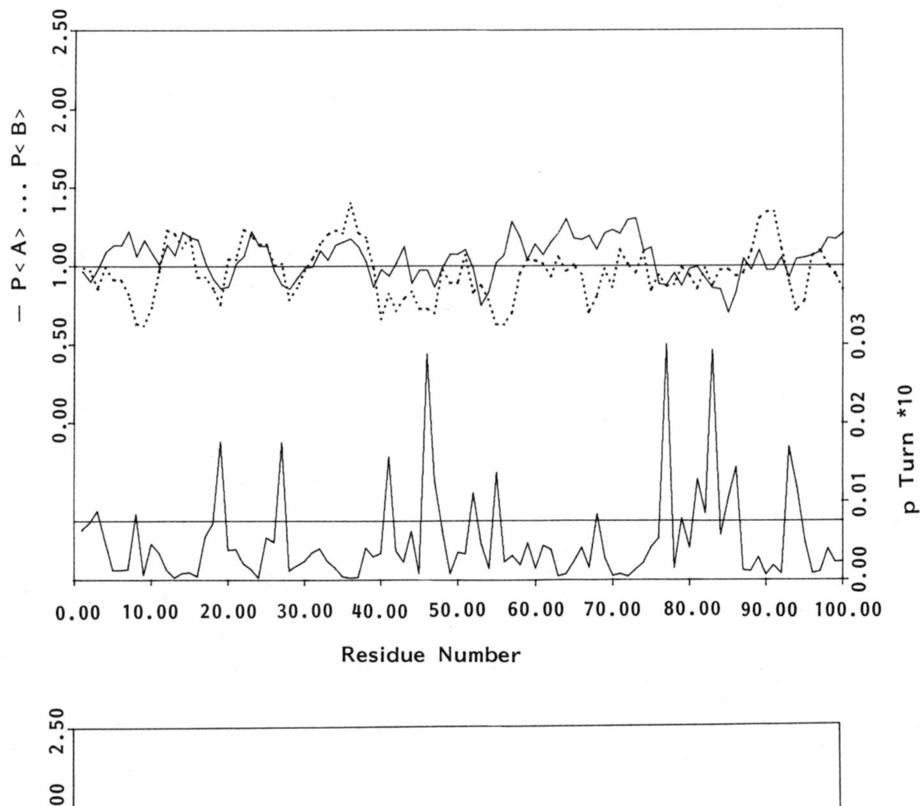

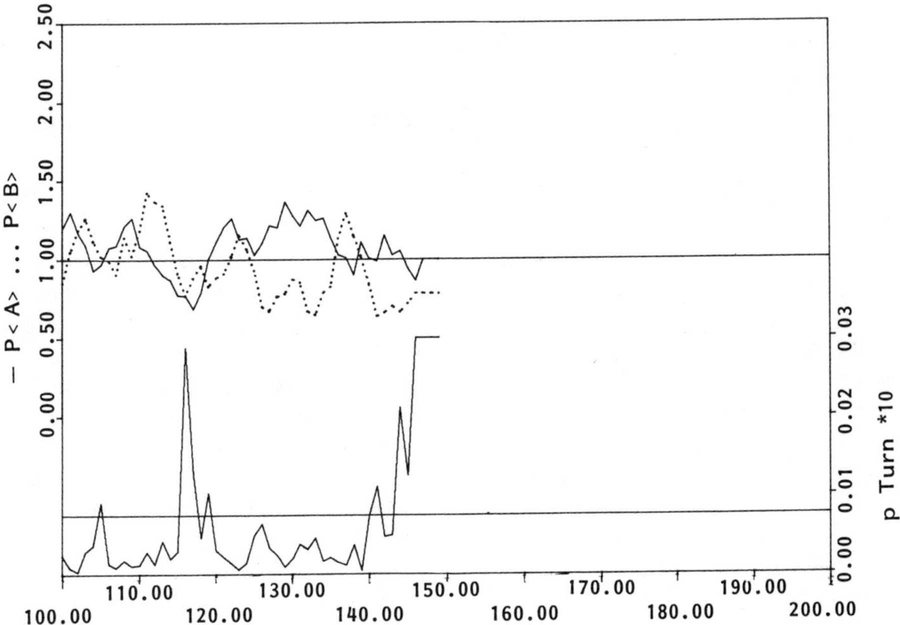

Figure 5. Predicted conformational profile of staphylococcal nuclease. The average helical and β-sheet potential of tetrapeptides i to $i + 3$ based on single-residue information. Data base is 64 proteins (see Chou, Chapter 12, this volume). $\langle P_\alpha \rangle$ and $\langle P_\beta \rangle$ are shown, respectively, by the solid line and dotted lines. The β-turn probability profile is shown as a solid line at the bottom of the profile. The horizontal line corresponds to the cutoff value of 0.75×10^{-5}.

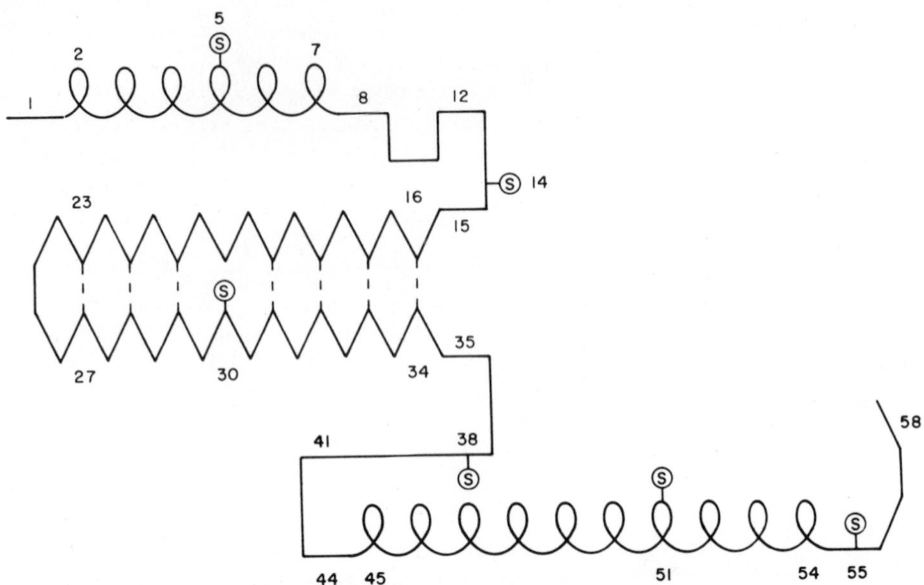

Figure 6. Schematic diagram of helical, β-sheet, and reverse β-turn regions predicted in pancreatic trypsin inhibitor. Residues are represented in their respective conformational states: helical, β-sheet, and coil. Chain reversals denote β-turn tetrapeptides. Hydrogen bonding between antiparallel β sheets is represented by dashed lines. Conformational boundary residues are numbered, as well as the six Cys residues indicated by S. It should be noted that in the present scale each helical loop represents a single helical residue and not a single turn consisting of 3.6 residues (Chou and Fasman, 1974b).

IX. ACCURACY OF PREDICTION

The prediction of the secondary structure of staphylococcal nuclease is compared to the x-ray diffraction results of Huber *et al.* (1971) in Table III. All regions of α and β structure were correctly predicted: four α-helical regions and two β-sheet regions. Within each region the numbers of overpredicted residues and missed residues are also listed in Table III. A total of 44 residues were either overpredicted or missed. Therefore, a total of 71% (subtracting both missed and overpredicted residues) of the residues were correctly predicted. The point to be emphasized is that all secondary structural regions were correctly predicted, and thus a fairly accurate picture of the secondary structure was obtained. The predicted β turns are also listed in Table III but cannot be compared, as they are not identified as such in the original paper (Huber *et al.*, 1971).

These results are an improvement over the original prediction of staphylococcal nuclease (Chou and Fasman, 1974b), and the improvement can be attributed to the larger data base (64 versus 29 proteins) used to obtain the conformational parameters P_α and P_β.

X. PREDICTION OF STAPHYLOCOCCAL NUCLEASE

This is an analysis of the prediction for staphylococcal nuclease. The output file is reproduced in Fig. 7.

Table III. The Conformational Prediction[a] of Staphylococcal Nuclease Compared to X-Ray Results[b]

Predicted	X ray	Residues overpredicted	Residues missed
7–18 α	12–19 α	5	1
23–26 β	21–27 β	0	3
31–40 β	30–36 β	4	1
59–76 α	54–67 α	10	8
97–104 α	99–106 α	2	2
120–140 α	122–134 α	8	0
Total		29	15
β turns[c]			
3–6			
19–22			
27–30			
41–44			
46–49			
55–58			
77–80			
83–86			
93–96			
105–108			
116–119			
141–144			
146–149			

[a]Based on the predictive analysis of Fig. 7.
[b]Huber *et al.* (1971).
[c]Percentage correctly predicted = (149 − 44)/149 × 100 = 71%.

Residue	Structural assignment
3–6	β turn: $\langle P_t \rangle > \langle P_\alpha \rangle$ and $\langle P_t \rangle > \langle P_\beta \rangle$; $\langle p_t \rangle >$ threshold.
7–18	α helix: $\langle P_\alpha \rangle = 1.13$; starts at residue 7; $\langle P_\alpha \rangle > \langle P_t \rangle$ for the region of 12–15; $\langle P_\alpha \rangle = 1.14$ and $\langle P_\beta \rangle = 1.18$; however, this is short for a β-sheet region, so it is not assigned as β. The proline is at position 5 of the helix. Therefore, it might cause a bend in the helix.
19–22	β turn.
23–26	β sheet: $\langle P_\beta \rangle = 1.20$ and $\langle P_\alpha \rangle = 1.19$.
27–30	β turn.
31–40	β sheet: $\langle P_\beta \rangle = 1.12$ and $\langle P_\alpha \rangle = 1.06$.
41–44	β turn.
46–49	β turn. The possible turn at residues 47–50, $\langle p_t \rangle = 1.28 \times 10^{-4}$, is precluded by the stronger turn at 46–49, $\langle p_t \rangle = 2.88 \times 10^{-4}$.
55–58	β turn. The presence of this turn precludes the one that is possible at residues 52–55.
59–76	α helix. The possible turn at residues 68–71 does not have $\langle P_t \rangle > \langle P_\alpha \rangle$. There is no region of β-sheet potential that is greater than α helix throughout this region. Notice that the region has been extended to include the helix-forming residue at 76, $\langle P_\alpha \rangle = 1.17$.

		Pa	Pb	Pt	a	b	<Pa>	<Pb>	<Pt>	<pt>	
1	A	142	83	66	H	i	96	99	100	6.40e-005	
2	T	83	119	96	i	h	89	96	109	7.38e-005	
3	S	77	75	143	i	b	98	85	110	8.86e-005	*
4	T	83	119	96	i	h	109 *	99	89	4.98e-005	
5	K	116	74	101	h	b	113 *	91	89	1.23e-005	
6	K	116	74	101	h	b	113 *	91	89	1.21e-005	
7	L	121	130	59	H	H	122 *	82	82	1.32e-005	
8	H	100	87	95	I	i	106 *	63	105	8.43e-005	*
9	K	116	74	101	h	b	116 *	62	98	6.51e-006	
10	E	151	37	74	H	B	108 *	73	97	4.66e-005	
11	P	57	55	152	B	B	100 *	96	93	3.53e-005	
12	A	142	83	66	H	i	113 *	123 *	67	1.31e-005	
13	T	83	119	96	i	h	107 *	120 *	75	2.66e-006	
14	L	121	130	59	H	H	121 *	111 *	68	8.66e-006	
15	I	108	160	47	h	H	118 *	119 *	65	9.69e-006	
16	K	116	74	101	h	b	116 *	92	90	4.40e-006	
17	A	142	83	66	H	i	102 *	93	103	5.55e-005	
18	I	108	160	47	h	H	91	85	123	7.28e-005	
19	D	101	54	146	I	B	85	75	136	1.77e-004	*
20	G	57	75	156	B	b	86	104 *	112	3.87e-005	
21	D	101	54	146	I	B	101 *	104 *	98	4.22e-005	
22	T	83	119	96	i	h	106 *	123 *	76	2.08e-005	
23	V	106	170	50	h	H	122 *	119 *	67	1.41e-005	
24	K	116	74	101	h	b	112 *	114 *	83	2.41e-006	
25	L	121	130	59	H	H	112 *	114 *	83	5.42e-005	
26	M	145	105	60	H	h	96	100 *	107	4.84e-005	
27	Y	69	147	114	b	H	88	101 *	117	1.76e-004	*
28	K	116	74	101	h	b	85	78	126	1.18e-005	
29	G	57	75	156	B	b	92	86	116	1.87e-005	
30	Q	111	110	98	h	h	99	97	101	2.46e-005	
31	P	57	55	152	B	B	99	104 *	92	3.53e-005	
32	M	145	105	60	H	h	109 *	113 *	77	4.06e-005	
33	T	83	119	96	i	h	103 *	120 *	77	2.44e-005	
34	F	113	138	60	h	h	113 *	122 *	68	1.58e-005	
35	R	98	93	95	i	i	115 *	120 *	68	4.41e-006	
36	L	121	130	59	H	H	117 *	140 *	56	2.91e-006	
37	L	121	130	59	H	H	112 *	121 *	78	3.46e-006	
38	L	121	130	59	H	H	102 *	118 *	87	4.14e-005	
39	V	106	170	50	h	H	86	99	111	3.01e-005	
40	D	101	54	146	I	B	98	66	117	3.45e-005	
41	T	83	119	96	i	h	93	82	104	1.57e-004	*
42	P	57	55	152	B	B	101 *	71	105	3.78e-005	
43	E	151	37	74	H	B	112 *	79	91	2.35e-005	
44	T	83	119	96	i	h	89	83	111	6.25e-005	
45	K	116	74	101	h	b	97	72	112	8.35e-006	
46	H	100	87	95	I	i	97	72	112	2.88e-004	*
47	P	57	55	152	B	B	86	69	127	1.28e-004	*
48	K	116	74	101	h	b	98	98	102	6.37e-005	
49	K	116	74	101	h	b	107 *	89	95	8.38e-006	
50	G	57	75	156	B	b	107 *	89	95	3.58e-005	
51	V	106	170	50	h	H	110 *	107 *	84	3.35e-005	
52	E	151	37	74	H	B	98	83	111	1.12e-004	*
53	K	116	74	101	h	b	74	87	130	4.62e-005	

Figure 7. Output file for staphylococcal nuclease. Note that a represents α, and b β. Chou–Fasman algorithm: input file, a:NUCLEASE; protein name, STAPH; data base used was 29 proteins.

54	Y	69	147	114	b	H	83		78		124	1.52e-005	
55	G	57	75	156	B	b	101	*	62		112	1.37e-004	*
56	P	57	55	152	B	B	106	*	62		108	2.27e-005	
57	E	151	37	74	H	B	128	*	69		87	3.09e-005	
58	A	142	83	66	H	i	118	*	94		83	1.90e-005	
59	S	77	75	143	i	b	103	*	103	*	91	4.68e-005	
60	A	142	83	66	H	i	113	*	103	*	80	1.52e-005	
61	F	113	138	60	h	h	107	*	101	*	89	4.36e-005	
62	T	83	119	96	i	h	115	*	93		89	3.92e-005	
63	K	116	74	101	h	b	120	*	105	*	78	4.69e-006	
64	K	116	74	101	h	b	129	*	96		71	8.08e-006	
65	M	145	105	60	H	h	117	*	100	*	85	2.29e-005	
66	V	106	170	50	h	H	116	*	94		86	4.12e-005	
67	E	151	37	74	H	B	119	*	70		99	1.55e-005	
68	N	67	89	156	b	i	110	*	80		106	8.37e-005	*
69	A	142	83	66	H	i	120	*	97		78	2.78e-005	
70	K	116	74	101	h	b	122	*	86		80	5.26e-006	
71	K	116	74	101	h	b	120	*	110	*	68	7.63e-006	
72	I	108	160	47	h	H	129	*	101	*	61	4.62e-006	
73	E	151	37	74	H	B	130	*	95		64	1.35e-005	
74	V	106	170	50	h	H	109	*	108	*	85	2.20e-005	
75	E	151	37	74	H	B	111	*	84		97	4.17e-005	
76	F	113	138	60	h	h	88		94		118	5.36e-005	
77	N	67	89	156	b	i	87		87		127	3.45e-004	*
78	K	116	74	101	h	b	95		88		112	1.47e-005	
79	G	57	75	156	B	b	87		99		111	7.82e-005	*
80	Q	111	110	98	h	h	98		94		108	4.13e-005	
81	R	98	93	95	i	i	99		85		109	1.29e-004	*
82	T	83	119	96	i	h	92		98		114	8.51e-005	*
83	D	101	54	146	I	B	85		87		129	2.93e-004	*
84	K	116	74	101	h	b	85		97		116	5.77e-005	
85	Y	69	147	114	b	H	70		97		130	1.05e-004	*
86	G	57	75	156	B	b	83		93		116	1.44e-004	*
87	R	98	93	95	i	i	104	*	95		94	1.24e-005	
88	G	57	75	156	B	b	97		108	*	98	1.12e-005	
89	L	121	130	59	H	H	110	*	130	*	71	2.96e-005	
90	A	142	83	66	H	i	97		134	*	85	6.34e-006	
91	Y	69	147	114	b	H	97		134	*	85	1.84e-005	
92	I	108	160	47	h	H	105	*	111	*	93	7.92e-006	
93	Y	69	147	114	b	H	92		89		120	1.70e-004	*
94	A	142	83	66	H	i	104	*	71		117	1.19e-004	*
95	D	101	54	146	I	B	104	*	77		115	4.95e-005	
96	G	57	75	156	B	b	106	*	106	*	91	8.70e-006	
97	K	116	74	101	h	b	108	*	109	*	91	1.15e-005	
98	M	145	105	60	H	h	117	*	100	*	85	3.99e-005	
99	V	106	170	50	h	H	116	*	94		86	2.30e-005	
100	N	67	89	156	b	i	120	*	84		88	2.37e-005	
101	E	151	37	74	H	B	130	*	105	*	62	8.12e-006	
102	A	142	83	66	H	i	116	*	119	*	67	3.57e-006	
103	L	121	130	59	H	H	109	*	125	*	75	2.84e-005	
104	V	106	170	50	h	H	93		112	*	99	3.70e-005	
105	R	98	93	95	i	i	96		102	*	102	9.12e-005	*
106	Q	111	110	98	h	h	107	*	99		94	1.31e-005	
107	G	57	75	156	B	b	109	*	90		95	8.48e-006	
108	L	121	130	59	H	H	121	*	114	*	69	1.77e-005	
109	A	142	83	66	H	i	126	*	102	*	70	1.12e-005	
110	K	116	74	101	h	b	108	*	118	*	82	1.15e-005	

Figure 7 (cont.)

111	V	106	170	50	h	H	105 *	142 *	70	2.85e-005
112	A	142	83	66	H	i	96	136 *	86	1.36e-005
113	Y	69	147	114	b	H	90	134 *	94	4.26e-005
114	V	106	170	50	h	H	87	111 *	104	1.97e-005
115	Y	69	147	114	b	H	77	91	130	2.92e-005
116	K	116	74	101	h	b	76	76	141	2.88e-004 *
117	P	57	55	152	B	B	68	88	140	1.28e-004 *
118	N	67	89	156	b	i	79	96	125	4.69e-005
119	N	67	89	156	b	i	100 *	83	105	1.03e-004 *
120	T	83	119	96	i	h	111 *	88	90	3.05e-005
121	H	100	87	95	I	i	120 *	91	81	2.18e-005
122	E	151	37	74	H	B	126 *	101 *	72	1.38e-005
123	Q	111	110	98	h	h	112 *	115 *	77	5.66e-006
124	L	121	130	59	H	H	114 *	106 *	78	1.43e-005
125	L	121	130	59	H	H	103 *	93	99	4.93e-005
126	R	98	93	95	i	i	110 *	69	103	6.44e-005
127	K	116	74	101	h	b	121 *	67	96	3.41e-005
128	S	77	75	143	i	b	120 *	76	95	2.47e-005
129	E	151	37	74	H	B	136 *	78	76	9.13e-006
130	A	142	83	66	H	i	127 *	87	82	1.96e-005
131	Q	111	110	98	h	h	121 *	85	91	3.85e-005
132	A	142	83	66	H	i	131 *	67	85	3.18e-005
133	K	116	74	101	h	b	124 *	64	94	4.63e-005
134	K	116	74	101	h	b	126 *	78	83	1.66e-005
135	E	151	37	74	H	B	113 *	82	97	2.11e-005
136	K	116	74	101	h	b	103 *	113 *	90	1.47e-005
137	L	121	130	59	H	H	101 *	129 *	89	1.10e-005
138	N	67	89	156	b	i	90	115 *	110	3.71e-005
139	I	108	160	47	h	H	111 *	102 *	90	4.47e-006
140	W	108	137	96	h	h	100 *	84	117	7.50e-005
141	S	77	75	143	i	b	99	63	129	1.11e-004 *
142	E	151	37	74	H	B	115 *	65	110	4.83e-005
143	N	67	89	156	b	i	102 *	70	128	5.02e-005
144	D	101	54	146	I	B	105 *	66	125	2.12e-004 *
145	A	142	83	66	H	i	94	71	127	1.25e-004 *
146	D	101	54	146	I	B	86	78	135	3.80e-004 *
147	S	77	75	143	i	b	0	0	0	0.00e+000
148	G	57	75	156	B	b	0	0	0	0.00e+000
149	Q	111	110	98	h	h	0	0	0	0.00e+000
150	^@	0	0	0	^@	^@	0	0	0	0.00e+000

Figure 7 (cont.)

77–80 β turn. This has been assigned as a turn because it has the largest local value of $\langle p_t \rangle$.

83–86 β turn. This assignment has been made for the same reason as above, the strongest local value of $\langle p_t \rangle$.

88–92 β sheet: $\langle P_\beta \rangle = 1.19$, $\langle P_\alpha \rangle = 0.97$.

93–96 β turn.

[97–115] The region from residue 97 to residue 115 has several possible predictions. This may be taken to mean that there is the possibility of conformational changes taking place (see Fasman, Chapter 6, this volume). A β-turn assignment could possibly be made at residues 105–108. The $\langle P_t \rangle$ is equal to the $\langle P_\beta \rangle$ there. If this turn is not predicted, then either α helix or β sheet could run through to about residue 112 or 115, respectively. For residues 97–104, $\langle P_\alpha \rangle = 1.19$ and $\langle P_\beta \rangle = 1.07$, so the assignment is made to α helix.

97–104	α helix. Since $\langle P_t \rangle > \langle P_\alpha \rangle$ at residue 105, the turn assignment was made for residues 105–108.
105–108	β turn.
109–115	β sheet. Residues 109–115 were assigned as β sheet because the most stable helix of six residues that could be made (residues 109–114) has $\langle P_\alpha \rangle = 1.13$, significantly less than the value obtained for $\langle P_\beta \rangle = 1.21$.
116–119	β turn.
120–143	α helix: $\langle P_\alpha \rangle = 1.13$.
[136–140]	Residues 136–140 have $\langle P_\beta \rangle = 1.08$ versus $\langle P_\alpha \rangle = 1.01$ and could be assigned as a β sheet, but as it is a short region within a much longer α region, it is not assigned.
141–144	β turn. This turn precludes the turn possible at residues 141–144. The turn at 144–147 has a higher $\langle p_t \rangle$.
146–149	β turn.

XI. PREDICTION OF SUBTILISIN

This is an analysis of the prediction for subtilisin. The output file is reproduced in Fig. 8.

Residue	Structural assignment
1–3	β sheet.
4–7	β turn.
8–10	β sheet: $\langle P_\beta \rangle = 1.18$, $\langle P_\alpha \rangle = 0.98$.
11–17	α helix. Note inclusion of indifferent residue, $\langle P_\alpha \rangle = 1.12$.
18–21	β turn.
23–26	β turn.
[11–26]*	α helix. Note these two β turns could be included in the α-helical region.
27–31	β sheet. Note inclusion of residue 31 because it is of class h, and the value $\langle P_\beta \rangle$ is 1.31.
32–35	β turn.
36–39	β turn.
44–47	β turn.
51–54	β turn.
55–58	β turn.
60–63	β turn. Highest local value of $\langle p_t \rangle$.
[32–63]	α helix + 3_{10} helix.
65–74	β sheet; $\langle P_\beta \rangle = 1.06$. $\langle P_\alpha \rangle = 1.02$ for residues 66–74. This region is difficult to assign and is a candidate for a region displaying an inducible change in secondary structure.
75–78	β turn.
79–84	β sheet, $\langle P_\beta \rangle = 1.30$, $\langle P_\alpha \rangle = 0.93$.
85–88	β turn.
89–96	β sheet, $\langle P_\beta \rangle = 1.23$. There is helical potential from residues 89–96; $\langle P_\alpha \rangle = 1.15$.
100–103	β turn.
104–122	β sheet, $\langle P_\beta \rangle = 1.14$, $\langle P_\beta \rangle = 1.03$. There may be a β turn located at residues 118–121.
123–126	β turn.

*The β-turn, ϕ,ψ angles are the same as those found in the 3_{10} helix. Therefore, several consecutive β turns can be added onto an α-helical region. Frequently terminal residues of an α helix have the 3_{10} geometry.

		Pa	Pb	Pt	a	b	<Pa>		<Pb>		<Pt>	<pt>	
1	A	142	83	66	H	i	109	*	109	*	89	3.90e-005	
2	Q	111	110	98	h	h	87		102	*	110	1.96e-005	
3	S	77	75	143	i	b	77		111	*	114	2.45e-005	
4	V	106	170	50	h	H	72		111	*	118	3.23e-004	*
5	P	57	55	152	B	B	72		111	*	118	6.68e-005	
6	Y	69	147	114	b	H	77		116	*	115	2.07e-005	
7	G	57	75	156	B	b	87		107	*	111	6.00e-005	
8	V	106	170	50	h	H	100	*	128	*	84	1.79e-005	
9	S	77	75	143	i	b	103	*	104	*	97	1.45e-005	
10	Q	111	110	98	h	h	119	*	106	*	78	1.05e-005	
11	I	108	160	47	h	H	105	*	93		91	1.18e-005	
12	K	116	74	101	h	b	114	*	73		96	8.24e-006	
13	A	142	83	66	H	i	115	*	87		85	4.42e-005	
14	P	57	55	152	B	B	105	*	88		93	1.51e-005	
15	A	142	83	66	H	i	110	*	93		90	1.48e-005	
16	L	121	130	59	H	H	102	*	100	*	98	3.51e-005	
17	H	100	87	95	I	i	86		86		123	1.09e-004	*
18	S	77	75	143	i	b	78		101	*	127	2.79e-004	*
19	Q	111	110	98	h	h	80		112	*	116	5.66e-005	
20	G	57	75	156	B	b	66		104	*	130	6.55e-005	
21	Y	69	147	114	b	H	71		104	*	127	1.78e-004	*
22	T	83	119	96	i	h	71		89		137	8.32e-005	*
23	G	57	75	156	B	b	76		102	*	126	1.44e-004	*
24	S	77	75	143	i	b	91		102	*	112	2.65e-005	
25	N	67	89	156	b	i	98		125	*	89	2.95e-005	
26	V	106	170	50	h	H	117	*	124	*	66	1.16e-005	
27	K	116	74	101	h	b	117	*	124	*	66	4.90e-006	
28	V	106	170	50	h	H	115	*	145	*	53	7.39e-006	
29	A	142	83	66	H	i	114	*	116	*	77	3.03e-006	
30	V	106	170	50	h	H	98		114	*	96	4.00e-005	
31	I	108	160	47	h	H	85		91		123	8.99e-005	*
32	D	101	54	146	I	B	85		91		123	2.17e-004	*
33	S	77	75	143	i	b	85		91		123	1.07e-005	
34	G	57	75	156	B	b	85		91		123	6.58e-005	
35	I	108	160	47	h	H	90		91		119	6.27e-005	
36	D	101	54	146	I	B	88		72		131	1.38e-004	*
37	S	77	75	143	i	b	77		73		133	1.05e-004	*
38	S	77	75	143	i	b	83		67		134	1.55e-005	
39	H	100	87	95	I	i	94		81		113	5.28e-004	*
40	P	57	55	152	B	B	98		78		114	3.84e-005	
41	D	101	54	146	I	B	111	*	107	*	89	1.40e-005	
42	L	121	130	59	H	H	121	*	114	*	69	1.14e-005	
43	K	116	74	101	h	b	105	*	100	*	93	1.40e-005	
44	V	106	170	50	h	H	90		100	*	107	1.36e-004	*
45	A	142	83	66	H	i	99		79		111	5.62e-005	
46	G	57	75	156	B	b	83		77		130	3.22e-005	
47	G	57	75	156	B	b	105	*	84		106	5.33e-005	
48	A	142	83	66	H	i	117	*	108	*	79	6.19e-006	
49	S	77	75	143	i	b	96		101	*	101	1.87e-005	
50	M	145	105	60	H	h	96		101	*	101	1.18e-005	
51	V	106	170	50	h	H	97		84		104	1.49e-004	*
52	P	57	55	152	B	B	92		71		116	8.62e-005	*
53	S	77	75	143	i	b	92		71		116	3.18e-005	
54	E	151	37	74	H	B	89		75		119	1.87e-005	

Figure 8. Output file for subtilisin. Note that a represents α, and b β. Chou–Fasman algorithm: input file, a:SUBTIL; protein name, SUBTILISIN; data base used was 29 proteins.

55	T	83	119	96	i	h	80		100	*	116	3.21e-004	*
56	P	57	55	152	B	B	87		98		116	5.39e-005	
57	N	67	89	156	b	i	98		97		115	1.98e-005	
58	F	113	138	60	h	h	106	*	89		112	8.38e-005	*
59	Q	111	110	98	h	h	95		76		136	1.33e-004	*
60	D	101	54	146	I	B	86		68		147	3.27e-004	*
61	D	101	54	146	I	B	86		76		135	8.24e-005	*
62	N	67	89	156	b	i	75		81		137	3.16e-004	*
63	S	77	75	143	i	b	79		89		122	8.47e-005	*
64	H	100	87	95	I	i	85		92		110	4.18e-005	
65	G	57	75	156	B	b	86		112	*	99	5.43e-005	
66	T	83	119	96	i	h	107	*	114	*	76	6.56e-006	
67	H	100	87	95	I	i	101	*	103	*	91	3.58e-005	
68	V	106	170	50	h	H	97		111	*	92	7.07e-005	
69	A	142	83	66	H	i	97		111	*	92	1.76e-005	
70	G	57	75	156	B	b	97		111	*	92	1.79e-005	
71	T	83	119	96	i	h	118	*	113	*	69	8.38e-006	
72	V	106	170	50	h	H	127	*	116	*	60	1.15e-005	
73	A	142	83	66	H	i	118	*	96		86	1.49e-005	
74	A	142	83	66	H	i	99		97		109	2.61e-005	
75	L	121	130	59	H	H	83		95		128	1.03e-004	*
76	N	67	89	156	b	i	79		103	*	125	9.35e-005	*
77	N	67	89	156	b	i	77		99		125	4.42e-005	
78	S	77	75	143	i	b	87		120	*	99	4.11e-005	
79	I	108	160	47	h	H	98		133	*	78	7.16e-006	
80	G	57	75	156	B	b	85		112	*	105	2.68e-005	
81	V	106	170	50	h	H	97		136	*	78	1.56e-005	
82	L	121	130	59	H	H	106	*	114	*	82	8.42e-006	
83	G	57	75	156	B	b	90		95		106	1.17e-005	
84	V	106	170	50	h	H	95		95		102	1.70e-005	
85	A	142	83	66	H	i	88		72		126	2.39e-004	*
86	P	57	55	152	B	B	88		72		126	1.03e-004	*
87	S	77	75	143	i	b	104	*	90		102	4.09e-005	
88	S	77	75	143	i	b	102	*	108	*	95	4.10e-005	
89	A	142	83	66	H	i	118	*	110	*	76	9.92e-006	
90	L	121	130	59	H	H	109	*	132	*	72	7.36e-006	
91	Y	69	147	114	b	H	108	*	118	*	82	1.66e-005	
92	A	142	83	66	H	i	117	*	124	*	66	1.10e-005	
93	V	106	170	50	h	H	112	*	136	*	65	1.40e-005	
94	K	116	74	101	h	b	100	*	112	*	91	1.44e-005	
95	V	106	170	50	h	H	96		107	*	102	2.39e-005	
96	L	121	130	59	H	H	105	*	85		106	5.38e-005	
97	G	57	75	156	B	b	89		71		131	5.97e-005	
98	D	101	54	146	I	B	94		71		127	2.25e-004	*
99	A	142	83	66	H	i	83		77		130	9.69e-005	*
100	G	57	75	156	B	b	75		83		138	2.64e-004	*
101	S	77	75	143	i	b	78		101	*	127	4.72e-005	
102	G	57	75	156	B	b	78		101	*	127	1.21e-004	*
103	Q	111	110	98	h	h	91		117	*	112	1.00e-004	*
104	Y	69	147	114	b	H	90		129	*	100	4.09e-005	
105	S	77	75	143	i	b	100	*	133	*	83	1.14e-006	
106	W	108	137	96	h	h	97		136	*	86	3.10e-006	
107	I	108	160	47	h	H	85		121	*	101	4.24e-005	
108	I	108	160	47	h	H	85		121	*	101	3.80e-005	
109	N	67	89	156	b	i	95		90		108	1.14e-005	
110	G	57	75	156	B	b	106	*	102	*	93	4.46e-005	
111	I	108	160	47	h	H	127	*	104	*	70	9.58e-006	

Figure 8 (cont.)

112	E	151	37	74	H	B	127	*	104	*	70	1.43e-006	
113	W	108	137	96	h	h	125	*	115	*	68	4.41e-006	
114	A	142	83	66	H	i	114	*	103	*	83	6.50e-006	
115	I	108	160	47	h	H	96		105	*	106	5.68e-005	
116	A	142	83	66	H	i	105	*	91		109	5.23e-005	
117	N	67	89	156	b	i	95		84		129	1.52e-005	
118	N	67	89	156	b	i	104	*	104	*	103	1.25e-004	*
119	M	145	105	60	H	h	115	*	122	*	75	1.17e-005	
120	D	101	54	146	I	B	95		118	*	99	8.35e-006	
121	V	106	170	50	h	H	106	*	131	*	78	2.21e-005	
122	I	108	160	47	h	H	99		107	*	101	5.30e-006	
123	N	67	89	156	b	i	102	*	99		104	1.16e-004	*
124	M	145	105	60	H	h	100	*	96		104	5.17e-005	
125	S	77	75	143	i	b	78		88		128	8.66e-005	*
126	L	121	130	59	H	H	73		83		130	6.70e-005	
127	G	57	75	156	B	b	62		70		151	3.12e-005	
128	G	57	75	156	B	b	62		70		151	5.83e-004	*
129	P	57	55	152	B	B	67		70		148	2.86e-004	*
130	S	77	75	143	i	b	88		77		127	7.39e-005	
131	G	57	75	156	B	b	104	*	79		107	2.88e-005	
132	S	77	75	143	i	b	120	*	92		83	2.23e-005	
133	A	142	83	66	H	i	130	*	92		73	1.56e-005	
134	A	142	83	66	H	i	130	*	92		73	6.26e-006	
135	L	121	130	59	H	H	130	*	92		73	1.42e-005	
136	K	116	74	101	h	b	126	*	102	*	70	7.75e-006	
137	A	142	83	66	H	i	122	*	97		82	1.03e-005	
138	A	142	83	66	H	i	116	*	95		90	4.90e-005	
139	V	106	170	50	h	H	116	*	95		90	2.85e-005	
140	D	101	54	146	I	B	116	*	95		90	3.14e-005	
141	K	116	74	101	h	b	126	*	102	*	70	6.79e-006	
142	A	142	83	66	H	i	116	*	102	*	81	1.07e-005	
143	V	106	170	50	h	H	95		100	*	103	8.95e-005	*
144	A	142	83	66	H	i	95		100	*	103	8.40e-005	*
145	S	77	75	143	i	b	86		122	*	99	1.51e-005	
146	G	57	75	156	B	b	93		146	*	76	7.27e-006	
147	V	106	170	50	h	H	106	*	170	*	50	4.42e-006	
148	V	106	170	50	h	H	115	*	148	*	54	4.83e-006	
149	V	106	170	50	h	H	124	*	126	*	58	6.04e-006	
150	V	106	170	50	h	H	133	*	104	*	62	9.57e-006	
151	A	142	83	66	H	i	120	*	81		88	2.43e-005	
152	A	142	83	66	H	i	102	*	82		111	7.88e-005	*
153	A	142	83	66	H	i	104	*	71		113	6.23e-005	
154	G	57	75	156	B	b	83		69		135	9.91e-005	*
155	N	67	89	156	b	i	88		69		132	1.95e-004	*
156	E	151	37	74	H	B	92		76		117	4.70e-005	
157	G	57	75	156	B	b	68		86		137	1.40e-004	*
158	S	77	75	143	i	b	73		86		134	2.61e-004	*
159	T	83	119	96	i	h	73		86		134	9.69e-005	*
160	G	57	75	156	B	b	72		75		146	1.88e-004	*
161	S	77	75	143	i	b	78		86		131	1.65e-004	*
162	S	77	75	143	i	b	85		109	*	108	5.75e-005	
163	S	77	75	143	i	b	80		109	*	111	5.52e-005	
164	T	83	119	96	i	h	78		127	*	104	9.80e-005	*
165	V	106	170	50	h	H	72		111	*	118	4.09e-005	
166	G	57	75	156	B	b	60		88		144	3.43e-005	
167	Y	69	147	114	b	H	74		87		130	4.46e-004	*
168	P	57	55	152	B	B	74		87		130	7.80e-005	*

Figure 8 (cont.)

169	G	57	75	156	B	b	74		87		130	9.09e-005	*
170	K	116	74	101	h	b	79		87		127	1.29e-005	
171	Y	69	147	114	b	H	77		111	*	114	1.64e-004	*
172	P	57	55	152	B	B	87		115	*	98	2.22e-005	
173	S	77	75	143	i	b	108	*	122	*	76	4.34e-006	
174	V	106	170	50	h	H	115	*	145	*	53	3.91e-006	
175	I	108	160	47	h	H	103	*	122	*	79	1.39e-005	
176	A	142	83	66	H	i	111	*	102	*	84	3.17e-005	
177	V	106	170	50	h	H	102	*	124	*	80	9.78e-006	
178	G	57	75	156	B	b	90		100	*	107	3.30e-005	
179	A	142	83	66	H	i	93		104	*	107	4.98e-005	
180	V	106	170	50	h	H	86		102	*	115	9.56e-005	*
181	G	57	75	156	B	b	77		96		131	7.62e-005	*
182	N	67	89	156	b	i	77		96		131	3.21e-004	*
183	K	116	74	101	h	b	96		94		109	3.94e-005	
184	Y	69	147	114	b	H	84		113	*	112	3.05e-005	
185	G	57	75	156	B	b	83		98		123	8.04e-005	*
186	A	142	83	66	H	i	83		98		123	1.13e-004	*
187	Y	69	147	114	b	H	69		107	*	130	1.02e-004	*
188	N	67	89	156	b	i	71		89		137	9.43e-005	*
189	G	57	75	156	B	b	90		93		113	7.57e-005	*
190	T	83	119	96	i	h	111	*	95		91	9.71e-006	
191	S	77	75	143	i	b	110	*	84		103	3.65e-005	
192	M	145	105	60	H	h	105	*	79		105	4.39e-005	
193	A	142	83	66	H	i	94		75		114	1.53e-005	
194	S	77	75	143	i	b	85		96		110	1.78e-004	*
195	P	57	55	152	B	B	101	*	98		90	7.79e-006	
196	H	100	87	95	I	i	101	*	103	*	91	3.58e-005	
197	V	106	170	50	h	H	111	*	102	*	84	5.19e-005	
198	A	142	83	66	H	i	120	*	81		88	1.04e-005	
199	G	57	75	156	B	b	120	*	81		88	1.57e-005	
200	A	142	83	66	H	i	136	*	94		64	1.12e-005	
201	A	142	83	66	H	i	128	*	114	*	59	9.19e-006	
202	A	142	83	66	H	i	123	*	125	*	57	1.36e-006	
203	L	121	130	59	H	H	106	*	123	*	77	7.91e-006	
204	I	108	160	47	h	H	105	*	109	*	87	1.28e-005	
205	L	121	130	59	H	H	103	*	91		99	3.30e-005	
206	S	77	75	143	i	b	87		72		122	8.73e-005	*
207	K	116	74	101	h	b	85		76		126	8.00e-006	
208	H	100	87	95	I	i	83		92		124	1.34e-003	*
209	P	57	55	152	B	B	78		100	*	125	4.28e-005	
210	N	67	89	156	b	i	81		108	*	126	1.24e-005	
211	W	108	137	96	h	h	85		116	*	111	1.25e-004	*
212	T	83	119	96	i	h	86		109	*	111	4.55e-005	
213	N	67	89	156	b	i	91		122	*	100	3.41e-005	
214	T	83	119	96	i	h	99		123	*	84	2.01e-005	
215	Q	111	110	98	h	h	98		112	*	96	3.73e-005	
216	V	106	170	50	h	H	89		103	*	107	8.71e-005	*
217	R	98	93	95	i	i	93		93		110	8.51e-005	*
218	S	77	75	143	i	b	96		97		110	5.88e-005	
219	S	77	75	143	i	b	94		101	*	114	1.01e-005	
220	L	121	130	59	H	H	95		112	*	102	9.02e-005	*
221	Q	111	110	98	h	h	86		109	*	111	3.15e-005	
222	N	67	89	156	b	i	79		111	*	111	8.93e-005	*
223	T	83	119	96	i	h	91		107	*	97	5.74e-005	
224	T	83	119	96	i	h	100	*	110	*	88	4.68e-005	
225	T	83	119	96	i	h	94		99		103	5.41e-005	

Figure 8 (cont.)

226	K	116	74	101	h	b	98		83		115	2.12e-005	
227	L	121	130	59	H	H	89		83		126	9.84e-005	*
228	G	57	75	156	B	b	87		85		126	9.12e-005	*
229	D	101	54	146	I	B	90		103	*	115	1.66e-004	*
230	S	77	75	143	i	b	82		126	*	107	7.01e-005	
231	F	113	138	60	h	h	77		126	*	111	6.65e-005	
232	Y	69	147	114	b	H	77		110	*	121	9.62e-005	*
233	Y	69	147	114	b	H	74		92		131	7.63e-005	*
234	G	57	75	156	B	b	87		88		118	1.56e-004	*
235	K	116	74	101	h	b	100	*	109	*	90	9.42e-006	
236	G	57	75	156	B	b	88		113	*	104	3.02e-006	
237	L	121	130	59	H	H	100	*	137	*	78	2.10e-005	
238	I	108	160	47	h	H	98		132	*	87	9.79e-006	
239	N	67	89	156	b	i	106	*	113	*	92	1.66e-005	
240	V	106	170	50	h	H	125	*	111	*	70	1.23e-005	
241	Q	111	110	98	h	h	134	*	89		74	1.14e-005	
242	A	142	83	66	H	i	134	*	89		74	1.56e-005	
243	A	142	83	66	H	i	0		0		0	0.00e+000	
244	A	142	83	66	H	i	0		0		0	0.00e+000	
245	Q	111	110	98	h	h	0		0		0	0.00e+000	
246	^@	0	0	0	^@	^@	0		0		0	0.00e+000	

Figure 8 (cont.)

128–131 β turn.
132–142 α helix.
143–146 β turn.
147–153 β sheet, $\langle P_\beta \rangle = 1.33$, $\langle P_\alpha \rangle = 1.21$.
154–157 β turn.
158–161 β turn.
162–166 β sheet.
167–170 β turn.
171–181 β sheet.
182–185 β turn.
186–189 β turn.
190–193 α helix.
194–197 β turn.
198–207 α helix.
208–211 β turn.
212–227 β sheet.
228–231 β turn.
234–237 β turn.
238–245 α helix, $\langle P_\alpha \rangle = 1.19$, $\langle P_\beta \rangle = 1.06$.

XII. APPENDIXES

Appendix 1: C Language Source Code for Program PREDICT

```
#include <stdio.h>
#include <ctype.h>
#define MAXLENGTH 1000
struct protein_data {
     int code;
     int p_alpha;
     int p_beta;
     int p_turn;
     float bend[4];
     int alpha_class;
     int beta_class;
     int p4_alpha;
     int p4_beta;
     int p4_turn;
     float turn_prod;
} sequence[MAXLENGTH];


#include <protein.dat>

char infile[128], outfile[128], prot_name[64];
main()
{
    int c,length=1;
    char ans[6];
    FILE *fopen(), *fpi, *fpo;
    int fclose(FILE *fpo);
         printf("\n\n     Chou-Fasman-Prevelige Algorithm\n\n\n\n\n");
         printf("\n\n   (C) copyright 1988, Peter Prevelige, all rights reserved\n\n\n");
    do {
        printf("Enter name of input file:  ");
        scanf("%s",infile);
        if ((fpi = fopen(infile, "r")) == NULL)
          printf("No such file exists.\n");
        }while (fpi == NULL);

    do {
        printf("Enter name of output file:  ");
        scanf("%s",outfile);
        if ((fpo = fopen(outfile, "w")) == NULL)
          printf("Cannot open the file for output.\nWill output to the screen.\n");
        }while (fpi == NULL);

        printf("Enter name of protein:   ");
        scanf("%s",prot_name);


   while ((c=getc(fpi)) != EOF)
       if (!isspace(c)) sequence[length++].code = toupper(c);


   get_probability(length);

   tetra_ave(length) ;

   print_it(length,fpo);
```

```
    fclose(fpo);
}


char d_base[5];

get_probability(length)
    int length;
{
    int i,j,k,dbase;

    do{
        printf("Use database of 29 proteins (Chou & Fasman '78)---> enter 29\n");
        printf("Use database of 64 proteins    (Chou '79)---> enter 64\n");

        dbase = 2;

        scanf("%s",d_base);

        if(d_base[0]=='2')dbase=0;

        if(d_base[0]=='6')dbase=1;

    } while (dbase > 1);


    for (i=1,j=0;i<length;i++){
        while((sequence[i].code != data[j].c) && j < 20) j++;
        if (j == 20) {printf("Illegal data point #%d is%d\n",i,sequence[i].code);exit(1);};
        sequence[i].p_alpha = data[j].p_a[dbase];
        sequence[i].p_beta  = data[j].p_b[dbase];
        sequence[i].p_turn  = data[j].p_t;
        for(k=0;k<=3;k++) sequence[i].bend[k]  = data[j].b[k];
        sequence[i].alpha_class = data[j].a_class[dbase];
        sequence[i].beta_class  = data[j].b_class[dbase];
        j=0;
        }

}

#define TURN_PRODUCT 0.75E-4


tetra_ave(length)          /* calculates tetrapeptide averages of protein */
int length;
{
    int i=1, j=0;
    int asum, bsum, tsum;
    float tprod;

    for (i=1; i < length - 3; i++){
        asum=bsum=tsum=0;
        tprod=1;
        for (j=0; j<=3; j++){
          asum += sequence[i+j].p_alpha;
          bsum += sequence[i+j].p_beta;
          tsum += sequence[i+j].p_turn;
```

```
          tprod*= sequence[i+j].bend[j];
    }
        sequence[i].p4_alpha = asum/4;
        sequence[i].p4_beta  = bsum/4;
        sequence[i].p4_turn  = tsum/4;
        sequence[i].turn_prod= tprod;
    }
}
#define ALPHA_CUT 100
#define BETA_CUT 100
#define TURN_CUT 0.75e-4

print_it(length,fpo)

FILE *fpo;
int length;

{
        int i;
        char format, *forma, *forml, *form;

        forma="%4d %c  %3d %3d %3d  %c  %c   %3d %c  %3d %c  %3d  %.2e %c\n";
        forml="%4d \"%c\" %3d %3d %3d  \"%c\"  \"%c\"   %3d \"%c\"  %3d \"%c\"  %3d
%.2e \"%c\"\n";

        format=getchar();
        do {printf("ASCII format enter A, or LOTUS format enter L:\n " );
        format=toupper(getchar());}

        while ((format != 'A') && (format != 'L'));
                fprintf(fpo," Chou-Fasman-Algorithm\n\n");
                fprintf(fpo,"Input file: %s\n",infile);
                fprintf(fpo,"Protein name: %s\n",prot_name);
                fprintf(fpo,"Database used was %s proteins\n\n",d_base);
                fprintf(fpo,"       Pa  Pb  Pt   a  b    <Pa>   <Pb>   <Pt>  <pt>\n");
                fprintf(fpo,"       --------------------------------------------\n");

        if (format == 'L') form=forml;
        else form=forma;

        for(i=1;i<=length;i++)
                fprintf(fpo,form,i,
                    sequence[i].code,
                    sequence[i].p_alpha,
                    sequence[i].p_beta,
                    sequence[i].p_turn,
                    sequence[i].alpha_class,
                    sequence[i].beta_class,
                    sequence[i].p4_alpha, (sequence[i].p4_alpha >= ALPHA_CUT) ? '*' : ' ',
                    sequence[i].p4_beta, (sequence[i].p4_beta >= BETA_CUT) ? '*' : ' ',
                    sequence[i].p4_turn,
                    sequence[i].turn_prod, (sequence[i].turn_prod >= TURN_CUT) ? '*' : ' ')
                    ;
```

Appendix 2: Include File "Protein.Dat"

```
struct p_data {
     int c;
     int p_a[2];
     int p_b[2];
     int p_t;
     float b[4];
     int a_class[2];
     int b_class[2];
     int p4_a;
     int p4_b;
     int p4_t;
};
struct p_data data[] ={
     'A',142,139, 83, 79, 66,0.060,0.076,0.035,0.058,'H','H','i','i',0,0,0,
     'R', 98,100, 93, 94, 95,0.070,0.106,0.099,0.085,'i','h','i','i',0,0,0,
     'N', 67, 78, 89, 66,156,0.161,0.083,0.191,0.091,'b','i','i','b',0,0,0,
     'D',101,106, 54, 66,146,0.147,0.110,0.179,0.081,'I','h','B','b',0,0,0,
     'C', 70, 95,119,107,119,0.149,0.053,0.117,0.128,'i','i','h','h',0,0,0,
     'Q',111,112,110,100, 98,0.074,0.098,0.037,0.098,'h','h','h','I',0,0,0,
     'E',151,144, 37, 51, 74,0.056,0.060,0.077,0.064,'H','H','B','b',0,0,0,
     'G', 57, 64, 75, 87,156,0.102,0.085,0.190,0.152,'B','B','b','i',0,0,0,
     'H',100,112, 87, 83, 95,0.140,0.047,0.093,0.054,'I','h','i','i',0,0,0,
     'I',108, 99,160,157, 47,0.043,0.034,0.013,0.056,'h','i','H','H',0,0,0,
     'L',121,130,130,117, 59,0.061,0.025,0.036,0.070,'H','H','H','h',0,0,0,
     'K',116,121, 74, 73,101,0.055,0.115,0.072,0.095,'h','h','b','b',0,0,0,
     'M',145,132,105,101, 60,0.068,0.082,0.014,0.055,'H','H','h','I',0,0,0,
     'F',113,111,138,123, 60,0.059,0.041,0.065,0.065,'h','h','h','h',0,0,0,
     'P', 57, 55, 55, 62,152,0.102,0.301,0.034,0.068,'B','B','B','B',0,0,0,
     'S', 77, 72, 75, 94,143,0.120,0.139,0.125,0.106,'i','b','b','i',0,0,0,
     'T', 83, 78,119,133, 96,0.086,0.108,0.065,0.079,'i','i','h','h',0,0,0,
     'W',108,103,137,124, 96,0.077,0.013,0.064,0.167,'h','I','h','h',0,0,0,
     'Y', 69, 73,147,131,114,0.082,0.065,0.114,0.125,'b','b','H','h',0,0,0,
     'V',106, 97,170,164, 50,0.062,0.048,0.028,0.053,'h','i','H','H',0,0,0
};
```

XIII. REFERENCES

Chou, P. Y., and Fasman, G. D., 1974a, Conformational parameters for amino acids in helical, β-sheet and random coil, regions calculated from proteins, *Biochemistry* **13:**211–222.

Chou, P. Y., and Fasman, G. D., 1974b, Prediction of protein conformation, *Biochemistry* **13:**222–245.

Chou, P. Y., and Fasman, G. D., 1977, β-Turns in proteins, *J. Mol. Biol.* **115:**135–175.

Chou, P. Y., and Fasman, G. D., 1978, Prediction of the secondary structure of proteins from their amino acid sequence, *Adv. Enzymol.* **47:**45–148.

Chou, P. Y., and Fasman, G. D., 1979, Prediction of β-turns, *Biophys. J.* **26:**367–384.

Fasman, G. D., 1985, A critique of the utility of the prediction of protein secondary structure, *J. Biosci.* **8:**15–23.

Huber, R., Kulka, D., Rühlmann, A., and Steigmann, W., 1971, Pancreatic trypsin inhibitor (Kunitz). Part I: Structure and function, *Cold Spring Harbor Symp. Quant. Biol.* **36:**141–150.

10

The GOR Method for Predicting Secondary Structures in Proteins

J. Garnier and B. Robson

J. Garnier • Laboratory of Physical Biochemistry, University of Paris-Sud, 91405 Orsay, France. *B. Robson* • Proteus Biotechnology Ltd., Marple, Cheshire SK6 6AB, England; and Epistron Peptide and Protein Engineering Research Unit, University of Manchester, Manchester M13 9PT, England. *Present address for JG:* Protein Engineering Unit, INRA, 78350 Jouy-en-Josas, France.

I. DEFINITION AND DATA BASES FOR THE CONFORMATIONAL STATES OF AN AMINO ACID RESIDUE

The widely used term "secondary structure" implies that it is of value to consider the structure of a protein as organized hierarchically. "Hierarchic" relates to the idea that the structure can be considered on at least two levels; there are, in fact, three levels of interest here, namely, the covalent structure (primary), the structural organization of stereoregular regions as specific backbone conformations (secondary), and the way in which they are assembled in a three-dimensional conformation (tertiary structure) to make a protein. The recently introduced concept of "supersecondary" structure intermediate to secondary and tertiary levels describes the interactions between secondary structures in space. It should also be stated that the more recently recommended definition of secondary structure covers all backbone conformations, stereoirregular as well as stereoregular.

The term secondary structure was rapidly favored by biochemists for several reasons. The most important secondary structures—the α helix (H) and the β pleated sheet (E, for extended chain of the sheet)—were first found theoretically to be most energetically stable through the pioneering work of Pauling and Corey and were shown to be consistent with the data obtained by Astbury from the x-ray diffraction patterns of hair keratin and silk fibroin. Biochemists have long been able to measure the approximate content of these structures by ORD and CD spectroscopy in protein solutions. In contrast, techniques such as viscometry, centrifugation, or light scattering gave information about the overall shape of the protein, and information of this kind was said to relate to the tertiary structure.

The accumulation of data from x-ray diffraction on protein crystals later revealed other types of recurrent folds of the polypeptide backbone, particularly reverse or β turns (T). Abundant regions of the backbone conformation with lack of stereoregularity were named aperiodic or more often if less precisely coil structure (C). Nonetheless, the first recognized secondary structures, α helix and β pleated sheet, are well represented in known protein structures. In a recent tabulation of 12,433 amino acid residues from the dictionary of proteins of known conformation edited by Kabsch and Sander (1983a), more than half of them are in a stereoregular conformation, with 30.4% in α helix, 19.9% in β sheet, 12.3% in β turns, and 37.4% in coil. The β turns may show periodic or aperiodic character and may represent turns of distorted or "nonclassical" helix.

These abundances justify the various attempts made to predict as exactly as possible the secondary structures. Another reason relates to the fact that α helix and β sheet have a well-defined stereoregularity, although as hinted above there are many examples from x-ray data of some distortion. This allows their convenient use as a good starting conformation for model building and energy-minimization procedures. Model building will help to identify supersecondary structures from the series of secondary structures. Supersecondary structure is a level of description between secondary and tertiary structures. It includes, for instance, the sequence β sheet, α helix, β sheet, α helix, β sheet (*babab*), or the Rossman fold, known to be a binding region for nucleotides. Here *a* stands for a sequence of H, and *b* for a sequence of E (Richardson, 1981; Robson and Garnier, 1986), which is fairly standard nomenclature for describing sequence segments as opposed to residues.

A. Criteria for Observed Conformational States

Methods for prediction of secondary structures rely on a data set of observed conformations. These data permit workers to develop statistical analyses to calculate prediction parameters or to establish prediction rules depending on the method. They also serve as the "experimental result" with which to test the quality of the prediction method. It is pertinent to mention, however, that the protein for which prediction is performed should be removed from the data base used to establish the rules and parameters (Levin *et al.*, 1986; Gibrat *et al.*, 1987).

Although two-dimensional NMR spectroscopy is entering the scene of protein structure determination, the existing data bases come from x-ray diffraction studies. These allow assignment of a set of coordinates for each atom of the protein, with the exception of hydrogen atoms, from an electron-density map. This is with the reservation that neutron diffraction has been applied in a very few cases and can determine hydrogen positions. Accuracy depends on various factors: the quality of the crystals and of their isomorphous replacements, the number of diffracted spots taken into account or resolution, and the rigidity of the polypeptide chain. Outside loops and N and C termini are typically flexible to an extent that depends on temperature. From the crystallographic point of view, the flexibility is measured by the *B* factor. By and large the coordinates of a protein atom may be determined within 0.5 Å.

These experimental limitations imply that not all published x-ray structures are suitable for a reliable data base of secondary structures. This data base has been extended gradually over the years (recently by Robson and Suzuki, 1976; Levitt and Greer, 1977; Kabsch and Sander, 1983a). In every case the authors have limited their data base to the best-refined x-ray structures available.

Any practical data base implies use of representations that are tidied, refined, or more drastically processed. The precise nature and content of a data base thus depend on other factors than simply the proteins it represents. In particular, secondary structure conformational state assignments will vary with the criteria used to recognize the secondary structure. First, calculated values of ϕ and ψ angles may be used. They are defined as the dihedral angles between the bonded atoms C′-N-C-C′ and N-C-C′-N, respectively, for each residue. Alternatively, the hydrogen bond network connecting the residues is used to identify the secondary structures. In practice, a balanced mixture of the two types of criteria may be applied with the inclusion for some purposes of the torsion angle defined by the path of four consecutive C_α atoms of the backbone. In all cases, there is also the question of defining how distorted a helix or other state need be before it ceases to be definable as that state.

The use of different refined x-ray structures and different criteria of secondary structure explains the discrepancies between authors in defining secondary structure in the same protein. Bovine pancreatic trypsin inhibitor (BPTI) is an example in which 12 residues out of 58 are differently assigned by Robson and Suzuki (1976) and Levitt and Greer (1977). Often the discrepancies occur at the ends of the secondary structure. In this instance, Levitt and Greer may define extra helical residues that are "not helical," at least from the viewpoint of ϕ,ψ angles. It is interesting to know how sensitive a prediction method will be to these different assignments. With BPTI, and using the method of Garnier *et al.* (1978) or the GOR method, Garnier and Robson (1979) found 62% of correctly predicted residues for the four states H, E, T, and C with their own conformational assignment and only 53% with the assignments of Levitt and Greer (1977). Bovine PTI is admittedly an extreme case, since in a set of 23 proteins, 56.4% of amino acid residues were correctly predicted in four states with the data base of Robson and Suzuki (1976) and 55% correct with the data base of Levitt and Greer (1977). The uncertainty in accuracy of prediction arising from the interpretation of the data base for a same set of proteins is here of the order of 1.5% or, in relative value, 2.5% of the total number of the correctly predicted residues. It is perfectly reasonable that a data base used to test predictions should yield better results for a prediction method based on the same basic definitions of secondary structure as are employed in the data base. The variation in the results obtained against data bases with different reasonable definitions is an important question of "robustness," however.

B. An Extended Set of Conformational States

The definition of only four states—H, E, T, and C—to describe the conformational state of a residue represents an oversimplification of the real situation in protein structure. Several attempts have been made to define other conformational states, which largely implies a higher resolution and a more detailed description.

Kabsch and Sander (1983a) define eight states according the hydrogen-bond pattern. First, a hydrogen bond is considered to be established when the electrostatic interaction energy between two hydrogen-bonding groups is less than −0.5 kcal/mole. The value of this energy depends on the distance *d* between the acceptor and the donor group (i.e., O of C═O and N of NH) and the angle θ between the N—H bond and O═C bond. An ideal value of the energy of the hydrogen bond is −3.0 kcal/mole for a distance of 2.9 Å and $\theta = 0$, and its energy is zero for a distance of 5 Å and also $\theta = 0$. An elementary hydrogen bond pattern or *n*-turn at residue *i* is defined as a pattern of $n + 1$ consecutive residues with hydrogen bond from C═O (*i*) to N—H ($i + n$) with $n = 3$, 4, or 5. A turn (T) is an isolated *n*-turn. Two or more consecutive n-turns are helix, namely, a 3_{10} helix for $n = 3$ (G), an α helix for $n = 4$ (H), and a π helix for $n = 5$ (I).

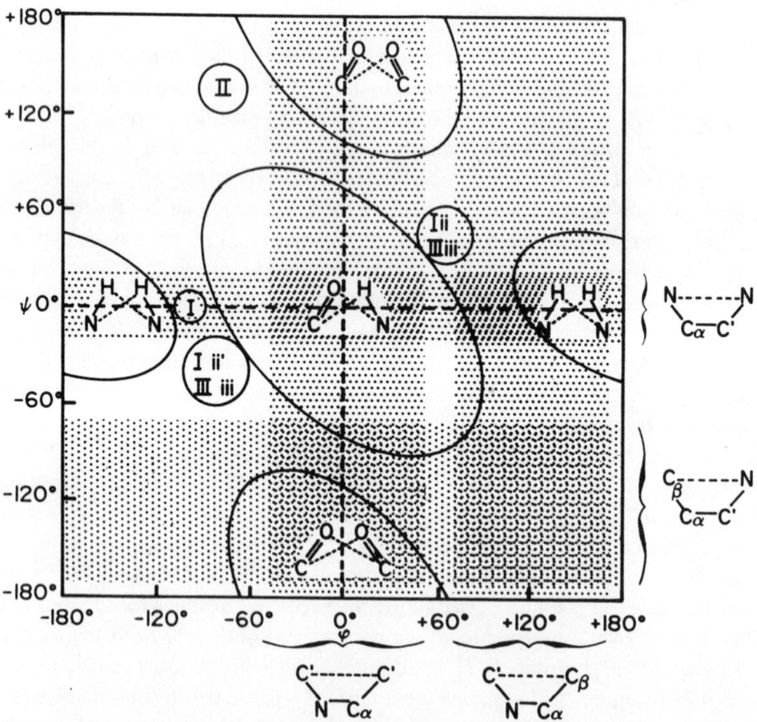

Figure 1. Steric clashes between atom pairs for N-acetylalanyl-N′-methylamide. The gray areas and the interiors of the ellipses, are "sterically forbidden" by at least one clash. Such a diagram is at the origin of the "Ramachandran plot." Interactions are represented by dotted lines. The circles are allowed regions favored by β turns. Reprinted by permission from Robson and Garnier (1986).

Another elementary hydrogen-bond pattern is the bridge where two non-overlapping stretches of three residues $i-1$, i, $i+1$ and $j-1$, j, $j+1$ form either a parallel or an antiparallel bridge with two hydrogen bonds characteristic of a β pleated sheet structure. Isolated they are named B at i and j, and a set of consecutive bridges of identical type are named extended strands (E).

Regions with high curvature or bends (S) are defined by a curvature of at least 70° at residue i when the vector joining the carbon atoms $C_{\alpha i}$ to $C_{\alpha i-2}$ makes an angle greater than 70° with the vector joining $C_{\alpha i+2}$ to $C_{\alpha i}$. Residues not classified in any of these categories are considered as coil (C).

It has been observed (J. Levin, J. A. Jenkins, & J. Garnier, unpublished data) that the Kabsch and Sander (1983a) algorithm is very sensitive to the coordinates of the main-chain atoms. We are currently retaining in our data set only proteins with resolution better than 2.8 Å with a crystallographic R factor less than 0.25 for correct assignments of observed states from the Brookhaven data bank (Bernstein *et al.*, 1977).

Another important classification of conformational states for a residue is to consider the various regions of the "Ramachandran map" of ϕ and ψ angles, which are related to the energy map of interactions of individual amino acid side chains (Fig. 1). A residue would be assigned a specific state, say helical, if its observed ϕ and ψ angles are within a defined range from a standard value of −55° and −55°, respectively; 40° is a fairly typical range despite a preference for 20° expressed by some crystallographers.

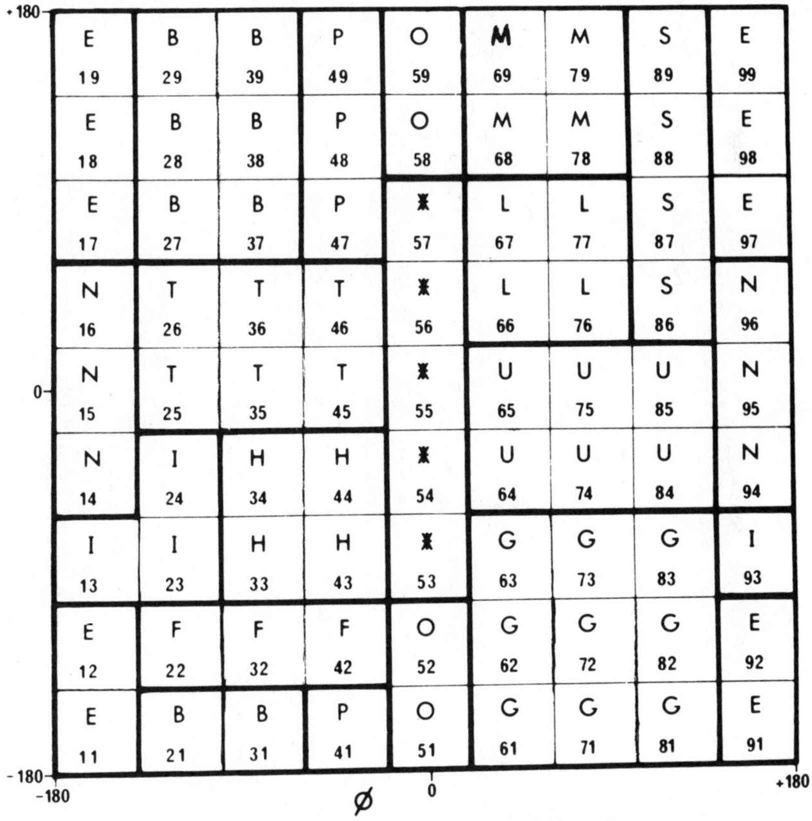

ψ \ ϕ (−180 … 0 … +180)								
E 19	B 29	B 39	P 49	O 59	M 69	M 79	S 89	E 99
E 18	B 28	B 38	P 48	O 58	M 68	M 78	S 88	E 98
E 17	B 27	B 37	P 47	* 57	L 67	L 77	S 87	E 97
N 16	T 26	T 36	T 46	* 56	L 66	L 76	S 86	N 96
N 15	T 25	T 35	T 45	* 55	U 65	U 75	U 85	N 95
N 14	I 24	H 34	H 44	* 54	U 64	U 74	U 84	N 94
I 13	I 23	H 33	H 43	* 53	G 63	G 73	G 83	I 93
E 12	F 22	F 32	F 42	O 52	G 62	G 72	G 82	E 92
E 11	B 21	B 31	P 41	O 51	G 61	G 71	G 81	E 91

ψ axis: +180 (top), 0, −180 (bottom); ϕ axis: −180, 0, +180

Figure 2. Definition of conformational states in terms of the ranges of backbone dihedral angles ϕ and ψ. For definitions of E, B, P, H, L, and T, see text. Only glycine residues tend to be found in the G region. N, O, *, and S are poorly populated. Other codes are arbitrary. Reprinted by permission from Robson and Garnier (1986).

Table I. Standard Values of Dihedral Angles for Secondary Structures in Globular Proteins[a]

Type[b]	Occurrence	φ	ψ	Rise per hydrogen bond: Residue (Å) (projection on helix axis)	Rise per hydrogen bond: Between residues i and $i + m$
3, 6_{13} Right-handed α helix	Globular and α-fibrous proteins	−55 ± 20	−55 ± 20	1.5	$CO_i \ldots NH_{i+4}$
Left handed α helix	Polyglycine short sections in globular proteins rich in Gly, Asn	+55 ± 20	+55 ± 20	1.5	$CO_i \ldots NH_{i+4}$
3_{10} Three-ten helix	Short sections in some globular proteins	−55 ± 20	−15 ± 20	2.0	$CO_i \ldots NH_{i+3}$
2_7 (eq) Ribbon with side-chain of residue equatorial to 7 atom hydrogen-bonded ring	Short sections in some globular proteins	−75 ± 20	+70 ± 20	2.8	$CO_i \ldots NH_{i+2}$
-3_x Right-handed collagen-type helix	Collagen, polyproline, short sections in few globular proteins (avian pancreatic polypeptide)	−65 ± 10	+145 ± 30	3 ± 0.1	Three cross-hydrogen-bonded strands in polyproline, collagen
$-2_{5,x}$ β-Sheet strand	β-Fibrous proteins, globular proteins	−120 ± 45	+130 ± 30	3.3	Weak hydrogen bond within strand; strong hydrogen bond to the other strands

Type[c]	Occurrence	φ	ψ	$C_{\alpha(i-1)} \ldots C_{\alpha(i+2)}$ (Å)
		Aperiodic secondary structure		
Type III turn (3_{10}-like)	Very common as distorted α helix, distorted 3_{10} helix Occasionaly used to invert β strand to form antiparallel β sheet Also often in coil regions	$\varphi_i = -75 \pm 65$ $\varphi_{i+1} = -60 \pm 40$	$\psi_i = -30 \pm 40$ $\psi_{i+1} = -15$ to -70	<6
Type I turn	As above	$\varphi_i = -75 \pm 65$ $\varphi_{i+1} = -90 \pm 40$	$\psi_i = -30 \pm 40$ $\psi_{i+1} = -15$ to $+40$	<6
Type II turn	Less common; inverted β strand, occurs in 'coil' regions Second residue usually glycine	$\varphi_i = -60 \pm 40$ $\varphi_{i+1} = 90 \pm 40$	$\psi_i = 120 \pm 40$ $\psi_{i+1} = 0 \pm 40$	<6
Type IV turn	Miscellaneous class of general occurrence	φ, ψ outside above ranges		<6
Type V kink	Occasional in globular proteins	$\varphi_i = -80$[d] $\varphi_{i+1} = +80$[d]	$\psi_i = +80$[d] $\psi_{i+1} = -80$[d]	<7
Type VI	Well characterized in Bence Jones protein	Involves *cis*-proline in residue $i + 1$		<7
Type VIIa kink	Occasional in globular proteins	$\varphi_{i+1} = +180 \pm 30$	$\psi_{i+1} = +30 \pm 30$	<7
Type VIIb kink	Occasional in globular proteins	$\varphi_{i+1} = +30 \pm 30$	$\psi_{i+1} = +180 \pm 30$	<7
Type X, or coil	Common in globular proteins	Any other value of φ and ψ		>7

[a]Reprinted by permission from Robson and Garnier (1986).
[b]Type is expressed as n_a, where n is the number of residues per turn, and a is the number of atoms in a hydrogen-bonded ring; x denotes a hydrogen bond between strands.
[c]*Note:* turns and kinds I′, II′, III′, IV′, V′ are also defined as approximately mirror images of I, II, III, IV, V, i.e., change signs of φ and ψ. Less common, except in tight hairpin loops of antiparallel β sheet.
[d]Wide variation, type not well defined.

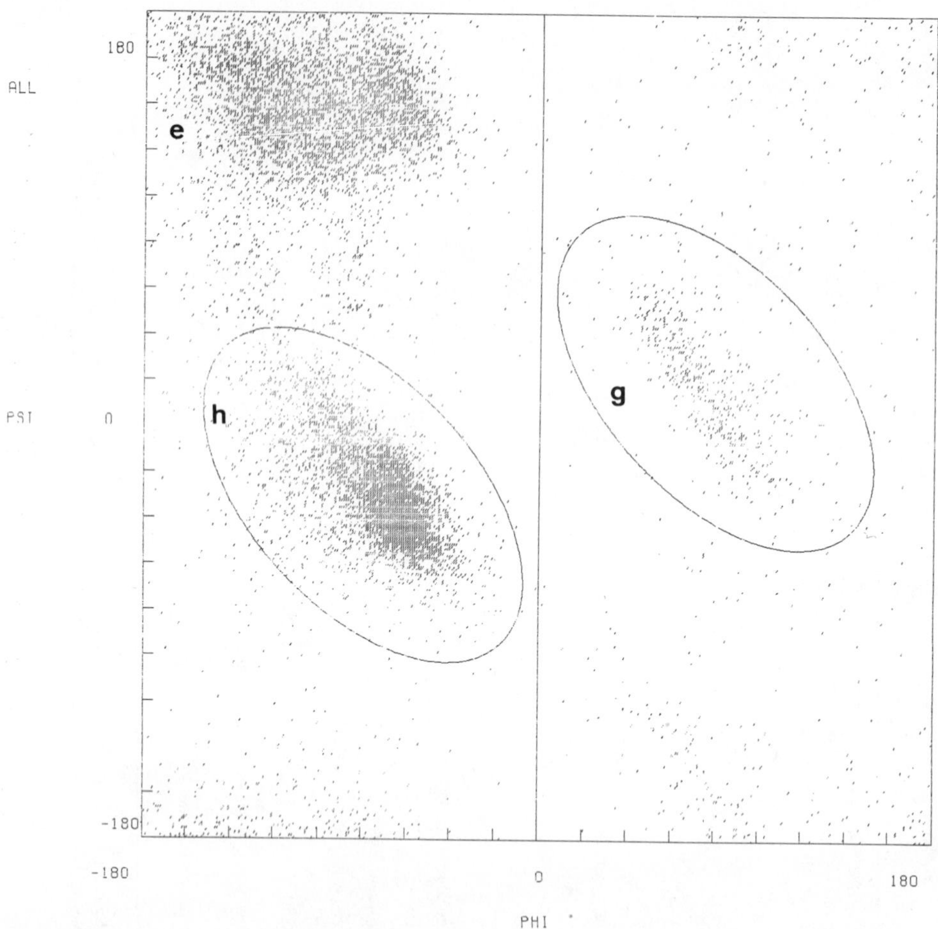

Figure 3. Ramachandran plot of observed φ and ψ angles of 11,237 amino acid residues of 68 proteins with refined x-ray coordinates from the Brookhaven data bank. The h region is defined by an ellipse centered at (−80°,−30°) of long axis $a = 90°$ and short axis $b = 50°$. The g region is the same ellipse symmetrical around the point (0°, 0°), and the e region is defined by (φ < 0°, ψ > 60° or <−140°) or (φ > 140°, ψ > 140°) or (φ > 140°, ψ < −140°).

Figure 2 exemplifies the way in which conformational states can be defined in terms of the ranges of backbone dihedral angles φ and ψ. The small square cells are 40° × 40°, and E represents extended chain, B β pleated sheet, P a proline-rich region, H the region characteristic of residues in the right-handed helical conformation, L the rather rare left-handed α helix, T the turn conformation, and so on. All the square cells can then define 81 different states. A compressed data base of format typified by ...A34E43G28..., where the digits relate to the corresponding 40° × 40° cell in Fig. 2, is given by Robson and Garnier (1986), Appendix 2. Standard values of φ,ψ angles for secondary structures are given in Table I.

All these states are not equally populated. Figure 3 represents a recent compilation made by J. F. Gibrat and J. Garnier (unpublished data). Among 11,237 amino acid residues, 90.7% are confined to the e and h regions, and 3.3% in the g region where Gly and Asn residues are mainly found. The Kabsch and Sander (1983a) algorithm identifies 38% of the residues as coil, or 51% if one includes the turns. This implies that those regions defined as coil structure include regions of unordered sequences of regular α helix and β sheet or slightly distorted α-helix and β-sheet structures.

C. Evaluation of the Quality of a Prediction

This question is of importance for users of prediction methods. However, there are many personalized evaluations that, even for a specialist in this field, make it difficult to appreciate the quality of predictions of protein secondary structure (Kabsch and Sander, 1983b).

The simplest and best way to evaluate prediction power is the fraction of residues correctly assigned, f

$$f = (\Sigma_S F_S^{+})/(\Sigma_S F_S) = \Sigma_S F_S^{+}/N \tag{1}$$

where F_S is the number of residues observed in a conformational state S, and its sum over all states S is equal to the total number of residues, N. F_S^{+} is the number of residues that are both predicted and observed in the state S. The fraction f is conveniently transformed on a percentage basis ($f \times 100$). If a method predicts only the helical conformational state, this prediction method is in fact a two-state prediction method, predicting helix and, by default, not helix ($\bar{\mathrm{H}}$). One should compute F_{H}^{+} and $F_{\bar{\mathrm{H}}}^{+}$ to calculate f from all the residues N. Such a method should be judged not only on its ability to predict residues in helix but also on its ability not to predict helix for residues not in helix, avoiding so-called overprediction. As an example, suppose a simple predictive rule predicts each residue to be H and the protein contains 30% of its residues in H conformation and 70% in the $\bar{\mathrm{H}}$ conformation. In this case $f = 0.30$ or 30%, but some authors have used instead the ratio of $F_{\mathrm{H}}^{+}/F_{\mathrm{H}}$. In the present example this ratio is equal to 1 or 100%, meaning that 100% of the helical residues are correctly predicted. But the use of this ratio does not take into account that the $\bar{\mathrm{H}}$ conformation was badly predicted (0%). Nonetheless, the ratio F_S^{+}/F_S is of some interest since it detects how well a specific secondary structure is predicted.

Equation 1 takes overprediction into consideration. Another equation has been used by Chou and Fasman (1974) and Matthews (1975).

$$Q_S = \tfrac{1}{2}\,[(F_S^{+}/F_S) + (F_{\bar{S}}^{+}/F_{\bar{S}})] \tag{2}$$

Except in the limiting case where $F_S = F_{\bar{S}}$, the value of Q_S differs from f. In the example of the simple rule above, "all is helix," $Q_{\mathrm{H}} = 0.5$ or 50% when $f = 0.3$. In the evaluation made by Matthews (1975) on T_4 phage lysozyme for the prediction of helical residues with the Chou and Fasman method (1974), $Q_{\mathrm{H}} = 70\%$ when f_{H} ($\times$ 100) is 67.7% (two states) and for extended conformation $Q_{\mathrm{E}} = 58\%$ when $f_{\mathrm{E}} \times 100 = 71.3\%$. Not only do Q values differ from f values, they vary here in opposite directions.

In a purely random assignment, the expected value of f (equation 1) for n states is $f = 1/n$; that is, for a random-assignment two-state prediction, $f = 0.50$, and for four states, $f = 0.25$ assuming an equal partition of the residues in each state. Comparison of prediction methods should be done on the basis of the same number of predicted states. A method that correctly predicts 50% of the residues for two states does not do better than random and is much less effective than a prediction method that correctly predicts the same percentage of residues, 50%, for four states.

Matthews (1975) has also proposed the use of a correlation coefficient C. For the state H:

$$C = (F_{\mathrm{H}}^{+}/N - F_{\mathrm{H}}\bar{P}/N^2)/[(F_{\mathrm{H}}\bar{P}/N^2)(1 - F_{\mathrm{H}}/N)(1 - \bar{P}/N)]^{1/2} \tag{3}$$

where $\bar{P}$ stands for $\bar{P} = F^{+}_{H} + (F_{\bar{H}} - F^{+}_{\bar{H}})$, i.e., the number of residues predicted as H. For a perfect prediction, $F^{+}_{H} = F_{H}$, $F^{+}_{\bar{H}} = F_{\bar{H}}$, and $C = 1$.

Garnier *et al.* (1978) proposed the use of an accuracy index:

$$\alpha = \Sigma_S(F^{+}_{S}/F_S) - 1 \tag{4}$$

This avoids the production of a false impression of predictive power when states are not equally distributed. The information measure log $(1 + \alpha)$ can be shown to have a value of log n if all residues are correctly predicted in the n different states and a value of 0 with $\alpha = 0$ if a purely random assignment is made.

Other evaluations have been used, such as a weighted evaluation when two predicted states are considered possible (Busetta and Hospital, 1982). Recently, Cohen *et al.* (1986) evaluated the correctness of their prediction of turns defined as $\bar{H}$ or $\bar{E}$ if they were predicted not more than three residues from either end of an observed secondary structure element (H or E).

Kabsch and Sander (1983b) proposed, in addition to f from equation 1, a quality index P defined as

$$P = F^{+}_{S}/(F^{+}_{S} + F_{\bar{S}} - F^{+}_{\bar{S}}) \tag{5}$$

or the ratio of correctly predicted residues in a given state S to the total number of residues predicted in that state. This ratio estimates the chance one has to be correct in assigning a residue in a state S from a given number of residues predicted in that state.

Although any evaluation method has its own logic and value, we recommend the use of at least the value of f (or $f \times 100$) on all predicted states, i.e., for the total number of residues in the protein or in the data base, as did Kabsch and Sander (1983b) in evaluating the prediction methods of Lim (1974), Chou and Fasman (1974), and Garnier *et al.* (1978).

II. INFORMATION THEORY AS A BASIS OF THE GOR METHOD

Information theory has been developed to handle very practical problems such as measurement of the amount of information communicated by a system of telegraphic signals or comparison of efficiency of coding devices (Brillouin, 1956). The theory is based on probability considerations and has found wide applications: telecommunication, computing, and pure physics. Pain and Robson (1970) and Robson and Pain (1971), in their pioneering work, recognized that a string of amino acids, R, of a polypeptide chain could be considered a message that was translated by the folding mechanism into another message of a string of conformational states, S. This work considered for the first time quantitatively the tendency of all types of residue, individually and as part of a pattern, to adopt different conformations. They used, to evaluate the information value, the relationships discussed by Fano (1961) and in a series of subsequent papers (Robson, 1974; Robson and Suzuki, 1976; Garnier *et al.*, 1978), this led to the GOR prediction method.

A. Principles of Information Theory

Consider two events, x and y, where x will be the conformational state S of an amino acid residue and y a specific amino acid residue of type R in a sequence of amino acids $R_1,\ldots,R_j,\ldots,R_N$.

The conditional probability $P(x|y)$ is the probability of occurrence of x knowing that y has happened. $P(x)$ is the probability of occurrence of x. The information or the statistical constraint of y on the occurrence of x is defined as

$$I(x;y) = \log\,[P(x|y)P(x)] \tag{6}$$

If y has no effect on the occurrence of the event x, then $P(x|y) = P(x)$, and $I(x;y) = 0$. If y favors the occurrence of x, then $P(x|y) > P(x)$, and $I(x;y) > 0$. Conversely, if y prevents x from occurring, $P(x|y) < P(x)$, and $I(x;y) < 0$.

The use of a logarithm is justified by the fact that one wants the information to have an additive property (see the following equations). Usually the natural logarithm is used, and the unit of information is the natural unit or nat, with subdivisions such as the hundredth of nat, i.e., centinat or cnat. In computer work, the binary unit or bit is used for a log to base 2, but the natural log confers useful properties when this approach is extended to finite data and the use of the "hash" function (see below). In trying to express the information from equation 6, one does not seek to understand how the event y modifies the occurrence of x, but one seeks to know how much it does. The mechanism of interaction between y and x is a different interesting problem, about which information theory tells nothing directly, but it helps us to determine which interactions are more important than others. This is one step toward an understanding of the mechanism.

If the event y can be decomposed into two events, y_1 and y_2, the information that y brings on the occurrence of x is equal to the information that y_1 and y_2 jointly bring to the event x:

$$I(x;y) = I(x;y_1,y_2) = \log\,[P(x|y_1,y_2)/P(x)] \tag{7}$$

Using the property of a logarithm function,

$$I(x;y_1,y_2) = \log\,[P(x|y_1,y_2)/P(x|y_1)] + \log\,[P(x|y_1)/P(x)] \tag{8}$$

The first term of the right-hand side of equation 8 expresses the effect of y_2 on the event x knowing that y_1 has occurred but without considering the contribution of y_1 to x. The second term expresses the influence of y_1 on the event x. Then:

$$I(x;y_1,\, y_2) = I(x;y_2|y_1) + I(x;y_1) \tag{9}$$

This relationship demonstrates that it is always possible to decompose the information brought by a complex event into a sum of information brought about by more simple events. By analogy with conditional probability, $I(x;y_2|y_1)$ is referred to as conditional information. Equation 9 can be extended to any number n of events y_i:

$$\begin{aligned} I(x;y_1,y_2,\ldots,y_n) = I(x;y_1) + I(x;y_2|y_1) + I(x;y_3|y_1,y_2) + \ldots \\ + I(x;y_n|y_1,y_2,\ldots,y_{n-1}) \end{aligned} \tag{10}$$

A Venn diagram of information decomposition for four nonindependent events is presented in Fig. 4, where each surface represents an element of information with the possible overlaps expressing the dependence between the various events and the need to avoid counting the same information (overlapping regions) twice by using conditional information.

However, even if y_1 and y_2 are independent events,

$$I(x;y_2|y_1) \neq I(x;y_2) \tag{11}$$

If one considers that the event x, which is the conformational state S, can occur in only two ways, for instance, helical (H) or nonhelical ($\bar{H}$) or more generally X or $\bar{X}$, then one defines the relationships

$$I(S = X{:}\bar{X};y) = I(S = X;y) - I(S = \bar{X};y) \tag{12}$$

where y can be a complex event such as the whole amino acid sequence. Equation 12 quantifies the preference of y for the occurrence of the conformational state X as opposed to the occurrence of state $\bar{X}$. Then:

$$I(S = X{:}\bar{X};y) = \log\,[P(S = X|y)/P(S = X)] - \log\,[P(S = \bar{X}|y)/P(S = \bar{X})] \tag{13}$$

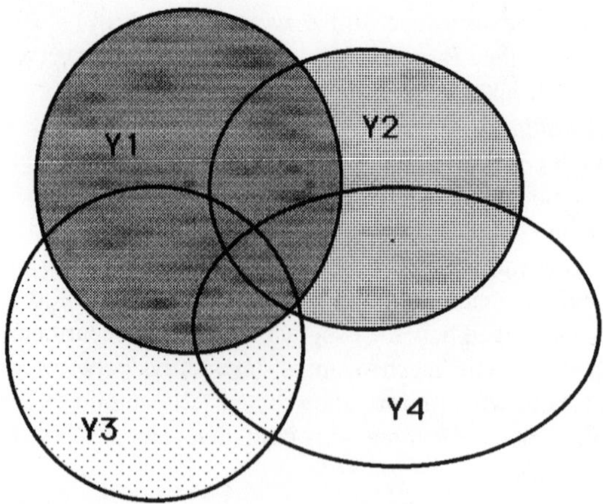

Figure 4. A Venn diagram representing the expansion of the information that the complex event y_1, y_2, y_3, and y_4 or $I(x; y_1, y_2, y_3, y_4)$ carries about the event x. $I(x;y_1)$, hatched area; $I(x;y_2|y_1)$, cross-hatched area; $I(x;y_3|y_1y_2)$, stippled area; $I(x;y_4|y_1y_2y_3)$, open area; and $I(x;y_1,y_2,y_3,y_4) = I(x;y_1) + I(x;y_2|y_1) + I(x;y_3|y_1y_2) + I(x;y_4|y_1y_2y_3)$. Reprinted by permission from Gibrat *et al.* (1987).

$$I(S = X{:}\bar{X};y) = \log\,[P(S = X|y)/P(S = \bar{X}|y)] - \log\,[P(S = \bar{X})/P(S = \bar{X})] \quad (14)$$

Since X and $\bar{X}$ are exhaustive and mutually exclusive, $P(S = \bar{X}|y) = 1 - P(S = X|y)$, and $P(S = \bar{X}) = 1 - P(S = X)$. If we define, as did Robson (1974),

$$*(S = X{:}\bar{X};y) = \log\,\{P(S = X|y)/[1 - P(S = X|y)]\} \quad (15)$$

$$*(S = X{:}\bar{X}) = \log\,\{P(S = X)/[1 - P(S = X)]\} \quad (16)$$

Equation 14 becomes

$$I(S = X{:}\bar{X};y) = *(S = X{:}\bar{X};y) - *(S = X{:}\bar{X}) \quad (17)$$

Suppose a data base includes 1830 amino acid residues, of which 390 are alanine residues, and that 240 of them are observed in helix and 150 are observed in nonhelix. In this data base 780 amino acid residues are observed in helix and the other 1050 residues are not in helix.

Assuming that there are enough observed frequencies to allow estimation of probability values from relative frequencies, and remembering that the conditional probability $P(S|y) = P(S,y)/P(y)$, then $P(\mathrm{H}|\mathrm{A}) = P(\mathrm{H,A})/P(\mathrm{A}) = (240/1830)/(390/1830) = 0.615$, where $S = \mathrm{H}$ for helical conformation and $y = \mathrm{A}$ for alanine. In addition, $*(S = \mathrm{H{:}\bar{H};A}) = \log\,(0.615/0.385) = 0.4684$, and $*(S = \mathrm{H{:}\bar{H}}) = \log\,(780/1830)/(1050/1830) = -0.2972$. When these results are put into equation 17, $I(S = \mathrm{H{:}\bar{H};A}) = 0.4684 + 0.2972 = 0.7656$ nat. The positive value indicates that alanine prefers to be helical rather than nonhelical with an information level of 76.56 cnat.

The real situation is more complex. The conformation of an amino acid residue will depend on the surrounding amino acids in the sequence, and the probability values are not always equal to the observed relative frequencies if their number is too small.

The left-hand side of equation 17 can be expended to the complete amino acid sequence

$$I(S_j = X{:}\bar{X};R_1,R_2,\ldots,R_n) \quad (18)$$

where S_j is the conformational state at the position j in the sequence of residues R_1, $R_2,\ldots,R_n$.

Some simplifications have to be made to develop equation 18. One consists of considering the local sequence $\pm m$ residues away from position j; typically $m = 8$. Then one can develop equation 18 using this approximation and equation 10:

$$I(S_j = X{:}\bar{X};R_1,\ldots,R_n) \simeq I(S_j = X{:}\bar{X};R_{j-m},\ldots,R_j,\ldots,R_{j+m}) \tag{19}$$

and

$$\begin{aligned} I(S_j = X{:}\bar{X};R_{j-m},\ldots,R_j,\ldots,R_{j+m}) = {} & I(S_j = X{:}\bar{X};R_j) \\ & + I(S_j = X{:}\bar{X};R_{j-1}|R_j) \\ & + I(S_j = X{:}\bar{X};R_{j+1}|R_j,R_{j-1}) \\ & + \ldots \\ & + I(S_j = X{:}\bar{X};R_{j+m}|R_{j-m},\ldots,R_j,\ldots,R_{j+m-1}) \end{aligned} \tag{20}$$

On the right-hand side of equation 20, there are $(2m + 1)$ terms: the first term is the single-residue information; the second term involves a pair of amino acids or pair information. The third term involves a triplet of amino acids, and so on.

At this stage one should consider that for four different conformational states with 20 amino acids, i.e., 400 different pairs of amino acids, the data base for pair information will have 1600 entries, whereas simple information will have only $4 \times 20 = 80$ entries. The present available data base for Kabsch and Sander (1983a) and C. Sander (personal communication) containing 12,757 residues corresponds to an average of 159 observed frequencies for single-residue information and only eight observed frequencies for pair information. Clearly triplet informations cannot be taken into consideration with the present data base.

In fact, two approximations to equation 20 have been made:

$$I(S_j = X{:}\bar{X};R_{j-m},\ldots,R_j,\ldots,R_{j+m}) \simeq \Sigma_m I(S_j = X{:}\bar{X};R_{j+m}) \tag{21}$$

and

$$I(S_j = X{:}\bar{X};R_{j-m},\ldots,R_j,\ldots,R_{j+m}) \simeq I(S_j = X{:}\bar{X};R_j) + \Sigma_{m,m \neq 0} I(S_j = X{:}\bar{X};R_{j+m}|R_j) \tag{22}$$

The approximation in equation 21, proposed by Robson and Suzuki (1976), is named directional information and was used by Garnier *et al.* (1978) to establish the GOR method. It is the information that a residue at the position $j + m$ (with $m \leqslant 8$) brings to bear on the conformational state of a residue at position j whatever this residue is.

The approximation in equation 22 involves the single-residue information and the pair information or the information brought to bear at position j by a residue of type R_{j+m} at position $j + m$ given that a residue of type R_j occurs at position j. The approximation in equation 21 neglects consideration of the dependence of information from neighbors on the nature of the amino acid residue at position j, whereas equation 22 does not. Equation 22 localizes information at $\pm m$ (with $m = 8$ in the GOR III method, see below) and replaces triplet information and higher-order information by residue pair information, neglecting possible overlap values. For instance, on the Venn diagram of Fig. 4, the triplet information $I(x;y_3|y_1,y_2)$ would be replaced by the pair information $I(x;y_3|y_1)$, neglecting the overlap between the surfaces y_2 and y_3, which is then counted twice, and so on for y_4 through y_8.

B. Experimental Determination of Information Values

When the observed frequencies are sufficiently large, as in the case of directional information, the probability terms can be estimated from the observed frequencies. The information values used in equation 21 are calculated from the frequencies observed in the data base for each of the 20 amino acids R and each of the conformations X in scanning the data base for every j position. Thus, for a given value of m and for directional information, one gets from equation 14 and the approximation that $P(S|y) = f(S,y)/f(y)$ the frequency-dependent equation

$$I(S_j = X{:}\bar{X};R_{j+m}) = \log\,[f(X_j,R_{j+m})/f(\bar{X}_j,R_{j+m})] - \log\,[f(X)/f(\bar{X})] \tag{23}$$

where $f(X_j,R_{j+m})$ and $f(\bar{X}_j,R_{j+m})$ are the number of times the amino acid R is at position $j + m$ of a residue observed in conformations X and $\bar{X}$, respectively, at a position j for the whole data base; $f(X)$ and $f(\bar{X})$ represent the total number of residues in conformations X and $\bar{X}$.

For pair information, the observed frequencies are on average less than 10, and so the information values used in the second term on the right-hand side of equation 22 have to be calculated with an estimate of the probability terms. Robson (1974) proposed an estimate, E, based on Bayesian reasoning, which can be written

$$\begin{aligned} E[I(S_j = X{:}\bar{X},R_{j+m}|R_j)|D(S_j,R_j,R_{j+m})] = \; & \#[f(X_j,R_j,R_{j+m})] \\ & - \#[f(\bar{X}_j,R_j,R_{j+m})] \\ & - \#[e(X_j,R_{j+m}|R_j)] \\ & + \#[e(\bar{X}_j,R_{j+m}|R_j)] \end{aligned} \tag{24}$$

where $f(X_j,R_j,R_{j+m})$ and $f(\bar{X}_j,R_j,R_{j+m})$ are the numbers of times the pair R_j and R_{j+m} is observed with R_j in the conformation X and $\bar{X}$ in the data base $D(S_j,R_j,R_{j+m})$, and

$$e(X_j,R_{j+m}|R_j) = f(R_j,R_{j+m})\,f(X_j,R_j)/f(R_j) \quad (25$$

and

$$e(\bar{X}_j,R_{j+m}|R_j) = f(R_j,R_{j+m})\,f(\bar{X}_j,R_j)/f(R_j) \tag{26}$$

where $f(R_j,R_{j+m})$ is the number of the specific pairs of amino acids separated by $m - 1$ residues in all the sequences of the data base, $f(R_j)$ is the number of amino acids of type R_j, and $f(X_j,R_j)$ and $f(\bar{X}_j,R_j)$ are the numbers of the amino acid type R_j in the conformation X and $\bar{X}$. Values of e are taken as the nearest integer, and $\#(f)$ or $\#(e)$ is the hash function, more usually defined as $\#(f) = 1 + 1/2 + 1/3 + \cdots + 1/(f-1)$, with $\#(0) = 1$, $\#(1) = 1$, and $\#(2) = 1$.

It is worth noting that $\#[f(X)] - \#[f(\bar{X})]$ converges to $\log[f(X)/f(\bar{X})]$ for high frequencies, which in practice occurs for frequencies greater than 10, i.e., for directional information values with the present data base.

C. Refinement of the Directional Information Values from the Kabsch and Sander Data Base: The GOR II Method

The GOR method, published in 1978 (Garnier *et al.*, 1978), used directional information values calculated from 26 proteins with about 4500 residues (Robson and Suzuki, 1976). Gibrat (1986) recalculated directional information values from 75 proteins or polypeptide chains from the data base of Kabsch and Sander (1983a) and C. Sander (personal communication) containing 12,757 residues.

This data base defines eight secondary structures (see Section I.B), which Gibrat (1986) has combined into four secondary structures: H, G, and I are considered as H or α helix, E is considered as E or β sheet, T is considered as T or β turn, and S, C, and B are considered as C or coil. This data base contains 3790 residues as H (29.7%), 2515 residues as E (19.7%), 1563 as T (12.2%), and 4889 residues as C (38.3%). The directional information values have been calculated using equation 23, and they are listed in Tables II, III, IV, and V for the four secondary structures and the 20 amino acids. Plots of the information values for all the amino acids are presented in Fig. 5. Only three conformational states—H, E, and T—are represented for clarity; m varies from -8 to $+8$; for $m = 0$, the information value is the information that the residue brings to its own conformation or single-residue information. For $m \neq 0$, the information value is the information that the residue at position $j + m$ brings to any residue at

Table II. Directional Information Values: $I(S_j = X{:}\bar{X};R_{j+m})$ for Helical Conformation $X = H$

Amino acid	Residue position $j-8$		$j-6$		$j-4$		$j-2$		j		$j+2$		$j+4$		$j+6$		$j+8$
Gly	0	0	−5	−10	−20	−30	−50	−70	−100	−70	−50	−30	−20	−10	−5	0	0
Ala	0	5	15	20	30	40	45	55	60	55	45	40	30	20	15	5	0
Val	0	0	0	0	−5	−15	−10	−5	0	−5	−10	−10	−5	0	0	0	0
Leu	0	0	5	5	5	10	15	20	25	20	15	10	5	5	5	0	0
Ile	0	0	0	0	0	0	5	10	15	10	5	0	0	0	0	0	0
Ser	0	0	−5	−10	−15	−20	−25	−30	−35	−33	−30	−25	−20	−15	−10	−5	−5
Thr	0	−5	−10	−15	−20	−25	−35	−40	−45	−40	−35	−25	−20	−15	−10	−5	0
Asp	0	0	0	0	10	15	10	5	0	−10	−25	−35	−25	−10	−5	0	0
Glu	25	28	32	35	44	50	60	65	70	55	40	20	10	5	5	0	0
Asn	0	0	0	0	−5	−10	−20	−30	−42	−30	−20	−10	−5	0	0	0	0
Gln	0	0	0	0	0	0	0	0	15	20	25	25	20	10	0	0	0
Lys	5	5	5	5	5	5	10	15	28	40	55	58	50	45	42	40	40
His	0	0	0	0	0	0	0	0	5	15	25	32	36	40	42	43	44
Arg	0	0	−10	−18	−20	−7	10	20	30	36	36	30	20	10	0	0	0
Phe	0	0	0	0	5	15	22	28	30	28	22	15	5	0	0	0	0
Tyr	0	−3	−10	−15	−22	−30	−35	−47	−35	−20	−10	−10	−10	−12	−17	−25	−45
Trp	25	40	30	20	10	42	42	33	20	12	5	3	0	0	0	0	0
Cys	−15	−15	−15	−15	−15	−12	−10	−5	0	0	−10	−25	−35	−43	−46	−47	−37
Met	5	5	8	10	20	30	38	45	50	48	45	42	35	30	20	15	10
Pro	0	−5	−12	−20	−25	−35	−50	−66	−96	−177	−118	−90	−70	−52	−40	−30	−20

Table III. Directional Information Values $I(S_j = X{:}\bar{X};R_{j+m})$ for Extended Conformation $X = E$

Amino acid	Residue position																
	$j-8$		$j-6$		$j-4$		$j-2$		j		$j+2$		$j+4$		$j+6$		$j+8$
Gly	5	12	22	35	40	33	22	−10	−45	−10	22	33	40	35	22	12	5
Ala	0	0	0	−5	−10	−15	−20	−30	−40	−30	−20	−15	−10	−5	0	0	0
Val	0	0	0	5	10	20	45	70	90	70	45	20	10	5	0	0	0
Leu	0	0	0	0	0	0	10	25	33	25	5	−15	−25	−25	−20	−10	−5
Ile	−15	−25	−40	−15	0	15	35	60	70	60	35	15	0	−15	−40	−25	−15
Ser	25	25	25	22	18	15	5	0	−5	0	5	15	18	22	25	25	25
Thr	5	5	5	7	10	13	21	35	28	22	18	15	15	15	15	15	15
Asp	0	0	0	0	0	−15	−55	−105	−55	−15	0	0	0	0	0	0	0
Glu	−10	−10	−15	−20	−27	−35	−45	−55	−65	−77	−63	−50	−40	−30	−20	−15	−10
Asn	5	12	20	20	15	−10	−45	−90	−60	−30	−5	10	25	30	35	30	20
Gln	15	20	20	10	5	0	−5	−15	−20	−35	−20	−15	0	10	15	20	20
Lys	0	0	0	0	−5	−12	−23	−35	−53	−70	−58	−45	−37	−30	−25	−20	−15
His	0	−5	−15	−5	0	0	0	0	0	0	5	15	25	15	5	0	0
Arg	0	0	0	0	0	0	0	0	−5	−10	−20	−25	−28	−25	−10	−5	0
Phe	−20	−35	−60	−60	−45	−30	0	25	40	25	0	−30	−45	−60	−60	−35	−20
Tyr	0	0	0	0	7	15	27	40	40	27	15	0	0	0	0	0	0
Trp	−15	−25	−40	−45	−80	−15	0	10	15	17	20	20	20	20	20	20	20
Cys	0	0	−20	−60	−55	−40	−20	5	15	17	10	5	0	0	0	0	0
Met	−20	−65	−90	−80	−60	−30	−5	15	30	15	−15	−45	−50	−45	−40	−30	−25
Pro	20	20	20	20	10	0	−30	−65	−110	−65	−30	0	10	20	20	20	20

Table IV. Directional Information Values $I(S_j = X{:}\bar{X};R_{j+m})$ for β Turn $X = T$

Amino acid	Residue position $j-8$		$j-6$		$j-4$		$j-2$		j		$j+2$		$j+4$		$j+6$		$j+8$
Gly	0	0	0	−5	−10	−20	−30	0	70	80	10	0	0	0	0	0	0
Ala	0	0	0	0	0	−3	−5	−10	−15	−22	−28	−32	−20	−7	−3	0	0
Val	0	0	0	0	−5	−10	−30	−50	−95	−50	−30	−10	−5	0	0	0	0
Leu	0	0	0	0	0	−3	−10	−15	−50	−15	−10	−3	0	0	0	0	0
Ile	0	0	0	−5	−20	−30	−50	−80	−140	−80	−30	−15	0	0	0	0	0
Ser	0	0	0	0	0	0	15	30	40	30	5	−15	−30	−15	−5	0	0
Thr	0	0	−5	−8	−10	−15	−18	−25	−42	−15	10	20	10	5	0	0	0
Asp	0	0	0	0	5	10	30	35	30	15	5	0	0	0	0	0	0
Glu	0	0	0	0	−5	−10	−10	25	20	−5	−10	20	20	5	0	0	0
Asn	0	0	−5	−20	−5	10	30	45	55	55	30	15	0	−10	−20	−5	0
Gln	0	0	0	0	0	0	0	0	0	0	0	−25	−60	−25	−15	−10	−15
Lys	0	0	0	0	0	3	5	18	30	25	10	−10	−25	−20	−12	−5	0
His	0	0	0	3	10	25	38	5	−20	15	15	−5	10	35	15	0	0
Arg	0	0	0	5	10	25	35	20	0	−35	−5	0	0	0	0	0	0
Phe	0	10	20	25	20	10	−15	−55	−50	−40	−30	−15	0	10	20	25	0
Tyr	0	0	0	5	10	20	30	5	−25	−10	0	15	30	40	30	15	0
Trp	0	−15	−45	15	35	25	10	−15	−57	−20	−10	−5	0	0	0	0	0
Cys	0	0	5	30	50	60	40	5	10	25	25	10	45	30	10	0	0
Met	0	0	0	0	0	−5	−25	−55	−85	−55	−25	−5	0	0	0	0	0
Pro	0	0	0	0	0	0	20	54	40	−154	40	50	20	0	0	0	0

Table V. Directional Information Values $I(S_j = X{:}\bar{X};R_{j+m})$ for Coil $X = C$

Amino acid	Residue position																
	$j-8$		$j-6$		$j-4$		$j-2$		j		$j+2$		$j+4$		$j+6$		$j+8$
Gly	0	0	0	0	0	10	35	70	45	20	10	5	0	0	0	0	0
Ala	0	0	0	0	−5	−10	−15	−20	−25	−20	−15	−10	−5	0	0	0	0
Val	0	0	0	0	−5	−10	−15	−25	−45	−25	−15	−10	−5	0	0	0	0
Leu	0	0	0	0	−5	−20	−30	−37	−35	−25	−15	−5	0	0	0	0	0
Ile	0	0	0	0	0	−10	−25	−32	−32	−25	−10	−5	0	0	0	0	0
Ser	0	0	0	0	0	0	5	10	20	10	5	0	0	0	0	0	0
Thr	0	0	0	0	0	0	5	15	25	15	5	0	0	0	0	0	0
Asp	0	0	0	0	0	0	5	20	35	45	40	30	15	5	0	0	0
Glu	−5	−7	−10	−20	−25	−35	−40	−45	−40	−15	−5	0	0	0	0	0	0
Asn	0	0	0	0	0	5	20	35	40	35	20	5	0	0	0	0	0
Gln	0	0	0	0	5	10	20	0	0	0	0	0	0	0	0	0	0
Lys	0	−7	−12	−18	−7	0	8	0	−8	−15	−30	−20	−12	−10	−8	−5	0
His	0	0	0	0	0	0	0	−5	−10	−15	−25	−40	−55	−40	−30	−25	−20
Arg	0	0	0	0	−5	−10	−20	−30	−35	−30	−20	−10	−5	0	0	0	0
Phe	0	0	0	0	−5	−10	−20	−30	−35	−30	−20	−10	−5	0	0	0	0
Tyr	0	0	10	23	10	0	0	0	0	0	0	0	5	18	5	0	0
Trp	0	0	−5	−20	−35	−43	−40	−35	−30	−23	−15	−5	0	0	0	0	0
Cys	15	15	20	30	20	10	5	−5	−15	−25	−10	0	10	20	30	40	25
Met	0	0	0	0	−5	−10	−15	−25	−40	−55	−30	−20	−10	−5	0	0	0
Pro	0	0	0	5	15	25	40	60	90	202	90	40	25	15	10	5	0

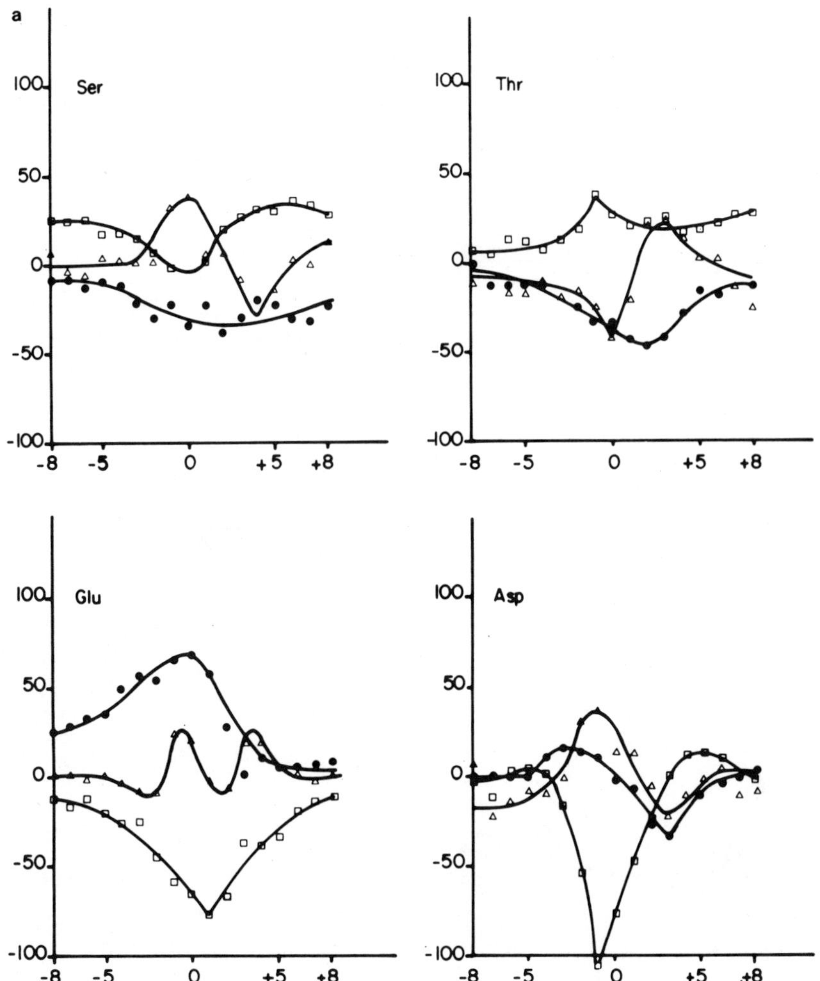

Figure 5a–e. Directional information values for the 20 amino acids, $I(S = X{:}\bar{x};R_{j+m})$ in centinats with $m > 0$ for residue R on the C-terminal side of the position j: ●, α helix; □, β sheet; △ or ▲, β turn. From Gibrat, 1986.

position j, m being positive for the residue on the C-terminal side of the position j. Note that the sign of m for a C-terminal position is opposite to the sign used by Robson and Suzuki (1976) and Garnier *et al.* (1978). Some workers have shown a preference for this modification as an easier way of memorizing the directional effect, since the sign of m that locates the residue exerting this effect along the sequence is the same as the sign (or order) of numbering of the amino acids in the sequence.

From Fig. 5 one can define four groups of amino acids:

1. Group I. The directional information is approximately symmetrical, with a maximum for $m = 0$ for the conformation—H, E, or T—that they favor. For instance, A for H, I and V for E, etc.

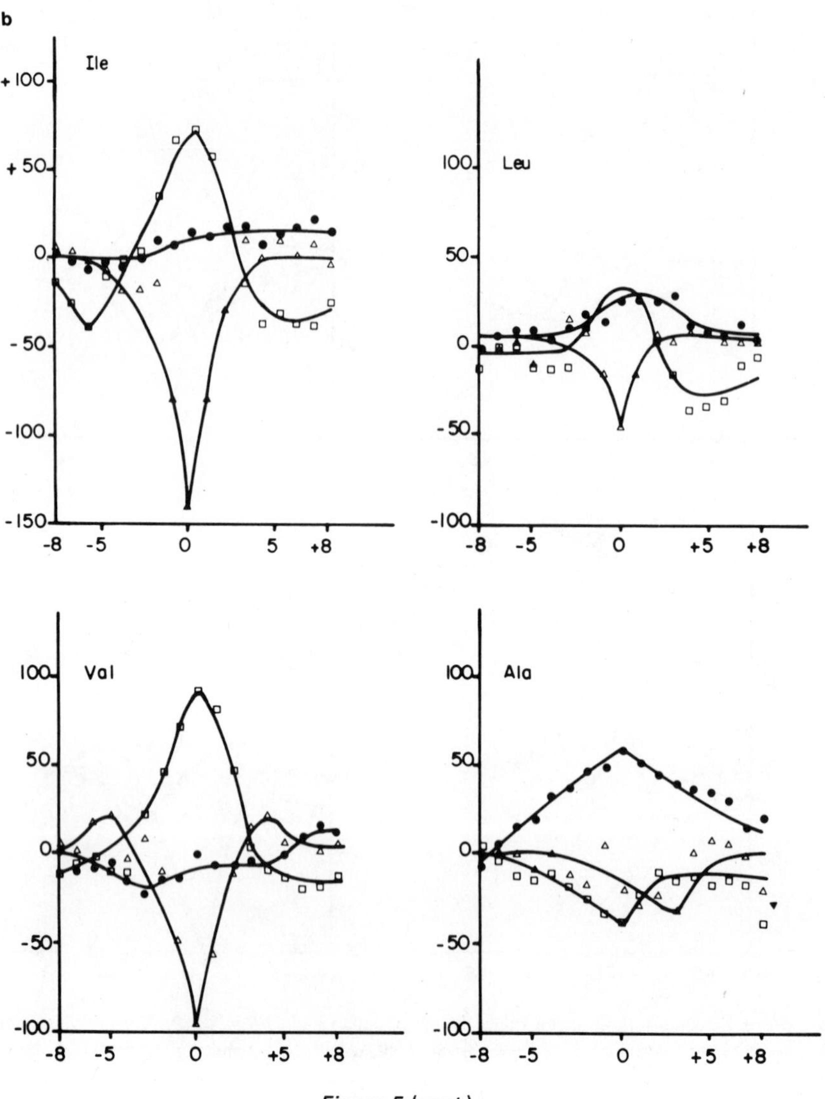

Figure 5 (*cont.*)

2. Group II. The directional information is nearly symmetrical, with a minimum at $m = 0$. For instance, G disfavors H, L and V disfavor T, etc.
3. Group III. Amino acids that exhibit asymmetric information, positive for $m < 0$ and negative for $m > 0$. These residues favor the conformational state when located towards the N-terminal end of the secondary structure, for instance, E and D for H, which can receive a hydrogen to make a hydrogen bond with the NH of an N-terminal residue of an α helix and stabilize its beginning.
4. Group IV. The opposite of group III: amino acids with negative values of directional

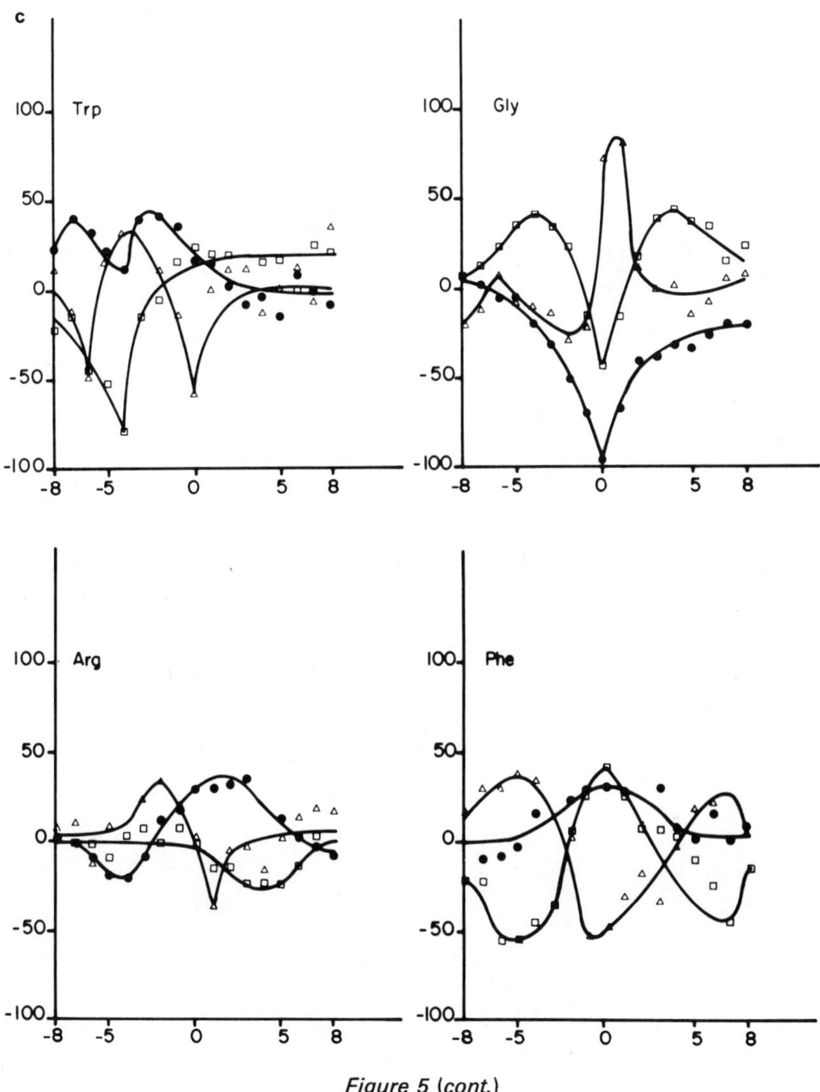

Figure 5 (cont.)

information for $m < 0$ and positive values for $m > 0$. For instance, K favors conformation H when located on the C-terminal end of a helix, stopping its propagation by making a hydrogen bond with a C-terminal carbonyl; the same holds true to a lesser degree for residues H and R.

Single-residue information (for $m = 0$) in Tables II–V identifies the strong helix formers such as E (70 cnat), A (60 cnat), and M (50 cnat), the strong β-sheet formers such as V (90 cnat), I (70 cnat), Y (40 cnat), and F (40 cnat), strong β-turn formers such as G (80 cnat at +1), N (55 cnat), S (40 cnat), C (60 cnat at −3 and 45 at +4), and P (54 cnat at −1 and +50 at +2), and strong coil formers such as P (202 cnat at +1), G (70 at −1), N (40 cnat), and C

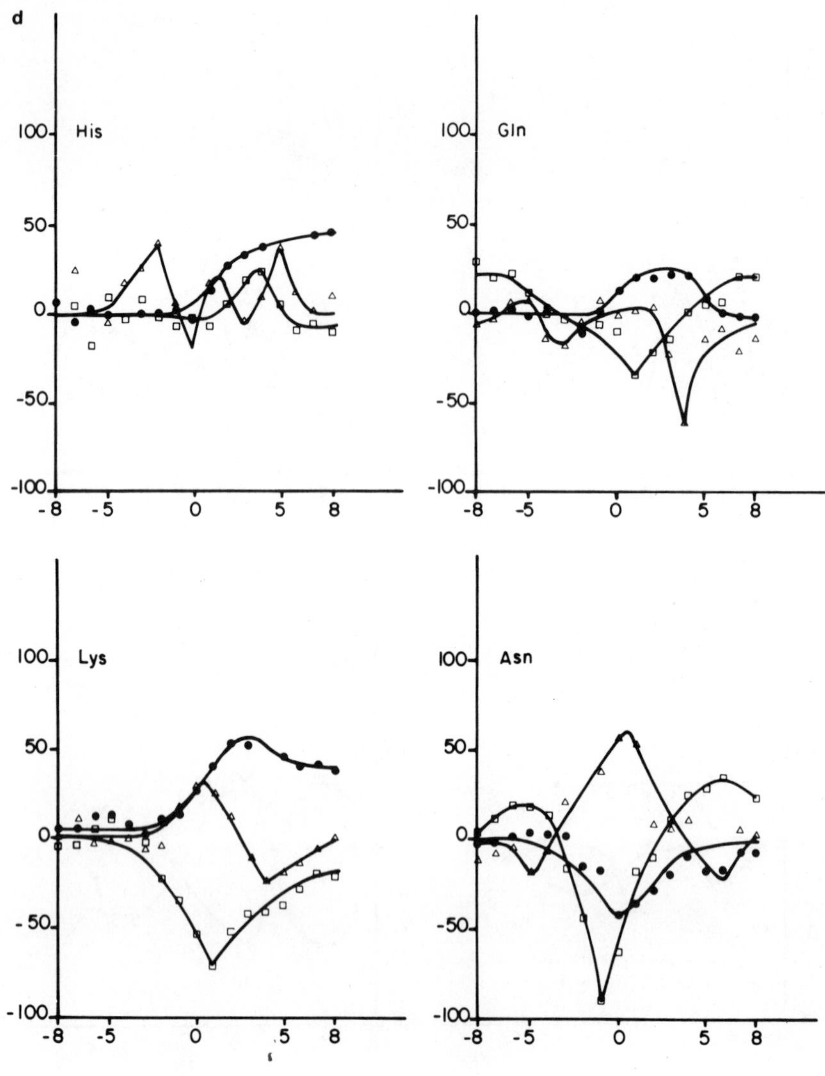

Figure 5 (cont.)

(40 cnat at +7). Note that P, which favors β turns, would prevent the residue next to its N terminal to be in a β turn (−154 cnat) or α helical (−177 cnat) or extended (−65 cnat) but will favor a coil conformation for it (+202). Other residues favorable to β-turn conformation (G, N, S, and C) are also helix and β-sheet breakers.

There are residues such as L and I that favor both H and E conformations and charged residues that favor H and break E.

From these directional information values one can calculate for a sequence of amino acids of unknown conformation the information value for each position j, summing up from $m = -8$ to $m = +8$ according to equation 21 and for each of four conformational states:

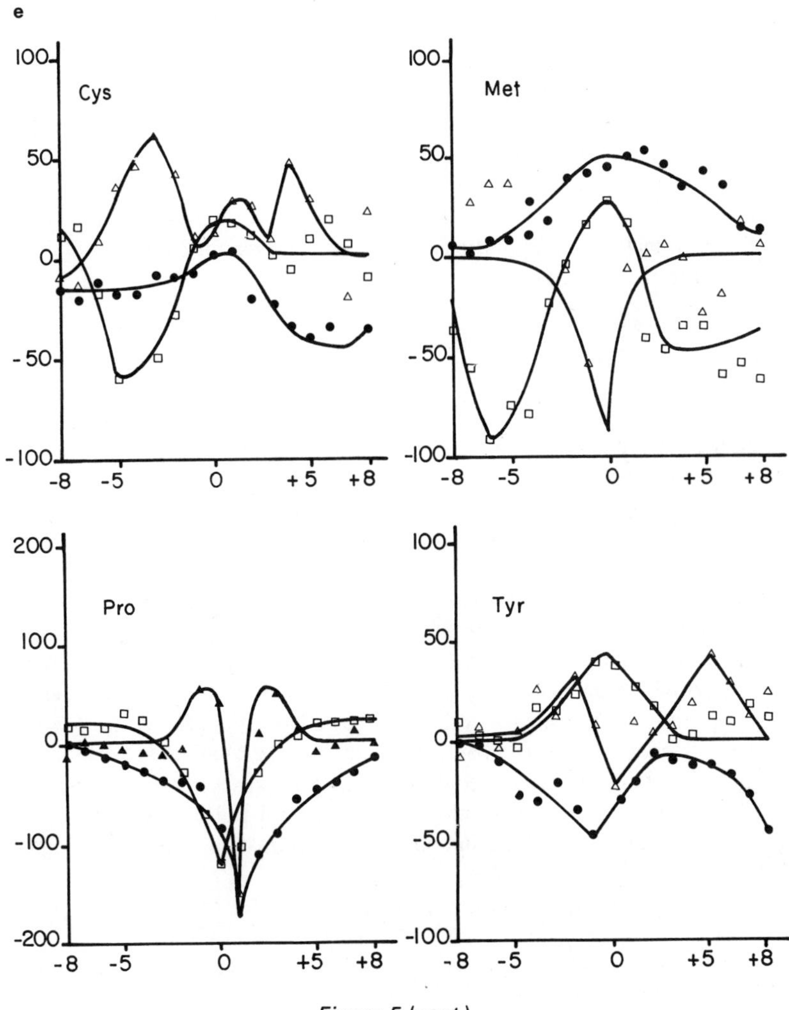

Figure 5 (cont.)

$$I(S = \text{H}:\bar{\text{H}}; R_{j-8},\ldots,R_j,\ldots,R_{j+8}) \tag{27}$$

$$I(S = \text{E}:\bar{\text{E}}; R_{j-8},\ldots,R_j,\ldots,R_{j+8}) \tag{28}$$

$$I(S = \text{T}:\bar{\text{T}}; R_{j-8},\ldots,R_j,\ldots,R_{j+8}) \tag{29}$$

$$I(S = \text{C}:\bar{\text{C}}; R_{j-8},\ldots,R_j,\ldots,R_{j+8}) \tag{30}$$

Residue j is predicted in the conformational state for which the values of equations 27 to 30 are highest. This process is pursued for all positions j in the sequence to obtain a prediction of the secondary structure. This algorithm is identical to the one used by Garnier *et al.* (1978) for the GOR method, but because new refined information values are used, this method of prediction is referred to as the GOR II method (see results in Section IV).

III. PAIR INFORMATION AND THE GOR III METHOD

A. Principles

The GOR III method is an attempt to use the pair information of equation 22 instead of directional information as in the GOR II method, but with the same algorithm as presented above; i.e., the predicted conformational state (of four states) will be the state with the highest positive information value. This method requires, as above, extracting from the data base single-residue information values and pair information values for each possible pair and for each value of m from -8 to $+8$.

B. Level of Significance

As mentioned above, if observed frequencies can be used for calculating single-residue information, the hash function has to be used for calculating pair information, since observed pair frequencies are small.

Robson (1974) suggested that the particular hash functions were specific solutions of the Bayesian information theory approach for specific values of a nondefined term g. Indeed, Gibrat *et al.* (1987) have demonstrated that direct use of equation 22 yields a prediction of a lower quality than the GOR method of Garnier *et al.* (1978): 53.2% of correctly predicted residues in three states or 41.1% in four states. It is the analysis of the origins of inaccuracy in measuring the information values that helps us to improve the prediction. One may reasonably assume that the origin of the inaccuracy is the low frequencies of some amino acid pairs in the data base compared with frequencies available for evaluating the other terms. For instance, in the Kabsch and Sander (1983a) data base, there are three methionines in position $j - 8$ when the residue in j is a valine in helical conformation, and seven methionines in $j - 8$ when a valine in j is in a nonhelical conformation. In the whole data base there are 294 valines in helical conformation and 640 valines in nonhelical conformation. From use of equations 24, 25, and 26:

$$E[I(S_j = \mathrm{H}:\bar{\mathrm{H}}, \mathrm{M}_{j-8}|\mathrm{V}_j)|D(S_j, \mathrm{V}_j, \mathrm{M}_{j-8})] = 0 \text{ cnat}$$

If four methionines had been observed at $j - 8$ when the residue in j is a valine in helical conformation, the information value would have been 48 cnat. Thus, a single additional observation can change the information by 48 cnat when the single-residue information, which is calculated with a large number of observed frequencies, only varies from 100 cnat for glycine to 70 cnat for glutamate for the helical conformation (Table II).

One way of avoiding this problem is to measure the level of significance for each pair interaction and to retain for prediction only those that are above a defined level. This can be done by using a statistical test on 2 × 2 contingency tables (Fisher, 1934). Such an approach was done for a similar problem by Maxfield and Scheraga (1975). A theoretical contingency table is given below, where R_{j+m} stands for a specific residue R at position $j + m$, $\bar{R}_{j+m}$ for any other residue than R at position $j + m$, and n_{11}, n_{12}, n_{21}, and n_{22} are the observed frequencies of occurrence of a specific residue R_j in the conformation H or $\bar{\mathrm{H}}$ according to the occurrence of R_{j+m} or $\bar{R}_{j+m}$ at position $j + m$, respectively.

	H	$\bar{\mathrm{H}}$
R_{j+m}	n_{11}	n_{12}
$\bar{R}_{j+m}$	n_{21}	n_{22}

In the example given above, $n_{11} = 3$, $n_{21} = 291$, $n_{12} = 7$, and $n_{22} = 633$. The Fisher test evaluates whether the difference between the two ratios n_{11}/n_{21} and n_{12}/n_{22} could be observed by chance by estimating the probability P (Fisher, 1934) of observing these two ratios when the four frequencies are independent events. The lower the value of P, the smaller is the chance that the difference between the two ratios is related to the sampling procedure of independent events and, consequently, the chance that R_{j+m} has no effect on the conformation of R_j at the j position. Usually a level of P is chosen so that when P is lower than 0.1, 0.05, or 0.01 (level of significance), one can consider the pair interaction to be significantly effective in influencing a conformational state. For a given value of the ratios n_{11}/n_{21} and n_{12}/n_{22}, the calculated value of P is very sensitive to the number of frequencies, decreasing rapidly when the number of frequencies increases.

C. Analysis of the Significant Pair Interactions

We have listed in Tables VI to IX the significant pair interactions for a level of $P \leqslant 0.001$ (Gibrat *et al.*, 1987) for the four conformational states and to a separation value m from -8 to $+8$ residues. This probability level is somewhat arbitrarily low, but it limits the discussion to a reasonable number of pairs, the most significant ones, instead of the 1600 possible pairs for each of the 16 values of m. The observed frequencies of pairs for a given conformation vary from 0 to about 50, with a significant number below 10, the average being 8. Since the information value is very sensitive to one additional frequency, as in the example given in Section III.B, we list in Tables VI to IX only the sign of the information, not its actual value.

Two results are apparent. Local pairs (m up to ± 3) are more frequent than remote ones ($m = \pm 6$ to ± 8). The most significant pairs are those in the vicinity of a residue, and this effect fades away after a dislocation of eight residues. This correlates well with the directional information shown in Fig. 5. Second, there are more significant pairs for $m > 0$; that is, more significant interactions of residues are observed toward the direction of the N terminal than toward the C terminal. Does this correspond to the biosynthesis direction from N to C terminal? When a residue is added to the growing polypeptide chain, it can at first exert a conformational influence only on the already synthesized N terminal end. The same seems to hold for the greater accessibility of the C-terminal part of a polypeptide chain to reagents such as antibodies, as if the C-terminal sequence is less involved in the rest of the folded part of the polypeptide chain (see also results in Section IV).

1. Significant Pairs for the Helical Structure

Pairs containing charged residues, mainly lysine and glutamate, appear favorable for helical structure, as does alanine. Interactions between hydrophobic residues such as leucine and valine are favorable for the helical structure for values of $m = 4$ or 3, corresponding to the pitch of an α helix, but they oppose the helical structure for $m = \pm 2$. One sees here the importance of using pairs, because these details are hidden when one uses directional information. Some favorable interactions are more difficult to explain such as L $\rightarrow$ P for $m = -8$, L $\rightarrow$ P for $m = -2$, A $\rightarrow$ G for $m = -4$, and K $\rightarrow$ P for $m = +8$.

2. Significant Pairs for the Extended Structure

Residues favorable for turn or aperiodic structure, such as glycine, serine, proline, and threonine, are favorable to the extended structure, but only for absolute values of m equal to 2, 4, and greater than 4. These residues belong primarily to a β turn or loop, often located

Table VI. Significant Pairs: Conformation in j:H $R_{j+m} \rightarrow R_j$ (+/−)[a]

$m = -8$	L → P (+)		
$m = -7$			
$m = -6$			
$m = -5$	P → V (−)		
$m = -4$	A → G (+)	E → A (+)	E → R (+)
	L → I (+)		
$m = -3$	A → K (+)	T → I (−)	F → L (+)
$m = -2$	A → V (+)	V → L (−)	L → P (+)
	S → S (−)	E → L (+)	E → I (+)
	P → L (−)		
$m = -1$	G → A (−)	G → L (−)	G → D (−)
	A → L (+)	E → A (+)	E → T (+)
	H → C (+)		
$m = +1$	G → A (−)	G → D (−)	A → V (+)
	A → L (+)	L → A (+)	E → G (+)
	E → A (+)	E → L (+)	P → A (−)
	P → L (−)	P → T (−)	P → D (−)
	P → E (−)	P → Q (−)	P → H (−)
$M = +2$	A → V (+)	L → V (−)	L → S (+)
	L → D (+)	S → S (−)	S → D (−)
	K → G (+)	K → Y (+)	F → D (+)
	P → A (−)	P → V (−)	
$M = +3$	G → K (+)	A → V (+)	L → G (+)
	L → V (+)	L → F (+)	K → A (+)
	K → S (+)	K → D (+)	K → H (+)
	R → W (+)	Q → P (+)	
$M = +4$	A → G (+)	A → P (+)	V → K (−)
	L → V (+)	L → L (+)	S → V (−)
	S → L (−)	K → A (+)	P → E (−)
$M = +5$	S → E (−)	K → I (+)	H → S (+)
	H → P (+)		
$m = +6$			
$m = +7$	L → G (+)	S → V (−)	T → T (−)
$m = +8$	A → K (+)	K → P (+)	Q → V (−)

[a]Influence of the residue located at position $j + m$ (m ranging from −8 to +8) on the conformation of the residue at j. The criterion of significance is the value of P (see text). Here only pairs with a value of P less than 0.1% are considered significant; (+) indicates that the pair is favorable to the conformation, and (−) indicates that the pair is unfavorable.

between two antiparallel β sheets. Valine and leucine appear in a number of favorable pairs, with the exception of leucine at $m = +4$ to valine and leucine at $m = +4$ to leucine. This value $m = +4$ corresponds to the pitch of a helix, and these two interactions are found favorable to the helical structure.

3. Significant Pairs for the β-Turn Structure

The β turn is a very localized structure, and significant pairs exist only for small values of m (Table VIII). Glycine plays the main role in position $m = +1$. Some pairs involve residues that are not abundant in the data base, such as cysteine at $m = -3$ from cysteine or histidine at $m = +1$ from cysteine.

Table VII. Significant Pairs: Conformation in j:E $R_{j+m} \rightarrow R_j$ (+/−)[a]

$m = -8$			
$m = -7$			
$m = -6$	G → H (+)		
$m = -5$	G → L (+)	N → Y (+)	P → K (+)
$m = -4$	G → K (+)	T → T (+)	
$m = -3$	F → L (−)		
$m = -2$	V → V (+)	V → K (+)	S → S (+)
	E → L (−)	D → V (−)	
$m = -1$	A → G (−)	V → G (+)	V → V (+)
	V → D (+)	V → S (+)	V → T (+)
	I → V (+)	D → V (−)	N → S (−)
$m = +1$	V → A (+)	V → V (+)	V → I (+)
	V → S (+)	V → T (+)	V → E (+)
	I → L (+)	E → S (−)	C → T (+)
	Q → S (−)	P → V (−)	P → F (−)
$m = +2$	V → F (+)	S → S (+)	T → S (+)
	E → L (−)		
$m = +3$	E → T (−)	K → A (−)	
$m = +4$	G → V (+)	L → V (−)	L → L (−)
	S → S (+)	S → F (+)	G → I (+)
$m = +5$	G → G (+)	G → A (+)	
$m = +6$	I → L (−)		
$m = +7$	T → S (+)		
$m = +8$	Q → V (+)	T → L (+)	

[a]Influence of the residue located at position $j + m$ (m ranging from −8 to +8) on the conformation of the residue at j. The criterion of significance is the value of P (see text). Here only pairs with a value of P less than 0.1% are considered significant; (+) indicates that the pair is favorable to the conformation, and (−) indicates that the pair is unfavorable.

4. Significant Pairs for the Aperiodic Structure

One can recognize in Table IX the effect of a proline one residue away on the C-terminal side of a residue to be coil, and this has to be compared with the strong value of the directional information $I(S = \text{C}:\bar{\text{C}};\text{P}) = 202$ cnat. Only P/M and P/W pairs are missing, most likely because of the low probability of occurrence in the data base, which contains 1.5% M, 1.4% W, and 4.2% P. Thus, their probability of occurrence by chance will be 6.3×10^{-4} for P/M and 5.9×10^{-4} for P/W. The probabilities of occurrence of the pairs A/A and G/G, which are the greatest, are 8.1×10^{-3} and 7.9×10^{-3}, respectively; that is, these pairs are 13–14 times more frequent than the pairs P/M and P/W. However, they are significant pairs although they correspond to pairs between less abundant residues; for instance, at $m = -4$, the pairs G/M and V/Y have a probability of occurrence of 1.36×10^{-3} and 2.8×10^{-3}, respectively, although their levels of significance are 0.0001 and 0.01, respectively.

Charged residues such as arginine, glutamate, and aspartate have a tendency to adopt aperiodic structure under the influence of threonine, aspartate, and serine, respectively, at position $m = +2$. Pairs that are unfavorable for coil can be found favorable for other structures; for example, the pair lysine/proline at $m = +8$ favors H and opposes C.

Table VIII. Significant Pairs: Conformation in j:T $R_{j+m} \rightarrow R_j$ (+/−)[a]

$m = -8$			
$m = -7$			
$m = -6$			
$m = -5$			
$m = -4$			
$m = -3$	N → T (+)	A → K (−)	C → C (+)
$m = -2$			
$m = -1$	P → G (+)	D → S (+)	D → P (+)
$m = +1$	G → A (+)	G → L (+)	G → I (+)
	G → S (+)	G → N (+)	G → Q (+)
	G → P (+)	F → G (−)	S → T (+)
	H → C (+)		
$m = +2$			
$m = +3$	P → G (+)		
$m = +4$			
$m = +5$			
$m = +6$			
$m = +7$			
$m = +8$			

[a]Influence of the residue located at position $j + m$ (m ranging from −8 to +8) on the conformation of the residue at j. The criterion of significance is the value of P (see text). Here only pairs with a value of P less than 0.1% are considered significant; (+) indicates that the pair is favorable to the conformation, and (−) indicates that the pair is unfavorable.

D. Use of Dummy Observations for Prediction Based on Pair Information

A low level of significance of a pair of amino acids for a given conformation may arise in two ways. Either the pair has no influence on that conformation or the number of observed frequencies is too small and the difference between the two ratios in the contingency table can be explained by a statistical variation of the sampling because of the data base. Robson (1974) considered in general terms the formal validity of a parameter g to resolve these difficulties but did not consider the method of calibration of g. Maxfield and Scheraga (1976) described a procedure for that purpose that is fully consistent with the formal definition of g and consists of increasing artificially the number of observations by adding dummy observations. Gibrat *et al.* (1987) chose to add at each position j a set of dummy frequencies, D, which reflect both the information that a residue has on its own conformation and the directional information of the pair residue for the same displacement. If a and b, dummy frequencies, are added to the observed pair frequencies of R_j,R_{j+m} for the conformations X and $\bar{X}$, respectively, then we may write

$$I(S_j = X{:}\bar{X};D) = I(S_j = X;a) - I(S_j = \bar{X};b)$$

$$I(S_j = X{:}\bar{X};D) = I(S_j = X{:}\bar{X};R_j) + I(S_j = X{:}\bar{X};R_{j+m}) \qquad (31)$$

Expressing the information functions in the two right-hand sides of equation 31 and rearranging produces:

$$\log(a/b) = \log[f(X_j,R_j)/f(\bar{X}_j,R_j)] + \log[f(X_j,R_{j+m})/f(\bar{X}_j,R_{j+m})] - \log[f(X)/f(\bar{X})] \qquad (32)$$

where $a + b = M$.

Table IX. Significant Pairs: Conformation in j:C $R_{j+m} \rightarrow R_j$ (+/−)[a]

$m = -8$	A → R (−)		
$m = -7$			
$m = -6$			
$m = -5$			
$m = -4$	E → K (−)	V → Y (−)	G → M (+)
$m = -3$			
$m = -2$	A → V (−)	G → K (+)	
$m = -1$	G → A (+)	G → V (+)	G → L (+)
	G → D (+)	G → Q (+)	G → K (+)
	A → L (−)	F → R (−)	
$m = +1$	P → G (+)	P → A (+)	P → V (+)
	P → L (+)	P → I (+)	P → S (+)
	P → T (+)	P → D (+)	P → E (+)
	P → N (+)	P → Q (+)	P → K (+)
	P → H (+)	P → R (+)	P → F (+)
	P → Y (+)	P → C (+)	P → P (+)
	V → V (−)	L → A (−)	G → P (−)
	T → L (+)		
$m = +2$	A → G (−)	S → D (+)	T → R (+)
	D → E (+)	P → A (+)	P → V (+)
	P → T (+)	L → D (−)	
$m = +3$	L → G (−)	S → I (+)	P → L (+)
$m = +4$			
$m = +5$	A → K (−)	V → G (−)	
$m = +6$			
$m = +7$			
$m = +8$	K → P (−)		

[a]Influence of the residue located at position $j + m$ (m ranging from −8 to +8) on the conformation of the residue at j. The criterion of significance is the value of P (see text). Here only pairs with a value of P less than 0.1% are considered significant; (+) indicates that the pair is favorable to the conformation, and (−) indicates that the pair is unfavorable.

In equation 24, $f(X_j,R_j,R_{j+m})$ is then replaced by $f(X_j,R_j,R_{j+m}) + a$, and $f(\bar{X}_j,R_j,R_{j+m})$ by $f(\bar{X}_j,R_j,R_{j+m}) + b$, and in equations 25 and 26, $f(R_j,R_{j+m})$ is replaced by $f(R_j,R_{j+m}) + M$, $f(X_j,R_j)$ by $f(X_j,R_j) + a$, $f(\bar{X}_j,R_j)$ by $f(\bar{X}_j,R_j) + b$, and $f(R_j)$ by $f(R_j) + M$.

This method is empirical, and hence the value of M was chosen to obtain the best prediction. An optimum value for M is 225 (see Table X). Greater values of M decreased the prediction accuracy to the level of the directional method (GOR II). Lower values of M (below 20) give a prediction accuracy lower than the GOR II method.

E. Decision Constants

Garnier *et al.* (1978) have shown that the predictions may be significantly improved by introducing specific decision constants for each conformational state. They determined the optimal values of the decision constants for each of the 25 proteins in the data base and proposed to use a decision constant based on the total content in α helix and β sheet of the protein to be predicted. This secondary structure content can be measured experimentally from the circular dichroism of the protein solution. Three classes of protein were distinguished: proteins with less than 20%, between 20% and 50%, and over 50% α helix or β sheet. The use

Table X. Influence of the Number of Dummy Observations M on the Accuracy of the GOR III Method

	Percentage of residues correctly predicted[a]	
M	Three states	Four states
0	53.2	41.1
20	57.1	45.9
100	59.7	49.9
225	61.0	51.7
300	60.5	51.1
950	57.1	46.6

[a]GOR II method (directional information) will predict correctly 56.8% (three states) or 46.5% (four states) of the residues.

of these decision constants came from the observation that on average proteins rich in α helix or in β sheet tended to have that conformation underpredicted by the GOR method, suggesting that this kind of protein has added information to form more α helix or β sheet. This added information is the decision constant. However, this method requires a prior knowledge of the circular dichroism data, which are not always available and may not be very accurate (Garnier *et al.*, 1976). Tentative attempts have been made to determine the folding type from the amino acid composition, either from the relative content of α-helix- or β-sheet-forming amino acids (Garnier *et al.*, 1976) or from a global amino acid composition taken from a set of proteins of known x-ray structure (Nakashima *et al.*, 1986; Levin and Garnier, 1988).

Levin and Garnier classified the proteins in the data base into four classes: α-rich, greater than 35% helix and less than or equal to 10% β sheet; β-rich, greater than 30% β sheet and less than or equal to 10% α helix; low secondary structure, regular secondary structure (helix and β sheet) less than 25%; and mixed, all proteins not in the other three categories. By comparing the global amino acid composition of a protein to the average global amino acid composition of the four protein classes (j) the folding type is predicted for the lowest value of the function C_j

$$C_j = \sum_{i=1}^{20} [\text{abs}(Av_{ij} - Per_i)/Sd_{ij}]W_i$$

Per is the percentage content of amino acid i of the unknown protein, *Av* and *Sd* are, respectively, class average percentage content and standard deviation for each amino acid. By optimizing the weighting factor W, 84% of the proteins of the data base were correctly predicted for their folding type. With the use of specific decision constants for each class, the accuracy of the prediction was increased by an additional 2% to 3%, typically going from 63% to 65.7% for a three-state prediction (Levin and Garnier, 1988).

Another way of using the decision constant was to adjust as closely as possible the number of predicted residues in a conformational state to the number of residues observed in that state in the data base overall (Gibrat *et al.*, 1987). The decision constants are the same for any protein to be predicted, whatever its secondary structure content, and they significantly improve the prediction. Table XI shows that a relatively small decision constant has a marked effect on the number of predicted residues. For example, the subtraction of 35 cnat from the information value of the E state for each residue reduced by 755 residues those being predicted

Table XI. Decision Constants[a] for the GOR III Method

State	Number of amino acid residues			
	Predicted			
	DC = 0	$DC_E = 35$	$DC_H = 5$, $DC_E = 25$	Observed
H	3760	4008	3795	3776
E	2930	2175	2409	2476
C	5743	6250	6229	6181
	Percentage correctly predicted residues[b]			
3 states	61.0%	61.7%	61.6%	
4 states	51.7%	51.9%	51.7%	

[a]Decision constants in centinats to be subtracted from the information values obtained from equations 22, 24–26, 31, and 32 with $M = 225$.
[b]The protein to be predicted is removed from the data base for calculating the pair information.

in E and added 248 and 507 residues predicted in H and C, respectively. That also means that a small number of regions in the amino acid sequences, 2% to 5% of the residues, exhibit information values within a few tens of cnats for several conformational states.

F. Predictions on Large Numbers of States

Such studies have been carried out on the 81 states of Fig. 2, and programs have been used routinely, particularly to obtain more detailed predictions of ϕ,ψ angles as a starting point for energy minimization.

Various methods have been explored, and the final method used pairs as described above without dummy observations but with the limitations of a too-small data base (Garnier and Robson, 1979). It is most important to note that when results by such methods are converted back to three- or four-state predictions by pooling predicted states, they yield the same results as the three- or four-state methods *per se*. The advantage is not, with some reservations, that predictions of secondary structure in the usual sense are improved. Rather, more detail is given, which is to say there is more information of other types. Even when these data are less reliable, they nonetheless assign a "best-bet" description in ϕ,ψ angle terms in the absence of other information and make it possible to start a folding simulation by energy minimization.

IV. ACCURACY OF THE GOR METHOD

It has already been mentioned in Section I.A that the criteria chosen to define the observed conformational states have an effect on the quality of the prediction. Other factors affecting this are also worth considering.

A. Influence of the Data Base

Different findings for quality of prediction for three and four states, expressed as $f \times 100$ as in equation 1, are presented in Table XII. Two prediction methods, GOR I and GOR II, are based on directional information. GOR I makes use of the parameters published by Garnier *et al.* (1978) with a decision constant equal to zero. GOR II uses the parameters calculated from the Kabsch and Sander data base (Tables II to V), and its detailed results for each protein in the

Table XII. Accuracy of the GOR Methods

GOR method	Data base	Quality of prediction ($f \times 100$)	
		4 states[a]	3 states[b]
GOR I[c]	R.S.[d]	49	56
	L.G.[e]	46.1	54.7
	K.S.[f]	42.6	56.0
GOR II[g]	L.G.[e]	46.1	53.8
	K.S.[f]	46.7	57.7
	−P[h]	44.4	56.9
	+P[h]	49.9	60.7
GOR III	K.S.[i]	51.9	61.7
	K.S.[j]		
	−P[k]		63.0
	+P[k]		69.7

[a]Value of $f \times 100$ (see text) for H, E, T, and C.
[b]Value of $f \times 100$ for H, E, and (T + C).
[c]GOR I refers to the directional information prediction of Garnier *et al.* (1978).
[d]Data base of Robson and Suzuki (1976) of 25 proteins or about 4500 amino acid residues.
[e]Data base of Levitt and Greer (1977) of 51 proteins containing 8898 residues, of which 36% are H, 33% are E, and 31% are (T + C).
[f]Data base of Kabsch and Sander (1983a) containing 55 proteins or 9472 residues, of which 28% are H, 19% are E, and 53% are (T + C).
[g]GOR II refers to the directional information prediction with values of Tables II to V (Gibrat, 1987) except for −P and +P.
[h]The directional information values are not averaged (see text), and the protein to be predicted is removed from the data base (−P) or left in the data base. (+P).
[i]GOR III refers to pair information prediction with $M = 225$ and DC = 35 cnat added to the information for the extended conformation E. The data base is made of 74 polypeptides chains (12,433 residues), and each protein is removed from the data base before prediction.
[j]GOR III with $M = 225$, $DC_H = 25$, and $DC_E = 30$ cnat. The data base is made of 68 polypeptides chains (11,237 residues) of refined x-ray coordinates (resolution better than 2.8 Å, R factor less than 0.25).
[k]−P: The protein to be predicted is removed from the data base to calculate at each prediction the pair information values. +P is the prediction with the protein left in the data base; i.e., the pair information values are identical for each predicted protein and are calculated with the whole data base.

data base are presented in Table XIII with a decision constant of zero. The third prediction method is based on pair information and is referred here as GOR III, the detailed resúlts of which are presented in Table XIV.

There are differences in prediction accuracy depending on which data base was used to make the test: the data base of Levitt and Greer (1977), that of Kabsch and Sander (1983a), and a data base made of proteins of x-ray-refined coordinates. This was found independently of the method of prediction used: GOR I, GOR II, or GOR III (see Table XII). The accuracies vary from 54.7% to 56%, 53.8% to 57.7%, and 61.7% to 63% for three states, or, in relative values, ($\Delta f/f \times 100$) from 2.4% to 7%. We mentioned in Section I.A that one component of the variation is the criteria used to define the conformational state; this accounts for about 1.5% to 2% in relative value, or half of the observed variation. A second factor is the accuracy of the x-ray data, accounting for about the other half.

The composition of the protein data base has its own influence. This was checked by removing from the data base the protein to be predicted and recalculating a new set of directional information values, which was used to predict that protein (−P in Table XII). This was done for each protein in the data base. The overall accuracy of prediction was 56.9% for

Table XIII. GOR II Method Result for the Whole Data Base: New Set of Parameters[a]

Proteins	Brookhaven identification	Results	
		3 states	4 states
Calcium-binding parvalbumin B	1CPV	57.4	51.9
L-Arabinose-binding protein	1ABP	54.6	48.7
Oxidized high-potential iron protein	1HIP	54.1	44.7
Cytochrome B_5	2B5C	49.4	49.4
Cytochrome B_{562}	156B	74.8	67.0
Cytochrome C	3CYT	43.7	34.0
Ferricytochrome C_2	3C2C	56.3	51.8
Cytochrome C_{550}	155C	47.8	35.1
Cytochrome C_{551}	351C	80.5	74.4
Ferredoxin	1FDX	48.1	22.2
Ferredoxin	3FXC	60.2	46.9
Flavodoxin	3FXN	63.0	57.2
Azurin	1AZU	47.2	28.8
Plastocyanin	1PCY	59.6	46.5
Avian pancreatic polypeptide	1PPT	72.2	61.1
Glucagon	1GCN	51.7	34.5
Insulin (chain 1)	1INS	19.0	4.8
Insulin (chain 2)	1INS	56.7	43.3
Phospholipase A_2	1BP2	48.0	38.2
Lysozyme bacteriophage T_4	1LZM	52.4	46.3
Lysosyme hen egg white	7LYZ	58.1	47.3
Staphylococcal nuclease	2SNS	49.6	41.1
Ribonuclease-S	1RNS	60.0	50.0
Ribonuclease-S (chain 2)	1RNS	59.6	41.3
Carboxypeptidase A	5CPA	51.5	40.4
Acid protease *Rhizopus chinensis*	1APR	50.9	35.2
Acid proteinase penicillopepsin	2APP	57.3	48.6
Thermolysin	3TLN	53.2	43.0
γ-Chymotrypsin A	2GCH	52.5	41.1
γ-Chymotrypsin A (chain 2)	2GCH	55.8	50.5
α-Lytic protease	2ALP	60.6	51.0
β-Trypsin	1PTN	56.1	42.2
Proteinase A	2SGA	60.8	54.1
Subtilisin BPN′	1SBT	57.8	45.1
Tosyl elastase	1EST	55.4	42.5
Actinidin	2ACT	54.1	41.3
Papain	8PAP	54.7	38.2
λ-Immunoglobulin Fab	1FAB	63.3	49.3
λ-Immunoglobulin (chain 2)	1FAB	60.7	50.2
Bence Jones immunoglobulin	1REI	57.9	46.7
Phosphoglycerate mutase	3PGM	57.8	45.7
Triose phosphate isomerase	1TIM	63.6	55.1
Concanavalin A	3CNA	59.5	43.9
Carbonic anhydrase form C	1CAC	55.1	41.4
Dihydrofolate reductase	3DFR	50.6	43.8
D-Glyceraldehyde-3-phosphate dehydrogenase	1GPD	57.7	46.8
Apo-liver alcohol dehydrogenase	4ADH	49.2	38.5
Lactate dehydrogenase	4LDH	55.0	44.4
Glutathione reductase	2GRS	57.3	43.4
Cu,Zn-Superoxide dismutase	2SOD	65.6	47.7

(*continued*)

Table XIII. (Continued)

Proteins	Brookhaven identification	Results	
		3 states	4 states
Myoglobin	1MBN	72.5	69.3
Hemoglobin (erythrocruorin)	1ECD	55.9	51.5
Hemoglobin (aquo Met)	2MHB	61.7	57.4
Hemoglobin (aquo Met) chain 2	2MHB	64.4	62.3
Hemoglobin (Met)-cyanide V	1LHB	56.1	53.4
Leghemoglobin (acetate Met)	1HBL	76.5	73.2
Crambin	1CRN	47.8	32.6
Ovomucoic third domain	1OVO	55.4	37.5
Streptomyces subtilisin inhibitor	2SSI	64.5	54.2
Trypsin inhibitor	4PTI	67.2	46.6
α-Cobratoxin	1CTX	74.6	62.0
Melittin	1MLT	42.3	34.6
Neurotoxin B	1NXB	72.6	58.1
Adenylate kinase	2ADK	68.6	56.2
Rhodanese	1RHD	60.1	49.5
Prealbumin	2PAB	45.6	33.3
Acid proteinase, *Endothia parasitica*	2APE	49.1	35.2
Aspartate carbamoyl transferase (chain 1)	4ATC	51.3	45.5
Aspartate carbamoyl transferase (chain 2)	4ATC	48.4	35.9
Catalase	7CAT	57.2	47.8
Hemoglobin (human deoxy) chain 1	2HHB	68.1	66.7
Hemoglobin (human deoxy) chain 2	2HHB	65.8	65.1
Malate dehydrogenase (chain 1)	2MDH	46.9	38.3
Immunoglobulin Bence Jones fragment	2RHE	64.0	52.6
Rubredoxin	4RXN	59.3	42.6
Results for the whole data base		56.9	46.5

[a]Parameters computed with a data base of 75 proteins (12,000 residues).

three states. In contrast, if the protein to be predicted was kept in the data base to calculate the directional information values, the overall quality of prediction went up to 60.7% for three states (+P in Table XII). Inspection of the information values obtained for each predicted residue with and without the protein to be predicted in the data base showed that they differ by 20 or 30 cnat; the contribution of the protein in the data base amounts to 1–2 cnat for each value of the displacement *m*. These small differences were enough to increase the quality of prediction by 3.8%, or 6% in relative value. To avoid these fluctuations and take advantage of the whole data base, the directional information was smoothed according to the solid lines in Fig. 5. In that case, the overall accuracy of prediction was 57.7% for three states, an absolute loss of 3% but an improvement of 0.8% compared with the prediction made by removing the protein to be predicted from the data base. The same was found with the GOR III method (Table XII), where the effect is even more pronounced. No predictive method is exempt from this effect; an example was studied by Kabsch and Sander (1983b) with Lim's method. Therefore, one should assess the accuracy of prediction without the protein included in the data base; this would indicate the real predictive value of the method for an unknown protein.

B. Influence of the Nature of the Protein on the Accuracy of Prediction: Its Size and Secondary Structure Content

Following the observation by Garnier *et al.* (1978) that a prediction can be improved by taking into account the total content of secondary structures, Garnier and Robson (1979) analyzed the quality of the prediction for different classes of proteins: those that are rich in helix, containing more than 20% α helix—α proteins; those that are poor in α helix (less than 20%) but rich in β sheets (more than 20%)—β proteins; and those that contain more than 20% α helix and 20% of β sheets—αβ proteins. With the Levitt and Greer (1977) data base, they found that the accuracy of prediction for four states was 67.3% for the α proteins (19 proteins or 2474 amino acid residues), 50% for the β proteins (16 proteins or 3022 residues), and 48.6% for the αβ proteins (16 proteins or 3402 residues). One reason was better prediction by the GOR I method of the helical conformational state. For instance, for the whole data base with a decision constant equal to zero, 57.9% of residues are correctly predicted in helix ($f_H \times 100$ for two states, H and $\bar{H}$) compared with 45.1% in E, 36% in T, and 27.9% in C conformational states. The GOR I method also predicted the residues in the H conformation better for the α proteins ($f_H \times 100 = 65.8\%$) than for the αβ proteins ($f_H \times 100 = 51.8\%$) and the β proteins ($f_H \times 100 = 29.5\%$). No such variation was found for the other three conformational states, E, T, and C, the prediction of which was 42–47%, 32–38%, and 27–29%, respectively, for the three different classes of proteins.

Proteins of fewer than 50 amino acids, five proteins of the Kabsch and Sander data base, are predicted by GOR II with 52.9% accuracy for three states. This contrasts with the higher prediction rate of proteins of larger size: 62.5% (three states) for nine proteins of 50 to 100 residues, 61.2% (three states) for 22 proteins of 100 to 200 residues, and 55.9% (three states) for ten proteins having more than 200 amino acid residues. It seems that the small proteins are not so well predicted. The bit pattern recognition method is more successful for small proteins (see below), perhaps because accessibility to the solvent plays a greater role in this method and in small proteins. For larger proteins the drop in accuracy either reflects the data base composition, as for the small proteins, or the folding becomes more dependent on long-range interactions for larger proteins.

C. Influence of the Postion along the Polypeptide Chain: N- and C-Terminal Sequences

Results concerning the N- and C-terminal sequences, defined as the first and last 40 amino acid residues in the sequences, are presented in Table XV. There is a significant increase in the accuracy of prediction for the C-terminal sequences. Their known frequent accessibility to antibodies suggests that this part of the sequence is less often buried in the folding of the whole chain and may be less sensitive to tertiary folding.

D. Relationship between the Value of the Information and the Quality of Prediction

The increase in accuracy of prediction brought about by the presence in the data base of the protein to be predicted, as shown in Section IV.A, has been related to the fact that the information values for at least two conformational states can be very close, within 10–20 cnat (directional information, GOR II method). Consequently, the addition to the data base of the protein to be predicted to calculate the directional information parameters leads to a significative improvement. Conversely, one can raise the following question: How different do the information values have to be to yield a prediction less dependent on the composition of the

Table XIV. GOR III Method Result for the Whole Data Base[a]

Proteins	Brookhaven identification	Results	
		3 states	4 states
Calcium-binding parvalbumin B	1CPV	61.1	53.7
L-Arabinose-binding protein	1ABP	53.9	47.4
Oxidized high-potential iron protein	1HIP	58.8	50.6
Cytochrome B_5	2B5C	52.9	50.6
Cytochrome B_{562}	156B	75.7	68.9
Cytochrome C	3CYT	62.1	54.4
Ferricytochrome C_2	3C2C	64.3	60.7
Cytochrome C_{550}	155C	58.2	45.5
Cytochrome C_{551}	351C	84.1	79.3
Ferredoxin	1FDX	68.5	46.3
Ferredoxin	3FXC	65.3	54.1
Flavodoxin	3FXN	66.7	61.6
Azurin	1AZU	53.6	40.8
Plastocyanin	1PCY	61.6	47.5
Avian pancreatic polypeptide	1PPT	83.3	75.0
Glucagon	1GCN	41.4	17.2
Insulin (chain 1)	1INS	23.8	9.5
Insulin (chain 2)	1INS	70.0	56.7
Phospholipase A_2	1BP2	52.0	39.8
Lysozyme bacteriophage T_4	1LZM	56.7	49.4
Lysosyme hen egg white	7LYZ	58.1	48.8
Staphylococcal nuclease	2SNS	54.6	45.4
Ribonuclease-S	1RNS	70.0	55.0
Ribonuclease-S (chain 2)	1RNS	58.7	49.0
Carboxypeptidase A	5CPA	58.6	45.9
Acid protease *Rhizopus chinensis*	1APR	—[b]	—[b]
Acid proteinase penicillopepsin	2APP	57.0	48.9
Thermolysin	3TLN	60.8	51.3
γ-chymotrypsin A	2GCH	69.7	58.9
γ-Chymotrypsin A (chain 2)	2GCH	61.1	54.7
α-Lytic protease	2ALP	58.6	47.5
β-Trypsin	1PTN	63.7	50.7
Proteinase A	2SGA	61.9	55.8
Subtilisin BPN′	1SBT	61.5	49.8
Tosyl elastase	1EST	63.3	49.6
Actinidin	2ACT	61.0	50.0
Papain	8PAP	62.3	47.6
λ-Immunoglobulin Fab	1FAB	67.6	56.0
λ-Immunoglobulin (chain 2)	1FAB	68.0	58.4
Bence Jones immunoglobulin	1REI	58.9	54.2
Phosphoglycerate mutase	3PGM	64.3	52.2
Triose phosphate isomerase	1TIM	68.8	60.7
Concanavalin A	3CNA	59.9	46.0
Carbonic anhydrase form C	1CAC	57.4	43.8
Dihydrofolate reductase	3DFR	55.6	46.9
D-Glyceraldehyde-3-phosphate dehydrogenase	1GPD	60.7	49.8
Apo-liver alcohol dehydrogenase	4ADH	56.4	46.5
Lactate dehydrogenase	4LDH	57.1	48.9
Glutathione reductase	2GRS	58.6	46.2
Cu,Zn-Superoxide dismutase	2SOD	66.2	49.7
Myoglobin	1MBN	75.8	72.5

Table XIV. (Continued)

Proteins	Brookhaven identification	Results 3 states	4 states
Hemoglobin (erythrocruorin)	1ECD	64.7	58.8
Hemoglobin (aquo Met)	2MHB	75.2	68.8
Hemoglobin (aquo Met) chain 2	2MHB	72.6	69.9
Hemoglobin (Met)-cyanide V	1LHB	56.8	52.0
Leghemoglobin (acetate Met)	1HBL	75.8	72.5
Crambin	1CRN	47.8	37.0
Ovomucoic third domain	1OVO	62.5	44.6
Streptomyces subtilisin inhibitor	2SSI	71.0	57.9
Trypsin inhibitor	4PTI	72.4	58.6
α-Cobratoxin	1CTX	71.8	64.8
Melittin	1MLT	46.2	38.5
Neurotoxin B	1NXB	74.2	58.1
Adenylate kinase	2ADK	71.1	60.3
Rhodanese	1RHD	61.4	51.2
Prealbumin	2PAB	48.2	36.8
Acid proteinase, *Endothia parasitica*	2APE	58.2	43.4
Aspartate carbamoyl transferase (chain 1)	4ATC	53.9	47.7
Aspartate carbamoyl transferase (chain 2)	4ATC	56.9	46.4
Catalase	7CAT	65.1	55.8
Hemoglobin (human deoxy) chain 1	2HHB	82.3	79.4
Hemoglobin (human deoxy) chain 2	2HHB	67.8	67.8
Malate dehydrogenase (chain 1)	2MDH	50.9	42.0
Immunoglobulin Bence Jones fragment	2RHE	66.7	54.4
Rubredoxin	4RXN	61.1	42.6
Results for the whole data base		61.7	51.9

[a]Parameters are computed with a data base of 74 polypeptide chains (12,433 residues). These parameters are biased by addition of 225 dummy observations. Only pair interactions are used (P_{max} = 1). The decision constant used for the extended conformation is DC_E = 35 and is subtracted from the information values.
[b]The secondary structure given by Kabsch and Sander for this protein is doubtful. Therefore, it was removed from the data base.

Table XV. Quality of Prediction along the Amino Acid Sequences with the GOR II Method[a]

	3 states	4 states
First 40 amino acid residues (N-terminal sequence)	57.9	47.6
Last 40 amino acid residues (C-terminal sequence)	60.6	49.0
Kabsch and Sander (1983a) data base	57.7	46.7

[a]With smoothed values of the directional information in Tables II to V. The results are expressed as percentage of correctly predicted residues for three states (H, E, and C) and four states (H, E, T, and C).

Table XVI. Number of Residues Correctly Predicted as a Function of the Threshold (V)[a]

Prediction using directional information					
Threshold (V) (cnat)	0	50	100	200	300
Percentage H correctly pred.	52.5	55.2	57.8	66.6	74.7
Percentage E correctly pred.	42.7	47.5	52.6	64.4	75.7
Percentage C correctly pred.	74.9	78.0	80.9	87.2	92.6
Percentage res. correctly pred.	58.0	61.8	65.3	74.3	82.6
Number of res. involved.	11,237	8756	6468	3140	1300
Prediction using pair information					
Threshold (V) (cnats)	0	50	100	200	300
Percentage H correctly pred.	57.6	62.2	67.8	78.8	84.5
Percentage E correctly pred.	51.1	57.9	64.9	76.4	93.9
Percentage C correctly pred.	71.1	75.6	80.9	86.6	93.0
Percentage res. correctly pred.	63.0	68.7	74.6	83.9	91.7
Number of res. involved.	11,237	8200	5710	2531	900

[a]The data base used contains 11,237 residues: 3226 helical residues, 2355 extended residues, and 5656 coil residues. The prediction using pair information was done with $P_{max} = 1.0$ and $M = 225$. The decision constants were $DC_H = 25$ and $DC_E = 30$. The column $V = 0$ gives the result of the prediction; i.e., the number of residues involved is simply the number of residues in the data base. The percentages represent the quality index P of equation 5 (Gibrat *et al.*, 1987).

data base? The answer to this question was pursued by Gibrat *et al.* (1987) in a slightly different way by examining the variation of the accuracy of prediction with a threshold defined by considering only regions of the sequences for which the information values (equations 27 to 30) are greater than a threshold of 50, 100, 200, or 300 cnat for one conformational state compared to any three other conformational states. These regions are better predicted than with a threshold of zero; for example, with a threshold of 100 cnat about half of the residues have a 75% probability of being predicted correctly with the GOR III method (Table XVI). About 25% of the residues have a probability of about 84% of being predicted correctly. It is then possible to print with each prediction along the amino acid sequence the range of probability for that residue to be predicted correctly (Gibrat *et al.*, 1987).

V. ASSOCIATION OF THE GOR METHOD WITH OTHER PREDICTION METHODS

Can different prediction methods based on different algorithms or properties of amino acids help to improve secondary structure prediction? An attempt has been made in our laboratory to combine the GOR III method with two other methods that we recently developed: the homologue method of Levin *et al.* (1986) and the bit pattern recognition method of Biou *et al.* (1988). These two methods are described before we discuss the algorithm of combination and its results.

A. The Homologue Method

The algorithm of this method is based on the hypothesis that short similar sequences of amino acids must have the same secondary structure tendencies, if not actually the same observed conformations, whether they come from homologous or nonhomologous proteins.

The main feature of this method relies on the definition of similar sequences. From observations of the amino acid sequences of proteins performing the same function in different species, it soon appeared that there are conserved amino acid residues; that is to say identical residues are found either at the same position in the sequence or slightly displaced in the sequence. These can be restored to identical positions if one allows a small number of amino acid deletions or insertions along the sequence. It also soon became evident that identity of amino acids was a too severe criterion for asserting family relationships between proteins from different species and that some amino acids in the sequence can be replaced by others without affecting the function or the tertiary structure. This means that there are amino acids that are similar although not identical, and the homology between two proteins can be expressed by two figures, the percentage of identical residues or the percentage of similar amino acids residues. This last figure includes the degeneracy of the code relating amino acid sequence to the tertiary structure and function of a polypeptide chain. The similarity between amino acids can be based on their physicochemical properties such as charge, solubility in water, and size or more empirically from the observed rate of amino acid substitution within a family of proteins leading to a "relatedness odds matrix" or similarity matrix such as the one set up by Schwartz and Dayhoff (1978). These matters and related subjects are examined in Chapter 4 of this volume.

A similarity matrix has been empirically developed (Levin *et al.*, 1986) to optimize secondary structure prediction. This matrix, shown in Fig. 6, assigns a similarity score between amino acids, and it differs in many respects from that of Schwartz and Dayhoff (1978). It can be considered a secondary structure similarity matrix. The amino acids G, P, T, H, C, and W are considered to have unique secondary structure properties, their replacement score is never greater than zero. There are several pairs of amino acids considered to have secondary structure properties in common: D and E, D and N, E and A, A and S, Q and N, Q and E, N and K, K and R, V and I, V and L, M and L, I and F, and F and Y. Their replacement score is 1 except for M and L, which have a score of 2. Other amino acid replacements considered dissimilar have a score of −1. All amino acids replaced by themselves have a score of 2 except P and N, with a score of 3.

	G	P	D	E	A	N	Q	S	T	K	R	H	V	I	M	C	L	F	Y	W
G	2																			
P	1	3																		
D	0	0	2																	
E	0	-1	1	2																
A	0	-1	0	1	2															
N	0	0	1	0	0	3														
Q	0	0	0	1	0	1	2													
S	0	0	0	0	1	0	0	2												
T	0	0	0	0	0	0	0	0	2											
K	0	0	0	0	0	1	0	0	0	2										
R	0	0	0	0	0	0	0	0	0	1	2									
H	0	0	0	0	0	0	0	0	0	0	0	2								
V	-1	-1	-1	-1	0	-1	-1	-1	0	-1	-1	-1	2							
I	-1	-1	-1	-1	0	-1	-1	-1	0	-1	-1	-1	1	2						
M	-1	-1	-1	-1	0	-1	-1	-1	0	-1	-1	-1	0	0	2					
C	0	0	0	0	0	0	0	0	0	0	0	0	0	0	0	2				
L	-1	-1	-1	-1	0	-1	-1	-1	0	-1	-1	-1	1	0	2	0	2			
F	-1	-1	-1	-1	-1	-1	-1	-1	-1	-1	-1	-1	0	1	0	-1	0	2		
Y	-1	-1	-1	-1	-1	-1	-1	-1	-1	-1	-1	0	0	0	0	-1	0	1	2	
W	-1	-1	-1	-1	-1	-1	-1	-1	-1	-1	0	-1	0	0	0	-1	0	0	0	2

Figure 6. The secondary structure similarity matrix used by Levin *et al.* (1986) with the scores assigned to each amino acid replacement (monoletter code).

The algorithm makes a comparison between every sequence n residues in length in the test protein and every n-residue fragment in the Kabsch and Sander data base. Each time a comparison is made, the similarity score between the two sequences is calculated. For instance, for $n = 7$ the test sequence STNGIYW will have a similarity score of 7 with the sequence ATSGVFL of the data base according to the replacement scores of Fig. 6. Suppose that the sequence ATSGVFL has an observed conformation CHHHHHT; the residue S of the test sequence will be assigned a score of 7 for the coil conformation and 0 for the others. The second residue, T, of the test sequence will be assigned the conformation H with a score of 7 and zero for the other conformations, and so on for the other residues. Every time a sequence of seven residues is found in the data base whose score is equal to or greater than a cutoff value of 7, its observed conformation is assigned to the test sequence with its similarity score as explained above and added to the previous score for each conformation. For instance, if a second peptide is found with a score of 8 and the observed conformation CEHHHHT, the residue S of the test sequence will be assigned a score of 8 for the coil conformation, which becomes now $7 + 8 = 15$, and zero for the other conformations. The residue T will have a score of 7 for H and 8 for E, and so on.

Once every fragment in the test protein has been compared in an iterative procedure—in the example given above, first residues 1–7 are compared against the data base, then residues 2–8—the secondary structure attributed to each residue is that which has the highest score value. In our example S will be predicted coil, and T will be predicted as E. In order to avoid overprediction of helix and underprediction of aperiodic structure, decision constants are used; i.e., the score values for the conformation H are reduced by 20%, the score values for the conformation C are increased by 5%, the score values for conformation T by 40%, and those for conformation S by 30%. This method predicts eight states as defined in the Kabsch and Sander dictionary (1983a). On a three-state basis with a sequence length n of 17 residues, this method has a prediction accuracy of 63% for 72 polypeptide chains or 12,058 residues if homologous proteins with a homology greater than 22% identical residues are removed from the data base. This method is very sensitive to the presence of homologous proteins in the data base. When this is properly taken into account, the accuracy gets up to 87% per chain with an identity greater than 30% (Levin and Garnier, 1988).

B. The Bit Pattern Recognition Method

This method is based on the recognition of a pattern of hydrophobic/hydrophilic residues along the sequence according to the known stereoregular features of helices and sheets, as was done by Schiffer and Edmundson (1967) for the helix.

1. A Hydrophobicity Scale Based on Information Theory

Many hydrophobicity scales have been proposed (see details and treatment of hydrophobicity profiles in Chapter 17 of this volume). However, it seemed worthwhile to develop a statistical scale of hydrophobicity based on the observed solvent accessibility of residues in proteins of known structure and using the formalism of information theory (Gibrat, 1986). The Kabsch and Sander (1983a) data base provides for each residue of the protein data base the number of water molecules in contact to its side chain according to the algorithm of Lee and Richards (1971). This number is proportional to the area of the side chain in contact with the solvent. By comparing this area to the total surface area of the amino acid side chain (Rose *et al.*, 1985), one can determine the percentage of total surface accessible to the solvent for each amino acid residue of the data base. A particular amino acid is considered to be exposed (e) when its surface accessible to the solvent is greater than or equal to 50% of its total surface.

Table XVII. Scale of Hydrophobicity (+) of the Amino Acid Residues in Proteins[a]

Amino acid	*I*
Cys	224
Ile	178
Leu	173
Phe	172
Met	135
Val	129
Trp	122
His	44
Tyr	36
Ala	6
Gly	−2
Pro	−30
Thr	−31
Asn	−38
Asp	−53
Gln	−59
Ser	−67
Arg	−81
Glu	−88
Lys	−147

[a]When more than 50% of their surface area is buried. From Gibrat (1986).

Otherwise it is considered buried in the interior of the protein (i). From this, the following information value was calculated from the data base:

$$I(S = \text{i:e};R) = \log \{f(\text{i},R)/[1 - f(\text{i},R)]\} - \log \{f(\text{i})/[1 - f(\text{i})]\} \qquad (33)$$

where $f(\text{i},R)$ is the number of times an amino acid residue of type R is found inside a protein, and $f(\text{i})$ is the number of amino acid residues found inside a protein whatever the amino acid type.

The first term on the right-hand side of equation 33 corresponds to the partition coefficient of the amino acid residue R between the interior and the exterior of a protein. This quantity multiplied by RT represents the free energy of transfer from the interior to the exterior of a protein of an amino acid residue R. The second term on the right-hand side takes into account the relative abundance of interior and exterior residues of the data base. It accounts for the average size of the proteins in the data base, since a residue in a small protein has more chance to be exposed than one in a larger protein.

The information values calculated from the data base of Kabsch and Sander (1983a) are presented in Table XVII and represent a hydrophobicity scale: the greater the information value, the more often this amino acid residue is found in the interior of a protein. One may notice that this scale gives a specific role to Ala and Gly with an information value close to zero. However, the ranking of the amino acid in the scale varies with the definition of exposure (Biou *et al.*, 1988).

2. Bit Pattern of Amino Acid Residues for α-Helix and β-Sheet Structures

Biou *et al.* (1988), making use of the hydrophobicity scale of Table XVII, designated the various patterns of hydrophobic and hydrophilic residues for stereoregular secondary structures: α helix and β sheet. The hydrophobic residues—C, I, L, F, M, V, W, H, and Y—were given a value of 1; the hydrophilic residues—K, E, R, S, Q, D, N, T, and P—were given a value of 0; and A and G were given a value of 0 or 1. In a sequence of eight amino acid residues, this gives a sequence of 0 and 1 numbers that can be considered as a binary number, where the binary value of the first N-terminal amino acid residue of the eight-amino-acid sequence corresponds to the unit value (2^0) of the binary number made by the eight-digit numbers. For instance, the amino acid sequence VLEACMST is written 11001100 or 11011100 because A can be either 0 or 1. These correspond to the binary numbers 110011 and 111011 or, respectively, to the decimal numbers 51 and 59. These arrangements manage a hydrophobic face to the α helix and are considered favorable to the α helix. These arrangements form hydrophobic arcs similar to the helical wheel rule of Schiffer and Edmundson (1967), but with two turns of helix rather than five. The decimal numbers corresponding to bit patterns considered favorable to α helix are the following: 9, 12, 13, 17, 18, 19, 25, 27, 29, 31, 34, 36, 38, 44, 45, 46, 47, 50, 51, 54, 55, 59, 61, 62, 77, 201, 205, 217, 219, and 237. For β sheet a series of hydrophobic residues $(1)_n$ or a series of alternate hydrophobic and hydrophilic residues $(10)_n$ were considered favorable.

3. Algorithm of the Bit Pattern Recognition Method and Its Results

A sequence of eight residues in the test protein is given a decimal number according to the hydrophobic scale and numbering proposed above. If this number is compatible with an α-helical structure (see the decimal list in Section V.B.2), the eight residues of the test sequence are considered as h. Then the test sequence is shifted one residue, e.g., from residues 1–8 to residues 2–9, and a new numbering is made. If this number is compatible with an α helix, the residues 2–9 are considered as h; if not, residues 2–9 are considered as $\bar{h}$. This is continued to the end of the sequence. When an $\bar{h}$ region contains Gly as first or last residue or a Pro as the first two residues, this region is considered as $\bar{h}$. Overlapping sequences of h are considered H; that is, all residues are predicted H from the first residue on the N terminal of the first overlapping sequence to the last C terminal of the last overlapping sequence. Then β-sheet regions are determined as in Section V.B.2, and the residues are considered as β sheet or E when there are more than two consecutive residues as E. In cases in which residues have both H or $\bar{h}$ and E conformational states, the assignments of H or h are conserved. Residues not assigned as H, h, or E are considered coil. These simple algorithms correctly predict 59% of the amino acid residues in three states—(H + h), E, and C—with a ratio $F_S^{+}/F_S \times 100$ of 54.4% for (H + h), 23.3% for E, and 76.5% for C. The prediction of the conformational state E is rather low and underpredicted in contrast to H and to a greater degree C, which are overpredicted. It is for the conformational state C that the quality index P (equation 5) is the highest: 0.626 against 0.54 and 0.515 for H and E, respectively. For proteins of less than 50 residues (five proteins and 208 residues), 73.1% of residues were correctly predicted in three states.

C. Algorithm of Association of GOR with Homologue and Bit Pattern Recognition Methods: The Combined Expert System

A tentative assay to develop an expert system to improve secondary structure has been undertaken by Biou *et al.* (1988) in our laboratory with three prediction methods: GOR III (Gibrat *et al.*, 1987), the homologue method (Levin and Garnier, 1988), and the bit pattern recognition method.

The algorithm takes advantage of some features specific to each method: the bit pattern is

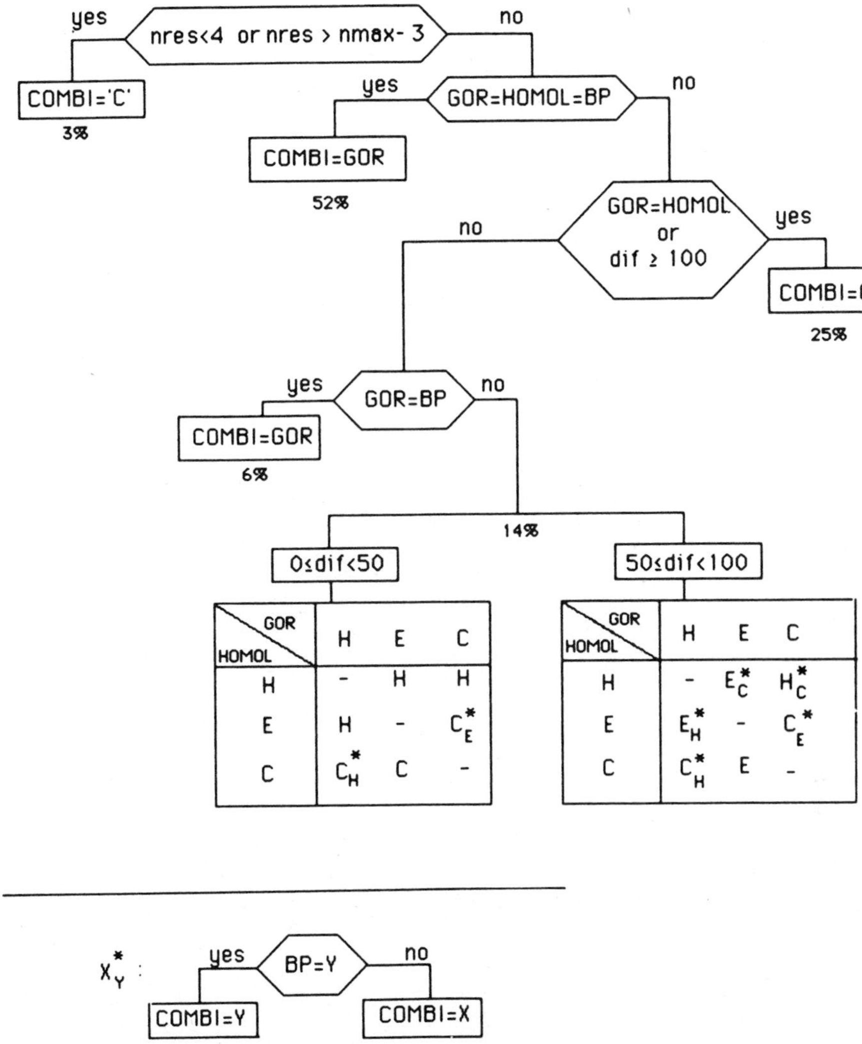

yes nres<4 or nres > nmax- 3 no

COMBI='C'

3%

yes GOR=HOMOL=BP no

COMBI=GOR

52%

no GOR=HOMOL or dif ≥ 100 yes

COMBI=GOR

25%

yes GOR=BP no

COMBI=GOR

6%

14%

0≤dif<50

HOMOL \ GOR	H	E	C
H	-	H	H
E	H	-	C^*_E
C	C^*_H	C	-

50≤dif<100

HOMOL \ GOR	H	E	C
H	-	E^*_C	H^*_C
E	E^*_H	-	C^*_E
C	C^*_H	E	-

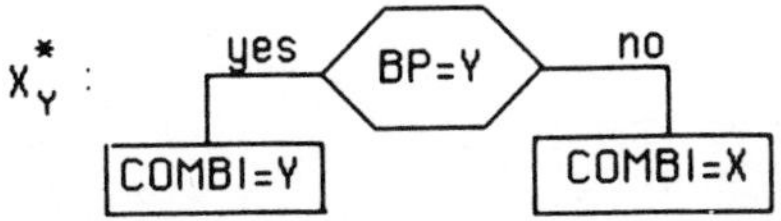

Figure 7. Flow chart of the expert program COMBINE. GOR, prediction obtained with the GOR III method (pair information) of Gibrat *et al.* (1987); HOMOL, prediction obtained with the homologue method of Levin and Garnier (1988); BP, bit pattern recognition method (Biou *et al.*, 1988). Reprinted by permission from Biou *et al.* (1988). The percentages under the assignments of COMBI indicate which proportion of the residues in the data base are involved in this assignment. dif, Difference between the information value of the predicted state (the largest one), and the second largest value; nmax, number of residues in the polypeptide chain; nres, position of the current residue in the polypeptide chain to be predicted.

more effective for small proteins (fewer than 50 residues); the GOR is most effective when the threshold is equal to or greater than 200 for H and E or 50 for T or C. By and large it was observed that when the three methods agreed, the accuracy of prediction was 75% for three states. When one of the methods disagreed with the other two, the residue was usually assigned the conformation predicted by the method for which it was found that the accuracy of prediction is the greatest; the same is done when the three methods disagree. The final algorithm of the expert program is given in Fig. 7. Isolated residues $\bar{H}$, $\bar{E}$, $\bar{T}$, and $\bar{C}$ in a sequence of H or E are made H or E, respectively.

Table XVIII. *Comparative Results for the Whole Data Base*[a]

Protein	File name	GOR III	Homologue	Bit pattern	Combine
Acid proteinase (Penicillium J)	2APP	59.4	56.3	56.0	58.5
Actinidin	2ACT	63.8	53.7	55.5	61.0
Agglutinin (wheat germ)	3WGA	76.5	74.1	81.2	76.5
Alcohol dehydrogenase (Apo)	4ADH	56.1	54.8	53.2	57.2
Alpha lytic protease	2ALP	65.7	50.0	48.5	59.6
Aspartate carbamoyltransferase	4ATC(1)	61.0	59.7	52.6	63.2
Aspartate carbamoyltransferase	4ATC(2)	54.2	55.6	51.0	56.2
Azurin (Alcaligenes denitrif.)	1AZA	45.7	45.7	45.7	41.9
Alpha-bungarotoxin	2ABX	81.1	81.1	66.2	86.5
Ca-binding parvalbumin	1CPV	57.4	58.3	55.6	62.0
Ca-binding protein (intestinal)	3ICB	88.0	94.7	66.7	92.0
Carbonic anhydrase B (human)	2CAB	62.9	64.8	57.4	63.7
Carboxipeptidase A	5CPA	59.6	66.7	59.9	67.4
Catalase (beef liver)	8CAT	66.5	65.3	57.0	67.3
Alpha chymotrypsin A (bovine)	5CHA(1)	66.4	65.6	53.4	64.9
Alpha chymotrypsin A (bovine)	5CHA(2)	59.8	60.8	52.6	57.7
Citrate synthase (pig)	2CTS	67.3	65.2	68.6	74.6
Crambin	1CRN	54.3	56.5	67.4	54.3
Gamma-II crystallin (calf)	1GCR	60.9	54.6	47.7	60.9
Cytochrome C (rice)	1CCR	72.1	61.3	57.7	74.8
Cytochrome C (prime)	2CCY	77.2	71.7	63.0	79.5
Cytochrome C perioxidase (yeast)	2CYP	63.5	66.6	57.0	67.9
Cytochrome C2 (reduced)	3C2C	63.4	64.3	64.3	69.6
Cytochrome C3 (D. vulgaris)	2CDV	68.2	73.8	55.1	71.0
Cytochrome C551 (oxidized)	351C	81.7	80.5	78.0	79.3
Dihydrofolate reducatase (L. casei)	3DFR	51.2	61.7	60.5	53.7
Elastase (porcine)	2EST	64.6	62.1	64.2	69.6
Erabutoxin B (sea snake)	2EBX	82.3	71.0	59.7	71.0
Erythrocruonin (reduced deoxy)	1ECD	58.8	74.3	58.1	71.3
Ferredoxin (P. aerogenes)	1FDX	79.6	70.4	85.2	81.5
Ferredoxin (S. platensis)	3FXC	61.2	67.3	68.4	67.3
Flavodoxin (clos. MP, oxidized)	3FXN	65.9	68.1	64.5	71.0

Ferredoxin (azobacter vinel)	2FD1	62.3	64.2	58.5	67.3
Glutathione peroxidase (bovine)	1GP1	62.0	62.5	64.7	62.0
Hemerythrin (Met)	1HMQ	69.9	52.2	66.4	66.4
Hemoglobin (human, deoxy)	2HHB(1)	68.1	61.7	56.7	70.8
Hemoglobin (human, deoxy)	2HHB(2)	64.4	56.8	44.5	68.5
Hemoglobin V (cyano, met, lamprey)	2LHB	65.8	73.2	64.4	75.2
High potential iron protein	1HIP	60.0	56.6	54.1	60.0
IGG FAB (kappa) MCPC603	1MCP(1)	64.1	67.3	55.5	66.4
IGG FAB (kappa) MCPC603	1MCP(2)	62.6	67.6	50.0	58.1
Kallikrein (porcine)	2PKA(1)	58.7	63.7	53.7	66.2
Kallikrein (porcine)	2PKA(2)	62.5	60.5	69.1	65.8
Lactate dehydrogenase	4LDH	59.6	52.9	62.3	61.7
Lehemoglobin	1LH1	75.2	76.5	60.8	79.7
Lysozyme (bacteriophage T4)	2LZM	67.7	69.5	70.7	70.7
Lysozyme (human)	1LZ1	60.8	62.3	60.8	63.1
Myoglobin (sperm whale, met)	1MBN	77.1	80.4	68.6	80.4
Melittin	1MLT	61.5	76.9	69.2	76.9
Scorpion neurotoxin (variant)	1SN3	67.7	73.8	67.7	75.4
Ovomucoid third domain (quail)	10V0	57.1	60.7	64.3	64.3
Papain D	1PPD	62.7	63.7	67.0	63.2
Phopholipase A2 (bovine)	1BP2	48.8	50.4	65.0	49.6
Plastocyanin	1PCY	67.7	72.7	70.7	70.7
Prealbumin (human plasma)	2PAB	56.1	55.3	46.5	56.1
Proteinase A (Strep. griseus)	2SGA	69.1	54.1	47.5	67.4
Proteinase II (rat mast cell)	3RP2	54.9	56.7	56.7	55.4
Ribonuclease A	1RN3	56.5	60.5	57.3	51.3
Rubredoxin	5RXN	75.9	74.1	74.1	77.8
Staphylococcal nuclease	2SNS	52.5	56.7	56.7	55.3
Subtilisin BPN prime	1SBT	63.3	65.8	58.9	65.1
Superoxide dismutase	2SOD	73.5	73.5	72.6	75.5
Thermolysin	3TLN	56.6	61.4	55.1	64.6
Trypsin (orthorhombic)	1TPO	67.3	51.4	62.3	67.3
Trypsin Inhibitor (bovine)	4PTI	70.7	79.3	70.7	82.8
Virus (satellite tobacco necrs)	2STV	57.6	57.1	46.2	54.3
Virus coat protein (SBMV, T=3)	4SBV	56.3	51.8	44.6	53.6

(continued)

Table XVIII. *(Continued)*

Protein	File name	GOR III	Homologue	Bit pattern	Combine
Global percentage[b]		63.3	62.9	59.0	65.5
Mean percentage for a protein[c]		63.9	63.9	59.9	66.3
Mean deviation for a protein[d]		6.4	7.2	7.0	7.2

[a]The percentage of correctly predicted residues in three states, H, E, and C, for the methods GOR III, homologue, bit pattern, and combine for each protein in the data base is given. File name refers to the Brookhaven Protein Data Bank name.

[b]Global percentage of correctly predicted residues = NCOR / NRES with NCOR = total number of correctly predicted residues in three states. NRES = total number of residues in the data base.

[c]Mean percentage: $MP = [\Sigma_{prot} (per_{prot})] / NPROT$.

[d]Mean absolute deviation of these percentages around the mean value: $MD = [\Sigma_{prot} |per_{prot} - MP|] / NPROT$, with per_{prot} = % of correctly predicted residues for each protein "prot"; NPROT = number of proteins in the data base.

The results of this expert system are shown in Table XVIII for three states, and the results obtained by GOR/bit pattern and the homologue methods are also listed for comparison. Clearly combining three prediction methods leads to an overall improvement of about 2.5%, reaching the level of 65.5% correctly predicted residues in three states for the whole data base. A notable observation is the regular decrease in accuracy of prediction with increased size of the proteins: from 73.1% for proteins of 50 residues or fewer to 66.2% for proteins of 100–200 amino acid residues (4270 residues) down to 61.2% for proteins with over 300 residues (4522 amino acid residues).

VI. PRACTICAL USE AND APPLICATIONS OF THE GOR APPROACH

It is fully acknowledged that the prediction of secondary structure gives imperfect results, and indeed this is inherent in the statistical approach that is employed. It also explains our increased interest over many years in the use of energy modeling of peptide and protein structures (Robson and Garnier, 1986). Indeed, one major application of secondary structure predictions is to provide a starting point and possibly a bias for such calculations but never to propose a secondary structure that is subsequently "locked" throughout the entire simulation.

Other important applications emphasized in the development of our studies are as follows. First, prediction of secondary structure provides a probabilistic template for identifying weakly homologous domains in the data base of known protein conformation. A domain or protein of known conformation is sought that has a similar pattern of secondary structure to that predicted. Second, the prediction of epitopic sites provides artificial peptides to serve as a basis for vaccines or diagnostics or biochemical probes. The "classic" Hopp–Woods procedure emphasizes the average polar character of segments of sequence, the most polar being the most promising as a sequence to be synthesized, linked to an immunogenic carrier, and used to raise antibodies. Unfortunately, there are many surface features that are not polar on average that can be very strong epitopes. The α helix, for example, is normally a surface feature yet has a strongly hydrophobic face against the protein core. We have developed methods based primarily on bit pattern recognition and secondary structure prediction methods that allow for this. They have turned out to be remarkably successful in studies to be reported elsewhere, yet the epitopes involved were not always highly polar. Finally, circular dichroism data are used to aid predictions or rather the use of predictions to assign the locations of secondary structures once composition of secondary content in the protein is experimentally established (Garnier *et al.*, 1978).

In all the above there is one important technique that we have often used in real applied studies. This is the identification of sequences homologous to that of interest and the use of a consensus prediction carried out on all the homologues. This is very useful for simply comparing results on each sequence, combining H, E, T, and C information values from all sequences prior to the final prediction, or introducing further rules for prediction based on homology arguments. For example, glutamate or alanine can replace proline or glycine in a strongly polar or loop segment. Insertion/deletion-rich regions are generally surface loops, in which case the homology alignment aids prediction. These are in turn also epitopic sites. It also follows that an ideal route in prediction of tertiary structure proceeds with such alignment and prediction followed by testing on the protein with antibodies raised against the predicted epitopes. That is, all aspects of the above come into play in an interactive, even recursive, way. Of course, if the three-dimensional structure of a homologue is known, this helps greatly, but it does not mean that secondary structure aspects are outmoded in such a case. Secondary structure can be different except in the very closest homologues. Even so, the prediction method available from us includes a method to boost secondary structure on homologues substantially if the conformation of one of them is known. Briefly, the "tertiary information" that would be required to give a perfect result for the protein of known conformation is calculated, and this is added over the proteins of unknown conformation. Finally, the complex algebra of supersecondary structure (Robson and Garnier, 1986) can be used to show what differences are permissible between secondary structures of weak homologues.

This potpourri of interrelated techniques well justifies the notion of an "expert" for the full suite of techniques being developed.

VII. CONCLUDING REMARKS

Within a 10-year period, secondary structure prediction have improved from 55% to 65% (three states). Some data such as the threshold values or agreement between different methods open the hope of reaching a 75% accuracy of prediction. One may wonder why this still seems to be a limit. Besides the shortage of x-ray structures, an obvious reason is the influence of the tertiary folding on the realization of a secondary structure. What are the hopes of overcoming this difficulty in the near future?

Many examples exist to show that secondary structures are organized in specific motifs named supersecondary structures. A careful analysis of these has been done by Richardson (1981), and some aspects of these features are presented in Chapter 15 of this volume. Attempts have been made to improve the prediction of some supersecondary structures such as the sequence β sheet–α helix–β sheet (Taylor and Thornton, 1983). Possibly sequence homologies can also be brought to improvement to identify these supersecondary structures and then propose a most probable short-range association of secondary structures.

Energy minimization has not been convincing when applied to large proteins, although it is giving valuable results for peptides and peptide loops (Robson and Garnier, 1986). However, with the improvement of potential functions and the advent of fast computers, the increase of accuracy of secondary structure and supersecondary structure prediction should bring in the near future a clue to the tertiary folding of protein from the amino acid sequence alone. Then the route will be open to a true function–amino-acid-sequence relationship to assist protein engineers and to direct genetic engineers in making new proteins as better catalysts, more thermostable enzymes, or new agonists or antagonists of neuropeptides or hormones. At present, secondary structure predictions are currently an aid in molecular modeling of a protein on a weakly similar protein of known structure, and should be one of the aids in choosing the residues to be changed by site-directed mutagenesis.

VIII. REFERENCES

Bernstein, F. C., Koetzle, T. F., Williams, G. J. B., Meyer, E. F., Brice, M. D., Rodgers, J. R., Kennard, O., Shimanouchi, T., and Tasumi, M., 1977, The protein data bank: A computer-based archival file for macromolecular structures, *J. Mol. Biol.* **112**:535–543.

Biou, V., Gibrat, J. F., Levin, J., Robson, B., and Garnier, J., 1988, Secondary structure prediction: Combination of three different methods, *Protein Engineering* **2**:185–191.

Brillouin, L., 1956, *Science and Information Theory,* Academic Press, New York.

Busetta, B., and Hospital, M., 1982, An analysis of the prediction of secondary structures, *Biochim. Biophys. Acta* **701**:111–118.

Chou, P. Y., and Fasman, G. D., 1974, Prediction of protein conformation, *Biochemistry* **13**:222–245.

Cohen, F. E., Abarbanel, R. M., Kuntz, I. D., and Fletterick, R. J., 1986, Turn prediction in proteins using a pattern matching approach, *Biochemistry* **25**:266–275.

Fano, R., 1961, *Transmission of Information,* John Wiley & Sons, New York.

Fisher, R. A., 1934, *Statistical Methods for Research Workers,* Oliver and Boyd, London, pp. 99–101.

Garnier, J., and Robson, B., 1979, Classes of protein folding and accuracy of prediction, in: *Workshop on Protein Structure,* CECAM, Orsay, France, pp. 147–148.

Garnier, J., Salesse, R., Rerat, B., Rerat, C., and Blake, C., 1976, Comparison of x-ray data to estimated secondary structures from amino acid sequence and circular dichroism of human prealbumin, *J. Chimie Phys.* **73:**1019–1023.

Garnier, J., Osguthorpe, D. J., and Robson, B., 1978, Analysis of the accuracy and implications of simple method for predicting the secondary structure of globular proteins, *J. Mol. Biol.* **120:**97–120.

Gibrat, J. F., 1986, *Modelisation by Computers of the 3-D Structure of Proteins,* Ph.D. thesis, University of Paris VI, Paris.

Gibrat, J. F., Garnier, J., and Robson, B., 1987, Further developments of protein secondary structure prediction using information theory. New parameters and consideration of residue pairs, *J. Mol. Biol.* **198:**425–443.

Kabsch, W., and Sander, C., 1983a, Dictionary of protein secondary structure: Pattern recognition of hydrogen-bonded and geometrical features, *Biopolymers* **22:**2577–2637.

Kabsch, W., and Sander, C., 1983b, How good are predictions of protein secondary structure? *FEBS Lett.* **155:** 179–182.

Lee, B., and Richards, F. M., 1971, The interpretation of protein structures: Estimation of static accessibility, *J. Mol. Biol.* **55:**379–400.

Levin, J. M., and Garnier, J., 1988, Improvements in a secondary structure prediction method based on a search for local sequence homologies and its use as a model building tool, *Biochim. Biophys. Acta.* **955:**283–295.

Levin, J. M., Robson, B., and Garnier, J., 1986, An algorithm for secondary structure determination in proteins based on sequence similarity, *FEBS Lett.* **205:**303–308.

Levitt, M., and Greer, J., 1977, Automatic identification of secondary structure in globular proteins, *J. Mol. Biol.* **114:**181–293.

Lim, V. I., 1974, Algorithm for prediction of α-helical and β-structural regions in globular proteins, *J. Mol. Biol.* **88:**873–894.

Matthews, B. B., 1975, Comparison of the predicted and observed secondary structure of T_4 phage lysozyme, *Biochim. Biophys. Acta* **405:**442–451.

Maxfield, F. R., and Scheraga, H. A., 1975, The effect of neighboring charges on the helix forming ability of charged amino acids in proteins, *Macromolecules* **8:**491–493.

Maxfield, F. R., and Scheraga, H. A., 1976, Status of empirical methods for the prediction of protein backbone topography, *Biochemistry* **15:**5138–5153.

Nakashima, H., Nishikawa, K., and Ooi, T., 1986, The folding type of a protein is relevant to the amino acid composition, *J. Biochem.* **99:**153–162.

Pain, R. H., and Robson, B., 1970, Analysis of the code relating sequence to secondary structure in proteins, *Nature* **227:**62–63.

Richardson, J. S., 1981, The anatomy and taxonomy of protein structure, *Adv. Protein Chem.* **34:**167–339.

Robson, B., 1974, Analysis of the code relating sequence to conformation in globular proteins, *Biochem. J.* **141:**853–867.

Robson, B., and Garnier, J., 1986, *Introduction to Proteins and Protein Engineering,* Elsevier, Amsterdam.

Robson, B., and Pain, R. H., 1971, Analysis of the code relating sequence to conformation in proteins: Possible implications for the mechanism of formation of helical regions, *J. Mol. Biol.* **58:**237–259.

Robson, B., and Suzuki, E., 1976, Conformational properties of amino acids residues in globular proteins, *J. Mol. Biol.* **107:**327–356.

Rose, G. D., Geselowitz, A. R., Lesser, G. J., Lee, R. H., and Zehfus, M. H., 1985, Hydrophobicity of amino acid residues in globular proteins, *Science* **229:**834–838.

Schiffer, M., and Edmundson, A. B., 1967, Use of helical wheels to represent the structures of protein and to identify segments with helical potential, *Biophys. J.* **7:**121–135.

Schwartz, R. M., and Dayhoff, M. O., 1978, Matrices for detecting distant relationships in: *Atlas of Protein Sequence and Structure,* Volume 5, Suppl. 3 (M. O. Dayhoff, ed.), National Biochemical Research Foundation, Washington, pp. 353–358.

Taylor, W. R., and Thornton, J. M., 1983, Prediction of supersecondary structure in proteins, *Nature* **301:**540–542.

11

Prediction of Packing of Secondary Structure

Kozo Nagano

I. INTRODUCTION

It is believed that a unique tertiary structure of a protein is dictated by its amino acid sequence. The folding process is finished within a limited time, possibly by the testing of a reasonable number of candidate conformations and not by searching a myriad of conceivable structures. The role of the side chains of hydrophobic amino acids such as Leu, Ile, Val, and Phe in the nucleation must be very important for a stable tertiary structure to be attained within a limited period of time. On the other hand, the methods proposed so far to simulate the folding process have met with insurmountable difficulties: energy-minimization approaches lead to the multiple-minima problem, and combinatorial approaches sometimes have to handle an enormous number of conformations to be tested.

The aim of this chapter is to describe a method of reducing the number to a testable one by use of a present-day computer and to show the systematic and stepwise prediction of tertiary structure by numerical interpretation of the sequence information and incorporating a set of empirical laws. It is a learning process about how a protein folds itself. Starting from a simple-minded folding mechanism that does not consume much computer time, we can develop more

Kozo Nagano • Faculty of Pharmaceutical Sciences, University of Tokyo, Tokyo 113, Japan.

sophisticated hypothetical rules from the deviation of the simulated conformational results from the experimental data.

I have reported a simplified dynamics method for detecting α/β topology, which was called a packing game simulation (Nagano, 1980). Geometric analyses of packing of α helices onto β sheets in α/β proteins have been presented (Janin and Chothia, 1980). Subsequently, Cohen *et al.* (1982, 1983) extended the analysis and developed an algorithm of secondary structure assignment for α/β proteins by a combinatorial approach. Taylor and Thornton (1983, 1984) applied a method of fitting templates to the sequences for recognition of βαβ units in α/β proteins by incorporating the secondary structure prediction method of Garnier *et al.* (1978). Also, a method of predicting the ADP-binding βαβ fold from its sequence fingerprint has been shown (Wierenga *et al.*, 1986). However, this chapter deals only with the generalization of my method to finding all types of chain topology and shows how far we can approach prediction of any tertiary structure of a protein using as many examples as possible of proteins of known sequence and conformation. The identification codes (or abbreviated names) of 130 proteins used in the present analyses are listed in Table I.

II. DEFINITION OF PREDICTION FUNCTIONS IN THE SECONDARY STRUCTURE PREDICTION METHOD OF NAGANO

The numerical interpretation of the sequence information began with the secondary structure prediction method of Nagano (1973, 1977b). A brief description of the method follows.

A doublet d is defined as a pair of amino acid residues m residues apart ($m = 0, 1, 2, \ldots 6$) by $d \equiv d(r,r';m)$, where r and r' represent the amino acid residues, and a triplet t is defined by $t \equiv t(x,x',x'';m,m')$, where the residues x and x' are separated by m residues, and x' and x'' by m' residues. The x, x', and x'' represent seven groups of amino acid residues: (1) Ala, (2) Arg, His, Lys, and Tyr, (3) Asn, Asp, Gln, and Glu, (4) Gly and Pro, (5) Cys and Thr, (6) Ser, and (7) Ile, Leu, Met, Phe, Trp, and Val.

The total numbers of weighted samples for doublets and triplets, N_m and $N_{mm'}$, are given by

$$N_m = \sum_{p=1}^{P} u(p)N_{\mathrm{cm}}(p) \quad \text{and} \quad N_{\mathrm{mm'}} = \sum_{p=1}^{P} u(p)N_{\mathrm{cmm'}}(p) \qquad (1)$$

where P is the total number of proteins used in the analysis, $N_{\mathrm{cm}}(p)$ is the number of residue pairs separated by m residues, $d(m)$, contained in the pth protein, $N_{\mathrm{cmm'}}(p)$ is the number of triplets of residues separated by m and m' residues, $t(mm')$, in the pth protein, and $u(p)$ is the statistical weight for the pth protein for the purpose of making the statistical significance equal between protein families as far as possible. The weighted totals of particular d and t terms are

$$N_d = \sum_{p=1}^{P} u(p)N_d(p) \quad \text{and} \quad N_t = \sum_{p=1}^{P} u(p)N_t(p) \qquad (2)$$

where $N_d(p)$ and $N_t(p)$ are the numbers of doublets d and triplets t, respectively, in the pth protein. The weighted totals of d and t in a conformation type s are given by

$$n_{ds} = \sum_{p=1}^{P} u(p)N_{ds}(p) \quad \text{and} \quad n_{ts} = \sum_{p=1}^{P} u(p)N_{ts}(p) \qquad (3)$$

Here $N_d(p)$ and $N_t(p)$ are the numbers of d and t, respectively, in a type s region in the pth protein; $s = 1$ denotes regions of one type of secondary structure, and $s = 2$ denotes the rest.

Table I. Protein Identification Codes or Abbreviated Protein Names Common to Those of Brookhaven Protein Data Bank for Proteins of Known Sequence and Conformation[a]

AAT	Chicken heart cytosolic aspartate aminotransferase
ABP	*E. coli* L-arabinose-binding protein
ACT	Kiwifruit actinidin
ADH	Horse liver alcohol dehydrogenase
ADK	Porcine muscle adenylate kinase
ALKP	*E. coli* alkaline phosphatase [Bradshaw *et al.*, 1981 (seq.); Sowadski *et al.*, 1985 (con.)]
ALP	Myxobacter 495 α-lytic protease
API	Human modified α_1-proteinase inhibitor
APP	Fungus penicillopepsin
ATCC	*E. coli* aspartate carbamoyltransferase C chain
ATCR	*E. coli* aspartate carbamoyltransferase R chain
AZU	*Pseudomonas aeruginosa* azurin
B2M	Bovin β_2-microglobulin
B5C	Bovine cytochrome b_5
CAB	Human erythrocyte carbonic anhydrase B
CAC	Human erythrocyte carbonic anhydrase C
CAT	Beef liver catalase
CDV	*Desulfovibrio vulgaris* cytochrome c_3
CHA	Bovine tosyl-α-chymotrypsin A
CLN	Chicken calmodulin
CNA	Jackbean lectin concanavalin A
CPA	Bovine carboxypeptidase A_α
CPB	Bovine carboxypeptidase B
CPP	*Pseudomonas putida* cytochrome P-450 (Poulos *et al.*, 1985)
CPV	Carp muscle calcium-binding parvalbumin
CRN	Abyssinian cabbage seed crambin
CRO	Bacteriophage λ gene-regulating protein (*cro* repressor)
CTS	Porcine heart citrate synthase
CTX	Cobra α-cobratoxin
CYC	Bonito heart cytochrome *c*
CYP	Yeast cytochrome *c* peroxidase
C2C	*Rhodospirillum rubrum* cytochrome c_2
C3A	Human C3a anaphylatoxin
DBP2	*Bacillus stearothermophilus* DNA-binding protein II (Tanaka *et al.*, 1984)
DFR	*Lactobacillus casei* dihydrofolate reductase
DPI	Klenow fragment of *E. coli* DNA polymerase I
EBX	Sea snake erabutoxin b
ECA	Insect erythrocruorin
EFM	GDP domain of *E. coli* elongation factor Tu
EST	Porcine tosyl-elastase
FABH	H part of λ immunoglobulin Fab′
FABL	L chain of λ immunoglobulin Fab′
FDX	Bacterial ferredoxin
FXC	Plant ferredoxin
FXN	*Clostridium MP* flavodoxin
GAP	*E. coli* catabolite gene activator protein
GCN	Porcine pancreas glucagon
GCR	Bovine γ-II crystallin

(*continued*)

Table I. (Continued)

GD1	*Bacillus stearothermophilus* D-glyceraldehyde-3-phosphate dehydrogenase
GN5	Bacteriophage fd gene-5 (DNA-binding) protein
GP	Bovine erythrocyte glutathione perioxidase (selenoenzyme) (Epp *et al.*, 1983)
GPD	Lobster glyceraldehyde-3-phosphate dehydrogenase
GRS	Human erythrocyte glutathione reductase
HBA	Horse hemoglobin α
HBB	Horse hemoglobin β
HIP	*Chromatium vinosum* high-potential iron protein
HPI	*Panulirus interruptus* hemocyanin (undetermined part of the sequence is assumed to be the same as the corresponding part of *Eurypelma californicum* hemocyanin subunit d)
HRB	Marine worm hemerythrin b
INSA	Porcine insulin A
INSB	Porcine insulin B
KGA	*Pseudomonas putida* KDPG-aldolase
LDH	Dogfish lactate dehydrogenase
LH	Yellow lupin root nodule leghemoglobin
LHB	Sea lamprey hemoglobin V
LRP	N-terminal domain of λ repressor
LYZ	Hen egg-white lysozyme
LZM	Bacteriophage T_4 lysozyme
MBN	Sperm-whale myoglobin
MDH	Porcine heart cytoplasmic malate dehydrogenase
MHR	Marine worm myohemerythrin
MLT	Honeybee venom melittin
MT2	Rat liver Cd,Zn-metallothionein II
OVO	Japanese quail ovomucoid third domain
PAB	Human plasma prealbumin
PAP	Papaya fruit latex papain
PCY	Poplar leaf plastocyanin
PEP	Porcine pepsin
PFK	*Bacillus stearothermophilus* phosphofructokinase
PFKE	*E. coli* phosphofructokinase [Hellinga and Evans, 1985 (seq.); Y. Shirokibara, personal communication (con.)]
PGKH	Horse phosphoglycerate kinase (Blake and Rice, 1981)
PGKY	Yeast phosphoglycerate kinase
PGL	Rabbit muscle phosphoglucomutase
PGM	Yeast phosphoglycerate mutase
PHH	*Pseudomonas fluorescens p*-hydroxybenzoate hydroxylase (Weijer *et al.*, 1983)
PKA	Porcine pancreas kallikrein
PPA	Rabbit muscle glycogen phosphorylase A
PPT	Avian pancreatic polypeptide
PP2	Rattlesnake venom Ca-free phospholipase A_2
PTI	Bovine pancreatic trypsin inhibitor
PTN	Bovine pancreatic trypsin
PYP	Yeast inorganic pyrophosphatase
P2P	Porcine phospholipase A_2
RHD	Bovine liver rhodanese

Table I. (Continued)

RHV1–RHV4	Human rhinovirus 14 coat protein VP1–VP4
RNB	*Bacillus amyloliquefaciens* barnase
RNS	Bovine pancreatic ribonuclease S
RNT	*Aspergillus oryzae* ribonuclease T1
RP2	Rat mast cell proteinase II
RST	*Streptomyces erythreus* ribonuclease St
RXN	*Clostridium pasteurianum* rubredoxin
SBP	*Salmonella typhimurium* sulfate-binding protein
SBT	*Bacillus amyloliquefaciens* subtilisin
SBV	Southern bean mosaic virus coat protein
SGB	*Streptomyces griseus* protease b
SNS	*Staphylococcus aureus* nuclease
SN3	Scorpion neurotoxin (variant 3)
SOD	Bovine erythrocyte Cu,Zn-superoxide dismutase
SRX	*E. coli* thioredoxin
SSI	*Streptomyces albogriseolus* subtilisin inhibitor
STI	Soybean trypsin inhibitor
STV	Satellite tobacco necrosis virus coat protein
TAA	*Aspergillus oryzae* Taka-amylase
TBV	Tomato bushy stunt virus coat protein
THI	Sweat-tasting protein thaumatin I from *Thaumatococcus danielli*
TIM	Chicken muscle triose phosphate isomerase
TLN	*Bacillus thermoproteolyticus* thermolysin
TMA	Rabbit muscle α-tropomyosin [Stone and Smillie, 1978 (seq.); McLachlan and Stewart, 1975 (con.)]
TMV	Tobacco mosaic virus coat protein
TNC	Turkey muscle troponin C
TS	N-terminal domain of *Bacillus stearothermophilus* tyrosyl-tRNA synthetase (Winter *et al.*, 1983)
TT4	Bacteriophage T_4 thioredoxin
UBQ	Human erythrocyte ubiquitin
UTG	Rabbit uteroglobin
WGA	Wheat germ agglutinin
YHX	Yeast hexokinase [Stachelek *et al.*, 1986 (seq.); Anderson *et al.*, 1978 (con.)]
ZGP	*Streptomyces albus* Zn-containing D-alanyl-D-alanine-cleaving carboxypeptidase
51C	*Pseudomonas aeruginosa* cytochrome c_{551}

[a]In this list of protein names, the references for amino acid sequence and tertiary structure should be obtained from the Brookhaven Protein Data Bank (Bernstein *et al.*, 1977) as of May 14, 1986, unless otherwise stated. If the sequence is referred to, "(seq.)" follows the reference, whereas "(con.)" follows it when the conformational data come from outside of the Protein Data Bank; without these notations both types of information are referred to in the reference. The protein identification codes adopted here are very similar to those used in the Protein Data Bank.

The statistical analyses are done separately for helix, loop (or β turn), and β structure. These analyses are distinguished by adding symbols (* and **) for loop and β structure, respectively. N_{ms} is the sum of n_{ds} over 20 × 20 = 400 combinations of doublets, $N_{mm's}$ is the sum of n_{ts} over 7 × 7 × 7 = 343 combinations of triplets; i.e.,

$$N_{ms} = \sum_{d = d(m)}^{\text{all}} n_{ds} \quad \text{and} \quad N_{mm's} = \sum_{t = t(mm')}^{\text{all}} n_{ts} \tag{4}$$

The overall proportions of doublets, q_{ms}, and triplets, $q_{mm's}$, in a type s region are $q_{ms} = N_{ms}/N_m$ and $q_{mm's} = N_{mm's}/N_{mm'}$, whereas the local proportions of d and t in the s region are $q_{ds} = n_{ds}/N_d$ and $q_{ts} = n_{ts}/N_t$. Those proportions are probabilities of expecting a conformation type s empirically for a particular doublet d or for a particular triplet t. The distribution of q_{ds} for various doublets is binomial and approaches a Gaussian curve when N_m becomes large enough. Since the standard deviation of the overall distribution having the proportion q_{ms} is given by

$$\sigma_{ms} = [q_{ms}(1 - q_{ms})]^{\frac{1}{2}}$$

the deviation of proportion, $q_{ds} - q_{ms}$, can be standardized as

$$z_{ds} = (q_{ds} - q_{ms})/\ (q_{ms} - q_{ms}{}^2)^{\frac{1}{2}}$$

We are going to use a modification of z_{ds} as a substitute for a measure representing how much conformation type s is stabilized by the interactions among the side chains and the main chain of a region including the residue pair d if the sample size N_d is always large enough and if q_{ms} is not too close to either 0 or 1. However, because N_d values are not always large enough,

$$y_{ds} = (q_{ds} - q_{ms})[N_d/(q_{ms} - q_{ms}{}^2)^2 N_m]^{\frac{1}{4}}$$

and

$$y_{ts} = (q_{ts} - q_{mm's})[N_t/(q_{mm's} - q_{mm's}{}^2)^2 N_{mm'}]^{\frac{1}{4}}$$

are used as such information functions for a doublet d and a triplet t.

When the triplet terms of the helical wheel effect are included in the helix prediction functions, the functions $g_s(p,i)$ for predicting a conformation type s for the residue number i of the pth protein become:

$$g_s(p,i) = -\sum_{m=0}^{4} c_{ms}(y_{ds} + y_{d's} + x_{ms} \sum_{d''}^{\text{all}} y_{d''s}) - \sum_{m,m'=0,2}^{2,0} c_{mm's} \sum_{t}^{\text{all}} y_{ts} \qquad (5)$$

$s = 1$ for helix;

$$g_s{}^*(p,i) = -\sum_{m=0}^{4} c_{ms}{}^*(y_{ds}{}^* + y_{d's}{}^* + x_{ms}{}^* \sum_{d''}^{\text{all}} y_{d's}) \qquad (6)$$

$s = 1$ for loop or turn;

$$g_s{}^{**}(p,i) = -\sum_{m=0}^{4} c_{ms}{}^{**}(y_{ds}{}^{**} + y_{d's}{}^{**} + x_{ms}{}^{**} \sum_{d''}^{\text{all}} y_{d''s}) \qquad (7)$$

$s = 1$ for β structure. Here, $d = d[r(p,i-m-1),r'(p,i);m]$, $d' = d[r(p,i),r'(p,i+m+1);m]$, $d'' = d[r(p,i'),r'(p,i'+m+1);m]$, $t = t[x(p,i''),x'(p,i''+m+1),x''(p,i''+m+m'+2);m,m']$, $1 \le i' < i < i'+m+1 \le N_{cm}(p)+m+1$, and $1 \le i'' < i < i''+m+m'+2 \le N_{cmm'}(p)+m+m'+2$.

A statistical data base was made by collecting sequences and assigned secondary structures of 133 proteins (Nagano, 1977b). The parameters defined above were adjusted to obtain the best predictions without using the data of the protein to be predicted. The values of the parameters used in this chapter are shown in Table II (same as those in Table 1 of Nagano, 1977b).

Plotter outputs of three prediction functions for LZM (phage lysozyme) (Matthews and

Table II. The Values of Parameters Used in the Calculation of Helix and β-Structure Prediction Functions[a]

Parameters	m: 0	1	2	3	4
For helix					
c_{m1}	−0.012	0.379	0.196	0.170	0.246
$c_{m1}x_{m1}$		0.521	0.334	0.324	0.110
$c_{021} = c_{201}$	0.413				
For β structure					
c^{**}_{m1}	0.484	0.545	0.117	0.518	−0.040
$c^{**}_{m1}x^{**}_{m1}$		0.614	0.332	0.532	0.081

[a]See Table 1 of Nagano (1977b).

Remington, 1974) and TIM (triose phosphate isomerase) (Banner *et al.*, 1975) are shown in Fig. 1.

III. DEFINITION OF A AND B FUNCTIONS

It was observed that long-range interactions occur between regions having strong potentials for both α helix and β structure (Nagano, 1974; Nagano and Hasegawa, 1975). The strength orders of the helical and β-structural predictabilities were defined as a pair of ranking numbers, one within a protein and the other over a range of 36 representative proteins, based on the prediction fucntions described in the preceding section; these represent the mode of long-range interactions in supersecondary structure quantitatively and schematically (Nagano, 1977a). It was suggested that one of the requirements for a right-handed βαβ unit is for well-balanced and fairly strong potentials of the β, α, and β segments. For the purpose of predicting all kinds of topological patterns in the chain folds of proteins, the general algorithm starts with the recognition of α/β proteins using the following functions.

The *A* and *B* functions are defined to measure approximately the long-range interaction potential dependent on whether the conformation is α helical or β structural:

$$A(p,i) = \sum_{i=j-2}^{j+2} [g_1^0 - g_1(p,i)]^2[g_1^{0**} - g_1^{**}(p,i)] \quad (8)$$

$$B(p,i) = \sum_{i=j-2}^{j+2} [g_1^0 - g_1(p,i)][g_1^{0**} - g_1^{**}(p,i)]^2 \quad (9)$$

Five consecutive amino acid residues are treated as a group (five-residue region) because the shortest length of a central β strand in a large β sheet is about five residues. This choice, however, is too small to represent the central α helices flanking such a large β sheet or constituting a domain of all-α protein. g_1^0 is the highest value, i.e., the weakest α potential, found so far for the helix prediction function $g_1(p,i)$, and g_1^{0**} is the highest value, i.e., the weakest β potential, found so far for the β-structure prediction function $g_1^{**}(p,i)$ (Nagano, 1977b). Values *p* and *i* denote a protein (or its primary sequence) and a residue number (sequence position counting from the N terminus), respectively. Local maxima of the *A* and *B* functions are searched for as loci separated from the nearest one by at least five residues. This

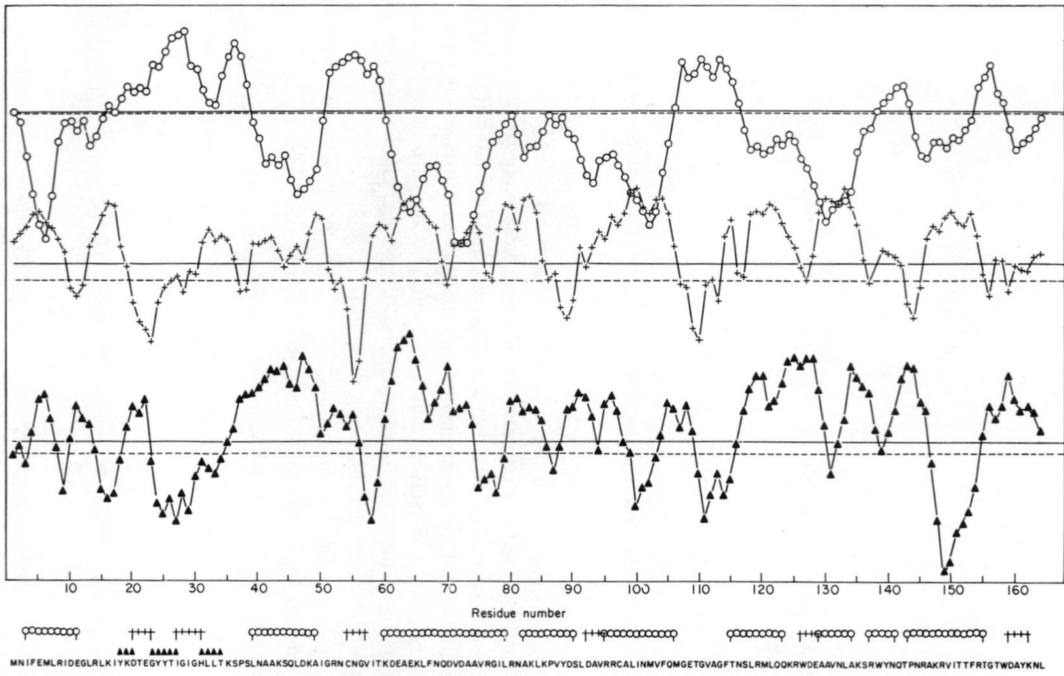

Figure 1. (a) Three prediction functions for LZM (bacteriophage T_4 lysozyme) plotted against residue number: helix prediction function (open circles), loop prediction function (crosses), and β-structure prediction function (triangles). The vertical scale is adjusted so that the difference between the values of the highest and lowest found is equal to half of the full range of the ordinate. Three horizontal lines represent the zero values of the three prediction functions. Three horizontal broken lines represent the thresholds, h_1, l_1, and b_1, given in Table 1 of Nagano (1977b). The primary structure is given by a sequence of one-letter codes. (b) Three prediction functions for TIM (chicken muscle triose phosphate isomerase) plotted against residue number. Symbols as in a. (Reproduced from Nagano, 1977b, with permission from *Journal of Molecular Biology*. Copyright 1977, Academic Press, Limited.)

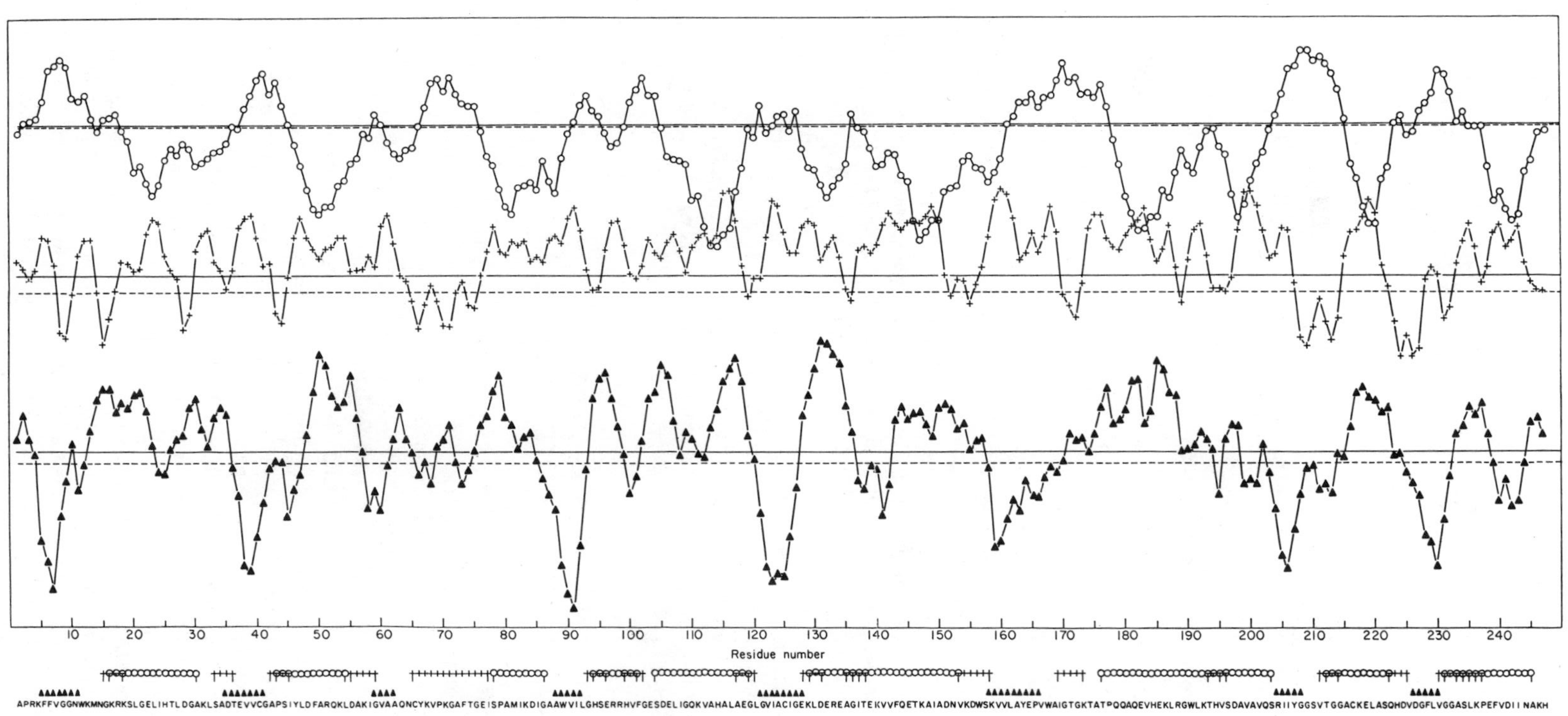

Figure 1 (cont.)

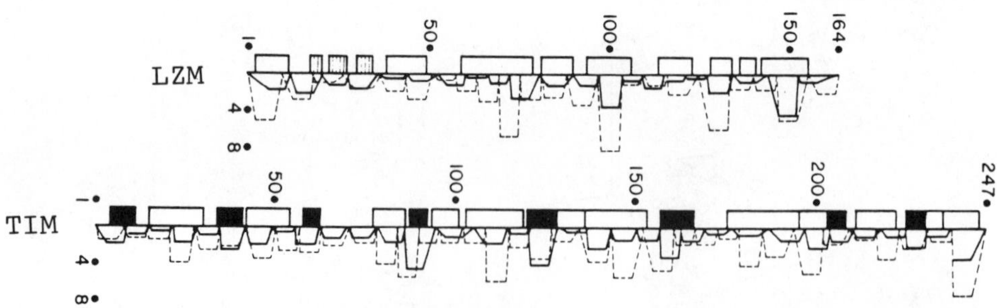

Figure 2. The *A* and *B* functions drawn in broken and solid lines, respectively, under the horizontal lines for LZM and TIM. These graphs correspond to Fig. 1. White boxes, striated boxes, and black boxes on the horizontal lines represent the observed ranges of α helices, antiparallel β strands, and β strands parallel to at least one adjacent strand, respectively. Residue numbers are given above. (Reproduced from Nagano, 1980, with permission from *Journal of Molecular Biology*. Copyright 1980, Academic Press, Limited.)

is because the shortest separation between central residues of two consecutive β strands is five residues. Both functions are drawn in broken and solid lines, respectively, under the horizontal lines of Fig. 2 for LZM and TIM, corresponding to Fig. 1. White boxes, striated boxes, and black boxes on the horizontal lines in Fig. 2 respectively represent the observed ranges of α helices, antiparallel β strands, and β strands parallel to at least one adjacent strand. The regions where the values of the *A* and *B* functions are weak very often coincide with outside regions where looping potential is strong. My algorithm for detecting α/β domains begins with locating maxima of hydrophobicity in terms of α-helical and β-structural predictabilities, whereas the algorithm of Cohen *et al.* (1983) starts by finding well-separated turns as maxima of hydrophilicity. However, residues 51 to 55 of LDH, the sequence VDVME, separates a β strand and an α helix but has a very low turn potential. This is why I consider my approach to be better.

IV. IMPROVED SCORING METHOD FOR CHOOSING THE BEST CANDIDATES FOR βαβ UNITS

If we treat the folding problem of a purely α/β protein, the algorithm of the folding simulation becomes a simple packing game of α and β candidates into a two-dimensional lattice consisting of 3 × 11 boxes (Nagano, 1980). Such a treatment was almost satisfactory for a relatively small typical α/β protein such as ADK, FXN, or TIM. It is most important in this approach to locate all the candidates for parallel β strands. The scoring method for such candidates was described in the Appendix of Nagano (1980) in a complicated way. The basic idea is to identify the strongest local β candidate between two α candidates and the strongest local α candidate between two β candidates, although one of the two might be missing.

Nagano (1980) first adjusted the parameters so that the program for calculating the central residue numbers of β candidates for all α/β domains, including TIM but without ABP, could be successful. The x-ray structure of ABP was subsequently published, and it became clear that the β-structure predictability for ABP was not satisfactory because the β strand 104–109 (sequence PLVMMA) contained a very rare Met-Met sequence. This was the main reason the statistical data base was updated at that time. After that the β candidate at residue 105 was satisfactorily picked up on one side, but a false β candidate for TIM, residue 140, was picked up on the other. The sequence around residue 140 is involved in a helix, 138–154, the

Table III. Class I and Class II Proteins[a]

Class I proteins[b]			Class II proteins[c]			
ABP	FDX	PCY	AAT	ECA	PFKE	SBV
ADH	FXN	PEP	ACT	EFM	PGKY	SN3
ADK	GCN	PGKH	ALKP	FABH	PGL	STI
ALP	GPD	PTI	API	FABL	PGM	STV
APP	HBA	PTN	ATCC	FXC	PHH	TAA
AZU	HBB	RHD	ATCR	GAP	PKA	TBV
B5C	HIP	RNS	B2M	GCR	PPA	THI
CAB	HRB	RXN	CAT	GD1	PPT	TMA
CAC	INSA	SBT	CDV	GN5	PP2	TNC
CHA	INSB	SGB	CLN	GP	PYP	TS
CNA	LDH	SNS	CPB	GRS	P2P	TT4
CPA	LHB	SOD	CPP	HPI	RHV1	UBQ
CPV	LYZ	SRX	CRN	KGA	RHV2	UTG
CYC	LZM	SSI	CRO	LH	RHV3	WGA
C2C	MBN	TIM	CTS	LRP	RHV4	YHX
DFR	MHR	TLN	CTX	MDH	RNB	ZGP
EBX	PAB	TMV	CYP	MLT	RNT	51C
EST	PAP		C3A	MT2	RP2	
			DBP2	OVO	RST	
			DPI	PFK	SBP	

[a]Protein identification codes and abbreviated names are defined in Table I.
[b]Class I proteins are those whose sequence and conformation are included in the statistical data base.
[c]Class II proteins are those whose sequence and conformation are not included in the statistical data base.

sequence of which is TEKVVFQETKAIADNVK. This made me aware of a characteristic hydrophobicity pattern of the typical parallel β strand. Since then I have not updated the statistical data base further and have tried to draw empirical rules based on characteristic patterns of amino acid sequences. The primary and secondary structural data of the representative 53 proteins were determined at an early stage (before 1979) and were included in the statistical data base used in this chapter. These 53 proteins are referred to as class I proteins. The other proteins of known sequence and conformation are called class II proteins. Table III lists the identification codes (protein names) of 53 class I proteins and 77 class II proteins. The secondary structure prediction for the class I proteins can be expected to be somewhat better than that for the class II proteins.

The improvement that was introduced by Nagano and Ponnuswamy (1984) is as follows. We define a β candidate of PAR as the most powerful parallel-type β strand by the following conditions, where n represents the amino acid type of nth residue count from the N terminus and $(\beta_P)_n$ the conformational preference for parallel β strand of n, as given by Lifson and Sander (1979). These preferences are shown in Table IV along with the conformational preferences for antiparallel β strand. The antiparallel β-strand conformational preference parameters are not used in this chapter.

The procedure is as follows:

1. Six consecutive amino acid residues are designated with residue numbers $n, n + 1, \ldots, n + 5$.
2. $n + 2 \neq$ P and $n + 3 \neq$ P; i.e., residues $n + 2$ and $n + 3$ should not be Pro.
3. Either $(\beta_P)_{n+2} \geqq 1.30$ or $(\beta_P)_{n+3} \geqq 1.30$ or both.
4. Unless 3 is true, both $(\beta_P)_{n+2} \geqq 0.79$ and $(\beta_P)_{n+3} \geqq 0.79$.

Table IV. Conformational Preference of Amino Acid Residues: β_P and β_A for Parallel and Antiparallel β Strands, Respectively[a]

Amino acid		β_P	β_A	Amino acid		β_P	β_A
Ala	(A)	1.00	0.90	Leu	(L)	1.42	1.26
Arg	(R)	0.68	1.02	Lys	(K)	0.59	0.74
Asn	(N)	0.54	0.62	Met	(M)	1.49	1.09
Asp	(D)	0.50	0.47	Phe	(F)	1.30	1.23
Cys	(C)	0.91	1.24	Pro	(P)	0.35	0.42
Gln	(Q)	0.28	1.18	Ser	(S)	0.70	0.87
Glu	(E)	0.59	0.62	Thr	(T)	0.59	1.30
Gly	(G)	0.79	0.56	Trp	(W)	0.89	1.75
His	(H)	0.38	1.12	Tyr	(Y)	1.08	1.68
Ile	(I)	2.60	1.54	Val	(V)	2.63	1.53

[a]From Lifson and Sander (1979).

5. If $(\beta_P)_{n+2} \geqq 1.30$ and $(\beta_P)_{n+3} \geqq 1.30$, then $(\beta_P)_{n+1} \geqq 0.50$ or $(\beta_P)_{n+4} \geqq 0.50$.*
6. Unless $(\beta_P)_{n+2} \geqq 1.30$ and $(\beta_P)_{n+3} \geqq 1.30$, then $(\beta_P)_{n+1} \geqq 0.68$ or $(\beta_P)_{n+4} \geqq 0.68$.
7. I_u is the number of residues $(n + 2k)$ $(k = 0, 1, 2)$ with $(\beta_P)_{n+2k} \geqq 1.30$, and I_d is the number of residues $(n + 2k + 1)$ $(k = 0, 1, 2)$ with $(\beta_P)_{n+2k+1} \geqq 1.30$. Then, $I_u \geqq 1$, $I_d \geqq 1$, and $I_u + I_d \geqq 3$.

When the above seven conditions are met for a strand of six consecutive residues, the strand is qualified to be a PAR type. The sequences of both ABP and TIM, which are found to be PAR type, are shown in Table V. The five-residue regions that are identified by the previous packing game simulation program and overlap the PAR-type sequences are shown by black balls and thick circles on the α-carbon backbone models of ABP and TIM tertiary structures in Fig. 3. The five-residue regions centering around residues 7, 36, 62, 86, 105, (110,) 138, 200, (211,) 227, and 287 for ABP and 6, 90, 123, 160, 228, (and 241) for TIM are shown.† The numbers in parentheses are the central residue numbers of those found to be in α-helical conformation, and the others are the numbers of those close to the central residues of the respective overlapping observed β strands and cluster in the separate domains of ABP and in the β-barrel domain of TIM. The underlined residues in one-letter amino acid codes in Table V are actually involved in the range of parallel β strands except for an antiparallel one at 287–291 of ABP as observed by x-ray crystallographic methods.

*This condition excludes the possibility of $(n + 1)$ = Pro and $(n + 4)$ = Pro being a β candidate. However, such an exception will be permitted later because a parallel β strand of KGA, 121–123, PLL, is found to be a part of 119–127 (EIPLLPGIS). See the discussion about the result for KGA below. In addition, two other cases, a part of 307–312 (YPLIPV) of CAT and a part of 62–67 (IPMIPG) of YHX are also found to be antiparallel β strands. So, at present, such a special sequence has a high β-structure-forming propensity in 100% probability. $(\beta_P)_{n+2} \geqq 1.30$ means that the residue at the $n + 2$ position counting from the N terminus is one of the following: Ile, Leu, Met, Phe, and Val. $(\beta_P)_{n+1} \geqq 0.50$ means that the residue at the $n + 1$ position is one of the following: Ala, Arg, Asn, Asp, Cys, Glu, Gly, Ile, Leu, Lys, Met, Phe, Ser, Thr, Trp, Tyr, and Val (Gln, His, and Pro are omitted).

†The central residue numbers of the five-residue regions of PAR or HLF types for the 130 proteins in Table I are listed in Table IX.

Table V. PAR-Type and HLF-Type Sequences of Two Proteins of α/β Type, ABP (L-Arabinose-Binding Protein) and TIM (Triose Phosphate Isomerase)

PAR-type sequences				HLF-type sequences			
ABP		TIM		ABP		TIM	
3- 10	LKLGFLVK[a]	4- 10	KFFVGGN	30- 35	DLGFEV	6- 11	FVGGNW
33- 38	FEVIKI	87- 94	AAWVILGH	36- 41	IKIAVP	27- 32	LDGAKL
58- 65	AKGFVICT	118-126	EGLGVIACI	55- 60	ASGAKG	37- 43	EVVCGAP
83- 89	MKVIAVD	158-164	KVVLAYE	71- 78	GSAIVAKA	44- 49	SIYLDF
103-111	VPLVMMAAT	225-233	VDGFLVGGA	78- 83	ARGYDM	56- 63	AKIGVAAQ
135-142	SAVMAITA	240-245	VDIINA	156-161	MDALKA	70- 75	KGAFTG
198-205	HWLIVGMN			160-165	KAAGFP	84- 89	DIGAAW
208-213	TVLGGV			179-185	IPGAFDA	86- 91	GAAWVI
224-231	ADIIGIGI			209-215	VLGGVRA	113-119	AHALAEG
285-290	DVVLIT *			218-223	GQGFKA	117-122	AEGLGV
				227-233	IGIGING	122-128	VIACIGE
				231-236	INGVDA	165-172	PVWAIGTG
				245-252	ATGFYGSL	187-192	LRGWLK
				265-270	EMLYNW	197-202	DAVAVQ
				268-273	YNWVAK	204-211	RIIYGGSV
				301-306	KGLGGK	211-216	VTGGNC
						229-235	LVGGASL

[a]The underline represents the range of parallel β strand. The underlines marked with an asterisk represents the range of antiparallel β strand, and the blank indicates that the β strand extends outsidethe defined region.

When the above two conditions are satisfied with the modifications that rules 5′ and 7′ are replaced by

5′. $(\beta_P)_{n+1} \geqq 0.50$ or $(\beta_P)_{n+4} \geqq 0.50$.
7′. $I_u \geqq 0$, $I_d \geqq 0$, and $I_u + I_d \geqq 0$.

and an additional condition that the strand does not meet the conditions for PAR type, the strand is called HLF type. The regions of sequences of HLF type for ABP and TIM are shown in Table V along with the underlined regions actually found to be parallel β strands.

The scoring method for detecting the best set of βαβ candidates, which was described in the Appendix of Nagano (1980), was improved by adding a score of 4* to a β candidate of PAR type if the central residue number is contained in the central four† residues of a PAR strand and by adding a score of 2* (i.e., half of the score to PAR type) to a β candidate of HLF type if the central residue number is in the central four† residues of HLF type (Nagano and Ponnuswamy, 1984).

The scores adopted first by Nagano (1980) and then by Nagano and Ponnuswamy (1984) were chosen to produce the best results for proteins of well-known α/β type. This is exactly the process of learning how to predict protein tertiary structure from sequence as well as possible. That is to say, the set of parameters and scores adopted earlier were adjusted to fit the observed data of class I proteins. As the number of class II proteins increases, it has turned out that the classification into PAR or HLF is more useful than had been expected. This is why the scores of PAR and HLF have been doubled recently.

*The score of 2 to a PAR-type β candidate and 1 to an HLF-type β candidate was first defined by Nagano and Ponnuswamy (1984).

†The central two residues instead of four residues was adopted when introduced by Nagano and Ponnuswamy (1984).

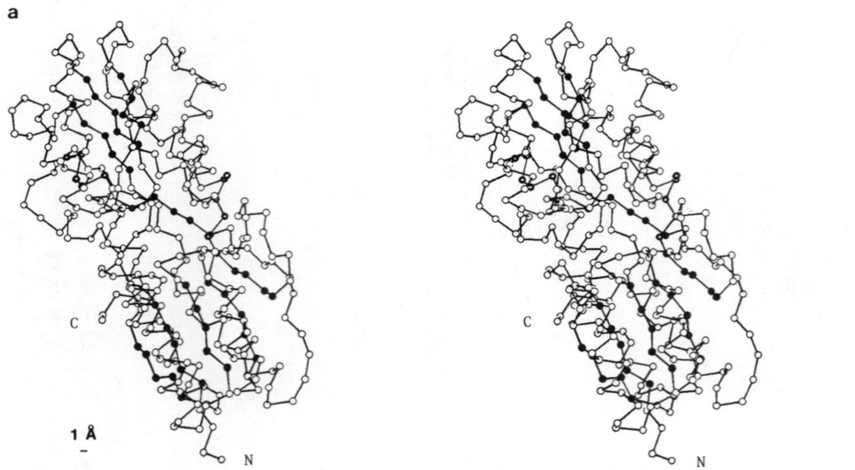

Figure 3. Stereoscopic α-carbon backbone models of two α/β proteins, (a) ABP and (b) TIM, representing the locations of amino acid residues on the PAR-type sequences. Black-ball α-carbon atoms represent β-structural five-residue regions of PAR type, and thick circle α-carbon atoms are helical ones. The central residue numbers of the regions are (a) 7, 36, 62, 86, 105, (110,) 138, 200, (211,) 227, and 287 for ABP and (b) 6, 90, 123, 160, 228, (and 241) for TIM. The numbers in parentheses are by thick circles. The figures are drawn by the program PLUTO78 (Motherwell, 1978). The characters N and C indicate the respective termini. The terminal characters and α-carbon symbols were drawn manually.

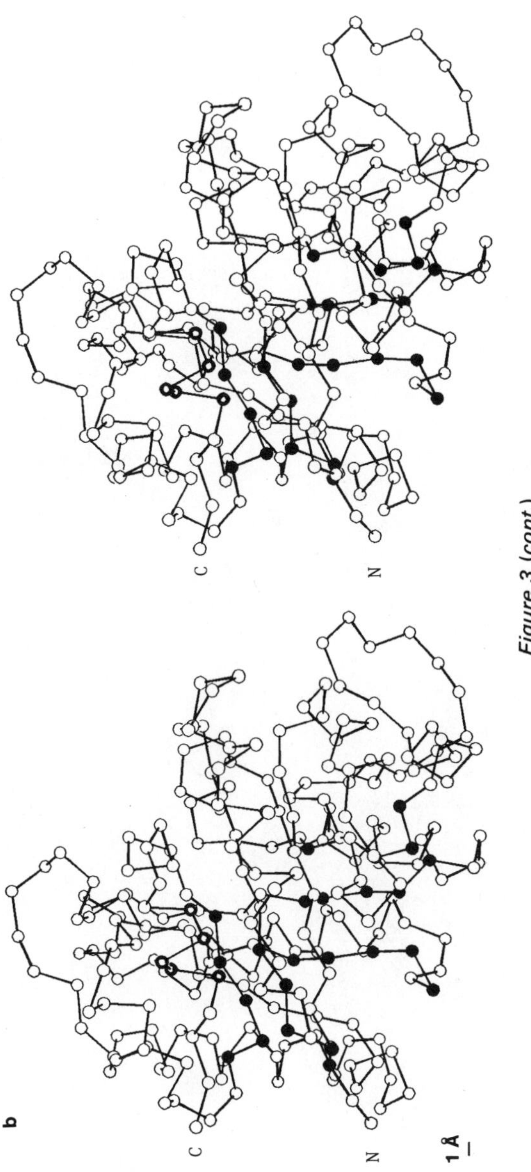

Figure 3 (cont.)

Table VI. Scores Given to β Candidates in ABP, LZM, PTI, and TIM[a]

ABP				LZM				PTI				TIM			
Central residue	Type	Score 1	Score 2	Central residue	Type	Score 1	Score 2	Central residue	Type	Score 1	Score 2	Central residue	Type	Score 1	Score 2
7	PAR	5	9	4	PAR	2	6	5		4	4	6	PAR	7	11
14		0	0	10		0	0	13		0	0	12		0	0
20		0	0	16		8	8	19	HLF	8	10	19		0	0
30		0	0	25	HLF	5	7	27	HLF	0	2	25		4	4
36	PAR	6	10	31	PAR	5	9	33	HLF	7	9	32		0	0
42		0	0	40		0	0	39	HLF	2	4	38	HLF	8	10
50		0	0	50		0	0	45		0	0	47	HLF	2	4
56	HLF	0	2	57	HLF	9	11	55		4	4	54		0	0
62	PAR	6	10	65		0	0					60	HLF	2	4
68		0	0	71		0	0					66		0	0
74	HLF	2	4	77	PAR	3	7					75		1	1
80	HLF	0	2	88		1	1					84		0	0
86	PAR	9	13	95		0	0					90	PAR	8	12
92		0	0	101	PAR	2	6					99		2	2
99		0	0	107		0	0					110		0	0
105	PAR	5	9	113	HLF	4	6					117	HLF	0	2
111		0	0	119		0	0					123	PAR	7	11
120		0	0	125		0	0					134		0	0
126		0	0	131	HLF	0	2					140		3	3
132		0	0	138		0	0					148		0	0
138	PAR	5	9	144		0	0					154		1	1
144		0	0	150		6	6					160	PAR	6	10
153		0	0	156		0	0					166	HLF	5	7
162	HLF	0	2									172		0	0
170		4	4									181		0	0
181	HLF	0	2									189	HLF	0	2
190		0	0									195		1	1
200	PAR	8	12									201	HLF	0	2
211	PAR	3	7									207	HLF	6	8
221	HLF	0	2									213	HLF	1	3
227	PAR	7	11									222		0	0
235	HLF	5	7									228	PAR	6	10
241		0	0									234	HLF	0	0
247	HLF	1	3									241	PAR	1	5
253		0	0												
260		0	0												
269	HLF	0	2												
275		0	0												
281		0	0												
287	PAR	8	12												
293		0	0												
303	HLF	0	2												

[a]*Central residue* refers to the central residue number of a β candidate composed of five consecutive residues. *Type* indicates if the central residue is one of the interior four residues of a six-residue region of PAR or HLF type as described in the text (as opposed to the middle two residues as described by Nagano and Ponnuswamy, 1984). *Score 1* is according to the Appendix of Nagano (1980). *Score 2* is Score 1 plus 4 for PAR-type β candidates and plus 2 for HLF-type β candidates (twice the additions proposed by Nagano and Ponnuswamy, 1984).

One of the themes of this chapter is how to detect α/β domains as accurately as possible. For that purpose it would be very important to judge whether a candidate is predicted to be helical (an α candidate) or β structural (a β candidate). The result of applying this scoring method to the sequences of four representative proteins, ABP, LZM, PTI, and TIM, is shown in Table VI. These scores are used to divide sequences into β-candidate regions and α-candidate regions as shown in Table VII for the above four proteins.

The β-candidate regions are first determined according to the scores and the residue distances between β candidates. The α candidates are searched for as those having the largest A values and the appropriate distances from the nearest β candidates. Furthermore, up to three more candidates are supplemented as local maxima of B values if there are any candidates having larger B and A values than threshold values. Both A and B values are calculated for individual candidates. A five-residue region of the PAR type is defined to be PAR(α) when the ratio of its A value to its B value, $A/B > 1.840$ and to be PAR(β) when the ratio $A/B \leqq 1.840$. Two PAR(α)-type regions, 4 and 77, are found in the LZM sequence.* A five-residue region of HLF type is HLF(α) when the ratio $A/B > 1.770$ and $B < -0.79$ or when $B \geqq -0.79$ and $A < -1.40$; it is HLF(β) when the ratio $A/B \leqq 1.770$ and $B \leqq -0.79$. A five-residue region of HLF type having $A \geqq -1.40$ and $B \geqq -0.79$ is only HLF and is neglected in the prediction.† A five-residue region of neither PAR type nor HLF type is α when the ratio $A/B > 1.590$ and $B > -0.79$ or when $A < -1.40$ and $B \geqq -0.79$; it is β when the ratio $A/B \leqq 1.590$ and $B \geqq -0.79$. A five-residue region of neither PAR type nor HLF type having $A \geqq -1.40$ and $B \geqq -0.79$ is not classified and is also neglected in the following analysis.†

Five-residue regions of either PAR(α) or HLF(α) or α are referred to as candidates of type $|\alpha|$, whereas those of either PAR(β) or HLF(β) or β are candidates of type $|\beta|$. In Table VIII, candidates of type $|\alpha|$ detected in the β line, e.g., residues 4 and 77 of LZM in Table VII, are replaced by asterisks in the β lines and appear in the α lines. This shows a compact assembly of both β and α candidates among the $\beta\alpha\beta$-triple candidates picked up by the most recent version of the packing game simulation program for the 130 proteins listed in Table I. The domains of some proteins are divided into separate positions according to their protein types. The protein types are (I) α/β type, 39 proteins or domains; (II) $\alpha + \beta$ type, 51 proteins or domains; (III) all-β type, 30 proteins or domains; (IV) all-α type, 36 proteins or domains; and (V) coil type, four proteins. It is very often found that small proteins such as PTI have only HLF-type regions, as shown in Table VIII (II.33). However, even such regions seem to play an important role in the folding because the three HLF(β) regions, 19, 33, and 38 of PTI, are situated very close to one another as shown in Fig. 4a. It is also understandable for PTI to form a disulfide bond between Cys^{14} and Cys^{38}.

PAR(β)-type sequences are also found in domains other than the α/β type. Table IX classifies many representative protein domains into seven groups according to the number of PAR(β) sequences. The protein domains in each group are further divided into five protein types: (1) α/β type, (2) $\alpha + \beta$ type, (3) all-β type, (4) all-α type, and (5) coil type. One $\alpha + \beta$ protein, LYZ, contains one PAR(β), 57, which is found near the center of the tertiary structure as shown in Fig. 4b, although it is not a part of parallel β sheet. Figure 4c shows two PAR(β) sequences of SNS, 24 and 36, which are found to be adjacent antiparallel β strands. Figure 4d shows an example of two PAR-type sequences, 144 and 209 of CAB, arranged on adjacent parallel β strands. SSI has three PAR(β)-type sequences, 14, 72, and 80, of which only 14 and 80 are detected as β candidates and appear on the β line in Table VIII (II.46). The two regions are observed very close to each other, as shown in Fig. 4e, and are also responsible for dimerization at the β sheet. Figure 4f shows another example of one PAR(β) sequence, 94 of

*The locations of the five-residue regions around fourth and 77th residues of LZM are shown in Fig. 5 on the α-carbon backbone.

†Asterisks in the α lines of Table VIII.

Table VII. Characteristic Patterns for Calculating Topology Packing Diagrams of ABP, LZM, PTI, and TIM

ABP: L-Arabinose-binding protein

Central residue[a]	7	36	62	86	105	138	170	200	227	247	287
Type[a]	PAR(β)	PAR(β)	PAR(β)	PAR(β)	PAR(β)	PAR(β)	β	PAR(β)	PAR(β)	HLF(β)	PAR(β)
-A[b]	2.83	6.45	1.35	3.42	3.91	4.68	2.46	4.95	2.56	1.56	3.89
-B[c]	2.35	4.15	2.17	3.14	2.84	2.79	2.10	3.64	2.47	1.27	4.07
A/B[d]	1.204	1.554	0.622	1.089	1.377	1.677	1.171	1.360	1.036	1.228	0.956
Central residue		41	70		110	143		205	233		
Type		β	α		PAR(α)	α		β	HLF(β)		
-A		1.37	1.43		3.42	2.12		0.98	1.11		
-B		1.12	0.87		1.34	1.13		1.08	1.63		
A/B		1.223	1.644		2.552	1.876		0.907	0.681		
Central residue						153					
Type						β					
-A						1.17					
-B						0.80					
A/B						1.463					
Central residue	21	51	75	92	123	160	184	211	238	269	296
Type	α	α	HLF(α)	α	α	HLF(α)	HLF(α)	PAR(β)	α	HLF(α)	α
-A	2.93	3.84	5.31	3.14	2.73	2.73	3.05	2.72	2.51	4.97	1.98
-B	0.85	1.44	2.27	1.40	0.61	0.46	0.72	2.26	1.23	1.90	0.31
A/B	3.447	2.667	2.339	2.243	4.475	5.935	4.236	1.204	2.041	2.616	6.387
Central residue	31	57	81		133		190	216		282	
Type	HLF(β)	HLF(α)	HLF(α)		α		β	β		β	
-A	1.48	2.11	2.25		2.08		2.11	1.16		2.26	
-B	1.18	0.80	0.95		0.95		1.42	0.81		1.71	
A/B	1.254	2.638	2.368		2.189		1.486	1.432		1.322	
Central residue								222			
Type								HLF(α)			
-A								2.09			
-B								0.80			
A/B								2.613			

LZM: Bacteriophage T_4 lysozyme

Central residue	4	16	31	57	77	113	150
Type	PAR(α)	β	PAR(β)	HLF(β)	PAR(α)	HLF(β)	β
$-A$	4.30	3.35	1.85	1.14	4.91	1.26	3.99
$-\overline{B}$	1.54	2.36	1.70	1.61	2.56	1.48	3.63
$A/\overline{B}$	2.792	1.419	1.088	0.708	1.918	0.851	1.099
Central residue			42	62	82	118	155
Type			α		α	α	β
$-A$			2.22	1.72	2.45	2.28	1.36
$-\overline{B}$			0.57	0.37	0.87	1.15	1.33
$A/\overline{B}$			3.895	4.649	2.816	1.983	1.023
Central residue					88	124	
Type					α		
$-A$					2.42	1.79	
$-\overline{B}$					1.36	0.49	
$A/\overline{B}$					1.779	3.653	
Central residue	10	22	48	67	95	131	161
Type	α	β	α	α	α	α	
$-A$	2.14	1.51	2.84	2.93	3.80	5.48	1.96
$-\overline{B}$	1.04	1.17	0.78	0.97	1.43	1.71	0.63
$A/\overline{B}$	2.058	1.291	3.641	3.021	2.657	3.205	3.111
Central residue				72	101	138	
Type				α	PAR(α)	α	
$-A$				4.97	8.00	1.78	
$-\overline{B}$				1.38	3.64	0.94	
$A/\overline{B}$				3.601	2.198	1.894	
Central residue					108	145	
Type					β		
$-A$					0.95	1.84	
$-\overline{B}$					0.89	0.70	
$A/\overline{B}$					1.067	2.629	

PTI: Pancreatic trypsin inhibitor

Central residue	3	19	33	55
Type	β	HLF(β)	HLF(β)	β
$-A$	1.09	5.02	1.80	1.67
$-\overline{B}$	0.80	3.25	2.08	1.44
$A/\overline{B}$	1.363	1.545	0.865	1.160
Central residue			38	
Type			HLF(β)	
$-A$			1.25	
$-\overline{B}$			0.87	
$A/\overline{B}$			1.437	
Central residue			46	
Type			α	
$-A$			1.85	
$-\overline{B}$			0.80	
$A/\overline{B}$			2.313	

(*continued*)

Table VII. *(Continued)*

TIM: Triose phosphate isomerase

Central residue	6	38	60	90	123	160	207	228	241
Type	PAR(β)	HLF(β)	HLF(β)	PAR(β)	PAR(β)	PAR(β)	HLF(β)	PAR(β)	PAR(α)
-A	1.38	1.94	2.56	4.75	3.76	4.20	0.92	2.03	6.05
-B	1.75	2.28	1.86	4.13	3.67	2.94	1.63	2.11	2.79
A/B	0.789	0.851	1.376	1.150	1.025	1.429	0.564	0.962	2.168
Central residue	11	43	66	99	128	165			
Type	β	β	α	β	α	β			
-A	1.25	1.16	2.39	1.96	2.16	1.70			
-B	1.01	1.15	1.46	1.27	1.01	1.64			
A/B	1.238	1.009	1.637	1.543	2.139	1.037			
Central residue	19		75	107	134	171			
Type			β	α		HLF(β)			
-A	1.87		1.62	2.46	1.49	0.74			
-B	0.71		1.13	1.16	0.52	0.91			
A/B	2.634		1.434	2.121	2.865	0.813			
Central residue	25				140	193			
Type	α				β	α			
-A	4.38				3.32	2.89			
-B	2.45				2.51	1.49			
A/B	1.788				1.323	1.940			
Central residue	33	48	83	112	148	199	220	235	
Type	α	HLF(α)	α	α	α	HLF(α)	α		
-A	2.27	3.33	3.90	5.23	4.23	5.18	2.61	1.43	
-B	0.95	1.48	1.34	1.63	1.13	2.33	0.68	0.77	
A/B	2.389	2.250	2.910	3.209	3.743	2.223	3.838	1.857	
Central residue		55		118	155				
Type		α		HLF(α)	α				
-A		2.34		2.46	3.22				
-B		0.79		0.80	1.35				
A/B		2.962		3.075	2.385				

[a]*Central residue* and *type* are defined as in Table VI.
[b]*A* is defined so that a large negative value represents a strong helical long-range interaction potential (Nagano, 1980).
[c]*B* is defined so that a large negative value represents a strong β-structural long-range interaction potential.
[d]The *A/B* ratio is used to judge whether the candidate is helical or β structural. See text.

Table VIII. *β and α Candidates Detected by the Packing Game Simulation Program for Predicting Patterns of the α/β-Type Domains*[a]

(I) α/β proteins

1. AAT III (98-295)

β 100 132 145|171|186 *** *** *** 260 270
α 117 140 157|181|208 218|233 240 *** 281

2. ABP I-II (1-306)

β 7 36 62 86 105 138 170 200 227 247 287
α 21 51 75 92 123 160 184 211 238 269 296

3. ADH II-III (166-375)

β 171|*** 196 207|220 238 268 290 308|*** *** 347 371
α 178 185 *** 213 233 253 276 301|316|332 340 359 ***

4. ADK 194 res.

β *** 12 35 91 116 161|
α 4 28 81 106 151 184

5. ALKP 449 res.

β 16 47 85 122 147 *** *** 204 236 *** 266 284 319 341 365 398 430
α 31 60 107 134 154 161 196 212 244 257|*** 307 325|351 374|419|440

6. ATCC I-II (1-310)

β *** 11 *** *** 44 70 101 123 141|*** *** 186 220 *** *** *** *** *** ***
α 6 18 23 30 64 90 113 *** 152|159|174 209|226 247|261|275 295 301 307

7. ATCR I (1-99)

β 3 *** *** 31| 44 61 84
α 10 18| 25 *** *** 73 91|

8. B5C 93 res.

β 6 25 75
α 12 31| 85

9. CPA 307 res.

β 3 *** *** 49 64 *** ***107 139 *** *** 202
α 10 20 34|*** 75 81 99 118 166 178 191|222

[a] The underlined abbreviated protein names represent the proteins whose sequence and conformational data are contained in the statistical data base (Class I proteins as defined in Table III). The underlined β and α candidates are found to be correctly predicted to be β structural and α helical, respectively, within certain limits of error in residue numbers (±1 for α helix and ±2 for β strand). Asterisks indicate that the candidate assigned to the particular location is not found. The β and α candidates followed by vertical lines to the right are found to be contrarily predicted, in other words, α helical and β structural, respectively. The upper-dotted candidates are in the sequence type |α|, which is one of the following three types: PAR(α), HLF(α), and α. See text for the definition of the sequence types. All candidates of type |α| are aligned in the α positions in this table.

Table VIII. (Continued)

245 267 297|
261 287 ***

10. CPB 306 res.
β 3 45 60 106 149 *** *** 191 244 265 300|
α 20 *** 82 112 172 178 184 219 257 295 ***

11. CPP I (1-67)
β 3 24 64
α 8 43

12. DFR I (1-136)
β 4 39 60 76 96 114
α 29 *** 69 81 104 131

13. EFM I (1-190)
β *** 16 33 62 92 104 131 *** *** 171
α 6 27 47 76 97|118 141 147 155 189

14. FXN 138 res.
β 4 *** *** 34 50 83 109
α 17 22 28 42 71 101 135

15. GD1 I (1-163)
β 4 29 58 72 91 117 129 146
α 18 42 65 85 109 122 *** 161

16. GP 182 res.
β 4 27 59 *** 105 *** *** 137 154 171|
α 17| 48 80 97|*** 113 127 145 164|179

17. GPD I (1-160)
β 5 34| 64 91 116 129 142
α 16 39 72|101 *** 135 156

18. GRS I-II (1-294)
β 23 59 97|125 153 192 216 245 267
α 47| 72 111 131|182 204 224 253|286|

19. KGA 225 res.
β *** *** 32 56 83 101 124 138 168 190 ***
α 3 19 42 68 93 112 131 156 181 202 221

20. LDH I (1-181)
β *** 25 50 71 93 *** *** 134 157
α 9 35 55 76|109 115 124 146 169

21. MDH I (1-166)
β 7 38 62 82 102 120 133
α 17 52 76 89 111 *** 142

22. PFK I-II (1-255)
β 5 34 59| 97 118 139|166 *** *** 218 246
α 22 44 88 111 134 148 188|202 213 235

23. PFKE I-II (1-255)
β 6 23| 36 58|***
α 17 29 51| 70 85

*** 98 121 139|166 189 218 248
90 111 128 148 180 203 231

24. PGKH I-II (1-416)
β *** *** 20 58 93 116 160 176|210 238 285 301 342 368
α 3 11 40 82 105 122 169 195 224 244 293 328 357 379

*** ***
400 413

25. PGKY I-II (1-415)
β 8 26 47|92 115 157 174 208 239|277 330 366 387 409|
α 20|41 58|104 120 164 193 222 265 324 355 376 398 ***

26. PGL I-II-III
β 4 26 56 85
α 21 40 72 103

(1-418)
111 128 166 186 226 249 311 328 *** *** *** ***
*** 147 176|202 234 272 318 348 362 390 400 410

27. PGM 230 res.
β 4 23 49 76 112 *** *** *** 210
α 18 35 66 100 138 167 176|198 ***

28. PHH I (1-160)
β 6 29 *** *** *** *** 98 140 156
α 19 47 53 63 77|87|126|147|

29. PPA I-II-III (1-767)
β *** 15 66|85 115|150 172 201 239 258 299|***
α 8 58 73 102 142 161 181 215| 245|265 333|351

*** 366|401|443|455 492|*** *** 564 583|605 641 687 710
356 380 411 450 462 501 522 531 578 590 626 653 697 762

30. RHD I-II (1-293)
β 10 31 58 95 124 186|*** ***
α 17 45 82 115 153 208|214 231

254|270
264 278

31. SBP I-II (1-214)
β 6 37 *** *** 100 122 *** 167 189 210
α 26 55 64|72 115 137 162 182 198

32. SBT 275 res.
β 8|28 46 81 93 121 149 176
α 16 34|71 87 113 137 *** 198|

(continued)

Table VIII. (Continued)

β 205 *** 252|269
α 226 233 *** ***

33. SRX 108 res.
β 5 26 55 79
α 16 43 73 89|

34. TAA I (1-375)
β *** 13 44| 60 82 117 151 171 193|226 *** ***
α 7 19 53 73 105 123|159 182 208|254 260 274

β 290 326 348
α 316 333 362

35. TIM 247 res.
β 6 38 60 90 123 160 207 228 ***
α 33 48 83 112 148 199 220 235 241

36. TS I (1-321)
β *** 33 52| 68 *** *** *** ***
α 5 45 57 87 95 102 110 120|

β 134|164|189 221 253|270|
α 147 171 205 *** 265 303

37. TT4 87 res.
β 4 24| 63 78
α 13 50 70| 83

38. YHX I-II (1-267)
β 3 *** *** 68 94 *** 155 206 229 265
α 24 32 43 82|128 138 171 217 248

39. YHX IV (409-485)
β 413 439|*** *** 482
α 429 451|465 472 ***

(II) α + β proteins

1. AAT I (1-45)
β *** 16|
α 6| 23

2. AAT III (296-411)
β 305|*** *** 357 379
α 316 329 338 371 405

3. ACT II (111-220)
β 135 165 178 207
α 125 144 *** 196|***

4. API (12-393)
β 25| 51 *** *** 111 144 *** *** *** *** 215 240 292 310 338 *** *** 372 386
α 35 59 64 99 128 158 171 180|200 208 222|273 301 329|347|353|365|378|***

5. CAB 260
β *** 60
α 17 66|

res.
β 87 110 144 *** 185|191 216 257
α 93|119|156 165 *** 209|228 ***

6. CAC 259 res.
β 4 48 66 86 120 *** 149 *** 189 207 241
α 33 56|76 92|131 143|159 166 201 222 256|

7. CAT
β ***
α 8

I-II (1-320)
β *** 40 53|82 110 133 153 *** 196|220 244 280 314
α 19 *** 61 99 123 143|162 183 210 229|266 299

8. CDV 107 res.
β 18 48 78
α 26 55 96

9. CPP II (68-378)
β 97 123|***
α 72 113 130 137

β *** *** 221|246|258|294 357 369|
α 162 209 237 252 273 342 363 375

10. CRN 46 res.
β 3 15 33
α 8 26 ***

11. CRO 66 res.
β 6 26|42
α 12 31 53|

12. CTS I (1-85)
β *** *** *** 34
α 5 11 21 43

β 52 66
α *** 76

13. DBP2 90 res.
β *** *** *** *** 43
α 3 8 30 36 87

14. DPI (324-928)
β 332 352 371 438 472 535|*** *** 628 646|679|701 ***
α 343 364 403 463 508 561 588 615 635 657|691 707|713

β *** 742 766|815 *** *** 886 914
α 725 760 772 857 863 878|898 925

15. FXC 98 res.
β 5 25|51 78|
α 16|33 72 89|

16. GD1 II (164-334)
β 169 *** *** 241 271 307 ***
α 175|216 229|258 298|319 329

17. GPD II (161-333)
β 166 178 201|240 268 305
α 172|*** 214 256 292|322

18. GRS III-IV (295-478)
β 314 328 369 381 420 432 *** ***
α 300|322|343 *** 408 426|443 449 464

19. HIP 85 res.
β *** *** *** ***
α 3|15 48|59|

(continued)

Table VIII. (Continued)

71
81

20. HPI II-III (176-659)
β 194|221|*** 276 285 313 349 410 447 481 493 529 *** *** 585 609 649
α 182 203 235 250|*** 298 338 389 426|463|*** 508|545 557 569 603 643|655|

21. INSA 21 res.
β 4|
α 17

22. INSB 30 res.
β *** *** 24
α 3 14 ***

23. LDH II (182-329)
β 188 202 *** *** 272 292
α *** 232 238 252 286|301|

24. LYZ 129 res.
β 4 29| 57 91|
α 10 35 82 109

25. MDH II (167-324)
β 176 191 225|*** *** 262 288
α 185|218 233 243 251 280|308

26. OVO 56 res.
β 5 41|
α 31| 51|

27. PAP II (109-212)
β 111 132 147 172 187 201
α 123 141 161|*** *** ***

28. PFK III (256-319)
β 285
α 273 310

29. PFKE III (256-340)
β 284 320
α 274 298 335

30. PGL IV (419-561)
β *** *** 465 501 515 *** *** 558
α 449 459 491|*** 535 540 547 ***

31. PHH II (161-292)
β *** 187 223 271
α 177|209|252 288|

32. PP2 122 res.
β *** *** 20| 63 85 108
α 4 10 41 71 94 ***

33. PTI 58 res.
β 3| 19 33 55|
α *** *** 46 ***

34. P2P 124 res.
β *** *** 27 63 93|
α 4 11 53 83|104

35. RHV1 289 res.
β *** 17 33 57 89 120 *** *** 178 192 217 *** 252
α 10 *** 40| 69 110 131|148|167 185|211 227|241|***

36. RHV2 262 res.
β *** 15 33 62| 80 109 124 142|171|199 220
α 5 27| 48 71| 98 116 132 157 189|205|226|

242 258
252 ***

37. RHV3 236 res.
β 36 54| 83 102|128 162 189 207 230
α 45 65| 93 113|152|172|194|219|***

38. RNB 110 res.
β 5 35| 55 96
α 12 43 76|107|

39. RNS 124
β *** 45 80
α 6 55 100|

res.
107
121|

40. RNT 104 res.
β 16| 50 78 90
α 23 61| 84|101

41. RST 100 res.
β *** *** 43 76 97
α 12 26 59| 91 ***

42. SBP III (215-310)
β 226 *** *** 281 306|
α 218|234 243 255 291 ***

43. SBV 260 res.
β *** *** 53 84 *** 119 132 155 170|*** *** 209|237 249
α 3 11 76|102 113|*** 141 *** *** 184|192 225|243|256

44. SNS 149 res.
β *** *** 24 36 91 *** ***
α 3 16|*** 73|101 112|123

139|
***

45. SN3 65 res.
β *** 16 44
α 4| 29 ***

46. SSI 113 res.
β *** 14 33 80 110
α 4 24 54 103 ***

47. STV 195 res.
β *** *** 31 53 67 87 103|133 154
α 3 14 43| 58 80| 93 123|145|173|

182
191|

48. TBV 387 res.
β *** *** 69 105 123 *** *** 161 177 201 228 241 262 295 334 375
α 11 27 83 112|132 143 152 169|194 210|234 256|282|304|342|384

49. TLN I
β 4 28 54
α 19| 45| 62|

(1-136)
77|102 120
86 114|129

50. UBQ 76 res.
β 4 43 69
α 26 60 ***

51. ZGP II (77-212)
β 85 *** *** *** *** 157 182 196
α 106 114 121 129|146 171 *** ***

(*continued*)

Table VIII. (Continued)

(III) all β proteins

1. ADH I (1-165)
β 7 39 63 80 132 157
α 13| 52 71| 91|151|***

2. ALP I-II (1-198)
β 5 21 49 68 87 116 149 195
α 11| 44| 55| 77 98|127|182|***

3. APP I-II (1-323)
β 5 41 72 91 124 166 193 219 245 258 296
α 30| 52 86|103|155|179|212|227 *** 285|306|

4. ATCR II (100-152)
β 106 *** *** 148|
α 127|135|142|***

5. AZU 128 res.
β 3 30 49 81 97 110 125
α 18| 43| 58 89| *** 120|***

6. B2M 98 res.
β 9 26 62 92
α *** 38| 82|***

7. CHA I-II (1-245)
β 3 15 41 66 *** 99 105 119 137 157 181 211 *** ***
α 8 31| 53| 80| 86|*** 112|130 148|170 204 229|235 241

8. CNA 237 res.
β 6 29 52 77 90 107 128 141 157 181 *** 212 230
α 11| 39| 62| 83 99|113|134 *** 174|190|197|218|***

9. CPP III (379-414)
β 397
α 411

10. CTX 71 res.
β 4 31 52
α 10 39|***

11. DFR II (137-162)
β 139 155
α *** ***

12. EBX 62 res.
β 4 35 51
α 29|*** ***

13. EST I-II (1-240)
β 4 30 55 76 *** 107 130 155 172 194 224
α 17| 42| 70| 90 97|*** 146|*** *** 201| 235

14. FABH I-II-III (1-220)
β *** 22 50 69 113 144 171 184 200
α 4| 34| 63 92|138 156|177|*** 213|

15. FABL I-II (1-208)
β 4 20 50 69 82 100 112 *** *** 174
α *** 36| 57|*** 90|*** 124|142|168|181

189
200|

res.
22 30 70 94 120
*** 36| 80 109|***

169 187 217 251 265 292 307
*** 196|234 258|270|302|318|

(1-229)
41 62 94 124 148 167 195 220|
49 71|104 142|161 189|215 ***

16. GAP I (1-98)
β 3 29 41 71
α 14 35| 49| 95|

17. GCR I-II (1-174)
β 4 39 55 92 132 165
α 23| 45| 73 123|153 ***

18. GN5 87 res.
β 4 45 74
α 32| 66| 80|

19. PAB 127
β *** ***
α 10| 15|

20. PCY 99 res.
β *** 13 39 *** 78 95
α 4| 28| 62| 72|*** ***

21. PEP I-II (1-331)
β *** *** 20 33 45 86 105 126 156
α 4| 10|*** *** 69| 93|120|143 ***

22. PKA I-II (1-232)
β 4 29 49 67 *** 99 123 147 168 190 219|
α 22| 38 55 86 93|*** 141|153 *** 195|225

23. PTN I-II
β 8 29
α 20| 35|

24. PYP 285 res.
β 4 28 *** 58 77 89 121 155 *** 204 242 279
α 15| 36 45|*** 83|110 127|172 189 216 271 ***

25. RP2 I-II (1-224)
β 4 28 53 76 106 164 194 ***
α 19| 42| 71| 99 146|176 210|221

26. RXN 53 res.
β 7 39
α *** 50

27. SGB I-II (1-186)
β 5 28 51 67 96 128 152 170
α 16| 45|*** 88|116|135|*** 176

28. SOD 151 res.
β 6 28 84 96 115 147
α 20| 45| 90|102|142|***

29. TAA II (376-478)
β 379 406 442
α 394|435|467|

30. THI 207 res.
β 4 39 75 101 123 151 183 201
α 15| 57| 89|110|128|173 192|***

(continued)

Table VIII. (Continued)

(IV) all α proteins

1. AAT II (46-97)

β 50 72
α 59 84

2. ACT I (1-110)

β *** 14 46 76 102|
α 5| 41 53 87

3. CAT III (321-506)

β *** *** 372 443| *** *** ***
α 329 353 361 383 463 472 478 494

4. CLN 148 res.

β *** *** *** *** 63 *** *** 136
α 3 11 34 52 87 102 107 143

5. CPV 108 res.

β *** *** *** *** 57 98
α 4 13 31 45 66 104

6. CTS

β 95|
α 116

II (86-437)

β 125|*** *** 170|202 232|251|283|313|*** *** 382|397|432|
α 139 146 156 187 214 244 264 289 337 349 357 388 408 ***

7. CYC 103 res.

β *** *** *** 34 58 81 ***
α 5 11 17 51 65 90 96

8. CYP 294 res.

β 6 47| 72|114|*** *** 198 247|
α 24 54 94 156 170 176 204 264

9. C2C 112 res.

β *** *** 18 61 *** *** ***
α 3 10 39 68 78 99 105

10. C3A 77 res.

β 4 43 ***
α 10 61 74

11. ECA 136 res.

β 7|*** *** 51| 64| 82|105|***
α 12 25 36 58 70 99 121 133

12. GAP II (99-208)

β *** *** *** 135 *** 165 174|204
α 106 114 125 *** 148 *** 185 ***

13. GCN 29 res.

β *** *** ***
α 13 18 24

14. HBA 141 res.

β *** *** 17|*** *** 68|*** *** 134|
α 3 12 25 34 55 87 109 122 ***

15. HBB 146 res.

β *** *** *** 33|*** *** *** *** 135|
α 6 12 18 67 76 106 112 130 142

16. HPI

β 3
α 16

I-II (1-175)
45 *** *** 99|121|168
54 72 81 105 152

17. HRB 113 res.
β 19|*** *** *** *** ***
α 28 34 61 68 82 104

18. LH 153 res.
β *** *** 32 *** *** 83 *** ***
α 3 10 62 69 76 114 120 135

144
150

19. LHB 149 res.
β 4 *** *** *** *** *** *** *** *** 142|
α 15 25 50 59 79 86 115 122 127 ***

20. LRP I (1-92)
β *** *** 49| 70|***
α 9 20 60 80 89

21. LZM 164 res.
β *** ***
α 4 10

16 31 57 *** *** 113|150|
22| 48 67 77 95 131 161

22. MBN 153 res.
β *** *** *** *** *** 68|100|*** *** 150|
α 5 11 17 30 53 76 106 112 135 ***

23. MHR 118 res.
β 16 48|***
α 22 82 96

*** 109|
103 115

24. MLT 26 res.
β *** ***
α 6 18

25. PAP I (1-108)
β *** 14 38| 80
α 6| 29 74 92

26. PHH III (293-394)
β *** 306|*** *** 380|
α 300 *** 312 335 ***

27. PPA IV (768-841)
β 771|802 824
α 787 818 ***

28. PPT 36 res.
β *** ***
α 13 18

29. TLN II (137-316)
β 139|155 173|191 231 254 *** *** *** *** 313|
α 146 166 180 *** 241 268 273 281 291 305 ***

30. TMA 284 res.
β *** *** *** *** ***
α 4 11 22 45 72

*** *** 128|170|*** ***
93 108 153 183 199 261

31. TMV 158 res.
β 12 *** *** 70 94 ***
α 23 43 48 82 120 131

32. TNC 162 res.
β *** *** *** *** *** *** 74|114|
α 3 12 21 27 45 61 80 120

(continued)

Table VIII. (Continued)

```
                33. UTG  70 res.    34. YHX  III (268-408)         35. ZGP  I (1-76)   36. 51C  82 res.
128 ***         β  *** ***  14|     β  *** 286|*** *** 372 398|     β   7  49|          β  ***  13|
153 159         α    3   9  43      α  275 301 315 321 381 405      α  22  73           α    6  76

(V)  coil proteins    1. FDX  55 res.   2. MT2  61 res.   3. RHV4  68 res.    4. WGA  I-II (1-171)
                      β  ***  10  30    β   9  40         β  ***  28  45      β   8 140
                      α    4  22  37    α  31  54         α   4 ***  58       α  86 145
```

Table IX. Classification of Proteins of Known Sequence and Conformation According to the Number of Detected PAR(β)-Type Sequences[a]

Group I. Proteins that have no PAR(β)-type sequence

I.1. α/β proteins

1. ADH III (345-375)
2. RHD II (151-290) 2 PAR(α) 214(c) 231(α); 5 HLF(β) 186(α) 198(c) 245(β) 254(α) 270(β); 3 HLF(α) 153(c) 259(α) 264(α)
3. TT4 87 res. 1 PAR(α) 34(β); 3 HLF(β) 4(β) 63(c) 78(β)

I.2. α + β proteins

1. CAT I (1-70)
2. CDV 107 res. 1 HLF(β) 48(c); 2 HLF(α) 66(α) 83(α)
3. FXC 98 res. 4 HLF(β) 25(α) 46(c) 51(β) 78(α); 2 HLF(α) 33(α) 72(c)
4. INSA 21 res. 1 HLF(β) 12(α)
5. PFK III (256-319) 1 PAR(α) 273(α); 1 HLF(β) 285(β); 2 HLF(α) 267(α) 310(α)
6. PTI 58 res. 3 HLF(β) 19(β) 33(β) 38(β)
7. RNB 110 res. 2 HLF(β) 35(α) 96(β); 2 HLF(α) 12(c) 43(c)
8. TLN I (1-136) 8 HLF(β) 9(β) 28(β) 40(β) 54(β) 77(α) 114(β) 120(β) 134(c); 2 HLF(α) 62(β) 72(α)
9. PP2 122 res. 1 PAR(α) 10(α); 4 HLF(β) 20(α) 25(c) 46(α) 108(c); 1 HLF(α) 94(α)
10. SN3 65 res. 5 HLF(β) 4(β) 16(c) 39(β) 44(β) 49(β)
11. ZGP II (77-212) 1 PAR(α) 114(α); 3 HLF(β) 101(c) 171(α) 203(β); 2 HLF(α) 129(β) 146(c)

I.3. All-β proteins

1. DFR II (137-162)
2. EBX 62 res.
3. GCR I (1-84) 1 HLF(β) 45(β)
4. RXN 53 res. 3 HLF(β) 7(β) 12(β) 39(c)
5. SGB I (1-85) 5 HLF(β) 5(β) 16(β) 28(β) 67(β) 83(c)
6. RP2 II (119-224) 1 PAR(α) 221(α); 4 HLF(β) 155(α) 169(β) 188(β) 194(β); 2 HLF(α) 126(β) 176(c)
7. TAA II (376-478) 1 HLF(β) 379(c)

I.4. All-α proteins

1. CAT III (321-506) 2 PAR(α) 472(α) 494(α); 1 HLF(β) 324(α); 4 HLF(α) 353(α) 383(c) 448(α) 463(α)

(*continued*)

[a]Class I proteins (see Table III) are underlined.

Table IX. (Continued)

2. CLN 148 res. 1 PAR(α) 143(α)
3. CPV 108 res. 5 PAR(α) 4(c) 31(α) 46(α) 66(α) 104(α);
   1 HLF(β) 98(β); 2 HLF(α) 13(α) 18(α)
4. HBA 141 res. 1 PAR(α) 34(α); 2 HLF(β) 17(α) 68(α); 2 HLF(α)
   25(α) 109(α)
5. HPI I (1-175) 2 PAR(α) 87(α) 116(α); 3 HLF(β) 3(c) 99(α)
   121(α); 1 HLF(α) 105(α)
6. HRB 113 res. 2 PAR(α) 34(α) 61(α); 1 HLF(α) 82(α)
7. GCN 29 res. 1 PAR(α) 24(α)
8. MLT 26 res. 1 PAR(α) 6(α); 1 HLF(α) 18(α)
9. PHH III (293-394) 1 PAR(α) 312(α); 3 HLF(β) 330(α) 380(α) 385(c);
   1 HLF(α) 300(α)
10. PPT 36 res.
11. ZGP I (1-76) 1 HLF(β) 49(α); 3 HLF(α) 32(c) 38(β) 62(c)
12. 51C 82 res. 2 HLF(β) 13(α) 24(c); 1 HLF(α) 76(α)

I.5. Coil proteins

1. MT2 61 res. 2 HLF(β) 15(c) 40(c)
2. WGA I-II (1-171) 3 HLF(β) 53(c) 145(c) 150(c)

Group II. Proteins that have one PAR(β)-type sequence

II.1. α/β proteins

1. ATCR I (1-99) 1 PAR(β) 31(α); 2 HLF(β) 44(β) 91(β); 3 HLF(α)
   10(c) 25(α) 78(α)
2. CPP I (1-67) 1 PAR(β) 24(c); 1 HLF(β) 64(β); 1 HLF(α) 43(α)
3. PGL III (316-418) 1 PAR(β) 318(c); 3 PAR(α) 362(α) 395(α) 400(α);
   1 HLF(β) 328(β); 1 HLF(α) 390(α)
4. SBP I (1-87) 1 PAR(β) 6(β); 1 HLF(β) 55(α); 2 HLF(α) 64(β)
   72(α)

II.2. α + β proteins

1. AAT I (1-45) 1 PAR(β) 16(α); 1 PAR(α) 6(β); 1 HLF(β) 34(β)
2. ACT II (111-220) 1 PAR(β) 165(β); 5 HLF(β) 135(β) 149(c) 178(β)
   201(c) 207(c)
3. CRO 66 res. 1 PAR(β) 42(β); 1 HLF(β) 26(α); 1 HLF(α) 12(α)
4. DBP2 90 res. 1 PAR(β) 43(β); 1 HLF(α) 8(α)
5. GRS III (295-362) 1 PAR(β) 314(β); 1 HLF(β) 328(β); 2 HLF(α) 338(α)
   343(α)
6. HIP 85 res. 1 PAR(β) 71(β); 1 HLF(β) 76(β); 3 HLF(α) 8(β)
   15(α) 53(c)
7. INSB 30 res. 1 PAR(β) 19(α); 1 PAR(α) 14(α); 1 HLF(β) 24(β)

Table IX. (Continued)

8. LYZ 129 res. 1 PAR(β) 57(β); 4 HLF(β) 4(β) 24(α) 29(α) 82(α); 3 HLF(α) 10(α) 109(α) 122(c)
9. OVO 56 res. 1 PAR(β) 5(c); 1 HLF(β) 41(α); 1 HLF(α) 36(α)
10. PFKE III (256-340) 1 PAR(β) 320(c); 1 PAR(α) 274(α); 1 HLF(β) 284(β); 2 HLF(α) 269(α) 312(α)
11. P2P 124 res. 1 PAR(β) 63(c); 2 HLF(β) 27(c) 93(α); 2 HLF(α) 11(α) 104(α)
12. RNS 124 res. 1 PAR(β) 107(β); 1 HLF(α) 55(α)
13. RNT 104 res. 1 PAR(β) 90(β); 1 HLF(α) 23(α)
14. RST 100 res. 1 PAR(β) 76(β); 1 HLF(β) 85(β)
15. SBP III (215-310) 1 PAR(β) 281(β); 1 HLF(β) 286(c); 2 HLF(α) 243(α) 291(α)

II.3. All-β proteins

1. ALP I (1-84) 1 PAR(β) 5(c); 7 HLF(β) 16(β) 21(β) 32(β) 39(α) 49(β) 60(β) 68(β); 1 HLF(α) 55(β)
2. ALP II (85-198) 1 PAR(β) 195(c); 3 HLF(β) 111(β) 138(c) 155(β); 2 HLF(α) 98(β) 127(β)
3. B2M 98 res. 1 PAR(β) 62(β); 1 HLF(β) 26(β)
4. CHA II (127-245) 1 PAR(β) 211(β); 1 PAR(α) 235(α); 4 HLF(β) 142(β) 181(β) 199(β) 229(β)
5. CPP III (379-414) 1 PAR(β) 404(c); 2 HLF(β) 382(c) 397(β)
6. CTX 71 res. 1 PAR(β) 4(c); 2 HLF(β) 31(c) 39(β)
7. EST II (117-240) 1 PAR(β) 194(β); 7 HLF(β) 120(c) 130(β) 155(α) 172(β) 201(β) 213(β) 229(c)
8. FABL I (1-104) 1 PAR(β) 50(β); 3 HLF(β) 31(α) 69(β) 82(β)
9. FABL II (105-208) 1 PAR(β) 129(β); 1 HLF(α) 142(β)
10. GCR II (85-174) 1 PAR(β) 132(β); 1 HLF(β) 165(β); 1 HLF(α) 158(α)
11. PAB 127 res. 1 PAR(β) 94(β); 2 PAR(α) 15(β) 109(β); 2 HLF(β) 30(β) 70(β); 1 HLF(α) 80(α)
12. PKA I (103-232) 1 PAR(β) 219(α); 3 HLF(β) 168(β) 190(β) 195(β)
13. PTN II (114-229) 1 PAR(β) 195(β); 4 HLF(β) 124(β) 167(β) 189(β) 220(α); 1 HLF(α) 142(β)
14. SGB II (86-186) 1 PAR(β) 128(β); 4 HLF(β) 96(β) 121(β) 152(β) 181(c); 1 HLF(α) 116(β)
15. THI 207 res. 1 PAR(β) 183(β); 4 HLF(β) 52(β) 75(c) 123(β) 133(c); 3 HLF(α) 15(β) 22(β) 128(β)
16. YHX II (224-267) 1 PAR(β) 229(β); 3 HLF(β) 234(β) 240(β) 265(β)

II.4. All-α proteins

1. C2C 112 res. 1 PAR(β) 34(c); 2 PAR(α) 78(α) 105(α); 1 HLF(β) 61(β); 1 HLF(α) 68(α)

(continued)

Table IX. (Continued)

2. C3A 77 res. 1 PAR(β) 53(α); 1 HLF(α) 74(α)
3. GCN 29 res. 1 PAR(β) 26(α)
4. LHB 149 res. 1 PAR(β) 142(α); 2 PAR(α) 37(α) 122(α); 1 HLF(β) 9(c); 2 HLF(α) 25(α) 86(α)
5. LZM 164 res. 1 PAR(β) 31(β); 3 PAR(α) 4(α) 77(α) 101(α); 2 HLF(β) 57(β) 113(α); 1 HLF(α) 131(α)
6. MBN 153 res. 1 PAR(β) 68(α); 4 PAR(α) 11(α) 30(α) 76(α) 112(α); 1 HLF(β) 150(α); 1 HLF(α) 130(α)
7. PAP I (1-108) 1 PAR(β) 38(α); 2 PAR(α) 29(α) 74(α); 1 HLF(β) 80(c)
8. PPA IV (768-841) 1 PAR(β) 771(α); 1 HLF(β) 824(c); 1 HLF(α) 787(α)
9. TMA 284 res. 1 PAR(β) 170(α); 1 PAR(α) 11(α); 1 HLF(α) 87(α)
10. TMV 158 res. 1 PAR(β) 12(c); 3 PAR(α) 23(α) 82(α) 131(α); 2 HLF(β) 70(β) 150(c); 3 HLF(α) 75(α) 87(α) 120(α)
11. UTG 70 res. 1 PAR(β) 14(α); 1 HLF(α) 38(α)

II.5. Coil proteins

1. FDX 55 res. 1 PAR(β) 30(c); 2 HLF(β) 10(c) 47(α)
2. RHV4 68 res. 1 PAR(β) 28(c)

Group III. Proteins that have two PAR(β)-type sequences

III.1. α/β proteins

1. B5C 93 res. 2 PAR(β) 25(β) 75(β); 3 HLF(β) 6(β) 31(β) 53(β)
2. EFM I (1-190) 2 PAR(β) 104(β) 131(β); 2 PAR(α) 147(α) 189(α); 1 HLF(β) 16β); 1 HLF(α) 176(α)
3. PPA I (1-281) 2 PAR(β) 85(β) 220(β); 1 PAR(α) 265(α); 8 HLF(β) 66(α) 90(β) 115(α) 137(α) 150(β) 172(β) 201(β) 258(c); 6 HLF(α) 53(α) 58(α) 102(α) 142(α) 155(β) 245(β)
4. SBP II (88-214) 2 PAR(β) 100(β) 189(β); 1 PAR(α) 162(α); 3 HLF(α) 137(α) 143(α) 198(α)
5. YHX IV (407-485) 2 PAR(β) 451(β) 482(c); 1 PAR(α) 465(α); 2 HLF(β) 413(β) 439(α)

III.2. α + β proteins

1. CAC 259 res. 2 PAR(β) 138(α) 207(β); 1 PAR(α) 143(β); 5 HLF(β) 4(c) 66(β) 97(β) 149(β) 189(β); 2 HLF(α) 76(c) 131(α)
2. CRN 46 res. 2 PAR(β) 15(α) 33(β); 1 HLF(β) 26(α)
3. CTS I (1-85) 2 PAR(β) 34(c) 52(c); 1 PAR(α) 11(α); 1 HLF(β) 43(α)
4. GRS IV (363-478) 2 PAR(β) 420(β) 432(β); 2 HLF(β) 381(β) 437(β); 2 HLF(α) 449(α) 454(α)

Table IX. (Continued)

5. HPI II (176-398) 2 PAR(β) 221(α) 349(α); 1 PAR(α) 326(α); 5 HLF(β) 189(α) 203(α) 211(c) 285(c) 338(c); 1 HLF(α) 182(c)
6. LDH II (182-329) 2 PAR(β) 286(β) 292(β); 2 PAR(α) 238(α) 252(α); 3 HLF(β) 188(β) 202(β) 247(α)
7. PHH II (161-292) 2 PAR(β) 187(β) 282(β); 4 HLF(β) 182(c) 199(β) 209(β) 223(β)
8. SNS 149 res. 2 PAR(β) 24(β) 36(β); 1 HLF(β) 91(β); 3 HLF(α) 16(β) 107(α) 112(β)
9. UBQ 76 res. 2 PAR(β) 4(β) 43(β); 1 PAR(α) 69(β)

III.3. All-β proteins
1. ATCR II (100-152) 2 PAR(β) 106(β) 148(α); 1 HLF(α) 135(β)
2. AZU 128 res. 2 PAR(β) 58(α) 110(β); 2 HLF(β) 63(α) 89(β); 1 HLF(α) 120(β)
3. EST I (1-116) 2 PAR(β) 55(β) 112(c); 1 PAR(α) 97(β); 2 HLF(β) 30(β) 42(β); 1 HLF(α) 90(c)
4. FABH II (119-220) 2 PAR(β) 144(β) 171(β); 1 HLF(β) 200(β)
5. GN5 87 res. 2 PAR(β) 45(β) 85(β); 3 HLF(β) 32(β) 61(β) 74(β)
6. PKA I (1-102) 2 PAR(β) 29(β) 67(β); 1 PAR(α) 93(β); 3 HLF(β) 17(β) 38(β) 49(β)
7. PTN I (1-113) 2 PAR(β) 62(c) 94(β); 4 HLF(β) 8(β) 13(β) 29(β) 41(β)

III.4. All-α proteins
1. AAT II (46-97) 2 PAR(β) 50(c) 72(c); 1 HLF(β) 59(α)
2. CYC 103 res. 2 PAR(β) 34(c) 81(c); 1 PAR(α) 96(α); 2 HLF(β) 46(c) 58(c)
3. ECA 136 res. 2 PAR(β) 64(α) 70(α); 3 PAR(α) 25(α) 36(α) 130(α); 1 HLF(β) 105(α); 1 HLF(α) 121(α)
4. GAP II (99-208) 2 PAR(β) 135(c) 204(c); 2 PAR(α) 114(α) 185(α); 1 HLF(β) 174(α)
5. HBB 146 res. 2 PAR(β) 33(α) 135(α); 3 PAR(α) 12(α) 106(α) 112(α); 3 HLF(α) 28(α) 86(α) 142(α)
6. LRP I (1-92) 2 PAR(β) 49(α) 70(α); 1 PAR(α) 65(α); 4 HLF(α) 20(α) 55(c) 60(α) 89(c)
7. MHR 118 res. 2 PAR(β) 16(c) 48(α); 1 PAR(α) 31(α) 1 HLF(β) 87(α)
8. TNC 162 res. 2 PAR(β) 50(α) 114(α); 6 PAR(α) 21(α) 27(α) 45(α) 61(α) 80(α) 159(α); 1 HLF(β) 103(α); 2 HLF(α) 12(α) 120(α)
9. YHX III (268-406) 2 PAR(β) 326(c) 398(α); 1 PAR(α) 315(α); 1 HLF(β) 270(c); 2 HLF(α) 310(α) 386(α)

III.5. Coil proteins

(*continued*)

Table IX. (Continued)

Group IV. Proteins that have three PAR(β)-type sequences

IV.1. α/β proteins

1. ATCC II (151-310) 3 PAR(β) 186(β) 220(c) 226(α); 4 PAR(α) 159(β) 198(α) 247(β) 301(α); 2 HLF(β) 209(β) 285(α); 2 HLF(α) 192(c) 295(α)
2. DFR I (1-136) 3 PAR(β) 39(β) 76(β) 96(β); 1 PAR(α) 81(α); 1 HLF(β) 4(β); 3 HLF(α) 29(α) 104(α) 119(β)
3. GP 182 res. 3 PAR(β) 27(β) 59(β) 154(β); 2 PAR(α) 97(β) 113(α); 2 HLF(β) 4(β) 105(c); 1 HLF(α) 22(c)
4. GRS I (1-165) 3 PAR(β) 23(β) 125(β) 153(β); 4 HLF(β) 54(c) 59(α) 84(c) 92(c); 4 HLF(α) 33(α) 42(α) 47(β) 111(α)
5. PFK II (136-255) 3 PAR(β) 166(β) 218(β) 246(β); 1 PAR(α) 202(α); 3 HLF(β) 139(α) 178(α) 223(β)
6. PGM 230 res. 3 PAR(β) 4(β) 49(β) 210(β); 3 PAR(α) 154(α) 167(α) 185(α); 2 HLF(β) 23(c) 76(c); 3 HLF(α) 66(α) 176(β) 198(c)
7. PHH I (1-160) 3 PAR(β) 6(β) 14(α) 41(c); 2 PAR(α) 19(α) 53(α); 2 HLF(β) 98(c) 156(β); 3 HLF(α) 47(c) 77(β) 115(α)
8. PPA II (282-546) 3 PAR(β) 299(α) 401(α) 492(α); 1 PAR(α) 351(α); 2 HLF(β) 443(α) 455(β); 3 HLF(α) 333(β) 501(α) 545(α)
9. SRX 108 res. 3 PAR(β) 21(β) 26(β) 79(β); 1 PAR(α) 38(α); 2 HLF(β) 55(β) 73(c); 2 HLF(α) 31(c) 94(c)

IV.2. α + β proteins

1. AAT IV (296-411) 3 PAR(β) 357(β) 379(β) 387(β); 2 PAR(α) 329(α) 371(α); 1 HLF(β) 305(α); 2 HLF(α) 316(α) 398(α)
2. CAB 260 res. 3 PAR(β) 144(β) 149(β) 209(β); 2 HLF(β) 110(c) 191(β); 2 HLF(α) 139(β) 165(α)
3. HPI III (399-659) 3 PAR(β) 410(β) 481(β) 649(β); 2 HLF(β) 585(β) 609(c); 2 HLF(α) 503(β) 557(c)
4. PAP II (109-212) 3 PAR(β) 132(β) 147(β) 172(β); 1 PAR(α) 123(α); 3 HLF(β) 166(β) 187(β) 201(c); 1 HLF(α) 161(β)
5. PGL IV (418-561) 3 PAR(β) 491(β) 501(β) 535(α); 2 PAR(α) 449(c) 540(α); 1 HLF(β) 515(β)
6. SSI 113 res. 3 PAR(β) 14(β) 72(c) 80(β); 3 HLF(β) 19(c) 59(c) 85(c); 2 HLF(α) 49(α) 54(α)

IV.3. All-β proteins

1. ADH I (1-165) 3 PAR(β) 39(β) 63(β) 157(β); 2 HLF(β) 7(c) 44(β); 3 HLF(α) 13(β) 71(β) 111(c)
2. APP II (174-323) 3 PAR(β) 219(β) 296(β) 301(α); 5 HLF(β) 212(β) 245(β) 271(α) 285(β) 318(β)
3. CHA I (1-126) 3 PAR(β) 66(β) 105(β) 119(β); 3 HLF(β) 41(β) 53(β) 59(α); 1 HLF(α) 86(β)

Table IX. (Continued)

4. FABH I (1-118) 3 PAR(β) 50(β) 69(β) 113(β); 4 HLF(β) 45(β) 92(β) 102(β) 108(β)
5. PCY 99 res. 3 PAR(β) 4(β) 13(β) 95(β); 2 HLF(β) 83(β) 90(β); 1 HLF(α) 72(β)

IV.4. All-α proteins

1. ACT I (1-110) 3 PAR(β) 14(c) 46(c) 76(α); 3 HLF(β) 26(α) 32(α) 102(α)
2. CYP 294 res. 3 PAR(β) 6(c) 47(α) 109(α); 1 PAR(α) 170(α); 2 HLF(β) 42(α) 114(α); 5 HLF(α) 24(α) 29(α) 176(α) 193(c) 204(α)
3. LH 153 res. 3 PAR(β) 32(α) 83(α) 144(α); 2 PAR(α) 45(c) 69(α); 5 HLF(α) 10(α) 37(α) 76(α) 125(c) 135(α)

IV.5. Coil proteins

Group V. Proteins that have four PAR(β)-type sequences

V.1. α/β proteins

1. AAT III (98-295) 4 PAR(β) 145(α) 171(α) 186(β) 270(β); 3 PAR(α) 208(α) 218(β) 240(α); 4 HLF(β) 112(α) 226(c) 252(β) 260(c); 1 HLF(α) 191(β)
2. ADK 194 res. 4 PAR(β) 12(β) 91(β) 116(β) 161(α); 1 PAR(α) 81(α)
3. ATCC I (1-150) 4 PAR(β) 70(β) 101(β) 123(β) 141(α); 1 PAR(α) 23(α); 2 HLF(β) 44(β) 49(β)
4. FXN 138 res. 4 PAR(β) 4(β) 50(β) 83(β) 109(β); 2 HLF(β) 55(β) 88(β); 2 HLF(α) 17(α) 22(α)
5. KGA 225 res. 4 PAR(β) 32(β) 101(β) 124(β) 190(β); 3 PAR(α) 27(c) 93(α) 221(α); 7 HLF(β) 51(α) 68(α) 78(β) 83(β) 138(β) 151(α) 195(β); 4 HLF(α) 42(c) 156(α) 209(α) 215(α)
6. PFK I (1-135) 4 PAR(β) 5(β) 34(β) 97(β) 118(β); 6 HLF(β) 39(β) 44(α) 54(α) 59(α) 68(c) 123(β); 1 HLF(α) 27(α)
7. PGKH II (191-416) 4 PAR(β) 210(β) 285(β) 342(c) 368(β); 2 PAR(α) 244(α) 400(α); 5 HLF(β) 238(α) 301(β) 308(β) 333(β) 408(α); 3 HLF(α) 195(α) 328(α) 379(α)
8. PGL II (189-315) 4 PAR(β) 226(α) 295(β) 301(c) 311(α); 1 HLF(β) 249(β); 2 HLF(α) 267(c) 284(β)
9. PPA III (547-767) 4 PAR(β) 583(α) 605(β) 641(β) 710(β); 3 PAR(α) 620(α) 680(c) 762(α); 3 HLF(β) 626(α) 687(β) 692(β); 2 HLF(α) 578(α) 614(α)
10. TS I (1-321) 4 PAR(β) 52(α) 68(β) 134(α) 147(α); 4 PAR(α) 5(α) 110(c) 171(α) 177(α); 8 HLF(β) 33(β) 73(α) 129(α) 189(β) 200(α) 239(c) 253(α) 270(α); 4 HLF(α) 11(c) 95(α) 216(β) 265(α)
11. YHX I (1-223) 4 PAR(β) 63(β) 68(c) 94(β) 206(β); 2 PAR(α) 82(β) 138(α); 3 HLF(β) 99(β) 155(β) 217(α); 2 HLF(α) 16(c) 128(α)

(continued)

Table IX. (Continued)

V.2. α + β proteins

1. GD1 II (164-334) 4 PAR(β) 169(β) 241(β) 271(β) 307(β); 3 PAR(α) 216(α) 229(β) 246(β); 3 HLF(α) 258(α) 324(α) 329(α)
2. GPD II (161-333) 4 PAR(β) 166(β) 240(β) 268(c) 305(β); 2 HLF(β) 178(β) 229(β); 1 HLF(α) 214(α)
3. MDH II (167-324) 4 PAR(β) 176(β) 191(β) 225(α) 262(β); 1 HLF(β) 288(β); 2 HLF(α) 202(β) 243(α)
4. RHV1 289 res. 4 PAR(β) 33(c) 120(β) 178(β) 192(β); 2 PAR(α) 69(α) 148(β); 5 HLF(β) 77(β) 131(β) 153(β) 199(c) 252(β); 1 HLF(α) 227(β)
5. STV 195 res. 4 PAR(β) 31(β) 87(β) 173(β) 182(β); 5 HLF(β) 43(β) 53(c) 80(β) 154(β) 191(β); 1 HLF(α) 161(α)

V.3. All-β proteins

1. APP I (1-173) 4 PAR(β) 41(β) 91(β) 124(β) 166(β); 1 PAR(α) 155(β); 4 HLF(β) 5(β) 25(β) 67(β) 96(β)
2. GAP I (1-98) 4 PAR(β) 41(β) 49(β) 61(β) 71(β); 1 PAR(α) 14(α); 2 HLF(α) 86(β) 95(β)
3. SOD 151 res. 4 PAR(β) 6(β) 28(β) 96(β) 147(β); 3 HLF(β) 84(β) 110(β) 142(β)
4. PYP 285 res. 4 PAR(β) 121(β) 155(β) 179(α) 279(β); 3 PAR(α) 45(β) 142(β) 189(α); 3 HLF(β) 127(β) 204(β) 228(α); 2 HLF(α) 36(c) 216(α)
5. RP2 I (1-118) 4 PAR(β) 28(β) 33(β) 53(β) 106(c); 2 PAR(α) 42(β) 92(β); 1 HLF(β) 4(c); 1 HLF(α) 14(β)

V.4. All-α proteins

1. TLN II (137-316) 4 PAR(β) 139(α) 173(α) 254(β) 313(α); 2 HLF(β) 155(c) 231(c); 3 HLF(α) 180(α) 241(α) 268(α)

V.5. Coil proteins

Group VI. Proteins that have five PAR(β)-type sequences

VI.1. α/β proteins

1. ABP I (1-107) 5 PAR(β) 7(β) 36(β) 62(β) 86(β) 105(β); 1 HLF(β) 31(c); 3 HLF(α) 57(β) 75(α) 81(α)
2. ABP II (108-306) 5 PAR(β) 138(β) 200(β) 211(α) 227(c) 287(β); 1 PAR(α) 110(α); 2 HLF(β) 233(β) 247(β); 4 HLF(α) 160(α) 184(α) 222(c) 269(α)
3. ALKP 449 res. 5 PAR(β) 47(β) 319(β) 365(β) 398(β) 430(β); 3 PAR(α) 161(c) 257(β) 440(α); 6 HLF(β) 22(c) 73(β) 122(c) 142(β) 204(β) 419(β); 4 HLF(α) 134(α) 217(c) 325(β) 446(α)

Table IX. (Continued)

4. CPA 307 res. 5 PAR(β) 49(β) 107(β) 202(β) 278(c) 297(α);
4 PAR(α) 20(α) 25(α) 99(α) 191(β); 5 HLF(β) 10(c) 64(β) 118(α)
139(c) 240(β); 2 HLF(α) 81(α) 228(α)

5. PFKE I (1-135) 5 PAR(β) 6(β) 23(α) 36(β) 98(β) 121(β);
2 PAR(α) 85(α) 111(α); 3 HLF(β) 41(α) 70(c) 106(α); 1 HLF(α) 17(α)

6. PFKE II (136-255) 5 PAR(β) 139(α) 166(β) 189(β) 218(β) 248(β);
3 HLF(α) 171(c) 180(α) 208(α)

7. PGL I (1-188) 5 PAR(β) 4(c) 56(β) 85(β) 111(β) 166(c);
1 PAR(α) 72(α); 2 HLF(β) 63(c) 128(c); 1 HLF(α) 77(α)

8. RHD I (1-150) 5 PAR(β) 31(β) 58(β) 72(c) 95(β) 124(β);
1 PAR(α) 115(α); 4 HLF(β) 10(β) 22(α) 87(α) 129(α); 1 HLF(α) 82(α)

9. TIM 247 res. 5 PAR(β) 6(β) 90(β) 123(β) 160(β) 228(β);
1 PAR(α) 241(α); 4 HLF(β) 38(β) 60(β) 171(c) 207(β); 3 HLF(α) 48(α)
118(α) 199(α)

VI.2. α + β proteins

1. SBV 260 res. 5 PAR(β) 43(β) 53(c) 84(β) 170(α) 218(α);
5 HLF(β) 96(c) 132(β) 155(β) 209(α) 243(β); 5 HLF(α) 102(α) 113(β)
184(β) 225(β) 256(c)

VI.3. All-β proteins

1. PEP I (1-177) 5 PAR(β) 33(β) 86(β) 105(β) 126(β) 169(β);
5 HLF(β) 20(β) 98(β) 115(α) 120(c) 156(β); 1 HLF(α) 4(β)

2. PEP II (178-331) 5 PAR(β) 251(β) 265(β) 302(β) 307(β) 327(β);
5 HLF(β) 187(β) 217(β) 270(β) 280(β) 292(β); 1 HLF(α) 234(α)

3. STI 181 res. 5 PAR(β) 66(c) 83(c) 88(c) 133(β) 171(β);
6 HLF(β) 18(β) 25(β) 30(β) 56(β) 105(c) 150(β); 1 HLF(α) 115(β)

VI.4. All-α proteins

1. CTS II (86-437) 5 PAR(β) 95(α) 283(α) 382(α) 397(α) 432(α);
3 PAR(α) 349(α) 403(α) 408(α); 6 HLF(β) 170(α) 202(c) 219(α) 232(α)
251(α) 313(α); 7 HLF(α) 146(α) 156(α) 187(α) 264(α) 269(α) 275(α)
321(c)

VI.5. Coil proteins

Group VII. Proteins that have 6-11 PAR(β)-type sequences

VII.1. α/β proteins

1. ADH II (166-344) 9 PAR(β) 171(α) 196(β) 202(α) 207(α) 220(β) 268(β)
290(β) 308(α) 321(c); 1 PAR(α) 332(α); 2 HLF(β) 263(β) 316(β);
2 HLF(α) 185(α) 213(α)

2. CPB 306 res. 6 PAR(β) 45(c) 60(β) 101(α) 106(β) 202(β)

(continued)

Table IX. (Continued)

276(c); 1 PAR(α) 295(α); 3 HLF(β) 112(α) 149(c) 191(β); 2 HLF(α) 77(α) 178(α)

3. GD1 I (1-163) 6 PAR(β) 29(β) 58(c) 72(β) 91(β) 117(β) 129(β); 2 HLF(β) 4(β) 85(α); 3 HLF(α) 42(α) 155(α) 161(α)
4. GPD I (1-160) 6 PAR(β) 27(β) 34(α) 44(α) 64(β) 116(β) 129(β); 2 PAR(α) 16(α) 39(α); 3 HLF(β) 5(β) 11(α) 91(β); 2 HLF(α) 21(α) 156(α)
5. GRS II (166-294) 6 PAR(β) 182(α) 192(β) 245(β) 262(β) 267(β) 277(β); 2 PAR(α) 204(α) 286(β); 3 HLF(β) 174(β) 199(α) 291(c); 1 HLF(α) 209(α)
6. LDH I (1-181) 8 PAR(β) 25(β) 30(α) 50(β) 71(c) 93(β) 134(β) 141(α) 151(α); 3 PAR(α) 9(α) 40(α) 115(α); 1 HLF(β) 76(β); 3 HLF(α) 35(α) 98(c) 146(α)
7. MDH I (1-166) 6 PAR(β) 7(β) 22(α) 38(β) 82(β) 102(α) 120(β); 1 HLF(β) 62(β); 2 HLF(α) 12(β) 17(α)
8. PGKH I (1-190) 6 PAR(β) 20(β) 58(β) 93(β) 116(β) 176(α) 187(β); 1 HLF(β) 160(β); 1 HLF(α) 82(α)
9. PGKY I (1-188) 6 PAR(β) 20(β) 41(α) 47(α) 58(β) 115(β) 174(c); 1 PAR(α) 184(β); 1 HLF(β) 157(β); 3 HLF(α) 76(α) 104(α) 164(c)
10. PGKY II (189-415) 7 PAR(β) 208(β) 277(β) 283(β) 330(β) 340(c) 366(β) 409(α); 1 PAR(α) 398(α); 2 HLF(β) 239(α) 392(α); 3 HLF(α) 193(α) 289(c) 324(α)
11. SBT 275 res. 7 PAR(β) 28(β) 81(c) 93(β) 121(β) 149(β) 176(β) 269(β); 2 PAR(α) 198(β) 233(α); 4 HLF(β) 8(α) 46(β) 108(α) 216(β); 3 HLF(α) 71(α) 113(α) 144(α)
12. TAA I (1-375) 10 PAR(β) 13(β) 53(α) 60(β) 112(β) 117(β) 151(c) 203(β) 248(β) 290(β) 326(β); 1 PAR(α) 316(α); 7 HLF(β) 44(α) 82(c) 193(α) 226(β) 231(β) 254(α) 348(c); 6 HLF(α) 73(c) 123(β) 164(β) 260(α) 310(α) 362(α)

VII.2. α + β proteins

1. API (12-393) 7 PAR(β) 118(β) 227(β) 240(β) 250(β) 338(β) 372(β) 386(β); 4 PAR(α) 64(α) 171(c) 186(β) 353(β); 3 HLF(α) 35(α) 59(α) 347(β)
2. CAT II (71-320) 7 PAR(β) 82(β) 153(c) 196(α) 220(β) 291(c) 309(β) 314(β); 1 PAR(α) 183(α); 4 HLF(β) 110(β) 133(β) 143(β) 280(β); 2 HLF(α) 229(β) 266(α)
3. CPP II (68-378) 6 PAR(β) 118(α) 221(α) 246(α) 258(α) 294(c) 369(α); 4 PAR(α) 162(c) 167(c) 203(α) 263(α); 2 HLF(β) 97(c) 357(c); 2 HLF(α) 113(α) 137(α)
4. DPI (324-928) 11 PAR(β) 352(β) 371(β) 438(β) 535(α) 628(α) 646(α) 701(β) 742(c) 766(α) 886(β) 914(β); 4 PAR(α) 713(α) 772(c) 863(α)

Table IX. (Continued)

878(β); 3 HLF(β) 381(β) 427(α) 919(β); 10 HLF(α) 343(α) 433(c) 477(c) 620(c) 635(α) 725(α) 760(α) 790(c) 799(α) 844(α)

5. RHV2 262 res. 6 PAR(β) 93(α) 124(β) 171(α) 181(β) 199(β) 220(β); 5 HLF(β) 5(c) 33(c) 62(α) 80(β) 162(c)

6. RHV3 236 res. 7 PAR(β) 70(β) 88(c) 128(β) 152(β) 162(β) 207(β) 212(β); 1 PAR(α) 45(α); 5 HLF(β) 102(α) 147(β) 157(β) 184(β) 189(β)

7. TBV 387 res. 7 PAR(β) 78(c) 123(β) 161(β) 201(β) 295(β) 334(β) 375(β); 5 PAR(α) 11(c) 20(c) 210(β) 256(β) 384(c); 6 HLF(β) 83(c) 177(β) 241(β) 262(β) 329(β) 370(β); 4 HLF(α) 27(c) 143(α) 194(α) 251(β)

VII.3. All-β proteins

1. <u>CNA</u> 237 res. 6 PAR(β) 6(β) 62(β) 128(β) 141(β) 212(β) 230(c); 1 PAR(α) 197(β); 5 HLF(β) 47(β) 52(β) 90(β) 174(β) 190(β)

VII.4. All-α proteins

VII.5. Coil proteins

PAB, appearing on the β line in Table VIII (III.19) and making a dimer as an antiparallel double-layer β sheet. It can be seen in this figure that two PAR(α) sequences, 15 and 109 (*A*/*B* ratios 1.884 and 1.891, respectively), which are separated by a relatively long range on the sequence, make an adjacent parallel β sheet. The above PAR(α) sequences are marginally |α| and often express their strong ability to be β strands and to play a role in the long-range interactions. Figure 4g shows two clusters of PAR-type sequences of CNA, the first composed of 6 [PAR(β)], 197 [PAR(α), *A*/*B* ratio 1.865], 212 [PAR(β)], and 230 [PAR(β)], and the second composed of 128 [PAR(β)] and 141 [PAR(β)]. The first cluster acts as a nucleus of the protein monomer, whereas the second works as the source of binding force for the dimerization by forming an extra antiparallel β sheet, as is also the case for PAB.

The results of Table VIII are statistically summarized in Table X. They show that pattern recognition of all-β proteins and all-α proteins is rather easy from the ratios, 0.32 and 1.84, respectively, of the number of |α| candidates to the number of |β| candidates.

Definition of the candidates of PAR and HLF types was introduced to improve the predictability of parallel β sheets in α/β proteins. As shown in Fig. 4, such candidates appear in key positions in the tertiary structure of various types of proteins. Inspection of Table IX suggests that the larger the number of residues in PAR(β) sequences in a domain, the bigger is the probability of its being an α/β type. The tendency can quite clearly be seen from the statistical summary of Table IX, as shown in Table XI.

On the other hand, the larger the number of residues in PAR(α) sequences, the bigger the probability is of the domain being an all-α type, as in the cases of CPV, LZM, and MBN. The five-residue regions around the central residues appear on both the β lines and the α lines of the three proteins in Table VIII. They are shown in the stereo views of the α-carbon backbone models in Fig. 5. The figure shows that PAR-type regions in helical conformations are widely distributed compared with those in the tertiary structure of proteins other than all-α type. It is

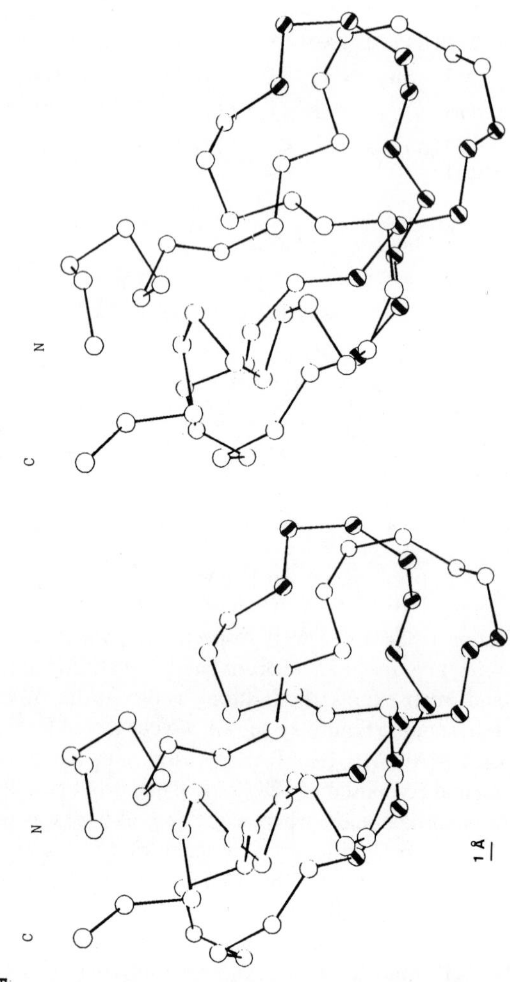
C
N
C
N
a
1 Å

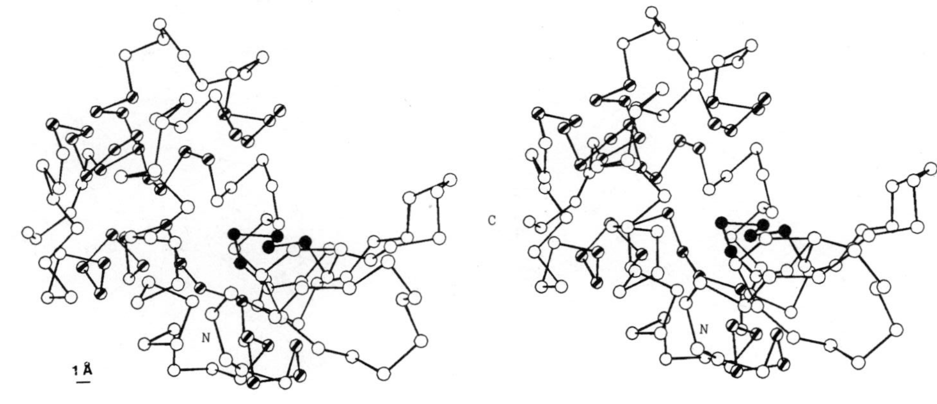

Figure 4. Stereoscopic α-carbon backbone models of five α + β proteins and two all-β proteins: (a) PTI, (b) LYZ, (c) SNS, (d) CAB, (e) SSI, (f) PAB, and (g) CNA. The models are drawn by PLUTO78. The N and C termini are also shown. (a) Two five-residue regions of HLF type, 19 and 33, of PTI are represented by ⦰, and one more HLF type, 38, is indicated by ○. It is clear that an antiparallel β sheet is formed between two HLF-type regions and that a disulfide bond between 38 and 14 is easily made. (b) One five-residue region of PAR type, 57, of LYZ is represented by ● and is situated near the center of the model. In addition, seven HLF-type regions, 4, 10, 24, 29, 82, 109, and 122, are shown by ∅.

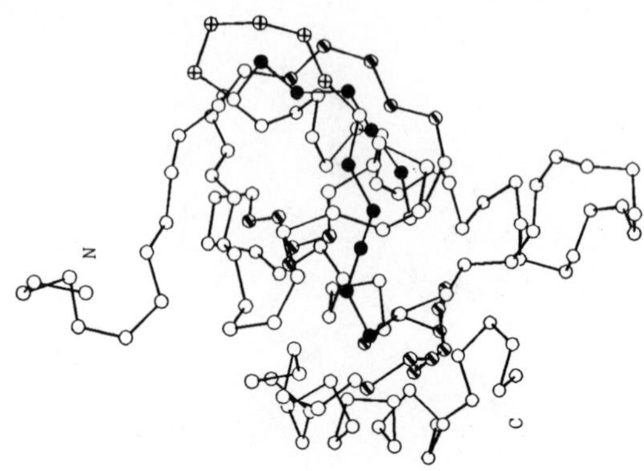
N
C

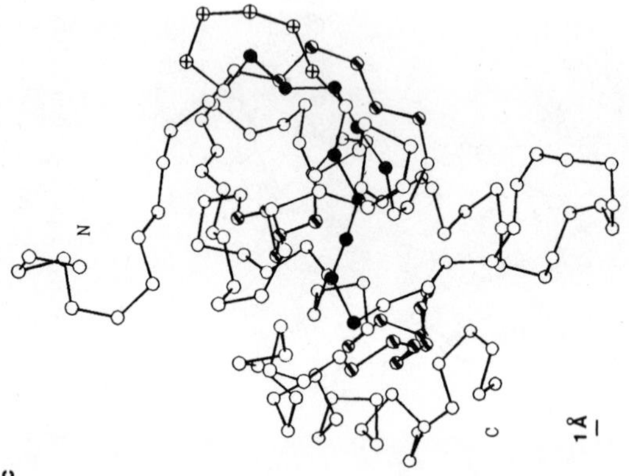
N
C
1Å
c

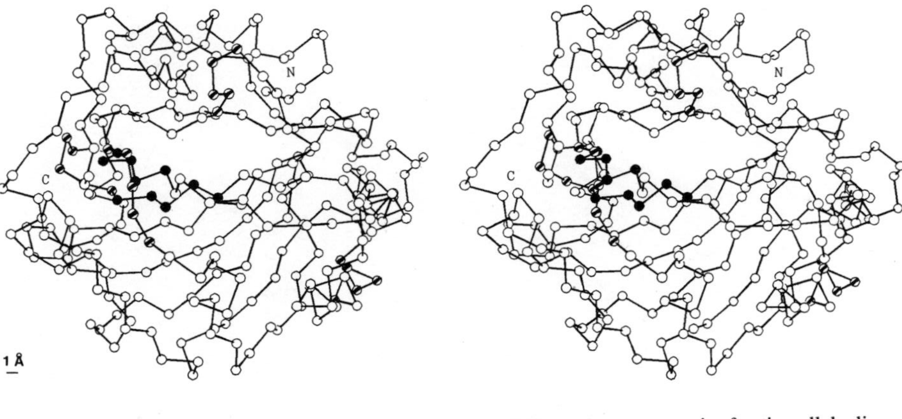

Figure 4 (*cont.*) (c) Two PAR-type regions, 24 and 36, of SNS are shown on a pair of antiparallel adjacent β strands by ●, and a connecting α-candidate region, 30, which is replaced by *** in Table VIII (II-44) because of its negligible *A* and *B* values, is shown by ⊕. Four HLF-type regions, 16, 91, 107, and 122, are shown by ∅. (d) Two PAR-type regions, 144 and 209, of CAB on a pair of parallel adjacent β strands near the center of its domain are shown by ●, and four HLF-type regions, 110, 139, 165, and 191, are shown by ∅.

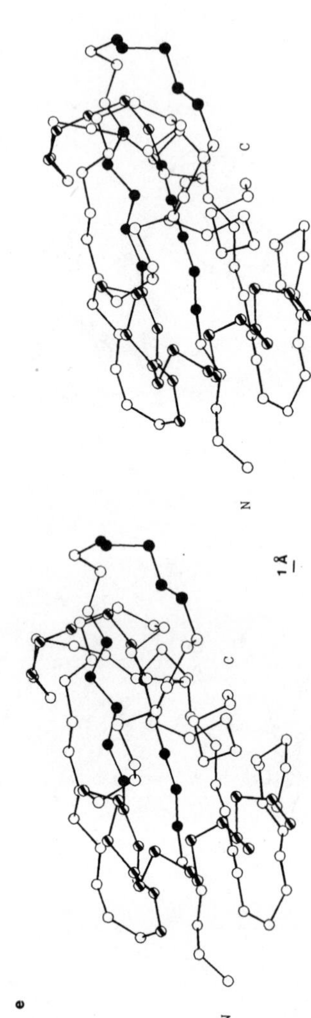
e
N
C
N
C
1 Å

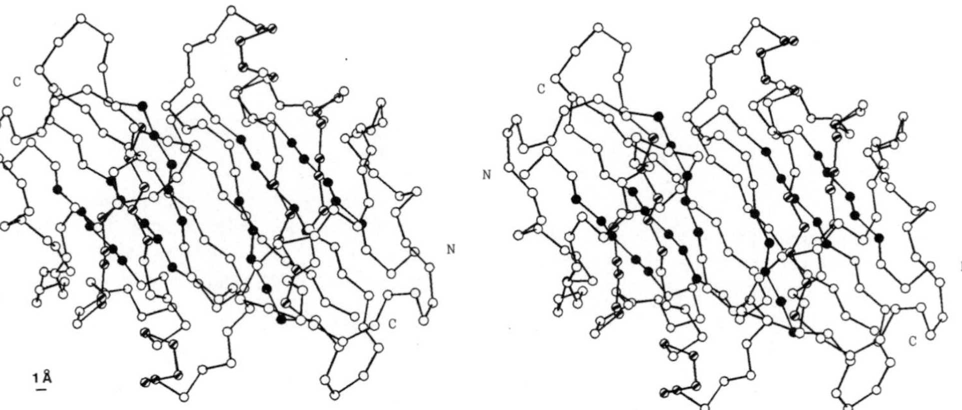

Figure 4 (cont.) (e) Two PAR-type regions, 14 and 80, of SSI form an antiparallel adjacent β sheet, and another PAR-type region, 72, functions to inhibit the activity of subtilisin. Five HLF-type regions are centered at 19, 49, 54, 59, and 85. The central strands of PAR type are responsible for forming a double-layered β-sheet dimer. (f) The PAB dimer is composed of an extended antiparallel double-layered β sheet on which three PAR-type regions, 15, 94, and 109, and three HLF-type regions, 30, 70, and 80, are shown. A pair of parallel β strands of PAR type, 15 and 109, is similar to that at 144 and 209 of CAB, and also is responsible for forming a quadruple-layered β-sheet teramer similar to the SSI dimer.

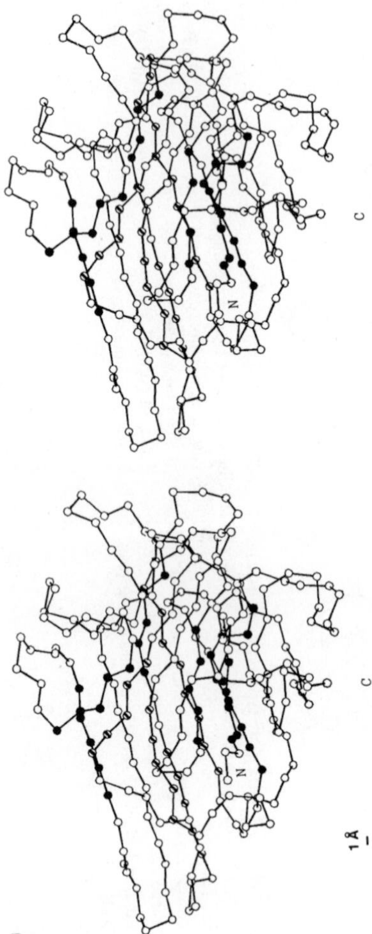

Figure 4 (cont.) (g) One cluster of five PAR-type regions, 6, 62, 197, 212, and 230, constitutes the hydrophobic core of the monomer of CNA, and another cluster of two PAR-type regions, 128 and 141, forms the core at the edge, which is capable of extending the β sheet by dimerization in an antiparallel fashion. Five HLF-type regions are 47, 52, 90, 174, and 190.

Table X. Statistical Evaluation of Secondary Structure Prediction in Table VIII

	Total		α helix		β structure		Coil	
(I) α/β proteins								
|β| candidates on β-line	317		43	(14%)	228	(72%)	46	(15%)
candidates on α-line	376	(100%)	256	(68%)	53	(14%)	67	(18%)
|α| on α-line	315	(84%)	233	(74%)	31	(10%)	51	(16%)
|β| on α-line	61	(16%)	23	(38%)	22	(36%)	16	(26%)
ratio of |α| candidates / |β| candidates	0.83							
(II) α + β proteins								
|β| candidates on β-line	258		44	(17%)	167	(65%)	47	(18%)
candidates on α-line	320	(100%)	147	(46%)	106	(33%)	67	(21%)
|α| on α-line	246	(77%)	134	(54%)	59	(24%)	53	(22%)
|β| on α-line	74	(23%)	13	(18%)	47	(64%)	14	(19%)
ratio of |α| candidates / |β| candidates	0.74							
(III) All-β proteins								
|β| candidates on β-line	196		3	(2%)	171	(87%)	22	(11%)
candidates on α-line	175	(100%)	28	(16%)	129	(74%)	18	(10%)
|α| on α-line	89	(51%)	20	(22%)	55	(62%)	14	(16%)
|β| on α-line	86	(49%)	8	(9%)	74	(86%)	4	(5%)
ratio of |α| candidates / |β| candidates	0.32							
(IV) All-α proteins								
|β| candidates on β-line	104		56	(54%)	10	(10%)	38	(37%)
candidates on α-line	220	(100%)	203	(92%)	3	(1%)	14	(6%)
|α| on α-line	210	(95%)	199	(95%)	1	(0%)	10	(5%)
|β| on α-line	10	(5%)	4	(40%)	2	(20%)	4	(40%)
ratio of |α| candidates / |β| candidates	1.84							
(V) Coil proteins								
|β| candidates on β-line	8		0	(0%)	0	(0%)	8	(100%)
candidates on α-line	9	(100%)	0	(0%)	0	(0%)	9	(100%)
|α| on α-line	4	(44%)	0	(0%)	0	(0%)	4	(100%)
|β| on α-line	5	(56%)	0	(0%)	0	(0%)	5	(100%)
ratio of |α| candidates / |β| candidates	0.31							

Table XI. Statistical Evaluation of Secondary Structure Prediction by Assignment of PAR(β), PAR(α), HLF(β), and HLF(α) Candidates in Table IX

	Total	Number per 100 residues	α helix	β structure	Coil
Group I: Proteins having 0 PAR(β), total 3218 residues					
(I.1) α/β proteins, 258 res. (8%), chain length 86 ± 55 res.					
PAR(β)	0	0.0	0	0	0
PAR(α)	3	1.2 ± 0.8	1 (33%)	1 (33%)	1 (33%)
HLF(β)	8	3.1 ± 2.0	2 (25%)	4 (50%)	2 (25%)
HLF(α)	3	1.0 ± 1.2	2 (67%)	0 (0%)	1 (33%)
(I.2) α + β proteins, 987 res. (31%), chain length 90 ± 37 res.					
PAR(β)	0	0.0	0	0	0
PAR(α)	3	0.3 ± 0.5	3 (100%)	0 (0%)	0 (0%)
HLF(β)	32	3.2 ± 2.3	8 (25%)	17 (53%)	7 (22%)
HLF(α)	13	1.3 ± 1.1	7 (54%)	2 (15%)	4 (31%)
(I.3) All-β proteins, 519 res. (16%), chain length 74 ± 29 res.					
PAR(β)	0	0.0	0	0	0
PAR(α)	1	0.2 ± 0.4	1 (100%)	0 (0%)	0 (0%)
HLF(β)	14	2.7 ± 2.6	1 (7%)	10 (71%)	3 (21%)
HLF(α)	2	0.4 ± 0.7	0 (0%)	1 (50%)	1 (50%)
(I.4) All-α proteins, 1222 res. (38%), chain length 102 ± 55 res.					
PAR(β)	0	0.0	0	0	0
PAR(α)	16	1.3 ± 1.6	15 (94%)	0 (0%)	1 (6%)
HLF(β)	14	1.2 ± 1.0	9 (64%)	1 (7%)	4 (29%)
HLF(α)	16	1.3 ± 1.4	13 (81%)	0 (0%)	3 (19%)
(I.5) Coil proteins, 232 res. (7%), chain length 116 ± 78 res.					
PAR(β)	0	0.0	0	0	0
PAR(α)	0	0.0	0	0	0
HLF(β)	5	2.2 ± 1.1	0 (0%)	0 (0%)	5 (100%)
HLF(α)	0	0.0	0	0	0
(I.1-5) All group I proteins					
PAR(β)	0	0.0	0	0	0
PAR(α)	23	0.7 ± 1.2	20 (87%)	1 (4%)	2 (9%)
HLF(β)	79	2.5 ± 2.1	26 (33%)	32 (41%)	21 (27%)
HLF(α)	34	1.1 ± 1.2	22 (65%)	3 (9%)	9 (26%)

Table XI. (Continued)

Group II: Proteins having 1 PAR(β), total 4838 residues								
(II.1) α/β proteins, 356 res. (7%), chain length 89 ± 16 res.								
PAR(β)	4	1.1 ± 0.2	1	(25%)	2	(50%)	1	(25%)
PAR(α)	3	0.8 ± 1.5	3	(100%)	0	(0%)	0	(0%)
HLF(β)	5	1.4 ± 0.5	1	(20%)	4	(80%)	0	(0%)
HLF(α)	7	2.0 ± 0.9	5	(71%)	1	(14%)	1	(14%)
(II.2) α + β proteins, 1312 res. (27%), chain length 87 ± 30 res.								
PAR(β)	15	1.1 ± 0.7	2	(13%)	10	(67%)	3	(20%)
PAR(α)	3	0.2 ± 1.0	2	(67%)	1	(33%)	0	(0%)
HLF(β)	17	1.3 ± 1.3	6	(35%)	8	(47%)	3	(18%)
HLF(α)	19	1.4 ± 1.1	17	(89%)	0	(0%)	2	(11%)
(II.3) All-β proteins, 1669 res. (34%), chain length 104 ± 39 res.								
PAR(β)	16	1.0 ± 0.6	1	(6%)	12	(75%)	3	(19%)
PAR(α)	3	0.2 ± 0.4	1	(33%)	2	(67%)	0	(0%)
HLF(β)	50	3.0 ± 2.5	4	(8%)	39	(78%)	7	(14%)
HLF(α)	11	0.7 ± 1.7	2	(18%)	9	(82%)	0	(0%)
(II.4) All-α proteins, 1378 res. (28%), chain length 125 ± 68 res.								
PAR(β)	11	0.7 ± 0.8	8	(73%)	1	(9%)	2	(18%)
PAR(α)	17	1.1 ± 1.0	17	(100%)	0	(0%)	0	(0%)
HLF(β)	9	0.6 ± 0.5	2	(22%)	3	(33%)	4	(44%)
HLF(α)	12	0.8 ± 0.6	12	(100%)	0	(0%)	0	(0%)
(II.5) Coil proteins, 123 res. (2%), chain length 62 ± 9 res.								
PAR(β)	2	1.6 ± 0.2	0	(0%)	0	(0%)	2	(100%)
PAR(α)	0	0.0	0		0		0	
HLF(β)	2	1.6 ± 2.6	1	(50%)	0	(0%)	1	(50%)
HLF(α)	0	0.0	0		0		0	
(II.1-5) All group II proteins								
PAR(β)	48	1.0 ± 0.7	12	(25%)	25	(52%)	11	(23%)
PAR(α)	26	0.5 ± 0.9	23	(88%)	3	(12%)	0	(0%)
HLF(β)	83	1.7 ± 2.0	14	(17%)	54	(65%)	15	(18%)
HLF(α)	49	1.0 ± 1.2	36	(73%)	10	(20%)	3	(6%)
Group III: Proteins having 2 PAR(β), total 3763 residues								
(III.1) α/β proteins, 770 res. (20%), chain length 154 ± 83 res.								
PAR(β)	10	1.3 ± 0.8	0	(0%)	9	(90%)	1	(10%)
PAR(α)	5	0.7 ± 0.5	5	(100%)	0	(0%)	0	(0%)

(*continued*)

Table XI. (Continued)

HLF(β)	14	1.8 ± 1.5	4	(29%)	9	(64%)	1	(7%)
HLF(α)	10	1.3 ± 1.2	8	(80%)	2	(20%)	0	(0%)
(III.2) α + β proteins, 1234 res. (33%), chain length 137 ± 69 res.								
PAR(β)	18	1.5 ± 1.1	4	(22%)	12	(67%)	2	(11%)
PAR(α)	6	0.5 ± 0.6	4	(67%)	2	(33%)	0	(0%)
HLF(β)	22	1.8 ± 0.9	5	(23%)	12	(55%)	5	(23%)
HLF(α)	8	0.7 ± 0.8	4	(50%)	2	(25%)	2	(25%)
(III.3) All-β proteins, 701 res. (19%), chain length 100 ± 24 res.								
PAR(β)	14	2.0 ± 0.8	2	(14%)	10	(71%)	2	(14%)
PAR(α)	2	0.3 ± 0.5	0	(0%)	2	(100%)	0	(0%)
HLF(β)	15	2.1 ± 1.3	1	(7%)	14	(93%)	0	(0%)
HLF(α)	3	0.4 ± 0.7	0	(0%)	2	(67%)	1	(33%)
(III.4) All-α proteins, 1058 res. (28%), chain length 118 ± 33 res.								
PAR(β)	18	1.7 ± 0.8	10	(56%)	0	(0%)	8	(44%)
PAR(α)	18	1.7 ± 1.1	18	(100%)	0	(0%)	0	(0%)
HLF(β)	8	0.8 ± 0.7	5	(63%)	0	(0%)	3	(38%)
HLF(α)	12	1.1 ± 1.4	10	(83%)	0	(0%)	2	(17%)
(III.5) Coil proteins, 0 res. (0%)								
(III.1-4) All group III proteins								
PAR(β)	60	1.6 ± 0.9	16	(27%)	31	(52%)	13	(22%)
PAR(α)	31	0.8 ± 0.9	27	(87%)	4	(13%)	0	(0%)
HLF(β)	59	1.6 ± 1.1	15	(25%)	35	(59%)	9	(15%)
HLF(α)	33	0.9 ± 1.1	22	(67%)	6	(18%)	5	(15%)
Group IV: Proteins having 3 PAR(β), total 3739 residues								
(IV.1) α/β proteins, 1526 res. (41%), chain length 170 ± 51 res.								
PAR(β)	27	1.8 ± 0.5	5	(19%)	20	(74%)	2	(7%)
PAR(α)	15	1.0 ± 0.7	12	(80%)	3	(20%)	0	(0%)
HLF(β)	20	1.3 ± 0.7	5	(25%)	7	(35%)	8	(40%)
HLF(α)	21	1.4 ± 0.8	10	(48%)	5	(24%)	6	(29%)
(IV.2) α + β proteins, 998 res. (27%), chain length 166 ± 74 res.								
PAR(β)	18	1.8 ± 0.8	1	(6%)	16	(89%)	1	(6%)
PAR(α)	4	0.4 ± 0.6	3	(75%)	0	(0%)	1	(25%)
HLF(β)	12	1.2 ± 1.0	1	(8%)	5	(42%)	6	(50%)
HLF(α)	9	0.9 ± 0.7	5	(56%)	3	(33%)	1	(11%)

Table XI. (Continued)

(IV.3) All-β proteins, 658 res. (18%), chain length 132 ± 26 res.					
PAR(β)	15	2.3 ± 0.5	1 (7%)	14 (93%)	0 (0%)
PAR(α)	0	0.0	0	0	0
HLF(β)	16	2.4 ± 0.9	2 (13%)	13 (81%)	1 (6%)
HLF(α)	5	0.8 ± 0.8	0 (0%)	4 (80%)	1 (20%)
(IV.4) All-α proteins, 557 res. (15%), chain length 186 ± 96 res.					
PAR(β)	9	1.6 ± 0.9	6 (67%)	0 (0%)	3 (33%)
PAR(β)	3	0.5 ± 0.7	2 (67%)	0 (0%)	1 (33%)
HLF(β)	5	0.9 ± 1.4	5 (100%)	0 (0%)	0 (0%)
HLF(α)	10	1.8 ± 1.6	8 (80%)	0 (0%)	2 (20%)
(IV.5) Coil proteins, 0 res. (0%)					
(IV.1-4) All group IV proteins					
PAR(β)	72	1.9 ± 0.6	13 (18%)	53 (74%)	6 (8%)
PAR(α)	22	0.6 ± 0.8	17 (77%)	3 (14%)	2 (9%)
HLF(β)	53	1.4 ± 1.0	13 (25%)	25 (47%)	15 (28%)
HLF(α)	45	1.2 ± 0.9	23 (51%)	12 (27%)	10 (22%)
Group V: Proteins having 4 PAR(β), total 4147 residues					
(V.1) α/β proteins, 2156 res. (52%), chain length 196 ± 57 res.					
PAR(β)	44	2.0 ± 0.6	10 (23%)	31 (70%)	3 (7%)
PAR(α)	19	0.9 ± 0.6	14 (74%)	2 (11%)	3 (16%)
HLF(β)	41	1.9 ± 1.2	16 (39%)	21 (51%)	4 (10%)
HLF(α)	21	1.0 ± 0.6	14 (67%)	3 (14%)	4 (19%)
(V.2) α + β proteins, 986 res. (24%), chain length 197 ± 53 res.					
PAR(β)	20	2.0 ± 0.4	1 (5%)	17 (85%)	2 (10%)
PAR(α)	5	0.5 ± 0.8	2 (40%)	3 (60%)	0 (0%)
HLF(β)	13	1.3 ± 1.0	0 (0%)	11 (85%)	2 (15%)
HLF(α)	8	0.8 ± 0.6	6 (75%)	2 (25%)	0 (0%)
(V.3) All-β proteins, 825 res. (20%), chain length 165 ± 73 res.					
PAR(β)	20	2.4 ± 1.0	1 (5%)	18 (90%)	1 (5%)
PAR(α)	7	0.9 ± 0.6	2 (29%)	5 (71%)	0 (0%)
HLF(β)	11	1.3 ± 0.7	1 (9%)	9 (82%)	1 (9%)
HLF(α)	5	0.6 ± 0.4	1 (20%)	3 (60%)	1 (20%)
(V.4) All-α proteins, 180 res. (4%), chain length 180 res.					

Table XI. (Continued)

PAR(β)	4	2.2	3 (75%)	1 (25%)	0 (0%)
PAR(α)	0	0.0	0	0	0
HLF(β)	2	1.1	0 (0%)	0 (0%)	2 (100%)
HLF(α)	3	1.7	3 (100%)	0 (0%)	0 (0%)
(V.5) Coil proteins, 0 res.					
(V.1-4) All group V proteins					
PAR(β)	83	2.0 ± 0.7	15 (18%)	62 (75%)	6 (7%)
PAR(α)	31	0.7 ± 0.6	18 (58%)	10 (32%)	3 (10%)
HLF(β)	62	1.5 ± 1.0	12 (19%)	41 (66%)	9 (15%)
HLF(α)	37	0.9 ± 0.8	24 (65%)	8 (22%)	5 (14%)
Group VI: Proteins having 5 PAR(β), total 3026 residues					
(VI.1) α/β proteins, 1902 res. (63%), chain length 211 ± 109 res.					
PAR(β)	45	2.4 ± 1.2	4 (9%)	36 (80%)	5 (11%)
PAR(α)	13	0.7 ± 0.5	10 (77%)	2 (15%)	1 (8%)
HLF(β)	27	1.4 ± 0.8	6 (22%)	12 (44%)	9 (33%)
HLF(α)	22	1.2 ± 0.9	17 (77%)	2 (9%)	3 (14%)
(VI.2) α + β proteins, 260 res. (9%), chain length 260 res.					
PAR(β)	5	1.9	2 (40%)	2 (40%)	1 (20%)
PAR(α)	0	0.0	0	0	0
HLF(β)	5	1.9	1 (20%)	3 (60%)	1 (20%)
HLF(α)	5	1.9	1 (20%)	3 (60%)	1 (20%)
(VI.3) All-β proteins, 512 res. (17%), chain length 171 ± 15 res.					
PAR(β)	15	2.9 ± 0.3	0 (0%)	12 (80%)	3 (20%)
PAR(α)	0	0.0	0	0	0
HLF(β)	16	3.1 ± 0.3	1 (6%)	13 (81%)	2 (13%)
HLF(α)	3	0.6 ± 0.1	1 (33%)	2 (67%)	0 (0%)
(VI.4) All-α proteins, 352 res. (12%), chain length 352 res.					
PAR(β)	5	1.4	5 (100%)	0 (0%)	0 (0%)
PAR(α)	3	0.9	3 (100%)	0 (0%)	0 (0%)
HLF(β)	6	1.7	5 (83%)	0 (0%)	1 (17%)
HLF(α)	7	2.0	6 (86%)	0 (0%)	1 (14%)
(VI.5) Coil proteins, 0 res.					

Table XI. (Continued)

(VI.1-4) All group VI proteins					
PAR(β)	70	2.3 ± 1.0	11 (16%)	50 (71%)	9 (13%)
PAR(α)	16	0.5 ± 0.5	13 (81%)	2 (13%)	1 (6%)
HLF(β)	54	1.8 ± 1.0	13 (24%)	28 (52%)	13 (24%)
HLF(α)	37	1.2 ± 0.8	25 (68%)	7 (19%)	5 (14%)
Group VII: Proteins having 6-11 PAR(β), total 5209 residues					
(VII.1) α/β proteins,			2539 res. (49%),	chain length	212 ± 72 res.
PAR(β)	83	3.3 ± 0.9	18 (22%)	56 (67%)	9 (11%)
PAR(α)	14	0.6 ± 0.6	11 (79%)	3 (21%)	0 (0%)
HLF(β)	30	1.2 ± 0.6	11 (37%)	15 (50%)	4 (13%)
HLF(α)	30	1.2 ± 0.4	23 (77%)	3 (10%)	4 (13%)
(VII.2) α + β proteins,			2433 res. (47%),	chain length	348 ± 129 res.
PAR(β)	51	2.1 ± 0.5	12 (24%)	34 (67%)	5 (10%)
PAR(α)	19	0.8 ± 0.5	9 (47%)	5 (26%)	5 (26%)
HLF(β)	25	1.0 ± 0.8	4 (16%)	17 (68%)	4 (16%)
HLF(α)	21	0.9 ± 0.6	13 (62%)	3 (14%)	5 (24%)
(VII.3) All-β proteins,			237 res. (5%),	chain length	237 res.
PAR(β)	6	2.5	0 (0%)	5 (83%)	1 (17%)
PAR(α)	1	0.4	0 (0%)	1 (100%)	0 (0%)
HLF(β)	5	2.1	0 (0%)	5 (100%)	0 (0%)
HLF(α)	0	0.0	0	0	0
(VII.4) All-α proteins,			0 res.		
(VII.5) Coil proteins,			0 res.		
(VII.1-3) All group VII proteins					
PAR(β)	140	2.7 ± 1.0	30 (21%)	95 (68%)	15 (11%)
PAR(α)	34	0.7 ± 0.5	20 (59%)	9 (26%)	5 (15%)
HLF(β)	60	1.2 ± 0.7	15 (25%)	37 (62%)	8 (13%)
HLF(α)	51	1.0 ± 0.6	36 (71%)	6 (12%)	9 (18%)

also clear from Fig. 5 that the candidate regions of α and β on the α line and β line, respectively, are important in the packing of α helices to hold the tertiary structure.

Statistical characteristics of α + β proteins (Tables X and XI) are not clear. It is rather difficult to distinguish a domain of α + β from α/β type. The former looks more like the all-β type, whereas some of the latter seem to be an intermediate type between α/β and one of the other types.

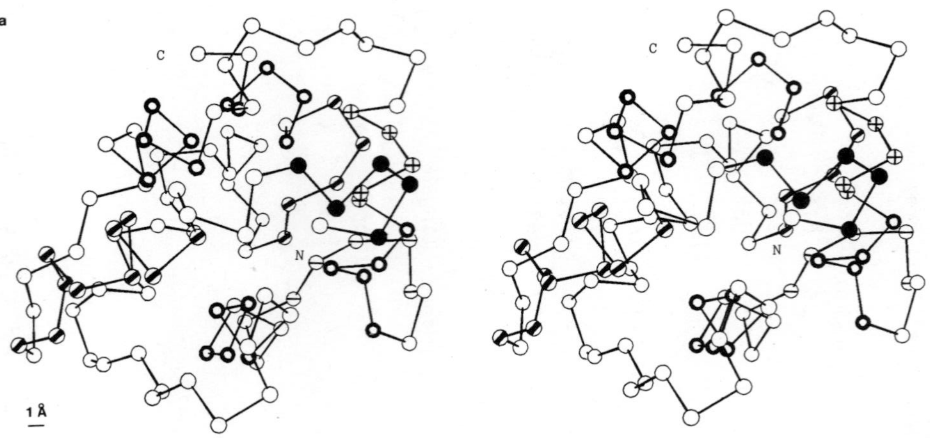

Figure 5. Stereoscopic α-carbon backbone models of three all-α proteins: (a) CPV, (b) LZM, and (c) MBN. The models are drawn by PLUTO78. Nonhelical PAR-type regions are represented by ●, and helical PAR-type regions are shown by **O**. HLF-type regions and the N and C termini are as in Figs. 3 and 4. Also, α regions, i.e., |α| regions other than PAR and HLF types, are represented by ⊕, and β regions, i.e., |β| regions other than PAR and HLF types, are by ⊖. (a) Five PAR-type regions, 4, 31, 50, 66, and 104, of CPV do not directly contact each other, but two HLF-type regions, 13 and 18, and α region 45 are also important in the packing of the α helices. An HLF(β) region, 98, and a β region, 57, make an antiparallel β sheet. (b) One PAR(β) region, 31, of LZM is very similar to 57 of LYZ. An HLF(β), 57, and a β, 16, form an antiparallel β sheet similar to 98 and 57 of CPV. Three PAR-type regions, 4, 77, and 101, and two HLF-type regions, 113 and 131, are dispersed. The helical packing arrangement is completed by adding five α regions, 10, 48, 67, 95, and 161, as well as one β region, 150, in a helical conformation. (c) Five PAR-type regions, 11, 30, 68, 76, and 112, and two HLF-type regions, 130 and 150, are widely dispersed, but a little more interaction among those is observed than in LZM. However, five α regions, 5, 17, 53, 106, and 135, are important in the packing arrangement. One β region, 100, is in an extended conformation.

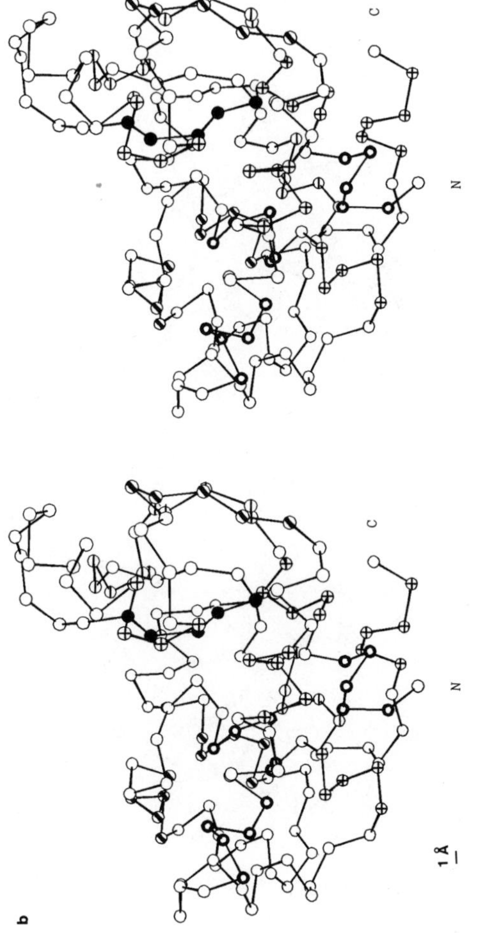

Figure 5 (cont.)

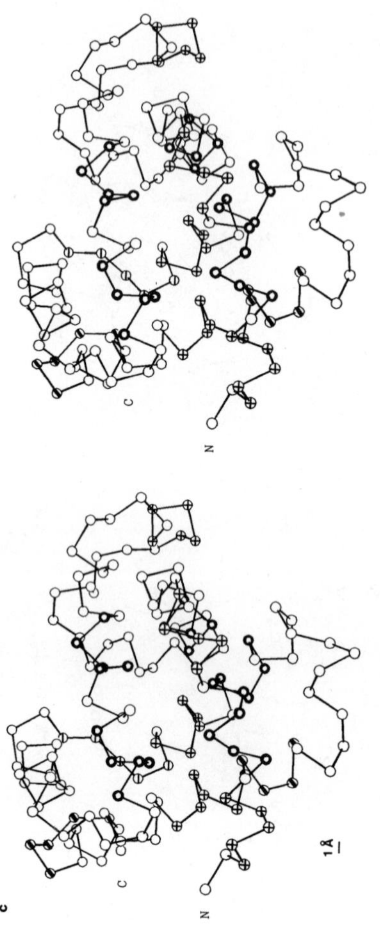

Figure 5 (cont.)

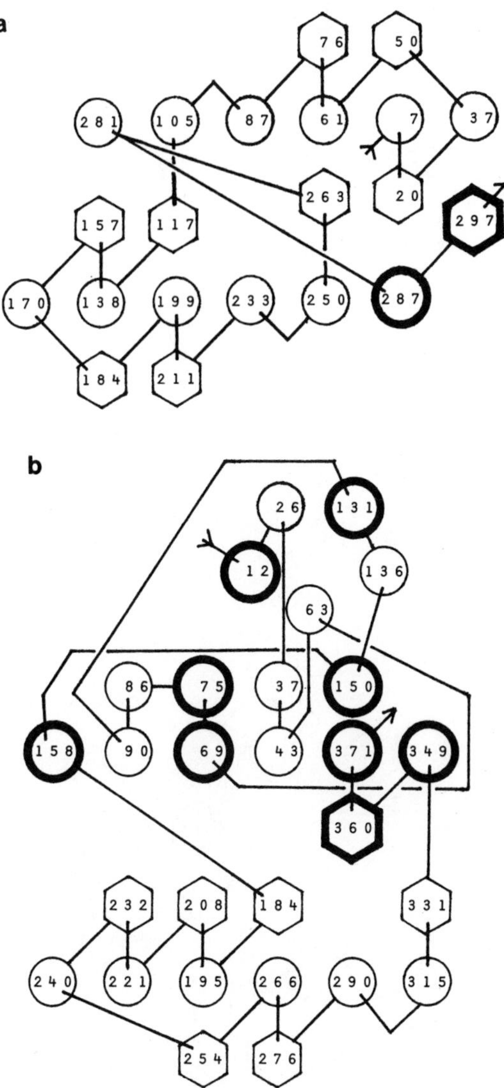

Figure 6. The observed packing arrangements of α helices and β strands are illustrated schematically for the proteins of α/β type: (a) ABP, (b) ADH, (c) ADK, (d) B5C, (e) CPA, (f) DFR, (g) FXN, (h) GPD, (i) KGA, (j) LDH, (k) PGKH, (l) RHD, (m) SBT, (n) SRX, (o) TAA, (p) TIM, and (q) YHX. Hexagons and circles represent α helices and β strands, respectively. Thin circles and thick hexagons are viewed from the C termini; thick circles and thin hexagons from the N termini. Arrow-tails and arrowheads represent N terminus and C terminus, respectively. Numerals in hexagons and circles represent central residue numbers of α helices and β strands, respectively. Broken circles represent a conformation very similar to that of β strand. The domains other than the α/β type are also shown schematically, but the spatial packing arrangements are not accurate enough. Such domains are (b) ADH I (all β), (f) DFR II (all β), (h) GPD II (α + β), (j) LDH II (α + β), (k) two excursions in PGKH (all β), (m) two excursions in SBT (all β), (o) one excursion in TAA I (all β) and TAA II (all β), and (q) YHX III (all α).

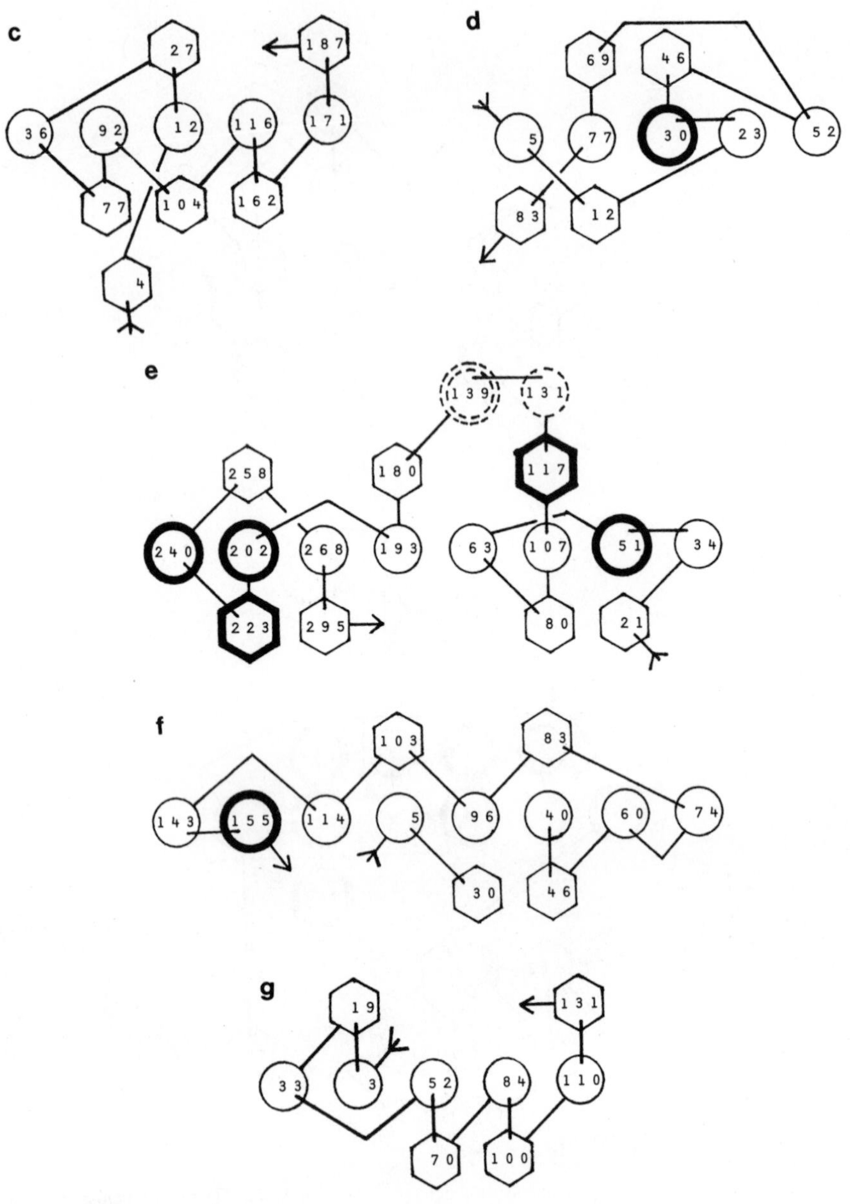

Figure 6 (cont.)

V. *PREDICTION OF PACKING OF SECONDARY STRUCTURES IN* α/β *PROTEINS*

In order to compare the predicted patterns of α/β proteins in Table VIII (I) with the topologies observed by x-ray crystallography, some of the schematic topological patterns are shown in Fig. 6, in which the central residue numbers represent individual α helices and β strands. We can see the best agreement between the calculated and observed secondary

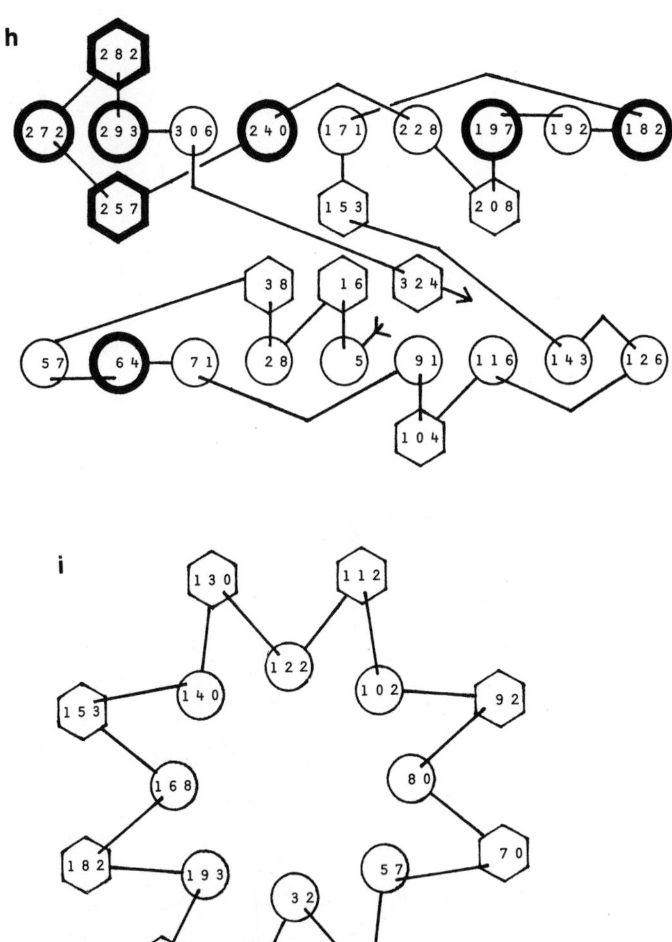

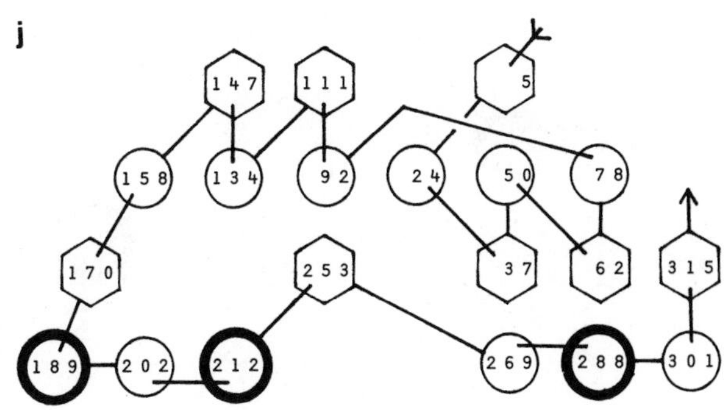

Figure 6 (cont.)

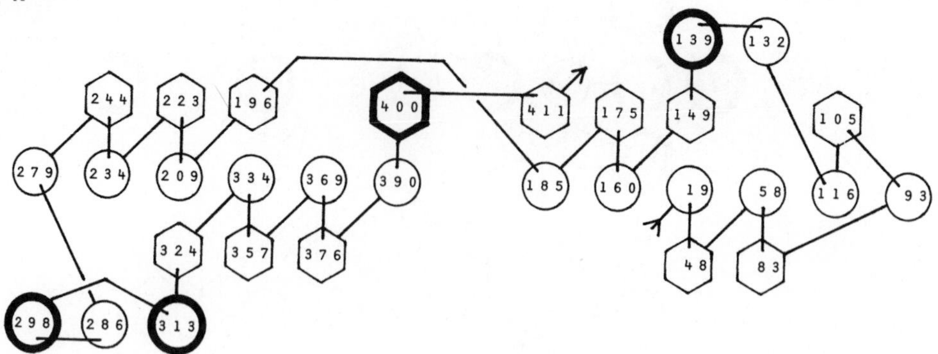

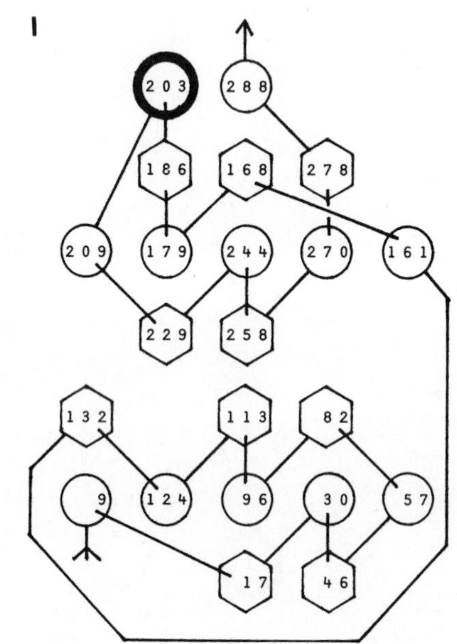

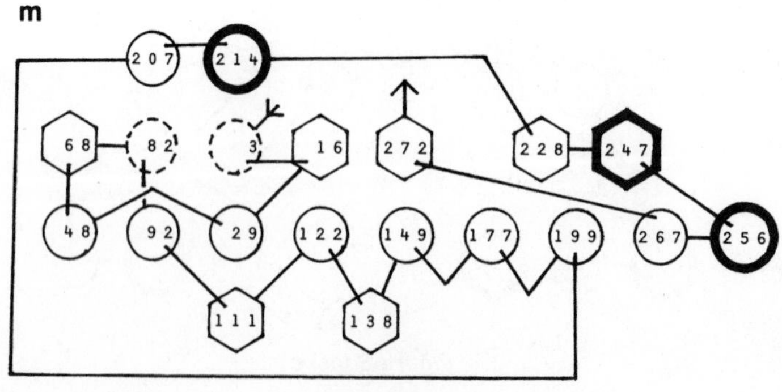

Figure 6 (cont.)

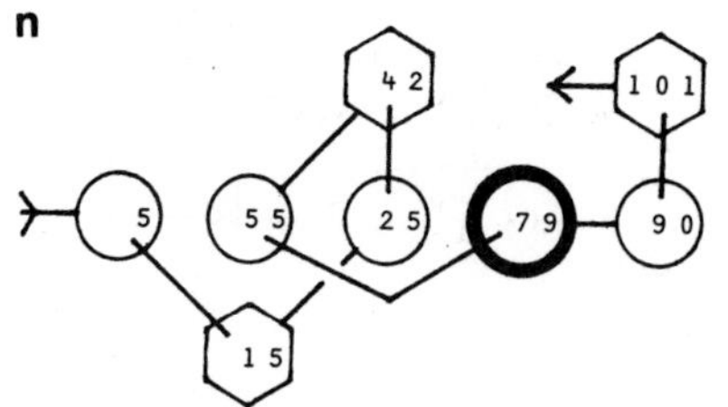

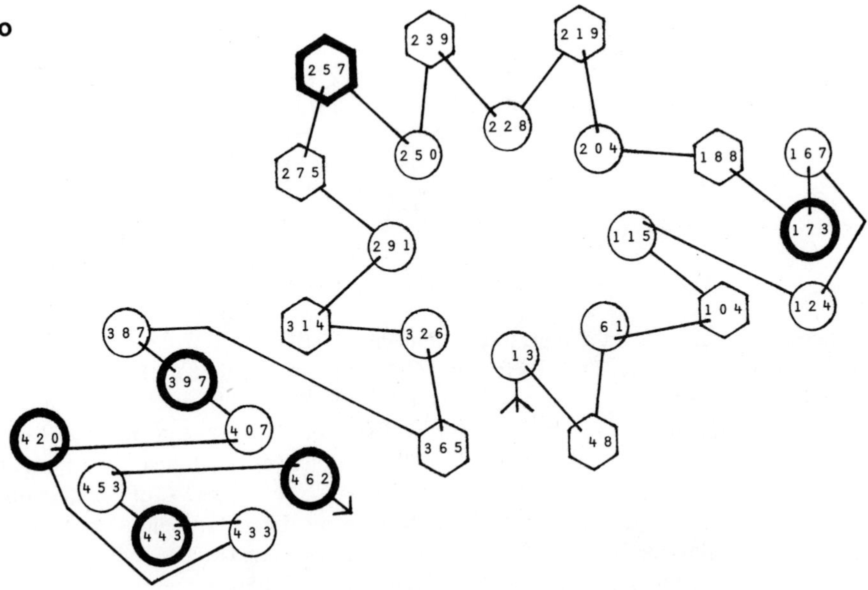

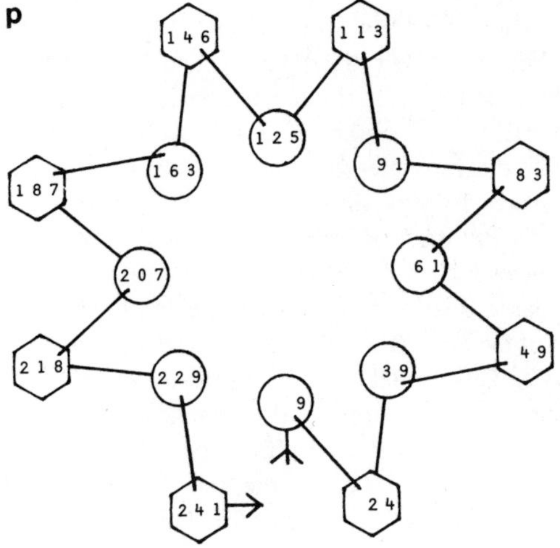

Figure 6 (cont.)

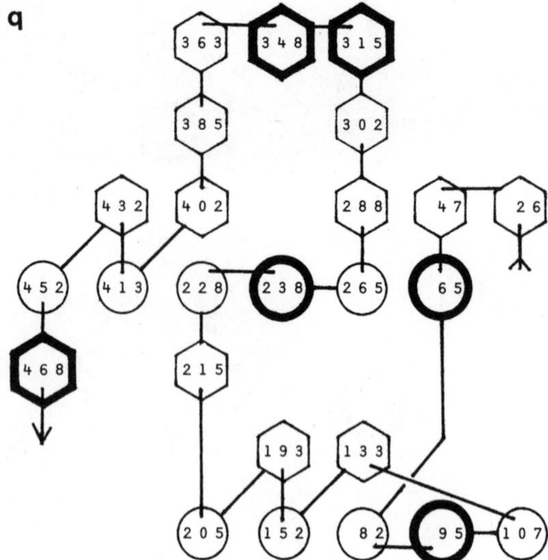

Figure 6 (cont.)

structures at the regions of the typical α/β domains of ABP I–II (227 instead of 233), ADK (162 instead of 171), FXN, TIM, DFR I, and GPD I (34 instead of 28), LDH I (71 instead of 78), RHD I and PGKH I (176 instead of 185). It is clear that the agreement is generally better for the class I proteins than the class II proteins. Even for class I proteins listed in Table III, discrepancies between the calculated and observed structures are found. The most minor discrepancies are nonhelical regions such as 92 of ABP I, 69 of DFR I, 42 of FXN, 76 of LDH I, 122 of PGKH, 34 of SBT, and 33 of TIM, which do not affect the predicted topology at all. Residues 31 of B5C and 89 of SRX are of greater concern because they actually form antiparallel β strands in mixed α/β domains. These could be only minor modifications of the predicted topology, but we must identify the exact locations of the modifications. More serious difficulties arise from misassignment of β candidates: 227 of ABP is not a major misassignment because it is a substitute for the nearby β strand, 233. This is also true for residues 245 of CPA (241), 176 and 342 of PGKH (185 and 334), and 254 of RHD II (244) (the numbers in parentheses are the observed β-central residue numbers). Some misassigned β candidates (176 of PGKH and 254 of RHD) are found to be α helical, but others are random coil.

The reasons for the discrepancies are not clear. The most serious difficulty in predicting the correct topology arises from a misassignment of a parallel β strand. Domains around residues 49 and 202 of CPA must be antiparallel to the central parallel β sheet because the residue number differences from 49 to 64 and from 191 to 202 are too small, but they are still situated in a common large β sheet. On the other hand, 139 of CPA, 285 and 301 of PGKH, and 81 of SBT are out of the large β sheet. The expulsion of 139 of CPA could be explained by a disulfide bond between Cys^{138} and Cys^{161}, which cannot be accommodated in a parallel β sheet in an α/β protein. Some typical parallel β strands in α/β proteins, such as 196 of ADH II, have Cys residues but do not form a disulfide bond. In some cases, there are no adequate partner Cys residues nearby. In other cases, the other Cys residue in its neighborhood could not pair, e.g., because it is involved in a large helix. However, on some occasions the absence of pairing between Cys residues arises from very delicate geometric constraints. Accordingly, it would be more feasible to design the algorithm to eliminate a β candidate from the members joining into a parallel β sheet if any chemical information is available about a disulfide bond involving a Cys residue in the center of the β candidate. It could be found in the future that a disulfide bond between a β strand and a flanking α helix is compatible with the α/β conformation. However, it would be wise to search for a best-fit conformation among α + β, all-β, coil, all-α, and α/β proteins, in that order.

The conformation of region 81 of SBT looks like a β strand, as schematically shown in Fig. 6m. The excursion of 81 of SBT is also related to a very rare case of left-handed βαβ unit (Nagano, 1977a), because it contributes to the formation of a hydrophobic core with region 68–75, VAGTVAAL, at the edge of the helix in the left-handed βαβ, with the antiparallel β sheet composed of 206–209, QSTL, and 216–221, AYNGTS, and also with the N-terminal 1–5, AQSVP, to form a seemingly parallel β sheet. It is very difficult to predict this kind of excursion at this stage of prediction.

In the case of ADH II, 207 is incorrectly detected as a β candidate, which would make the residue distance between the neighboring candidates (196 and 220) too small. This result permits many interpretations if the conformation is unknown. For all three to be β structural, the β sheet would have to be antiparallel. One, in particular 207, could be expelled, as is the case for 81 for SBT. Or 207 could be α helical, and this turns out to be the correct answer. Even if the residue number distances between neighboring β candidates are all larger than 16, multiple interpretations would be necessary for unknown proteins. Cohen *et al.* (1982, 1983) started their algorithm for seeking β candidates by finding hydrophilic turns. Even their method would have difficulty in defining a unique conformation for the region 196–207 of ADH II, the sequence of which is AVFGLGGVGLS. It is the least hydrophilic (in other words, HLF-type) sequence. The conformation of the region would be highly dependent on the conformations in its neighborhood.

Prediction of parallel β strands of KGA was almost complete except for 124. The sequence of its neighborhood is (119–127) EIPLLPGIS. As stated in the footnote defining PAR type, the PLLP sequence was first omitted because of low predictability of Pro residues for β structure. This sequence is very special because it cannot form a shortest α helix. When two residues between two Pro residues are both hydrophobic, along with two more residues outside the Pro residues, satisfying the condition of PAR type, such a sequence is highly β structural. Although the definition of PAR type was thus modified, the candidate at 124 was not detected as a β candidate because it has a marginally low *B* value. I decided to give such a situation a score of 8, and the predicted result of KGA in Table VIII (I.19) was obtained. The distance between 124 and 138 is 14 residues, which is the shortest distance in the allowed parallel β sheet.

TIM and KGA have an eight-stranded parallel β barrel, which will be referred to as TIM-like barrel. The topologies of both of these proteins are satisfactorily predicted by the results in Tables VIII and IX in the sense that eight parallel β candidates with two PAR(β)-type candidates at the N and C termini are predicted. Such a β barrel is also found in TAA. The schematic drawing of its topology is shown in Fig. 6o. Its N-terminal 375 residues, TAA I, have a TIM-like barrel with an excursion of residues 122–175 as a three-stranded antiparallel β sheet. The C-terminal domain, TAA II, is all-β type. The result for TAA in Table VIII (I.34) is not sufficient to establish the possibility of TIM-like barrel.

Figure 7 shows the computer output for the TAA sequence. It is similar to those already explained in Table VI. Although 248 is not detected because of very weak α-helical predictability in the region of 231–240, the chemically demonstrated steric constraint that four disulfide bonds, 30–38, 150–164, 240–283, and 439–474, would help align one more α helix near 240 in addition to the predicted one at 274 and would make the β candidate of PAR(β), 248, join the parallel β sheet made by 193 (or 203), 226, 290, and 326 flanked by large α helices at 182, 208, 260, 274, 316, and 333 (or 362). Note that two terminal β candidates, 13 and 326, are PAR(β) candidates to satisfy one of the seemingly necessary conditions, as is also

```
TAA   TAKA-AMYLASE

THE RANGE OF ALPHA HELIX

  44  53    99 110    182 195    215 222    237 242    253 261    270 280    308 320    357 374

THE RANGE OF BETA SHEET

  10  15    58  64    112 118    122 127    165 169    170 175    202 206    226 231    248 252    289 293    323 329    384 390
 392 398   404 410    417 424    430 436    440 445    450 456    458 465

PAR   CANDIDATES ARE AS FOLLOWS.

  1  10  15 SIYFLL     2  11  16 IYFLLT     3  52  57 IQGMGF     4  57  62 FTAIWI     5 110 115 RGMYLM     6 111 116 GMYLMV
  7 112 117 MYLMVD     8 113 118 YLMVDV     9 116 121 VDVVAN    10 148 153 PFCFIQ    11 200 205 IDGLRI    12 246 251 MDGVLN
 13 287 292 TLLGTF    14 314 319 VAAFII    15 315 320 AAFIIL    16 316 321 AFIILN    17 324 329 LPIIYA

HLF   CANDIDATES ARE AS FOLLOWS.

  1  36  41 KYCGGT     2  37  42 YCGGTW     3  42  47 WQGIID     4  49  54 LDYIQG     5  53  58 QGMGFT     6  54  59 GMGFTA
  7  58  63 TAIWIT     8  72  77 DCAYGD     9  73  78 CAYGDA    10  76  81 GDAYTG    11  79  84 YTGYWQ    12 122 127 HMGYDG
 13 125 130 YDGAGS    14 163 168 DCWLGD    15 188 193 YDWVGS    16 189 194 DWVGSL    17 221 226 KAAGVY    18 222 227 AAGVYC
 19 223 228 AGVYCI    20 224 229 GVYCIG    21 225 230 VYCIGE    22 226 231 YCIGEV    23 229 234 GEVLDG    24 247 252 DGVLNY
 25 252 257 YPIYYP    26 253 258 PIYYPL    27 256 261 YPLLNA    28 307 312 DIALAK    29 325 330 PIIYAG    30 326 331 IIYAGQ
 31 334 339 HYAGGN    32 346 351 ATWLSG    33 359 364 YKLIAS    34 360 365 KLIASA    35 377 382 TGFVTY    36 422 427 LSGASY
 37 435 440 EVIGCT    38 452 457 VPMAGG    39 453 458 PMAGGL    40 454 459 MAGGLP

  ****      13      44      60      82     117     151     171     193     226     260     290     326     348     379     406     442
  ****     PAR     HLF     PAR     HLF     PAR     PAR             HLF     HLF     HLF     PAR     PAR     HLF     HLF
          BETA    BETA    BETA    BETA    BETA    BETA    BETA    BETA    BETA    ALPH    BETA    BETA    BETA    BETA    BETA    BETA
******    3.25    3.06    3.34    1.81    8.12    1.90    0.84    3.76    2.46    4.51    3.69    2.05    2.08    1.07    4.58    0.70
******    2.40    2.03    2.88    1.82    4.64    1.71    1.21    2.47    2.70    1.69    2.55    2.19    1.35    1.39    3.70    1.53
  ****    ****    ****      65      88    ****    ****     177     203     231     265     295    ****     357     389     411     452
  ****    ****    ****                    ****    ****             PAR     HLF                    ****
                          BETA    BETA                    BETA    BETA    BETA    BETA    ALPH            ALPH    BETA            BETA
******************        2.22    2.36**************      1.33    2.48    1.70    1.13    1.47*******     2.02    1.42    1.17    0.89
******************        1.46    2.05**************      0.87    1.88    1.22    1.02    0.85*******     0.81    0.94    0.71    1.04
  ****    ****    ****    ****      95    ****    ****    ****    ****     240    ****     302    ****    ****    ****     420     460
  ****    ****    ****    ****            ****    ****    ****    ****            ****            ****    ****    ****
                                  BETA                                    BETA            BETA                            BETA    BETA
**************************        1.67****************************        0.97*******     1.82********************        0.81    1.77
**************************        1.15****************************        0.93*******     1.26********************        1.48    1.26
  ****    ****    ****    ****    ****    ****    ****    ****    ****     248    ****     310    ****    ****    ****     430    ****
  ****    ****    ****    ****    ****    ****    ****    ****    ****     PAR    ****     HLF    ****    ****    ****            ****
                                                                          BETA            ALPH                            BETA
************************************************************              3.18*******     4.88********************        1.81*******
************************************************************              2.12*******     1.31********************        1.38*******

     7      19      53      73     105     123     159     182     208     254     274     316     333     362     394     435     467
                   PAR     HLF             HLF                             HLF             PAR             HLF
  ALPH    ALPH    BETA    ALPH    ALPH    ALPH    ALPH    ALPH    ALPH    BETA    ALPH    ALPH    ALPH    ALPH    ALPH    ALPH    ALPH
  2.15    2.40    1.43    2.18    3.39    1.82    2.27    3.32    3.90    1.23    4.38    6.53    2.27    4.30    3.43    4.19    1.88
  1.33    0.97    1.43    0.76    0.97    0.83    1.01    2.01    2.27    1.11    1.78    3.12    0.85    1.87    1.58    2.38    0.68
  ****      28    ****    ****     112     128     164     188     213    ****     279     321     338     369     401    ****     472
  ****            ****    ****     PAR     HLF     HLF                    ****                     HLF                    ****
          BETA                    BETA            ALPH    ALPH    ALPH            BETA    BETA            ALPH    BETA            BETA
******    1.20**************      3.19    0.62    1.45    3.87    1.90*******     2.31    1.80    0.68    4.17    0.96*******     1.25
******    1.25**************      1.81    0.38    0.80    1.86    1.02*******     1.78    1.87    0.49    1.62    1.00*******     1.21
  ****      37    ****    ****    ****     134    ****    ****     221    ****     285    ****    ****    ****    ****    ****    ****
  ****     HLF    ****    ****    ****            ****    ****            ****            ****    ****    ****    ****    ****    ****
                                          BETA                    ALPH
******    1.30**********************      2.02**************      1.42*******     0.72****************************************
******    0.72**********************      1.98**************      0.53*******     0.58****************************************
  ****    ****    ****    ****    ****     145    ****    ****    ****    ****    ****    ****    ****    ****    ****    ****    ****
  ****    ****    ****    ****    ****            ****    ****    ****    ****    ****    ****    ****    ****    ****    ****    ****
                                          BETA
**********************************        1.56************************************************************************
**********************************        1.24************************************************************************
```

Figure 7. Computer output for TAA corresponding to the contents of Table VII for (1) ABP, (2) LZM, (3) PTI, and (4) TIM. The combinations of (PAR,ALPH), (PAR,BETA), (HLF,ALPH), (HLF,BETA), (□,ALPH), and (□,BETA) represent PAR(α), PAR(β), HLF(α), HLF(β), α, and β, respectively. Two subsequent numbers represent $-A$ and $-B$, respectively, but the A/B ratio is not shown. The candidates having (□,□) are also shown as far as they can be detected to help prediction. Asterisks indicate that no candidate is detected.

true for TIM and KGA. As possible domains to occupy the rest of the eight β-strand sites with PAR(β) candidates, 60 and 117 would be likely even though a combinatorial approach must be introduced; candidate 151 could be easily excluded on the basis of information about the disulfide bonds. A similar algorithm would be required to get the correct topology about the C-terminal sequence by fitting the candidates to the patterns obtained from GCR or other proteins like it. It is not so easy to determine how the possibility of an α-helix is affected by the disulfide bond 240–283. Very many examples are found in the protein data bank of various kinds of antiparallel β sheets. One example of a pair of antiparallel α helices is the α helices 39–58 and 89–108 of P2P, held by two disulfide bonds, 44–105 and 51–98. Residues 240–283 of TAA provide the first example of parallel α helices bound with a disulfide bond, although Cys^{283} is at the turn of the α helix 270–280. The excursion of residues from 122 to 175 is also an important problem in predicting the correct packing of secondary structure. Similar phenomena also occur in the case of PGKH. One of the biggest excursions is an all-α domain of YHX III inserted between two β sheets, which make the second domain, which is very similar to the YHX I domain shown schematically in Fig. 6q. It has not yet been determined how to detect such regions efficiently and correctly.

VI. CONCLUSION

I must admit that no unique interpretation is possible at this stage in predicting the packing of the secondary structure of a protein of unknown conformation on the basis of calculated results such as those in Tables VIII and IX. If we could choose a correct set of β and α candidates, we would try to predict the strand order by the method presented by Nagano (1980). When we obtain an amino acid sequence belonging to group V or VI in Table IX, the probability of the protein being of the α/β type is greatest. For group VII, containing very large proteins, the probability of predicting parallel β strands from PAR(β) or HLF(β) candidates seems to be smaller, as is the case for TAA. In this case, the packing game should be played with a limited number of parallel β strands, and weaker candidates such as HLF(β) sequences would turn into parts of α helices.

If the protein to be predicted is not an α/β type, the packing game would be more complicated, and the atomic coordinates supplied by the Protein Data Bank would be the packing playground instead of a two-dimensional lattice of 3 × 11 boxes in the method of Nagano (1980). If we try to fit two PAR(β) sequences into a set of atomic coordinates of SNS, for example, the amino acid residues outside the candidates would become very important. Figure 4c suggests that Val^{39} of SNS, which just follows one of the two PAR(β) candidates, 36, is important for making a turn covered by an HLF-type region. Accordingly, it would be necessary in the next stage to check the sequence homology inside and outside the PAR- or HLF-type candidates to find both extra hydrophobic core and hydrophilic turns. In this way the problem of finding the correct strand order in an α/β protein will also be solved. In order to reduce the total number of conformations to be tested, it would be very important to start checking amino acid sequence homology from hydrophobic regions such as PAR and HLF domains. The basic idea of the packing game simulation can be generalized to the three-dimensional space of orderly packed atoms of all kinds of proteins solved previously by x-ray methods and to finding a similar hydrophobic core or a similar motif of packing of secondary structure to that of the protein in question.

VII. RECENT RESEARCH

Chothia (1988) has reported that 14 TIM-like α/β barrels were determined by crystallography; presently, their number is 16. They are:

1. TIM
2. KGA
3. TAA
4. Cat muscle pyruvate kinase (PK) (Stuart *et al.*, 1979)
5. Rabbit skeletal muscle aldolase (ALD) (Sygusch *et al.*, 1987)
6. Yeast enolase (ENL) (Lebioda and Stec, 1988)
7. Yeast flavocytochrome b_2 (FCB) (Xia *et al.*, 1987)
8. Spinach glycolate oxidase (GO) (Bränden *et al.*, 1987)

9.,10. *E. coli* N-(5′-phosphoribosyl)anthranilate isomerase-indole-3-glycerol-phosphate synthase (PRA-IGP) which is a bifunctional enzyme and has two α/β barrels with different catalytic activities (Priestle *et al.*, 1987)

11. Trimethylamine dehydrogenase (TMD) (Lim *et al.*, 1986)
12. The α subunit of *S. typhimirium* tryptophan synthase (WSA) (D. R. Davies and E. W. Miles, unpublished results)
13. *R. rubrum* RuBisCO (RBCR) (Schneider *et al.*, 1986)
14. Tobacco RuBisCo (RBCT) large subunit (Eisenberg *et al.*, 1987)

a) PK   CAT MUSCLE PYRUVATE KINASE

```
THE RANGE OF ALPHA HELIX

  22   29    57   67    80   99   125  128   140  145   149  154   161  173   227  234   247  258   273  277   279  284   295  299
 303  319   340  351   370  387   391  403   408  420   435  441   455  465   481  499

THE RANGE OF BETA SHEET

  44   51    70   75   106  112   117  123   130  133   135  138   176  182   185  191   201  205   207  209   212  215   221  223
 238  241   265  270   289  292   322  327   356  361   425  429   449  452   467  471   509  513   519  524


   10     48     71    108    140    157    193    238    266    291    324    358    428    451    470    509    525
 HLF    PAR    PAR    HLF           PAR    PAR    PAR           PAR    HLF    PAR    PAR    PAR    PAR    PAR    PAR
 BETA   BETA   BETA   ALPH   BETA   BETA   BETA   ALPH   BETA   BETA   BETA   BETA   BETA   BETA   BETA   BETA   BETA
 3.09   2.16   3.72   4.90   3.67   3.04   3.32   5.36   4.19   3.21   2.52   1.88   5.84   3.44   2.64   6.01   1.89
 2.62   2.31   2.09   2.52   2.87   2.52   2.52   2.30   2.73   2.52   2.31   1.34   3.68   2.79   2.77   5.74   2.16
 ****     55     76    113   ****   ****    198    243    276    296    329    363    433    456    478   ****   ****
 ****                        ****   ****                        HLF                                      ****   ****
        BETA                               BETA   ALPH   BETA   BETA   ALPH          BETA   BETA   ALPH
******  1.71   0.86   1.38**************   0.87   4.10   2.09   1.25   3.00   1.17   1.14   1.71   3.17**************
******  1.34   0.76   0.72**************   0.86   1.55   1.34   1.25   0.96   0.76   0.89   1.08   0.98**************
 ****   ****     84    121   ****   ****    208    249   ****    302    335    373   ****   ****    484   ****   ****
 ****   ****          HLF    ****   ****                 ****                 PAR    ****   ****          ****   ****
               ALPH   BETA                 BETA   ALPH          ALPH                 ALPH                 ALPH
*************  2.53   1.60**************   0.96   2.94*******   1.82   1.11   2.80**************   3.43**************
*************  0.80   1.51**************   0.99   1.29*******   0.79   0.45   0.88**************   1.26**************
 ****   ****   ****    127   ****   ****    218   ****   ****   ****    344    380   ****   ****   ****   ****   ****
 ****   ****   ****          ****   ****          ****   ****   ****                 ****   ****   ****   ****   ****
                      BETA                 ALPH                               ALPH   ALPH
********************  1.10**************   2.50*********************   1.51   3.79**************************************
********************  1.10**************   1.27*********************   0.61   1.40**************************************

   18     60     89    132    145    165    226    254    281    307    349    388    441    463    489    517   ****
                                    PAR                         PAR    HLF    HLF           HLF                  ****
 ALPH   ALPH   ALPH   ALPH   ALPH   BETA   ALPH   ALPH   ALPH   ALPH   ALPH   ALPH   ALPH   ALPH   ALPH
 4.60   4.70   4.66   3.34   2.68   3.86   2.77   4.36   3.90   3.80   4.55   4.39   3.76   3.59   5.55   0.24*******
 1.24   1.33   2.43   1.31   1.12   2.82   1.26   2.04   1.55   1.71   1.62   1.06   1.38   1.40   2.57   0.41*******
   24     66     94   ****    152    173    231    261    286    313   ****    398    446   ****    494   ****   ****
                      ****                 HLF                  PAR    ****                 ****   HLF    ****   ****
 ALPH   ALPH   ALPH          ALPH   BETA   BETA   ALPH   ALPH   BETA                 ALPH                 ALPH
 3.92   2.57   3.22*******   1.62   1.98   1.96   1.58   1.45   1.81*******   2.38   1.35*******   3.42**************
 1.48   1.44   1.23*******   0.66   2.03   1.38   0.59   0.77   1.61*******   1.05   0.67*******   1.59**************
   30   ****    103   ****   ****    181   ****   ****   ****    319   ****    416   ****   ****    504   ****   ****
        ****          ****   ****   PAR    ****   ****   ****          ****          ****   ****          ****   ****
 ALPH          BETA                 ALPH                                              ALPH                 BETA
 2.89*******   1.57**************   5.17*********************   1.17*******   2.01**************   1.11**************
 1.12*******   1.15**************   2.73*********************   0.71*******   0.96**************   0.97**************
   42   ****   ****   ****   ****    188   ****   ****   ****   ****   ****    422   ****   ****   ****   ****   ****
        ****   ****   ****   ****          ****   ****   ****   ****   ****   HLF    ****   ****   ****   ****   ****
 BETA                               ALPH                                              ALPH
 1.44*****************************  1.82***********************************   3.08**************************************
 1.16*****************************  0.69***********************************   1.11**************************************
```

Figure 8. Predicted packing arrangements of secondary structure. (a) Cat muscle pyruvate kinase (PK). The ranges of parallel β strands in the α/β barrel are: 44–51, 70–75, 106–112, 238–241, 265–270, 289–292, 322–327, and 356–361.

```
b) ALD         RABBIT MUSCLE ALDOLASE A

THE RANGE OF ALPHA HELIX

    10     23     38     46     53     64     80     84     95    100    131    140    161    180    198    218    245    258    277    289    304    313    319    339

THE RANGE OF BETA SHEET

    27     34     73     78    102    109    144    153    183    191    226    231    263    271    296    301

      3     21     62     77    105    151    184    221    254    266    297    357
               PAR    PAR    PAR    PAR    PAR    HLF           HLF    HLF
        ALPH   BETA   BETA   BETA   BETA   BETA   BETA   ALPH   BETA   BETA   BETA
   0.83   4.04   3.06   3.47   2.17   3.43   2.49   2.61   3.94   1.52   3.30   2.04
   0.50   2.39   1.85   3.72   2.48   2.31   2.59   1.61   2.29   1.84   2.15   1.81
      9     30   ****   ****    110    157    189    226   ****    271    306   ****
        HLF      ****   ****                               ****  PAR    HLF      ****
        BETA                 BETA                 ALPH          BETA   ALPH
   1.13   1.94**************   1.70   0.94   0.98   1.71*******   0.45   3.79*******
   0.40   1.23**************   1.55   0.58   0.66   1.04*******   0.82   1.34*******
   ****     36   ****   ****    116    163    200    232   ****    279    312   ****
   ****          ****   ****  HLF    PAR                   ****         HLF      ****
        BETA                 BETA   ALPH   ALPH   BETA          ALPH   ALPH
 ******   1.08**************   0.86   3.59   2.02   0.85*******   2.22   2.75*******
 ******   0.82**************   0.88   1.44   0.90   0.88*******   0.89   0.84*******
   ****   ****   ****   ****    123   ****    206    239   ****   ****    325   ****
   ****   ****   ****   ****          ****                 ****   ****          ****
                             BETA          ALPH   BETA                 ALPH
****************************   0.80*******   4.59   1.84**************   2.72*******
****************************   0.93*******   2.15   1.24**************   0.58*******

     14     41     69     84    135    170    211    244    260    284    330   ****
                                    HLF    PAR                                 ****
 ALPH   ALPH   ALPH   ALPH   ALPH   ALPH   ALPH   ALPH   BETA   ALPH   ALPH
   2.23   2.45   1.34   3.65   2.58   5.45   8.98   2.12   1.34   3.36   4.35*******
   0.59   1.03   0.80   1.35   0.93   1.53   2.62   1.10   0.99   1.50   1.05*******
   ****     46   ****     89    140    175    216    249   ****    292    337   ****
   ****          ****               HLF           PAR      ****         HLF      ****
        BETA                        BETA   ALPH   ALPH          ALPH   ALPH
 ******   1.16*******   0.85   1.35   2.24   2.81   4.44*******   1.84   2.72*******
 ******   1.01*******   0.32   0.30   1.53   0.73   1.46*******   0.97   1.27*******
   ****     57   ****     97    146   ****   ****   ****   ****   ****    348   ****
   ****          ****         HLF      ****   ****   ****   ****   ****  HLF      ****
        ALPH          ALPH   ALPH                                      ALPH
 ******   2.60*******   2.60   3.64***********************************   2.14*******
 ******   0.86*******   1.57   1.06***********************************   0.65*******
   ****   ****   ****   ****   ****   ****   ****   ****   ****   ****   ****   ****
   ****   ****   ****   ****   ****   ****   ****   ****   ****   ****   ****   ****

************************************************************************************
************************************************************************************
```

Figure 8 (cont.) (b) Rabbit skeletal muscle aldolase (ALD). All β strands listed in the output are parallel.

```
c) ENL        YEAST ENOLASE

THE RANGE OF ALPHA HELIX

   61   72    74   80    88   96   110  125   129  136   172  181   220  234   276  288   303  312   327  336   353  364   379  387
  404  419

THE RANGE OF BETA SHEET

    4   11    17   24    29   35   148  153   189  193   244  248   292  296   316  320   341  345   368  373   394  397

   ****     21     94    112    152    170    228    243    292    317    344    369    386    424
   ****           PAR    PAR    PAR    PAR    PAR    HLF    PAR    PAR    PAR    PAR    PAR    HLF
          BETA   BETA   BETA   BETA   ALPH   ALPH   BETA   BETA   BETA   ALPH   BETA   ALPH   BETA
 ******   2.91   3.35   3.26   2.41   2.99   8.24   2.15   1.34   3.84   5.56   3.13   5.47   2.33
 ******   2.38   2.32   2.85   1.72   1.25   3.56   1.61   1.91   3.16   2.82   2.71   2.95   1.52
   ****     26     99   ****    157    176   ****    251    302    322    351    374    391   ****
   ****                 ****                 ****                                              ****
          BETA                        ALPH                 ALPH   BETA   BETA          BETA
 ******   1.66   1.08*******   0.57   1.95*******   1.39   2.13   1.89   1.93   0.95   1.27*******
 ******   1.09   0.69*******   0.64   0.96*******   0.66   0.57   1.30   1.51   0.73   1.36*******
   ****     32   ****   ****   ****   ****   ****    260   ****   ****    357   ****    397   ****
   ****          ****   ****   ****   ****   ****          ****   ****          ****          ****
          BETA                                      BETA                 ALPH
 ******   2.71***********************************   1.54**************   2.64*******   0.96*******
 ******   2.31***********************************   1.01**************   0.86*******   0.72*******
   ****     41   ****   ****   ****   ****   ****    275   ****   ****   ****   ****    406   ****
   ****          ****   ****   ****   ****   ****          ****   ****   ****   ****          ****
          BETA                                                                         ALPH
 ******   1.85***********************************   0.55****************************   2.55*******
 ******   1.31***********************************   0.53****************************   0.51*******

      4     63    106    118    162    181    235    280    307    332    363    379    411    430
           HLF    HLF           HLF    HLF
   ALPH   ALPH   ALPH   ALPH   ALPH   ALPH   ALPH   ALPH   ALPH   ALPH   ALPH          ALPH
   3.88   5.72   1.77   2.94   1.90   3.74   2.98   3.33   3.50   4.80   3.23   1.27   3.72   1.38
   2.17   2.58   0.76   0.65   0.95   0.89   0.89   1.24   1.28   1.57   1.23   0.76   1.30   0.70
      9     70   ****    128   ****    187   ****    287    312    339   ****   ****    416   ****
                 ****          ****          ****                        ****   ****          ****
   BETA   ALPH          ALPH          BETA          ALPH   BETA   ALPH                 ALPH
   1.90   4.56*******   2.18*******   2.39*******   1.73   1.81   2.60**************   1.87*******
   1.34   2.43*******   1.25*******   1.55*******   1.06   1.21   0.44**************   0.64*******
     16     81   ****    135   ****    195   ****   ****   ****   ****   ****   ****   ****   ****
                 ****          ****          ****   ****   ****   ****   ****   ****   ****   ****
          ALPH          ALPH          ALPH
   0.39   2.95*******   2.42*******   2.22********************************************************
   0.48   1.72*******   0.74*******   1.02********************************************************
   ****     89   ****    146   ****    219   ****   ****   ****   ****   ****   ****   ****   ****
   ****          ****    PAR   ****          ****   ****   ****   ****   ****   ****   ****   ****
          ALPH          BETA          ALPH
 ******   3.60*******   0.83*******   2.43********************************************************
 ******   1.36*******   1.08*******   1.14********************************************************
```

Figure 8 (cont.) (c) Yeast enolase (ENL). The ranges of parallel β strands in the α/β barrel are: 148–153, 189–193, 244–248, 292–296, 316–320, 341–345, 368–373, and 394–397.

```
d) FCB        FLAVOCYTOCHROME B2  FROM  BAKER'S YEAST

THE RANGE OF ALPHA HELIX


THE RANGE OF BETA SHEET


    3     28     60    124    172    194    228    251    279    333    348    367    386    422    463    488
         PAR    PAR    HLF    PAR    HLF                  PAR           PAR    PAR    PAR    PAR    PAR
 ALPH   BETA   BETA   ALPH   BETA   BETA   ALPH   BETA   BETA   ALPH   BETA   BETA   BETA   ALPH   ALPH   BETA
 1.66   2.21   3.74   4.06   5.08   3.43   3.63   4.11   4.46   1.93   3.72   1.90   3.30   4.38   3.86   1.40
 0.69   2.75   2.09   2.22   3.25   2.78   2.17   2.92   3.11   1.05   3.40   2.56   1.94   1.83   2.01   1.55
   10   ****     65    129   ****    199    233    256    297   ****    353    372    393    427   ****   ****
        ****          HLF    ****   HLF                         ****                        HLF    ****   ****
 ALPH          ALPH   ALPH          ALPH   BETA   BETA   ALPH          BETA   BETA   BETA   BETA
 1.72*******   1.96   3.01*******   3.22   0.97   1.98   2.12*******   2.11   1.08   1.95   1.30**************
 1.07*******   0.68   1.25*******   1.58   0.80   1.42   1.14*******   1.59   0.93   1.27   0.88**************
 ****   ****   ****    135   ****    207   ****    262    310   ****   ****   ****   ****    437   ****   ****
 ****   ****   ****          ****          ****                 ****   ****   ****   ****   HLF    ****   ****
                      BETA                        BETA   ALPH                               ALPH
******************    2.22*******   0.70*******   2.02   1.95***************************   1.70**************
******************    1.45*******   0.26*******   1.31   1.21***************************   0.81**************
 ****   ****   ****   ****   ****   ****   ****   ****    317   ****   ****   ****   ****    447   ****   ****
 ****   ****   ****   ****   ****   ****   ****   ****          ****   ****   ****   ****          ****   ****
                                                                                            ALPH
*************************************************   1.28***************************   2.86**************
*************************************************   0.64***************************   1.17**************

   16     37     71    141    178    212    239    267    322    340    359    381    398    452    471    495
                      HLF
 ALPH   ALPH   ALPH   ALPH   BETA   ALPH   ALPH   ALPH   ALPH   ALPH   ALPH   ALPH   ALPH   ALPH   ALPH   ALPH
 3.51   3.10   3.62   4.32   2.13   1.66   3.55   4.27   2.92   2.49   5.03   2.06   2.21   6.21   3.51   2.84
 1.01   1.61   1.85   1.83   1.81   0.68   1.40   1.78   0.96   0.55   2.19   0.92   0.89   2.61   1.80   1.36
   23     42     76    146    183    217    246    274    327   ****   ****   ****    406    458    476    500
                                    HLF           HLF                  ****   ****   ****   PAR
               ALPH          ALPH          ALPH   ALPH   BETA                        BETA   ALPH   ALPH   BETA
 0.80   0.67   2.05   1.23   1.84   0.58   2.30   3.28   1.28*********************   4.35   3.41   2.05   1.37
 0.61   0.38   0.72   0.73   0.97   0.60   1.14   1.60   0.87*********************   2.69   1.24   0.92   1.00
 ****     50    106    152    189    223   ****   ****   ****   ****   ****   ****    417   ****    482    506
 ****                                      ****   ****   ****   ****   ****   ****          ****
        BETA   ALPH   ALPH   BETA   BETA                                              ALPH          ALPH
******  3.04   2.82   2.28   2.25   1.54****************************************   3.49*******   3.37   1.39
******  2.13   0.69   1.03   1.71   1.27****************************************   1.65*******   1.94   0.71
 ****   ****    114    163   ****   ****   ****   ****   ****   ****   ****   ****   ****   ****   ****   ****
 ****   ****                 ****   ****   ****   ****   ****   ****   ****   ****   ****   ****   ****   ****
               ALPH   BETA
************   2.48   3.19*****************************************************************************
************   0.83   2.02*****************************************************************************
```

Figure 8 (cont.) (d) Yeast flavocytochrome b_2 (FCB). The ranges of secondary structure are not reported.

e) GO GLYCOLATE OXIDASE FROM SPINACH

THE RANGE OF ALPHA HELIX

THE RANGE OF BETA SHEET

```
4        33       53       103      129      154      205      231      249      282      312      354
                  PAR      PAR      HLF      HLF      HLF      PAR      PAR      PAR      PAR
BETA     ALPH     BETA     BETA     BETA     BETA     BETA     BETA     BETA     BETA     BETA     ALPH
2.60     2.81     3.75     2.23     3.83     3.92     2.88     4.05     2.92     1.91     4.35     3.04
2.36     1.74     3.35     1.90     2.56     2.55     2.00     3.39     3.08     2.21     3.29     1.63
****     39       58       108      134      159      213      ****     254      291      317      359
****                                                           ****                       HLF
         ALPH     BETA     BETA     ALPH              BETA              ALPH     BETA     ALPH
******   2.08     2.09     1.79     2.51     1.03     2.27*******       1.52     1.58     2.54     1.02
******   0.54     1.75     1.34     1.52     0.79     1.44*******       0.81     1.38     0.98     0.64
****     ****     64       ****     ****     ****     ****     ****     260      ****     ****     ****
****     ****     PAR      ****     ****     ****     ****     ****     HLF      ****     ****     ****
                  BETA                                                  BETA
*************     2.80*********************************************     2.34***************************
*************     2.52*********************************************     1.48***************************
****     ****     71       ****     ****     ****     ****     ****     266      ****     ****     ****
****     ****              ****     ****     ****     ****     ****     PAR      ****     ****     ****
                  BETA                                                  ALPH
*************     2.32*********************************************     3.35***************************
*************     1.47*********************************************     1.76***************************

11       44       92       114      139      169      218      240      272      296      327      366
HLF               HLF               PAR               HLF      HLF               PAR               HLF
ALPH     ALPH     ALPH     BETA     ALPH     ALPH     ALPH     ALPH     ALPH     ALPH     ALPH     ALPH
3.32     3.70     3.29     3.36     6.60     3.44     4.72     4.07     6.45     5.56     5.44     3.31
0.82     1.87     1.28     2.31     2.20     1.26     2.62     1.32     2.15     1.81     1.78     1.49
16       ****     98       124      149      179      223      ****     277      305      332      ****
         ****     HLF      PAR      HLF                        ****              PAR               ****
ALPH              ALPH     BETA     ALPH     ALPH     BETA              ALPH     BETA     ALPH
1.95*******       1.95     2.22     2.92     2.02     2.29*******       2.36     1.73     2.31*******
0.61*******       0.70     1.75     1.14     1.12     2.36*******       1.14     1.62     0.56*******
22       ****     ****     ****     ****     186      ****     ****     ****     ****     338      ****
HLF      ****     ****     ****     ****     HLF      ****     ****     ****     ****     HLF      ****
BETA                                         BETA                                         ALPH
3.09************************************     1.46************************************     4.30*******
1.78************************************     0.95************************************     1.74*******
28       ****     ****     ****     ****     200      ****     ****     ****     ****     349      ****
         ****     ****     ****     ****              ****     ****     ****     ****              ****
                                             BETA                                         ALPH
1.29************************************     1.31************************************     2.43*******
0.57************************************     1.06************************************     1.24*******
```

Figure 8 (cont.) (e) Spinach glycolate oxidase (GO). The ranges of secondary structure are not reported.

```
f) PRA-IGP  PRAI-IGPS FROM E. COLI  BIFUNCTIONAL ENZYME

THE RANGE OF ALPHA HELIX

   4   23    68   77    98  107   126  132   147  156   168  177   196  206   221  230   237  242   248  255   266  274   295  302
 317  327   343  350   367  375   387  395   414  419   441  451

THE RANGE OF BETA SHEET

  49   55    82   87   113  119   138  143   159  165   181  186   210  215   233  236   257  261   279  285   306  313   331  335
 354  362   377  383   402  407   423  427

****     16     51     87    102    139    182    211    236    261    283    310    330    361    403    423    444
****    HLF    PAR           PAR    PAR    PAR           PAR    HLF    PAR    PAR    PAR    HLF    PAR    HLF    PAR
       BETA   BETA   BETA   BETA   BETA   BETA   BETA   BETA   BETA   BETA   BETA   BETA   ALPH   BETA   BETA   ALPH
****** 4.64   3.46   2.89   3.69   5.83   2.83   1.71   2.10   1.52   3.34   2.33   6.41   2.17   3.22   1.71   5.71
****** 2.78   2.16   2.50   2.64   3.33   2.50   2.39   2.06   1.34   3.49   2.41   3.96   1.21   2.13   1.27   2.34
****     24     56   ****    107    147    187    219    241    266    288   ****    335    371    408   ****   ****
****                 ****                               HLF                  ****                 HLF    ****   ****
       ALPH   ALPH          ALPH   ALPH   ALPH   BETA   ALPH          BETA                 ALPH   ALPH   BETA
****** 1.49   2.16*******   2.27   3.43   1.53   1.74   2.96   0.96   1.06*******   1.69   2.08   1.26**************
****** 0.76   0.78*******   1.27   1.48   0.91   1.46   1.14   0.51   0.98*******   0.95   0.77   0.88**************
****     30     65   ****    117   ****    194   ****   ****   ****    294   ****    343   ****   ****   ****   ****
****                 ****          ****          ****   ****   ****          ****          ****   ****   ****   ****
       ALPH   BETA          BETA          ALPH                               BETA          BETA
****** 1.93   1.45*******   1.85*******   2.45*********************          1.90*******   3.71****************************
****** 1.15   1.43*******   1.48*******   1.36*********************          1.32*******   2.45****************************
****     38     71   ****    124   ****   ****   ****   ****   ****   ****   ****    351   ****   ****   ****   ****
****                 ****   HLF    ****   ****   ****   ****   ****   ****   ****          ****   ****   ****   ****
       BETA   ALPH          BETA                                                       ALPH
****** 1.75   2.32*******   3.37************************************************       3.46****************************
****** 1.25   0.74*******   2.38************************************************       0.94****************************

     7     43     76     93    132    152    200    224    250    273    300    319    356    376    415    429    450
   PAR    HLF    HLF                  HLF                         HLF    HLF                  HLF
  ALPH   ALPH   ALPH   ALPH   ALPH   ALPH   ALPH   ALPH   ALPH   ALPH   ALPH   ALPH   ALPH   ALPH   ALPH   ALPH   ALPH
  9.23   3.62   5.60   1.91   2.52   5.23   2.62   3.81   5.88   2.62   4.60   6.58   7.53   3.20   2.71   2.71   3.52
  3.85   1.40   1.90   1.04   1.06   1.39   0.74   1.82   1.67   0.57   1.16   2.91   2.94   1.72   1.40   1.40   1.30
  ****   ****     82   ****   ****    157    206    231    255    278    305    325   ****    381   ****    439   ****
  ****   ****          ****   ****                         PAR    HLF           PAR    ****          ****          ****
                ALPH                 ALPH   ALPH   ALPH   BETA   BETA   ALPH   ALPH                        ALPH
*************   2.33**************   2.90   1.54   2.14   2.25   2.05   2.45   4.80*******   0.45*******   3.10*******
*************   0.95**************   1.20   0.73   1.32   1.54   1.49   1.14   2.00*******   0.51*******   1.39*******
  ****   ****   ****   ****   ****    163   ****   ****   ****   ****   ****   ****   ****    387   ****   ****   ****
  ****   ****   ****   ****   ****          ****   ****   ****   ****   ****   ****   ****          ****   ****   ****
                                     BETA
**********************************   2.90**************************************************   0.50*********************
**********************************   2.11**************************************************   0.43*********************
  ****   ****   ****   ****   ****    177   ****   ****   ****   ****   ****   ****   ****    393   ****   ****   ****
  ****   ****   ****   ****   ****   HLF    ****   ****   ****   ****   ****   ****   ****   HLF    ****   ****   ****
                                     ALPH                                                      BETA
**********************************   3.93**************************************************   2.16*********************
**********************************   1.39**************************************************   1.80*********************
```

Figure 8 (*cont.*) (f) *E. coli N*-(57-phosphoribosyl) anthranilate isomerase-indole-3-glycerol-phosphate synthase (PRA-IGP). The ranges of parallel β strands in the PRA domain are: 49–55, 82–87, 113–119, 138–143, 159–165, 181–186, 210–215, and 233–236. Those in the IGP domain are: 257– 261, 279–285, 306–313, 331–335, 354–362, 377–383, 402–407, and 423–427.

```
g) WSA          TRYPTOPHAN SYNTHASE ALPHA CHAIN   E. COLI K-12

THE RANGE OF ALPHA HELIX

    1   14    30   40    79   89   112  124   134  144   160  167   188  201   217  227   248  268

THE RANGE OF BETA SHEET

   19   24    46   50    64   68    95  104   125  130   149  154   173  179   208  212   237  241

     8     21     50     81    106    127    150    176    214    238    257
  PAR    PAR    HLF    HLF    PAR    PAR    PAR    HLF    HLF    PAR    PAR
  ALPH   BETA   BETA   ALPH   BETA   BETA   BETA   BETA   BETA   ALPH   ALPH
  4.01   2.82   2.05   2.83   3.14   3.78   2.38   2.20   0.65   6.22   6.82
  1.29   2.26   1.26   1.25   2.30   3.15   1.73   2.10   0.92   3.35   3.48
  ****     26     55   ****   ****    132    155    181   ****   ****   ****
  ****                 ****   ****                 HLF    ****   ****   ****
                                     BETA          BETA
******   0.20   0.62**************   2.01   0.34   1.51*********************
******   0.49   0.44**************   1.79   0.49   1.24*********************
  ****   ****     64   ****   ****    138   ****   ****   ****   ****   ****
  ****   ****          ****   ****          ****   ****   ****   ****   ****
                BETA                 ALPH
*************   1.71**************   2.16***********************************
*************   1.13**************   0.81***********************************
  ****   ****   ****   ****   ****   ****   ****   ****   ****   ****   ****
  ****   ****   ****   ****   ****   ****   ****   ****   ****   ****   ****

****************************************************************************
****************************************************************************

    15     32     69     86    114    144    164    188    222    245   ****
                       PAR    HLF                         HLF           ****
  ALPH   ALPH   ALPH   ALPH   ALPH   ALPH   ALPH   ALPH   ALPH   ALPH
  1.55   2.43   3.94   5.80   3.57   3.64   4.95   2.84   5.09   2.04*******
  0.44   1.20   1.48   1.45   1.14   1.09   2.25   0.57   1.48   1.01*******
  ****     37     74     94    122   ****    169    197    227    252   ****
  ****          HLF                  ****   HLF    PAR    HLF    PAR    ****
         ALPH   ALPH          BETA          BETA   ALPH   ALPH   ALPH
******   4.13   2.52   0.61   1.87*******   1.78   5.20   2.39   4.58*******
******   2.54   1.19   0.70   1.87*******   1.51   1.72   0.60   1.06*******
  ****     43   ****    100   ****   ****   ****    203    233   ****   ****
  ****   HLF    ****   PAR    ****   ****   ****          HLF    ****   ****
         ALPH          BETA                        ALPH   BETA
******   2.18*******   3.72*********************   2.10   2.06**************
******   0.77*******   2.56*********************   0.76   1.56**************
  ****   ****   ****   ****   ****   ****   ****    209   ****   ****   ****
  ****   ****   ****   ****   ****   ****   ****          ****   ****   ****

************************************************   0.74*********************
************************************************   0.54*********************
```

Figure 8 (*cont.*) (g) *E. coli* α subunit of tryptophan synthase (WSA). The ranges of parallel β strands in the α/β barrel are: 19–24, 46–50, 95–104, 125–130, 149–154, 173–179, 208–212, and 237–241.

```
h) RBCR        RUBISCO ....      HODOSPIRILLUM RUBRUM

THE RANGE OF ALPHA HELIX


THE RANGE OF BETA SHEET


   3     20     39     57     72     89    126    162    189    210    241    260    306    341    365    389    423
        HLF    HLF                  PAR    PAR           PAR           PAR    PAR    PAR    HLF    PAR           PAR    PAR
 ALPH   BETA   BETA   BETA   BETA   BETA   BETA   BETA   BETA   ALPH   BETA   BETA   BETA   ALPH   BETA   BETA   BETA
 1.56   1.69   2.35   1.92   5.41   3.37   1.73   2.64   1.49   6.11   3.73   5.34   2.96   5.61   1.05   2.77   2.40
 0.91   1.23   1.75   2.16   3.17   2.36   1.43   3.03   1.07   2.98   2.23   3.26   2.19   2.80   1.26   2.76   2.06
 ****   ****     44   ****   ****     97    134    167    199    215   ****    265   ****   ****    370    394    428
 ****   ****   HLF    ****   ****          HLF                         ****   HLF    ****   ****   HLF    HLF
               ALPH                 BETA   ALPH   ALPH          ALPH          BETA                 BETA   BETA   ALPH
************   4.11*************    1.80   2.75   1.41   0.82   3.75*******   2.42*************    1.19   1.47   3.13
************   1.08*************    1.16   1.34   0.85   0.62   0.71*******   1.52*************    0.84   1.30   0.93
 ****   ****   ****   ****   ****    103    142    173   ****    222   ****   ****   ****   ****    376    402   ****
 ****   ****   ****   ****   ****   PAR                  ****          ****   ****   ****   ****                 ****
                                    ALPH   BETA                                                     ALPH   BETA
********************************    5.23   2.23   0.90*******   1.28*************************    1.82   1.18*******
********************************    2.37   1.92   0.46*******   0.45*************************    0.89   0.85*******
 ****   ****   ****   ****   ****    109   ****   ****   ****   ****   ****   ****   ****   ****   ****    409   ****
 ****   ****   ****   ****   ****          ****   ****   ****   ****   ****   ****   ****   ****   ****          ****
                                    BETA                                                                   ALPH
********************************    0.93************************************************************      2.50*******
********************************    1.04************************************************************      0.94*******

  10     26     49     66     81    120    147    181    204    227    247    272    313    346    381    414    435
 HLF    HLF                         HLF    HLF    HLF           HLF    HLF           HLF
 ALPH   ALPH          ALPH   ALPH   ALPH   ALPH   ALPH   ALPH   ALPH   ALPH   ALPH   ALPH          ALPH   ALPH   ALPH
 3.52   3.53   1.28   2.17   3.94   3.07   6.69   3.31   1.90   3.21   3.41   5.06   2.51   1.25   2.68   4.78   3.75
 1.40   1.74   0.68   1.35   1.64   0.92   2.80   1.31   0.90   1.43   1.83   2.31   0.87   0.57   1.09   1.46   0.88
   15     34   ****   ****   ****   ****    157   ****   ****    232    255    277    322    355   ****   ****    446
        HLF    ****   ****   ****   ****   PAR    ****   ****                                      ****   ****
 ALPH                                      BETA                 BETA   ALPH   ALPH   BETA   BETA                 BETA
 1.79   0.99***************************    1.14*************    2.44   2.39   2.04   0.90   1.06*************    1.95
 0.54   0.58***************************    1.32*************    1.54   0.92   0.84   1.13   1.10*************    1.36
 ****   ****   ****   ****   ****   ****   ****   ****   ****   ****   ****    284    328   ****   ****   ****    452
 ****   ****   ****   ****   ****   ****   ****   ****   ****   ****   ****          PAR    ****   ****   ****
                                                                              ALPH                               ALPH
*****************************************************************************  3.00   0.68******************    2.64
*****************************************************************************  1.24   0.50******************    1.03
 ****   ****   ****   ****   ****   ****   ****   ****   ****   ****   ****    292    336   ****   ****   ****    458
 ****   ****   ****   ****   ****   ****   ****   ****   ****   ****   ****   HLF           ****   ****   ****
                                                                                     ALPH                        BETA
*****************************************************************************  1.10   2.11******************    1.70
*****************************************************************************  0.77   0.74******************    1.22
```

Figure 8 (*cont.*) (h) *R. rubrum* RuBisCO (RBCR). The ranges of secondary structure are not reported.

```
i) RBCS      RUBISCO OF LARGE SUBUNIT  FROM  SPINACH

THE RANGE OF ALPHA HELIX


THE RANGE OF BETA SHEET


I1    137 I2    137

     3     23     38     68     99    125    144    172    197    220    264    291    312    331    364    383    444
                 PAR           HLF    PAR    PAR    HLF    HLF    PAR    PAR    PAR    PAR    PAR    HLF    HLF
          BETA   ALPH   BETA   BETA   BETA   BETA   BETA   BETA   BETA   BETA   BETA   BETA   BETA   BETA   BETA   BETA
   1.09   1.71   7.07   2.31   3.46   2.86   2.93   1.64   1.49   2.95   2.81   5.85   4.97   2.31   1.80   1.77   2.73
   0.64   2.03   2.13   2.42   2.86   1.87   1.77   1.49   1.48   1.88   2.73   3.31   2.83   2.52   2.24   2.05   2.00
   ****   ****     43     73    104   ****    153    179    206    226    269   ****   ****    336   ****    392    449
   ****   ****                 HLF   ****          HLF           HLF                 ****   ****          ****
                 BETA   BETA   BETA          BETA   BETA   ALPH   ALPH   BETA                 ALPH          ALPH   ALPH
 *************   1.17   1.21   2.01*******   0.63   1.38   1.75   3.30   1.53**************   1.76*******   3.08   2.47
 *************   1.03   1.19   1.17*******   0.84   0.90   0.99   0.60   1.04**************   0.71*******   1.54   1.01
   ****   ****     52   ****    114   ****   ****    185   ****    235    276   ****   ****    344   ****    399    455
   ****   ****          ****          ****   ****          ****                 ****   ****   PAR    ****
                               BETA                 ALPH          ALPH   BETA                 BETA          BETA   ALPH
 *************   0.96*******   2.62**************   2.59*******   1.93   1.27**************   2.90*******   1.07   2.42
 *************   0.13*******   2.08**************   1.26*******   0.97   1.26**************   2.28*******   1.23   0.48
   ****   ****   ****   ****   ****   ****   ****   ****   ****    245   ****   ****   ****    350   ****    417   ****
   ****   ****   ****   ****   ****   ****   ****   ****   ****   HLF    ****   ****   ****          ****   HLF    ****
                                                                  BETA                        ALPH          ALPH
 *****************************************************************1.44******************   2.22*******   2.70*******
 *****************************************************************0.88******************   1.03*******   0.90*******

    11     32     57     78    119    131    158    190    214    254    281    299    319    356    373    422    460
                 HLF                  PAR           HLF           HLF           PAR    HLF           PAR    HLF    HLF
   ALPH   ALPH   ALPH   ALPH   BETA   ALPH   ALPH   ALPH   ALPH   ALPH   BETA   ALPH   ALPH   ALPH          ALPH   ALPH
   3.51   1.45   4.55   2.21   4.06   3.06   2.79   3.35   2.69   4.72   2.31   5.64   2.97   1.62   0.34   5.93   3.95
   1.80   0.77   1.19   0.84   2.92   0.89   1.41   1.69   1.01   1.79   1.79   2.43   1.36   0.61   0.70   2.04   1.47
     16   ****     63     88   ****    139    163   ****   ****    259    286    304    326   ****    378    427    465
   HLF    ****                 ****                 ****   ****                               ****   HLF    HLF
   ALPH          BETA   ALPH          ALPH   ALPH                 ALPH   BETA   ALPH   BETA          BETA   ALPH   ALPH
   2.31*******   0.62   1.72*******   2.19   1.80**************   2.70   1.85   2.41   1.55*******   0.68   4.26   3.56
   1.00*******   0.84   0.96*******   1.29   0.77**************   1.25   1.50   1.10   1.51*******   0.89   1.46   1.56
   ****   ****   ****     94   ****   ****   ****   ****   ****   ****   ****   ****   ****   ****   ****    433    473
   ****   ****   ****          ****   ****   ****   ****   ****   ****   ****   ****   ****   ****   ****
                                                                                                             ALPH   ALPH
                                                                                                             1.73   2.23
 ********************   1.32*******************************************************************************   0.42   1.19
 ********************   0.69*******************************************************************************    439   ****
   ****   ****   ****   ****   ****   ****   ****   ****   ****   ****   ****   ****   ****   ****   ****           ****
   ****   ****   ****   ****   ****   ****   ****   ****   ****   ****   ****   ****   ****   ****   ****
                                                                                                             ALPH
                                                                                                             1.52*******
 ***********************************************************************************************************   0.63*******
 ***********************************************************************************************************
```

Figure 8 (*cont.*) (i) Spinach RuBisCO (RBCS). Prediction was made assuming that the conformation of RBCS is quite the same as that of RBCT. The ranges of secondary structure are not reported.

```
j) KHGA          2-KETO-4-HYDROXYGLUTARATE ALDOLASE FROM E. COLI

THE RANGE OF ALPHA HELIX

   1   14    33   41    56   62    78   84    97  105   117  122   139  145   168  174   198  210

THE RANGE OF BETA SHEET

  19   22    44   49    66   72    89   93   110  112   127  131   154  159   178  187


     3     21     44     69    114    126    156    179
          PAR    PAR    HLF           PAR           PAR
  ALPH   BETA   BETA   BETA   BETA   BETA   BETA   BETA
  1.90   5.42   3.62   2.78   0.94   1.65   1.47   2.66
  0.96   5.05   2.67   2.04   1.20   1.30   1.23   2.46
  ****     26     49     74   ****    133    165    184
  ****                        ****   PAR           HLF
         ALPH   BETA   BETA          BETA          BETA
******   3.85   2.48   0.79*******   2.17   0.82   1.09
******   1.42   1.65   0.97*******   1.23   0.73   1.40
  ****   ****   ****   ****   ****    142   ****    195
  ****   ****   ****   ****   ****   HLF    ****   HLF
                                     ALPH          ALPH
**********************************   1.61*******   1.59
**********************************   0.81*******   0.77
  ****   ****   ****   ****   ****   ****   ****    201
  ****   ****   ****   ****   ****   ****   ****
                                                   ALPH
************************************************   3.11
************************************************   1.44

     8     31     58     81    121    147    170    208
          HLF    HLF           PAR    HLF
  ALPH   ALPH   ALPH   ALPH   BETA   ALPH   ALPH   ALPH
  2.37   3.87   6.04   3.38   3.66   3.72   2.13   4.07
  1.16   1.06   2.05   1.24   2.07   1.65   0.88   1.30
    16     39     63     90   ****   ****   ****   ****
  PAR    HLF           HLF    ****   ****   ****   ****
  BETA   BETA   ALPH   ALPH
  0.95   2.37   3.33   2.58****************************
  1.42   1.95   0.97   1.37****************************
  ****   ****   ****    101   ****   ****   ****   ****
  ****   ****   ****          ****   ****   ****   ****
                       ALPH
********************   2.87****************************
********************   0.80****************************
  ****   ****   ****    107   ****   ****   ****   ****
  ****   ****   ****          ****   ****   ****   ****
                       BETA
********************   1.29****************************
********************   1.10****************************
```

Figure 8 (cont.) (j) *E. coli* 2-keto-4-hydroxyglutarate aldolase (KHGA). Prediction was made assuming that the conformation of KHGA is quite the same as that of KGA except the N- and C-terminal regions. The assigned ranges of β strands are listed in the output and all parallel.

15. Xylose isomerase (XIM) (Farber *et al.*, 1987)
16. Muconate lactonizing enzyme (MLE) (Goldman *et al.*, 1987).

TIM, KGA, and TAA were discussed in Section V. The sequence of PK (Muirhead *et al.*, 1986) appeared shortly after the submission of the manuscript, and the sequence of ALD has been updated (Tolan *et al.*, 1984). Also, the sequences of yeast ENL (Chin *et al.*, 1981), yeast FCB (Lederer *et al.*, 1985), and spinach GO (Cederlund *et al.*, 1988) have been reported. A comparison between the sequences of the *R. rubrum* RuBisCO (RBCR) and the spinach large subunit RuBisCO (RBCS) was made (Nargang *et al.*, 1984). The sequences from various sources and the observed ranges of secondary structure for PRA-IGP (Priestle *et al.*, 1987) and for WSA (Crawford *et al.*, 1987) were reported. The observed ranges of secondary structure for PK (Muirhead *et al.*, 1986), ALD (Sygusch *et al.*, 1987), and ENL (Lebioda and Stec, 1988) were given. Besides, the sequence of *E. coli* 2-keto-4-hydroxy-glutarate aldolase (KHGA) was also recently discovered (Vlahos and Dekker, 1988). It has a 49% homology with KGA and the same polypeptide length not including both termini. As a test of the present approach the computed output for the ten proteins of known sequence with α/β barrel(s) are shown in Fig. 8.

ACKNOWLEDGMENTS. This work was supported by scientific research grants 57580160, 59570917, and 61490009 from the Ministry of Education, Science and Culture of Japan. The program is written in FORTRAN IV and is available on request.

VIII. REFERENCES

Anderson, C. M., McDonald, R. C., and Steitz, T. A., 1978, Sequencing a protein by x-ray crystallography. I. Interpretation of yeast hexokinase B at 2.5 Å resolution by model building, *J. Mol. Biol.* **123:**1–13.

Banner, D. W., Bloomer, A. C., Petsko, G. A., Phillips, D. C., Pogson, C. I., Wilson, I. A., Corran, P. H., Furth, A. J., Milman, J. D., Offord, R. E., Priddle, J. D., and Waley, S. G., 1975, Structure of chicken muscle triose phosphate isomerase determined crystallographically at 2.5 Å resolution using amino acid sequence data, *Nature (London)* **255:**609–614.

Bernstein, F. C., Koetzle, T. F., Meyer, E. F., Jr., Brice, M. D., Rodgers, J. R., Kennard, O., Shimanouchi, T., and Tasumi, M., 1977, The protein data bank: A computer-based archival file for macromolecular structures, *J. Mol. Biol.* **112:**535–542.

Blake, C. C. F., and Rice, D. W., 1981, Phosphoglycerate kinase, *Phil. Trans. R. Soc. Lond. [A]* **293:**93–104.

Bradshaw, R. A., Cancedda, F., Ericsson, L. H., Neumann, P. A., Piccoli, S. P., Schlesinger, M. J., Schriefer, K., and Walsh, K. A., 1981, Amino acid sequence of *Escherichia coli* alkaline phosphatase, *Proc. Natl. Acad. Sci. U.S.A.* **78:**3473–3477.

Bränden, C.-I., Schneider, G., Lindqvist, Y., Andersson, I., Knight, S., and Lorimer, G., 1987, Structural and evolutionary aspects of the key enzymes in photorespiration: RuBisCO and glycolate oxidase, *Cold Spring Harb. Symp. Quant. Biol.* **52:**491–498.

Cederlund, E., Lindqvist, Y., Söderlund, G., Bränden, C.-I., and Jörnvall, H., 1988, Primary structure of glycolate oxidase from spinach, *Eur. J. Biochem.* **173:**523–530.

Chin, C. C. Q., Brewer, J. M., and Wold, F., 1981, The amino acid sequence of yeast enolase, *J. Biol. Chem.* **256:**1377–1384.

Chothia, C., 1988, The 14th barrel rolls out, *Nature (London)* **333:**598–599.

Cohen, F. E., Sternberg, M. J. E., and Taylor, W. R., 1982, Analysis and prediction of the packing of α-helices against a β-sheet in the tertiary structure of globular proteins, *J. Mol. Biol.* **156:**821–862.

Cohen, F. E., Abarbanel, R. M., Kuntz, I. D., and Fletterick, R. J., 1983, Secondary structure assignment for α/β proteins by a combinatorial approach, *Biochemistry* **22:**4894–4904.

Crawford, I. P., Niermann, T., and Kirschner, K., 1987, Prediction of secondary structure by evolutionary comparison: Application to the α subunit of tryptophan synthase, *Proteins* **2:**118–129.

Eisenberg, D., Almassy, R. J., Janson, C. A., Chapman, M. S., Suh, S. W., Casio, D., and Smith, W. W., 1987, Some evolutionary relationships of the primary biological catalysts glutamine synthase and RuBisCO, *Cold Spring Harb. Symp. Quant. Biol.* **52:**483–490.

Epp, O., Ladenstein, R., and Wendel, A., 1983, The refined structure of the selenoenzyme glutathione peroxidase at 0.2-nm resolution, *Eur. J. Biochem.* **133:**51–59.

Farber, G. K., Petsko, G. A., and Ringe, D., 1987, The 3.0 Å crystal structure of xylose isomerase from *Streptomyces olivochromogenes, Protein Engineering* **1:**459–466.

Garnier, J., Osguthorp, D. J., and Robson, B., 1978, Analysis of the accuracy and implications of simple methods for predicting the secondary structure of globular proteins, *J. Mol. Biol.* **120:**97–120.

Goldman, A., Ollis, D. L., and Steitz, T. A., 1987, Crystal structure of muconate lactonizing enzyme at 3 Å resolution, *J. Mol. Biol.* **194:**143–153.

Hellinga, H. W., and Evans, P. R., 1985, Nucleotide sequence and high-level expression of the major *Escherichia coli* phosphofructokinase, *Eur. J. Biochem.* **149:**363–373.

Janin, J., and Chothia, C., 1980, Packing of α-helices onto β-pleated sheets and the anatomy of α/β proteins, *J. Mol. Biol.* **143:**95–128.

Lebioda, L., and Stec, B., 1988, Crystal structure of enolase indicates that enolase and pyruvate kinase evolved from a common ancestor, *Nature (London)* **333:**683–686.

Lederer, F., Cortial, S., Becam, A.-M., Haumont, P.-Y., and Perez, L., 1985, Complete amino acid sequence of flavocytochrome b_2 from baker's yeast, *Eur. J. Biochem.* **152:**419–428.

Lifson, S., and Sander, C., 1979, Antiparallel and parallel β-strands differ in amino acid residue preferences, *Nature (London)* **282:**109–111.

Lim, L. W., Shamala, N., Mathews, F. S., Steenkamp, D. J., Hamlin, R., and Xuong, N.-h., 1986, Three-dimensional structure of the iron-sulfur flavoprotein trimethylamine dehydrogenase at 2.4 Å resolution, *J. Biol. Chem.* **261:**15140–15146.

Matthews, B. W., and Remington, S. J., 1974, The three dimensional structure of the lysozyme from bacteriophage T4, *Proc. Natl. Acad. Sci. U.S.A.* **71:**4178–4182.

McLachlan, A. D., and Stewart, M., 1975, Tropomyosin coiled-coil interactions: Evidence for an unstaggered structure, *J. Mol. Biol.* **98:**293–304.

Motherwell, S., 1978, PLUTO78, *Cambridge Crystallographic Database User's Manual,* pp. 56–66, Cambridge, Crystallographic Data Centre, Cambridge, England.

Muirhead, H., Clayden, D. A., Barford, D., Lorimer, C. G., Fothergill-Gilmore, L. A., Schiltz, E., and Schmitt, W., 1986, The structure of cat muscle pyruvate kinase, *EMBO J.* **5:**475–481.

Nagano, K., 1973, Logical analysis of the mechanism of protein folding. I. Prediction of helices, loops and β-structures from primary structure, *J. Mol. Biol.* **75:**401–420.

Nagano, K., 1974, Logical analysis of the mechanism of protein folding. II. The nucleation process, *J. Mol. Biol.* **84:**337–372.

Nagano, K., 1977a, Logical analysis of the mechanism of protein folding. IV. Super-secondary structures, *J. Mol. Biol.* **109:**235–250.

Nagano, K., 1977b, Triplet information in helix prediction applied to the analysis of super-secondary structures, *J. Mol. Biol.* **109:**251–274.

Nagano, K., 1980, Logical analysis of the mechanism of protein folding. V. Packing game simulation of α/β proteins, *J. Mol. Biol.* **138:**797–832.

Nagano, K., and Hasegawa, K., 1975, Logical analysis of the mechanism of protein folding. III. Prediction of the strong long-range interactions, *J. Mol. Biol.* **94:**257–281.

Nagano, K., and Ponnuswamy, P. K., 1984, Prediction of packing of secondary structure, *Adv. Biophys.* **18:** 115–148.

Nargang, F., McIntosh, L., and Somerville, C., 1984, Nucleotide sequence of the ribulosebisphosphate carboxylase gene from *Rhodospirillum rubrum, Mol. Gen. Genet.* **193:**220–224.

Poulos, T. L., Finzel, B. C., Gunsalus, I. C., Wagner, G. C., and Kraut, J., 1985, The 2.6-Å crystal structure of *Pseudomonas putida* cytochrome P-450, *J. Biol. Chem.* **260:**16122–16130.

Priestle, J. P., Grütter, M. G., White, J. L., Vincent, M. G., Kania, M., Wilson, E., Jardetzky, T. S., Kirschner, K., and Jansonius, J. N., 1987, Three-dimensional structure of the bifunctional enzyme *N*-(5′-phosphoribosyl)anthranilate isomerase-indole-3-glycerol-phosphate synthase from *Escherichia coli, Proc. Natl. Acad. Sci. U.S.A.* **84:**5690–5694.

Schneider, G., Lindqvist, Y., Bränden, C.-I., and Lorimer, G., 1986, Three-dimensional structure of ribulose-1,5-bisphosphate carboxylase/oxygenase from *Rhodospirillum rubrum* at 2.9 Å resolution, *EMBO J.* **5:**3409–3415.

Sowadski, J. M., Handschumacher, M. D., Murthy, H. M. K., Foster, B. A., and Wyckoff, H. W., 1985, Refined structure of alkaline phosphatase from *Escherichia coli* at 2.8 Å resolution, *J. Mol. Biol.* **186:**417–433.

Stachelek, C., Stachelek, J., Swan, J., Botstein, D., and Konigsberg, W., 1986, Identification, cloning and sequence determination of the genes specifying hexokinase A and B from yeast, *Nucleic Acids Res.* **14:** 945–962.

Stone, D., and Smillie, L. B., 1978, The amino acid sequence of rabbit skeletal α-tropomyosin. The NH_2-terminal half and complete sequence, *J. Biol. Chem.* **253:**1137–1148.

Stuart, D. I., Levine, M., Muirhead, H., and Stammers, D. K., 1979, Crystal structure of cat muscle pyruvate kinase at a resolution of 2.6 Å, *J. Mol. Biol.* **134:**109–142.

Sygusch, J., Beaudry, D., and Allaire, M., 1987, Molecular architecture of rabbit skeletal muscle aldolase at 2.7 Å resolution, *Proc. Natl. Acad. Sci. U.S.A.* **84:**7846–7850.

Tanaka, I., Appelt, K., Dijk, J., White, S. W., and Wilson, K. S., 1984, 3-Å resolution structure of a protein with histone-like properties in prokaryotes, *Nature (London)* **310:**376–381.

Taylor, W. R., and Thornton, J. M., 1983, Prediction of supersecondary structure in proteins, *Nature (London)* **301:**540–542.

Taylor, W. R., and Thornton, J. M., 1984, Recognition of super-secondary structure in proteins, *J. Mol. Biol.* **173:**487–514.

Tolan, D. R., Amsden, A. B., Putney, S. D., Urdea, M. S., and Penhoet, E. E., 1984, The complete nucleotide sequence for rabbit muscle aldolase A messenger RNA, *J. Biol. Chem.* **259:**1127–1131.

Vlahos, C. J., and Dekker, E. E., 1988, The complete amino acid sequence and identification of the active-site arginine peptide of *Escherichia coli* 2-keto-4-hydroxyglutarate aldolase, *J. Biol. Chem.* **263:**11683–11691.

Weijer, W. J., Hofsteenge, J., Beintema, J. J., Wierenga, R. K., and Drenth, J., 1983, *p*-Hydroxybenzoate hydroxylase from *Pseudomonas fluorescens*. 2. Fitting of the amino-acid sequence to the tertiary structure, *Eur. J. Biochem.* **133:**109–118.

Wierenga, R. K., Terpstra, P., and Hol, W. G. J., 1986, Prediction of the occurrence of the ADP-binding βαβ-fold in proteins, using an amino acid sequence fingerprint, *J. Mol. Biol.* **187:**101–107.

Winter, G., Koch, G. L. E., Hartley, B. S., and Barker, D. G., 1983, The amino acid sequence of the tyrosyl-tRNA synthetase from *Bacillus stearothermophilus, Eur. J. Biochem.* **132:**383–387.

Xia, Z.-x., Shamala, N., Bethge, P. H., Lim, L. W., Bellamy, H. D., Xuong, N.-h., Lederer, F., and Mathews, F. S., 1987, Three-dimensional structure of flavocytochrome b_2 from baker's yeast at 3.0-Å resolution, *Proc. Natl. Acad. Sci. U.S.A.* **84:**2629–2633.

12

Prediction of Protein Structural Classes from Amino Acid Compositions

Peter Y. Chou

I. INTRODUCTION

Proteins are simple and yet quite complex biological macromolecules. They are simple because their building blocks are the 20 naturally occurring amino acids. They are complex because these different amino acid sequences can assume an infinite variety of spatial conformations. Fortunately, there is a link between the simple and the complex in that the protein's conformation is coded in its amino acid sequence. This was demonstrated by the classical experiments on ribonuclease refolding (Anfinsen *et al.*, 1961) that showed that the information for the protein's three-dimensional architecture resides in the primary structure alone.

Since proteins play an essential role in the biological processes of plants and animals, the amino acids are nature's alphabet of life. In this analogy, we may say that the secondary structures of proteins (α helices, β sheets, β turns) are the words of nature. Structural domains

Peter Y. Chou • 3893 Ross Road, Palo Alto, California 94303.

such as βαβ folds or crossover β connections (Sternberg and Thornton, 1976; Richardson, 1976, 1977, 1981), s-helix globules (Lim, 1978), multiple bends (Isogai *et al.*, 1980), and omega loops (Leszczynski and Rose, 1986) may be comparable to sentences. Likewise, reverse turns (Lewis *et al.*, 1971; Chou and Fasman, 1977; Rose *et al.*, 1985) and β bulges (Richardson *et al.*, 1978) may also serve as punctuation marks or signal points for termination of helix and β-sheet propagation. On the other hand, supersecondary structures such as the nucleotide-binding domain (Rossmann *et al.*, 1974), heme pockets of the globins and cytochromes (Rossmann and Argos, 1975), as well as the immunoglobulin fold (Poljak, 1975), may be likened to paragraphs. Finally, the tertiary fold of the protein unfolds an adventurous and fascinating story that describes its enzymatic, hormonal, or structural function within the biological system.

Predicting the three-dimensional structure of a protein from its amino acid sequence is one of the great quests of molecular biology. Current predictive methods of protein conformation have been reviewed (Schulz, 1977; Sternberg and Thornton, 1978; Chou and Fasman, 1978a,b) and updated elsewhere in this volume (Fasman, Chapter 6). In this chapter, we focus on predicting protein structural classes from amino acid compositions as an aid in improving conformational predictions and providing some insights on protein folding (Richardson, 1981; Rossman and Argos, 1981).

The helix, β-sheet, and β-turn conformational parameters of 20 amino acids, which were calculated from the x-ray structure of 15 proteins (Chou and Fasman, 1974a,b), were updated to 29 proteins (Chou and Fasman, 1977, 1978a). Levitt (1978) has published the amino acid conformational preferences in 49 different proteins using an automated algorithm (Levitt and Greer, 1977) for identifying the α, β, and turn regions. Despite the increased Levitt data base (normalized frequencies from 5523 residues) and the considerable shifts of amino acids as α and β formers from the Chou and Fasman data on 15 proteins (1939 effective residues), the Levitt parameters did not improve the predictive quality in 24 proteins (Argos *et al.*, 1978) and gave less accuracy in four proteins not included in either the Chou–Fasman or Levitt data bases. These results seem to indicate that increasing the data sampling did not necessarily increase the predictive accuracy or that the Levitt parameters are somewhat less reliable than the earlier Chou–Fasman parameters in predicting secondary structures of proteins. In order to compare the two data sets more objectively, the Chou–Fasman P_α and P_β parameters were expanded from 29 to 64 proteins and calculated for different protein classes.

Just as a zoological survey will be more meaningful when different species are grouped together as amphibians, birds, mammals, and reptiles, computing the helical and sheet potentials P_α and P_β for different protein classes may yield conformational parameters that are useful in predicting proteins in that class. The topology-packing diagrams of Levitt and Chothia (1976) revealed that proteins can be classified as α (most helical and little or no β structures), β (mostly β sheets and little or no helical), α + β (α-helical and β structures that do not mix but tend to segregate in different domains), and α/β (approximately alternating α and β strands). Examples of α proteins include hemoglobins and cytochromes, whereas β proteins are represented by immunoglobulins and the serine proteases. Ferredoxin and ribonuclease are typical α + β proteins, and the dehydrogenases and kinases are representative of α/β proteins.

The P_α and P_β values for the 20 amino acids from 64 proteins were calculated based on four distinct protein classes and for the sample as a whole (Chou, 1979). Statistics were collected on the amino acid compositions for these four protein classes so that these data may aid in assigning proteins to their proper conformational classes when their x-ray structure is still unknown. A computerized algorithm is presented herein that can assign the correct structural class to proteins based on their amino acid compositions with 80% accuracy (Chou, 1980). Several previously erroneously predicted regions (Chou and Fasman, 1974b) may be correctly assigned based on the P_α and P_β parameters for distinct protein classes. These analyses will provide additional clues in elucidating and understanding the code for protein folding and nature's alphabet of life.

II. METHODS

A. Examination of Proteins with Known X-Ray Structure and Sequence

The identification of secondary structural regions was outlined according to the criteria of Chou and Fasman (1978a), where 29 proteins were analyzed in deriving the conformational parameters. Of the 35 new proteins examined, the following 19 were obtained from the Protein Data Bank (Bernstein *et al.*, 1977) tape updated to March 1978 and used as a source for the x-ray atomic coordinates of proteins, amino acid sequence data, and secondary structural information: actinidin (Baker, 1977), alcohol dehydrogenase (Eklund *et al.*, 1976), carbonic anhydrase B (Kannan *et al.*, 1975), carboxypeptidase B (Schmid and Herriott, 1976; Titani *et al.*, 1975), cytochrome c_{550} (Timkovich and Dickerson, 1976), lobster D-glyceraldehyde 3-phosphate dehydrogenase (Moras *et al.*, 1975), α and β chains of human deoxyhemoglobin A (Fermi, 1975), γ chain of human deoxyhemoglobin F (Frier and Perutz, 1977), immunoglobulin FAB′ V_H-C_H, V_L-C_L chains (Poljak *et al.*, 1974; Saul *et al.*, 1978), immunoglobulin REI (Epp *et al.*,1974), prealbumin (Blake *et al.*, 1978), protease B (Delbaere *et al.*, 1979), rhodanese (Ploegman *et al.*, 1978a,b), superoxide dismutase (Richardson *et al.*, 1975a,b), thioredoxin (Holmgren *et al.*, 1975), triose phosphate isomerase (Banner *et al.*, 1975), and trypsin (Chambers and Stroud, 1977).

The secondary structures and sequence data for the remaining 16 proteins were obtained from the literature: cytochrome b_{562} (Mathews *et al.*, 1979), cytochrome c_{555} (Korszun and Salemme, 1977; Stone *et al.*, 1977), erabutoxin b (Low *et al.*, 1976), D-glyceraldehyde 3-phosphate dehydrogenase, *B. stearothermophilus* (Biesecker, 1977), hemerythrin, *G. gouldii* (Hendrickson and Ward, 1975; Klippenstein *et al.*, 1972), methemerythrin, *T. dyscritum* (Stenkamp *et al.*, 1978; Loehr *et al.*, 1978), myohemerythrin, *T. pyroides* (Hendrickson *et al.*, 1975; Hendrickson and Ward, 1975; Klippenstein *et al.*, 1976), immunoglobulin Mcg (Edmundson *et al.*, 1975; Fett and Deutsch, 1974), bacteriophage T4 lysozyme (Remington *et al.*, 1978), seal myoglobin (Scouloudi and Baker, 1978), penicillopepsin (Hsu *et al.*, 1977), horse phosphoglycerate kinase (Banks *et al.*, 1979), bovine phospholipase A (Dijkstra *et al.*, 1978; Fleer *et al.*, 1978), protease B (Brayer *et al.*, 1978), and subtilisin inhibitor (Mitsui *et al.*, 1979; Ikenaka *et al.*, 1974).

In addition, the following revisions were made on the previous data base of 29 proteins (Chou and Fasman, 1978b): carp myogen has been renamed as calcium-binding parvalbumin, and residues 57–59, 96–97 assigned as β strand (Kretsinger and Nockolds, 1973); the cytochrome *c* sequence of horse (104 residues) was replaced by tuna (103 residues with 18 positions different from horse) since the 2.0-Å x-ray data for tuna (Swanson *et al.*, 1977; Takano *et al.*, 1977) are more refined than the horse (2.8 Å); Asn^{122} is replaced by Asp^{122} in sperm whale myoglobin (Romero-Herrera and Lehmann, 1974; Takano, 1977); amino acid revisions in concanavalin A were made at positions 78 (Asn to Asp), 187–190 (Gln-Ser-Ala-Val to Thr-Val-Ser-Ala), and 196 (Gly to Ala) (Cunningham *et al.*, 1975) as well as updated revisions for β regions 3–11, 23–30, 35–40, 45–56, 59–67, 73–79, 87–97, 103–116, 124–134, 139–143, 147–149, 152–160, 163–166, 169–181, 187–200, and 208–215 based on ϕ,ψ angles and hydrogen bonding (Reeke *et al.*, 1975); two small helical segments were located in ferredoxin 13–18 and 39–45, and the β regions expanded to 1–5, 22–25, 28–31, and 49–52 on the basis of more refined data (Adman *et al.*, 1976); similar refinements of

Table I. Helix and Sheet Regions, Helical and β-Sheet Residues, and %α and %β in 19 α Proteins[a]

α Protein	α	β	n	n_α	n_β	%α	%β
Ca^{2+}-binding parvalbumin	6	2	108	61	5	56.5	4.6
Cytochrome b_{562}	4	0	103	81	0	78.6	0.0
Cytochrome *c*	5	2	103	44	6	42.7	5.8
Cytochrome c_2	5	2	112	49	4	43.8	3.6
Cytochrome c_{550}	5	2	134	53	6	39.6	4.5
Cytochrome c_{555}	3	0	86	31	0	36.1	0.0
Hemerythrin (Met-)	4	0	113	73	0	64.6	0.0
Hemerythrin (Myo-)	4	0	118	81	0	68.6	0.0
Hemerythrin (*G. gouldi*)	4	0	113	81	0	71.7	0.0
Hemoglobin, α (human)	7	0	141	109	0	77.3	0.0
Hemoglobin, β (human)	8	0	146	112	0	76.7	0.0
Hemoglobin, α (horse)	7	0	141	109	0	77.3	0.0
Hemoglobin, β (horse)	8	0	146	115	0	78.8	0.0
Hemoglobin, (glycera)	7	0	147	112	0	76.2	0.0
Hemoglobin, (lamprey)	8	0	148	117	0	79.1	0.0
Hemoglobin, (midge larva)	8	0	136	113	0	83.1	0.0
Hemoglobin, γ (human)	8	0	146	113	0	77.4	0.0
Myoglobin (seal)	8	0	153	121	0	79.1	0.0
Myoglobin (sperm whale)	8	0	153	121	0	79.1	0.0
Total	117	8	2447	1696	21	69.3	0.9

[a]α is the number of α-helical regions in the protein; β is the number of β-sheet regions; n is the number of amino acid residues in the protein; n_α is the number of helical residues; n_β is the number of β-sheet residues; $\%\alpha = (n_\alpha/n) \times 100$; $\%\beta = (n_\beta/n) \times 100$.

rubredoxin *C. pasteruianum* to 1.2 Å showed β regions at 3–7, 10–14, and 49–52 (Watenpaugh *et al.*, 1978).

B. Classification of Proteins in Four Structural Classes

Topological packing diagrams of proteins similar to those of Levitt and Chothia (1976) were used to assign proteins to four distinct classes. The 19 α proteins (69% α, 1% β) surveyed in Table I consist of predominantly helical regions with little or no β sheets (117 α and eight β regions). The average percentage helicity in the ten globins and three hemerythrins were 78.4% and 68.3%, respectively, with no β regions, although the number of helices in the globins (seven or eight) is twice that of the hemerythrins (four). Calcium-binding protein with six α and two β regions (56.5% α, 4.6% β) was assigned as an α protein as previously (Levitt and Chothia, 1976). However, the cytochromes were classified as α instead of as α + β proteins in contrast to Levitt and Greer (1977), since they have an average of 48% α and 3% β residues. There are no β residues in cytochrome b_{562} and cytochrome c_{555}, and the four β residues in cytochrome c_2 as well as the six β residues in cytochrome *c* and cytochrome c_{550} are really not enough to class them as α + β proteins.

Proteins assigned to the β class have mainly β sheets with minimal or no helical content (Table II). The 15 β proteins (51% β, 5% α) examined have 172 β and only 20 α regions or an average 11.5 : 1 ratio of β : α regions. Comparatively, the 19 α proteins (69% α, 1% β) have an average 6 : 0.4 ratio of α : β regions.

Proteins classified as α + β generally have helices and β sheets clustered in separate domains. The 14 α + β proteins (34% α, 21% β) surveyed in Table III have 63 α and 72 β

Table II. Helix and Sheet Regions, Helical and β-Sheet Residues, and %α and %β in 15 β Proteins[a]

β Protein	α	β	n	n_α	n_β	%α	%β
α-Chymotrypsin	2	13	241	20	97	8.3	40.2
Concanavalin A	1	6	237	5	137	2.1	57.8
Elastase	2	14	240	18	118	7.5	49.2
Erabutoxin B	0	5	62	0	31	0.0	50.0
Immunoglobulin Fab′ (V&C,H)	1	17	220	5	130	2.3	59.1
Immunoglobulin Fab′ (V&C,L)	1	17	208	5	122	2.4	58.7
Immunoglobulin Mcg	3	16	216	23	142	10.7	65.7
Immunoglobulin REI	1	9	108	5	61	4.6	56.5
Penicillopepsin	4	18	323	28	135	8.7	41.8
Prealbumin	1	8	127	8	58	6.3	45.7
Protease A	2	13	181	15	94	8.3	51.9
Protease B	1	12	185	9	104	4.9	56.2
Rubredoxin	0	3	54	0	14	0.0	25.9
Superoxide dismutase	1	8	151	7	76	4.6	50.3
Trypsin	0	13	223	0	92	0.0	41.3
Total	20	172	2776	148	1411	5.3	50.8

[a]α is the number of α-helical regions in the protein; β is the number of β-sheet regions; n is the number of amino acid residues in the protein; n_α is the number of helical residues; n_β is the number of B-sheet residues; $\%\alpha = (n_\alpha/n) \times 100$; $\%\beta = (n_\beta/n) \times 100$.

regions or an average 4.5 : 5 ratio of α : β regions. In contrast, α/β proteins have alternating α helix and β strand such as the discernible repeating βαβ folds found in the dehydrogenases (Holbrook *et al.*, 1975; Moras *et al.*, 1975; Eklund *et al.*, 1976). The 16 α/β proteins (35% α, 23% β) surveyed in Table IV have 135 α and 169 β regions or an average of 8.5 α : 10.5 β regions.

Table III. Helix and Sheet Regions, Helical and β-Sheet Residues, and %α and %β in 14 α + β Proteins[a]

α + β Protein	α	β	n	n_α	n_β	%α	%β
Actinidin	6	4	218	61	31	28.0	14.2
Cytochrome b_5	6	5	93	43	26	46.2	28.0
Ferredoxin	2	4	54	13	17	24.1	31.5
High-potential iron protein	2	3	85	10	13	11.8	15.3
Insulin	3	2	51	25	12	49.0	23.5
Lysozyme (bacteriophage T4)	10	4	164	107	20	65.2	12.2
Lysozyme (chicken)	6	4	129	54	22	41.9	17.1
Papain	5	17	212	56	30	26.4	14.2
Phospholipase A_2	5	2	123	61	12	49.6	9.8
Ribonuclease S	3	6	124	31	55	25.0	44.4
Staphylococcal nuclease	4	6	149	38	43	25.5	28.9
Subtilisin inhibitor	2	5	113	20	38	17.7	33.6
Thermolysin	7	8	316	112	65	35.4	20.6
Trypsin inhibitor	2	2	58	11	16	19.0	27.6
Total	63	72	1889	642	400	34.0	21.2

[a]α is the number of α-helical regions in the protein; β is the number of β-sheet regions; n is the number of amino acid residues in the protein; n_α is the number of helical residues; n_β is the number of β-sheet residues; $\%\alpha = (n_\alpha/n) \times 100$; $\%\beta = (n_\beta/n) \times 100$.

Table IV. Helix and Sheet Regions, Helical and β-Sheet Residues, and %α and %β in 16 α/β Proteins[a]

α/β Protein	α	β	n	n_α	n_β	%α	%β
Adenylate kinase	10	5	194	105	24	54.1	12.4
Alcohol dehydrogenase	10	21	374	106	115	28.3	30.8
Carbonic anhydrase B	7	10	260	49	72	18.9	27.7
Carbonic anhydrase C	7	10	259	53	69	20.5	26.6
Carboxypeptidase A	8	8	307	108	45	35.2	14.7
Carboxypeptidase B	8	8	306	96	45	31.4	14.7
Dihydrofolate reductase	3	8	159	28	49	17.6	30.8
Flavodoxin	4	5	138	50	37	36.2	26.8
Glyceraldehyde 3-phosphate dehydrogenase (lobster)	9	19	333	109	115	32.7	34.5
Glyceraldehyde 3-phosphate dehydrogenase (*B. stearotherm.*)	9	15	334	104	89	31.1	26.7
Lactate dehydrogenase	10	11	329	133	79	40.4	24.0
Phosphoglycerate kinase	15	17	416	170	99	40.9	23.8
Rhodanese	12	10	293	120	42	41.0	14.3
Subtilisin BPN′	8	9	275	80	55	29.1	20.0
Thioredoxin	4	5	108	52	30	48.2	27.8
Triose phosphate isomerase	11	8	248	134	50	54.0	20.2
Total	135	169	4333	1497	1015	34.5	23.4

[a] α is the number of α-helical regions in the protein; β is the number of β-sheet regions; n is the number of amino acid residues in the protein; n_α is the number of helical residues; n_β is the number of β-sheet residues; $\%\alpha = (n_\alpha/n) \times 100$; $\%\beta = (n_\beta/n) \times 100$.

C. Calculations of Protein Conformational Parameters

The number of α residues ($N_\alpha = 3983$) and β residues ($N_\beta = 2846$) in the 64 proteins (total residues, $N = 11{,}445$) is, respectively, more than double and triple that of the previous analysis (Chou and Fasman 1978a) on 29 proteins ($N_\alpha = 1798$, $N_\beta = 930$, $N = 4741$). The frequency of residues j in α and β regions may be easily computed by dividing its occurrence in these respective regions by its total occurrence: $f_{\alpha,j} = n_{\alpha,j}/n_j$ and $f_{\beta,j} = n_{\beta,j}/n_j$. When these frequencies are divided by their average frequency, respectively, their conformational parameters are obtained: $P_{\alpha,j} = f_{\alpha,j}/\langle f_\alpha \rangle$ and $P_{\beta,j} = f_{\beta,j}/\langle f_\beta \rangle$ (Chou and Fasman, 1974a,b). The standard errors $\sigma_{P\alpha}$ and $\sigma_{P\beta}$ were calculated in the same manner as σ_{P_t} (Chou and Fasman, 1977).

D. Prediction of Protein Classes from Amino Acid Compositions

The percentage of amino acids in a protein is compared with the percentages found in the four structural classes, α, β, α + β, and α/β. Two methods are used to assign a protein to a structural class based on its amino acid composition. The first method sums the differences between $\%_i$ and $(\%_i)_\alpha$, $(\%_i)_\beta$, $(\%_i)_{\alpha+\beta}$, and $(\%_i)_{\alpha/\beta}$ for all 20 amino acids. The lowest overall percentage differences between a protein and a structural class would indicate closest percentage similarities in amino acid composition, and are assigned automatically by a computer algorithm to that protein class. Hence,

$$(\Delta\%_i)_\beta = \Sigma|\%_i - (\%_i)_\alpha| \quad (1)$$

$$(\Delta\%_i)_\beta = \Sigma|\%_i - (\%_i)_\beta| \quad (2)$$

$$(\Delta\%_i)_{\alpha+\beta} = \Sigma|\%_i - (\%_i)_{\alpha+\beta}| \quad (3)$$

$$(\Delta\%_i)_{\alpha/\beta} = \Sigma|\%_i - (\%_i)_{\alpha/\beta}| \quad (4)$$

After the total percentages from the above four equations are compared, a protein is assigned as belonging to the α class if equation 1 is the lowest, to the β class if equation 2 is the lowest, to the α + β class if equation 3 is the lowest, and to the α/β class if equation 4 is the lowest.

The second method looks for the highest correspondences in $\%_i$ by comparing the $|\%_i - (\%_i)_j|$ of the 4 j classes against each other for each of the 20 amino acids. The lowest value of $|\%_i - (\%_i)_j|$ indicates a correspondence value of 1 for that protein class, with a value of 0 for the other three structural classes. The value of 0.5 is assigned to classes tied for the lowest value of $|\%_i - (\%_i)_j|$. These correspondence values are then summed by a computer algorithm, which automatically assigns a protein to the structural class having the greatest number of correspondences. Thus, a total correspondence value of 20 would indicate that the protein's amino acid composition fits the structural class j ideally, although its percentages for the 20 amino acids are not identical to the average percentage composition for that class. Both of these methods, the "minimum sum of differences" and the "greatest correspondences," are utilized in predicting protein structural classes based on amino acid compositions.

III. RESULTS AND DISCUSSIONS

A. Amino Acid Compositions in Four Different Protein Classes

The amino acid compositions for the 64 proteins of known x-ray structure are grouped in four protein structural classes. The 19 α, 15 β, 14 α + β, and 16 α/β proteins are presented in Tables V–VIII, respectively. It is immediately apparent that the protein size fluctuates more between different classes than within the same class. The average chain lengths for the four protein classes in decreasing order are $\langle N_{\alpha/\beta}\rangle = 271$, $\langle N_{\beta}\rangle = 185$, $\langle N_{\alpha+\beta}\rangle = 135$, $\langle N_{\alpha}\rangle = 129$. To discern more relationships among the residue compositions in these protein classes, their amino acid frequencies were calculated and arranged in hierarchic order by percentages in Table IX. The average chain lengths and structural characteristics in the four protein classes (Tables I–IV) are summarized in Table X. The characteristic features of α, β, α + β, and α/β proteins are enumerated below.

1. α Proteins

Table I shows that α proteins are predominantly helical with practically no β sheets (69% α helix and 1% β sheet). Only four of the 19 α proteins were found to contain β strands: calcium-binding parvalbumin 57–59, 96–97; cytochrome c 18–21, 31–32; cytochrome c_2 38–39, 61–62; cytochrome c_{550} 18–20, 35–37. Thus, the 21 β residues out of 2447 total residues amount to less than 1% β. The α proteins have the smallest average chain length, $\langle N_{\alpha}\rangle = 129$, of the four protein classes (Table X). The ten globin chains ($\langle N_{\alpha}\rangle = 146$) and three hemerythrins ($\langle N_{\alpha}\rangle = 115$) are oxygen transport proteins and are completely devoid of β sheets. The cytochromes with $\langle N_{\alpha}\rangle = 108$ serve as electron carriers. These transport proteins are in general of smaller size than average ($\langle N\rangle = 179$ for the 64 proteins surveyed), thus providing them with greater mobility and accessibility within the biological system. The predominant helical character in transport proteins reflects the greater flexibility of helices than β sheets in accommodating oxygen to the heme ring in globins and cytochromes and to the two iron atoms in the hemerythrins. If appreciable β sheets were present in transport proteins, they might be less efficient as oxygen or electron carriers, as β sheets form interchain hydrogen bonds, thus making the protein chains more rigid.

Table V. Amino Acid Composition in 19 α Proteins

α Proteins	Ala	Arg	Asn	Asp	Cys	Gln	Glu	Gly	His	Ile	Leu	Lys	Met	Phe	Pro	Ser	Thr	Trp	Tyr	Val	Total
Calcium-binding parvalbumin (carp)	20	1	3	14	1	2	6	8	1	5	9	13	0	10	0	5	5	0	0	5	108
Cytochrome b_{562} (*E. coli*)	17	4	9	10	0	8	7	3	2	3	9	12	3	2	3	1	4	0	2	4	103
Cytochrome *c* (tuna)	7	2	6	4	2	4	5	13	2	4	6	16	2	3	3	4	7	2	5	6	103
Cytochrome c_2 (*R. rubrum*)	15	0	7	6	2	1	9	8	2	2	8	17	2	5	3	5	8	1	5	6	112
Cytochrome c_{550} (*P. denitrificans*)	15	1	8	10	2	6	10	17	1	5	6	17	4	4	6	3	8	1	3	7	134
Cytochrome c_{555} (*C. thiosulfatophilum*)	16	0	4	4	2	2	0	12	2	2	1	11	8	0	4	3	4	1	4	6	86
Hemerythrin B (*G. gouldii*)	6	3	7	12	1	3	6	7	6	9	8	11	1	9	4	3	4	4	5	4	113
Methemerythrin (*T. dyscritum*)	7	4	6	12	2	5	4	6	7	9	10	9	1	7	3	3	7	3	6	2	113
Myohemerythrin (*T. pyroides*)	7	2	5	9	2	1	11	6	6	6	7	15	3	7	5	4	5	3	5	9	118
α-Methemoglobin (horse)	16	3	4	9	1	1	3	10	10	0	21	11	1	7	6	13	9	1	3	12	141
β-Methemoglobin (horse)	15	4	6	8	1	4	10	14	9	0	19	11	1	8	5	6	3	2	3	17	146
α-Deoxyhemoglobin (human)	21	3	4	8	1	1	4	7	10	0	18	11	2	7	7	11	9	1	3	13	141
β-Deoxyhemoglobin (human)	15	3	6	7	2	3	8	13	9	0	18	11	1	8	7	5	7	2	3	18	146
γ-Deoxyhemoglobin (human fetal)	11	3	5	8	1	4	8	13	7	4	17	12	2	8	4	11	10	3	2	13	146
Hemoglobin (*glycera*)	28	3	4	8	1	6	4	20	6	8	11	11	5	3	3	10	1	2	3	10	147
Hemoglobin (lamprey)	21	5	4	10	1	4	6	6	2	8	10	13	4	8	6	13	9	2	4	12	148
Hemoglobin (midge larva)	17	3	5	9	0	4	5	11	4	9	6	10	4	14	5	9	9	1	2	9	136
Myoglobin (seal)	14	5	3	8	0	3	14	12	13	8	19	19	2	7	4	7	5	2	2	6	153
Myoglobin (sperm whale)	17	4	1	7	0	5	14	11	12	9	18	19	2	6	4	6	5	2	3	8	153
Total	285	53	97	163	22	67	134	197	111	91	221	249	48	123	82	122	119	33	63	167	2447

Table VI. Amino Acid Composition in 15 β Proteins

β Proteins	Ala	Arg	Asn	Asp	Cys	Gln	Glu	Gly	His	Ile	Leu	Lys	Met	Phe	Pro	Ser	Thr	Trp	Tyr	Val	Total
α-Chymotrypsin (bovine)	22	3	13	9	10	10	5	23	2	10	19	14	2	6	9	27	22	8	4	23	241
Concanavalin A (jack bean)	19	6	12	20	0	5	7	16	6	15	18	12	2	11	11	31	19	4	7	16	237
Elastase (porcine)	17	12	18	6	8	15	4	25	6	10	18	3	2	3	7	22	19	7	11	27	240
Erabutoxin B (sea snake)	0	3	3	1	8	4	4	5	2	4	1	4	0	2	4	8	5	1	1	2	62
Immunoglobulin Fab′ (V_H and C_H) (human)	11	7	7	8	6	8	5	18	3	4	19	9	1	6	14	34	23	4	9	24	220
Immunoglobulin Fab′ (V_L and C_L) (human)	19	5	7	5	5	11	10	14	4	5	15	13	0	5	14	30	19	3	8	16	208
Immunoglobulin B-J MCG (human)	17	3	9	7	5	9	12	18	3	4	11	14	0	5	14	32	20	3	11	19	216
Immunoglobulin B-J REI (human)	6	3	2	5	2	13	2	8	0	8	8	4	1	3	6	14	11	1	8	3	108
Penicillopepsin (*P. janthinellum*)	24	0	18	19	2	25	4	40	3	13	21	5	0	20	12	47	29	3	14	24	323
Prealbumin (human)	12	4	3	5	1	2	10	10	4	5	7	8	1	5	8	11	12	2	5	12	127
Protease A (*S. griseus*)	19	7	13	3	5	6	2	31	3	9	10	0	1	5	4	21	21	1	8	12	181
Protease B (*S. griseus*)	14	8	10	7	4	2	2	33	2	7	7	1	2	5	5	22	28	2	10	14	185
Rubredoxin (*C. pasteurianum*)	0	0	1	10	4	0	6	6	0	2	1	4	1	2	5	0	3	1	3	5	54
Superoxide dismutase (bovine)	9	4	6	11	3	2	9	25	8	9	8	10	1	4	6	8	12	0	1	15	151
Trypsin (bovine)	14	2	17	5	12	10	4	25	3	15	14	14	2	3	8	34	10	4	10	17	223
Total	203	67	139	121	75	122	86	297	49	120	177	115	16	85	127	341	253	44	110	229	2776

Table VII. Amino Acid Composition in 14 α + β Proteins

α + β Proteins	Ala	Arg	Asn	Asp	Cys	Gln	Glu	Gly	His	Ile	Leu	Lys	Met	Phe	Pro	Ser	Thr	Trp	Tyr	Val	Total
Actinidin (kiwi fruit)	18	5	11	16	7	10	10	28	1	17	8	6	2	5	7	12	18	6	14	17	218
Cytochrome b_5 (bovine)	4	3	2	7	0	2	12	6	5	5	8	9	0	3	3	8	7	1	4	4	93
Ferredoxin (*P. aerogenes*)	7	0	3	5	8	2	2	4	0	6	0	1	0	0	5	5	0	0	2	4	54
High-potential iron protein (chromatium)	19	2	5	5	4	5	4	6	1	2	5	5	1	2	5	3	4	3	1	3	85
Insulin (A and B chains) (porcine)	2	1	3	0	6	3	4	4	2	2	6	1	0	3	1	3	2	0	4	4	51
Lysozyme (bacteriophage T4)	15	13	12	10	2	5	8	11	1	10	16	13	5	5	3	6	11	3	6	9	164
Lysozyme (chicken)	12	11	13	8	8	3	2	12	1	6	8	6	2	3	2	10	7	6	3	6	129
Papain (papaya)	14	12	13	6	7	13	7	28	2	12	11	10	0	4	10	13	8	5	19	18	212
Phospholipase A_2 (bovine)	6	2	16	9	14	3	5	6	2	5	8	11	1	4	5	10	4	1	7	4	123
Ribonuclease S (bovine)	12	4	10	5	8	7	5	3	4	3	2	10	4	3	4	15	10	0	6	9	124
Staphylococcal nuclease (*S. aureus*)	14	5	7	7	0	6	12	10	4	5	11	23	4	3	6	5	10	1	7	9	149
Subtilisin inhibitor (streptomyces)	18	4	3	6	4	1	5	11	2	0	9	2	3	3	8	9	8	1	3	13	113
Thermolysin (*B. thermoproteolyticus*)	28	10	19	25	0	13	8	36	8	18	16	11	2	10	8	26	25	3	28	22	316
Trypsin inhibitor (bovine)	6	6	3	2	6	1	2	6	0	2	2	4	1	4	4	1	3	0	4	1	58
Total	175	78	120	111	74	74	86	171	33	93	110	112	25	52	71	126	117	30	108	123	1889

Table VIII. Amino Acid Composition in 16 α/β Proteins

α/β Proteins	Ala	Arg	Asn	Asp	Cys	Gln	Glu	Gly	His	Ile	Leu	Lys	Met	Phe	Pro	Ser	Thr	Trp	Tyr	Val	Total
Adenylate kinase (porcine)	8	11	2	11	2	6	19	19	2	9	18	21	6	5	6	11	14	0	7	17	194
Alcohol dehydrogenase (horse)	28	12	8	17	14	8	21	38	7	24	25	30	9	18	20	26	24	2	4	39	374
Carbonic anhydrase B (human)	19	7	17	14	1	9	13	16	11	10	20	18	2	11	17	30	14	6	8	17	260
Carbonic anhydrase C (human)	13	7	8	21	1	10	14	22	12	9	26	24	1	12	17	18	12	7	8	17	259
Carboxypeptidase A (bovine)	21	11	17	12	2	11	14	23	8	21	23	15	3	16	10	32	26	7	19	16	307
Carboxypeptidase B (bovine)	22	13	15	13	7	6	19	22	7	16	21	17	6	12	12	27	27	8	22	14	306
Dihydrofolate reductase (*E. coli*)	13	9	6	13	2	5	11	10	5	12	11	6	5	6	10	9	6	5	4	11	159
Flavodoxin (*Clostridium* MP)	6	2	8	9	3	3	18	14	0	15	8	10	5	5	3	8	5	3	3	10	138
Glyceraldehyde 3-P dehydrogenase (lobster)	32	9	10	22	5	7	17	30	5	18	18	28	10	15	12	25	20	3	9	38	333
Glyceraldehyde 3-P dehydrogenase (*B. stearo-therm.*)	39	14	22	18	2	3	23	25	9	19	26	23	7	5	11	17	18	2	8	43	334
Lactate dehydrogenase (dogfish)	19	9	13	22	7	10	14	24	12	22	35	29	11	7	10	26	12	7	7	33	329
Phosphoglycerate kinase (horse)	41	11	22	23	7	7	26	40	6	18	38	42	13	16	16	24	18	4	4	40	416
Rhodanese (bovine)	23	20	8	14	4	6	22	25	8	7	25	15	5	15	18	21	13	8	11	25	293
Subtilisin BPN′ (*B. amylo-liquefaciens*)	37	2	17	11	0	11	4	33	6	13	15	11	5	3	14	37	13	3	10	30	275
Thioredoxin (*E. coli*)	12	1	4	11	2	3	5	9	1	9	13	10	1	4	5	3	6	2	2	5	108
Triose phosphate isomerase (chicken)	28	8	6	13	4	9	17	27	8	17	17	22	2	8	7	13	10	5	4	23	248
Total	361	146	183	244	63	114	257	377	107	239	339	321	91	158	188	327	238	72	130	378	4333

Table IX. Amino Acid Compositions and Frequencies in Four Protein Structural Classes[a]

19 α Proteins			15 β Proteins			14 α + β Proteins			16 α/β Proteins		
	n_j	$\%_j$		n_j	$\%_j$		n_j	$\%_j$		n_j	$\%_j$
Ala	285	11.6	Ser	341	12.3	Ala	175	9.3	Val	378	8.7
Lys^+	249	10.2	Gly	297	10.7	Gly	171	9.1	Gly	377	8.7
Leu	221	9.0	Thr	253	9.1	Ser	126	6.7	Ala	361	8.3
Gly	197	8.1	Val	229	8.2	Val	123	6.5	Leu	339	7.8
Val	167	6.8	Ala	203	7.3	Asn	120	6.4	Ser	327	7.5
Asp^-	163	6.7	Leu	177	6.4	Thr	117	6.2	Lys^+	321	7.4
Glu^-	134	5.5	Asn	139	5.0	Lys^+	112	5.9	Glu^-	257	5.9
Phe	123	5.0	Pro	127	4.6	Asp^-	111	5.9	Asp^-	244	5.6
Ser	122	5.0	Gln	122	4.4	Leu	110	5.8	Ile	239	5.5
Thr	119	4.9	Asp^-	121	4.4	Tyr	108	5.7	Thr	238	5.5
His^+	111	4.5	Ile	120	4.3	Ile	93	4.9	Pro	188	4.3
Asn	97	4.0	Lys^+	115	4.1	Glu^-	86	4.6	Asn	183	4.2
Ile	91	3.7	Tyr	110	4.0	Arg^+	78	4.1	Phe	158	3.6
Pro	82	3.4	Glu^-	86	3.1	Cys	74	3.9	Arg^+	146	3.4
Gln	67	2.7	Phe	85	3.1	Gln	74	3.9	Tyr	130	3.0
Tyr	63	2.6	Cys	75	2.7	Pro	71	3.8	Gln	114	2.6
Arg^+	53	2.2	Arg^+	67	2.4	Phe	52	2.8	His^+	107	2.5
Met	48	2.0	His^+	49	1.8	His^+	33	1.7	Met	91	2.1
Trp	33	1.3	Trp	44	1.6	Trp	30	1.6	Trp	72	1.7
Cys	22	0.9	Met	16	0.6	Met	25	1.3	Cys	63	1.5
$N =$	2447		$N =$	2776		$N =$	1889		$N =$	4333	

[a] n_j = number of amino acid j found in the four protein classes from Tables V–VIII. $\%_j = (n_j/N) \times 100$, where N = total number of amino acids for the protein class from the bottom rows of Tables V–VIII for α, β, α + β, and α/β proteins, respectively.

A glance at Table V shows that in the α proteins, there is a high Lys content (10%) that is fairly constant, with $\langle n_{\text{Lys}} \rangle = 13 \pm 2.5$ Lys residues/α protein. The small Cys content (1%) in α proteins, averaging 1.2 ± 0.6 residues, is compensated for by greater hydrophobicity (42%) and helicity (69%). The greater abundance of Ala ($\langle n_{\text{Ala}} \rangle = 14.9$) and Leu ($\langle n_{\text{Leu}} \rangle = 11.6$) in the α proteins is not surprising since they are strong helix formers (Chou and Fasman, 1974a,b). Greater solubility of fetal deoxyhemoglobin (deoxyHb F) than adult deoxyhemoglobin (deoxyHb A) was observed by Frier and Perutz (1977). This may be attributed to the increase of Ser + Thr (12 to 21) or +9 polar residues with a corresponding decrease of hydrophobic residues (−7) when the γ and β chains are compared (Table V).

Table X. Chain Length and Structural Characteristics in Four Different Classes of Proteins[a]

Protein class	Number of proteins	Number of α res.	Number of β res.	Average no. of α : β regions	Total res.	Avg. length	%α	%β
α	19	1696	21	6 : 0.4	2,447	129	69.3	0.9
β	15	148	1411	1 : 11.5	2,776	185	5.3	50.8
α + β	14	642	400	4.5 : 5	1,889	135	34.0	21.2
α/β	16	1497	1015	8.5 : 10.5	4,333	271	34.5	23.4
Total	64	3983	2847	5.2 : 6.6	11,445	179	34.8	24.9

[a] The helical and β-sheet regions in 19 α proteins are 117 : 8, in 15 β proteins 20 : 172, in 14 α + β proteins 63 : 72, and in 16 α/β proteins 135 : 169 (see Tables I–IV).

2. β Proteins

The 15 β proteins listed in Table II have a majority of β sheets with little helicity (51% β, 5% α). No helix region was found in the smaller β proteins erabutoxin B (62 residues) and rubredoxin (54 residues). The average chain length, $\langle N_\beta \rangle = 185$, for β proteins is the second largest size of the four protein classes (Table X). The serine proteases (chymotrypsin, elastase, trypsin, proteases A and B) and the immunoglobulins all have sequence length around 200 (Table II). Although immunoglobulin REI is listed with 108 amino acids, it should be noted that only the variable (V) domain was analyzed by x-ray studies, since the constant (C) portion of the chain did not crystallize (Epp *et al.*, 1974). Because immunoglobulins are antibody proteins that recognize and bind specifically with antigens, it is not surprising that β sheets are utilized as their major structural components. Interchain hydrogen bonds of β sheets are more effective than intrachain hydrogen bonds of helices in the intermolecular binding of antigens. Furthermore, β sheets are generally more stable than α helices at higher temperatures (Snell and Fasman, 1973); hence, they may function more efficiently in the immunoglobulin's defense system as antibodies.

There is also a tendency for β proteins to form dimers (e.g., α-chymotrysin, concanavalin A, prealbumin, and immunoglobulins). High Cys content is found in the smallest β proteins: erabutoxin b (eight Cys) and rubredoxin (four Cys) have an average of 10.2% Cys residues (Table VI). That these two β proteins with high Cys content are devoid of helices strengthens the correlation of low Cys content (1%) found in predominantly helical α proteins (Table V). A glance at the amino acid composition in 15 β proteins (Table VI) shows that Ser is the most abundant residue. An average of 25 Ser residues (11.7%) is found in the five serine proteases (chymotrypsin, elastase, proteases A and B, and trypsin). The highest Ser content was also present (14.9%) in the three full immunoglobulins (FAB′ heavy and light chains and Mcg) and penicillopepsin (14.6%), thus indicating the importance of serine in antibody proteins.

3. α + β Proteins

Helices and β sheets are clustered in separate domains in α + β proteins. The 14 α + β proteins in Table III are 34% α helical and 21% β sheet and are generally smaller than proteins from the other three classes. Only three proteins have sequences longer than 200 residues (actinidin, papain, tharmolysin), and five have sequences below 100 residues (cytochrome b_5, ferredoxin, high-potential iron protein, insulin, and trypsin inhibitor). The average chain length is $\langle N_{\alpha+\beta} \rangle = 135$ (Table X).

Two of the larger α + β proteins in Table VII, actinidin (218 residues) and papain (212 residues), are from tropical fruits. They have nearly the same number of amino acids and the same Cys content (seven residues). However, actinidin has three times more acidic residues (16 Asp, 28 Glu) than papain (six Asp, seven Glu), whereas the latter has double the number of basic residues (12 Arg, 10 Lys) than the former (five Arg, six Lys). Nevertheless, their structural folding appears similar, with six α, four β, and 14 turn regions in actinidin (28% α, 14% β, 26% turn residues) and five α, seven β, and 18 turn regions in papain (26% α, 17% β, 27% turn residues). The disulfide bridges (22–65, 56–98, 156–206) in actinidin resemble the S-S linkages in papain (22–63, 56–95, 153–200). Other similarities in tertiary folding between these two proteins were discussed by Baker (1977, 1980).

It is interesting to note that the largest α + β protein, thermolysin (316 residues), has no Cys residues, although a greater abundance occurs in the three smallest α + β proteins, ferredoxin (eight Cys), insulin (six Cys), and trypsin inhibitor (six Cys) with an average of 12.3% Cys composition. The same trend was noted earlier in the two smallest β proteins, erabutoxin b and rubredoxin, which have an average of 10.2% Cys residues. This suggests that

S-S bridges play an important role in tertiary stabilization of smaller proteins. In contrast, the absence of Cys residues in thermolysin is compensated for by the 40.2% hydrophobic residues and an abundance of secondary structures: seven helical, eight β-sheet, and 32 β-turn regions (35% α, 21% β, 37% turn residues). On the other hand, the reduced number of hydrophobic residues in phospholipase A (29.3%) is compensated for by an increase in Cys content (11.4%), as evident in Table VII.

4. α/β *Proteins*

The α/β proteins are characterized by alternating α-helical and β-strand topological packing such as the βαβ fold. They are generally divided into two domains (e.g., dehydrogenases, kinases, rhodanes). The 16 α/β proteins in Table IV are 35% α helical and 23% β sheet. The α/β proteins have the longest average chain length, $\langle N_{\alpha/\beta} \rangle = 271$, of the four protein classes (Table X). It is interesting to note that although the percentage helical and β-sheet content for α/β and α + β proteins are almost identical, there are more α and β regions in the α/β proteins than in the α + β proteins, since the average sequence length for the former (271 residues) is double that for the latter class (135 residues). There is a greater occurrence of Val, Ala, and Leu (25%) in α/β proteins, so that their hydrophobic residue content (41%) is higher than that in the α + β proteins (38%). This may account for the lower Cys content (1.5%) in α/β proteins than in α + β proteins (3.9% Cys). Another factor may be that S-S bridges stabilize the smaller α + β proteins, whereas the larger α/β proteins are stabilized by interdomain associations, thus requiring fewer Cys residues.

5. *Residue Compositions in the Four Protein Classes*

The amino acid compositions of the four protein classes (bottom rows of Tables V–VIII) are summarized in Table IX, with their percentages arranged in hierarchic order. There is a greater distribution of percentage residue composition in the α and β proteins than in the mixed classes α + β and α/β. The differences between the highest and lowest percentage compositions are: β proteins (12.3% Ser − 0.6% Met = 11.7%), α proteins (11.6% Ala − 0.9% Cys = 10.7%), α + β proteins (9.3% Ala − 1.3% Met = 8.0%), and α/β proteins (8.7% Val − 1.5% Cys = 7.2%). This clearly indicates that there is a greater residue preference in pure α and β proteins than in mixed structural classes. When the three highest-occurring amino acids in the α class are summed for both α and β proteins, their averages approach that of the mixed classes as expected. Thus, the total Ala content in α and β proteins (285 + 203 = 488) yields a percentage of 9.3% (488/5223), which compares favorably with the 9.3% Ala in α + β and 8.3% Ala in α/β proteins. Likewise, the 7.0% Lys content (364/5223) and 7.6% Leu content (398/5223) in α and β proteins are in the range of 5.9% Lys (α + β), 7.4% Lys (α/β), 5.8% Leu (α + β), and 7.8% Leu (α/β). A similar analysis for the three highest-occurring residues in β proteins shows that the 8.9% Ser (463/5223) in α and β proteins is still higher than the 6.7% Ser (α + β) and 7.5% Ser (α/β). Higher percentages are also found in the 9.5% Gly (494/5223) and 7.1% Thr (372/5223) content in α and β proteins when compared with the mixed classes: 9.1% Gly (α + β), 8.7% Gly (α/β), 6.2% Thr (α + β), and 5.5% Thr (α/β).

Another apparent distinction between structural classes is that total charged residues (Asp, Glu, Arg, His, Lys) occur with greater than average frequency in the α proteins and less than average in the β proteins. There are 29.0% charged residues in α proteins, 15.8% in β proteins, 22.2% in α + β proteins, and 24.8% in α/β proteins. The greater abundance of charged residues in α proteins is not surprising since they play an important role at both terminals of helical boundaries (Chou and Fasman, 1974a,b). Likewise, each of the five charged residues occurs with less than 5% frequency in the β proteins since they are unfavorable to β-sheet formation (Chou and Fasman, 1974a,b).

It is interesting to find Ser and Gly as the most abundant residues in β proteins with 12.3% and 10.7%, respectively, even though they were classified as β-sheet breakers, both having $P_\beta = 0.75$ (Chou and Fasman, 1978a,b). However, these two small amino acids have strong bend conformational potential, with $(P_t)_{Gly} = 1.54$ and $(P_t)_{Ser} = 1.35$ (Chou and Fasman, 1977), and are therefore utilized at the corners of antiparallel β sheets. Of the 845 middle-positioned (i.e., $i + 1$, $i + 2$) β-turn residues found in 29 proteins, 118 Gly (14%) and 92 Ser residues (11%) occurred with the greatest frequency (Chou and Fasman, 1977). Because the middle bend residues are not counted in antiparallel β-sheet regions, most of the Ser and Gly residues found between β strands were instead assigned to the coil or turn conformation (Chou and Fasman, 1978a). The predominance of Gly (44.6%), Ala (29.4%), and Ser (12.2%) in the fibrous β-sheet protein silk fibroin was noted earlier (Dickerson and Geis, 1969), so it is not surprising that Ser and Gly also have the greatest occurrence in globular proteins that are mostly β structured.

Comparing the residue percentages in the α and β proteins in Table IX, one can postulate the following guidelines in distinguishing these two protein classes on the basis of their amino acid composition. The α proteins have a greater abundance of the following residues than β proteins: Lys (10.2% versus 4.1%), Asp (6.7% versus 4.4%), Glu (5.5% versus 3.1%), His (4.5% versus 1.8%), Ala (11.6% versus 7.3%), and Leu (9.0% versus 6.4%). The β proteins have more of the following residues than α proteins: Ser (12.3% versus 5.0%), Thr (9.1% versus 4.9%), Gly (10.7% versus 8.1%), and Cys (2.7% versus 0.9%).

When the α + β proteins and the α/β proteins are compared with respect to their amino acid frequencies, the differences are not as sharp as for the α and β proteins. This is to be expected, since both α + β and α/β proteins contain almost identical amounts of helices and β sheets (Table X). The only difference between these two classes is the topological packing of their α and β regions (Levitt and Chothia, 1976). The compositions of 15 out of 20 amino acids between these two classes are within 2%, thus making it difficult to predict α + β proteins apart from α/β proteins. Nevertheless, the following differences may be helpful in distinguishing them. The α + β proteins have greater amounts of the following amino acids than α/β proteins: Asn (6.4% versus 4.2%), Tyr (5.7% versus 3.0%), and Cys (3.9% versus 1.5%). The α/β proteins have greater amounts of the following residues than α + β proteins: Val (8.7% versus 6.5%), Leu (7.8% versus 5.8%), and Lys + Glu (13.3% versus 10.5%).

Perhaps the best way of predicting a protein to be in the α/β class is its size. If an arbitrary cutoff sequence length of 248 is used, only four of the 16 α/β proteins (Table IV) will be assigned incorrectly (adenylate kinase, dihydrofolate reductase, flavodoxin, and thioredoxin), resulting in 75% accuracy. More impressive is the fact that only two out of 48 proteins in the other classes (Tables I–III) will be erroneously identified as α/β based on this length criterion alone (penicillopepsin and thermolysin, with 323 and 316 residues, respectively). These two proteins may next be examined according to their amino acid composition. For thermolysin, we find 6.0% Asn, 8.9% Tyr, 0% Cys, 7.0% Val, 5.1% Leu, and 6.0% Glu + Lys. With the exception of its Cys content, which is closer to α/β (1.5%) than α + β (3.9%), the other percentage compositions are closer to the α + β class, which is the correct assignment for thermolysin. In the case of penicillopepsin, we find 14.6% Ser, 12.4% Gly, 9.0% Thr, and 7.7% Gln, which are indicative of its β-class assignment, since these percentages are too high for typical α/β proteins. Furthermore, penicillopepsin has only 2.8% Glu + Lys, which is more like the β proteins than α/β proteins, which have 13.3% Glu + Lys.

Hence, it is possible to assign proteins according to their structural class based on their sequence length and amino acid compositions. This information will be useful in assigning a protein of unknown x-ray structure to a given class on the basis of its residue composition. Then the conformational parameters derived from known proteins of that class will be utilized in predicting the unknown protein conformation based on its sequence.

B. Protein Conformational Parameters

1. Helical and β-Sheet Parameters of 64 Proteins

The number of α-helical and β-sheet residues elucidated in each of the four different protein classes is given in Table X. The overall secondary structural content in the 64 proteins (11,445 residues) is 35% helix and 25% β sheet. These percentages are similar to the α and β content found in α + β (34% α, 21% β) as well as α/β proteins (35% α, 23% β). An earlier analysis based on 29 proteins (4741 residues) showed an average of 38% helicity and 20% β (Chou and Fasman, 1978a), whereas Levitt (1978) found 31% α and 28% β in 66 proteins (effective residues 5507). It should be noted that of the 66 proteins analyzed by Levitt (1978), 17 were redundant in the sense that some proteins were independently solved halves of a dimer, and some proteins were the same but solved by different laboratories. This reduces the sampling of Levitt to 49 different proteins with 8084 residues before homologous proteins were deleted from the data pool. The present analysis contains 17 proteins that were not included in the surveys of Levitt (1978) and Levitt and Greer (1977). A preliminary attempt was made to delete homologous proteins from our 64-protein data set. When 17 α proteins with 1513 α residues out of 2200 total residues (69% α) were corrected for sequence repeats, the remaining 1181 α residues out of 1724 residues (69% α) gave almost identical P_α values within ±0.03 except for Asn (0.86 → 0.81), His (0.90 → 0.99), and Pro (0.79 → 0.74). In this treatment, the conformation of horse cytochrome *c* was analyzed at 18 positions where its sequence differed from the tuna sequence. Likewise, 73 residues of hemerythrin differed from myohemerythrin, 18 and 25 residues of α and β chains of horse hemoglobin differed from human hemoglobin, and 39 residues of γ-chain human fetal hemoglobin differed from β-chain human adult hemoglobin. Because of the small deviation of P_α values from the two treatments, the few homologous proteins were not deleted for easier statistical analysis.

Table X shows the numbers of helical and β-sheet residues along with their respective percentages in the four different protein classes. Although the α-helical class contains the largest protein sampling (19), their 2447 total residues are smaller than the 2776 residues for 15 β proteins and the 4333 residues for 16 α/β proteins. The smallest sampling is the α + β class, having 14 proteins containing 1889 residues. It is worthy of mention that most of the earlier proteins determined to high x-ray resolution were α + β proteins (e.g., lysozyme, ribonuclease, papain, staphylococcal nuclease, insulin, and trypsin inhibitor). Hence, α + β proteins occurred more than the other structural classes in the Chou–Fasman (1974a,b) analysis of 15 proteins (α : β : α + β : α/β = 5 : 2 : 6 : 2) and their later analysis (Chou and Fasman, 1978a) of 29 proteins (α : β : α + β : α/β = 8 : 4 : 10 : 6). Levitt's (1978) survey of 49 different proteins is better balanced by protein classes (α : β : α + β : α/β = 15 : 12 : 10 : 12), as is the case with Chou's (1979) analysis of 64 proteins (α : β : α + β : α/β = 19 : 15 : 14 : 16). Since calculations of protein conformational parameters depend on the number as well as types of proteins analyzed in the data set, it is important that the sampling be taken from a wide variety of proteins and that the statistical data be large enough to be meaningful.

The frequency of amino acids in all 64 proteins ranged from Gly (n_j = 1042, $\%_j$ = 9.1%) to Trp (n_j = 179, $\%_j$ = 1.6%); for helical residues, the range is from Ala ($n_{\alpha,j}$ = 496, $\%_{\alpha,j}$ = 12.5%) to Trp ($n_{\alpha,j}$ = 64, $\%_{\alpha,j}$ = 1.6%); and for β-sheet residues, the range is from Val ($n_{\beta,j}$ = 365, $\%_{\beta,j}$ = 12.8%) to Met ($n_{\beta,j}$ = 45, $\%_{\beta,j}$ = 1.6%). Corresponding statistics based on 29 proteins (Chou and Fasman, 1978a) are: 434 Ala (9.2%) to 73 Met (1.5%) for all residues: 234 α Ala (13.0% α) to 25 α Cys (1.4% α) for α-helical residues; and 119 β Val (12.8% β) to 15 β Met (1.6% β) for β residues. The 49-protein data set of Levitt (1978) gave the following ranges: 519 Gly (9.4%) to 84 Met or 84 Trp (1.5%) for all residues; 186 α Ala (10.8%) to 26 Trp (1.5%) for α residues; 185 Val (11.9% β) to 23 Met (1.5%) for β residues.

Table XI. Conformational Parameters for α-Helical and β-Sheet Residues on 64 Proteins[a]

α Residues[b]	P_α		β Residues[c]	P_β	
Glu(−)	1.44 ± 0.06	H_α	Val	1.64 ± 0.07	H_β
Ala	1.39 ± 0.05		Ile	1.57 ± 0.08	
Met	1.32 ± 0.11		Thr	1.33 ± 0.07	h_β
Leu	1.30 ± 0.05		Tyr	1.31 ± 0.09	
Lys(+)	1.21 ± 0.05	h_α	Trp	1.24 ± 0.14	
His(+)	1.12 ± 0.08		Phe	1.23 ± 0.09	
Gln	1.12 ± 0.07		Leu	1.17 ± 0.06	
Phe	1.11 ± 0.07		Cys	1.07 ± 0.12	
Asp(−)	1.06 ± 0.06		Met	1.01 ± 0.13	I_β
Trp	1.03 ± 0.10	I_α	Gln	1.00 ± 0.09	
Arg(+)	1.00 ± 0.07		Ser	0.94 ± 0.06	i_β
Ile	0.99 ± 0.06	i_α	Arg(+)	0.94 ± 0.09	
Val	0.97 ± 0.05		Gly	0.87 ± 0.05	
Cys	0.95 ± 0.09		His(+)	0.83 ± 0.09	
Thr	0.78 ± 0.05		Ala	0.79 ± 0.05	
Asn	0.78 ± 0.06		Lys(+)	0.73 ± 0.06	b_β
Tyr	0.73 ± 0.06	b_α	Asp(−)	0.66 ± 0.06	
Ser	0.72 ± 0.04		Asn	0.66 ± 0.06	
Gly	0.63 ± 0.04	B_α	Pro	0.62 ± 0.07	B_β
Pro	0.55 ± 0.05		Glu(−)	0.51 ± 0.06	

[a]Chou (1979), updated from Chou and Fasman (1974b, 1978a,b).
[b]Helical assignments: H_α, strong α former; h_α, α former; I_α, weak α former; i_α, α indifferent; b_α, α breaker; B_α, strong α breaker.
[c]β-Sheet assignments: H_β, strong β former; h_β, β former; I_β, weak β former; i_β, β indifferent; b_β, β breaker; B_β, strong β breaker.

Hence, identical residues are seen to represent the extreme ranges in the Levitt sampling (1978) as well as the updated analysis by Chou (1979).

The conformational parameters P_α and P_β for the 20 amino acids are arranged in their hierarchic order in Table XI based on an analysis of 64 proteins (Tables I–VIII). The standard error σ given represents a 68.27% confidence level in the P_α and P_β values and was calculated in the same manner as σ_{P_t} (Chou and Fasman, 1977). For a 95.45% confidence level, $P_\alpha \pm 2\sigma_{P_\alpha}$ and $P_\beta \pm 2\sigma_{P_\beta}$ may be obtained by doubling the number after the ± sign in Table XI. As can be seen, σ is approximately 10% of the P_α or P_β value for the less frequently occurring amino acids (e.g., Cys, Met, Trp) and decreases to approximately 4% for the more abundant residues.

A comparison of the conformational assignments in Table XI with the P_α and P_β values based on 29 proteins (Chou and Fasman, 1978a) shows the following changes: Trp, Ile, and Val, which were α formers with P_α = 1.08, 1.08, and 1.06, respectively, are now weakly helical or helix indifferent with P_α = 1.03, 0.99, and 0.97. The largest change in P_α is 30% for Cys (0.70 → 0.95), which also has a decrease in P_β = 1.19 → 1.07. Aspartate and His, which were considered weak helix formers with P_α close to unity, are now classified as helix formers with P_α = 1.06 and 1.12, respectively. Other changes in assignments include: Ser [P_α = 0.77 (i_α) → 0.72 (b_α); P_β = 0.75 (b_β) → 0.94 (i_β)], Asn [P_α = 0.67 (b_α) → 0.78 (i_α); P_β = 0.89 (i_β) → 0.66 (i_β)], Gln [P_β = 1.10 (h_β) → 1.00 (I_β)], Gly [P_β = 0.75 (b_β) → 0.87 (i_β)].

The similarity of P_α values calculated from proteins to the Zimm–Bragg helix growth parameters *s* evaluated from poly(α-amino acids) was mentioned earlier (Chou and Fasman,

1974a; Suzuki and Robson, 1976). We note that the new assignment for Val as helix indifferent with $P_\alpha = 0.97 \pm 0.05$ agrees with the experimental data on the helix stability constants of Val in random copolymer studies in water with $s = 0.93$ and 0.97 at 20°C and 30°C, respectively (Alter *et al.*, 1973). An updated investigation for Val gave $s = 0.95, 0.96, 0.97$ at 20°C, 25°C, 30°C, respectively (Chang *et al.*, 1981). More recently determined experimental s values at 20°C based on the guest–host copolymer technique show close similarity to the P_α values derived from proteins: Asn with $s = 0.78$ (Matheson *et al.*, 1977) and $P_\alpha = 0.78 \pm 0.06$, Ser with $s = 0.76$ (Van Nispen *et al.*, 1977) and $P_\alpha = 0.72 \pm 0.04$, Met with $s = 1.20$ (Hill *et al.*, 1977) and $P_\alpha = 1.32 \pm 0.11$, Arg with $s = 1.03$ (Konishi *et al.*, 1977) and $P_\alpha = 1.00 \pm 0.07$, Thr with $s = 0.82$ (Hecht *et al.*, 1978) and $P_\alpha = 0.78 \pm 0.05$, and Trp with $s = 1.11$ (Nagy *et al.*, 1980) and $P_\alpha = 1.03 \pm 0.10$. The average difference between the s values and P_α parameters of the seven amino acids is 4.6%.

It is interesting to note that all five charged residues have values of $P_\alpha > 1.00$ and values of $P_\beta < 1.00$, thus indicating their preferences for helices instead of β sheets. The conformational parameters derived in Table XI are in general agreement with the Levitt (1978) data set. However, several differences should be noted: Cys with $P_\alpha = 1.11$ was assigned as a helix former with 56% confidence (Levitt, 1978) based on 121 Cys residues of which 42 were helical. Cysteine is listed here as helix indifferent with $P_\alpha = 0.95 \pm 0.09$ (Chou, 1979) based on 234 Cys residues of which 77 were helical. Earlier analysis based on 29 proteins showed Cys as helix indifferent, $P_\alpha = 0.70$, based on 94 Cys residues of which 25 were helical (Chou and Fasman, 1978a). Although the helix stability constant s for Cys has not been determined experimentally by the guest–host technique (Sueki *et al.*, 1984), an extrapolated value of $s = 0.92$ has been obtained recently (Kiders *et al.*, 1985), which is in close agreement with the $P_\alpha = 0.95$ for Cys based on 64 proteins (Chou, 1979b). Levitt assigned Phe ($P_\alpha = 1.07$) and Asp ($P_\alpha = 1.04$) as helix indifferent, but these residues are given helical-forming propensities here. Threonine ($P_\alpha = 0.82$) was assigned as helix breaking by Levitt (1978) instead of helix indifferent in Table XI. It should be noted that Levitt's P_α numerical values in these three residues are within 5% of the values given in Table XI despite the differences in conformational assignments.

Levitt's assignments of Cys ($P_\beta = 0.74$) and Gln ($P_\beta = 0.80$) as β breakers and Leu as β indifferent ($P_\beta = 1.02$) are in contrast with the β-forming ability of Leu ($P_\beta = 1.17 \pm 0.06$) and Cys ($P_\beta = 1.07 \pm 0.12$) and the β neutrality of Gln ($P_\beta = 1.00 \pm 0.09$) shown in Table XI. This explains the poorer β-sheet predictions observed by Argos *et al.* (1978) using the Levitt parameters and better results with the Chou–Fasman parameters, since Cys, Gln, and Leu were assigned as β formers based on 15 as well as 29 proteins (Chou and Fasman, 1974b, 1978a). We note that residues 26–30 (Thr-Thr-Leu-Asn-Leu), 100–105 (Gln-Ala-Val-Gln-Ala-Ala), 121–125 (Leu-Leu-Gly-Leu-Ala), and 216–222 (Thr-Thr-Leu-Leu-Leu-Leu-Asn) in penicillopepsin (Hsu *et al.*, 1977); residues 27–31 (Gly-Leu-Arg-Val-Leu) and 177–181 (Gln-Leu-Val-Asp-Ser) in rhodanese (Ploegman *et al.*, 1978a,b); and residues 108–112 (Gln-Lys-Leu-Tyr-Leu) in dihydrofolate reductase (Matthews *et al.*, 1977) are correctly predicted as β sheet by the Chou and Fasman (CAF) parameters (1974b, 1978a) but missed by the Levitt (1978) (LEV) assignments. These differences account for the higher correlation coefficients for β predictions using the CAF parameters ($C_\beta = 0.29$) than the LEV parameters ($C_\beta = 0.10$) on four proteins not included in either the CAF or LEV data bases (Argos *et al.*, 1978). Several of the β regions correctly predicted by Chou and Fasman (1974b) were missed by Levitt's assignment of Cys as a β breaker. These regions include 39–43 (Phe-His-Phe-Cys-Gly), 118–122 (Val-Ser-Ala-Val-Cys), and 179–183 (Ala-Met-Ile-Cys-Gly) in α-chymotrypsin (Birktoft and Blow, 1970); residues 26–30 (Trp-Ala-His-Thr-Cys), 172–174 (Met-Val-Cys-Ala-Gly), and 192–196 (Leu-His-Cys-Leu-Val) in elastase (Sawyer *et al.*, 1978); and residues 2–7 (Val-Asn-Gln-His-Leu-Cys) in the β chain of insulin (Blundell *et al.*, 1972). Furthermore, experimental studies showed the lack of helix formation in poly-S-methylcysteine (Fasman and Potter, 1967) and predominantly β structures in S-methylcysteine oligopeptides (Toniolo *et al.*, 1975). Perhaps Levitt's set II assignments derived without weights for Cys, $P_\beta = 1.04 > P_\alpha = 0.92$, are more accurate than the set I data determined with weights for Cys, $P_\alpha = 1.11 > P_\beta = 0.74$ (Levitt, 1978). The former parameters showing Cys as more of a β former than α former is in closer agreement with experimental results in proteins and polypeptides, where Cys is found to prefer the β structure.

2. α-Helical Parameters in Four Classes of Proteins

The helical conformational parameters for the four different protein classes (Tables I–VIII) are given in Table XII as $(P_\alpha)_\alpha$ for α proteins, $(P_\alpha)_\beta$ for β proteins, $(P_\alpha)_{\alpha+\beta}$ for α + β proteins, and $(P_\alpha)_{\alpha/\beta}$ for α/β proteins. The helical residues in β proteins are quite small in number (5%), and so the derived $(P_\alpha)_\beta$ parameters are not statistically significant. They are nevertheless listed in the last column of Table XII for sake of reference. As can be seen, their standard errors, σ, are much larger than those found in the three other classes. When the P_α values for 64 proteins (Table XI) are compared to the $(P_\alpha)_\alpha$ values based on 19 helical proteins, it is striking that the range of P_α from Glu to Pro has decreased about one-half from 0.89 to 0.45. This may be attributed to the 69.3% helicity for the α proteins, which is double that for all 64 proteins (%α = 34.8%). That is, for predominantly α proteins, most of the residues are utilized in forming helices so that there is not a great discrepancy between the $(P_\alpha)_\alpha$ values among the 20 amino acids.

Decreases in helical potential are noted for Met [$P_\alpha = 1.32 \rightarrow (P_\alpha)_\alpha = 1.02$], Lys [$P_\alpha = 1.21 \rightarrow (P_\alpha)_\alpha = 0.94$], and Phe [$P_\alpha = 1.11 \rightarrow (P_\alpha)_\alpha = 0.84$] while increases in α potential are observed for Val [$P_\alpha = 0.95 \rightarrow (P_\alpha)_\alpha = 1.19$], Ser [$P_\alpha = 0.72 \rightarrow (P_\alpha)_\alpha = 1.01$], and Thr [$P_\alpha = 0.77 \rightarrow (P_\alpha)_\alpha = 0.97$]. The above trends may be rationalized in terms of the bulkiness of the side chains, since α proteins with their 70% helicity are probably more densely packed than the other classes. Hence, the more bulky Phe, Met, and Lys residues (i.e., the fourth, sixth, and eighth largest amino acids according to molecular weight) are less favored in a helical environment in all α proteins. For proteins with mixed α and β regions we would expect their P_α values to increase, since their bulky side chains will not perturb adjacent helices as often as in α proteins. This is indeed the case, as we find greater than unity α potentials for Phe [$(P_\alpha)_{\alpha+\beta} = 1.47$, $(P_\alpha)_{\alpha/\beta} = 1.10$], Met [$(P_\alpha)_{\alpha+\beta} = 1.41$, $(P_\alpha)_{\alpha/\beta} = 1.18$], and Lys [$(P_\alpha)_{\alpha+\beta} = 1.02$, $(P_\alpha)_{\alpha/\beta} = 1.16$]. In contrast, the smaller residues Ser, Val, and Thr (i.e., the third, fifth, and sixth smallest amino acids) should be more favored in helices in α proteins, thus accounting for their increases in P_α values. We note that these three residues have lower α potentials in non-α proteins: Ser [$(P_\alpha)_{\alpha+\beta} = 0.75$, $(P_\alpha)_{\alpha/\beta} = 0.84$], Val [$(P_\alpha)_{\alpha+\beta} = 1.05$, $(P_\alpha)_{\alpha/\beta} = 0.83$], and Thr [$(P_\alpha)_{\alpha+\beta} = 0.73$, $(P_\alpha)_{\alpha/\beta} = 0.89$].

Although Cys appears as helix indifferent in α proteins with $(P_\alpha)_\alpha = 0.92$, it assumes helical character in α + β and α/β proteins with values of $(P_\alpha)_{\alpha+\beta} = 1.27$ and $(P_\alpha)_{\alpha/\beta} = 1.29$, respectively. Hence, the earlier statement that Cys is more favored in β regions than helices should be qualified for different protein classes. Tables XII and XIII show that Cys has higher α than β potential in α + β proteins [$(P_\alpha)_{\alpha/\beta} = 1.29 > (P_\beta)_{\alpha/\beta} = 0.88$] but possesses a stronger β than α potential in β proteins [$(P_\beta)_\beta = 1.00 > (P_\alpha)_\beta = 0.75$]. It is likely that disulfide linkages in the Cys residues are primarily responsible for these trends observed in the different protein classes. Thus, the interchain hydrogen bonding of β sheets would favor Cys adopting the β conformation during S-S formation. Of the 75 Cys residues in β proteins, 38 are found in β regions, and only three in α helices. In this respect, Cys is helix breaking as its $(P_\alpha)_\beta = 0.75$ is near the bottom of the α-potential list for β proteins (Table XII). It is interesting to note that Phe is helix breaking in all α and all β proteins with $(P_\alpha)_\alpha = 0.84$ and (

Table XII. α-Helical Conformational Parameters for Amino Acids in Four Different Classes of Proteins[a]

	$(P_\alpha)_\alpha$		$(P_\alpha)_{\alpha+\beta}$		$(P_\alpha)_{\alpha/\beta}$		$(P_\alpha)_\beta$[b]
Glu(−)	1.21 ± 0.05	Glu(−)	1.57 ± 0.16	Gln	1.38 ± 0.14	Glu(−)	2.18 ± 0.65
Val	1.19 ± 0.04	Phe	1.47 ± 0.20	Ala	1.35 ± 0.08	Trp	2.13 ± 0.90
Ala	1.15 ± 0.03	Ala	1.46 ± 0.11	Leu	1.29 ± 0.08	Asp(−)	1.56 ± 0.48
Leu	1.12 ± 0.04	Met	1.41 ± 0.29	Cys	1.29 ± 0.18	Tyr	1.53 ± 0.49
Gln	1.08 ± 0.08	Gln	1.39 ± 0.17	Glu(−)	1.28 ± 0.09	Asn	1.48 ± 0.43
Ile	1.06 ± 0.07	Leu	1.39 ± 0.14	Met	1.18 ± 0.15	Gln	1.22 ± 0.42
Arg(+)	1.03 ± 0.09	Cys	1.27 ± 0.17	Lys(+)	1.16 ± 0.08	Ala	1.20 ± 0.32
Met	1.02 ± 0.10	Trp	1.18 ± 0.26	Phe	1.10 ± 0.11	Lys(+)	1.14 ± 0.42
Trp	1.01 ± 0.12	Ile	1.17 ± 0.15	Arg(+)	1.05 ± 0.12	Thr	0.96 ± 0.26
Ser	1.01 ± 0.06	Arg(+)	1.09 ± 0.16	Trp	0.96 ± 0.16	Ile	0.94 ± 0.38
Thr	0.99 ± 0.06	Asp(−)	1.09 ± 0.14	Ile	0.94 ± 0.09	Val	0.90 ± 0.27
Lys(+)	0.94 ± 0.04	Val	1.05 ± 0.13	Asp(−)	0.94 ± 0.09	Ser	0.88 ± 0.22
His(+)	0.94 ± 0.07	Lys(+)	1.02 ± 0.13	His(+)	0.89 ± 0.13	Leu	0.85 ± 0.29
Asp(−)	0.92 ± 0.05	His(+)	0.89 ± 0.24	Thr	0.89 ± 0.09	His(+)	0.77 ± 0.53
Cys	0.92 ± 0.15	Ser	0.75 ± 0.11	Tyr	0.87 ± 0.12	Cys	0.75 ± 0.43
Asn	0.91 ± 0.07	Thr	0.73 ± 0.12	Ser	0.84 ± 0.07	Pro	0.74 ± 0.32
Phe	0.84 ± 0.06	Asn	0.69 ± 0.11	Val	0.83 ± 0.07	Phe	0.66 ± 0.38
Tyr	0.82 ± 0.09	Tyr	0.54 ± 0.11	Asn	0.72 ± 0.09	Gly	0.51 ± 0.18
Gly	0.76 ± 0.05	Gly	0.40 ± 0.08	Gly	0.71 ± 0.06	Arg(+)	0.00 ± 0.00
Pro	0.76 ± 0.08	Pro	0.33 ± 0.11	Pro	0.52 ± 0.03	Met	0.00 ± 0.00

[a]The 64 proteins of known x-ray structure were divided into four distinct classes consisting of 19 α, 15 β, 14 α + β, and 16 α/β proteins as listed in Tables V–VIII, from which these P_α values were calculated. The standard error, σ, after the ± sign represents a 68% confidence level in the P_α values.

[b]Because of the small amount of α residues in β proteins (5.3%), the $(P_\alpha)_\beta$ values are statistically insignificant and are listed here for reference only.

Table XIII. β-Sheet Conformational Parameters for Amino Acids in Four Different Classes of Proteins[a]

	$(P_\beta)_\beta$		$(P_\beta)_{\alpha+\beta}$		$(P_\beta)_{\alpha/\beta}$		$(P_\beta)_\alpha$[b]
Phe	1.41 ± 0.10	Ile	1.62 ± 0.23	Val	1.89 ± 0.11	Cys	5.30 ± 5.15
Leu	1.41 ± 0.07	Tyr	1.62 ± 0.22	Ile	1.81 ± 0.14	Asn	3.60 ± 2.05
Met	1.35 ± 0.23	Val	1.54 ± 0.20	Phe	1.35 ± 0.16	Trp	3.53 ± 3.46
Val	1.35 ± 0.06	Phe	1.45 ± 0.30	Met	1.27 ± 0.21	Thr	2.94 ± 1.67
Ile	1.26 ± 0.09	Thr	1.25 ± 0.19	Trp	1.19 ± 0.23	Ile	2.56 ± 1.79
Trp	1.21 ± 0.14	Met	1.13 ± 0.41	Leu	1.18 ± 0.10	Met	2.43 ± 2.41
Tyr	1.18 ± 0.09	Leu	1.12 ± 0.19	His(+)	1.04 ± 0.18	His(+)	2.10 ± 1.47
Thr	1.17 ± 0.06	Trp	1.10 ± 0.37	Tyr	1.02 ± 0.16	Glu(−)	1.74 ± 1.22
Arg(+)	1.15 ± 0.12	Gln	1.08 ± 0.23	Thr	1.01 ± 0.12	Pro	1.42 ± 1.40
Lys(+)	1.11 ± 0.09	Arg(+)	1.03 ± 0.22	Cys	0.88 ± 0.22	Phe	0.95 ± 0.94
His(+)	1.08 ± 0.14	His(+)	1.00 ± 0.34	Ala	0.83 ± 0.09	Val	0.70 ± 0.70
Gln	1.04 ± 0.09	Lys(+)	0.89 ± 0.18	Gly	0.83 ± 0.09	Gly	0.59 ± 0.59
Cys	1.00 ± 0.11	Ala	0.84 ± 0.14	Asp(−)	0.77 ± 0.11	Leu	0.53 ± 0.55
Ala	0.98 ± 0.07	Asn	0.83 ± 0.16	Lys(+)	0.77 ± 0.09	Lys(+)	0.47 ± 0.47
Gly	0.83 ± 0.06	Ser	0.79 ± 0.16	Ser	0.77 ± 0.09	Ala	0.00 ± 0.00
Ser	0.77 ± 0.05	Asp(−)	0.72 ± 0.16	Asn	0.72 ± 0.12	Arg(+)	0.00 ± 0.00
Asp(−)	0.70 ± 0.09	Gly	0.69 ± 0.13	Arg(+)	0.70 ± 0.13	Asp(−)	0.00 ± 0.00
Pro	0.59 ± 0.08	Cys	0.64 ± 0.19	Pro	0.57 ± 0.11	Gln	0.00 ± 0.00
Glu(−)	0.57 ± 0.10	Glu(−)	0.55 ± 0.16	Glu(−)	0.56 ± 0.09	Ser	0.00 ± 0.00
Asn	0.47 ± 0.07	Pro	0.53 ± 0.18	Gln	0.45 ± 0.12	Tyr	0.00 ± 0.00

[a]The 64 proteins of known x-ray structure were divided into four distinct classes consisting of 19 α, 15 β, 14 α + β, and 16 α/β proteins as listed in Tables V–VIII, from which these P_β values were calculated. The standard error, σ, after the ± sign represents a 68% confidence level in the P_β values.

[b]Because of the negligible amount of β residues in α proteins (0.9%), the $(P_\beta)_\alpha$ values are statistically insignificant and are listed here for reference only.

$P_\alpha)_\beta = 0.66$, respectively, but is helix forming in proteins containing mixtures of α and β regions with $(P_\alpha)_{\alpha+\beta} = 1.47$ and $(P_\alpha)_{\alpha/\beta} = 1.10$. On the other hand, Val appears to be helix breaking in α/β proteins $[(P_\alpha)_{\alpha/\beta} = 0.83]$ but helix forming in α and α + β proteins with $(P_\alpha)_\alpha = 1.19$ and $(P_\alpha)_{\alpha+\beta} = 1.05$, respectively. Of the four protein classes, only the α + β proteins retain the same assignments for helix breakers (Asn, Tyr, Gly, Pro) as was found in 15 and 29 proteins (Chou and Fasman, 1974b, 1978a). This is not surprising since these earlier surveys contained a greater number of α + β proteins in the data base.

We have discussed earlier that although the α and β contents in α + β and α/β proteins are almost identical (Table VI), these two protein classes may be distinguished on the basis of their chain length and amino acid compositions. We may now ask whether there are differences in the conformational parameters in these two protein classes. A glance at Table XII shows that the hierarchic order in the P_α values is nearly the same for α + β and α/β proteins. Notable differences arise in the larger $(P_\alpha)_{\alpha+\beta}$ than $(P_\alpha)_{\alpha/\beta}$ values for all the hydrophobic residues: Phe $[(P_\alpha)_{\alpha+\beta} = 1.47 > (P_\alpha)_{\alpha/\beta} = 1.10]$, Met $[(P_\alpha)_{\alpha+\beta} = 1.41 > (P_\alpha)_{\alpha/\beta} = 1.18]$, Leu $[(P_\alpha)_{\alpha+\beta} = 1.39 > (P_\alpha)_{\alpha/\beta} = 1.29]$, Trp $[(P_\alpha)_{\alpha+\beta} = 1.18 > (P_\alpha)_{\alpha/\beta} = 0.96]$, Ile $[(P_\alpha)_{\alpha+\beta} = 1.17 > (P_\alpha)_{\alpha/\beta} = 0.94]$, and Val $[(P_\alpha)_{\alpha+\beta} = 1.05 > (P_\alpha)_{\alpha/\beta} = 0.83]$. Of the helix formers, only Lys showed an increase in α potential in the α/β proteins

Table XIV. Sample Computer Output for Automatic Prediction of Protein Class (α) Based on Amino Acid Composition[a]

Calcium-binding parvalbumin—carp (108 residues)
Differences in percentage composition from four protein classes

A.A.	Count	%	α	β	α + β	α/β
Ala	20	18.5	6.9*	11.2	9.2	10.2
Arg	1	0.9	1.3*	1.5	3.2	2.5
Asn	3	2.8	1.2*	2.2	3.6	1.4
Asp	14	13.0	6.3*	8.6	7.1	7.4
Cys	1	0.9	0.0*	1.8	3.0	0.6
Gln	2	1.9	0.8	2.5	2.0	0.7*
Glu	6	5.6	0.1*	2.5	1.0	0.3
Gly	8	7.4	0.7*	3.3	1.7	1.3
His	1	0.9	3.6	0.9	0.8*	1.6
Ile	5	4.6	0.9	0.3*	0.3*	0.9
Leu	9	8.3	0.7	1.9	2.5	0.5*
Lys	13	12.0	1.8*	7.9	6.1	4.6
Met	0	0.0	2.0	0.6*	1.3	2.1
Phe	10	9.3	4.3*	6.2	6.5	5.7
Pro	0	0.0	3.4*	4.6	3.8	4.3
Ser	5	4.6	0.4*	7.7	2.1	2.9
Thr	5	4.6	0.3*	4.5	1.6	0.9
Trp	0	0.0	1.3*	1.6	1.6	1.7
Tyr	0	0.0	2.6*	4.0	5.7	3.0
Val	5	4.6	2.2	3.6	1.9*	4.1
Total	108	100.0	40.8*	77.4	65.0	56.7
Correspondence	—	—	14.0*	1.5	2.5	2.0

[a]An asterisk (*) next to the figure indicates that the percentage of amino acid *i* of this protein has the least difference or the greatest similarity to the average percentage of amino acid *i* (see Table V) in the protein structural class under which the asterisk appears. The "total" and "correspondence" figures are, respectively, the sum of percentage differences and the sum of asterisks listed under each column.

Table XV. Sample Computer Output for Automatic Prediction of Protein Class (β) Based on Amino Acid Composition[a]

Penicillopepsin—*P. janthinellum* (323 residues)
Differences in percentage composition from four protein classes

A.A.	Count	%	α	β	α + β	α/β
Ala	24	7.4	4.2	0.1*	1.9	0.9
Arg	0	0.0	2.2*	2.4	4.1	3.4
Asn	18	5.6	1.6	0.6*	0.8	1.4
Asp	19	5.9	0.8	1.5	0.0*	0.3
Cys	2	0.6	0.3*	2.1	3.3	0.9
Gln	25	7.7	5.0	3.3*	3.8	5.1
Glu	4	1.2	4.3	1.9*	3.4	4.7
Gly	40	12.4	4.3	1.7*	3.3	3.7
His	3	0.9	3.6	0.9	0.8*	1.6
Ile	13	4.0	0.3*	0.3*	0.9	1.5
Leu	21	6.5	2.5	0.1*	0.7	1.3
Lys	5	1.5	8.7	2.6*	4.4	5.9
Met	0	0.0	2.0	0.6*	1.3	2.1
Phe	20	6.2	1.2*	3.1	3.4	2.6
Pro	12	3.7	0.3	0.9	0.1*	0.6
Ser	47	14.6	9.6	2.3*	7.9	7.1
Thr	29	9.0	4.1	0.1*	2.8	3.5
Trp	3	0.9	0.4*	0.7	0.7	0.8
Tyr	14	4.3	1.7	0.3*	1.4	1.3
Val	24	7.4	0.6*	0.8	0.9	1.3
Total	323	100.0	57.7	26.3*	45.9	50.0
Correspondence	—	—	5.5	11.5*	3.0	0.0

[a]See footnote to Table XIV.

$[(P_\alpha)_{\alpha/\beta} = 1.16 > (P_\alpha)_{\alpha+\beta} = 1.02]$, while substantial increases in P_α values are noted for all the helix breakers: Pro $[(P_\alpha)_{\alpha/\beta} = 0.52 > (P_\alpha)_{\alpha+\beta} = 0.33]$, Gly $[(P_\alpha)_{\alpha/\beta} = 0.71 > (P_\alpha)_{\alpha+\beta} = 0.40]$, and Tyr $[(P_\alpha)_{\alpha/\beta} = 0.87 > (P_\alpha)_{\alpha+\beta} = 0.54]$. These differences in P_α values between the α + β and α/β classes may lead to more accurate helical predictions in these mixed protein classes.

3. β-Sheet Parameters in Four Classes of Proteins

In examining the β potentials on the 64 proteins (Table XI) and the data set based on 15 β proteins (Table XIII), we find only a few changes in P_β values. Notable increases are seen in Leu $[P_\beta = 1.17 \rightarrow (P_\beta)_\beta = 1.41]$, Met $[P_\beta = 1.01 \rightarrow (P_\beta)_\beta = 1.35]$, Arg $[P_\beta = 0.94 \rightarrow (P_\beta)_\beta = 1.16]$, His $[P_\beta = 0.83 \rightarrow (P_\beta)_\beta = 1.08]$, and Lys $[P_\beta = 0.73 \rightarrow (P_\beta)_\beta = 1.11]$, while decreases in P_β are observed for the strongest β formers (i.e., Val, Ile, Thr, and Tyr) in all β proteins. It may be noted that the values of $P_\beta = 1.64$ and 1.57 for the strongest β formers Val and Ile are higher than the $P_\alpha = 1.44$ and 1.39 values for the strongest α formers Glu and Ala (Table XII). Similar high P_β values are observed in the α + β and α/β protein classes, where $(P_\beta)_{\alpha+\beta} = 1.62$ for both Ile and Tyr and $(P_\beta)_{\alpha/\beta} = 1.89$ and 1.81 for Val and Ile (Table XIII). These higher P_β values for the strong β formers were observed in the analysis of 15 and 29 proteins (Chou and Fasman, 1974b, 1978a) and are a major factor in the overpredictions of β regions in proteins using the Chou–Fasman predictive methods.

Table XVI. Sample Computer Output for Automatic Prediction of Protein Class (α + β) Based on Amino Acid Composition[a]

Ribonuclease S—bovine (124 residues)
Differences in percentage composition from four protein classes

A.A.	Count	%	α	β	α + β	α/β
Ala	12	9.7	1.9	2.4	0.4*	1.4
Arg	4	3.2	1.0	0.8	0.9	0.2*
Asn	10	8.1	4.1	3.1	1.7*	3.9
Asp	5	4.0	2.7	0.4*	1.9	1.6
Cys	8	6.5	5.6	3.8	2.6*	5.0
Gln	7	5.6	2.9	1.2*	1.7	3.0
Glu	5	4.0	1.5	0.9	0.6*	1.9
Gly	3	2.4	5.7*	8.3	6.7	6.3
His	4	3.2	1.3	1.4	1.5	0.7*
Ile	3	2.4	1.3*	1.9	2.5	3.1
Leu	2	1.6	7.4	4.8	4.2*	6.2
Lys	10	8.1	2.1	4.0	2.2	0.7*
Met	4	3.2	1.2	2.6	1.9	1.1*
Phe	3	2.4	2.6	0.7	0.4*	1.2
Pro	4	3.2	0.2*	1.4	0.6	1.1
Ser	15	12.1	7.1	0.2*	5.4	4.6
Thr	10	8.1	3.2	1.0*	1.9	2.6
Trp	0	0.0	1.3*	1.6	1.6	1.7
Tyr	6	4.8	2.2	0.8*	0.9	1.8
Val	9	7.3	0.5*	0.9	0.8	1.4
Total	124	100.0	55.8	42.2	40.4*	49.5
Correspondence	—	—	5.0	5.0	6.0*	4.0

[a]See footnote to Table XIV.

Some normalizing factor may be used in bringing the two sets of P_α and P_β values within the same order of magnitude so that overpredictions may be minimized. The $(P_\beta)_\beta$ values in Table XIII may fit this description, since the strongest β formers Phe and Leu with $(P_\beta)_\beta = 1.41$ are similar to the $P_\alpha = 1.44$ and 1.39 values for the strongest α formers (Table XII). Trials along these lines, when incorporated into the Chou–Fasman algorithm, may lead to better predictions of secondary structures. It should be noted that the standard error, σ, for $(P_\beta)_\beta$ values is lower for the β proteins than the other protein classes since approximately half of all residues (1411 out of 2847) in the 64 proteins analyzed were found in the all-β class (Table X). On the other hand, the minuscule amount of β residues (0.9%) in α proteins (21 out of 2447 residues) makes the $(P_\beta)_\beta$ values listed in Table XIII statistically insignificant. Because of these large standard errors, attempts were not made to correlate the $(P_\beta)_\alpha$ values to the other conformational parameters derived in Table XIII (Chou, 1979).

In comparing the β potentials of β proteins to those in the α + β and α/β classes, we note the following P_β decreases: Leu $[(P_\beta)_\beta = 1.41 > (P_\beta)_{\alpha/\beta} = 1.18 > (P_\beta)_{\alpha+\beta} = 1.12]$, Arg $[(P_\beta)_\beta = 1.15 > (P_\beta)_{\alpha+\beta} = 1.03 > (P_\beta)_{\alpha/\beta} = 0.70]$, Lys $[(P_\beta)_\beta = 1.15 > (P_\beta)_{\alpha+\beta} = 0.89 > (P_\beta)_{\alpha/\beta} = 0.77]$, and Cys $[(P_\beta)_\beta = 1.00 > (P_\beta)_{\alpha/\beta} = 0.88 > (P_\beta)_{\alpha+\beta} = 0.64]$. Increases in P_β values were observed between protein classes in the following residues: Asn $[(P_\beta)_\beta = 0.47 < (P_\beta)_{\alpha/\beta} = 0.72 < (P_\beta)_{\alpha+\beta} = 0.83]$, Tyr $[(P_\beta)_{\alpha/\beta} = 1.02 < (P_\beta)_\beta = 1.18 < (P_\beta)_{\alpha+\beta} = 1.62]$, and Ile $[(P_\beta)_\beta = 1.26 < (P_\beta)_{\alpha+\beta} = 1.54 < (P_\beta)_{\alpha/\beta} = 1.89]$. In addition to the P_β differences noted above between α + β and α/β proteins, we find a decrease in β potential for

Table XVII. Sample Computer Output for Automatic Prediction of Protein Class (α/β) Based on Amino Acid Composition[a]

Triose phosphate isomerase—chicken (248 residues)
Differences in percentage composition from four protein classes

A.A.	Count	%	α	β	α + β	α/β
Ala	28	11.3	0.3*	4.0	2.0	3.0
Arg	8	3.2	1.0	0.8	0.9	0.2*
Asn	6	2.4	1.6*	2.6	4.0	1.8
Asp	13	5.2	1.5	0.8	0.7	0.4*
Cys	4	1.6	0.7	1.1	2.3	0.1*
Gln	9	3.6	0.9	0.8	0.3*	1.0
Glu	17	6.9	1.4	3.8	2.3	1.0*
Gly	27	10.9	2.8	0.2*	1.8	2.2
His	8	3.2	1.3	1.4	1.5	0.7*
Ile	17	6.9	3.2	2.6	2.0	1.4*
Leu	17	6.9	2.1	0.5*	1.1	0.9
Lys	22	8.9	1.3*	4.8	3.0	1.5
Met	2	0.8	1.2	0.2*	0.5	1.3
Phe	8	3.2	1.8	0.1*	0.4	0.4
Pro	7	2.8	0.6*	1.8	1.0	1.5
Ser	13	5.2	0.2*	7.1	1.5	2.3
Thr	10	4.0	0.9*	5.1	2.2	1.5
Trp	5	2.0	0.7	0.4	0.4	0.3*
Tyr	4	1.6	1.0*	2.4	4.1	1.4
Val	23	9.3	2.5	1.1	2.8	0.6*
Total	248	100.0	27.0	41.6	34.8	23.5*
Correspondence	—	—	7.0	4.0	1.0	8.0*

[a]See footnote to Table XIV.

Thr $[(P_\beta)_{\alpha+\beta} = 1.25 > (P_\beta)_{\alpha/\beta} = 1.01]$ and Gln $[(P_\beta)_{\alpha+\beta} = 1.08 > (P_\beta)_{\alpha/\beta} = 0.45]$. The last case is particularly striking, since Gln was found to be a weak β-sheet former in 64 proteins with $P_\beta = 1.00 \pm 0.09$, though it has slightly higher β potential in β and α + β proteins with $P_\beta = 1.04$ and 1.08, respectively. However, in α/β proteins, Gln is the strongest β-sheet breaker with $(P_\beta)_{\alpha/\beta} = 0.45 \pm 0.12$ as well as the strongest helix former with $(P_\alpha)_{\alpha/\beta} = 1.38 \pm 0.14$. Thus, Levitt's (1978) assignment of Gln as β breaking for proteins in general should be particularized: Gln is a β breaker in α/β proteins but a β former in β and α + β proteins. Of the 377 Gln residues in the 64 proteins surveyed, 94 were found in β regions (24.9% β). In β proteins, 65 of the 123 Gln residues were in β sheets (52.8% β), whereas in α + β proteins, 17 of the 74 Gln were β (23.1% β). In the α/β proteins, however, only 12 of the 113 Gln were β (10.6% β) compared to 54 helical Gln residues (47.8% α). It is hoped that by using the conformational potentials derived for a particular protein class in elucidating the secondary structures of proteins within that class, improvements will be made in protein conformational predictions.

C. Evaluation of the Prediction of the Four Classes of Proteins from the Amino Acid Composition

Specific examples in Tables XIV through XXVII illustrate the computer output for the automated prediction of the four classes of proteins based on their amino acid composition.

Table XVIII. Differences in Percentage Composition of Four Protein Classes and Correspondences for 19 α Proteins[a]

Protein	α	β	α + β	α/β
Calcium-binding parvalbumin—carp	40.8*	77.4	65.0	56.7
	14.0*	1.5	2.5	2.0
Cytochrome b_{562}—*E. coli*	46.4*	78.4	60.0	58.7
	12.0*	2.0	4.0	2.0
Cytochrome *c*—tuna	39.8	37.4	30.2*	39.3
	5.0	4.0	9.0*	2.0
Cytochrome c_2—*R. rubrum*	33.1*	56.3	42.5	42.4
	10.0*	2.5	3.0	4.5
Cytochrome c_{550}—*P. denitrificans*	35.1*	51.1	40.1	41.8
	8.0*	4.0	5.0	3.0
Cytochrome c_{555}—*C. thiosulfatoph.*	56.8*	67.0	62.0	62.7
	8.0*	7.0	2.0	3.0
Hemerythrin (Met)—*T. dyscritium*	43.4*	56.4	44.8	45.1
	7.0*	5.0	3.0	5.0
Hemerythrin (Myo)—*T. pyroides*	32.5*	56.3	47.7	36.6
	8.0*	3.0	2.0	7.0
Hemerythrin—*G. gouldi*	37.7*	61.9	48.7	44.4
	11.0*	4.5	2.0	2.5
Hemoglobin (deoxy-, α)—human	36.1*	56.9	54.9	40.8
	11.0*	2.0	2.0	5.0
Hemoglobin (deoxy-, β)—human	27.7*	52.5	47.9	32.6
	10.5*	3.0	1.5	5.0
Hemoglobin (Met-, α)—horse	33.8*	51.8	52.8	39.7
	12.0*	2.0	1.0	5.0
Hemoglobin (Met-, β)—horse	29.1*	56.9	49.7	35.2
	11.5*	2.0	2.0	4.5
Hemoglobin—glycera (bloodworm)	38.9*	52.9	46.5	40.2
	10.0*	2.0	3.0	5.0
Hemoglobin—lamprey	27.5	43.5	36.1	25.2*
	9.5*	2.0	3.0	5.5
Hemoglobin—midge larva	29.6*	54.6	40.2	32.1
	9.0*	0.0	5.0	6.0
Hemoglobin fetal (γ chain)—human	23.2	41.0	38.6	21.9*
	11.0*	1.0	2.5	5.5
Myoglobin—seal	31.8*	67.6	53.2	41.3
	12.0*	0.0	3.5	4.5
Myoglobin—sperm whale	30.4*	66.2	54.2	43.3
	12.5*	2.0	2.5	3.0

[a]The computer prediction of the structural class for each protein is indicated by an asterisk (*) using two methods. The first row represents "total" differences in percentage composition from the average percentage in the four structural classes; the second row represents the number of "correspondences" (see details in Table XIV).

Table XIX. Differences in Percentage Composition from Four Protein Classes and Correspondences for 15 β Proteins[a]

Protein	α	β	α + β	α/β
α-Chymotrypsin—bovine	46.9	20.7*	30.5	33.2
	2.0	7.0	8.0*	3.0
Concanavalin A—jack bean	37.9	24.9*	35.5	27.6
	5.0	7.0*	1.0	7.0*
Elastase—porcine	60.7	29.1*	36.9	42.8
	1.0	10.5*	3.5	5.0
Erabutoxin B—sea snake	75.0	53.4*	58.2	59.3
	2.0	9.5*	5.5	3.0
Immunoglobulin Fab′(V&C,H)—human	58.0	24.8*	46.2	42.7
	4.0	10.0*	3.0	3.0
Immunoglobulin Fab′(V&C,L)—human	47.1	23.7*	36.9	36.2
	4.0	11.0*	4.0	1.0
Immunoglobulin Mcg—human	51.0	24.0*	37.0	36.3
	6.0	7.0*	5.0	2.0
Immunoglobulin REI—human	69.7	37.5*	51.9	54.8
	3.0	10.0*	4.0	3.0
Penicillopepsin—*P. janthinellum*	57.7	26.3*	45.9	50.0
	5.5	11.5*	3.0	0.0
Prealbumin—human	37.9	30.9	34.1	25.4*
	4.0	5.5	3.5	7.0*
Protease A—*S. griseus*	65.5	33.9*	40.9	55.0
	3.5	9.0*	7.5	0.0
Protease B—*S. griseus*	70.6	35.2*	48.2	56.3
	3.0	10.0*	6.0	1.0
Rubredoxin—*C. pasteurianum*	79.0	80.0	72.2	69.5*
	6.0	3.0	4.0	7.0*
Superoxide dismutase—bovine	42.0	42.6	41.6	33.7*
	6.0*	5.0	4.0	5.0
Trypsin—bovine	52.2	30.0*	36.4	40.5
	4.5	9.0*	4.5	2.0

[a]See footnote to Table XVIII and details in Table XV.

Presented are the differences in percentage composition of the four protein classes and correspondence for α proteins, β proteins, α and β, and α/β proteins. Also, a comparison of computer prediction of protein classes based on minimum total differences and maximum correspondences in amino acid composition is given, as is a summary of computerized predictions of the four protein classes based on amino acid compositional differences.

An automated prediction of the four protein classes, based on amino acid composition of phosphorylase a, which is not included in the data base, was developed (not included herein). The computerized protein class assignments of 12 proteins of known x-ray structure not included in the data base, and differences in percentages from four protein classes and correspondence are listed. Finally the improvements on erroneously predicted structural regions are listed, using P_α and P_β conformational parameters from the specific classes, rather than from all of the proteins.

Table XX. Differences in Percentage Composition from Four Protein Classes and Correspondences for 14 α + β Proteins[a]

Protein	α	β	α + β	α/β
Actinidin—kiwi fruit	49.8	31.2	28.0*	39.5
	3.5	8.5*	5.0	3.0
Cytochrome b_5—bovine	39.4	53.2	48.6	38.3*
	9.0*	4.0	2.0	5.0
Ferredoxin—*P. aerogenes*	79.5	70.7	66.3*	74.2
	7.0*	6.0	5.0	2.0
High-potential iron protein	49.0	56.2	40.0*	51.5
	6.0	3.0	10.0*	1.0
Insulin—porcine	58.7	60.7	55.7*	59.2
	9.0*	8.0	4.0	1.0
Lysozyme—bacteriophage T4	34.7	49.7	30.3*	30.6
	5.5	2.0	8.0*	4.5
Lysozyme—chicken	53.0	47.0	31.4*	41.9
	2.0	3.0	11.0*	4.0
Papain—papaya	60.3	35.3	31.5*	42.4
	1.0	8.5*	7.5	3.0
Phospholipase A_2—bovine	54.8	55.2	42.8*	50.1
	6.0*	5.0	6.0*	3.0
Ribonuclease S—bovine	55.8	42.2	40.4*	49.5
	5.0	5.0	6.0*	4.0
Staphylococcal nuclease	32.4*	48.8	35.8	33.5
	6.0*	3.0	6.0*	5.0
Subtilisin inhibitor—*Streptomyces*	53.6	45.0	48.0	43.3*
	5.0	5.0	3.0	7.0*
Thermolysin—*B. thermoproteolyt.*	47.3	29.5	26.9*	36.0
	5.0	6.0*	5.0	4.0
Trypsin inhibitor—bovine	62.3	64.3	49.7*	60.4
	5.5	5.0	7.0*	2.5

[a]See footnote to Table XVIII and details in Table XVI.

Table XXI. Differences in Percentage Composition from Four Protein Classes and Correspondences for 16 α/β Proteins[a]

Protein	α	β	α + β	α/β
Adenylate kinase—porcine	37.9	48.7	42.1	32.2*
	8.0*	2.5	5.5	4.0
Alcohol dehydrogenase—horse	36.5	35.1	31.9	20.6*
	5.0	6.0*	3.0	6.0*
Carbonic anhydrase B—human	31.2	30.4	31.2	24.7*
	4.0	5.0	4.0	7.0*
Carbonic anhydrase C—human	22.5*	44.3	38.3	27.8*
	10.0*	4.0	3.0	3.0
Carboxypeptidase A—bovine	41.4	27.6*	28.2	30.1
	4.0	6.0*	5.0	5.0
Carboxypeptidase B—bovine	41.3	29.5	25.7	25.0*
	2.0	6.0	5.5	6.5*
Dihydrofolate reductase—*E. coli*	35.3	46.5	37.1	27.0*
	8.0*	3.0	1.0	8.0*
Flavodoxin—*Clostridium* MP	48.9	55.1	45.3	39.8*
	7.0*	3.0	3.0	7.0*
Glyceraldehyde 3-P dehydrogenase	26.6	38.6	27.4	17.1*
—lobster	6.5	1.0	5.5	7.0*
Glyceraldehyde 3-P dehydrogenase	29.7	50.5	33.9	24.0*
(*B. stearotherm.*	7.0	0.0	3.0	10.0*
Lactate dehydrogenase—dogfish	30.0	44.2	39.2	23.9*
	9.5*	2.5	2.0	6.0
Phosphoglycerate kinase—horse	21.4	41.6	32.6	20.5*
	6.0	3.0	4.0	7.0*
Rhodanese—bovine	34.0	39.8	36.2	22.1*
	6.0	3.5	2.0	8.5*
Subtilisin BPN′—*B. amyloliquef.*	51.2	30.2*	41.4	42.7
	6.0	7.0*	5.0	2.0
Thioredoxin—*E. coli*	29.9*	54.3	42.3	35.0
	10.0	2.0	3.0	5.0
Triose phosphate isomerase—chicken	27.0	41.6	34.8	23.5*
	7.0	4.0	1.0	8.0*

[a]See footnote to Table XVIII and details in Table XVII.

Table XXII. Comparison of Computer Predictions of Protein Classes Based on Minimum Total Differences and Maximum Correspondences in Amino Acid Compositions[a]

Protein class	Total				Correspondence			
	α	β	α + β	α/β	α	β	α + β	α/β
α	16	0	1	2	18	0	1	0
β	0	12	0	3	1	10.5[b]	1	2.5[b]
α + β	1	0	11	2	4[c]	3	6[c]	1
α/β	2	2	0	12	5[c]	2.5[b]	0	8.5[d]

[a]The computer prediction of protein classes denoted by (*) in Tables XVIII–XXI are summarized above for "total" differences in percentage composition and maximum "correspondence." Proteins tied for maximum "correspondence" are given a value of 0.5 instead of 1. Superscripts *b*, *c*, and *d* refer to 1, 2, and 3 ties, respectively.

Table XXIII. Summary of Computerized Predictions of Four Protein Classes Based on Amino Acid Compositional Differences

α Proteins predicted as		β Proteins predicted as		α + β Proteins predicted as		α/β Proteins predicted as	
Calcium-binding parvalbumin	α	α-Chymotrypsin	β	Actinidin	α + β	Adenylate kinase	α/β
Cytochrome b_{562}	α	Concanavalin A	β	High-potential iron protein	α + β	Alcohol dehydrogenase	α/β
Cytochrome *c*	α + β	Elastase	β	Cytochrome b_5	α/β	Carbonic anhydrase B	α/β
Cytochrome c_2	α	Erabutoxin B	β	Ferredoxin	α + β	Carbonic anhydrase C	α
Cytochrome c_{550}	α	Immunoglobulin Fab′(V_H	β	Insulin	α + β	Carboxypeptidase A	β
Cytochrome c_{555}	α	and C_H)		Lysozyme—phage T_4	α + β	Dihydrofolate reductase	α/β
Hemerythrin B	α	Immunoglobulin Fab′(V_L	β	Lysozyme—chicken	α + β	Flavodoxin	α/β
Methemerythrin	α	and C_L)		Papain	α + β	Glyceraldehyde 3-P dehy-	α/β
Myohemerythrin	α	Immunoglobulin MCG	β	Phospholipase A_2	α + β	drogenase (lobster)	
α-Methemoglobin—horse	α	Immunoglobulin REI	β	Ribonuclease S	α + β	Glyceraldehyde 3-P dehy-	α/β
β-Methemoglobin—horse	α	Penicillopepsin	β	Staphylococcal nuclease	α	drogenase (*B. stearo-*	
α-Deoxyhemoglobin—human	α	Prealbumin	α/β	Subtilisin inhibitor	α/β	*thermophilus*)	
Hemoglobin—glycera	α	Protease A	β	Thermolysin	α + β	Lactate dehydrogenase	α/β
Hemoglobin—lamprey	α/β	Protease B	β	Trypsin inhibitor	α + β	Phosphoglycerate kinase	α/β
γ-Hemoglobin—fetal	α/β	Rubredoxin	α/β			Carboxypeptidase B	α/β
Hemoglobin—midge larva	α	Superoxide dismutase	α/β			Rhodanese	α/β
Myoglobin—seal	α	Trypsin	β			Subtilisin BPN′	β
Myoglobin—sperm whale	α					Thioredoxin	α
β-Deoxyhemoglobin—human	α					Triose phosphate isomer-	α/β
						ase	
Predicted correctly 84%		Predicted correctly 80%		Predicted correctly 79%		Predicted correctly 75%	

Table XXIV. Sample Computer Output for Automatic Prediction of Protein Class Based on Amino Acid Composition of Phosphorylase a Not Included in the Data Base[a]

Phosphorylase *a*—rabbit muscle (841 residues)
Differences in percentage composition from four protein classes

A.A.	Count	%	α	β	α + β	α/β
Ala	63	7.5	4.1	0.2*	1.8	0.8
Arg	65	7.7	5.5	5.3	3.6*	4.3
Asn	45	5.4	1.4	0.4*	1.0	1.2
Asp	51	6.1	0.6	1.7	0.2*	0.5
Cys	9	1.1	0.2*	1.6	2.8	0.4
Gln	31	3.7	1.0	0.7	0.2*	1.1
Glu	64	7.6	2.1	4.5	3.0	1.7*
Gly	48	5.7	2.4*	5.0	3.4	3.0
His	22	2.6	1.9	0.8	0.9	0.1*
Ile	49	5.8	2.1	1.5	0.9	0.3*
Leu	78	9.3	0.3*	2.9	3.5	1.5
Lys	47	5.6	4.6	1.5	0.3*	1.8
Met	21	2.5	0.5	1.9	1.2	0.4*
Phe	38	4.5	0.5*	1.4	1.7	0.9
Pro	36	4.3	0.9	0.3	0.5	0.0*
Ser	29	3.4	1.6*	8.9	3.3	4.1
Thr	35	4.2	0.7*	4.9	2.0	1.3
Trp	12	1.4	0.1*	0.2	0.2	0.3
Tyr	36	4.3	1.7	0.3*	1.4	1.3
Val	62	7.4	0.6*	0.8	0.9	1.3
Total	841	100.0	32.8	44.8	32.8	26.3*
Correspondence	—	—	8.0*	3.0	4.0	5.0

[a]See footnote to Table X.

Table XXV. Computerized Protein Class Assignments of 12 Proteins of Known X-Ray Structure Not Included in the Data Base Based on Amino Acid Compositional Differences from α, β, α + β, and α/β Proteins[a]

Protein	Source	Number of residues	α	β	α + β	α/β	X-ray	Predicted
α-Lytic protease	Myxobacter 495	198	63.1	35.9*	44.9	53.8	β	β
L-Arabinose binding protein	*E. coli*	306	21.8	41.0	32.0	20.5*	α/β	α/β
Azurin	*P. aeruginosa*	128	34.7	34.9	37.5	30.0*	β	β
Cytochrome c_{551}	*P. aeruginosa*	82	34.8*	52.8	43.8	38.1	α	α
Dihydrofolate reductase	Mouse L1210	186	39.2	52.4	44.4	31.5*	α/β	α/β
Neurotoxin A	Sea snake	62	78.3	56.7*	61.5	62.6	β	β
Immunoglobulin RHE	Human	113	57.1	33.3*	42.9	49.0	β	β
Insulin	Hagfish	52	57.5	54.5	39.5*	49.6	α + β	α + β
Insulin-like growth factor	Human	70	51.7	50.5	42.1*	45.2	α + β	α + β
Leghemoglobin A	Soybean	140	28.9*	56.5	50.9	36.2	α	α
Phosphorylase B	Rabbit muscle	841	32.8	44.8	32.8	26.3*	α/β	α/β
Uteroglobin	Rabbit	70	49.9	54.7	50.9	43.2*	α	α/β

[a]The percentage composition of amino acids in each protein was compared with the $\%_j$ values in the four protein classes (Table IX). The sum of the differences in percentage compositions for 20 amino acids are given under the columns α, β, α + β, and α/β. The smallest difference is indicated by an asterisk (*), thus predicting the protein class according to closest percentage compositional resemblance. The predictive accuracy for the above 12 proteins not included in the data base is 83%.

Table XXVI. Differences in Percentage Composition from Four Protein Classes and Correspondences for 12 Proteins of Known X-Ray Structure Not Included in Data Base[a,c]

Protein	X-ray[b]	α	β	α + β	α/β
α-Lytic protease (myxobacter 495)	β	63.1	35.9*	44.9	53.8
		4.5	9.5*	5.0	1.0
L-Arabinose-binding protein (*E. coli*)	α/β	21.8	41.0	32.0	20.5*
		6.5*	3.5	5.0	5.0
Azurin (*P. aeruginosa*)	β	34.7	34.9	37.5	30.0*
		6.0	7.0*	0.0	7.0*
Cytochrome c_{551} (*P. aeruginosa*)	α	34.8*	52.8	43.8	38.1
		8.0*	3.0	4.0	5.0
Dihydrofolate reductase (mouse L1210)	α/β	39.2	52.4	44.4	31.5*
		5.5	4.0	4.5	6.0*
Neurotoxin A (sea snake)	β	78.3	56.7*	61.5	62.6
		2.0	9.5*	5.5	3.0
Immunoglobulin RHE (human)	β	57.1	33.3*	42.9	49.0
		3.0	9.0*	5.0	3.0
Insulin (hagfish)	α + β	57.5	54.5	39.5*	49.6
		5.0	3.0	8.0*	4.0
Insulinlike growth factor (human)	α + β	51.7	50.5	42.1*	45.2
		7.0*	4.0	6.5	2.5
Leghemoglobin A (soybean)	α	28.9*	56.5	50.9	36.2
		12.5*	1.5	1.5	4.5
Phosphorylase A (rabbit muscle)	α/β	32.8	44.8	32.8	26.3*
		8.0*	3.0	4.0	5.0
Uteroglobin (rabbit)	α	49.9	54.7	50.9	43.2*
		6.0*	5.5	3.5	5.0

[a]See footnote to Table XVIII and details in Table XXIV.
[b]The protein class was assigned from the three-dimensional x-ray structure. "Total" structural class prediction: 10/12 = 83%. "Correspondence" structural class prediction: 8.5/12 = 71%.
[c]See footnote *a*, Table XVIII.

Table XXVII. Improvements on Erroneously Predicted Structural Regions[a] Using P_α and P_β Conformational Parameters from Specific Protein Classes Instead of from All the Proteins

Protein	Unpredicted α region	Conformational parameters based on 64 Proteins	Protein class
Cytochrome b_5	82–87	$<P_\alpha> = 1.01$	$<P_\alpha>_{\alpha+\beta} = 1.04$
Subtilisin BPN′	245–251	$<P_\alpha> = 0.99$	$<P_\alpha>_{\alpha/\beta} = 1.09$
Elastase	237–245	$<P_\alpha> = 0.94$	$<P_\alpha>_\beta = 1.27$
	Overpredicted α region		
Concanavalin A	154–159	$<P_\alpha> = 1.13$	$<P_\alpha>_\beta = 0.96$
α-Chymotrypsin	111–116	$<P_\alpha> = 1.08$	$<P_\alpha>_\beta = 1.00$
Ca^{2+}-binding parvalbumin	1–6	$<P_\alpha> = 1.13$	$<P_\alpha>_\alpha = 1.04$
Elastase	56–61	$<P_\alpha> = 1.08$	$<P_\alpha>_\beta = 0.86$
	Overpredicted β region		
Subtilisin BPN′	205–209	$<P_\beta> = 1.20$	$<P_\beta>_{\alpha/\beta} = 1.04$
Subtilisin BPN′	250–255	$<P_\beta> = 1.14$	$<P_\beta>_{\alpha/\beta} = 0.90$
Carboxypeptidase A	206–211	$<P_\beta> = 1.19$	$<P_\beta>_{\alpha/\beta} = 0.89$
Carboxypeptidase A	234–238	$<P_\beta> = 1.15$	$<P_\beta>_{\alpha/\beta} = 0.93$

[a]These erroneously predicted regions based on 15 proteins (Chou and Fasman, 1974a,b) are still incorrectly predicted based on 64 proteins but correctly predicted based on protein class P_α and P_β parameters.

IV. REFERENCES

Adman, E. T., Sieker, L. C., and Jensen, L. H., 1976, Structure of *Peptococcus aerogenes* ferredoxin, *J. Biol. Chem.* **251:**3801–3806.

Alter, J. E., Andreatta, R. H., Taylor, G. T., and Scheraga, H. A., 1973, Helix–coil stability constants for the naturally occurring amino acids in water. VIII. Valine parameters from random poly (hydroxypropylglutamine-co-L-valine) and poly(hydroxybutylglutamine-co-L-valine), *Macromolecules* **6:**564–570.

Anfinsen, C. B., Haber, E., Sela, M., and White, F. H., Jr., 1961, The kinetics of formation of the native ribonuclease during oxidation of the reduced polypeptide chain, *Proc. Natl. Acad. Sci. U.S.A.* **47:**1309–1314.

Argos, P., Hanei, M., and Garavito, R. M., 1978, The Chou–Fasman secondary prediction method with an extended data base, *FEBS Lett.* **93:**19–24.

Baker, E. N., 1977, Structure of actinidin: Details of the polypeptide chain conformation and active site from an electron density map at 2.8 Å resolution, *J. Mol. Biol.* **115:**263–277.

Baker, E. N., 1980, Structure of actinidin, after refinement at 1.7 Å resolution, *J. Mol. Biol.* **141:**441–484.

Banks, R. D., Blake, C. C. F., Evans, P. R., Haser, R., Rice, D. W., Hardy, G. W., Merrett, M., and Phillips, A. W., 1979, Sequence, structure and activity of phosphoglycerate kinase: A possible hinge-bending enzyme, *Nature* **279:**773–777.

Banner, D. W., Bloomer, A. C., Petsko, G. A., Phillips, D. C., Pegron, C. I., Wilson, I. A., Corran, P. H., Furth, A. J., Milman, J. D., Offord, R. E., Priddle, J. D., and Waley, S. G., 1975, Structure of chicken muscle triose phosphate isomerase determined crystallographically at 2.5 Å resolution using amino acid sequence data, *Nature* **255:**609–614.

Bernstein, F. C., Koetzle, T. F., Williams, G. J. B., Meyer, E. F., Jr., Brice, M. D., Rodgers, J. R., Kennard, O., Shimanouchi, T., and Tasumi, M., 1977, The Protein Data Bank: A computer-based archival file for macromolecular structures, *J. Mol. Biol.* **112:**535–542.

Biesecker, G., Harris, J. I., Thierry, J. C., Walker, J. E., and Wonacott, A. J., 1977, Sequence and structure of D-glyceraldehyde 3-phosphate dehydrogenase from *Bacillus stearothermophilus, Nature* **266:**328–334.

Birktoft, J. J., and Blow, D. M., 1970, Structure of crystalline α-chymotrypsin. V. The atomic structure of tosyl-C-chymotrypsin at 2 Å resolution, *J. Mol. Biol.* **68:**187–240.

Blake, C. C. F., Geisow, M. J., Oatley, S. J., Rerat, B., and Rerat, C., 1978, Structure of prealbumin: Secondary, tertiary and quarternary interactions determined by Fourier refinement at 1.8 Å, *J. Mol. Biol.* **121:**339–356.

Blundell, T., Dodson, G., Hodgkin, D., and Mercola, D., 1972, Insulin: The structure in the crystal and its reflection in chemistry and biology, *Adv. Protein Chem.* **26:**279–402.

Brayer, G. D., Delbaere, L. T. J., and James, M. N. G., 1978, Molecular structure of crystalline *Streptomyces griseus* protease A at 2.8 Å resolution. II. Molecular conformation, comparison with α-chymotrypsin and active-site geometry, *J. Mol. Biol.* **124:**261–283.

Chambers, J. L., and Stroud, R. M., 1977, Difference Fourier refinement of the structure of DIP-trypsin at 1.5 Å with a minicomputer technique, *Acta Crystallogr.* **B33:**1824–1837.

Chang, M. C., Fredrickson, R. A., Powers, S. P., and Scheraga, H., 1981, Helix-coil stability constants for the naturally occurring amino acids in water. 20. Reinvestigation of valine parameters from random poly[(hydroxypropyl) glutamine-co-L-valine], *Macromolecules* **14:**633–634.

Chou, P. Y., 1979, New approaches to protein structural analysis and conformational predictions, in: *Rapport d'Activité Scientifique du CECAM,* Orsay, p. 149–216.

Chou, P. Y., 1980, Amino acid compositions of four structural classes of proteins, in: *Abstracts of Papers, Part I, Second Chemical Congress of the North American Continent,* Las Vegas.

Chou, P. Y., and Fasman, G. D., 1974a, Conformational parameters for amino acids in helical, β-sheet, and random coil regions calculated from proteins, *Biochemistry* **13:**211–222.

Chou, P. Y., and Fasman, G. D., 1974b, Prediction of protein conformation, *Biochemistry* **13:**222–245.

Chou, P. Y., and Fasman, G. D., 1975, The conformation of glucagon: Prediction and consequences, *Biochemistry* **14:**2536–2541.

Chou, P. Y., and Fasman, G. D., 1977, β-Turns in proteins. *J. Mol. Biol.* **115:**135–175.

Chou, P. Y., and Fasman, G. D., 1978a, Prediction of the secondary structure of proteins from their amino acid sequence, *Adv. Enzymol.* **47:**45–148.

Chou, P. Y., and Fasman, G. D., 1978b, Empirical prediction of protein conformation, *Annu. Rev. Biochem.* **47:**251–276.

Cunningham, B. A., Wang, J. L., Waxdal, M. J., and Edelman, G. M., 1975, The covalent and three-dimensional structure of concanavalin A. II. Amino acid sequence of cyanogen bromide fragment F_3, *J. Biol. Chem.* **250:**1503–1512.

Delbaere, L. T. J., Brayer, G. D., and James, M. N. G., 1979, The 2.8 Å resolution structure of *Streptomyces griseus* protease B and its homology with α-chymotrypsin and *Streptomyces griseus* protease A, *Can. J. Biochem.* **57:**135–144.

Dickerson, R. E., and Geis, I., 1969, *The Structure and Function of Proteins,* Harper & Row, New York, pp. 34–37.

Dijkstra, B. W., Drenth, J., Kalk, K. H., and Vandermaelen, P. J., 1978, Three-dimensional structure and disulfide bond connections in bovine pancreatic phospholipase A_2, *J. Mol. Biol.* **124:**53–60.

Edmundson, A. B., Ely, K. R., Abola, E. E., Schiffer, M., and Panagiotopoulos, N., 1975, Rotational allomerism and divergent evolution of domains in immunoglobulin light chains, *Biochemistry* **14:**3953–3961.

Eklund, H., Nordstrom, B., Zeppezauer, E., Soderlund, G., Ohlsson, I., Boiwe, T., Soderberg, B.-O., Tapia, O., Brandern, C.-I., and Akeson, A., 1976, Three-dimensional structure of horse liver alcohol dehydrogenase at 2.4 Å resolution, *J. Mol. Biol.* **102:**27–59.

Epp, O., Colman, P., Fehlhammer, H., Bode, W., Schiffer, M., Huber, R., and Palm, W., 1974, Crystal and molecular structure of a dimer composed of the variable portions of the Bence-Jones protein REI, *Eur. J. Biochem.* **45:**513–524.

Fasman, G. D., and Potter, J., 1967, The optical rotary dispersion of two beta structures, *Biochem. Biophys. Res. Commun.* **27:**209–215.

Fermi, G., 1975, Three-dimensional Fourier synthesis of human deoxyhemoglobin at 2.5 Å resolution: Refinement of the atomic model, *J. Mol. Biol.* **97:**237–256.

Fett, J. W., and Deutsch, H. F., 1974, Primary Structure of the Mcg λ chain, *Biochemistry* **13:**4102–4114.

Fleer, E. A. M., Verheij, H. M., and De Hass, G. H., 1978, The primary structure of bovine pancreatic phospholipase A, *Eur. J. Biochem.* **82:**261–270.

Frier, J. A., and Perutz, M. F., 1977, Structure of human foetal deoxyhemoglobin, *J. Mol. Biol.* **112:**97–112.

Hecht, M. H., Zweifel, B. O., and Scheraga, H. A., 1978, Helix-coil stability constants for the naturally occurring amino acids in water. 17. Threonine parameters from random poly[(hydroxypropyl) glutamine-co-L threonine], *Macromolecules* **11:**545–551.

Hendrickson, W. A., and Ward, K. B., 1975, Atomic models for the polypeptide backbones of myohemerythrin and hemerythrin, *Biochem. Biophys. Res. Commun.* **66:**1349–1356.

Hendrickson, W. A., Klippenstein, G. L., and Ward, K. B., 1975, Tertiary structure of myohemerythrin at low resolution, *Proc. Natl. Acad. Sci. U.S.A.* **72:**2160–2164.

Hill, D. J. T., Cardinaux, F., and Scheraga, H., 1977, Helix-coil stability constants in the naturally occurring amino acids in water. XIV. Methionine parameters from random poly(hydroxypropylglutamine, L-methionine), *Biopolymers* **16:**2447–2467.

Holbrook, J. J., Liljas, A., Steindel, S. J., and Rossmann, M. G., 1975, Lactate dehydrogenase, in: *The Enzymes,* 3rd ed., Volume XI (P. D. Boyer, ed.), Academic Press, New York, pp. 191–292.

Holmgren, A., Söderberg, B.-O., Eklund, H., and Bränden, C.-I., 1975, Three-dimensional structure of *Escherichia coli* thioredoxin-S_2 to 2.8 Å resolution, *Proc. Natl. Acad. Sci. U.S.A.* **72:**2305–2309.

Hsu, I.-N., Delbaere, L. T. J., James, M. N. G., and Hofmann, T., 1977, Penicillopepsin from *Penicillium janthinellum* crystal structure at 2.8 Å and sequence homology with porcine pepsin, *Nature* **266:**140–145.

Ikenaka, T., Odani, S., Sakai, M., Nabeshima, Y., Sata, S., and Murao, S., 1974, Amino acid sequence of an alkaline proteinase inhibitor (*Streptomyces* subtilisin inhibitor) from *Streptomyces albogriseolus* S-3253, *J. Biochem.* **76:**1191–1209.

Isogai, Y., Nemethy, G., Rackovsky, S., Leach, S. J., and Scheraga, H. A., 1980, Characterization of multiple bends in proteins, *Biopolymers* **19:**1183–1210.

Kannan, K. K., Nostrand, B., Fridborg, K., Lovgren, S., Ohlsson, A., and Peter, M., 1975, Crystal structure of human erythrocyte carbonic anhydrase B. Three-dimensional structure at a nominal 2.2 Å resolution, *Proc. Natl. Acad. Sci. U.S.A.* **72:**51–55.

Kidera, A., Konishi, Y., Oka, M., Ooi, T., and Scheraga, H. A., 1985, Statistical analysis of the physical properties of the 20 naturally occurring amino acids, *J. Protein Chem.* **4:**23–55.

Klipperstein, G. L., Van Riper, D. A., and Oosterom, E. A., 1972, A comparative study of the oxygen transport proteins of *Dendrostomum pyroides, J. Biol. Chem.* **247:**5959–5963.

Klipperstein, G. L., Cote, J. L., and Ludlam, S. E., 1976, The primary structure of myohemerythrin, *Biochemistry* **15:**1128–1136.

Konishi, Y., van Nispen, J. W., Davenport, G., and Scheraga, H. A., 1977, Helix-coil stability constants for the naturally occurring amino acids in water. 15. Arginine parameters from random poly(hydroxybutylglutamine-co-L-arginine), *Macromolecules* **10:**1264–1271.

Korszun, Z. R., and Salemme, F. R., 1977, Structure of cytochrome c_{555} of *Chlorobium thiosulfatophilum:* Primitive low-potential cytochrome c, *Proc. Natl. Acad. Sci. U.S.A.* **74:**5244–5247.

Kretsinger, R. H., and Nockolds, C. E., 1973, Carp muscle calcium-binding protein. II. Structure determination and general description, *J. Biol. Chem.* **248:**3313–3326.

Leszczynski, J. F., and Rose, G. D., 1986, Loops in globular proteins: A novel category of secondary structure, *Science* **234:**849–855.

Levitt, M., 1978, Conformational preferences of amino acids in globular proteins, *Biochemistry* **17:**4277–4285.

Levitt, M., and Chothia, C., 1976, Structural patterns in globular proteins, *Nature* **261:**552–558.

Levitt, M., and Greer, J., 1977, Automatic identification of secondary structure in globular proteins, *J. Mol. Biol.* **114:**181–239.

Lewis, P. N., Momany, F. A., and Scheraga, H. A., 1971, Folding of polypeptide chains in proteins: A proposed mechanism of folding, *Proc. Natl. Acad. Sci. U.S.A.* **68:**2293–2297.

Lim, V. I., 1978, Polypeptide chain folding through a highly intermediate as a general principle of globular protein structure formation, *FEBS Lett.* **89:**10–14.

Loehr, J. S., Lammers, P. J., Brimhall, B., and Hermodson, M. A., 1978, Amino acid sequence of hemerythrin from *Themiste dyscritum, J. Biol. Chem.* **253:**5726–5731.

Low, B. W., Preston, H. S., Sato, A., Rosen, L. S., Searl, J. E., Rudko, A. D., and Richardson, J. S., 1976, Three dimensional structure of erabutoxin b neurotoxic protein: Inhibitor of acetylcholine receptor, *Proc. Natl. Acad. Sci. U.S.A.* **73:**2291–2294.

Matheson, R. R., Jr., Nemenoff, R. A., Cardinaux, F., and Scheraga, H. A., 1977, Helix–coil stability constants for the naturally occurring amino acids in water. XII. Asparagine parameters from random poly(hydroxybutylglutamine-co-L-asparagine), *Biopolymers* **16:**1567–1585.

Mathews, F. S., Bethge, P. H., and Czerwinski, E. W., 1979, The structure of cytochrome b_{562} from *Escherichia coli* at 2.5 Å resolution, *J. Biol. Chem.* **254:**1699–1706.

Matthews, D. A., Alden, R. A., Bolin, J. T., Freer, S. T., Hamlin, R., Xuong, N., Kraut, J., Poe, M.,

Williams, M., and Hoogsteen, K., 1977, Dihydrofolate reductase: X-ray structure of the binary complex with methotrexate, *Science* **197:**452–455.

Mitsui, Y., Satow, Y., Watanabe, Y., and Iitaka, Y., 1979, Crystal structure of a bacterial protein proteinase inhibitor (*Streptomyces* subtilisin inhibitor) at 2.6 Å resolution, *J. Mol. Biol.* **131:**697–724.

Moras, D., Olson, K. W., Sabesan, M. N., Buehner, M., Ford, G. C., and Rossman, M. G., 1975, Studies of asymmetry in the three-dimensional structure of lobster D-glyceraldehyde 3-phosphate dehydrogenase, *J. Biol. Chem.* **250:**9137–9162.

Nagy, J. A., Powers, S. P., Zweifel, B. O., and Scheraga, H. A., 1980, Helix-coil stability constants for the naturally occurring amino acids in water. 18. Tryptophan parameters from random poly[hydroxypropyl glutamine-co-L-tryptophan], *Macromolecules* **13:**1428–1440.

Ploegman, J. H., Drent, G., Kalk, K. H., Hol, W. G. J., Heinrikson, R. L., Keim, P., Weng, L., and Russell, J., 1978a, The covalent and tertiary structure of bovine liver rhodanese, *Nature* **273:**124–129.

Ploegman, J. H., Drent, G., Kalk, K. H., and Hol, W. G. J., 1978b, Structure of bovine liver rhodanese. I. Structure determination at 2.5 Å resolution and a comparison of the conformation and sequence of its two domains, *J. Mol. Biol.* **123:**557–594.

Poljak, R. J., 1975, Three-dimensional structure, function and genetic control of immunoglobulins, *Nature* **256:** 373–376.

Poljak, R. J., Amzel, L. M., Chen, B. L., Phizackerley, R. P., and Saul, F., 1974, The three-dimensional structure of the Fab′ fragment of a human myeloma immunoglobulin at 2.0 Å resolution, *Proc. Natl. Acad. Sci. U.S.A.* **71:**3440–3444.

Reeke, G. N., Jr., Becker, J. W., and Edelman, G. M., 1975, The covalent and three-dimensional structure of concanavalin A. IV. Atomic coordinates, hydrogen bonding, and quaternary structure, *J. Biol. Chem.* **250:** 1525–1547.

Remington, S. J., Anderson, W. F., Owen, J., TenEyck, L. F., Grainger, C. T., and Matthews, B. W., 1978, Structure of the lysozyme from bacteriophage T4: An electron density map at 2.4 Å resolution, *J. Mol. Biol.* **118:**81–98.

Richardson, J. S., 1976, Handedness of crossover connections in β-sheets, *Proc. Natl. Acad. Sci. U.S.A.* **73:** 2619–2623.

Richardson, J. S., 1977, β-sheet topology and the relatedness of proteins, *Nature* **268:**495–500.

Richardson, J. S., 1981, The anatomy and taxonomy of protein structure, *Adv. Protein Chem.* **34:**167–339.

Richardson, J. S., Thomas, K. A., Rubin, B. H., and Richardson, D. C., 1975a, Crystal structure of bovine Cu, Zn superoxide dismutase at 3 Å resolution: Chain tracing and metal ligands, *Proc. Natl. Acad. Sci. U.S.A.* **72:**1349–1353.

Richardson, J. S., Thomas, K. A., and Richardson, D. C., 1975b, Alpha-carbon coordinates for bovine Cu,Zn superoxide dismutase, *Biochem. Biophys. Res. Commun.* **63:**986–992.

Richardson, J. S., Getzoff, E. D., and Richardson, D. C., 1978, The β-bulge: A common small unit of nonrepetitive protein structure, *Proc. Natl. Acad. Sci. U.S.A.* **75:**2574–2578.

Romero-Herrera, A. E., and Lehmann, H., 1974, Residue 122 of sperm whale and horse myoglobin, *Biochim. Biophys. Acta* **336:**318–323.

Rose, G. D., Gierasch, L. M., and Smith, J. A., 1985, Turns in peptides and proteins, *Adv. Protein Chem.* **37:** 1–109.

Rossmann, M. G., and Argos, P., 1975, A comparison of the heme binding pocket in globulins and cytochrome b-5, *J. Biol. Chem.* **250:**7525–7532.

Rossmann, M. G., and Argos, P., 1981, Protein folding, *Annu. Rev. Biochem.* **50:**497–532.

Rossmann, M. G., Moras, D., and Olsen, K., 1974, Chemical and biological evolution of a nucleotide-binding protein, *Nature* **250:**194–199.

Saul, F. A., Amzel, L. M., and Poljak, R. J., 1978, Preliminary refinement and structural analysis of the Fab′ fragment from human immunoglobulin New at 2.0 Å resolution, *J. Biol. Chem.* **253:**585–597.

Sawyer, L., Shotton, D. M., Campbell, J. W., Wendell, P. L., Muirhead, H., Watson, H. C., Diamond, R., and Ladner, R. C., 1978, The atomic structure of crystalline porcine pancreatic elastase at 2.5 Å resolution: Comparisons with the structure of α-chymotrypsin, *J. Mol. Biol.* **118:**137–208.

Schmid, M. F., and Herriott, J. R., 1976, Structure of carboxypeptidase B at 2.8 Å resolution, *J. Mol. Biol.* **103:**175–190.

Schulz, G. E., 1977, Structural rules for globular proteins, *Angew. Chem. [Engl.]* **16:**23–32.

Scouloudi, H., and Baker, E. N., 1978, X-ray crystallographic studies of seal myoglobin: The molecule at 2.5 Å resolution, *J. Mol. Biol.* **126:**637–660.

Snell, C. R., and Fasman, G. D., 1973, Kinetics and thermodynamics of the α helix $\rightleftharpoons$ β transconformation of poly(L-lysine) and L-leucine copolymers. A compensation phenomenon, *Biochemistry* **12:**1017–1025.

Stenkamp, R. E., Sieker, L. C., Jensen, L. H., and McQueen, J. E., Jr., 1978, Structure of methemerythrin at 2.8 Å resolution: Computer graphics fit of an averaged electron density map, *Biochemistry* **17:**2499–2504.

Sternberg, M. J. E., and Thornton, J. M., 1976, On the conformation of proteins: The handedness of the β-strand-α-helix-β-strand unit, *J. Mol. Biol.* **105:**367–382.

Sternberg, M. J. E., and Thornton, J. M., 1978, Prediction of protein structure from amino acid sequence, *Nature* **271:**15–20.

Stone, D., Phillips, A. W., and Burchall, J. J., 1977, The amino acid sequence of the dihydrofolate reductase of a trimethoprin-resistant strain of *Escherichia coli, Eur. J. Biochem.* **72:**613–624.

Sueki, M., Lee, S., Powers, S. P., Denton, J. B., Konishi, Y., and Scheraga, H. A., 1984, Helix-coil stability constants for the naturally occurring amino acids in water. 22. Histidine parameters from random poly(hydroxybutyl) glutamine-co-L-histidine, *Macromolecules* **17:**148–155.

Suzuki, E., and Robson, B., 1976, Relationship between helix–coil transition parameters for synthetic polypeptides and helix conformation parameters for globular proteins. A simple model, *J. Mol. Biol.* **107:**357–367.

Swanson, R., Trus, B. L., Mandel, N., Mandel, G., Kallai, O. B., and Dickerson, R. E., 1977, Tuna cytochrome c at 2.0 Å resolution. I. Ferricytochrome structure analysis, *J. Biol. Chem.* **252:**759–775.

Takano, T., 1977, Structure of myoglobin refined at 2.0 Å resolution, *J. Mol. Biol.* **110:**537–568.

Takano, T., Trus, B. L., Mandel, N., Mandel, G., Kallai, O. B., Swanson, R., and Dickerson, R. E., 1977, Tuna cytochrome c at 2.0 Å resolution. I. Ferrocytochrome structure analysis, *J. Biol. Chem.* **252:**776–785.

Timkovich, R., and Dickerson, R. E., 1976, The structure of *Paracoccus denitrificans* cytochrome c_{550}, *J. Biol. Chem.* **251:**4033–4046.

Titani, K., Ericsson, L. H., Walsh, K. A., and Neurath, H., 1975, Amino-acid sequence of bovine carboxypeptidase B, *Proc. Natl. Acad. U.S.A.* **72:**1666–1670.

Toniolo, C., Bonora, G. M., and Scatturin, A., 1975, Linear oligopeptides. XXV. The effect of a sulphur atom in γ-position in the secondary structure of homo-oligopeptides, *Gaz. Chim. Ital.* **105:**1063–1071.

Van Nispen, J. W., Hill, D. J., and Scheraga, H. A., 1977, Helix–coil stability constants for the naturally occurring amino acids in water. XIII. The presence of by-products in amino-acid analysis of copolymers and their effect on the guest parameters: Recomputed values of σ and s for L-serine, *Biopolymers* **16:**1587–1592.

Watenpaugh, K. D., Margulis, T. N., Sieker, L. C., and Jensen, L. H., 1978, Water structure in a protein crystal: Rubredoxin at 1.2 Å resolution, *J. Mol. Biol.* **122:**175–190.

13

Use of Class Prediction to Improve Protein Secondary Structure Prediction

Joint Prediction with Methods Based on Sequence Homology

Gilbert Deléage and Bernard Roux

I. INTRODUCTION

In the absence of crystallographic data, many structural features of proteins can be deduced from the analysis of protein sequences. One of the most promising tools for the near future relates to the prediction of antigenic sites (for a review, see Berzofsky, 1985; Delisi and Berzofsky, 1985; Margalit *et al.*, 1987) for the engineering of synthetic vaccines. In addition, with the increasing number of protein sequences known from DNA cloning and sequencing, the need for a theoretical treatment of protein sequences has never been greater. In this context, many different methods for predicting the secondary structure of proteins have been

Gilbert Deléage and Bernard Roux • Laboratory of Biological Chemistry, Claude Bernard University of Lyon I, 69622 Villeurbanne Cedex, France.

developed (Finkelstein and Ptitsyn, 1971; Robson and Pain, 1971; Kabat and Wu, 1973; Burgess *et al.*, 1974; Chou and Fasman, 1974, 1978; Lim, 1974; Nagano, 1977; Garnier *et al.*, 1978; Cid *et al.*, 1982). Several of these methods are statistical; i.e., they are based on the observed frequency with which individual residues are found in given structural states.

In 1974, a hierarchic classification of all 20 amino acids for their helix and sheet conformational potential was established by Chou and Fasman (1974). A set of empirical rules has also been formulated to use these parameters (determined for 29 proteins) to predict the secondary structure of proteins (Chou and Fasman, 1978). Several computer programs that use this method have been described (Corrigan and Huang, 1982; Rawlings *et al.*, 1983; Nishikawa, 1983; Parilla *et al.*, 1986; Deléage *et al.*, 1987; Ralph *et al.*, 1987). Another well-known method, that of Garnier *et al.* (1978), is based on the effect of the "directional information" carried by residues on the conformation of neighboring amino acids. A method of modifying predictions from the α-helix and β-sheet content discovered from preliminary calculations has been explored (Robson and Pain, 1971) and has led to the introduction of decision constants, which improve the accuracy of the final predicted secondary structure (Garnier *et al.*, 1978). Recently, a comparison carried out by Kabsch and Sander (1983a) of three of the most widely used methods has revealed that the average accuracy of a three-state description (α, β, and coil) of secondary structure ranges between 50 and 59%.

More recently, several groups (Sweet, 1986; Nishikawa and Ooi, 1986; Levin *et al.*, 1986; Zvelebil *et al.*, 1987) have explored a different approach. They developed methods based on the sequence homology between the protein to be predicted and proteins belonging to a secondary structure database. Their success was at the 60%–66% level, depending on the number of proteins considered.

In this context, the development of methods based on a different concept may provide a potential way to make combined predictions. Therefore, we have examined the possibility of including the information taken from the prediction of the structural class (i.e., all-α, all-β, α/β, α + β, and irregular) of proteins (Levitt and Chothia, 1976; Richardson, 1981; Nakashima *et al.*, 1986) to improve the rate of success in secondary structure prediction (Deléage and Roux, 1987). The improvement in the accuracy in secondary structure prediction was 6%, with an average level of success of 61%, when this method was checked on the entire secondary structure data base of Kabsch and Sander (1983b). The success in class prediction was 72% on these proteins. We report here the main steps of this "double prediction method" as well as the promising results obtained by the joint prediction with the "homology" method of Levin *et al.* (1986). The success rate can be improved by about 10% in residues for 56% of residues predicted in the same conformational state by these two methods.

II. METHODS

A. Double Prediction Method

The "double prediction method" consists of three successive steps.

1. Protein Class Prediction

The class of the protein is predicted from its amino acid composition. This is carried out with the algorithm of Nakashima *et al.* (1986) except for the α/β class of proteins, which has been divided into two subclasses, α/β-α and α/β-β, according to the smallest distance from the centers of the α and β classes, respectively.

Table I. Conformational Parameters for the 20 Amino Acids[a]

Amino acid	P_α	P_β	P_c	P_t
E	1.504	0.567	0.761	1.149
A	1.489	0.709	0.66	0.788
L	1.236	1.261	0.810	0.670
H	1.003	0.863	1.068	0.970
M	1.363	1.210	0.810	0.436
Q	1.164	0.840	0.947	0.997
W	1.090	1.306	0.941	0.546
V	0.990	1.965	0.772	0.387
F	1.195	1.393	0.797	0.624
K	1.172	0.721	0.897	1.302
I	1.003	1.799	0.886	0.240
D	0.924	0.541	1.197	1.197
T	0.785	1.221	1.148	0.739
S	0.739	0.928	1.130	1.316
R	1.224	0.920	0.983	0.912
C	0.966	1.191	0.953	0.965
N	0.772	0.604	1.167	1.572
Y	0.787	1.266	1.109	0.795
P	0.492	0.354	1.540	1.415
G	0.510	0.657	1.251	1.860

[a]The conformational potentials for helix, sheet, coil, and turn are designated by P_α, P_β, P_c, and P_t, respectively. All the parameters are calculated as described by Chou and Fasman (1974).

2. Secondary Structure Prediction

The set of statistical parameters has been calculated over the Kabsch and Sander (1983b) secondary structure data base (helix is H, G, and I; sheet is E; turn is T; and coil is S and B) from the frequency of occurrence of each amino acid in the four states (helix, sheet, coil, turn). The P_{coil} has been added to make a true three-state (α, β, coil) prediction and to minimize the overprediction of ordered structures. These new parameters (Table I) of the Chou and Fasman type were used in a simple algorithm for the unambiguous prediction of each amino acid in one of the three states (α, β, and coil). A window width was chosen as a function of the predicted class of protein. The window width was 11 ($v = 5$) for all α proteins, 7 ($v = 3$) for all β proteins, and 9 ($v = 4$) for the other classes:

$$S(k,i) = \sum_{-v}^{v} P(k,i+v) \,/\, (|v| + 1)$$

where i is the predicted residue, k is the conformational state (helix, sheet, turn, coil), v determines the window ($v = 3$, $v = 4$, or $v = 5$, see above), and $|v|$ is the absolute value of v. The highest $S(k,i)$ determines the conformation k for the ith residue.

3. Decision on "Reprediction" with Optimized Parameters

The parameters (P_α, P_β, P_c) have been optimized independently for all proteins of each predicted class using the following three criteria: (1) maximal accuracy in secondary structure

Table II. Optimization of Parameters as a Function of Predictions of Class and Secondary Structure[a]

Predicted class of proteins	α% > β%[b] Changes	α% > β%[b] Total number of run	β% > α% Changes	β% > α% Total number of run
α	$P_\beta = P_\beta - \|(\beta_f/5) \times P_\beta\|$[c] $P_\alpha = P_\alpha + \|(\alpha_f/6.5) \times P_\alpha\|$ $P_c = P_c + \|(C_f/6.5) \times P_c\|$	Until $\beta\% < 5$	—	1
β	—	1	$P_\alpha = P_\alpha - \|(\beta_f/4) \times P_\alpha\|$ $P_c = P_c + \|(\alpha_f/4) \times P_c\|$	2
α/β-α	$P_c = P_c + \|(\beta_f/3) \times P_c\|$ $P_\beta = P_\beta - \|(\alpha_f/3.5) \times P_\beta\|$	2	$P_\alpha = P_\alpha - \|(\alpha_f/4) \times P_\alpha\|$ $P_\beta = P_\beta - \|(\beta_f/3) \times P_\beta\|$	2
α/β-β	—[d]	1	$P_\alpha = P_\alpha - \|(\beta_f/4) \times P_\alpha\|$ $P_c = P_c + \|(\beta_f/4) \times P_c\|$	2
α + β	$P_\beta = P_\beta - \|(\beta_f/6) \times P_\beta\|$	2	$P_\beta = P_\beta - \|(\beta_f/6) \times P_\beta\|$	2
R	$P_c = P_c + \|(C_f) \times P_c\|$	2	$P_c = P_c + \|(C_f) \times P_c\|$	2

[a]From Deléage and Roux, 1987.
[b]α% and β% represent, respectively, the percentage of helix and sheet found by the single prediction method (see Section II). The P_α, P_β, and P_c parameters are the parameters of Table I calculated over the data base.
[c]α_f, β_f, and C_f are the fractions of residues predicted in, respectively, helix, sheet, and coil states, i.e., the number of predicted residues in a state divided by the total number of residues of the protein. For α + β and R proteins, the change in parameters is not dependent on α% and β%.
[d]No example in the database.

prediction, (2) maximal agreement between predicted and observed content in secondary structure, and (3) no loss of accuracy for the majority of the proteins in a given class. A summary of the conditions and of the changes in the conformational parameters is supplied in Table II. These new parameters are then used in the algorithm described above to generate the final predicted structure. But, if the results of the class prediction and of the secondary structure prediction are not in agreement, the prediction is limited to the preliminary prediction since at least one of the two methods may be wrong.

B. Other Predictive Methods

The computerized Chou and Fasman method has been used as previously described (Deléage *et al.*, 1987) except for the thresholds for the nucleation of helix (1.03) and sheet (1.05), which have been changed to 1.13 and 1.22, respectively. This program (with these minor modifications) yields an average accuracy in secondary structure prediction of 57% instead of 51% and a good agreement between the predicted and observed structure contents (see Table IV).

The GOR program using the directional information (Garnier *et al.*, 1978) has been keyed in as described by the authors. All decision constants were taken as equal to zero to avoid the overpredictions of ordered secondary structures.

The "homology" method of Levin *et al.* (1986), using the secondary structure data base of Kabsch and Sander (1983b) except for rubredoxin and cytochrome c_{550}, has been implemented on a microcomputer.

All the predictions were obtained from a computer package available for microcomputer (Deléage *et al.*, 1988) and displayed on a Hewlett Packard (HP 7470A) plotter.

III. RESULTS AND DISCUSSION

A. Predictive Success of the Double Prediction Method

The main idea of the double prediction method was to utilize an independent prediction of the structural class of a protein to bias the prediction of its secondary structure. Generally, the approach was the reverse, since algorithms have been described to predict the class of a protein from isolated (Garnier *et al.*, 1978; Busetta, 1986) or combined (Klein and Delisi, 1986) methods of secondary structure prediction. The reliability of these methods was in the range 60–80%. In other respects, the structural class of a protein can be predicted from its amino acid composition with a success rate of 75% (Nakashima *et al.*, 1986). When this latter method has been applied to allocate 60 proteins of the secondary structure data base (Kabsch and Sander, 1983b) into one of the five classes (α, β, α/β, α + β, and irregular), the allocation matrix (Klein and Delisi, 1986) is:

$$\mathbf{M} = \begin{pmatrix} 9 & 2 & 0 & 2 & 0 \\ 0 & 11 & 1 & 3 & 0 \\ 0 & 2 & 12 & 1 & 0 \\ 1 & 1 & 1 & 6 & 1 \\ 0 & 0 & 0 & 0 & 1 \end{pmatrix}$$

where the i,j element is the number of times a protein observed in class i is predicted in class j. In other words, the number on the diagonal represents the proteins for which the prediction is consistent with experimental data. The reliability of the prediction of the class of proteins is 72% (39/54) instead of the score of 75% reported by Nakashima *et al.* (1986), indicating that the secondary structure data base could be a representative sample of a larger folding-type data base.

In order to obtain a gain in prediction reliability for most proteins of a given (predicted) class, the optimization of the parameters has been carried out on each isolated class including the proteins wrongly assigned.

The direction of the changes has been deduced from a comparison between the global predicted content found in the preliminary prediction with that found in observed classes (Fig. 1). The aperiodic structure is always underpredicted, whereas the sheet content is over-predicted in all classes except for the β class. As expected, for α and β proteins, the helix content is underestimated and overestimated, respectively, whereas the contrary is observed for the sheet content. Thus, the contents in predicted secondary structures found after the optimization of the parameters are in better agreement with those observed than the ones previously predicted.

On the other hand, no universal changes of parameters have proved to be satisfactory for all the proteins of a given class, since changes adapted to correctly assigned proteins will make little improvement in the predictions for incorrectly classified proteins. That is why the fraction of each state found in the first run has been used to modify the parameters (see Table II).

The double prediction method has been tested on 60 of the 62 proteins constituting the data base of Kabsch and Sander (1983b). The prediction accuracy is 72% for the class determination and reaches 61% (6,471/10,569) for a three-state (helix, sheet, coil) description of the secondary structure of proteins (Table III, columns A). In addition, the removal of each protein from the secondary structure data base before its own prediction is made leads to a loss in accuracy as little as 58 amino acids over 10,569 (0.55%), indicating a rather poor sensitivity of the method toward the data base. Similar decreases in accuracy have been reported for

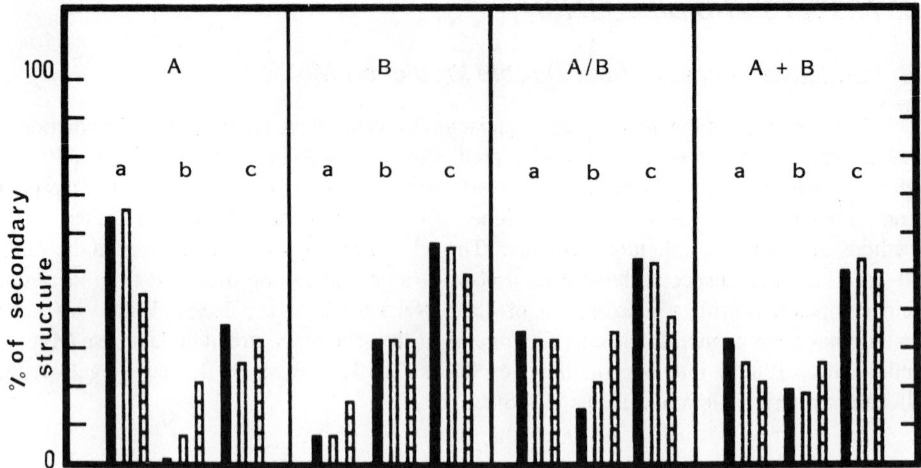

Figure 1. The contents of secondary structure of proteins of observed (solid bars) and predicted classes (α, β, α/β, α + β) are plotted for the helix (a), sheet (b), and coil (c) states. Hatched and open rectangles designate the contents found by a single run and by the double prediction method (see Section II), respectively. From Deléage and Roux (1987).

statistical methods (Kabsch and Sander, 1983a). However, if the prediction of the class of proteins is omitted, the level of success drops to 55% (Table III, columns B). Thus, a substantial increase of 6% in residues is brought by the double prediction method.

B. Comparison of Accuracy with Existing Methods

Comparison of the double prediction method with other predictive methods (Chou and Fasman, 1974; Garnier *et al.*, 1978; Levin *et al.*, 1986) has been made among observed and predicted conformations for four proteins belonging to the α, β, α/β, and α + β classes (Fig. 2). For all four proteins (deoxyhemoglobin, lambda immunoglobulin Fab, rhodanese, and ribonuclease S), the best agreement with x-ray data was obtained from the double prediction method or the homology method. Moreover, for mixed α and β proteins, the discrimination

Table III. Prediction Success of the Double Prediction Method and the Preliminary Prediction

Conformation	Number of observed residues[a]	Number of predicted residues		Number of correctly predicted residues		Percentage correct of observed residues[b]	
		A	B	A	B	A	B
Helix	3,005	2,799	2,795	1,624	1,386	54	46.1
Sheet	2,080	2,400	3,243	1,075	1,197	51.6	57.5
Coil	5,484	5,370	4,531	3,772	3,297	68.8	60.1
Total	10,569	10,569	10,569	6,471	5,880	61.2	55.6

[a]The proteins are those listed by Kabsch and Sander (1983b) except for cytochrome C_{550} and rubredoxin, for which some doubts persist in sequence or in structure.

[b]The percentage correct for a given conformation is the number of correctly predicted residues in that conformation divided by the number of residues observed in that state. Columns A and B are for the results obtained with the double prediction method and the prediction by a single run, respectively.

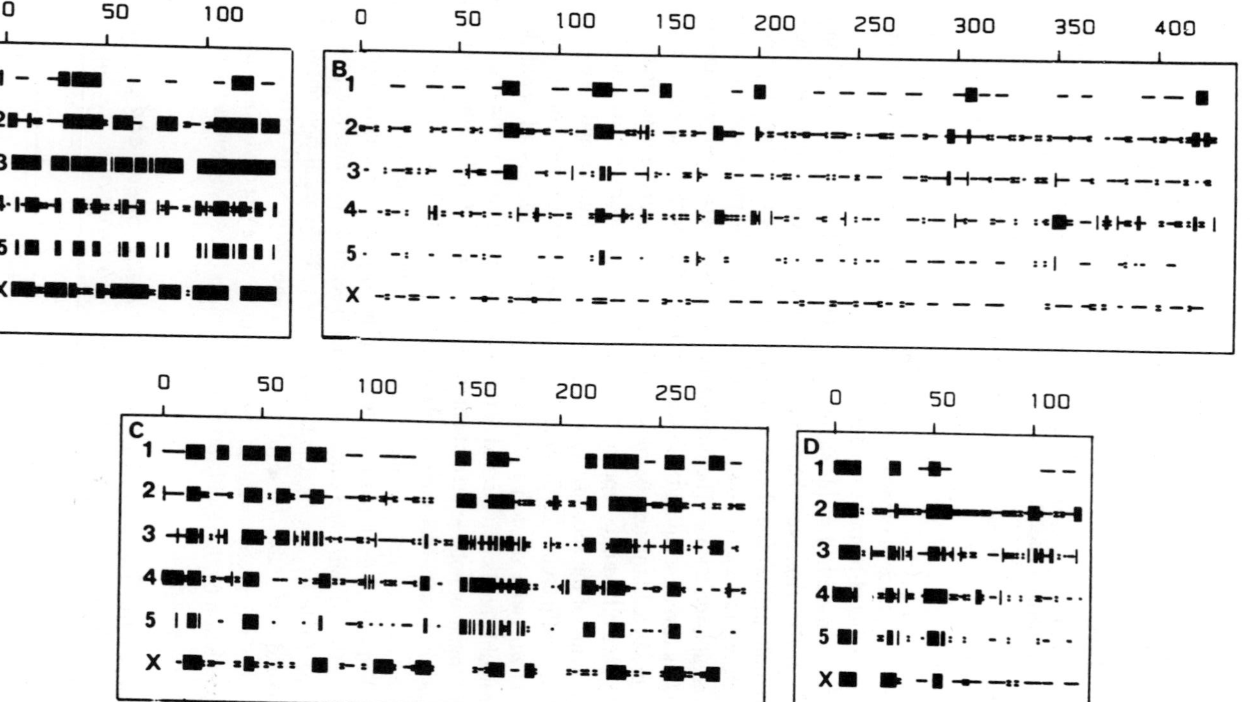

Figure 2. The predicted secondary structures are given for deoxyhemoglobin (A), lambda immunoglobulin Fab (B), rhodanese (C), and ribonuclease S (D) according to (1) Chou and Fasman (1978), (2) Garnier *et al.* (1978), (3) Deléage and Roux (1987), and (4) Levin *et al.* (1986) methods. The structures observed from x rays and predicted in common by the last two methods (see the text) are in lines X and 5, respectively. Heavy bars and single and double lines designate the regions in helices, sheets, and turns, respectively. The blanks in the sequences represent coil segments. The numbered scales indicate the positions of residues in the sequences.

Table IV. Comparison of the Prediction Success of Four Different Methods: Details of Helix, Sheet, and Coil Prediction Averaged on 60 Proteins

		Chou		Garnier		Deléage		Levin	
	Observed[a]	P	CP	P	CP	P	CP	P	CP
Helix	3,005	2,992	1,510	3,199	1,693	2,799	1,624	2,823	1,550
Sheet	2,080	2,026	812	2,929	1,167	2,400	1,075	2,025	925
Coil	5,484	5,551	3,666	4,441	3,118	5,370	3,772	5,721	3,843
Total	10,569	10,569	5,988	10,569	5,978	10,569	6,471	10,569	6,318

[a]The first column indicates the number of residues observed in each conformational state (helix, sheet, or aperiodic) in 60 proteins of the secondary structure dictionary described by Kabsch and Sander (1983b). The number of amino acids predicted (P) and correctly predicted (CP) are given from left to right for the computerized version (Deléage *et al.*, 1987) of the method of Chou and Fasman (1978), the algorithm of Garnier *et al.* (1978) using the directional information, our double prediction method, and the homology method of Levin *et al.* (1986).

between α/β and $\alpha + \beta$ classes is obvious from the respective alternation and separation of ordered structures as predicted by most methods.

The double prediction method yields slightly different results for helix (54%) and sheet (52%) prediction (Table III). However, the accuracy of the predictions for the aperiodic state is much better than for both helices and sheets. This leads to the conclusion that the less regular structure is also the easiest to predict. This is possibly because of the rather low conformational constraints in such an aperiodic state. The optimization has been carried out with the aim that most proteins of a class (even including those wrongly predicted) take advantage of it and that the total content in predicted structure be most consistent with observed structures.

In order to compare different computerized methods, the prediction procedure has been carried out on 60 proteins of the secondary structure data base of Kabsch and Sander (1983b) (Table IV). The statistical methods of Chou and Fasman (with raised thresholds, see Section II.B) and Garnier *et al.* (with no decision constants) yield an accuracy of about 56–57%. When applying the double prediction method or the homology method of Levin *et al.* (1986) to an unknown protein, one can hope to predict correctly about 60% of the residues in the actual conformation. Thus, it appears that these methods are two of the best techniques currently available. In addition, the root-mean-square deviation between predicted contents in secondary structure and observed ones, which has been found to be a significant criterion of reliability (Kabsch and Sander, 1983a), has been calculated on a subset of 45 proteins (i.e., those longer than 100 residues; see Table V). The largest deviations (14 to 17%) are always obtained for the helix state. This could reasonably be a result of the rather large scale of variation in the observed content of the helical conformation. In addition, the smallest deviation is obtained for the sheet content prediction by the double prediction method (9.6%), whereas the second-best

Table V. Average and Standard Deviations for Helix, Sheet, and Coil Contents Predicted by Four Methods on 45 Proteins[a]

Method	Helix	Sheet	Coil
Chou–Fasman	−0.5 (17)	−1.3 (12.8)	+1.7 (14.8)
Garnier *et al.*	2.8 (15.6)	7 (10.9)	−9.8 (13.4)
Double prediction	−1.2 (14)	1.7 (9.6)	−0.6 (12.2)
Levin *et al.*	−2 (14)	−1 (12)	3 (10.7)

[a]The difference between observed and predicted secondary structure contents has been calculated over the 45 proteins of the secondary structure data base that are more than 100 residues in length. The average percentage of deviation is given as well as the root-mean-square average deviation (in parentheses) for the four indicated methods.

score (10.7%) is obtained for the coil content by the homology method. Nevertheless, the uncertainty is high when compared with that of circular dichroism, and a potential way of reducing this error rate is to make a joint prediction.

C. Joint Prediction

For anyone who needs to use methods of secondary structure prediction, at least two questions arise: which predictive method has to be chosen, and is there any advantage in using different methods? The answer to the first question mainly depends on the aim of the predictor and on his knowledge of his protein. For example, if the protein is known to be homologous with a protein of the secondary structure data base, the homology methods are required. On the other hand, the double prediction method will be useful in almost all other cases. In order to answer the second question, we have investigated the possibility of increasing the accuracy by considering the amino acids predicted in the same state by different methods. This information is particularly useful for anyone who is interested in knowledge of the secondary structure of precisely limited regions as potential antigenic sites.

In making combined predictions, one has to make a compromise between the number of residues jointly predicted and the number of correctly predicted residues, because the combination of methods based on a similar approach (say, statistical) will yield a large number of jointly predicted residues and a small improvement in accuracy. In contrast, if two methods that are based on different approaches are combined, the level of coincidence can be low, but the resulting success rate will be increased. Therefore, the joint prediction must deal with methods based on dissimilar principles.

Since the double prediction method is based on a different approach from that of the homology method (Levin *et al.*, 1986) and exhibits comparable accuracy, we report in Table VI the accuracy and the percentage of residues predicted in the same conformational state by these two methods. For comparison of isolated or combined predictions, the accuracies are given, as recommended by Kabsch and Sander (1983a), as the "percentage correct of predicted." Although these two methods yield a mean accuracy of 60%, the joint prediction gives 70% accuracy for the 56% amino acids predicted in the same conformational state by the two methods. Thus, an improvement in prediction accuracy of 10% can be expected for more than half of the residues from the combination of statistical and homologous methods.

Table VI. Prediction Success of Isolated Methods and of the Joint Prediction

				Correctly predicted			
		Predicted[a]			Percentage correct of predicted[b]		
	Observed Number	Number	Percentage	Number	A	B	C
Helix	3,005	1,587	52.8	1,095	58	54.9	69
Sheet	2,080	1,182	56.8	650	44.8	45.7	55
Coil	5,484	3,195	58.3	2,413	70.2	67.2	75.5
Total	10,569	5,964	56.4	4,158	61.2	59.8	69.7

[a]The "predicted" columns are for results concerning the residues that have been predicted in the same conformational state by the double prediction method and the homology method of Levin *et al.* (1986).

[b]Columns A and B give the fraction correct of predicted residues (i.e., the number of correctly predicted residues divided by the number of predicted residues) for the double prediction and the homology methods, respectively. Column C is the percentage of amino acids correctly predicted by the joint prediction.

IV. PERSPECTIVES

A possible evolution of the double prediction method will consist of dividing a long protein into domains (Klein and Delisi, 1986; Vonderviszt and Simon, 1986; Taylor and Thornton, 1983) and proceeding to the prediction of class on isolated domains, allowing the utilization of optimized parameters on separated strands of proteins. Another complementary direction would be to include hydrophobic effects (Garnier *et al.*, 1978; Cid *et al.*, 1982; Eisenberg *et al.*, 1982; Busetta and Barrans, 1984; Garatt *et al.*, 1985). A method to ensure the retrieval of the information given by many isolated predictive methods is currently under investigation.

V. REFERENCES

Berzofsky, J. A., 1985, Intrinsic and extrinsic factors in protein antigenic structure, *Science* **229:**932–940.

Burgess, A. W., Ponnuswamy, P. K., and Scheraga, H. A., 1974, Analysis of conformations of amino acid residues and prediction of backbone topography in proteins, *Isr. J. Chem.* **12:**239–286.

Busetta, B., 1986, Examination of folding patterns for predicting protein topologies, *Biochim. Biophys. Acta* **870:**327–338.

Busetta, B., and Barrans, Y., 1984, The prediction of protein domains, *Biochim. Biophys. Acta* **790:**117–124.

Chou, P. Y., and Fasman, G. D., 1974, Prediction of protein conformation, *Biochemistry* **13:**222–244.

Chou, P. Y., and Fasman, G. D., 1978, Prediction of secondary structure of proteins from amino acid sequence, *Adv. Enzymol. Relat. Subj. Biochem.* **47:**45–148.

Cid, H., Bunster, M., Arriagada, E., and Campos, M., 1982, Prediction of secondary structure of proteins by means of hydrophobicity profiles, *FEBS Lett.* **150:**247–254.

Corrigan, A. J., and Huang, P. C., 1982, A BASIC microcomputer program for plotting the secondary structure of proteins, *Comput. Prog. Biomed.* **15:**163–168.

Deléage, G., and Roux, B., 1987, An algorithm for protein secondary structure prediction based on class prediction, *Protein Eng.* **1:**289–294.

Deléage, G., Tinland, B., and Roux, B., 1987, A computerized version of the Chou and Fasman method for predicting the secondary structure of proteins, *Anal. Biochem.* **163:**292–297.

Deléage, G., Clerc, F. F., Roux, B., and Gautheron, D. C., 1988, ANTHEPROT: A package for protein sequence analysis using a microcomputer, *CABIOS* **4:**351–356.

Delisi, C., and Berzofsky, J. A., 1985, T-cell antigenic sites tend to be amphipathic structures, *Proc. Natl. Acad. Sci. U.S.A.* **82:**7048–7052.

Eisenberg, D., Weiss, R. M., and Terwilliger, T. C., 1982, The helical hydrophobic moment: A measure of the amphiphilicity of a helix, *Nature* **299:**371–374.

Finkelstein, A. V., and Ptitsyn, O. B., 1971, Statistical analysis of the correlation among amino acid residues in helical, β-structural and non-regular regions of globular proteins, *J. Mol. Biol.* **62:**613–624.

Garatt, R. C., Taylor, W. R., and Thornton, J. M., 1985, The influence of tertiary structure on secondary structure prediction, *FEBS Lett.* **188:**59–62.

Garnier, J., Osguthorpe, D. J., and Robson, B., 1978, Analysis of the accuracy and implications of simple methods for predicting the secondary structure of globular proteins, *J. Mol. Biol.* **120:**97–120.

Kabat, E. A., and Wu, T. T., 1973, The influence of nearest-neighbor amino acids on the conformation of the middle amino acid in proteins: Comparison of predicted and experimental determination of β sheets in concanavalin A, *Proc. Natl. Acad. Sci. U.S.A.* **70:**1473–1477.

Kabsch, W., and Sander, C., 1983a, How good are predictions of protein secondary structure? *FEBS Lett.* **155:** 179–182.

Kabsch, W., and Sander, C., 1983b, Dictionary of protein secondary structure: Pattern recognition of hydrogen-bonded and geometrical features, *Biopolymers* **22:**2577–2637.

Klein, P., and Delisi, C., 1986, Prediction of protein structural class from the amino acid sequence, *Biopolymers* **25:**1659–1672.

Levin, J. M., Robson, B., and Garnier, J., 1986, An algorithm for secondary structure determination in proteins based on sequence similarity, *FEBS Lett.* **205:**303–308.

Levitt, M., and Chothia, C., 1976, Structural patterns in globular proteins, *Nature* **261:**552–558.

Lim, V. I., 1974, Structural principles of the globular organization of protein chains. A stereochemical theory of globular protein secondary structure, *J. Mol. Biol.* **88:**857–872.

Margalit, H., Spouge, J. L., Cornette, J. L., Cease, K. P., Delisi, C., and Berzofsky, J. A., 1987, Prediction of immunodominant helper T cell antigenic sites from the primary sequence, *J. Immunol.* **138:**2213–2229.

Nagano, K., 1977, Triplet information in helix prediction applied to the analysis of super-secondary structures, *J. Mol. Biol.* **109:**251–274.

Nakashima, H., Nishikawa, K., and Ooi, T., 1986, The folding type of a protein is relevant to the amino acid composition, *J. Biochem. (Tokyo)* **99:**153–162.

Nishikawa, K., 1983, Assessment of secondary-structure prediction of proteins. Comparison of computerized Chou–Fasman method with others, *Biochim. Biophys. Acta* **748:**285–299.

Nishikawa, K., and Ooi, T., 1986, Amino acid sequence homology applied to the prediction of protein secondary structure, and joint prediction with existing methods, *Biochim. Biophys. Acta* **871:**45–54.

Parilla, A., Domenech, A., and Querol, E., 1986, A pascal microcomputer program for prediction of protein secondary structure and hydropathic segments, *Cabios* **2:**211–215.

Ralph, W. W., Webster, T., and Smith, T. F., 1987, A modified Chou and Fasman protein structure algorithm, *Cabois* **3:**211–216.

Rawlings, N., Ashman, K., and Wittmann-Liebold, B., 1983, Computerized version of the Chou and Fasman protein secondary structure predictive method, *Int. J. Peptide Res.* **22:**515–524.

Richardson, J. S., 1981, The anatomy and taxonomy of protein structure, *Adv. Prot. Chem.* **34:**167–339.

Robson, B., and Pain, R. H., 1971, Analysis of the code relating sequence to conformation in proteins: Possible implications for the mechanism of formation of helical regions, *J. Mol. Biol.* **58:**237–259.

Sweet, R. M., 1986, Evolutionary similarity among peptide segments is a basis for prediction of protein folding, *Biopolymers* **25:**1565–1577.

Taylor, W. R., and Thornton, J. M., 1983, Prediction of super-secondary structure in proteins, *Nature* **301:** 540–542.

Vonderviszt, F., and Simon, I., 1986, A possible way for prediction of domain boundaries in globular proteins from amino acid sequence, *Biochem. Biophys. Res. Commun.* **139:**11–17.

Zvelebil, M. J., Barton, G. J., Taylor, W. R., and Sternberg, M. J. E., 1987, Prediction of protein secondary structure and active sites using the alignment of homologous sequences, *J. Mol. Biol.* **195:**957–961.

14

Redundancies in Protein Sequences

Russell F. Doolittle

I. INTRODUCTION

At one point in the history of protein chemistry it was thought that—because of the vast number of possible sequences—proteins ought to exist in a countless number of forms befitting any conceivable function or structure. In fact, as the number of known amino acid sequences continues to mount, it is becoming abundantly clear that there is a practical limit to the number of types of protein structures that exist in living systems on Earth. Thus, although the number of possible sequences for 20 amino acids arranged randomly in strings of 350 units

Russell F. Doolittle • Center for Molecular Genetics, University of California, San Diego, La Jolla, California 92093.

is a superastronomical 20^{350}, certainly nowhere near that number of protein sequences has ever or will ever exist. Instead, a small number of genetically encoded protein structures has been expanded by the general route of "duplication and modification." The duplications come in various degrees, from the very short to the supragenic or chromosomal. Postduplication modification mostly takes the form of base substitutions leading to amino acid replacement, and, theoretically, the history of any protein ought to be evident by appropriate comparison of the diverging sequences.

Depending on the recency of a duplication and the rate of modification, however, it may be easy or difficult to recognize the preduplication heritage of a protein. Given enough time, relationships can be blurred to the point where recognition is no longer possible. Furthermore, other genetic events may conspire to confuse the interpretation. Deletions or insertions of all sizes may occur, for example. Gene conversions—by which we usually mean the replacement of an older version of a member of a multigene family by a new copy of some other member of the same family—can distort our view of the history of the system. Moreover, similar sequences may result from convergence as a result of selection for particular structures. Finally, distinguishing either divergent or convergent similarities from the vagaries of random sequence fluctuation is always a challenge.

In spite of these complicating factors, it should be possible to reconstruct most of the history of the vast majority of extant proteins on the basis of sequence similarities. It is also likely that an appreciation of this evolutionary expansion will allow the three-dimensional structures of many proteins to be inferred on the basis of their amino acid sequences alone whenever a structure from a related protein is available (Greer, 1985).

In this chapter I review the various kinds of sequence redundancy one finds in proteins, without paying much heed to the genetic mechanisms that lead to their existence. It is enough to say that nucleic acid replication by its very nature is delinquent. Beyond that, it should be kept in mind that the mispairing of similar nucleic acid sequences lies at the heart of many of these duplications, and as a result, duplication tends to beget more duplication (Doolittle, 1979).

The variety of redundant sequences in proteins defies simple categorization, ranging as it does from simple repeats of short sequences to the duplication of entire proteins. Our plan is to begin with the simplest kinds of short repeats and simple redundancies and then to move progressively in the direction of very large-scale duplications, emphasizing as we go the possible impact of the former on the latter.

II. ABOUT THE STRUCTURES OF PROTEINS

It is a part of the dogma of protein chemistry that three-dimensional structures are entirely dictated by primary structures, and in all of this discussion we must constantly remind ourselves of the structural implications of any redundancies we may find. Given the same environment, identical amino acid sequences must fold into identical three-dimensional structures. Implicit in this truth is the thought that divergent proteins with similar sequence ought to have similar structures. What we must be alert to, nonetheless, are those situations where similar sequences are not the result of a divergence but instead are the result of either chance or convergence. It has been shown, for example, that the same pentapeptidyl sequence can exist in very different secondary structure patterns in different proteins (Kabsch and Sander, 1984), which tells us that these pentapeptidyl sequences have different environments in the different proteins and, in all likelihood, different histories.

Regularities in protein sequences have been sought almost from the moment the first data appeared (Gamow *et al.*, 1956; Sorm, 1959). In that era, also, attempts were initiated to predict folding patterns and three-dimensional structures on the basis of amino acid sequences alone. These efforts have ranged from first-principle energy calculations (Scott *et al.*, 1967), on the one hand, to the empirical correlation of sequences observed in x-ray structures, on the other (Chou and Fasman, 1974). Numerous other studies have sought to establish departures from randomicity among amino acid sequences (Holmquist and Moise, 1975; Black *et al.*, 1976; Ivanov and Ivanov, 1980, *inter alia*), and reports of various kinds of repeats are too common even to list. Some studies have tried to find whether particular constellations of amino acids are preferentially involved in certain geometries. For example, α helices are well known to invoke a rhythmic repetition of polar and nonpolar residues. It remains to be shown, nonetheless, whether of all the permutations of polar and nonpolar residues, some are more likely to exist in helices than others. And, if some short sequences do indeed show up more often than others, we must determine whether they are the result of independent selection or are merely the progeny of earlier duplications.

Put another way, when we find a similar sequence in different proteins, can we distinguish gradual selection for advantageous structures from divergence from a common ancestor? As we shall see, given the large number of combinations and permutations for 20 amino acids, proteins that exhibit similarities over the full course of their lengths must be the result of divergence. In the cases of short sequences, however, distinguishing divergence from convergence can be troublesome. In general, the greater the number of residues over which a resemblance extends, the less likely it is to be the result of either chance or convergence.

III. ABOUT DATA BASES

One of the most stimulating aspects of the recent accrual of so many new protein sequences has been the availability of data banks that can be rapidly searched through by computer. Today, the first thing an investigator should do upon determining a new sequence is to compare it with all other known sequences to see if there are similarities (Doolittle, 1986b). Interpreting the results of such a search is something that ought not be dealt with lightly, however. For example, what are we to suppose if the sequence doesn't resemble any other? In such a case, we must keep in mind that the sizes of the collections are still somewhat modest, and the uniqueness of the new sequence may merely be a matter of a current limitation in data collection. On the other hand, suppose the sequence is absolutely identical to one already in the collection. In this case, two groups—knowingly or unknowingly—obviously have been working on the same protein. In many other cases, however, a newly determined sequence will be found to resemble one in the collection. Indeed, these are the findings that are revealing the general truth of a relatively small number of protein families.

A great deal can be inferred about a protein on the basis of sequence similarities with other proteins, and naturally investigators are eager to find related sequences. There are many marginal situations where investigators try to read too much into chance similarities, however, and one of the minor themes of this chapter is to encourage a cautious if not skeptical attitude about low-level resemblances.

Only a small fraction of contemporary proteins have had their sequences determined, although reckoning how many different extant sequences there are is not a trivial task (Doolittle, 1981). The present count shows that about 5000 protein sequences have been determined, most now being derived from DNA sequences. The representation in the available data banks is far from a random sampling, of course. Many of the entries are actually the same protein from different species, and although such comparative data are very valuable, we must be aware that they introduce a redundancy that must be taken into account in any analysis of the data, as is borne out by the following example.

Some years ago, I began a study of small-scale structural redundancies in protein sequences in a search for examples of structural convergences in proteins. For a data base, I obtained a tape of the *Atlas of Protein Sequence and Structure* (Dayhoff, 1978). The first question addressed was, do any dipeptidyl sequences occur more or less frequently than expected by chance? The data base contained about 120,000 residues, and therefore it was expected that each dipeptidyl sequence ought to occur often enough that a significant measure could be obtained (the average dipeptide ought to have occurred about 300 times, depending on the amino acids involved). Of the 400 possible dipeptidyl sequences, one exhibited a frequency that was dramatically above expectation. Thus, cysteinyl-histidine (Cys-His) occurred 255 times, even though the overall frequency expected on the basis of the composition of the data base indicated that it ought to occur only 99 times. Indeed, the inverted sequence His-Cys occurred 113 times, quite within the range of expectation.

Was there something special about Cys-His? As it happened, there was, but not in the way originally supposed. The sequence Cys-His occurs in every known cytochrome *c* as a part of the structure that binds the porphyrin. Since there were over a hundred cytochromes in the data base (almost 10% of the entries), they accounted for most of the excess. It was clear that this was not an appropriate data base for studying structural convergence, if only because of the great species redundancy.

It might be asked, why not just use all the known proteins from a single species? The sequences of many human proteins are known, and in fact some attempts along these lines have been undertaken (Saroff and Pretorius, 1981). As already implied, however, redundancy extends well beyond that observed from species to species. As an example, the 1978 *Atlas of Protein Sequence and Structure* data base contained over 100 immunoglobulins, mostly human and mouse, and all descended from a common ancestral stock.

For this and other reasons, I began a protein sequence collection called NEWAT (''new atlas''), originally as an accessory to the larger professional collections. The intention was to make the collection representative rather than comprehensive. From the start, however, the impossibility of its being truly nonredundant was recognized, if only because most proteins are evolved from other proteins. In the end, it came to contain many related sequences. Still, the collection proved very useful, and in much of this chapter I have leaned heavily on the 1985 version to sort out certain simple redundancies.

IV. SAMPLING AND STRUCTURAL TRENDS

Since most protein evolution is the result of duplication and modification, how are we to identify structurally convergent situations if they occur? For example, if we find that a certain grouping of amino acids occurs more often than expected by chance, how can we tell whether it was ''selected for'' on different occasions or is merely the radiative result of common ancestry? Moreover, given the large variance in many of the frequency distributions we will be considering, how can we distinguish between a ''trend'' and a ''drift''? One way to cope with the latter problem is to divide the sequence collection into biological groups. If an excess or deficiency shows up only in one group, it is not likely to be a universal feature of protein structure. Contrarily, if the same trend occurs in all groups, then there is a connotation of structural benefit strong enough to be selected for on independent occasions. We use this argument mainly in our discussion of the relative frequencies of the 20 amino acids, in one instance, and short peptidyl sequences, in another.

Table I. Amino Acid Compositions[a] of Six Groups in NEWAT '85

	Pro	Col	Vir	Euk	Ver	Hum	All
Leucine	8.9	9.5	9.6	8.9	8.7	9.2	9.14
Alanine	9.5	10.1	6.8	7.5	7.2	6.3	7.77
Glycine	8.0	7.5	6.5	6.9	7.6	7.1	7.21
Serine	6.0	5.5	7.4	7.1	7.0	7.5	6.83
Valine	7.1	7.5	5.9	6.1	6.8	6.5	6.61
Glutamic acid	6.1	6.6	5.5	5.8	6.8	6.8	6.27
Lysine	5.7	5.7	5.2	6.5	6.5	5.8	5.91
Threonine	5.9	5.3	6.6	5.8	5.6	6.1	5.90
Aspartic acid	5.8	5.6	4.8	5.4	5.4	5.1	5.30
Isoleucine	5.9	5.8	5.0	6.1	5.0	4.3	5.29
Proline	4.1	3.9	7.1	4.6	4.9	5.6	5.18
Arginine	5.0	5.9	5.3	4.3	4.9	4.9	5.11
Asparagine	4.1	4.1	4.2	5.4	4.1	4.1	4.29
Glutamine	3.9	4.3	4.5	3.9	3.9	4.3	4.18
Phenylalanine	3.8	3.5	3.6	4.3	4.1	4.0	3.87
Tyrosine	3.3	2.7	3.2	3.8	3.2	3.3	3.23
Histidine	2.0	2.0	2.5	2.4	2.2	2.7	2.33
Methionine	2.5	2.6	2.2	2.0	2.3	2.0	2.25
Cysteine	1.0	0.9	2.1	1.7	2.5	3.0	1.94
Tryptophan	1.3	1.0	1.7	1.4	1.3	1.5	1.40

[a]Values are residue-percents. Pro, prokaryotes except *E. coli*; Col, *E. coli*; Vir, viruses; Euk, eukaryotes but not vertebrates; Ver, vertebrates but not human; Hum, Human (taken from Doolittle, 1986b).

V. AMINO ACID COMPOSITIONS

Fundamental to any discussion of the distribution of amino acids in proteins is a consideration of their bulk composition. It is common knowledge that the 20 amino acids are not equally distributed in nature (Table I). In fact, the four most abundant amino acids (Leu, Ala, Gly, Ser) occur about four times more often than the least frequent four (Met, His, Cys, Trp).

There has been much discussion over the years as to whether the frequency of amino acids in proteins is a function of the number of codons available, or natural selection for an appropriate mix of side-chain properties or some combination of both. It is certainly true that there are six codons for the very common amino acids leucine and serine and only one each for the infrequent amino acids methionine and tryptophan (King and Jukes, 1969). Beyond that, however, the correlation is less than weak, and selection for a characteristic mix seems likely (Jukes *et al.*, 1975). The situation is confounded by variations in the numbers of tRNAs with different anticodons in various species, especially with regard to a tendency for high G and C in the third positions of many codons. What is clear, nonetheless, is that in all groups of organisms, certain amino acids are used in proteins more than others (Table I). Beyond that, the small differences that occur in different groups are sometimes attributable to sampling differences. For example, cysteine occurs more often among human proteins than among *E. coli* proteins (Table I), but this is at least partly because so many extracellular proteins—from blood and other fluids—have been studied in the case of humans.

VI. THE DISTRIBUTION OF DI- AND TRIPEPTIDYL SEQUENCES IN PROTEINS

The NEWAT '85 data base was analyzed with regard to the occurrences of all possible dipeptides and tripeptides. The aim was to find if any of these were selected for on independent

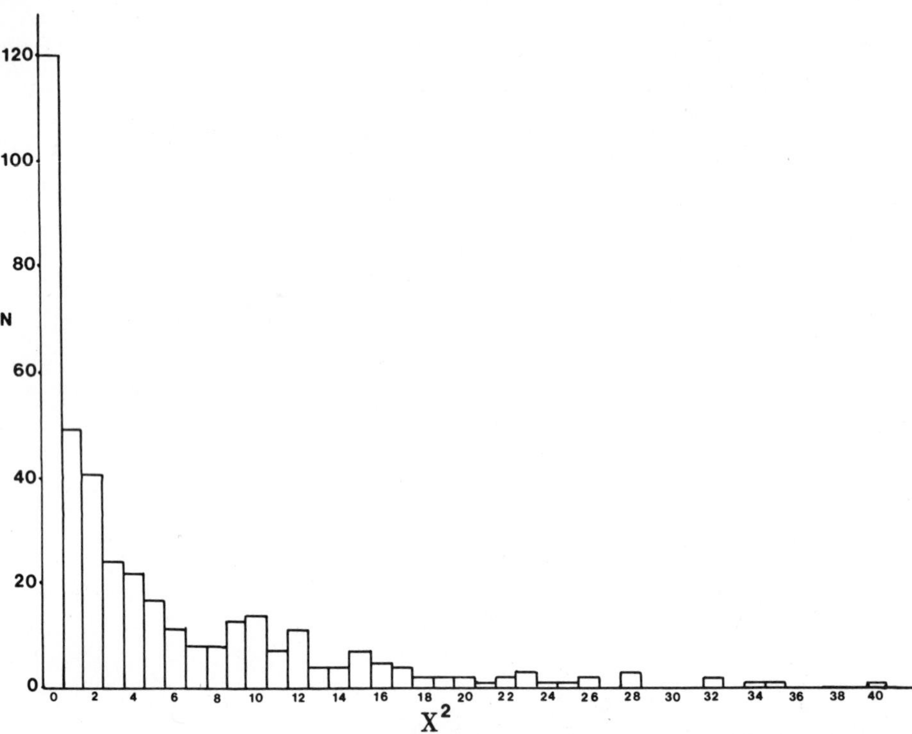

Figure 1. Distribution of χ^2 values for the occurrence of all 800 dipeptidyl sequences in NEWAT '85. The ten highest values (54, 62, 70, 72, 86, 92, 93, 94, 159, and 162) are not shown on the plot. Of these, the eight highest were for the homodipeptides of Glu, Ala, Lys, Pro, Cys, Gln, Arg, and Ser (see also Tables II and III).

occasions because of particularly useful properties. At the same time, we could test to see if any of them were subject to natural rejection. Are there any "forbidden" constellations of amino acids? At this microlevel, apparently not. Virtually every one of the 8000 possible tripeptidyl sequences has been identified in some protein, even those made exclusively of the least frequent amino acids. Of the 16 tripeptides not found among the 300,688 residues in NEWAT '85, all were composed of cysteine, histidine, and/or tryptophan. Most of these have now been identified in more recent and larger collections.

As for the relative frequencies of the 400 dipeptides, most fall reasonably within expectations based on the overall frequencies of the individual amino acids (Fig. 1). Of those that were overrepresented, the most significant were homodipeptides, a situation we will focus on momentarily. The most overrepresented heterodipeptide was Pro-Glu, which occurred 1157 times, even though only 977 were expected. The heterodipeptide Glu-Pro, on the other hand, was—in what may only be a curious coincidence—the most underrepresented dipeptidyl sequence, occurring only 730 times for the 977 expected.

There are other dipeptidyl sequences in which both peptides of a pair occur much less often than expected on the basis of the overall amino acid composition. These include Glu-Ser and Ser-Glu, Ser-Lys and Lys-Ser, Asp-Thr and Thr-Asp, Pro-Lys and Lys-Pro, and finally, Ala-Tyr and Tyr-Ala (Table II). With the exception of Thr-Asp, which is the least underrepresented of the ten sequences, the deficits are evident in all the biological divisions of the collection: prokaryotic, virus, and eukaryotic. The data strongly imply that there is a small but significant incompatibility between the two amino acids in each of these pairs.

Table II. Frequencies of Some Underrepresented Combinations of Amino Acids in NEWAT '85

	Observed	Expected	χ^{2a}
Glu-Ser	1022	1286	54
Ser-Glu	1116	1286	22
Lys-Ser	992	1213	40
Ser-Lys	1027	1213	28
Lys-Pro	811	921	13
Pro-Lys	776	921	23
Ala-Tyr	608	753	28
Tyr-Ala	616	753	25
Asp-Thr	786	938	24
Thr-Asp	880	938	3

[a] χ^2 = (expected − observed)^2/expected. See Fig. 1 for overall distribution of χ^2 values.

Another group of short sequences does seem to be overrepresented: homodipeptides (e.g., Asp-Asp, Thr-Thr, etc.) and homotripeptides (Asp-Asp-Asp, etc.) occur significantly more often than expected by chance (Table III). At first the thought was that this was an artifact created by long runs of particular amino acids, and so pains were taken to have all "long runs" represented only once. Even then, many of the excesses persisted, especially among the homotripeptides. Moreover, the overrepresentation occurs in all groups of proteins: prokaryotic, viral, and eukaryotic.

Table III. Occurrences of Homodipeptidyl and Homotripeptidyl Sequences in a Representative Data Base (NEWAT '85)[a]

	Homodipeptidyl[a]			Homotripeptidyl[a]		
Amino acid	Observed	Expected	Obs/exp	Observed	Expected	Obs/exp
Leucine	2649 (2360)	2514	1.05	281 (229)	229	1.23
Alanine	2353 (2052)	1815	1.30	287 (241)	141	2.04
Glycine	1702 (1499)	1563	1.09	186 (147)	112	1.66
Serine	1714 (1519)	1400	1.22	192 (158)	95	2.02
Valine	1441 (1337)	1314	1.10	98 (90)	86	1.14
Glutamic acid	1621 (1431)	1182	1.37	188 (134)	74	2.54
Lysine	1365 (1242)	1051	1.30	120 (101)	62	1.94
Threonine	1125 (1039)	1045	1.08	82 (75)	61	1.41
Aspartic acid	866 (790)	842	1.03	72 (49)	44	1.64
Isoleucine	874 (829)	841	1.04	40 (37)	44	0.91
Proline	1083 (916)	808	1.34	156 (129)	41	3.80
Arginine	1022 (925)	784	1.30	94 (78)	40	2.35
Asparagine	649 (601)	553	1.17	41 (37)	23	1.78
Glutamine	738 (647)	525	1.41	71 (45)	21	3.38
Phenylalanine	461 (434)	450	1.02	25 (21)	17	1.47
Tyrosine	372 (357)	312	1.19	12 (12)	10	1.20
Histidine	239 (228)	163	1.47	11 (9)	3	3.67
Methionine	192 (185)	152	1.26	7 (6)	3	2.33
Cysteine	215 (211)	112	1.92	3 (3)	2	1.50
Tryptophan	75 (73)	59	1.27	1 (1)	0	—

[a] NEWAT '85 contained 1169 entries and 300,688 residues. The di- or tripeptidyl sequences are listed in accordance with the relative abundances of the amino acids in the data base. Values in parentheses represent the frequencies when a run of the same amino acid, no matter how long, was only counted once (see text for fuller explanation).

The excess does not seem to be an adaptation to a need for similar side-chain properties in similar settings (even if adjacent positions in a polypeptide chain should ever be considered "similar settings") in that there is no corresponding excess of those heterodipeptidyl sequences composed of amino acids with similar side chains. Both Lys-Lys and Arg-Arg are in significant excess (Table III), but Lys-Arg and Arg-Lys are not. The basis for these apparent excesses remains mysterious.

VII. OTHER EXCESSES AND DEFICIENCIES

The survey of all possible 800 tripeptidyl sequences in NEWAT '85 uncovered some apparent excesses and deficiencies over and beyond the homotripeptides. A number of the alleged excesses were tracked further, and most turned out to be the result of individual proteins that contained large numbers of tandem repeats, a category we will deal with shortly. Even in a "representative sequence collection," then, there is always a threat that excesses will result from genetically clustered redundancy, and claims of overrepresentation that imply structural convergence must be judged very cautiously.

Deficiencies are another matter. If a tripeptidyl sequence is observed at a much lower frequency than expected—given the general amino acid frequencies—it cannot be attributed to excessive tandem duplication, nor can it reflect overrepresentation of proteins from a particular family. Accordingly, attention was given to identifying "rarer than expected" tripeptidyl sequences, in the event there were genuine structural reasons discouraging their existence.

Some of the more significant deficiencies in this category are listed in Table IV. They have been divided into two groups: those that are minimally represented in "both directions," which is to say the inverted sequence is also deficient, and those that are deficient but whose inverted sequences are present in the expected numbers. As an example of the former, both Glu-Ser-Lys and Lys-Ser-Glu are present well below expectation. This is in accord with all four constituent dipeptidyl sequences being underrepresented (Table II). In contrast, the tripeptidyl sequence Glu-Pro-Asp is significantly underrepresented, but Asp-Pro-Glu is present in excess (Table IV).

Table IV. Some Underrepresented Tripeptidyl Sequences and Their Inverted Counterparts[a]

	Sequence			Inverted sequence		
	Observed	Expected	χ^2	Observed	Expected	χ^2
Underrepresented one way only						
Glu-Pro-Asp	16	51	24	78	51	
Pro-Leu-Gly	52	102	23	95	102	
Ala-Glu-Ser	57	99	18	86	99	
Gln-Asn-Glu	11	33	15	36	33	
Asp-Arg-Gly	28	58	15	57	58	
Glu-Pro-Asn	16	41	15	56	41	
Glu-Pro-Lys	24	57	14	65	57	
Underrepresented both directions						
Ile-Lys-Leu	46	85	18	66	85	5
Arg-Phe-Glu	13	37	15	18	37	9
Arg-Ala-Thr	37	70	15	55	70	3
Glu-Ser-Lys	45	76	12	48	76	10
Arg-Glu-Ser	37	65	12	54	65	2

[a]The inverted sequence of Glu-Pro-Asp is Asp-Pro-Glu, etc. χ^2 = (expected − observed)^2/expected.

Table V. Occurrence of the 18 "Most Favorable" Tripeptidyl Sequences in β Turns in the NEWAT '85 Data Base[a]

Tripeptidyl	$(f \times f \times f)^b$	Observed	Expected	Obs/exp
1. Pro-Asn-Trp	9.60	11	9	1.22
2. Pro-Gly-Trp	9.55	16	15	1.07
3. Asn-Pro-Asn	9.26	32	28	1.14
4. Asn-Pro-Gly	9.21	52	48	1.08
5. Pro-Asp-Trp	9.00	16	11	1.46
6. Pro-Asn-Gly	8.74	58	48	1.21
7. Pro-Gly-Gly	8.69	66	81	0.81
8. Asn-Pro-Asp	8.67	67	35	1.91
9. Cys-Pro-Asn	8.58	21	12	1.75
10. Cys-Pro-Gly	8.52	25	21	1.19
11. Asp-Pro-Asn	8.45	67	35	1.91
12. Asp-Pro-Gly	8.41	30	59	0.51
13. Pro-Asp-Gly	8.19	90	59	1.53
14. His-Pro-Asn	8.05	22	15	1.47
15. Cys-Pro-Asp	8.03	26	15	1.73
16. His-Pro-Gly	8.01	36	26	1.39
17. Asp-Pro-Asp	7.92	55	43	1.28
18. His-Pro-Asp	7.54	46	19	2.42
Total		736	579	

[a]As judged by combining the probabilities at successive positions in the Chou and Fasman (1978) survey of 408 observed turns.
[b]All values multiplied by 10^3.

VIII. TURNS

The "turn" is a common structural feature of proteins. Indeed, many proteins are largely composed of chain segments that fold back on themselves to yield a series of β strands punctuated by turns. Chou and Fasman (1978) found that such turns tend to be composed of a definite subset of amino acids. Prominent among these is the frequent occurrence of a proline residue at the second position of a four-residue turn sequence. Given this general observation, we could further ask if the same short peptidyl sequences are used repeatedly in particular settings, as opposed to merely the same kinds of amino acids. For example, is the exact sequence Asp-Pro-Asn-Gly used more often in turns than the sequence Asn-Pro-Gly-Asp?

This problem can be addressed indirectly by examining the distribution of tripeptides involving these amino acids to see if—quite independent of known three-dimensional structures—they occur more or less often than predicted for a random distribution (the data base is not yet large enough to afford meaningful numbers for the 160,000 possible tetrapeptides). In this regard, I examined the frequency in NEWAT '85 of the 18 tripeptides most expected in turns as judged by the Chou and Fasman (1978) data. As a group, these tripeptidyl sequences do indeed occur more often than expected (Table V), although this does not mean that they exist as turns in all these situations.

Surprisingly, one of the sequences, Asp-Pro-Gly, occurs only half as often as it should, however, only 30 being observed for the 59 expected. Is there something uncomfortable about this arrangement? The inverted sequence, Gly-Pro-Asp, more closely meets expectations, there being 47 in the data base for the same 59 expected. In another vein, one of these tripeptidyl sequences, His-Pro-Asp, occurs more than twice as often as expected (Table V). A breakdown of the data revealed that the bulk of the overrepresentation occurs among the viral proteins, however, and it was not a universal feature of proteins from all groups of organisms.

IX. ACTIVE SITES

It is well known that certain short sequences are often particularly important for the function of particular enzymes. For example, Asp-Ser-Gly has long been associated with the "active sites" of serine proteases, just as Asp-Thr-Gly is characteristic of acid proteases. The question arises, do these short sequences occur more or less frequently than one would expect, and, in either case, is it because of selection or merely because the data base has a disproportionate number of those kinds of enzymes? As a test, the NEWAT '85 data base was examined, and the numbers of these sequence occurrences were counted. For example, Asp-Ser-Gly occurred 113 times, even though only 78 were expected. The collection contained 18 sequences that could be positively identified as serine proteases. As for Asp-Thr-Gly, 65 were observed, and 67 expected; only four acid proteases were identified in the collection. We may conclude from this that the vast majority of these sequences occur in nonprotease settings and that, far from being discriminated against, in at least one instance an "active site" sequence occurs in significant excess.

The same situation unfolds for most consensus sequences used to search for various "binding sites." Thus, the tripeptidyl sequence Arg-Gly-Asp, allegedly involved in binding certain proteins to particular cells (Ruoslahti and Pierschbacher, 1987), occurs 70 times in NEWAT'85, and the vast majority of these proteins clearly are not involved in attachments to cell surfaces. Similarly, the loose consensus sequence used for ATP-binding sites (Walker *et al.*, 1982) occurs in many proteins that do not bind nucleotides (Doolittle, 1986a). The rule that emerges is: most consensus sequences are not quite essential and almost never sufficient to identify a structure–function relationship.

X. PROTEASE ACTIVATION TARGETS

It is well known that many eukaryotic hormones and growth factors are synthesized as parts of precursor proteins and are liberated by the proteolysis of characteristic target sequences, among the most common of which are "double-basic" combinations of lysine and arginine residues. A few years ago Jornvall and Persson (1983) reported that proline is overrepresented in the position immediately following double-basic sequences, the presumed reason being to curb wanton proteolysis of nontarget proteins.

In the survey of the NEWAT '85 collection, no significant excess of those particular tripeptidyl sequences was observed, however. Indeed, the agreement with expectation was remarkable:

Sequence	Observed	Expected
Lys-Lys-Pro	67	54
Lys-Arg-Pro	29	46
Arg-Lys-Pro	33	46
Arg-Arg-Pro	56	40
Total	185	186

Jornvall and Persson (1983), working with a less representative data base, also concluded that all other tripeptidyl sequences beginning with the double-basic dipeptidyl combination were underrepresented. That this is not the case is apparent from the surplus already noted above for Arg-Arg and Lys-Lys.

Table VI. Some Long Runs of Identical Residues

Amino acid	Length of run	Protein	Reference
Glutamine	31	Notch protein, *Drosophila*	Wharton *et al.* (1985)
Glutamic acid	25	High-mobility protein, trout	Watson and Dixon (1981)
Asparagine	15	*Plasmodium falciparum* antigen	Wahlgren *et al.* (1986)
Alanine	14	Body pattern development protein, *Drosophila*	Poole *et al.* (1985)
Aspartic acid	13	RAD 6 protein, *S. cerevisiae*	Reynolds *et al.* (1985)
Serine	14	Phosvitin, chicken	Byrne *et al.* (1984)
Threonine	10	Salivary glue protein, *Drosophila*	Garfinkel *et al.* (1983)
Leucine	10	Kinase-related protein, feline sarcoma virus	Hampe *et al.* (1984)
Histidine	9	Histidine-rich protein, *Plasmodium lophurae*	Ravetch *et al.* (1984)
Glycine	9	Homeotic gene, *Drosophila*	Wharton *et al.* (1985)
Arginine	7	Sperm histone, bovine	Coelingh *et al.* (1972)
Proline	7	*gag* protein, T-cell leukemia virus	Shimotohno *et al.* (1985)
Lysine	6	Transforming protein, mouse	George *et al.* (1985)
Isoleucine	6	URF F, *Aspergillus* mitochondria	Netzker *et al.* (1982)
Valine	5	ILVB leader peptide, *E. coli*	Friden *et al.* (1982)
Tyrosine	5	Ig Heavy chain V-II region, human	Putnam *et al.* (1973)
Phenylalanine	4	Lactose permease gene, *E. coli*	Büchel *et al.* (1980)
Methionine	4	Endochitinase, *Phaseolus vulgaris*	Broglie *et al.* (1986)
Tryptophan	3	Capsid protein VP2 Poliovirus	Kitamura *et al.* (1981)
Cysteine	3	Neurotoxin, funnel-web spider	Sheumack *et al.* (1985)

XI. LONG RUNS OF IDENTICAL RESIDUES

Many proteins have long runs of a single amino acid, and many more have even longer runs of two amino acids with similar properties, such as glutamic and aspartic acids. The record repetition of a single residue occurs in the *Drosophila* notch protein, where there is a string of 31 glutamines in a row (Table VI). Other long strings include 25 glutamic acids in a row in the calf high-mobility-group protein and 14 consecutive serines in chicken phosvitin. How were these single-residue redundancies generated, and what are the ensuing structural consequences? Neither question has been answered satisfactorily.

Sometimes these long strings abruptly appear as intrusions in otherwise prosaic sequences. For example, the high-mobility-group protein from calf thymus can be aligned with the homologous sequence from a trout protein. The calf sequence, however, has a string of 41 glutamic and aspartic acids that the trout protein does not have (Fig. 2). As should become clear shortly, it is unlikely that the long run of negatively charged residues came about by single residue-by-residue replacement.

XII. SHORT TANDEM REPEATS

The evolution of periodic proteins with short tandem repeats has been much discussed in the past (Black and Dixon, 1967; Ycas, 1972; Doolittle, 1979; Ohno, 1984; *inter alia*). In many cases structural or functional benefits of short peptidyl repeats appear obvious. For example, regularity of structure and high tensile strength often go together, and the structural benefits of the Gly-Xxx-Pro tripeptidyl repeat in collagens are well known. Many other

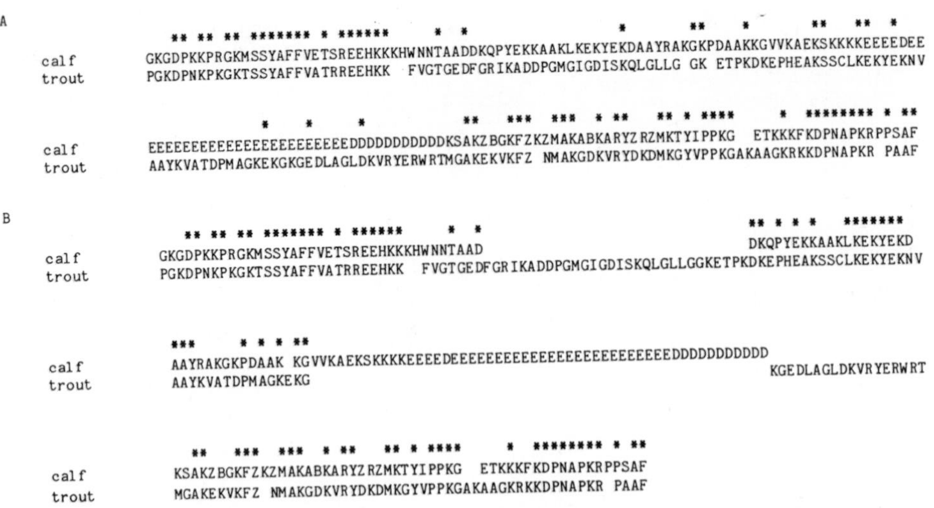

Figure 2. Two different alignments of high-mobility-group proteins from trout and calf (Watson and Dixon, 1981). The first (A) is the result of an alignment by a conventional computer program where no allowance is made for "shuffling." The second (B) shows that various segments have been rearranged. The long string of negatively charged amino acids in the calf sequence has no counterpart in the trout protein and was likely acquired from some other part of the genome. Asterisks (*) denote identical residues.

structural proteins, including insect egg casings and chorion proteins, have periodic sequences, often involving four or five residues per repeat. For example, the eggshell protein of schistosomes is mainly a repeated pentapeptide, Gly-Tyr-Asp-Lys-Tyr (Johnson *et al.*, 1987). The cytoskeletal protein involucrin has a long series of ten-residue repeats (Eckert and Green, 1986), and keratins in general are renowned for repeating structures.

Myosin, the principal extended protein of muscle, has an α-helical structure that consists mainly of heptad repeats (McLachlan *et al.*, 1975). For the most part, these are amphopathic structures with polar and nonpolar faces; the intertwined helices ("coiled coils") in myosin wrap together with their nonpolar residues inwards. Apoliproteins, which provide the topical covering for extracellular lipid transport, usually have an 11-residue repeat beginning with a proline residue. These are also amphopathic helices, the difference between this rendition of an α-helical protein and myosin being that the apoliproteins are able to kink and change direction every 11 residues (McLachlan, 1977), whereas the myosin rod continues on unswervingly.

Small-scale repeats also occur in nonstructural proteins, of course, often with a function that seems in line with the repetition. Thus, a negatively charged residue often alternates with a nonpolar one to give a periodic structure for binding cations, as in the case of the repetitive Pro-Glu structure in the *E. coli* iron-binding protein (Postle and Good, 1983). Similarly, a clam shell protein for seeding calcium deposition has an Xxx-Asp repeat (Weiner and Hood, 1975). The antifreeze protein found in the blood plasmas of some fish consists of a number of repeats of the tripeptide Ala-Ala-Thr (DeVries *et al.*, 1972). Evidently, the periodic but flexible structure prevents the seeding of ice crystals. In an interesting counterpoint, the ice nucleation protein of certain bacteria also has a highly repetitive structure, the sequence from the *Pseudomonas syringiae* protein containing more than 40 (imperfect) repeats of an octapeptidyl sequence. In this case the repeated structure is thought to play a role in causing rather than preventing seeding (Green and Warren, 1985).

There are many other examples of tandemly repeated short sequences. The most highly exposed portions of mammalian fibrinogen molecules, for example, have many repeats of a very hydrophilic 13-residue sequence (Doolittle *et al.*, 1979). Interestingly, the corresponding structure in the fibrinogen from a primitive fish has multiple repeats of an 18-residue repeat made up of the same kinds of residues, implying that these two duplication lines have independent origins and are the result of parallel evolution.

Other examples of relatively short tandem repeats with recognizable functional advantages are the antigenic proteins found in certain trypanosomes and malaria parasites (Van der Ploeg, 1987). An interesting feature is that these organisms appear to have a repertoire of different repeat segments that they can draw upon in different situations. Another fascinating example is afforded by some neuropeptide precursors found among invertebrates, in which cases peptide repeats occur that are subsequently released by a limited proteolysis that leads to a fusillade of biologically active agents (Scheller *et al.*, 1983).

Finally, there may be an added benefit to some tandem repeats that transcends ordinary structure–function considerations. In this regard, Ohno (1984) has made the point that repetitive oligonucleotide sequences have built-in advantages over and beyond their obvious ability to encode periodic amino acid sequences. Thus, oligonucleotide repeats that are not multiples of three automatically repeat themselves in all three coding frames, thereby rendering such sequences relatively impervious to deletions or insertions.

Consider a repeating sequence of a randomly chosen pentanucleotide, G-C-C-A-T:

```
             GCC  ATG   CCA   TGC  CAT   GCC  ATG   CCAT
frame 1:     Ala  Met   Pro   Cys  His   Ala  Met   Pro      ..
  frame 2:    Pro   Cys  His   Ala   Met   Pro  Cys   His    ..
    frame 3:   His   Ala   Met   Pro   Cys   His  Ala   Met  ..
```

This DNA sequence not only encodes a repeating pentapeptide: it does so in all three frames. Deletion of any base leaves the sequence more or less intact, no matter which frame is the initial coding frame.

Although this is an intriguing observation, the fact that so many highly repetitive sequences—including our collagen and fish antifreeze protein examples—involve repeats that are in fact multiples of three tends to vitiate the evolutionary importance of such oligonucleotide repeats being refractory to insertions and deletions.

XIII. PARTIAL GENE DUPLICATIONS

I am arbitrarily using the term "partial gene duplication" for those situations where the duplicated portion of the protein amounts to at least 20 amino acid residues, in contrast to either the short repeats we have been dealing with heretofore or those cases where all or most of a protein is duplicated, a subject treated in the next section. The intention is to limit this section to modular or domainal repeats, the inference being that these sequences are usually sufficiently long that their folding patterns are self-contained.

Probably the first example of an internal duplication of this sort was that found in bacterial ferredoxin (Jukes, 1966; Eck and Dayhoff, 1966). The entire ferredoxin sequence is composed of only 55 residues, and it was observed that residues 30–55 can be readily aligned with residues 1–26. Since that initial observation, many examples of subdomainal repeats of this sort have been identified in both prokaryotic and eukaryotic proteins. There may be multiple repeats of a relatively short segment, or a single repeat of a very long one. In *E. coli*, the ribosomal protein S1 contains six readily detected repeats of an 88-residue segment

Table VII. Some Proteins Known to Contain the EGF Domain[a]

Protein	Reference
Thyroid peroxidase	Magnusson *et al.* (1986)
Complement C9	DiScipio *et al.* (1984)
Factor VII	F. S. Hagen *et al.* (1986)
Factor IX	Kurachi and Davie (1982)
Factor X	Enfield *et al.* (1980)
Factor XII	McMullen and Fujikawa (1985)
Protein C	Fernlund *et al.* (1978)
Protein S	Lundwall *et al.* (1986)
Urokinase	Nagamine *et al.* (1984)
Plasminogen activator	Pennica *et al.* (1983)
Thrombomodulin	Jackman *et al.* (1986)
Protein Z	Hojrup *et al.* (1985)
Cartilage matrix protein	Argraves *et al.* (1987)
Vaccinia virus	Venkatesan *et al.* (1982)
LDL receptor	Yamamoto *et al.* (1984)
Sea urchin regulatory protein	Hursh *et al.* (1987)
EGF precursor	Scott *et al.* (1983)
Drosophila notch protein	Wharton *et al.* (1985)
C. elegans Lin-12	Greenwald (1985)

[a]The EGF unit, first observed in epidermal growth factor (EGF), consists of a segment 40–45 residues long containing six cysteines involved in three disulfide bonds.

(Doolittle *et al.*, 1982; Wittmann-Liebold *et al.*, 1983), for example, whereas β-galactosidase contains a twice-repeated structure of about 400 residues (Hood *et al.*, 1978).

One particular repeat found in a wide variety of animal proteins is the so-called "EGF domain." Epidermal growth factor (EGF) is a small polypeptide agent, originally purified from mammalian sources, that interacts with receptors on certain cells and stimulates them to grow and divide. The mammalian growth factor itself contains 54 amino acids and has a compact structure consistent with three disulfide bonds provided by six cysteine residues. The arrangement of disulfides is such that the polypeptide chain is constrained to turn back on itself three times. Interestingly, when the precursor of the growth factor was cloned and characterized, it was found to contain more than 1200 amino acids (Scott *et al.*, 1983), including eight or nine additional segments built on the same general plan as the genuine EGF. These units, which lacked several residues at the carboxy-terminal end of the active factor, were 40–45 residues long.

Surprisingly, units that were similar to these "EGF domains" were identified in several other mammalian proteins, including a number of blood-clotting factors (Banyai *et al.*, 1983; Doolittle *et al.*, 1984), and then, along with other portions of the EGF-precursor, in the low-density lipoprotein (LDL) receptor (Südhoff *et al.*, 1985). After that came the deluge, the same structure being found in many different settings (Table VII) and sometimes in long series of tandem repeats (Fig. 3). So far, all of the occurrences are in animal proteins, implying that the invention occurred after the divergence of plants and animals. On the other hand, the number of plant proteins whose sequences are known is still relatively small, and the structures may yet show up among them. In any event, the EGF domain appears to be a particular apt structure for an extracellular existence in an oxygenic world.

There are a number of other disulfide sequence motifs found in animal extracellular proteins. The β_2 glycoprotein repeat has about 60 amino acids, four of which are cysteines

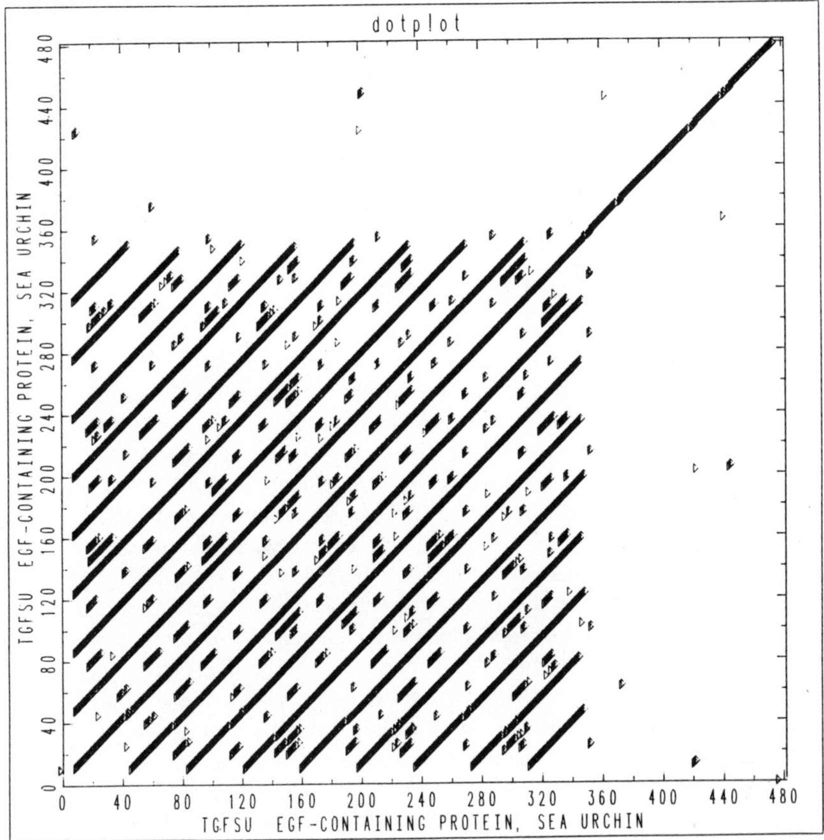

Figure 3. Self-diagonal plot of the sea urchin regulatory protein (Hursh *et al.*, 1987) showing a high degree of periodicity corresponding to segments about 40 residues in length (see also Table VII).

Table VIII. Assorted Proteins Containing the "β_2GP Repeat" Unit[a]

Protein	Number of repeats	Reference
Cartilage proteoglycan core protein	2	Sai *et al.* (1986)
Factor B regulatory unit	3	Mole *et al.* (1984)
Complement C2	3	Bentley (1986)
Decay-accelerating factor	4	Medof *et al.* (1987)
β_2 Glycoprotein	5	Lozier *et al.* (1984)
Complement C4-binding protein	8	Chung *et al.* (1985)
Factor XIII b chain	10	Ichinose *et al.* (1986)
Complement factor H	20	Kristensen and Tack (1986)

[a]The "β_2GP repeat" unit, first observed in β_2 glycoprotein, averages 60 residues and contains four characteristic cysteines.

bound in two disulfide bonds. So far this stable segment has been reported in eight otherwise unrelated mammalian proteins (Table VIII). Its invention is likely to be fairly recent, perhaps in a primitive vertebrate animal. Like the EGF domain, its widespread occurrence is attributable to "exon shuffling" (Doolittle, 1985; Davie *et al.*, 1986). Thus, many of these modular units have been found to be encoded in DNA such that they are bounded by intervening sequences (introns). The increased possibilities for crossing over, as well as other possible genetic peccadillos provided by intervening sequences, have resulted in these genetic units being widely dispersed and exchanged. It is likely that the abrupt segmental differences in the high-mobility-group proteins (Fig. 2) are the result of exon shuffling.

XIV. TOTAL GENE DUPLICATIONS

Many families of proteins are known, the members of which are the result of an extended series of total gene duplications. By "total," we are implying that all the attendant wherewithal for gene expression was included in the duplication event, including—usually—promoters, initiators, terminators, and any other accessory signals required. As a result, some of these events may involve the duplication of hundreds of kilobases of DNA.

Some of the families are very old and date back several billion years, considerably antedating the prokaryote—eukaryote divergence. Indeed, most if not all enzymes have roots that can be traced to these primordial times. There are hordes—literally thousands—of serine proteases, for example, that can be rooted back to a common ancestor in preprokaryotic times. The same is true for many other enzyme families (Doolittle *et al.*, 1986).

Sometimes the divergence is so deep and the rate of change so marked that the trail blurs. The acyl-amino acid-tRNA synthetases are a case in point. Given their common function, it has long been assumed that all of these enzymes descended from a common ancestor (Doolittle, 1979). The determination of the first dozen sequences, however, gave rise to a surprisingly unrelated set. Only a few icebergs of resemblance existed, and most of these could easily have been the result of chance or convergence. It was not until the recent unraveling of the valine-tRNA synthetase sequence that a set of relationships could be drawn that had any real statistical validity (Heck and Hatfield, 1988). The point to be made here is that many of the earliest and most interesting relationships are subtle, and many good cases are still to be made.

In the meantime, reports of family connections among nonenzyme proteins are pouring in at an amazing rate, particularly for proteins from eukaryotes. Some of these are now so routine as to be prosaic; the data bases already contain large families of hormones, growth factors, globins, histones, immunoglobulins, cell adhesion proteins, protease inhibitors, blood-clotting and complement factors, and interferons, *inter alia.*

XV. UNEXPECTED RELATIONSHIPS

Although reports of unexpected common ancestry for proteins are no longer greeted with the astonishment they once evoked, it is worth our while to review some of them, if only for the lesson they provide in how focused the expansion of protein structures can be. Probably the earliest of the "sequence surprises" was the finding that the milk protein lactalbumin is decended from an ancestor of the enzyme lysozyme (Brew *et al.*, 1967). Shortly thereafter, haptoglobin, a constituent of vertebrate blood plasma, was found to be related to serine proteases (Barnett *et al.*, 1972). Angiotensinogen, a large-molecular-weight precursor of a small polypeptide hormone, was found to be a member of the serine protease inhibitor family (Doolittle, 1983), and blood-clotting factor VIII, the protein affected in classical hemophilia, is clearly related to the copper-binding protein ceruloplasmin (Vehar *et al.*, 1984). More recently, the eukaryotic β-adrenergic receptor was found to be remarkably similar to rhodopsin (Dixon *et al.*, 1986). In none of these cases was there any indication in advance of the sequence determination that these proteins had anything in common. We can anticipate that many more such surprises are in store.

At the same time, we must be cautious not to push the data too far. In their zeal to establish common ancestry, some investigators attach too much significance to similarities that could easily have resulted from chance or convergence. Indeed, during the course of searching thousands of sequences during the past 10 years, I have happened on a number of similarities that had all the trappings of genuine homology, and yet common sense indicated that they were merely spurious similarities. The tails of any distribution can stretch out very far, of course. Consider, for example, a "match" recently picked up during a routine search of a newly published sequence against the data base:

.. F R Ⓛ Ⓐ Ⓐ S P Ⓛ T N Ⓚ Ⓟ Ⓠ T Ⓟ S Ⓡ Ⓟ Ⓐ Ⓢ E Q P P A Ⓠ Ⓟ R ..
.. L L Ⓛ Ⓐ Ⓐ G S Ⓛ G Q Ⓚ Ⓟ Ⓠ R Ⓟ R Ⓡ Ⓟ Ⓐ Ⓢ P I S T I Ⓠ Ⓟ K ..

Do you think this is a significant relationship? Half of the residues are identical (14 of 28), and there are several cases of similar amino acids (Ser/Gly, Asn/Gln, Arg/Lys). As it happens, the upper sequence is from the carboxy-terminal portion of a bacterial protein involved in chemoreception (Krikos *et al.*, 1983), while the lower sequence is from the amino terminus of the γ chain of the C8 component of human complement (Ng *et al.*, 1987)! As we will see momentarily, the complement protein belongs to a family of eukaryotic proteins, none of the others of which have any detectable similarity to the bacterial protein. Accordingly, we would be hard pressed to think there is any evolutionary connection between these two sequences. The question remains, then, does the similarity result from convergence for a particular structure, or is it just the tail end of a chance distribution?

XVI. HOW MANY SUPERFAMILIES?

At this point, the number of known amino acid sequences is such that the chances are about even that a search of any newly determined sequence will be found to resemble—at least, in part—one already known. Although this is a broad statement requiring some qualification—in line with our previous comments about the biased nature of current sequence collections—it is made on the basis of actual observation. The question arises, can we use this information to estimate the total number of families or superfamilies in existence? Several such attempts have been made previously (Zuckerkandl, 1975; Dayhoff, 1978; Doolittle, 1981), those early estimates falling diffusely between 100 and 1000. Now that we have considerably more data, we might be able to gauge this figure more accurately, especially in the light of our awareness of modular shuffling among modern proteins. As predicted (Doolittle, 1981), as more sequences are determined, more relationships are uncovered among already known sequences. Thus, low-level sequence resemblances become more significant as more members are added to the alignment. Recently, for example, an extended relationship has been developed connecting a wide variety of cellular receptors and immunoglobulins (Hayashida *et al.*, 1988).

There are many other examples, but let me bring attention to only one. The α_1 acid glycoprotein (a.k.a. orosomucoid) sequence has been known for a very long time (Schmid *et al.*, 1973), but it was not until the publication of a number of other sequences a dozen years later that its position in an evolutionary network became known (Pervaiz and Brew, 1987).

```
                                         *
rbph             ERDCRVSSFRV KENFDKARFSGTWYAMAKKDPEGLFLQDN IVAEFSVDETGQMSATAKGRV
purc              QTCAVDSFSV KDNFDPKRYAGKWYALAKKDPEGLFLQDN ISAEYTVEEDGTMTASSKGRV
c8gh      QKPQRPRRPASPISTIQP KANFDAQQFAGTWLLVAV GSACRFLQEQGHRAEAT TLHVAPQGTAMAVS
hchu           GPVPTPP DNIQV QENFNISRIYGKWYNLAI GSTCPWLKIMDRMTVSTLVLGEGATEAEISMT
blgb                  LIVTQT MKGLDIQKVAGTWYSLAMAASDISLLDAQSAPLRVYVEELKPTPEGDLEIL
agpr            QNPEPANITLGIPITNGTLRWLSDKWFYMGAAFRDPVFKQAVQTIQTEYFYLTPNLINDTIELR

                  *
rbph      RLLNNWDVCADMVGTFT  DTEDPAKFKMKYWGVASFLQKGNDDHWIVDTDYDTYAVQYSCRLLNLDGT
purc      KLFGFWVICADMAAQYTVPDPTTPAKMYMTYQGLASYLSSGGDNYWVIDTDYDNYAITYACRSLKEDGS
c8gh      TFRKLDGICWQVRQLYG  DTGVLGRFL     LQARGARGAVHVVVAETDYQSFAVL    YLERAGQ
hchu      STRWRKGVCEETSGAYE  KTDTDGKFL     YHKSKWNITMESYVVHTNYDEYAIFLTKKFSRHHGPT
blgb      LQKWENGECAQKKIIAE  KTKIPAVFK     IDALN   ENKVLVLDTDYKKYLLF CMENSAEPEQS
agpr      EFQTTDDQCVYNFTHLG  VQRENGTL      SKCAGAVKIFAHLIVLKKHGTFMLAF   NLTDENR

rbph       CADSYSFVFSRDPNGLPPEAQKI  VRQRQ EELCLARQYRLIVHNGYCDGRSERNLL
purc       CDDGYSLIFSRNPRGLPPAIQRI  VRQKQ EEICMSGQFQPVLQSGAC
c8gh      LSVKLYARSLPVSDSVLSGFEQRVQEAHLTE DQIFYFPKYGFCEAADQFHVLDEVRR
hchu      ITAKLYGRAPQLRETLLQDFRVVAQGVGIPE DSIFTMADRGECVPGEQEPEPILIPR
blgb      LACQCLVRTPEVDDEALEKFDKALKALPMHI RLSF         NPTQLEEQCHI
agpr       GLSFYAKKPDLSPELRKIFQQAVKDVGMDESEIVFVDWTKDKCSEQQKQQLELEKETKKETKKDP
```

Figure 4. Multiple alignment showing that the γ chain of the C8 component of complement (Ng *et al.*, 1987) belongs to the recently described "lipocalin" family (Pervaiz and Brew, 1987). Abbreviations: rbph, retinol-binding protein, human (Laurent *et al.*, 1985); purc, porpurin, chicken (Schubert *et al.*, 1986); c8gh, complement C8 γ chain, human (Ng *et al.*, 1987); hchu, α_1-microglobulin, human (López Otin *et al.*, 1984); blgb, β-lactoglobulin, bovine (Godovac-Zimmermann *et al.*, 1985); agpr, α_1 acid glycoprotein, rat (Ricca and Taylor, 1981). Asterisks (*) denote residues identical in all six sequences.

Thus, although the sequence resemblance is low, when α_1 acid glycoprotein is aligned with several other proteins, including retinol-binding protein (Laurent *et al.*, 1985), β-lactoglobulin (Godovac-Zimmermann *et al.*, 1985), α_2-microglobulin (López Otin *et al.*, 1984), and purpurin (Schubert *et al.*, 1986), the significance becomes obvious. The family has been dubbed "lipocalins" to reflect the binding of lipophilic substances that these proteins enact (Pervaiz and Brew, 1987). Interestingly, while I was in the final stages of writing this chapter, the sequence of the γ chain of the C8 component of human complement was published (Ng *et al.*, 1987). When I searched its sequence against the data base, the computer reported that this protein, also, clearly belongs to this family (Fig. 4).

In the meantime, the membership of some other families is growing rapidly. The rhodopsin–hormone–receptor family, mentioned above, now has at least six known members, including the α-adrenergic receptor (Kobilka *et al.*, 1987), the muscarinic acetylcholine receptor (Kubo *et al.*, 1986), the yeast mating-type receptor (D. C. Hagen *et al.*, 1986), and a receptor for substance K (Masu *et al.*, 1987).

What will the final number of families be? In the shadowy light of the still incomplete data set, let me make a prediction. Of the 50,000 to 100,000 gene products that are expected to be revealed when the human genome is sequenced, it will not be surprising if as many as 90% are identifiable on the basis of their sequences alone, and of these a full 90% may belong to as few as 100 superfamilies. The 10% that will not be recognizable amounts to a large number, of course (5,000–10,000). That they are not obviously related to other proteins may be because they represent unique and singular materials ("one-member families") or—more likely—that sequence resemblances have been blurred beyond the point of recognition by a long course of amino acid replacement.

Table IX. Some Members of the "Serpin" Family

Known protease inhibitors	
α_1-Antitrypsin	Carrell *et al.* (1982)
α_1-Antichymotrypsin	Chandra *et al.* (1983a)
α_2-Antiplasmin	Sumi *et al.* (1986)
Antithrombin III	Chandra *et al.* (1983b)
Protein C inhibitor	Suzuki *et al.* (1987)
C1 inhibitor	Bock *et al.* (1986)
Others	
Leuserpin-2	Ragg (1986)
Ovalbumin	McReynolds *et al.* (1978)
Angiotensinogen	Kageyama *et al.* (1984)
Thyroxine-binding globulin	Flink *et al.* (1986)
Corticosteroid-binding globulin	Hammond *et al.* (1987)
Endosperm protein Z, barley	Hejgaard *et al.* (1985)

XVII. CLOSING COMMENTS

The vast majority of extant proteins are the direct result of a continuous series of genetic duplications and subsequent modification. As a result, redundancy is a built-in characteristic of protein sequences. Already, backward trails through the course of the expansion can be traced to ancient roots in many cases, and eventually it should be possible to reconstruct a large part of the hierarchy, even though some of the most ancient relationships may have been blurred beyond the point of certainty.

Particular structures are found repeatedly in certain kinds of proteins, and in most cases the evidence favors a single invention and propagation by duplication and genetic shuffling rather than coincident evolution by selective convergence. Thus, the many nucleotide binding sites found in diverse settings appear to have enough in common at both the primary and tertiary structure levels to preclude separate *de novo* invention. Similarly, the widespread occurrence of the "EGF domain" among extracellular animal proteins reflects a particularly stable modular unit being born once and then being shaped and modified for different recognition roles in different settings. And on a grander scale, the chance evolution of a particularly stable structure found in certain protease inhibitors seems to be at the heart of this same full-sized protein being employed in diverse roles that can be very different from protease inhibition (Table IX).

Given the nature of the expansion, proteins at the microsequence level—di- and tripeptidyl—are remarkably random. Thus, virtually all 8000 tripeptidyl sequences are observed in current data bases in numbers consistent with a random distribution for the amino acid composition of the data bases as a whole. Given a genetic expansion from a small number of starter types, this might not have been predicted, and what it implies to me is that there is a very large number of ways to assemble 20 amino acids into equivalent structures and that there has been a very thorough churning over the eons. This will not be news to crystallographers, of course, who often observe very similar three-dimensional structures that have very dissimilar amino acid sequences.

There are a few significant departures from expectation at the dipeptidyl and tripeptidyl levels, but these do not appear to be great enough to have any significant impact on the results of searching and aligning distantly related sequences. With rare exception, similarities that meet the test of conventional matching procedures that depend on randomized sequences as controls (e.g., Feng *et al.*, 1985) are the results of a divergence from a common ancestral sequence; they are not likely to be the result of chance matching of overrepresented short sequences. On the other hand, some of the sequence differences between related proteins are so bizarre that even the crystallographers must be surprised. There appear to be intrusions in many sequences—long strings of identical residues—that apparently do not disrupt other elements of the protein's structure. It is as though many proteins are constantly "blebbing"; if these structural warts do no harm, they have as good a chance of passing through the filter of natural selection as any other neutral mutation.

The full extent of redundancies in protein sequences may be revealed if current efforts to sequence the human genome are enacted. I have already commented that it is likely that the vast majority of the potential gene products will be readily assigned to known protein families. As a result, it should be possible in many cases to model three-dimensional structures on the basis of known relatives. Beyond that, enough data should be forthcoming to resolve questions about whether or not certain constellations of amino acids really are favored or discouraged, even though we already can see that the distinction between short-range convergent tendencies and longer-range genuine homology is not a significant problem in determining most family relationships. In the limit, the human inventory may make it possible to estimate the number of starter protein types.

ACKNOWLEDGMENTS. I am grateful to Karen Anderson for considerable help in compiling materials for this manuscript and to Da-Fei Feng for programming assistance.

XVIII. REFERENCES

Argraves, W. S., Deak, F., Sparks, K. J., Kiss, I., and Goetinck, P. F., 1987, Structural features of cartilage matrix protein deduced from cDNA, *Proc. Natl. Acad. Sci. U.S.A.* **84:**464–468.

Banyai, L., Varadi, A., and Patthy, L., 1983, Common evolutionary origin of the fibrin-binding structures of fibronectin and tissue-type plasminogen activator, *FEBS Lett.* **163:**37–41.

Barnett, D. R., Lee, T. H., and Bowman, B. H., 1972, Amino acid sequence of the human haptoglobin β chain. Amino- and carboxyl-terminal sequences, *Biochemistry* **11:**1189–1194.

Bentley, D. R., 1986, Primary structure of human complement component C2, *Biochem. J.* **239:**339–345.

Black, J. A., and Dixon, G.H., 1967, Evolution of protamine: A further example of partial gene duplication, *Nature* **216:**152–154.

Black, J. A., Harkins, R. N., and Stenzel, P., 1976, Non-random relationships among amino acids in protein sequences, *Int. J. Peptide Protein Res.* **8:**125–130.

Bock, S. C., Shriver, K., Nielson, E., Thogersen, H.-C., Wiman, B., Donaldson, V. H., Eddy, R. L., Marrinan, J., Radziejewska, E., Huber, R., Shows, T. B., and Magnusson, S., 1986, Human C1 inhibitor: Primary structure, cDNA cloning, and chromosomal localization, *Biochemistry* **25:**4292–4301.

Brew, K., Vanaman, T. C., and Hill, R. L., 1967, Comparison of the amino acid sequence of bovine α-lactalbumin and hens eggwhite lysozyme, *J. Biol. Chem.* **242:**3747–3749.

Broglie, K. E., Gaynor, J. J., and Broglie, R. M., 1986, Ethylene-regulated gene expression: Molecular cloning of the genes encoding an endochitinase from *Phaseolus vulgaris, Proc. Natl. Acad. Sci. U.S.A.* **83:**6820–6824.

Büchel, D. E., Gronenborn, B., and Müller-Hill, B., 1980, Sequence of the lactose permease gene, *Nature* **283:** 541–545.

Byrne, B. M., van het Schip, A. D., van de Klundert, J. A. M., Arnberg, A. C., Gruber, M., and Geert, A. B., 1984, Amino acid sequence of phosvitin derived from the nucleotide sequence of part of the chicken vitellogenin gene, *Biochemistry* **23:**4275–4279.

Carrell, R. W., Jeppsson, J.-O., Laurell, C.-B., Brennan, S. O., Owen, M. C., Vaughan, L., and Boswell, D. R., 1982, Structure and variation of human α_1-antitrypsin, *Nature* **298:**329–334.

Chandra, T., Stackhouse, R., Kidd, V. J., Robson, K. J. H., and Woo, S. L. C., 1983a, Sequence homology between human α_1-antichymotrypsin, α_1-antitrypsin, and antithrombin III, *Biochemistry* **22:**5055–5061.

Chandra, T., Stackhouse, R., Kidd, V. J., and Woo, S. L. C., 1983b, Isolation and sequence characterization of a cDNA clone of human antithrombin III, *Proc. Natl. Acad. Sci. U.S.A.* **80:**1845–1848.

Chou, P. Y., and Fasman, G. D. 1974, Prediction of protein conformation, *Biochemistry,* **13:**222–245.

Chou, P. Y., and Fasman, G. D., 1978, Empirical prediction of protein conformation, *Annu. Rev. Biochem.* **47:** 251–276.

Chung, L. P., Bentley, D. R., and Reid, K. B. M., 1985, Molecular cloning and characterization of the cDNA coding for C4b-binding protein, a regulatory protein of the classical pathway of the human complement system, *Biochem. J.* **230:**133–141.

Coelingh, J. P., Monfoort, C. H., Rozijn, T. H., Gevers Leuven, J. A., Schiphof, R., Steyn-Parvé, E. P., Braunitzer, G., Schrank, B., and Ruhfus, A., 1972, The complete amino acid sequence of the basic nuclear protein of bull spermatozoa, *Biochim. Biophys. Acta* **285:**1–14.

Davie, E. W., Ichinose, A., and Leytus, S. P., 1986, Structural features of the proteins participating in blood coagulation and fibrinolysis, *Cold Spring Harbor Symp. Quant. Biol.* **51:**509–514.

Dayhoff, M. O., 1978, *Atlas of Protein Sequence and Structure,* Volume 5, Suppl. 3, National Biomedical Research Foundation, Washington.

DeVries, A. L., Vandenheede, J., and Feeney, R. E., 1972, Primary structure of freezing point-depressing glycoproteins, *J. Biol. Chem.* **246:**305–308.

DiScipio, R. G., Gehring, M. R., Podack, E. R., Kan, C. C., Hugli, T. E., and Fey, G. H., 1984, Nucleotide sequence of cDNA and derived amino acid sequence of human complement component C9, *Proc. Natl. Acad. Sci. U.S.A.* **81:**7298–7302.

Dixon, R. A. F., Kobilka, B. K., Strader, D. J., Benovic, J. L., Dohlman, H. G., Frielle, T., Bolanowski, M. A., Bennett, C. D., Rands, E., Diehl, R. E., Mumford, R. A., Slater, E. E., Sigal, I. S., Caron, M. G., Lefkowitz, R. J., and Strader, C. D., 1986, Cloning of the gene and cDNA for mammalian β-adrenergic receptor and homology with rhodopsin, *Nature* **321:**75–79.

Doolittle, R. F., 1979, Protein evolution, in: *The Proteins* (H. Neurath and R. L. Hill, eds.), 2nd ed., Volume 4, Academic Press, New York, pp. 1–118.

Doolittle, R. F., 1981, Similar amino acid sequences: Chance or common ancestry? *Science* **214:**149–159.

Doolittle, R. F., 1983, Angiotensinogen is related to the antitrypsin–antithrombin–ovalbumin family, *Science* **222:**417–419.

Doolittle, R. F., 1985, The genealogy of some recently evolved vertebrate proteins, *Trends Biochem. Sci.* **10:** 233–237.

Doolittle, R. F., 1986a, Protein sequence data banks: The continuing search for related structures, in: *Protein Engineering* (M. Inouye, ed.), Academic Press, Orlando, FL, 15–27.

Doolittle, R. F., 1986b, *Of URFs and ORFs: A Primer on How to Analyze Derived Amino Acid Sequences,* University Science Books, Mill Valley, CA.

Doolittle, R. F., Watt, K. W. K., Cottrell, B. A., Strong, D., and Riley, M., 1979, Amino acid sequence of the α-chain of human fibrinogen, *Nature* **280:**464–468.

Doolittle, R. F., Woodbury, N. W., and Jue, R. A., 1982, Ribosomal protein S1 is the product of a series of contiguous duplications, *Biosci. Rep.* **2:**405–412.

Doolittle, R. F., Feng, D.-F., and Johnson, M. S., 1984, Computer-based characterization of epidermal growth factor precursor, *Nature* **307:**558–560.

Doolittle, R. F., Feng, D.-F., Johnson, M. S., and McClure, M. A., 1986, Relationships of human protein sequences to those of other organisms, *Cold Spring Harbor Symp. Quant. Biol.* **51:**447–455.

Eck, R. V., and Dayhoff, M. O., 1966, Evolution of the structure of ferredoxin based on living relics of primitive amino acid sequences, *Science* **152:**363–366.

Eckert, R. L., and Green, H., 1986, Structure and evolution of the human involucrin gene, *Cell* **46:**583–589.

Enfield, D. L., Ericsson, L. H., Fujikawa, K., Walsh, K. A., Neurath, H., and Titani, K., 1980, Amino acid sequence of the light chain of bovine factor X, *Biochemistry* **19:**659–667.

Feng, D. F., Johnson, M. S., and Doolittle, R. F., 1985, Aligning amino acid sequences: Comparison of commonly used methods, *J. Mol. Evol.* **21:**112–125.

Fernlund, P., Stenflo, J., and Tufvesson, A., 1978, Bovine protein C: Amino acid sequence of the light chain, *Proc. Natl. Acad. Sci. U.S.A.* **75:**5889–5892.

Flink, I. L., Bailey, T. J., Gustafson, T. A., Markham, B. E., and Morkin, E., 1986, Complete amino acid sequence of human thyroxine-binding globulin deduced from closed DNA: Close homology to the serine antiproteases, *Proc. Natl. Acad. Sci. U.S.A.* **83:**7708–7712.

Friden, P., Newman, T., and Freundlich, M., 1982, Nucleotide sequence of the *ilvB* promoter–regulatory region: A biosynthetic operon controlled by attenuation and cyclic AMP, *Proc. Natl. Acad. Sci. U.S.A.* **79:** 6156–6160.

Gamow, G., Rich, A., and Ycas, M., 1956, The problem of information transfer from the nucleic acids to proteins, *Adv. Biol. Med. Phys.* **4:**23–68.

Garfinkel, M. D., Pruitt, R. E., and Meyerowitz, E. M., 1983, DNA sequence, gene regulation and modular protein evolution in the *Drosophila* 68C glue gene cluster, *J. Mol. Biol.* **168:**765–789.

George, D. L., Scott, A. F., Trusko, S., Glick, B., Ford, E., and Dorney, D., 1985, Structure and expression of amplified cKi-*ras* gene sequences in Y1 mouse adrenal tumor cells, *EMBO J.* **4:**1199–1203.

Godovac-Zimmermann, J., Conti, A., Liberatori, J., and Braunitzer, G., 1985, The amino-acid sequence of β-lactoglobulin II from horse colostrum (*Equus caballus,* Perissodactyla): β-Lactoglobulins are retinol-binding proteins, *Biol. Chem. Hoppe-Seyler* **366:**601–608.

Green, R. L., and Warren, G. J., 1985, Physical and functional repetition in a bacterial ice nucleation gene, *Nature* **317:**645–648.

Greenwald, I., 1985, *lin*-12, A nematode homeotic gene, is homologous to a set of mammalian proteins that includes epidermal growth factor, *Cell* **43:**583–590.

Greer, J., 1985, Protein structure and function by comparative model building, *Ann. N.Y. Acad. Sci.* **439:**144–163.

Hagen, D. C., McCaffrey, G., and Sprague, G. F., 1986, Evidence the yeast STE3 gene encodes a receptor for the peptide pheromone a factor: Gene sequence and implications for the structure of the presumed receptor, *Proc. Natl. Acad. Sci. U.S.A.* **83:**1418–1422.

Hagen, F. S., Gray, C. L., O'Hara, P., Grant, F. J., Saari, G. C., Woodbury, R. G., Hart, C. E., Insley, M., Kisiel, W., Kurachi, K., and Davie, E. W., 1986, Characterization of a cDNA coding for human factor VII, *Proc. Natl. Acad. Sci. U.S.A.* **83:**2412–2416.

Hammond, G. L., Smith, C. L., Goping, I. S., Underhill, D. A., Harley, M. J., Reventos, J., Musto, N. A., Gunsalus, G. L., and Bardin, C. W., 1987, Primary structure of human corticosteroid binding globulin, deduced from hepatic and pulmonary cDNAs, exhibits homology with serine protease inhibitors, *Proc. Natl. Acad. Sci. U.S.A.* **84:**5153–5157.

Hampe, A., Gobet, M., Sherr, C. J., and Galibert, F., 1984, Nucleotide sequence of the feline retroviral oncogene v-*fms* shows unexpected homology with oncogenes encoding tyrosine-specific protein kinases, *Proc. Natl. Acad. Sci. U.S.A.* **81:**85–89.

Hayashida, H., Kuma, K., and Miyata, T., 1988, Immunoglobulin-like sequences in the extracellular domains of proto-oncogene *fms* and platelet-derived growth factor receptor, *Proc. Japan. Acad.* **64**(Suppl. B):113–118.

Heck, J. D., and Hatfield, G. W., 1988, Valyl-tRNA synthetase gene of *Escherichia coli* K12: Primary structure of the *valS* gene and homology with other aminoacyl-tRNA synthetases, *J. Biol. Chem.* **263:**857–867.

Hejgaard, J., Rasmussen, S. K., Brandt, A., and Svendsen, I., 1985, Sequence homology between barley endosperm protein Z and protease inhibitors of the α_1-antitrypsin family, *FEBS Lett.* **180:**89–94.

Hojrup, P., Jensen, M. S., and Petersen, T. E., 1985, Amino acid sequence of bovine protein Z: A vitamin K-dependent serine protease homolog, *FEBS Lett.* **184:**333–338.

Holmquist, R., and Moise, H., 1975, Compositional nonrandomness: A quantitatively conserved evolutionary invariant, *J. Mol. Evol.* **6:**1–14.

Hood, J. M., Fowler, A. V., and Zabin, I., 1978, On the evolution of β-galactosidase, *Proc. Natl. Acad. Sci. U.S.A.* **75:**113–116.

Hursh, D. A., Andrews, M. E., and Raff, R. A., 1987, A sea urchin gene encodes a polypeptide homologous to epidermal growth factor, *Science* **237:**1487–1490.

Ichinose, A., McMullen, B. A., Fujikawa, K., and Davie, E. W., 1986, Amino acid sequence of the b subunit of human factor XIII, a protein composed of ten repetitive segments, *Biochemistry* **25:**4633–4638.

Ivanov, O. C., and Ivanov, C. P., 1980, Some evidence for the universality of structural periodicity in proteins, *J. Mol. Evol.* **16:**47–68.

Jackman, R. W., Beeler, D. L., VanDeWater, L., and Rosenberg, R. D., 1986, Characterization of a thrombomodulin cDNA reveals structural similarity to the low density lipoprotein receptor, *Proc. Natl. Acad. Sci. U.S.A.* **83:**8834–8838.

Johnson, K. S., Taylor, D. W., and Cordingley, J. S., 1987, Possible eggshell protein gene from *Schistosoma mansoni, Mol. Biochem. Parasitol.* **22:**89–100.

Jornvall, H., and Persson, B., 1983, Amino acid sequence restriction in relation to proteolysis, *Biosci. Rep.* **3:** 225–232.

Jukes, T. H., 1966, *Molecules and Evolution,* Columbia University Press, New York.

Jukes, T. H., Holmquist, R., and Moise, H., 1975, Amino acid composition of proteins: Selection against the genetic code, *Science* **189:**50–51.

Kabsch, W., and Sander, C., 1984, On the use of sequence homologies to predict protein structure: Identical pentapeptides can have completely different conformations, *Proc. Natl. Acad. Sci. U.S.A.* **81:**1075–1078.

Kageyama, R., Ohkubo, H., and Nakanishi, S., 1984, Primary structure of human preangiotensinogen deduced from the cloned cDNA sequence, *Biochemistry* **23:**3603–3609.

King, J. L., and Jukes, T. H., 1969, Non-Darwinian evolution, *Science* **164:**788–798.

Kitamura, N., Semler, B. L., Rothberg, P. G., Larsen, G. R., Adler, C. J., Dorner, A. J., Emini, E. A., Hanecak, R., Lee, J. J., van der Werf, S., Anderson, W. W., and Wimmer, E., 1981, Primary structure, gene organization and polypeptide expression of poliovirus RNA, *Nature* **291:**547–553.

Kobilka, B. K., Matsui, H., Kobilka, T. S., Yang-Feng, T. L., Francke, U., Caron, M. G., Lefkowitz, R. J., and Regan, J. W., 1987, Cloning, sequencing, and expression of the gene coding for the human platelet α_2-adrenergic receptor, *Science* **238:**650–656.

Krikos, A., Mutoh, N., Boyd, A., and Simon, M. I., 1983, Sensory transducers of *E. coli* are composed of discrete structural and functional domains, *Cell* **33:**615–622.

Kristensen, T., and Tack, B. F., 1986, Murine protein H is comprised of 20 repeating units, 61 amino acids in length, *Proc. Natl. Acad. Sci. U.S.A.* **83:**3963–3967.

Kubo, T., Fukuda, K., Mikami, A., Maeda, A., Takahashi, H., Mishina, M., Haga, T., Haga, K., Ichiyama, A., Kangawa, K., Kojima, M., Matsuo, H., Hirose, T., and Numa, S., 1986, Cloning, sequencing and expression of complementary DNA encoding the muscarinic acetylcholine receptor, *Nature* **323:**411–416.

Kurachi, K., and Davie, E. W., 1982, Isolation and characterization of a cDNA coding for human factor IX, *Proc. Natl. Acad. Sci. U.S.A.* **79:**6461–6464.

Laurent, B. C., Nilsson, M. H. L., Bavik, C. O., Jones, T. A., Sundelin, J., and Peterson, P. A., 1985, Characterization of the rat retinol-binding protein gene and comparison to the three-dimensional structure of the protein, *J. Biol. Chem.* **260:**11476–11480.

López Otin, C., Grubb, A. O., and Méndez, E., 1984, The complete amino acid sequence of human complex-forming glycoprotein heterogeneous in charge (protein HC) from one individual, *Arch. Biochem. Biophys.* **228:**544–554.

Lozier, J., Takahashi, N., and Putnam, F. W., 1984, Complete amino acid sequence of human plasma β_2-glycoprotein I, *Proc. Natl. Acad. Sci. U.S.A.* **81:**3640–3644.

Lundwall, A., Dackowski, W., Cohen, E., Shaffer, M., Mahr, A., Dahlback, B., Stenflo, J., and Wydro, R., 1986, Isolation and sequence of the cDNA for human protein S, a regulator of blood coagulation, *Proc. Natl. Acad. Sci. U.S.A.* **83:**6716–6720.

Magnusson, R. P., Gestautas, J., Seto, P., Taurog, A., and Rapoport, B., 1986, Isolation and characterization of a cDNA clone for porcine thyroid peroxidase, *FEBS Lett.* **208:**391–396.

Masu, Y., Nakayama, K., Tamaki, H., Harada, Y., Kuno, M., and Nakanishi, S., 1987, cDNA cloning of bovine substance-K receptor through oocyte expression system, *Nature* **329:**836–838.

McLachlan, A. D., 1977, Repeated helical pattern in apolipoprotein-A-I, *Nature* **267:**465–466.

McLachlan, A. D., Stewart, M., and Smillie, L. B., 1975, Sequence repeats in α-tropomyosin, *J. Mol. Biol.* **98:**281–291.

McMullen, B. A., and Fujikawa, K., 1985, Amino acid sequence of the heavy chain of human α-factor XIIa (activated Hageman factor), *J. Biol. Chem.* **260:**5328–5341.

McReynolds, L., O'Malley, B. W., Nisbet, A. D., Fothergill, J. E., Givol, D., Fields, S., Robertson, M., and Brownlee, G. G., 1978, Sequence of chicken ovalbumin mRNA, *Nature* **273:**723–728.

Medof, M. E., Lublin, D. M., Holers, V. M., Ayers, D. J., Getty, R. R., Leykam, J. F., Atkinson, J. P., and Tykocinski, M. L., 1987, Cloning and characterization of cDNAs encoding the complete sequence of decay-accelerating factor of human complement, *Proc. Natl. Acad. Sci. U.S.A.* **84:**2007–2011.

Mole, J. E., Anderson, J. K., Davison, E. A., and Woods, D. E., 1984, Complete primary structure for the zymogen of human complement factor B, *J. Biol. Chem.* **259:**3407–3412.

Nagamine, Y., Pearson, D., Altus, M. S., and Reich, E., 1984, cDNA and gene nucleotide sequence of porcine plasminogen activator, *Nucleic Acids Res.* **12:**9525–9541.

Netzker, R., Kchel, H. G., Basak, N., and Kntzel, H., 1982, Nucleotide sequence of *Aspergillus nidulans* mitochondrial genes coding for ATPase subunit 6, cytochrome oxidase subunit 3, seven unidentified proteins, four tRNAs and L-rRNA, *Nucleic Acids Res.* **10:**4783–4794.

Ng, S. C., Rao, A. G., Zack Howard, O. M., and Sodetz, J. M., 1987, The eighth component of human complement: Evidence that it is an oligomeric serum protein assembled from products of three different genes, *Biochemistry* **26:**5229–5233.

Ohno, S., 1984, Repeats of base oligomers as the primordial coding sequences of the primeval earth and their vestiges in modern genes, *J. Mol. Evol.* **20:**313–321.

Pennica, D., Holmes, W. E., Kohr, W. J., Harkins, R. N., Vehar, G. A., Ward, C. A., Bennett, W. F., Yelverton, E., Seeburg, P. H., Heyneker, H. L., Goeddel, D. V., and Collen, D., 1983, Cloning and expression of human tissue-type plasminogen activator cDNA in *E. coli, Nature* **301:**214–221.

Pervaiz, S., and Brew, K., 1987, Homology and structure-function correlations between α_1 acid glycoprotein and serum retinol-binding protein and its relatives, *FASEB J.* **1:**209–214.

Poole, S. J., Kauvar, L. M., Drees, B., and Kornberg, T., 1985, The *engrailed* locus of drosophila: Structural analysis of an embryonic transcript, *Cell* **40:**37–43.

Postle, K., and Good, R. F., 1983, DNA sequence of the *Escherichia coli tonB* gene, *Proc. Natl. Acad. Sci. U.S.A.* **80:**5235–5239.

Putnam, F. W., Florent, G., Paul, C., Shinoda, T., and Shizizu, A., 1973, Complete amino acid sequence of the mu heavy chain of a human IgM immunoglobulin, *Science* **182:**287–291.

Ragg, H., 1986, A new member of the plasma protease inhibitor gene family, *Nucleic Acids Res.* **14:**1073–1088.

Ravetch, J. V., Feder, R., Pavlovec, A., and Blobel, G., 1984, Primary structure and genomic organization of the histidine-rich protein of the malaria parasite *Plasmodium lophurae, Nature* **312:**616–620.

Reynolds, P., Prakash, L., Dumais, D., Perozzi, G., and Prakash, S., 1985, Nucleotide sequence of the *RAD10* gene of *Saccharomyces cerevisiae, EMBO J.* **4:**3549–3552.

Ricca, G. A., and Taylor, J. M., 1981, Nucleotide sequence of rat α-acid glycoprotein messenger RNA, *J. Biol. Chem.* **256:**11199–11202.

Ruoslahti, E., and Pierschbacher, M. D., 1987, New perspectives in cell adhesion: RGD and integrins, *Science* **238:**491–497.

Sai, S., Tanaka, T., Kosher, R. A., and Tanzer, M. L., 1986, Cloning and sequence analysis of a partial cDNA for chicken cartilage proteoglycan core protein, *Proc. Natl. Acad. Sci. U.S.A.* **83:**5081–5085.

Saroff, H. A., and Pretorius, H. T., 1981, The uniqueness of protein sequences, *Fed. Proc.* **40:**1676.

Scheller, R. H., Jackson, J. F., McAllister, L. B., Rothman, B. S., Mayeri, E., and Axel, R., 1983, A single gene encodes multiple neuropeptides mediating a stereotyped behavior, *Cell* **32:**7–22.

Schmid, K., Kaufman, H., Isemura, S., Bauer, F., Emura, J., Motoyarma, T., Ishiguro, M., and Nanno, S., 1973, Structure of α_1-acid glycoprotein. The complete amino acid sequence, multiple amino acid substitutions, and homology with the immunoglobulins, *Biochemistry* **12:**2711–2724.

Schubert, D., LaCorbiere, M., and Esch, F., 1986, A chick neural retina adhesion and survival molecule is a retinol-binding protein, *J. Cell Biol.* **102:**2295–2301.

Scott, J., Urdea, M., Quiroga, M., Sanchez-Pescador, R., Fong, N., Selby, M., Rutter, W. J., and Bell, G. I., 1983, Structure of a mouse submaxillary messenger RNA encoding epidermal growth factor and seven related proteins, *Science* **221:**236–240.

Scott, R. A., Vanderkooi, G., Tuttle, R. W., Shames, P. M., and Scheraga, H. A., 1967, Minimization of polypeptide energy, III. Application of a rapid energy minimization technique to the calculation of preliminary structures of gramicidin-S, *Proc. Natl. Acad. Sci. U.S.A.* **58:**2204–2211.

Sheumack, D. D., Claassens, R., Whiteley, N. M., and Howden, M. E. H., 1985, Complete amino acid sequence of a new type of lethal neurotoxin from the venom of the funnel-web spider *Atrax robustus, FEBS Lett.* **181:**154–156.

Shimotohno, K., Takahashi, Y., Shimizu, N., Gojobori, T., Golde, D. W., Chen, I. S. Y., Miwa, M., and Sugimura, T., 1985, Complete nucleotide sequence of an infectious clone of human T-cell leukemia virus type II: An open reading frame for the protease gene, *Proc. Natl. Acad. Sci. U.S.A.* **82:**3101–3105.

Sorm, F., 1959, Ähnlichkeitder struktur bei eiweißstoffen, in: *1st International Symposium on The Origin of Life* (A. Oparin *et al.*, eds.), Pergamon Press, Oxford, pp. 231–240.

Südhoff, T. C., Russell, D. W., Goldstein, J. L., Brown, M. S., Sanchez-Pescador, R., and Bell, G. I., 1985, Cassette of eight exons shared by genes for LDL receptor and EGF precursor, *Science* **228:**893–895.

Sumi, Y., Nakamura, Y., Aoki, N., Sakai, M., and Muramatsu, M., 1986, Structure of the carboxyl-terminal half of human α_2-plasmin inhibitor deduced from that of cDNA, *J. Biochem. (Tokyo)* **100:**1399–1402.

Suzuki, K., Dehayashiki, Y., Nishioka, J., Kurachi, K., Akira, M., Yamamoto, S., and Hashimoto, S., 1987, Characterization of a cDNA for human protein C inhibitor, *J. Biol. Chem.* **262:**611–615.

Van der Ploeg, L. H. T., 1987, Control of variant surface antigen switching in trypanosomes, *Cell* **51:**159–161.

Vehar, G. A., Keyt, B., Eaton, D., Rodriguez, H., O'Brien, D. P., Rotblat, F., Oppermann, H., Keck, R., Wood, W. I., Harkins, R. N., Tuddenham, E. G. D., Lawn, M., and Capon, D. J., 1984, Structure of human factor VIII, *Nature* **312:**337–342.

Venkatesan, S., Gershowitz, A., and Moss, B., 1982, Complete nucleotide sequences of two adjacent early vaccinia virus genes located within the inverted terminal repetition, *J. Virol.* **44:**637–646.

Wahlgren, M., Aslund, L., Franźen, L., Sundvall, M., Wahlin, B., Berzins, K., McNicol, L. A., Björkman, A., Wigzell, H., Perlmann, P., and Pettersson, U., 1986, A *Plasmodium falciparum* antigen containing clusters of asparagine residues, *Proc. Natl. Acad. Sci. U.S.A.* **83:**2677–2681.

Walker, J. E., Saraste, M., Runswick, M. J., and Gay, N. J., 1982, Distantly related sequences in the α- and β-subunits of ATP synthase, myosin, kinases and other ATP-requiring enzymes and a common nucleotide binding fold, *EMBO J.* **1:**945–951.

Watson, D. C., and Dixon, G. H., 1981, Amino acid sequence homologies between the high-mobility-group proteins, HMB-T from trout testis and HMG-1 and -2 from calf thymus: Is the poly-aspartic-glutamic acid polypeptide within the main chain? *Biosci. Reports* **1:**167–175.

Weiner, S., and Hood, L., 1975, Soluble protein of the organic matrix of mollusk shells: A potential template for shell formation, *Science* **190:**987–989.

Wharton, K. A., Johansen, K. M., Xu, T., and Artavanis-Tsaknoas, S., 1985, Nucleotide sequence from the neurogenic locus notch implies a gene product that shares homology with proteins containing EGF-like repeats, *Cell* **43:**567–581.

Wittmann-Liebold, B., Ashman, K., and Dzionara, M., 1983, Multiple internal repeats within protein S1 from the *Escherichia coli* ribosome, *FEBS Lett.* **154:**31–41.

Yamamoto, T., Davis, C. G., Brown, M. S., Schneider, W. J., Casey, M. L., Goldstein, J. L., and Russell, D. W., 1984, The human LDL receptor: A cysteine-rich protein with multiple Alu sequences in its mRNA, *Cell* **39:**27–38.

Ycas, M., 1972, *De novo* origin of periodic proteins, *J. Mol. Evol.* **2:**17–27.

Zuckerkandl, E., 1975, The appearance of new structures and functions in proteins during evolution, *J. Mol. Evol.* **7:**1–57.

15

The Hydrophobicity Profile

George D. Rose and Jonathan E. Dworkin

I. HYDROPHOBICITY AND PROTEIN FOLDING

The hydrophobic effect has been a topic of chemical interest for more than a century. This phenomenon, in its most general terms, is concerned with the solubility of molecules in water and in nonpolar solvents (Tanford, 1980). Liquid water dissolves polar substances readily but nonpolar substances only sparingly. On mixing, water ''squeezes out'' hydrophobic molecules, resulting in a separation into polar and nonpolar phases. The spontaneous segregation of oil after mixing with water serves as a familiar example.

Hydrophobicity is believed to play a major role in organizing the self-assembly of protein molecules because some amino acid residues are abundantly water soluble whereas others are not (Tanford, 1978). The 20 commonly occurring residues can be classified with respect to their side chains as polar, nonpolar, or amphipathic. However, even residues with nonpolar side chains should be viewed as amphipathic because backbone N—H and C=O groups are themselves highly polar.

According to the popular oil-drop model, the protein interior is expected to be enriched in nonpolar (''oily'') residues, which cluster together to form, in effect, an organic phase

George D. Rose and Jonathan E. Dworkin • Department of Biological Chemistry, Milton S. Hershey Medical Center, Pennsylvania State University, Hershey, Pennsylvania 17033.

(Kauzmann, 1959). The effect is analogous to the spontaneous segregation of oil in water, with the important distinction that residues in proteins are covalently bound to their chain neighbors and cannot partition independently. This appealing model, if correct, should predict the relative tendencies of residues to partition between the inside and outside of water-soluble proteins. The model is complicated by the fact that the protein interior is more heterogeneous than a simple hydrocarbon solvent, and the solvation of groups on the surface may be hindered by intramolecular neighbors.

The oil-drop concept of a protein molecule stems from the solution thermodynamics of simple model compounds (Cohn and Edsall, 1943). The solubility of hydrocarbons and amino acids has been studied extensively both in water and in organic solvents such as ethanol (Nozaki and Tanford, 1971) or octanol (Fauchère and Pliška, 1983). Adoption of these measurements to explain larger composites, such as proteins, would seem to be a valid extrapolation.

However, quantitative assessment of the actual situation in proteins of known structure reveals unsuspected complexity. Richards noted that

> of the accessible areas of native structures, roughly half represent polar atoms and half nonpolar atoms. Thus, the "grease" is by no means all "buried." In the folding process there are roughly equivalent decreases in the accessibility of both the polar and nonpolar groups. The relevant forces and the final structure require more careful definition than is implied by the common feeling that inside equals nonpolar and outside equals polar (Richards, 1977).

Can the thermodynamics of simple model compounds account for the observed heterogeneity of actual molecules? Ideally, the distribution of residues between the inside and outside of a protein molecule would be governed solely by side-chain solubility, since backbone atoms remain invariant from residue to residue. Practically, significant sources of nonideality arise because, as mentioned previously, the residues are covalently linked and cannot distribute independently. In particular, a chain site with a polar residue adjacent in sequence to a nonpolar residue may not be able to satisfy both tendencies simultaneously.

Despite these complications, it is conceivable that segments of the polypeptide chain distribute according to their aggregate solubility. In the simplest case, the hydrophobicity of a composite segment would be reckoned by summing the contributions of constituent chemical groups. Such group additivity obtains, for example, in the case of straight-chain aliphatic hydrocarbons (Tanford, 1980).

II. THE HYDROPHOBICITY PROFILE

The hydrophobicity profile is a simple way to quantify the concentration of hydrophobic residues along the linear polypeptide chain. Local extrema in hydrophobicity are then used predictively. The method is based on the arguments, presented in the preceding section, that residues with hydrophobic side chains will tend to bury themselves within proteins, away from solvent water. The effect is similar to the spontaneous self segregation of oil in water with the important distinction that residues in proteins are covalently bound to their chain neighbors and cannot partition independently.

Specifically, a hydrophobicity profile for a protein is a graph of the average hydrophobicity per residue against position in the sequence. Plotting the curve reveals the loci of minima and maxima in hydrophobicity along the linear polypeptide chain.

The topic of hydrophobicity profiles has been reviewed recently by Rose *et al.* (1985b) and Engelman *et al.* (1986). The reader is also referred to Doolittle's excellent monograph (Doolittle, 1986). Given the sequence of a protein of unknown structure, hydrophobicity profiles have been used to predict peptide chain turns, interior/exterior regions, antigenic sites, and membrane-spanning segments.

Rose (1978) used a hydrophobicity profile to predict turns, and Rose and Roy (1980) used the method to predict interior and exterior regions in the molecule. Both and Sleigh (1980) compared the hydrophobicity of hemagglutinins from different viral strains in order to identify antigenic sites. Hopp and Woods (1981) emphasized use of the approach to predict antigenic determinants, referring to their running average as a hydrophilicity value. Kyte and Doolittle (1982) predicted molecular segmentation into interior and exterior regions for both soluble and membrane-bound proteins. Novotny and Auffray (1984) applied the method to homologous sequences of histocompatibility antigens. Eisenberg (1984) showed that a related analysis by hydrophobic moments can distinguish transmembrane from globular proteins. Engelman *et al.* (1986) use the method to identify transbilayer helices in membrane proteins.

III. UNDERLYING ASSUMPTIONS

The interpretation of hydrophobicity profiles of globular proteins is based on the assumption that the protein and its aqueous surrounding can be viewed as distinct solvent "phases." The tendency of any type of residue to occur at the surface or in the interior can be described by a partition coefficient between these phases. The distribution of the residues should be influenced by the partitioning tendencies of their side chains, because backbone atoms remain the same from residue to residue. However, the residues cannot be treated independently, because they are covalently linked. (For example, a chain site with a polar residue adjacent in sequence to a nonpolar residue may not be able to satisfy the partitioning tendencies of both residues simultaneously.) In addition, the simple two-solvent model is a gross simplification; the interior of the protein is an anisotropic phase, unlike a bulk solvent. Furthermore, the packing between segments of secondary structure is cooperative, and longer-range forces may predominate (Richardson, 1981; Chothia, 1984). These complexities notwithstanding, trends in partitioning can be discerned (Rose *et al.*, 1985a).

The hydrophobicity profile for an amino acid sequence is in principle simple to construct. It depends on (1) the choice of hydrophobicity scale and (2) the degree of averaging (the number of consecutive residues considered as a unit). A hydrophobicity scale assigns a hydrophobicity value to each of the 20 amino acids. The profile is then computed by averaging the hydrophobicity within a moving window that is stepped along the sequence. The appropriate window size depends on the size of the structural component one wants to resolve. To distinguish secondary structural elements, the window must be larger than a single turn (i.e., greater than four residues) but smaller than a complex segment such as a turn-helix (i.e., less than approximately 12 residues). To identify membrane-spanning segments, the window should approximate the expected length of a segment (i.e., about 19 residues). Smaller windows yield profiles with greater detail, but most of the minor fluctuations in hydrophobicity appear to have little relation to overall segmentation of the molecule.

IV. CHOICE OF HYDROPHOBICITY SCALE

To test the idea that chain segments behave as predictable composites, the hydrophobicity of individual residues must be known. In such studies, hydrophobicity is measured as the free energy required to transfer a standard quantity of substance from water to a nonaqueous solvent. This standard-state free energy, $\Delta G^0_{\text{water}\rightarrow\text{organic}}$, is derived by measuring the solubility in either phase. The free energy of transfer, $\Delta G^0_{p\rightarrow p'}$, between phases p and p' is related to the partition coefficient between the phases, $K_{p\rightarrow p'}$, by the equation $\Delta G^0_{p\rightarrow p'} = -RT\ln K_{p\rightarrow p'}$ (Cohn and Edsall, 1943).

Many scales of hydrophobicity have been determined for the natural amino acids and their analogues. Such scales can be divided into two general classes. Solution scales are based on the solubilities of amino acids in aqueous and nonaqueous solvents. These measured solubilities are used to calculate partition coefficients and free energies of transfer. Empirical scales are derived by statistical analysis of the observed distributions of the residues between the solvent-accessible surface (aqueous phase) and the buried interior (nonaqueous phase) in proteins of known structure.

Hydrophobicity scales from different sources do not agree in all respects. There are even qualitative discrepancies about which amino acids are considered hydrophobic and which hydrophilic. For example, on two scales based on solution studies, tryptophan and tyrosine are very hydrophobic in the scale of Nozaki and Tanford (1971) but quite hydrophilic in the scale of Wolfenden *et al.* (1981). Qualitative disagreement occurs in the empirical scales also, where different investigators have used differing criteria to determine residue "buriedness." The issues were analyzed by Rose *et al.* (1985b).

Some hydrophobicity scales lack values for some residue types. Although the scale of Nozaki and Tanford (1971) covers only 11 of the residues, all of the hydrophobic amino acids are included except isoleucine. Both and Sleigh (1980) used this scale directly. Rose (1978) and Rose and Roy (1980) assigned isoleucine a value equal to that of leucine and values of zero to all hydrophilic residues. Although this strategy does not affect the positions of local minima in hydrophobicity, it does preclude assigning a quantitative significance to the values. In their prediction of antigenic determinants, Hopp and Woods (1981) used the Levitt (1976) scale with adjustment of the values of acidic residues and proline to maximize predictive success. Kyte and Doolittle (1982) supplemented and modified the Wolfenden scale on the basis of chemical arguments (Wolfenden *et al.*, 1981). Eisenberg *et al.* (1982) averaged several scales to achieve a "consensus" scale. Novotny and Auffrey (1984) supplemented the parameters of Rose and Roy (1980) by positions of the charged residues. Engelman *et al.* (1986) developed a new scale calibrated specifically for transmembrane helices.

V. A SCALE BASED ON SURFACE AREA

A recent empirical scale developed by Rose *et al.* (1985a) measures the mean solvent accessibility (Lee and Richards, 1971) of residues in a data base of x-ray-elucidated proteins. The mean solvent accessibility, $\langle A_{\mathrm{res}} \rangle$, is the residue surface area that remains accessible to solvent, on average, within folded proteins. In work of this sort, it is presumed that surface area is related to hydrophobicity in a linear way and, in particular, that the more hydrophobic the residue, the more completely buried it will be on average.

Mean accessibilities must be normalized before valid comparisons can be made. For example, although $\langle A_{\mathrm{val}} \rangle$ and $\langle A_{\mathrm{gly}} \rangle$ are approximately equal, an isolated valyl residue has almost twice the area of an isolated glycyl residue, and hence it must bury a larger fraction of its available surface on folding.

To normalize $\langle A_{\mathrm{res}} \rangle$, a standard-state accessible surface area is needed. For a given residue type, *X*, this standard-state accessibility is defined as the average surface area that residue has in Gly-*X*-Gly tripeptides. The tripeptide surface area depends on the values of its dihedral angles. Values are chosen to reflect the actual ensemble of dihedral angles observed in the data base, and these are used to construct a representative set of tripeptides. The standard-state accessibility, A^{0}_{res}, is taken as the average over this representative set. For example, if there are *N* phenylalanine residues in the data base, then *N* tripeptides, Gly-Phe-Gly, are constructed with angles $(\phi_i, \psi_i, \chi_i^1)$ $(i = 1, 2, \ldots, N)$. A^{0}_{phe} is the average of this ensemble.

The fractional accessibility of a residue can now be calculated as its mean accessibility in proteins divided by the standard-state accessibility, $\langle A_{\mathrm{res}} \rangle / A^{0}_{\mathrm{res}}$. This fractional accessibility is

Table I. Standard State and Mean Surface Areas for the Amino Acid Residues[a]

Residue	A^0	$\langle A \rangle$	$A^0 - \langle A \rangle$	$(A^0 - \langle A \rangle)/A^0$	σ
Ala	118.1	31.5	86.6	0.74	0.28
Arg	256.0	93.8	162.2	0.64	0.23
Asn	165.5	62.2	103.3	0.63	0.25
Asp	158.7	60.9	97.8	0.62	0.26
Cys	146.1	13.9	132.3	0.91	0.12
Gln	193.2	74.0	119.2	0.62	0.22
Glu	186.2	72.3	113.9	0.62	0.25
Gly	88.1	25.2	62.9	0.72	0.27
His	202.5	46.7	155.8	0.78	0.23
Ile	181.0	23.0	158.0	0.88	0.18
Leu	193.1	29.0	164.1	0.85	0.20
Lys	225.8	110.3	115.5	0.52	0.20
Met	203.4	30.5	172.9	0.85	0.19
Phe	222.8	28.7	194.1	0.88	0.17
Pro	146.8	53.7	92.9	0.64	0.28
Ser	129.8	44.2	85.6	0.66	0.29
Thr	152.5	46.0	106.5	0.70	0.25
Trp	266.3	41.7	224.6	0.85	0.19
Tyr	236.8	59.1	177.7	0.76	0.19
Val	164.5	23.5	141.0	0.86	0.19

[a]Abbreviations used: A^0, area in the standard state, in square Angstroms—for a given residue type, X, the standard-state accessibility is defined as the average surface area that residue has in a representative ensemble of Gly-X-Gly tripeptides; $\langle A \rangle$, average accessible area in proteins, in square Angstroms; $A^0 - \langle A \rangle$, average area buried on transfer from the standard state to the folded protein, in square Angstroms; $(A^0 - \langle A \rangle)/A^0$, mean fractional area loss, equal to the average area buried normalized by the standard-state area; σ, the standard deviation of the mean fractional area loss. From Rose *et al.* (1985a) and Dworkin and Rose (1987).

an intrinsic measure of residue hydrophobicity. In addition, the difference between the standard-state accessibility and the mean accessibility, $A^0_{res} - \langle A_{res} \rangle$, measures the area the residue buries, on average, on folding. This difference is proportional to the residue's hydrophobic contribution to the conformational free energy of the protein.

For each of the 20 residues, Table I lists the standard-state (A^0) and mean accessibilities ($\langle A \rangle$) together with the mean fraction buried $[(A^0 - \langle A \rangle)/A^0]$ and its standard deviation (σ).

VI. CONSTRUCTION OF THE PROFILE

The mean fractions buried, from Table I, can be used to construct a hydrophobicity profile that includes limits of probable error (Dworkin and Rose, 1987). The usual practice in calculating the average hydrophobicity within a window of N residues has been to sum individual residue hydrophobicities and divide by N. However, a size correction can be introduced to compensate for the fact that, for a given value of the mean fractional accessibility, large residues make a greater contribution than small residues to the conformational free energy. Thus, the weight, P_i, assigned to the ith residue of an N-residue window would be

$$P_i = A_i^0/\Sigma A_i^0 \tag{1}$$

Error bars are assigned to each residue by calculating the standard error of the mean, σ_μ, using the standard deviation and number of residues in Table I. For an individual residue, $\sigma_\mu(\text{res}) = \sigma/(N)^{\frac{1}{2}}$. The standard error for each residue of an N-residue window is given by the square root of the summed weighted squared individual values; i.e.,

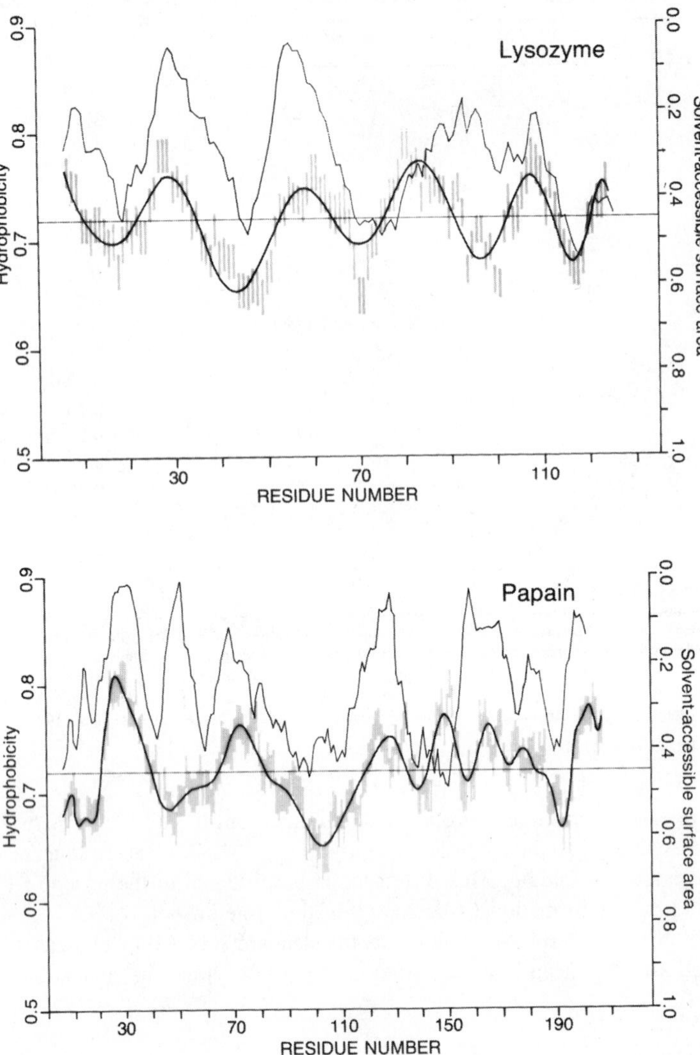

Figure 1. Hydrophobicity and accessibility profiles for flavodoxin, lysozyme, and papain (Bernstein *et al.*, 1977) using a nine-residue window. The hydrophobicity profile is based on sequence information alone, whereas the accessibility profile requires knowledge of the three-dimensional structure. The histogram is the hydrophobicity profile proper; the midpoint of each solid bar represents the mean fraction buried for the central residue, calculated as a weighted average of all nine residues within the window using methods described in Section VI. The length of the bar represents the standard error, as given by equation 2. The smooth curve (bold line) superimposed on the histogram is another representation of the hydrophobicity profile, achieved by graphing the *N*th-degree polynomial that accounts for at least 95% of the variance in these data. This curve, which underscores the central tendency of the histogram, was obtained from curvilinear regression analysis with orthogonal polynomials, using the IMSL routine RLFOR (International Mathematical Subroutine Library). The remaining smooth curve (faint line) is an accessibility profile; it plots the mean solvent-accessible surface area averaged within the same window, using the method of Lee and Richards (1971). The accessibility profile is used to assess predictive success of the hydrophobicity profile.

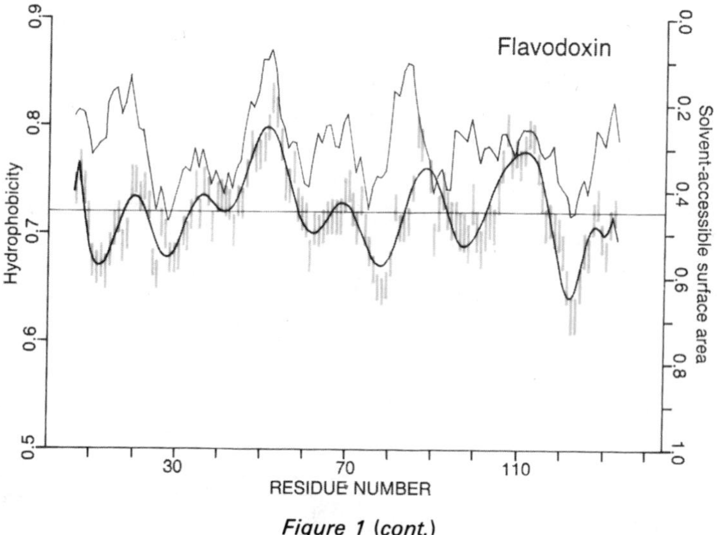

Figure 1 (cont.)

$$\sigma_{\mu}(\text{window}) = [\Sigma P_i(\sigma_{\mu}(i))^2]^{\frac{1}{2}} \tag{2}$$

Figure 1 shows hydrophobicity profiles for flavodoxin, lysozyme, and papain using a nine-residue window. The histogram is the hydrophobicity profile proper; the midpoint of each solid bar represents the mean fraction buried for the central residue, calculated as a weighted average of all nine residues within the window. The length of the bar represents the standard error, as given by equation 2.

The smooth curve (bold line) superimposed on the histogram is an alternative representation of the hydrophobicity profile, achieved by graphing the *N*th degree polynomial that accounts for at least 95% of the variance in these data. This optional curve, which underscores the central tendency of the histogram, was obtained from curvilinear regression analysis with orthogonal polynomials, using the International Mathematical Subroutine Library (IMSL) routine RLFOR.*

The remaining curve (faint line) is the accessibility profile; it plots the mean solvent-accessible surface area averaged within the same window. The solvent-accessible surface was calculated using the method of Lee and Richards (1971). The accessibility profile requires atomic coordinates, and it is used in these three cases to assess the predictive quality of the hydrophobicity profile.

VII. INTERPRETATION OF THE PROFILE

The histograms in Fig. 1 are derived from sequence information alone, using the values in Table I. The nine-residue window employed in these examples is suitable for predicting partitioning of the chain into interior/exterior regions. The flavodoxin profile, for example, predicts that segments centered at residues 21, 36, 51, 70, 89, and 110 will be buried, while segments centered at 13, 29, 62, 77, 97, and 122 will be accessible.

A comparison of the two curves in Fig. 1 is instructive. The hydrophobicity profile (bold line) is derived from the sequence, whereas the accessibility profile (faint line) requires full

*International Mathematical Subroutine Library, 7500 Bellaire Blvd., Houston, TX 77036.

three-dimensional information. It can be seen that peaks and valleys in the hydrophobicity profile predict corresponding peaks and valleys in measured accessibility to a large extent, but not perfectly. Significant failures are apparent, for example, in flavodoxin residues 95–105, in lysozyme residues 90–100, and in papain residues 45–65 and 145–155.

The degree of predictive success in Fig. 1 is representative of what can be expected for globular proteins in general. A discussion of predictive error for membrane proteins can be found in Engelman *et al.* (1986) and in Wallace *et al.* (1986). The implied caution for those using these methods is apparent.

ACKNOWLEDGMENTS. We thank Barry Dworkin and Marshall Jones for helpful suggestions. This work was supported by grants from the National Institutes of Health.

VIII. REFERENCES

Bernstein, F. C., Koetzle, T. G., Williams, G. J. B., Meyer, E. F., Jr., Brice, M. D., Rogers, J. R., Kennard, O., Shimanouchi, T., and Tasumi, M., 1977, The protein data bank: A computer-based archival file for macromolecular structures, *J. Mol. Biol.* **112:**535–542.

Both, G. W., and Sleigh, M. J., 1980, Complete nucleotide sequence of the haemagglutinin gene from a human influenza virus of the Hong Kong subtype, *Nucleic Acids Res.* **8:**2561–2575.

Chothia, C., 1984, Principles that determine the structure of proteins, *Annu. Rev. Biochem.* **53:**537–572.

Cohn, E. J., and Edsall, J. T., 1943, *Proteins, Amino Acids and Peptides as Ions and Dipolar Ions,* Reinhold, New York.

Doolittle, R. F., 1986, *Of Urfs and Orfs,* University Science Books, Mill Valley, CA.

Dworkin, J. E., and Rose, G. D., 1987, Hydrophobicity profiles revisited, in: *Methods in Protein Sequence Analysis—1986* (K. A. Walsh, ed.), Humana Press, Clinton, NJ, pp. 573–586.

Eisenberg, D., 1984, Three-dimensional structure of membrane and surface proteins, *Annu. Rev. Biochem.* **53:** 595–623.

Eisenberg, D., Weiss, R. M., Terwilliger, T. C., and Wilcox, W., 1982, Hydrophobic moments and protein structure, *Faraday Symp. Chem. Soc.* **17:**109–120.

Engelman, D. M., Steitz, T. A., and Goldman, A., 1986, Identifying nonpolar transbilayer helices in amino acid sequences of membrane proteins, *Annu. Rev. Biophys. Biophys. Chem.* **15:**321–353.

Fauchère, J.-L., and Pliška, V., 1983, Hydrophobic parameters π of amino-acid side chains from the partitioning of N-acetyl-amino-acid amides, *Eur. J. Med. Chem. Chim. Ther.* **18:**369–375.

Hopp, T. P., and Woods, K. R., 1981, Prediction of protein antigenic determinants from amino acid sequences, *Proc. Natl. Acad. Sci. U.S.A.* **78:**3824–3828.

Kauzmann, W., 1959, Some factors in the interpretation of protein denaturation, *Adv. Protein Chem.* **14:**1–63.

Kyte, J., and Doolittle, R. F., 1982, A simple method for displaying the hydropathic character of a protein, *J. Mol. Biol.* **157:**105–132.

Lee, B. K., and Richards, F. M., 1971, The interpretation of protein structures: Estimation of static accessibility, *J. Mol. Biol.* **55:**379–400.

Levitt, M., 1976, A simplified representation of protein conformations for rapid simulation of protein folding, *J. Mol. Biol.* **104:**59–107.

Novotny, J., and Auffray, C., 1984, A program for prediction of protein secondary structure from nucleotide sequence data: Application to histocompatibility antigens, *Nucleic Acids Res.* **12:**243–255.

Nozaki, Y., and Tanford, C., 1971, The solubility of amino acids and two glycine peptides in aqueous ethanol and dioxane solutions—Establishment of a hydrophobicity scale, *J. Biol. Chem.* **246:**2211–221.

Richards, F. M., 1977, Areas, volumes, packing, and protein structure, *Annu. Rev. Biophys. Bioeng.* **6:**151–176.

Richardson, J. S., 1981, The anatomy and taxonomy of protein structure, *Adv. Protein Chem.* **34:**167–339.

Rose, G. D., 1978, Prediction of chain turns in globular proteins on a hydrophobic basis, *Nature* **272:**586–590.

Rose, G. D., and Roy, S., 1980, Hydrophobic basis of packing in globular proteins, *Proc. Natl. Acad. Sci. U.S.A.* **77:**4643–4647.

Rose, G. D., Geselowitz, A. R., Lesser, G. J., Lee, R. H., and Zehfus, M. H., 1985a, Hydrophobicity of amino acids in globular proteins, *Science* **229:**834–838.

Rose, G. D., Gierasch, L. M., and Smith, J. A., 1985b, Turns in peptides and proteins, *Adv. Protein Chem.* **37:** 1–109.

Tanford, C., 1978, The hydrophobic effect and the organization of living matter, *Science* **200:**1012–1018.

Tanford, C., 1980, *The Hydrophobic Effect,* John Wiley & Sons, New York.

Wallace, B. A., Cascio, M., and Mielke, D. L., 1986, Evaluation of methods for prediction of membrane protein secondary structures, *Proc. Natl. Acad. Sci. U.S.A.* **83:**9423–9427.

Wolfenden, R., Andersson, L., Cullis, P. M., and Southgate, C. C. B., 1981, Affinities of amino acid side chains for solvent water, *Biochemistry* **20:**849–855.

16

Hydrophobic Moments as Tools for Analyzing Protein Sequences and Structures

David Eisenberg, Morgan Wesson, and William Wilcox

I. INTRODUCTION

This chapter deals with the two main themes of this book—prediction of protein structure from sequence and principles of protein conformation—but from a narrow point of view. The point of view is that the interaction with water of the amino acid side chains is a major determinant of protein structure. Charged and polar side chains seek contact with water, and apolar side chains avoid water, preferring to cluster together beneath the surface of the protein. This avoidance of water is the hydrophobic interaction, first described in detail by Kauzmann (1959). In this point of view, it is the simultaneous attraction of some amino acid side chains to water and the avoidance of water by others that is a major factor in dictating the conformation taken by the polypeptide backbone.

To compute these attractive and repulsive forces for proteins from first principles is a

David Eisenberg, Morgan Wesson, and William Wilcox • Molecular Biology Institute, and Departments of Chemistry and Biochemistry, University of California, Los Angeles, Los Angeles, California 90024.

problem at the frontier of quantum chemical research. So for the present we may have to settle for semiempirical computer methods of learning about the hydrophobic effect. The hydrophobic moments discussed in this chapter are semiempirical quantities based on both computation and experimental measurements that describe the distribution of water-attracting and water-avoiding groups within a protein structure. In essence, they combine free energy and geometry.

The concepts of residue, hydrophobicities, and hydrophobic moments are coarse-grained compared to the concepts of the quantum chemist; they describe amino acid side chains rather than atoms. Also their numerical values, based on measurements of free energies of transfer, are only roughly known compared to the parameters from atomic physics. However, the concepts are helpful in analyzing biochemical structures because they deal with the free energy of folding and with structure. The free energy is the physical property that determines the structure at constant temperature and pressure, and the structure is often what we want to learn about the protein. So hydrophobic moments are a useful supplement to conventional energetics in understanding protein folding.

Hydrophobic moments can aid in understanding protein folding in two situations, each frequently encountered by the working biochemist. The first is when an amino acid sequence is known but little or nothing is known about the conformation of the protein that the sequence describes. Then the sequence hydrophobic moment can be used to detect segments of the sequence that are capable of forming strongly amphiphilic α helices or β sheets. The second situation is when a hypothetical folding model has been developed, perhaps by analogy to a similar sequence of known structure or from some partial experimental information. Then the structural hydrophobic moments of the segments of secondary structure can suggest whether the model is energetically reasonable or not. In real proteins, the hydrophobic moments of segments of secondary structure tend to oppose one another in their directions; in incorrect models, the directions of hydrophobic moments do not have simple patterns. Also, structural hydrophobic moments provide a picture of the hydrophobic forces at work in a folded protein.

II. HYDROPHOBICITIES

Biophysical chemists, at least since Cohn and Edsall (1943), have recognized that the free energy change for transfer of a chemical group from an apolar solvent to water is to some extent independent of the larger molecule to which the group is attached. For amino acid residues, these free energies of transfer have been termed "hydrophobicities." An early determination of values for some of the amino acid residues was made by Nozaki and Tanford (1971), who measured the free energy of transfer of amino acids from ethanol and other apolar solvents to water. These hydrophobicity values are given with others in Table I.

Each hydrophobicity represents the free energy of transfer of one mole of residue at a defined concentration from an apolar solvent to water. Apolar residues, such as Trp or Leu, have positive values. Polar residues, such as Glu and Lys, have negative values. The magnitude of the hydrophobicity is a measure of how much the side chain seeks or avoids association with water. Because the spontaneous folding of a chain of amino acids into a compact globular protein must depend to some extent on association with water or avoidance of it by its residues, it might be expected that the sequence of amino acid hydrophobicities would be a determinant of protein structure. In fact, the extent of correlation of residue hydrophobicities in two protein sequences seems to reflect the similarity of their three-dimensional structures (Sweet and Eisenberg, 1983).

There are now numerous determinations of residue hydrophobicities by a variety of methods (including those of Wolfenden *et al.*, 1981; Kyte and Doolittle, 1982; Hopp and

Table I. Amino Acid Hydrophobicities (Free Energies of Transfer)[a]

Residue	Normalized consensus[b]	ΔG_{FP}[c]	ΔG_{ATOM}[d]	ΔG_{TAN}[e]
Gly	0.48	(0)	(0)	(0)
Ala	0.62	0.42	0.67	0.50
Val	1.08	1.66	1.5	1.50
Leu	1.06	2.32	1.9	1.80
Ile	1.38	2.46	1.9	
Pro	0.12	0.98	1.2	
Cys	0.29	1.34	0.38	
Met	0.64	1.68	2.4	1.30
Thr	−0.05	0.35	0.52	0.40
Ser	−0.18	−0.05	0.01	−0.30
Phe	1.19	2.44	2.3	2.50
Trp	0.81	3.07	2.6	3.40
Tyr	0.26	1.31	1.6	2.30
Gln	−0.85	−0.30	−0.22	
Asn	−0.78	−0.82	−0.60	
Glu	−0.74	−0.87	−0.76	
Asp	−0.90	−1.05	−1.2	
His	−0.40	0.18	0.64	0.50
Lys	−1.50	−1.35	−0.57	
Arg	−2.53	−1.37	−2.1	

[a]Observed and calculated free energies of transfer are given in units of kcal $mole^{-1}$. ΔG_{FP} and ΔG_{TAN} are experimental free energies of transfer relative to glycine. ΔG_{ATOM} is a calculated value.

[b]An average of five hydrophobicity scales (Eisenberg *et al.*, 1984a, 1982b).

[c]Free energy of transfer for amino acid analogues from 1-octanol to water determined by Fauchère and Pliška (1983).

[d]A calculated, atom-based free energy of transfer (Eisenberg and McLachlan, 1986), based on data from Fauchère and Pliška (1983).

[e]Free energy of transfer for amino acid side chains determined by Nozaki and Tanford (1971).

Woods, 1981; among many others) and much discussion about their proper interpretation (see especially Rose *et al.*, 1985; Guy, 1985; Frömmel, 1984; Edsall and McKenzie, 1978). Rose and Dworkin (Chapter 15) discuss these matters. With the variety of hydrophobicity scales available, it remains difficult to know exactly which scale of hydrophobicities to work with for each application. The applications described in this chapter use our normalized consensus scale (Table I; Eisenberg *et al.*, 1984a), which was synthesized from five earlier scales (those of Nozaki and Tanford, 1971; Chothia, 1976; Janin, 1979; von Heijne and Blomberg, 1979; Wolfenden *et al.*, 1981). For comparison, Table I also contains a scale determined from the transfer of amino acid analogues from octanol to water (Fauchère and Pliška, 1983). The derivation and use of this atomic scale have been described elsewhere (Eisenberg and McLachlan, 1986).

III. HYDROPHOBIC MOMENTS

Hydrophobic moments describe asymmetry of hydrophobicity or amphiphilicity (Eisenberg *et al.*, 1982a,b). The hydrophobic dipole moment, or simply hydrophobic moment, is the exact analogue of the electric dipole moment except that it measures the asymmetry of the hydrophobicity in a structure rather than the asymmetry of electric charge. An α helix having apolar side chains protruding from one cylindrical surface and polar and charged side chains protruding from the opposite cylindrical surface has a large hydrophobic moment (Fig. 1). In this case, the hydrophobic moment is simply a quantitative expression of the familiar idea of

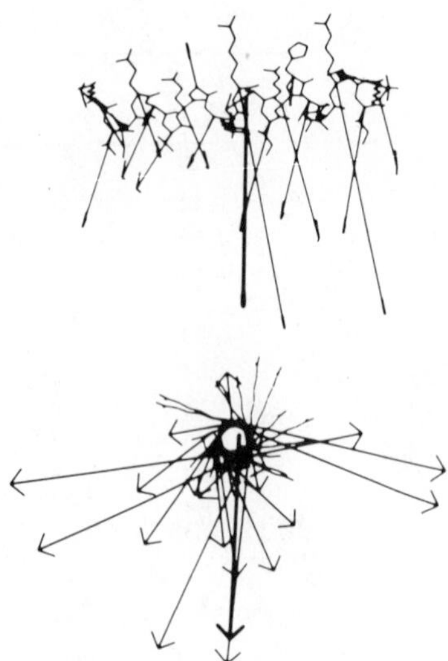

Figure 1. An α helix with very large hydrophobic moment, viewed down the helix axis (below) and perpendicular to the axis (above). The apolar amino acid side chains extend generally downward, and the polar side chains extend generally upward. The amino acid sequence of the helix is RVIEVVQGAYRAIRHIPRRIR from the C terminus of the gp41 protein of HIV. Each apolar side chain (that is, each residue with a positive hydrophobicity) is represented by an arrow extending from its α carbon in the direction of the center of the side chain. The length of the arrow is proportional to the numerical hydrophobicity of the side chain. Similarly, each polar side chain (that is, each residue with a negative hydrophobicity) has an arrow on this line, but because of the negative sign of its hydrophobicity, the arrow points from the α carbon away from the center of the side chain. Because the helix shown here is so highly amphiphilic, arrows from both polar and apolar side chains tend to point in the same direction. The vector sum of all arrows is the hydrophobic moment. If drawn to scale, it would extend off the bottom of the page. It is shown here, much scaled down, as the heavy arrow, which represents the direction of the hydrophobic moment: from the hydrophilic side of the helix toward the hydrophobic side.

α-helix amphiphilicity (Perutz *et al.*, 1965; Schiffer and Edmundson, 1967; Kaiser and Kezdy, 1983). Hydrophobic moments can also describe amphiphilicity in β structures, irregular structure, entire molecules, or even small groups of atoms (Eisenberg *et al.*, 1982b, 1984b; Eisenberg and McLachlan, 1986).

The hydrophobic moment of a protein, or of a segment of it, can be calculated from the atomic coordinates in cases where they are known. This is called the *structural hydrophobic moment* and is given by

$$\boldsymbol{\mu}_s = \sum_{\text{residues } n} H_n \mathbf{s}_n \qquad (1)$$

in which H_n is the hydrophobicity of the nth residue of the segment and s_n is a unit vector pointing from its α carbon atom to the center of the residue's side chain. Thus, the moment is a vector sum of the directions of the side chains, with each being weighted by its hydrophobicity. A very hydrophobic residue contributes strongly to the moment, whereas a very hydrophilic residue (having a negative value for H) decreases the moment if it has the same direction in space as the hydrophobic residue. However, if the hydrophilic side chain is directed 180° away from the hydrophobic one, it also increases the moment. Thus, the largest possible moment for a given segment is achieved when hydrophobic residues protrude from one side and hydrophilic residues protrude from the other, because then both sets contribute to the sum with the same sign. The meaning of equation 1 is illustrated in Fig. 1.

Even if the atomic coordinates of a protein are not known, it is still possible to estimate a hydrophobic moment, called the *sequence hydrophobic moment*. Normally it is assumed that the segment has some particular regular secondary structure and that the side chains protrude perpendicular to the axis of the segment. The magnitude of the sequence hydrophobic moment is then given by

$$\mu_H = \left\{\left[\sum_{\text{residues } n} H_n \sin(\delta n)\right]^2 + \left[\sum_{\text{residues } n} H_n \cos(\delta n)\right]^2\right\}^{\frac{1}{2}} \tag{2}$$

As in equation 1, the sum is over all residues of the segment, and H_n is the hydrophobicity of the nth residue; δ is the angle, measured in radians, at which successive side chains emerge from the central axis of the structure. For an α helix, δ is 100°; for a strand of β sheet, δ is about 160°. For a flat β sheet, δ is 180°.

A limitation of the above definitions of the hydrophobic moment is that they emphasize the amphiphilicity of a helix perpendicular to its axis. An alternative definition of the structural hydrophobic moment may be useful when it is necessary to describe amphiphilicity both parallel and perpendicular to its axis. It is:

$$\boldsymbol{\mu}_{s2} = \sum_{\text{residues } n} (H_n \boldsymbol{R}_n - \langle H \rangle \boldsymbol{R}_n) \tag{3}$$

where $\boldsymbol{R}_n$ is a vector from any origin to the center of the side chain of the nth residue. The first term of this equation gives the distribution in space of the residue hydrophobicities. The second term makes the definition independent of origin.

IV. THE HYDROPHOBIC MOMENT PLOT

The hydrophobic properties of an α helix can be classified on a hydrophobic moment plot (Eisenberg, 1984), an example of which is given in Fig. 2. The horizontal coordinate represents the average hydrophobicity of the residues in the helix, using the normalized consensus scale from Table I. The vertical coordinate registers the magnitude of the hydrophobic moment per residue. The curved boundary shows the largest moment per residue that any α helix can have for a given average hydrophobicity. The curve intersects the horizontal axis at a value of 1.38, the hydrophobicity per residue of a hypothetical polyisoleucine α helix. Other points on the curve represent hypothetical helical copolymers of isoleucine and arginine. In general, hydrophobic helices plot toward the right; amphiphilic helices plot toward the top.

Alpha helices of different types tend to cluster in different regions of the hydrophobic moment plot. Helices from globular proteins tend to plot in the region of Fig. 2 labeled GLOB, whereas probable transmembrane helices tend to plot in the two triangular regions at the lower right. The α helix of the lytic protein melittin and related molecules plot in the region labeled SURF, as discussed in the following section.

V. SURFACE-SEEKING HELICES

In earlier work it was found that several hemolytic peptide toxins all have hydrophobic properties that plot in the "SURF" region of the hydrophobic moment plot, assuming the

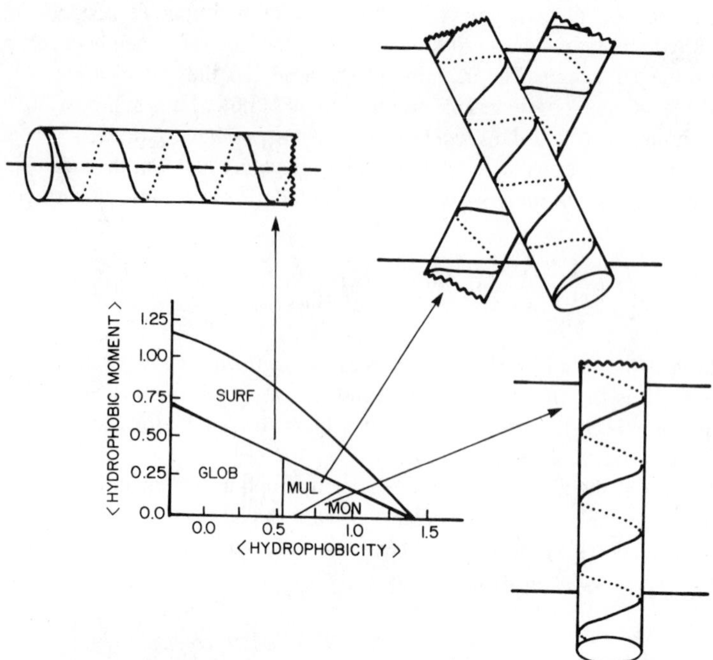

Figure 2. The hydrophobic moment plot. Each point on the plot represents the hydrophobic properties of an α helix. The horizontal axis gives the hydrophobicity per residue of the helix, and the vertical axis gives the hydrophobic moment per residue. Helices that are characterized by the surface (SURF) region tend to be amphiphilic and to align themselves parallel to the boundary between a polar and an apolar phase. Helices that are highly hydrophobic (MON region) tend to be immersed in an apolar phase (membrane interior) and are evenly hydrophobic around the helix (small hydrophobic moment). Helices that have somewhat larger hydrophobic moments may tend to form multimeric bundles with their hydrophilic faces together (MUL region). In all cases an 11-residue window is considered in the sum of equation 2, and for a given protein, the particular 11-residue helix that is plotted is the one with the largest moment.

peptides are helical (Eisenberg *et al.*, 1982a). Two of these are melittins from the venoms of two species of bee. Melittin is known to be an α helix from crystal structure studies (Terwilliger and Eisenberg, 1982a,b) and is known to accumulate at an air–water surface (Habermann, 1972). The origin of melittin's surface-seeking activity is almost certainly that it forms a highly amphiphilic α helix with an apolar side and a polar side. The helix aligns itself parallel to the air–liquid surface, with its apolar side toward air and its polar side in water. In other words, a protein segment with large hydrophobic moment tends to seek the surface between an aqueous and an apolar phase. This same principle is probably involved in melittin's lytic action, with the apolar phase now being lipid rather than air (Terwilliger *et al.*, 1982).

More recently a class of 17-residue lytic peptides was discovered in the venom of the bumblebee, the bombolitins (Argiolas and Pisano, 1985). Five such peptides lyse erythrocytes and liposomes, release histamine from rat peritoneal mast cells, and stimulate phospholipase A_2. All five peptides, if assumed to be α helical, have large hydrophobic moments that plot in the "SURF" region of the hydrophobic moment plot. Similarly, mastoparan (from wasps, hornets, and yellowjackets) and crabrolin (from European hornets) plot in the same region (see Table II and Fig. 3). These seven additional peptides bring to 15 the total of apparently amphiphilic peptides whose hydrophobic properties are similar. This strengthens the hypothesis that lytic, highly surface-seeking peptides can be recognized from their amino acid sequences by their positions on the hydrophobic moment plot.

Table II. Lytic Peptides Whose Hydrophobic Properties Are Represented by the Triangular Points of Fig. 3[a]

Peptide	Starting residue	Plot coordinates ⟨*H*⟩	⟨μ⟩	Reference
Cecropin A	3 (Lys)	−0.22	0.80	Steiner *et al.* (1981)
Cecropin B	3 (Lys)	−0.41	0.89	Steiner *et al.* (1981)
Delta-hemolysin				
S. aureus	16 (Ile)	−0.11	0.70	Fitton *et al.* (1980)
S. aureus (canine strain)	16 (Ile)	−0.17	0.67	Fitton *et al.* (1980)
Mellitin				
A. mellifera	12 (Gly)	0.25	0.57	Habermann (1972)
A. florea	12 (Gly)	0.34	0.47	Kriel (1973)
Artificial cytotoxin	12 (Leu)	0.19	0.62	DeGrado *et al.* (1981)
Artificial cytotoxin	2 (Leu)	−0.10	0.84	Eisenberg *et al.* (1986b)
Bombolitin I	2 (Lys)	0.06	0.60	Argiolas and Pisano (1985)
Bombolitin II	2 (Lys)	0.05	0.68	Argiolas and Pisano (1985)
Bombolitin III	2 (Lys)	0.11	0.70	Argiolas and Pisano (1985)
Bombolitin IV	2 (Asn)	0.04	0.66	Argiolas and Pisano (1985)
Bombolitin V	6 (Ile)	0.47	0.44	Argiolas and Pisano (1985)
Mastoparan	2 (Leu)	0.23	0.58	Argiolas and Pisano (1985)
Crabrolin	2 (Leu)	0.33	0.54	Argiolas and Pisano (1985)

[a]Each point refers to an 11-residue helix starting at the indicated residue, having a hydrophobicity per residue of ⟨*H*⟩ and a hydrophobic moment per residue of ⟨μ⟩.

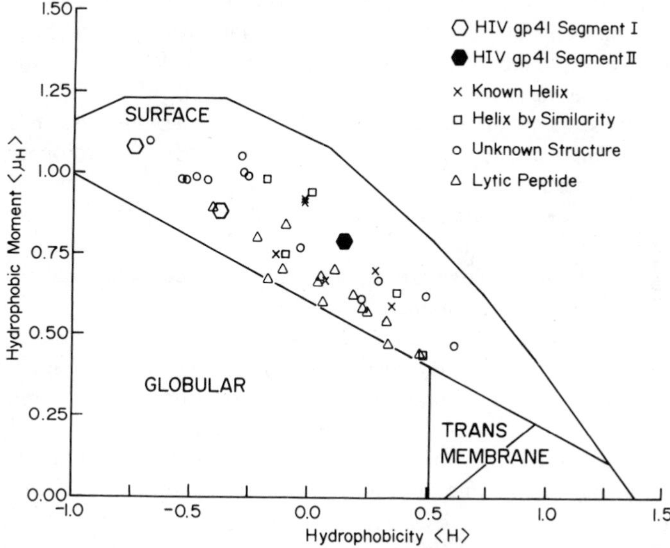

Figure 3. Hydrophobic moment plot for highly amphiphilic segments. The triangular points represent lytic peptides of Table II. Each point is the 11-residue segment of the sequence having the largest hydrophobic moment. The other points are 11-residue segments from proteins selected from the PIR data base of amino acid sequences, using equation 2.

Also plotting in the surface region of the hydrophobic moment plot of Fig. 3 are possible helical segments from larger proteins. These segments were identified by applying equation 2 to the PIR database and selecting the most amphiphilic regions, assuming all proteins to be entirely α helical (Eisenberg and Wesson, 1989). Among these segments with very large hydrophobic moments, all segments from proteins of known structure are in fact α helices that lie at the protein surface. These are indicated in Fig. 3. This finding suggests that if a segment of a protein is capable of forming a highly amphiphilic α helix, it often does so at the surface of the protein. If the segment is not attached to a larger protein, the helix can bind to a membrane. This is the case for the small lytic peptides.

Two intriguing segments with large hydrophobic moments are from the envelope glycoprotein gp41 of HIV, the etiologic agent of AIDS. One of these, segment I, is at the C terminus of gp41 and could conceivably function in one of the membrane-related activities of the virus.

Transmembrane protein pores may also contain highly amphiphilic helices. Finer-Moore and Stroud (1984) and Eisenberg *et al.* (1984a) have proposed algorithms for detecting these in amino acid sequences using hydrophobic moments.

VI. STRUCTURAL HYDROPHOBIC MOMENTS

The structural hydrophobic moment, defined by equation 1 above, can help in visualizing the hydrophobic component of the free energy of protein folding. For each segment (α helix, strand of β structure, or irregular stretch of sequence) of a known structure, the magnitude and direction of the hydrophobic moment is computed from coordinates. Then the relative positions and directions of the moments are examined by computer graphics. For a model structure, the same type of display can be generated.

Two general features are found in globular proteins (Eisenberg *et al.*, 1982b): (1) hydrophobic moments of segments at the aqueous surface point inward (toward greater hydrophobicity); and (2) hydrophobic moments of neighboring segments tend to oppose each other. The result is that in smaller proteins formed from two layers of structure (Richardson, 1981), the hydrophobic moments of virtually all segments point inward toward the center.

These two features can be seen in the hydrophobic moments of the melittin tetramer. The melittin monomer consists of 26 amino acid residues folded as an α helix bent by 60° near the center. In aqueous salt solution and in the crystalline state, melittin exists as a tetramer. The four monomers within the tetramer are held together mainly by the hydrophobic interaction. There are no hydrogen bonds among the four monomers of the tetramer, and electrostatic charge interactions between melittin side chains are all repulsive. This is so because there are no negative charges in melittin: the C terminus is amidated, and there are no acidic residues. Thus, the tendency of the 40 large apolar side chains to avoid water must account in large measure for the stability of the tetramer.

A visual representation of the hydrophobic energy of melittin is provided by the structural hydrophobic moments in Fig. 4. Hydrophobic moments are shown by straight lines drawn from the center of each melittin monomer toward the direction of increasing hydrophobicity. Because melittin is more hydrophobic on the inside than the outside, the moments point inward. Furthermore, their directions oppose each other. This opposition of moments from neighboring segments reflects the directional aspect of the hydrophobic energy of folding.

A quantitative measure of the extent of opposition of moments of different protein segments in the folded structure is given by C.

$$C = 1 - \left| \left(\sum_{\text{segments } i} \boldsymbol{\mu}_i \Big/ \sum_{\text{segments } i} |\boldsymbol{\mu}_i| \right) \right| \tag{4}$$

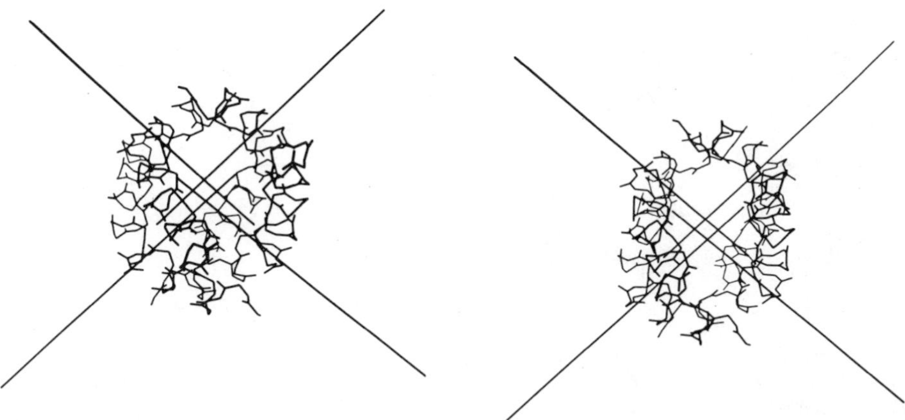

Figure 4. The melittin tetramer, showing the hydrophobic dipole moment (μ_s) of each of the four α-helical polypeptide chains. As in the following stereo pairs, the origin of each hydrophobic moment vector is placed at the center of the corresponding segment of secondary structure, and the moment extends in the direction of greater hydrophobicity (Eisenberg *et al.*, 1982b).

We can evaluate *C* using values of the hydrophobicity from the atomic scale of Table I. With this scale we find that the hydrophobic moments of melittin monomers 1 and 3 are 13.2 kcal mole^{-1} Å, and those of monomers 2 and 4 are 12.0 kcal mole^{-1} Å. This difference results from the slightly different conformation these monomers adapt. The vector sum of the moments for the four chains is 2.4 kĉal mole^{-1} Å, showing that four moments are in nearly perfect opposition, as suggested by Fig. 4. The fractional cancellation determined from equation 4 is 0.95.

It is interesting to consider the melittin monomers in pairs (Table III). Monomers 1 and 2 run nearly antiparallel and are bonded by only a few hydrophobic residues. Most of their apolar side chains extend toward monomers 3 and 4. The moments form an angle of 62° with each other and therefore only partly oppose each other; the fractional cancellation is 0.14 (Table III). In contrast, monomer pairs 1–3 and 1–4 have moments that more effectively oppose each other. They form angles nearer to 180° and thus have much more substantial fractional cancellation. In other words, the hydrophobic forces are most directed between chains 1 and 3 and between chains 1 and 4.

Table III. Opposition of Hydrophobic Moments in Melittin

Monomer(s)	Moment length (kcalÅ)[a]	Angle (deg)[b]	*C* (cancellation)[c]
1	13.2		
2	12.0		
3	13.2		
4	12.0		
1+2+3+4	2.4		0.95
1+2	21.6	62.	0.14
1+3	11.8	127.	0.55
1+4	7.6	145.	0.70

[a]Vector sum.
[b]Angle between two moments.
[c]From equation 4.

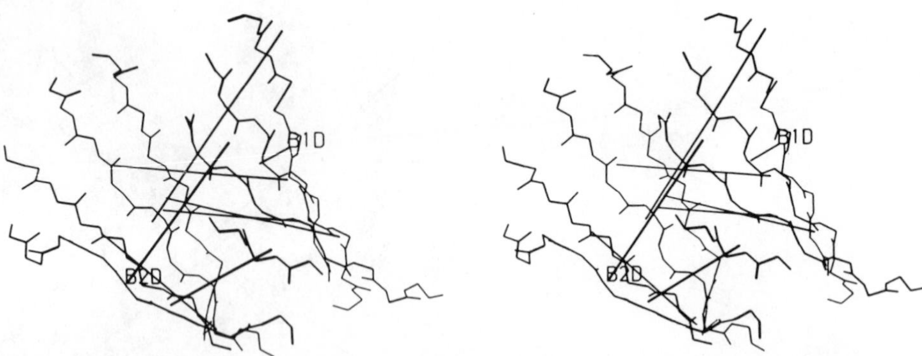

Figure 5. Hydrophobic moments of the eight strands of β sheet forming the β barrel of prealbumin. The region inside the barrel is hydrophobic; nearly all of the moments point into this region. To give the orientation of the barrel, the outer strands have been labeled in accord with the Brookhaven Data Bank. (From Eisenberg *et al.*, 1982b.)

In structures more complicated than melittin, there are more interesting patterns of hydrophobic moments. An example is the antiparallel β barrel of prealbumin (Blake *et al.*, 1978; See Fig. 5). Of the eight strands of the barrel, seven have hydrophobic moments that point toward the interior of the barrel, although one of the moments is small. Other examples are proteins of the α/β classification of Richardson (1981). In these, the sheet often has one polar side and one nonpolar side, as judged by the directions of the hydrophobic moments of the individual strands. This is illustrated in Fig. 6 for the β sheet of carboxypeptidase A (Rees *et al.*, 1983), in which six of the eight strands have their moments pointing toward the same side.

A still more involved pattern is found in triosephosphate isomerase (Banner *et al.*, 1976), which has four layers of polypeptide chains. The moments of the β barrel inside tend to point outward, opposing the moments of the surrounding α helices, which themselves point inward (Eisenberg *et al.*, 1982b). In other words, the very center of the enzyme is less apolar on average than the region midway between the center and the outer aqueous surface. This information is presented visually by the opposed pattern of hydrophobic moments.

VII. STRUCTURAL HYDROPHOBIC MOMENTS IN MODEL ASSESSMENT AND PROTEIN DESIGN

In concluding our discussion of structural hydrophobic moments, we note that these quantities may find use in assessing proposed protein models and also in protein design. They

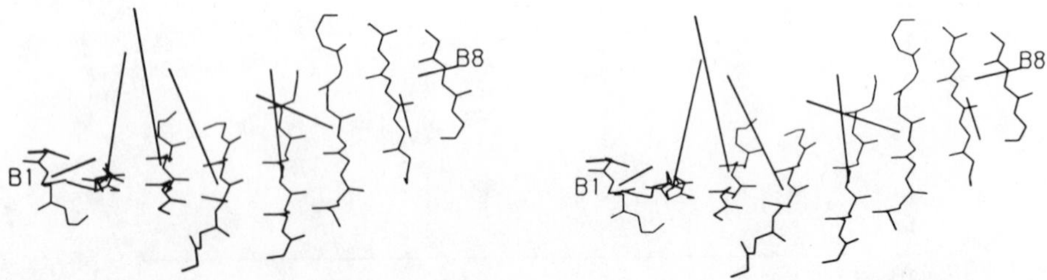

Figure 6. Hydrophobic moments in the β sheet of carboxypeptidase A. Two strands are labeled according to the nomenclature of the Brookhaven Protein Data Bank. (From Eisenberg *et al.*, 1982b.)

may be useful for assessing the correctness of proposed models because the coherent pattern of opposing hydrophobic moments found in melittin and in other real proteins is not found in hypothetical, misfolded protein structures (Novotny *et al.*, 1984; Eisenberg *et al.*, 1986b). In improperly folded proteins, segment moments are small and do not point inward in opposition to each other. Similarly, in designing new proteins, it is important to build in realistic hydrophobic forces. A visual check of the directions of hydrophobic moments may aid in such work. This criterion was used in the design, synthesis, and crystallization of a simple helical peptide (Eisenberg *et al.*, 1986a).

VIII. PROGRAM

The program MOMENT is available for calculation of sequence hydrophobic moments according to equation 2. It is written in FORTRAN for the VAX with VMS operating system. It requests a window length, a file of hydrophobicity values, and a file listing amino acid sequences to process. The list file contains the names of sequence files and their formats. MOMENT accepts sequences in a three-letter format, the Protein Identification Resource (PIR) format, and the University of Wisconsin format.

ACKNOWLEDGMENTS. The authors gratefully acknowledge support from NIH research (USPHS GM 31299) and NSF grants DMB 85-05867 and CHE 85-09657.

IX. REFERENCES

Argiolas, A., and Pisano, J. J., 1985, Bombolitins: A new class of mast cell degranulating peptides from the venom of the bumblebee *Megabombus pennsylvanicus*, *J. Biol. Chem.* **260**(3):1437–444.

Banner, D. W., Bloomer, A. C., Petsko, G. A., Phillips, D. C., and Wilson, 1976, Atomic coordinates for triose phosphate isomerase from chicken muscle, *Biochem. Biophys. Res. Commun.* **72**(1): 146–155.

Blake, C. C. F., Geisow, M. J., Oatley, S. J., Rerat, B., Rerat, C., 1978, Structure of prealbumin: Secondary, tertiary, and quaternary interactions determined by Fourier refinement at 1.8 Å, *J. Mol. Biol.* **121**(3):339–356.

Chothia, C., 1976, The nature of the accessible and buried surfaces in proteins, *J. Mol. Biol.* **105**(1):1–12.

Cohn, E. J., and Edsall, J. T., 1943, *Proteins, Amino Acids, and Peptides as Ions and Dipolar Ions*, Reinhold, New York, pp. 206–212.

DeGrado, W. F., Kezdy, F. J., and Kaiser, E. T., 1981, Design, synthesis, and characterization of a cytotoxic peptide with melittin-like activity, *J. Am. Chem. Soc.* **103**(3):679–681.

Edsall, J. T., and McKenzie, H. A., 1978, Water and proteins: I. The significance and structure of water: Its interaction with electrolytes and non-electrolytes, *Adv. Biophys.* **10:**137–207.

Eisenberg, D., 1984, Three-dimensional structure of membrane and surface proteins, *Annu. Rev. Biochem.* **53:** 595–623.

Eisenberg, D., and McLachlan, A. D., 1986, Solvation energy in protein folding and binding, *Nature (London)* **319**(6050):199–203.

Eisenberg, D., and Wesson, M., Hydrophobic moment analysis identifies two possible surface-seeking alpha helices in HIV gp41, in preparation.

Eisenberg, D., Weiss, R. M., and Terwilliger, T. C., 1982a, The helical hydrophobic moment: A measure of the amphiphilicity of a helix, *Nature (London)* **299**(5881):371–374.

Eisenberg, D., Weiss, R. M., Terwilliger, T. C., and Wilcox, W., 1982b, Hydrophobic moments and protein structure, *Faraday Symp. Chem. Soc.* **17:**109–120.

Eisenberg, D., Schwarz, E., Komaromy, M., and Wall, R., 1984a, Analysis of membrane and surface protein sequences with the hydrophobic moment plot, *J. Mol. Biol.* **179**(1):125–142.

Eisenberg, D., Weiss, R. M., and Terwilliger, T. C., 1984b, The hydrophobic moment detects periodicity in protein hydrophobicity, *Proc. Natl. Acad. Sci. U.S.A.* **81**(1):140–144.

Eisenberg, D., Wilcox, W., Eshita, S. M., Pryciak, P. M., Ho, S. P., and DeGrado, W. F., 1986a, The design, synthesis, and crystallization of an alphahelical peptide, *Proteins: Struct. Funct. Genet.* **1**(1):16–22.

Eisenberg, D., Wilcox, W., and McLachlan, A. D., 1986b, Hydrophobicity and amphiphilicity in protein structure, *J. Cell. Biochem.* **31**(1):11–17.

Fauchère, J. L., and Pliška, V., 1983, Hydrophobic parameters π of amino acid side chains from the partitioning of N-acetyl-amino acid amides, *Eur. J. Med. Chem.-Chim. Ther.* **18**(4):369–375.

Finer-Moore, J., and Stroud, R. M., 1984, Amphipathic analysis and possible formation of the ion channel in an acetylcholine receptor, *Proc. Natl. Acad. Sci. U.S.A.* **81**(1):155–159.

Fitton, J. E., Dell, A., and Shaw, W. V., 1980, The amino acid sequence of the delta hemolysin of *Staphylococcus aureus, FEBS Lett.* **115**(2):209–212.

Frömmel, C., 1984, The apolar surface area of amino acids and its empirical correlation with hydrophobic free energy, *J. Theor. Biol.* **111**(2):247–260.

Guy, H. R., 1985, Amino acid side-chain partition energies and distribution of residues in soluble proteins, *Biophys. J.* **47**(1):61–70.

Habermann, E., 1972, Bee and wasp venoms, *Science* **177**(4046):314–322.

Hopp, T. P., and Woods, K. R., 1981, Prediction of protein antigenic determinants from amino acid sequences, *Proc. Natl. Acad. Sci. U.S.A.* **78**(6):3824–3828.

Janin, J., 1979, Surface and inside volumes in globular proteins, *Nature (London)* **277**(5696): 491–492.

Kaiser, E. T., and Kezdy, F. J., 1983, Secondary structures of proteins and peptides in amphiphilic environments (A review), *Proc. Natl. Acad. Sci. U.S.A.* **80**(4):1137–1143.

Kauzmann, W., 1959, Some factors in the interpretation of protein denaturation, *Adv. Protein Chem.* **14:**1–63.

Kreil, G., 1973, Structure of melittin isolated from two species of honey bees, *FEBS Lett.* **33**(2):241–244.

Kyte, J., and Doolittle, R. F., 1982, A simple method for displaying the hydropathic character of a protein, *J. Mol. Biol.* **157**(1):105–132.

Novotny, J., Bruccoleri, R., and Karplus, M., 1984, An analysis of incorrectly folded protein models: Implications for structure predictions, *J. Mol. Biol.* **177**(4):787–818.

Nozaki, Y., and Tanford, C., 1971, Solubility of amino acids and two glycine peptides in aqueous ethanol and dioxane solutions: Establishment of a hydrophobicity scale, *J. Biol. Chem.* **246**(7):2211–2217.

Perutz, M. F., Kendrew, J. C., and Watson, H. C., 1965, Structure and function of hemoglobin: II. Some relations between polypeptide chain configuration and amino acid sequence, *J. Mol. Biol.* **13:**669–678.

Rees, D. C., Lewis, M., and Lipscomb, W. N., 1983, Refined crystal structure of carboxypeptidase A at 1.54 Å resolution, *J. Mol. Biol.* **168**(2): 367–387.

Richardson, J. S., 1981, The anatomy and taxonomy of protein structure, *Adv. Protein Chem.* **34:**167–339.

Rose, G. D., Geselowitz, A. R., Lesser, G. J., Lee, R. H., and Zehfus, M. H., 1985, Hydrophobicity of amino acid residues in globular proteins *Science* **229**(4716):834–838.

Schiffer, M., and Edmundson, A. B., 1967, Use of helical wheels to represent the structures of proteins and to identify segments with helical potential, *Biophys. J.* **7**(2):121–135.

Steiner, H., Hultmark, D., Engstroem, A., Bennich, H., and Boman, H. G., 1981, Sequence and specificity of two antibacterial proteins involved in insect immunity, *Nature (London)* **292**(5820):246–248.

Sweet, R. M., and Eisenberg, D., 1983, Correlation of sequence hydrophobicities measures similarity in three-dimensional protein structure, *J. Mol. Biol.* **171**(4):479–488.

Terwilliger, T. C., and Eisenberg, D., 1982a, The structure of melittin: I. Structure determination and partial refinement, *J. Biol. Chem.* **257**(11):6010–6015.

Terwilliger, T. C., and Eisenberg, D., 1982b, The structure of melittin: II. Interpretation of the structure, *J. Biol. Chem.* **257**(11):6016–6022.

Terwilliger, T. C., Weissman, L., and Eisenberg, D., 1982, The structure of melittin in the form I crystals and its implication for melittin's lytic and surface activities, *Biophys. J.* **37**(1):353–361.

Von Heijne, G., and Blomberg, C., 1979, Trans-membrane translocation of proteins: The direct transfer model, *Eur. J. Biochem.* **97**(1):175–181.

Wolfenden, R., Andersson, L., Cullis, P. M., and Southgate, C. C. B., 1981, Affinities of amino acid side chains for solvent water, *Biochemistry* **20**(4):849–855.

17

Tertiary Structure Prediction

Fred E. Cohen and Irwin D. Kuntz

I. INTRODUCTION

The polypeptide chain of a globular protein is linear, but the three-dimensional or tertiary structure is quite contorted. This was apparent from the first crystallographic determination of the structure of a protein by Kendrew *et al.* (1960). The contortion arises because of the need to satisfy a multitude of conflicting interactions: the hydrogen-bonding requirements of buried nitrogen and oxygen atoms, placement of the remaining polar groups near the protein–solvent interface, and internalization of most hydrophobic residues. Anfinsen *et al.* (1961) demonstrated that the amino acid sequence contained enough information to define a protein tertiary

Fred E. Cohen and Irwin D. Kuntz • Department of Pharmaceutical Chemistry, School of Pharmacy, University of California, San Francisco, California 94143–04460.

structure. These experiments defined the protein-folding problem: by computation determine the precise tertiary structure of a protein from its amino acid sequence. In the 25 years that have followed, much regularity and order have been recognized in protein tertiary structure (for reviews, see Richards, 1977; Nemethy and Scheraga, 1977; Schulz and Schirmer, 1979; Richardson, 1981; Sternberg, 1983; Chothia, 1984). This chapter describes some of the theoretical methods designed to predict protein structure.

X-ray crystallographers have solved the structures of more than 100 proteins (see Brookhaven Data Bank, Bernstein *et al.*, 1977). This provides an enormous structural set by which predictive schemes for secondary or tertiary structure can be tested. Although the atomic accuracy of crystallographic data is limited by experimental difficulties, crystalline disorder (*R* values), and thermal motion (*B* factors), these limits of resolution are small compared to the accuracy of current theoretical methods for proteins. Protein structures have both static and dynamic properties. This review emphasizes the static features.

A protein in solution adopts the conformation of minimum free energy. In principle, proper application of known chemical interactions among all parts of the full covalent structure of the peptide and all aspects of solvation would define the native structure. Two general approaches rely on specific potential energy functions for the various interactions between atoms or groups of atoms. The first method involves minimization of the overall energy by adjustment of conformational parameters. It can be applied either to the full covalent structure of the protein (Nemethy and Scheraga, 1977) or to idealized representations using modified potential functions (Levitt and Warshel, 1975). The second method is based on calculations of molecular dynamics. It probes fluctuations in structure in the time domain (McCammon *et al.*, 1977). Both procedures follow physically realizable paths during the required small incremental changes in structure. At the present stages of development, neither approach has been able to fold an extended chain into its known compact native conformation using unconstrained potential functions.

In contrast to these fundamental methods, an alternative approach disregards any explicit consideration of energy terms and concentrates solely on probable geometry, based on examples from the known structures of peptides and proteins. Structural elements are simply placed in space, and the packing is optimized. Such procedures will usually have a computational "path," but this need bear no relation to a physically realizable process. These heuristic procedures depend heavily on the analyses of known protein structures. The problem is conveniently divided into three stages:

1. Determine secondary structure from amino acid sequence, e.g., Chou and Fasman (1974, 1977), Lim (1974), Garnier *et al.*, (1978), and Cohen *et al.* (1983, 1986a).

Abbreviations used in this chapter: α/α proteins: 562, cytochrome b_{562} (Mathews *et al.*, 1979); cpv, calcium-binding protein, carp muscle (Kretsinger and Nockolds, 1973); lzm, lysozyme, T_4 phage (Matthews and Remington, 1974); mbm, myoglobin (Takano, 1977); mhr, myohemerythrin (Hendrickson and Ward, 1977); thn, thermolysin (Holmes and Matthews, 1982); tmv, tobacco mosaic virus protein (Bloomer *et al.*, 1978).

β/β proteins: adh(1–179), alcohol dehydrogenase, liver (Eklund *et al.*, 1976); cna, concanavalin A (Reeke *et al.*, 1975); est(1–111), elastase (Sawyer *et al.*, 1978); ger, γ-crystallin (Blundell *et al.*, 1981); imnh, Fab fragment of human immunoglobulin; imnl, heavy and light chains (Saul *et al.*, 1978); pre, prealbumin (Blake *et al.*, 1978); rei, Bence Jones protein variable dimer (Epp *et al.*, 1974); sns, nuclease, staphylococcal (or micrococcal) (Arnone *et al.*, 1971); sod, superoxide dismutase, Cu,Zn (Tainer *et al.*, 1982); stnv, satellite tobacco necrosis virus (Jones and Liljas, 1984).

α/β proteins: adh(180–374), alcohol dehydrogenase (Eklund *et al.*, 1976); adk, adenylate kinase (Schulz *et al.*, 1974); fxn, flavodoxin (Smith *et al.*, 1977); ldh, lactate dehydrogenase (Holbrook *et al.*, 1975); pgk, phosphoglycerate kinase (Banks *et al.*, 1979); rhd, rhodanese (Ploegman *et al.*, 1978); sbt, subtilisin (Wright *et al.*, 1969); tim, triosephosphate isomerase (Banner *et al.*, 1975).

Others: cpa, carboxypeptidase A (Hartsuck and Lipscomb, 1971); srx, thioredoxin, *E. coli* (Holmgren *et al.*, 1975); utg, uteroglobin (Mornon *et al.*, 1980).

2. Determine approximate tertiary fold by packing secondary structure, e.g., Ptitsyn and Rashin (1975), Cohen *et al.* (1979, 1980, 1981a, 1982), Efimov (1979), Lifson and Sander (1980).
3. Calculate the native conformation by refining the tertiary fold.

We restrict this discussion to such heuristic techniques. The first issue is abstraction of structural elements. Pauling made the crucial observations that hydrogen-bonding patterns greatly favored two regular elements: helices and sheets. A second important step was the realization that proteins of more than 150–200 amino acids fold into compact subunits called domains (Wetlaufer and Ristow, 1973). An important paper of Levitt and Chothia (1976) described a taxonomic sorting of protein domains into four classes based roughly on the content and sequential arrangement of secondary structure. These classes were α/α domains, composed almost exclusively of α helices like myoglobin; β/β domains, composed almost exclusively of β strands as in the immunoglobulins; α/β domains constructed from a regular alternation of α helices and β strands to form a β sheet covered by α helices and β strands as in ribonuclease. The beauty of this simple observation lies in the organizational features that can be recognized in each class. For example, β strands in β/β domains tend to have hydrophobic and hydrophilic residues in alternation along the sequence, whereas β strands in α/β domains have short sequences containing predominantly hydrophobic residues. These concepts have proved useful not only in the analysis of protein structure but in the prediction of protein structure from amino acid sequence considerations.

II. PRIMARY → SECONDARY STRUCTURE

Methods for heuristic secondary structure prediction have divided along two avenues: numerical (statistical) and pattern-based (structural) methods. Chou and Fasman (1974) devised the most popular numerical algorithm for secondary structure. The propensity for an individual amino acid to adopt a local conformation (α helical, β strand, coil) was evaluated from the data base of known structures as the ratio of the occurrences in one local conformation to the number of examples not in that local conformation. This method had the advantage of being easy to use and relatively accurate (~60%). It suffers from the slow increase in accuracy with the increasing data base. The algorithm of Garnier *et al.* (1978) is similar in spirit and is easily programmed, but the upper limit on the accuracy of statistical schemes seems limited to ~65%.

The existence of an upper bound on the accuracy of schemes that depend on a single parameter determined by local effects can be rationalized as follows. Any numerical parameter p that defines some property of structure can be rank ordered. Those subsequences with the largest values of p will always adopt a p-like structure, whereas those with the smallest values of p never will. Unfortunately, the distribution of p in proteins tends to be more gaussian than bimodal. Thus, there is a range of values for p in which members of this subset are p-like only sometimes. Any attempt to divide this region will result in one of three results: all p-like structures with many non-p structures called p-like, no p-like structures with many p-like structures called non-p-like, or a hybrid of the two (see Fig. 1a). More poignant examples of this paradigm have been found by Kabsch and Sander (1984) and Cohen *et al.* (1983), who located pentapeptides and hexapeptides with identical or nearly identical sequences that adopt a β-strand configuration in one protein and an α-helical structure in another (see Fig. 1b). This is evidence of the well-known impact of long-range interactions in defining local conformation.

In an attempt to avoid these inherent limitations, Lim (1974) and later Cohen *et al.* (1983, 1986a) and Taylor and Thornton (1983) used a pattern-based approach to secondary structure prediction. Protein domains were divided into the four taxonomic classes of Levitt and

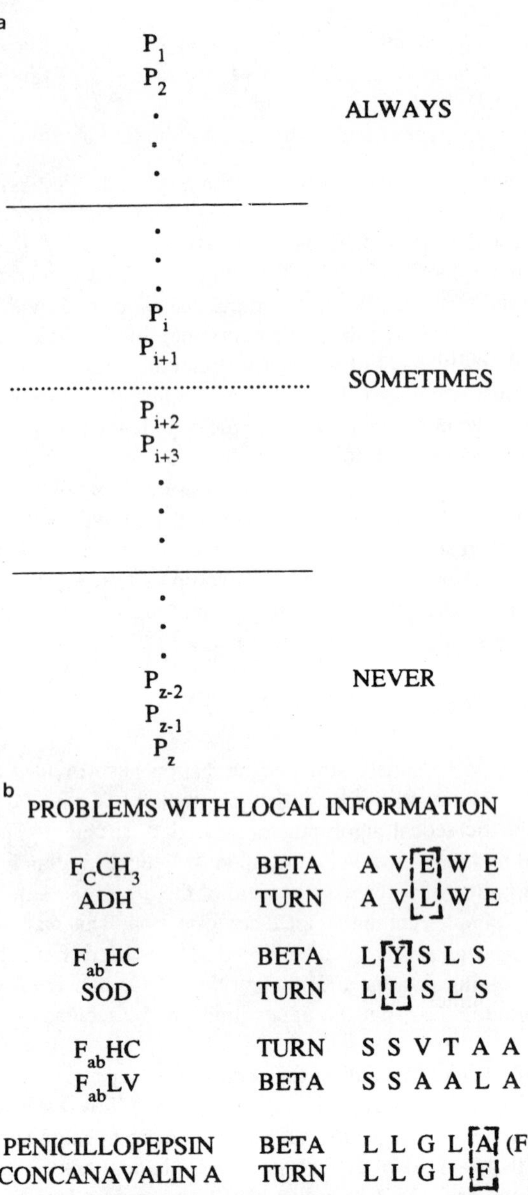

Figure 1. (a) Schematic rank ordering of protein subsequences by property *P*. This list can always be divided into three zones: the top, with all subsequences having the property *P;* the middle, with some subsequences having the property *P;* and the bottom, with no subsequences having the property *P*. Any attempt to divide the list into two parts (e.g., the dotted line) will result in assignment errors for the property *P*. This is the main reason why secondary structure algorithms parameterized for local features fail. (b) Local sequences from a variety of proteins. Highly homologous sequences that adopt very different local structures are paired. The importance of long-range interactions is evident.

Chothia, and pattern-based rules were developed to predict secondary structure. Each domain is assumed to be classifiable in terms of the Levitt–Chothia analysis. Such classifications can be based on experiments (circular dichroism, Raman, two-dimensional NMR), compositional arguments (Zav'yalov and Denesyuk, 1982; Nishikawa *et al.*, 1983; Sheridan *et al.*, 1985), or homology with domains of known structure. This method relies on first locating turns and subsequently assigning short sequence blocks to α-helix, β-strand, or coil conformation. Turns are defined broadly as irregular pieces of chain between regular secondary structure. Further subdivision of turns into specific conformational categories is avoided. Location of turns is a problem of considerable interest. Our procedures are developed from a simple model of protein structure: a chain adopts a modular structure and traverses space under the restrictions of a globular boundary, with hydrophobic interior and hydrophilic exterior. Turns logically follow as regions rich in hydrophilic residues or regions sequentially distant from other turns forced to be near the globule's boundary. Depending on the protein class, the modular structure is either α helices, β strands, or both. This distinction largely determines the spacing between turns.

We have chosen to illustrate the details of the pattern-based approach through a discussion of finding turns in globular protein domains. From the physical model, we associate turns with four sequence-related features: local maxima in hydrophilicity; avoidance of recognizable secondary structure; regions containing proline; and weakly hydrophobic segments sequentially distant from well-defined turns. The above order defines the hierarchic search. We consider each of these ideas in more detail to motivate the actual patterns we have used.

Subsequences rich in hydrophilic residues have been identified as turns by many investigators (Kuntz, 1972; Rose, 1978; Kyte and Doolittle, 1982). Cohen *et al.* (1983) employed a pattern-based version of this notion. We note, however, that there are certain sequences that are rich in hydrophilic side chains that are not turns; for example, glycine residues flanked by pairs of hydrophilic residues can play a special role in α/α proteins (see Table I). Such "glycine sites" are discussed further.

Turns are found in regions between secondary structural elements. Residues sequentially arrayed in secondary structure are, by definition, not in turns. Thus, a secondary structure identification algorithm is of obvious benefit for turn location. Moreover, since domains in globular proteins have diameters roughly related to the molecular weight, turns are generally found within a fixed number of residues of the center of secondary structure feature. Thus, some patterns match a high density of hydrophobic residues that corresponds to the center of an internal β strand. Since the typical β strand is no more than nine residues in length, turns are sought five residues before and five after such a site. In a variant of this approach, in the α/α set, the termini of α helices can sometimes be found by a nonhelical alternation of hydrophobic and hydrophilic residues. Turns are located between adjacent sets of these termini.

Proline occupies a special role in turns. It is given a high turn propensity in empirical prediction schemes (Lewis *et al.*, 1971; Kuntz, 1972; Chou and Fasman, 1974, 1977). However, proline is occasionally observed in "kinks" in the middle of α helices or in "bulges" in β strands (Richardson *et al.*, 1978). In helical domains, we find that prolines are not in turns when strong hydrophilic turns are among neighboring residues on the N-terminal side. We speculate that the longer helix with a kink at the proline is favored over two short helices divided by an internal turn. The occurrence of proline in β bulges limits its usefulness in turn identification in the β/β class.

Weakly hydrophilic segments in the sequence are also candidates for turns. Although not all such "weak spots" are turns, those sequentially distant from other "stronger" turns frequently are. Specific patterns of amino acids with reduced requirements for local hydrophilicity are searched for at the desired turn spacings.

The term "swing residue" implies that certain residues occupy both hydrophobic and hydrophilic roles. Swing residues are not a fixed set of amino acids. Instead, they vary with protein class. We have learned that they are GKPSTY in β/β, AKY in α/α, and KY in α/β. Swing residues generally occur in physical locations where a small change in torsional bond angles of the main chain or side chain can move a side chain from a buried to an exposed position without altering the surrounding protein architecture. In β/β proteins, each strand has

Table Ia. α/β Turn Patterns

```
Pattern Name       Offset  Pattern

gly-ala-site1         4 "[ACFIKLMTVWY][ABDEGHKNPQRSTZ]%2.2
                          [AG][ABDEGHKNPQRSTZ]%2.2[ACFIKLMTVWY]"
gly-ala-site2         5 "[ACFIKLMTVWY]%2.2[ABDEGHKNPQRSTZ]
                          [ABDEGHKNPQRSTZ][AG][ABDEGHKNPQRSTZ]%2.2
                          .[ACFIKLMTVWY]"
gly-ala-site3         5 "[ACFIKLMTVWY].[ABDEGHKNPQRSTZ]%2.2
                          [AG][ABDEGHKNPQRSTZ]%2.2[ACFIKLMTVWY]%2.2"
```

CLUSTER OF HYDROPHILIC RESIDUES BOUNDED BY HYDROPHOBIC RESIDUES, CENTERED ON A GLYCINE OR ALANINE.
TYPICAL OF HELICAL AREAS PARTICIPATING IN TIGHTLY PACKED HELIX-HELIX INTERACTIONS.

```
many_ala              5 (density(>=,3,9,1,"A"))
```

ALTHOUGH ALANINE MAY OCCUPY MANY POSITIONS IN THIS PATTERN, WHEN 3 OR MORE ARE SEEN, THE GEOMETRIC ASSYMMETRY SUGGESTS THAT HELIX-HELIX PACKING IS UNLIKELY. THIS PATTERN MATCHES WHEREVER THERE ARE GREATER THAN 2 ALANINES OUT OF 9 RESIDUES.

```
gly-ala-site          0 ((gly-ala-site1 or gly-ala-site2 or
                          gly-ala-site3) and not many_ala)
```

A GLY-ALA TYPE HELICAL SITE WITHOUT TOO MANY ALANINES.

```
Proline              -4 "P"
```

A PROLINE RESIDUE WITH OFFSET -4.

```
alpha_philic          0 "[BDEGKNQRSZ]"
```

HYDROPHILIC RESIDUES.

```
alpha_begin          -3 "[BDEGHKNQRSZ][ACFIKLMPTVWY][BDEGHKNQRSZ]
                          [ACFIKLMPTVWY]"
```

PATTERN FREQUENTLY SEEN AT THE BEGINNING OF α-HELICES.

```
alpha_end             0 "[ACFIKLMPTVWY][BDEGHKNQRSZ][ACFIKLMPTVWY]
                          [BDEGHKNQRSZ]"
```

PATTERN FREQUENTLY SEEN AT THE END OF α-HELICES.

```
alpha_phobic               0 "[ACFIKLMPTVWY]"    HYDROPHOBIC RESIDUES.
alpha_strong_phobic        0 "[ACFILMVW]"        STRONG HYDROPHOBICS.

charge_a              2 "[DE]...[HKR]"   CHARGE INTERACTION FAVORED
                                         BY HELICAL GEOMETRY.
charge_b              1 "[DE]..[HKR]"    SAME WITH DIFFERENT OFFSET
                                         AND SEPARATION OF CHARGES.
                                         (KIM AND BALDWIN 1984)

charge_c              2 "[HKR]...[DE]"   SAME AS "CHARGE_A" BUT NOT
                                         FAVORED BY HELICAL GEOMETRY.

alpha1charges         4 ((charge_c{1,1} and charge_a{2,2}) or
                          (charge_c{0,0} and charge_a{3,3}) or
                          (charge_c{0,0} and charge_a{4,4}) or
                          (charge_c{-1,1} and charge_a{4,4}))

alpha2charges         2 ((charge_a{1,1} and charge_c{2,2}) or
                          (charge_a{0,0} and charge_c{3,3}) or
                          (charge_a{0,0} and charge_c{4,4}) or
                          (charge_a{-1,1} and charge_c{4,4}))
```

TWO HELICAL PATTERNS WHICH FAVORABLY ARRAYS CHARGED RESIDUES ALONG A HELICAL FACE EITHER +...-...+ OR -...+...-

```
Pattern Name          Offset  Pattern

AA                       0 (gly-ala-site or alpha1charges or alpha2charges)
```

REGIONS LIKELY TO BE HELICAL WITH SPECIAL PATTERNS OF CHARGED OR HYDROPHILIC RESIDUES WHICH ARE SUBSEQUENTLY MASKED TO PREVENT THEIR ASSIGNMENT AS TURN REGIONS.

```
alpha_phob               1 (not density(>=,2,3,1,alpha_philic))
```

REGIONS UNLIKELY TO BE GOOD TURNS.

```
AA_Turn1_no-phobics      2 (density(=,0,5,1,alpha_strong_phobic)
                            and not charge_a{-2,2}
                            and not charge_b{-2,2}
                            and not gly-ala-site)
```

FIVE RESIDUES IN SEQUENCE WITHOUT ANY STRONG HYDROPHOBICS.
THIS PATTERN AVOIDS KIM/BALDWIN CHARGE PAIRS AND GLYCINE INTERACTION SITES.

```
AA_T_philic              1 (density(=,4,4,1,alpha_philic))
```

STRONG TURN WITH FOUR HYDROPHILIC RESIDUES IN SEQUENCE.

```
AA_Turn2_4-philics       0 (T_philic and not AA{-5,5}
                            and not AA_Turn1_no-phobics{-11,11})
```

WEAKER TURN PATTERNS WITH APPROPRIATE MASKING OF WHAT HAS COME BEFORE.

```
AA_Turn3_proline         0 (Proline and not AA_Turn1_no-phobics{-11,0}
                            and not AA_Turn2_4-philics{-11,0})
```

PROLINE SITE STRONGLY SUGGESTIVE OF TURN IN α/α PROTEINS.

```
AA_Turn4_helix-ends      1 (alpha_end and
                            alpha_begin{-2,0}
                            and not (AA_Turn1_no-phobics{-11,11} or
                                     AA_Turn2_4-philics{-11,11} or
                                     AA_Turn3_proline))
```

TURN SUGGESTED BY ADJACENT HELICAL BEGINNING AND END IN REGION AWAY FROM REGULAR TURN.

```
AA_Turn5_weak            0 (not AA{-5,5} and not AA_T_possible{-11,11}
                            and not alpha_phob{-1,1})
```

NOT A TYPICAL CENTRAL HELIX OR PHOBIC, MASKING OUT MORE LIKELY AA_T_POSSIBLE TURN PATTERN.

```
AA_Turn5_group           0 (group(AA_Turn5_weak,7))
   Merged AA_Turn5_weak hits for up to 7 adjacent hits.

AA_T_possible            0 (   group(AA_Turn1_no-phobics,7)
                            or group(AA_Turn2_4-philics,7)
                            or group(AA_Turn4_helix-ends,7)
                            or group(AA_Turn3_proline,7)  )
```

GROUPED POSSIBLE TURN SITES.

```
AlphaTurns               0 (AA_T_possible or AA_turn5_group)
```

FINAL CONSENSUS TURN PATTERN.

Table Ib. α/α *Turn Patterns*

Pattern Name	Offset	Pattern
yphil1	0	"[BDEGHKNPQRSTYZ][BDEGHKNPQRSTZ]%3,3"
yphil2	0	"[BDEGHKNPQRSTZ][BDEGHKNPQRSTYZ][BDEGHKNPQRSTZ]%2,2"
yphil3	0	"[BDEGHKNPQRSTZ]%2,2[BDEGHKNPQRSTYZ][BDEGHKNPQRSTZ]"
yphil4	0	"[BDEGHKNPQRSTZ]%3,3[BDEGHKNPQRSTYZ]"
yphilic	1	(yphil1 or yphil2 or yphil3 or yphil4) 4 HYDROPHILIC RESIDUES WITH AT MOST 1 SWING RESIDUE Y.
5of7	3	(density(>=,5,7,1,"[BDEGHKNPQRSTZ]")) 5 OF 7 RESIDUES ARE HYDROPHILIC.
charge_a	2	"[DE]...[KR]" CHARGE PAIR WITH FAVORABLE INTERACTION
charge_b	1	"[DE]..[KR]" IN THE KIM/BALDWIN (1984) SENSE.
charge_c	2	"[KR]...[DE]" UNFAVORABLE INTERACTION.
charge_d	2	"[KR]..[DE]"
charges	0	(charge_a or charge_b or charge_c or charge_d)
beta_middle	2	(density(>=,4,5,1,"[CFILMVW]")) HIGH DENSITY OF HYDROPHOBIC RESIDUE PATHOGNOMONIC FOR AN INTERNAL β-STRAND.
ab_phob	1	(density(>=,2,3,1,"[ACFILMTVW]")) PATTERN NOT SUITABLE FOR TURNS.
AB_Turn1_4-philics	0	(group(yphilic,5)) MOST DEFINITE TURNS.
AB_Turn2_Proline	0	("P" and not AB_Turn1_4-philics{-7,7}) PROLINES MASKING TURNS ABOVE.
AB_Turn3_5of7	0	(5of7 and not charges{-2,2} and not AB_Turn1_4-philics{-7,7} and not AB_Turn2_Proline{-7,7})
AB_Turn3_group	0	(group(AB_Turn3_5of7,5)) GROUPED REGIONS WITH 5 OF 7 RESIDUES HYDROPHILIC, MASKING TURNS ABOVE.
AB_Turn4_not-beta	0	((beta_middle{7,7} or beta_middle{-3,-3}) and not AB_Turn1_4-philics{-7,7} and not AB_Turn2_Proline{-7,7} and not AB_Turn3_group{-7,7}) BORDERS OF SEQUENCE MOSTLY LIKELY A MIDDLE OF β-STRAND.
AB_Turn5_weak	1	(not AB_Turn1_4-philics{-7,7} and not AB_Turn2_Proline{-7,7} and not AB_Turn3_group{-7,7} and not AB_Turn4_not-beta{-7,7} and not ab_phob{0,2})
AB_Turn5_group	0	(group(AB_Turn5_weak,5)) GROUPED WEAK HYDROPHILIC REGIONS, MASKING OTHER TURNS.
ABTurns	0	(AB_Turn1_4-philics or AB_Turn2_Proline or AB_Turn3_group or AB_Turn4_not-beta or AB_Turn5_group) FINAL CONSENSUS TURN PATTERN.

Table Ic. β/β Turn Patterns

```
Pattern Name       Offset   Pattern

ypphil1             0 "[BDEGHKNPQRSTYZ][BDEGHKNQRSZ]%3,3"
ypphil2             0 "[BDEGHKNQRSZ][BDEGHKNPQRSTYZ][BDEGHKNQRSZ]%2,2""
ypphil3             0 "[BDEGHKNQRSZ]%2,2[BDEGHKNPQRSTYZ][BDEGHKNQRSZ]"
ypphil4             0 "[BDEGHKNQRSZ]%3,3[BDEGHKNPQRSTYZ]"
        4 HYDROPHILIC RESIDUES WITH AT MOST 1 Y OR P SWING RESIDUE.

loc_min1+           2 "[ACFIKLMTVWY][BDEGHKNPQRSTYZ]%3,4[ACFIKLMTVWY]"
loc_min3+           3 "[ACFIKLMTVWY][BDEGHKNPQRSTYZ]%5,5[ACFIKLMTVWY]"
        LOCAL "MINIMUM" IN HYDROPHOBICITY.

loc_min1-           1 (density(>=,2,3,1,beta_swing1))
loc_min2-           1 (density(>=,3,4,1,beta_swing1))
loc_min3-           2 (density(>=,3,5,1,beta_swing1))
        USED IN PATTERN BB_LOC-MIN TO AVOID HYDROPHILIC
        LOCAL MINIMA WHERE THE MAJORITY OF THE HYDROPHILIC RESIDUES
        ARE ACTUALLY "SWING RESIDUES."

bulge_res           0 "[GSY]"       ANIMO ACIDS FREQUENTLY SEEN IN \(*B BULGES.
beta_swing1         0 "[GKSTY]"  SWING RESIDUES.
beta_swing2         0 "[AY]"     ANOTHER SET OF SWING RESIDUES.

many_bulge          1 (density(>=,3,4,1,bulge_res))
        LOCAL CONCENTRATION OF TYPICAL "BULGE" RESIDUES.

betaphilic          0 "[BDEGHKNPQRSTZ]"      HYDROPHILIC RESIDUES.
beta_strong_phobic  0 "[ACFILMVW]"         STRONG HYDROPHOBICS.
beta_weak_phobic    0 "[ACFGIKLMSTVWY]"            WEAK HYDROPHOBICS.

BB_4-philic         1 (ypphil4 or ypphil3 or ypphil2 or ypphil1)
        STRONGEST TURN PATTERN.

BB_loc-min   0 ((loc_min1+ and not (loc_min1- or loc_min2-))
                   or (loc_min3+ and not loc_min3-) )
        USING LOCAL HYDROPHOBICITY MINIMA.

BB_no-phobic        2 (density(=,0,5,1,beta_strong_phobic))
        ABSENCE OF STRONG HYDROPHOBIC RESIDUES.

BB_3-philic         1 (density(>=,3,4,1,betaphilic) and
                       not density(>=,4,5,1,beta_weak_phobic))
        HIGH DENSITY OF HYDROPHILIC RESIDUES WITH LIMITED NUMBER
        OF SWING RESIDUES.
```

```
Pattern Name       Offset   Pattern

BB_weak-philic      3 (( density(>=,4,7,1,betaphilic) and
                           density(>=,2,7,1,beta_swing2) )
                        or
                         ( density(>=,3,7,1,betaphilic) and
                           density(>=,4,7,1,beta_swing2) ))
        MINIMUM DENSITY OF HYDROPHILIC RESIDUES SEEN IN TURNS.

BB_4-philic_group     0 (group(BB_4-philic,7))
BB_loc-min_group      0 (group(BB_loc-min,7))
BB_no-phobic_group    0 (group(BB_no-phobic,7))
BB_3-philic_group     0 (group(BB_3-philic,7))
BB_weak-philic_group  0 (group(BB_weak-philic,7))
        GROUPING OF BB_TURN PATTERNS ABOVE.

BB_Turn1_4-philic     0 (BB_4-philic_group and not many_bulge{0,1})
        MOST DEFINITE TURN PATTERN, MASKING TYPICAL BULGE RESIDUES.

BB_Turn2_loc-min      0 (BB_loc-min_group and not BB_Turn1_4-philic{-4,4})
        LOCAL MINIMUM TYPE TURNS MASKING ONES ABOVE.

BB_Turn3_no-phobics   0 (BB_no-phobic_group
                           and not BB_Turn1_4-philic{-8,8}
                           and not BB_Turn2_loc-min{-4,4})
        NO PHOBIC TYPE TURNS MASKING ONES ABOVE.

BB_Turn4_3-philic     0 (BB_3-philic_group
                           and not BB_Turn1_4-philic{-4,4}
                           and not BB_Turn2_loc-min{-4,4}
                           and not BB_Turn3_no-phobics{-4,4})
        THREE PHILIC TYPE TURNS MASKING ONES ABOVE.

BB_Turn5_weak-philic  0 (BB_weak-philic_group
                           and not BB_Turn1_4-philic{-8,8}
                           and not BB_Turn2_loc-min{-8,8}
                           and not BB_Turn3_no-phobics{-8,8}
                           and not BB_Turn4_3-philic{-8,8})
        WEAK PHILIC TYPE TURNS MASKING ONES ABOVE.

BetaTurns             0 (BB_Turn1_4-philic    or
                           BB_Turn2_loc-min     or
                           BB_Turn3_no-phobics  or
                           BB_Turn4_3-philic    or
                           BB_Turn5_weak-philic)
        FINAL CONSENSUS TURN PATTERN.
```

Table II. Link Lengths and Protein Classes

Class	Approximate domain size	r[a]	Pitch (Å/ residue)	Link length Å[b]	Residues[c]	Used[e]
α/α	150	17.5	1.5	31.0	20.7	22
α/β	200	19.3	2.25[d]	34.6	15.4	14
β/β	100	15.3	3.0	26.6	8.9	8

[a]Computed from $(3 \times 110n/0.602 \times 4\pi\bar{v})^{1/3}$ where n is the number of residues and $\bar{v}$ is the partial specific volume (0.75 cm).
[b]Computed as 2(r − 4) Å, where 4 Å is taken as the width of a turn.
[c]Computed as (2r − 8Å)/pitch.
[d]Assumed to be the average of the pitch for α and β structure.
[e]To obtain the average length between turns, add 4 residues to the value used for the secondary structure length. Only even values appear in this column as ½ of this value appears as a natural parameter in many pattern expressions and non-integral values are not tolerated. In Cohen *et al.* (1983) a value of 14 ± 4 was used for α/β turn separation equivalent to the value chosen here.

at least two such positions, one at either end (Cohen *et al.*, 1981b). When β bulges are included, the number of potential positions for swing residues increases by two per bulge. In α/β proteins, the side-chain packing tends to be more regular, and β bulges are rarely observed (Cohen *et al.*, 1982). Whatever the precise structural, kinetic, or evolutionary reasons for the existence of swing residues, it is clear, empirically, that these residues can fit both hydrophobic and hydrophilic environments.

In summary, we find that the local maxima in hydrophilicity identify 50% of the turns in the domains studied here. Patterns relating to proline and the avoidance of secondary structure account for another 30%, and weakly hydrophilic regions that are sequentially distant from well-defined turns account for about 10% more. These categories are made mutually exclusive through the use of sequence masking.

Masking means that certain regions of the chain are removed from consideration as possible turns. Regions of a sequence are masked because they are considered likely to be α helices or β strands or because a turn has already been located in their immediate vicinity. The length of sequence that is masked follows from the rules presented below and in Table II.

Table III. Symbols Used in Pattern Specifications

Special symbol	Meaning
^	Beginning of sequence
$	End of sequence
*	Zero or more repeats of the preceding symbol, equivalent to %0*,
%*n*,*m*%	Between *n* and *m* repeats of previous symbol or parenthesized set of symbols; *m* may be "*" to indicate *n* or more repetitions
[ ]	Logical OR of symbols in brackets, as in [ABC]
-	"Through", used in [. . .] to indicate a range of values; e.g., [A-CG-K] means [ABCGHIJK]
{*m*,*u*}	Spreading of previous symbol that hits at position *i*, to all sites between *i* + *m* and *i* + *u*
( )	Used for grouping of characters for repetitions or logical combinations of pattern expressions involving AND, OR, or NOT
Group	Explained in the text
Density	Explained in the text

In work on α/β proteins (Cohen *et al.*, 1983), we noted that a knowledge of the expected spacing between turns offered an excellent means for exploiting strongly signaled turns to aid in the location of more weakly signaled turns. Though the path of the polypeptide chain is complex, the important turns occur at fairly regular spatial and sequential intervals. To approximate the average length of secondary structure or link length, the following calculation is useful. The average domain size for each protein class is known empirically. Precision is not needed since the link length varies as the one-third power of the domain size. If the proteins are assumed to be spherical and the turns taken to occupy the outermost 2 Å of the sphere, then link length as a function of protein class can be quickly calculated (see Table II). For example, in an α/α protein, if residues 20 and 40 were identified as strong turns, the program would not allow additional turns within the 18 residues from 21 to 39, as the link length for α-helices is 22 residues. If the same strong turns were found in a β/β protein, an attempt to find a weaker turn pattern between residues 28 and 32 would be made. This avoidance or masking of the sequence close to strong turns improves the accuracy of the algorithm.

Of course, some proteins show large axial ratios and occasionally long α helices. This is the case in influenza hemagglutinin (Wilson *et al.*, 1981). Such structures would be outside the limits of our current models, but we anticipate that they could be included by developing rules for subclasses.

A. Pattern Language

A formal pattern language to search sequences has been developed (Abarbanel, 1984). It permits direct residue-for-residue matches and allows residues to be grouped in various ways. Table III explains the special characters used in pattern definitions. All patterns also have an associated offset, an integer that is added to the actual sequence location of a match. This is useful for bringing related patterns together to obtain new patterns. Several examples of patterns from Table I are discussed below.

The simplest pattern is just a quoted string that must match exactly, for example,

Proline −4 "P"

A sequence will be marked with pattern "Proline" at offset −4 residues, that is, 4 residues before each P residue.

PLANS (Pattern Language for Amino acids and Nucleic acid Sequences) allows for single-residue variability at a given site. These residues may be specified to be either IN a restricted set, NOT IN a set, or completely free. More than one residue may take the role of "P" in the pattern above by using the notation "[. . .]." For example,

alpha_strong_phobic 0 "[ACFILMVW]"

would match a single residue at any of the residues in the group "ACFILMVW" with zero offset.

The "." character may be used to represent any residue and is therefore a place marker for specific separations or where any residue may match. For example,

charge_a 2 "[DE]...[HKR]"

will mark a site two residues after the D or E whenever there is an H, K, or R at residue 4 downstream. The offset of 2 here places the "hit" location for this pattern at the center of the charged pair. In an α helix, this would constitute a reasonable stabilizing charge pair (Kim and Baldwin, 1984).

Logical operators are provided to allow one to specify that more than one pattern matches at a given site (AND), at least one pattern matches there (OR), or that some pattern not match

(NOT). Expressions using these connective words are evaluated using the matches of the patterns included in the expression. Expressions may also include parentheses.

An example of the use of logical connectives and parentheses is

gly-ala-site 0 ((gly-ala-site1 OR gly-ala-site2 OR gly-ala-site3) AND NOT many_ala)

which would match wherever any one or more of the three OR'd patterns match, as long as the many_ala pattern is not also found there. PLANS both notes the match with pattern gly-ala-site and remembers which of the four subpatterns was found.

It is convenient to define densities of matches within the pattern syntax. One can specify an exact number (using =) or some relational operator (>, >=, <, <=, !=) applied to a number of patterns. For example,

many_ala 5 (density(>=,3,9,1,"A"))

would match wherever the sequence contained at least three alanines out of nine contiguous residues. The "A" here could be replaced by the name of any other pattern.

PLANS also provides "merging" and "spreading" functions. The "group" function merges up to a specified number of matches into a single mark. In this way, local signals spread over adjacent residues may be combined in a single hit. For example,

AA_Turn5_group 0 (group(AA_Turn5_weak,7)

will merge up to seven AA_Turn5_weak matches that are adjacent. These groups will be marked at the beginning of such a group in sequences matched.

Spreading allows the location of a match to be marked over a number of adjacent residues. This allows patterns to act as masks so that regions of a sequence may be effectively excluded from matching during the evaluation of another, possibly less specific, pattern. This masking idea is used powerfully in the turn assignment patterns, where a precedence of turn types is brought about by each stronger or more definite type of turn masking a region that then will not be hit by weaker patterns.

The symbol for spreading is incorporated in the pattern as $\{m,n\}$

AA_Turn2_4-philics 0 (AA_T_philic AND NOT AA{−5,5} AND NOT AA_Turn1_no-phobics{−11,11})

In this case, matches will be found where the pattern AA_T_philic is found provided that those matches are NOT also within five residues of a match with pattern AA AND also NOT within 11 residues of a match with pattern AA_Turn1_no-phobics. In the most recent version of the pattern-matching algorithm, the offset function is replaced by the spread function. For example, offset −4 is equivalent to spread {−4,−4.}

Since patterns are defined in terms of each other, PLANS must keep track of related patterns and note any missing pattern definitions. In addition, these objects keep the results of previous matches so that a given pattern needs to be run only once on a sequence. Any pattern defined in terms of another will cause the other pattern(s) to be evaluated. This process may be monitored so that complex matches may be decomposed and understood. A complete example is worked out in Fig. 2 for the turns in part of the myoglobin sequence.

Table 1 contains the complete list of α/α turn patterns. The hierarchy or order of precedence of turn expressions is AA_Turn1_no-phobics, AA_Turn2_4-philics, AA_Turn3_proline, AA_Turn4_helix-ends, AA_Turn5_weak. AlphaTurns is the aggregate final turn expression. Each of the turn expressions is a combination of patterns that are also listed in the table. Many of these terms are defined by other terms, possibly by using the spread notation and masks. As an example, see the pattern AA_Turn2_4-philics explained above. The "not

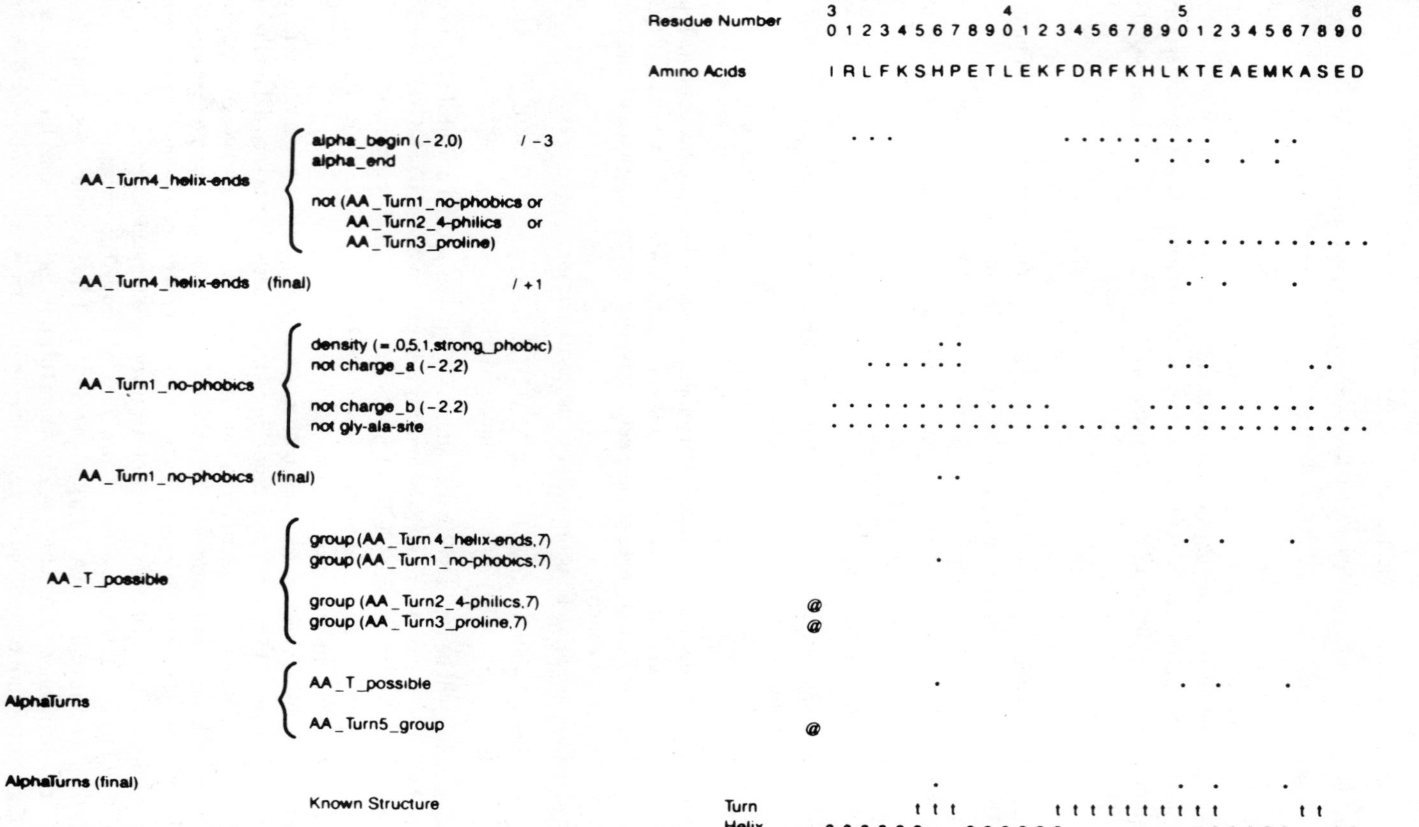

Figure 2. Myoglobin residues 30–60. Pattern hits are marked ●, and @ indicates no match for this pattern over the region. Reprinted with permission from Cohen *et al.* (1986a).

AA'' part of this expression avoids gly-ala-site, alpha1charges, and alpha2charges patterns. The ''not AA_Turn1_no-phobics'' logical expression prevents looking for turns in regions where a stronger turn has already been noted. By continued substitution of terms, one finally arrives at a list of specific amino acid sequences. This list is then checked against the sequence under study.

In a similar fashion, patterns for the β/β and α/β-class turns were developed. In the β/β work (see Table I), initial emphasis was placed on local peaks in hydrophilicity (BB_4-philic, BB_loc-min, BB_no-phobic). Because β bulges can create high densities of certain hydrophilic residues in the centers of β strands, these combinations are searched for (many_bulge) and avoided (e.g., BB_4-philic). Less hydrophilic areas are labeled as turns if adequately separated from stronger turns. In the case of β structure, an eight-residue lower bound is used on the separation between turns (BB_Turn4-3philic, BB_Turn5_weak-philic). A consensus pattern, BetaTurns, is formed.

The α/β turns patterns (see Table I) combine features seen in α/α and β/β patterns. A reliance is placed on hydrophilicity (AB_Turn4_not-Beta). The consensus pattern, ABTurns, groups several turn concepts under one heading.

B. Implementation

A version of this program called MATCH written in C and running under VMS or UNIX is available. Patterns exist for locating turns in each protein class. A scheme for assigning the secondary structure to the regions between the turns is under development.

C. Successes and Failures

A test set consisting of 7 α/α proteins with 45 turns, 11 β/β proteins with 127 turns, and eight α/β proteins with 145 turns was studied. The errors were α/α 4% (2/45), β/β 8% (10/127), α/β 5% (7/145) when the appropriate turn algorithm was applied to each class. Turns were judged correct if they occurred between secondary structure elements or if they were not more than three residues from either end of a secondary structure element. An incorrectly predicted turn divides a secondary element. We do not assign actual ''turn'' residues for these purposes, merely partitions for secondary structures. We also applied the turn algorithm for each class to sequences in other classes. The expected and predicted number of turns usually agreed best when the appropriate class algorithm was applied, but the data were not predictive.

Although the overall error rate is low, both overprediction and errors of underprediction occurred. Overprediction implies that a turn was predicted in the middle of a secondary structure segment. Errors of this variety are difficult to recover from if one is intent on predicting tertiary structure from sequence. Fortunately, only three overpredictions occur in the test set. These are cataloged in Table IV. The two sequences have in common hydrophilicity and adequate separation from neighboring turns. Both features lead to failures in the present algorithms. We were unable to develop any reasonable rule to recognize these regions and subsequently avoid the errors. The error rate for overprediction is 0.9%.

Errors of omission—the failures to identify a turn—are more common (Table V) but are more easily addressed. For instance, a more determined search for increasingly weaker turns in segments too long to remain as a single link would yield a suitable turn in most cases. A few of the turns in β/β proteins deserve further mention. Some β strands have N-terminal residues hydrogen-bonded to residues in one β sheet while the C-terminal residues are hydrogen-bonded to residues in the second sheet. Since there is a 90° kink between the N-terminal and C-terminal residues, this strand is sometimes labeled as two separate strands (for a complete

Table IV. False Positive Assignments of Turns

Protein class	Protein name	Overpredicted turn: Sequence no.	Overpredicted turn: Amino acids	Secondary structure split
α/α	None			
α/β	LDH	36	ADA	[a]
	SBT	111	NGIE	α 103–117[b]
	TIM	145	QETK	α 137–153[b]
β/β	STNV	187	DSSYE	β 183–192[c]

[a]Helix longer than allowed by model, weak turn is sought.
[b]Strong turn pattern. However, crystallographically, a good helix. Potential interactions of K 147 with D 151 to form salt bridge.
[c]Unusual collection of residues in the center of an internal β strand in a β sandwich.

discussion see Chothia and Janin, 1982). For example, trypsin can be thought of as being composed of domains that are either six- or eight-stranded barrels. The separations between some of the neighboring turns are short (0–3 residues), and these turns are hydrophobic. In the tertiary structure, these turns are frequently less accessible to solvent. We miss these turns with the patterns described here but could retrieve them fairly easily with a pattern designed to recognize the situation.

In general, the protein class—α/α, β/β, or α/β—is not known, and yet information about turns is desired. If the goal was to locate a subset of the turns that had a very high likelihood of being correct, it would be possible to run all three prediction schemes and take the common answers. These triple hits could prove useful to those interested in antigenicity.

In a pilot study to assess the value of a pattern-based approach to secondary structure

Table V. True Turns Not Assigned

Protein class	Protein name	Missed: Sequence no.	Missed: Amino acids	Comment
α/α	LZM	91–92	LD	[a]
	MBN	18–20	EAD	[b]
α/β	PGK	330–332	AKQ	[b]
	RHD	10–11	VS	[b]
		247–250	CRKG	[b]
	TIM	86–88	GAA	[c]
		203–204	SR	[d]
β/β	ADH	40–41	MV	[e]
	EST	46–47	GG	[e]
		88–89	GV	[e]
	GCR	127–130	LEGS	perhaps [a]
	IMMH	100–102	IAG	[a]
		176–178	SSG	[a]
	SNS	11–12	PA	[e]
		90–91	AY	[e]
		95–97	DGK	[a]

[a]Masked from consideration since turn separation close to desired helix limit (22 residues). Error of model assumption.
[b]Spacing between neighboring turns is long enough to suggest additional turns. Length-based parse of sequence would locate turn.
[c]Sequence not hydrophilic enough to be recognized as a potential turn. Perhaps footnote *a* applies.
[d]Masked from consideration sincd 208 is too close (< 7 residues) to 203–204. Fortunately, the α-helix is non-core.
[e]Single β strand hydrogen bonding with both sheets with 90° kink in strand direction between turn residues. Note relative hydrophobicity.

prediction, Cohen *et al.* (1983) examined α/β proteins in detail. Similar methods were used to locate turns. The protein was then divided into subsegments based on turn location, and then each subsegment was examined for possible α or β structure. Pathognomonic patterns for α structure (residues i, $i+3$, $i+4$, $i+7$ hydrophobic) and β structure (residues i, $i+1$, $i+2$, $i+3$ hydrophobic) were located. The general alternation of helix–strand–helix was used to facilitate the categorization of unassigned segments. When ambiguity remained in a particular subsegment, alternative secondary structure schemes were allowed. This scheme proved accurate in locating secondary structure and in distinguishing α/β from non-α/β proteins. Future extensions of the current turn algorithms along these lines are planned. Additional mechanisms for introducing tertiary structure features into secondary structure assignment will be required.

D. Implications and Future Developments

When the protein sequence can be placed into one of three classes (αα, ββ, αβ) from composition or spectroscopic data or by homology, the turn algorithm described achieves a >90% accuracy. It can locate approximately half the turns without prior classification. Immediately, this procedure should be useful to the immunologist interested in predicting antigenic features of a sequence to isolate, characterize, or neutralize proteins.

It is our long-term hope that the accuracy of this turn prediction algorithm can be enhanced and that it will form the basis for a complete secondary structure assignment. Clearly, the ability to incorporate tertiary features into secondary structure assignment was vital to improving the accuracy of this method. Future advances will depend on the ability to add additional tertiary features directly into a secondary structure prediction algorithm or to create a procedure for the feedback of information from the tertiary assembly methods described below.

III. SECONDARY → TERTIARY STRUCTURE

Although an accurate secondary structure prediction has inherent utility, its value is significantly enhanced to the extent that it defines tertiary structure. Numerous investigators (e.g., Havel *et al.*, 1979; Momany *et al.*, 1975) have shown that a complete knowledge of secondary structure in the absence of other structural details restricts predictive schemes little. Yet, it is clear from an examination of a protein's three-dimensional structure that recognizing secondary structure aids greatly in following the peptide chain.

Cohen, Sternberg, and Richards developed a constructive approach to tertiary structure prediction that relied on the geometries observed in secondary structure interactions. Instead of searching a path through conformation space from the denatured to the native state, they created combinatorial schemes for generating large numbers of structures that sampled the majority of conformation space. Structures were eliminated that failed to meet simple criteria pertaining to the connectivity of the chain or the steric properties of the molecule. In a series of studies, it was shown that given the amino acid sequence and secondary structure, a small list of tertiary structures (usually 1–100 but occasionally as large as 4000 for large proteins) can be selected from the larger starting list (10^5–10^{12} structures). At least one member of the list closely resembled the native structure (r.m.s. deviation 1.7–4.5 Å). The details of these methods are described below.

A. α/α Proteins

1. Description of an Idealized α Helix

In a cylindrical coordinate system (z,r,θ), the description of an ideal α helix is straightforward. The z axis is coincident with the helix axis. In a simplified model, residues can be

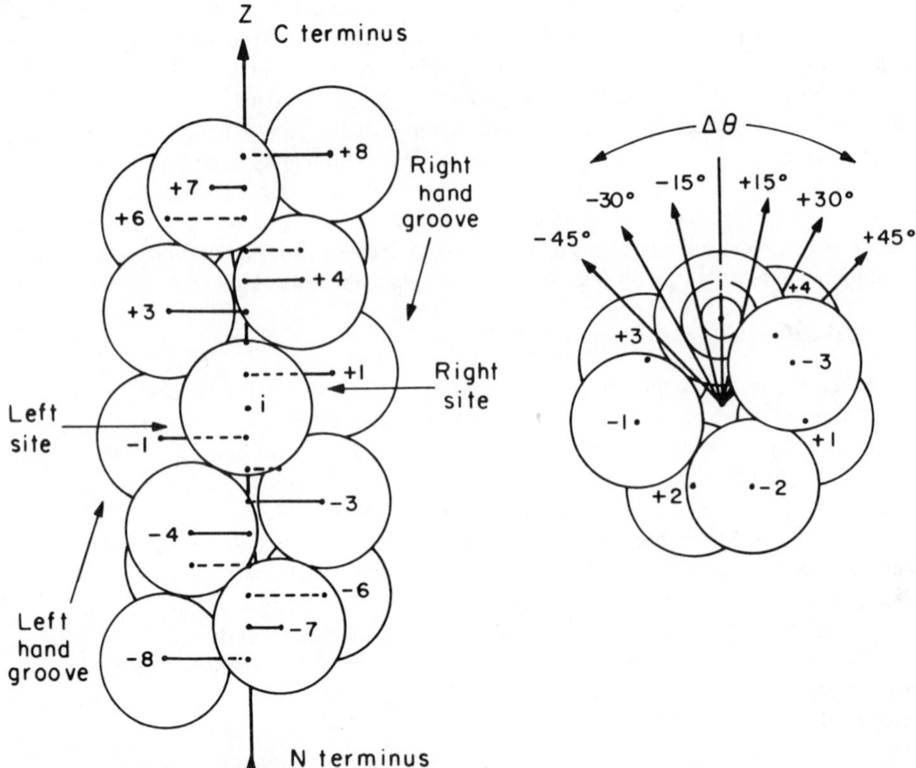

Figure 3. Sketch of a helix of close-packed spheres approximating an α helix. The left panel is the view perpendicular to the helix axis along the radius through the center of a central residue *i*. The helix axis points for the other spheres are shown at the end of the projected radii. The right and left grooves and the probable packing sites for central residues of a second helix are shown. The right panel is the view along the helix axis and shows the directions of the contact normal of residue *i* for the various interaction classes. The positive and negative skew angles shown are related to the assumed size of the central residue as it affects packing in the left or right site. Reprinted with permission from Cohen *et al.* (1979).

represented by a single sphere of radius 6.39 Å. Residue centers are located 4.12 Å from the z axis with a pitch of 1.5 Å and a periodicity of 3.6 residues per turn. Sphere centers take the coordinate form ($1.5n$, 4.12, $100°n$), where n is the residue number. Points of the form ($1.5n$, 0.00, $100°n$) are the helix axis points. No other parameter of the actual residues enters into the geometric description of a single helix, but residue types will affect the packing of two helices.

2. Helix Pairs

The geometric relationship of two helices is specified by four parameters: the perpendicular distance between the two helix axes, referred to subsequently as the contact distance; the dihedral angle between the two helix axis directions, the helix axis angle; and two angles relating the internal angle coordinates for each helix and the contact normal, the skew angles.

When two helices pack together, the residues of one helix will tend to fit between the residues of the other helix in one of several characteristic arrangements (Crick, 1953; Chothia *et al.*, 1977). In the representation used here, the radii simulating the two closest residues on the two helices, the central residues, will not be colinear with the contact normal but will have

Table VI. Assumed Parameters for the Three Helix–Helix Interaction Classes[a]

Interaction class	Central residue types	Assumed length of contact normal (Å)	Acute helix axis angle (°)	Skew angles (°)
I	Gly only	7.5	−80	±15
II	Ala, Val, Ile, Ser, Thr, Cys	8.5	−60	±30
III	Class II + Leu, Met	10.5	+19	±45

[a]Adapted from Chothia *et al.* (1977).

angular displacements that are related to the approximate size of the residues, the displacement being larger for the larger residues. This is shown schematically in Fig. 3 for the skew angles assumed in this study. Following Richmond and Richards (1978), the types of helix–helix interactions are considered to fall into three classes, which are related both to the size of the central residues and to the helix axis angles. The assumed mean values are listed in Table VI.

3. Prediction of Potential Interaction Sites

A helix–helix interaction site is a collection of residues from two helices that participate in the packing of the two helices. Packing is operationally defined as the loss of substantial surface area on taking the two helices from infinite separation to the native conformation. Typically, the residues involved in helix–helix packing are hydrophobic and are arrayed along one face of each helix. The prediction scheme for strong helix–helix interaction sites proposed by Richmond and Richards (1978) is based on the probable magnitude of the change in solvent-accessible area (Lee and Richards, 1971) in going from the separated helices to the helix pair. Given the primary structure and information (given or assumed) that the helix actually exists, the regions of the helix that might be involved in such contacts can be rank ordered on potential area change. Each site consists of a central residue i and a subset of the nearest neighbors (±4, ±3, ±1 for class II, for example; see Fig. 3). Prediction lists are prepared independently for each class. One can decide on the basis of other criteria how far down these rank-ordered lists to go in setting up the search procedure. The helix data and list of potential sites used for myoglobin are given in Table VII.

Nine sites were eliminated from the complete list of predictions because of steric conflicts when several sites occurred on one helix. Two rules were developed to handle these complications, based on the geometry of the interaction patch.

1. If i is a central residue in a type I or type III interaction, residues $i + 3$ and $i + 4$ or $i - 3$ and $i - 4$ cannot also be used as type I or type III sites.
2. If i is a central residue in a type II interaction, residues $i + 4$ and $i - 4$ cannot also be used as type II sites.

4. The Assembly Algorithm

The first program, ALPHAFOLD, produces lists of all possible site pairs based solely on combinatorial sums subject only to the following criteria:

1. Separate sites on the same helix cannot be paired.
2. No two helices can form more than one pair with each other.
3. All helices must appear in each list.
4. Site pairs are made only between sites of the same class.

Table VII. Interaction Site Prediction Data from the Myoglobin Sequence[a]

Helix designation	Residues in helices: Sequence numbers of termini N	C	Number of residues In helix	Between helices	Predicted number of possible strong sites	Potential central residue[b]	Interaction class[c]
A	3	18	16		2	10	II or III*
				1		13	II
B	20	35	16		2	25	I
				0		28	II
C	36	42	7		0	—	—
				8			
D	51	57	7		0	—	—
				0			
E	58	77	20		3	65	I
				8		68	II or III*
						69	III*
						71	II
F	86	94	9		1	90	II
				5			
G	100	118	19		4	107	II or III*
				6		108	II* or III*
						110	II
						111	II* or III
						112	II
						114	II*
H	125	147	23		3	134	II or III*
						135	III
						142	II

[a]See Richmond and Richards (1978).
[b]The following potential sites were not included in the list given above. One or more of the surrounding residues in each of these cases is not in an acutal helical segment. if secondary structure were unknown or uncertain, some or all of these would have to be tested: 17(II or III), 21(III), 29(III), 72(III), 75(II or III), 101(II), 104(III), 114(III), 127(II or III), 131(III), 142(III).
[c]Entries marked with an asterisk were not used in the site list for the calculations reported here. These deletions are explained in the text.

From the site data in Table VII, and always including the B25–E65 class I pair, ALPHAFOLD produced 10,370 lists of pairings acceptable by these criteria and required 10 min of VAX-780 CPU time. (If the B–E pair was omitted and the sites were left open, ALPHAFOLD produced an additional 9600 lists of pairs.) Each list showed five pairs between helices A, B, E, F, G, and H and five unused sites. Helices C and D do not appear in the lists, as they contain no predicted sites. Fifteen sites randomly paired without regard to class or helix position and five sites left unpaired would have given 135,135 combinations.

The second program, ALPHABUILD, is designed to test which of the lists from ALPHAFOLD can lead to stereochemically acceptable structures. As described earlier, each site pair actually represents eight possible structures. The acute helix angle is defined by the class, but there are two possible N → C directions for the second helix axis. Each central residue has two possible skew angles, whose numerical value is again defined by class. The total number of permutations to be tested is 8^5 or 32,768. Each of the 10,370 lists, in principle, must be processed through this number of trial structures, for a grand total of 3.4×10^8, unless early termination of the search occurs.

Only two selection criteria, or filters, have been used. (1) The end-to-end distance of helices consecutive in the sequence had to be less than a maximum permissible number; (2) only a limited number of close contacts were permitted in an accepted structure.

The permissible end-to-end distance between the C terminus of one helix and the N terminus of the next helix in the chain was set equal to or less than $3(m + 1) + 3$ Å, where m is the number of nonhelical residues between the two helices. The unit distance of 3 Å is added to represent the distances from the helix axis to the inner edge of the helix cylinder. The further processing of a particular branch in the search was discontinued when any distance failed this test.

Distances between axis points of neighboring helices were checked. Close contacts (bad contacts) were scored when this distance was less than 7.5 Å. When the number of such contacts exceeded 50, the structure was discarded. Fifty bad contacts correspond to ten misplaced axis points (triangular matrix check). This rejection number is, of course, a variable parameter to be adjusted on the basis of experience. Helices that are paired are automatically placed so that there are no bad contacts between them. Helices not actually paired may collide, and it is these that are checked for collision. For example, if helix A packs with helix B and helix B packs with helix C, helices A and C could collide, since their spatial proximity is determined solely by interactions with helix B. No test was made for collision of the connecting residues, although this could occur, in principle, even with no bad contact between the helices themselves.

In this particular study, a list from ALPHAFOLD consisted of five pairs of residues in an ordered sequence. The first residue and the contact normal defined a coordinate system. The other helix of the pair was then placed in this reference frame with the correct distance and helix angle for that site class. The third helix was then placed in a vacant site of one of the first two as specified by the list. The end-to-end distance check on segment lengths between consecutive helices was then performed. This procedure was then repeated with succeeding helices in the list. The calculation terminated whenever the distance check failed. All subsequent lists with the same entries up to that level were also discarded. The collision check was made only on structures satisfying all endpoint restrictions. A schematic drawing of a predicted myoglobin structure with helix axes and contact normals linking central residues is shown in Fig. 4.

B. β/β *Proteins*

1. *Overview*

A flow diagram of the algorithm to predict the structure of β-sandwich proteins is shown in Fig. 5. The input consists of the amino acid sequence and an actual (or postulated) strand assignment. The algorithm considers each β strand as a unit and constructs strand alignment diagrams. These are filtered to obtain a reduced set that is then converted into a Cα representation. The details of the various steps are now described, classified according to whether they involve the construction of the structures (BETABUILD I, II, AND III) or the filtering of the structures (BETABUILD I, II, AND III). Some aspects of the algorithm resemble those used by Ptitsyn *et al.* (1979), who report a nonautomated procedure to construct β-sandwich topologies for several proteins.

2. *BETABUILD I: Locate Hydrophobic Patches on the* β *Strands*

From the input of sequence and strand assignments, one first predicts not only which side of each β strand will point toward the other sheet but also the location along each strand of the few residues that mediate the hydrophobic sheet–sheet interaction. We postulate that the site

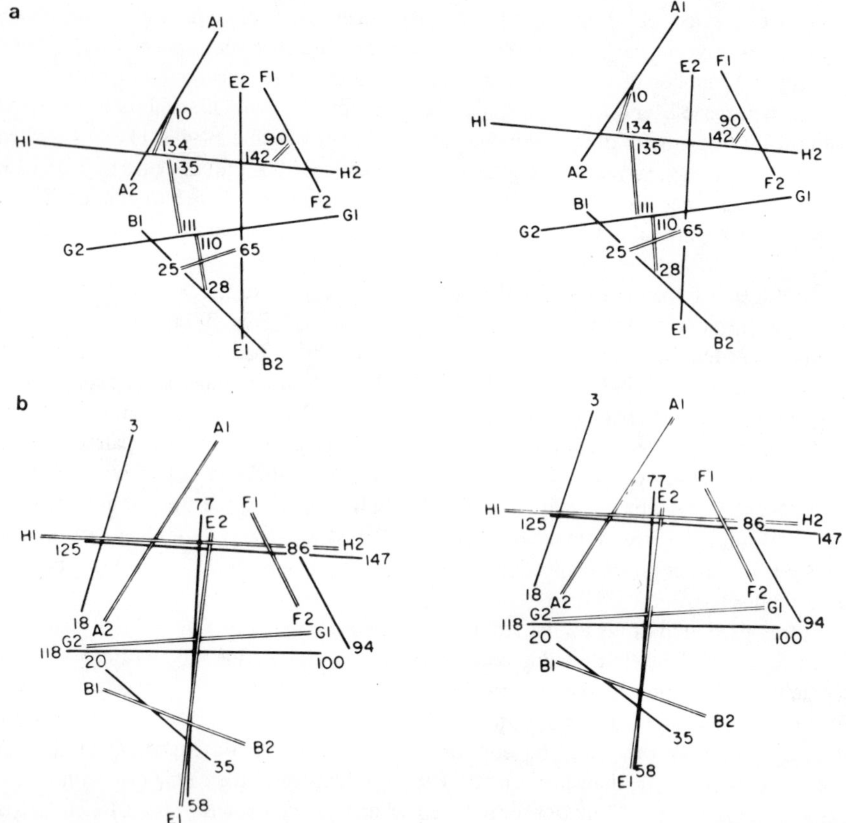

Figure 4. (a) Stereo diagram of the predicted structure of MBN showing contact normals (double lines) joining central residues (numbered). Helix axes are labelled A–H with 1 and 2 for the N and C termini. (b) Predicted structure of MBN overlaid on the helix axes of the crystal structure. The helix termini of the crystal structure are numbered. Reprinted with permission from Cohen *et al.* (1979).

will be the region on the strand that, when sandwiched, will screen from the solvent the maximum number of strongly hydrophobic side chains.

Accordingly, each β strand was scanned for a continuous site of nonpolar residues (see below) on one face of the strand. The location of the site was represented by one or two central residues. The site with largest number of nonpolar residues was chosen preferentially. In a strand, the first two and last two residues are not allowed to be a central residue. The sites were:

4 residues at positions $i-2, i, i+2, i+4$	central residues $i, i+2$
3 residues at positions $i-2, i, i+2$	central residues i
2 residues at positions $i, i+2$	central residues $i, i+2$

When two or more sites with the same number of residues were located, one selected the site with the largest potential φ-area (accessible surface area) change on sheet–sheet packing. For example, the φ-area of a three-residue site would be the sum of the areas of the residues at positions $i-2$, i, $i+2$. The φ-area of a residue type X was the exposed area in the model

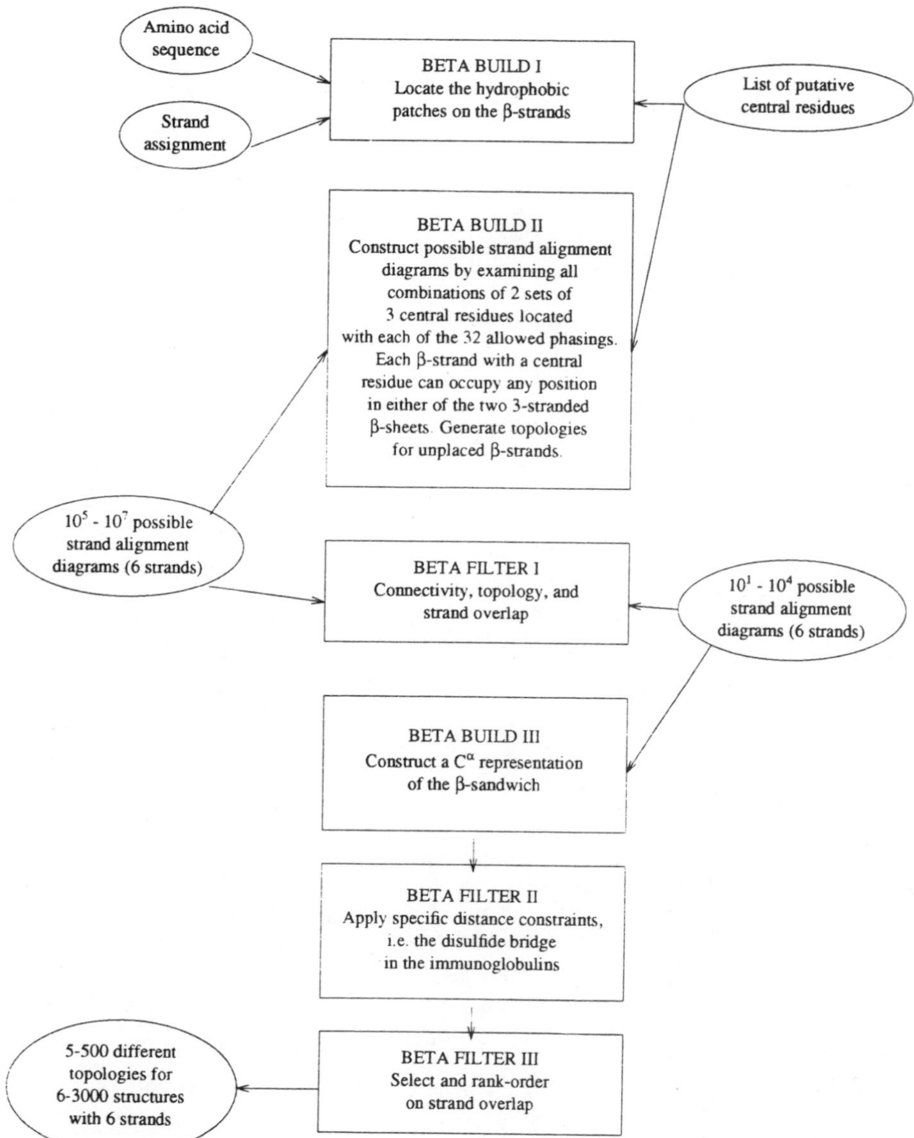

Figure 5. Flow diagram of the β-sheet sandwich combinational algorithm for packing secondary structure. Program names appear in capital letters.

three-stranded β sheet $(Cys)_9$:$(Cys)_4$ X $(Cys)_4$:$(Cys)_9$. These calculated areas agreed well with the observed area changes on sheet stacking (see Tables VIII and IX; Cohen *et al.*, 1981b). The nonpolar residues with the area (Å^2) are: Ala (8), Cys (11), Ile (18), Leu (19), Met (26), Phe (24), Pro (9), Trp (34), Tyr (17), Val (13), and Lys (12) (see below). Other residues (e.g., Gly) were designated polar and assigned zero area. The role of Lys is special. It can only contribute to a site if that β strand is later placed on the edge of a β sheet, and accordingly an alternative site is searched for so the β strand can be buried within the sheet. Generally the

Table VIII. Contact Area Changes on β-Sandwich Formation[a]

Sandwich	No. of residues in β sheet	Area change (Å²) β strands → β sheet			Area change (Å²) β sheet → β sandwich			Ratio φ-area change strands → sheets/sheets → sandwich
		Nonpolar	Polar		Nonpolar	Polar		
		(φ area)	Main chain	Side chain	(φ area)	Main chain	Side chain	
CONA	118	1710	764	50	796	222	44	2.1
PRE	64	924	363	23	303	87	8	3.0
FAHV	63	963	334	25	370	138	46	2.6
REI	60	743	336	35	336	139	64	2.2
FALV	57	692	319	11	230	102	11	3.0
FCH3	57	806	301	24	306	89	29	2.6
FAHC	56	803	278	21	323	93	29	2.5
FALC	56	658	317	18	312	113	16	2.1
FCH2	36	492	201	17	243	48	8	2.0

[a]The two processes considered are (1) β strands → isolated β sheet, and (2) β sheets → β sandwich. The contact area change is dissected into non-polar atoms (C^{α}, side chain C and S), polar main chain atoms (N, C′, O), and polar side chain atoms.

Table IX. Nonpolar Area for Residues in Actual and Model β Sandwiches[a]

Residue	Mean and standard deviation of φ area (Å^2): Isolated β strand	β Strand → GB sheet	Isolated β sheet	β Sheet → β sandwich	φ-Area model structures (Å^2): β Strand *b*	β Sheet *c*	β Strand → β sheet *d*
Ala	21.5 ± 3.4	15.2 ± 3.6	9.8 ± 5.4	8.0 ± 5.0	22.7	8.2	14.5
Arg	23.8 ± 2.5	18.4 ± 2.4	11.7 ± 6.0	2.3	22.9	5.7	17.2
Asn	8.1 ± 4.8	10.4	3.1 ± 2.7	5.4	12.4	2.9	9.5
Asp	13.4 ± 3.2	12.8	7.5 ± 4.6	3.7	13.4	2.8	10.6
Cys	30.9 ± 2.0	19.8 ± 3.7	14.0 ± 8.4	11.2 ± 3.7	29.6	11.2	18.4
Gln	17.1 ± 4.1	12.9 ± 2.9	10.3 ± 6.4	7.1 ± 3.0	14.5	0.6	13.9
Glu	17.4 ± 3.0	11.7 ± 1.1	11.6 ± 4.7	6.4	16.9	1.1	15.8
Gly	11.3 ± 2.1	6.2 ± 2.7	6.7 ± 4.2	4.5 ± 2.0	11.9	4.2	7.5
His	31.0 ± 4.2	12.1 ± 3.4	22.0 ± 7.0	8.3 ± 6.9	32.6	13.2	19.4
Ile	43.1 ± 3.7	22.7 ± 4.1	24.9 ± 10.2	16.0 ± 7.1	41.8	18.2	23.6
Leu	42.2 ± 4.2	23.7 ± 4.1	26.1 ± 8.7	17.9 ± 7.2	45.3	19.4	25.9
Lys	32.3 ± 4.8	18.5 ± 6.4	23.4 ± 9.9	16.0 ± 1.5	33.3	12.4	20.9
Met	45.2 ± 3.3	—	33.4 ± 6.5	29.4 ± 4.4	47.9	25.8	22.1
Phe	51.0 ± 5.3	31.1 ± 7.2	30.4 ± 11.1	20.8 ± 7.3	52.2	24.2	28.0
Pro	35.0 ± 3.0	—	29.4 ± 9.4	10.6 ± 4.2	36.7	8.9	27.8
Ser	14.8 ± 2.6	10.1 ± 2.6	9.4 ± 5.6	4.1 ± 2.7	14.3	4.7	9.6
Thr	23.3 ± 3.3	14.7 ± 3.8	16.0 ± 7.7	9.9 ± 4.7	22.5	6.7	15.8
Trp	59.9 ± 4.7	25.8 ± 4.4	35.2 ± 9.0	32.5 ± 7.4	63.0	34.1	28.9
Tyr	44.5 ± 2.0	31.3 ± 5.7	22.8 ± 11.5	12.6 ± 4.5	44.9	16.7	28.2
Val	34.4 ± 3.9	20.9 ± 4.3	21.3 ± 7.7	14.3 ± 5.7	37.2	13.4	23.8

[a]Only residues with φ-area changes > 1 Å^2 are included in the average.
[b]Gives the φ area of residue *X* in a model β strand $(Cys)_4X(Cys)_4$.
[c]Gives the φ area of residue *X* in a model anti-parallel β sheet $(Cys)_9{:}(Cys)_4X(Cys)_4{:}(Cys)_9$.
[d]Gives the φ area of residue *X* in the model transition β strand → β sheet.

predicted sites correspond to the residues that were found to mediate the sheet–sheet interaction (see Cohen *et al.*, 1981b).

The predicted sites for FALC are shown in Fig. 6. For example, in strand E there are three possible two-residue sites: Lys-Ala (φ-area = 20 Å^2), Tyr-Ala (25 Å^2), Leu-Leu (38 Å^2). The last pair is chosen, as it has the largest area. These two residues have the observed φ-area changes on sheet–sheet packing.

To summarize, BETABUILD I predicts between zero and four central residues on each β strand that delineate the position of the sheet–sheet interaction.

3. BETABUILD II: Construct Possible Strand-Alignment Diagrams

Next the β strands are placed into the two β sheets so as to construct a two-dimensional strand-alignment diagram. Inspection of the observed area changes on sheet packing (Fig. 7) shows that not all the β strands have sizable φ-area changes on sheet–sheet interaction. Accordingly, we constructed a strand-alignment diagram of only a central core of two three-stranded β sheets that is found in all β sandwiches. The additional strands will have to be placed once the central core has been predicted. Clearly a different approach is required for the 14 β strands in concanavalin A.

First, all possible topologies for the two three-stranded sheets are constructed. Each β

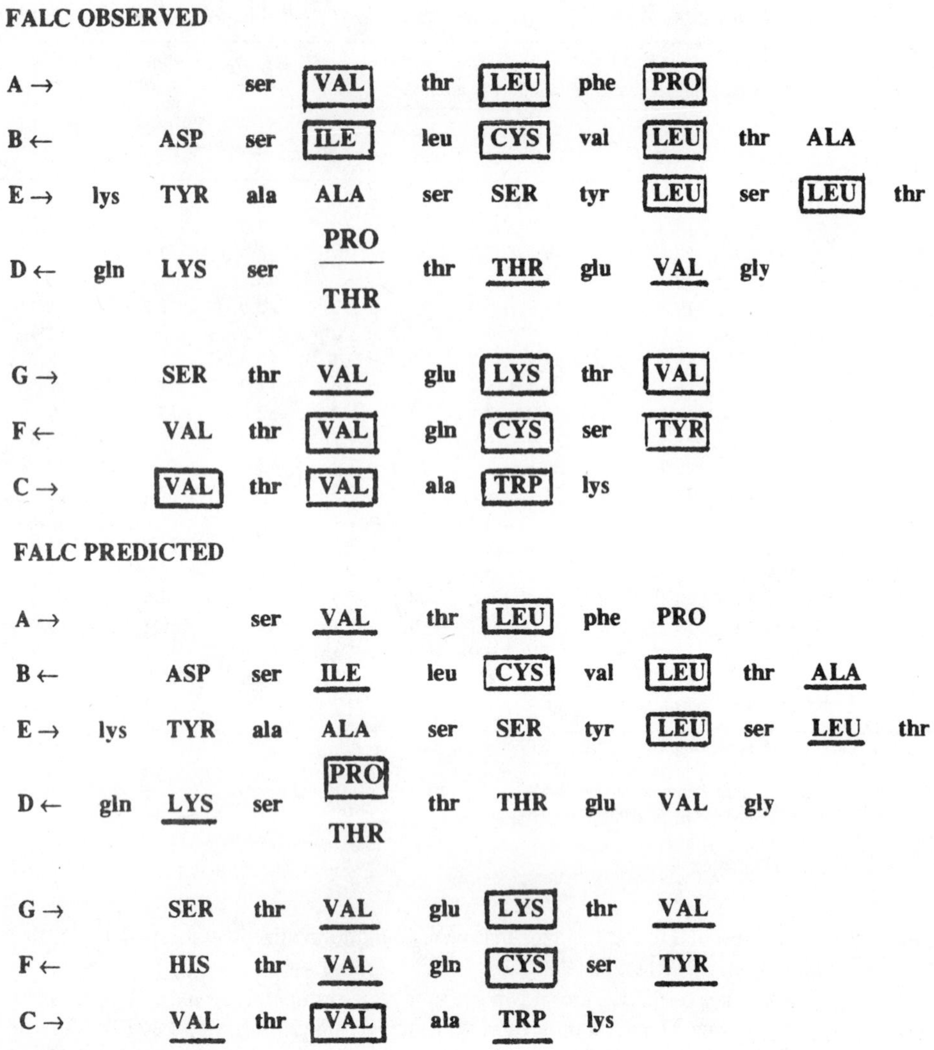

FALC OBSERVED

A →			ser	VAL	thr	LEU	phe	PRO		
B ←		ASP	ser	ILE	leu	CYS	val	LEU	thr	ALA
E →	lys	TYR	ala	ALA	ser	SER	tyr	LEU	ser	LEU	thr
D ←	gln	LYS	ser	PRO / THR	thr	THR	glu	VAL	gly	
G →		SER	thr	VAL	glu	LYS	thr	VAL		
F ←		VAL	thr	VAL	gln	CYS	ser	TYR		
C →		VAL	thr	VAL	ala	TRP	lys			

FALC PREDICTED

A →			ser	VAL	thr	LEU	phe	PRO			
B ←		ASP	ser	ILE	leu	CYS	val	LEU	thr	ALA	
E →	lys	TYR	ala	ALA	ser	SER	tyr	LEU	ser	LEU	thr
D ←	gln	LYS	ser	PRO / THR	thr	THR	glu	VAL	gly		
G →		SER	thr	VAL	glu	LYS	thr	VAL			
F ←		HIS	thr	VAL	gln	CYS	ser	TYR			
C →		VAL	thr	VAL	ala	TRP	lys				

Figure 6. Observed and predicted strand-alignment diagram for FALC. The relative positions of the strand residues are shown; β bulges (Richardson *et al.*, 1978) are indicated as Pro-Thr in strand D. Residues with side chains pointing toward the other sheet are in capital letters. The observed area changes on sheet sandwiching are indicated by enclosing the residue in a rectangle if the area change is >10 Å² and by underlining if the area change is between 5 and 10 Å². In the predicted structure, the central residues are in rectangles, and the other residues that form the site are underlined.

strand with a site can occupy any position or any direction in either of the sheets.* For n β strands with sites, there are in general $({}^{n-3}C_3)\ 3!3!2^3 2^3/4$ topologies.† There are $({}^{n}C_3)$ ways of sorting the remaining strands into a second sheet, 3! cyclic permutations of strand order

*The exception is that if there is a B strand that only has a site with Lys (e.g., strand G in FALC, see Fig. 6), then this strand can only be placed at the edge of the sheet.

†${}^{n}C_r = n!/(n-r)!r!$

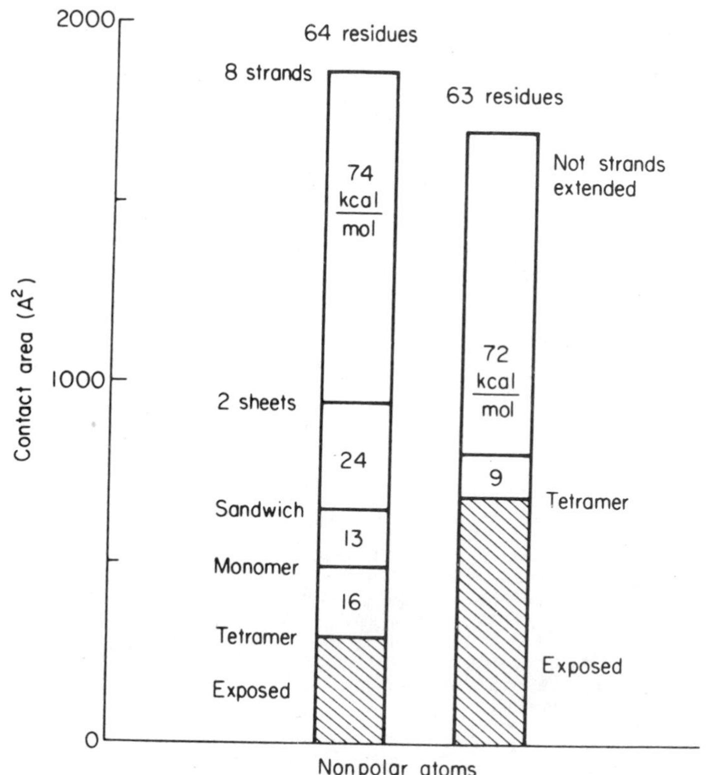

Figure 7. Nonpolar area changes for different structural organizations in prealbumin. The exposed φ-area for different levels of structural organization of prealbumin are dissected according to whether the residues form β strands (eight strands) or intervening loops (not strands). The condensation model starts with the observed φ-area for the β strands and the hypothetical φ-area of the remaining residues in a β-strand conformation (Table III, column 5). The remaining φ-areas are calculated from the relevant crystal structure. The hydrophobic effect is quantified by converting the area changes (1 Å^2 ≡ 80 cal/mole: Richmond and Richards, 1978). Reprinted with permission from Cohen *et al.* (1981b).

within each sheet, 2^3 possible antiparallel and parallel arrangements of three strands, and a fourfold degeneracy arising from symmetry in this construction.

Next, the strands are phased relative to each other in each β sheet. We wish to place the hydrophobic sites on the sheet surfaces so as to reconstruct the two anticomplementary patterns observed on sheet–sheet packing. Through this construction, the predicted nonpolar site of FALC corresponds closely to the observed changes for strands ABE and GFC (see Fig. 6). Formally, the relative positions of the central residues are specified. For example, in FALC (Fig. 6), the two three-stranded sheets with sites are A(Leu)B(Cys)E(Leu) and G(Lys)F(Cys)-C(Val). The positions of the strands in Fig. 6 can therefore be expressed as 00+2, 00−2. To provide the required variation in hydrophobic surfaces compatible with sheet–sheet packing, 32 different phasings are allowed for any set of six central residues (Table X). In strand B of FALC there are two central residues, and each will be taken to generate 32 strand-alignment diagrams (often the same diagram will be constructed from the alternative central residues using different phasings). This feature and the special treatment of Lys at the edge of sheets makes it difficult to evaluate accurately the number of sheet diagrams constructed, but we estimate this as 32 times the number of different topologies. Typically this number is between

Table X. Combinatorics of β-Sheet/β-Sheet Packing

Top sheet	Bottom sheet	Top sheet	Bottom sheet
−2,0,+2[a]	+2,0,−2	0,0,+2	+2,0,−2
	+2,0,0		+2,0,0
	0,0,−2		0,0,−2
	0,0,0		0,0,0
	−2,0,0		−2,0,0
	0,0,+2		0,0,+2
−2,0,0	+2,0,−2	0,0,0	+2,0,−2
	+2,0,0		+2,0,0
	0,0,−2		0,0,−2
	0,0,0		0,0,0
	−2,0,0		−2,0,0
	0,0,+2		0,0,+2
0,0,−2	+2,0,−2	+2,0,0	+2,0,−2
	+2,0,0		+2,0,0
	0,0,−2		0,0,−2
	0,0,0		0,0,0

[a]The relative positions for the three strands in a sheet. For example +2,0,−2 is $i+2,j,k-2$ where i,j,k are the positions of residues vertically below one another in the strand-alignment diagrams. −2,0,+2; −2,0,0; 0,0,+2; 0,0,0 have the correct positive shift (i.e., to the right) for the top sheet; 0,0,−2 and +2,0,0 have the wrong (i.e., negative shift). Similarly +2,0,−2; +2,0,0; 0,0,−2 and 0,0,0 have the correct shift for the lower sheet, whilst −2,0,0 and +2,0,−2 the wrong shift. Allowed pairings are: top correct and bottom correct; top incorrect and bottom correct; top correct and bottom incorrect.

10^5 and 10^7. The list of 32 allowed phasings is a substantial reduction in the possible ways (say 15,625 = $5^2 \times 5^2$) of aligning the β strands in the core while still having a sensible sheet structure with hydrogen bonds.

Finally, from each central core a set of complete structures was constructed with the previously unplaced strands being located in every possible unoccupied position with every direction. This was required to implement some topological filters (see below), although no attempt was made to phase the unplaced strands with respect to the central core.

4. BETABUILD III: *Construction of a Cα Representation*

From a predicted sheet diagram for the central core, suitable Cα co-ordinates can be obtained for the six β strands. The structure is constructed based on idealized geometry typical for sheet sandwiches (Fig. 8). Each β strand was generated by the model-building program of Scheraga and co-workers (Momany *et al.*, 1975) with the main-chain dihedral angles (ϕ = 120°, ψ = +140°) set to a typical antiparallel conformation with a right-handed twist (Chothia, 1973). Each sheet was constructed based on the position of the central residues. The β strands were first placed parallel or antiparallel to the x axis in the x–y plane with an interstrand separation of 4.25 Å between the x coordinates and the Cα–Cβ bond of the central residue normal to the strand axis and in the x–z plane. Each strand was then rotated by 20° about a line parallel to the y axis that passed through its central residue. The rotation produced a β sheet with adjacent β strand having the usual (left-handed) interstrand dihedral angle of −20° (Chothia, 1973). The two sheets were then stacked with an intersheet separation (along the z axis) of 10 Å. The central residues in the central strands of the two sheets were placed on top of each other (same x and y coordinates). The top sheet was then rotated by −30° about the line joining these two central residues to acknowledge the intersheet dihedral angle (see Cohen *et al.*, 1981b) typically observed between sheets in a β sandwich.

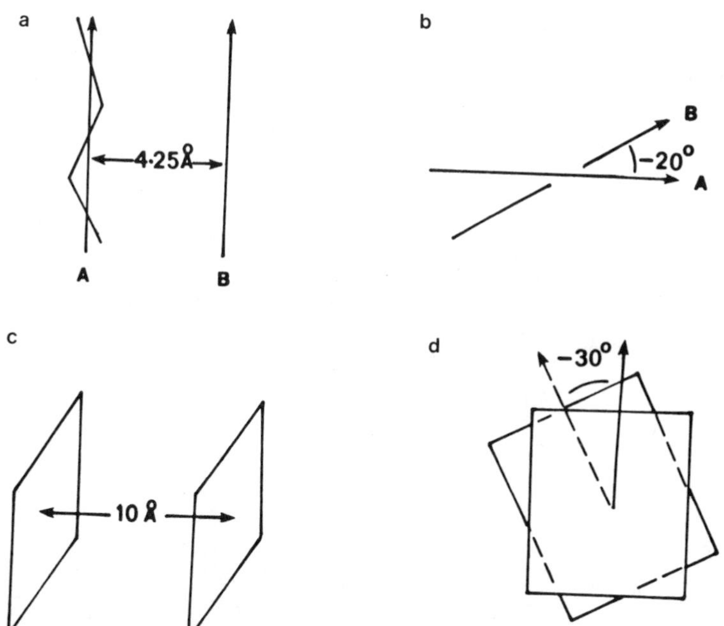

Figure 8. Standard geometry of β-sheet sandwich. The geometry used to construct a β sandwich from the strand-alignment diagram. (a) Interstrand separation: typically 4.25 Å, although variations are expected for parallel and antiparallel hydrogen bonding. (b) Interstrand angle: this quantifies the characteristic twist between adjacent strands (Chothia, 1973). (c) Intersheet separation: this is the perpendicular distance between the two sheet planes for two close-packed sheets in a sandwich. (d) Intersheet angle: the packing between β sheets exhibits a characteristic rotation of −30° in a sandwich.

5. BETAFILTER I: General Constraints

The first set of filters applied to the list of generated structures imposed restrictions that quantify observed topological and steric features of known β sandwiches. Each central core (six β strands) must be allowed by all the filters. In addition, from each central core, a set of possible topologies was constructed by locating the previously unplaced β strands (BETABUILD II). At least one member of this set had to be allowed.

The connection between two parallel β strands in the same sheet is right-handed (Richardson, 1976; Sternberg and Thornton, 1977a). This almost invariant feature of protein structure is believed to be the consequence of the twist of the β sheet. Thus, for example, in FALC strands B and D are in the top sheet and C in the bottom. B and D are parallel and out of the page, so B must be placed to the right of D (see Fig. 9).

6. BETAFILTER II: Specific Distance Constraints

At this stage, we incorporated information about the proximity of specific residues as an additional filter. We considered only the restriction imposed by the intradomain disulfide bridge that occurs in the immunoglobulins. Knowledge of disulfide bridges is of course available from chemical studies. The coordinates obtained from BETABUILD III were examined to ensure that the C^{α} atoms of the two cysteine residues were within 15 Å of each other. The choice of 15 Å is a conservative estimate to allow for errors in the construction of the coordinates. In addition, this disulfide bridge is not allowed between cysteine residues in

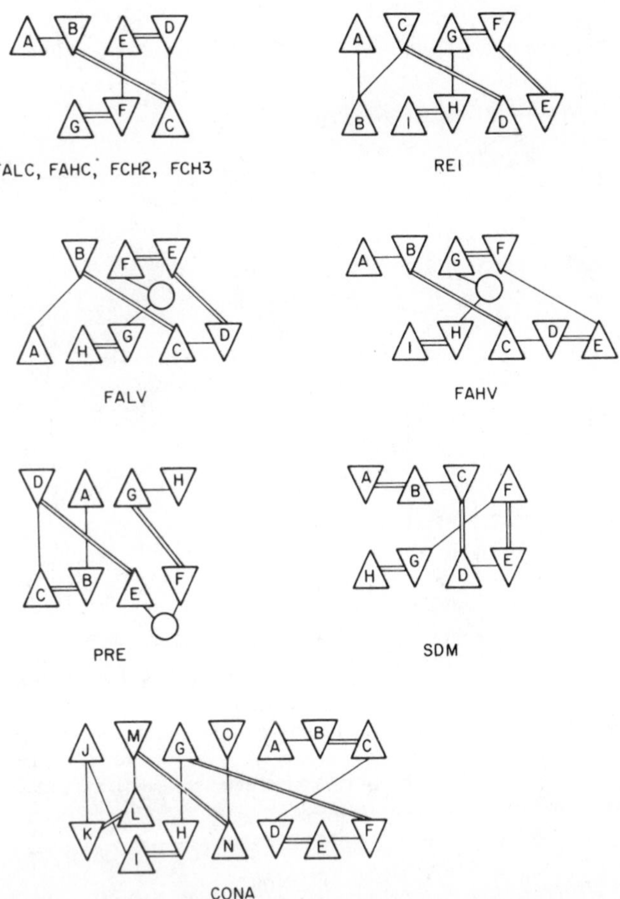

Figure 9. The topology of sandwiches. Each sheet is viewed along the strand direction, and each strand is represented by a triangle whose apex points up or down according to whether the strand is viewed from the N or the C end. The strands are lettered sequentially. Connections between strands are shown in double lines if they start close to the viewer; otherwise they are in single lines. Each strand was located by first delineating the ends of the correct interstrand hydrogen bonding (amino-N–carbonyl-O distance <3.3 Å). The β strand was then extended sequentially if the main chain dihedral angles of a residue were in the range $-170° < \phi < -90°$ and $100° < \psi < 170°$. Reprinted with permission from Cohen *et al.* (1981b).

nonadjacent β strands in the same β sheet (Thornton, 1982). Other specific distance constraints could easily be incorporated at this stage (cf. Cohen and Sternberg, 1980a). The information would be obtained from chemical studies such as cross linking or from spectroscopic measurements, e.g., nuclear magnetic resonance.

7. BETAFILTER III: *Strand Overlap*

We now assess whether each β sheet has a favorable juxtaposition of β strands in terms of the interactions between the β strands. In a folded protein, the β sheet provides a convenient way of enabling buried main-chain nitrogen and oxygen atoms to form the required hydrogen bonds. In addition, β sheets shield from the solvent a substantial fraction of the surface area of the nonpolar atoms in a β strand. We have attempted to quantify these requirements in terms of strand overlap and strand length.

The observed strand overlap of a sheet is defined as the number of pairs of residues in adjacent strands that are opposite each other (i.e., vertically below one another in the strand-alignment diagram). For example, in the core of FALC, the strand overlap in ABE (IOVER1) is 15, and that in GFC (IOVER2) is 13, making a total of 28. We required that the observed strand overlap for both sheets be the maximum or one less than the maximum that could be formed given the strand composition but not the strand location of a sheet. The maximum overlap of a sheet is the sum of the lengths of the two shorter strands. In ABE this is 15, and in GFC 13. Thus, the core of FALC has maximum overlap. The margin of one between the observed and maximum overlap allows the algorithm to accommodate the particular delineation of the strand termini or a distortion in a strand caused by a β bulge (Richardson *et al.*, 1978).

Finally, the allowed structures are rank-ordered on a measure of the extent of interactions between the β strands, fraction strand overlap (F). This is defined by

$$F = \frac{\text{IOVER1} + \text{IOVER2}}{\tfrac{1}{2}[(N_1{}^1 + 2N_2{}^1 + N_3{}^1) + (N_1{}^2 + 2N_2{}^2 + N_3{}^2)]} \times 100\%$$

IOVER1 and IOVER2 are the observed strand overlaps for the two sheets. For sheet 1, the number of half side chains available to be buried is $N_1^1+2N_2^1+N_3^1$ where N_1^1 and N_3^1 are the number of residues in the two edge strands and N_2^1 is the number in the buried strand. Similarly for sheet 2. For example, in FALC, $F=28/\tfrac{1}{2}(35 + 27)x100\% = 90\%$. We consider the position of the native structure in the rank-ordered list.

8. *Implementation*

The computer algorithm was applied uniformly to nine β sandwiches: FCH2, FCH3, FALC, FAHC, FAHV, FALV, REI, PRE, and SDM. All the structural data, including the strand assignments, are as described in the analysis of β sandwiches (see Cohen *et al.*, 1981b). The programs are written in FORTRAN and are available from the authors.

For each β sandwich we need to examine between 10^5 and 10^7 possible six-stranded cores, each of which would generate several (10^1 to $\sim 10^3$) topologies for the unplaced strands. In practice not all these structures need to be examined. The β sandwiches are generated in a tree structure, and filters are employed at the nodes to establish whether each subtree needs to be examined. The efficiency of the algorithm depends on the structure of the tree and the order of implementing the filters. The present computer program used 20 hr of central processor time (IBM 360/195) to consider the nine sandwiches. However, from the experience gained from these trials, the efficiency of the algorithm could probably be improved further.

C. α/β *Proteins*

A flow diagram of the computer algorithm to predict the tertiary structure of α/β proteins by a combinatorial approach is presented in Fig. 10. For each protein domain, the inputs to the algorithm are the amino acid sequence, an assignment of the secondary structures, and the segregation of the secondary structures to the domain. In this study, the assignment and segregation of the secondary structures were taken from the known crystal structures.

The algorithm bifurcates to consider separately structures with pure parallel and mixed β-sheet hydrogen bonding. The mixed β-sheet alternative is chosen only if the connection between any sequential β strands is too short to form a parallel connection (fewer than seven residues). Different sets of topological restrictions are used to reduce the number of allowed structures for pure parallel and mixed β sheets. The relative stagger of the strands in the sheets is constrained by the requirements for specific patterns in the patches of nonpolar residues on

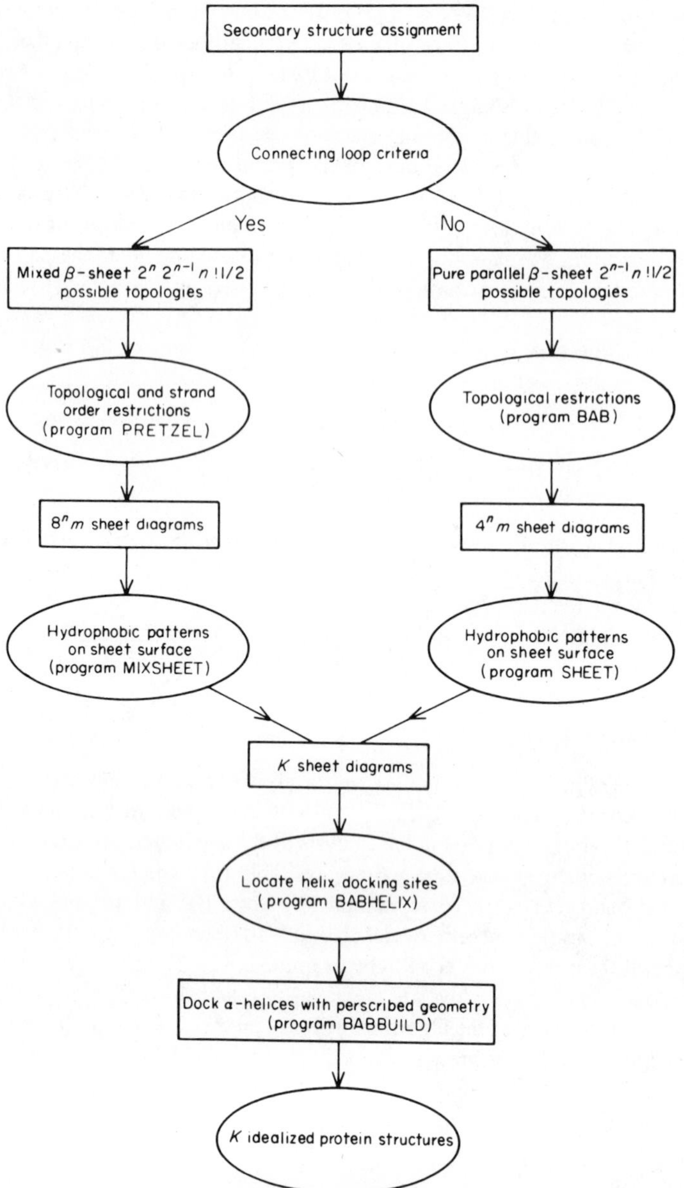

Figure 10. Overview of the combinatorial approach to α/β proteins: flow diagram. *n* is the number of strands, and *m* the number of allowed topologies. Reprinted with permission from Cohen *et al.* (1982).

the two surfaces of the twisted β sheets, the absence of polar residues inside these patches, and the formation of most of the possible hydrogen bonds. Different formalizations of these requirements are used for the pure parallel and mixed β sheets. Thus, a list of possible strand alignments is constructed.

The two paths then converge for the placement of α helices. Which α helices are to be docked on the β sheet and which residues on the α helices mediate the interaction are predicted by a search for a constellation of nonpolar residues on the surfaces of the α helix.

Finally, an α-carbon model for the α/β structure is constructed from the standard geometry of α-helix/β-sheet packing. The predicted structures are compared to the crystal structure, and the agreement expressed by the r.m.s. deviation between equivalent α-carbon atoms.

1. Prediction of Topology

Choice between Pure Parallel and Mixed β Sheets. A secondary structure prediction algorithm will produce a list of N- and C-terminal residues for every secondary structure segment. If the proximity of two β strands along the chain is such that the C terminus of the first is within seven residues of the N terminus of the second, this protein has a mixed β sheet. If no such example exists, then the β sheet is pure parallel.

Pure Parallel β Sheets. These can be divided into the more common planar sheets with two edge strands (e.g., LDH) and the closed β barrel (e.g., TIM). A pure parallel "planar" sheet with n β strands could adopt theoretically $\frac{1}{2}n!2^{n-1}$ topologies; n strands could have $n!$ strand orders, but rotational symmetry reduces this by one half. Each of the $n - 1$ connecting loops could be left- or right-handed, so there are 2^{n-1} possibilities. For sheets with seven or more β strands, the option to form a closed β barrel as in TIM or PK must be allowed, so the number of possible topologies doubles. Two topological restrictions have been reported.

P1. The connection between two sequential parallel β strands is right-handed (Sternberg and Thornton 1976, 1977a; Richardson, 1977). Out of 58 structures, Sternberg and Thornton (1977a) found 57 right-handed connections, the only left-handed unit being in subtilisin. A left-handed connection joining the two domains of phosphofructokinase has been found (Evans and Hudson, 1979). This preference in chirality has been attributed to the fact that the right-handed path joining sequential parallel β strands is shorter than the left-handed alternative (Sternberg and Thornton, 1976, 1977a). The right-handed rule reduces the number of possible topologies by a factor of 2^{n-1}.

P2. Pure parallel sheets have no more than one chain reversal (Richardson, 1977). A reversal in the order of the strands in the sheets (i.e., chain reversal) might well be kinetically less favorable than simply continuing in the same direction but would be more favorable than multiple reversals. There are no reversals in β barrels (e.g., TIM), and always one in "planar" sheets (e.g., LDH). Thus, when sequential strands are labeled alphabetically, ABEDC is an acceptable planar topology, but ABDEC is not allowed.

A survey of 11 parallel β sheets, nine "planar" and two barrels (ADH, ADK, FXN, LDH, PGK_N, PGK_C, RHO_N, RHO_C, SBT, PK, TIM), suggests two additional restrictions.

P3. Pure parallel sheets are shielded from solvent by roughly an even covering of α helices. The connecting segment between two consecutive β strands may contain no intervening helical segments (e.g, FXN B→C), one helical segment (e.g., FXN C→D), or several helical segments (e.g., ADK B→C). When more than one helical segment is present, only one interacts with the β sheet. For purposes of requiring an even covering of a β sheet by helices, the issue is whether or not a segment connecting consecutive β strands contains a helix. In "planar" sheets, the difference in the number of α helices on either side of the sheet is never more than one. This "even" covering of α helices shields both hydrophobic faces of the β sheet from the solvent. Thus, the N-terminal domain of PGK with the form $(\beta\alpha)_5\beta$ could have the topology

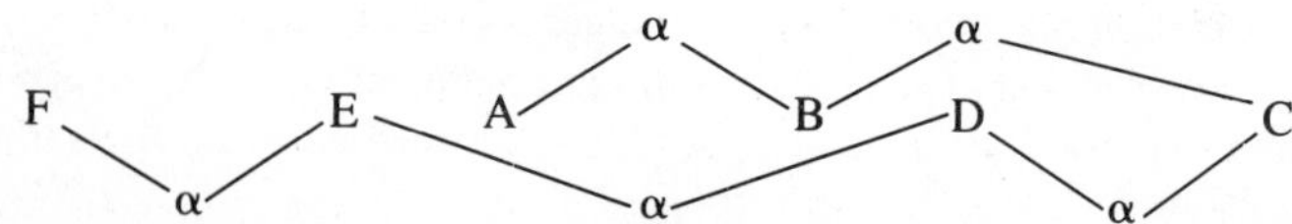

which puts two α helices on the top of the sheet and three on the bottom, if right-handed connections are assumed. However,

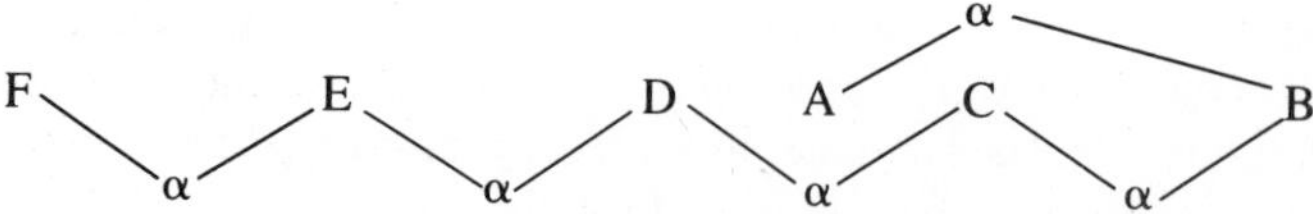

which forces four of the five helices to be on the bottom of the sheet, would not be allowed. In β barrels, the sheet is shielded from solvent by placing all the α helices on the outside.

P4. Certain topologies with the C-terminal strand not on the edge of the sheet are disallowed, as they produce steric conflicts between α helices. If the top of the β sheet is the side on which the first helix resides, then an α helix trailing the final β strand lies on top of the β sheet and always to the left of the first strand when the first strand is placed to the left of the second strand. In ADK, with the form $(\beta\alpha)_5$, the topologies

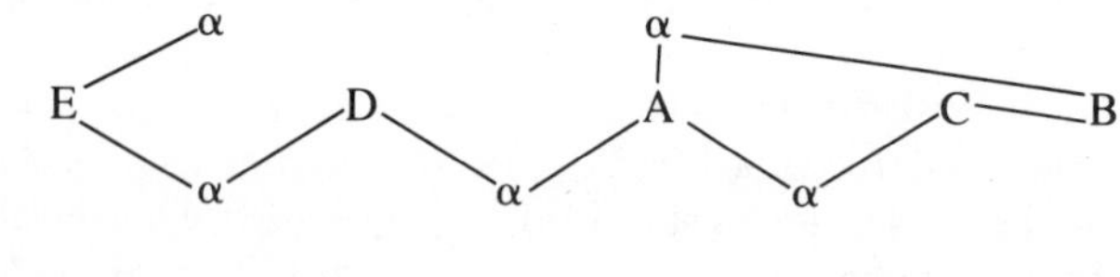

and

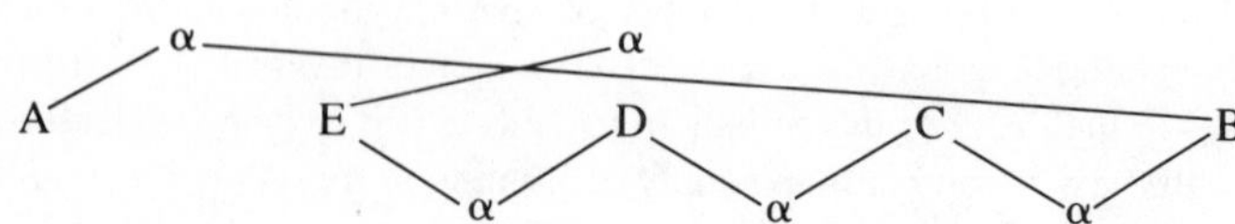

have two helices on the top and three on the bottom. However, the second would not be allowed because of the steric problems created by the α-helices and their connecting loops that pack against the top of the β-sheet. This is analogous to the suggestion of Ptitsyn *et al.* (1979) that there are no crossover connections in all-β proteins.

Although the number of possible topologies is vast, the topologies allowed by these four rules forms a computationally manageable subset (see Table XI). Clearly, the handedness restriction (rule P1) introduces the most severe constraint on the plausible topologies. However, rules P2, P3, and P4 are successful in reducing the problem by an additional order of magnitude. Subtilisin presents a problem since the structure contains a left-handed βαβ connection in a long loop joining strands B and C. Because the original explanation of the preference for right-handed chirality rested on economical connection paths, it is reasonable to assume that all sufficiently long links could make left-handed connections if necessary.

Mixed β Sheets. In contrast to the pure parallel β sheets where four topological rules severely restrict the number of paths for the polypeptide chain, much more diversity seems possible for mixed β sheets. Although the connection between parallel β strands remains right-handed (Sternberg and Thorton, 1976, 1977a; Richardson, 1977), there can be more than one

Table XI. Allowed β-Sheet Topologies

Sheet	No. of strands	No. of possible topologies	No. of topologies allowed by rule 1	No. of topologies allowed by rules 1 to 4
Flavodoxin	5	960	60	6
Adenyl kinase	5	960	60	6
Rhodanese N- or C-domain	5	960	60	6
Alcohol dehydrogenase NAD binding domain	6	11,520	360	15
Phosphoglycerate kinase N-domain	6	11,520	360	10
Phosphoglycerate kinase C-domain	6	11,520	360	10
Lactate dehydrogenase NAD binding domain	6	11,520	360	13
Subtilisin	6	11,520	360[a]	6[a]
Triose phosphate isomerase	8	5×10^6	40,321[b]	21[b]

[a]The correct topology would not be in this list, since subtilisin has a left-handed βαβ connection. If this one left-handed connection were allowed, an additional 360 topologies would be allowed by rule 1 and an additional six by rules 1 to 4.
[b]An additional right-handed barrel topology would be allowed for sheets with seven or more strands. Left-handed barrels have been shown by Sternberg and Thornton (1978) to be topologically unreasonable.

chain reversal (see Table XII). Moreover, the helices joining the strands do not form an equal covering of the sheet. CPA has four helices on one side and two helices on the other side of its β sheet; GPD also has this asymmetry, but here the poorly covered face is involved in the protein–protein interface of the tetramer. This decrease in regularity, coupled with the 2^{n-1}-fold increase in the number of possible topologies (since antiparallel and parallel strand pairs are allowed) dramatically magnifies the computational problems.

Table XII. A Survey of Some Topological Properties of Mixed β Sheets

Protein	Number of strands	Number of helices between strands	Number of reversals[a]	Number of antiparallel connections	Adjacency[b]	Mixed interfaces
Cytochrome b_5	5	3	3	2	1	2
Thioredoxin	5	3	2	2	2	1
Ribonuclease S	5[c]	1	0	3	3	0
Phosphoglycerate mutase	6	4	1	2	2	1
Hexokinase I[d]	6	3	2	3	3	1
II	5	2	1	2	3	1
Papain	7[c]	3	2	3	3	0
Carboxypeptidase A	8	4	3	4	3	2
Glyceraldehyde-3-phosphate dehydrogenase						
C-domain	7	2	2	4	3	3
N-domain	9	3	2	2	5	1
Carbonic anhydrase	10	4	6	6	5	2

[a]A reversal occurs when there is a change in the sense of the strand order, e.g., CAB′, starting at A. B is to the right of A, but C is to the left of B. For pure parallel β sheets, this number was, at most, 1.
[b]Computed as the number of sequential strands that are nearest neighbors in a sheet.
[c]Two sequentially distinct segments form one effective strand.
[d]I: sheet with strands AKJINO; II: sheet with DCBEH.

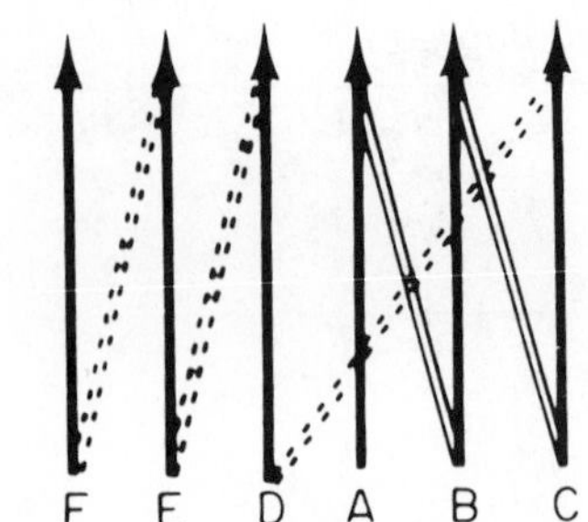

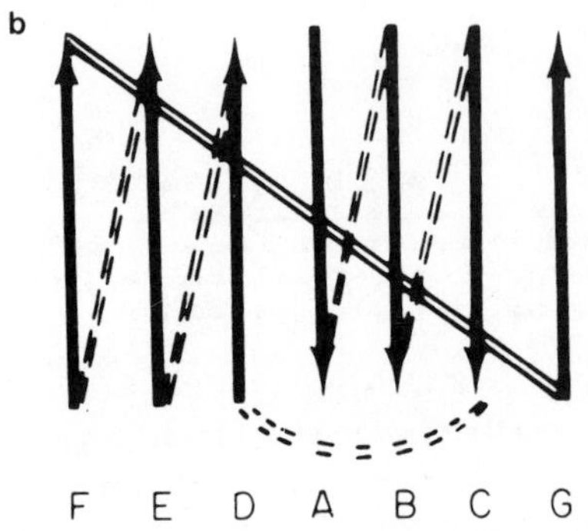

Figure 11. (a) A six-stranded pure parallel β sheet with an adjacency of 4 (AB, BC, DE, EF) and one chain reversal between C and D. (b) A seven-stranded mixed β sheet with an adjacency of four (AB, BC, DE, EF), two chain reversals (between C and D and F and G), and two points of mixture (AD and CG). Reprinted with permission from Cohen *et al.* (1982).

A survey of some other topological properties suggests that adjacency is directly related to the size of the β sheet (see Table XII). Adjacency is defined as the number of sequential consecutive β strands that are spatially adjacent to one another in the β sheet (see Fig. 11). This is consistent with the earlier work on adjacency by Sternberg and Thorton (1977b).

Richardson (1977) has noted the tendency for β sheets to minimize the number of mixed strands, strands that run parallel to one neighbor and antiparallel to the other (see Fig. 11). The number of possible mixed strands increases as the number of strands in the sheet. Although the maximum number of mixed strands is $n-2$, where n is the number of strands, the number of mixed strands observed is always less than $n/2$ and frequently much smaller (see Table XII).

An analysis of strand orders in all known β sheets reveals one further restriction: if $i,j,k,$ and l are sequential strands in a sheet, then the strand order is never k,i,l,j or j,l,i,k. These convoluted strand orders will be known as pretzels or reverse pretzels. In the two extreme situations, it is easy to understand why this highly convoluted strand order is never seen: if all strands are parallel, a knot is created (Richardson, 1977), and if all strands are antiparallel, a crossover of connecting loops is forced (Ptitsyn *et al.*, 1979). Moreover, the fact that this order is never observed is very significant. All possible strand orders for β sheets with four to ten strands were generated, and the fraction that did not contain a pretzel or a reverse pretzel was counted. This allowed fraction shows sigmoid variation as a function of number of strands in the sheet (see Fig. 12). Approximately one-half of all possible six-stranded sheets contain pretzels or reverse pretzels. As the number of strands in the sheet increases, the fraction of strand orders that are allowed by this rule decreases rapidly.

These observations can be formulated into three rules that restrict the number of allowed strand orders and topologies:

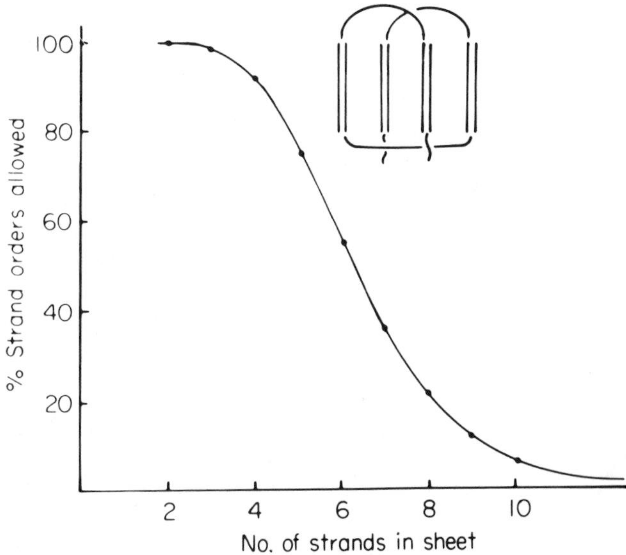

Figure 12. All possible strand orders were generated for β sheets with two to ten strands. The percentage of sheets that did not have a pretzel strand order is plotted as a function of sheet size. A pretzel occurs when four sequential strands, a,b,c,d, have the sheet order b,d,a,c or c,a,d,b (see the schematic β sheet in the upper right hand corner of the graph). Although pretzels are potentially very frequent in sheets with six or more strands, they have never been observed in protein structures. Reprinted with permission from Cohen *et al.* (1982).

M1. The connection between parallel strands is right-handed.

M2. The number of mixed strands is minimal. In the prediction of an unknown structure, this value would be increased from 1 to 2 and so on until a reasonable structure is found. In practice, this value is set to the actual number of mixed interfaces in a particular β sheet in the protein under study.

M3. No pretzels or reverse pretzels are permitted.

2. Prediction of Residue Alignment in β Sheets

Pure Parallel. Although a restricted set of topologies for pure parallel β sheets provides a reasonable reduction of the problem of predicting the structure of αβ proteins, the arrangement of residues along one strand relative to neighboring strands must also be established. A survey of nine pure parallel β sheets (RHO (N and C), PGK (N and C), ADK, ADH, SBT, LDH, and FXN) guided by the conclusions of the analysis of ϕ-area loss in revealed three additional restrictions for allowed strand alignment (see Fig. 13). Hydrophobic residues were Ala, Cys, Ile, Leu, Met, Phe, Pro, Trp, Tyr, and Val, and hydrophilic residues were Arg, Asn, Asp, Gln, Glu, Gly, Lys, His, Ser, and Thr.

P5. Pure parallel β sheets have a central hydrophobic stripe on both faces. Let i and $i + 1$ be sequential residues on β strand β_i, where i or $i + 1$ is within one residue of the middle residue of the strand. (For strands with an even number of resides, the middle residue was taken as that closer to the N terminus of the strand.) For all strands (β_i, β_j, β_k) not on the edge of the sheet, positions i,j,k will be adjacent. Thus, in FXN (Fig. 14), the pair Ala-Leu in strand D aligns with Ile-Leu in strand C and with Ile-Val in strand A. The number of hydrophilic residues in this stripe is small, typically none or one.

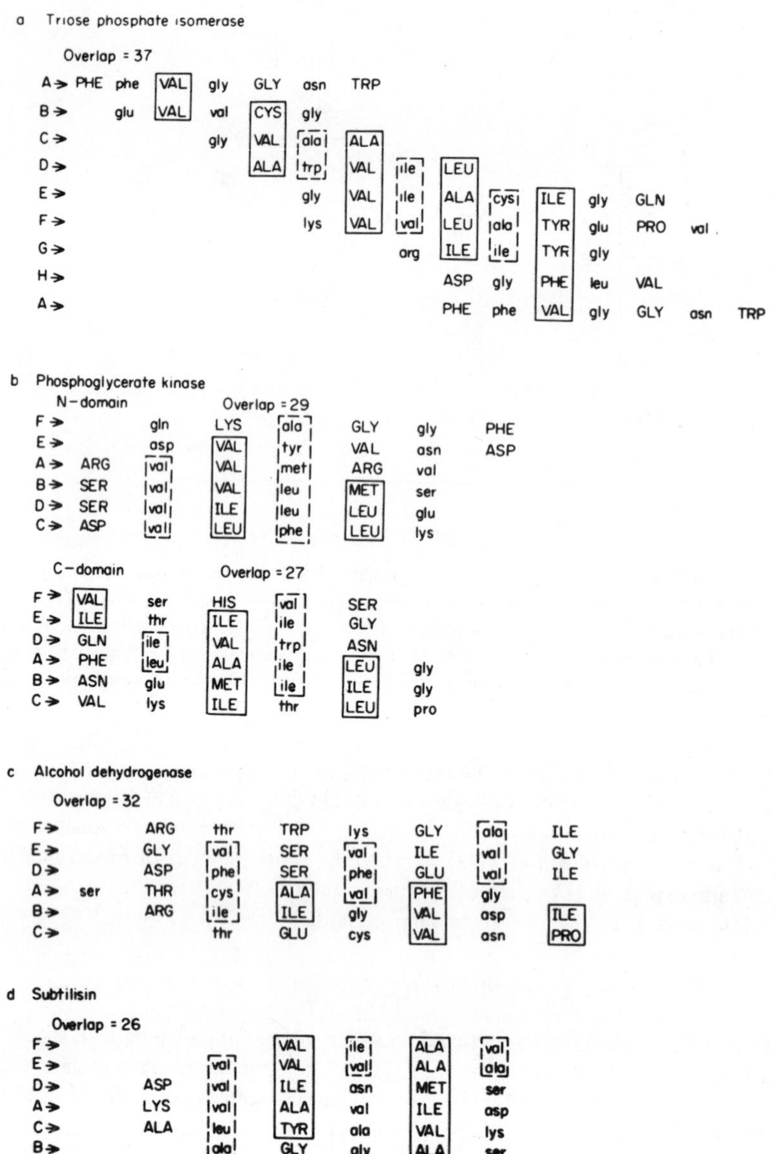

Figure 13. Strand alignment diagrams for five pure parallel β sheets derived from crystallographic data. Residues in capital letters are on the top face of the sheet, and those in lower case are on the bottom. Strands are labeled alphabetically, and their N → C direction is indicated. The unbroken boxes around clusters of residues define a hydrophobic island on the top of the sheet, and the broken boxes delineate the hydrophobic islands on the bottom of the sheet. The hydrophobic residues on the top of the sheet cluster around a line joining the upper left-hand corner to the lower right-hand corner of the β sheet. On the bottom of the sheet, the hydrophobic pattern is from lower left to upper right. The size of carbonic anhydrase and the nucleotide-binding domain of glyceraldehyde-3-phosphate dehydrogenase and the strand discontinuities in ribonuclease S and papain make their analysis by MIXSHEET impossible at present. These four sheets still fit the topological and hydrophobic organization rules M1–M6. Because triose phosphate isomerase is a barrel, it would not be reasonable to expect the same patterns of hydrophobic residues as seen in other βα proteins. Reprinted with permission from Cohen *et al.* (1982).

	1	2	3	4	5	6	7	8	9	10	11	12
E (1)	C	V	V	V	E	T	P	L	I	V	Q	N
D (2)			K	K	V	A	L	F	G	S		
C (3)				I	L	I	L	G	C	S	A	
A (4)				M	K	I	V	Y	W			
B (5)			D	V	N	T	I	N	V			

Figure 14. A strand-alignment diagram for flavodoxin with the strands labeled sequentially A to E and running left to right in rows from N to C. Columns are marked with numbers 1 to 12. Hydrophobic islands on the top of the sheet are boxed with unbroken lines, and those on the bottom in broken lines. Reprinted with permission from Cohen *et al.* (1982).

P6. Hydrophobic residues in β sheets have a characteristic alignment because of the twist of the β strands. The twist of the β sheet (Chothia, 1973) causes the side chains to vary in their displacement from the strand axis when the sheet surface is viewed in projection (see Fig. 15). Certain positions are more likely to be occupied by hydrophobic residues as the side chains point toward the center of the sheet and will be shielded from solvent by the surrounding α helices. This tendency can be formulated explicitly.

Figure 14 displays the alignment of the five strands (rows) of flavodoxin dictated by the hydrogen-bonding pattern. In this format, a column (even numbered) (e.g., column 6,

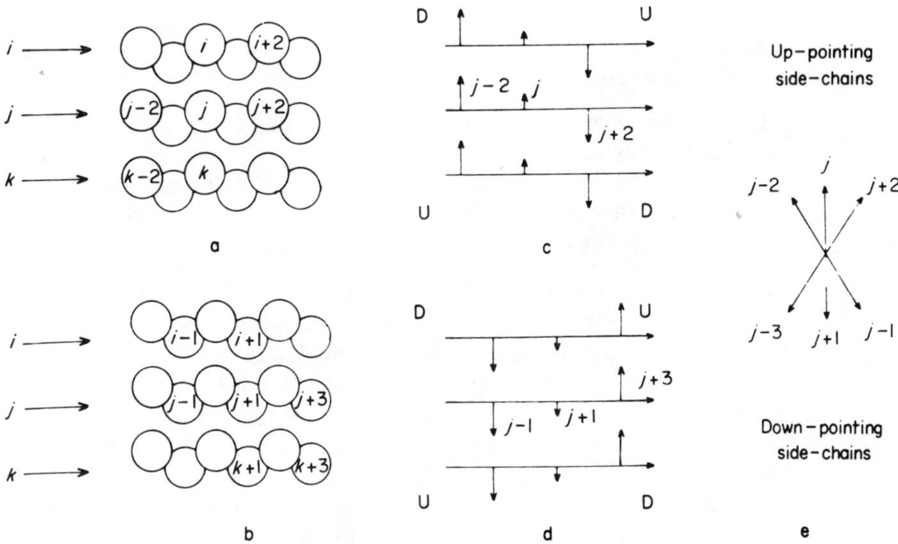

Figure 15. β-Sheet geometry and φ-area change: the explanation for the observed φ-area change when one α helix packs onto a β sheet. a and c consider the up-pointed side chains that interact with an α-helix that lies above the sheet. b and d consider down-pointing side chains. (a and b) Three β strands (*i,j,k*) that run horizontally, with each residue that generally interacts with a docked α helix being labeled. (c and d) The twisted nature of β sheets. The vertical arrows indicate the rough orientation of the up-pointing (c) or down-pointing (d) side chains. U and D denote the raised and lowered corners of the β sheet. (e) The rotation of side-chain positions as one progresses along a β strand. Reprinted with permission from Cohen *et al.* (1982).

Thr/Ala/Ile/Ile/Thr) contains residues that contribute to the top face of the β sheet, and the odd numbered columns constitute the bottom face of the β sheet. A hydrophobic island is defined as a subset of a column of residues containing two or more adjacent hydrophobic amino acids. Thus, Ala/Ile/Ile in column 6 would be a hydrophobic island, but in column 4, Val would not contribute to the hydrophobic island formed by Ile/Met/Val. The midpoints of these hydrophobic islands, h_i [e.g., Met (4) in column 4 or between Val/Leu ($2\frac{1}{2}$) in column 5] must have $h_i \geq h_{i+2} \geq ... \geq h_k$ and $h_{i'} \leq h_{i'+2} \leq ... \leq h_{k'}$, with $|i - i'| = 1$, $|k - k'| = 1$, and $\mathrm{Max}(k,k') - \mathrm{Min}(i,i') \geq 3$ for some column i. In the flavodoxin example, $i = 5, i' = 4, k = 9$, and $k' = 8$. The structural implication of this formulation is that groups of hydrophobic residues in alternating columns progress from the lower left to the upper right corner of the sheet while the intervening rows proceed from the upper left to the lower right-hand corner. The residues involved in the lower-left-to-upper-right patch are on the top of the sheet when the strands run from right to left across the page and the first strand is above the second strand. Moreover, a stretch of hydrophobic islands must occupy at least four consecutive columns, and no intervening columns are seen.

In the set of nine α/β proteins with pure parallel β sheets that we examined, this rule was satisfied by all of the strand-alignment diagrams (see Fig. 13) derived from the crystallographic data with the exception of ADK. In ADK, a suitable solution can be found if strand D is shifted one residue to the right (see Fig. 13), although the phase of strand D is now incorrect with respect to the β sheet. The coordinate displacement is small.

P7. Hydrogen bonding is maximized. In practice, potential hydrogen bonding between strands is always within one of the maximum number possible. This is quantified by strand overlap. Overlap is calculated as one-half the sum over all residues in the sheet of the number of residues on neighboring strands that lie directly across from each residue. Thus, the hypothetical sheet:

Strand A→ Gly-Ala-Ala
Strand B→ Leu-Leu
Strand C→ Lys-Leu

would have a strand overlap of four. This concept succeeds in unifying two contributions to the stability of β sheets: a favorable hydrophobic effect following the burial of nonpolar atoms on sheet formation and a favorable electrostatic effect from the hydrogen bonds. Calorimetric experiments reviewed by Privalov (1979) suggest the extreme importance of hydrogen bonds in stabilizing proteins.

A computer program, BABSHEET (see Fig. 10), was written to apply constraints P5–P7. As input, this program uses the set of allowed topologies supplied by BAB together with the amino acid sequence and strand assignments. For each strand, four possible positions for the strand midpoints of each strand are sampled. This produces a list of 4^n strand-alignment diagrams, where n is the number of strands for each topology. Increasing the potential separation of the midpoints would increase the number of possible strand-alignment diagrams, but all of these structures would have a smaller strand overlap. Each structure is tested against rules P5–P7 and rank ordered on degree of overlap (hydrogen bonding) (see Fig. 14). Structures with overlap comparable to or greater than the native alignment together with a number of hydrophilics in the central stripe (rule P5) less than or equal to that observed in the native alignment are output. The computation time is short, ranging from less than 30 sec for FXN to about 50 min for TIM on an ICL2980. The possibility of a β-barrel-type structure is allowed for proteins with seven or more strands and built explicitly by a segment of the BABBUILD algorithm, the coordinate construction algorithm based on ideal helix and β-sheet geometry.

The time requirements are small compared to those for α-helical proteins (Cohen *et al.*, 1979) or β-sandwich proteins (Cohen *et al.*, 1980, 1981a). This disparity in time is related to the power of the topological constraints, which are easily evaluated and exclude a large number of structures.

Mixed β Sheets. Guided by the rules for hydrophobic organization in pure parallel β sheets, the distribution of hydrophobic residues in mixed sheets was examined (see Fig. 13). Three similar though slightly more general restrictions are suggested:

M4. Mixed β sheets have a central hydrophobic stripe on both faces. An analogous hydrophobic stripe to that found in pure parallel β sheets occurs for all nonedge strands. The difference between mixed and pure β sheets is that mixed sheets have more hydrophilics in this stripe, although the absolute number is still small. Typically, it is one or two but can be as large as six in the ten-stranded sheet of CAC, which is covered by only five α helices.

M5. Hydrophobic residues in β sheets have a characteristic alignment because of the twist of the β strands. Hydrophobic islands have been defined and discussed for pure parallel β sheet in Section III.C.2 (P6). For mixed β sheets, the following relationship must hold:

$$h_i \geq h_{i+2n} \ldots \geq h_{i+2m} \text{ and } h_{i'} \leq h_{i'+2n'} 5 \ldots \leq h_{i'+2m'}$$

with $|i - i'| = 2k + 1$ and $\text{Max}(m,m') \geq 1$ where n, m, n', m', and k are positive integers with $n < m$ and $n' < m'$. As with pure parallel sheets, the structural result is that groups of hydrophobic residues in alternating columns cluster along the diagonal joining the lower left to the upper right corner of the sheet while the intervening rows organize hydrophics along the diagonal joining the upper left and lower right corners of the sheet. The residues involved in the lower-left-to-upper-right patch are on the top of the sheet when the strands run from right to left across the page and the first strand is above the second. The formulation of the hydrophobic organization constraint for mixed β sheets is a generalized version of its counterpart for pure parallel β sheets (P6). This generalized form allows columns with no hydrophobic islands to insert between columns with hydrophobic islands while still maintaining the anticomplementary pattern of organization for hydrophobic residues on opposite faces of β sheets. The anticomplementary organization of hydrophobic residues is central to this work as well as previous studies of all-β proteins (Cohen *et al.*, 1980, 1981b).

M6. Hydrogen bonding is maximized. In practice, potential hydrogen bonding between strands is always within two of the maximum number possible. This is quantified by strand overlap.

A computer program MIXSHEET (see Fig. 10) was written to apply constraints M1–M6. As input, the program uses the set of all strand orders that could have an overlap consistent with rule M2 together with the amino acid sequence and strand assignments.

For each strand, four possible positions for the strand midpoint and two N→C directions are allowed. This produces a list of 8^n hydrogen-bonding diagrams, where n is the number of strands for each strand order. Each structure is tested against rules M1–M6 and sorted by degree of overlap. Structures with overlap comparable to or greater than the native together with a number of hydrophobics in the central stripe (rule M4) less than or equal to that observed in the native strand alignment are output. The computation time ranges from 3 min on TRX to 15 hr for CPA of VAX-780 CPU time.

Table XIII. Distribution and Φ-Area Change for α and β Residues

	α-Helical residues					β-Sheet residues		
	Observed change[a]					Observed change[d]		
	Fraction out of 174	φ-Area change		φ Area in a model α helix[b]	Prediction of potential φ area	Fraction out of 180	φ-Area change	
Residue type	(%)	Mean (Å²)	Max (Å²)	(Å²)	(Å²)	(%)	Mean (Å²)	Max (Å²)
Ala	11.5	14	21	18	21	3.8	11	15
Arg	1.1	7	8	19	0	1.3	11	16
Asn	0.6	8	8	11	0	0.6	5	5
Asp	1.1	7	7	11	0	1.3	6	7
Cys	4.6	16	26	24	24	3.8	10	17
Gln	0.6	6	7	14	0	0.0	0	5
Glu	1.1	12	12	15	0	1.9	17	21
Gly	2.3	7	9	11	0	0.0	0	5
His	1.7	17	22	27	11	1.9	11	21
Ile	12.1	19	35	39	35	16.9	15	34
Leu	18.4	17	31	33	34	13.8	19	32
Lys	5.7	11	21	27	23	1.3	9	10
Met	7.5	24	35	41	35	3.8	15	21
Phe	6.9	23	39	45	35	6.9	22	37
Pro	1.1	14	15	20	18	1.3	20	25
Ser	3.4	10	13	12	0	5.0	10	14
Thr	5.2	14	25	23	0	3.1	9	18
Trp	0.6	35	35	60	35	1.9	34	51
Tyr	2.9	19	24	38	35	5.0	20	26
Val	11.5	15	25	33	33	26.9	15	14

[a]Only α residues with a φ area change of greater than 5 Å² when one α helix docks to a β sheet are included, as they mediate the α/β interaction. The total number of α helical residues (i.e., φ area ≥ 0 Å²) is 517.

[b]The φ areas of a residue type X exposed in a model α helix Ala_4XAla_4. Taken from Richards and Richmond (1978).

[c]These values are used to predict the site on an α helix that packs against a β sheet. The exact values were determined by trial and error.

[d]Only β residues with a φ area change of greater than 5 Å² when all the α helices dock onto the β sheet are included. The total number of β sheet residues (i.e., φ area ≥ Å²) is 306.

3. *Prediction of Helix–Sheet Packing*

Having predicted the hydrogen-bonding pattern of the β sheet in an α/β protein and ensured that this sheet structure will have clusters of hydrophobic residues suitable for helix–sheet packing, we still have to position the interacting α helices. An algorithm predicting helix–helix interaction sites in myoglobin from a sequence and known secondary structure was introduced by Richmond and Richards (1978). They searched the helical regions of the myoglobin sequence for clusters of hydrophobic residues that could bury large quantities of nonpolar surface area on helix–helix packing. Special constellations of residues (e.g., i - 4, i - 3, i - 1, i + 4) were associated with different interaction geometries. A similar approach allowed us to located the α-helical residues involved in helix–sheet packing.

Analysis of α/β packing (Janin and Chothia, 1980; Cohen *et al.*, 1982) show that residues i, i + 1, i + 4, i + 5, i + 8, and i + 9, which cluster on the face of an α helix, have sizable φ-area loss on α/β association. The average and maximum φ-area lost by α-helical residue types on α/β packing provides the basis for φ-area scale used in the prediction (Table XIII). The values were amended in the light of trials; in particular, most hydrophilic residues were set to zero, and His and Lys were decremented slightly. The desired hydrophobic patch included residues $i + 1, i + 4, i + 5, i + 8$, and $i + 9$. To increase the prominence of certain positions, weights of 0.5, 1.5, 1.0, 1.0, 1.5, and 0.5 were used. This follows from the geometry of an α helix, where a repeat of 3.6 residues per turn induces the phase relationship with respect to the α/β contact normal of −90°, 10°, −50°, +50°, −10°, +90°. The predicted hydrophobic patch was that which maximized the sum $H\phi$:

$$H\phi = 0.5(A_i + A_{i+9}) + 1.0(A_{i+4} + A_{i+5}) + 1.5(A_{i+1} + A_{i+7})$$

where A_i is the φ-area for residue type at position j except if j is not in the helix, when $A_j = 0$. Occasionally, there will be several helices joining two consecutive β strands only one of which will interact with the β sheet. Thus, in ADK, helices 3a, b, and c link strands B and C, but only helix 3c interacts with the sheet. Two rules can be developed to deal with this additional complication.

P8. The first or last helix is involved in the interaction.

P9. This ambiguity can be resolved by choosing the helix with the maximum potential φ-area loss.

While helix 3*a* in ADK could lose 101 Å^2, helix 3*c* could lose 174 Å^2 (see Table XIV). On several occasions, the helix leading into the first strand of trailing from the last strand of a β sheet does not participate in the helix–sheet interaction. This occurs when the potential area loss is less than 147 Å^2.

A computer program, BABHELIX, was written to apply these rules for locating potential helix–sheet interaction sites on the surface of α helices. BABHELIX was applied uniformly to 58 α helices in six α/β proteins. The hydrophobic patch actually involved in the helix–sheet interactions had been located previously from an analysis of φ-area changes on packing and interaction geometry. The prediction scheme succeeded in identifying all the 17 of the helices that were not involved in helix–sheet interactions and located 22 of the 41 actual hydrophobic patches on the remaining 41 α helices. This translates into an overall accuracy of 67% (39 out of 58). When the correct hydrophobic patches are not found, the error is always a shift of ±1 residue (9 times), ±3 residues (5 times), or ±4 residues (5 times). These shifts do not imply dramatic shifts of hydrophobic residues into hydrophilic environments but instead preserve the correct phasing of the α helices with respect to the β sheet. The result is a small translation of 1.5–6.0 Å.

4. Construction of C_α Coordinates

The final task in the prediction of the structure of α/β proteins is to combine the strand alignments generated by SHEET or MIXSHEET with the hydrophobic patches predicted by BABHELIX and produce a list of C_α coordinates for the residues involved in secondary structure. This construction rests heavily on the average geometry observed in α/β proteins.

The construction is done in two phases: a twisted β sheet is built, and then α helices are docked to this sheet. To build the β sheet, each strand is built and placed along the x axis. The hydrophobic core of the predicted strand alignment determines the origin for each strand. The strands are twisted relative to one another about the strand midpoints. Strands are ordered by the appropriate topology, separated by 4.25 Å, and twisted about the y axis so that there is a −20° twist between adjacent strands. Helices are placed by a similar construction. Each helix axis is placed on the x axis such that residues $i + 3$ and $i + 4$ symmetrically straddle the z axis. Each helix is translated by ±10 Å in the z direction to maintain the correct handedness of the connection between the preceding and following β strands. The helix is then displaced along the y axis so as to lie between the preceding and following strands and rotated in the x–y plane to make an angle with the x axis equal to the average angle these two strands made with the x

Table XIV. α/β Helix Site Prediction

Protein	Helix number	Actual helix site	Predicted helix site	Potential area change (Å²)	Site prediction error
ADHase	1a	—	—	147	
	1b	187	184	148	+3
	2	208	209	155	−1
	3	233	233	96	
	4	254	254	171	
	5	275	276	170	−1
	6a	—	—	86	
	6b	—	—	196	
	6c	—	—	87	
ADKase	1	—	—	51	
	2	29	29	104	
	3a	—	—	101	
	3b	—	—	138	
	3c	80	80	174	
	4	102	102	52	
	5a	—	—	103	
	5b	—	—	155	
	5c	157	157	120	
	6	190	187	176	+3
FXN	1	20	20	173	
	2	70	70	174	
	3	100	100	156	
	4	135	135	162	
GDPHase	1	16	15	160	+1
	2	40	40	192	
	3	108	105	136	+3
	4	153	157	173	−4
	5	210	207	151	+3
	6	259	255	173	+4
	7	324	324	169	
PGKase	1a	—	—	73	
	1b	50	47	168	+3
	2	81	81	167	
	3	107	107	94	
	4	154	150	149	+4
	5a	—	—	84	
	5b	175	179	85	−4
	6	196	197	158	−1
	7	223	223	172	
	8a	244	244	168	
	8b	—	—	166	
	9	327	327	148	
	10	359	355	144	+4
	11	375	376	57	−1
	12a	394	398	89	−4
	12b	—	—	84	
TIMase	1	24	24	171	
	2	51	50	176	+1
	3	83	82	122	+1
	4a	—	—	68	
	4b	112	112	173	

(*continued*)

Table XIV. (*Continued*)

Protein	Helix number	Actual helix site	Predicted helix site	Potential area change (Å²)	Site prediction error
	5a	—	—	103	
	5b	146	146	127	
	6a	188	187	175	+1
	6b	—	—	70	
	7	220	220	107	
	8a	—	—	57	
	8b	243	243	156	

axis. Each helix predicted to participate in helix–sheet interactions was placed and phased by the predicted interacting diamond.

The separation between the C and N termini of consecutive units of secondary structure is required to be less than $(n + 1)3.5$ Å + 4.0 Å, where n is the number of residues in the connecting segment, 3.5 Å is the effective length of a polypeptide chain in an extended conformation, and 4.0 Å is allowed for errors in the model. All of the nativelike predicted structures satisfy this connectivity constraint, but this filter has yet to be applied to the entire set of predicted structures.

D. Successes and Failures

The combinatorial algorithms for tertiary structure prediction have been applied to a large number of proteins in three of the four major taxonomic classes. These data are presented in Table XV. Stereo diagrams comparing the predicted and crystal structures are shown in Fig. 16. In several cases, the crystal structure has not been determined. These should provide direct tests of the method.

Beginning with the amino acid sequence, structures were predicted for flavodoxin and interleukin-2 using the method described for secondary structure prediction and tertiary structure assembly. Since the structure of flavodoxin is known, a sense of the strengths and weaknesses of existing procedures can be evaluated. The αβ secondary structure prediction suggests three possible secondary structure assignments, which differ as to whether residues 30–35 or 47–55 are on the edge of the β sheet and if the α helix 65–79 is present or absent. Six possible topologies are generated by BABTOPS. BABSHEET was applied and suggested that the first strand should begin with residue Met^1 instead of Lys^2. Eight sheet diagrams were generated, and 16 sets of coordinates were generated, half of which contained a helix at residues 65–79. This process is shown schematically in Fig. 17. One of the structures closely resembles the native structure, with others having increasingly larger r.m.s. deviations. Perhaps the most exciting feature of these algorithms rests in their ability to sort through very large numbers of structures and arrive at a much smaller number of acceptable structures. The common themes in this work are: secondary structure packing provides a useful way of sampling all of conformation space; empirical topological constraints can be recognized that facilitate efficient searches of the possible structures; and connectivity and steric constraints place severe restrictions on allowable structures. Clearly, the size of conformation space for a polypeptide constrained by simple properties is much smaller than the initial figures proposed by Levinthal.

The distressing side of this work is that the answer is hardly unique, and the family of

Table XV. Results of a Combinatorial Approach to Protein Structure Prediction

Protein class	Protein	No. of structures generated	No. of acceptable structures	Best r.m.s. deviation
α/α	Myoglobin (Cohen *et al.*, 1979)	3.4×10^7	20	4.4
	Tobacco mosaic virus coat protein	9.2×10^5	62	4.45
	Interferon α (Sternberg and Cohen, 1982)	3.0×10^6	77	—
	Interleukin 2 (Cohen *et al.*, 1986)	3.9×10^4	27	—
β/β	Immunoglobulin			
	Heavy chain (Cohen *et al.*, 1980)			
	Variable	6×10^8	509	4.9
	Constant 1	5×10^7	6	3.2
	Constant 2	5×10^7	327	2.3
	Constant 3	5×10^7	36	4.9
	Light chain (Cohen *et al.*, 1980)			
	Variable	2×10^8	3343	4.9
	Constant	5×10^7	10	1.4
	REI	6×10^8	988	3.6
	Super oxide dismutase	2×10^8	1135	5.1
	Prealbumin	2×10^8	12	2.0
	Thy-1 (Cohen *et al.*, 1981a)	5×10^7	327	—
	HLA-B_2 (Cohen *et al.*, 1980)	5×10^7	93	—
	B_2 Microglobulin (Cohen *et al.*, 1980)	5×10^7	34	—
α/β	Alcohol dehydrogenase (Cohen *et al.*, 1982)	4.7×10^7	148	4.6
	Adenylate kinase	1.0×10^6	11	5.6
	Cytochrome B_5	6.3×10^7	510	7.3
	Carboxypeptidase A	8.7×10^{13}	2388	7.3
	Flavodoxin	1.0×10^6	8	4.1
	Glyceradehyde 3-phosphate dehydrogenase			
	N-Domain	1.2×10^{16}	3	4.7
	C-Domain	6.8×10^{11}	3768	7.1
	Lactate dehydrogenase	4.7×10^7	96	4.8
	Phosphoglycerate kinase			
	N-Domain	4.7×10^7	47	6.0
	C-Domain	4.7×10^7	2	8.4
	Rhodenese			
	N-Domain	1.0×10^6	67	7.0
	C-Domain	1.0×10^6	32	5.9
	Triose phosphate isomerase	1.7×10^{11}	1	4.5
	Thioredoxin	6.3×10^7	63	4.8

structures generated is difficult to distinguish. In effect, the multiple-minima problem that has crippled energy calculations has been exchanged for a combinatorial problem. If a complete theoretical solution to the protein-folding problem is required, a significant improvement in our understanding of residue–residue interactions will be necessary to sort among empirical structures. If a hybrid solution is allowed, then experimental methods for measuring a limited set of interatomic distances could be very effective in eliminating certain alternatives. This issue is developed in the following section.

IV. ROLE OF HYBRIDIZED EXPERIMENTAL/THEORETICAL SCHEMES FOR OBTAINING STRUCTURAL DATA

Existing theoretical methods fall short of solving the protein-folding problem in that they fail to produce a unique tertiary structure. Although enhancing existing schemes is obviously

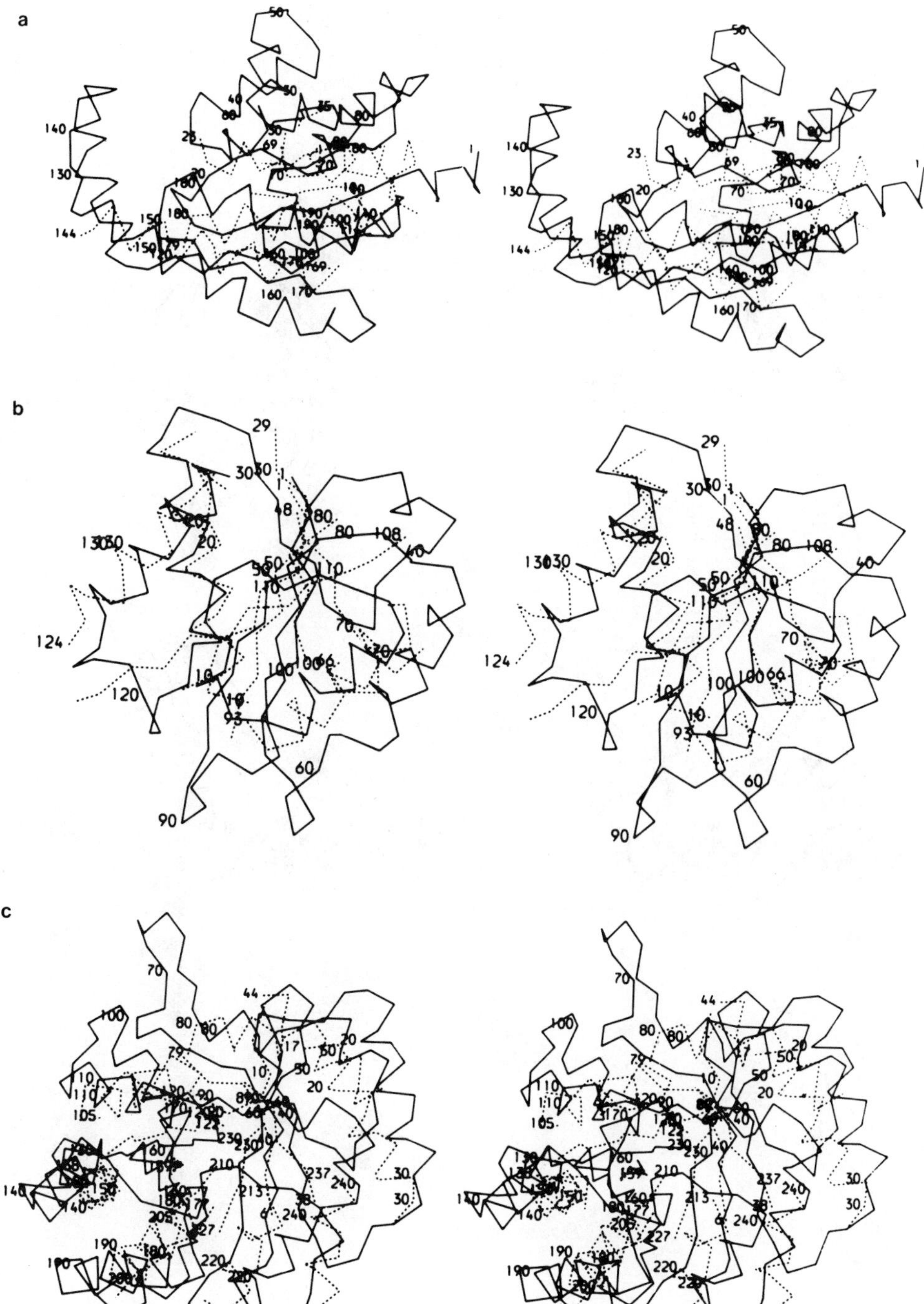

Figure 16. α-Carbon diagrams of the crystal structures of (a) ADK, (b) FXN, (c) TIM, (d) MBN, (e) TMV, and (f) FALC as continuous lines with the predicted structures superimposed as dotted lines. Residues in the crystal-predicted structures are numbered every 10 along the sequence. The N termini of the α and β segments of the predicted structure are also numbered. Reprinted with permission from Cohen and Sternberg (1980a) and Cohen *et al.* (1982).

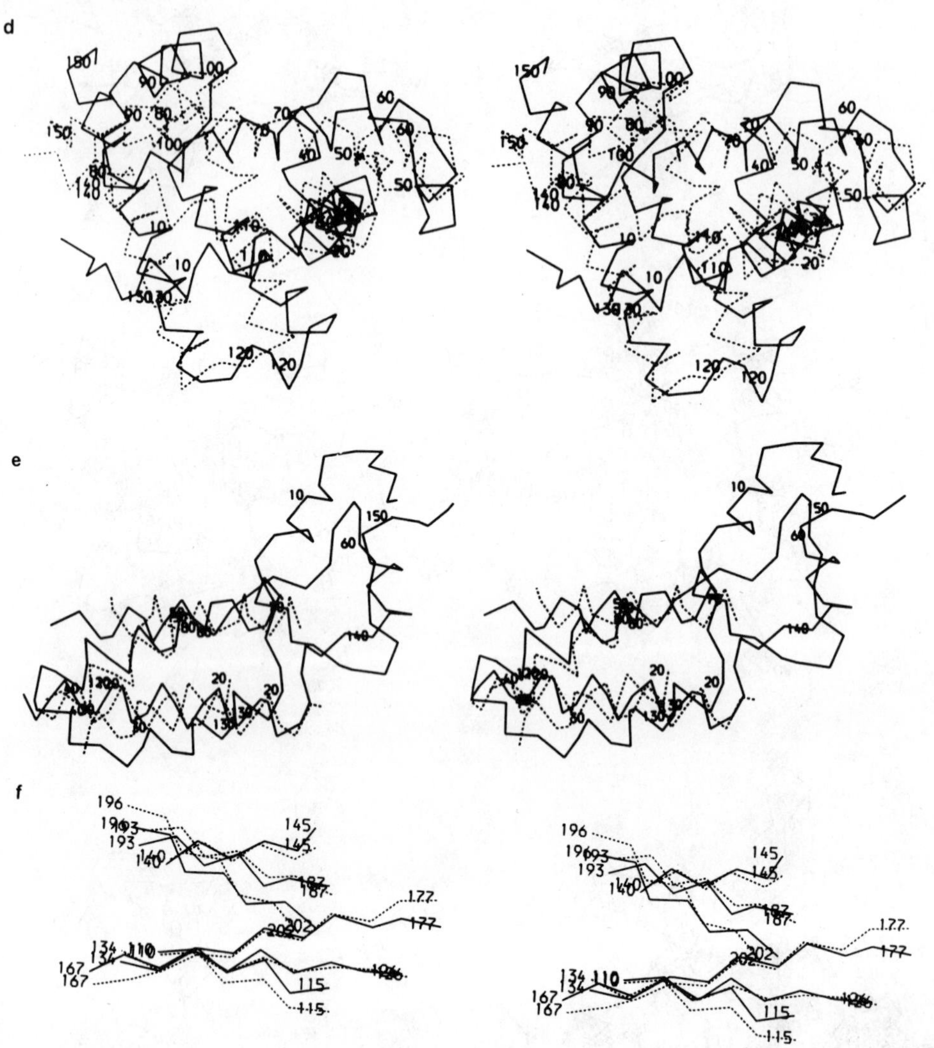

Figure 16 (cont.)

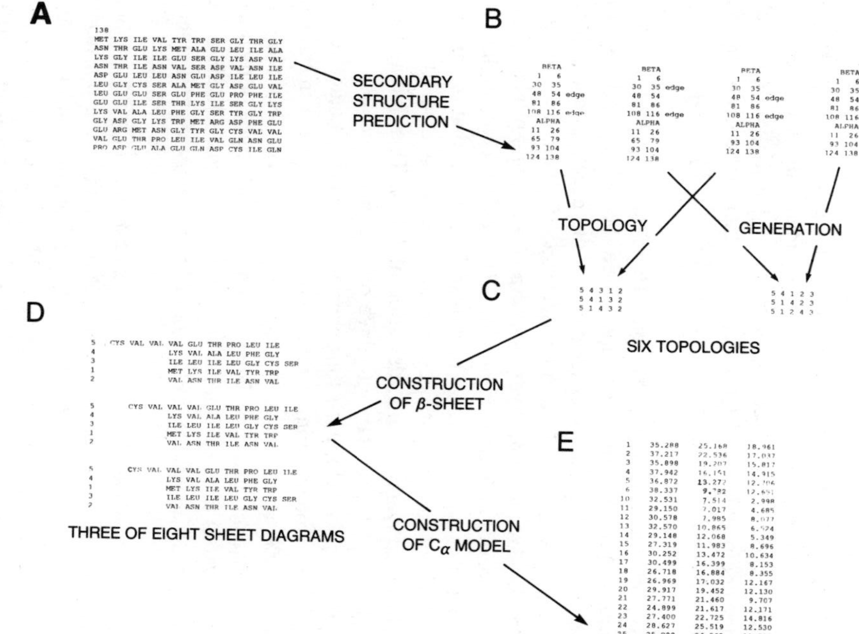

Figure 17. Flow diagram of the prediction of atomic coordinates (E) from amino acid sequence (A) for flavodoxin. Secondary structure (B) is predicted with some ambiguities using the method of Cohen *et al.* (1983). β-Sheet topologies (C) are generated from the secondary structure assignments, and β-sheet hydrogen-bonding diagrams (D) are constructed using the method of Cohen *et al.* (1982). Finally, α-carbon coordinates are constructed from β-sheet hydrogen-bonding patterns using ideal β geometry, and α helices are docked to the β sheet using ideal packing geometries as described by Cohen *et al.* (1982). The combinatorial nature of the structure prediction problem is evident. One sequence gives rise to four possible secondary structures, six topologies, eight sheet hydrogen-bonding diagrams, and 16 lists of coordinates. Of course, many share common features. For example, the top two sheet H-bonding diagrams differ only in the register of of the H-bonding of the fifth and fourth strands. In the bottom two sheet H-bonding diagrams, strands 1 and 3 exchange positions.

desirable, useful results can be obtained from existing methods if they can be combined with experimental constraints. In the simplest case, specific distance information can be gained from the residue pairing in disulfide bridges. This can restrict the set of possible models. With the new techniques for DNA manipulation, site-directed mutagenesis experiments could be designed to sort between alternative structures. Distance geometry techniques developed by Crippen, Kuntz, and co-workers offer a formalism for incorporating this information (e.g., Havel *et al.*, 1979; Goel and Ycas, 1979).

Distance geometry provides an unbiased mathematic formalism for incorporating information on interatomic distances and the accuracy of these distances into three dimensional models. Two matrices are required for input: $[l_{ij}]$ and $[u_{ij}]$, the lower and upper-bounds matrices. Each l_{ij} is large enough to prevent steric conflict, and each u_{ij} is no greater than the diameter of the molecule. Specific changes can be made to incorporate secondary structure assignments, disulfide bridges, chemical cross links, or NMR distance data. For example, if Cys^{50} and Cys^{100} are bridged, then $l_{50,100} = 6.0$ Å and $u_{50,100} = 8.0$ Å reflect the high degree of spatial coupling of these two C_α atoms while allowing some conformational freedom.

Once $[l_{ij}]$ and $[u_{ij}]$ have been determined, the triangle inequality is applied to restrict further these matrices, and the algorithm selects $[d_{ij}]$ such that $[l_{ij}] \leq [d_{ij}] \leq [u_{ij}]$ for all i,j. A multidimensional structure compatible with $[d_{ij}]$ is created and then embedded in three dimensions. This structure is then refined to satisfy the constraints imposed by the upper- and lower-bound matrices.

With pancreatic trypsin inhibitor as an example, Havel *et al.* (1979) showed that a random structure would have Δd r.m.s. deviation (see below) of 7.1 Å from the crystallographic coordinates. Adding the exact locations of the α helices and β strand still produced structures that averaged $\Delta d = 7.0$ Å. A complete specification of the distance of all C_α atoms from the centroid, which a simple hydrophobic potential might produce, resulted in $\Delta d = 5.0$ Å. When the number of specified interatomic distances was equal to the number of residues, the r.m.s. deviation averaged 3.0 Å. Their final finding was that the extensive but qualitative assignment of all interatomic distances being greater or less than 10 Å forced the generated structures to have an average r.m.s. deviation of 1.1 Å. This finding has proved useful in translating nuclear Overhauser data into three-dimensional structure.

Circular dichroism (CD) spectroscopy has long been advocated as a method for assessing secondary structure content in proteins. Fitting procedures (e.g., Provencher and Glockner, 1981) cite accuracies of ±10% across a set of known structures. The secondary structure assignment scheme of Garnier *et al.* used CD data to improve the accuracy of their algorithm marginally. Cohen *et al.* (1986b) used CD on interleukin-2 to test qualitatively a secondary structure assignment (see Fig. 18 and Table XVI). Both CD and theory suggested that helices dominated the structure.

Mutagenesis offers several avenues for gathering distance or proximity information at low concentrations of protein and in nonordered samples. Some of the more practicable ideas involve double revertants, introduction of disulfide linkages, introduction of metal-binding sites, and construction of antigenic features. The use of double revertants to infer proximity of two amino acids has been suggested (Craik *et al.*, 1985). There is a fundamental logical limitation in using the restoration of activity to deduce structure because false-positive and false-negative results are readily obtained. Thus, as Shortle has shown, some natural revertants of staphylococcal nuclease involve amino acids that are not close together in the crystal structure. The converse situation, that alterations in proximal amino acids might not restore activity, should also be a common occurrence. We propose the use of sequentially related revertants to overcome some of these difficulties. For instance, if modification at sequence position i causes loss of activity that can be restored by revertant at position j and modification at $i+2$ can be restored by changes at $j+2$, the inference that i, j, $i+2$, $j+2$ are close and in a

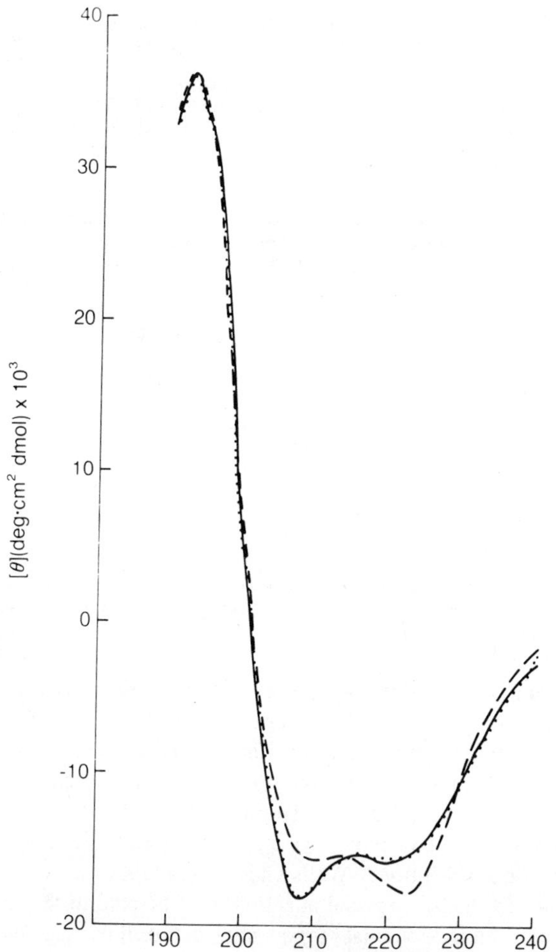

Figure 18. Far-ultraviolet circular dichroism spectrum of recombinant human IL-2 (——) in 5 mM ammonium acetate (pH 5) at room temperature, recorded with a Jasco 500A spectropolarimeter. The spectrum is the average of three scans with protein concentrations of 3.4×10^{-5} M and 2.2×10^{-5} M determined with a molar extinction coefficient of 9.6×10^{-3} M^{-1} cm^{-1} at 280 nm. $[\theta]$ is expressed as mean residue elipticity. Also shown are the least-squares fits to the experimental spectrum by the procedures of Chang *et al.* (1978) (---) and Provencher and Glockner (1981) (···). Predicted fractions of secondary structure are $f_{\text{helix}} = 0.495$, $f_{\text{sheet}} = 0.235$, and $f_{\text{remainder}} = 0.27$; and $f_{\text{helix}} = 0.65$ and $f_{\text{sheet}} = 0.35$, with $f_{\text{remainder}} = 0.0$, respectively. Reprinted with permission from Cohen *et al.* (1986b).

parallel β sheet is quite plausible. As a control, mutating residue $i+1$ should offer nò advantage. Given the geometry of β sheets, this distortion should prove counterproductive.

Extensions to other β-sheet geometries are straightforward, and helix–β interactions can also be studied. Helix–helix crossings are more difficult to probe by this technique. What types of sequence modification have the most general applicability? If the first change is significantly destabilizing (e.g., replacing a small side chain with a large one or replacing a hydrophobic residue with a charged one), and if a compensating alteration is made at appropriate positions (i.e., large to small or hydrophobic to oppositely charged side chain), measurable modulation of activity should occur. Only positive results (loss and restoration of activity) with suitable controls are reliable.

Table XVI. Mouse and Human Interleukin-2 Sequences[a]

```
                       @@@@@@@@@            AAAAAAAAAAAAAA
     t                       aaaaaaaaaaaaaaat              t
APTSSSTSSSTAEAQQQQQQQQQQQQHLEQLLMDLQELLSRMENYRNLKLPRMLTFKFYLP
.......KK-.-------------QLQLEH..L...MI.NGIN..K.P..T........M.
          1                     2         3        4
          0                     0         0        0

AAAAAAAA         AAAAAAAAAAAAAA    AAAAAAAAAAAAAAAAAAAA
    t       t  aaaaaaaaaa   tt     aaaaaaaaaaaaaaaa t
KQATELKDLQCLEDELGPLRHVLDLTQSKSFQLEDAENFISNIRVTVVKKLKGSDNTFECQ
.K.....H.....E..K..EE..N.A..LN.H.R-PRDL....N.I.LG-....ET..M.E
  5        6        7         8         9         1
  0        0        0         0         0         0
                                                  0

@@@@@@@   AAAAAAAAAAAAAAAAA
         aaaaaaaaaaaaaaaaa
FDDESATVVDFLRRWIAFCQSIISTSPQ
YA..T..I.E..N...T.........L
   1         1         1
   1         2         3
   0         0         0
```

[a]Numbers pertain to the human interleukin-2 sequence: (.), homology; (-), deletion/insertion; (aaaaa), helical region predicted; (AAAAA), helical region in x-ray structure of human IL-2; (@@@@@), non-core helix in x-ray structure of human IL-2; (t), turn identified by pattern matching algorithm.

The introduction of Cys residues to form new disulfide bridges is a straightforward way to improve the thermodynamic stability of proteins. Such modifications can also carry structural information, since successful formation of the disulfide bond means that two α carbons must lie within 6.5 ± 1 Å of each other. Negative results (the bond does not form) are not useful, since the bond angle constraints are severe (Thornton, 1982; Richardson, 1981). If the protein of interest already contains one or more Cys residues, special care must be taken to identify firmly the partners in each S-S bond. Wells and Powers (1986) have engineered a disulfide bridge into subtilisin. Obviously, special attention must be paid to the sensitivity of enzyme activity to the solution redox potential. Frequently, extracellular enzymes are stabilized by disulfide bridges, whereas intracellular proteins are not.

Disulfide bond formation is a limiting case of a cross-linking experiment. Since endogenous reagents are used, concentration is not a problem. In fact, the reaction will proceed more cleanly at low protein concentration since bimolecular events will be reduced. Distance information is also obtained by cross linking any suitable pair of residues with conventional bifunctional agents, but the laboratory effort involved is quite substantial, and if a mixture of products is obtained, current separation technology is not promising. Thus, efforts to put Tyr, Lys, or His residues into a sequence solely for cross-linking purposes is probably premature unless the proteins are quite small.

The use of immunologic probes of structure has a long history. The high sensitivity and selectivity to features as subtle as single amino acid substitutions make the general approach quite attractive. The stimulating experiments of Atassi (Twining *et al.*, 1981) and Lerner (1984) have renewed interest in elucidating the molecular basis of antigencity. From our perspective, the important question is what molecular features can be reliably identified. At the primary level, antibodies attack a molecular surface, and positive reactivity provides unambiguous evidence that the sites involved are exposed to solvent. If the commitment is made to raise antibodies to many peptides (preferably derived synthetically for a given protein), it should be possible to map out in a systematic way which peptides are involved in the major loops and turns. At the next level would be efforts to measure the proximity of pairs of loops or

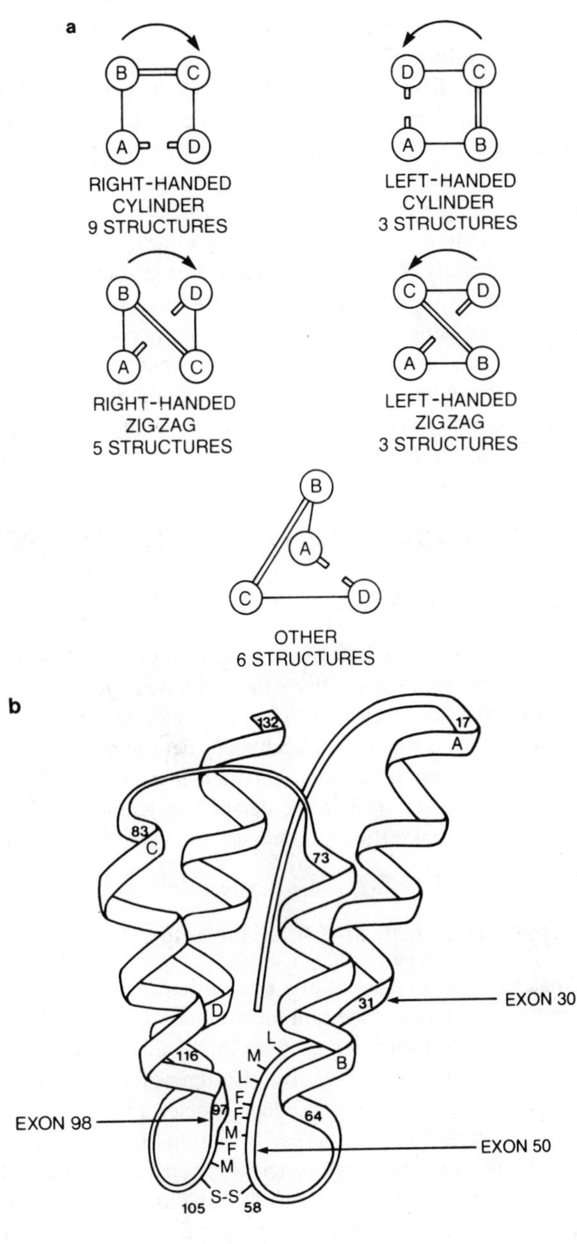

Figure 19. (a) Schematic drawings of the helical topologies for the five families of predicted IL-2 structures. The α helices are represented by circles and labeled sequentially A through D. They are shown end on. Connecting loops are shown as single or double lines: double lines represent connections in front of the page, and single connections behind the page. Arrows highlight the sequential rotation of the helices. (b) Ribbon diagram of a member of the right-handed cylinder family of predicted IL-2 structures. Helical boundaries are located by a pattern-based turn scheme (Cohen *et al.*, 1986a), with most turns located by the patterns *pppp*, *pxppp*, *ppxpp*, or *pppxp*, where *p* is a hydrophilic residue and *x* is any amino acid. Helices are segments bounded by turns with the patterns *hqqhhqqh*, *hhqqhh*, *hqqqhhqqh*, or *hqqhhqqqh*, where *h* is a hydrophobic residue and *q* tends to be hydrophilic. Helical boundary predictions have implied errors of ±3 residues. Helices are labeled

adjacency of nonsequentially related residues. We note that experiments can also be designed to seek secondary structure information by taking advantage of the $i,i + 2$ periodicity in β structure and the $i,i + 4$ periodicity in α helices using immunologic probes.

In summary, experiments can be designed to probe several geometric features of proteins—surface exposure, secondary structure, and residue–residue proximity—that might lead to inferences concerning the ordering of strands in β sheets or helix–helix contacts. Clearly, it is not feasible to start such experiments on a random basis. Fortunately, some of the newer techniques for predicting protein secondary and tertiary structures offer specific structural models that can be tested directly. Cohen *et al.* (1986b) used some of these approaches with interleukin-2. After generation of a secondary structure from the amino acid sequence and building of tertiary structures using the secondary structure packing algorithms (see Fig. 19), experiments were reviewed, and new experiments were designed to sort through the alternative structures (see Table XVII). A combination of immunologic and mutant data was used to eliminate alternative predictions and to enhance confidence in the remaining family of models (see Cohen *et al.*, 1986b, and references therein).

V. CRITERIA FOR JUDGING SUCCESS AND FAILURE OF METHODS

A. Secondary Structure Prediction Type I and II Errors

The critical evaluation of secondary structure prediction schemes has been complicated by the fact that no single quantity accurately reflects the assignment goals: location of all secondary structures and their boundaries. Measures that simply score the correctness of residue-by-residue assignments overestimate the role of accurately determining the secondary structure boundaries relative to locating all structures. Although it is possible to imagine a tertiary packing algorithm recovering from errors in the detailed assignments of N terminus of an α helix (type I error), it is much more difficult to imagine recovery if a core helix is missed (type II error).

B. Tertiary Structure Prediction and the Meaning of the R.M.S. Deviation

A list of coordinates is the usual result of a protein-folding simulation. As a measure of their success, most investigators report the root mean square (r.m.s.) deviation of the atomic positions in their model(s) from those in the crystallographically observed native structure. This deviation offers some measure of the relative effectiveness of two procedures for folding the same protein. However, a standard is needed to determine the absolute success of a procedure or to relate quantitatively the success of various methods on different proteins.

Havel *et al.* (1979), in their studies of pancreatic trypsin inhibitor (PTI) and carp parvalbumin (CPV), have obtained r.m.s. deviations for randomized structures using a distance geometry approach. Hagler and Honig (1978) assessed the quality of energy-minimization

---

A (17–31), B (64–73), C (83–97), D (116–132), and the amino and carboxy termini are numbered. Strong sites for helix–helix packing were determined by the method of Richmond and Richards (1978). Interaction classes pertain to helix–helix docking angles II = −60° (residues 86, 128), III = 19° (residues 23, 24, 25, 69, 92, 93, 118, 122). The disulfide bridge between Cys^{58} and Cys^{105} is indicated. A collection of hydrophobic residues in the loop between helices A and B, which may be important in receptor binding, is marked. Exon boundaries are marked with arrows. The region of binding for monoclonal antibodies to 23–41 (dots), 27–41 (fine dots), and 42–60 (cross hatch), which neutralize activity, are shown. Lys^{76}, a site of limited proteolysis, is also marked. Reprinted with permission from Cohen *et al.* (1986b).

Table XVII. Structure–Activity Changes in Interleukin-2

Structural change	Activity[a]	Comment based on model	Reference
Modification			
Recombinant IL-2	→	Glycosylation not required for activity	Navro *et al.* (1985)
N-Met IL-2	→	N-terminus not important to activity	Yamada *et al.* (1986)
Des Ala^1 IL-2	→	N-terminus not important to activity	Ralph *et al.* (1986)
Mutation			
$Cys^{125} \rightarrow$ Ser	→ ↑	May enhance stability of helix D	Lifson *et al.* (1984)
$Cys^{125} \rightarrow$ Ala	↑	Enhances stability of helix D	Amgen Technical Bulletin (1985)
$Cys^{105} \rightarrow$ Ala	↓	Precludes native disulfide	Liang *et al.* (1986)
$Cys^{105} \rightarrow$ Ser	↓ ↓ ↓	Precludes native disulfide and places hydrophilic Ser near hydrophobic active loop	Lifson *et al.* (1986)
$Cys^{58} \rightarrow$ Ser	↓ ↓	Precludes native disulfide	Lifson *et al.* (1986)
$Cys^{58} \rightarrow$ Ala	↓ ↓	Precludes native disulfide	Liang *et al.* (1984)
$Met^{104} \rightarrow$ Ala	→	Loop between C and D unimportant to activity	Ralph *et al.* (1986)
$Phe^{42\ \&\ 44} \rightarrow$ Tyr	↓ ↓	Disruption of potential binding site	Butler *et al.* (1986)
$Phe^{42\ \&\ 44} \rightarrow$ Trp	↓ ↓ ↓	Disruption of potential binding site	Butler *et al.* (1986)
Deletions			
Del 1-29	0	Loss of helix A	Chizzonite *et al.* (1986)
Del 30-49	0	Loss of binding site loop	Chizzonite *et al.* (1986)
Del 100-133	0	Loss of helix D	Chizzonite *et al.* (1986)
Del 114-133	0	Loss of helix D	Butler *et al.* (1986)
Del 100-104, Cys^{125}Acm[b]	↓ ↓	Possible to divide the chain creating 2 units, "ABC" and "D" which can pack preserving low level of activity	Ciardelli *et al.* (1985)
Dell 100-104, Cys^{125}Acm, Trp^{121}N Formyl	0	Disruption of helix D packing site eliminates binding loop between helix B and C	Ciardelli *et al.* (1985)
Del 76-77, His $^{79} \rightarrow$ Asp	→	Unimportant to activity	Butler *et al.* (1986)
Del Cys^{105}	↓	Loss of native disulfide	Liang *et al.* (1986)
Del Cys^{58}	↓ ↓ ↓	Loss of native disulfide and deletion near active site	Liang *et al.* (1986)
Del Cys^{125}	↓ ↓ ↓	Destabilizes helix-helix interaction for D helix	Liang *et al.* (1986)
Del Phe^{125}	↓ ↓ ↓	Destabilizes helix-helix interaction for D helix	Liang *et al.* (1986)
Del Gln^{126}	↓ ↓ ↓	Destabilizes helix-helix interaction for D helix	Liang *et al.* (1986)

[a]Summary of reported IL-2 structure modifications: (→) no significant change in biologic activity compared to recombinant IL-2 as determined and reported by the authors cited; (↓) an approximate decrease in activity of one order of magnitude; (↑) an approximate increase in activity of one order of magnitude; and (0) no biologic activity.
[b]ACM is the acetamidomethyl derivative of cysteine.

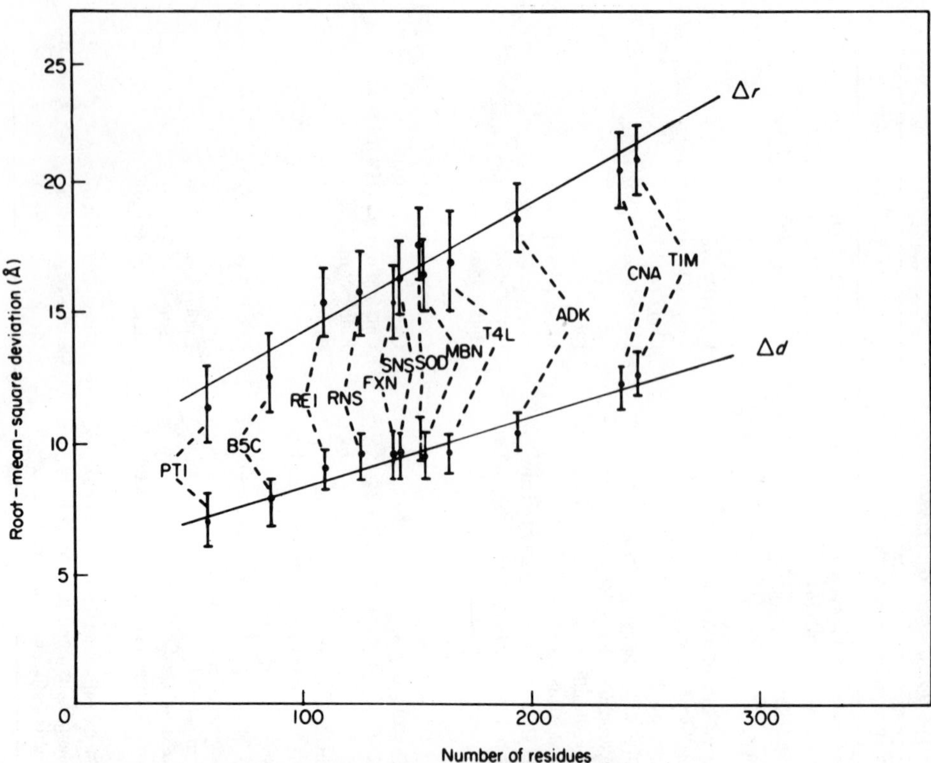

Figure 20. The linear behavior of Δr and Δd with molecular weight. The least-squares equations are Δr = 0.0468 (number of residues) + 9.25 and Δr = 0.0271 (number of residues) + 5.56. Reprinted with permission from Cohen *et al.* (1980).

studies of PTI by folding a chain composed solely of alanines and glycines. In a more general study based on random walks, Cohen and Sternberg (1980b) developed an equation that related the number of residues in the polypeptide chain (N) to the r.m.s. deviation (ΔR) for a random structure from the native structure (see Fig. 20).

$$\Delta R = 0.0468\, N + 9.25$$

with ΔR computed using the formula

$$\Delta R = \text{Min}[\Sigma\, (x_i - y_i)^2]^{1/2}$$

When the r.m.s. deviation is computed on the basis of interatomic distances ΔD, a method that can potentially obscure errors in handedness, the following equations apply:

$$\Delta D_{\text{random}} = 0.0271\, N + 5.56$$

with ΔD computed using the formula

$$\Delta D = [\Sigma\Sigma\, (d_{ij} - e_{ij})^2/n^2]^{1/2}$$

d_{ij} and e_{ij} being the distances between atoms i and j in the native and predicted structures. Table XVIII contains a list of predictive studies and the r.m.s. deviations reported. Clearly, the results obtained in the earlier simulations were little better than random. More recent attempts have met with some success but still remain, on a per-atom basis, several angstroms away from the native structure.

Table XVIII. An Evaluation of Past Protein Folding Studies

Study	Number of residues	R.m.s. deviation reported for predicted structure (Å)	R.m.s. deviation for a random structure with *n* residues (Å)
Levitt (1976) PTI with helices	58	6.2 ($\vec{\Delta d}$)	6.86 ($\vec{\Delta d}$)
Levitt (1976) PTI	58	8.5 ($\vec{\Delta d}$)	7.13 ($\vec{\Delta d}$)
Warshel and Levitt (1976) CPV with helices	108	7.4 ($\vec{\Delta d}$)	8.15 ($\vec{\Delta d}$)
Warshel and Levitt (1976)	108	8.15 ($\vec{\Delta d}$)	8.48 ($\vec{\Delta d}$)
Kuntz et al. (1976) PTI with disulphides	58	4.7–6.5 ($\vec{\Delta d}$)	6.86 ($\vec{\Delta d}$)
Cohen et al. (1979) MBN with helices	99	3.6 ($\vec{\Delta d}$)	7.97 ($\vec{\Delta d}$)
Robson and Osguthorpe	58	6.0 ($\vec{\Delta d}$)	6.86 ($\vec{\Delta d}$)
Cohen et al. (1980)			
All β Proteins			
FCH2	34	2.3 (Δr)	10.84 (Δr)
FCH3	50	4.9 (Δr)	11.59 (Δr)
FALC	46	1.4 (Δr)	11.40 (Δr)
FAHC	48	3.2 (Δr)	11.50 (Δr)
PRE	51	2.0 (Δr)	11.63 (Δr)
FALV	48	4.9 (Δr)	11.50 (Δr)
SDM	51	5.1 (Δr)	11.63 (Δr)
FAHV	46	3.5 (Δr)	11.40 (Δr)
REI	48	3.6 (Δr)	11.50 (Δr)
Cohen et al. (1982)			
α/β Proteins			
ADH	105	4.6 (Δr)	14.16 (Δr)
ADK	92	5.6 (Δr)	13.56 (Δr)
B5C	36	7.3 (Δr)	10.93 (Δr)
CP A	122	7.3 (Δr)	14.95 (Δr)
FXN	97	4.1 (Δr)	13.79 (Δr)
$GPDH_N$	81	4.7 (Δr)	13.64 (Δr)
$GPDH_C$	95	7.1 (Δr)	13.70 (Δr)
LDH	105	4.8 (Δr)	14.16 (Δr)
PGK_N	88	6.0 (Δr)	13.37 (Δr)
PGK_C	108	8.4 (Δr)	14.30 (Δr)
RHO_N	66	7.0 (Δr)	12.34 (Δr)
RHO_C	73	5.9 (Δr)	12.67 (Δr)
TIM	158	4.5 (Δr)	16.64 (Δr)

VI. CONCLUSION

Methods for the prediction of protein tertiary structure from amino acid sequence are under development. Through the use of empirical pattern-based schemes, the ability to locate secondary structure has improved. However, many details remain unresolved. Secondary structure provides a logical starting point for generating approximate tertiary structures. Although unique answers are generated rarely, a manageable set of structures is produced that contains at least one good approximation of the native structure. These methods have led to the prediction of the three-dimensional structure of proteins currently undergoing crystallorgraphic study.

ACKNOWLEDGMENTS. We are deeply indebted to our colleagues of many years standing who made fundamental contributions to the work described here: Robert Abarbanel, Robert Fletterick, Adam Janin, Sir David Phillips, Frederic Richards, Timothy Richmond, Michael Sternberg, and William Taylor. Interactions with members of the UCSF Computer Graphics Laboratory, especially Donald Kneller and Robert Langridge, were most helpful. This work was supported by the National Institutes of Health (GM-31497). The Computer Graphics Laboratory is supported by the Division of Research Resources (RR-1695).

VII. REFERENCES

Abarbanel, R. A., 1984, *Protein Structural Knowledge Engineering,* Ph.D. Thesis, University of California, San Francisco.

Amgen Technical Bulletin (1985).

Anfinsen, C. B., Haber, E., Sela, M., and White, F. H., 1961, The kinetics of formation of native ribonuclease during oxidation of the reduced polypeptide chain, *Proc. Natl. Acad. Sci. U.S.A.* **47:**1309–1314.

Arnone, A., Bier, C. J., Cotton, F. A., Day, V. W., Hazen, E. E., Jr., Richardson, D. C., Richardson, J. S., and Yonath, A., 1971, A high resolution structure of an inhibitor complex of the extracellular nuclease of *Staphylococcus aureus.* I. Experimental procedures and chain tracing, *J. Biol. Chem.* **246:**2302–2316.

Banks, R. D., Blake, C. C. F., Evans, P. R., Haser, R., Rice, D. W., Hardy, G. W., Merrett, M., and Phillips, A. W., 1979, Sequence, structure and activity of phosphoglycerate kinase: A possible hinge-bending enzyme, *Nature* **279:**773–777.

Banner, D. W., Bloomer, A. C., Petsko, G. A., Phillips, D. C., Pogson, C. I., and Wilson, I. A., 1975, Structure of chicken muscle triose phosphate isomerase determined crystallographically at 2.5 Å resolution using amino acid sequence data, *Nature* **255:**609–614.

Bernstein, F. C., Koetzle, T. F., Williams, G. J. D., Meyer, E. F., Brice, M. D., Rodgers, J. R., Kennard, O., Shimanochi, T., and Tasumi, M., 1977, The Protein Data Bank: A computer-basee archival file for macromolecular structures, *J. Mol. Biol.* **112:**535–542.

Blake, C. C. F., Geisow, M. J., Oatley, S. J., Rerat, B., and Rerat, C., 1978, Structure of prealbumin: Secondary, tertiary and quaternary interactions determined by Fourier refinement at 1.8 Å, *J. Mol. Biol.* **121:**339–356.

Bloomer, A. C., Champness, J. N., Bricogne, G., Staden, R., and Klug, A., 1978, Protein disk of tobacco mosaic virus at 2.8 Å resolution showing the interactions within and between subunits, *Nature* **276:**362–368.

Blundell, T., Lindley, P., Miller, L., Moss, D., Slingsby, C., Tickle, I., Turnell, B., and Wistow, G., 1981, The molecular structure and stability of the eye lens: X-ray analysis of gamma-crystallin II, *Nature* **289:** 771–777.

Chang, C. T., Wu, C. S. C., and Yang, J. T., 1978, Circular dichroic analysis of protein conformation: Inclusion of the β-turns, *Anal. Biochem.* **91:**13–31.

Chizzonite, R., Truitt, T., Danho, W., Kilian, P., Gately, M., Tsien, W. H., Moschera, J., Collins, L., and Ju, G., 1986, Analysis of the functional domains of human and mouse interleukin-2 with monoclonal antibodies, *J. Cell. Biochem. Suppl.* **10A:**73.

Chothia, C., 1973, Conformation of twisted beta-pleated sheets in proteins, *J. Mol. Biol.* **75:**295–302.

Chothia, C., 1984, Principles that determine the structure of proteins, *Annu. Rev. Biochem.* **53:**537–572.

Chothia, C., and Janin, J., 1982, Orthogonal packing of beta-placed sheets in proteins, *Biochemistry* **21:**3955–3965.

Chothia, C., Levitt, M., and Richardson, D., 1977, Structure of proteins: Packing of alpha-helices and pleated sheets, *Proc. Natl. Acad. Sci. U.S.A.* **74:**4130–4134.

Chou, P. Y., and Fasman, G. D., 1974, Prediction of protein conformation, *Biochemistry* **13:**211–245.

Chou, P. Y., and Fasman, G. D., 1977, Beta-turns in proteins, *J. Mol. Biol.* **115:**135–175.

Ciardelli, T. L., Smith, K. A., Gadski, R., Butler, L., Strnad, J., and Cohen, F. E., 1985, Structural studies of interleukin-2, in *Peptides: Structure and Function* (C. M. Deber, ed.), Pierce Chemical Co., Rockford, IL, p. 75.

Cohen, F. E., and Sternberg, M. J. E., 1980a, On the use of chemically derived distance constraints in the prediction of protein structure with myoglobin as an example, *J. Mol. Biol.* **137**:9–22.

Cohen, F. E., and Sternberg, M. J. E., 1980b, On the prediction of protein structure: The significance of the root-mean-square deviation, *J. Mol. Biol.* **138**:321–333.

Cohen, F. E., Richmond, T. J., and Richards, F. M., 1979, Protein folding: Evaluation of some simple rules for the assembly of helices into tertiary structure with myoglobin as an example, *J. Mol. Biol.* **132**:275–288.

Cohen, F. E., Sternberg, M. J. E., and Taylor, W. R., 1980, Analysis and prediction of protein beta-sheet structures by a combinatorial approach, *Nature* **285**:378–382.

Cohen, F. E., Novotny, J. N., Sternberg, M. J. E., Campbell, D. G., and Williams, A. F., 1981a, Analysis of structural similarities between brain Thy-1 antigen and immunoglobulin domains: Evidence for an evolutionary relationship and an hypothesis for its functional significance, *Biochem. J.* **195**:31–40.

Cohen, F. E., Sternberg, M. J. E., and Taylor, W. R., 1981b, The analysis of the tertiary structure of protein beta-sheet sandwiches, *J. Mol. Biol.* **148**:253–272.

Cohen, F. E., Sternberg, M. J. E., and Taylor, W. R., 1982, The analysis and prediction of the tertiary structure of globular proteins involving the packing of α-helices against a β-sheet: A combinatorial approach, *J. Mol. Biol.* **156**:821–862.

Cohen, F. E., Abarbanel, R. A., Kuntz, I. D., and Fletterick, R. J., 1983, A combinatorial approach to secondary structure prediction: α/β proteins, *Biochemistry* **22**:4894–4904.

Cohen, F. E., Abarbanel, R. A., Kuntz, I. D., and Fletterick, R. J., 1986a, Turn prediction in proteins using a pattern matching approach, *Biochemistry* **25**:266–275.

Cohen, F. E., Kosen, P. A., Kuntz, I. D., Epstein, L., Ciardelli, T., and Smith, K. A., 1986b, Structure activity studies of interleukin-2, *Science* **234**:349–352.

Craik, C. S., Langman, C., Fletcher, T., Roczniak, S., Barr, P. J., Fletterick, R. J., and Rutter, W. J., 1985, Redesigning trypsin: Alteration of substrate specificity, *Science* **228**:291–297.

Crick, F. H. C., 1953, The packing of α-helices: Simple coiled-coils, *Acta. Crystallogr.* **6**:689–697.

Efimov, A. V., 1979, Packing of alpha-helices in globular proteins. Layer-structure of globin hydrophobic cores, *J. Mol. Biol.* **134**:23–40.

Eklund, H., Nordstrom, B., Zeppezauer, E., Soderlund, G., Ohlsson, I., Boiwe, T., Soderberg, B.-O., Tapia, O., Branden, C.-I., and Akeson, A., 1976, Three-dimensional structure of horse liver alcohol dehydrogenase at 2.4 Å resolution, *J. Mol. Biol.* **102**:27–59.

Epp, O., Colman, P., Fehlhammer, H., Bode, W., Schiffer, M., Huber, R., and Palm, W., 1974, Crystal and molecular structure of a dimer composed of the variable portions of the Bence-Jones protein REI, *Eur. J. Biochem.* **45**:513–524.

Evans, P. R., and Hudson, P. J., 1979, Structure and control of phosphofructokinase from *Bacillus stearothemophilus, Nature* **279**:500–504.

Garnier, J., Osguthorpe, D. J., and Robson, B., 1978, Analysis of the accuracy and implications of simple methods for predicting the secondary structure of globular proteins, *J. Mol. Biol.* **120**:97–120.

Goel, N. S., and Ycas, M., 1979, On the computation of the tertiary structure of globular proteins, *J. Theor. Biol.* **77**:253–305.

Hagler, A. T., and Honig, B., 1978, On the formation of protein tertiary structure on a computer, *Proc. Natl. Acad. Sci. U.S.A.* **75**:554–558.

Hartsuck, J. A., and Lipscomb, W. N., 1971, Carboxypeptidase A, in: *The Enzymes* (P. D. Boyer, ed.), Academic Press, New York, pp. 1–56.

Havel, T. F., Crippen, G. M., and Kuntz, I. D., 1979, Effects of distance constraints on macromolecular conformation. II Simulation of experimental results and theoretical predictions, *Biopolymers* **18**:73–81.

Hendrickson, W. A., and Ward, K. B., 1977, Pseudosymmetry in the structure of myohemerythrin, *J. Biol. Chem.* **252**:3012–3018.

Holbrook, J. J., Liljas, A., Steindel, J., and Rossman, M. G., 1975, Lactate dehydrogenase, in: *The Enzymes* (P. D. Boyer, ed.), Academic Press, New York, pp. 191–292.

Holmes, M. A., and Matthews, B. W., 1982, Structure of thermolysin refined at 1.6 Å resolution, *J. Mol. Biol.* **160**:623–639.

Holmgren, A., Soderberg, B.-O., Eklund, H., and Branden, C.-I., 1975, Three-dimensional structure of *Escherichia coli* thioredoxin-S2 to 2.8 Å resolution, *Proc. Natl. Acad. Sci. U.S.A.* **72**:2305–2309.

Janin, J., and Chothia, C., 1980, Packing of alpha-helices onto beta-pleated sheets and the anatomy of alpha/beta proteins, *J. Mol. Biol.* **143**:95–128.

Jones, T. A., and Liljas, L., 1984, Structure of satellite tobacco necrosis virus after crystallographic refinement at 2.5 Å resolution, *J. Mol. Biol.* **177**:753.

Kabsch, W., and Sander, C., 1984, On the use of sequence homologies to predict protein structure: Identical pentapeptides can have completely different conformations, *Proc. Natl. Acad. Sci. U.S.A.* **81:**1075–1078.

Kendrew, J. C., Dickerson, R. E., Strandberg, B. E., Hart, R. G., Davies, D. R., Phillips, D. C., and Shore, V. C., 1960, Structure of myoglobin, a three-dimensional Fourier synthesis at 2 Å resolution, *Nature* **185:** 422–427.

Kim, P. S., and Baldwin, R. L., 1984, A helix stop signal in the isolated S-peptide of ribonuclease A, *Nature* **307:**329–334.

Kretsinger, R. H., and Nockolds, C. E., 1973, Carp muscle calcium-binding protein. II. Structure determination and general description, *J. Biol. Chem.* **248:**3313–3326.

Kuntz, I. E., 1972, Protein folding, *J. Am. Chem. Soc.* **94:**4009–4012.

Kyte, J., and Doolittle, R. F., 1982, A simple method for displaying the hydropathic character of a protein, *J. Mol. Biol.* **157:**105–132.

Lee, B. K., and Richards, F. M., 1971, The interpretation of protein structures: Estimation of static accessibility, *J. Mol. Biol.* **55:**379–400.

Lerner, R. A., 1984, Antibodies of predetermined specificity in biology and medicine, *Adv. Immunol.* **36:**1–44.

Levitt, M., and Chothia, C., 1976, Structural patterns in globular proteins, *Nature* **261:**552–558.

Levitt, M., and Warshel, A., 1975, Computer simulation of protein folding, *Nature* **253:**694–698.

Lewis, P. N., Momany, F. A., and Scheraga, H. A., 1971, Folding of polypeptide chains in proteins: A proposed mechanism for folding, *Proc. Natl. Acad. Sci. U.S.A.* **68:**2293–2297.

Liang, S. M., Thatcher, D. R., Liang, C. M., and Allet, B., 1986, Studies of structure–activity relationships of human interleukin-2, *J. Biol. Chem.* **261:**334–337.

Lifson, S., and Sander, C., 1980, Specific recognition in the tertiary structure of beta-sheets of proteins, *J. Mol. Biol.* **139:**627–639.

Lifson, J. D., Benike, C. J., Mark, D. F., Koth, S. K., and Engleman, E. B., 1984, Human recombinant interleukin-2 partly reconstitutes deficient *in-vitro* immune responses of lymphocytes from patients with AIDS, *Lancet* **1:**698.

Lim, V. I., 1974, Algorithms for prediction of alpha-helical and beta-structural regions in globular proteins, *J. Mol. Biol.* **88:**857–894.

Mathews, F. S., Bethge, P. H., and Czerwinski, E. W., 1979, The structure of cytochrome b_{562} from *Escherichia coli* at 2.5 Å resolution, *J. Biol. Chem.* **254:**1699–1706.

Matthews, B. W., and Remington, S. J., 1974, The three dimensional structure of the lysozyme from bacteriophage T4, *Proc. Natl. Acad. Sci. U.S.A.* **71:**4178–4182.

McCammon, J. A., Gelin, B. R., and Karplus, M., 1977, Dynamics of folded proteins, *Nature* **267:**585–590.

Momany, F. A., McGuire, R. F., Burgess, A. W., and Scheraga, H. A., 1975, Energy parameters in polypeptides. VII. Geometric parameters, partial atomic charges, nonbonded interactions, hydrogen bond interactions, and intrinsic torsional potentials for the naturally occurring amino acids, *J. Phys. Chem.* **79:**2361–

Mornon, J. P., Fridlansky, F., Bally, R., and Milgrom, E., 1980, X-ray crystallographic analysis of a progesterone-binding protein. The C2221 crystal form of oxidized uteroglobin at 2.2 Å resolution, *J. Mol. Biol.* **137:**415–429.

Nemethy, G., and Scheraga, H. A., 1977, Protein folding, *Q. Rev. Biophys.* **10:**239–352.

Nishikawa, K., Kubota, Y., and Ooi, T., 1983, Classification of proteins into groups based on amino acid composition and other characters. I. Angular distribution, *J. Biochem.* **94:**981–995.

Ploegman, J. H., Drenth, G., Kalk, K. H., and Hol, W. G. J., 1978, Structure of bovine liver rhodanese. I. Structure determination at 2.5 Å resolution and a comparison of the conformation and sequence of its two domains, *J. Mol. Biol.* **123:**557–594.

Privalov, P. L., 1979, Stability of proteins: Small globular proteins, *Adv. Protein Chem.* **33:**167–241.

Provencher, S. W., and Glockner, J., 1981, Estimation of globular protein secondary structure from circular dichroism, *Biochemistry* **20:**33–37.

Ptitsyn, O. B., and Rashin, A. A., 1975, A model of myoglobin self-organization, *Biophys. Chem.* **3:**1–20.

Ptitsyn, O. B., Finklestein, A. B., and Falk, P., 1979, Principal folding pathway and topology of all-beta proteins, *FEBS Lett.* **101:**1–5.

Ralph, P., Nakoinz, I., Doyle, M., Lee, M. T., Koths, K., Halenback, R., and Mark, D. F., 1986, Human B and T lymphocytes stimulating properties of interleukin-2, *J. Cell. Biochem.* **10A:**71.

Reeke, G. N., Becker, J. W., and Edelman, G. M., 1975, The covalent and three-dimensional structure of concanavalin A. IV. Atomic coordinates, hydrogen bonding, and quaternary structure, *J. Biol. Chem.* **250:** 1525–1547.

Richards, F. M., 1977, Areas, volumes, packing and protein structure, *Annu. Rev. Biophys. Bioeng.* **6:**151–176.

Richards, F. M., and Richmond, T., 1978, Solvents, interfaces, and protein structure, CIBA Symposium on Molecular Interactions and Activity in Proteins, Exerpta Medica., Amsterdam, Oxford: New York, pp. 23–45.

Richardson, J. S., 1976, Handedness of crossover connections in beta sheets, *Proc. Natl. Acad. Sci. U.S.A.* **73:** 2619–2623.

Richardson, J. S., 1977, β-sheet topology and the relatedness of proteins, *Nature* **268:**495–500.

Richardson, J. S., 1981, The anatomy and taxonomy of protein structure, *Adv. Protein Chem.* **34:**167–339.

Richardson, J. S., Getzoff, E. D., and Richardson, D. C., 1978, The beta bulge: A common small unit of nonrepetitive protein structure, *Proc. Natl. Acad. Sci. U.S.A.* **75:**2574–2578.

Richmond, T. J., and Richards, F. M., 1978, Packing of alpha-helices: Geometrical constraints and contact areas, *J. Mol. Biol.* **119:**537–555.

Rose, G. D., 1978, Prediction of chain turns in globular proteins on a hydrophobic basis, *Nature* **272:**586–590.

Saul, F. A., Amzel, L. M., and Poljak, R. J., 1978, Preliminary refinement and structural analysis of the Fab fragment from human immunoglobulin new at 2.0 Å resolution, *J. Biol. Chem.* **253:**585–597.

Sawyer, L., Shotton, D. M., Campbell, J. W., Wendell, P. L., Muirhead, H., Watson, H. C., Diamond, R., and Ladner, R. C., 1978, The atomic structure of crystalline porcine pancreatic elastase at 2.5 Å resolution: Comparisons with the structure of alpha-chymotrypsin, *J. Mol. Biol.* **118:**137–208.

Schulz, G. E., and Schirmer, R. H., 1979, *Principles of Protein Structure,* Springer-Verlag, New York.

Schulz, G. E., Elzinga, M., Marx, F., and Schirmer, R. H., 1974, Three dimensional structure of adenyl kinase, *Nature* **250:**120–123.

Sheridan, R. P., Dixon, J. S., Venkataraghavan, R., Kuntz, I. D., and Scott, K. P., 1985, Amino acid composition and hydrophobicity patterns of protein domains correlate with their structures, *Biopolymers* **24:**1995–2023.

Smith, W. W., Burnett, R. M., Darling, G. D., and Ludwig, M. L., 1977, Structure of the semiquinone form of flavodoxin from *Clostridium* MP. Extension of 1.8 Å resolution and some comparisons with the oxidized state, *J. Mol. Biol.* **117:**195.

Sternberg, M. J. E., 1983, The analysis and prediction of protein structure, in: *Computing in Biological Science* (M. Geisow and A. Barrett, eds.), Elsevier, Amsterdam, pp. 143–177.

Sternberg, M. J. E., and Thorton, J. M., 1976, On the conformation of proteins: The handedness of the beta-strand–alpha-helix–beta-strand unit, *J. Mol. Biol.* **105:**367–382.

Sternberg, M. J. E., and Thorton, J. M., 1977a, On the conformation of proteins: An analysis of beta-pleated sheets, *J. Mol. Biol.* **110:**285–296.

Sternberg, M. J. E., and Thorton, J. M., 1977b, On the conformation of proteins: Towards the prediction of strand arrangements in beta-pleated sheets, *J. Mol. Biol.* **113:**401–418.

Sternberg, M. J. E., and Thorton, J. M., 1978, Prediction of protein structure from amino acid sequence, *Nature (London)* **271:**15–20.

Tainer, J. A., Getzoff, E. D., Beem, K. M., Richardson, J. S., and Richardson, D. C., 1982, Determination and analysis of the 2 Å-structure of copper, zinc superoxide dismutase, *J. Mol. Biol.* **160:**181–217.

Takano, T., 1977, Structure of myoglobin refined at 2.0 Å resolution. II. Structure of deoxymyoglobin from sperm whale, *J. Mol. Biol.* **110:**569–584.

Taylor, W. R., and Thorton, J. M., 1983, Prediction of super-secondary structure in proteins, *Nature* **301:**540–542.

Thornton, J. M., 1982, Disulphide bridges in globular proteins, *J. Mol. Biol.* **151:**261–287.

Twining, S. S., David, C. S., and Atassi, M. Z., 1981, Genetic control of the immune response to myoglobin. IV. Mouse antibodies in outbred and congenic strains against sperm-whale myoglobin recognize the same antigenic sites that are recognized by antibodies raised in other species, *Mol. Immunol.* **18:**447–450.

Wells, J. A., and Powers, D. B., 1986, *In vivo* formation and stability of engineered disulfide bonds in subtilisin, *J. Biol. Chem.* **261:**6564–6570.

Wetlaufer, D. B., and Ristow, S. S., 1973, Acquisition of three-dimensional structure of proteins, *Annu. Rev. Biochem.* **42:**135–158.

Wilson, I. A., Skehel, J. J., and Wiley, D. C., 1981, Structure of the haemagglutinin membrane glycoprotein of influenza virus at 3 Å resolution, *Nature* **289:**366–373.

Wright, C. S., Alden, R. A., and Kraut, J., 1969, Structure of subtilisin BPN at 2.5 Å resolution, *Nature* **221:** 235–242.

Yamada, T., Kato, K., Kawahara, K., and Nishimura, O., 1986, Separation of recombinant human interleukin-2 and methionyl interleukin-2 produced in *Escherichia coli, Biochem. Biophys. Res. Commun.* **135:**837–843.

Zav'yalov, V. P., and Denesyuk, A. I., 1982, Possible conformation of interferons: A prediction based on amino acid composition and sequence, *Immunol. Lett.* **4:**7–14.

18

Structure Prediction for Membrane Proteins

Fritz Jähnig

I. INTRODUCTION

To predict the secondary structure of soluble proteins several schemes are available (see other chapters in this book). Most of them are variations and extensions of the same principle developed by Chou and Fasman (1974). To each amino acid residue one attributes a potential for α-helix, β-strand, and β-turn conformation, which has been deduced from proteins of known three-dimensional structure. For a protein of unknown structure, the profiles of the three potentials along the amino acid sequence are compared to predict the secondary structure. The predictive power of such an analysis is limited, but in many cases this is the sole piece of structural information available.

Even this moderate reliability is lost when such an analysis is applied to membrane proteins (Green and Flanagan, 1976; Wallace *et al.*, 1986). Roughly speaking all membrane proteins, or more precisely their membrane-incorporated parts, are predicted to be in β-strand conformation, which is not true. The reason for this failure is the following: The Chou–Fasman potential for β-strand conformation is highest for Val, Ile, Tyr, Phe, Trp, Leu. These are the most hydrophobic residues and, therefore, are found predominantly in the interior of soluble proteins (Janin, 1979). Hence, on the average soluble proteins have a hydrophobic core of β-strand conformation. One may speculate about the reason for that, but it is likely that a β

Fritz Jähnig • Max Planck Institute for Biology, D-7400 Tübingen, West Germany.

sheet in the interior of a soluble protein provides the necessary stability. Since the integral parts of membrane proteins are always hydrophobic, they are predicted to form β-strands.

In fact, the membrane-incorporated segments that are in contact with lipids must be in an ordered conformation, but they may be either in α-helix or in β-strand conformation (Kennedy, 1978). The reason for this requirement lies in the necessity to saturate the hydrogen bonds intramolecularly, since the lipid hydrocarbon chains cannot participate in hydrogen bonding. In the case of a helix, the hydrogen bonds can be formed along the helix as intrachain bonds; hence, a single helix represents a stable structure in a membrane. In the case of β strands, the hydrogen bonds must be formed between neighboring strands as interchain bonds; hence, each β strand must have two neighbors, leading to a closed arrangement, a β barrel. A single β strand, therefore, is not a stable structure in a membrane; at least six to eight β strands are required to form a β barrel.

A single helix in a membrane often serves as an anchor to expose a protein on the cell surface. Examples are glycophorin of erythrocytes (Tomita and Marchesi, 1975) and hemagglutinin of influenza virus (Wilson *et al.*, 1981). A helix may also act as a pore for the permeation of other particles through the membrane. For an α helix with 3.6 residues per turn, the diameter of the pore is too small, but a π helix with 4.3 residues per turn is wide enough to permit permeation of at least protons. The relevant example is gramicidin A (Urry, 1971). For the permeation of larger particles several helices are required to form a ring with a hole in the middle. In the simplest case, the helices may all be identical. Examples are the melittin pore, probably formed by four monomers (Vogel and Jähnig, 1986b), or alamethicin, involving 8–14 monomers (Fox and Richards, 1982). Proteins with more complex functions such as pumps usually consist of several different helices on a single polypeptide chain. Examples are bacteriorhodopsin, made of probably seven helices (Henderson and Unwin, 1975), lactose permease of *E. coli,* comprising 10–14 helices (Vogel *et al.*, 1985), or the reaction center of *S. viridis,* comprising 11 helices on three subunits (Deisenhofer *et al.*, 1985). Proteins of the β-barrel type may act as pores, because even the smallest barrel made of six to eight β strands has a hole of several Å diameter in the middle. Examples are the porins and possibly also OmpA protein of the outer membrane of *E. coli* (Vogel and Jähnig, 1986a).

The question is how to predict such structures from the amino acid sequence in view of the inapplicability of a Chou–Fasman analysis. A first step was made by Kyte and Doolittle (1982) and Argos *et al.* (1982), who applied the concept of hydrophobicity in a quantitative manner to analyze membrane protein sequences. As a measure for the hydrophobicity of an amino acid residue, Kyte and Doolittle used the mean value of two quantities: the transfer free energy of a residue between water and vapor phase and the interior/exterior distribution of the residue in soluble proteins. Since a membrane-spanning helix requires about 20 residues, one searches for hydrophobic stretches of about 20 residues along the sequence. These stretches are then predicted to be membrane-spanning helices.

The actual search is done by averaging the hydrophobicities $h(i)$ of the residues i over seven neighbors on the sequence and plotting the mean hydrophobicity

$$H(i) = \tfrac{1}{7}[h(i \pm 3) + h(i \pm 2) + h(i \pm 1) + h(i)] \qquad (1)$$

along the sequence. This averaging leads to a smoother curve. From the case of bacteriorhodopsin, a value of $H(i) = 1.6$ was found to be the lower limit for a membrane-spanning α helix. This value has been verified for the reaction center (Michel *et al.*, 1986). Hence, a membrane-spanning α helix is predicted if $H(i) \geq 1.6$ over a region of about 20 residues. Averaging over seven neighbors on the sequence corresponds to averaging over two turns on the helix. Therefore, helices predicted in this way are hydrophobic on the entire surface. If a protein has only one membrane-spanning helix, e.g., an anchor, the entire helix must be hydrophobic, and, consequently, a membrane anchor can be identified by such a Kyte–Doolittle analysis.

If a membrane protein consists of several helices that, e.g., form a pore, the helices will not be hydrophobic on the entire surface. On the side facing the lipids they will be hydrophobic, but on the other side facing the pore they will be hydrophilic to provide a pore with a hydrophilic surface. Such helices are called amphipathic. They cannot be identified in a Kyte–Doolittle plot because of the averaging over two turns of the helices. The oldest recipe to demonstrate the amphipathy of a helix is to draw a helical wheel (Schiffer and Edmundson, 1967). A similar but more quantitative plot was proposed by Eisenberg *et al.* (1982), who introduced the hydrophobic moment. These plots are well suited to demonstrate the amphipathy of an individual helix but not appropriate to identify amphipathic helices on an amino acid sequence of several hundred residues. Therefore, other plots have been proposed.

Finer-Moore and Stroud (1984) and Eisenberg *et al.* (1984) performed a Fourier transformation of the hydrophobicity to identify any periodic variation such as the amphipathy of helices or β-strands. However, the entire hydrophobicity curve $h(i)$ is not subjected to Fourier transformation, but only a small region of about 20 residues around each residue i; the result is given by the set of Fourier spectra of all such regions along the sequence. Actually, one is not interested in the full Fourier spectra but only in the components specifying amphipathic helices and β strands. Therefore, in our studies we have restricted ourselves to these two components and plotted only these two along the sequence. To facilitate the analysis even more, we replaced the Fourier transformation of a region of 20 residues by a summation over one side of an α helix or β strand.

II. PREDICTION METHOD

In our analysis, residue i is first assumed to lie in the middle of one side of a membrane-spanning α-helix, and the mean hydrophobicity of this side is calculated as

$$H_\alpha(i) = \tfrac{1}{10}[h(i \pm 8) + h(i \pm 7) + \tfrac{1}{4}h(i \pm 5) + h(i \pm 4) + \tfrac{3}{4}h(i \pm 3) + \tfrac{1}{2}h(i \pm 1) + h(i)] \tag{2}$$

The helix is amphipathic if $H_\alpha(i)$ is high on one side and $H_\alpha(i \pm 2)$ on the other side is low. This corresponds to an oscillation of $H_\alpha(i)$ with a period of four residues (sometimes three, to yield an average of 3.6). Thus, an amphipathic helix is predicted if $H_\alpha(i) \geq 1.6$ and $H_\alpha(i \pm 2) << 1.6$. Next, residue i is assumed to lie in the middle of one side of a membrane-spanning β-strand, and the mean hydrophobicity of this side is calculated as

$$H_\beta(i) = \tfrac{1}{5}[h(i \pm 4) + h(i \pm 2) + h(i)]. \tag{3}$$

An amphipathic β strand corresponds to an oscillation of $H_\beta(i)$ with a period of two residues and, therefore, is predicted if $H_\beta(i) \geq 1.6$ and $H_\beta(i \pm 1) << 1.6$. To perform this analysis for a protein, $H_\alpha(i)$ and $H_\beta(i)$ are plotted along the sequence. For this purpose, a computer program called AMPHI has been written which is available upon request.

An additional aid in predicting amphipathic α-helices and β-strands is the condition that membrane-spanning α-helices must comprise about 20 residues, whereas membrane-spanning β-strands are about ten residues long. Finally, one may profit from the restriction that the minimum number of amphipathic helices is two, whereas six to eight β-strands are required to form a β-barrel.

The secondary structure of residues outside the membrane and therefore in water may be predicted employing the Chou–Fasman rules. Since the connecting loops between α-helices or between β-strands must contain at least one β-turn, the search for β-turns seems to be especially useful. The β-turn potential is calculated as

$$P_t(i) = f_1(i) \cdot f_2(i + 1) \cdot f_3(i + 2) \cdot f_4(i + 3) \tag{4}$$

where the f_i values are given by Chou and Fasman (1978).

III. TWO EXAMPLES

We now present two examples for the prediction of the structure of membrane proteins: lactose permease of the inner membrane of *E. coli* (Vogel *et al.*, 1985) and OmpA protein of the outer membrane (Vogel and Jähnig, 1986a). In both cases, we first discuss the prediction plots and their results and then adjust the prediction to experimental results on the secondary structure. On that basis, models for the folding of the two proteins across the membrane will be developed.

The prediction plots for lactose permease are shown in Fig. 1. From the plot of the hydrophobicity $H(i)$, eight membrane-spanning α-helices may be predicted, six at the N terminus and two at the C terminus. However, this prediction is not very strong, since some of the helices would be shorter than 20 residues and the hydrophobicity of others is not constantly above the critical value of 1.6. A more reliable prediction is obtained from the plot of the α-helix side hydrophobicity $H_\alpha(i)$. Ten membrane-spanning α-helices are predicted, half of them being hydrophobic (those numbered 1, 2, 3, 6, 14) and the other half amphipathic (numbered 4, 5, 10, 11, 13). The difference between a hydrophobic and an amphipathic helix is clearly seen in comparing helices 13 and 14. In the case of helix 14, the α-helix side hydrophobicity $H_\alpha(i)$ remains high over the entire helix, whereas for helix 13, $H_\alpha(i)$ oscillates with a period of 4 residues between values above and below the critical value of 1.6. Hence, helix 14 is hydrophobic on the entire surface, whereas helix 13 is hydrophobic on one side and hydrophilic on the other. This explains the different profiles of the hydrophobicity $H(i)$ across the two helices. The dip of $H(i)$ in the middle of helix 13 is a consequence of its amphipathic character.

The β-strand side hydrophobicity $H_\beta(i)$ necessarily reflects the hydrophobic helices as regions of high hydrophobicity. Amphipathic β-strands showing up as strong oscillations of $H_\beta(i)$ are found mainly in regions outside the predicted α-helices. Only those that fulfill the condition $H_\beta(i) > 1.6$ on one side are candidates for membrane-spanning β-strands in contact with lipids. They are located around residues 116, 212, 279, and 347. The number of β strands, however, is four, too small for the formation of a β-barrel, and therefore the four possible membrane-spanning β-strands are not strongly predicted. This is especially true for the two β-strands around residues 116 and 279 which overlap with predicted α-helices and are abandoned in favor of the helices.

The β-turns predicted from the turn potential $P_t(i)$ in most cases fall into regions outside the predicted α-helices.

These structural predictions may be used to develop a model for the folding of lactose permease across the membrane. The reliability of such a model is improved considerably if experimental data on the secondary structure are taken into account in devising it. Raman measurements indicate that the total α-helix content of lactose permease is 70% (Vogel *et al.*, 1985). Since ten membrane-spanning α-helices correspond to an α-helix content of only 50%, about 80 additional residues must be in α-helix conformation. In the absence of other criteria, they were assigned to the four helices numbered 7–9 and 12 in Fig. 1B, which were chosen to provide the sides of the next most hydrophobic membrane-spanning α-helices still possible. This means that these helices are not predicted but assumed as membrane-spanning.

Combination of our structure prediction and the Raman result leads to the model for the folding of lactose permease presented in Fig. 2. It agrees with experimental results obtained by surface iodination, proteases, and antibodies raised against synthetic polypeptides, which indicate that the N terminus, the two loops between helices 4 and 5 and between helices 6 and 7, and the C terminus are on the same side, the cytoplasmic side, of the membrane (see Vogel *et al.*, 1985).

In the last step of model building for lactose permease, the ten membrane-spanning

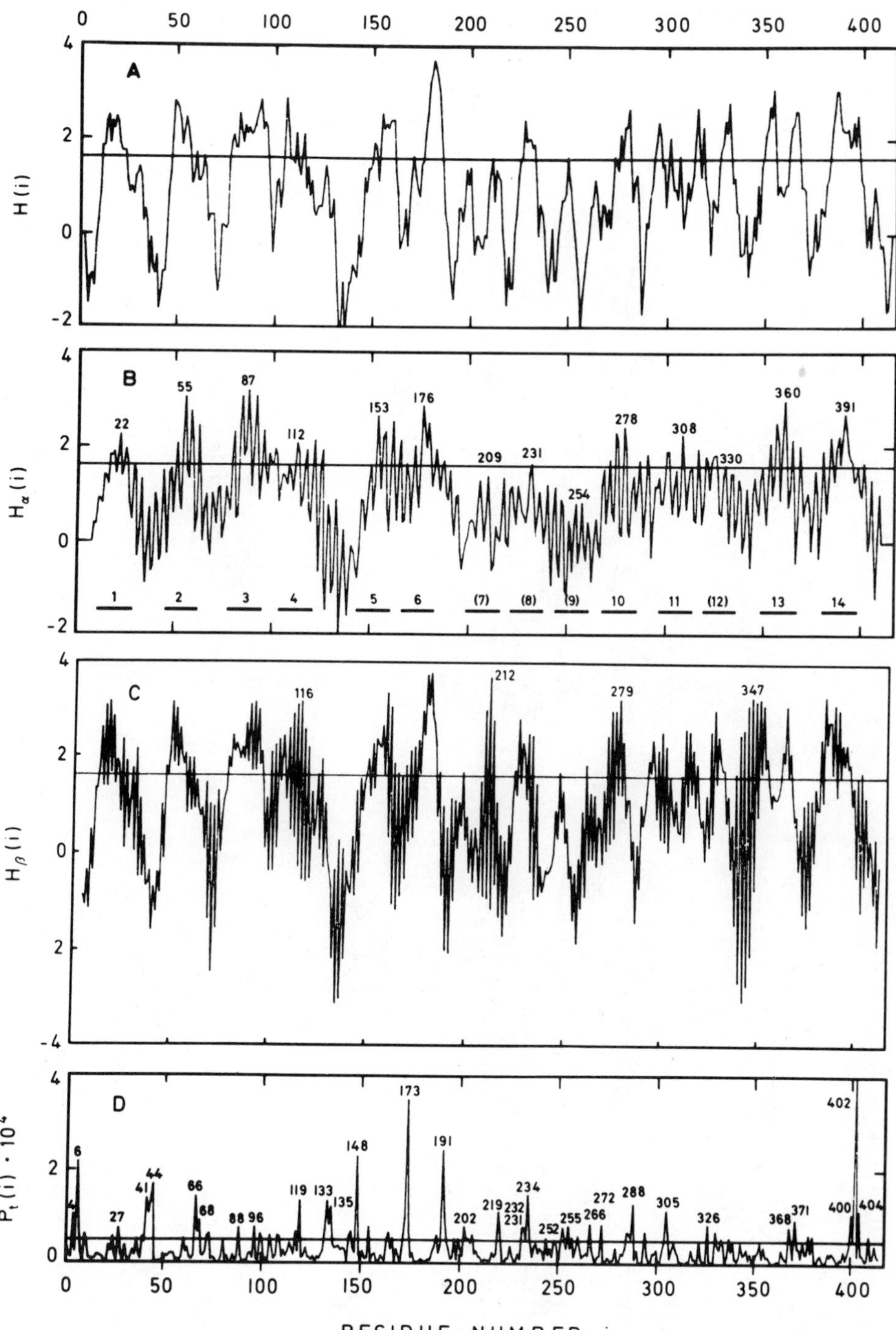

Figure 1. Structure prediction plots for the lactose permease: (A) hydrophobicity $H(i)$, (B) α-helix side hydrophobicity $H_\alpha(i)$, (C) β-strand side hydrophobicity, and (D) β-turn potential $P_t(i)$. Numbers in B specify the residues in the middle of the most hydrophobic sides; the bars specify the extension of the corresponding helices.

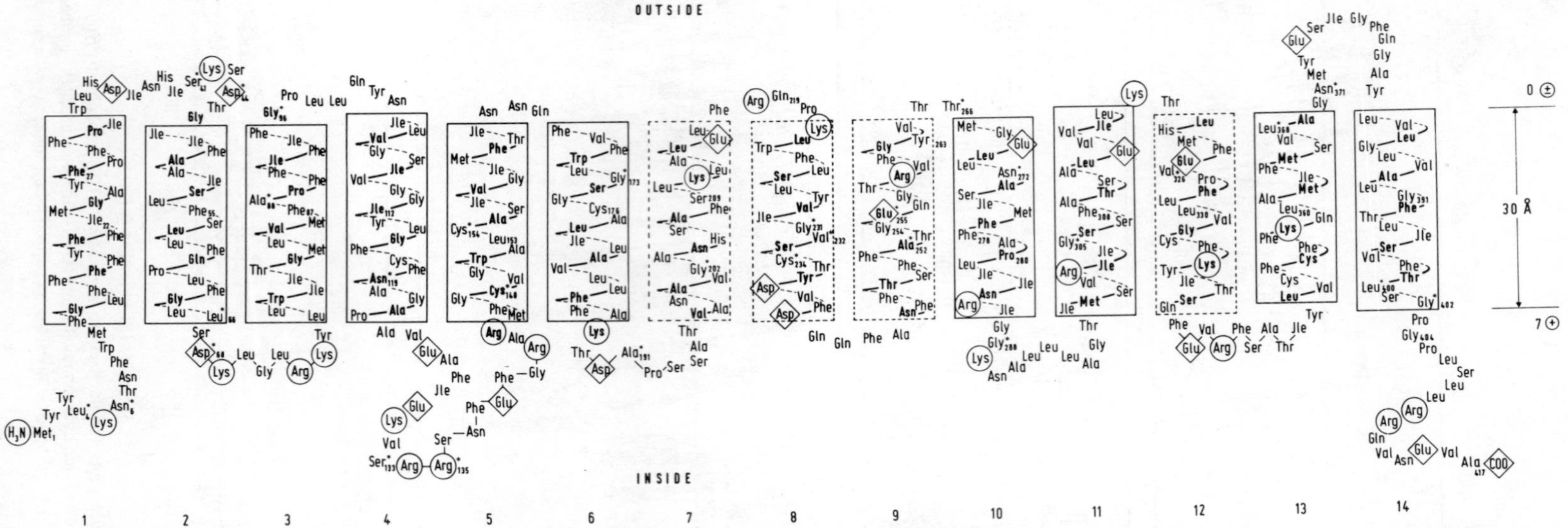

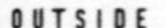

Figure 2. Folding model for the lactose permease. Solid rectangles represent predicted membrane-spanning α-helices, with boldface letters denoting the residues on the more hydrophilic sides. Dashed rectangles represent hydrophilic α-helix regions that are not predicted as membrane-spanning. Zigzags symbolize β-strands, and asterisks denote β-turns. Circles and squares indicate positively and negatively charged residues, respectively.

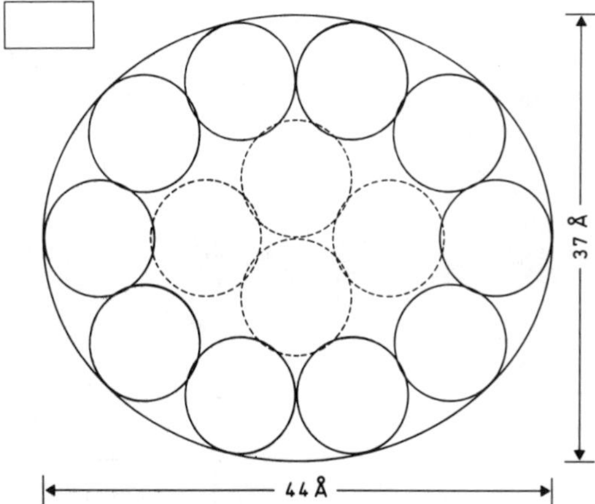

Figure 3. Model for the packing of lactose permease in the membrane plane. Solid circles represent the ten predicted membrane-spanning α helices, and dashed circles four further membrane-spanning α helices that are not predicted as such. The radius of an α helix is assumed to be 5 Å. The rectangle illustrates the approximate size of a lactose molecule of dimensions 4 Å × 8 Å × 12 Å.

helices predicted are folded into a ring in the membrane plane as shown in Fig. 3. The hydrophobic sides of the helices face the lipids, and their more hydrophilic sides face the interior of the ring. The interior is filled with the remaining four hydrophilic helices, which were not predicted but assumed to be membrane-spanning. If their packing in the membrane plane is optimal, the cross section of the protein must be slightly elliptical. The dimensions of the ellipse derived in that way agree with experimental results on the size and shape of lactose permease provided by rotational diffusion measurements (Dornmair *et al.*, 1985). Point mutations that lead to altered substrate binding are located mainly on the four hydrophilic helices (see Vogel *et al.*, 1985; Carrasco *et al.*, 1986). This supports the suggestion that these hydrophilic helices in the middle of the protein provide the sugar and proton binding sites and the pathway to reach them from either side of the membrane. Thus, our model may be helpful to develop ideas about the molecular mechanism of sugar transport by lactose permease.

Figure 4 shows the prediction plots for the OmpA protein. Special attention is paid to the N-terminal fragment of residues 1–177, which is the membrane-incorporated part of the protein. The plot of the hydrophobicity $H(i)$ indicates the absence of any hydrophobic α-helix. The two narrow peaks around residues 230 and 300 closely resemble the pattern of hydrophobic β-strands in soluble proteins, e.g., the arabinose-binding protein of *E. coli*. From the plot of the α-helix side hydrophobicity $H_\alpha(i)$, one amphipathic membrane-spanning α-helix around residue 127 may be predicted. One such helix, however, cannot form a stable structure in a membrane. The plot of the β-strand side hydrophobicity $H_\beta(i)$ permits the prediction of six amphipathic membrane-spanning β-strands (numbered 3–8) for the OmpA fragment. This number is just sufficient for the formation of a β-barrel; hence, the OmpA fragment is predicted to form a β-barrel in the membrane.

The six membrane-spanning β-strands correspond to a β-strand content of 35% for the OmpA fragment. The experimental result, obtained again by Raman spectroscopy, is 70% with a strong predominance of antiparallel β-strands (Vogel and Jähnig, 1986a). Thus, in addition to the six β-strands predicted, about 60 residues must be assigned to β-strand confor-

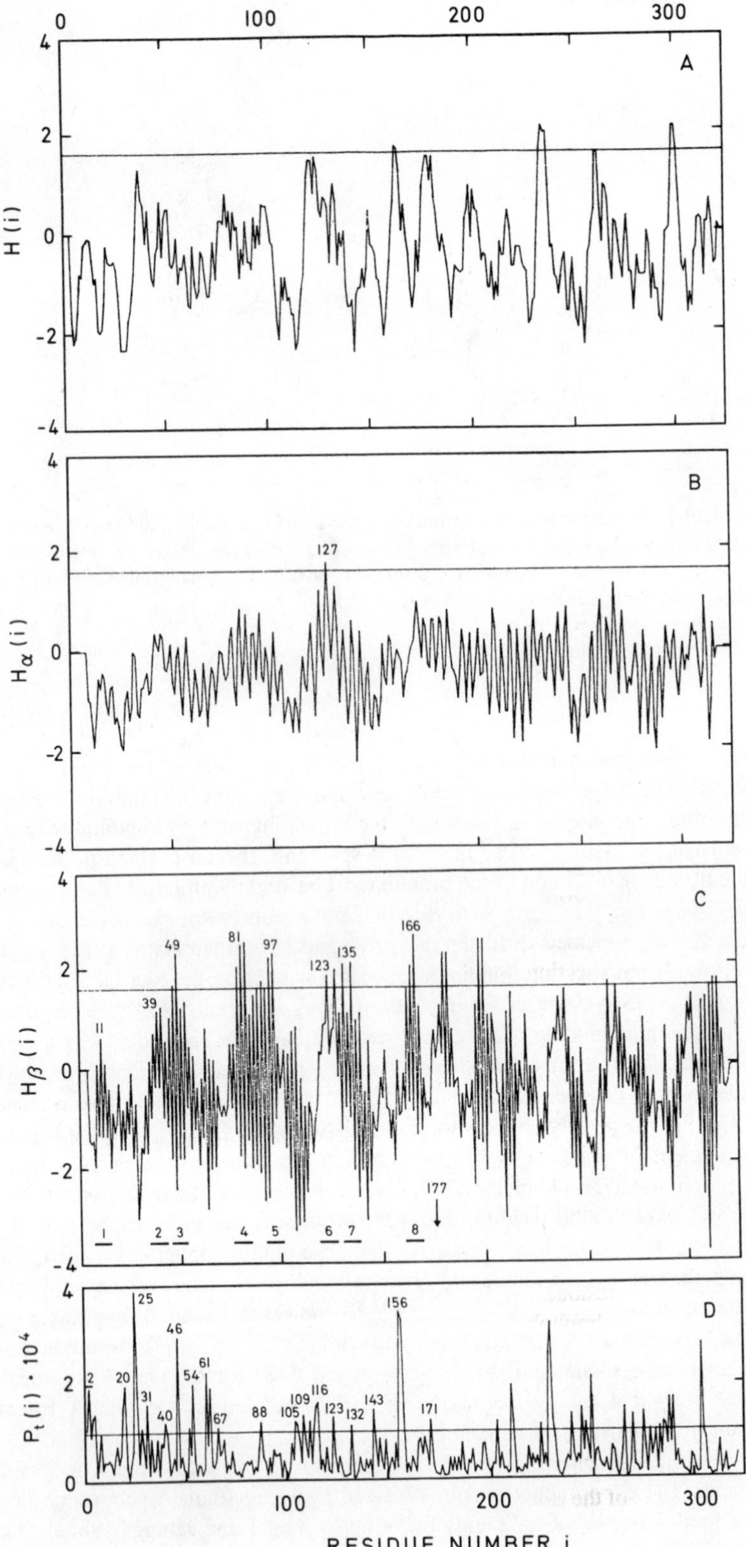

0
100
200
300
4
2
0
-2
-4
A
H (i)
B
127
H_α (i)
C
11
39
49
81
97
123
135
166
177
1
2
3
4
5
6
7
8
H_β (i)
D
2
20
25
31
40
46
54
61
67
88
105
109
116
123
132
143
156
171
P_t (i) · 10^4
RESIDUE NUMBER i

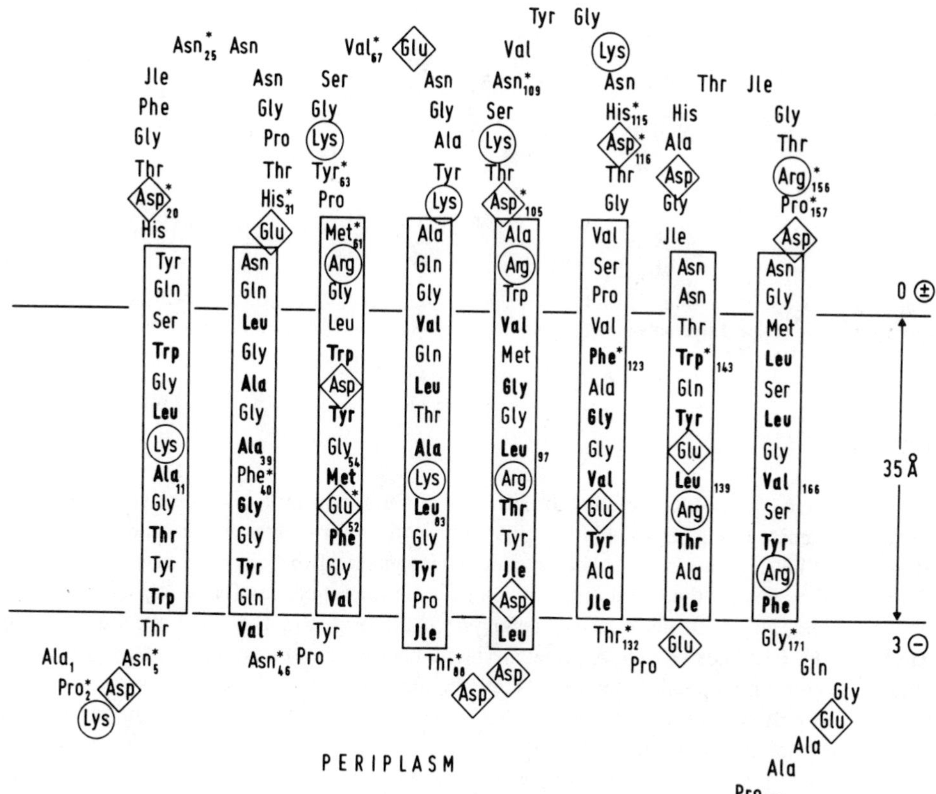

Figure 5. Folding model for the OmpA fragment. Rectangles represent membrane-spanning β strands, with boldface letters denoting the residues on the hydrophobic sides. Asterisks denote β turns, and circles and squares positively and negatively charged residues, respectively.

mation. Guided by the experimental results of Morona *et al.* (1985) on phage binding, two further membrane-spanning β-strands were specified (numbered 1 and 2) that are slightly less hydrophobic. The remaining residues were assigned by extending the predicted β-strands on the side of the large loops to an average length of 13 residues per strand.

The plot of the β-turn potential indicates that most of the predicted β-turns lie between predicted β-strands.

The resulting model for the folding of the OmpA fragment is shown in Fig. 5. It agrees with the results of Morona *et al.* (1985) on phage binding, which indicate that the four regions around residues 25, 70, 110, and 154 are exposed on the cell surface. The eight membrane-spanning β-strands are supposed to be folded into a β-barrel with their hydrophobic sides facing the lipids and their hydrophilic sides forming the polar surface of a pore of about 5 Å diameter, as shown in Fig. 6. This would permit the OmpA protein to function as a membrane pore, but this has not yet been demonstrated.

Figure 4. Structure prediction plots for the OmpA protein: (A) hydrophobicity $H(i)$, (B) α-helix side hydrophobicity $H_\alpha(i)$, (C) β-strand side hydrophobicity $H_\beta(i)$, (D) β-turn potential $P_t(i)$. Numbers in C specify the residues in the middle of the most hydrophobic sides; the bars specify the extension of the corresponding β strands.

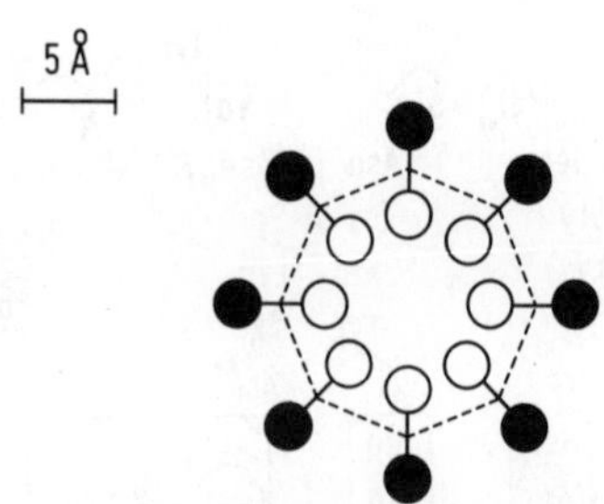

Figure 6. Model for the packing of the OmpA fragment in the membrane plane. Dumbbells denote β strands with hydrophilic (○) and hydrophobic (●) residues, and broken lines indicate hydrogen bonds.

IV. OUTLOOK

The scheme proposed to predict the structure of membrane proteins leads to rather distinct models for their folding, especially in conjunction with experimental results on the secondary structure. What remains open is the detailed arrangement of the predicted α-helices and β-strands in the plane of the membrane. Which helix on the sequence of lactose permease (Fig. 2) corresponds to which helix in the membrane plane (Fig. 3)? The same question may be raised for the β strands of OmpA protein. The number of possible assignments, however, is limited by the amphipathic character of the helices as discussed as well as by the short connections between the helices or strands (Trewhella *et al.*, 1983). Hence, the problem is reduced to finding, among a limited number of possible arrangements, that with the lowest energy. This problem should be solved by molecular dynamics simulations (see other chapters in this book). The possible arrangements would be used as input configurations, and, after some time of simulation, their energies compared to determine the structure of lowest energy. Thus, the structure of a membrane protein may become calculable in the near future.

Although the situation with soluble proteins is not as favorable, progress may also be achieved here by adopting some of the concepts developed for membrane proteins. The concept of hydrophobicity has already been employed by Hopp and Woods (1983) to predict antigenic sites on the surface of soluble proteins. Lim (1974) has made use of the amphipathy of α-helices and β-strands for secondary structure prediction. The above prediction scheme for membrane proteins may be considered as a quantification of Lim's more qualitative rules. An attempt to apply this scheme to soluble proteins has already been made in predicting the structure of a fiber protein of phages on the basis of its amphipathy (Riede *et al.*, 1987).

V. REFERENCES

Argos, P., Rao, J. K. M., and Hargrave, P. A., 1982, Structural prediction of membrane-bound proteins, *Eur. J. Biochem.* **128:**565–575.

Carrasco, N., Antes, L. M., Poonian, M. S., and Kaback, H. R., 1986, *Lac* permease of *Escherichia coli:* Histidine-322 and glutamic acid-325 may be components of a charge-relay system, *Biochemistry* **25:**4486–4488.

Chou, P. Y., and Fasman, G. D., 1974, Prediction of protein conformation, *Biochemistry* **13:**222–245.

Chou, P. Y., and Fasman, G. D., 1978, Empirical predictions of protein conformation, *Annu. Rev. Biochem.* **47:**251–276.

Deisenhofer, J., Epp, O., Miki, K., Huber, R., and Michel, H., 1985, Structure of the protein subunit in the photosynthetic reaction centre of *Rhodopseudomonas viridis* at 3 Å resolution, *Nature* **318:**618–624.

Dornmair, K., Corin, A. F., Wright, J. K., and Jähnig, F., 1985, The size of the lactose permease derived from rotational diffusion measurements, *EMBO J.* **4:**3633–3638.

Eisenberg, D., Weiss, R. M., and Terwilliger, T. C., 1982, The helical hydrophobic moment: A measure of the amphiphilicity of a helix, *Nature* **299:**371–374.

Eisenberg, D., Weiss, R. M., and Terwilliger, T. C., 1984, The hydrophobic moment detects periodicity in protein hydrophobicity, *Proc. Natl. Acad. Sci. U.S.A.* **81:**140–144.

Finer-Moore, J., and Stroud, R. M., 1984, Amphipathic analysis and possible formation of the ion channel in an acetylcholine receptor, *Proc. Natl. Acad. Sci. U.S.A.* **81:**155–159.

Fox, R. O., and Richards, F. M., 1982, A voltage-gated ion channel model inferred from the crystal structure of alamethicin at 1.5-Å resolution, *Nature* **300:**325–330.

Green, N. M., and Flanagan, M. T., 1976, The prediction of the conformation of membrane proteins from the sequence of amino acids, *Biochem. J.* **153:**729–732.

Henderson, R., and Unwin, P. N. T., 1975, Three-dimensional model of purple membrane obtained by electron microscopy, *Nature* **257:**28–32.

Hopp, T. P., and Woods, K. R., 1983, Prediction of protein antigenic determinants from amino acid sequences, *Proc. Natl. Acad. Sci. U.S.A.* **78:**3824–3828.

Janin, J., 1979, Surface and inside volumes in globular proteins, *Nature* **277:**491–492.

Kennedy, S. J., 1978, Structures of membrane proteins, *J. Membr. Biol.* **42:**265–279.

Kyte, J., and Doolittle, R. F., 1982, A simple method for displaying the hydropathic character of a protein, *J. Mol. Biol.* **157:**105–132.

Lim, V. I., 1974, Structural principles of the globular organization of protein chains. A stereochemical theory of globular protein secondary structure, *J. Mol. Biol.* **88:**857–872.

Michel, H., Weyer, K. A., Gruenberg, H., Dunger, I., Oesterhelt, D., and Lottspeich, F., 1986, The "light" and "medium" subunits of the photosynthetic reaction centre from *Rhodopseudomonas viridis:* Isolation of the genes, nucleotide and amino acid sequence, *EMBO J.* **5:**1149–1158.

Morona, R., Krämer, C., and Henning, U., 1985, Bacteriophage receptor area of outer membrane protein OmpA of *Escherichia coli* K-12, *J. Bacteriol.* **164:**539–543.

Riede, I., Schwarz, H., and Jähnig, F., 1987, Predicted structure of tail-fiber proteins of T-even type phages, *FEBS Lett.* **215:**145–150.

Schiffer, M., and Edmundson, A. B., 1967, Use of helical wheels to represent the structures of protein and to identify segments with helical potential, *Biophys. J.* **7:**121–135.

Tomita, M., and Marchesi, V. T., 1975, Amino-acid sequence and oligosaccharide attachment sites of human erythrocyte glycophorin, *Proc. Natl. Acad. Sci. U.S.A.* **72:**2964–2968.

Trewhella, J., Anderson, S., Fox, R., Gogol, E., Khan, S., and Engelman, D., 1983, Assignment of segments of the bacteriorhodopsin sequence to positions in the structural map, *Biophys. J.,* **42:**233–241.

Urry, D. W., 1971, The gramicidin A transmembrane channel: A proposed $\pi_{L,D}$ helix, *Proc. Natl. Acad. Sci. U.S.A.* **68:**672–676.

Vogel, H., and Jähnig, F., 1986a, Models for the structure of outer-membrane proteins of *Escherichia coli* derived from Raman spectroscopy and prediction methods, *J. Mol. Biol.* **190:**191–199.

Vogel, H., and Jähnig, F., 1986b, The structure of melittin in membranes, *Biophys. J.* **50:**573–582.

Vogel, H., Wright, J. K., and Jähnig, F., 1985, The structure of the lactose permease derived from Raman spectroscopy and prediction methods, *EMBO J.* **4:**3625–3631.

Wallace, B. A., Cascio, M., and Mielke, D. L., 1986, Evaluation of methods for the prediction of membrane protein secondary structure, *Proc. Natl. Acad. Sci. U.S.A.* **83:**9423–9427.

Wilson, I. A., Skehel, J. J., and Wiley, D. C., 1981, Structure of the haemagglutinin membrane glycoprotein of influenza virus at 3 Å resolution, *Nature* **289:**366–373.

19

Identification of Membrane Proteins and Soluble Protein Secondary Structural Elements, Domain Structure, and Packing Arrangements by Fourier-Transform Amphipathic Analysis

Janet Finer-Moore, J. Fernando Bazan, John Rubin, and Robert M. Stroud

Janet Finer-Moore, J. Fernando Bazan, and Robert M. Stroud • Department of Biochemistry and Biophysics, University of California, San Francisco, California 94143. *John Rubin* • Genentech, Inc., South San Francisco, California 94080.

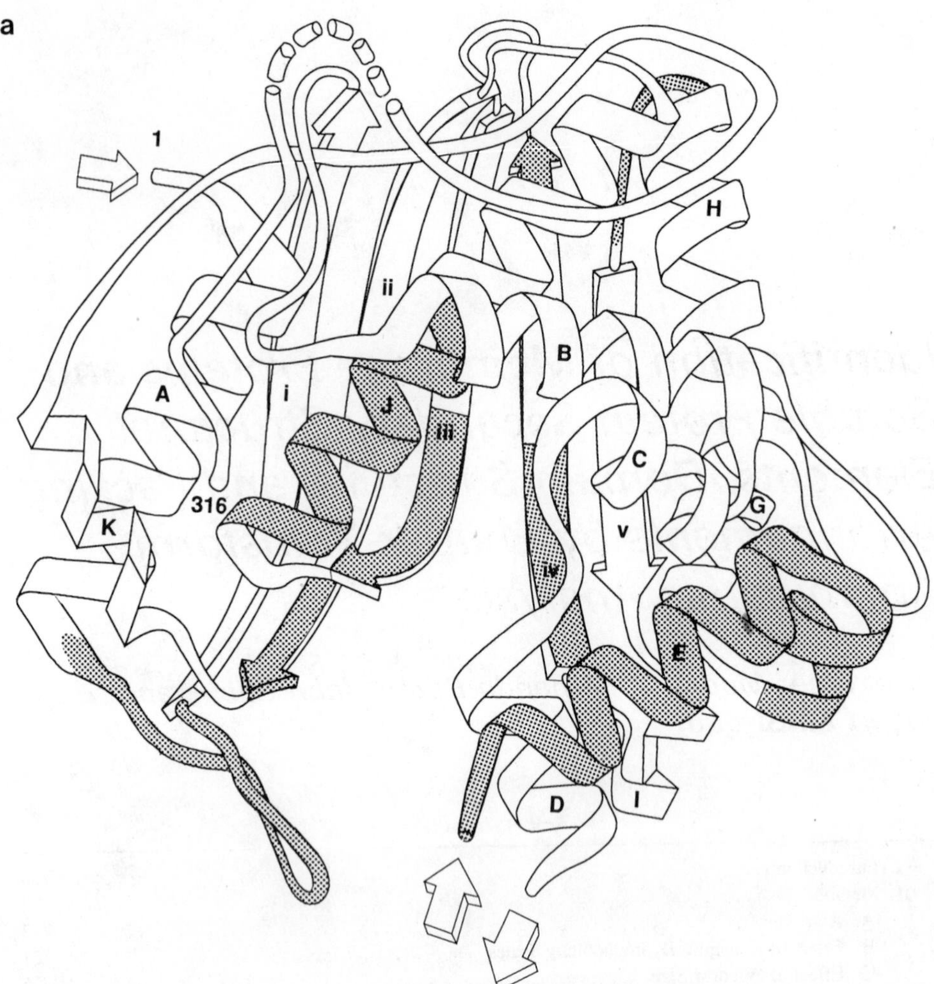

Figure 1. (a) Ribbon drawing of thymidylate synthase (Hardy *et al.*, 1987), a typical globular protein. Amphipathic helices A and H are located at the interface between protein and solvent in the three-dimensional structure of the enzyme. Hydrophilic residues in helix A occur with a period of four residues, as shown in the helical net diagram (b). Helix J, in contrast, is almost completely buried in the three-dimensional structure and is composed primarily of hydrophobic residues, as shown in c.

I. INTRODUCTION

The value of a model is that it simplifies the description of a system: it focuses attention onto potentially critical features, and it will be superseded by better models that explain more. Above all it suggests critical and focused experimental tests designed to refine the model, eliminate inconsistency, and uncover function.

Early in the study of protein structure, the pattern of hydrophobicity in a protein's sequence was identified as a major determinant of its three-dimensional structure. Perutz *et al.* (1965) noted that although the myoglobins and hemoglobins from different species were structurally conserved, they had little sequence homology. However, the hydrophobic character of 30 interior amino acid sites was preserved. Several of these amino acids were on the

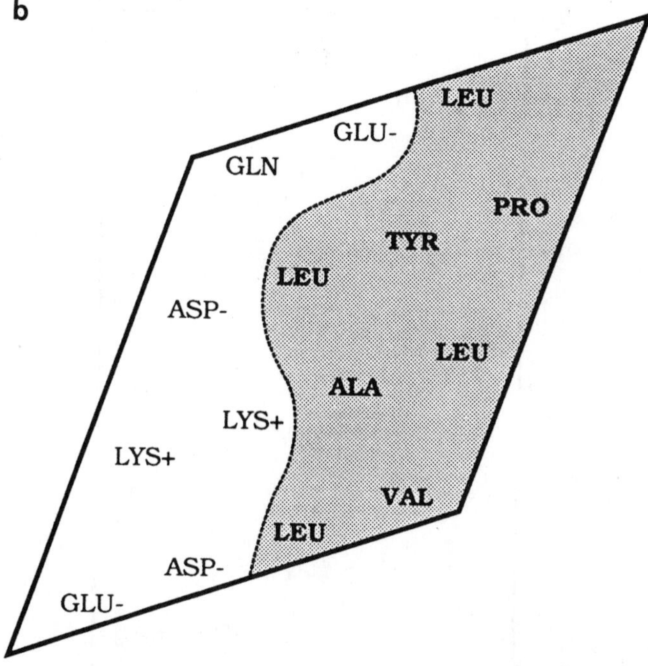

Figure 1 (cont.)

buried sides of amphipathic α helices. Many more crystal structures of proteins are now known, and as many as half of the helices in these proteins are amphipathic (Cornette *et al.*, 1987) (Fig. 1). The tendency for helices to be amphipathic in globular proteins has been exploited in the prediction of helical secondary structure from protein sequences (Lim, 1974; Garnier *et al.*, 1978).

Several methods have been published for predicting amphipathic structures from protein sequences. The earliest methods employed two-dimensional representations of helices. Two complementary approaches, the helical wheel (Schiffer and Edmundson, 1967) and the helical net (Dunnill, 1968), allowed assessment of overall amphipathicity and identification of hydrophobic patches on a helix by simple inspection. The overall amphipathicity of a helix has been quantitated in the hydrophobic moment, defined as the mean vector sum of hydrophobicities of the side chains of the helix (Eisenberg *et al.*, 1982a). This calculation has been generalized to quantitate the amphipathicity of any periodic structure and shown to be equivalent to the modulus of the Fourier transform of the hydrophobicities of the structure calculated at the relevant frequency, which would be $1/w$ where w is the number of residues between residues that face the same side and direction in the putative structure (Eisenberg *et al.*, 1984a).

Teleologically hydrophobicity seems important in, for example, correctly orienting residues and secondary structural elements on the surface of a protein. But it is not necessary that the two sides of a feature be most distinguished by difference in hydrophobicity. A surface that binds to DNA may be more distinguishable by its preponderance of positive charge, whereas sequences that lie in contact with lipids will be distinguished by features other than hydrophobicity that account for structural fit or lipophilicity.

We have used two-dimensional plots of Fourier transforms of hydrophobicities, which we refer to as amphipathic plots, to identify possible amphipathic structures (Finer-Moore and Stroud, 1984). These plots, which are a function of residue number in the sequence as well as

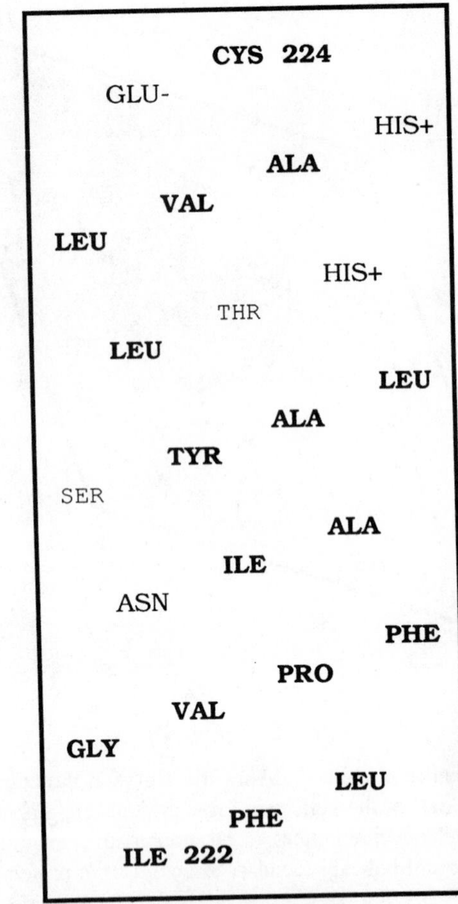

Figure 1 (cont.)

frequency, contain a wealth of information, not all of which is readily interpretable in terms of secondary structure prediction. The complexity of the plots is an advantage. As described in many of the following examples, the plots constitute a ''fingerprint'' of the sequence that can be used not only for identification of common amphipathic structures but also for recognition of domain structure of a protein and recognition of the structural and functional similarity of different proteins.

II. METHOD

A. Algorithm

The premise for our algorithm is that regular secondary structural features located at an interface between a milieu of one kind and one of another will show a variation in affinity for each kind of milieu that has a periodicity characteristic of the particular secondary structure.

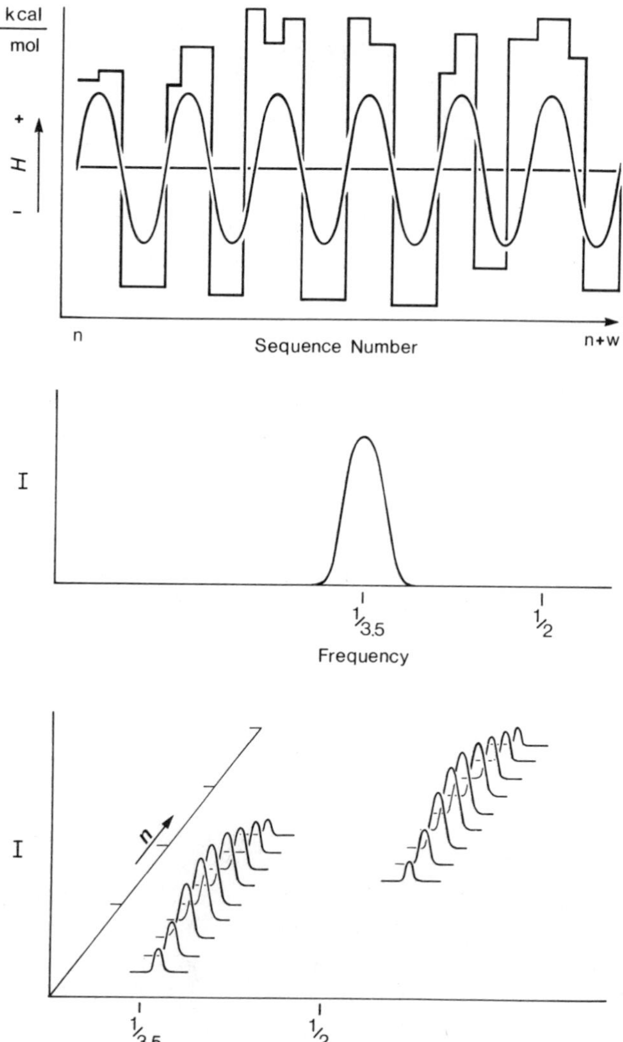

Figure 2. Illustration of the algorithm for detecting periodicity in the hydrophobicities of a sequence. In the top figure hydrophobicities for a stretch of 20 amino acids are plotted as a bar graph. The graph is overlaid by a sine wave that best matches the periodicity, p, in residues, and magnitudes of the hydrophobicities. The Fourier transform of the hydrophobicities will then have a peak at frequency $\nu = 1/p$ and a height equal to the amplitude of the sine wave (central figure). The 20-residue window is moved through the sequence in one-residue steps, and the Fourier transform is calculated at each step and referred to the central residue. The transforms are plotted sequentially along the residue axis (n) to produce a topographical map (lower figure); which we project onto a plane and contour.

As discussed by McLachlan and Karn (1983) and illustrated in Fig. 2, the Fourier transform of amino acid hydrophobicities as a function of sequence will then have peaks at the frequency given by the inverse of the period in residues. Simply, it defines the amplitude of a sine wave whose period best matches the hydrophobicities of a sequence of amino acid residues within a chosen window. We calculate the square of this amplitude of the Fourier transform (power spectrum) of the hydrophobicities within a certain length, or window, of sequence according to:

$$I(k,\nu) = \left| \sum_{j = k - n/2}^{k + n/2} [h_j - \bar{h}(k)]\exp(2\pi i \cdot j \cdot \nu) \right|^2 \tag{1}$$

where k is the residue number in the sequence, ν is the frequency, and h_j is the hydrophobicity of residue j. The average hydrophobicity, $\bar{h}(k)$, of the $w = n + 1$ residues in the window is subtracted, $[h_j - \bar{h}(k)]$ in the equation to remove the origin peak $I(0)$ from the spectra. $I(0)$ corresponds to the square of the average hydrophobicity, $\bar{h}(k)^2$, and contains no information relevant to periodic change.

The spectra may be scaled by a normalization factor, $1/I_m$, where I_m is the mean intensity in the power spectrum of a random arrangement of the amino acids in the window (McLachlan and Stewart, 1976). Normalization thus permits the predominant underlying frequency to be identified with equal signal, irrespective of the absolute difference in hydrophobicity on one side or the other of the secondary structure. If one wished to identify the most amphipathic structures (say, at a lipid interface), one would not use the normalizer. One would normalize to identify features at the surface of a protein, since periodicity rather than absolute range of property is the characteristic sought. The power spectra are plotted as a two-dimensional contour map with frequency ranging from 0.0 to 0.5 along the abscissa and residue number, k, along the ordinate. Frequencies less than $1/w$ have no meaning since they correspond to a period longer than the window.

B. Effect of Changing Hydrophobicity Scales

Amphiphilic α and β structures in proteins are identified by the Fourier transform of hydrophobicities of the primary sequence, represented as local patterns of characteristic periodicity. Clearly the choice of hydrophobicity scales for the amino acids in the protein chain could strongly affect this periodic analysis. There are a great number of numerically distinct yet qualitatively similar hydrophobic/hydrophilic scales for the 20 gene-coded amino acids. These scales reflect the different behavior of amino acid residues in nonisotropic environments as a result of variations in their structural and chemical properties and are generated by a variety of either semiempirical, or statistical techniques (for review see Rose *et al.*, 1985; Hopp, 1986; Appendix II of Cornette *et al.*, 1987). They are widely used in the generation of hydrophobicity profiles for protein sequences, which suggest, by the peaks and valleys of a moving-window average, regions of the molecule that are either hidden from or exposed to solvent. Much of the published work on amphiphilic analysis and the examples in this chapter utilize the Eisenberg scale of hydrophobicities, which is a consensus set derived by combining both experimental and statistical scales (Eisenberg *et al.*, 1982b).

Cornette *et al.*, (1987), in a comparative study of the use of hydrophobicity scales in detecting amphipathic structures in proteins, concluded that some scales and computational techniques are better at detecting amphiphilicity than others. The authors derive amphipathic scales using an eigenvector method that seeks to optimize the periodic signal strength in soluble proteins. Although the description of the mathematical techniques used to calculate these new scales is best left to the original paper by Cornette *et al.* (1987), it is interesting to compare the Fourier maps calculated using these optimized scales to those calculated with the established Eisenberg scale. Our experience in general has been that one of the new scales (termed PRIFT) is somewhat better at picking out α helices in α/α and α/β proteins but not measurably better at identifying β amphipathicity. As an example we can use the protein adenylate kinase (using the porcine sequence from Heil *et al.*, 1974; referred to here as

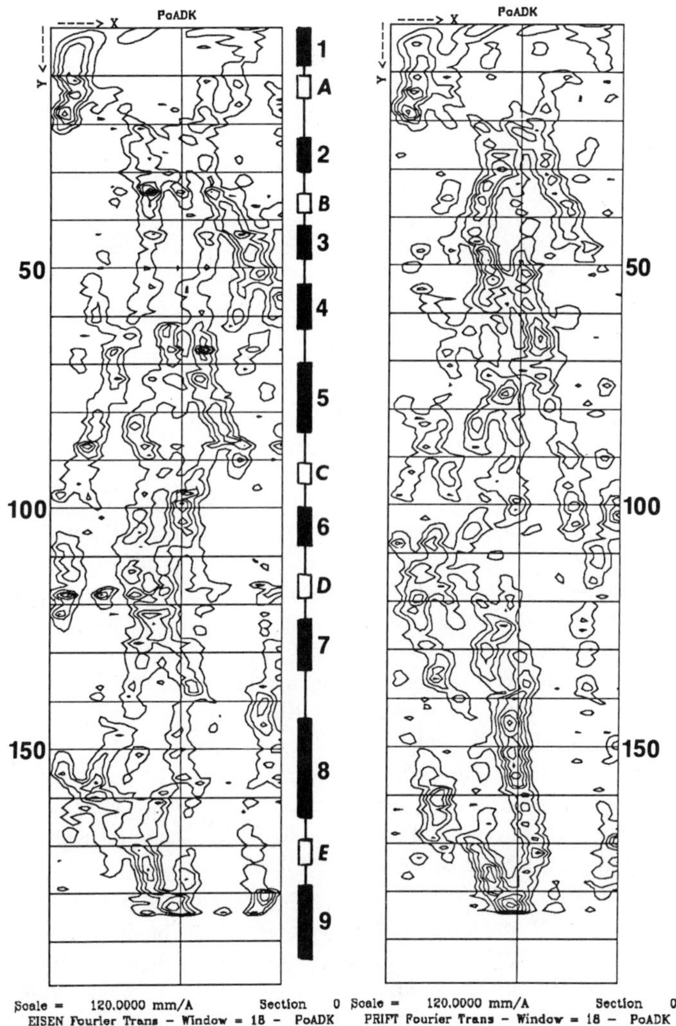

Figure 3. Amphipathic plots for porcine adenylate kinase. The plot on the left was calculated with the Eisenberg consensus hydrophobicity scale (Eisenberg *et al.*, 1982b). The plot on the right was calculated with a hydrophobicity scale designed to optimize the amphipathic signal in helices of globular proteins (Cornette *et al.*, 1987) and shows a stronger signal for the α helices at the C-terminal end of the protein. Positions of α helices (black rectangles) and β strands (white rectangles) in the crystal structures are marked along the residue axis.

PoADK), whose structure has been crystallographically determined by Schulz *et al.* (1974). It is an ATP-binding protein of α/β type and contains a typical nucleotide-binding fold (Branden, 1980) with several additional helical excursions. Figure 3 shows that although qualitatively the PRIFT and Eisenberg Fourier contour maps are very similar, the PRIFT map is better able to identify the series of α helices in the C-terminal half of the PoADK sequence that are seen in the x-ray structure.

Different scales or scales of different properties, such as charge, propensity, or the probability of residues lying in contact with the lipids in membrane proteins, are often more appropriate to particular interfaces. So far, most statistical scales derive from the existing data bank of globular soluble protein structures. Newer scales, such as those statistical scales

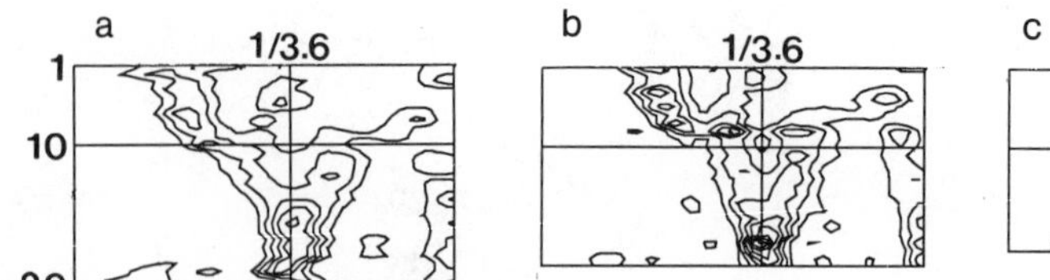

Figure 4. Amphipathic plots for avian pancreatic polypeptide calculated with window sizes of (a) 15 residues, (b) 20 residues, and (c) 25 residues. The α-helix frequency at 1/3.6 is marked by a vertical line.

utilizing structural information now arising from crystallized membrane proteins, may be better at interpreting transmembrane and amphiphilic regions of integral membrane proteins (Kuhn and Leigh, 1985; Rao and Argos, 1986).

C. Effect of Window Size

Amphipathic plots for a given sequence that are calculated with different window sizes are qualitatively similar. Prediction of amphipathic structures does not usually change with a small change in window size. However, adjusting the window size may make the plot easier to interpret. When window size is precisely tuned to the length of a periodic sequence, the peak in the Fourier transform is maximized, as illustrated in Fig. 4. In this example the Fourier transforms of avian pancreatic polypeptide are calculated with window sizes of 15, 20, and 25 residues. The protein contains a 26-amino-acid amphipathic helix at its C terminus, the first 20 amino acids of which have a regular periodicity of 3.6 residues (Blundell *et al.*, 1981). When the window size is 20 amino acids, the peak in the amphipathic plot is maximized, and its center is located in the center of the periodic sequence. This is because the best-fit periodic wave through the hydrophobicity-versus-sequence graph is most perfectly fitted over the longest windowed stretch when it exactly matches the feature length. Thus, varying window length can be used to determine feature length and location.

Figure 5 illustrates how a proper choice of window can help resolve peaks so that the positions and the lengths of amphipathic secondary structures can be more precisely determined. Concanavalin A contains several stretches of amphipathic extended structure that form a β barrel (Reeke *et al.*, 1975). The β strands are on average ten residues long. When a 25-amino-acid window is used, peaks for the individual strands are merged into long continuous features at frequencies between 1/2.0 and 1/3.0 (Fig. 5a). In addition, for some regions of sequence, strong intensities at more than one frequency are observed. When a window size of 11 is used, peaks for the individual β strands are resolved, and a short helix, marked next to the residue axis of Fig. 5b, is accurately located. Thus, the highest peaks and clearest peak separations are obtained when the window matches the length of the feature.

D. Quantitative Comparison of Similarity

In order to recognize similar features in folded proteins it is advantageous to have some quantitative evaluation of the similarity of their amphipathic plots. Such an evaluation requires precise overlap of similar regions of the plots. Determining regions of overlap is simplified when there is enough sequence homology between the proteins to identify positions of insertions and deletions in the sequences. In the case of proteins that are not evolutionarily related but whose structures are of the same class, two procedures for automatic overlap of amphipathic plots are possible. In the first, regions of high correlation are determined by multiplying together values of the two power spectra at equivalent sequence numbers and frequencies. In determining such correlations, the Fourier transforms are first "floated" in their average

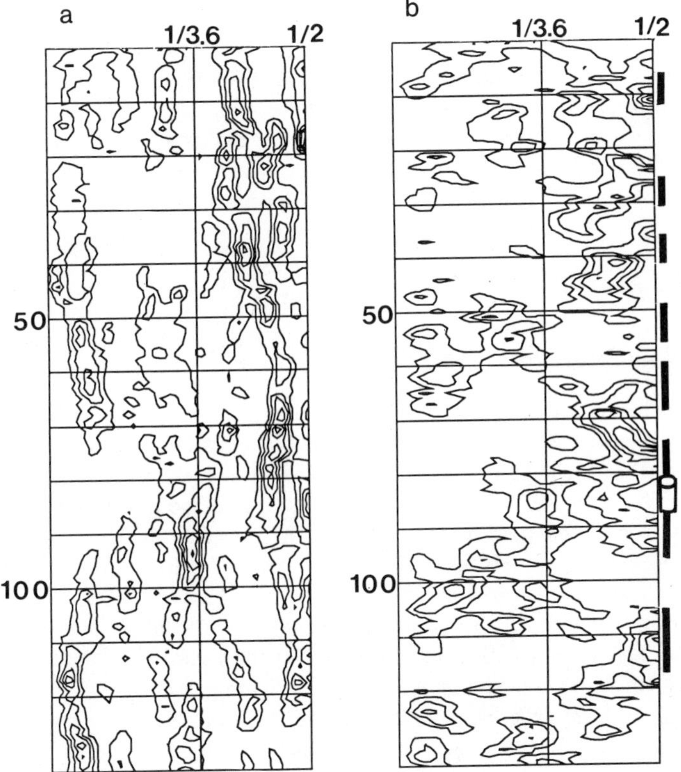

Figure 5. (a) Amphipathic plot for the first 130 residues of concanavalin A calculated with a 25-residue window. (b) Amphipathic plot for the same sequence calculated with an 11-residue window. Bars along the residue axis mark positions of β strands seen in the crystal structure, and a cylinder marks the position of an α helix.

value: the average value, $\bar{I}(k) = \sum_{\nu=0}^{\frac{1}{2}} I(\nu,k)/\Sigma\nu$, is subtracted from all individual $I(\nu,k)$ as indicated in equation 2, so that the average value in the plot is zero.

$$C(k) = \sum_{\nu=0}^{\frac{1}{2}} [I_l(\nu,k) - \bar{I}_l(k)]\,[I_2(\nu,k) - \bar{I}_2(k)] \tag{2}$$

A second method utilizes the phases from the initial Fourier transform calculation. The amphipathic plots display the power spectra $I(k,\nu) = F(k,\nu)^2$ derived from Fourier analysis. These spectra contain only positive numbers. However, the Fourier transform calculation has both amplitude $F(k,\nu)$ and phase $\phi(k,\nu)$ associated with its terms. These amplitudes and phases can be used to produce difference maps between vectors from two aligned sequences analogous to the difference maps used by crystallographers. The magnitudes $F(k,\nu)$ and phase angle $\phi(k,\nu)$, represented in polar coordinates for a region of one sequence are subtracted from those for the equivalent region of the second sequence and expressed as a residual R.

$$R = \frac{\sum_{\nu=0}^{\frac{1}{2}} |F_1(k,\nu)\, e^{2\pi i\phi_1(k,\nu)} - F_2(k,\nu)\, e^{2\pi i\phi_2(k,\nu)}|}{\sum_{\nu=0}^{\frac{1}{2}} |F_1(k,\nu)\, e^{2\pi i\phi_1(k,\nu)}| + |F_2(k,\nu)\, e^{2\pi i\phi_2(k,\nu)}|} \tag{3}$$

The phase encodes information related to the origin of a periodic wave through the hydrophobicity values. Thus, the minimum residual, R, as a function of trial alignments provides a means of high-resolution alignment of features in the sequence so that they will be in register for calculation of the difference maps: the alignment that produces the lowest peaks in the Fourier transform difference map will be the optimum alignment. The peak heights are a quantitative fingerprint of the difference in character of the folded proteins. More automatic procedures for recognition of similar tertiary structures by this means are possible.

III. EXAMPLES

A. Predicting Structures of Membrane Domains

Hydrophobicity analysis is a particularly useful method for interpreting membrane-associated protein sequences. The structures of two membrane proteins, bacteriorhodopsin (Henderson and Unwin, 1975; Agard and Stroud, 1982) and the reaction center complex from *R. viridis* (Deisenhofer *et al.*, 1985), are known. The membrane domains of both proteins are composed of hydrophobic transmembrane helices. The residues in these helices (Ovchinnikov *et al.*, 1979; Khorana *et al.*, 1979; Michel *et al.*, 1986), when in similar sequences in globular proteins, usually form extended structures rather than helices. Presumably the helix is a favorable conformation in a hydrophobic environment because all possible hydrogen bonds in the peptide backbone are satisfied, at least during synthesis and insertion if not in the folded protein, and there is no competition with water for hydrogen bonding.

Sequences of many other membrane proteins have similar characteristics to the sequences of these known protein structures, suggesting that they too may be composed of parallel hydrophobic transmembrane helices. The simplicity of this structural motif has made it possible to predict membrane-spanning regions by locating a range of residues with greater than average hydrophobicity that is long enough to span the lipid bilayer as an α helix (Engelman *et al.*, 1982; Kyte and Doolittle, 1982; Argos *et al.*, 1982). However, overall hydrophobicity alone does not always provide an unambiguous means of identifying membrane-spanning regions in sequences. In membrane proteins with several membrane crossings, such as bacteriorhodopsin, overall hydrophobicity of membrane-spanning helices is often only moderate. The helices may contain several charged groups, which tend to cluster on one side of the helix (Engelman *et al.*, 1980; Katre and Stroud, 1981). Amphipathicity is therefore an additional criterion that is useful for identifying membrane-spanning regions (Eisenberg *et al.*, 1984b). The distribution of hydrophilic groups on membrane-spanning helices is generally so sparse that the amphipathicity of the helix is not detected on amphipathic plots. However, for the special case of transmembranous channel-forming structures, it is possible that some membrane-spanning regions may show exaggerated amphipathicity and be readily detected by the plots, although in no sense should it be presumed without experimental test that such amphipathic regions do cross the membrane.

We discuss two examples of membrane protein structure prediction. HMG-CoA reductase has a membrane domain that may be similar in structure to bacteriorhodopsin. Several criteria, including amphipathicity, were used to predict membrane-spanning regions in this protein. The acetylcholine receptor contains an ion channel that spans the membrane. Possible channel-forming structures were identified in the receptor sequences using amphipathic analysis.

1. HMG-CoA Reductase

3-Hydroxy-3-methylglutaryl coenzyme A reductase (HMG-CoA reductase), the rate-controlling enzyme in cholesterol biosynthesis (Brown and Goldstein, 1980), is a 97-kDa

(Chin *et al.*, 1984) transmembrane glycoprotein (Liscum *et al.*, 1983) in the endoplasmic reticulum of animal cells. It has a large soluble cytoplasmic domain containing the enzyme active site (Liscum *et al.*, 1983; Faust *et al.*, 1982). The domain structure of the protein, which was established by proteolysis combined with antibody binding to specific sequences (Liscum *et al.*, 1985), can be roughly seen in the hydrophobicity plot (Kyte and Doolittle, 1982) (Fig. 6a). The N-terminal membrane-bound portion of the protein is generally more hydrophobic than the C-terminal, soluble region, but the soluble region contains large regions of hydrophobic residues that could be mistaken for membrane-spanning regions if hydrophobicity alone were used as the criterion. These regions are predicted to form a hydrophobic core of the soluble domain composed primarily of β structure (Liscum *et al.*, 1985) and have characteristics that distinguish them from what we believe are the membrane-spanning regions of the protein.

Several discrete peaks in the N-terminal portion of the hydrophobicity plot suggest that the membrane-bound portion of HMG-CoA reductase contains several membrane-spanning helices. The exact number and location of the membrane-spanning sequences cannot be unambiguously predicted from the hydrophobicity plot alone. Some of the peaks are less positive, in fact, than peaks in the soluble domain of the protein. The assignment of seven membrane-spanning regions as indicated in Fig. 6a was based on several additional criteria including experimental evidence for the protein topography, amino acid composition, and amphipathicity and was corroborated by the location of introns at the ends of the regions in the gene sequence (Liscum *et al.*, 1985).

The enzymatically active C-terminal portion of the enzyme is released from ER membrane vesicles by proteolytic cleavage in the region of residues 368–379 (Liscum *et al.*, 1985). The protein beyond residue 379 must therefore lie on the cytoplasmic side of the ER membrane. The protein is known to possess N-linked carbohydrate (Liscum *et al.*, 1983), and the sequence of the membrane-bound region contains only one N-glycosylation consensus sequence, at asparagine 281. Asparagine 281 therefore must lie on the lumen of the endoplasmic reticulum, and the sequence between residues 281 and 379 must cross the membrane at least one time. A single peak in the hydrophobicity plot between residues 281 and 379, labeled G, probably corresponds to the required membrane-spanning region, even though the sequence has at least three charged groups and only moderate overall hydrophobicity.

Like segment G, predicted transmembrane segment D has fairly low overall hydrophobicity and contains several charges. However, it has other properties typical of membrane-spanning helices. For example, it is similar in hydrophobicity, amphipathicity, and in the types of its amino acids (McLachlan, 1971) to the seventh transmembrane helix in bacteriorhodopsin (Fig. 7). Its similarity to typical membrane helices is shown most clearly in the membrane propensity plot (Fig. 6b). This plot is analogous to a hydrophobicity plot but uses a membrane propensity scale (Kuhn and Leigh, 1985) based on the frequency of occurrence of amino acids in transmembrane segments instead of a hydrophobicity scale. Segments D and G, the least hydrophobic segments, have prominent peaks in this plot. The membrane propensity plot more clearly distinguishes the membrane domain: only one peak in the soluble region of the protein, that at the C terminus, has a membrane propensity as high as any of the predicted membrane-spanning regions. The predicted β regions in the soluble domain are very hydrophobic but are rich in the common turn-forming residues glycine and proline and also have many cysteines. Proline and cysteine, although not hydrophilic, have low membrane propensities.

Several of the other predicted membrane-spanning regions contain two or more charged residues. The length of the hydrophobic stretches, 22–30 residues, suggests that they form parallel α helices. The helix net representation of these sequences shows that in their predicted α-helical structures, positive and negative charges would cluster on one side of the helix. As in bacteriorhodopsin (Engelman and Zaccai, 1980), the putative membrane-spanning helices

a

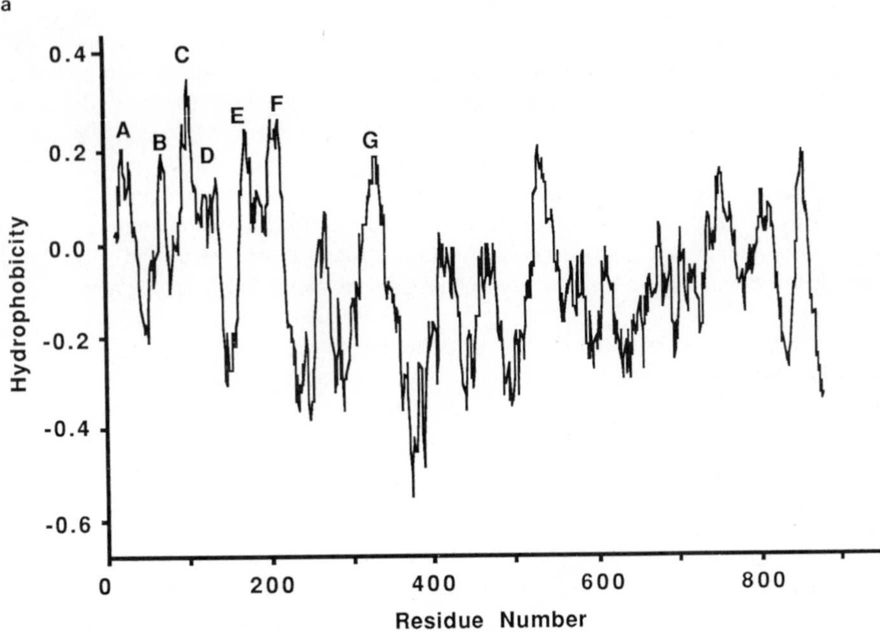

b

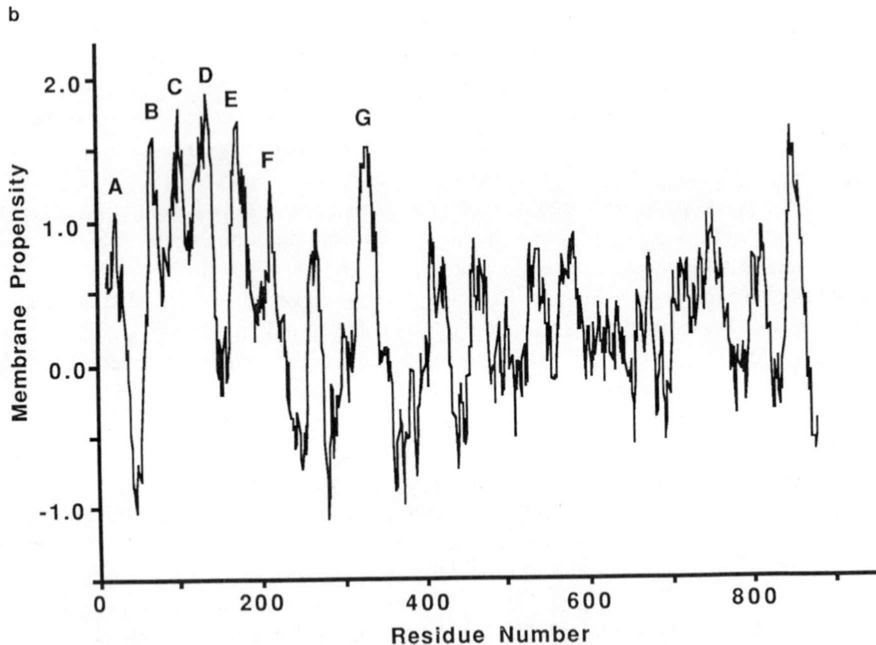

Figure 6. (a) Hydrophobicity plot (Kyte and Doolittle, 1982) for hamster HMG-CoA reductase calculated with a moving window of 21 residues using the consensus hydrophobicities of Eisenberg *et al.* (1982b). Peaks for the seven predicted membrane-spanning regions are labeled A–G. (b) Membrane propensity plot calculated as in (a) but using the membrane propensity scale of Kuhn and Leigh (1985) instead of hydrophobicity. Predicted membrane-spanning regions are labeled as in a.

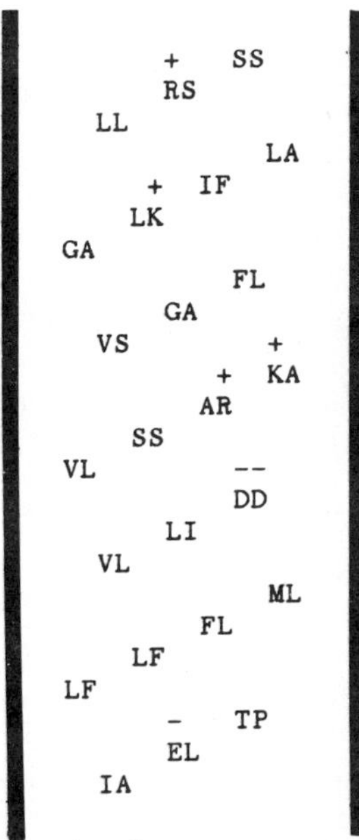

Figure 7. Helix net diagram of the seventh transmembrane segment of bacteriorhodopsin (Ovchinnikov *et al.*, 1979; Khorana *et al.*, 1979) aligned to the fourth predicted membrane-spanning region of hamster HMG-CoA reductase (Liscum *et al.*, 1985). The amino acids of the two segments are displayed in one-letter code with aligned residues adjacent to one another. Charges are marked above the residues. Both putative helices contain charged residues but also a completely hydrophobic face (on the left of the diagram). Differences between most of the aligned residues in the center of the helices are conservative.

could pack so as to isolate the charged groups on the interior of the membrane domain. The α-helical periodicity of the charged groups in these segments is therefore consistent with their assignment to a membrane domain.

The exon–intron structure of the HMG-CoA reductase gene from hamster has been determined (Reynolds *et al.*, 1984). Of eight introns in the membrane domain, seven lie at the boundaries of regions coding for the predicted membrane-spanning regions. All but the first predicted spanning region is bounded by at least one intron. Although this distribution of introns is not direct evidence that the membrane-spanning regions are correctly predicted, it is consistent with the pattern seen in other membrane proteins such as rhodopsin (Nathans and Hogness, 1983) and the acetylcholine receptor (reviewed in Stroud and Finer-Moore, 1985) with respect to their proposed transmembrane segments.

2. Ion Channel-Forming Proteins: AChR and the Colicins

The acetylcholine receptor is a cell surface receptor found at the neuromuscular junction of all vertebrates as well as in the electrocytes of electric fish (for recent reviews see Popot and Changeux, 1984; Stroud and Finer-Moore, 1985; McCarthy *et al.*, 1986). This heteropentameric complex of four similar subunits signals between nerve and muscle cells by means of an ion channel. The structure of acetylcholine receptor from *Torpedo californica* and *T. marmorata* has been much studied, and the amino acid sequences of all its subunits are now known (Sumikawa *et al.*, 1982; Noda *et al.*, 1982, 1983a,b; Claudio *et al.*, 1983; Devillers-Thiery *et al.*, 1983). The sequences of the four subunit types have two striking characteristics:

profound homology (Noda *et al.*, 1983b), indicating that the subunits have similar tertiary structures, and the presence in each of the subunits of four highly conserved very hydrophobic stretches, 19 to 27 amino acids long (Claudio *et al.*, 1983; Noda *et al.*, 1983b; Devillers-Thiery *et al.*, 1983). The hydrophobic stretches are generally more hydrophobic than those seen in bacteriorhodopsin (Ovchinnikov *et al.*, 1979; Khorana *et al.*, 1979) and, as in the transmembrane helices of the reaction center from *R. viridis* (Michel *et al.*, 1986), contain long regions uninterrupted by charged residues. These hydrophobic stretches are likely to form the helical (Ross *et al.*, 1977) membrane domain of the protein (Claudio *et al.*, 1983; Noda *et al.*, 1983b; Devillers-Thiery *et al.*, 1983; Finer-Moore and Stroud, 1984), although assignment of all four stretches to the membrane domain is still controversial (Ratnam *et al.*, 1986).

The extreme hydrophobicity of the four putative transmembrane helices posed a dilemma. The membrane domain is known to form the cation-conducting channel of the protein. This channel traverses the membrane (Kistler *et al.*, 1982) and provides a water-filled environment for the conducted ions (Lewis and Stevens, 1983). It seemed unlikely that hydrophobic residues within the lipid milieu would form such a water-containing volume. Amphipathic plots of the four sequences were used to search for amphipathic structures that could interface both with a waterlike environment (channel) and also with the hydrophobic exterior of the membrane domain (Finer-Moore and Stroud, 1984). Arithmetically averaging the aligned power spectra from the four subunits seemed a valid method for increasing signal to noise, given the likely structural homology of the four subunits. The averaged plot had two prominent features: a long peak with a frequency between 1/3.0 and 1/2.0 in a region of the sequence known to lie outside the membrane and a peak at frequency 1/3.6 near the C terminus corresponding to an amphipathic helix in each subunit (Fig. 8). Each subunit, it was proposed, contributed the hydrophilic face of its amphipathic helix to the wall of the ion channel (Finer-Moore and Stroud, 1984). A close-packed arrangement of the five parallel amphipathic helices, perhaps interspersed with others, would form a channel of approximately correct dimensions (Kistler *et al.*, 1982).

The proposed model (Finer-Moore and Stroud, 1984) for the ion channel in the acetylcholine receptor probably cannot be definitively evaluated without a higher-resolution receptor structure. There is inconsistency, for example, in the experimental evidence on the topology of the C-terminal residues. These residues are generally found to be cytoplasmic (Lindstrom *et al.*, 1984; Young *et al.*, 1985), although through indirect experiments, they are sometimes thought to be extracellular (McCrea *et al.*, 1986; Dunn *et al.*, 1986). To this date (August 1987) it remains an interesting and completely viable model. Sequences with amphipathic helical periodicity have been observed in the same relative positions of a wide variety of AChR sequences (Stroud and Finer-Moore, 1985). These sequences are poorly conserved, however, indicating that if they have a functional role in ion conductance, it is a very simple one, namely, to provide a water-loving environment for the channel. If they are not involved in conductance they must be involved in an unusual protein fold on the cytoplasmic surface that solubilizes five of the most hydrophobic sequences of 20 or more residues outside of the membrane. This would also represent an unusual and new scheme for protein folding.

Another group of channel-forming proteins, the colicins, have strong periodic features corresponding to potentially amphipathic helices in their channel-forming domains (S. Choe and R. M. Stroud, unpublished results) (Fig. 8). Signals for amphipathic helices have been seen in the amphipathic plots for colicins Ia, Ib, E1, and A. Circular dichroism measurements have verified that, at least in its soluble form, colicin Ia is largely α helical (S. Mel and R. M. Stroud, unpublished data). Each amphipathic sequence is flanked by a hydrophobic stretch of about 25 amino acids that is located 10 to 20 amino acids from the C terminus of the protein (S. Choe and R. M. Stroud, unpublished data). In spite of the apparent similarities in the structures of the channel-forming domains of these colicins, their sequence homology is fairly low: 40% between channel domains of Ia and Ib, and less between other colicins (Choe, 1987).

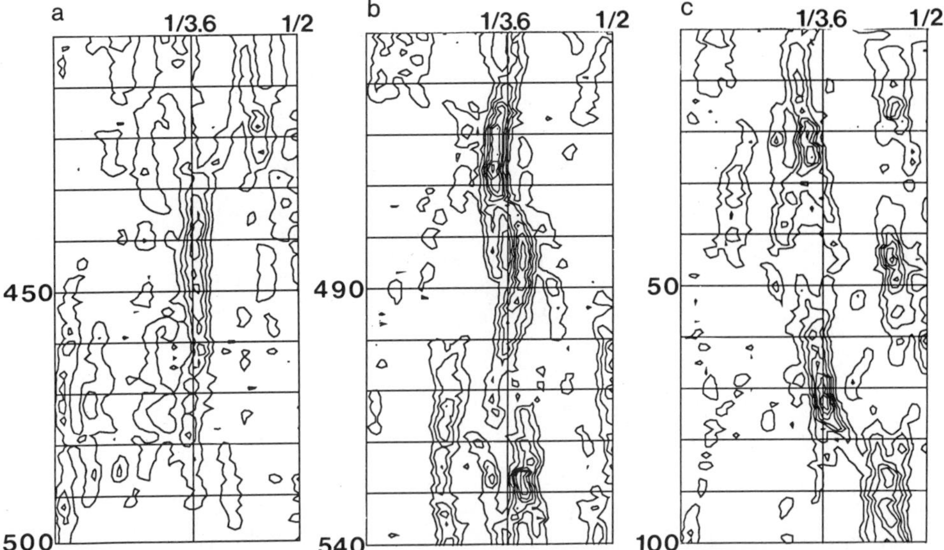

Figure 8. (a) Averaged amphipathic plot for the four aligned subunits of the acetylcholine receptor from *T. californica* in the region of the predicted channel-forming helix. The calculation was done with a 25-residue window. A 35-residue-long peak at the α-helix frequency, marked by a vertical line, identifies the predicted helices. Averaging the four subunits' plots increased the signal-to-noise ratio in the map but also reduced the signal of the helix relative to the signals in the individual maps. (b) Amphipathic plot for the first 100 residues of the channel-forming domain of colicin Ib calculated with a 25-residue window. (c) Amphipathic plot of the first 100 amino acids of hemoglobin, calculated with a 25-residue window. This plot serves as a control for a and b. Peaks centered at residues 25 and 70 correspond to amphipathic helices A/B and amphipathic helix E, respectively, of hemoglobin (Perutz *et al.*, 1965).

On the other hand, sequences for two other channel-forming proteins that conduct chloride anions rather than cations, the strychnine-binding subunit of the glycine receptor (Grenningloh *et al.*, 1987) and the GABA receptor (Schofield *et al.*, 1987), show homology to AChR sequences, particularly in the hydrophobic crossings, but neither contains an amphipathic sequence analogous to that seen in AChR sequences. Therefore, amphipathic secondary structure is apparently not a general feature of ion channel-forming proteins.

B. Amphipathic Structures in Small Peptides, Hormones, and Signaling Sequences

The biological activity of a hormone or other effector depends on correct spatial orientation of its amino acids. An amphipathic environment may induce correct conformation in the case of small proteins (Kaiser and Kezdy, 1983). Therefore, it is not surprising that the biologically active regions of small effector proteins and peptides are frequently amphipathic. One example of an effector that may have amphipathic structure is prolactin release-inhibiting factor found within the precursor for human gonadotropin-releasing hormones (Adelman *et al.*, 1986). This protein of 56 amino acids affects prolactin release *in vitro* (Nikolics *et al.*, 1985). Its amphipathic plot has several strong signals close to the α-helix frequency of 1/3.6 that are predicted to correspond to two intersecting helices connected by a flexible loop (Fig. 9b). The packing of apolar stripes on the helix surfaces, which determines the crossing angle of the helices, can be deduced from changes in the frequency of the signal versus sequence number and by examination of helical nets for regions identified using the plots (Fig. 9a). The

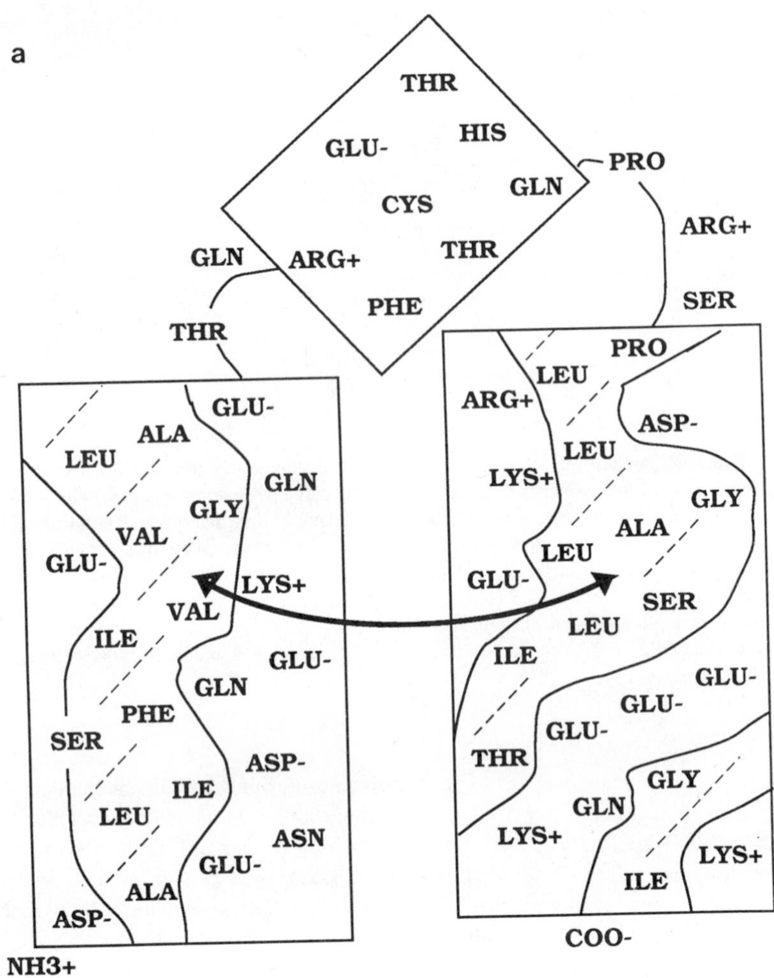

Figure 9. (a) Helix net diagrams of two predicted amphipathic helices in a prolactin release-inhibiting factor (Adelman *et al.*, 1986). The hydrophobic surfaces of each helix are marked off and shaded by dashed lines. The predicted helices are connected by a flexible loop and would pack together as in b with a crossing angle that best buries their hydrophobic surfaces. This crossing angle is deduced from the angles of the hydrophobic stripes on the two cylindrical helical surfaces. A detailed molecular model of the structure was built, shown in c. This model was useful in designing further tests using site-directed mutagenesis of the putative interface regions. These in turn give further evidence for the tertiary fold and hence for those regions that provide the functional receptor binding site on the hormone.

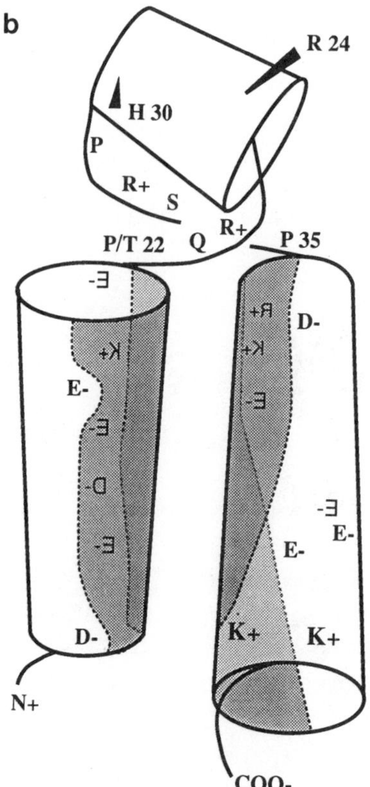

Figure 9 (cont.)

c

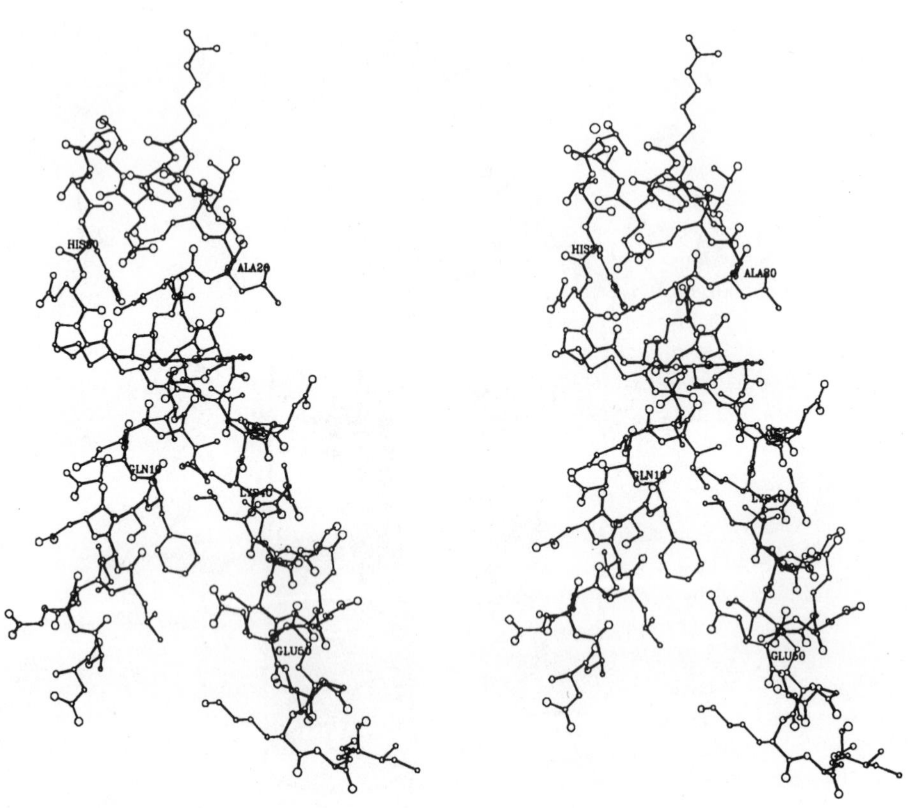

Figure 9 (cont.)

predicted structure suggests that an immune response to the hormone might best be generated to the loop between the two helices. It also suggests which surface residues could be presented to the hormone receptor protein.

Amphipathic helical regions are also found in the receptor binding domains at the C-terminal regions of the α chains of the G proteins (Masters *et al.*, 1986). The hydrophilic surfaces of these putative helices have charge distributions that may be responsible for the specificity of the α chains for their receptors (Fig. 10). The α chains with similar charge distributions cross react with each other's receptors.

Other evidence indicating a signal detector function for these residues is the high sequence homology between the C termini of the G-protein α chains and an internal region of arrestin, a retinal protein (Wistow *et al.*, 1986) that competes with transducin for binding to photorhodopsin and prevents light-dependent activation of α_{t1} (Fig. 10). Thus, the complementary pattern of a hydrophobic or hydrophilic stripe on the surface of a helix may be an important determinant of function in signaling peptides.

Another example where amphipathic helices seem to be effectors is in the mitochondrial targeting sequences. Douglas and co-workers theorize that such proteins are correctly transported into mitochondria using the surfaces of amphipathic helices rather than linear sequences as signals (Vassarotti *et al.*, 1987). The precursor of one mitochondrial protein, the F_1-ATPase β subunit, contains a 19-residue transient amino-terminal sequence that is required for import of the F_1-ATPase β subunit into the mitochondria (Emr *et al.*, 1986). Fourier analysis of the N-terminal 100 amino acids of the precursor indicated that several regions of the sequence, including the presequence, could form amphipathic helices (Vassarotti *et al.*, 1987) (Fig. 11a). Residues 4–37 of the precursor were predicted to form a basic amphipathic helix on interaction with a membrane (Fig. 11b). This helix was postulated to float in the membrane, as melitten does, and perhaps disrupt the membrane. Deletion of residues 4–37 completely eliminated transport of the F-ATPase β subunit to the mitochondria (Vassarotti *et al.*, 1987). Transport was partially restored by mutations in the first predicted amphipathic helix of the deletion construct (Fig. 11c) that would make this putative helix more basic while preserving its amphipathicity (Vassarotti *et al.*, 1987). These mutations replaced acidic amino acids with either hydrophobic or basic amino acids. Thus, the three essential properties of the mitochondrial targeting sequences seem to be basicity, accessibility, and the tendency to form amphipathic helices. Many of the other signal sequences that have membrane-active functions may likewise encode their target specificity on one surface of a helix.

C. Identification of Domains in Proteins

The amphipathic plot of a particular sequence represents an encoded image of the pattern of polarity or amphipathic propensity within a protein or a domain of a protein. Thus, such a plot can be considered a fingerprint that characterizes the structure of that domain. In certain cases, the amphipathic plot of a protein sequence may suggest the domain structure of the protein. Some amphipathic plots reveal regions with a high degree of internal secondary structure separated by regions with little or no secondary structure. These plots suggest boundaries between domains. In the case of membrane proteins, the amphipathic plot in the region of the membrane domain will have a distinctly different character than in the region of the soluble domain. Symmetry in a plot may also indicate domain structure. A domain can sometimes also be identified from the similarities of a portion of the plot to the plot of a structurally homologous domain in another protein. Three examples of the analysis of domain structure in proteins are discussed.

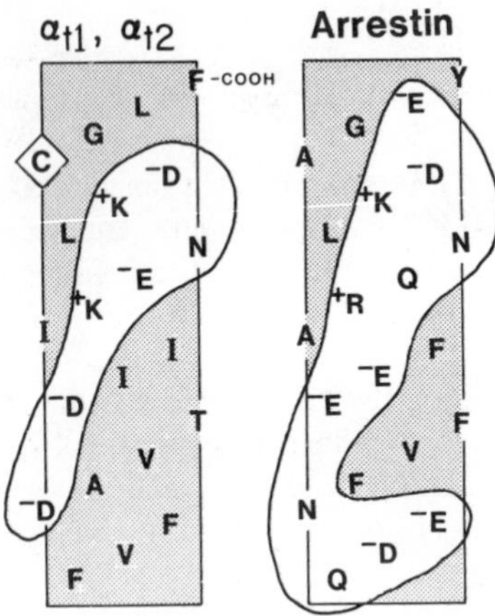

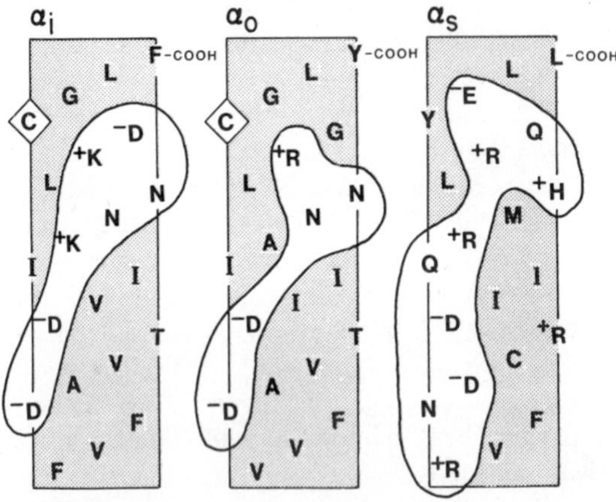

Figure 10. Helix nets for the carboxy termini of G protein α chains and an internal sequence (Wistow *et al.*, 1986) of arrestin. Shading indicates hydrophobic portions of the surfaces of the presumed α helices. The pattern of hydrophilic residues on the helices correlates with the specificities of the α chains for receptor proteins. Thus, GTP hydrolysis by both α_{t1} and α_i is effectively stimulated by photorhodopsin, whereas α_s responds poorly to this receptor. Arrestin competes with α_{t1} for binding to photorhodopsin.

1. HMG-CoA Reductase

As discussed in Section III.A.1, HMG-CoA reductase is a membrane protein with an established domain structure that is evident in the hydrophobicity plot of the protein (Fig. 6). The unnormalized Fourier transform of the hydrophobicities of the HMG-CoA reductase sequence (Fig. 12) shows the domain structure of the protein in greater detail than the hydrophobicity plot. The transmembrane region shows little structure, with only one feature of more than three contour levels, indicating a general uniformity in hydrophobicity over 20-residue stretches of amino acids. The membrane-bound domain terminates just before a region of about 100 amino acids whose most intense features have frequencies less than 1/4.0. This region was therefore predicted to correspond to a hydrophilic linker between domains that has little definite secondary structure. Subsequently, it was found that this linking region is poorly conserved, even between closely related species of HMG-CoA reductase (Luskey and Stevens, 1985). Thus, prediction of interdomain regions was consistent with genetic evidence for conservation within predicted domains and variation in the interdomain regions.

The Fourier transforms of the hydrophobic sequences in the soluble domain are nearly featureless and show a remarkable symmetry. Each hydrophobic sequence is 90–100 amino

a

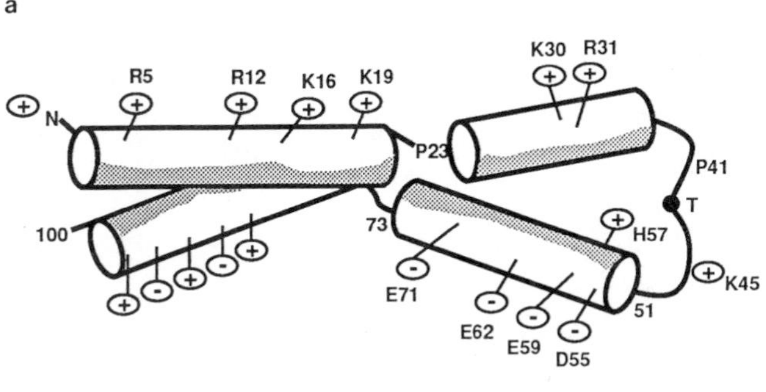

b

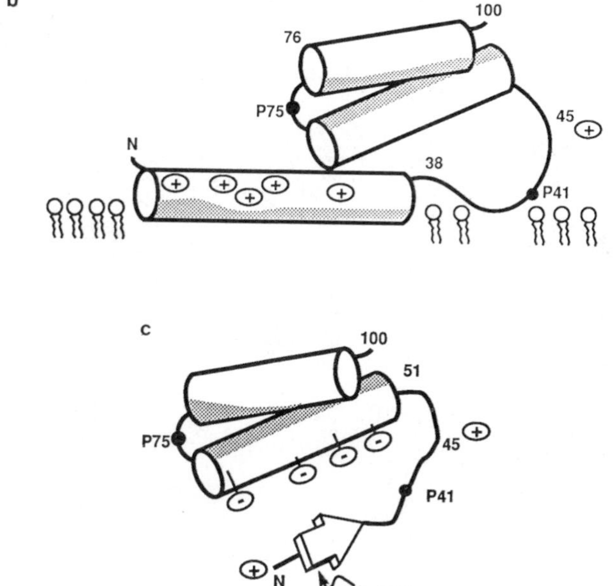

Figure 11. Sketches of the predicted structures of the first 100 residues of the F_1-ATPase β subunit precursor (a) in the cytoplasm and (b) in a lipid environment. In (a) four amphipathic helices, including the presequence, residues 1–19, pack so as to bury their hydrophobic surfaces (the shaded areas on the cylinders). In b, residues 4–37 form a single basic amphipathic helix, which associates with the membrane. When residues 4–37 are removed, as in (c) the N-terminal amphipathic helix is acidic. This helix acts as a targeting sequence when its negatively charged amino acids are mutated to neutral or basic amino acids.

acids long and flanked by strong features of frequency 1/3.0 to 1/4.0, which probably correspond to α helices (Liscum *et al.*, 1985). The hydrophobic sequences together with their adjacent helices may comprise domains with similar structure but with no sequence homology within the soluble portion of the enzyme. These two possible domains are highly conserved among all known HMG-CoA reductase sequences (Chin *et al.*, 1984; Luskey and Stevens, 1985; Skalnik and Simoni, 1985; D. J. Chin, personal communication; Basson *et al.*, 1988) and encode the active center of the enzyme.

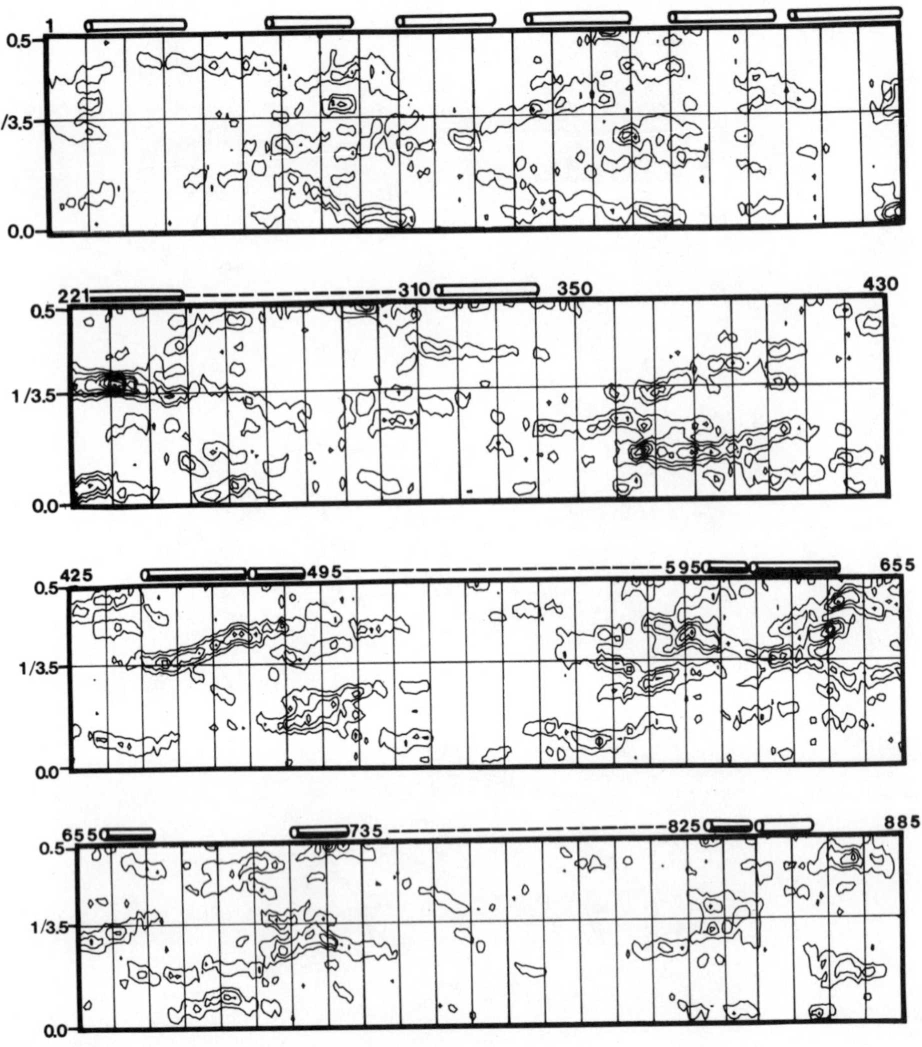

Figure 12. Unnormalized amphipathic plot for hamster HMG-CoA reductase calculated with a moving window of 20 residues. The residue axis is horizontal. Predicted β structure is denoted by dashed lines above this axis, and open and shaded cylinders mark positions of predicted hydrophobic and amphipathic helices, respectively. The membrane domain of the protein, residues 1–339, is connected to the soluble, enzymatically active domain by a linking region (residues 339 to 445) with little predicted, regular secondary structure. The soluble portion of the enzyme may itself be divided into two domains of similar structure but no sequence homology, residues 450 to 650 and residues 650 to 845.

Figure 13. (a) Schematic illustration of internal homologies within PDI. Numbers above boxed regions refer to residue numbers. The sequences within A and A′ are also homologous to thioredoxin and listed below. Below A and A′, the sequence homology within B and B′ is indicated. (b) Amphipathic Fourier transforms between frequencies 1/6.0 and 1/2.0 per residue for the sequence of mature PDI. Frequency is plotted on the vertical axis. Residue number is plotted on the horizontal axis. A frequency of 1/2.0 corresponds to the amphipathic β structure, and a frequency of 1/3.6 corresponds to the amphipathic α-helical structure. Above the maps is a schematic of the consensus secondary structure. Open cylinders, α helices; shaded cylinders, amphipathic α helices; dashed lines, β structures; T, reverse turn.

a

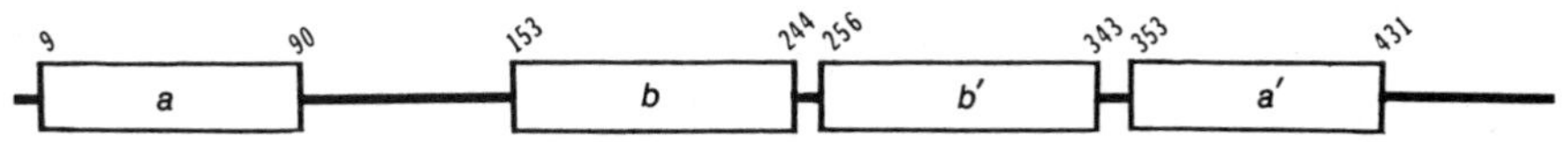
9
90
153
244
256
343
353
431
a
b
b′
a′

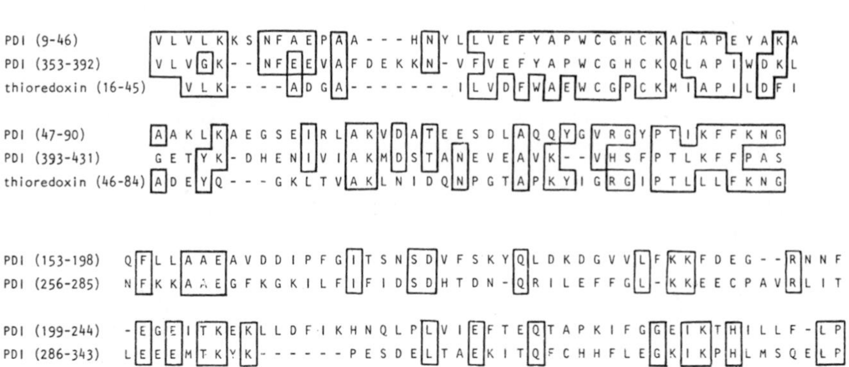
PDI (9-46) V L V L K K S N F A E P A A - - - H N Y L L V E F Y A P W C G H C K A L A P E Y A K A
PDI (353-392) V L V G K - - N F E E V A F D E K K N - V F V E F Y A P W C G H C K Q L A P I W D K L
thioredoxin (16-45) V L K - - - - A D G A - - - - - - - I L V D F W A E W C G P C K M I A P I L D F I
PDI (47-90) A A K L K A E G S E I R L A K V D A T E E S D L A Q Q Y G V R G Y P T I K F F K N G
PDI (393-431) G E T Y K - D H E N I V I A K M D S T A N E V E A V K - - V H S F P T L K F F P A S
thioredoxin (46-84) A D E Y Q - - - G K L T V A K L N I D Q N P G T A P K Y I G R G I P T L L L F K N G
PDI (153-198) Q F L L A A E A V D D I P F G I T S N S D V F S K Y Q L D K D G V V L F K K F D E G - - R N N F
PDI (256-285) N F K K A A E G F K G K I L F I F I D S D H T D N - Q R I L E F F G L - K K E E C P A V R L I T
PDI (199-244) - E G E I T K E K L L D F I K H N Q L P L V I E F T E Q T A P K I F G G E I K T H I L L F - L P
PDI (286-343) L E E E M T K Y K - - - - - - P E S D E L T A E K I T Q F C H H F L E G K I K P H L M S Q E L P

b

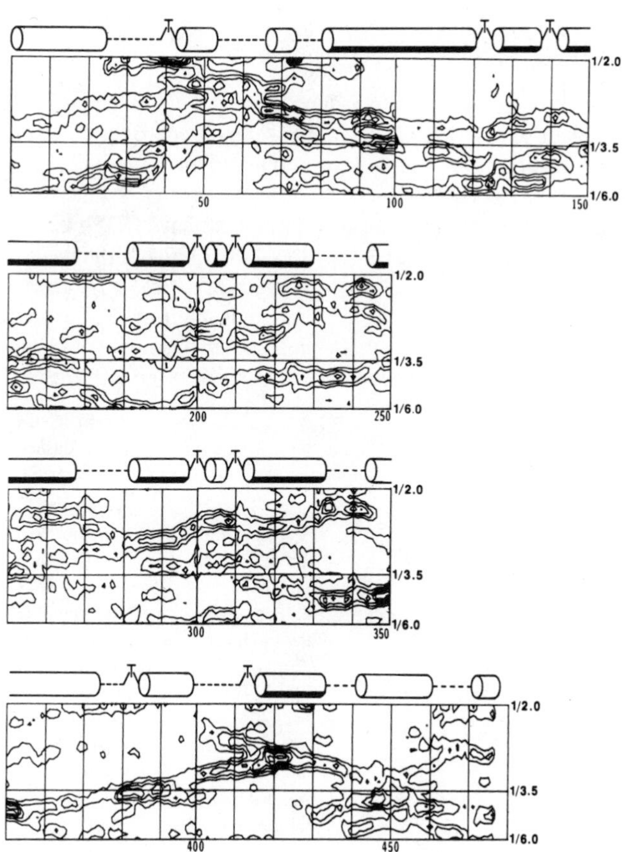
1/2.0
1/3.5
1/6.0
50
100
150
200
250
300
350
400
450

2. Protein Disulfide Isomerase

Protein disulfide isomerase (PDI) is located in the lumen of the endoplasmic reticulum and catalyzes isomerization of both intramolecular (Freedman *et al.*, 1984; Creighton *et al.*, 1980) and intermolecular (Roth and Koshland, 1981) disulfide bonds in proteins. The sequence consists of 508 amino acids, of which the first 19 correspond to a signal peptide (Edman *et al.*, 1985) (Fig. 13a). Region A (residues 9–90) and region A′ (residues 353–431) have 47% identical amino acids, and each shows considerable sequence homology to thioredoxin from *E. coli* (Holmgren *et al.*, 1975). Regions B (residues 153–254) and B′ (residues 256–343) (which lie between A and A′) are only 28% identical. Despite the considerable divergence in primary structure between B and B′, the "fingerprint" of the B and B′ domains suggests an identical arrangement of α and β amphipathic regions within the folded domains (Fig. 13b).

Both regions A and A′ and thioredoxin contain a highly conserved sequence Trp-Cys-Gly-His-Cys-Lys, which may be of functional importance. The two cysteines in thioredoxin function as an oxidation–reduction couple for insulin disulfides, which suggests a similar active site role for these sites in the A and A′ domains of PDI. A second highly conserved sequence downstream from this region has ten amino acids identical, four functionally conserved, and two different out of the subsequent 16 residues. Amphipathic analysis suggested that these highly conserved and possibly functionally important regions lie between a short stretch of β and a long α-helical structure (Edman *et al.*, 1985).

3. G-Protein α Chains

The intracellular G proteins that transduce hormonal and other signals into regulation of enzymes such as adenylate cyclase (Gilman, 1984; Stryer and Bourne, 1986) or retinal cGMP phosphodiesterase (Stryer, 1986) provide examples of proteins with domains that are not formed by contiguous sequence. The α chains of the G proteins and the GDP-binding domains of the 21-kDa products of the *ras* oncogenes (p21*ras*) all have GDP-binding domains that are homologous to that of elongation factor Tu (EF-Tu) (Halliday, 1984; Sullivan *et al.*, 1986; Bourne, 1986; Hurley *et al.*, 1984). The crystal structure of EF-Tu is known (Jurnak, 1985; La Cour *et al.*, 1985), but in the G proteins this domain is interrupted or extended relative to EF-Tu at three different sites by domains that encode other functions. Components of the GDP-binding domain were located in the sequences of the α chains by sequence homology to the so-

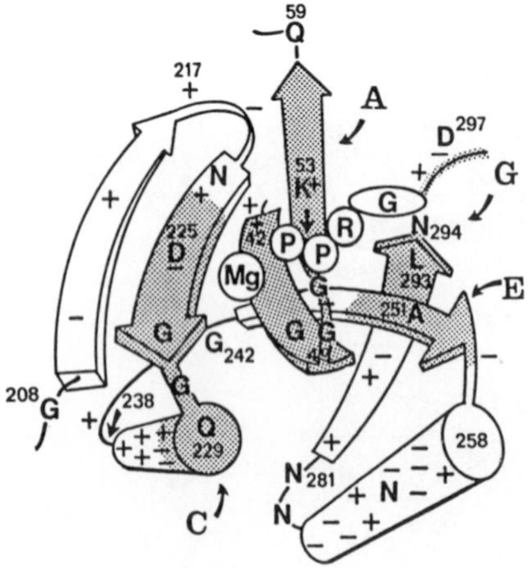

Figure 14. Schematic model for the proposed tertiary structure of the presumptive guanine-binding domain of α_{avg} residues 42–58 and 209–298. The position of the GDP molecule and the folding of α helices and β strands are modeled after the corresponding domain of EF-Tu (Jurnak, 1985). Stippled areas indicate the A (α_{avg} residues 39–56), C (α_{avg} residues 224–230), E (α_{avg} residues 247–255), and G (α_{avg} residues 289–298) regions; + and − symbols denote the presence of a charged residue in any one of the five α chains. The letters G, N, P, and Q are at positions where at least three of the five α chains contain a glycine, asparagine, proline, or glutamine, respectively. Several other important residues are marked with letters designating specific amino acids.

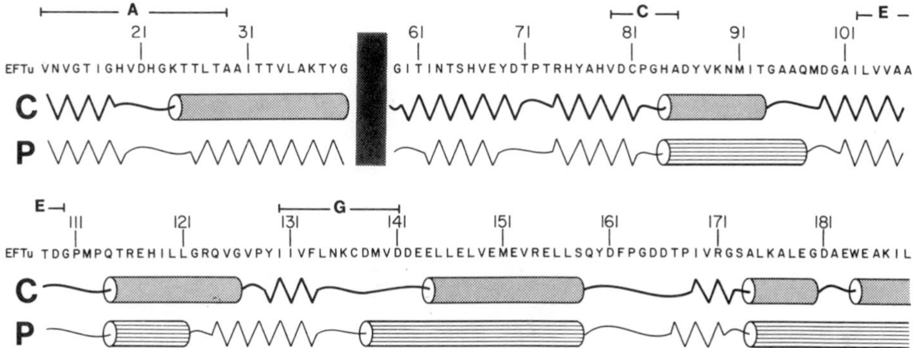

Figure 15. Comparison of the x-ray crystallographic prediction (Jurnak, 1985) of the secondary structure of the GDP-binding site of EF-Tu (C) and the secondary structure prediction for the same sequences (P). The filled bar interposed in the structure prediction and sequence represents a region that was missing from the trypsinized form of EF-Tu used for crystallography (Jurnak, 1985). Symbols used to denote secondary structure are as follows: cylinders, α helix; zig-zag lines, β sheet; wavy lines, turn or coil. The A, C, E, and G regions corresponding to the presumptive guanine nucleotide-binding site are indicated above the sequences.

called Halliday regions (Halliday, 1984) in the EF-Tu sequence. The four Halliday regions, designated A, C, E, and G, each consist of a turn between a β strand and an α helix. Residues in or near these turns interact with the bound GDP molecule and with an associated magnesium ion (Fig. 14). Sequence homology and analysis of amphipathic plots were then used to model the three-dimensional structure of this domain. This defining of the GDP-binding domain also located the other functional domains in the α-chain sequences.

Initially, the secondary structure for the GDP-binding domain of EF-Tu was predicted as a control. In comparison with the crystal structure, the prediction located nine out of ten turns, five out of six α helices, and all six β strands (Fig. 15). The general vicinities of most structures were correctly predicted, although some were of incorrect length by about three amino acids. In one case a 17-amino-acid stretch was predicted to have extended structure and was in fact helical. Nevertheless, the overall prediction correctly located the four Halliday regions.

Five highly homologous α chains from the G proteins (Robishaw *et al.*, 1986; Itoh *et al.*, 1986; Tanabe *et al.*, 1985; Yatsunami and Khorana, 1985; Medynski *et al.*, 1985; Sullivan *et al.*, 1986; Lochrie *et al.*, 1985) were aligned with each other, and their Halliday regions were aligned with the GDP-binding domain of EF-Tu. Amphipathic plots for the α chains were calculated, and the five plots were arithmetically averaged to increase signal to noise (Fig. 16). Based on the averaged plot, the secondary structure of the α chains was predicted to be very similar to the secondary structure of EF-Tu in the vicinity of the Halliday regions. Therefore, the complete three-dimensional structure of the GDP-binding domain of the α chains could be modeled on the known structure of EF-Tu (Fig. 17).

The predicted model for the GDP-binding domain of the G-protein α chains differs from that of EF-Tu in that an α helix (residues 24–30 of EF-Tu) is replaced by a β strand (residues 53–59 of the G proteins), leading into an effector binding domain, and the C-terminal helix–β-strand pair in the GDP-binding domain of the G proteins has extra amino acids with respect to the corresponding structures in EF-Tu. With these changes the GDP-binding domain appears to be structurally homologous in all GDP-binding proteins. Critical residues involved in GDP binding are conserved, and mutations of these residues affect activity. For example, replacement of Gly^{12} of $p21^{ras}$ produces a protein with enhanced transforming activity and diminished

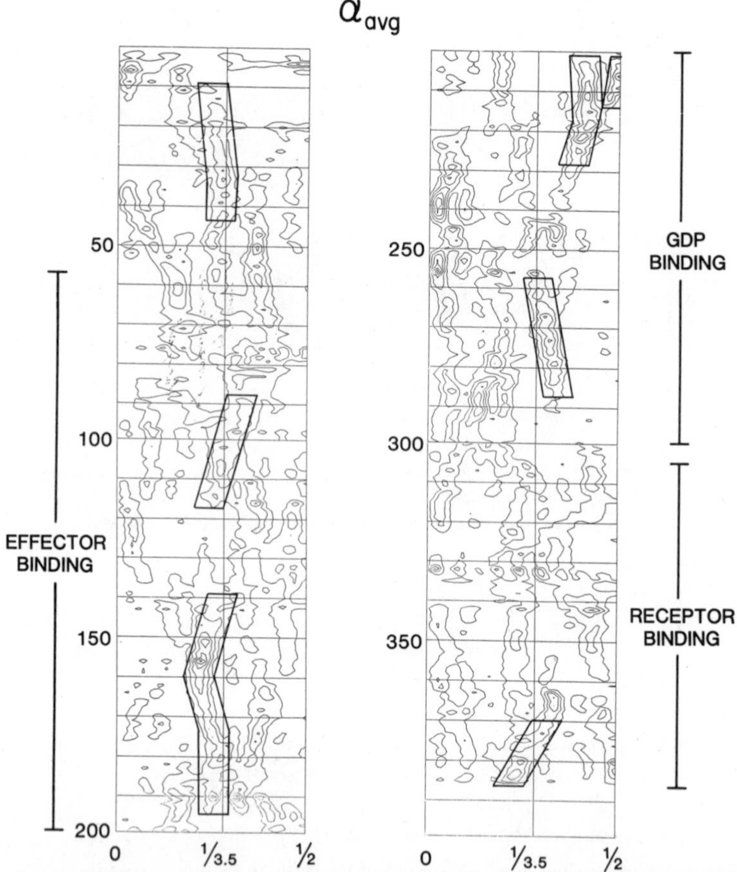

Figure 16. Amphipathic Fourier transform between frequencies 0 and 1/2 residues^{-1} for the sequence of α_{avg}. The transform was calculated with a moving window of 20 residues. The central residue number is plotted on the vertical axis, and horizontal lines are drawn every 10 residues on the plots. The frequency is plotted on the horizontal axis. A frequency of 1/2.0 corresponds to the characteristic amphipathic β-structure frequency. A vertical line on the plot is drawn near the frequency (1/3.6 residues^{-1}) characteristic of amphipathic α helices. Boxes indicate strong amphipathic α-helical and β-strand regions.

GTPase activity (Bishop and Varmus, 1985). The effects of these mutations can now be rationalized in terms of the prediction for domain organization and the known tertiary structure of EF-Tu.

D. Deduction of Structural and Functional Homology Based on Similar "Fingerprints" within the Plots

1. γ-Interferon and the Tryptophan Repressor

The amphipathic transform can be particularly useful in comparing structures of α-helical proteins. Comparison of the amphipathic plot for human γ-interferon (Gray *et al.*, 1982) (of unknown tertiary structure) with that of the tryptophan repressor from *E. coli* (Gunsalus and Yanofsky, 1980) [whose x-ray structure is determined (Schevitz *et al.*, 1985)] shows a similar "fingerprint" pattern. That is, the maps of peaks in the amphipathic plots have similar patterns

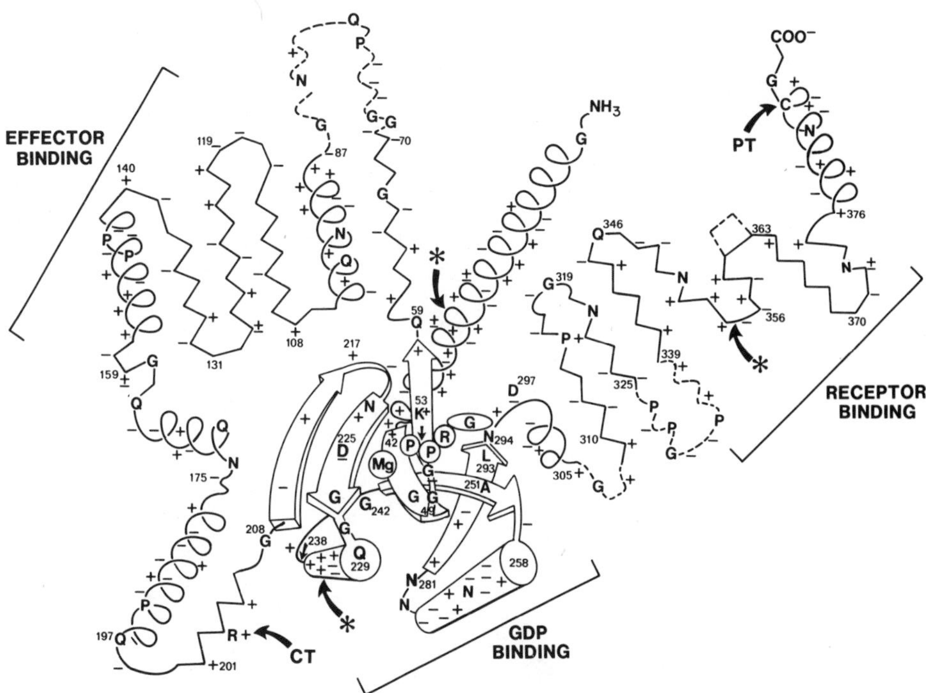

Figure 17. Schematic model for the predicted secondary structure of α_{avg}. The presumptive guanine nucleotide binding region in α_{avg} forms the core of the α chain. Dashed lines represent sequences present only in α_s.

of features (Fig. 18). This suggests that the relative orientation of α-helical segments in these two proteins may be similar. The sequence of human γ-interferon reveals a striking zig-zag pattern of peaks that cross the principal α-helical frequency from 1/3.0 to 1/4.0 residues^{-1}. This zig-zag pattern in α-helical proteins often correlates with the length and endpoints of helical segments. A sketch of the sequence in the region around residue 100 of bovine γ-interferon shows how this change in frequency can be related to the direction in which a polar or an apolar stripe is located on a putative α-helical region (Fig. 19). The frequency 1/3.0 corresponds to a left-handed twist to the stripe, and the frequency 1/4.0 to a right-handed stripe on the surface of the helix. These stripes often arise from the angle at which helices pack against one another. Thus, the zig-zag fingerprint pattern encodes information relating to the angle at which helices pack against each other.

A diagonal line peak, a "zig" in the zig-zag pattern, then corresponds to the intersection of a helix with two others, one of which crosses to the left and another to the right of the region at the center of the peaks. This is seen clearly in helical proteins whose structures are known. The similarity in the fingerprint zig-zag pattern for tryptophan repressor and interferon may thus represent some similarity in the packing of helices one against the other. There is no sequence homology between these proteins and therefore no implication from sequence comparisons that the tertiary structures are themselves similar; rather, the notion derived from the amphipathic plots is that similar angles of crossing between different helical regions are expected. Although the structure of γ-interferon is not yet known, the clear indication from the amphipathic plots is that it consists of helical bundles with predictable crossing angles. It should not be concluded that its tertiary structure (or function) is related to that of the trp repressor.

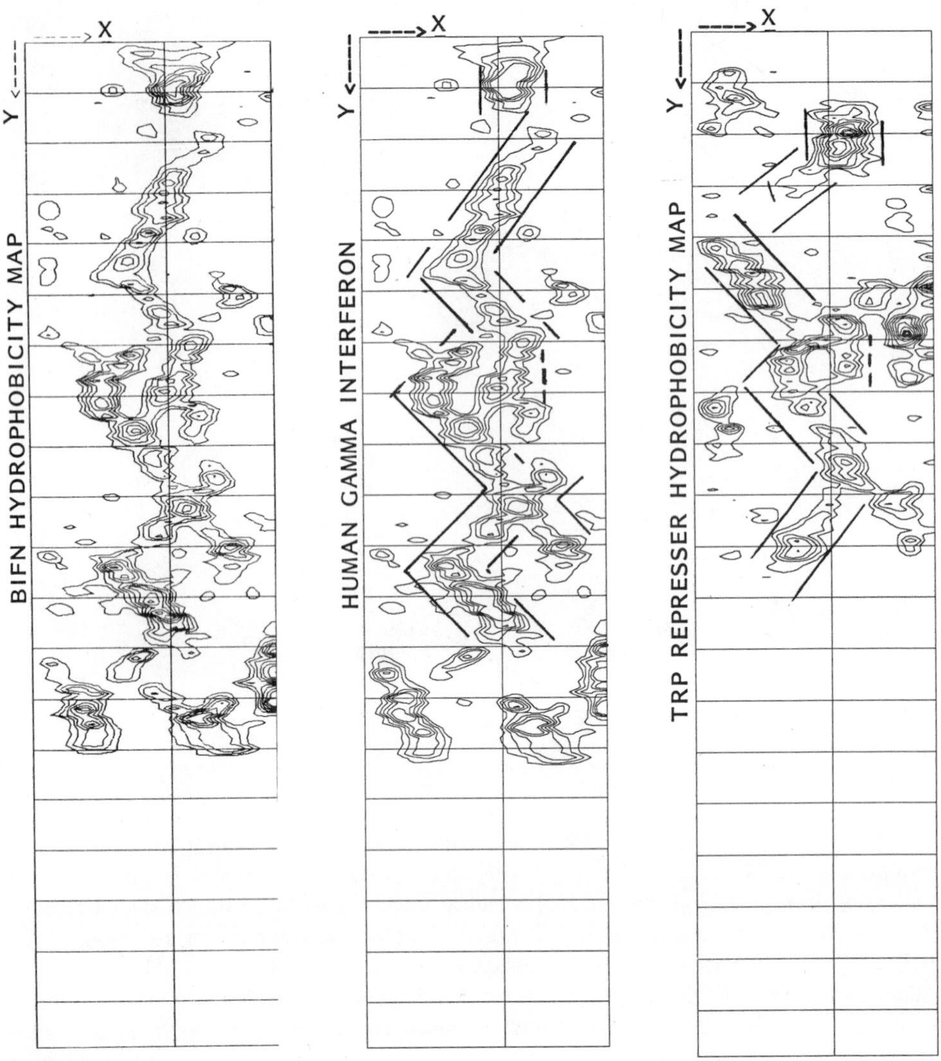

Figure 18. Amphipathic plots for bovine (left) and human (center) γ-interferon and for the tryptophan repressor. γ-Interferon and tryptophan repressor have no sequence homology, but both have a characteristic zigzag pattern in their amphipathic plots centered around the α-helix frequency of 1/3.6.

Human growth hormone is another sequence with features on the plots that strongly suggest that it is composed of a bundle of several parallel helices (Fig. 20). The putative helical regions deduced from the plots are indicated with letters (helix A, B, C, D) and bars indicating the length of the helical signal and therefore the probable positions of the helices. The window size is 12 residues, and the vertical line is at α-helical periodicity. The criss-cross pattern seen for trp repressor where helices cross at large angles, or in γ interferon, are not so apparent, and the angle of the feature is closer to being vertical on the plots. One "zig" is visible for putative helix D. Thus, this protein is probably composed mostly of parallel helices. The sequence in the region of helix D, 153–189, is shown in Fig. 21. Near residue 153, the periodicity of 1/4 seen in Fig. 20 is clearly seen on the helix net. At residues 165–180 the

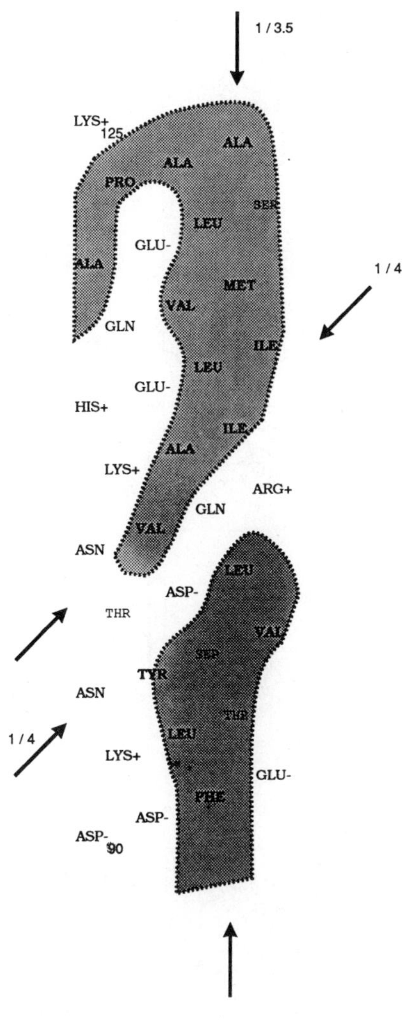

Figure 19. Sketch of a portion of the γ-interferon sequence showing how the periodicity of the hydrophobicities would be related to the direction of the apolar stripe on a helix in this region of the sequence. This could derive from the angles of helix–helix crossings between this helix and two or three others.

normal 1/3.5 period indicates an almost parallel interface between helical crossings. At 180–190, the periodicity of 1/3 seen in Fig. 20 is visible in the helix net of Fig. 21, indicative of helical crossings with a left-handed twist.

2. *An Extracellular Domain of the Acetylcholine Receptor*

During our analysis of the amphipathic secondary structure of the acetylcholine receptor from *T. californica* (Section III.A.2), we noted that the most prominent periodic feature common to all four subunits occurred in the extracellular region of the protein (Finer-Moore and Stroud, 1984). This feature was detected by a strong signal about 100 amino acids in length with a frequency that varied between 1/2.0 and 1/2.3 in amphipathic plots calculated with a 25-residue window. The frequency, length, and intensity of the signal suggested that it corresponded to several contiguous amphipathic β strands. This pattern of secondary structure

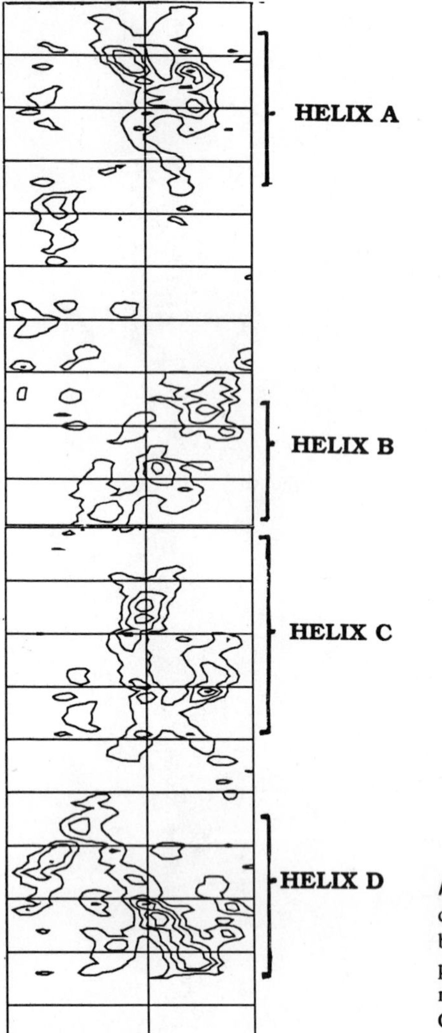

Figure 20. Fourier plot of human growth hormone sequence. Sequence numbers go downward. The horizontal bars are every 10 residues. Features close to the α-helical periodocity (vertical line) indicate where α-helical regions may be expected, as indicated at the right for helices A, B, C, and D. The window size is 12 residues, and the plot is unnormalized.

would be consistent with a β-barrel tertiary structure, where β strands interface both with each other (inside the barrel) and with solvent.

Amphipathic plots of several known β-barrel domains showed that in general they did not have amphipathic β periodicity as intense or over such a long region as seen in the acetylcholine receptor. Several reasons why the β strands in these domains are not always amphipathic have been noted (Turnell *et al.*, 1986). However, some β-barrel domains that are flattened into a sandwichlike structure in which the β strands are less twisted do show strong signals for amphipathic β structure even when a large window size is used. Figure 22 shows the amphipathic plots of regions of concanavalin A and IgG, two proteins that form β-sandwich structures (Reeke *et al.*, 1975; Amzel and Poljak, 1979), alongside the average amphipathic plot for a portion of the extracellular domains of the four subunits of the acetylcholine receptor. These results suggest that the acetylcholine receptor may have β-sandwich domains in its extracellular region. Supporting evidence for this suggestion is the prediction of

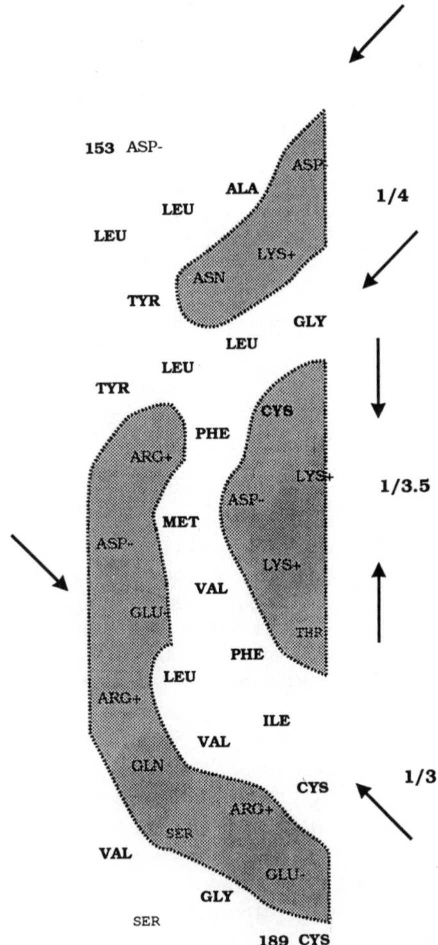

Figure 21. Helical net of the sequence for helix D of human growth hormone. Arrows and associated frequencies listed indicate the feature on the helix surface represented by the "zig" opposite helix D in Fig. 20.

β structure in the putative β-sandwich region by statistical methods (Garnier *et al.*, 1978) and the estimation of 30–35% β structure in the acetylcholine receptor by CD (Moore *et al.*, 1974) or Raman spectroscopy (Chang *et al.*, 1983). Our analysis led to a quite precise estimation of where the turns between the β stretches are likely to be.

3. *Structural Studies of a Bifunctional Enzyme: 6-Phosphofructo-2-kinase/Fructose-2,6-biphosphatase*

The glycolytic enzyme 6-phosphofructo-2-kinase/fructose-2,6-bisphosphatase (PF2K/FBP) catalyzes both the synthesis and degradation of fructose-2,6-bisphosphate (fru-2,6-P_2) (Pilkis *et al.*, 1986). Thus, the two opposing reactions catalyzed by PF2K/FBP are

$$\text{fru-6-P} + \text{ATP} \rightarrow \text{fru-2,6-P}_2 + \text{ADP}$$

$$\text{fru-2,6-P}_2 \rightarrow \text{fru-6-P} + \text{P}_i$$

They are believed to occur at two active sites (El-Maghrabi and Pilkis, 1984). Protein modification experiments (with selective chemical reagents or proteases) have different effects on the two reactions (El-Maghrabi *et al.*, 1984a,b, 1987), indicating that the two catalytic sites are

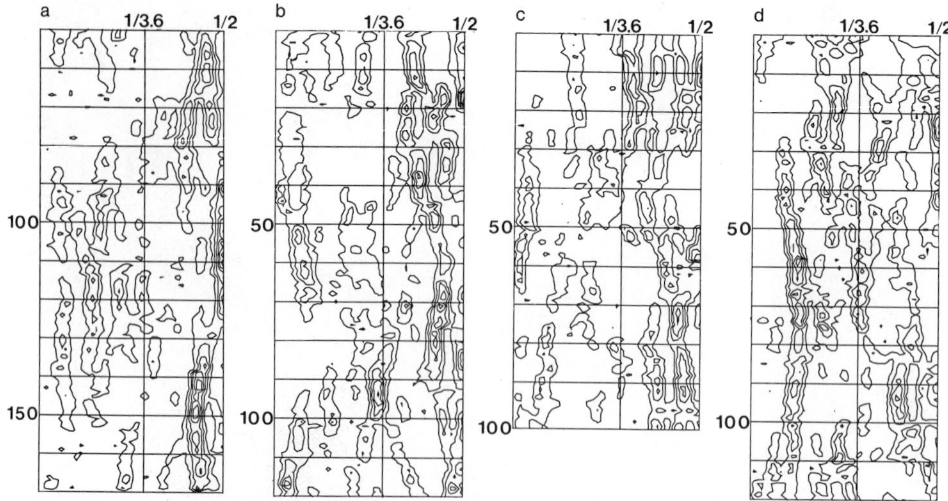

Figure 22. (a) Averaged amphipathic plot for the four subunits of the acetylcholine receptor from *T. californica* for a portion of the extracellular domain. The plot was calculated with a 25-residue window. (b and c) Plots for the first 130 residues of concanavalin A and for the variable regions of the human Ig heavy chains (Wang *et al.*, 1977) calculated with a 25-residue window. Both proteins have β-sandwich structures, and the strongest signals in their amphipathic plots fall to the right of the α-helix frequency of 1/3.6. (d) Amphipathic plot for the first β barrel in trypsin calculated with a 25-residue window. In contrast to plots a–c, the signals in this plot do not have primarily the amphipathic β frequency (1/2.0–1/3.0).

unique but still leaving open the question of whether the bifunctional enzyme consists of two catalytic domains. The amino acid sequence of PF2K/FBP from rat liver is composed of 470 amino acid residues (Colosia *et al.*, 1987; Lively *et al.*, 1988).

In the hope of identifying other better-characterized proteins that are either wholly homologous to PF2K/FBP or to one of its two postulated domains, computer searches were carried out and revealed that the amino acid sequence as a whole was not homologous to any other known sequence. Furthermore, the enzyme did not seem to be the product of a gene duplication event as assessed by diagonal matrix self-comparison (Maizel and Lenk, 1981). A more in-depth search of the protein sequence data bank with separate N- and C-terminal halves of the RtPF2K/FBP sequence revealed an intriguing local similarity between seven residues around histidine 258 in the C-terminal half of the enzyme and residues around active site histidine 8 of yeast phosphoglycerate mutase (YePGM; Fothergill and Harkins, 1982).

RtPF2K/FBP	(255)	LCRHGESELN
YePGM	(5)	LVRHGQSEWN

Phosphoglycerate mutase catalyzes the phosphotransfer reaction between glycerate-3-phosphate and glycerate-2-phosphate (reviewed in Rose, 1980), and His^{8} is known to be the site of phosphorylation during catalysis (Han and Rose, 1979; Hass *et al.*, 1980). Experiments verified that the homologous residue in RtPF2K/FBP, His^{258}, is the histidine that is phosphorylated in the PF2K/FBP biphosphatase reaction (Pilkis *et al.*, 1987).

Following this hint of possible homology between the active sites of the two enzymes, amphipathic plots gave striking evidence for structural relatedness of residues 250–470 of RtPF2K/FBP to both PGM and to bisphosphoglycerate mutase (BPGM), an enzyme that is

Figure 23. Amphipathic plot of yeast phosphoglycerate mutase aligned with the plot of the bisphosphatase domain of the bifunctional enzyme PF2K/FBP from rat. The structural homology of these two proteins was deduced from the similarity of their plots.

highly homologous to PGM (Joulin *et al.*, 1986) (Fig. 23). Based on this evidence the sequence of the second half of PF2K/FBP was modeled based on the crystal structure of PGM (Winn *et al.*, 1981) (Fig. 24). Phosphoglycerate mutase has a typical three-layer α/β structure, a parallel β-sheet core sandwiched by α helices. It can be classified as a nucleotide-binding fold (nbf) commonly seen in dehydrogeneases and kinases. It does not bind nucleotides but has found the nbf structure favorable for binding glycerates. The active site of the mutase lies in a crevice at the C-terminal end of the β sheet.

Alignment of the structurally homologous regions of PF2K/FBP and the two mutases was accomplished to first approximation by alignment of similar regions in the amphipathic plots. A more precise alignment was then obtained using traditional structure prediction algorithms (Chou and Fasman, 1974; Garnier *et al.*, 1978) as well as a new, highly accurate turn

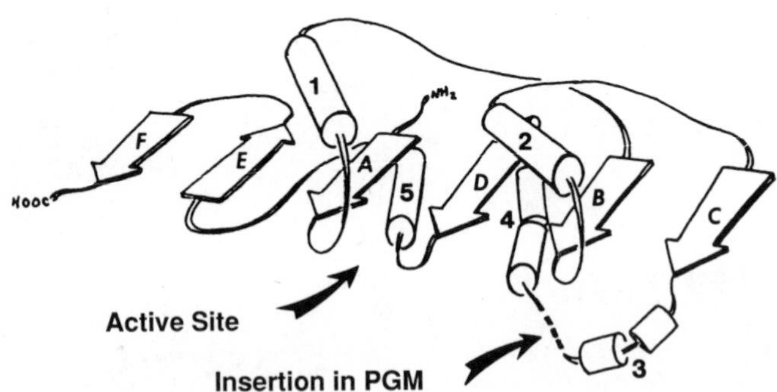

Figure 24. Structure of the bisphosphatase domain of rat PF2K/FBP modeled on the crystal structure of yeast phosphoglycerate mutase. The structure has a nucleotide-binding fold. The position of a 25-residue insertion in phosphoglycerate mutase is indicated by dotted lines.

prediction program specific for α/β proteins (Cohen *et al.*, 1983, 1986). After alignment, α and β secondary structure elements largely superimposed among the three sequences, although only about 25% of the residues in either of the mutase structures were conserved in PF2K/FBP.

Two histidines at the end of loops and two arginines are in the active site of YePGM (Winn *et al.*, 1981). Based on the alignment of the sequences, these residues are conserved and would occur in similar positions in the PF2K/FBP structure. This result suggested that PF2K/FBP might show an unexpected mutase activity toward glycerates, not its normal substrate. This prediction was tested and confirmed (Tauler *et al.*, 1987). Thus, the amphipathic plots permitted a very accurate alignment with an apparently unrelated protein and ultimately led to generation of a detailed three-dimensional model that predicted a new biological activity. This mutase activity was experimentally tested and found to be present, a remarkable vindication of the predictive power of this scheme.

The N-terminal half of PF2K/FBP must constitute the kinase domain that binds sugar-phosphate and ATP molecules. It contains a cAMP-dependent phosphorylation site (serine 32) that can modulate the activity of the kinase domain (Murray *et al.*, 1984; Lively *et al.*, 1988) as well as the signature sequence G-X_4-G-K-T, which typically occurs in the turn between the first β strand and helix in ATP/GTP nucleotide-binding folds (Moller and Amons, 1985; Dever *et al.*, 1987). The kinase domain is therefore being modeled on a nbf using the approach outlined for the bisphosphatase domain (Bazan *et al.*, 1988).

E. Applications to DNA Sequences

Just as the Fourier transform method is well suited to examining periodicities in the hydrophobic character of a protein chain, it can also be used to detect underlying patterns in nucleic acid sequences. These would not necessarily be reflections of amphiphilicity patterns but rather a measure of the local distribution of bases along the DNA (or RNA) chain. For proteins we usually apply the Fourier transform technique in order to predict secondary or tertiary structure. Nucleic acids, on the other hand, have a rather well understood molecular structure, the double helix, that we can consider as a constant "secondary" structure. By assigning the different nucleotides arbitrary values, we can tailor a Fourier transform analysis of a primary sequence to examine a particular property of the double-helix surface. Examples of the different scales that may be useful are:

1. Single-nucleotide (base) pattern. Assign nucleotide *N* a value of +1, the others 0.
2. Four-nucleotide pattern. Assign each nucleotide (A, T or U, G, C) a different value.
3. Purine/pyrimidine pattern. Assign A and G an equal value distinct from that of T (or U) and C.
4. AT/GC distribution pattern. Assign A and T (or U) an equal value distinct from that of G and C.

Patterns of bases in a nucleotide string become structural patterns when woven into the double helix. Bases that occur with a periodicity equal to integer (or half-integer) multiples of the characteristic 10.5-basepair repeat of the double-stranded molecule will be spaced by a turn (or half-turn) of the helix. Fourier analysis of a DNA sequence that reveals such repeats may be diagnostic of a structural pattern in the helix that can be "read" by a sequence-motif-specific DNA-binding protein. Another use is to detect sequence-dependent variations of the DNA strand (like local bending) marked by turn-periodic patches of certain nucleotides.

1. Analysis of DNA Regions That Interact with Regulatory Proteins

Several x-ray crystal structures of "helix–turn–helix" motif DNA-binding proteins exist (for review see Ohlendorf and Matthews, 1983). This group of regulatory proteins is able to recognize only one face of the DNA helix, interacting primarily with bases along the major groove. Another recent structure, the EcoR1 endonuclease complexed with a stretch of double-helical DNA that encodes the enzymes' restriction site, reveals as well that contacts are made only on one side of the helix (McClarin *et al.*, 1986). Rhodes and Klug (1986), applying Fourier analysis to the primary base sequence and DNAase I cleavage patterns of the double-stranded DNA region implicated in binding the transcription factor TFIIIA, discovered a 5.5-basepair periodicity in G residues. The TFIIIA protein has a ninefold repeated motif, a "zinc-binding" finger, that is responsible for binding to transcriptional control sequences in DNA (Miller *et al.*, 1985). The structural periodicity noted in the control region of approximately 50 basepairs inspired Rhodes and Klug to propose that pairs of TFIIIA fingers intercalate in major grooves on opposite sides of the DNA helix. Further analysis of regulatory sites that bound RNA pol III and transcription factor Sp1 revealed the same half-turn repeat motif, leading the authors to predict that both of these DNA-binding control factors would possess the finger repeats seen in TFIIIA.

2. Variations in the DNA Structure

Drew and Travers (1985) experimentally examined the phenomenon of DNA bending and its relationship to nucleosome positioning. They concluded that turn-periodic separation of AT or GC patches strongly affected the plane of bending. In particular, AT pairs in minor grooves would face inward; GC pairs in major grooves face outward. Fourier analysis of longer stretches of DNA can detect these bending regions by analyzing the periodicity of AT versus GC tracts.

IV. DISCUSSION

The amphipathic Fourier method has a strong advantage over data-base-oriented methods for predicting the structures of folded proteins. Data-base methods very rarely take advantage of any cooperativity in the folding of proteins. There are simply too many possible peptide sequences to store tertiary information on all of them. Furthermore, as Kabsch and Sander

(1984) have shown, it is possible that the same sequences of pentapeptides can occur in different secondary structures within different proteins. Thus, data-base methods will in general encode properties of individual imino acids for which some rationalization can sometimes be generated; proline breaks helices because it is an imino acid, whereas glycine offers many possibilities for angles unavailable to amino acids with side chains, etc.

The amphipathic method seeks to identify features of the folded structure that may be important in maintaining that structure. For example, an amphipathic β-sheet structure that lies on the surface of a protein interfaces between a polar outside and a less polar inside. It is perhaps inherent to the feature's order in that location that properties such as polarity or propensity for lying within a hydrophobic membrane will be represented with the periodicity expected for the secondary structure in which they lie. In this respect similarity in properties of amino acids rather than the probability of their substitution for one another, which may depend on frequencies of base changes during mutation as well as on natural selection, preserves the amphipathic signal. The footprint of the secondary structure remains within the protein while it evolves through individual amino acid sequence changes. The amphipathic plots can suggest similarities of structure between quite unrelated proteins that may lie in the same assembly of secondary structures. The plots are clearly sensitive to the angles at which helices cross one another and are powerful at identifying domains as well as regions of relatively featureless structure such as the linkers in HMG-CoA reductase. These predictions have been useful in a variety of ways. They permit some estimation of which regions may be immunogenically active, which regions may lie in channel-forming domains of membrane protein, or which regions may lie at the interfaces between secondary structural elements in the formation of a folded protein structure.

ACKNOWLEDGMENTS. This work was supported by PHS grant CA41323; NIH Grant GM 24485; and NSF grant PCM-8316401.

V. REFERENCES

Adelman, J. P., Mason, A. J., Hayflick, J. S., and Seeburg, P. H., 1986, Isolation of the gene and hypothalamic cDNA for the common precursor of gonadotropin-releasing hormone and prolactin release-inhibiting factor in human and rat, *Proc. Natl. Acad. Sci. U.S.A.* **83:**179–183.

Agard, D. A., and Stroud, R. M., 1982, Linking regions between helices in bacteriorhodopsin revealed, *Biophys. J.* **37:**589–602.

Amzel, L. M., and Poljak, R. J., 1979, Three-dimensional structure of immunoglobins, *Annu. Rev. Biochem.* **48:**961–997.

Argos, P., Rao, J. K. M., and Hargrave, P. A., 1982, Structural prediction of membrane-bound proteins, *Eur. J. Biochem.* **128:**565–575.

Basson, M. E., Thorsness, M., Finer-Moore, J., Stroud, R. M., and Rine, J., 1988, Structural and functional conservation between yeast and human 3-hydroxy-3-methylglutaryl coenzyme A reductases, the rate-limiting enzyme of sterol biosynthesis, *Molecular and Cellular Biology,* **8:**3797–3808.

Bazan, J. F., Fletterick, R. J., and Pilkis, S. J., 1988, Evolution of a bifunctional enzyme: 6-Phosphofructo-2,6-bisphosphate, *Proc. Natl. Acad. Sci.,* submitted.

Bishop, J. M., and Varmus, H. E., 1985, Functions and origins of retroviral transforming genes, in: *Molecular Biology of Tumor Viruses* (R. Weiss, R. Teich, H. Varmus, and J. Coffin, eds.), Cold Spring Harbor Laboratory Press, New York, pp. 249–356.

Blundell, T. L., Pitts, J. E., Tickle, I. J., Wood, S. P., and Wu, C.-W., 1981, X-ray analysis (1.4-Å resolution) of avian pancreatic polypeptide: Small globular protein hormone, *Proc. Natl. Acad. Sci. U.S.A.* **78:**4175–4179.

Bourne, H. R., 1986, GTP-binding proteins: One molecular machine can transduce diverse signals, *Nature* **321:** 814.

Branden, C.-I., 1980, Relation between structure and function of α/β-proteins, *Q. Rev. Biophys.* **13:**317–338.

Brown, M. S., and Goldstein, J. L., 1980, Multivalent feedback regulation of HMG CoA reductase, a control mechanism coordinating isoprenoid synthesis and cell growth, *J. Lipid Res.* **21:**505–517.

Chang, E. L., Yager, P., Williams, R. W., and Dalziel, A. W., 1983, The secondary structure of reconstituted acetylcholine receptor as determined by Raman spectroscopy, *Biophys. J.* **41:**65a.

Chin, D. J., Gil, G., Russell, D. W., Liscum, L., Luskey, K. L., Basu, S. K., Okayama, H., Berg, P., Goldstein, J. L., and Brown, M. S., 1984, Nucleotide sequence of 3-hydroxy-3-methyl-glutaryl coenzyme A reductase, a glycoprotein of endoplasmic reticulum, *Nature* **308:**613–617.

Choe, S., 1987, *Structural Study of Channel-Forming Proteins,* Ph.D. Thesis, University of California, Berkeley.

Chou, P. Y., and Fasman, G. D., 1974, Prediction of protein conformation, *Biochemistry* **13:**222–245.

Claudio, T., Ballivet, M., Patrick, J., and Heinemann, S., 1983, Nucleotide and deduced amino acid sequences of *Torpedo californica* acetylcholine receptor α subunit, *Proc. Natl. Acad. Sci. U.S.A.* **80:**1111–1115.

Cohen, F. E., Abarbanel, R. M., Kuntz, I. D., and Fletterick, R. J., 1983, Secondary structure assignment for α/β proteins by a combinatorial approach, *Biochemistry* **22:**4894–4904.

Cohen, F. E., Abarbanel, R. M., Kuntz, I. D., and Fletterick, R. J., 1986, Turn prediction in proteins using a pattern-matching approach, *Biochemistry* **25:**266–275.

Colosia, A. D., Lively, M., El-Maghrabi, M. R., and Pilkis, S. J., 1987, Isolation of a cDNA clone for rat liver 6-phosphofructo 2-kinase/fructose 2,6-bisphosphatase, *Biochem. Biophys. Res. Commun.* **143:**1092–1098.

Cornette, J. L., Cease, K. B., Margalit, H., Spouge, J. L., Berzofsky, J. A., and DeLisi, C., 1987, Hydrophobicity scales and computational techniques for detecting amphipathic structures in proteins, *J. Mol. Biol.* **195:**659–685.

Creighton, T. E., Hillson, D. A., and Freedman, R. B., 1980, Catalysis by protein-disulphide isomerase of the unfolding and refolding of proteins with disulphide bonds, *J. Mol. Biol.* **142:**43–62.

Deisenhofer, J., Epp, O., Miki, K., Huber, R., and Michel, H., 1985, Structure of the protein subunits in the photosynthetic reaction centre of *Rhodopseudomonas viridis* at 3 Å resolution, *Nature* **318:**618–624.

Dever, T. E., Glynias, M. J., and Merrick, W. C., 1987, GTP-binding domain: Three consensus sequence elements with distinct spacing, *Proc. Natl. Acad. Sci. U.S.A.* **84:**1814–1818.

Devillers-Thiery, A., Giraudat, J., Bentaboulet, M., and Changeux, J.-P., 1983, Complete mRNA coding sequence of the acetylcholine binding α-subunit from *Torpedo marmorata:* A model for the transmembrane organization of the polypeptide chain, *Proc. Natl. Acad. Sci. U.S.A.* **80:**2067–2071.

Drew, H. R., III, and Travers, A. A., 1985, Structural junctions in DNA: The influence of flanking sequence on nuclease digestion specificities, *Nucleic Acids Res.* **13:**4445–4467.

Dunn, S. M. J., Conti-Tronconi, B. M., and Raftery, M. A., 1986, Acetylcholine receptor dimers are stabilized by extracellular disulfide bonding, *Biochem. Biophys. Res. Commun.* **139:**830–837.

Dunnill, P., 1968, The use of helical net-diagrams to represent protein structures, *Biophys. J.* **8:**865–875.

Edman, J. C., Ellis, L., Blacher, R. W., Roth, R. A., and Rutter, W. J., 1985, Sequence of protein disulphide isomerase and implications of its relationship to thioredoxin, *Nature* **317:**267–270.

Eisenberg, D., Weiss, R. M., and Terwilliger, T. C., 1982a, The helical hydrophobic moment: A measure of the amphiphilicity of a helix, *Nature* **299:**371–374.

Eisenberg, D., Weiss, R. M., Terwilliger, T. C., and Wilcox, W., 1982b, Hydrophobic moments and protein structure, *Faraday Symp. Chem. Soc.* **17:**109–120.

Eisenberg, D., Weiss, R. M., and Terwilliger, T. C., 1984a, The hydrophobic moment detects periodicity in protein hydrophobicity, *Proc. Natl. Acad. Sci. U.S.A.* **81:**140–144.

Eisenberg, D., Schwarz, E., Komaromy, M., and Wall, R., 1984b, Analysis of membrane and surface protein sequences with the hydrophobic moment plot, *J. Mol. Biol.* **179:**125–142.

El-Maghrabi, M. R., and Pilkis, S. J., 1984, Rat liver 6-phosphofructo-2-kinase/fructose 2,6-bisphosphatase: A review of relationships between the two activities of the enzyme, *J. Cell. Biochem.* **26:**1–17.

El-Maghrabi, M. R., Pate, T. M., Murray, K. J., and Pilkis, S. J., 1984a, Differential effects of proteolysis and protein modification on the activities of 6-phosphofructo-2-kinase/fructose-2,6-bisphosphatase, *J. Biol. Chem.* **259:**13096–13103.

El-Maghrabi, M. R., Pate, T. M., Pilkis, J., and Pilkis, S. J., 1984b, Effect of sulfhydryl modification on the activities of rat liver 6-phosphofructo-2-kinase/fructose-2,6-bisphosphatase, *J. Biol. Chem.* **259:**13104–13110.

El-Maghrabi, M. R., Pate, T. M., D'Angelo, G., Correia, J. J., Lively, M. O., and Pilkis, S. J., 1987, Rat liver 6-phosphofructo-2-kinase/fructose-2,6-bisphosphatase: Identification of essential sulfhydryl residues in the primary sequence of the enzyme, *J. Biol. Chem.* **262:**11714–11720.

Emr, S., Vassarotti, A., Garrett, J., Geller, B., Takeda, M., and Douglas, M. G., 1986, The amino terminus of the yeast F1-ATPase β-subunit precursor functions as a mitochondrial import signal, *J. Cell Biol.* **102:**523–533.

Engelman, D. M., and Zaccai, G., 1980, Bacteriorhodopsin is an inside-out protein, *Proc. Natl. Acad. Sci. U.S.A.* **77:**5894–5898.

Engelman, D. M., Henderson, R., McLachlan, A. D., and Wallace, B. A., 1980, Path of the polypeptide in bacteriorhodopsin, *Proc. Natl. Acad. Sci. U.S.A.* **77:**2023–2027.

Engelman, D. M., Goldman, A., and Steitz, T. A., 1982, The identification of helical segments in the polypeptide chain of bacteriorhodopsin, *Methods Enzymol.* **88:**81–88.

Faust, J. R., Luskey, K. L., Chin, D. J., Goldstein, J. L., and Brown, M. S., 1982, Regulation of synthesis and degradation of 3-hydroxy-3-methylglutaryl-coenzyme A reductase by low density lipoprotein and 25-hydroxycholesterol in UT-1 cells, *Proc. Natl. Acad. Sci. U.S.A.* **79:**5205–5209.

Finer-Moore, J., and Stroud, R. M., 1984, Amphipathic analysis and possible formation of the ion channel in an acetylcholine receptor, *Proc. Natl. Acad. Sci. U.S.A.* **81:**155–159.

Fothergill, L. A., and Harkins, R. N., 1982, The amino acid sequence of yeast phosphoglycerate mutase, *Proc. R. Soc. London [Biol.]* **215:**19–44.

Freedman, R. B., Brockway, B. E., and Lambert, N., 1984, Protein disulphide-isomerase and the formation of native disulphide bonds, *Biochem. Soc. Trans.* **12:**929–932.

Garnier, J., Osguthorpe, D. J., and Robson, B., 1978, Analysis of the accuracy and implications of simple methods for predicting the secondary structure of globular proteins, *J. Mol. Biol.* **120:**97–120.

Gilman, A. G., 1984, G proteins and dual control of adenylate cyclase, *Cell* **36:**577–579.

Gray, P. W., Leung, D. W., Pennica, D., Yelverton, E., Najarian, R., Simonsen, C. C., Derynck, R., Sherwood, P. J., Wallace, D. M, Berger, S. L., Levinson, A. D., and Goeddel, D. V., 1982, Expression of human immune interferon cDNA in *E. coli* and monkey cells, *Nature* **295:**503–508.

Grenningloh, G., Rienitz, A., Schmitt, B., Methfessel, C., Zensen, M., Beyreuther, K., Gundelfinger, E. D., and Betz, H., 1987, The strychnine-binding subunit of the glycine receptor shows homology with nicotinic acetylcholine receptors, *Nature* **328:**215–220.

Gunsalus, R. P., and Yanofsky, C., 1980, Nucleotide sequence and expression of *Escherichia coli* trpR, the structural gene for the trp aporepressor, *Proc. Natl. Acad. Sci. U.S.A.* **77:**7117–7121.

Halliday, K., 1984, Regional homology in GTP-binding proto-oncogene products and elongation factors, *J. Cyclic Nucleotide Res.* **9:**435–448.

Han, C.-H., and Rose, Z. B., 1979, Active site phosphohistidine peptides from red cell bisphosphoglycerate synthase and yeast phosphoglycerate mutase, *J. Biol. Chem.* **254:**8836–8840.

Hardy, L. W., Finer-Moore, J. S., Montfort, W. R., Jones, M. O., Santi, D. V., and Stroud, R. M., 1987, Atomic structure of thymidylate synthase: Target for rational drug design, *Science* **235:**448–455.

Hass, L. T., Place, A. R., Miller, K. B., and Powers, D. A., 1980, The isolation and characterization of an active site phosphohistidine peptide from human erythrocyte bisphosphoglycerate synthase, *Biochem. Biophys. Res. Commun.* **95:**1570–1576.

Heil, A., Muller, G., Noda, L. H., Pinder, T., Schirmer, I., Schirmer, R. H., and von Zabern, I., 1974, The amino acid sequence of porcine adenylate kinase from skeletal muscle, *Eur. J. Biochem.* **43:**131–144.

Henderson, R., and Unwin, P. N. T., 1975, Three-dimensional model of purple membrane obtained by electron microscopy, *Nature* **257:**28–32.

Holmgren, A., Soderberg, B. O., Eklund, H., and Branden, C. I., 1975, Three-dimensional structure of *Escherichia coli* thioredoxin-S_2 to 2.8 Å resolution, *Proc. Natl. Acad. Sci. U.S.A.* **72:**2305–2309.

Hopp, T. P., 1986, Protein surface analysis. Methods for identifying antigenic determinants and other interaction sites, *J. Immunol. Methods* **88:**1–18.

Hurley, J. B., Simon, M. I., Teplow, D. B., Robishaw, J. D., and Gilman, A. G., 1984, Homologies between signal transducing G proteins and *ras* gene products, *Science* **226:**860–862.

Itoh, H., Kozasa, T., Nagata, S., Nakamura, S., Katada, T., Ui, M., Iwai, S., Ohtsuka, E., Kawasaki, H., Suzuki, K., and Kaziro, Y., 1986, Molecular cloning and sequence determination of cDNAs for α subunits of the guanine nucleotide-binding proteins G_s, G_i, and G_o from rat brain, *Proc. Natl. Acad. Sci. U.S.A.* **83:** 3776–3780.

Joulin, V., Peduzzi, J., Romeo, P.-H., Rosa, R., Valentin, C., Dubart, A., Lapeyre, B., Blouquit, Y., Garel, M.-C., Goossens, M., Rosa, J., and Cohen-Solal, M., 1986, Molecular cloning and sequencing of the human erythrocyte 2,3-bisphosphoglycerate mutase cDNA: Revised amino acid sequence, *EMBO J.* **5:** 2275–2283.

Jurnak, F., 1985, Structure of the GDP domain of EF-Tu and location of the amino acids homologous to *ras* oncogene products, *Science* **230:**32–36.

Kabsch, W., and Sander, C., 1984, On the use of sequence homologies to predict protein structure: Identical pentapeptides can have completely different conformations, *Proc. Natl. Acad. Sci. U.S.A.* **81:**1075–1078.

Kaiser, E. T., and Kezdy, F. J., 1983, Secondary structures of proteins and peptides in amphiphilic environments (a review), *Proc. Natl. Acad. Sci. U.S.A.* **80:**1137–1143.

Katre, N. V., and Stroud, R. M., 1981, A probable linking sequence between two transmembrane components of bacteriorhodopsin, *FEBS Lett.* **136:**170–174.

Khorana, H. G., Gerber, G. E., Herlihy, W. C., Gray, C. P., Anderegg, R. J., Nihei, K., and Biemann, K., 1979, Amino acid sequence of bacteriorhodopsin, *Proc. Natl. Acad. Sci. U.S.A.* **76:**5046–5050.

Kistler, J., Stroud, R. M., Klymkowsky, M. W., Lalancette, R. A., and Fairclough, R. H., 1982, Structure and function of an acetylcholine receptor, *Biophys. J.* **37:**371–381.

Kuhn, L. A., and Leigh, J. S., Jr., 1985, A statistical technique for predicting membrane protein structure, *Biochim. Biophys. Acta* **828:**351–361.

Kyte, J., and Doolittle, R. F., 1982, A simple method for displaying the hydropathic character of a protein, *J. Mol. Biol.* **157:**105–132.

La Cour, T. F. M., Nyborg, J., Thirup, S., and Clark, B. F. C., 1985, Structural details of the binding of guanosine diphosphate to elongation factor Tu from *E. coli* as studied by x-ray crystallography, *EMBO J.* **4:**2385–2388.

Lewis, C. A., and Stevens, C. F., 1983, Acetylcholine receptor channel ionic selectivity: Ions experience an aqueous environment, *Proc. Natl. Acad. Sci. U.S.A.* **80:**6110–6113.

Lim, V. I., 1974, Algorithms for prediction of α-helical and β structural regions in globular proteins, *J. Mol. Biol.* **88:**873–894.

Lindstrom, J., Criado, M., Hochschwender, S., Fox, J. L., and Sarin, V., 1984, Immunochemical tests of acetylcholine receptor subunit models, *Nature* **311:**573–575.

Liscum, L., Cummings, R. D., Anderson, R. G. W., DeMartino, G. N., Goldstein, J. L., and Brown, M. S., 1983, 3-Hydroxy-3-methylglutaryl-CoA reductase: A transmembrane glycoprotein of the endoplasmic reticulum with N-linked "high-mannose" oligosaccharides, *Proc. Natl. Acad. Sci. U.S.A.* **80:**7165–7169.

Liscum, L., Finer-Moore, J., Stroud, R. M., Luskey, K. L., Brown, M. S., and Goldstein, J. L., 1985, Domain structure of 3-hydroxy-3-methylglutaryl coenzyme A reductase, a glycoprotein of the endoplasmic reticulum, *J. Biol. Chem.* **260:**522–530.

Lively, M. O., El-Maghrabi, M. R., Pilkis, J., D'Angelo, G., Colosia, A. D., Ciavola, J., Fraser, B. A., and Pilkis, S. J., 1988, Complete amino acid sequence of rat liver 6-phosphofructo-2-kinase/fructose-2,6-bisphosphatase, *J. Biol. Chem.* **263:**839–849.

Lochrie, M. A., Hurley, J. B., and Simon, M. I., 1985, Sequence of the α subunit of photoreceptor G protein: Homologies between transducin, *ras,* and elongation factors, *Science* **228:**96–99.

Luskey, K. L., and Stevens, B., 1985, Human 3-hydroxy-3-methylglutaryl coenzyme A reductase. Conserved domains responsible for catalytic activity and sterol-regulated degradation, *J. Biol. Chem.* **260:**10271–10277.

Maizel, J. V., and Lenk, R. P., 1981, Enhanced graphic matrix analysis of nucleic acid and protein sequences, *Proc. Natl. Acad. Sci. U.S.A.* **78:**7665–7669.

Masters, S. B., Stroud, R. M., and Bourne, H. R., 1986, Family of G protein α chains: Amphipathic analysis and predicted structure of functional domains, *Protein Eng.* **1:**47–54.

McCarthy, M. P., Earnest, J. P., Young, E. F., Choe, S., and Stroud, R. M., 1986, The molecular neurobiology of the acetylcholine receptor, *Annu. Rev. Neurosci.* **9:**383–413.

McClarin, J. A., Frederick, C. A., Wang, B.-C., Greene, P., Boyer, H. W., Grable, J., and Rosenberg, J. M., 1986, Structure of the DNA-Eco R1 endonuclease recognition complex at 3 Å resolution, *Science* **234:** 1526–1541.

McCrea, P., Popot, J.-L., and Engelman, D., 1986, Accessibility of the acetylcholine receptor δ chain C-terminus to hydrophilic reagents in reconstituted vesicles, *Biophys. J.* **49:**355a.

McLachlan, A. D., 1971, Tests for comparing related amino-acid sequences. Cytochrome *c* and cytochrome c_{551}, *J. Mol. Biol.* **61:**409–424.

McLachlan, A. D., and Karn, J., 1983, Periodic features in the amino acid sequence of nematode myosin rod, *J. Mol. Biol.* **164:**605–626.

McLachlin, A. D., and Stewart, M., 1976, The 14-fold periodicity in α-tropomyosin and the interaction with actin, *J. Mol. Biol.* **103:**271–298.

Medynski, D. C., Sullivan, K., Smith, D., Van Dop, C., Chang, F. H., Fung, B. K. K., Seeburg, P. H., and Bourne, H. R., 1985, Amino acid sequence of the α subunit of transducin deduced from the cDNA sequence, *Proc. Natl. Acad. Sci. U.S.A.* **82:**4311–4315.

Michel, H., Weyer, K. A., Gruenberg, H., Dunger, I., Oesterhelt, D., and Lottspeich, F., 1986, The 'light' and 'medium' subunits of the photosynthetic reaction centre from *Rhodopseudomonas viridis:* Isolation of the genes, nucleotide and amino acid sequence, *EMBO J.* **5:**1149–1158.

Miller, J., McLachlan, A. D., and Klug, A., 1985, Repetitive zinc-binding domains in the protein transcription factor IIIA from *Xenopus* oocytes, *EMBO J.* **4:**1609–1614.

Moller, W., and Amons, R., 1985, Phosphate-binding sequences in nucleotide-binding proteins, *FEBS Lett.* **186:**1–7.

Moore, W. M., Holliday, L. A., Puett, D., and Brady, R. N., 1974, On the conformation of the acetylcholine receptor protein from *Torpedo nobiliana, FEBS Lett.* **45:**145–149.

Murray, K. J., El-Maghrabi, M. R., Kountz, P. D., Lukas, T. J., Soderling, T. R., and Pilkis, S. J., 1984, Amino acid sequence of the phosphorylation site of rat liver 6-phosphofructo-2-kinase/fructose-2,6-bisphosphatase, *J. Biol. Chem.* **259:**7673–7681.

Nathans, J., and Hogness, D. S., 1983, Isolation, sequence analysis, and intron–exon arrangement of the gene encoding bovine rhodopsin, *Cell* **34:**807–814.

Nikolics, K., Mason, A. J., Szonyi, E., Ramachandran, J., and Seeburg, P. H., 1985, A prolactin-inhibiting factor within the precursor for human gonadotropin-releasing hormone, *Nature* **316:**511–517.

Noda, M., Takahashi, H., Tanabe, T., Toyosato, M., Furutani, Y., Hirose, T., Asai, M., Inayama, S., Miyata, T., and Numa, S., 1982, Primary structure of α-subunit precursor of *Torpedo californica* acetylcholine receptor deduced from a cDNA sequence, *Nature* **299:**793–797.

Noda, M., Takahashi, H., Tanabe, T., Toyosato, M., Kikyotani, S., Hirose, T., Asai, M., Takashima, H., Inayama, S., Miyata, T., and Numa, S., 1983a, Primary structures of β and δ-subunit precursors of *Torpedo californica* acetylcholine receptor deduced from cDNA sequences, *Nature* **301:**251–255.

Noda, M., Takahashi, H., Tanabe, T., Toyosato, M., Kikyotani, S., Furutani, Y., Hirose, T., Takashima, H., Inayama, S., Miyata, T., and Numa, S., 1983b, Structural homology of *Torpedo californica* acetylcholine receptor subunits, *Nature* **302:**528–532.

Ohlendorf, D. H., and Matthews, B. W., 1983, Structural studies of protein–nucleic acid interactions, *Annu. Rev. Biophys. Bioeng.* **12:**259–284.

Ovchinnikov, Yu. A., Abdulaev, N. G., Feigina, M. Yu., Kiselev, A. V., and Lobanov, N. A., 1979, The structural basis of the functioning of bacteriorhodopsin: An overview, *FEBS Lett.* **100:**219–224.

Perutz, M. F., Kendrew, J. C., and Watson, H. C., 1965, Structure and function of haemoglobin. II. Some relations between polypeptide chain configuration and amino acid sequence, *J. Mol. Biol.* **13:**669–678.

Pilkis, S. J., Fox, E., Wolfe, L., Rothbarth, L., Colosia, A., Stewart, H. B., and El-Maghrabi, M. R., 1986, Hormonal modulation of key hepatic regulatory enzymes in the gluconeogenic/glycolytic pathway, *Ann. N.Y. Acad. Sci.* **478:**1–19.

Pilkis, S. J., Claus, T. H., Kountz, P. D., and Maghrabi, M. R., 1987, Enzymes of the fructose-6-phosphate fructose-1,6-bisphosphate substrate cycle, in: *The Enzymes,* Volume 28 (P. D. Boyer and E. G. Krebs, eds.), Academic Press, New York, pp. 3–45.

Popot, J.-L., and Changeux, J.-P., 1984, Nicotinic receptor of acetylcholine: Structure of an oligomeric integral membrane protein, *Physiol. Rev.* **64:**1162–1239.

Rao, J. K. M., and Argos, P., 1986, A conformational preference parameter to predict helices in integral membrane proteins, *Biochim. Biophys. Acta* **869:**197–214.

Ratnam, M., Sargent, P., Sarin, V., Fox, J. L., LeNguyen, D., Rivier, J., Criado, M., and Lindstrom, J., 1986, Location of antigenic determinants on primary sequences of subunits of nicotinic acetylcholine receptor by peptide mapping, *Biochemistry* **25:**2621–2632.

Reeke, G. N., Jr., Becker, J. W., and Edelman, G. M., 1975, The covalent and three-dimensional structure of concanavalin A. IV. Atomic coordinates, hydrogen bonding, and quaternary structure, *J. Biol. Chem.* **250:** 1525–1547.

Reynolds, G., Basu, S. K., Osborne, T. F., Chin, D. J., Gil, G., Brown, M. S., Goldstein, J. L., and Luskey, K. L., 1984, HMG CoA reductase: A negatively regulated gene with unusual promoter and 5′ untranslated regions, *Cell* **38:**275–286.

Rhodes, D., and Klug, A., 1986, An underlying repeat in some transcriptional control sequences corresponding to half a double helical turn of DNA, *Cell* **46:**123–132.

Robishaw, J. D., Russell, D. W., Harris, B. A., Smigel, M. D., and Gilman, A. G., 1986, Deduced primary structure of the α subunit of the GTP-binding stimulatory protein of adenylate cyclase, *Proc. Natl. Acad. Sci. U.S.A.* **83:**1251–1255.

Rose, G. D., Gierasch, L. M., and Smith, J. A., 1985, Turns in peptides and proteins, *Adv. Protein Chem.* **37:** 1–109.

Rose, Z. B., 1980, The enzymology of 2,3-bisphosphoglycerate, *Adv. Enzymol.* **51:**211–253.

Ross, M. J., Klymkowsky, M. W., Agard, D. A., and Stroud, R. M., 1977, Structural studies of a membrane-bound acetylcholine receptor from *Torpedo californica, J. Mol. Biol.* **116:**635–659.

Roth, R. A., and Koshland, M. E., 1981, Role of disulfide interchange enzyme in immunoglobulin synthesis, *Biochemistry* **20:**6594–6599.

Schevitz, R. W., Otwinowski, Z., Joachimiak, A., Lawson, C. L., and Sigler, P. B., 1985, The three-dimensional structure of trp repressor, *Nature* **317:**782–786.

Schiffer, M., and Edmundson, A. B., 1967, Use of helical wheels to represent the structures of proteins and to identify segments with helical potential, *Biophys. J.* **7:**121–135.

Schofield, P. R., Darlison, M. G., Fujita, N., Burt, D. R., Stephenson, F. A., Rodriguez, H., Rhee, L. M., Ramachandran, J., Reale, V., Glencorse, T. A., Seeburg, P. H., and Barnard, E. A., 1987, Sequence and functional expression of the GABA receptor shows a ligand-gated receptor super-family, *Nature* **328:**221–227.

Schulz, G. E., Elzinga, M., Marx, F., and Schirmer, R. H., 1974, Three dimensional structure of adenyl kinase, *Nature* **250:**120–123.

Skalnik, D. G., and Simoni, R. D., 1985, The nucleotide sequence of syrian hamster HMG-CoA reductase cDNA, *DNA* **4:**439–443.

Stroud, R. M., and Finer-Moore, J., 1985, Acetylcholine receptor structure, function, and evolution, *Annu. Rev. Cell Biol.* **1:**317–351.

Stryer, L., 1986, Cyclic GMP cascade of vision, *Annu. Rev. Neurosci.* **9:**87–119.

Stryer, L., and Bourne, H. R., 1986, G proteins: A family of signal transducers, *Annu. Rev. Cell. Biol.* **2:**391–419.

Sullivan, K. A., Liao, Y.-C., Alborzi, A., Beiderman, B., Chang, F.-H., Masters, S. B., Levinson, A. D., and Bourne, H. R., 1986, Inhibitory and stimulatory G proteins of adenylate cyclase: cDNA and amino acid sequences of the α chains, *Proc. Natl. Acad. Sci. U.S.A.* **83:**6687–6691.

Sumikawa, K., Houghton, J., Smith, J. G., Bell, L., Richards, B. M., and Barnard, E. A., 1982, The molecular cloning and characterization of cDNA coding for the α subunit of the acetylcholine receptor, *Nucleic Acids Res.* **10:**5809–5822.

Tanabe, T., Nukada, T., Nishikawa, Y., Sugimoto, K., Suzuki, H., Takahashi, H., Noda, M., Haga, T., Ichiyama, A., Kangawa, K., Minamino, N., Matsuo, H., and Numa, S., 1985, Primary structure of the α-subunit of transducin and its relationship to *ras* proteins, *Nature* **315:**242–245.

Tauler, A., El-Maghrabi, M. R., and Pilkis, S. J., 1987, Functional hemology of 6-phosphofructo-2-kinase/fructose-2,6-bisphosphatase, phosphoglycerate mutase, and 2,3-bisphosphoglycerate mutase, *J. Biol. Chem.* **262:**16808–16815.

Turnell, W., Sarra, R., Glover, I. D., Baum, J. O., Caspi, D., Baltz, M. L., and Pepys, M. B., 1986, Secondary structure prediction of human SAA1. Presumptive identification of calcium and lipid binding sites, *Mol. Biol. Med.* **3:**387–407.

Vassarotti, A., Stroud, R., and Douglas, M., 1987, Independent mutations at the amino terminus of a protein act as surrogate signals for mitochondrial import, *EMBO J.* **6:**705–711.

Wang, A. C., Wang, I. Y., and Fudenberg, H. H., 1977, Immunoglobulin structure and genetics, *J. Biol. Chem.* **252:**7192–7199.

Winn, S. I., Watson, H. C., Harkins, R. N., and Fothergill, L. A., 1981, Structure and activity of phosphoglycerate mutase, *Phil. Trans. R. Soc. Lond. [Biol.]* **293:**121–130.

Wistow, G. J., Katial, A., Craft, C., and Shinohara, T., 1986, Sequence analysis of bovine retinal S-antigen, *FEBS Lett.* **196:**23–28.

Yatsunami, K., and Khorana, G., 1985, GTPase of bovine rod outer segments: The amino acid sequence of the α subunit as derived from the cDNA sequence, *Proc. Natl. Acad. Sci. U.S.A.* **82:**4316–4320.

Young, E. F., Ralston, E., Blake, J., Ramachandran, J., Hall, Z. W., and Stroud, R. M., 1985, Topological mapping of acetylcholine receptor: Evidence for a model with five transmembrane segments and a cytoplasmic COOH-terminal peptide, *Proc. Natl. Acad. Sci. U.S.A.* **82:**626–630.

20

Guide for Studies on Structure and Function Employing Synthetic Polypeptides

Emil T. Kaiser

I. INTRODUCTION

The systematic design of protein tertiary structure poses a formidable challenge in protein engineering. Several years ago we undertook the design of models for those peptides and proteins for which to a first approximation tertiary structure can be neglected (Kroon *et al.*, 1978; Kaiser and Kezdy, 1983, 1984). Although the prediction of tertiary structure from primary sequence cannot as yet be made with confidence, it was our thesis that secondary structures could be built in a rational fashion. Furthermore, the reasonable hypothesis was proposed that in amphiphilic environments like that of a biological interface, complementary amphiphilic secondary structures (secondary structures having distinct hydrophobic and hydrophilic faces) may be induced in peptides and proteins binding on such interfaces. Therefore, as our initial objective we engaged in the modeling of peptides and proteins that have affinity for membranes.

Emil T. Kaiser • Late of the Laboratory of Bioorganic Chemistry and Biochemistry, Rockefeller University, New York, New York 10021.

As our first target we were drawn to apolipoprotein A-I (apo A-I), the principal protein constituent of high-density lipoprotein (HDL) (Kroon *et al.*, 1978; Fukushima *et al.*, 1979, 1981; Yokoyama *et al.*, 1980). On the basis of physical data (such as circular dichroism spectra) and of considerations of the known amino acid sequence, it had already been proposed that when apo A-I was bound to the surface of the HDL particle it had the potential to form amphiphilic α-helical regions (Segrest *et al.*, 1974). Further analysis suggested that the apo A-I molecule could form six to seven amphiphilic α-helical regions consisting of hydrophilic and hydrophobic residues punctuated by "hinges" containing Gly or Pro residues (Fitch, 1977; McLachlan, 1977).

To test the roles of the putative amphiphilic α-helical regions in determining the biological and physical properties of apo A-I, we decided to design and construct a model peptide that "idealized" this secondary structural feature of apo A-I (Fukushima *et al.*, 1979, 1980). Since the design principles we adopted in constructing the model have subsequently been applied in our modeling of many other biologically active peptides and proteins, some examples of which are discussed here, it is worthwhile to describe the principles briefly at the outset.

II. MODELS FOR APOLIPOPROTEINS

Since the amphiphilic secondary structure for which we were building an idealized model was an α helix, we employed amino acids with high helix-forming potential in our modeling (Fukushima *et al.*, 1979). Thus, we employed Leu as the hydrophobic aliphatic residue, Lys as the positively charged hydrophilic residue (near neutral pH), and Glu as the negatively charged residue. The first model we prepared, peptide 1, was 22 amino acids in length, since this is roughly the length of the typical amphiphilic helices postulated to be formed in apo A-I. For convenience in our solid-phase synthesis of the model, Ala was chosen as the C-terminal residue, and Pro was placed at the N terminus because this residue or Gly usually punctuates the helices in apo A-I. As illustrated in the helical wheel diagram of the model shown in Fig. 1,

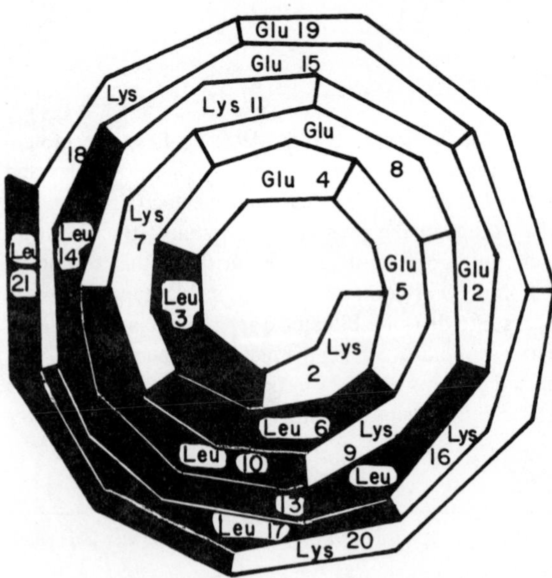

Figure 1. "Spiral" axial projection in two dimensions of the α-helical conformation of residues 2–21 of peptide 1, a model simulating the properties of apolipoprotein A-I (Kaiser and Kezdy, 1984).

the sequence of the designed peptide was chosen so that approximately one-third of the α-helical surface would be hydrophobic and the remainder hydrophilic. The charge distribution of the Lys and Glu residues in the model was chosen to approximate what we believe to be the most common distribution of these charged groups in the amphiphilic α-helical regions of apo A-I itself.

What should be stressed here is that in the case of the apo A-I model, as in most of our later modeling of surface-active peptides and proteins, we have tried to minimize the homology of the amino acid sequence of our designed structure to the sequence of the corresponding naturally occurring secondary structural region (Kaiser and Kezdy, 1983, 1984). The reason we have done this is that if the amino acid sequence of our model is very dissimilar to that of the natural sequence, having in common mainly the potential to form a similar amphiphilic α helix, and yet the model mimics the physical and biological properties of the natural system, this makes a strong case that it is the characteristics of this secondary structural feature and not the specific amino acid sequence that are crucial in determining these properties. Although in our later studies we moved to the use of unnatural amino acids (Rajashekhar and Kaiser, 1986), when we began our modeling work we generally employed natural amino acids. Therefore, in our attempts to minimize amino acid sequence homology between the models and the natural molecules we sometimes did not eliminate sequence homologies. For instance, if a Leu residue occurred in a given molecule, we did not generally change this residue in our model since another related aliphatic amino acid that we might use would have a lower α-helical potential, probably reducing the effectiveness of our model as a mimic for the natural peptide.

Although peptide I was only 22 amino acids in length, we found that this idealized model of the amphiphilic α helices of apo A-I mimicked effectively many of the properties of the 243-amino-acid-containing protein apo A-I in amphiphilic environments (Fukushima *et al.*, 1979, 1981; Yokoyama *et al.*, 1980). A monolayer of peptide 1 formed at the air–water interface was very stable, giving a collapse pressure of 21 dyne/cm, a value essentially identical to that seen for a monolayer of apo A-I itself (Edelstein *et al.*, 1979). Similarly, peptide 1 bound to egg lecithin unilamellar vesicles at pH 7.0 with a K_d value of 2×10^{-6} M, as compared to the K_d of 10^{-6} M seen for apo A-I. Reversible aggregation of peptide 1 to form a tetramer occurred at pH 7.0 with a dissociation constant of 9.7×10^{-16} M^3 (Fukushima *et al.*, 1979). This aggregation was reflected in the measurements of $\theta_{222\ nm}$ in the circular dichroism spectrum. At low concentrations of the peptide, measurement of $\theta_{222\ nm}$ indicated that about 30% α helix was formed, whereas at high concentration an estimate of 50% α helix was obtained. In the α-helix-inducing solvent trifluoroethanol roughly 60% α-helix formation was estimated for peptide 1, very similar to that calculated for apo A-I. Perhaps most importantly, in a biological system our model peptide 1 behaved very similarly to apo A-I. In particular, we found that peptide 1 was not far from being as effective as apo A-I in activating the esterification of egg lecithin–cholesterol vesicles catalyzed by lecithin:cholesterol acyltransferase.

In parallel with our studies on the idealized model peptide 1, we also synthesized naturally occurring segments of apo A-I that appeared to have a relatively good potential to form amphiphilic α helices. In essence what we found was that although a naturally occurring 22-amino-acid segment of apo A-I did not very effectively mimic the properties of apo A-I (Kroon *et al.*, 1978), a larger segment of apo A-I corresponding to residues 121–164 that contained two putative amphiphilic α-helical regions did show some of the characteristics of the whole protein (Fukushima *et al.*, 1980). For this reason we decided to construct a 44-amino-acid model, a dimer of peptide 1, peptide 2 (Fig. 2), which had the potential to form two idealized amphiphilic α helices connected by a linker containing a Pro residue (Nakagawa *et al.*, 1985). This peptide, which we prepared through a segment synthesis–condensation approach, proved

Pro–Lys–Leu–Glu–Glu–Leu–Lys–Glu–Lys–Leu–Lys–

Glu–Leu–Leu–Glu–Lys–Leu–Lys–Glu–Lys–Leu–Ala

1

Pro–Lys–Leu–Glu–Glu–Leu–Lys–Glu–Lys–Leu–Lys–

Glu–Leu–Leu–Glu–Lys–Leu–Lys–Glu–Lys–Leu–Ala–

Pro–Lys–Leu–Glu–Glu–Leu–Lys–Glu–Lys–Leu–Lys–

Glu–Leu–Leu–Glu–Lys–Leu–Lys–Glu–Lys–Leu–Ala

2

Figure 2. The amino acid sequences of peptides 1 and 2, which are models of human plasma apo A-I.

to have properties even more similar to those of apo A-I than did peptide 1. For instance, in our monolayer studies the limiting area per residue, A_0, was measured to be 23 Å^2 for peptide 1 (Fukushima *et al.*, 1979), indicating an appreciably less compact structure than that measured for apo A-I (A_0 = 16 Å^2), but the value found for peptide 2 (A_0 = 16 Å^2) corresponded exactly to that for apo A-I. We have concluded that the prototypic structural unit for apo A-I consists of two amphiphilic α-helical segments flanking a "hinge" region containing a helix-breaking residue like Pro.

Several years ago, on the basis of indirect evidence (Reynolds, 1980) that the dominant amphiphilic secondary structure affecting the properties of apolipoprotein B (apo B), the principal protein in low-density lipoprotein, might be the amphiphilic β strand, we undertook the design of model peptides possessing this type of structure (Osterman *et al.*, 1984).

Specifically, we designed peptide 3, a tridecapeptide containing three different amino acids, valine, glutamate, and ornithine. The sequence of the peptide was arranged so that alternating amino acids would be hydrophilic and lipophilic and that hydrophilic amino acids would alternate in having acidic and basic side chains, which would help in staggering the electric charges on self-association of the peptide. We chose valine as one of the constituent amino acids because it has a high β-forming potential and forms soluble β sheets in copolymers like poly(Val-Lys) (Brack and Orgel, 1975). Usually glutamic acid is considered to have a low β-forming potential, but poly(Glu) and copolymers including this residue such as poly(Glu-Val) are known to have a potential to form β structures under certain conditions (Rippon *et al.*, 1973; Itoh *et al.*, 1976). Our choice of ornithine was made to maximize the side-chain interactions, and we used this residue rather than lysine because the former residue can serve readily as a chemical marker, since apo B does not contain ornithine. The peptide was synthesized by the solid-phase method using the trifluoroacetyl group for protection of the δ-amino group of ornithine. At a concentration of 2.5×10^{-4} M, the peptide showed in aqueous solution a circular dichroic spectrum consistent with the β-strand structure with a minimum in the spectrum at 218 nm. Under these conditions we calculated that the peptide contained 55% β strand and 45% random coil in distilled water and 66% β strand and 34%

random coil in 0.16 M KCl. A film of the peptide obtained by evaporating its aqueous solution on a CaF_2 plate showed an infrared spectrum consistent with an antiparallel β-sheet structure characterized by a strong absorption band centered at 1628 cm^{-1} and a weak band at 1690 cm^{-1}. Analytical gel permeation chromatography on a Toyo Soda G2000 SW HPLC column showed that the peptide had a high tendency to self-associate.

NH_2-Val-Glu-Val-Orn-Val-Glu-Val-Orn-Val-Glu-Val-Orn-Val-COOH (peptide 3)

At a variety of interfaces the peptide showed amphiphilic properties. The peptide was found to form monolayers of great stability at the air–water interface. On a subphase with a rather low ionic strength, 0.16 M KCl, the monolayer showed ideal gas behavior at $\pi \leq 1$ dyne/cm and was characterized by an exclusion area of 16 Å^2/amino acid. Compression of the monolayer up to 44 dyne/cm did not show any sign of collapse. When a monolayer of human apo B was spread on a subphase of high ionic strength, the exclusion area was calculated to be 17 Å^2/amino acid, and the collapse pressure was 38 dyne/cm, indicating that the surface properties of peptide 3 are very similar to those of apo B.

Measurements with hyperlipemic LDL employing peptide 3 in concentrations ranging from 2.5×10^{-8} to 5.0×10^{-5} M at 21°C showed that the peptide binds to the hyperlipemic LDL according to a Langmuir isotherm with $K_d = 1.45 \times 10^{-7}$ M and with 88 sites per LDL. Again, the observation that peptide 3 binds quite well to LDL demonstrates the considerable similarity of the properties of this peptide with those of the whole apolipoprotein B molecule. The recent elucidation of the amino acid sequence of apo B from its DNA sequence (Knott *et al.*, 1986; Yang *et al.*, 1986) strongly suggests that, as anticipated from the results of our modeling work, considerable sections of the apolipoprotein consist of amphiphilic β-strand structures.

III. MODELING OF THE PEPTIDE TOXIN MELITTIN

In the cases of the apolipoproteins we have modeled, we have focused on purely secondary structural considerations. However, in moving toward our eventual goal of modeling tertiary structural systems, including enzymes, we sought to examine the possibility of developing models for peptides in which, in addition to the secondary structural region, an active site region exists. Accordingly, early in our work we were drawn to the study of the peptide toxin melittin (DeGrado *et al.*, 1981). In its principal version this peptide consists of 26 amino acids the sequence of which is shown in structure 4. When we began to model this peptide toxin, it seemed to us on the basis of data in the literature that melittin could be divided into two principal regions. One region consists of the sequence of residues 1–20, which has the potential to form a rather hydrophobic amphiphilic α helix with a kink at the proline residue. The other region corresponds to the C-terminal hexapeptide portion, which includes four basic residues that should be positively charged in the vicinity of neutral pH. A peptide corresponding to the 1–20 region will bind to the erythrocyte membrane without lysing it (Schröder *et al.*, 1971). The C-terminal hexapeptide is inactive by itself, but when it is covalently attached to the 1–20 region it acts in concert with that amphiphilic structure to cause lysis of the red cell.

H_2N-Gly-Ile-Gly-Ala-Val-Leu-Lys-Val-Leu-Thr-Thr-Gly-Leu-Pro-Ala-Leu-Ile-
Ser-Trp-Ile-Lys-Arg-Lys-Arg-Gln-Gln-$CONH_2$ (peptide 4)

In undertaking to model the melittin system, we considered the hexapeptide portion to be a kind of primitive "active site," and in our model we preserved this sequence. Indeed, in most of our modeling of surface-active peptides and proteins, when we have redesigned secondary structural regions, we have typically maintained the original active site residues of the peptide or protein we were modeling. In the case of the 1–20 region of melittin, however, we made global mutations. In particular, we used Gln residues as neutral hydrophilics that are also good α-helix formers, and we also used Ser residues, which are hydrophilic but not quite as good in terms of their α-helical potential. We employed Leu as a hydrophobic residue that has a strong tendency to go into α helices. No charges were included in the 1–20 region of the model peptide, and the kink present at position 14 in the natural sequence because of the Pro residue was eliminated by replacing this residue with a Ser residue.

It should be noted that although we attempted to reduce the homology to the sequence of melittin itself in the 1–20 region as much as possible, we did not completely eliminate this homology. For instance, where there were Leu residues in the natural sequence, we maintained these residues in our model. This was done because in the category of aliphatic hydrophobic α-helix formers Leu is a particularly good residue. In the early phases of modeling work we usually employed naturally occurring amino acids because we wished to develop modeling approaches that could be readily transferred to genetic engineering studies. We felt that if we replaced Leu with any other naturally occurring hydrophobic aliphatic amino acid, this would probably be deleterious to the activity of our model.

The sequence of the model melittin peptide is given in structure 5, and a Schiffer–Edmundson (1967) helical wheel projection of this sequence is shown in Fig. 3. This figure illustrates an α-helical projection of the 1–20 region in which one is looking down the helix axis. The design of the peptide is readily apparent from a consideration of the segregation of the residues present. The hydrophobic face of this helix covers close to two-thirds of the surface of the helix, and one can clearly see the segregation of the residues into hydrophobic and hydrophilic patches.

H_2N-Leu-Leu-Gln-Ser-Leu-Leu-Ser-Leu-Leu-Gln-Ser-Leu-Leu-Ser-Leu-Leu-
Leu-Gln-Trp-Leu-Lys-Arg-Lys-Arg-Gln-Gln-$CONH_2$ (peptide 5)

We have done extensive kinetic studies on lysis by melittin and our model. However, to keep within the scope of this review the details of these kinetic studies are not discussed here.

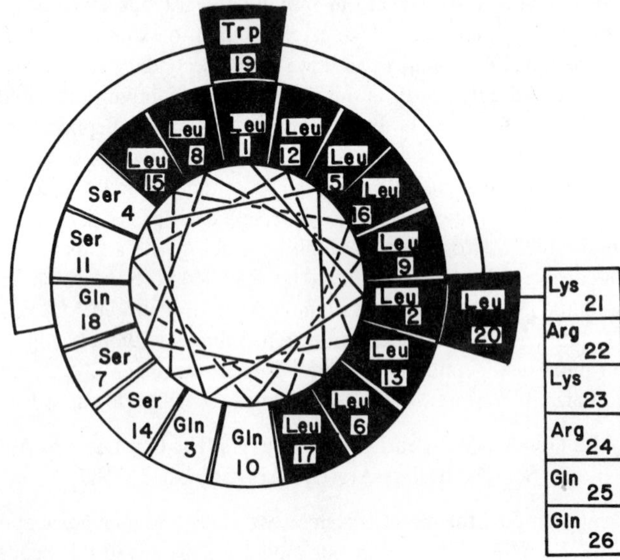

Figure 3. Axial projection of the α-helical conformation of a model peptide designed to have melittinlike activity.

One example comparing the model system and melittin can be described briefly. When both the model peptide and melittin were incubated with red cells for short times in an isotonic solution, we found that the rate of lysis, as evidenced by the release of hemoglobin, caused by the model toxin was appreciably greater than that seen for lysis by melittin itself. We have found that it is an aggregate of melittin that causes lysis of the red cell. The reasons why the model toxin is more effective in this assay than is melittin are not completely clear. However, measurements on the model peptide show that it undergoes association to form a tetramer in aqueous solution with a more favorable dissociation constant than has been observed for the corresponding tetramerization of melittin, and it is possible that the more facile lysis by the model peptide reflects this aggregation behavior. Since the model peptide lacks a Pro residue in the amphiphilic α-helical region, if aggregation of the peptide occurs with the hydrophobic faces coming together, there should be a more extended contact between the hydrophobic faces in the case of the model peptide than in the case of melittin, which shows a kink because of the Pro residue.

Both the model peptide and melittin bind readily to unilamellar egg lecithin vesicles, and we have found that both peptides readily disrupt the phospholipid bilayers. To get an assessment of the strength of binding of melittin to phospholipid, we carried out studies on the binding of melittin to phospholipid-coated polystyrene beads. The buffer used for these experiments was 0.025 M Tris-HCl, pH 7.4, containing 0.15 M KCl. Under these conditions melittin bound with a dissociation constant of 0.51×10^{-6} M (Retzinger *et al.*, 1985).

IV. THE DESIGN OF PEPTIDE HORMONES

In view of our success in modeling the apolipoproteins and the peptide toxin melittin, we decided to go a step further in modeling peptides of greater complexity. Consideration of the possible structures for many peptide hormones that are in the range of about 10–50 amino acids in length and are not extensively cross linked by disulfide linkages suggested to us that many of these peptides might have potentially amphiphilic secondary structural regions (Kaiser and Kezdy, 1983, 1984; Kaiser *et al.*, 1985). If the hormones possess such regions, there would be a number of possible advantages to their having these characteristics. In particular, the secondary structural regions might play a role in assisting peptide hormones to bind to their receptors. Another possible function of such a region would be to hold the active site of the hormone in just the right geometry relative to the hormonal receptor to allow the activity of the receptor to be turned on. A further possible role for amphiphilic secondary structure in the hormones would be a stabilization of the hormonal molecule against enzymatic degradation. Because the design of various models for peptide hormones has been described in detail in other recent articles, only some selected cases of our modeling work is described here.

Perhaps the most thoroughly studied hormone in our laboratory has been β-endorphin (Taylor *et al.*, 1981, 1982, 1983; Blanc *et al.*, 1983; Blanc and Kaiser, 1984; Rajashekhar and Kaiser, 1986). According to our analysis of the biologically active structure of this hormone, there are three important regions (Taylor and Kaiser, 1986). The N-terminal region corresponds to the sequence of [Met^5]-enkephalin, and this is what we believe to be the "active site" of the hormone. The sequence of residues 6–12 is quite hydrophilic and is postulated to be a spacer region. Finally, the region of residues 13–29, nearly to the end of the C-terminal region, is thought to be a potential amphiphilic helix. In all of the model peptides we have built, the amino-terminal segment of β-endorphin was retained because we expected it to have highly specific interactions with opiate receptors as indicated by the considerable literature on single-residue deletion and substitution analogues.

Analysis of the propensity for formation of secondary structure using several algorithms suggested that the hydrophilic linker region had little such structure. Model building suggests that the amphiphilic helical region at the C terminus of β-endorphin corresponds either to an α helix in which the hydrophobic face twists along the surface of the helix as it moves down the helix axis or to an amphiphilic π helix in which the hydrophobic face lies along the helix axis. Because π helices are not commonly encountered in globular proteins, not much is known about such structures, and the question of whether or not such structures occur in surface-active proteins remains an open one. Since we had learned quite a bit in earlier studies about the construction of the α helices, we elected to focus on the α-helical models of β-endorphin in designing our analogues. In the various models built for the amphiphilic helical region, the naturally occurring amino acids were typically replaced by sequences involving Leu, Lys, and Gln residues.

The pharmacological effects of β-endorphin are diverse, and it is clear that different aspects of its activity rely to varying extents on the presence of the C-terminal putative amphiphilic helical region. On the basis of our structural hypothesis we could systematically vary the sequences of our analogues in such a way as to provide stringent tests of the importance of the various proposed structural regions for the biological and physical properties of β-endorphin. In our initial efforts in analogue design we asked whether idealization of the helical region together with straightening of the hydrophobic face of the helix would give active β-endorphin models. In the first model we prepared we carried out an idealization of only part of the proposed amphiphilic helical sequence, but shortly afterwards we also prepared a more extended idealized structure for the helix that also had a simplified hydrophilic linkage in the region of residues 6–12 (a repeating Ser-Gly sequence). Earlier literature data had suggested that the μ- and δ-receptor-binding activities of β-endorphin were largely attributable to the N-terminal region of the hormone, although the carboxy-terminal region does play a role in the δ/μ selectivity of the interaction of β-endorphin with opioid receptors (Taylor and Kaiser, 1986). Therefore, it was not very surprising that both of the analogues we prepared were effective in binding to these receptors. More interestingly, however, both analogues were quite effective as inhibitors of the electrically stimulated contractions of rat vas deferens, a property of β-endorphin that shows a strong dependence on the nature of the C-terminal region. Furthermore, both analogues showed considerable stability when exposed to the enzymes in a rat brain homogenate, stability even greater than that of β-endorphin, which survives such treatment considerably more effectively than does the short peptide [Met^5]-enkephalin. Indeed, both of the first two analogues that we prepared were so effective, even though the hydrophobic faces of their α-helical regions lay along the helix axis, that we wondered how crucial the shape of the hydrophobic face of the amphiphilic helix was for obtaining β-endorphin activity.

Accordingly, in a third model we prepared a sequence in which the amphiphilic helical region was constructed to give not only an idealized structure but also a hydrophobic face that would twist along the α-helical axis in a fashion similar to that proposed for β-endorphin. Once again the properties of the analogue in μ- and δ-receptor binding as well as in the inhibition of the contractions of the electrically stimulated rat vas deferens and in enzymatic stability were closely related to the properties of β-endorphin. In one respect, however, this third model, which was designed to have a hydrophobic face with a shape similar to that proposed for the amphiphilic α-helical model of β-endorphin, resembled the hormone better than did the first two analogues. In particular, neither of the first two analogues that had their hydrophobic faces lying along the helix axis showed analgesic activity, whereas the third analogue did exhibit potent analgesic effects. This finding illustrates nicely the use of systematic structural variation in probing the dependence of the various pharmacological properties of a hormone on its secondary structural features.

In a fourth model the C-terminal region was constructed with an amino acid composition corresponding to that of the second model including the repeating Ser-Gly sequence in residues 6–12, except that the arrangement of the residues in the C-terminal part of the fourth model was such that a continuous hydrophobic face could not be formed in the α-helical conformation. Although this peptide showed tight binding to the μ- and δ-receptors, it was rapidly degraded by the enzymes in rat brain homogenates, and in the receptor assays it became rapidly inactivated. In the rat vas deferens assay, inhibition of electrically stimulated contractions was seen only at very high concentrations and was exhibited in a nonopioid manner. Thus, the inhibitory effects of this peptide seen at very high concentration could not be reversed by the addition of naloxone but were slowly reversed with time by another mechanism, which may be proteolytic degradation. This result, taken together with the relatively high activities in the rat vas deferens assay seen for the other peptide models, all of which exhibited naloxone reversibility and resistance to proteolytic degradation, strongly supports the argument that an amphiphilic helical structure is necessary in the C-terminal segment of β-endorphin.

In our fifth model we used only D-amino acid residues in positions 13–31, arranged in a sequence that allows formation of a left-handed α helix with characteristics similar to those of the natural right-handed helical structure (Blanc and Kaiser, 1984). In effect, the structure of this model is diastereomeric with the β-endorphin structure, since the N-terminal portion of the sequence remains in the original configuration while the helix itself has an opposite chirality to that of β-endorphin. The peptide showed considerable binding potency in δ- and μ-receptor binding assays. In the rat vas deferens assay the fifth peptide model was approximately equipotent to β-endorphin, although it exhibited a mixed agonist–antagonist behavior. The peptide also produced potent and long-lasting antinociception.

The most recent model that we have constructed, a sixth analogue, is one in which we have utilized the idealized helix of the third model with the twisting hydrophobic face in addition to a very considerable simplification of the hydrophilic spacer region. In this peptide model we replaced the natural linker in the region 6–12 by a repeating unit of γ-hydroxymethyl-γ-aminobutyric acid (Rajashekhar and Kaiser, 1986). Assuming that the linker region does not possess a defined secondary structure, the length of the naturally occurring linker can be approximated by four residues of the γ-hydroxymethyl-γ-aminobutyric acid species linked in amide bonds where the carboxyl of one molecule is covalently bound to the amino of the next. This sixth analogue of β-endorphin has physical properties similar to those of the third analogue, which has the natural spacer in it, and this has been shown by surface monolayer and circular dichroism studies on the two peptides. Furthermore, the binding affinities of the two peptides to δ and μ receptors are also similar. In assays in the rat vas deferens, model 6 is equipotent to the third model. Most strikingly, the potent analgesic activity displayed by the sixth peptide model when injected intracerebroventricularly into mice is comparable to that of the third model. Thus, the rather drastic structural change made in the sixth model in the linker region did not seem to affect in a significant way either the solution or the *in vitro* and *in vivo* biological properties. Our findings with this model show that one can make mutations in structural regions of biologically active peptides (and probably proteins) using building blocks other than α-amino acids. In view of our success with the sixth model, in the next stages of our research we plan to construct the secondary structural regions of biologically active peptides from nonpeptidic building blocks. It is worth pointing out here that the approach of designing peptide hormones with important structural regions that do not contain typical amide bonds is one that cannot be pursued by the use of recombinant DNA techniques.

The hormone that we have studied most extensively other than β-endorphin is calcitonin (Moe *et al.*, 1983; Moe and Kaiser, 1985). A brief description of the current status of work on this hormone is now given because it illustrates how we have been able to proceed from the construction of fairly extreme models in which we have greatly reduced homology to the corresponding natural sequences in the secondary structural region to the design of analogues having increased potencies, which can result at least in part from reintroduction of certain specific features of the naturally occurring hormones. We have considered the calcitonin molecule to be divided into three structurally distinct regions. These are (1) a loop formed by linking the cysteine residues in positions 1 and 7, (2) a potential amphiphilic α-helical segment in residues 8–22, and (3) a linker region connecting the helix to the carboxy-terminal proline amide covering the region residues 23–31.

In the early phases of our work two peptide models of calcitonin were designed and constructed based on this analysis of the hormone structure. In both of the peptide models we concentrated on design of the amphiphilic α-helical structure in residues 8–22. In each instance we used leucine, lysine, and glutamine residues to maximize the α-helix-forming potential of the structure, and we tried to reduce homology with the natural sequences of the various calcitonins. We did retain the length and amphiphilicity of the helix, the size, shape, and orientation of its hydrophobic domain relative to the rest of the peptide, and the occurrence of the mostly neutral or basic residues on the hydrophilic face. However, in many of the naturally occurring calcitonins a single acidic residue (Asp or Glu) occurs in the hydrophobic domain, typically in position 15 (Taylor and Kaiser, 1986), and in the first two models we prepared this residue was replaced by a leucine residue to idealize the amphiphilicity. Furthermore, in the first model we tried to reduce the sequence homology to salmon calcitonin, the most active form of the naturally occurring hormone, by incorporating a tryptophan residue in place of the naturally occurring Leu. In our second model we went back to using Leu at this position and we also introduced a Tyr residue at position 22 where in the first model we had had a Leu residue and where aromatic residues are common to all of the natural structures.

The amphiphilic helical region of the second peptide model of calcitonin is compared to the α-helical structure proposed for salmon calcitonin I, residues 8–22, in Fig. 4. Comparison of the physicochemical properties of the two model peptides with those of salmon calcitonin I suggested that the models form more stable helical structures with increased amphiphilic properties, as expected from their design. The biological activities were compared to that of salmon calcitonin I by carrying out assays of binding to rat brain membranes, using ^{125}I-labeled salmon calcitonin I to radiolabel calcitonin receptors and employing assays of the hypocalcemic effects in rats 1 hr after subcutaneous injection. The second peptide model behaved roughly comparably to salmon calcitonin in the binding assays and showed approximate equipotency in the *in vivo* assay. The first peptide model was 10–20 times less potent in each assay than was salmon calcitonin, corresponding to potency similar to that of porcine calcitonin, the most active of the mammalian structures that have been tested (Nakamuta *et al.*, 1981). Also, in binding assays to rat kidney cortical membranes and in assays of adenylate cyclase activation in the same tissue, the second peptide model was equipotent to salmon calcitonin I. Thus, it can be seen that the pharmacological assays of the two model peptides provide strong evidence that peptides designed to have idealized α-helical structures in residues 8–22 of the calcitonin structure can reproduce the most potent and specific actions of the natural hormones. The fact that in both models considerable reduction in sequence homology in the amphiphilic region could be achieved as compared to the natural sequence also provides strong evidence that residues 8–22 are in the α-helical conformation on the calcitonin receptor surface.

The relatively lower activity of the first model in which a Trp residue is present in position 12 is consistent with older observations of Maier *et al.*, (1976), who suggested that aromatic residues in positions 12, 16, and 19 were responsible for the lower activities of the mammalian calcitonins in the hypocalcemic assay and who showed that sequential substitution of leucine residues in all three positions increased the potency of human calcitonin in this assay significantly.

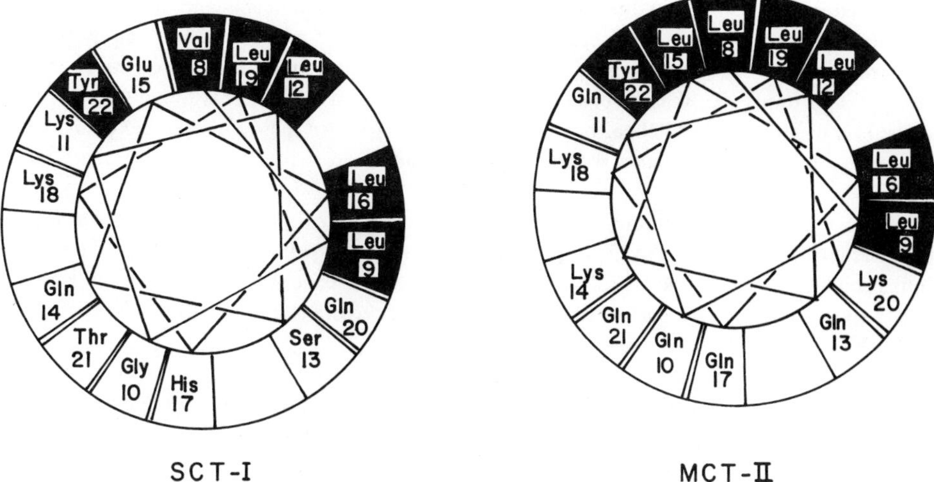

Figure 4. Axial projections of the potential α-helical conformation of residues 8 to 22 in salmon calcitonin I and our second model calcitonin, MCT-II.

In the most recent calcitonin analogue that we designed, we have proceeded further in restoring the sequence homology between our model and the naturally occurring salmon calcitonin on the hydrophobic face of the amphiphilic helical portion of the structure (Green *et al.*, 1987). Specifically, we reintroduced the carboxylate-containing Glu residue at position 15 in the model. Perhaps the most interesting finding made with this third model is that the *in vivo* hypocalcemic effect seen is actually appreciably higher than that for the very active naturally occurring salmon calcitonin. In other words, in the case of the calcitonin models, by retaining the residues on the hydrophobic face including the Glu residue in position 15, which is the one residue that violates the segregation of the amphiphilic region of salmon calcitonin into hydrophobic and hydrophilic faces, and by idealizing just the hydrophilic face of the amphiphilic helix, we can prepare a hormone even more effective than any of the naturally occurring ones. In further experiments on the design of calcitonin analogues, we hope to explore, as in the case of β-endorphin, the possibility that at least part of the structure may be replaced by nonpeptidic units.

V. SIGNAL REGIONS

In concluding this chapter it is worthwhile to touch briefly on signal peptides or the signal regions of various proteins. For either eukaryotic or prokaryotic systems, the secretion of many proteins occurs through a precursor form containing an amino-terminal extension sequence or signal (leader) peptide region (Kaiser and Kezdy, 1987). In the present review we do not go into the details of the many proposals for possible mechanisms by which the signal peptides interact with membranes. There is a great deal of sequence variation among signal peptides, but most of them contain a large number of nonpolar amino acids. For this reason it has been considered that when they interact with membranes they may have similar lipophilic secondary structures (Emr and Silhavy, 1983; Engelman and Steitz, 1981). Generally speaking, the amino-terminal portion of a signal region contains at least one charged amino acid residue. Usually a hydrophobic core follows the N-terminal charged region and precedes the more polar region where the signal peptidase cleavage site is located. The value of predictive algorithms for conformations may be somewhat questionable in the cases of signal regions because the signal sequences operate in hydrophobic (membrane) environments, and most of the algorithms that are available have been developed for globular proteins rather than those that primarily interact with membranes. If, nevertheless, one considers the predictions made through algorithms like that proposed by Chou and Fasman (1978), it is thought that signal sequences usually have α-helical or β-strand conformations in their hydrophobic cores. Further, they are predicted to have β turns in the regions close to the signal peptidase cleavage site.

One approach to examining the potential secondary structural requirements for signal peptides has been through physical studies on isolated signal peptides obtained synthetically. In particular, studies on the signal sequence of phage M13 coat protein (Shinnar and Kaiser, 1984) and on the signal sequence of *E. coli* λ-receptor protein (LamB) (Briggs and Gierasch, 1984) by circular dichroism spectroscopy have been interpreted to indicate that in aqueous solution these peptides are mostly in the random conformation. However, in solvents like trifluoroethanol and hexafluoroisopropanol these peptides, as well as a 29-residue peptide consisting of the prepro region of parathyroid hormone (Rosenblatt *et al.*, 1980), became at least partially α helical.

When fragments of *E. coli* λ-receptor protein wild-type and mutant signal sequences containing hydrophobic cores were examined, they had primarily random conformations in aqueous solution (Briggs and Gierasch, 1984). In this series, however, the peptides corresponding to signal regions that were found to be functional in transport in the whole-protein systems showed a strong tendency to form α helices, although they also showed the ability to form significant amounts of β structure in some solvents. Examination of a signal peptide corresponding to a nonfunctional deletion mutant signal of the LamB system showed that it had a considerably lower tendency to form an α helix than the functional wild-type or revertant peptides. The results obtained have been interpreted as indicating that the functional signal sequences adopt α-helical conformations during at least part of the secretion process. It has been proposed from the circular dichroism spectra of functional signal peptide fragments in lysolecithin micelles that these peptides can adopt a significant amount of β structure, whereas the corresponding nonfunctional deletion mutant peptide cannot (Briggs and Gierasch, 1984; Briggs *et al.*, 1986). For this reason it has been suggested that the signal sequence may have to take on a β-structure arrangement or another non-α-helical secondary structure during secretion.

Returning, however, from the simple peptides to the whole-protein systems, it has been shown by Kendall *et al.* (1986) that idealization of the hydrophobic core of the *E. coli* alkaline phosphatase signal region by site-directed mutagenesis to give a mutant signal sequence containing nine consecutive leucines results in a very effective signal system. This signal sequence, which has a maximized potential for hydrophobic α-helix formation in the core region, shows properties in transport studies that indicate that the mature alkaline phosphatase is correctly targeted to the *E. coli* periplasm and that processing of the precursor to the mature form of the enzyme occurs extremely rapidly. Had β-structure formation been obligatory at some point in the secretion process, this finding might not have been expected.

VI. CONCLUSION

The studies on the design, construction, and determination of the biological and physical properties of surface-active peptides as described in this chapter demonstrate that it is now possible to predict and to prepare in a systematic fashion new peptide sequences that can act analogously to apolipoproteins, toxins, or hormones. The peptides and proteins that we have modeled have been, for the most part, ones for which tertiary structural considerations can be neglected, at least in the early phases of model design. In further research that is underway, we are attempting to replace in a similar manner secondary structural regions of folded proteins where x-ray structural information is available. Our objective is to determine whether the introduction of appropriately designed secondary structural units with new amino acid sequences in these proteins will permit proper folding of the altered proteins to occur. Among the systems where this approach is being employed are ribonuclease T_1 and a methotrexate-resistant dihydrofolate reductase. We intend to determine whether the introduction of designed secondary structural regions in these enzymes still permits the modified enzymes to show significant binding and catalytic activity.

ACKNOWLEDGMENT. Much of the work from our laboratory that has been described here was supported by NIH Program Project 18577.

VII. REFERENCES

Blanc, J. P., and Kaiser, E. T., 1984, Biological and physical properties of a β-endorphin analog containing only D-amino acids in the amphiphilic helical segment 13–31, *J. Biol. Chem.* **259:**9549–9556.

Blanc, J. P., Taylor, J. W., and Kaiser, E. T., 1983, Examination of the requirement for an amphiphilic helical structure in β-endorphin through the design, synthesis and study of model peptides, *J. Biol. Chem.* **258:** 8277–8284.

Brack, A., and Orgel, L. E., 1975, β-Structures of alternating polypeptides and their possible prebiotic significance, *Nature* **256:**383–387.

Briggs, M. S., and Gierasch, L. M., 1984, Exploring the conformational roles of signal sequences: Synthesis and conformation analysis of receptor protein wild-type and mutant signal peptides, *Biochemistry* **23:**3111–3114.

Briggs, M. S., Cornell, D. G., Dluhy, R. A., and Gierasch, L. M., 1986, Spectroscopic studies of signal peptides in phospholipid monolayers: Conformations induced by lipids suggest possible initial steps in protein export, *Science* **233:**206–208.

Chou, P. Y., and Fasman, G. D., 1978, Empirical predictions of protein conformation, *Annu. Rev. Biochem.* **47:**251–276.

DeGrado, W. F., Kézdy, F. J., and Kaiser, E. T., 1981, Design, synthesis and characterization of a cytotoxic peptide with melittin-like activity, *J. Am. Chem. Soc.* **103:**679–681.

Edelstein, C. Kézdy, F. J., Scanu, A. M., and Shen, B. W., 1979, Apolipoproteins and the structural organization of plasma lipoproteins: Human plasma high density lipoprotein-3, *J. Lipid Res.* **20:**143–153.

Emr, S. D., and Silhavy, T. J., 1983, Importance of secondary structure in the signal sequence for protein secretion, *Proc. Natl. Acad. Sci. U.S.A.* **80:**4599–4603.

Engelman, D. M., and Steitz, T. A., 1981, The spontaneous insertion of proteins into and across membranes. The helical hairpin hypothesis, *Cell* **23:**411–422.

Fitch, W. M., 1977, Phylogenies constrained by the crossover process as illustrated by human hemoglobins and a thirteen cycle, eleven amino acid repeat in human apolipoprotein A-I, *Genetics* **86:**623–644.

Fukushima, D., Kupferberg, J. P., Yokoyama, D., Kroon, D., Kaiser, E. T., and Kézdy, F. J., 1979, A synthetic amphiphilic helical docosapeptide with the surface properties of plasma apolipoprotein A-I, *J. Am. Chem. Soc.* **101:**3703–3704.

Fukushima, D., Kaiser, E. T., Kézdy, F. J., Kroon, D. J., Kupferberg, J. P., and Yokoyama, S., 1980a, Rational design of synthetic models for lipoproteins, *Ann. N.Y. Acad. Sci.* **348:**365–373.

Fukushima, D., Yokoyama, S., Kroon, D. J., Kezdy, F. J., and Kaiser, E. T., 1980b, Chain length-function correlation of amphiphilic peptides, *J. Biol. Chem.* **255:**10651–10657.

Fukushima, D., Yokoyama, S., Kezdy, F. J., and Kaiser, E. T., 1981, Binding of amphiphilic peptides to phospholipid/cholesterol unilamellar vesicles: A model for protein–cholesterol interactions, *Proc. Natl. Acad. Sci. U.S.A.* **78:**2732–2736.

Green, F. R., Lynch, B., and Kaiser, E. T., 1987, Biological and physical properties of a model calcitonin containing a glutamate residue interrupting the hydrophobic face of the idealized amphiphilic α-helical region, *Proc. Natl. Acad. Sci. U.S.A.* **84:**8340–8344.

Itoh, K., Foxman, B. M., and Fasman, G. D., 1976, The two β-forms of poly(L-GLUTAMIC ACID), *Biopolymers* **15:**419–455.

Kaiser, E. T., and Kezdy, F. J., 1983, Secondary structures of proteins and peptide in amphiphilic environments (a review), *Proc. Natl. Acad. Sci. U.S.A.* **80:**1137–1143.

Kaiser, E. T., and Kezdy, F. J., 1984, Amphiphilic secondary structure: Design of peptide hormones, *Science* **223:**249–255.

Kaiser, E. T., and Kezdy, F. J., 1987, Peptides with affinity for membranes, *Annu. Rev. Biophys. Biophys. Chem.* **16:**561–581.

Kaiser, E. T., Lynch, B., and Rajashekhar, B., 1985, The design and construction of peptides and proteins with amphiphilic secondary structures, in: *Proceedings 9th Annual Peptide Symposium, Toronto,* Pierce Chemical Co., Rockford, IL, p. 855–864.

Kendall, D. A., Bock, S. C., and Kaiser, E. T., 1986, Idealization of the hydrophobic segment of the alkaline phosphatase signal peptide, *Nature* **321:**706–708.

Knott, T. J., Pease, R. J., Powell, L. M., Wallis, S. C., Rall, S. C., Jr., Innearity, T. L., Blackhart, B., Taylor, W. H., Marcel, Y., Milne, R., Johnson, D., Fuller, M., Lusis, A. J., McCarthy, B. J., Mahley, R. W., Levy-Wilson, B., and Scott, J., 1986, Complete protein sequence and identification of structural domains of human apolipoprotein B, *Nature* **323:**734–738.

Kroon, D. J., Kupferberg, J. P., Kaiser, E. T., and Kezdy, F. J., 1978, Mechanism of lipid–protein interactions in lipoproteins. A synthetic peptide–lecithin vesicle model system, *J. Am. Chem. Soc.* **100:**5975–5977.

Maier, R., Kamber, B., Riniker, B., and Rittel, W., 1976, Analogs of human calcitonin IV. Influence of leucine substitutions in positions 12, 16 and 19 on hypocalcemic activity in the rat, *Clin. Endocrinol.* **5:** 327s–332s.

McLachlan, A. D., 1977, Repeated helical pattern in apolipoprotein A-I, *Nature* **267:**465–466.

Moe, G. R., and Kaiser, E. T., 1985, Design, synthesis and characterization of a model peptide having potent calcitonin-like biological activity: Implications for calcitonin structure/activity, *Biochemistry* **24:**1971–1976.

Moe, G. R., Miller, R. J., and Kaiser, E. T., 1983, Design of a peptide hormone: Synthesis and characterization of a model peptide with calcitonin-like activity, *J. Am. Chem. Soc.* **105:**4100–4102.

Nakagawa, S. H., Lau, H. S. H., Kezdy, F. J., and Kaiser, E. T., 1985, The use of polymer-bound oxime for the synthesis of large peptides usable in segment condensation: Synthesis of a 44 amino acid amphiphilic peptide model of apolipoprotein A-I, *J. Am. Chem. Soc.* **107:**7087–7092.

Nakamuta, H., Furukawa, S., Koida, M., Yajima, H., Orlowski, R. C., and Schleuter, R., 1981, Specific binding of ^{125}I-salmon calcitonin to rat brain: Regional variation and calcitonin specificity, *Jpn. J. Pharmacol.* **31:**53–60.

Osterman, D., Mora, R., Kezdy, F. J., Kaiser, E. T., and Meredith, S., 1984, A synthetic amphiphilic β-strand tridecapeptide: A model for apolipoprotein B, *J. Am. Chem. Soc.* **106:**6845–6847.

Rajashekhar, B., and Kaiser, E. T., 1986, Design of biologically active peptides with non-peptidic structural elements: Biological and physical properties of a synthetic analogue of β-endorphin with unnatural amino acids in the region 6–12, *J. Biol. Chem.* **261:**13617–13623.

Retzinger, G. S., Meredith, S. C., Lau, S. H., Kaiser, E. T., and Kezdy, F. J., 1985, A method for probing the affinity of peptides for amphiphilic surfaces, *Anal. Biochem.* **150:**131–140.

Reynolds, J. A., 1980, Binding studies with apolipoproteins, *Ann. N.Y. Acad. Sci.* **348:**174–183.

Rippon, W. B., Chen, H. H., and Walton, A. G., 1973, Spectroscopic characterization of poly(Glu-Ala), *J. Mol. Biol.* **75:**369–375.

Rosenblatt, M., Beaudette, N. V., and Fasman, G. D., 1980, Conformation studies of the synthetic precursor-specific region of preproparathyroid hormone, *Proc. Natl. Acad. Sci. U.S.A.* **77:**3983–3987.

Schiffer, M., and Edmundson, A. B., 1967, Use of helical wheels to represent the structures of proteins and to identify segments with helical potential, *Biophys. J.* **7:**121–135.

Schröder, E., Lubke, K., Lehmann, M., and Beetz, I., 1971, Hemolytic activity and action on the surface tension of aqueous solutions of synthetic melittins and their derivatives, *Experentia* **26:**764–768.

Segrest, J. P., Jackson, R. L., Morrisett, J. D., and Gotto, A. M., Jr., 1974, A molecular theory of lipid–protein interactions in the plasma lipoproteins, *FEBS Lett.* **38:**247–253.

Shinnar, A. E., and Kaiser, E. T., 1984, Physical and conformation properties of a synthetic leader peptide from M13 coat protein, *J. Am. Chem. Soc.* **106:**5006–5007.

Taylor, J. W., and Kaiser, E. T., 1986, The structural characterization of β-endorphin and related peptide hormones and neurotransmitters, *Pharmacol. Rev.* **38:**291–319.

Taylor, J. W., Osterman, D. G., Miller, R. J., and Kaiser, E. T., 1981, Design and synthesis of a model peptide with β-endorphin-like properties, *J. Am. Chem. Soc.* **103:**6965–6966.

Taylor, J. W., Miller, R. J., and Kaiser, E. T., 1982, Structural characterization of β-endorphin through the design, synthesis and study of model peptides, *Mol. Pharmacol.* **22:**657–666.

Taylor, J. W., Miller, R. J., and Kaiser, E. T., 1983, Characterization of an amphiphilic helical structure in β-endorphin through the design, synthesis and study of model peptides, *J. Biol. Chem.* **258:**4464–4471.

Yang, C.-Y., Chan, S.-H., Gianturco, S. H., Bradley, W. A., Sparrow, J. T., Tanimura, M., Li, W.-H., Sparrow, D. A., DeLoof, H., Rosseneu, M., Lee, F.-S., Gu, Z.-W., Gotto, A. M., Jr., and Chan, L., 1986, Sequence, structure, receptor-binding domains and internal repeats of human apolipoprotein B-100, *Nature* **323:**738–742.

Yokoyama, S., Fukushima, D., Kupferberg, J. P., Kezdy, F. J., and Kaiser, E. T., 1980, The mechanism of activation of lecithin:cholesterol acyltransferase by apolipoprotein A-I and an amphiphilic peptide, *J. Biol. Chem.* **255:**7333–7337.

Index